Prealgebra

Fourth Edition

K. Elayn Martin-Gay

University of New Orleans

PEARSON

Prentice Hall

Upper Saddle River, New Jersey 07458

Library of Congress Cataloging-in-Publication Data
Martin-Gay, K. Elayn
  Prealgebra/K. Elayn Martin-Gay.–4th ed.
  p. cm.
  Includes index.
  ISBN 0-13-144447-6

  1. Arithmetic. I. Title
QA107.2.M37 2004
513'.1-dc   2003056484

Editor in Chief: *Christine Hoag*
Project Manager: *Mary Beckwith*
Production Editor: *Lynn Savino Wendel*
Vice President/Director of Production and Manufacturing: *David W. Riccardi*
Senior Managing Editor: *Linda Mihatov Behrens*
Executive Managing Editor: *Kathleen Schiaparelli*
Assistant Manufacturing Manager/Buyer: *Michael Bell*
Manufacturing Manager: *Trudy Pisciotti*
Executive Marketing Manager: *Eilish Collins Main*
Marketing Assistant: *Annett Uebel*
Marketing Project Manager: *Barbara Herbst*
Development Editor: *Elka Block*
Editor in Chief, Development: *Carol Trueheart*
Media Project Manager, Developmental Math: *Audra J. Walsh*
Assistant Managing Editor, Math Media Production: *John Matthews*
Assistant Editor: *Christina Simoneau*
Art Directors: *Geoffrey Cassar/Maureen Eide*
Interior Designer: *Circa 86*
Cover Designer: *Jack Robol; Dina Curro*
Art Editor: *Tom Benfatti*
Creative Director: *Carole Anson*
Director of Creative Services: *Paul Belfanti*
Director, Image Resource Center: *Melinda Reo*
Manager, Rights and Permissions: *Zina Arabia*
Interior Image Specialist: *Beth Brenzel*
Image Permission Coordinator: *Charles Morris*
Photo Researcher: *Elaine Soares*
Cover Art: ©Dale Chihuly, Purple & Pink Persian with Process Yellow Lip Wrap. 1992 13 x 23 x 18".
       Photo: Claire Garoutte 92/92.738.p1
Art Studio: Artworks
  Managing Editor, AV Production & Management: *Patty Burns*
  Production Manager: *Ronda Whitson*
  Production Technologies Manager: *Matthew Haas*
  Illustrators: *Dan Knopsnyder, Scott Wieber, Stacy Smith, Nate Storck, Mark Landis, Ryan Currier, Audrey Simonetti*
  Quality Assurance: *Pamela Taylor, Ken Mooney, Tim Nguyen*
Formatting Manager: *Jim Sullivan*
Electronic Production Specialists: *Karen Noferi, Joanne Del Ben, Karen Stephens, Jackie Ambrosius, Vicki Croghan, Julita Nazario*

 ©2004, 2001, 1997, 1993 Pearson Education, Inc.
Pearson Prentice Hall
Pearson Education, Inc., Upper Saddle River, New Jersey 07458

10 9 8 7 6 5

Photo Credits appear on page I105, which constitutes a continuation of the copyright page.
ISBN 0-13-144447-6 (paperback) 0-13-144520-0 (casebound)

Pearson Education Ltd., London
Pearson Education Australia Pty. Limited, Sydney
Pearson EducationSingapore Pte. Ltd.
Pearson Education North Asia, Ltd, Hong Kong
Pearson Education Canada, Ltd., Toronto
Pearson Educacion de Mexico, S.A.,de C.V.
Pearson Education, Japan, Tokyo
Pearson Education Malaysia, Pte. Ltd.

To Jewett B. Gay, and to the memory of her husband, Jack Gay

CONTENTS

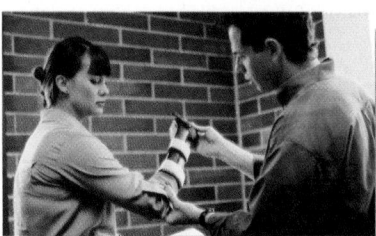

PREFACE

About the Book

Prealgebra, Fourth Edition was written to help students make the transition from arithmetic to algebra. To reach this goal, I introduce algebraic concepts early and repeat them as I treat traditional arithmetic topics, thus laying the groundwork for the next algebra course your students will take. A second goal was to show students the relevancy of the mathematics in everyday life and in the workplace.

In preparing this Fourth Edition, I considered the comments and suggestions of colleagues throughout the country and of the many users of previous editions. The numerous features that contributed to the success of the third edition have been retained. This updated revision includes an increased emphasis on study and test preparation skills, increased attention to geometric concepts, and real-life applications. I have carefully chosen pedagogical features to help students understand and retain concepts. As suggested by the AMATYC Crossroads Document and the NCTM Standards (plus Addenda), real-life and real-data applications, data interpretation, conceptual understanding, problem solving, writing, cooperative learning, appropriate use of technology, mental mathematics, number sense, estimation, critical thinking, and geometric concepts are emphasized and integrated throughout the book. In addition, the Fourth Edition now includes a new resource—the Chapter Test Prep Video CD. With this CD/video, students have instant access to video solutions for each of the chapter test questions contained in the text. It's designed to help them study efficiently.

Prealgebra, Fourth Edition is part of a series of texts that can include *Introductory Algebra*, Second Edition; *Basic Mathematics*, Second Edition; *Intermediate Algebra*, Second Edition; and a combined text, *Algebra A Combined Approach*, Second Edition. Throughout the series pedagogical features are designed to develop student proficiency in algebra and problem solving, and to prepare students for future courses. Key pedagogical features and resources are described on the following pages.

Key Pedagogical Features and Changes in the Fourth Edition

Readability and Connections I have tried to make the writing style as clear as possible while still retaining the mathematical integrity of the content. When a new topic is presented, an effort has been made to relate the new ideas to those that students may already know. Constant reinforcement and connections within problem-solving strategies, data interpretation, geometry, patterns, graphs, and situations from everyday life can help students gradually master both new and old information. In addition, each section begins with a list of objectives covered in the section. Clear organization of section material based on objectives further enhances readability.

Problem-Solving Process This is formally introduced in Chapter 3 with a four-step process that is integrated throughout the text. The four steps are **Understand, Translate, Solve**, and **Interpret**. The repeated use of these steps in a variety of examples shows their wide applicability. Reinforcing the steps can increase students' comfort level and confidence in tackling problems.

Applications and Connections Every effort was made to include as many interesting and relevant real-life applications as possible throughout the text in both worked-out examples and exercise sets. In the Fourth

Edition, the applications have been thoroughly revised and updated, and the number of applications has increased. The applications help to motivate students and strengthen their understanding of mathematics in the real world. They show connections to a wide range of fields including agriculture, allied health, anthropology, art, astronomy, biology, business, chemistry, construction, consumer affairs, earth science, education, entertainment, environmental issues, finance, geography, government, history, medicine, music, nutrition, physics, sports, travel, and weather. Many of the applications are based on recent real data. Sources for data include newspapers, magazines, publicly held companies, government agencies, special-interest groups, research organizations, and reference books. Opportunities for obtaining your own real data are also included. See the Applications Index on page xviii.

Practice Problems   Throughout the text, each worked-out example has a parallel Practice Problem placed next to the example in the margin. These invite students to be actively involved in the learning process before beginning the end-of-section exercise set. Practice Problems immediately reinforce a skill after it is developed. Answers appear at the bottom of the page for quick reference.

Concept Checks   These margin exercises are appropriately placed throughout the text. They allow students to gauge their grasp of an idea as it is being explained in the text. Concept Checks stress conceptual understanding at the point of use and help suppress misconceived notions before they start. Answers appear at the bottom of the page.

Increased Integration of Geometry Concepts   In addition to the traditional topics in prealgebra courses, this text contains a strong emphasis on problem solving and geometric concepts, which are integrated throughout. The geometry concepts presented are those most important to a student's understanding, and I have included many applications and exercises devoted to this topic. These are marked with the the geometry icon △. Also, Chapter 9, Geometry and Measurement, provides a focused treatment of the topics.

Helpful Hints   Helpful Hints contain practical advice on applying mathematical concepts. These are found throughout the text and strategically placed where students are most likely to need immediate reinforcement. Helpful Hints are highlighted for quick reference.

Visual Reinforcement of Concepts   The Fourth Edition contains a wealth of graphics, models, photographs, and illustrations to visually clarify and reinforce concepts. These include new and updated bar graphs, line graphs, calculator screens, application illustrations, and geometric figures.

Calculator Explorations   These optional explorations offer point-of-use intruction, through examples and exercises, on the proper use of scientific calculators as tools in the mathematical problem-solving process. Placed appropriately throughout the text, Calculator Explorations also reinforce concepts learned in the corresponding section.

Additional exercises building on the skill developed in the Explorations may be found in exercise sets throughout the text. Exercises requiring a calculator are marked with the ▦ icon.

Study Skills Reminders   New Study Skills Reminder boxes are integrated throughout the text. They are strategically placed to constantly remind and encourage students as they hone their study skills. A new **Section 1.1**, Tips on Success in Mathematics, provides an overview of the Study Skills needed to succeed in math. These are reinforced by the Study Skills Reminder boxes throughout the text.

Focus On Appropriately placed throughout each chapter, these are divided into Focus on Mathematical Connections, Focus on Business and Career, Focus on the Real World, and Focus on History. They are written to help students develop effective habits for engaging in investigations of other branches of mathematics, understanding the importance of mathematics in various careers and in the world of business, and seeing the relevance of mathematics in both the present and past through critical thinking exercises and group activities.

Chapter Highlights Found at the end of each chapter, these contain key definitions, concepts, and examples to help students understand and retain what they have learned and help them organize their notes and study for tests.

Chapter Activity This feature occurs once per chapter at the end of the chapter, often serving as a chapter wrap-up. For individual or group completion, the Chapter Activity, usually hands-on or data-based, complements and extends to concepts of the chapter, allowing students to make decisions and interpretations and to think and write about algebra.

Integrated Reviews These "midchapter reviews" are appropriately placed once per chapter. Integrated Reviews allows students to review and assimilate the many different skills learned separately over several sections before moving on to related material in the chapter.

Pretests Each chapter begins with a Pretest that is designed to help students identify areas where they need to pay special attention in the upcoming chapter.

Chapter Review and Test The end of each chapter contains a review of topics introduced in the chapter. The Chapter Review offers exercises that are keyed to sections of the chapter. The Chapter Test is a practice test and is not keyed to sections of the chapter. This text is accompanied by the Chapter Test Prep Video CD, which gives students instant access to a video solution to each chapter test question.

Cumulative Review These features are found at the end of each chapter (except Chapter 1). Each odd problem contained in the Cumulative Review is an earlier worked example in the text that is referenced in the back of the book along with the answer. Students who need to see a complete worked-out solution, with explanation, can do so by turning to the appropriate example in the text. The evens are not keyed to examples.

Student Resource Icons At the beginning of each section, videotape and CD, tutorial software, Prentice Hall Tutor Center, and solutions manual icons are displayed. These icons help reinforce that these learning aids are available should students wish to use them to help them review concepts and skills at their own pace. These items have direct correlation to the text and emphasize the text's methods of solution.

SSM
TUTOR CENTER   SG   CD & VIDEO   MATH PRO   WEB

Functional Use of Color and New Design Elements of this text are highlighted with color or design to make it easier for students to read and study. Special care has been taken to use color within solutions to examples or in the art to **help clarify, distinguish, or connect concepts**.

Exercise Sets Each text section ends with an Exercise Set. Each exercise in the set, except those found in parts labeled Review and Preview or Combining Concepts, is keyed to one of the objectives of the section. Wherever possible, a specific example is also referenced. In addition to the approximately 4400 exercises in end-of-section exercise sets, exercises may also be found in the Pretests, Integrated Reviews, Chapter Reviews, Chapter Tests, and Cumulative Reviews.

 Exercises and examples marked with a video icon have been worked out step-by-step by the author in the lecture videos that accompany this text.

Throughout the exercises in the text there is an emphasis on data and graphical interpretation via tables, charts, and graphs. The ability to interpret data and read and create a variety of types of graphs is developed gradually so students become comfortable with it. Geometric concepts—such as perimeter and area—are integrated throughout the text. Exercises and examples marked with a geometry icon have been identified for convenience. In addition, Chapter 9 provides a focused treatment of the topic.

Each exercise set contains one or more of the following features.

Mental Math Found at the beginning of an exercise set, these mental warmups reinforce concepts found in the accompanying section and increase students' confidence before they tackle an exercise set. By relying on their own mental skills, students increase not only their confidence in themselves but also their number sense and estimation ability.

Review and Preview These exercises occur in each exercise set (except for those in Chapter 1) after the exercises keyed to the objectives of the section. Review and Preview problems are keyed to earlier sections and review concepts learned earlier in the text that are needed in the next section or in the next chapter. These exercises show the links between earlier topics and later material.

 Combining Concepts These exercises are found at the end of each exercise set after the Review and Preview exercises. Combining Concepts exercises require students to combine several concepts from that section or to take the concepts of the section a step further by combining them with concepts learned in previous sections. For instance, sometimes students are required to combine the concepts of the section with the problem-solving process they learned in Chapter 1 to try their hand at solving an application problem.

 **Writing Exercises** These exercises occur in almost every exercise set and are marked with an icon. They require students to assimilate information and provide a written response to explain concepts or justify their thinking. Guidelines recommended by the American Mathematical Association of Two Year Colleges (AMATYC) and other professional groups recommend incorporating writing in mathematics courses to reinforce concepts.

Vocabulary Checks Vocabulary Checks, **new to this edition**, provide an opportunity for students to become more familiar with the use of mathematical terms as they strengthen their verbal skills. These appear at the end of the chapter before the Chapter Highlights.

Data and Graphical Interpretation There is an emphasis on data interpretation in exercises via tables and graphs. The ability to interpret data and read and create a variety of types of graphs is developed gradually so students become comfortable with it.

 **Internet Excursions** These exercises occur once per chapter. Internet Excursions require students to use the Internet as a data-collection tool to complete the exercises, allowing students first-hand experience with manipulating and working with real data.

Key Content Features in the Fourth Edition

Overview This new edition retains many of the factors that have contributed to its success. Even so, **every section of the text was carefully reexamined**. Throughout the new edition you will find numerous new applications, examples, and many real-life applications and exercises. Overall, many new applications have been added from the nursing and allied health fields. For example, look at the exercise sets of Sections 6.4, 9.6, or 9.7. Some sections have internal reorganization to better clarify and enhance the presentation.

Chapter 1 now begins with Tips for Success in Mathematics (Section 1.1). New Study Skills Reminder boxes have been inserted throughout the text. These boxes reinforce the tips from Section 1.1. They are placed strategically to encourage students to hone their study skills.

Increased Integration of Geometry Concepts In addition to the traditional topics in prealgebra courses, this text contains a strong emphasis on problem solving, and geometric concepts are integrated throughout. The geometry concepts presented are those most important to a student's understanding, and I have included many **applications and exercises** devoted to this topic. These are marked with a geometry icon. Also, Chapter 9 focuses on geometry and measurement.

New Examples Detailed step-by-step examples were added, deleted, replaced, or updated as needed. Many of these reflect real life.

Exercise Sets Revised and Updated The exercise sets have been carefully examined and extensively revised. The **real-world and real-data applications** have been thoroughly updated and many new applications are included. In addition, an **increased number of challenging problems** have been included in the new edition. **Writing exercises** are now included in most exercise sets and new **Vocabulary Checks** have been added to the end of the chapter to help students become proficient in the language of mathematics.

Additional Content Changes

New Section 9.7 Conversions between the U.S. and Metric Systems.

Section 3.3 now contains more types of equations to solve using the multiplication property of equality.

Section 4.8 now also contains operations on negative mixed numbers.

Section 5.1 now also contains comparing negative decimals.

Section 5.5 provides expanded content with estimation.

Section 6.5 now also contains congruent triangles.

Section 8.1 now also contains histograms.

Cumulative Reviews now include twice as many exercises.

New Student Resource the Chapter Test Prep Video CD. This Video CD gives students instant access to video solutions to each chapter test question in the text. See the supplements description page xv for details. It is also described on p. ix of the Preface, page xxx of the walkthrough and in the Note to Students. If your text does not include the Video CD, it is available through your campus bookstore.

Resources for the Instructor

Printed Supplements

Annotated Instructor's Edition (0-13-144448-4)

- Answers to all exercises printed on the same text page.
- Teaching Tips throughout the text placed at key points in the margin.

Instructor's Solution Manual (0-13-144527-8)

- Solutions to even-numbered section exercises.
- Solutions to every (even and odd) Mental Math exercise.
- Solutions to every (even and odd) Practice Problem (margin exercise).
- Solutions to every (even and odd) exercise found in the Pretests, Integrated Reviews (mid-chapter reviews), Chapter Reviews, Chapter Tests, and Cumulative Reviews.

Instructor's Resource Manual with Tests (0-13-144529-4)

- Notes to the Instructor including suggested homework assignments for each section, guidelines for course pacing, and much more.
- Two free-response Pretests per chapter.
- Eight Chapter Tests per chapter (3 multiple-choice, 5 free-response).
- Two Cumulative Review Tests (one multiple-choice, one free-response) every two chapters (after chapters 2, 4, 6, 8, 10).
- Eight Final Exams (4 multiple-choice, 4 free-response).
- Twenty additional exercises per section for added test exercises if needed.
- New Additional examples for each section for use in the classroom.
- Group Activities (an average of two per chapter; providing short group activities in a convenient, ready-to-use format).
- Answers to all items.

Media Supplements

TestGen with QuizMaster CD-ROM (Windows/Macintosh) (0-13-144533-2)

- Algorithmically driven, text-specific testing program.
- Networkable for administering tests and capturing grades on-line.
- Edit and add your own questions to create a nearly unlimited number of tests and worksheets.
- Use the new "Function Plotter" to create graphs.
- Tests can be easily exported to HTML so they can be posted to the Web for student practice.
- Includes an email function for network users, enabling instructors to send a message to a specific student or an entire group.
- Network-based reports and summaries for a class or student and for cumulative or selected scores are available.

MathPro Explorer 4.0

- Network Version IBM/Mac 0-13-144532-4
- Enables instructors to create either customized or algorithmically generated practice quizzes from any section of a chapter.

- Includes e-mail function for network users, enabling instructors to send a message to a specific student or to an entire group.
- Network-based reports and summaries for a class or student and for cumulative or selected scores.

Instructor's CD Series

- Written and presented by K. Elayn Martin-Gay.
- Contains suggestions for presenting course material, utilizing the integrated resource package, time-saving tips, and much more.

Resources for the Student

Printed Supplements

Student's Solution Manual 0-13-144539-1

- Solutions to odd-numbered section exercises.
- Solutions to every (even and odd) Mental Math exercise.
- Solutions to every (even and odd) Practice Problem (margin exercise).
- Solutions to every (even and odd) exercise found in the Pretests, Integrated Reviews (mid-chapter reviews), Chapter Reviews, Chapter Tests, and Cumulative Reviews.

Student's Study Guide: 0-13-144538-3

- Additional step by step worked out examples and exercises
- Practice Tests and Final Exam
- Includes study skill and note taking suggestions
- Includes hints and warnings section
- Solutions to the exercises, tests and final exams contained in this study guide.

Media Supplements

- New **Chapter Test Prep Video CD** provides students with instant access to video solutions for each of the chapter test questions contained in the text. The Video CD is packaged in each new student text. It is also available for purchase through your campus bookstore.
- The Video CD's easy navigation is designed to help students study efficiently. Students select the exact chapter test questions they missed or need help with and instantly view the solution worked by K. Elayn Martin-Gay.
- Start making the most of your study time today. Turn to the back of the text to access the Video CD. You'll also find a Note to the Students with steps to success for studying for tests and using the video.

MathPro 4.0 Explorer Student Version 0-13-144541-3

- Available on CD-ROM for stand alone use or can be networked in the school laboratory.
- Text specific tutorial exercises and instructions at the objective level.
- Algorithmically generated Practice Problems.
- "Watch" screen videoclips by K. Elayn Martin-Gay.

Videotape Series 0-13-144531-6

■ Written and presented by K. Elayn Martin-Gay.

■ Keyed to each section of the text.

■ Step-by-step solutions to exercises from each section of the text. Exercises that are worked in the videos are marked with a video icon.

Digitized Lecture Videos on CD-ROM 0-13-144534-0

■ The entire set of *Prealgebra, Fourth Edition* lecture videotapes in digital form.

■ Convenient access anytime to video tutorial support from a computer at home or on campus.

■ Available shrinkwrapped with the text or stand-alone.

SSM
TUTOR CENTER

Prentice Hall Tutor Center

■ Staffed with qualified math instructors and open 5 days a week, 7 hours per day.

■ Obtain help for examples and exercises in Martin-Gay, *Prealgebra, Fourth Edition*, via toll-free telephone, fax, or e-mail.

■ The Prentice Hall Tutor Center is accessed through a registration number that may be bundled with a new text or purchased separately with a used book. Visit http://www.prenhall.com/tutorcenter to learn more.

Acknowledgments

First, as usual, I would like to thank my husband, Clayton, for his constant encouragement. I would also like to thank my children, Eric and Bryan, for their sense of humor and especially for asking Dad to cook the bacon that I always used to burn.

I would also like to thank my extended family for their invaluable help and also their sense of humor. Their contributions are too numerous to list. They are Rod, Karen, and Adam Pasch; Michael, Christopher, Matthew, and Jessica Callac; Stuart, Earline, Melissa, Mandy, Bailey, and Ethan Martin; Mark, Sabrina, and Madison Martin; Leo and Barbara Miller; and Jewett Gay.

I would like to thank the following reviewers for their input and suggestions:

Reviewers

Bob Angus, Northeastern University
Nanci Barker, Caroll Community College
Rob Dengler, Southeastern Community College
Kathryn Gunderson, Three Rivers Community College
Kelly Houlton, University of Alaska–Fairbanks
Pat Kline, Southwestern Community College
Rose Turn, Westchester Community College
Aretha Vernette, Indian River Community College
Kinley Alston, Trident Technical College
John Close, Salt Lake Community College
Kay Haralson, Austin Peay State Community College
Barbara Hughes, San Jacinto Community College

Karen Pagel, Dona Ana Branch
Jonathan Weissman, Essex Community College

There were many people who helped me develop this text and I will attempt to thank some of them here. Ellen Sawyer and Cindy Trimble were invaluable for contributing to the overall accuracy of this text. Elka Block, Chris Callac, and Miriam Daunis were invaluable for their many suggestions and contributions during the development and writing of this fourth edition. Lynn Savino Wendel provided guidance throughout the production process. I thank Richard Semmler and Carrie Green for all their work on the solutions, text, and accuracy.

Sadly, executive editor of this project, Karin Wagner, passed away this year. She will be dearly remembered by me and the rest of the staff at Prentice Hall for her integrity, wisdom, commitment, and especially her sense of humor and infectious laugh.

A very special thank you to my project manager, Mary Beckwith, for taking over during a difficult period for all of us.

Lastly, my thanks to the staff at Prentice Hall for all their support: Linda Behrens, Mike Bell, Patty Burns, Tom Benfatti, Paul Belfanti, Geoffrey Cassar, Eilish Main, Patrice Jones, John Tweeddale, Chris Hoag, Paul Corey, and Tim Bozik.

K. Elayn Martin-Gay

About the Author

K. Elayn Martin-Gay has taught mathematics at the University of New Orleans for more than 20 years and has received numerous teaching awards including the local University Alumni Association's Award for Excellence in Teaching.

Over the years, Elayn has developed a videotaped lecture series to help her students understand algebra better. This highly successful video material is the basis for her books: *Basic College Mathematics*, Second Edition; *Prealgebra*, Fourth Edition: *Introductory Algebra*, Second Edition; *Intermediate Algebra*, Second Edition; *Algebra A Combined Approach*, Second Edition; and her hardback series: *Beginning Algebra*, Fourth Edition; *Intermediate Algebra*, Fourth Edition; *Beginning and Intermediate Algebra*, Third Edition; and *Intermediate Algebra: A Graphing Approach*, Third Edition.

APPLICATIONS INDEX

HIGHLIGHTS OF PREALGEBRA, FOURTH EDITION

Prealgebra, Fourth Edition is the primary learning tool in a fully integrated learning package to help you succeed in this course. Author K. Elayn Martin-Gay focuses on enhancing the traditional emphasis of mastering the basics with innovative pedagogy and a meaningful learning program. There are three goals that drive her authorship:

▲ **Master and apply skills and concepts**

▲ **Build confidence**

▲ **Increase motivation**

Take a few moments now to examine some of the features that have been incorporated into *Prealgebra, Fourth Edition* to help students excel.

Solving Equations and Problem Solving

CHAPTER 3

Throughout this text, we have been making the transition from arithmetic to algebra. We have said that in algebra letters called variables represent numbers. Using variables is a very powerful method for solving problems that cannot be solved with arithmetic alone. This chapter introduces operations on algebraic expressions and solving variable equations.

Americans are in love with traveling the open roads. Throughout the country, roads and highways lead travelers on an exciting journey through every period of U.S. history. The *Santa Fe Trail Scenic and Historic Byway* in Colorado retraces the famous road taken by wagon trains of the Old West. The *Seward Highway* in Alaska is named for the U.S. Secretary of State who arranged the purchase of Alaska from Russia in 1867. The *Selma to Montgomery March Byway* in Alabama traces the historic route taken by Dr. Martin Luther King, Jr. in 1965. Whether by scenic byways or interstate highways, touring the country by automobile remains a favorite pastime of vacationers today. Before you venture out on the open road, check out Exercises 107–110 on page 198 and see how we can use equations to relate the distance, rate, and time of an automobile trip.

169

Page 169

◀ **REAL-WORLD APPLICATIONS**

Chapter-opening real-world applications introduce you to everyday situations that are applicable to the mathematics you will learn in the upcoming chapter, showing the relevance of mathematics in daily life.

xxiv

Become a Confident Problem Solver

A goal of this text is to help you develop problem-solving abilities.

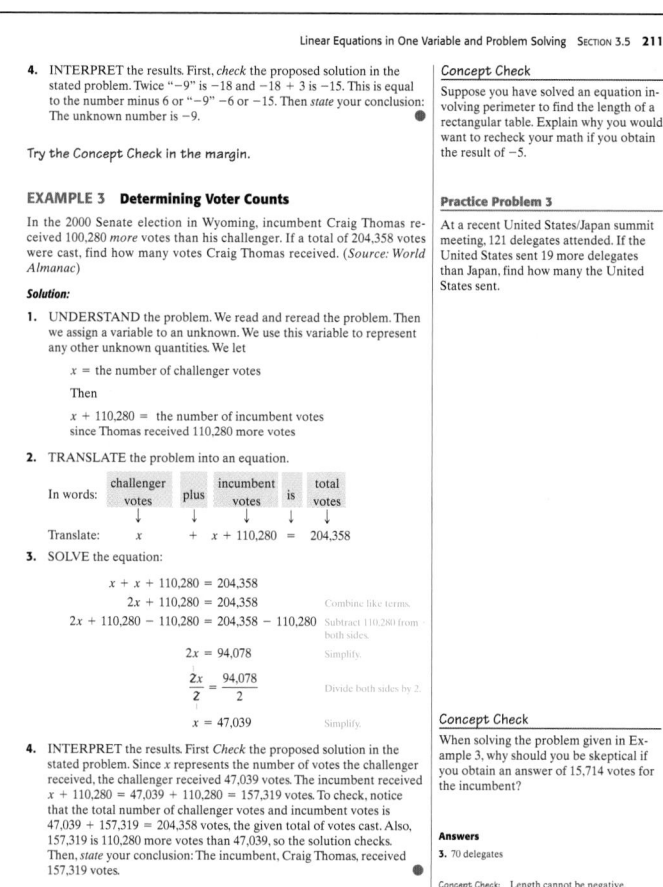

4. INTERPRET the results. First, *check* the proposed solution in the stated problem. Twice "−9" is −18 and −18 + 3 is −15. This is equal to the number minus 6 or "−9" −6 or −15. Then *state* your conclusion: The unknown number is −9. ●

Try the *Concept Check* in the margin.

EXAMPLE 3   Determining Voter Counts

In the 2000 Senate election in Wyoming, incumbent Craig Thomas received 100,280 *more* votes than his challenger. If a total of 204,358 votes were cast, find how many votes Craig Thomas received. (*Source: World Almanac*)

Solution:

1. UNDERSTAND the problem. We read and reread the problem. Then we assign a variable to an unknown. We use this variable to represent any other unknown quantities. We let

x = the number of challenger votes

Then

$x + 110,280$ = the number of incumbent votes
since Thomas received 110,280 more votes

2. TRANSLATE the problem into an equation.

In words: challenger votes | plus | incumbent votes | is | total votes
Translate: x + $x + 110,280$ = $204,358$

3. SOLVE the equation:

$$x + x + 110,280 = 204,358$$
$$2x + 110,280 = 204,358 \quad \text{Combine like terms.}$$
$$2x + 110,280 - 110,280 = 204,358 - 110,280 \quad \text{Subtract 110,280 from both sides.}$$
$$2x = 94,078 \quad \text{Simplify.}$$
$$\frac{2x}{2} = \frac{94,078}{2} \quad \text{Divide both sides by 2.}$$
$$x = 47,039 \quad \text{Simplify.}$$

4. INTERPRET the results. First *Check* the proposed solution in the stated problem. Since x represents the number of votes the challenger received, the challenger received 47,039 votes. The incumbent received $x + 110,280 = 47,039 + 110,280 = 157,319$ votes. To check, notice that the total number of challenger votes and incumbent votes is $47,039 + 157,319 = 204,358$ votes, the given total of votes cast. Also, 157,319 is 110,280 more votes than 47,039, so the solution checks. Then, *state* your conclusion: The incumbent, Craig Thomas, received 157,319 votes. ●

Try the *Concept Check* in the margin.

Concept Check

Suppose you have solved an equation involving perimeter to find the length of a rectangular table. Explain why you would want to recheck your math if you obtain the result of −5.

Practice Problem 3

At a recent United States/Japan summit meeting, 121 delegates attended. If the United States sent 19 more delegates than Japan, find how many the United States sent.

Concept Check

When solving the problem given in Example 3, why should you be skeptical if you obtain an answer of 15,714 votes for the incumbent?

Answers

3. 70 delegates

Concept Check: Length cannot be negative.

Concept Check: Answers may vary.

Page 211

◀ **GENERAL STRATEGY FOR PROBLEM-SOLVING**

Save time by having a plan. This text's organization can help you. Note the outlined problem-solving steps, *Understand, Translate, Solve,* and *Interpret.*

Problem solving is introduced early, emphasized and integrated throughout the book. The author provides patient explanations and illustrates how to apply the problem-solving procedure to the in-text examples.

GEOMETRY ▶

Geometric concepts are integrated throughout the text. Examples and exercises involving geometric concepts are identified with a triangle icon △. This text includes chapter 9 on Geometry and Measurement as well as dedicated Appendices.

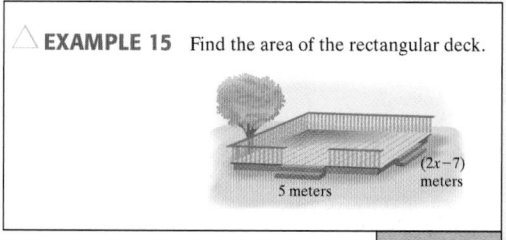

△ **EXAMPLE 15**   Find the area of the rectangular deck.

$(2x−7)$ meters
5 meters

Page 175

Master and Apply Basic Skills and Concepts

K. Elayn Martin-Gay provides thorough explanations of key concepts and enlivens the content by integrating successful and innovative pedagogy. *Prealgebra, Fourth Edition* integrates skill building throughout the text and provides problem-solving strategies and hints along the way. These features have been included to enhance your understanding of algebraic concepts.

◄ CONCEPT CHECKS

Concept Checks are special margin exercises found in most sections. Work these to help gauge your grasp of the concept being developed in the text.

Concept Check

What number should be added to or subtracted from both sides of the equation in order to solve the equation $-3 = y + 2$?

Page 184

COMBINING CONCEPTS ▶

Combining Concepts exercises are found at the end of each exercise set. Solving these exercises will expose you to the way mathematical ideas build upon each other.

Page 78

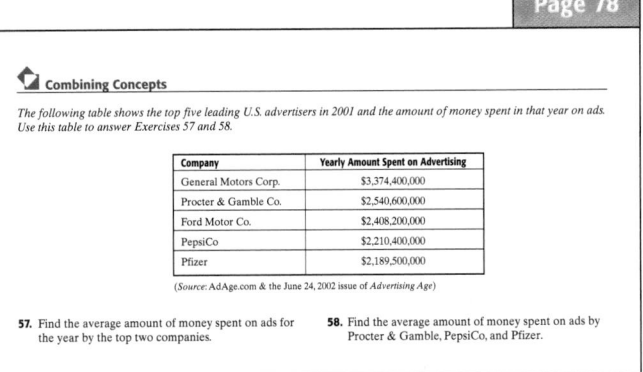

Combining Concepts

The following table shows the top five leading U.S. advertisers in 2001 and the amount of money spent in that year on ads. Use this table to answer Exercises 57 and 58.

Company	Yearly Amount Spent on Advertising
General Motors Corp.	$3,374,400,000
Procter & Gamble Co.	$2,540,600,000
Ford Motor Co.	$2,408,200,000
PepsiCo	$2,210,400,000
Pfizer	$2,189,500,000

(*Source*: AdAge.com & the June 24, 2002 issue of *Advertising Age*)

57. Find the average amount of money spent on ads for the year by the top two companies.

58. Find the average amount of money spent on ads by Procter & Gamble, PepsiCo, and Pfizer.

◄ PRACTICE PROBLEMS

Practice Problems occur in the margins next to every Example. Work these problems after an example to immediately reinforce your understanding.

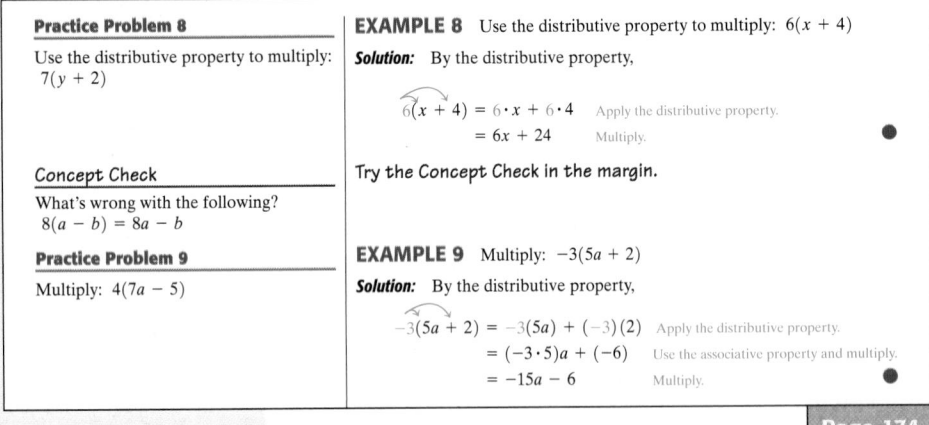

Practice Problem 8

Use the distributive property to multiply:
$7(y + 2)$

Concept Check

What's wrong with the following?
$8(a - b) = 8a - b$

Practice Problem 9

Multiply: $4(7a - 5)$

EXAMPLE 8 Use the distributive property to multiply: $6(x + 4)$

Solution: By the distributive property,

$$6(x + 4) = 6 \cdot x + 6 \cdot 4 \quad \text{Apply the distributive property.}$$
$$= 6x + 24 \quad \text{Multiply.} \qquad \bullet$$

Try the Concept Check in the margin.

EXAMPLE 9 Multiply: $-3(5a + 2)$

Solution: By the distributive property,

$$-3(5a + 2) = -3(5a) + (-3)(2) \quad \text{Apply the distributive property.}$$
$$= (-3 \cdot 5)a + (-6) \quad \text{Use the associative property and multiply.}$$
$$= -15a - 6 \quad \text{Multiply.} \qquad \bullet$$

Page 174

Page 207

WRITING EXERCISES ▶

Writing Exercises, marked by an icon, ✎ are found in most practice sets.

85. A classmate tries to solve $3x = 39$ by subtracting 3 from both sides of the equation. Will this step solve the equation for x? Why or why not?

Test Yourself and Check Your Understanding

Good exercise sets and an abundance of worked-out examples are essential for building student confidence. The exercises you will find in this worktext are intended to help you build skills and understand concepts as well as motivate and challenge you. In addition, features like Chapter Highlights, Chapter Reviews, Chapter Tests, and Cumulative Reviews are found at the end of each chapter to help you study and organize your notes.

Chapter **3** Pretest

Simplify.
1. $9x - 4 + 6x + 8$ **2.** $3(2x - 1) - (x - 8)$

Multiply.
3. $8(7b)$ **4.** $-5(2y - 7)$

△ **5.** Find the perimeter of the triangle.   △ **6.** Find the area of the rectangle.

Page 170

◀ PRETESTS

Pretests open each chapter. Take a Pretest to evaluate where you need the most help before beginning a new chapter.

Page 199

INTEGRATED REVIEWS ▶

Integrated Reviews serve as mid-chapter reviews and help you to assimilate the new skills you have learned separately over several sections.

Name _____ Section _____ Date _____

Integrated Review—Expressions and Equations

Simplify each expression by combining like terms.
1. $7x + x$ **2.** $6y - 10y$

3. $2a + 5a - 9a - 2$ **4.** $6a - 12 - a - 14$

Review and Preview

Perform each indicated operation. See Sections 2.2 and 2.3.

81. $-13 + 10$ **82.** $-7 - (-4)$ **83.** $-4 - (-12)$

84. $-15 + 23$ **85.** $-4 + 4$ **86.** $8 + (-8)$

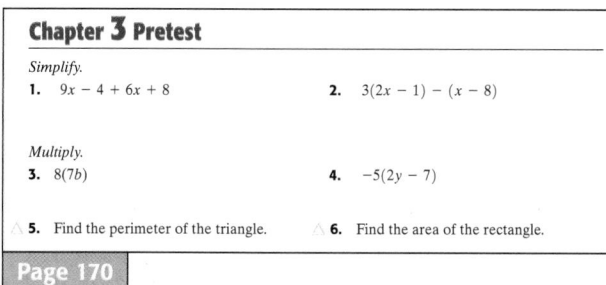 **Combining Concepts**

Simplify.

87. $9684q - 686 - 4860q + 12,960$ **88**

Page 180

◀ REVIEW AND PREVIEW

Review and Preview exercises review concepts learned earlier in the text that are needed in the next section or chapters.

Page 157

CHAPTER HIGHLIGHTS ▶

Found at the end of every chapter, the Chapter Highlights contain key definitions, concepts, and examples to help students understand and retain what they have learned.

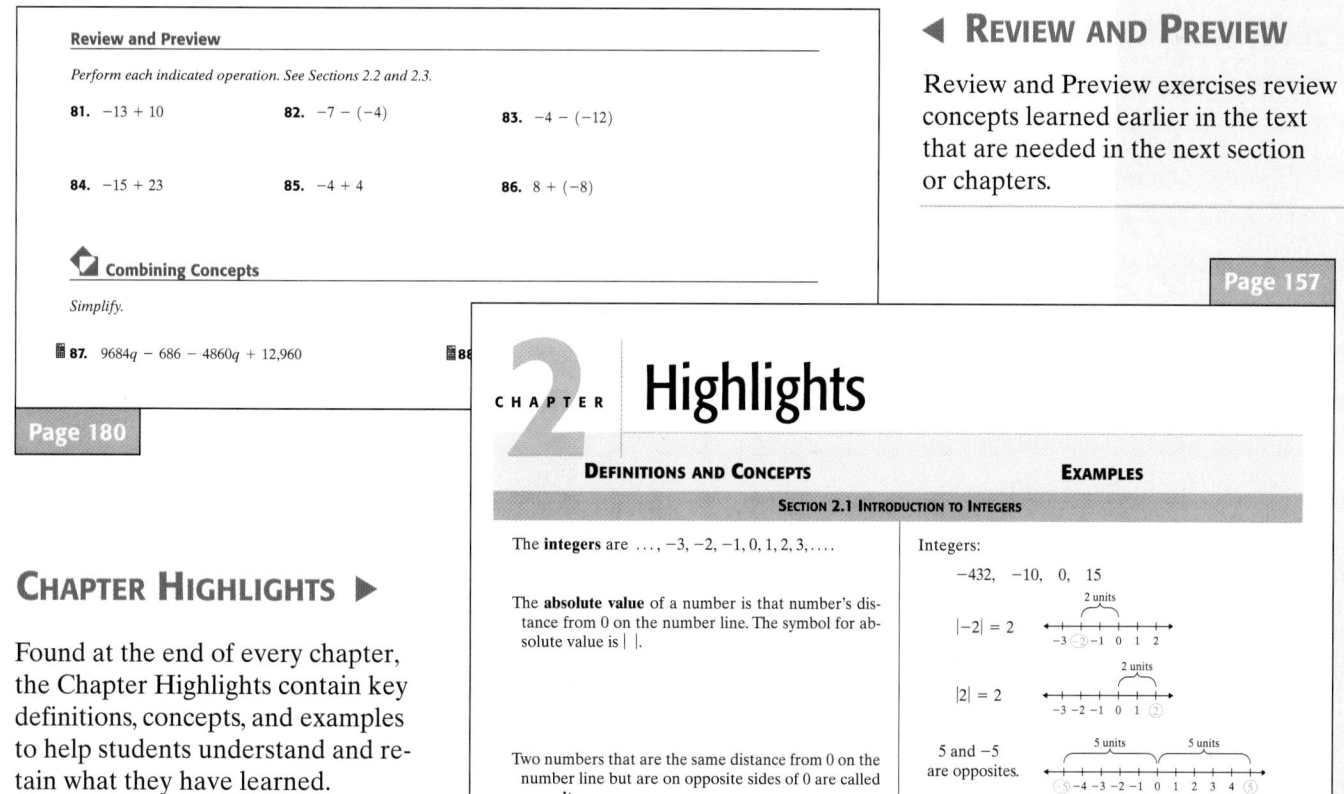

CHAPTER **2** Highlights

DEFINITIONS AND CONCEPTS	EXAMPLES
SECTION 2.1 INTRODUCTION TO INTEGERS	
The **integers** are $\ldots, -3, -2, -1, 0, 1, 2, 3, \ldots$.	Integers: $-432, \quad -10, \quad 0, \quad 15$
The **absolute value** of a number is that number's distance from 0 on the number line. The symbol for absolute value is $|\ \ |$.	$|-2| = 2$
	$|2| = 2$
Two numbers that are the same distance from 0 on the number line but are on opposite sides of 0 are called **opposites**.	5 and -5 are opposites.

Increase Motivation

Throughout *Prealgebra, Fourth Edition*, K. Elayn Martin-Gay provides interesting real-world applications to strengthen your understanding of the relevance of math in everyday life. When a new topic is presented, an effort has been made to relate the new ideas to those that students may already know. The Fourth Edition increases emphasis on visualization to clarify and reinforce key concepts.

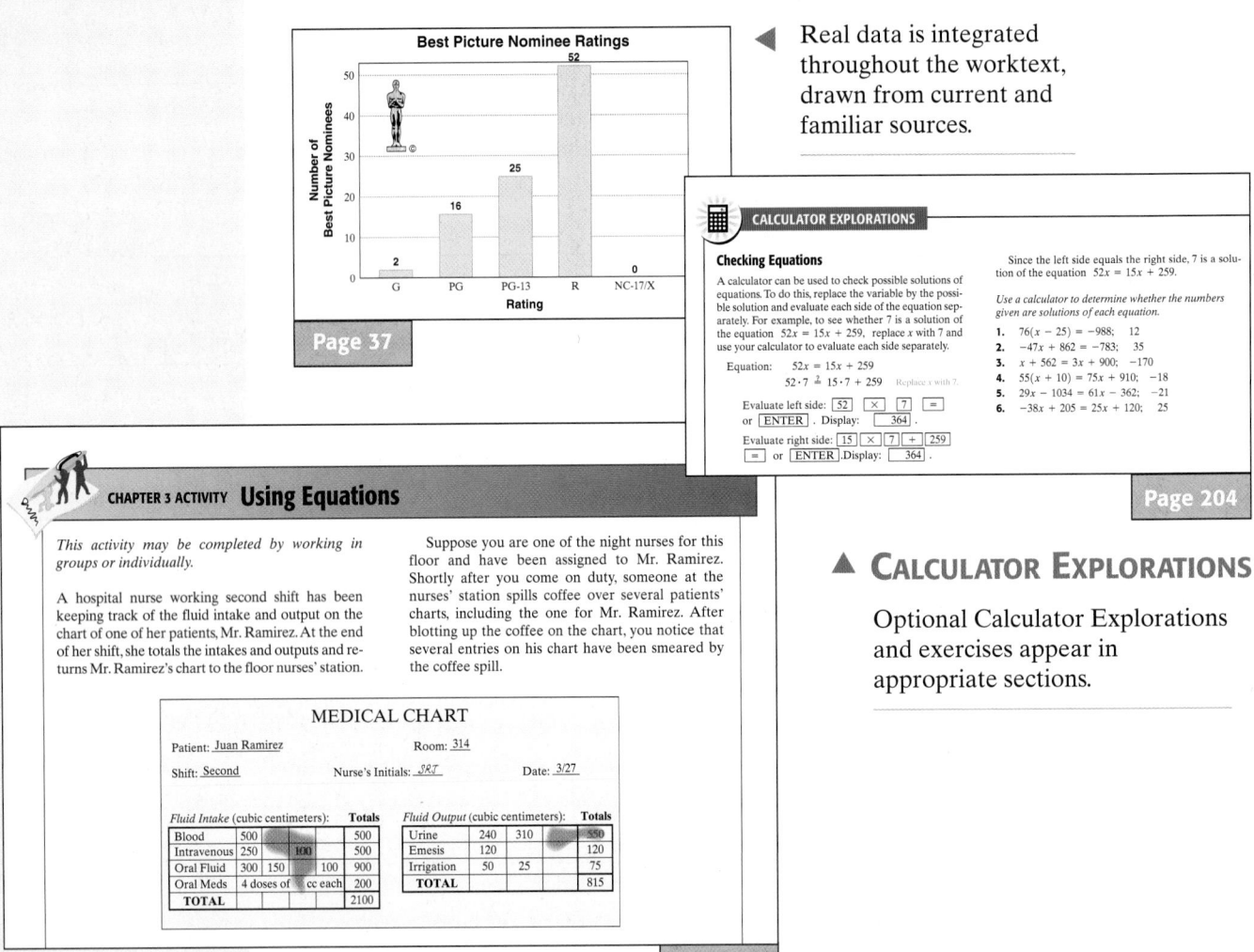

◀ Real data is integrated throughout the worktext, drawn from current and familiar sources.

Page 37

Page 204

▲ **CALCULATOR EXPLORATIONS**

Optional Calculator Explorations and exercises appear in appropriate sections.

Page 218

▲ Graphics, models, and illustrations provide visual reinforcement.

FOCUS ON BOXES ▶

Focus On boxes found throughout each chapter help you see the relevance of math through critical-thinking exercises and group activities. Try these on your own or with a classmate. Focus On covers the areas of: History, Mathematical Connections, Real World, and Business and Career.

Page 148

Build Confidence

Several features of this text can be helpful in building your confidence and mathematical competence. As you study, also notice the connections the author makes to relate new material to ideas that you may already know.

1.1 Tips for Success in Mathematics

Before reading this section, remember that your instructor is your best source for information. Please see your instructor for any additional help or information.

A Getting Ready for This Course

Now that you have decided to take this course, remember that a *positive attitude* will make all the difference in the world. Your belief that yo̶ ceed is just as important as your commitment to this course. Make

Page 3

◀ **TIPS FOR SUCCESS**

New coverage of study skills in Section 1.1 reinforces this important component to success in this course.

Page 148

STUDY SKILLS REMINDER ▶

New Study Skills Reminders are integrated throughout the book to reinforce section 1.1 and encourage the development of strong study skills.

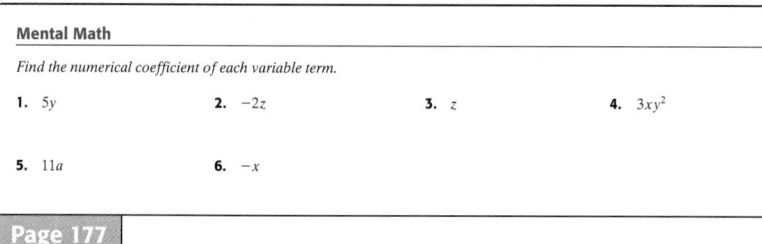

STUDY SKILLS REMINDER

How are your homework assignments going?

It is so important in mathematics to keep up with homework. Why? Many concepts build on each other. Often, your understanding of a day's lecture in mathematics depends on an understanding of the previous day's material.

Remember that completing your homework assignment involves a lot more than attempting a few of the problems assigned.

To complete a homework assignment, remember these four things:

1. Attempt all of it.   3. Correct it.
2. Check it.   4. If needed, ask questions about it.

Mental Math

Find the numerical coefficient of each variable term.

1. $5y$ 2. $-2z$ 3. z 4. $3xy^2$

5. $11a$ 6. $-x$

Page 177

◀ **MENTAL MATH**

Mental Math warm-up exercises reinforce concepts found in the accompanying section and can increase your confidence before beginning an exercise set.

HELPFUL HINTS ▶

Found throughout the text, these contain practical advice on applying mathematical concepts. They are strategically placed where you are most likely to need immediate reinforcement.

Page 113

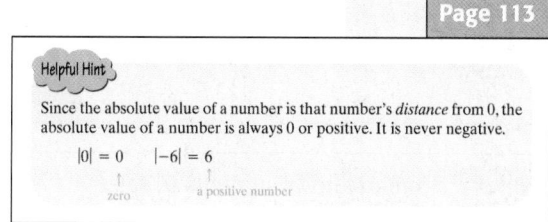

Helpful Hint

Since the absolute value of a number is that number's *distance* from 0, the absolute value of a number is always 0 or positive. It is never negative.

$$|0| = 0 \qquad |-6| = 6$$
$$\quad\uparrow \qquad\qquad \uparrow$$
$$\text{zero} \qquad \text{a positive number}$$

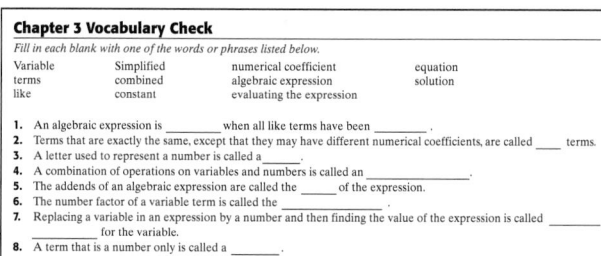

Chapter 3 Vocabulary Check

Fill in each blank with one of the words or phrases listed below.

Variable	Simplified	numerical coefficient	equation
terms	combined	algebraic expression	solution
like	constant	evaluating the expression	

1. An algebraic expression is _____ when all like terms have been _____ .
2. Terms that are exactly the same, except that they may have different numerical coefficients, are called _____ terms.
3. A letter used to represent a number is called a _____ .
4. A combination of operations on variables and numbers is called an _____ .
5. The addends of an algebraic expression are called the _____ of the expression.
6. The number factor of a variable term is called the _____ .
7. Replacing a variable in an expression by a number and then finding the value of the expression is called _____ _____ for the variable.
8. A term that is a number only is called a _____ .
9. An _____ is of the form expression = expression.

Page 219

◀ **VOCABULARY CHECKS**

New Vocabulary Checks allow you to write your answers to questions about chapter content and strengthen verbal skills.

Enrich Your Learning

Good study skills are essential for success in mathematics. Use the student resources that accompany *Prealgebra, Fourth Edition* to make the most of your valuable study time.

New—Chapter Test Prep Video CD provides students with instant access to video solutions for each of the chapter test questions contained in the text.

Step-by-step solutions, presented by the textbook author, are included for every problem on the Chapter Tests in *Prealgebra, Fourth Edition*. To make the most of this resource, when studying for a test, follow these three steps:

1. Take the Chapter Test found at the end of the chapter.

2. Check your answers in the back of the text.

3. Use this video to review *every step* of the worked out solution for those specific questions you answered in correctly or need help with

The Video CD's easy navigation is designed to help students study efficiently. Students select the exact chapter test questions they missed or need help with and instantly view the solution worked by K. Elayn Martin-Gay.

Previous and Next buttons allow for easy navigation within tests.

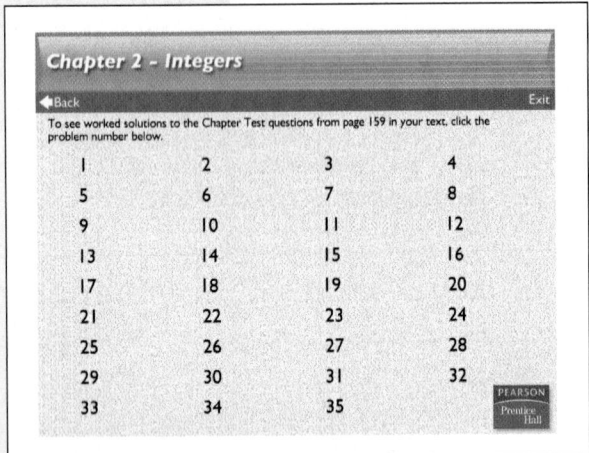

Chapter test menu allows students to select just those test questions they wish to view.

Rewind, Pause, and Play buttons let students view the solutions at their own pace.

The Video CD includes an Introduction with instructions on how to use the Chapter Test Prep Video CD as well as test taking study skills tips. To begin preparing for your test today, see the Video CD and Note to Students in the back of your text. The Video CD is also available through the campus bookstore.

Also Available

- Math Pro 5 and Math Pro 4 tutorial software
- Student Solutions Manual
- Study Guide
- CD Lecture Series - For complete instruction, see the text specific videos available on CD. They are hosted by K. Elayn Martin-Gay and cover each objective in every chapter section of your text.

Prentice Hall Tutor Center

◀ Prentice Hall Tutor Center provides text-specific tutoring via phone, fax, and e-mail. Visit
http://prenhall.com/tutorcenter
for details.

Whole Numbers and Introduction to Algebra

Mathematics is an important tool for everyday life. Knowing basic mathematical skills can help simplify tasks such as creating a monthly budget. Whole numbers are the basic building blocks of mathematics. The whole numbers answer the question "How many?"

This chapter covers basic operations on whole numbers. Knowledge of these operations provides a good foundation on which to build further mathematical skills.

Yosemite National Park was established on October 1, 1890. With over 4 million visitors annually, it is a favorite tourist destination in the Sierra Nevada Mountains in central California. Its nearly 750,000 acres are home to many of nature's most beautiful sites, including rock formations, giant sequoias, and waterfalls. In Exercise 59 on page 30, we will see how whole numbers can be used to measure the height of Yosemite Falls, the highest waterfall in the United States.

Answers
1. _____
2. _____
3. _____
4. _____
5. _____
6. _____
7. _____
8. _____
9. _____
10. _____
11. _____
12. _____
13. _____
14. _____
15. _____
16. _____
17. _____
18. _____
19. _____
20. _____
21. _____
22. _____

Chapter **1** Pretest

1. Determine the place value of the digit 7 in the whole number 5732.

2. Write the whole number 23,490 in words.

3. Add: $58 + 29$

4. Multiply: $\begin{array}{r} 413 \\ \times\ \ 9 \\ \hline \end{array}$

Subtract. Check by adding.

5. $\begin{array}{r} 857 \\ -231 \\ \hline \end{array}$

6. $\begin{array}{r} 51 \\ -19 \\ \hline \end{array}$

Solve.

7. Karen Lewis is reading a 329-page novel. If she has just finished reading page 193, how many more pages must she read to finish the novel?

8. Round 9045 to the nearest ten.

9. Round each number to the nearest hundred to find an estimated sum.
$\begin{array}{r} 382 \\ 436 \\ 2084 \\ +\ \ 176 \\ \hline \end{array}$

10. Use the distributive property to rewrite the following expression: $9(3 + 11)$

△ **11.** Find the perimeter of the following figure:

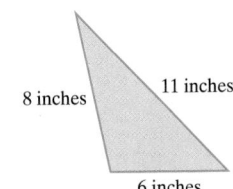

8 inches   11 inches

6 inches

△ **12.** Find the area of the following rectangle:

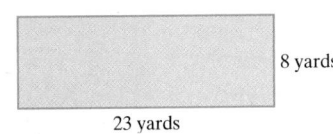

8 yards

23 yards

13. The seats in the history lecture hall are arranged in 32 rows with 18 seats in each row. Find how many seats are in this room.

Divide and then check by multiplying.

14. $2187 \div 9$

15. $\dfrac{5361}{12}$

Solve.

16. Find the average of the following list of numbers: $29, 36, 84, 41, 6, 12, 65$

17. Paul Crandall left $29,640 in his will to go to his three favorite nephews. If each boy was to receive the same amount of money, how much did each boy receive?

18. Write $9 \cdot 9 \cdot 9 \cdot 9 \cdot 9 \cdot 9 \cdot 9$ using exponential notation.

19. Evaluate: 7^4.

20. Simplify: $36 + 18 \div 6$

21. Evaluate
$$\dfrac{2x^2 + 3}{y} \text{ for } x = 3 \text{ and } y = 7.$$

22. Write the phrase as a variable expression. Use x to represent "a number." The sum of twice a number and 6

1.1   Tips for Success in Mathematics

Before reading this section, remember that your instructor is your best source for information. Please see your instructor for any additional help or information.

OBJECTIVES

(A) Get ready for this course.

(B) Understand some general tips for success.

(C) Understand how to use this text.

(D) Get help as soon as you need it.

(E) Learn how to prepare for and take a test.

(F) Develop good time management.

SSM TUTOR CENTER   SG   CD & VIDEO   MATH PRO   WEB

(A) Getting Ready for This Course

Now that you have decided to take this course, remember that a *positive attitude* will make all the difference in the world. Your belief that you can succeed is just as important as your commitment to this course. Make sure that you are ready for this course by having the time and positive attitude that it takes to succeed.

Next, make sure that you have scheduled your math course at a time that will give you the best chance for success. For example, if you are also working, you may want to check with your employer to make sure that your work hours will not conflict with your course schedule. Also, schedule your class during a time of day when you are more attentive and do your best work.

On the day of your first class period, double-check your schedule and allow yourself extra time to arrive in case of traffic problems or difficulty locating your classroom. Make sure that you bring at least your textbook, paper, and a writing instrument. Are you required to have a lab manual, graph paper, calculator, or other supplies besides this text? If so, also bring this material with you.

(B) General Tips for Success

Below are some general tips that will increase your chance for success in a mathematics class. Many of these tips will also help you in other courses you may be taking.

Exchange names and phone numbers with at least one other person in class. This contact person can be a great help if you miss an assignment or want to discuss math concepts or exercises that you find difficult.

Choose to attend all class periods and be on time. If possible, sit near the front of the classroom. This way, you will see and hear the presentation better. It may also be easier for you to participate in classroom activities.

Do your homework. You've probably heard the phrase "practice makes perfect" in relation to music and sports. It also applies to mathematics. You will find that the more time you spend solving mathematics problems, the easier the process becomes. Be sure to schedule enough time to complete your assignments before the next class period.

Check your work. Review the steps you made while working a problem. Learn to check your answers in the original problems. You may also compare your answers with the answers to selected exercises section in the back of the book. If you have made a mistake, try to figure out what went wrong. Then correct your mistake. If you can't find what went wrong don't erase your work or throw it away. Bring your work to your instructor, a tutor in a math lab, or a classmate. It is easier for someone to find where you had trouble if they look at your original work.

Learn from your mistakes and be patient with yourself. Everyone, even your instructor, makes mistakes. Use your errors to learn and to become a better math student. The key is finding and understanding your errors.

Was your mistake a careless one, or did you make it because you can't read your own math writing? If so, try to work more slowly or write more neatly and make a conscious effort to carefully check your work.

Did you make a mistake because you don't understand a concept? Take the time to review the concept or ask questions to better understand it.

Did you skip too many steps? Skipping steps or trying to do too many steps mentally may lead to preventable mistakes.

Know how to get help if you need it. It's all right to ask for help. In fact, it's a good idea to ask for help whenever there is something that you don't understand. Make sure you know when your instructor has office hours and how to find his or her office. Find out whether math tutoring services are available on your campus. Check out the hours, location, and requirements of the tutoring service. Videotapes and software are available with this text. Learn how to access these resources.

Organize your class materials, including homework assignments, graded quizzes and tests, and notes from your class or lab. All of these items will make valuable references throughout your course especially when studying for upcoming tests and the final exam. Make sure that you can locate these materials when you need them.

Read your textbook before class. Reading a mathematics textbook is unlike leisure reading such as reading a book or newspaper. Your pace will be much slower. It is helpful to have paper and a pencil with you when you read. Try to work out examples on your own as you encounter them in your text. You should also write down any questions that you want to ask in class. When you read a mathematics textbook, some of the information in a section may be unclear. But after you hear a lecture or watch a videotape on that section, you will understand it much more easily than if you had not read your text beforehand.

Don't be afraid to ask questions. Instructors are not mind readers. Many times we do not know a concept is unclear until a student asks a question. You are not the only person in class with questions. Other students are normally grateful that someone has spoken up.

Hand in assignments on time. This way you can be sure that you will not lose points for being late. Show every step of a problem and be neat and organized. Also be sure that you understand which problems are assigned for homework. You can always double-check the assignment with another student in your class.

C Using This Text

There are many helpful resources that are available to you in this text. It is important that you become familiar with and use these resources. They should increase your chances for success in this course.

- Each example in every section has a parallel Practice Problem. As you read a section, try each Practice Problem after you've finished the corresponding example. This "learn-by-doing" approach will help you grasp ideas before you move on to other concepts.

- The main section of exercises in each exercise set is referenced by an objective, such as Ⓐ or Ⓑ, and also via examples. Use this referencing if you have trouble completing an assignment from the exercise set.

- If you need extra help in a particular section, look at the beginning of the section to see what videotapes and software are available.

- Make sure that you understand the meaning of the icons that are beside many exercises. The video icon 📼 tells you that the corresponding exercise may be viewed on the videotape that corresponds to that section. The pencil icon ✎ tells you that this exercise is a writing exercise in which you should answer in complete sentences. The △ icon tells you that the exercise involves geometry.

- Integrated Reviews in each chapter offer you a chance to practice—in one place—the many concepts that you have learned separately over several sections.

- There are many opportunities at the end of each chapter to help you understand the concepts of the chapter.

  Chapter Highlights contain chapter summaries and examples.
  Chapter Reviews contain review problems organized by section.
  Chapter Tests are sample tests to help you prepare for an exam.
  Cumulative Reviews are reviews consisting of material from the beginning of the book to the end of that particular chapter.

See the preface at the beginning of this text for a more thorough explanation of the features of this text.

D Getting Help

If you have trouble completing assignments or understanding the mathematics, get help as soon as you need it! This tip is presented as an objective on its own because it is so important. In mathematics, usually the material presented in one section builds on your understanding of the previous section. This means that if you don't understand the concepts covered during a class period, there is a good chance that you will not understand the concepts covered during the next class period. If this happens to you, get help as soon as you can.

Where can you get help? Many suggestions have been made in the section on where to get help, and now it is up to you to do it. Try your instructor, a tutoring center, or a math lab, or you may want to form a study group with fellow classmates. If you do decide to see your instructor or go to a tutoring center, make sure that you have a neat notebook and are ready with your questions.

E Preparing for and Taking a Test

Make sure that you allow yourself plenty of time to prepare for a test. If you think that you are a little "math anxious," it may be that you are not preparing for a test in a way that will ensure success. The way that you prepare for a test in mathematics is important. To prepare for a test,

1. Review your previous homework assignments.

2. Review any notes from class and section-level quizzes you have taken. (If this is a final exam, also review chapter tests you have taken.)

3. Review concepts and definitions by reading the Highlights at the end of each chapter.

4. Practice working out exercises by completing the Chapter Review found at the end of each chapter. (If this is a final exam, go through a Cumulative Review. There is one found at the end of each chapter except Chapter 1. Choose the review found at the end of the latest chapter that you have covered in your course.) *Don't stop here!*

5. It is important that you place yourself in conditions similar to test conditions to find out how you will perform. In other words, as soon as you feel that you know the material, get a few blank sheets of paper and take a sample test. There is a Chapter Test available at the end of each chapter. During this sample test, do not use your notes or your textbook. Once you complete the Chapter Test, check your answers in the back of the book. If any answer is incorrect, there is a CD available with each exercise of each chapter test worked. Use this CD or your instructor to correct your sample test. Your instructor may also provide

you with a review sheet. If you are not satisfied with the results, study the areas that you are weak in and try again.

6. Get a good night's sleep before the exam.

7. On the day of the actual test, allow yourself plenty of time to arrive at where you will be taking your test.

When taking your test,

1. Read the directions on the test carefully.

2. Read each problem carefully as you take the test. Make sure that you answer the question asked.

3. Watch your time and pace yourself so that you can attempt each problem on your test.

4. If you have time, check your work and answers.

5. Do not turn your test in early. If you have extra time, spend it double-checking your work.

Ⓕ Managing Your Time

As a college student, you know the demands that classes, homework, work, and family place on your time. Some days you probably wonder how you'll ever get everything done. One key to managing your time is developing a schedule. Here are some hints for making a schedule:

1. Make a list of all of your weekly commitments for the term. Include classes, work, regular meetings, extracurricular activities, etc. You may also find it helpful to list such things as laundry, regular workouts, grocery shopping, etc.

2. Next, estimate the time needed for each item on the list. Also make a note of how often you will need to do each item. Don't forget to include time estimates for the reading, studying, and homework you do outside of your classes. You may want to ask your instructor for help estimating the time needed.

3. In the following Exercise Set, you are asked to block out a typical week on the schedule grid given. Start with items with fixed time slots like classes and work.

4. Next, include the items on your list with flexible time slots. Think carefully about how best to schedule some items such as study time.

5. Don't fill up every time slot on the schedule. Remember that you need to allow time for eating, sleeping, and relaxing! You should also allow a little extra time in case some items take longer than planned.

6. If you find that your weekly schedule is too full for you to handle, you may need to make some changes in your workload, classload, or in other areas of your life. You may want to talk to your advisor, manager or supervisor at work, or someone in your college's academic counseling center for help with such decisions.

Note: In this chapter, we begin a feature called Study Skills Reminder. The purpose of this feature is to remind you of some of the information given in this section and to further expand on some topics in this section.

EXERCISE SET 1.1

1. What is your instructor's name?

2. What are your instructor's office location and office hours?

3. What is the best way to contact your instructor?

4. What does the ✏ icon mean?

5. What does the ▦ icon mean?

6. What does the △ icon mean?

7. Where are answers located in this text?

8. What Exercise Set answers are available to you in the answers section?

9. What Chapter Review, Chapter Text, and Cumulative Text answers are available to you in the answer section?

10. Are there worked-out solutions to exercises in this text?

11. If the answer to Exercise 10 is yes, what worked-out solutions are available to you in this text?

12. Go to the Highlights Section at the end of this chapter. Describe how this section may be helpful to you when preparing for a test.

13. Do you have the name and contact information of at least one other student in class?

14. Will your instructor allow you to use a calculator in this class?

15. Are videotapes, CDs, and/or tutorial software available to you? If so, where?

16. Is there a tutoring service available? If so, what are its hours?

17. Have you attempted this course before? If so, write down ways that you might improve your chances of success during this second attempt.

18. List some steps that you can take if you begin having trouble understanding the material or completing an assignment.

19. Read or reread objective Ⓕ and fill out the schedule grid below.

	Monday	Tuesday	Wednesday	Thursday	Friday	Saturday	Sunday
7:00 a.m.							
8:00 a.m.							
9:00 a.m.							
10:00 a.m.							
11:00 a.m.							
12:00 a.m.							
1:00 p.m.							
2:00 p.m.							
3:00 p.m.							
4:00 p.m.							
5:00 p.m.							
6:00 p.m.							
7:00 p.m.							
8:00 p.m.							
9:00 p.m.							

20. Study your filled-out grid from Exercise 19. Decide whether you have the time necessary to successfully complete this course and any other courses you may be registered for.

1.2   Place Value and Names for Numbers

The **digits** 0, 1, 2, 3, 4, 5, 6, 7, 8, and 9 can be used to write numbers. For example, the **natural numbers** are

1, 2, 3, 4, 5, 6, 7, 8, 9, 10, 11, ...

The natural numbers are also called the counting numbers. The three dots (...) after the 11 mean that this list continues indefinitely. That is, there is no largest natural number. The smallest natural number is 1.

If 0 is included with the natural numbers, we have the whole numbers. The whole numbers are

0, 1, 2, 3, 4, 5, 6, 7, 8, 9, 10, 11, 12, ...

There is no largest whole number. The smallest whole number is 0.

(A) Finding the Place Value of a Digit in a Whole Number

The position of each digit in a number determines its **place value**. A place-value chart is shown next for the whole number 48,337,000.

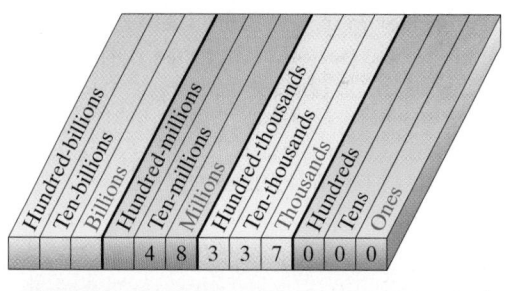

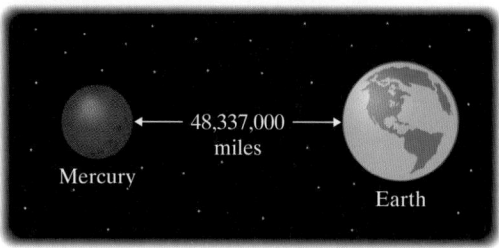

The two 3s in 48,337,000 represent different amounts because of their different placements. The place value of the 3 on the left is hundred-thousands. The place value of the 3 on the right is ten-thousands.

EXAMPLES   Find the place value of the digit 4 in each whole number.

1. 48,761
↑
ten-thousands

2. 249
↑
tens

3. 524,007,656
↑
millions

Practice Problems 1–3

Find the place value of the digit 7 in each whole number.

1. 72,589,620
2. 67,890
3. 50,722

Answers

1. ten-millions   **2.** thousands   **3.** hundreds

B Writing a Whole Number in Words and in Standard Form

A whole number such as 1,083,664,500 is written in **standard form**. Notice that commas separate the digits into groups of three, starting from the right. Each group of three digits is called a **period**. The names of the first four periods are shown in red and above the chart.

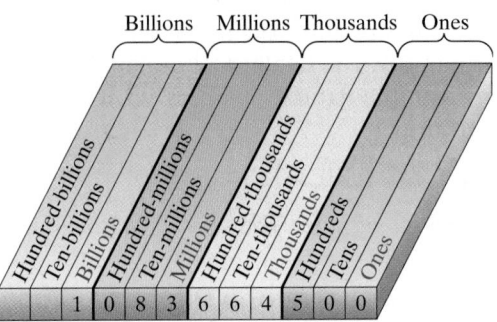

Writing a Whole Number in Words

To write a whole number in words, write the number in each period followed by the name of the period. (The ones period is usually not written.) This same procedure can be used to read a whole number.

For example, we write 1,083,664,500 as

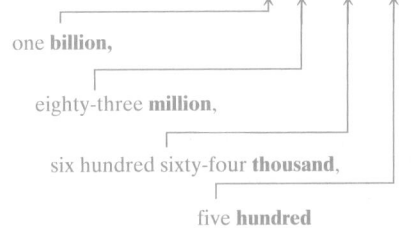

one **billion,**

eighty-three **million,**

six hundred sixty-four **thousand,**

five **hundred**

> **Helpful Hint**
>
> The name of the ones period is not used when reading and writing whole numbers. For example,
>
> 9,265
>
> is read as
>
> "nine **thousand**, two **hundred** sixty-five."

Practice Problems 4–5

Write each number in words.

4. 395
5. 2807

EXAMPLES Write each number in words.

4. 126 one hundred twenty-six
5. 3005 three thousand five

> **Helpful Hint**
>
> The word "and" is *not* used when reading and writing whole numbers. It is used when reading and writing mixed numbers and some decimal values, as shown later in this text.

Answers

4. three hundred ninety-five
5. two thousand eight hundred seven

EXAMPLE 6   Write 106,052,447 in words.

Solution: 106,052,447 is written as

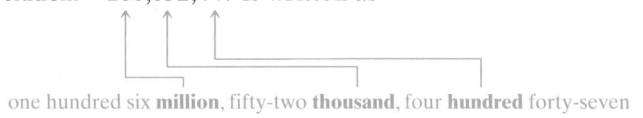

one hundred six **million**, fifty-two **thousand**, four **hundred** forty-seven

Try the Concept Check in the margin.

Writing a Whole Number in Standard Form

To write a whole number in standard form, write the number in each period, followed by a comma.

EXAMPLES   Write each number in standard form.

7.   sixty-one   61
8.   eight hundred five   805
9.   two million, five hundred sixty-four thousand, three hundred fifty

    2,564,350
10.   nine thousand, three hundred eighty-six

    9,386 or 9386

A comma may or may not be inserted in a four-digit number. For example, both

    9,386   and   9386

are acceptable ways of writing nine thousand, three hundred eighty-six.

(c)   Writing a Whole Number in Expanded Form

The place value of a digit can be used to write a number in expanded form. The **expanded form** of a number shows each digit of the number with its place value. For example, 5672 is written in expanded form as

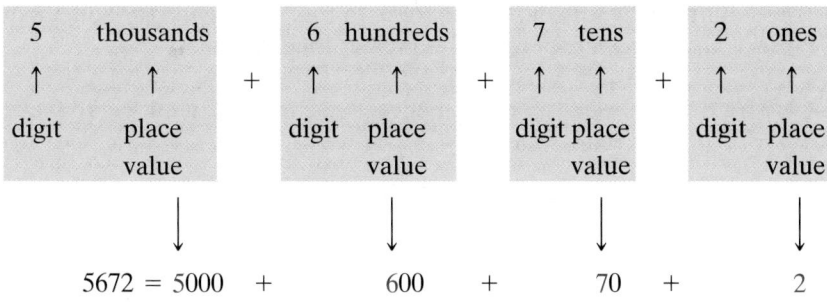

Practice Problem 6

Write 321,670,200 in words.

Concept Check

True or false? When writing a check for $2600, the word name we write for the dollar amount of the check would be "two thousand sixty." Explain your answer.

Practice Problems 7–10

Write each number in standard form.

7.   twenty-nine
8.   seven hundred ten
9.   twenty-six thousand seventy-one
10.   six thousand, five hundred seven

Concept Check: false

Answers

6.   three hundred twenty-one million, six hundred seventy thousand, two hundred

7. 29   **8.** 710   **9.** 26,071   **10.** 6507

Practice Problem 11

Write 1,047,608 in expanded form.

EXAMPLE 11 Write 706,449 in expanded form.

Solution: $700{,}000 + 6000 + 400 + 40 + 9$

STUDY SKILLS REMINDER

Many of the terms used in this text may be new to you. It will be helpful to make a list of new mathematical terms and symbols as you encounter them and to review them frequently. Placing these new terms (including page references) on 3×5 index cards might help you later when you're preparing for a quiz.

D Comparing Whole Numbers

We can picture whole numbers as equally spaced points on a line called the **number line**. The whole numbers are written in "counting" order on the number line beginning with 0. An arrow at the right end of the number line means that the whole numbers continue indefinitely.

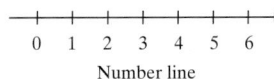

Number line

A whole number is **graphed** by placing a dot on the number line at the point corresponding to the number. For example, 5 is graphed on the number line as shown.

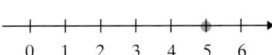

For any two numbers graphed on a number line, the number to the *right* is the **greater number**, and the number to the *left* is the **smaller number**. For example, 3 is *to the left* of 6, so 3 **is less than** 6. Also, 6 is *to the right* of 3, so 6 **is greater than** 3.

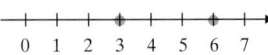

We use the symbol $<$ to mean "is less than" and the symbol $>$ to mean "is greater than." The symbols $<$ and $>$ are called **inequality** symbols. For example, the true statement "*3 is less than 6*" can be written in symbols as

$3 < 6$

Also, "*6 is greater than 3*" is written as

$6 > 3$

The statements $3 < 6$ and $6 > 3$ are both *true statements*. The statement $5 < 2$ is a *false statement* because 5 is greater than 2, not less than.

Practice Problem 12

Insert $<$ or $>$ between each pair of numbers to make a true statement.

a. 0   19   b. 18   32   c. 107   103

EXAMPLE 12 Insert $<$ or $>$ between each pair of numbers to make a true statement.

a. 5   50   **b.** 101   0   **c.** 29   27

Solution:

a. $5 < 50$ **b.** $101 > 0$ **c.** $29 > 27$

Answers

11. $1{,}000{,}000 + 40{,}000 + 7000 + 600 + 8$
12. a. $<$ **b.** $<$ **c.** $>$

Helpful Hint

One way to remember the meaning of the inequality symbols $<$ and $>$ is to think of them as arrowheads "pointing" toward the smaller number. For example, $5 < 11$ and $11 > 5$ are both true statements.

Reading Tables

Now that we know about place value and names for whole numbers, we introduce one way that whole number data may be presented. **Tables** are often used to organize and display facts that contain numbers. The table below shows the countries that have won the most medals during the Olympic winter games for the years 1924 through 2002. (Although the medals are truly won by athletes from the various countries, for simplicity we will state that countries have won the medals.)

Most Medals Olympic Winter (1924 – 2002) Games		Gold	Silver	Bronze	Total
	Germany[1]	107	104	86	297
	Russia	113	83	78	274
	Norway	94	92	74	260
	United States	69	71	51	191
	Austria	41	57	64	162
	Finland	41	51	49	141
	Sweden	39	30	39	108
	Switzerland	32	33	38	103
	Canada	30	28	36	94

(*Source:* The Sydney Morning Herald)
[1]Includes West Germany 1952, 1968–1988; East Germany 1968–1988

For example, by reading from left to right along the row marked "U.S." we find that the United States won 69 gold, 71 silver, and 51 bronze medals during the Olympic Winter Games 1924–2002.

Practice Problem 13

Use the winter Games table to answer the following questions:

a. How many gold medals did Russia win during the winter Games of the Olympics?

b. Which countries shown have won more than 100 gold medals?

EXAMPLE 13 Use the Winter Games table to answer each question.

a. How many silver medals did Sweden win during the Winter Games of the Olympics?

b. Which countries shown have won fewer bronze medals than Austria?

Solution:

a. Find "Sweden" in the left column. Then read from left to right until the "Silver" column is reached. We find that Sweden has won 30 silver medals.

b. Austria has won 64 bronze medals. United States, Finland, Sweden, Switzerland, and Canada have each won fewer than 64 bronze medals. ●

Answers

13. a. 113   **b.** Germany and Russia

EXERCISE SET 1.2

(A) *Determine the place value of the digit 5 in each whole number. See Examples 1 through 3.*

1. 352

2. 905

3. 5890

4. 6527

5. 62,500,000

6. 79,050,000

7. 5,070,099

8. 51,682,700

(B) *Write each whole number in words. See Examples 4 through 6.*

9. 5420

10. 3165

11. 26,990

12. 42,009

13. 1,620,000

14. 3,204,000

15. 53,520,170

16. 47,033,107

Write each number in the sentence in words. See Examples 4 through 6.

17. Bermuda is a British dependency consisting of many small islands located in the Atlantic Ocean east of North Carolina. At this writing, the population of Bermuda is 63,960. (*Source:* The World Almanac, 2003)

18. The Goodyear blimp *Eagle* holds 202,700 cubic feet of helium. (*Source:* The Goodyear Tire & Rubber Company)

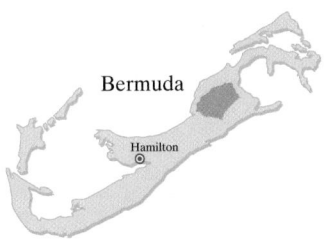

Bermuda

Hamilton

19. The world's tallest building, the PETRONAS Twin Towers in Kuala Lumpur, Malaysia, is 1483 feet tall. (*Source:* Council on Tall Buildings and Urban Habitat)

20. Liz Harold has the number 16,820,409 showing on her calculator display.

21. The highest point in Idaho is at Granite Peak, at an elevation of 12,662 feet. (*Source*: U.S. Geological Survey)

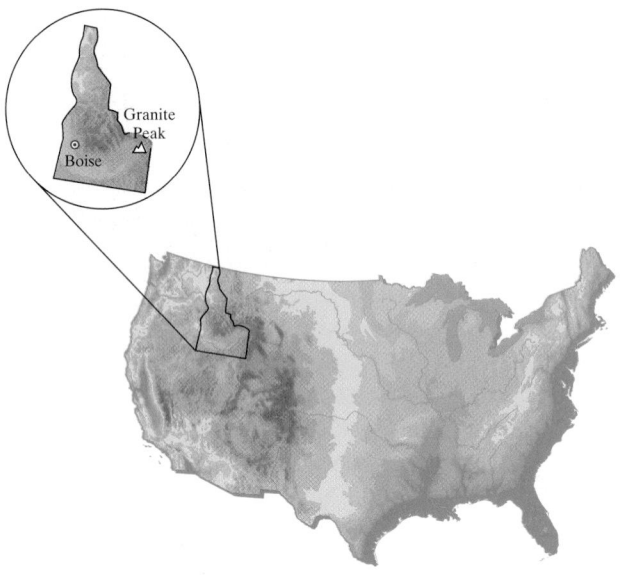

22. The highest point in New Mexico is Wheeler Peak, at an elevation of 13,161 feet. (*Source*: U.S. Geological Survey)

23. Each day, UPS delivers 13,600,000 packages and documents worldwide. (*Source*: United Parcel Service of America, Inc.)

24. In a recent year, zinc mines in the United States mined 799,000 metric tons of zinc. (*Source*: U.S. Dept. of Interior)

25. In a recent year, there were 3895 patients in the United States waiting for a heart transplant. (*Source*: United Network for Organ Sharing)

26. Each Home Depot store in the United States and Canada stocks at least 40,000 different kinds of building materials, home improvement supplies, and lawn and garden products. (*Source*: The Home Depot, Inc.)

Write each whole number in standard form. See Examples 7 through 10.

27. Six thousand, five hundred eight

28. Three thousand, three hundred seventy

29. Twenty-nine thousand, nine hundred

30. Forty-two thousand, six

31. Six million, five hundred four thousand, nineteen

32. Ten million, thirty-seven thousand, sixteen

33. Three million, fourteen

34. Seven million, twelve

Write the whole number in each sentence in standard form. See Examples 7 through 10.

35. The International Space station orbits above Earth at an altitude of two hundred twenty miles. (NASA)

36. The average distance between the surfaces of the Earth and the Moon is about 234 thousand miles.

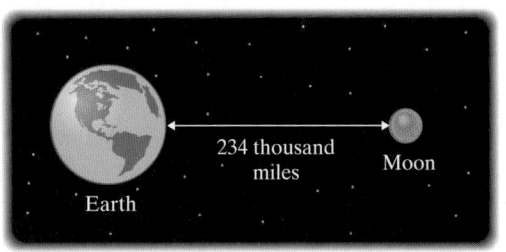

37. Hank Aaron holds the career record for home runs in Major League baseball, with a total of seven hundred fifty-five home runs. (*Source*: Major League Baseball)

38. The average annual salary for an NHL hockey player for the 2000–2001 season was one million, four hundred thousand dollars. (*Source*: NHL Players Association)

39. In 2002, there were seventy-three million, five hundred thousand U.S. households that have cable TV. (*Source*: Nielsen Media Research)

40. There were thirty thousand slices of pizza consumed at a recent Daytona 500 race. (*Source*: Americrown Service Corporation)

41. The world's tallest self-supporting structure is the CN Tower in Toronto, Canada. It is one thousand, eight hundred fifteen feet tall. (*Source*: The World Almanac, 2003)

42. At this writing, there are about 3 billion Web pages.

43. As of 2003, there were one thousand, two hundred sixty-two species classified as either threatened or endangered in the United States. (*Source*: U.S. Fish & Wildlife Service)

44. Grain storage facilities in Toledo, Ohio, have a capacity of sixty-three million, one hundred thousand bushels. (*Source*: Chicago Board of Trade Market Information Department)

45. FedEx employees retrieve packages from over forty-five thousand FedEx drop boxes around the world each business day. (*Source*: Federal Express Corporation)

46. One hundred million pounds of nonhazardous waste is recycled at Kodak Park each year. (*Source*: Eastman Kodak Company)

C *Write each whole number in expanded form. See Example 11.*

47. 406

48. 789

49. 5290

50. 6040

51. 62,407

52. 20,215

53. 30,680

54. 99,032

55. 39,680,000

56. 47,703,029

57. 1006

58. 20,304

D *Insert* < *or* > *to make a true statement. See Example 12.*

59. 3   8

60. 7   10

61. 9   0

62. 0   12

63. 6   2

64. 15   14

65. 22   0

66. 39   47

E *The table shows the five longest rivers in the world. Use this table to answer Exercises 67 through 72. See Example 13.*

River	Miles
Chang jiang-Yangtze (China)	3964
Amazon (Brazil)	4000
Tenisei-Angara (Russia)	3442
Mississippi-Missouri (U.S.)	3740
Nile (Egypt)	4145

67. Write the length of the Amazon River in words.

68. Write the length of the Tenisei-Angara River in words.

69. Write the length of the Nile River in expanded form.

70. Write the length of the Mississippi-Missouri River in expanded form.

71. Which river is the longest in the world?

72. Which river is the second longest in the world?

The table shows the top ten breeds of dogs in 2001 according to the American Kennel Club. Use this table to answer Exercises 73 through 76. See Example 13.

Top Ten American Kennel Club Registrations in 2001	
Breed	**Number of Registered Dogs**
Beagles	50,419
Boxers	37,035
Chihuahuas	36,627
Dachshunds	50,478
German Shepherd Dogs	51,625
Golden Retrievers	62,497
Labrador Retrievers	165,970
Poodles	40,550
Shih Tzu	33,240
Yorkshire Terriers	42,025

73. Which breed has the most American Kennel Club registrations? Write the number of registrations for this breed in words.

74. Which of the listed breeds has the fewest registrations? Write the number of registered dogs for this breed in words.

75. Which breed has more dogs registered, Chihuahua or Golden retriever?

76. Which breed has fewer dogs registered, Beagle or Yorkshire terrier?

◆ Combining Concepts

77. Write the largest four-digit number that can be made from the digits 3, 6, 7, and 2 if each digit must be used once. _____ _____ _____ _____

78. Write the largest five-digit number that can be made using the digits 4, 5, and 3 if each digit must be used at least once. _____ _____ , _____ _____ _____

79. If a number is given in words, describe the process used to write this number in standard form.

80. If a number is written in standard form, describe the process used to write this number in expanded form.

81. The Pro-Football Hall of Fame was established on September 7, 1963, in this town. Use the information and the accompanying diagram to find the name of the town.

- Alliance is east of Massillon.
- Dover is between Canton and New Philadelphia.
- Massillon is not next to Alliance.
- Canton is north of Dover.

1.3 Adding Whole Numbers and Perimeter

OBJECTIVES

Ⓐ Add whole numbers.

Ⓑ Find the perimeter of a polygon.

Ⓒ Solve problems by adding whole numbers.

SSM
TUTOR CENTER   SG   CD & VIDEO   MATH PRO   WEB

Ⓐ Adding Whole Numbers

If one computer in an office has a 2-megabyte memory and a second computer has a 4-megabyte memory, the total memory in the two computers can be found by adding 2 and 4.

$$2 \text{ megabytes} + 4 \text{ megabytes} = 6 \text{ megabytes}$$

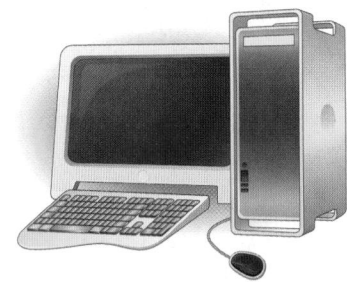

The **sum** is 6 megabytes of memory. Each of the numbers 2 and 4 is called an **addend**.

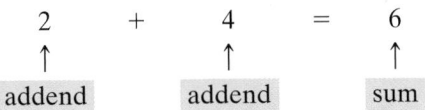

To add whole numbers, we add the digits in the ones place, then the tens place, then the hundreds place, and so on. For example, let's add $2236 + 160$.

$$
\begin{array}{r}
2236 \\
+\ 160 \\
\hline
2396
\end{array}
$$

     — sum of ones
     — sum of tens
     — sum of hundreds
     — sum of thousands

Line up numbers vertically so that the place values correspond. Then add digits in corresponding place values, starting with the ones place.

EXAMPLE 1   Add: $23 + 136$

Solution:
$$
\begin{array}{r}
23 \\
+136 \\
\hline
159
\end{array}
$$

When the sum of digits in corresponding place values is more than 9, "carrying" is necessary. For example, to add $365 + 89$, add the ones-place digits first.

$$
\begin{array}{r}
\overset{1}{3}65 \\
+\ 89 \\
\hline
4
\end{array}
$$

5 ones + 9 ones = **14 ones** or **1 ten** + **4 ones**

Write the 4 ones in the ones place and carry the 1 ten to the tens place.

Next, add the tens-place digits.

Practice Problem 1

Add: $7235 + 542$

●

Answer
1. 7777

$$
\begin{array}{r}
\overset{1\ 1}{3\,6\,5} \\
+\ 8\,9 \\
\hline
5\,4
\end{array}
$$
1 ten + 6 tens + 8 tens = **15 tens** or **1 hundred** + **5 tens**
Write the 5 tens in the tens place and carry the 1 hundred to the hundreds place.

Next, add the hundreds-place digits.

$$
\begin{array}{r}
\overset{1\ 1}{3\,6\,5} \\
+\ 8\,9 \\
\hline
4\,5\,4
\end{array}
$$
1 hundred + 3 hundreds = 4 hundreds
Write the 4 hundreds in the hundreds place.

Practice Problem 2

Add: $27{,}364 + 92{,}977$

Concept Check

What is wrong with the following computation?

$$
\begin{array}{r}
3\,9\,4 \\
+\ 2\,8\,3 \\
\hline
5\,7\,7
\end{array}
$$

EXAMPLE 2   Add: $34{,}285 + 149{,}761$

Solution:
$$
\begin{array}{r}
\overset{1\ 1\ \ 1}{34{,}285} \\
+\ 149{,}761 \\
\hline
184{,}046
\end{array}
$$

Try the Concept Check in the margin.

Before we continue adding whole numbers, let's review some properties of addition that you may have already discovered. The first property that we will review is the **addition property of 0**. This property reminds us that the sum of 0 and any number is that same number.

Addition Property of 0

The sum of 0 and any number is that number. For example,

$$7 + 0 = 7$$
$$0 + 7 = 7$$

Next, notice that we can add any two whole numbers in any order and the sum is the same. For example,

$$4 + 5 = 9 \quad \text{and} \quad 5 + 4 = 9$$

We call this special property of addition the **commutative property of addition**.

Commutative Property of Addition

Changing the **order** of two addends does not change their sum. For example,

$$2 + 3 = 5 \quad \text{and} \quad 3 + 2 = 5$$

Another property that can help us when adding numbers is the **associative property of addition**. This property states that when adding numbers, the grouping of the numbers can be changed without changing the sum. We use parentheses to group numbers. They indicate what numbers to add first. For example, let's use two different groupings to find the sum of $2 + 1 + 5$.

$$2 + (1 + 5) = 2 + 6 = 8$$

Also,

$$(2 + 1) + 5 = 3 + 5 = 8$$

Both groupings give a sum of 8.

Concept Check: forgot to carry 1 hundred to the hundreds place

Answer

2. 120,341

Associative Property of Addition

Changing the grouping of addends does not change their sum. For example,

$$3 + (5 + 7) = 3 + 12 = 15 \quad \text{and} \quad (3 + 5) + 7 = 8 + 7 = 15$$

The commutative and associative properties tell us that we can add whole numbers using any order and grouping that we want.

When adding several numbers, it is often helpful to look for two or three numbers whose sum is 10, 20, and so on.

EXAMPLE 3   Add: $13 + 2 + 7 + 8 + 9$

Solution:

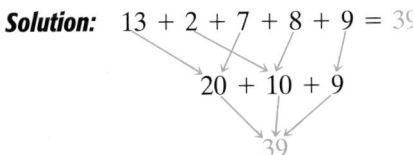

$$13 + 2 + 7 + 8 + 9 = 39$$
$$20 + 10 + 9$$
$$39$$

Remember these properties as we continue practicing addition of whole numbers.

EXAMPLE 4   Add: $1647 + 246 + 32 + 85$

Solution:
$$\begin{array}{r} \overset{1\,2\,2}{1647} \\ 246 \\ 32 \\ +\quad 85 \\ \hline 2010 \end{array}$$

Ⓑ **Finding the Perimeter of a Polygon**

A special application of addition is finding the perimeter of a polygon. A **polygon** can be described as a flat figure formed by line segments connected at their ends. (For more review, see Appendix D.) Geometric figures such as triangles, squares, and rectangles are called polygons.

Triangle   Square   Rectangle

The **perimeter** of a polygon is the *distance around* the polygon. This means that the perimeter of a polygon is the sum of the lengths of its sides.

EXAMPLE 5   Find the perimeter of the polygon shown.

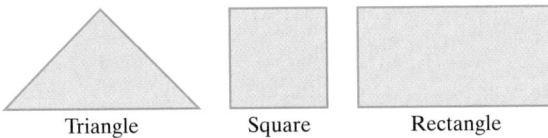

Solution: To find the perimeter (distance around), we add the lengths of the sides.

$$2 \text{ in.} + 3 \text{ in.} + 1 \text{ in.} + 3 \text{ in.} + 4 \text{ in.} = 13 \text{ in.}$$

The perimeter is 13 inches.

Practice Problem 3

Add: $11 + 7 + 8 + 9 + 13$

Practice Problem 4

Add: $19 + 5042 + 638 + 526$

Practice Problem 5   △

Find the perimeter of the polygon shown. (A centimeter is a unit of length in the metric system.)

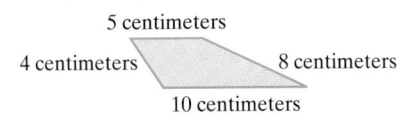

Answers

3. 48   **4.** 6225   **5.** 27 cm

Practice Problem 6

A new shopping mall has a floor plan in the shape of a triangle. Each of the mall's three sides is 532 feet. Find the perimeter of the building.

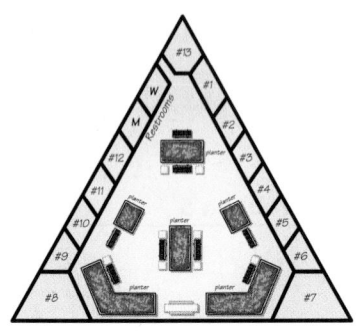

532 feet

EXAMPLE 6   Calculating the Perimeter of a Building

The largest commercial building in the world under one roof is the flower auction building of the cooperative VBA in Aalsmeer, Netherlands. The floor plan is a rectangle that measures 776 meters by 639 meters. Find the perimeter of this building. (A meter is a unit of length in the metric system.) (*Source: The Handy Science Answer Book*, Visible Ink Press)

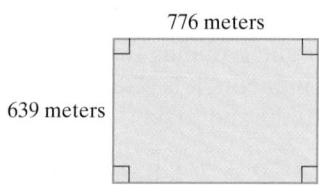

776 meters

639 meters

Solution: Recall that opposite sides of a rectangle have the same lengths. To find the perimeter of this building, we add the lengths of the sides. The sum of the lengths of its sides is

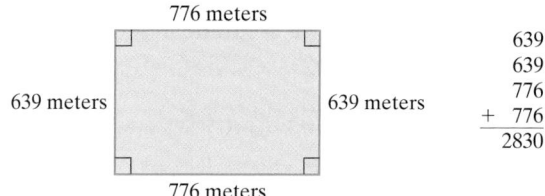

776 meters

639 meters   639 meters

776 meters

$$\begin{array}{r} 639 \\ 639 \\ 776 \\ +\ 776 \\ \hline 2830 \end{array}$$

The perimeter of the building is 2830 meters.

(C) Solving Problems by Adding

Often, real-life problems occur that can be solved by writing an addition statement. The first step in solving any word problem is to *understand* the problem by reading it carefully. Descriptions of problems solved through addition *may* include any of these key words or phrases:

Key Words or Phrases	Example	Symbols
added to	5 added to 7	$7 + 5$
plus	0 plus 78	$0 + 78$
increased by	12 increased by 6	$12 + 6$
more than	11 more than 25	$25 + 11$
total	the total of 8 and 1	$8 + 1$
sum	the sum of 4 and 133	$4 + 133$

To solve a word problem that involves addition, we first use the facts described to write an addition statement. Then we write the corresponding solution of the real-life problem. It is sometimes helpful to write the statement in words (brief phrases) and then translate to numbers.

Answer

6. 1596 ft

EXAMPLE 7 Finding a Salary

The governor's salary in the state of Florida was recently increased by $3004. If the old salary was $120,171, find the new salary.
(*Source: The World Almanac and Book of Facts*, 2003)

Solution: The key phrase here is "increased by," which suggests that we add. To find the new salary, we add the increase, $3004, to the old salary.

In Words		Translate to Numbers
old salary	→	120,171
increased by 3004	→	+ 3004
new salary	→	123,175

The Florida governor's salary is now $123,175.

EXAMPLE 8 Determining the Number of Baseball Cards in a Collection

Alan Mayfield collects baseball cards. He has 109 cards for the New York Yankees, 96 for the Chicago White Sox, 79 for the Kansas City Royals, 42 for the Seattle Mariners, 67 for the Oakland Athletics, and 52 for the California Angels. How many cards does he have in total?

Solution: The key word here is "total." To find the total number of Alan's baseball cards, we find the sum of the quantities for each team.

In Words		Translate to Numbers
		3 3
New York Yankees cards	→	109
Chicago White Sox cards	→	96
Kansas City Royals cards	→	79
Seattle Mariners cards	→	42
Oakland Athletics cards	→	67
+ California Angels cards	→	+ 52
Total cards	→	445

Alan has a total of 445 baseball cards.

Practice Problem 7

Texas produces 90 million pounds of pecans per year. Georgia is the world's top pecan producer and produces 15 million pounds more pecans than Texas. How much does Georgia produce?
(*Source:* Absolute Trivia.com)

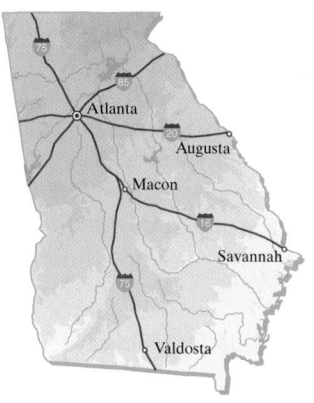

Practice Problem 8

Elham Abo-Zahrah collects thimbles. She has 42 glass thimbles, 17 steel thimbles, 37 porcelain thimbles, 9 silver thimbles, and 15 plastic thimbles. How many thimbles are in her collection?

Answers

7. 105 million lb   **8.** 120 thimbles

CALCULATOR EXPLORATIONS

Adding Numbers

To add numbers on a calculator, find the keys marked $+$ and $=$ or ENTER.

For example, to add 5 and 7 on a calculator, press the keys 5 $+$ 7 $=$ or ENTER.
The display will read [12].
Thus, $5 + 7 = 12$.
To add 687 and 981 on a calculator, press the keys 687 $+$ 981 $=$ or ENTER.
The display will read [1668].
Thus, $687 + 981 = 1668$. (Although entering 687, for example, requires pressing more than one key, here numbers are grouped together for easier reading.)

Use a calculator to add.

1. $89 + 45$

2. $76 + 97$

3. $285 + 55$

4. $8773 + 652$

5.
$$
\begin{array}{r}
985 \\
1210 \\
562 \\
+ \quad 77 \\
\hline
\end{array}
$$

6.
$$
\begin{array}{r}
465 \\
9888 \\
620 \\
+ \quad 1550 \\
\hline
\end{array}
$$

Name _____ Section _____ Date _____

Mental Math

Find each sum.

1. 5 + 7   **2.** 20 + 30   **3.** 5000 + 4000   **4.** 4300 + 26   **5.** 1620 + 0   **6.** 6 + 126 + 4

EXERCISE SET 1.3

A   *Add. See Examples 1 through 4.*

1.
```
  14
+ 22
```

2.
```
  27
+ 31
```

3.
```
  62
+ 30
```

4.
```
  37
+ 42
```

5.
```
  12
  13
+ 24
```

6.
```
  23
  45
+ 30
```

7.
```
  5267
+  132
```

8.
```
   236
+ 6243
```

9. 53 + 64   **10.** 41 + 74   **11.** 22 + 49   **12.** 35 + 47

13. 38 + 79   **14.** 92 + 37

15.
```
  8
  9
  2
  5
+ 1
```

16.
```
  3
  5
  8
  5
+ 7
```

17.
```
   6
  21
  14
   9
+ 12
```

18.
```
  12
   4
   8
  26
+ 10
```

19.
```
  81
  17
  23
  79
+ 12
```

20.
```
  64
  28
  56
  25
+ 32
```

21. 62 + 18 + 14   **22.** 23 + 49 + 18

23. 40 + 800 + 70   **24.** 30 + 900 + 20   **25.** 7542 + 49 + 682   **26.** 1624 + 1832 + 1976

27. 24 + 9006 + 489 + 2407   **28.** 16 + 748 + 1056 + 770

29.
```
  627
  628
+ 629
```

30.
```
  427
  383
+ 229
```

31.
```
  6820
  4271
+ 5626
```

32.
```
  6789
  4321
+ 5555
```

33.
```
  507
  593
+  10
```

34.
```
  864
  733
+ 356
```

35.
```
  4200
  2107
+ 2692
```

36.	37.	38.	39.	40.
5000	49	26	121,742	504,218
400	628	582	57,279	321,920
+ 3021	5762	4763	6586	38,507
	+ 29,462	+ 62,511	+ 426,782	+ 594,687

B *Find the perimeter of each figure. See Examples 5 and 6.*

△ **41.**

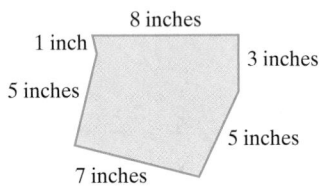

8 inches
1 inch
3 inches
5 inches
5 inches
7 inches

△ **42.**

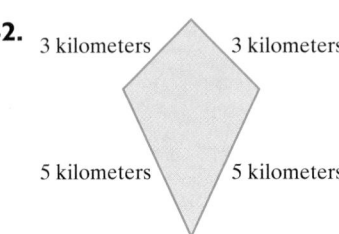

3 kilometers 3 kilometers
5 kilometers 5 kilometers

43.

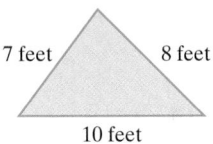

7 feet   8 feet
10 feet

△ **44.**

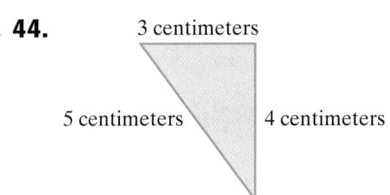

3 centimeters
5 centimeters   4 centimeters

△ **45.**

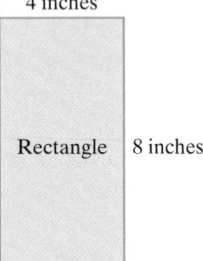

4 inches
Rectangle 8 inches

△ **46.**

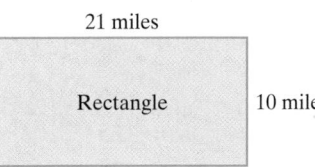

21 miles
Rectangle   10 mile

△ **47.**

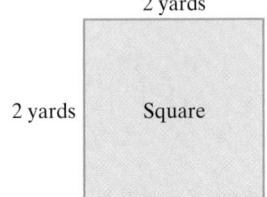

2 yards
2 yards   Square

△ **48.**

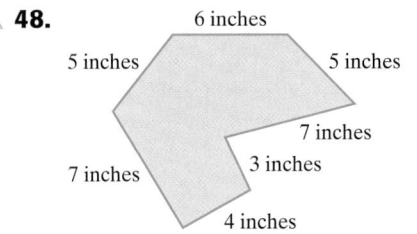

6 inches
5 inches   5 inches
7 inches
7 inches   3 inches
4 inches

Solve. See Examples 7 and 8.

49. The highest point in South Carolina is Sassafras Mountain at 3560 feet above sea level. The highest point in North Carolina is Mt. Mitchell, whose peak is 3124 feet higher than Sassafras Mountain. Find how high Mt. Mitchell is. (*Source:* U.S. Geological Survey)

50. The distance from Kansas City, Kansas, to Hays, Kansas, is 285 miles. Colby, Kansas, is 98 miles farther from Kansas City than Hays. Find how far it is from Kansas City to Colby.

△ **51.** Leo Callier is installing an invisible fence in his backyard. How many feet of wiring are needed to enclose the yard below?

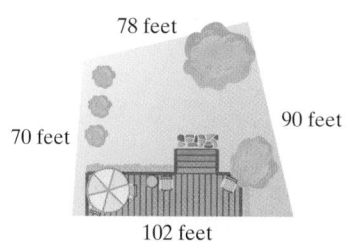

78 feet

70 feet   90 feet

102 feet

△ **52.** A homeowner is considering installing gutters around her home. To help her estimate the cost of gutters find the perimeter of the edge of her roof.

60 feet

45 feet

53. In 2001, Harley-Davidson shipped 234,500 motorcycles worldwide. In 2002, it shipped 29,200 more motorcycles. What was the total number of Harley-Davidson motorcycles shipped in 2002? (*Source:* Harley-Davidson, Inc.)

54. Dan Marino holds the NFL career record for most passes completed. He completed 2305 passes from the beginning of his NFL career in 1983 through 1989. He completed another 2662 passes from 1990 through 1999, his last season before retiring from professional football. How many passes did he complete during his NFL career? (*Source:* National Football League)

55. During the spring of 2001, Kellogg Company acquired Keebler Foods Company. Before the merger, Kellogg employed 15,000 people and Keebler employed 13,000 people. Assuming there were no layoffs after the merger, how many employees did the newly combined company have? (*Source:* Kellogg Company)

56. The DVD video format was introduced in March 1997. From 1997 through 1999, a total of 5,423,786 DVD players were sold in the United States. A total of 8,498,545 DVD players were sold in the United States in 2000. How many DVD players in all were sold from 1997 through the end of 2000? (*Source:* Consumer Electronics Association)

57. In 2001, there were 7981 Blockbuster video rental stores worldwide. In 2002, the Blockbuster store tally had increased by 564 stores. How many Blockbuster stores were there in 2002? (*Source:* Blockbuster Inc.)

58. Wilma Rudolph, who won three gold medals in track and field events in the 1960 Summer Olympics, was born in 1940. Marion Jones, who also won three gold medals in track and field events but in the 2000 Summer Olympics, was born 35 years later. In what year was Marion Jones born?

59. The highest waterfall in the United States is Yosemite Falls in Yosemite National Park in California. Yosemite Falls is made up of three sections, as shown in the graph. What is the total height of Yosemite Falls? (*Source:* U.S. Department of the Interior)

Highest U.S. Waterfalls

Section	Height (in feet)
Upper Yosemite Falls	1430
Cascades	675
Lower Yosemite Falls	320

60. Jordan White, a nurse at Mercy Hospital, is recording fluid intake in a patient's medical chart. During his shift, the patient had the following types and amounts of intake measured in cubic centimeters (cc). What amount should Jordan record as the total fluid intake for this patient?

Oral	Intravenous	Blood
240	500	500
100	200	
355		

61. The State of Alaska has 1795 miles of urban highways and 11,460 miles of rural highways. Find the total highway mileage in Alaska. (*Source:* U.S. Federal Highway Administration)

62. The state of Hawaii has 1851 miles of urban highways and 2291 miles of rural highways. Find the total highway mileage in Hawaii. (*Source:* U.S. Federal Highway Administration)

63. As of January 3, 2003, the Broadway show *Cats* had staged 953 more performances than the Broadway show *Les Miserables*. At that time, *Les Miserables* had staged 6532 performances. How many performances had *Cats* staged by January 3, 2003? (*Source: The League of American Theatres and Producers*)

64. Charles Schulz, the creator of the Peanuts comic strip, was born in 1922. Scott Adams, the creator of the Dilbert comic strip, was born 35 years later. In what year was Scott Adams born? (*Source: The World Almanac*)

65. There were 9821 kidney transplants and 592 kidney-pancreas transplants performed in the United States in 2002. What was the total number of organ transplants performed in 2002 that involved a kidney? (*Source:* Organ Procurement and Transplantation Network)

66. There were 1450 heart transplants and 23 heart-lung transplants performed in the United States in 2002. What was the total number of organ transplants performed in 2002 that involved a heart? (*Source:* Organ Procurement and Transplantation Network)

 Combining Concepts

The table shows the number of Wal-Mart stores in ten states. Use this table to answer Exercises 67 through 72.

67. Which state listed in the table has the most Wal-Mart stores?

68. Which state listed in the table has the fewest Wal-Mart stores?

The Top States for Wal-Mart Stores in 2002	
State	**Number of Stores**
Pennsylvania	102
California	154
Florida	182
Georgia	115
Illinois	140
North Carolina	112
Tennessee	104
Missouri	128
Ohio	116
Texas	342

(*Source:* Wal-Mart Corporation)

69. What is the total number of Wal-Mart stores located in the three states with the most Wal-Mart stores?

70. Which pair of neighboring states have more Wal-Mart stores, Pennsylvania and Ohio or Florida and Georgia?

71. How many Wal-Mart stores are located in the ten states listed in the table? Use a calculator to check your total.

72. Wal-Mart operates stores in all 50 states. There are 1287 Wal-Mart stores located in the states not listed in the table. What is the total number of Wal-Mart stores in the United States?

73. In your own words, explain the commutative property of addition.

74. In your own words, explain the associative property of addition.

75. Add: $78,962 + 129,968,350 + 36,462,880$

76. Add: $56,468,980 + 1,236,785 + 986,768,000$

77. A student shows you the following computation and asks you to check it for him. What is your response and why?

$$
\begin{array}{r}
1288 \\
+\ \ 377 \\
\hline
1555
\end{array}
$$

1.4 Subtracting Whole Numbers

OBJECTIVES

Ⓐ Subtract whole numbers.

Ⓑ Subtract whole numbers when borrowing is necessary.

Ⓒ Solve problems by subtracting whole numbers.

SSM
TUTOR CENTER   SG   CD & VIDEO   MATH PRO   WEB

Ⓐ Subtracting Whole Numbers

If you have $5 and someone gives you $3, you have a total of $8, since $5 + 3 = 8$. Similarly, if you have $8 and then someone borrows $3, you have $5 left. **Subtraction** is finding the **difference** of two numbers.

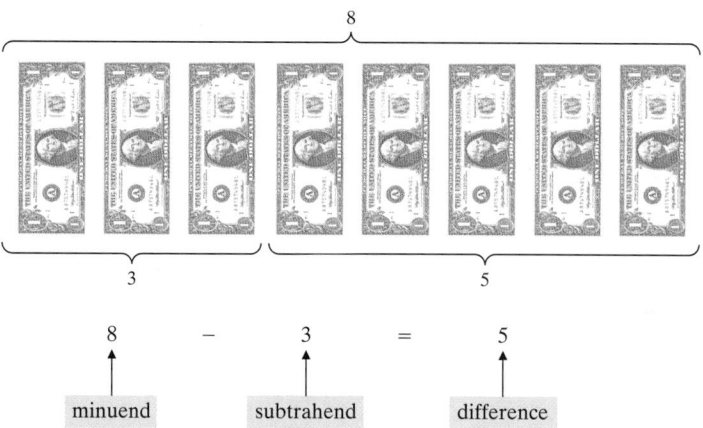

$$8 - 3 = 5$$

minuend   subtrahend   difference

Notice that addition and subtraction are very closely related. In fact, subtraction is defined in terms of addition.

$$8 - 3 = 5 \text{ because } 5 + 3 = 8$$

This means that subtraction can be *checked* by addition, and we say that addition and subtraction are reverse operations.

EXAMPLE 1 Subtract. Check each answer by adding.

a. $12 - 9$ **b.** $11 - 6$ **c.** $5 - 5$ **d.** $7 - 0$

Solution:

a. $12 - 9 = 3$ because $3 + 9 = 12$
b. $11 - 6 = 5$ because $5 + 6 = 11$
c. $5 - 5 = 0$ because $0 + 5 = 5$
d. $7 - 0 = 7$ because $7 + 0 = 7$ ●

Look again at Examples 1(c) and 1(d).

1(c) $5 - 5 = 0$ 1(d) $7 - 0 = 7$

same   difference   a number   difference is the
number   is 0   minus 0   same number

These two examples illustrate the subtraction properties of 0.

Subtraction Properties of 0

The difference of any number and that same number is 0. For example,

$$11 - 11 = 0$$

The difference of any number and 0 is that same number. For example,

$$45 - 0 = 45$$

Practice Problem 1

Subtract. Check each answer by adding.

a. $14 - 9$
b. $9 - 9$
c. $4 - 0$

Answers

1. a. 5   **b.** 0   **c.** 4

When subtraction involves numbers of two or more digits, it is more convenient to subtract vertically. For example, to subtract $893 - 52$,

$$
\begin{array}{r}
8\;9\;3 \\
-\;\;5\;2 \\
\hline
8\;4\;1
\end{array}
$$

← minuend
← subtrahend
← difference

$3 - 2$
$9 - 5$
$8 - 0$

Line up the numbers vertically so that the minuend is on top and the place values correspond. Subtract digits that are in corresponding places, starting with the ones place.

To check, add.

$$
\begin{array}{r}
\text{difference} \\
+\;\text{subtrahend} \\
\hline
\text{minuend}
\end{array}
\quad \text{or} \quad
\begin{array}{r}
841 \\
+\;\;\;52 \\
\hline
893
\end{array}
$$

← Since this is the original minuend, the problem checks.

Practice Problem 2

Subtract. Check by adding.

a. $4689 - 253$

b. $981 - 630$

EXAMPLE 2 Subtract: $7826 - 505$. Check by adding.

Solution:
$$
\begin{array}{r}
7826 \\
-\;\;505 \\
\hline
7321
\end{array}
$$

Check:
$$
\begin{array}{r}
7321 \\
+\;\;505 \\
\hline
7826
\end{array}
$$

B Subtracting with Borrowing

When a digit in the second number (subtrahend) is larger than the corresponding digit in the first number (minuend), **borrowing** is necessary. For example, consider

$$
\begin{array}{r}
81 \\
-\;63
\end{array}
$$

← We cannot take away 3 ones from 1 one, so we borrow.

Since the 3 in the ones place of 63 is larger than the 1 in the ones place of 81, borrowing is necessary. We borrow 1 ten from the tens place and add it to the ones place.

Borrowing

$$
\underset{\text{tens}}{8} - \underset{\text{ten}}{1} = \underset{\text{tens}}{7} \quad \rightarrow \quad
\begin{array}{r}
\overset{7\;\;11}{8\;\;\cancel{1}} \\
-\;6\;\;3
\end{array}
$$

← 1 ten + 1 one = 11 ones

Now we subtract the ones-place digits and then the tens-place digits.

$$
\begin{array}{r}
\overset{7\;\;11}{8\;\;\cancel{1}} \\
-6\;\;3 \\
\hline
1\;\;8
\end{array}
$$

← $11 - 3 = 8$
$7 - 6 = 1$

Check:

$$
\begin{array}{r}
18 \\
+\;63 \\
\hline
81
\end{array}
$$
The original minuend

Copyright 2004 Pearson Education, Inc.

Answers

2. a. 4436   **b.** 351

EXAMPLE 3 Subtract: 43 − 29. Check by adding.

Solution:

$$
\begin{array}{r}
\overset{3}{\cancel{4}}\,\overset{13}{\cancel{3}} \\
-2\ 9 \\
\hline
1\ 4
\end{array}
$$

Check:

$$
\begin{array}{r}
14 \\
+\ 29 \\
\hline
43
\end{array}
$$

Sometimes we may have to borrow from more than one place value. For example, to subtract 7631 − 152, we first borrow from the tens place.

$$
\begin{array}{r}
7\ 6\ \overset{2}{\cancel{3}}\ \overset{11}{\cancel{1}} \\
-\ \ 1\ 5\ 2 \\
\hline
9
\end{array}\quad \leftarrow 11 - 2 = 9
$$

In the tens place, 5 is greater than 2, so we borrow again. This time we borrow from the hundreds place

$$
\begin{array}{r}
\overset{12}{} \\
7\ \overset{5}{\cancel{6}}\ \overset{\cancel{2}}{\cancel{3}}\ \overset{11}{\cancel{1}} \\
-\ \ 1\ 5\ 2 \\
\hline
7\ 4\ 7\ 9
\end{array}
\quad
\begin{cases}
6 \text{ hundreds} - 1 \text{ hundred} = 5 \text{ hundreds} \\
\textbf{1 hundred} + 2 \text{ tens or} \\
10 \text{ tens} + 2 \text{ tens} = 12 \text{ tens}
\end{cases}
$$

Check:

$$
\begin{array}{r}
7479 \\
+\ \ 152 \\
\hline
7631
\end{array}\ \text{The original minuend}
$$

EXAMPLE 4 Subtract: 900 − 174. Check by adding.

Solution: In the ones place, 4 is larger than 0, so we borrow from the tens place. But the tens place of 900 is 0, so to borrow from the tens place we must first borrow from the hundreds place.

$$
\begin{array}{r}
\overset{8}{\cancel{9}}\ \overset{10}{\cancel{0}}\ 0 \\
-\ 1\ 7\ 4 \\
\end{array}
$$

Now borrow from the tens place.

$$
\begin{array}{r}
\overset{9}{} \\
\overset{8}{\cancel{9}}\ \overset{\cancel{10}}{\cancel{0}}\ \overset{10}{\cancel{0}} \\
-1\ 7\ 4 \\
\hline
7\ 2\ 6
\end{array}
$$

Check:

$$
\begin{array}{r}
\overset{1\ 1}{726} \\
-\ 174 \\
\hline
900
\end{array}
$$

C Solving Problems by Subtracting

Descriptions of real-life problems that suggest solving by subtraction include these key words or phrases:

Practice Problem 3

Subtract. Check by adding.

a. $\begin{array}{r} 227 \\ -\ 175 \\ \hline \end{array}$

b. $\begin{array}{r} 1136 \\ -\ 914 \\ \hline \end{array}$

c. $\begin{array}{r} 8627 \\ -\ 4119 \\ \hline \end{array}$

Practice Problem 4

Subtract. Check by adding.

a. $\begin{array}{r} 400 \\ -\ 164 \\ \hline \end{array}$

b. $\begin{array}{r} 200 \\ -\ 45 \\ \hline \end{array}$

c. $\begin{array}{r} 1000 \\ -\ 762 \\ \hline \end{array}$

Answers

3. a. 52 **b.** 222 **c.** 4508
4. a. 236 **b.** 155 **c.** 238

Key Words or Phrases	Examples	Symbols
subtract	subtract 5 from 8	$8 - 5$
difference	the difference of 10 and 2	$10 - 2$
less	17 less 3	$17 - 3$
take away	14 take away 9	$14 - 9$
decreased by	7 decreased by 5	$7 - 5$
subtracted from	9 subtracted from 12	$12 - 9$

Concept Check

In each of the following problems, identify which number is the minuend and which number is the subtrahend.

a. What is the result when 9 is subtracted from 20?

b. What is the difference of 15 and 8?

c. Find a number that is 15 fewer than 23.

Helpful Hint

Be careful when solving applications that suggest subtraction. Although order *does not* matter when adding, order *does* matter when subtracting. For example, $10 - 3$ and $3 - 10$ do *not* simplify to the same number.

Try the Concept Check in the margin.

Practice Problem 5

The radius of Earth is 6378 kilometers. The radius of Mars is 2981 kilometers less than the radius of Earth. What is the radius of Mars? (*Source:* National Space Science Data Center)

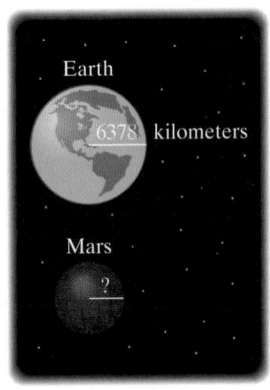

EXAMPLE 5 Finding the Radius of a Planet

The radius of Venus is 6052 kilometers. The radius of Mercury is 3612 kilometers less than the radius of Venus. Find the radius of Mercury. (*Source:* National Space Science Data Center)

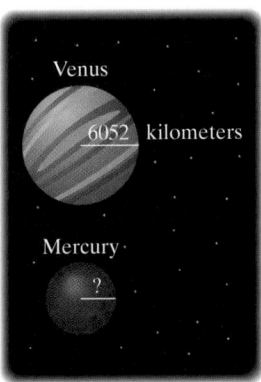

Solution:

In Words		Translate to Numbers
radius of Venus	$\rightarrow$	6052
less 3612	$\rightarrow$	-3612
radius of Mercury	$\rightarrow$	2440

The radius of Mercury is 2440 kilometers.

EXAMPLE 6 Calculating Miles Per Gallon

A subcompact car gets 42 miles per gallon of gas. A full-size car gets 17 miles per gallon of gas. How many more miles per gallon does the subcompact car get than the full-size car?

Solution:

In Words		Translate to Numbers
subcompact miles per gallon	$\rightarrow$	$\overset{3\ \ 12}{4\ 2}$
$-$ full-size miles per gallon	$\rightarrow$	$-\ 1\ 7$
more miles per gallon		$2\ 5$

The subcompact car gets 25 more miles per gallon than the full-size car.

Practice Problem 6

A new suit originally priced at $92 is now on sale for $47. How much money was taken off the original price?

Answers

5. 3397 km   **6.** $45

Concept Check: **a.** minuend: 20; subtrahend: 9
b. minuend: 15; subtrahend: 8
c. minuend: 23; subtrahend: 15

Helpful Hint

Since subtraction and addition are reverse operations, don't forget that a subtraction problem can be checked by adding.

Graphs can be used to visualize data. The graph shown next is called a **bar graph**. Notice that the height of each bar and each number above it correspond to the number's graph on the vertical number line, or axis.

EXAMPLE 7 Reading a Bar Graph

The graph below shows the ratings of Best Picture nominees since PG-13 was introduced in 1984. In this graph, each bar represents a different rating, and the height of each bar represents the number of Best Picture nominees for that rating.

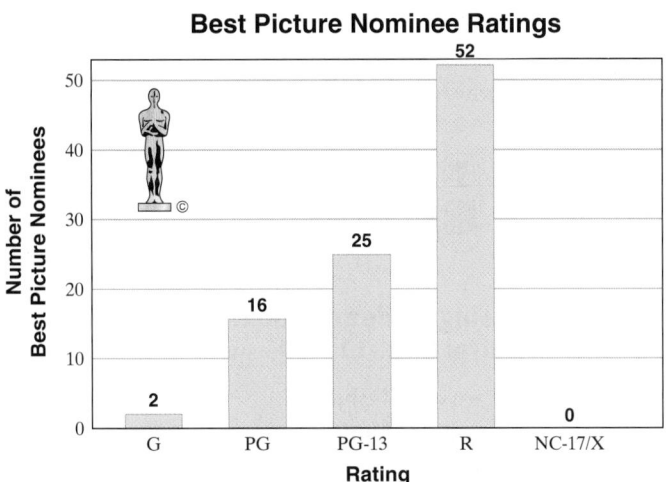

Best Picture Nominee Ratings

Source: The internet Movie Database and Cuadra Associates Movie Star Database, 2003.

a. Which rating did most Best Picture nominees have?

b. How many more Best Picture nominees were rated PG-13 than PG?

Solution:

a. The rating for most Best Picture nominees is the one corresponding to the highest bar, which is an R rating.

b. The number of Best Picture nominees rated PG-13 is 25. The number of Best Picture nominees rated PG is 16. To find how many more pictures were rated PG-13, we subtract.

$$25 - 16 = 9$$

Nine more Best Picture nominees were rated PG-13 than PG. ●

Practice Problem 7

Use the graph in Example 7 to answer the following:

a. Which rating had the least number of Best Picture nominees?

b. How many more Best Picture nominees were rated R than G?

Answers

7. a. NC-17/X   **b.** 50

CALCULATOR EXPLORATIONS

Subtracting Numbers

To subtract numbers on a calculator, find the keys marked $-$ and $=$ or ENTER.

For example, to find $83 - 49$ on a calculator, press the keys 83 $-$ 49 $=$ or ENTER.

The display will read [34]. Thus, $83 - 49 = 34$.

Use a calculator to subtract.

1. $865 - 95$

2. $76 - 27$

3. $147 - 38$

4. $366 - 87$

5. $9625 - 647$

6. $10,711 - 8925$

FOCUS ON Mathematical Connections

MODELING SUBTRACTION OF WHOLE NUMBERS

A mathematical concept can be represented or modeled in many different ways. For instance, subtraction can be represented by the following symbolic model:

$11 - 4$

The following verbal models can also represent subtraction of these same quantities:

"Four subtracted from eleven" or
"Eleven take away four"

Physical models can also represent mathematical concepts. In these models, a number is represented by that many objects. For example, the number 5 can be represented by five pennies, squares, paper clips, tiles, or bottle caps.

A physical representation of the number 5

Take-Away Model for Subtraction: 11 − 4

■ Start with 11 objects.
■ Take 4 objects away.
■ How many objects remain?

Comparison Model for Subtraction: 11 − 4

■ Start with a set of 11 of one type of object and a set of 4 of another type of object.

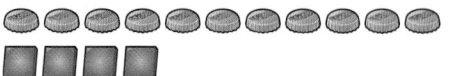

■ Make as many pairs that include one object of each type as possible.
■ How many more objects are in the larger set?

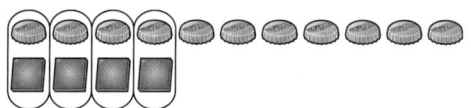

Missing Addend Model for Subtraction: 11 − 4

■ Start with 4 objects.
■ Continue adding objects until a total of 11 is reached.
■ How many more objects were needed to give a total of 11?

CRITICAL THINKING

Use an appropriate physical model for subtraction to solve each of the following problems. Explain your reasoning for choosing each model.

1. Sneha has assembled 12 computer components so far this shift. If his quota is 20 components, how many more components must he assemble to reach his quota?

2. Yuko wants to plant 14 daffodil bulbs in her yard. She planted 5 bulbs in the front yard. How many bulbs does she have left for planting in the backyard?

3. Todd is 19 years old and his sister Tanya is 13 years old. How much older is Todd than Tanya?

Name _____ Section _____ Date _____

Mental Math

Find each difference.

1. $9 - 2$ **2.** $6 - 6$ **3.** $5 - 0$ **4.** $44 - 22$ **5.** $93 - 93$

6. $700 - 400$ **7.** $700 - 300$ **8.** $700 - 700$ **9.** $600 - 100$ **10.** $600 - 0$

EXERCISE SET 1.4

A *Subtract. Check by adding. See Examples 1 and 2.*

1. $\begin{array}{r} 67 \\ -\ 23 \\ \hline \end{array}$ **2.** $\begin{array}{r} 72 \\ -\ 41 \\ \hline \end{array}$ **3.** $\begin{array}{r} 82 \\ -\ 22 \\ \hline \end{array}$ **4.** $\begin{array}{r} 27 \\ -\ 10 \\ \hline \end{array}$ **5.** $\begin{array}{r} 389 \\ -\ 124 \\ \hline \end{array}$ **6.** $\begin{array}{r} 572 \\ -\ 321 \\ \hline \end{array}$

7. $\begin{array}{r} 677 \\ -\ 423 \\ \hline \end{array}$ **8.** $\begin{array}{r} 766 \\ -\ 324 \\ \hline \end{array}$ **9.** $\begin{array}{r} 998 \\ -\ 453 \\ \hline \end{array}$ **10.** $\begin{array}{r} 912 \\ -\ 610 \\ \hline \end{array}$ **11.** $\begin{array}{r} 749 \\ -\ 149 \\ \hline \end{array}$ **12.** $\begin{array}{r} 257 \\ -\ 257 \\ \hline \end{array}$

A **B** *Subtract. Check by adding. See Examples 1 through 4.*

13. $\begin{array}{r} 62 \\ -\ 37 \\ \hline \end{array}$ **14.** $\begin{array}{r} 55 \\ -\ 29 \\ \hline \end{array}$ **15.** $\begin{array}{r} 70 \\ -\ 25 \\ \hline \end{array}$ **16.** $\begin{array}{r} 80 \\ -\ 37 \\ \hline \end{array}$ **17.** $\begin{array}{r} 938 \\ -\ 792 \\ \hline \end{array}$ **18.** $\begin{array}{r} 436 \\ -\ 275 \\ \hline \end{array}$

19. $\begin{array}{r} 922 \\ -\ 634 \\ \hline \end{array}$ **20.** $\begin{array}{r} 674 \\ -\ 299 \\ \hline \end{array}$ **21.** $\begin{array}{r} 600 \\ -\ 432 \\ \hline \end{array}$ **22.** $\begin{array}{r} 300 \\ -\ 149 \\ \hline \end{array}$ **23.** $\begin{array}{r} 42 \\ -\ 36 \\ \hline \end{array}$ **24.** $\begin{array}{r} 73 \\ -\ 29 \\ \hline \end{array}$

25. $\begin{array}{r} 923 \\ -\ 476 \\ \hline \end{array}$ **26.** $\begin{array}{r} 813 \\ -\ 227 \\ \hline \end{array}$ **27.** $\begin{array}{r} 6283 \\ -\ 560 \\ \hline \end{array}$ **28.** $\begin{array}{r} 5349 \\ -\ 720 \\ \hline \end{array}$ **29.** $\begin{array}{r} 533 \\ -\ 29 \\ \hline \end{array}$ **30.** $\begin{array}{r} 724 \\ -\ 16 \\ \hline \end{array}$

31. $\begin{array}{r} 200 \\ -\ 111 \\ \hline \end{array}$ **32.** $\begin{array}{r} 300 \\ -\ 211 \\ \hline \end{array}$ **33.** $\begin{array}{r} 1983 \\ -\ 1904 \\ \hline \end{array}$ **34.** $\begin{array}{r} 1983 \\ -\ 1914 \\ \hline \end{array}$ **35.** $\begin{array}{r} 56,422 \\ -\ 16,508 \\ \hline \end{array}$ **36.** $\begin{array}{r} 76,652 \\ -\ 29,498 \\ \hline \end{array}$

37. $50,000 - 17,289$ **38.** $40,000 - 23,582$ **39.** $7020 - 1979$ **40.** $6050 - 1878$

41. $51,111 - 19,898$ **42.** $62,222 - 39,898$ **43.** Subtract 5 from 9. **44.** Subtract 9 from 21.

45. Find the difference of 41 and 21.

46. Find the difference of 16 and 5.

47. Subtract 56 from 63.

48. Subtract 41 from 59.

49. Find 108 less 36.

50. Find 25 less 12.

51. Find 12 subtracted from 100.

52. Find 86 subtracted from 90.

Solve. See Examples 5 through 7.

53. Dyllis King is reading a 503-page book. If she has just finished reading page 239, how many more pages must she read to finish the book?

54. When Lou and Judy Zawislak began a trip, the odometer read 55,492. When the trip was over, the odometer read 59,320. How many miles did they drive on their trip?

55. In 1997, the hole in the Earth's ozone layer over Antartica was about 21 million square kilometers in size. In 2001, the hole had grown to 25 million square kilometers. By how much has the hole grown from 1997 to 2001? (*Source*: U.S. Environmental Protection Agency EPA: *http://www.epa.gov/ozone/ science/hole/sizedata.html#areatime*)

56. Bamboo can grow to 98 feet while Pacific giant kelp (a type of seaweed) can grow to 197 feet. How much taller is the kelp than the bamboo?

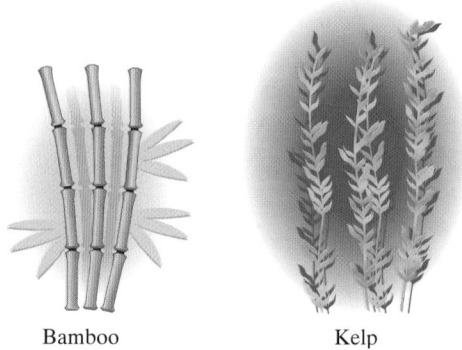

Bamboo   Kelp

57. The peak of Mt. McKinley in Alaska is 20,320 feet above sea level. The peak of Long's Peak in Colorado is 14,255 feet above sea level. How much higher is the peak of Mt. McKinley than Long's Peak? (*Source:* U.S. Geological Survey)

58. On one day in May the temperature in Paddin, Indiana, dropped 27 degrees from 2 p.m. to 4 p.m. If the temperature at 2 p.m. was 73° Fahrenheit, what was the temperature at 4 p.m.?

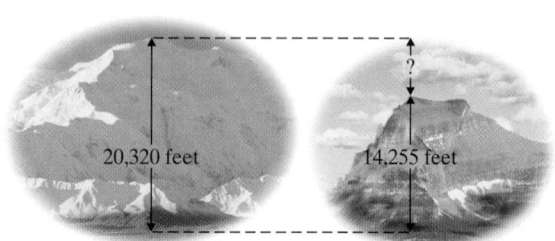

20,320 feet   14,255 feet

Mt. McKinley, Alaska   Long's Peak, Colorado

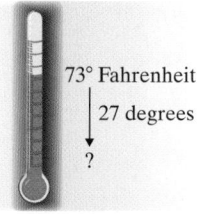

73° Fahrenheit

27 degrees

?

59. Buhler Gomez has a total of $539 in his checking account. If he writes a check for each of the items below, how much money will be left in his account?

Bell South $27
Cleco $101
Mellon Finance $236

60. Pat Salanki's blood cholesterol level is 243. The doctor tells him it should be decreased to 185. How much of a decrease is this?

61. The distance from Kansas City to Denver is 645 miles. Hays, Kansas, lies on the road between the two and is 287 miles from Kansas City. What is the distance between Hays and Denver?

62. Nhoc Tran is trading his car in on a new car. The new car costs $ 15,425. His car is worth $7998. How much more money does he need to buy the new car?

63. A new home theater system with DVD player costs $732. Pat Gomez has $971 in his checking account. How much will he have left in his checking account after he buys the home theater?

64. A stereo that regularly sells for $547 is discounted by $99 in a sale. What is the sale price?

65. During the 2000–2001 regular season, Jerry Stackhouse of the Detroit Pistons led the NBA in total points scored with 2380. The Philadelphia 76ers' Allen Iverson placed second for total points scored with 2207. How many more points did Stackhouse score than Iverson during the 2000–2001 regular season? (*Source:* National Basketball Association)

66. In 1999, Americans bought 233,125 Ford Expeditions. In 2000, 19,642 fewer Expeditions were sold in the United States. How many Expeditions were sold in the United States in 2000? (*Source:* Ford Motor Company)

67. In 2000, there were 38,803 boxers registered with the American Kennel Club. In 2001, there were 1768 fewer boxers registered. How many boxers were registered with the AKC in 2001? (*Source:* American Kennel Club)

68. In the United States, there were 41,589 tornadoes from 1950 through 2000. In all, 13,205 of these tornadoes occurred from 1990 through 2000. How many tornadoes occurred during the period prior to 1990? (*Source:* Storm Prediction Center, National Weather Service)

69. Jo Keen and Trudy Waterbury were candidates for student government president. Who won the election if the votes were cast as follows? By how many votes did the winner win?

| Class | Candidate | |
	Jo	Trudy
Freshman	276	295
Sophomore	362	122
Junior	201	312
Senior	179	18

70. Two students submitted advertising budgets for a student government fund-raiser. If $1200 is available for advertising, how much excess would each budget have?

	Student A	Student B
Radio ads	$600	$300
Newspaper ads	$200	$400
Posters	$150	$240
Handbills	$120	$170

71. The population of Florida grew from 12,937,926 in 1990 to 15,982,378 in 2000. What was Florida's population increase over this time period? (*Source:* U.S. Census Bureau)

72. The population of El Paso, Texas, was 515,342 in 1990 and 563,662 in 2000. By how much did the population of El Paso grow from 1990 to 2000? (*Source:* U.S. Census Bureau)

73. Until recently, the world's largest permanent maze was located in Ruurlo, Netherlands. This maze of beech hedges covers 94,080 square feet. A new hedge maze using hibiscus bushes at the Dole Plantation in Wahiawa, Hawaii, covers 100,000 square feet. How much larger is the Dole Plantation maze than the Ruurlo maze? (*Source*: The Guinness Book of Records)

74. There were only 27 California condors in the entire world in 1987. By 2002, the number of California condors had increased to 200. How much of an increase was this? (*Source:* California Department of Fish and Game)

75. The Mackinac Bridge is a suspension bridge that connects the lower and upper peninsulas of Michigan across the Straits of Mackinac. Its total length is 26,372 feet. The Lake Pontchartrain Bridge is a twin concrete trestle bridge in Slidell, Louisiana. Its total length is 28,547 feet. Which bridge is longer and by how much? (*Sources:* Mackinac Bridge Authority and Federal Highway Administration, Bridge Division)

76. Papa John's is the third largest pizza chain in the United States. In 1999, there were 2486 Papa John's restaurants worldwide. By 2000, the number of Papa John's restaurants had grown to 2817. How many new Papa John's restaurants were added during 2000? (*Source:* Papa John's International Inc.)

The bar graph shows the number of passenger arrivals and departures in 2001 (in thousands) for the top busiest airports in the U.S. Use this graph to answer Exercises 77 through 80. See Example 7.

77. Which airport is the busiest?

78. Which airports have fewer than 60 million passenger arrivals and departures per year?

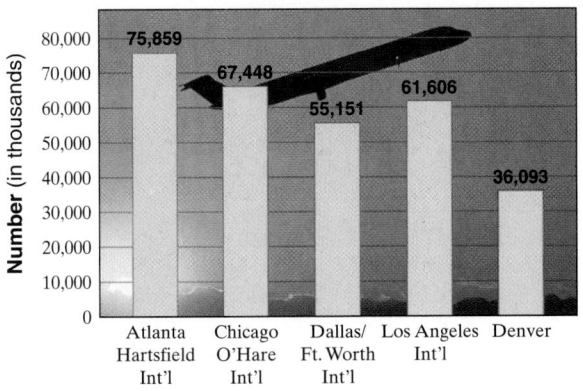

Passenger Arrivals and Departures 2001 (in thousands) **for the Busiest Airports in the U.S.**

Source: Airports Council International

79. How many more passenger arrivals and departures per year does the Chicago O'Hare International Airport have than the Los Angeles International Airport?

80. How many more passenger arrivals and departures per year does the Atlanta Hartsfield International Airport have than the Dallas/Ft. Worth International Airport?

 Combining Concepts

The table shows the top ten leading advertisers in the United States in 2001 and the amount of money spent in that year on advertising. Use this table to answer Exercises 81 through 85.

Company	Yearly Amount Spent on Advertising
General Motors Corp.	$3,374,400,000
Procter & Gamble Co.	$2,540,600,000
Ford Motor Co.	$2,408,200,000
PepsiCo	$2,210,400,000
Pfizer	$2,189,500,000
DaimlerChrysler	$1,985,300,000
AOL Time Warner.	$1,885,300,000
Phillip Morris Cos	$1,815,700,000
Walt Disney Co.	$1,757,300,000
Johnson & Johnson	$1,618,100,000

(*Source:* AdAge.com & the June 24, 2002 issue of *Advertising Age*)

81. Which company spent more than $3 billion on advertising?

82. Which companies spent fewer than $2 billion on advertising?

83. How much less money did DaimlerChrysler spend on advertising than General Motors Corp.?

84. How much more money did PepsiCo spend on advertising than Pfizer?

85. Find the total amount of money spent by these ten companies on advertising.

86. The local college library is having a Million Pages of Reading promotion. The freshmen have read a total of 289,462 pages; the sophomores have read a total of 369,477 pages; the juniors have read a total of 218,287 pages; and the seniors have read a total of 121,685 pages. Have they reached a goal of one million pages? If not, how many more pages need to be read?

Fill in the missing digits in each problem.

87.
```
   526_
 − 2_85
 ─────
  28_4

 ─────
  ──_

 ─────
  _
```

88.
```
  10,_4_
 − 8 5_4
 ─────
   _710

 ─────
  _ _

 ─────
  _
```

89. Is there a commutative property of subtraction? In other words, does order matter when subtracting? Why or why not?

90. The Gap, Inc., had net income of $13,847,873,000 in 2001. In 1997 The Gap's net income was only $6,507,825,000. How much did The Gap's net income increase from 1997 to 2001? (*Source:* The Gap, Inc.)

FOCUS ON **Business and Career**

THE EARNING POWER OF A COLLEGE DEGREE

According to data from the U.S. Bureau of the Census, average annual wages tend to increase with additional education. The following table shows the average annual earnings for both men and women in 1996 and 2001 for workers with only a high school diploma and for those who have earned a bachelor's degree.

Gender	Year	Average Annual Wages	
		High School Diploma	**Bachelor's Degree**
Men	1996	$24,814	$39,624
	2001	$28,343	$49,985
Women	1996	$12,702	$25,192
	2001	$15,665	$30,973

CRITICAL THINKING

Use the table to answer each question.

1. First, analyze men's earnings.
 a. Did the average annual wages for men with a high school diploma increase or decrease from 1996 to 2001? By how much?
 b. Did the average annual wages for men with a bachelor's degree increase or decrease from 1996 to 2001? By how much?

c. How much more could a man with a bachelor's degree earn than a man with a high school diploma in 1996? In 2001?

2. Now analyze women's earnings.
 a. Did the average annual wages for women with a high school diploma increase or decrease from 1996 to 2001? By how much?
 b. Did the average annual wages for women with a bachelor's degree increase or decrease from 1996 to 2001? By how much?
 c. How much more could a woman with a bachelor's degree earn than a woman with a high school diploma in 1996? In 2001?

3. Now compare men's and women's earnings.
 a. Find the difference between average annual wages (the "wage gap") for men and women with high school diplomas in 1996.
 b. Find the wage gap for men and women with high school diplomas in 2001.
 c. Find the wage gap for men and women with bachelor's degrees in 1996.
 d. Find the wage gap for men and women with bachelor's degrees in 2001.

4. Write a paragraph summarizing the conclusions that can be drawn from this table of data. Identify any apparent trends.

1.5   Rounding and Estimating

Ⓐ Rounding Whole Numbers

Rounding a whole number means approximating it. A rounded whole number is often easier to use, understand, and remember than the precise whole number. For example, instead of trying to remember the Iowa state population as 2,851,792, it is much easier to remember it rounded to the nearest million: 3 million people.

To understand rounding, let's look at the following illustrations. The whole number 36 is closer to 40 than 30, so 36 rounded to the nearest ten is 40.

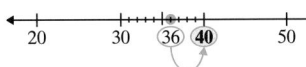

The whole number 52 rounded to the nearest ten is 50 because 52 is closer to 50 than to 60.

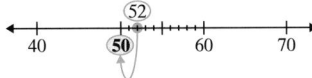

In trying to round 25 to the nearest ten, we see that 25 is halfway between 20 and 30. It is not closer to either number. In such a case, we round to the larger ten, that is, to 30.

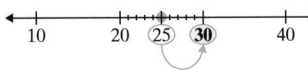

To round a whole number without using a number line, follow these steps:

Rounding Whole Numbers to a Given Place Value

Step 1. Locate the digit to the right of the given place value to be rounded.

Step 2. If this digit is 5 or greater, add 1 to the digit in the given place value and replace each digit to its right by 0.

Step 3. If this digit is less than 5, keep the digit in the given place value and replace each digit to its right by 0.

EXAMPLE 1   Round 568 to the nearest ten.

Solution: 5 6 ⑧
                    ↑
              tens place   The digit to the right of the tens place is the ones place, which is circled.

              5 6 ⑧
                 ↑
         Add 1. Replace   Since the circled digit is 5 or greater, add 1 to the 6 in the
              with 0.   tens place and replace the digit to the right by 0.

We find that 568 rounded to the nearest ten is 570.

Practice Problem 1

Round to the nearest ten.

a. 46

b. 731

c. 125

Answers

1. a. 50   **b.** 730   **c.** 130

Practice Problem 2

Round to the nearest thousand.

a. 56,702

b. 7444

c. 291,500

Practice Problem 3

Round to the nearest hundred.

a. 2777

b. 38,152

c. 762,955

Concept Check

Round each of the following numbers to the nearest *hundred*. Explain your reasoning.

a. 79

b. 33

Practice Problem 4

Round each number to the nearest ten to find an estimated sum.

```
     79
     35
     42
     21
  +  98
```

Answers

2. a. 57,000 **b.** 7000 **c.** 292,000   **3. a.** 2800

  b. 38,200 **c.** 763,000   **4.** 280

Concept Check:

a. 100   **b.** 0

EXAMPLE 2   Round 278,362 to the nearest thousand.

Solution: Thousands place

        ↓   ┌─ 3 is less than 5.

      278,③62

        ↑   ↑

      Do not add 1. Replace with zeros.

The number 278,362 rounded to the nearest thousand is 278,000.

EXAMPLE 3   Round 248,982 to the nearest hundred.

Solution: Hundreds place

        ↓   ┌─ 8 is greater than or equal to.5

      248,9⑧2

        ↑

      Add 1.   9 + 1 = 10, so replace the digit 9 by 0 and carry 1 to the place
                    value to the left.

```
              8+1   0
   2    4    8,  9   8   2
```

              ↑   ↑

        Add 1.   Replace with zeros.

The number 248,982 rounded to the nearest hundred is 249,000.

Try the Concept Check in the margin.

B   Estimating Sums and Differences

By rounding addends, we can estimate sums. An estimated sum is appropriate when an exact sum is not necessary. To estimate the sum shown, round each number to the nearest hundred and then add.

```
   768    rounds to        800
  1952    rounds to       2000
   225    rounds to        200
+  149    rounds to     +  100
                          3100
```

The estimated sum is 3100, which is close to the exact sum of 3094.

EXAMPLE 4   Round each number to the nearest hundred to find an estimated sum.

```
   294
   625
  1071
+  349
```

Solution:
```
    294    rounds to        300
    625    rounds to        600
   1071    rounds to       1100
+   349    rounds to     +  300
                           2300
```
The estimated sum is 2300. (The exact sum is 2339.)

EXAMPLE 5 Round each number to the nearest hundred to find an estimated difference.

          4725
      −   2879

Solution:

      4725   rounds to   4700
    − 2879   rounds to   − 2900
                                1800

The estimated difference is 1800. (The exact difference is 1846.) ●

C Solving Problems by Estimating

Making estimates is often the quickest way to solve real-life problems when their solutions do not need to be exact.

EXAMPLE 6 Estimating Distances

Jose Guillermo is trying to estimate quickly the distance from Temple, Texas, to Brenham, Texas. Round each distance given on the map to the nearest ten to estimate the total distance.

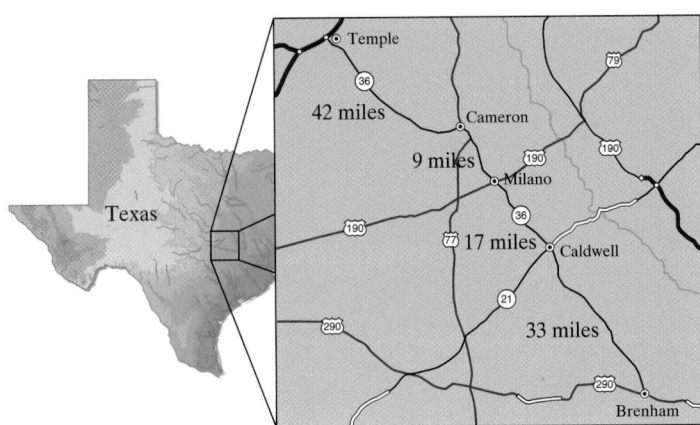

Solution:

Distance		Estimation
42	rounds to	40
9	rounds to	10
17	rounds to	20
+ 33	rounds to	+ 30
		100

It is approximately 100 miles from Temple to Brenham. (The exact distance is 101 miles.) ●

Practice Problem 5

Round each number to the nearest thousand to find an estimated difference.
          4725
      −   2879

Practice Problem 6

Tasha Kilbey is trying to estimate how far it is from Grove, Kansas, to Hays, Kansas. Round each given distance on the map to the nearest ten to estimate the total distance.

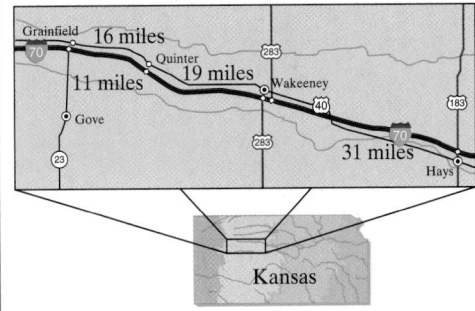

Answers

5. 2000   **6.** 80 mi

Practice Problem 7

In a recent year, there were 120,624 reported cases of chicken pox, 22,866 reported cases of tuberculosis, and 45,970 reported cases of salmonellosis in the United States. Round each number to the nearest ten-thousand to estimate the total number of cases reported for these diseases. (*Source*: Centers for Disease Control and Prevention)

EXAMPLE 7   Estimating Data

In three recent years the numbers of reported cases of mumps in the United States were 906, 1537, and 1692. Round each number to the nearest hundred to estimate the total number of cases reported over this period. (*Source*: Centers for Disease Control and Prevention)

Solution:

Number of Cases		Estimation
906	rounds to	900
1537	rounds to	1500
+ 1692	rounds to	+ 1700
		4100

The approximate number of cases reported over this period is 4100.   ●

Answer

7. 190,000

EXERCISE SET 1.5

(A) *Round each whole number to the given place. See Examples 1 through 3.*

1. 632 to the nearest ten

2. 273 to the nearest ten

3. 635 to the nearest ten

4. 275 to the nearest ten

5. 792 to the nearest ten

6. 394 to the nearest ten

7. 395 to the nearest ten

8. 582 to the nearest ten

9. 1096 to the nearest ten

10. 2198 to the nearest ten

11. 42,682 to the nearest thousand

12. 42,682 to the nearest ten-thousand

13. 248,695 to the nearest hundred

14. 179,406 to the nearest hundred

15. 36,499 to the nearest thousand

16. 96,501 to the nearest thousand

17. 99,995 to the nearest ten

18. 39,994 to the nearest ten

19. 59,725,642 to the nearest ten-million

20. 39,523,698 to the nearest million

Complete the table by rounding the given number to the given place value.

		Tens	Hundreds	Thousands
21.	5281			
22.	7619			
23.	9444			
24.	7777			
25.	14,876			
26.	85,049			

27. Round to the nearest thousand the 2001–2002 enrollment of East Tennessee State University: 11,331. (*Source:* Peterson's)

28. Round to the nearest hundred the hourly cost of operating a B747-400 aircraft: $6592. (*Source:* Air Transport Association of America)

Round each number to the indicated place.

29. Kareem Abdul-Jabbar holds the NBA record for points scored, a total of 38,387 over his NBA career. Round this number to the nearest thousand. (*Source:* National Basketball Association)

30. It takes 10,759 days for Saturn to make a complete orbit around the Sun. Round this number to the nearest hundred. (*Source:* National Space Science Data Center)

31. In 2001, U.S. farms produced 116,856,000 bushels of oats. Round the oat production figure to the nearest ten-million. (*Source:* U.S. Department of Agriculture)

32. In 2002, there were 485,536 U.S. Army personnel on active duty. Round this personnel figure to the nearest ten-thousand.

33. A total of 18,188 women participate in college soccer in the United States. Round this number to the nearest thousand. (*Source:* NCAA Sports)

34. The United States currently has 110,547,000 cellular mobile phone users (about 40% of population) while Austria has 6,150,000 users (about 75% of population). Round each of the user numbers to the nearest million. (*Note:* We will study percents in a later chapter.) (*Source:* Siemens AG, International Telecom Statistics, 2001)

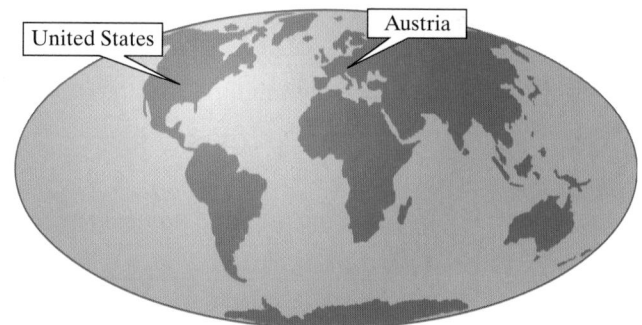

B *Estimate the sum or difference by rounding each number to the nearest ten. See Examples 4 and 5.*

35.
```
   29
   35
   42
 + 16
```

36.
```
   62
   72
   15
 + 19
```

37.
```
   649
 - 272
```

38.
```
   555
 - 235
```

Estimate the sum or difference by rounding each number to the nearest hundred. See Examples 4 and 5.

39.
```
   1812
   1776
 + 1945
```

40.
```
   2010
   2001
 + 1984
```

41.
```
   1774
 - 1492
```

42.
```
   1989
 - 1870
```

43.
```
   2995
   1649
 + 3940
```

44.
```
   799
   1655
 + 271
```

Two of the given calculator answers below are incorrect. Find them by estimating each sum.

45. 362 + 419   781

46. 522 + 785   1307

47. 432 + 679 + 198   1139

48. 229 + 443 + 606   1278

49. 7806 + 5150   12,956

50. 5233 + 4988   9011

51. 31,439 + 18,781   50,220

52. 68,721 + 52,335   121,056

Solve each problem by estimating. See Examples 6 and 7.

53. Campo Appliance Store advertises three refrigerators on sale at $799, $1299, and $999. Round each cost to the nearest hundred to estimate the total cost.

54. Jared Nuss scored 89, 92, 100, 67, 75, and 79 on his calculus tests. Round each score to the nearest ten to estimate his total score.

55. Arlene Neville wants to estimate quickly the distance from Stockton to LaCrosse. Round each distance given on the map to the nearest ten miles to estimate the total distance.

56. Carmelita Watkins is pricing new stereo systems. One system sells for $1895 and another system sells for $1524. Round each price to the nearest hundred dollars to estimate the difference in price of these systems.

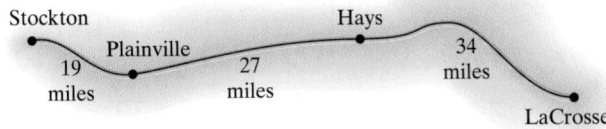

57. The peak of Mt. McKinley, in Alaska, is 20,320 feet above sea level. The top of Mt. Rainier, in Washington, is 14,410 feet above sea level. Round each height to the nearest thousand to estimate the difference in elevation of these two peaks. (*Source:* U.S. Geological Survey)

58. The Gonzales family took a trip and traveled 458, 489, 377, 243, 69, and 702 miles on six consecutive days. Round each distance to the nearest hundred to estimate the distance they traveled.

59. In 2000 the population of Chicago was 2,896,016, and the population of Philadelphia was 1,517,550. Round each population to the nearest hundred-thousand to estimate how much larger Chicago was than Philadelphia. (*Source:* U.S. Census Bureau, 2000 census)

60. The distance from Kansas City to Boston is 1429 miles and from Kansas City to Chicago, 530 miles. Round each distance to the nearest hundred to estimate how much farther Boston is from Kansas City than Chicago is.

61. In the 1964 presidential election, Lyndon Johnson received 41,126,233 votes and Barry Goldwater received 27,174,898 votes. Round each number of votes to the nearest million to estimate the number of votes by which Johnson won the election.

62. Enrollment figures at Normal State University showed an increase from 49,713 credit hours in 1988 to 51,746 credit hours in 1989. Round each number to the nearest thousand to estimate the increase.

63. Head Start is a national program that provides developmental and social services for America's low-income preschool children ages three to five. Enrollment figures in Head Start programs showed an increase from 750,696 children in 1995 to 905,235 children in 2000. Round each number of children to the nearest thousand to estimate this increase. (*Source:* Head Start Bureau)

64. In 2000, General Motors produced 271,800 Saturn cars. Similarly, in 1999 only 232,570 Saturns were produced. Round each number of cars to the nearest thousand to estimate the increase in Saturn production from 1999 to 2000. (*Source:* General Motors Corporation)

65. Jupiter has the largest planetary moon. It is called Ganymede and has a diameter of 3274 miles. In contrast, the diameter of our moon is 2159 miles. Round each of these numbers to the nearest hundred to estimate the difference in the diameters.

66. AOL Time Warner has the most visited website with 39,458,234 unique visitors per week. In second place is Yahoo! with 32,514,019 visitors per week. Round each number to the nearest million to estimate how many more visitors AOL Time Warner has than Yahoo!. (*Source*: Nielsen/Net Ratings Audience Measurement Service)

3,274 miles — Ganymede   2,159 miles — Moon

The following table (from Section 1.4) shows the top ten leading advertisers in the United States for 2001 and the amount of money spent in that year on advertising. Use this table to answer Exercises 67 through 70.

Company	Yearly Amount Spent on Advertising
General Motors Corp.	$3,374,400,000
Procter & Gamble Co.	$2,540,600,000
Ford Motor Co.	$2,408,200,000
PepsiCo	$2,210,400,000
Pfizer	$2,189,500,000
DaimlerChrysler	$1,985,300,000
AOL Time Warner.	$1,885,300,000
Phillip Morris Cos	$1,815,700,000
Walt Disney Co.	$1,757,300,000
Johnson & Johnson	$1,618,100,000

(*Source:* Ad Age.com & the June 24, 2002 issue of *Advertising*)

67. Approximate the amount of money spent on advertising by General Motors Corp. to the nearest hundred-million.

68. Approximate the amount of money spent on advertising by Johnson & Johnson Ltd. to the nearest hundred-million.

69. Approximate the amount of money spent on advertising by PepsiCo to the nearest billion.

70. Approximate the amount of money spent on advertising by Procter & Gamble Co. to the nearest billion.

◆ Combining Concepts

71. A number rounded to the nearest hundred is 8600. Determine the smallest possible number.

72. Determine the largest possible number.

73. On August 23, 1989, it was estimated that 1,500,000 people joined hands in a human chain stretching 370 miles to protest the fiftieth anniversary of the pact that allowed what was then the Soviet Union to annex the Baltic nations in 1939. If the estimate of the number of people is to the nearest hundred-thousand, determine the largest possible number of people in the chain.

74. In your own words, explain how to round a number to the nearest thousand.

75. Estimate the perimeter by first rounding each length to the nearest hundred.

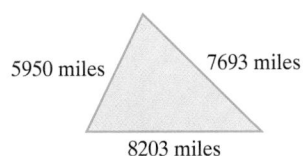

5950 miles   7693 miles

8203 miles

Internet Excursions

 Go To: http://www.prenhall.com/martin-gay_prealgebra What's Related

This World Wide Web site provides access to a site that allows the user to find the distance, as the crow flies, between two cities. It also gives population and elevation figures for each city. Visit this site and answer the following questions:

76. Choose three cities in your state: (A) _____, (B) _____, and (C) _____.

78. List the population for each city. Which city has the greatest population? How many more people does that city have than the city with the least population? Check your results by estimating.

77. Find the distance between city A and city B. Then find the distance from city B to city C. If a crow were to fly directly from city A to city B and then continue on to city C, how far would it fly?

1.6 Multiplying Whole Numbers and Area

Suppose that we wish to count the number of desks in a classroom. The desks are arranged in 5 rows, and each row has 6 desks.

6 desks in each row

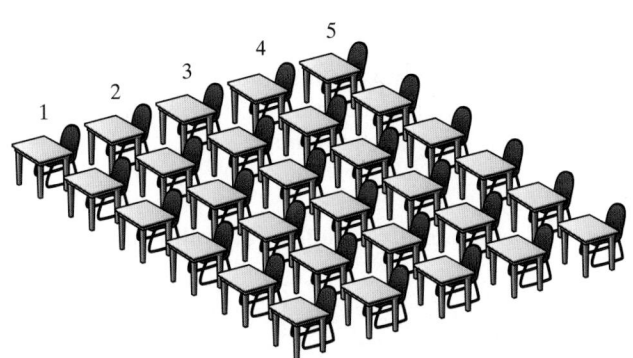

Adding 5 sixes gives the total number of desks: $6 + 6 + 6 + 6 + 6 = 30$ desks. When each addend is the same, we refer to this as **repeated addition**. **Multiplication** is repeated addition but with different notation.

$$\underbrace{6 + 6 + 6 + 6 + 6}_{\text{5 sixes}} = \underset{\text{factor}}{5} \quad \underset{}{\times} \quad \underset{\text{factor}}{6} = \underset{\text{product}}{30}$$

The $\times$ is called a **multiplication sign**. The numbers 5 and 6 are called **factors**. The number 30 is called the **product**. The notation 5×6 is read as "five times six." The symbols $\cdot$ and () can also be used to indicate multiplication.

$$5 \times 6 = 30, \quad 5 \cdot 6 = 30, \quad (5)(6) = 30, \quad \text{and} \quad 5(6) = 30$$

Try the Concept Check in the margin.

Ⓐ Using the Properties of Multiplication

As with addition, we memorize products of one-digit whole numbers and then use certain properties of multiplication to multiply larger numbers. Notice in the appendix that when any number is multiplied by 0, the result is always 0. This is called the **multiplication property of 0**.

Multiplication Property of 0

The product of 0 and any number is 0. For example,

$5 \cdot 0 = 0$

$0 \cdot 8 = 0$

Also notice in the appendix that when any number is multiplied by 1, the result is always the original number. We call this result the **multiplication property of 1**.

Multiplication Property of 1

The product of 1 and any number is that same number. For example,

$1 \cdot 9 = 9$

$6 \cdot 1 = 6$

Concept Check

a. Rewrite $4 + 4 + 4 + 4 + 4 + 4 + 4$ using multiplication.

b. Rewrite 3×16 as repeated addition. Is there more than one way to do this? If so, show all ways.

Answers

Concept Check:

a. $7 \times 4 = 28$

b. $16 + 16 + 16 = 48$; yes

$3 + 3 + 3 + 3 + 3 + 3 + 3 + 3$

$+ 3 + 3 + 3 + 3 + 3 + 3 = 48$

Practice Problem 1

Multiply.

a. 3×0

b. $4(1)$

c. $(0)(34)$

d. $1 \cdot 76$

EXAMPLE 1 Multiply.

a. 6×1 **b.** $0(8)$ **c.** $1 \cdot 45$ **d.** $(75)(0)$

Solution:

a. $6 \times 1 = 6$ **b.** $0(8) = 0$

c. $1 \cdot 45 = 45$ **d.** $(75)(0) = 0$ ●

Like addition, multiplication is commutative and associative. Notice that when multiplying two numbers, the order of these numbers can be changed without changing the product. For example,

$$3 \cdot 5 = 15 \quad \text{and} \quad 5 \cdot 3 = 15$$

This property is the **commutative property of multiplication**.

Commutative Property of Multiplication

Changing the **order** of two factors does not change their product. For example,

$$9 \cdot 2 = 18 \quad \text{and} \quad 2 \cdot 9 = 18$$

Another property that can help us when multiplying is the **associative property of multiplication**. This property states that when multiplying numbers, the grouping of the numbers can be changed without changing the product. For example,

$$2 \cdot (3 \cdot 4) = 2 \cdot 12 = 24$$

Also,

$$(2 \cdot 3) \cdot 4 = 6 \cdot 4 = 24$$

Both groupings give a product of 24.

Associative Property of Multiplication

Changing the **grouping** of factors does not change their product. For example,

$$5 \cdot (3 \cdot 2) = (5 \cdot 3) \cdot 2$$
$$5 \cdot (3 \cdot 2) = 5 \cdot 6 = 30 \quad \text{and} \quad (5 \cdot 3) \cdot 2 = 15 \cdot 2 = 30$$

With these properties, along with the **distributive property**, we can find the product of any whole numbers. The distributive property says that multiplication **distributes** over addition. For example, notice that $3(2 + 5)$ is the same as $3 \cdot 2 + 3 \cdot 5$.

$$3(2 + 5) = 3(7) = 21$$

$$3 \cdot 2 + 3 \cdot 5 = 6 + 15 = 21$$

Notice in $3(2 + 5) = 3 \cdot 2 + 3 \cdot 5$ that each number inside the parentheses is multiplied by 3.

Answers

1. a. 0   **b.** 4   **c.** 0   **d.** 76

Distributive Property

Multiplication distributes over addition. For example,

$$2(3 + 4) = 2\cdot 3 + 2\cdot 4$$

EXAMPLE 2   Rewrite each using the distributive property.

a. $3(4 + 5)$ **b.** $10(6 + 8)$ **c.** $2(7 + 3)$

Solution: Using the distributive property, we have

a. $3(4 + 5) = 3\cdot 4 + 3\cdot 5$
b. $10(6 + 8) = 10\cdot 6 + 10\cdot 8$
c. $2(7 + 3) = 2\cdot 7 + 2\cdot 3$

B Multiplying Whole Numbers

Let's use the distributive property to multiply 7(48). To do so, we begin by writing the expanded form of 48 and then applying the distributive property.

$$7(48) = 7(40 + 8)$$
$$= 7\cdot 40 + 7\cdot 8 \quad \text{Apply the distributive property.}$$
$$= 280 + 56 \quad \text{Multiply.}$$
$$= 336 \quad \text{Add.}$$

This is how we multiply whole numbers. When multiplying whole numbers, we will use the following notation.

$$\begin{array}{r} \overset{5}{48} \\ \times\ 7 \\ \hline 336 \end{array}$$ $\leftarrow 7\cdot 8 = 56$ Write 6 in the ones place and carry 5 to the tens place.

$7\cdot 4 = 28$ and $28 + 5 = 33$

EXAMPLE 3   Multiply: $\begin{array}{r} 25 \\ \times\ 8 \end{array}$

Solution: $\begin{array}{r} \overset{4}{25} \\ \times\ 8 \\ \hline 200 \end{array}$

To multiply larger whole numbers, use the following similar notation. Multiply 89 × 52.

Step 1
$$\begin{array}{r} \overset{1}{89} \\ \times\ 52 \\ \hline 178 \end{array}$$ $\leftarrow$ Multiply 89 × 2.

Step 2
$$\begin{array}{r} \overset{4}{89} \\ \times\ 52 \\ \hline 178 \\ 4450 \end{array}$$ $\leftarrow$ Multiply 89 × 50.

Step 3
$$\begin{array}{r} 89 \\ \times\ 52 \\ \hline 178 \\ 4450 \\ \hline 4628 \end{array}$$ Add.

The numbers 178 and 4450 are called **partial products**. The sum of the partial products, 4628, is the product of 89 and 52.

Rewrite each using the distributive property.

a. $5(2 + 3)$
b. $9(8 + 7)$
c. $3(6 + 1)$

Practice Problem 3

Multiply.

a. $\begin{array}{r} 36 \\ \times\ 4 \end{array}$ b. $\begin{array}{r} 92 \\ \times\ 9 \end{array}$

Answers

2. a. $5(2 + 3) = 5\cdot 2 + 5\cdot 3$
   b. $9(8 + 7) = 9\cdot 8 + 9\cdot 7$
   c. $3(6 + 1) = 3\cdot 6 + 3\cdot 1$
3. a. 144 **b.** 828

Practice Problem 4

Multiply.

a.   594   b.   306
  × 72   × 81

Practice Problem 5

Multiply.

a.   726   b.   4
  × 142   × 288

Concept Check

Find and explain the error in the following multiplication problem:

```
    102
  ×  33
    306
    306
    612
```

EXAMPLE 4   Multiply: 236 × 86

Solution:

```
      236    ← 6(236)
    ×  86
     1416
   18,880    ← 80(236)
   20,296    Add.
```

EXAMPLE 5   Multiply: 631 × 125

Solution:

```
      631
    × 125
     3155    ← 5(631)
   12,620    ← 20(631)
   63,100    ← 100(631)
   78,875    Add.
```

Try the Concept Check in the margin.

Ⓐ Finding the Area of a Rectangle

A special application of multiplication is finding the area of a region. Area measures the amount of surface of a region. For example, we measure a plot of land or the living space of a home by area. The figures show two examples of units of area measure. (A centimeter is a unit of length in the metric system.)

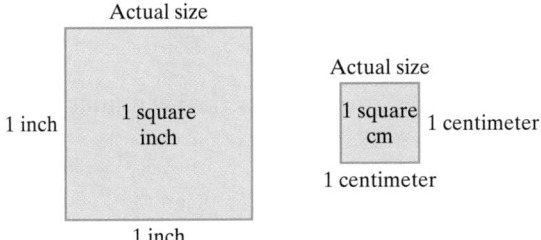

To measure the area of a geometric figure such as the rectangle shown, count the number of square units that cover the region.

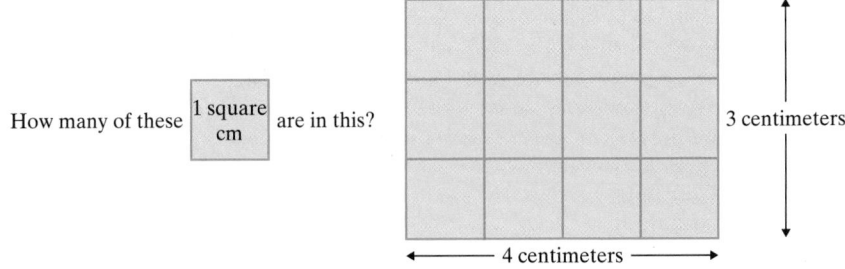

This rectangular region contains 12 square units, each 1 square centimeter. Thus, the area is 12 square centimeters. This total number of squares can be found by counting or by multiplying $4 \cdot 3$ (length · width).

$$\text{Area of a rectangle} = \text{length} \cdot \text{width}$$
$$= (4 \text{ centimeters})(3 \text{ centimeters})$$
$$= 12 \text{ square centimeters}$$

This example above shows why area is always measured in square units. In this section, we find the areas of rectangles only. In later sections, we find the areas of other geometric regions.

Answers

4. a. 42,768   **b.** 24,786
5. a. 103,092   **b.** 1152

Concept Check:
```
      102
    ×  33
      306
     3060
     3366
```

Remember that *perimeter* (distance around a plane figure) is measured in units. *Area* (space enclosed by a plane figure) is measured in square units.

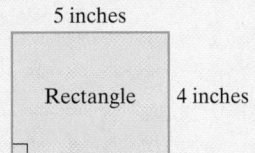

Perimeter = 5 inches + 4 inches + 5 inches
+ 4 inches = 18 inches

Area = (5 inches)(4 inches) = 20 square inches

 EXAMPLE 6   Finding the Area of a State

The state of Colorado is in the shape of a rectangle whose length is about 380 miles and whose width is about 280 miles. Find its area.

Solution: The area of a rectangle is the product of its length and its width.

Area = length · width

= (380 miles)(280 miles)

= 106,400 square miles

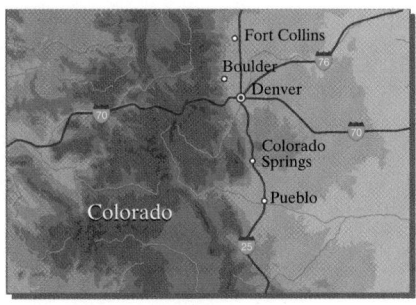

The area of Colorado is approximately 106,400 square miles.

[Note: The abbreviations sq in. and in.² both mean square inches. In this text, we will use the notation sq in.]

Solving Problems by Multiplying

There are several words or phrases that indicate the operation of multiplication. Some of these are as follows:

Key Words or Phrases	Example	Symbols
multiply	multiply 5 by 7	5 · 7
product	the product of 3 and 2	3 · 2
times	10 times 13	10 · 13

Many key words or phrases describing real-life problems that suggest addition might be better solved by multiplication instead. For example, to find the **total** cost of 8 shirts, each selling for $27, we can either add 27 + 27 + 27 + 27 + 27 + 27 + 27 + 27, or we can multiply 8(27).

Practice Problem 6

The state of Wyoming is in the shape of a rectangle whose length is 360 miles and whose width is 280 miles. Find its area.

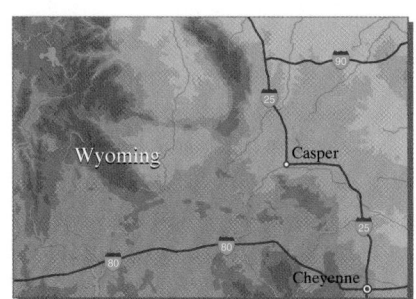

Answer

6. 100,800 square mile

Practice Problem 7

A computer printer can print 240 characters per second in draft mode. How many total characters can it print in 15 seconds?

EXAMPLE 7 Finding Disk Space

A certain computer disk can hold about 1510 thousand bytes of information. How many total bytes can 42 such disks hold?

Solution: Forty-two disks will hold 42 × 1510 thousand bytes.

In Words		Translate to Numbers
bytes per disk	→	1510
× disks	→	× 42
		3020
		60,400
total bytes		63,420

Forty-two disks will hold 63,420 thousand bytes.

Practice Problem 8

Softball T-shirts come in two styles: plain at $6 each and striped at $7 each. The team orders 4 plain shirts and 5 striped shirts. Find the total cost of the order.

EXAMPLE 8 Budgeting Money

Earline Martin agrees to take her children and their cousins to the San Antonio Zoo. The ticket price for each child is $4 and for each adult, $6. If 8 children and 1 adult plan to go, how much money is needed for admission?

Solution: If the price of one child's ticket is $4, the price for 8 children is $8 \cdot 4 = \$32$. The price of one adult ticket is $6, so the total cost is

In Words		Translate to Numbers
price of 8 children	→	32
+ price of adult	→	+ 6
total cost		38

The total cost is $38.

Practice Problem 9

If an average page in a book contains 259 words, estimate, rounding to the nearest hundred, the total number of words contained on 195 pages.

EXAMPLE 9 Estimating Word Count

The average page of a book contains 259 words. Estimate, rounding to the nearest ten, the total number of words contained on 22 pages.

Solution: The exact number of words is 259×22. Estimate this product by rounding each factor to the nearest ten.

$$
\begin{array}{r}
259 \quad \text{rounds to} \quad 260 \\
\times\, 22 \quad \text{rounds to} \quad \times\, 20 \\
\hline
5200
\end{array}
$$

There are approximately 5200 words contained on 22 pages.

Answers

7. 3600 characters   **8.** $59   **9.** 60,000 words

CALCULATOR EXPLORATIONS

Multiplying Numbers

To multiply numbers on a calculator, find the keys marked ⟨ × ⟩ and ⟨ = ⟩ or ⟨ ENTER ⟩. For example, to find $31 \cdot 66$ on a calculator, press the keys ⟨ 31 ⟩ ⟨ × ⟩ ⟨ 66 ⟩ ⟨ = ⟩ or ⟨ ENTER ⟩. The display will read ⟨ 2046 ⟩. Thus, $31 \cdot 66 = 2046$.

Use a calculator to multiply.

1. 72×48 **2.** 81×92

3. $163 \cdot 94$ **4.** $285 \cdot 144$

5. $983(277)$ **6.** $1562(843)$

Name _____ Section _____ Date _____

Mental Math

A *Multiply. See Example 1.*

1. $1 \cdot 24$

2. $55 \cdot 1$

3. $0 \cdot 19$

4. $27 \cdot 0$

5. $8 \cdot 0 \cdot 9$

6. $7 \cdot 6 \cdot 0$

7. $87 \cdot 1$

8. $1 \cdot 41$

EXERCISE SET 1.6

A *Use the distributive property to rewrite each expression. See Example 2.*

1. $4(3 + 9)$ **2.** $5(8 + 2)$ **3.** $2(4 + 6)$ **4.** $6(1 + 4)$ **5.** $10(11 + 7)$ **6.** $12(12 + 3)$

B *Multiply. See Example 3.*

7.
$$\begin{array}{r} 42 \\ \times 6 \\ \hline \end{array}$$

8.
$$\begin{array}{r} 79 \\ \times 3 \\ \hline \end{array}$$

9.
$$\begin{array}{r} 624 \\ \times 3 \\ \hline \end{array}$$

10.
$$\begin{array}{r} 638 \\ \times 5 \\ \hline \end{array}$$

 **11.**
$$\begin{array}{r} 227 \\ \times 6 \\ \hline \end{array}$$

12.
$$\begin{array}{r} 882 \\ \times 2 \\ \hline \end{array}$$

13.
$$\begin{array}{r} 1062 \\ \times 5 \\ \hline \end{array}$$

14.
$$\begin{array}{r} 9021 \\ \times 3 \\ \hline \end{array}$$

Multiply. See Examples 4 and 5.

15.
$$\begin{array}{r} 298 \\ \times 14 \\ \hline \end{array}$$

16.
$$\begin{array}{r} 591 \\ \times 72 \\ \hline \end{array}$$

17.
$$\begin{array}{r} 231 \\ \times 47 \\ \hline \end{array}$$

18.
$$\begin{array}{r} 526 \\ \times 23 \\ \hline \end{array}$$

19.
$$\begin{array}{r} 809 \\ \times 14 \\ \hline \end{array}$$

20.
$$\begin{array}{r} 307 \\ \times 16 \\ \hline \end{array}$$

21. $(620)(40)$

22. $(720)(80)$

23. $(998)(12)(0)$

24. $(593)(47)(0)$

25. $(590)(1)(10)$

26. $(240)(1)(20)$

27.
$$\begin{array}{r} 1234 \\ \times 48 \\ \hline \end{array}$$

28.
$$\begin{array}{r} 1357 \\ \times 79 \\ \hline \end{array}$$

**29.**
$$\begin{array}{r} 609 \\ \times 234 \\ \hline \end{array}$$

30.
$$\begin{array}{r} 505 \\ \times 127 \\ \hline \end{array}$$

31.
$$\begin{array}{r} 5621 \\ \times 324 \\ \hline \end{array}$$

32.
$$\begin{array}{r} 1234 \\ \times 567 \\ \hline \end{array}$$

33.
$$\begin{array}{r} 1941 \\ \times 235 \\ \hline \end{array}$$

34.
$$\begin{array}{r} 1876 \\ \times 437 \\ \hline \end{array}$$

35.
$$\begin{array}{r} 589 \\ \times 110 \\ \hline \end{array}$$

36.
$$\begin{array}{r} 426 \\ \times 110 \\ \hline \end{array}$$

37.
$$\begin{array}{r} 964 \\ \times 207 \\ \hline \end{array}$$

38.
$$\begin{array}{r} 462 \\ \times 305 \\ \hline \end{array}$$

Estimate the products by rounding each factor to the nearest hundred. See Example 9.

39. 576 × 354   **40.** 982 × 650   **41.** 604 × 451   **42.** 111 × 999

Estimate the products by rounding each factor to the nearest ten. See Example 9.

43. 872 × 27   **44.** 126 × 41   **45.** 36 × 87   **46.** 57 × 77

C *Find the area of each rectangle. See Example 6.*

 47.

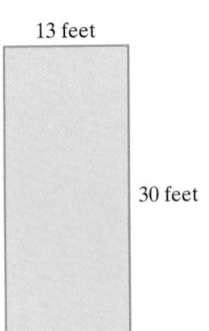

9 meters

7 meters

△ **48.** 4 inches

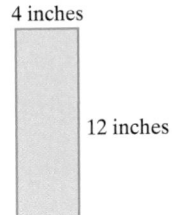

12 inches

△ **49.** 13 feet

30 feet

△ **50.** 25 centimeters

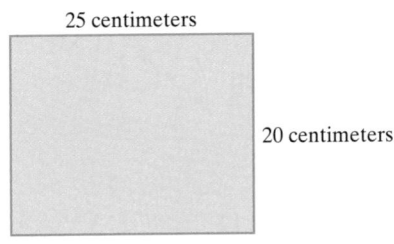

20 centimeters

C **D** *Solve. See Examples 6 through 9.*

 51. One tablespoon of olive oil contains 125 calories. How many calories are in 3 tablespoons of olive oil? (*Source: Home and Garden Bulletin No. 72*, U.S. Department of Agriculture).

52. One ounce of hulled sunflower seeds contains 14 grams of fat. How many grams of fat are in 6 ounces of hulled sunflower seeds? (*Source: Home and Garden Bulletin No. 72*, U.S. Department of Agriculture).

53. The textbook for a course in Civil War history costs $54. There are 35 students in the class. Find the total cost of the history books for the class.

54. The seats in the mathematics lecture hall are arranged in 12 rows with 6 seats in each row. Find how many seats are in this room.

55. A case of canned peas has *two layers* of cans. In each layer are 8 rows with 12 cans in each row. Find how many cans are in a case.

56. An apartment building has *three floors*. Each floor has five rows of apartments with four apartments in each row. Find how many apartments there are.

57. A plot of land measures 90 feet by 110 feet. Find its area.

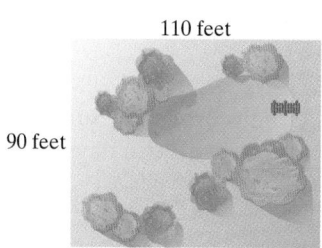

110 feet

90 feet

58. A house measures 45 feet by 60 feet. Find the floor area of the house.

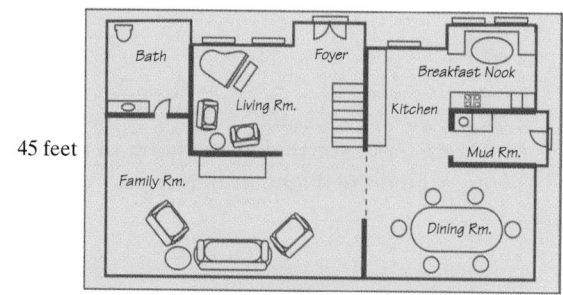

45 feet

60 feet

59. The largest lobby can be found at the Hyatt Regency in San Francisco, CA. It is in the shape of a rectangle that measures 350 feet by 160 feet. Find its area.

60. Recall from an earlier section that the largest commercial building in the world under one roof is the flower auction building of the cooperative VBA in Aalsmeer, Netherlands. The floor plan is a rectangle that measures 776 meters by 639 meters. Find the area of this building. (A meter is a unit of length in the metric system.) (*Source: The Handy Science Answer Book,* Visible Ink Press)

776 meters

639 meters

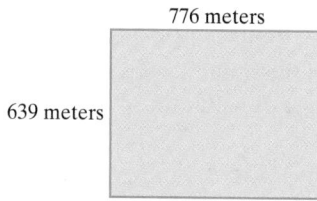

61. A pixel is a rectangular dot on a graphing calculator screen. If a graphing calculator screen contains 62 pixels in a row and 94 pixels in a column, find the total number of pixels on a screen.

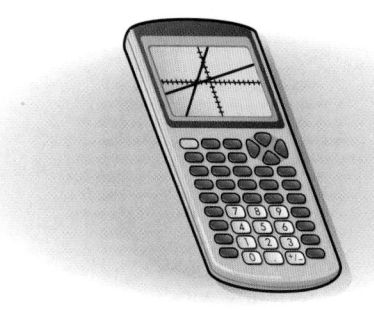

62. A CD (compact disk) can hold 650 megabytes (MB) of information. How many MBs can 17 disks hold?

63. A line of print on a computer contains 80 characters (letters, spaces, punctuation marks). Find how many characters there are in 25 lines.

64. An average cow eats 3 pounds of grain per day. Find how much grain a cow eats in a year. (Assume 365 days in 1 year.)

65. One ounce of Planters® Dry Roasted Peanuts has 160 calories. How many calories are in 8 ounces? (*Source:* RJR Nabisco, Inc.)

66. One ounce of Planters® Dry Roasted Peanuts has 13 grams of fat. How many grams of fat are in 8 ounces? (*Source:* RJR Nabisco, Inc.)

67. The diameter of the planet Saturn is 9 times as great as the diameter of Earth. The diameter of Earth is 7927 miles. Find the diameter of Saturn.

68. The planet Uranus orbits the Sun every 84 Earth years. Find how many Earth days two orbits take. (Assume 365 days in 1 year.)

69. A window washer in New York City is bidding for a contract to wash the windows of a 23-story building. To write a bid, the number of windows in the building is needed. If there are 7 windows in each row of windows on 2 sides of the building and 4 windows per row on the other 2 sides of the building, find the total number of windows.

70. In North America, the average toy expenditure per child is $328 per year. On average, how much is spent on toys for a child by the time he or she reaches age 18? (*Source:* The NPD Group Worldwide)

71. Hershey's main chocolate factory in Hershey, Pennsylvania, uses 700,000 quarts of milk each day. How many quarts of milk would be used during the month of March, assuming that chocolate is made at the factory every day of the month? (*Source:* Hershey Foods Corp.)

72. Among older Americans (age 65 years and older), there are 4 times as many widows as widowers. There were 1,994,000 widowers in 2000. How many widows were there in 2000? (*Sources:* Administration on Aging, U.S. Census Bureau)

73. In 2001, the average cost of the Head Start program was $6,633 per child. That year, 905,235 children participated in the program. Round each number to the nearest thousand and estimate the total cost of the Head Start program in 2001. (*Source:* Head Start Bureau)

74. American households currently spend an average of $4810 on food each year. A small community encompasses 1643 households. Round each number to the nearest hundred and estimate the total annual food expenditure for the residents of this community. (*Source:* U.S. Bureau of Labor Statistics)

 Combining Concepts

In a survey, college students were asked to name their favorite fruit. The results are shown in the picture graph. Use this graph to answer Exercises 75 through 78.

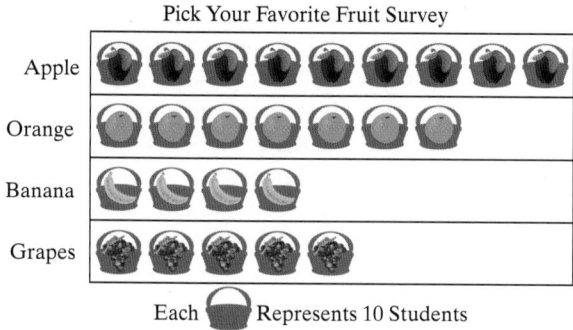

Pick Your Favorite Fruit Survey

Apple
Orange
Banana
Grapes

Each ⬤ Represents 10 Students

75. How many students chose grapes as their favorite fruit?

76. How many students chose bananas as their favorite fruit?

77. Which two fruits were the most popular?

78. Which two fruits were chosen by a total of 110 students?

Fill in the missing digits in each problem.

79.

$$\begin{array}{r} 4\_ \\ \times\ \underline{\ \ 3\_} \\ 126 \\ \underline{3780\ } \\ 3906 \end{array}$$

80.

$$\begin{array}{r} \_7 \\ \times\ \underline{\ 6\_} \\ 171 \\ \underline{3420\ } \\ 3591 \end{array}$$

81. Explain how to multiply two 2-digit numbers using partial products.

82. A slice of enriched white bread has 65 calories. One tablespoon of jam has 55 calories, and one tablespoon of peanut butter has 95 calories. Suppose a peanut butter and jelly sandwich is made with two slices of white bread, one tablespoon of jam, and one tablespoon of peanut butter. How many calories are in two such peanut butter and jelly sandwiches? (*Source: Home and Garden Bulletin No. 72,* U.S. Department of Agriculture)

83. During the NBA's 2000–2001 season, Kobe Bryant of the Los Angeles Lakers scored 61 three-point field goals, 640 two-point field goals, and 475 free throws (worth one point each). How many points did Kobe Bryant score during the 2000–2001 season? (*Source:* National Basketball Association)

FOCUS ON **The Real World**

JUDGING DISTANCES

Do you know how to estimate a distance without using a tape measure? One easy way to do this is to use the length of your own stride. First, measure the length of your stride in inches. You can do this with the following steps:

- Lay a yardstick on the floor.
- Stand next to the yardstick with feet together so that both heels line up with the 0-mark on the yardstick.
- Take a normal-sized step forward.
- Find the whole-inch mark nearest the toe of the foot that is farthest from the 0-mark. This is roughly the length of your stride.

To judge a distance, simply pace it off using normal-sized strides. Multiply the number of strides by the length of your stride to get a rough estimate of the distance.

Suppose you need to measure a distance that can't easily be paced, such as a pond or a busy street. You can easily "transfer" the distance to a more easily paced area by using a baseball cap.

- Stand at the edge of the pond or street while wearing a baseball cap.
- Bend your head until your chin rests on your chest.
- Pull the bill of the cap up or down until it appears to touch the other side of the pond or street.
- Without moving your head or the cap, pivot your body to the right until you have a clear path straight ahead for walking.
- Notice where the bill seems to be touching the ground now. The distance to this point is the same as the distance across the pond or street you are measuring.
- Pace off the distance to this point and find an estimate of the distance as before.

Pace off this distance

Suppose a distance estimate using this method is 1302 inches. To write the distance in terms of feet, divide the estimate by 12. The quotient is the number of whole feet, and the remainder is the number of inches. A distance of 1302 inches is the same as 108 feet 6 inches.

$$
\begin{array}{r}
108 \text{ R } 6 \\
12\overline{)1302} \\
\underline{12} \\
10 \\
\underline{0} \\
102 \\
\underline{96} \\
6
\end{array}
$$

GROUP ACTIVITY

Materials: yardstick, baseball cap

Use the baseball cap procedure to estimate the distance across a river, stream, pond, or busy road on or near your campus. Have each person in your group estimate the same distance using the length of his or her own stride. Compare your results. Write a brief report summarizing your findings. Be sure to include what distance your group estimated, the length of each member's stride, the number of strides each member paced off, and each member's distance estimate. Conclude by discussing reasons for any differences in estimates.

1.7   Dividing Whole Numbers

Suppose three people pooled their money and bought a raffle ticket at a local fund-raiser. Their ticket was the winner and they won a $60 cash prize. They then divided the prize into three equal parts so that each person received $20.

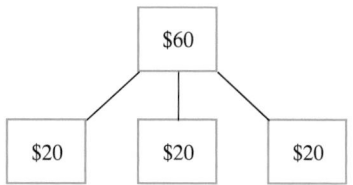

A Dividing Whole Numbers

The process of separating a quantity into equal parts is called **division**. Division can be symbolized by several notations.

quotient

$$3)\overline{60} \leftarrow \text{dividend}$$
divisor

$$\frac{60}{3} = 20 \leftarrow \text{quotient}$$
divisor

$$60 \div 3 = 20$$
dividend   divisor

(In the notation $\frac{60}{3}$, the bar separating 60 and 3 is called a **fraction bar**.) Just as subtraction is the reverse of addition, division is the reverse of multiplication. This means that division can be checked by multiplication.

$$20 \atop 3)\overline{60}$$ because $20 \cdot 3 = 60$

Since multiplication and division are related in this way, you can use the multiplication table in Appendix B to review quotients of one-digit divisors if necessary.

EXAMPLE 1   Find each quotient. Check by multiplying.

a. $42 \div 7$ **b.** $\frac{81}{9}$ **c.** $4)\overline{24}$

Solution:

a. $42 \div 7 = 6$ because $6 \cdot 7 = 42$

b. $\frac{81}{9} = 9$ because $9 \cdot 9 = 81$

c. $4)\overline{24}^{\,6}$ because $6 \cdot 4 = 24$

EXAMPLE 2   Find each quotient. Check by multiplying.

a. $1)\overline{8}$ **b.** $11 \div 1$ **c.** $\frac{9}{9}$ **d.** $7 \div 7$ **e.** $\frac{10}{1}$ **f.** $6)\overline{6}$

Solution:

a. $1)\overline{8}^{\,8}$ because $8 \cdot 1 = 8$

b. $11 \div 1 = 11$ because $11 \cdot 1 = 11$

Practice Problem 1

Find each quotient. Check by multiplying.

a. $8)\overline{48}$

b. $35 \div 5$

c. $\frac{49}{7}$

Practice Problem 2

Find each quotient. Check by multiplying.

a. $\frac{8}{8}$ b. $3 \div 1$ c. $1)\overline{12}$

d. $2 \div 1$ e. $\frac{5}{1}$

Answers

1. a. 6   **b.** 7   **c.** 7   **2. a.** 1   **b.** 3   **c.** 12
d. 2   **e.** 5

c. $\dfrac{9}{9} = 1$ because $1 \cdot 9 = 9$

d. $7 \div 7 = 1$ because $1 \cdot 7 = 7$

e. $\dfrac{10}{1} = 10$ because $10 \cdot 1 = 10$

f. $6\overline{)6}^{1}$ because $1 \cdot 6 = 6$

Example 2 illustrates important properties of division as described next:

Division Properties of 1

The quotient of any number and that same number is 1. For example,

$$8 \div 8 = 1 \qquad \dfrac{7}{7} = 1 \qquad 4\overline{)4}^{1}$$

The quotient of any number and 1 is that same number. For example,

$$9 \div 1 = 9 \qquad \dfrac{6}{1} = 6 \qquad 1\overline{)3}^{3} \qquad \dfrac{0}{1} = 0$$

Practice Problem 3

Find each quotient. Check by multiplying.

a. $\dfrac{0}{7}$ b. $5\overline{)0}$ c. $9 \div 0$

d. $0 \div 6$

EXAMPLE 3 Find each quotient. Check by multiplying.

a. $9\overline{)0}$ **b.** $0 \div 12$ **c.** $\dfrac{0}{5}$ **d.** $\dfrac{3}{0}$

Solution:

a. $9\overline{)0}^{0}$ because $0 \cdot 9 = 0$

b. $0 \div 12 = 0$ because $0 \cdot 12 = 0$

c. $\dfrac{0}{5} = 0$ because $0 \cdot 5 = 0$

d. If $\dfrac{3}{0} = $ a **number**, then the **number** times $0 = 3$. Recall that any number multiplied by 0 is 0 and not 3. We say, then, that $\dfrac{3}{0}$ is **undefined**.

Example 3 illustrates important division properties of 0.

Division Properties of 0

The quotient of 0 and any number (except 0) is 0. For example,

$$0 \div 9 = 0 \qquad \dfrac{0}{5} = 0 \qquad 14\overline{)0}$$

The quotient of any number and 0 is not a number. We say that

$$\dfrac{3}{0}, \qquad 0\overline{)3}, \qquad \text{and} \qquad 3 \div 0$$

are **undefined**.

Answers

1. a. 0 **b.** 0 **c.** undefined **d.** 0

B Performing Long Division

When dividends are larger, the quotient can be found by a process called **long division**. For example, let's divide 2541 by 3.

$$3\overline{)2541}$$

We can't divide 3 into 2, so we try dividing 3 into the first two digits.

$$\begin{array}{r} 8 \\ 3\overline{)2541} \end{array}$$ $25 \div 3 = 8$ with 1 left, so our best estimate is 8. We place 8 over the 5 in 25.

Next, multiply 8 and 3 and subtract this product from 25. Make sure that this difference is less than the divisor.

$$\begin{array}{r} 8 \\ 3\overline{)2541} \\ -\ 24 \\ \hline 1 \end{array}$$ $8(3) = 24$
$25 - 24 = 1$, and 1 is less than the divisor 3.

Bring down the next digit and go through the process again.

$$\begin{array}{r} 84 \\ 3\overline{)2541} \\ -\ 24\downarrow \\ \hline 14 \\ -\ 12 \\ \hline 2 \end{array}$$ $14 \div 3 = 4$ with 2 left

$4(3) = 12$
$14 - 12 = 2$

Once more, bring down the next digit and go through the process.

$$\begin{array}{r} 847 \\ 3\overline{)2541} \\ -\ 24 \\ \hline 14 \\ -\ 12 \\ \hline 21 \\ -\ 21 \\ \hline 0 \end{array}$$ $21 \div 3 = 7$

$7(3) = 21$
$21 - 21 = 0$

The quotient is 847. To check, see that $847 \times 3 = 2541$.

EXAMPLE 4   Divide: $3705 \div 5$. Check by multiplying.

Solution:

$$\begin{array}{r} 7 \\ 5\overline{)3705} \\ -\ 35\downarrow \\ \hline 20 \end{array}$$ $37 \div 5 = 7$ with 2 left. Place this estimate, 7, over the 7 in 37.

$7(5) = 35$
$37 - 35 = 2$, and 2 is less than the divisor 5.
└── Bring down the 0.

$$\begin{array}{r} 74 \\ 5\overline{)3705} \\ -\ 35 \\ \hline 20 \\ -\ 20\downarrow \\ \hline 05 \end{array}$$ $20 \div 5 = 4$

$4(5) = 20$
$20 - 20 = 0$, and 0 is less than the divisor 5.
└── Bring down the 5.

Practice Problem 4

Divide. Check by multiplying.

a. $6\overline{)5382}$

b. $4\overline{)2212}$

Answers

4. a. 897   **b.** 553

$$
\begin{array}{r}
741 \\
5{\overline{\smash{)}3705}} \\
-\ 35 \\
\hline
20 \\
-\ 20 \\
\hline
5 \\
-\ 5 \\
\hline
0
\end{array}
$$

$5 \div 5 = 1$

$1(5) = 5$

$5 - 5 = 0$

Check:

$$
\begin{array}{r}
741 \\
\times\ \ 5 \\
\hline
3705
\end{array}
$$

> **Helpful Hint**
>
> Since division and multiplication are reverse operations, don't forget that a division problem can be checked by multiplying.

Practice Problem 5

Divide and check.

a. $3{\overline{\smash{)}2397}}$

b. $7{\overline{\smash{)}2520}}$

EXAMPLE 5 Divide and check: $1872 \div 9$

Solution:

$$
\begin{array}{r}
208 \\
9{\overline{\smash{)}1872}} \\
-\ 18 \\
\hline
07 \\
-\ 0 \\
\hline
72 \\
-\ 72 \\
\hline
0
\end{array}
$$

$2(9) = 18$

$18 - 18 = 0$; bring down the 7.

$0(9) = 0$

$7 - 0 = 7$; bring down the 2.

$8(9) = 72$

$72 - 72 = 0$

Check: $208 \cdot 9 = 1872$

Naturally, quotients don't always "come out even." Making 4 rows out of 26 chairs, for example, isn't possible if each row is supposed to have exactly the same number of chairs. Each of 4 rows can have 6 chairs, but 2 chairs are still left over.

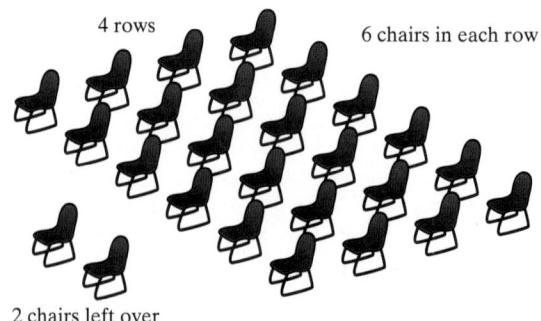

4 rows

6 chairs in each row

2 chairs left over

We signify "leftovers" or **remainders** in this way:

$$
\begin{array}{r}
6\ \text{R}\,2 \\
4{\overline{\smash{)}26}} \\
-24 \\
\hline
2
\end{array}
$$

The **whole number part of the quotient** is 6; the **remainder part of the quotient** is 2. Checking by multiplying,

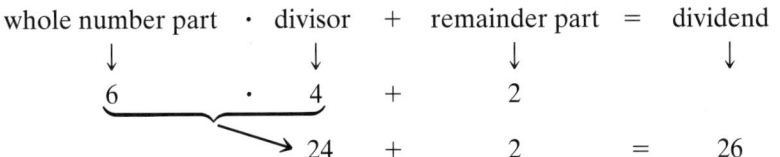

EXAMPLE 6   Divide and check: 2557 ÷ 7

Solution:

$$
\begin{array}{r}
365 \quad \text{R } 2 \\
7\overline{)2557} \\
\end{array}
$$

$$
\begin{array}{r}
-\ 21\!\downarrow \\
\hline
45 \\
-\ 42\!\downarrow \\
\hline
37 \\
-\ 35 \\
\hline
2 \\
\end{array}
$$

3(7) = 21
25 − 21 = 4; bring down the 5.
6(7) = 42
45 − 42 = 3; bring down the 7.
5(7) = 35
37 − 35 = 2; the remainder is 2.

Check: 365 · 7 + 2 = 2557
                ↑   ↑   ↑   ↑
            whole
            number · divisor + remainder = dividend
            part   part

Practice Problem 6

Divide and check.

a. 5)949

b. 6)4399

EXAMPLE 7   Divide and check: 56,717 ÷ 8

Solution:

$$
\begin{array}{r}
7089 \quad \text{R } 5 \\
8\overline{)56717} \\
\end{array}
$$

$$
\begin{array}{r}
-\ 56\!\downarrow \\
\hline
07 \\
-\ 0\!\downarrow \\
\hline
71 \\
-\ 64\!\downarrow \\
\hline
77 \\
-\ 72 \\
\hline
5 \\
\end{array}
$$

7(8) = 56
Subtract and bring down the 7.
0(8) = 0
Subtract and bring down the 1.
8(8) = 64
Subtract and bring down the 7.
9(8) = 72
Subtract. The remainder is 5.

Check: 7089 · 8 + 5 = 56,717
                ↑   ↑   ↑   ↑
            whole
            number · divisor + remainder = dividend
            part   part

Practice Problem 7

Divide and check.

a. 5)40,841

b. 7)22,430

When the divisor has more than one digit, the same pattern applies. For example, let's find 1358 ÷ 23.

$$
\begin{array}{r}
5 \\
23\overline{)1358} \\
115\!\downarrow \\
\hline
208 \\
\end{array}
$$

135 ÷ 23 = 5 with 20 left over. Our estimate is 5.
5(23) = 115
135 − 115 = 20. Bring down the 8.

Now we continue estimating.

Answers

6. a. 189 R 4   **b.** 733 R 1   **7. a.** 8168 R 1
b. 3204 R 2

$$\begin{array}{r} 59 \text{R } 1 \\ 23\overline{)1358} \\ -115 \\ \hline 208 \\ -207 \\ \hline 1 \end{array}$$

$208 \div 23 = 9$ with 1 left over.

$9(23) = 207$

$208 - 207 = 1.$ The remainder is 1.

To check, see that $59 \cdot 23 + 1 = 1358$.

Practice Problem 8

Divide: $5740 \div 19$

EXAMPLE 8   Divide: $6819 \div 17$

Solution:
$$\begin{array}{r} 401 \text{R } 2 \\ 17\overline{)6819} \\ -68 \\ \hline 01 \\ -0 \\ \hline 19 \\ -17 \\ \hline 2 \end{array}$$

$4(17) = 68$

Subtract and bring down the 1.

$0(17) = 0$

Subtract and bring down the 9.

$1(17) = 17$

Subtract. The remainder is 2.

To check, see that $401 \cdot 17 + 2 = 6819$.

Practice Problem 9

Divide: $16,589 \div 247$

EXAMPLE 9   Divide: $51,600 \div 403$

Solution:
$$\begin{array}{r} 128 \text{R } 16 \\ 403\overline{)51600} \\ -403 \\ \hline 1130 \\ -806 \\ \hline 3240 \\ -3224 \\ \hline 16 \end{array}$$

$1(403) = 403$

Subtract and bring down the 0.

$2(403) = 806$

Subtract and bring down the 0.

$8(403) = 3224$

Subtract. The remainder is 16.

To check, see that $128 \cdot 403 + 16 = 51,600$.

Concept Check

Which of the following is the correct way to represent "the quotient of 20 and 5"? Or are both correct? Explain your answer.

a. $5 \div 20$

b. $20 \div 5$

Try the Concept Check in the margin.

(C) **Solving Problems by Dividing**

Below are some key words and phrases that indicate the operation of division:

Key Words or Phrases	Examples	Symbols
divide	divide 10 by 5	$10 \div 5$ or $\dfrac{10}{5}$
quotient	the quotient of 64 and 4	$64 \div 4$ or $\dfrac{64}{4}$
divided by	9 divided by 3	$9 \div 3$ or $\dfrac{9}{3}$
divided or **shared equally among**	$100 divided equally among five people	$100 \div 5$ or $\dfrac{100}{5}$

Answers

8. 302 R 2

9. 67 R 40

Concept Check: **a.** incorrect   **b.** correct

EXAMPLE 10 Finding Shared Earnings

Zachary, Tyler, and Stephanie McMillan share a paper route to earn money for college expenses. The total in their fund after expenses was $2895. How much is each person's equal share? Use estimation to check your result.

Solution: Each person's equal share is (total) ÷ (number of people) or

$$2895 \div 3$$

Then
```
      965
  3)2895
  - 27
    ---
     19
   - 18
    ---
     15
   - 15
    ---
      0
```

Each person's share is $965.

Check: To check, we can round $2895 to $3000 and mentally divide $3000 by 3. The result is $1000, which is close to our actual answer of $965. ●

EXAMPLE 11 Calculating Shipping Needs

How many boxes are needed to ship 56 pairs of Nikes to a shoe store in Texarkana if 9 pairs of shoes will fit in each shipping box?

Solution:

number of boxes	=	total pairs of shoes	÷	how many pairs in a box
↓		↓		↓
number of boxes	=	56	÷	9

```
    6 R 2
  9)56
  - 54
    ---
     2
```

There are 6 full boxes with 2 pairs of shoes left over, so 7 boxes will be needed. ●

EXAMPLE 12 Dividing Holiday Favors among Students

Mary Schultz has 48 kindergarten students. She buys 260 stickers as Thanksgiving Day favors for her students. Can she divide the stickers up equally among her students? If not, how many stickers will be left over?

Solution:

number of stickers	÷	number of students
↓		↓
260	÷	48

```
     5 R 20
  48)260
   - 240
    ----
      20
```

No, the stickers cannot be divided equally among her students since there is a nonzero remainder. There will be 20 stickers left over. ●

Practice Problem 10

Marina, Manual, and Min bought 10 dozen high-density computer diskettes to share equally. How many diskettes did each person get?

Practice Problem 11

Peanut butter and cheese cracker sandwiches come in 6 sandwiches to a package. How many full packages are formed with 195 sandwiches?

Practice Problem 12

Calculators can be packed 24 to a box. If 497 calculators are to be packed but only full boxes are shipped, how many full boxes will be shipped? How many calculators are left over and not shipped?

Answers

10. 40 diskettes   **11.** 32 full packages
12. 20 full boxes; 17 calculators left over

(D) Finding Averages

A special application of division (and addition) is finding the average of a list of numbers. The **average** of a list of numbers is the sum of the numbers divided by the number of numbers.

$$\text{average} = \frac{\text{sum of numbers}}{\text{number of numbers}}$$

EXAMPLE 13   Averaging Scores

Liam Reilly's scores in his mathematics class so far are 93, 86, 71, and 82. Find his average score.

Solution: To find his average score, we find the sum of his scores and divide by 4, the number of scores.

$$
\begin{array}{r}
93 \\
86 \\
71 \\
+\ 82 \\
\hline
332 \ {\scriptstyle \text{sum}}
\end{array}
\qquad
\text{average} = \frac{332}{4} = 83
\qquad
\begin{array}{r}
83 \\
4\overline{)332} \\
-\ 32 \\
\hline
12 \\
-\ 12 \\
\hline
0
\end{array}
$$

His average score is 83.

Practice Problem 13

To compute a safe time to wait for reactions to occur after allergy shots are administered, a lab technician is given a list of elapsed times between administered shots and reactions. Find the average of the times 5 minutes, 7 minutes, 20 minutes, 6 minutes, 9 minutes, 3 minutes, and 48 minutes.

Answer

13. 14 min

🖩 CALCULATOR EXPLORATIONS

Dividing Numbers

To divide numbers on a calculator, find the keys marked $\boxed{\div}$ and $\boxed{=}$ or $\boxed{\text{ENTER}}$. For example, to find $435 \div 5$ on a calculator, press the keys $\boxed{435}$ $\boxed{\div}$ $\boxed{5}$ $\boxed{=}$ or $\boxed{\text{ENTER}}$. The display will read $\boxed{\qquad 87}$. Thus, $435 \div 5 = 87$.

Use a calculator to divide.

1. $848 \div 16$ **2.** $564 \div 12$

3. $95\overline{)5890}$ **4.** $27\overline{)1053}$

5. $\dfrac{32,886}{126}$ **6.** $\dfrac{143,088}{264}$

7. $0 \div 315$ **8.** $315 \div 0$

Mental Math

 Find each quotient. See Examples 1 through 3.

1. $40 \div 8$ **2.** $72 \div 9$ **3.** $45 \div 5$ **4.** $24 \div 3$ **5.** $0 \div 5$

6. $0 \div 8$ **7.** $9 \div 1$ **8.** $12 \div 1$ **9.** $\dfrac{16}{16}$ **10.** $\dfrac{49}{49}$

11. $\dfrac{25}{5}$ **12.** $\dfrac{45}{9}$ **13.** $6 \div 0$ **14.** $\dfrac{12}{0}$ **15.** $7 \div 1$

16. $6 \div 6$ **17.** $0 \div 4$ **18.** $7 \div 0$ **19.** $16 \div 2$ **20.** $18 \div 3$

EXERCISE SET 1.7

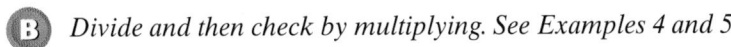 *Divide and then check by multiplying. See Examples 4 and 5.*

1. $9\overline{)108}$ **2.** $5\overline{)85}$ **3.** $6\overline{)222}$ **4.** $8\overline{)640}$ **5.** $3\overline{)1014}$ **6.** $4\overline{)504}$

Divide and then check by multiplying. See Examples 6 and 7.

7. $6\overline{)98}$ **8.** $7\overline{)422}$ **9.** $2\overline{)1127}$ **10.** $3\overline{)1240}$

11. $186 \div 5$ **12.** $167 \div 3$ **13.** $2121 \div 8$ **14.** $333 \div 4$

Divide and then check by multiplying. See Examples 8 and 9.

15. $23\overline{)1127}$ **16.** $42\overline{)2016}$ **17.** $55\overline{)715}$ **18.** $32\overline{)1856}$ **19.** $97\overline{)9449}$

20. $1938 \div 44$ **21.** $3708 \div 18$ **22.** $7224 \div 12$ **23.** $6578 \div 13$ **24.** $5670 \div 14$

25. $9299 \div 46$ **26.** $2539 \div 64$ **27.** $\dfrac{10{,}620}{236}$ **28.** $\dfrac{5781}{123}$ **29.** $\dfrac{10{,}194}{103}$

30. $\dfrac{23{,}048}{240}$ **31.** $20{,}619 \div 102$ **32.** $40{,}803 \div 203$ **33.** $45{,}046 \div 223$ **34.** $164{,}592 \div 543$

(c) *Solve. See Examples 10 through 12.*

35. Kathy Gomez teaches Spanish lessons for $85 per student for a 5-week session. From one group of students, she collects $4930. Find how many students are in the group.

36. Martin Thieme teaches American Sign Language classes for $55 per student for a 7-week session. He collects $1430 from the group of students. Find how many students are in the group.

37. Twenty-one people pooled their money and bought lottery tickets. One ticket won a prize of $5,292,000. Find how many dollars each person received.

38. The gravity of Jupiter is 318 times as strong as the gravity of Earth, so objects on Jupiter weigh 318 times as much as they weigh on Earth. If a person would weigh 52,470 pounds on Jupiter, find how much the person weighs on Earth.

39. A truck hauls wheat to a storage granary. It carries a total of 5810 bushels of wheat in 14 trips. How much does the truck haul each trip if each trip it hauls the same amount?

40. The white stripes dividing the lanes on a highway are 25 feet long, and the spaces between them are 25 feet long. Find how many whole stripes there are in 1 mile of highway. (A mile is 5280 feet.)

5810 bushels

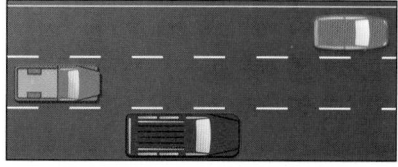

41. There is a bridge over highway I-35 every three miles. The first bridge is at the beginning of the 265-mile stretch of highway. Find how many bridges there are over 265 miles of I-35.

42. An 18-hole golf course is 5580 yards long. If the distance to each hole is the same, find the distance between holes.

43. Wendy Holladay has a piece of rope 185 feet long that she wants to cut into pieces for an experiment in her second-grade class. Each piece of rope is to be 8 feet long. Determine whether she has enough rope for her 22-student class. Determine the amount extra or the amount short.

44. Jesse White is in the requisitions department of Central Electric Lighting Company. Light poles along a highway are placed 492 feet apart. The first light pole is at the beginning of the 1-mile strip. Find how many poles he should order for a 1-mile strip of highway. (A mile is 5280 feet.)

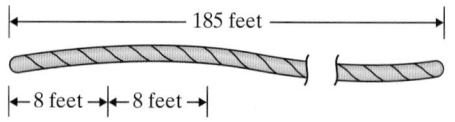

185 feet

8 feet   8 feet

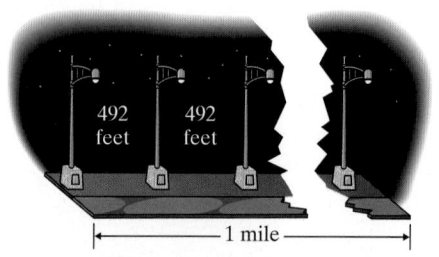

492 feet   492 feet

1 mile

45. Priest Holmes led the NFL in touchdowns during the 2002 football season, scoring a total of 144 points as touchdowns. If a touchdown is worth 6 points, how many touchdowns did Priest make during 2002? (*Source:* National Football League)

46. Broad Peak in Pakistan is the twelfth-tallest mountain in the world. Its elevation is 26,400 feet. A mile is 5280 feet. How many miles high is Broad Peak? (*Source:* National Geographic Society)

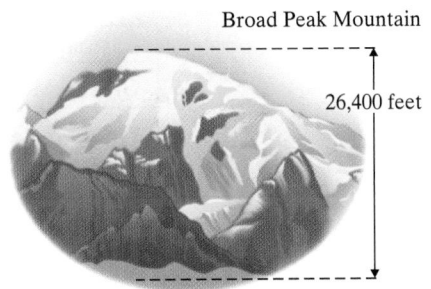

Broad Peak Mountain

26,400 feet

47. Find how many yards are in 1 mile. (A mile is 5280 feet; a yard is 3 feet.)

48. Find how many whole feet are in 1 rod. (A mile is 5280 feet; 1 mile is 320 rods.)

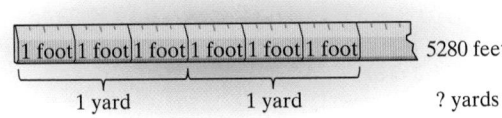

1 foot | 1 foot | 1 foot | 1 foot | 1 foot | 1 foot | 5280 feet

1 yard   1 yard   ? yards

D *Find the average of each list of numbers. See Example 13.*

49. 14, 22, 45, 18, 30, 27

50. 37, 26, 15, 29, 51, 22

51. 204, 968, 552, 268

52. 121, 200, 185, 176, 163

53. 86, 79, 81, 69, 80

54. 92, 96, 90, 85, 92, 79

The normal monthly temperature in degrees Fahrenheit for Minneapolis, Minnesota, is given in the graph. Use this graph to answer Exercises 55 and 56. (Source: National Climatic Data Center)

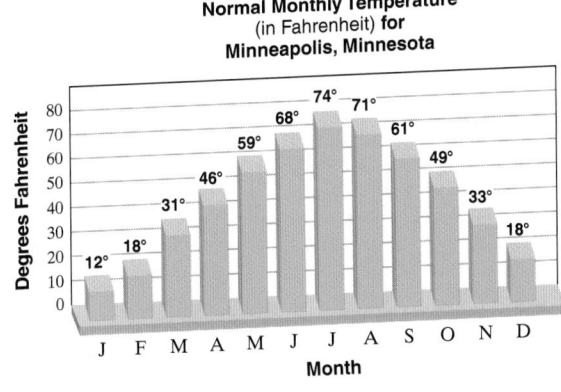

Normal Monthly Temperature
(in Fahrenheit) for
Minneapolis, Minnesota

55. Find the average temperature for December, January, and February.

56. Find the average temperature for the entire year.

The following table shows the top five leading U.S. advertisers in 2001 and the amount of money spent in that year on ads. Use this table to answer Exercises 57 and 58.

Company	Yearly Amount Spent on Advertising
General Motors Corp.	$3,374,400,000
Procter & Gamble Co.	$2,540,600,000
Ford Motor Co.	$2,408,200,000
PepsiCo	$2,210,400,000
Pfizer	$2,189,500,000

(*Source*: AdAge.com & the June 24, 2002 issue of *Advertising Age*)

57. Find the average amount of money spent on ads for the year by the top two companies.

58. Find the average amount of money spent on ads by Procter & Gamble, PepsiCo, and Pfizer.

In Example 13 in this section, we found that the average of 93, 86, 71, and 82 is 83. Use this information to answer Exercises 59 and 60.

59. If the number 71 is removed from the list of numbers, does the average increase or decrease? Explain why.

60. If the number 93 is removed from the list of numbers, does the average increase or decrease? Explain why.

61. Without computing it, tell whether the average of 126, 135, 198, 113 is 86. Explain why it is or why it is not.

62. If the area of a rectangle is 30 square feet and its width is 3 feet, what is its length?

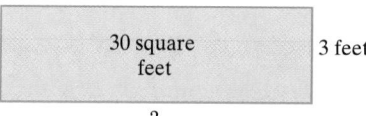

Integrated Review–Operations on Whole Numbers

Perform each indicated operation. Use estimation to see if your results are reasonable.

1. 23
      46
    + 79

2. 7006
     − 451

3. 36
     × 45

4. 8)4496

5. 1 · 79

6. $\frac{36}{0}$

7. 9 ÷ 1

8. 9 ÷ 9

9. 0 · 13

10. 7 · 0 · 8

11. 0 ÷ 2

12. 12 ÷ 4

13. 4219
      − 1786

14. 1861
      7965
       199
    + 2870

15. 5)1068

16. 1259
      × 63

17. 3 · 9

18. 45 ÷ 5

19. 207
      − 69

20. 207
      + 69

21. 7)7695

22. 9)1000

23. 32)21,222

24. 65)70,000

25. 4000 − 2976

26. 10,000 − 101

27. 303
      × 101

28. (475)(100)

29. 7)0

30. $\frac{14}{0}$

31. $\frac{0}{6}$

32. 0 ÷ 105

33. Subtract 14 from 100.

34. Find the difference of 43 and 21.

1. _____
2. _____
3. _____
4. _____
5. _____
6. _____
7. _____
8. _____
9. _____
10. _____
11. _____
12. _____
13. _____
14. _____
15. _____
16. _____
17. _____
18. _____
19. _____
20. _____
21. _____
22. _____
23. _____
24. _____
25. _____
26. _____
27. _____
28. _____
29. _____
30. _____
31. _____
32. _____
33. _____
34. _____

Complete the table by rounding the given number to the given place value.

		Tens	Hundreds	Thousands
35.	8625			
36.	1553			
37.	10,901			
38.	432,198			

Find the perimeter and area of each figure.

△**39.**

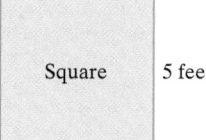

Square   5 feet

△**40.**

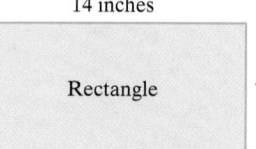

14 inches

Rectangle   7 inches

Find the perimeter of each figure.

△**41.**

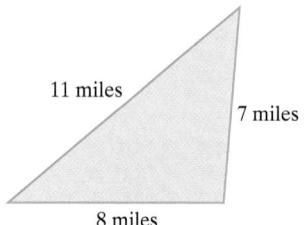

11 miles

7 miles

8 miles

△**42.**

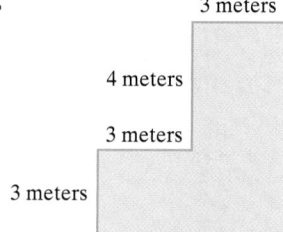

3 meters

4 meters

3 meters

3 meters

43. The length of the southern boundary of the conterminous United States is 1933 miles. The length of the northern boundary of the conterminous United States is 2054 miles longer than this. What is the length of the northern boundary? (*Source:* U.S. Geological Survey)

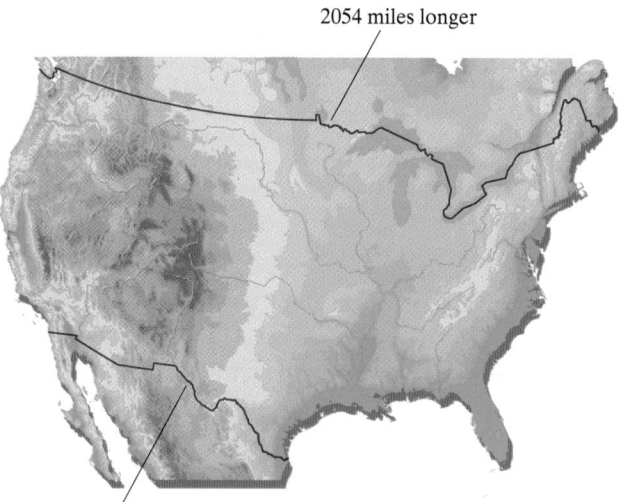

2054 miles longer

1933 miles

44. In humans, 14 muscles are required to smile. It takes 29 more muscles to frown. How many muscles does it take to frown?

45. How many cases can be filled with 9900 cans of jalapeños if each case holds 48 cans? How many cans will be left over? Will there be enough cases to fill an order for 200 cases?

46. The director of a learning lab at a local community college is working on next year's budget. Thirty-three new video players are needed at a cost of $540 each. What is the total cost of these video players?

1.8   Exponents and Order of Operations

(A) Using Exponential Notation

An **exponent** is a shorthand notation for repeated multiplication. When the same number is a factor several times, an exponent may be used. In the product

$$2 \cdot 2 \cdot 2 \cdot 2 \cdot 2$$

2 is a factor 5 times.

Using an exponent, this product can be written as

exponent

$$2^5$$ Read as "two to the fifth power."

base

Thus,

$$2 \cdot 2 \cdot 2 \cdot 2 \cdot 2 = 2^5$$

This is called **exponential notation**. The **exponent**, 5, indicates how many times the **base**, 2, is a factor.

Certain expressions are used when reading exponential notation.

$5 = 5^1$ is read as "five to the first power."

$5 \cdot 5 = 5^2$ is read as "five to the second power" or "five squared."

$5 \cdot 5 \cdot 5 = 5^3$ is read as "five to the third power" or "five cubed."

$5 \cdot 5 \cdot 5 \cdot 5 = 5^4$ is read as "five to the fourth power."

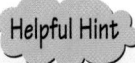

Helpful Hint

Usually, an exponent of 1 is not written, so when no exponent appears, we assume that the exponent is 1. For example, $2 = 2^1$ and $7 = 7^1$.

EXAMPLES   Write using exponential notation.

1. $4 \cdot 4 \cdot 4 = 4^3$
2. $7 \cdot 7 = 7^2$
3. $5 \cdot 5 \cdot 5 \cdot 5 = 5^4$
4. $6 \cdot 6 \cdot 6 \cdot 8 \cdot 8 \cdot 8 \cdot 8 \cdot 8 = 6^3 \cdot 8^5$

(B) Evaluating Exponential Expressions

To **evaluate** an exponential expression, we write the expression as a product and then find the value of the product.

EXAMPLES   Evaluate.

5. $8^2 = 8 \cdot 8 = 64$
6. $7^1 = 7$
7. $2^5 = 2 \cdot 2 \cdot 2 \cdot 2 \cdot 2 = 32$
8. $5 \cdot 6^2 = 5 \cdot 6 \cdot 6 = 180$

Example 8 illustrates an important property: An exponent applies only to its base. The exponent 2, in $5 \cdot 6^2$, applies only to its base, 6.

Practice Problems 1–4

Write using exponential notation.

1. $2 \cdot 2 \cdot 2$
2. $3 \cdot 3$
3. $10 \cdot 10 \cdot 10 \cdot 10 \cdot 10 \cdot 10$
4. $5 \cdot 5 \cdot 4 \cdot 4 \cdot 4$

Practice Problems 5–8

Evaluate.

5. 2^3
6. 5^2
7. 10^1
8. $4 \cdot 5^2$

Answers

1. 2^3 **2.** 3^2 **3.** 10^6 **4.** $5^2 \cdot 4^3$ **5.** 8   **6.** 25
7. 10   **8.** 100

> **Helpful Hint**
>
> An exponent applies only to its base. For example, $4 \cdot 2^3$ means $4 \cdot 2 \cdot 2 \cdot 2$.

> **Helpful Hint**
>
> Don't forget that 2^4, for example, is *not* $2 \cdot 4$. 2^4 means repeated multiplication of the same factor.
>
> $$2^4 = 2 \cdot 2 \cdot 2 \cdot 2 = 16, \text{ whereas } 2 \cdot 4 = 8$$

Concept Check

Which of the following statement(s) is correct?

a. 3^6 is the same as $6 \cdot 6 \cdot 6$.
b. "Eight to the fourth power" is the same as 8^4.
c. "Ten squared" is the same as 10^3.
d. 11^2 is the same as $11 \cdot 2$.
e. $5^2 = 10$

Try the Concept Check in the margin.

Ⓒ Using the Order of Operations

Suppose that you are in charge of taking inventory at a local bookstore. An employee has given you the number of a certain book in stock as the expression

$$3 + 2 \cdot 10$$

To calculate the value of this expression, do you add first or multiply first? If you add first, the answer is 50. If you multiply first, the answer is 23.

Mathematical symbols wouldn't be very useful if two values were possible for one expression. Thus, mathematicians have agreed that, given a choice, we multiply first.

$$3 + 2 \cdot 10 = 3 + 20 \qquad \text{Multiply.}$$
$$= 23 \qquad \text{Add.}$$

This agreement is one of several **order of operations** agreements.

Order of Operations

1. Perform all operations within grouping symbols such as parentheses or brackets.
2. Evaluate any expressions with exponents.
3. Multiply or divide in order from left to right.
4. Add or subtract in order from left to right.

For example, using the order of operations, let's evaluate $2^3 \cdot 4 - (10 \div 5)$.

$$2^3 \cdot 4 - (10 \div 5) = 2^3 \cdot 4 - 2 \qquad \text{Simplify inside parentheses.}$$
$$= 8 \cdot 4 - 2 \qquad \text{Write } 2^3 \text{ as 8.}$$
$$= 32 - 2 \qquad \text{Multiply } 8 \cdot 4.$$
$$= 30 \qquad \text{Subtract.}$$

ANSWER

Concept Check: b

EXAMPLE 9   Simplify: $2 \cdot 4 - 3 \div 3$

Solution: There are no parentheses and no exponents, so we start by multiplying and dividing, from left to right.

$$2 \cdot 4 - 3 \div 3 = 8 - 3 \div 3 \qquad \text{Multiply.}$$
$$= 8 - 1 \qquad \text{Divide.}$$
$$= 7 \qquad \text{Subtract.}$$

Practice Problem 9

Simplify: $16 \div 4 - 2$

EXAMPLE 10   Simplify: $4^2 \div 2 \cdot 4$

Solution: We start by evaluating 4^2.

$$4^2 \div 2 \cdot 4 = 16 \div 2 \cdot 4 \qquad \text{Write } 4^2 \text{ as 16.}$$

Next we multiply or divide *in order* from left to right. Since division appears before multiplication from left to right, we divide first, then multiply.

$$16 \div 2 \cdot 4 = 8 \cdot 4 \qquad \text{Divide.}$$
$$= 32 \qquad \text{Multiply.}$$

Practice Problem 10

Simplify: $18 \div 3^2 \cdot 2^2$

EXAMPLE 11   Simplify: $(8 - 6)^2 + 2^3 \cdot 3$

Solution: $(8 - 6)^2 + 2^3 \cdot 3 = 2^2 + 2^3 \cdot 3 \qquad$ Simplify inside parentheses.

$$= 4 + 8 \cdot 3 \qquad \text{Write } 2^2 \text{ as 4 and } 2^3 \text{ as 8.}$$
$$= 4 + 24 \qquad \text{Multiply.}$$
$$= 28 \qquad \text{Add.}$$

Practice Problem 11

Simplify: $(9 - 8)^3 + 3 \cdot 2^4$

EXAMPLE 12   Simplify: $4^3 + [3^2 - (10 \div 2)] - 7 \cdot 3$

Solution: Here we begin with the innermost set of parentheses.

$$4^3 + [3^2 - (10 \div 2)] - 7 \cdot 3 = 4^3 + [3^2 - 5] - 7 \cdot 3 \qquad \text{Simplify inside parentheses.}$$
$$= 4^3 + [9 - 5] - 7 \cdot 3 \qquad \text{Write } 3^2 \text{ as 9.}$$
$$= 4^3 + 4 - 7 \cdot 3 \qquad \text{Simplify inside brackets.}$$
$$= 64 + 4 - 7 \cdot 3 \qquad \text{Write } 4^3 \text{ as 64.}$$
$$= 64 + 4 - 21 \qquad \text{Multiply.}$$
$$= 47 \qquad \text{Add and subtract from left to right.}$$

Practice Problem 12

Simplify: $24 \div [20 - (3 \cdot 4)] + 2^3 - 5$

EXAMPLE 13   Simplify: $\dfrac{7 - 2 \cdot 3 + 3^2}{2^2 + 1}$

Solution: Here, the fraction bar is like a grouping symbol. We simplify above and below the fraction bar separately.

$$\frac{7 - 2 \cdot 3 + 3^2}{2^2 + 1} = \frac{7 - 2 \cdot 3 + 9}{4 + 1} \qquad \text{Write } 3^2 \text{ as 9 and } 2^2 \text{ as 4.}$$
$$= \frac{7 - 6 + 9}{5} \qquad \text{Multiply } 2 \cdot 3 \text{ in the numerator and add 4 and 1 in the denominator.}$$
$$= \frac{10}{5} \qquad \text{Add and subtract from left to right.}$$
$$= 2 \qquad \text{Divide.}$$

Practice Problem 13

Simplify: $\dfrac{60 - 5^2 + 1}{3(1 + 1)}$

Answers

9. 2   **10.** 8   **11.** 49   **12.** 6   **13.** 6

Ⓓ Finding the Area of a Square

Since a square is a special rectangle, we can find its area by finding the product of its length and its width.

Area of a rectangle = length · width

By recalling that each side of a square has the same measurement, we can use the following procedure to find its area:

Area of a square = length · width

= side · side

= (side)^2

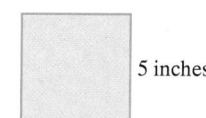

Square   Side

Side

Practice Problem 14

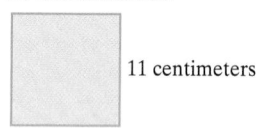

Find the area of a square whose side measures 11 centimeters.

11 centimeters

Answer

14. 121 sq cm

EXAMPLE 14   Find the area of a square whose side measures 5 inches.

Solution: Area of a square = (side)^2

= (5 inches)^2

= 25 square inches

5 inches

The area of the square is 25 square inches.

CALCULATOR EXPLORATIONS

Exponents

To evaluate an exponent such as 4^7 on a calculator, find the keys marked $\boxed{y^x}$ and $\boxed{=}$ or $\boxed{\text{ENTER}}$. To evaluate 4^7 press the keys $\boxed{4}$ $\boxed{y^x}$ $\boxed{7}$ $\boxed{=}$ or $\boxed{\text{ENTER}}$. The display will read $\boxed{\qquad 16384}$. Thus, $4^7 = 16{,}384$.

Use a calculator to evaluate.

1. 3^6 **2.** 5^6

3. 4^5 **4.** 7^6

5. 2^{11} **6.** 6^8

Order of Operations

To see whether your calculator has the order of operations built in, evaluate $5 + 2 \cdot 3$ by pressing the keys $\boxed{5}$ $\boxed{+}$ $\boxed{2}$ $\boxed{\times}$ $\boxed{3}$ $\boxed{=}$ or $\boxed{\text{ENTER}}$. If the display reads $\boxed{\quad 11}$, your calculator does have the order of operations built in. This means that most of the time you can key in a problem exactly as it is written and

the calculator will perform operations in the proper order. When evaluating an expression containing parentheses, key in the parentheses. (If an expression contains brackets, key in parentheses.) For example, to evaluate $2[25 - (8 + 4)] - 11$, press the keys

$\boxed{2}$ $\boxed{\times}$ $\boxed{(}$ $\boxed{25}$ $\boxed{-}$ $\boxed{(}$ $\boxed{8}$ $\boxed{+}$ $\boxed{4}$ $\boxed{)}$ $\boxed{)}$ $\boxed{-}$ $\boxed{11}$ $\boxed{=}$

or $\boxed{\text{ENTER}}$.

The display will read $\boxed{\qquad 15}$.

Use a calculator to evaluate.

7. $7^4 + 5^3$

8. $12^4 - 8^4$

9. $63 \cdot 75 - 43 \cdot 10$

10. $8 \cdot 22 + 7 \cdot 16$

11. $4(15 \div 3 + 2) - 10 \cdot 2$

12. $155 - 2(17 + 3) + 185$

EXERCISE SET 1.8

A *Write using exponential notation. See Examples 1 through 4.*

1. $3 \cdot 3 \cdot 3 \cdot 3$ **2.** $5 \cdot 5 \cdot 5$ **3.** $7 \cdot 7 \cdot 7 \cdot 7 \cdot 7 \cdot 7 \cdot 7 \cdot 7$ **4.** $6 \cdot 6 \cdot 6 \cdot 6 \cdot 6$

**5.** $12 \cdot 12 \cdot 12$ **6.** $10 \cdot 10$ **7.** $6 \cdot 6 \cdot 5 \cdot 5 \cdot 5$ **8.** $4 \cdot 4 \cdot 4 \cdot 3 \cdot 3$

9. $9 \cdot 9 \cdot 9 \cdot 8$ **10.** $7 \cdot 7 \cdot 7 \cdot 4$ **11.** $3 \cdot 2 \cdot 2 \cdot 2 \cdot 2 \cdot 2$ **12.** $4 \cdot 6 \cdot 6 \cdot 6 \cdot 6$

13. $3 \cdot 2 \cdot 2 \cdot 5 \cdot 5 \cdot 5$ **14.** $6 \cdot 6 \cdot 2 \cdot 9 \cdot 9 \cdot 9 \cdot 9$

B *Evaluate. See Examples 5 through 8.*

15. 5^2 **16.** 6^2 **17.** 5^3 **18.** 6^3 **19.** 2^6 **20.** 2^7

21. 2^{10} **22.** 1^{12} **23.** 7^1 **24.** 8^1 **25.** 3^5 **26.** 5^4

27. 2^8 **28.** 3^3 **29.** 4^3 **30.** 4^4 **31.** 9^2 **32.** 8^2

33. 9^3 **34.** 8^3 **35.** 10^2 **36.** 10^3 **37.** 10^4 **38.** 10^5

39. 10^1 **40.** 14^1 **41.** 1920^1 **42.** 6849^1 **43.** 3^6 **44.** 4^5

C *Simplify. See Examples 9 through 13.*

45. $15 + 3 \cdot 2$ **46.** $24 + 6 \cdot 3$ **47.** $20 - 4 \cdot 3$ **48.** $17 - 2 \cdot 6$ **49.** $5 \cdot 9 - 16$

50. $8 \cdot 4 - 10$ **51.** $28 \div 4 - 3$ **52.** $42 \div 7 - 6$ **53.** $14 + \dfrac{24}{8}$ **54.** $32 + \dfrac{8}{2}$

55. 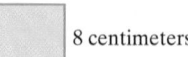 $6 \cdot 5 + 8 \cdot 2$ **56.** $3 \cdot 4 + 9 \cdot 1$ **57.** $0 \div 6 + 4 \cdot 7$ **58.** $0 \div 8 + 7 \cdot 6$ **59.** $6 + 8 \div 2$

60. $6 + 9 \div 3$ **61.** $(6 + 8) \div 2$ **62.** $(6 + 9) \div 3$ **63.** $(6^2 - 4) \div 8$ **64.** $(7^2 - 7) \div 7$

65. $(3 + 5^2) \div 2$ **66.** $(13 + 6^2) \div 7$ 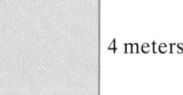**67.** $6^2 \cdot (10 - 8)$ **68.** $5^3 \div (10 + 15)$ **69.** $\dfrac{18 + 6}{2^4 - 4}$

70. $\dfrac{15 + 17}{5^2 - 3^2}$ **71.** $(2 + 5) \cdot (8 - 3)$ **72.** $(9 - 7) \cdot (12 + 18)$ 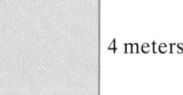**73.** $\dfrac{7(9 - 6) + 3}{3^2 - 3}$

74. $\dfrac{5(12 - 7) - 4}{5^2 - 2^3 - 10}$ **75.** $5 \div 0 + 24$ **76.** $18 - 7 \div 0$ **77.** $3^4 - [35 - (12 - 6)]$

78. $[40 - (8 - 2)] - 2^5$ 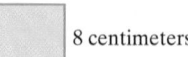**79.** $(7 \cdot 5) + [9 \div (3 \div 3)]$ **80.** $(18 \div 6) + [(3 + 5) \cdot 2]$

81. $8 \cdot [4 + (6 - 1) \cdot 2] - 50 \cdot 2$ **82.** $35 \div [3^2 + (9 - 7) - 2^2] + 10 \cdot 3$

83. $7^2 - \{18 - [40 \div (4 \cdot 2) + 2] + 5^2\}$ **84.** $29 - \{5 + 3[8 \cdot (10 - 8)] - 50\}$

D *Find the area of each square. See Example 14.*

△ **85.**

20 miles

△ **86.**

4 meters

△ **87.**

8 centimeters

△ **88.**

31 feet

89. The Eiffel Tower stands on a square base measuring 100 meters on each side. Find the area of the base.

90. A square lawn that measures 72 feet on each side is to be fertilized. If 5 bags of fertilizer are available and each bag can fertilize 1000 square feet, is there enough fertilizer to cover the lawn?

 Combining Concepts

Insert grouping symbols (parentheses) so that each given expression evaluates to the given number.

91. $2 + 3 \cdot 6 - 2$; evaluate to 28

92. $2 + 3 \cdot 6 - 2$; evaluate to 20

93. $24 \div 3 \cdot 2 + 2 \cdot 5$; evaluate to 14

94. $24 \div 3 \cdot 2 + 2 \cdot 5$; evaluate to 15

95. A building contractor is bidding on a contract to install gutters on seven homes in a retirement community, all in the shape shown. To estimate her cost of materials, she needs to know the total perimeter of all seven homes. Find the total perimeter.

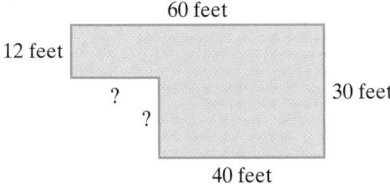

Simplify.

96. $25^3 \cdot (45 - 7 \cdot 5) \cdot 5$

97. $(7 + 2^4)^5 - (3^5 - 2^4)^2$

98. Explain why $2 \cdot 3^2$ is not the same as $(2 \cdot 3)^2$.

FOCUS ON **Mathematical Connections**

INVESTIGATING WHOLE NUMBERS AND ORDER OF OPERATIONS

This activity may be completed by working in groups or individually.

Try the following activity. You will need 30 index cards.

1. Label each index card with a number from 1 to 30, using each number once.

2. Shuffle the cards. Deal 5 cards face up to each player or team. Then deal a single card face down to each player or team. This single card is the goal card.

3. Play begins when the goal cards are turned over simultaneously. The object is for each player or team to use all 5 number cards (each only once) along with any of the operations +, −, ×, or ÷ to obtain the number on the goal card. The winner is the first team to correctly write out the se-

quence of numbers and operations equaling the number on the goal card, according to the order of operations. Parentheses or other grouping symbols may be used as needed.

4. If no player/team is able to obtain the number on the goal card, the player/team that is able to come closest to their goal wins.

Example:

Number Cards | Goal Cards

9 14 15 10 24   12

Sequence of numbers and operations to obtain goal:

$$24 \div [(15 - 14) + (10 - 9)] = 12$$

1.9 Introduction to Variables and Algebraic Expressions

A Evaluating Algebraic Expressions

Perhaps the most important quality of mathematics is that it is a science of patterns. Communicating about patterns is often made easier by using a letter to represent all the numbers fitting a pattern. We call such a letter a **variable**. For example, in Section 1.2 we presented the addition property of 0, which states that the sum of 0 and any number is that number. We might write

$$0 + 1 = 1$$
$$0 + 2 = 2$$
$$0 + 3 = 3$$
$$0 + 4 = 4$$
$$0 + 5 = 5$$
$$0 + 6 = 6$$
$$\vdots$$

continuing indefinitely. This is a pattern, and all whole numbers fit the pattern. We can communicate this pattern for all whole numbers by letting a letter, such as a, represent all whole numbers. We can then write

$$0 + a = a$$

Using variable notation is a primary goal of learning **algebra**. We now take some important first steps in beginning to use variable notation.

A combination of operations on letters (variables) and numbers is called an **algebraic expression** or simply an **expression**.

Algebraic Expressions

$$3 + x \qquad 5 \cdot y \qquad 2 \cdot z - 1 + x$$

If two variables or a number and a variable are next to each other, with no operation sign between them, the operation is multiplication. For example,

$$2x \quad \text{means} \quad 2 \cdot x$$

and

$$xy \text{ or } x(y) \quad \text{means} \quad x \cdot y$$

Also, the meaning of an exponent remains the same when the base is a variable. For example,

$$x^2 = \underbrace{x \cdot x}_{2 \text{ factors of } x} \quad \text{and} \quad y^5 = \underbrace{y \cdot y \cdot y \cdot y \cdot y}_{5 \text{ factors of } y}$$

Algebraic expressions such as $3x$ have different values depending on replacement values for x. For example, if x is 2, then $3x$ becomes

$$3x = 3 \cdot 2$$
$$= 6$$

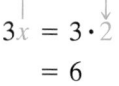

If x is 7, then $3x$ becomes

$$3x = 3 \cdot 7$$
$$= 21$$

Replacing a variable in an expression by a number and then finding the value of the expression is called **evaluating the expression** for the variable. When finding the value of an expression, remember to follow the order of operations given in Section 1.8.

Practice Problem 1

Evaluate $x - 2$ if x is 5.

EXAMPLE 1   Evaluate $x + 7$ if x is 8.

Solution: Replace x with 8 in the expression $x + 7$.

$$x + 7 = 8 + 7 \qquad \text{Replace } x \text{ with 8.}$$
$$= 15 \qquad \text{Add.}$$

When we write a statement such as "x is 5," we can use an equals symbol to represent "is" so that

$$x \text{ is } 5 \quad \text{can be written as} \quad x = 5.$$

Practice Problem 2

Evaluate $y(x - 3)$ for $x = 3$ and $y = 7$.

EXAMPLE 2   Evaluate $2(x - y)$ for $x = 8$ and $y = 4$.

Solution: $2(x - y) = 2(8 - 4) \qquad \text{Replace } x \text{ with 8 and } y \text{ with 4.}$
$$= 2(4) \qquad \text{Subtract.}$$
$$= 8 \qquad \text{Multiply.}$$

Practice Problem 3

Evaluate $\dfrac{y + 6}{x}$ for $x = 2$ and $y = 8$.

EXAMPLE 3   Evaluate $\dfrac{x - 5y}{y}$ for $x = 21$ and $y = 3$.

Solution: $\dfrac{x - 5y}{y} = \dfrac{21 - 5(3)}{3} \qquad \text{Replace } x \text{ with 21 and } y \text{ with 3.}$

$$= \dfrac{21 - 15}{3} \qquad \text{Multiply.}$$

$$= \dfrac{6}{3} \qquad \text{Subtract.}$$

$$= 2 \qquad \text{Divide.}$$

Practice Problem 4

Evaluate $25 - z^3 + x$ for $z = 2$ and $x = 1$.

EXAMPLE 4   Evaluate $x^2 + z - 3$ for $x = 5$ and $z = 4$.

Solution: $x^2 + z - 3 = 5^2 + 4 - 3 \qquad \text{Replace } x \text{ with 5 and } z \text{ with 4.}$
$$= 25 + 4 - 3 \qquad \text{Evaluate } 5^2.$$
$$= 26 \qquad \text{Add and subtract from left to right.}$$

Answers

1. 3   **2.** 0   **3.** 7   **4.** 18

Try the Concept Check in the margin.

EXAMPLE 5   The expression $\dfrac{5(F - 32)}{9}$ can be used to write degrees Fahrenheit F as degrees Celsius C. Find the value of this expression for $F = 86$.

Solution $\dfrac{5(F - 32)}{9} = \dfrac{5(86 - 32)}{9}$

$$= \dfrac{5(54)}{9}$$

$$= \dfrac{270}{9}$$

$$= 30$$

Thus 86°F is the same temperature as 30°C.

B   Translating Phrases into Variable Expressions

To aid us in solving problems later, we practice translating verbal phrases into algebraic expressions. Certain key words and phrases suggesting addition, subtraction, multiplication, or division are reviewed next.

Addition	Subtraction	Multiplication	Division
sum	difference	product	quotient
plus	minus	times	divided by
added to	subtracted from	multiply	into
more than	less than	twice	per
increased by	decreased by	of	
total	less	double	

EXAMPLE 6   Write as an algebraic expression. Use x to represent "a number."

a. 7 increased by a number
b. 15 decreased by a number
c. The product of 2 and a number
d. The quotient of a number and 5
e. 2 subtracted from a number

Concept Check

What's wrong with the solution to the following problem?

Evaluate $3x + 2y$ for $x = 2$ and $y = 3$.

Solution: $3x + 2y = 3(3) + 2(2)$
$$= 9 + 4$$
$$= 13$$

Practice Problem 5

Evaluate $\dfrac{5(F - 32)}{9}$ for $F = 41$.

Practice Problem 6

Write as an algebraic expression. Use x to represent "a number."

a.   Twice a number.
b.   8 increased by a number.
c.   10 minus a number.
d.   10 subtracted from a number.
e.   The quotient of 6 and a number.

Answers

5. 5   **6. a.** $2x$ **b.** $x + 8$ **c.** $10 - x$ **d.** $x - 10$

e. $6 \div x$ or $\dfrac{6}{x}$

Concept Check:
$3x + 2y = 3(2) + 2(3)$
$$= 6 + 6$$
$$= 12$$

Solution:

a. In words: $\boxed{7}$ $\boxed{\text{increased by}}$ $\boxed{\text{a number}}$

Translate: $\quad 7 \qquad\quad + \qquad\qquad x$

b. In words: $\boxed{15}$ $\boxed{\text{decreased by}}$ $\boxed{\text{a number}}$

Translate: $\quad 15 \qquad\quad - \qquad\qquad x$

c. In words: $\boxed{\begin{array}{c}\text{The product}\\ \text{of}\end{array}}$

$\boxed{2} \quad \boxed{\text{and}} \qquad \boxed{\text{a number}}$

Translate: $\quad 2 \qquad\cdot \qquad\qquad x \qquad \text{or } 2x$

d. In words: $\boxed{\begin{array}{c}\text{The quotient}\\ \text{of}\end{array}}$

$\boxed{\text{a number}} \quad \boxed{\text{and}} \quad \boxed{5}$

Translate: $\qquad x \qquad\quad \div \qquad 5 \quad \text{or } \dfrac{x}{5}$

e. In words: $\boxed{2}$ $\boxed{\begin{array}{c}\text{subtracted}\\ \text{from}\end{array}}$ $\boxed{\text{a number}}$

Translate: $\qquad x \quad\longleftarrow\quad - \quad\longrightarrow\quad 2$

Helpful Hint

Remember that order is important when subtracting. Study the order of numbers and variables below.

Phrase	Translation
a number *decreased by* 5	$x - 5$
a number *subtracted from* 5	$5 - x$

Name _____ Section _____ Date _____

EXERCISE SET 1.9

A *Evaluate each following expression for $x = 2$, $y = 5$, and $z = 3$. See Examples 1 through 5.*

 1. $3 + 2z$

2. $7 + 3z$

3. $6xz - 5x$

4. $4yz + 2x$

5. $z - x + y$

6. $x + 5y - z$

7. $3x - z$

8. $2y + 5z$

 9. $y^3 - 4x$

10. $y^3 - z$

11. $2xy^2 - 6$

12. $3yz^2 + 1$

13. $8 - (y - x)$

14. $5 + (2x - 1)$

15. $y^4 + (z - x)$

16. $x^4 - (y - z)$

17. $\dfrac{6xy}{z}$

18. $\dfrac{8yz}{15}$

19. $\dfrac{2y - 2}{x}$

20. $\dfrac{6 + 3x}{z}$

21. $\dfrac{x + 2y}{z}$

22. $\dfrac{2z + 6}{3}$

23. $\dfrac{5x}{y} - \dfrac{10}{y}$

24. $\dfrac{70}{2y} - \dfrac{15}{z}$

25. $2y^2 - 4y + 3$

26. $3z^2 - z + 10$

27. $(3y - 2x)^2$

28. $(4y + 3z)^2$

29. $(xy + 1)^2$

30. $(xz - 5)^4$

 **31.** $2y(4z - x)$

32. $3x(y + z)$

33. $xy(5 + z - x)$

34. $xz(2y + x - z)$

 **35.** $\dfrac{7x + 2y}{3x}$

36. $\dfrac{6z + 2y}{4}$

Introduction to Variables and Algebraic Expressions   Section 1.9   **93**

37. The expression $16t^2$ gives the distance in feet that an object falls after t seconds. Complete the table by evaluating $16t^2$ for each given value of t.
see table

t	1	2	3	4
$16t^2$				

38. The expression $\dfrac{5(F - 32)}{9}$ gives the equivalent degrees Celsius for F degrees Fahrenheit. Complete the table by evaluating this expression for each given value of F.

see table

F	50	59	68	77
$\dfrac{5(F - 32)}{9}$				

B *Write each phrase as a variable expression. Use x to represent "a number." See Example 6.*

39. The sum of a number and five

40. Ten plus a number

41. The total of a number and eight

42. The difference of a number and five hundred

43. Twenty decreased by a number

44. A number less thirty

45. The product of 512 and a number

46. A number times twenty

47. A number divided by 2

48. The quotient of six and a number

49. The sum of seventeen and a number added to the product of five and the number

50. The difference of twice a number, and four

51. The product of five and a number

52. The quotient of twenty and a number, decreased by three

53. A number subtracted from 11

54. Twelve subtracted from a number

55. A number less 5

56. The product of a number and 7

57. 6 divided by a number

58. The sum of a number and 7

59. Fifty decreased by eight times a number

60. Twenty decreased by twice a number

Combining Concepts

Use a calculator to evaluate each expression for x = 23 and y = 72.

61. $x^4 - y^2$

62. $2(x + y)^2$

63. $x^2 + 5y - 112$

64. $16y - 20x + x^3$

65. If x is a whole number, which expression is the largest: $2x$, $5x$, or $\frac{x}{3}$? Explain your answer.

66. If x is a whole number, which expression is the smallest: $2x$, $5x$, or $\frac{x}{3}$? Explain your answer.

67. In Exercise 37, what do you notice about the value of $16t^2$ as t gets larger?

68. In Exercise 38, what do you notice about the value of $\frac{5(F - 32)}{9}$ as F gets larger?

This activity may be completed by working in groups or individually.

An **endangered** species is one that is thought to be in danger of becoming extinct throughout all or a major part of its habitat. A **threatened** species is one that may become endangered. The Division of Endangered Species at the U.S. Fish and Wildlife Service keeps close tabs on the state of threatened and endangered wildlife in the United States and around the world. The table below was compiled from data in the Division of Endangered Species' box score published on July 1, 2003. The "Total Species" column gives the total number of endangered and threatened species for each group.

1. Round each number of *endangered animal species* to the nearest ten to estimate the Animal Total.

2. Round each number of *endangered plant species* to the nearest ten to estimate the Plant Total.

3. Add the exact numbers of endangered animal species to find the exact Animal Total and record it in the table in the Endangered Species column. Add the exact numbers of endangered plant species to find the Plant Total and record it in the table in the Endangered Species column. Then find the total number of endangered species (animals and plants combined) and record this number in the table as the Grand Total in the Endangered Species column.

4. Find the Animal Total, Plant Total, and Grand Total for the Total Species column. Record these values in the table.

5. Use the data in the table to complete the Threatened Species column.

6. Write a paragraph discussing the conclusions that can be drawn from the table.

7. (Optional) The Division of Endangered Species updates its endangered/threatened species box score monthly. Visit the current box score of endangered and threatened species at http://ecos.fws.gov/tess/html/boxscore.html on the World Wide Web. How have the figures changed since July, 2003? Write a paragraph summarizing the changes in the numbers of endangered and threatened animals and plants.

Endangered and Threatened Species Worldwide				
	Group	**Endangered Species**	**Threatened Species**	**Total Species**
Animals	Mammals	316		342
	Birds	253		273
	Reptiles	78		115
	Amphibians	20		30
	Fishes	82		126
	Clams	64		72
	Snails	22		33
	Insects	39		48
	Arachnids	12		12
	Crustaceans	18		21
	Animal Total			
Plants	Flowering Plants	578		716
	Conifers	2		5
	Ferns and Allies	24		26
	Lichens	2		2
	Plant Total			
	Grand Total			

(*Source:* U.S. Fish and Wildlife Service, Division of Endangered Species)

Chapter 1   Vocabulary Check

Fill in each blank with one of the words or phrases listed below.

place value	whole numbers	perimeter	variable
area	natural numbers	place value	exponent

1. The _____ are 0, 1, 2, 3, ...
2. The _____ of a polygon is its distance around or the sum of the lengths of its sides.
3. The position of each digit in a number determines its _____ .
4. An _____ is a shorthand notation for repeated multiplication of the same factor.
5. To find the _____ of a rectangle, multiply length times width.
6. The _____ are 1, 2, 3, 4, ...
7. The position of a digit in a number determines its _____ .
8. A letter used to represent a number is called a _____ .

CHAPTER 1 Highlights

DEFINITIONS AND CONCEPTS	EXAMPLES

SECTION 1.2 PLACE VALUE AND NAMES FOR NUMBERS

The **whole numbers** are 0, 1, 2, 3, 4, 5,...

The position of each digit in a number determines its **place value**. A place-value chart is shown next with the names of the periods given.

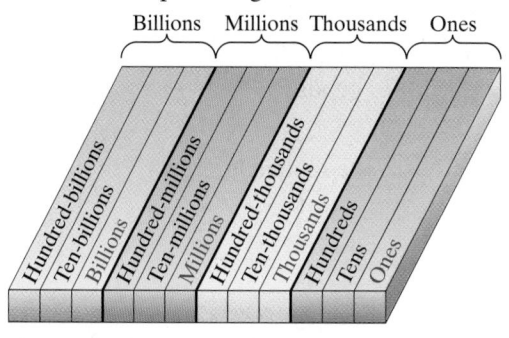

Billions   Millions   Thousands   Ones

Hundred-billions, Ten-billions, Billions, Hundred-millions, Ten-millions, Millions, Hundred-thousands, Ten-thousands, Thousands, Hundreds, Tens, Ones

0,   14,   968,   5,268,619

To write a whole number in words, write the number in the period followed by the name of the period. The name of the ones period is not included.

9,078,651,002 is written as nine billion, seventy-eight million, six hundred fifty-one thousand, two.

SECTION 1.3 ADDING WHOLE NUMBERS

To add whole numbers, add the digits in the ones place, then the tens place, then the hundreds place, and so on, carrying when necessary.

Find the sum:

```
 2 1 1
  2689    ←   addend
  1735    ←   addend
+  662    ←   addend
  5086    ←   sum
```

△ Find the perimeter of the polygon shown.

The **perimeter** of a polygon is its distance around or the sum of the lengths of its sides.

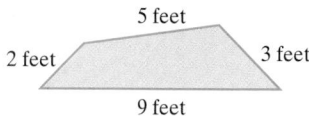

5 feet

2 feet   3 feet

9 feet

The perimeter is
5 feet + 3 feet + 9 feet + 2 feet = 19 feet.

DEFINITIONS AND CONCEPTS	EXAMPLES

SECTION 1.4 SUBTRACTING WHOLE NUMBERS

To subtract whole numbers, subtract the digits in the ones place, then the tens place, then the hundreds place, and so on, borrowing when necessary.

Subtract:

$$
\begin{array}{r}
{}^{8\,15}\\
79\cancel{8}4 \leftarrow \quad \text{minuend}\\
-\ 5673 \leftarrow \quad \text{subtrahend}\\
\hline
2281 \leftarrow \quad \text{difference}
\end{array}
$$

SECTION 1.5 ROUNDING AND ESTIMATING

ROUNDING WHOLE NUMBERS TO A GIVEN PLACE VALUE

Step 1. Locate the digit to the right of the given place value.

Step 2. If this digit is 5 or greater, add 1 to the digit in the given place value and replace each digit to its right with 0.

Step 3. If this digit is less than 5, replace it and each digit to its right with 0.

Round 15,721 to the nearest thousand.

15, ⑦ 21

Add 1 ⟶

Replace with zeros.

Since the circled digit is 5 or greater, add 1 to the given place value and replace digits to its right with zeros.

15,721 rounded to the nearest thousand is 16,000.

SECTION 1.6 MULTIPLYING WHOLE NUMBERS

To multiply 73 and 58, for example, multiply 73 and 8, then 73 and 50. The sum of these partial products is the product of 73 and 58. Use the notation to the right.

$$
\begin{array}{r}
73 \leftarrow \quad \text{factor}\\
\times\ 58 \leftarrow \quad \text{factor}\\
\hline
584 \leftarrow \quad 73 \times 8\\
3650 \leftarrow \quad 73 \times 50\\
\hline
4234 \leftarrow \quad \text{product}
\end{array}
$$

To find the **area** of a rectangle, multiply length times width.

Find the area of the rectangle shown.

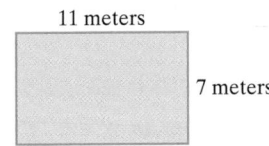

11 meters

7 meters

area of rectangle = length · width
    = (11 meters)(7 meters)
    = 77 square meters

DEFINITIONS AND CONCEPTS	**EXAMPLES**

SECTION 1.7 DIVIDING WHOLE NUMBERS

To divide larger whole numbers, use the process called **long division** as shown to the right.

$$
\begin{array}{r}
\text{divisor} \rightarrow \quad 507 \text{ R } 2 \quad \leftarrow \text{quotient} \\
14\overline{)7100} \quad \leftarrow \text{dividend} \\
-70\downarrow \qquad 5(14) = 70 \\
\overline{10} \qquad \text{Subtract and bring down the 0.} \\
-0\downarrow \qquad 0(14) = 0 \\
\overline{100} \qquad \text{Subtract and bring down the 0.} \\
98 \qquad 7(14) = 98 \\
\overline{2} \qquad \text{Subtract. The remainder is 2.}
\end{array}
$$

To check, see that $507 \cdot 14 + 2 = 7100$.

The **average** of a list of numbers is

$$\text{average} = \frac{\text{sum of numbers}}{\text{number of numbers}}$$

Find the average of 23, 35, and 38.

$$\text{average} = \frac{23 + 35 + 38}{3} = \frac{96}{3} = 32$$

SECTION 1.8 EXPONENTS AND ORDER OF OPERATIONS

An **exponent** is a shorthand notation for repeated multiplication of the same factor.

ORDER OF OPERATIONS

1. Do all operations within grouping symbols such as parentheses or brackets.

2. Evaluate any expressions with exponents.

3. Multiply or divide in order from left to right.

4. Add or subtract in order from left to right.

$$\underset{\text{base}}{\uparrow} 3^{\overset{\text{exponent}}{4}} = \underbrace{3 \cdot 3 \cdot 3 \cdot 3}_{\text{4 factors of 3}} = 81$$

Simplify: $\dfrac{5 + 3^2}{2(7 - 6)}$

Simplify above and below the fraction bar separately.

$$
\begin{aligned}
\frac{5 + 3^2}{2(7 - 6)} &= \frac{5 + 9}{2(1)} \qquad &&\text{Evaluate } 3^2. \\
&&&\text{Subtract: } 7 - 6. \\
&= \frac{14}{2} \qquad &&\text{Add.} \\
&&&\text{Multiply.} \\
&= 7 \qquad &&\text{Divide.}
\end{aligned}
$$

The **area of a square** is $(\text{side})^2$.

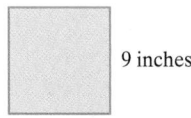

9 inches

Find the area of the square shown.

$$
\begin{aligned}
\text{area of the square} &= (\text{side})^2 \\
&= (9 \text{ inches})^2 \\
&= 81 \text{ square inches}
\end{aligned}
$$

DEFINITIONS AND CONCEPTS	**EXAMPLES**

A letter used to represent a number is called a **variable**.

A combination of operations on variables and numbers is called an **algebraic expression**.

Replacing a variable in an expression by a number, and then finding the value of the expression is called **evaluating the expression** for the variable.

Variables:

$$x, \quad y, \quad z, \quad a, \quad b$$

Algebraic Expressions:

$$3 + x, \quad 7y, \quad x^3 + y - 10$$

Evaluate $2x + y$ for $x = 22$ and $y = 4$.

$$
\begin{aligned}
2x + y &= 2 \cdot 22 + 4 \quad &\text{Replace x with 22 and y with 4.} \\
&= 44 + 4 \quad &\text{Multiply.} \\
&= 48 \quad &\text{Add.}
\end{aligned}
$$

Chapter 1 Review

(1.2) *Determine the place value of the digit 4 in each whole number.*

1. 5480

2. 46,200,120

Write each whole number in words.

3. 5480

4. 46,200,120

Write each whole number in expanded form.

6. 403,225,000

5. 6279

Write each whole number in standard form.

7. Fifty-nine thousand, eight hundred

8. Six billion, three hundred four million

The following table shows the populations of the ten largest cities in the United States. Use this table to answer Exercises 9 through 12.

Rank	City	2000	1990	1980
1	New York, NY	8,008,278	7,322,564	7,071,639
2	Los Angeles, CA	3,694,820	3,485,398	2,968,528
3	Chicago, IL	2,896,016	2,783,726	3,005,072
4	Houston, TX	1,953,631	1,630,553	1,595,138
5	Philadelphia, PA	1,517,550	1,585,577	1,688,210
6	Phoenix, AZ	1,321,045	983,403	789,704
7	San Diego, CA	1,223,400	1,110,549	875,538
8	Dallas, TX	1,188,580	1,006,877	904,599
9	San Antonio, TX	1,144,646	935,933	785,940
10	Detroit, MI	951,270	1,027,974	1,203,368

(*Source*: U.S. Census Bureau)

9. Find the population of Houston, Texas, in 1990.

10. Find the population of Los Angeles, California, in 1980.

11. Find the increase in population for Phoenix, Arizona, from 1980 to 2000.

12. Find the decrease in population for Detroit, Michigan, from 1990 to 2000.

(1.3) *Add.*

13. $7 + 6$

14. $8 + 9$

15. $3 + 0$

16. $0 + 10$

17. $25 + 8 + 5$

18. $27 + 41$

19. $32 + 24$

20. $19 + 21$

21. $47 + 63$

22. $77 + 43$

23. 567 + 383   **24.** 463 + 787   **25.** 591 + 623 + 497   **26.** 5982 + 1647 + 2238

27. The distance from Chicago to New York City is 714 miles. The distance from New York City to New Delhi, India, is 7318 miles. Find the total distance from Chicago to New Delhi if traveling through New York City.

28. Susan Summerline earned salaries of $62,589, $65,340, and $69,770 during the years 1998, 1999, and 2000, respectively. Find her total earnings during those three years.

Find the perimeter of each figure.

△ **29.**
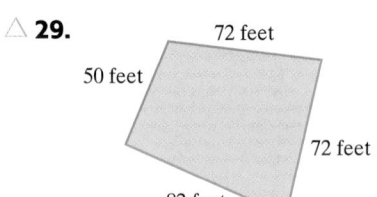
72 feet
50 feet
72 feet
82 feet

△ **30.**
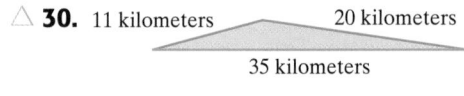
11 kilometers   20 kilometers
35 kilometers

(1.4) *Subtract and then check.*

31.
  42
− 9

32.
  67
− 24

33.
  93
− 79

34.
  60
− 27

35.
  599
− 237

36.
  462
− 397

37.
  583
− 279

38.
  600
− 124

39.
  4000
− 1886

40.
  4268
− 3947

41. Bob Roma is proofreading the Yellow Pages for his county. If he has finished 315 pages of the total 712 pages, how many pages does he have left to proofread?

42. Shelly Winters bought a new car listed at $28,425. She received a discount of $1599 and a factory rebate of $1200. Find how much she paid for the car.

The following bar graph shows the monthly savings account balance for a freshman attending a local community college. Use this graph to answer Exercises 43 through 46.

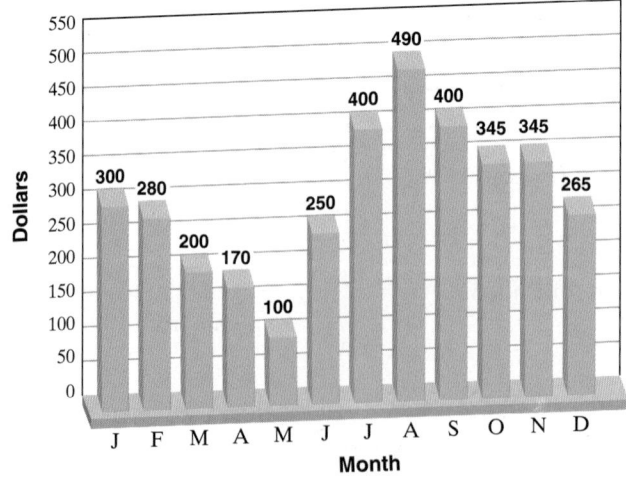

43. During what month was the balance the least?

44. During what month was the balance the greatest?

45. For what months was the balance greater than $350?

46. For what months was the balance less than $200?

(1.5) *Round to the given place.*

47. 93 to the nearest ten

48. 45 to the nearest ten

49. 467 to the nearest ten

50. 493 to the nearest hundred

51. 4832 to the nearest hundred

52. 57,534 to the nearest thousand

53. 49,683,712 to the nearest million

54. 768,542 to the nearest hundred-thousand

Estimate the sum or difference by rounding each number to the nearest hundred.

55. 4892 + 647 + 1876

56. 5925 − 1787

57. In 2001, there were 68,490,000 households in the United States subscribing to cable television services. Round this number to the nearest million.(*Source:* Nielsen Media Research-NTI)

58. In 2000, the total number of employees working for U.S. airlines was 679,967. Round this number to the nearest thousand.(*Source:* The Air Transport Association of America, Inc.)

(1.6) *Multiply.*

59. $6 \cdot 7$

60. $8 \cdot 3$

61. $5(0)$

62. $0(9)$

63. $\begin{array}{r} 47 \\ \times\ 30 \\ \hline \end{array}$

64. $\begin{array}{r} 69 \\ \times\ 42 \\ \hline \end{array}$

65. $20(8)(5)$

66. $25(9 \times 4)$

67. $\begin{array}{r} 48 \\ \times\ 77 \\ \hline \end{array}$

68. $\begin{array}{r} 77 \\ \times\ 22 \\ \hline \end{array}$

69. $49 \cdot 49 \cdot 0$

70. $62 \cdot 88 \cdot 0$

71. $\begin{array}{r} 586 \\ \times\ 29 \\ \hline \end{array}$

72. $\begin{array}{r} 242 \\ \times\ 37 \\ \hline \end{array}$

73. $\begin{array}{r} 642 \\ \times\ 177 \\ \hline \end{array}$

74. $\begin{array}{r} 347 \\ \times\ 129 \\ \hline \end{array}$

75. $\begin{array}{r} 1026 \\ \times\ 401 \\ \hline \end{array}$

76. $\begin{array}{r} 2107 \\ \times\ 302 \\ \hline \end{array}$

Estimate each product by rounding each factor to the given place.

77. 49 · 32; tens

78. 586 · 357; hundreds

79. 5231 · 243; hundreds

80. 7836 · 912; hundreds

81. One ounce of Swiss cheese contains 8 grams of fat. How many grams of fat are in 3 ounces of Swiss cheese? (*Source: Home and Garden Bulletin No. 72, U.S. Department of Agriculture*)

82. There were 5283 students enrolled at Weskan State University in the fall semester. Each paid $927 in tuition. Find the total tuition collected.

Find the area of each rectangle.

△ **83.**

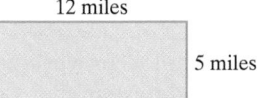

12 miles

5 miles

△ **84.**

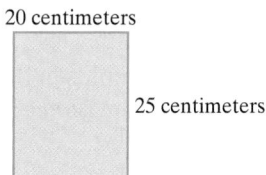

20 centimeters

25 centimeters

(1.7) *Divide and then check.*

85. 18 ÷ 6

86. 36 ÷ 9

87. 42 ÷ 7

88. 25 ÷ 5

89. 27 ÷ 5

90. 18 ÷ 4

91. $\dfrac{16}{0}$

92. $\dfrac{0}{8}$

93. $\dfrac{9}{9}$

94. $\dfrac{10}{1}$

95. 918 ÷ 0

96. 0 ÷ 668

97. $5\overline{)75}$

98. $8\overline{)159}$

99. $26\overline{)626}$

100. $6\overline{)336}$

101. $32\overline{)49}$

102. $19\overline{)680}$

103. $20\overline{)10,000}$

104. $43\overline{)909}$

105. $47\overline{)23,782}$

106. $30\overline{)480}$

107. $16\overline{)3192}$

108. $25\overline{)5000}$

109. One foot is 12 inches. Find how many feet there are in 5496 inches.

110. Find the average of the numbers 76, 49, 32, and 47.

(1.8) *Write using exponential notation.*

111. $7 \cdot 7 \cdot 7 \cdot 7$

112. $6 \cdot 6 \cdot 3 \cdot 3 \cdot 3$

113. $4 \cdot 2 \cdot 2 \cdot 2 \cdot 3 \cdot 3$

114. $5 \cdot 5 \cdot 7 \cdot 7 \cdot 7 \cdot 2 \cdot 2$

Simplify.

115. 7^2

116. 2^6

117. $5^3 \cdot 3^2$

118. $4^1 \cdot 10^2 \cdot 7^2$

119. $18 \div 3 + 7$

120. $12 - 8 \div 4$

121. $\dfrac{(6^2 - 3)}{3^2 + 2}$

122. $\dfrac{16 - 8}{2^3}$

123. $2 + 3[1 + (20 - 17) \cdot 3]$

124. $21 - [2^4 - (7 - 5) - 10] + 8 \cdot 2$

Find the area of each square.

△ **125.**

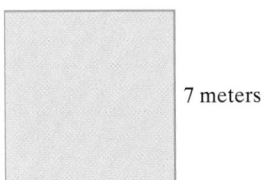

7 meters

△ **126.**

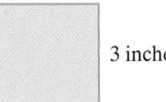

3 inches

(1.9) *Evaluate each expression for $x = 5$, $y = 0$, and $z = 2$.*

127. $\dfrac{2x}{z}$

128. $4x - 3$

129. $\dfrac{x + 7}{y}$

130. $\dfrac{y}{5x}$

131. $x^3 - 2z$

132. $\dfrac{7 + x}{3z}$

133. $(y + z)^2$

134. $\dfrac{100}{x} + \dfrac{y}{3}$

Translate each phrase into a variable expression.

135. Five subtracted from a number

136. Seven more than a number

137. Ten divided by a number

138. The product of 5 and a number

139. Complete the table by evaluating $8x^2$ for each value of x.

x	0	1	2	3
$8x^2$				

Name _____ Section _____ Date _____

Chapter 1 Test Remember to check your answers and use the Chapter Test Prep Video to view solutions.

Evaluate. Where it's helpful, use estimation to see if your results are reasonable.

1. $59 + 82$ **2.** $600 - 487$

3. $\begin{array}{r} 496 \\ \times\ 30 \\ \hline \end{array}$ **4.** $52{,}896 \div 69$

5. $2^3 \cdot 5^2$ **6.** $6^1 \cdot 2^3$ **7.** $98 \div 1$ **8.** $0 \div 49$

9. $62 \div 0$ **10.** $(2^4 - 5) \cdot 3$ **11.** $16 + 9 \div 3 \cdot 4 - 7$

12. $2[(6 - 4)^2 + (22 - 19)^2] + 10$ **13.** Round 52,369 to the nearest thousand.

Estimate each sum or difference by rounding each number to the nearest hundred.

14. $6289 + 5403 + 1957$ **15.** $4267 - 2738$

16. Twenty-nine cans of Sherwin-Williams paint cost $493. How much was each can?

17. Admission to a movie costs $7 per ticket. The Math Club has 17 members who are going to a movie together. What is the total cost of their tickets?

18. Jo McElory is looking at two new refrigerators for her apartment. One costs $599 and the other costs $725. How much more expensive is the higher-priced one?

19. Aspirin was 100 years old in 1997 and was the first U.S. drug made in tablet form. Today, people take 11 billion tablets a year for heart disease prevention and 4 billion tablets a year for headaches. How many more tablets are taken a year for heart disease prevention? (*Source:* Bayer Market Research)

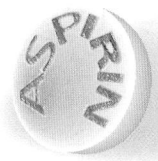

Answers
1. _____
2. _____
3. _____
4. _____
5. _____
6. _____
7. _____
8. _____
9. _____
10. _____
11. _____
12. _____
13. _____
14. _____
15. _____
16. _____
17. _____
18. _____
19. _____

20. _____

21. _____

22. _____

23. a. _____

b. _____

24. _____

25. _____

26. _____

27. _____

28. _____

20. The U.S. Federal Bureau of Investigation (FBI) maintains a collection of over 250,000,000 sets of fingerprint records. Each work day, the FBI Identification Division receives over 34,000 standard fingerprint cards to add to its collection. During a standard five-day work week, how many fingerprint cards does the FBI receive? (*Source:* Federal Bureau of Investigation)

21. Evaluate $5(x^3 - 2)$ for $x = 2$.

22. Evaluate $\dfrac{3x - 5}{2y}$ for $x = 7$ and $y = 8$.

23. Translate the following phrases into mathematical expressions. Use x to represent "a number."

   a. The product of a number and 17

   b. Twice a number subtracted from 20

Find the perimeter and the area of each figure.

△ **24.**

Square   5 centimeters

△ **25.**

20 yards

Rectangle   10 yards

The following table shows the top grossing movies for 2001 and 2002. Use this table to answer Exercises 26 through 28.

2001	
Movie	**Gross**
Harry Potter and the Sorcerer's Stone	$317,557,891
The Lord of the Rings: The Fellowship of the Ring	$313,364,114
Shrek	$267,652,016
Monsters, Inc.	$255,870,172
Rush Hour 2	$226,138,454

2002	
Movie	**Gross**
Spiderman	$403,706,375
The Lord of the Rings: The Two Towers	$339,554,276
Star Wars: Episode II— Attack of the Clones	$310,675,583
Harry Potter and the Chamber of Secrets	$261,970,615
My Big Fat Greek Wedding	$182,805,123

(*Source:* The Internet Movie Database Ltd.)

26. Find the amount earned by the movie *Spiderman* in 2002.

27. How much more did *Lord of the Rings: The Two Towers* in 2002 earn than *Lord of the Rings: The Fellowship of the Ring* in 2001?

28. How much more did the top-grossing movie of 2002 earn than the top-grossing movie in 2001?

Integers

Whole numbers are not sufficient for representing many situations in real life. For example, to express 5° below 0° or $100 in debt, numbers less than 0 are needed. This chapter is devoted to integers, which include numbers less than 0, and to operations on integers.

The weather on our planet is of great interest to its occupants. From snowstorms to tornadoes to hurricanes, torrential rain to severe drought, some sort of extreme weather affects nearly every portion of the Earth. But did you ever wonder about the weather on other planets? Scientists do! Information from satellites, probes, and telescopes tell of drastic atmospheric conditions of our Solar System neighbors. Pluto's distance from the Sun contributes to its frigid temperatures, some recorded as low as 338°F *below zero*. Compare that to the extreme heat of Venus, where temperatures on the surface are in excess of 800°F. In Exercises 63–66 on page 134, we will see how integers can be used to compare average daily temperatures of several planets in our Solar System.

Name _____ Section _____ Date_____

1. _____

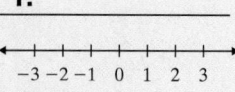

2. _____

3. _____

4. _____

5. _____

6. _____

7. _____

8. _____

9. _____

10. _____

11. _____

12. _____

13. _____

14. _____

15. _____

16. _____

17. _____

18. _____

19. _____

20. _____

21. _____

22. _____

110

Chapter **2** Pretest

1. The Washington Redskins lost 22 yards on a play. Represent this quantity using an integer.

2. Graph $-3, 2$, and 0 on the number line.

3. Insert $<$ or $>$ between the given pair of numbers to make a true statement.

$-31 \qquad -36$

4. Simplify: $|-8|$

5. Find the opposite of -12.

6. Add: $-19 + 8$

7. Evaluate $x + y$ if $x = -6$ and $y = -11$.

8. On February 2 the temperature in Manassas, Virginia, at 7 a.m. was $-9°$ Fahrenheit. By 8 a.m. the temperature had risen by $5°$, and then between 8 a.m. and 9 a.m. it rose another $7°$. What was the temperature at 9 a.m.?

9. Subtract: $-9 - (-14)$

10. Simplify: $4 - 6 - (-2) + 8$

11. Evaluate $m - n$ if $m = -5$ and $n = 10$.

12. Andrea Roberts has $88 in her checking account. She makes a deposit of $35 and writes a check for $72, and then writes another check for $55. Find the balance in her account. (Write the amount as an integer.)

13. Multiply: $(-8)(13)$

14. Divide: $\dfrac{-36}{-4}$

15. Evaluate $\dfrac{x}{y}$ if $x = -96$ and $y = -2$.

16. Evaluate xy if $x = 4$ and $y = -5$.

17. A card player had a score of -20 for each of three games. Find her total score.

Simplify.

18. -5^2

19. $\dfrac{9 - 17}{-6 + 2}$

20. $(-2) \cdot |-7| - (-6)$

21. Evaluate $x - y^2$ if $x = 2$ and $y = -6$.

22. Find the average of $-17, 10, -9, 5, -19$.

2.1   Introduction to Integers

Ⓐ   Representing Real-Life Situations

Thus far in this text, all numbers have been 0 or greater than 0. Numbers greater than 0 are called **positive numbers**. However, sometimes situations exist that cannot be represented by a number greater than 0. For example,

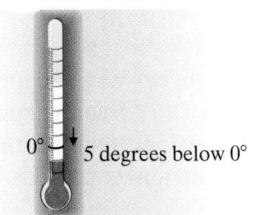

0°   5 degrees below 0°

Sea level

20 feet below sea level

OBJECTIVES

Ⓐ   Represent real-life situations with integers.

Ⓑ   Graph integers on a number line.

Ⓒ   Compare two integers.

Ⓓ   Find absolute value.

Ⓔ   Find the opposite of a number.

Ⓕ   Read bar graphs containing negative integers.

SSM
TUTOR CENTER   SG   CD & VIDEO   MATH PRO   WEB

To represent these situations, we need numbers less than 0.

Extending the number line to the left of 0 allows us to picture **negative numbers**, numbers that are less than 0.

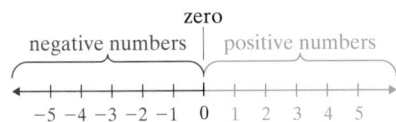

zero

negative numbers | positive numbers

−5 −4 −3 −2 −1 0 1 2 3 4 5

When a single + sign or no sign is in front of a number, the number is a positive number. When a single − sign is in front of a number, the number is a negative number. Together, we call positive numbers, negative numbers, and 0 the **signed numbers**.

−5 indicates "negative five."

5 and +5 both indicate "positive five."

The number 0 is neither positive nor negative. ──

Helpful Hint

Notice that 0 is neither positive nor negative.

Some signed numbers are integers. The **integers** consist of the numbers labeled on the number line above. The integers are

..., −3, −2, −1, 0, 1, 2, 3, ...

Now we have numbers to represent the situations previously mentioned.

5 degrees below 0°   −5°

20 feet below sea level   −20 feet

EXAMPLE 1   Representing Depth with an Integer

Jack Mayfield, a miner for the Molly Kathleen Gold Mine, is presently 150 feet below the surface of the earth. Represent this position using an integer.

Solution: If 0 represents the surface of the earth, then 150 feet below the surface can be represented by −150.

Practice Problem 1

a.   A deep-sea diver is 800 feet below the surface of the ocean. Represent this position using an integer.

b.   A company reports a $2 million loss for the year. Represent this amount using an integer.

Answers

1. a. −800   **b.** −2 million

Practice Problem 2

Graph -5, -4, 3, and -3 on the number line.

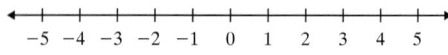

Concept Check

Is there a largest positive number? Is there a smallest negative number? Explain.

Practice Problem 3

Insert $<$ or $>$ between each pair of numbers to make a true statement.

a. 0 -3
b. -5 5
c. -8 -12

Answers

2.

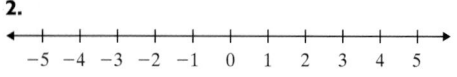

3. **a.** $>$ **b.** $<$ **c.** $>$

Concept Check: No.

B Graphing Integers

EXAMPLE 2   Graph 0, -3, 2, and -2 on the number line.

Solution:

```
<---+---+---+---+---+---+---+---+---+--->
   -4  -3  -2  -1   0   1   2   3   4
```

C Comparing Integers

We compare integers just as we compare whole numbers. For any two numbers graphed on a number line, the number to the **right** is the **greater number** and the number to the **left** is the **smaller number**. Recall that the inequality symbol $>$ means "is greater than" and the inequality symbol $<$ means "is less than."

Both -5 and -7 are graphed on the number line shown.

```
<---+---+---+---+---+---+---+---+---+--->
   -7  -6  -5  -4  -3  -2  -1   0   1
```

On the graph, -7 is **to the left of** -5, so -7 **is less than** -5, written as

$$-7 < -5$$

We can also write

$$-5 > -7$$

since -5 is **to the right** of -7, so -5 **is greater than** -7.

Try the Concept Check in the margin.

EXAMPLE 3   Insert $<$ or $>$ between each pair of numbers to make a true statement.

a. -7 7
b. 0 -4
c. -9 -11

Solution:

a. -7 is to the left of 7 on a number line, so $-7 < 7$.
b. 0 is to the right of -4 on a number line, so $0 > -4$.
c. -9 is to the right of -11 on a number line, so $-9 > -11$.

> **Helpful Hint**
>
> If you think of $<$ and $>$ as arrowheads, notice that in a true statement the arrow always points to the smaller number.
>
> $$5 > -4 \qquad -3 < -1$$
>
>   ↑   ↑
> smaller smaller
> number number

D Finding the Absolute Value of a Number

The **absolute value** of a number is the number's distance from 0 on the number line. The symbol for absolute value is | |. For example, |3| is read as "the absolute value of 3."

|3| = 3 because 3 is 3 units from 0.

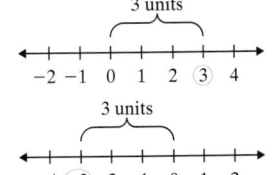

|−3| = 3 because −3 is 3 units from 0.

EXAMPLE 4   Simplify.

a. |−2| **b.** |5| **c.** |0|

Solution:

a. |−2| = 2 because −2 is 2 units from 0.

b. |5| = 5 because 5 is 5 units from 0.

c. |0| = 0 because 0 is 0 units from 0.

Helpful Hint

Since the absolute value of a number is that number's *distance* from 0, the absolute value of a number is always 0 or positive. It is never negative.

|0| = 0 |−6| = 6
  ↑   ↑
 zero   a positive number

E Finding Opposites

Two numbers that are the same distance from 0 on the number line but are on opposite sides of 0 are called **opposites**.

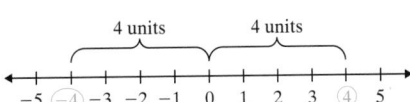

When two numbers are opposites, we say that each is the opposite of the other. Thus **4 is the opposite of** −4 and −4 **is the opposite of 4**.

The phrase "the opposite of" is written in symbols as "−." For example,

The opposite of   5   is   −5
        ↓   ↓   ↓   ↓
        −   (5) =   −5

The opposite of   −3   is   3
        ↓   ↓   ↓   ↓
        −   (−3) =   3

Notice we just stated that

−(−3) = 3

If a is a number, then $-(-a) = a$.

Practice Problem 4

Simplify.

a. |−4| b. |2|

c. |−8|

Practice Problem 5

Find the opposite of each number.

a. 7

b. −17

Concept Check

True or false? The number 0 is the only number that is its own opposite.

Practice Problem 6

Simplify.

a. $-|-2|$

b. $-|5|$

c. $-(-11)$

Practice Problem 7

Evaluate $-|x|$ if $x = -9$.

EXAMPLE 5 Find the opposite of each number.

a. 11 **b.** −2 **c.** 0

Solution:

a. The opposite of 11 is −11.

b. The opposite of −2 is −(−2) or 2.

c. The opposite of 0 is 0.

> **Helpful Hint**
> Remember that 0 is neither positive nor negative.

Try the Concept Check in the margin.

EXAMPLE 6 Simplify.

a. $-(-4)$ **b.** $-|-5|$ **c.** $-|6|$

Solution:

a. $-(-4) = 4$ The opposite of negative 4 is 4.

b. $-|-5| = -5$ The opposite of the absolute value of −5 is the opposite of 5, or −5.

c. $-|6| = -6$ The opposite of the absolute value of 6 is the opposite of 6, or −6.

EXAMPLE 7 Evaluate $-|x|$ if $x = -2$.

Solution: Carefully replace x with −2; then simplify.

$$-|-x| = -|-(-2)|$$ Replace x with −2.

Then $-|-(-2)| = -|2| = -2$.

Answers

5. a. −7 **b.** 17

6. a. −2 **b.** −5 **c.** 11

7. −9

Concept Check: True.

Reading Bar Graphs Containing Negative Numbers

The bar graph below shows the average temperature (in Fahrenheit) of known planets. Notice that a negative temperature is illustrated by a bar below the horizontal line representing 0°F, and a positive temperature is illustrated by a bar above the horizontal line representing 0°F.

Average Surface Temperature of Planets*

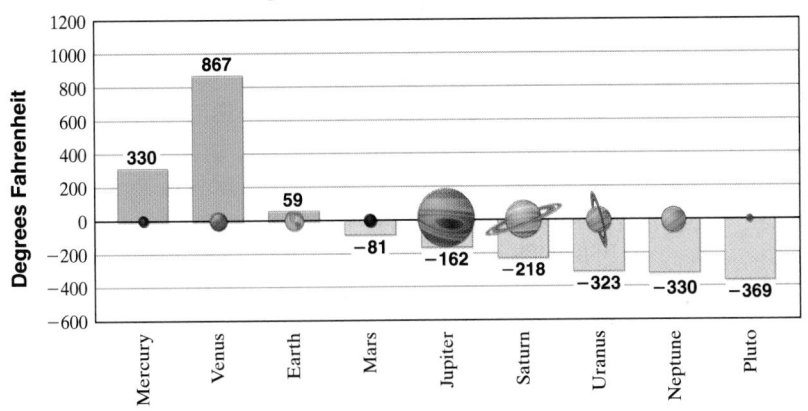

*(For some planets, the temperature given is the temperature where the atmosphere pressure equals 1 Earth atmosphere; Source: *The World Almanac*, 2003)

EXAMPLE 8.   Which planet has the lowest temperature?

Solution   The planet with the lowest temperature is the one that corresponds to the bar that extends the furthest in the negative direction (downward.) Pluto has the lowest temperature of −369°F.   ●

Later, we will be reading graphs without labels on bars. Notice that the height of the bar can be checked or estimated by looking horizontally to the left and checking the vertical axis scale.

Practice Problem 8

Which planet has the highest temperature?

Answer

8. Venus; 867°F

 **STUDY SKILLS REMINDER**

Are you reviewing new terms?

Remember that many of the terms used in this text may be new to you. It will be helpful to make a list of new mathematical terms and symbols as you encounter them and to review them frequently.

Focus on The Real World

USING NEGATIVE NUMBERS

Did you know that many countries around the world label the underground floors in a tall building with negative numbers? As we have seen in this section and will see throughout this chapter, negative numbers are very useful in many different situations where numbers less than 0 are needed. Another use of negative numbers is in pregnancy and childbirth. Just prior to labor, a baby will drop, or engage, in the mother's pelvis. Doctors and midwives use "stations," ranging from −5 to +5, to describe how much a baby has engaged compared to the 0 station at the entrance to the pelvic canal. A negative station means that a baby is still above the pelvic bones used for reference. A positive station describes how far the baby has moved into the pelvic canal.

CRITICAL THINKING

Make a list of as many real-world situations as you can think of that use negative numbers in some way. Consider flipping through a newspaper or news magazine for inspiration.

EXERCISE SET 2.1

A *Represent each quantity by an integer. See Example 1.*

1. A worker in a silver mine in Nevada works 1445 feet underground.

2. A scuba diver is swimming 35 feet below the surface of the water in the Gulf of Mexico.

3. The peak of Mount Elbert in Colorado is 14,433 feet above sea level. (*Source:* U.S. Geological Survey)

4. The lowest elevation in the United States is found at Death Valley, California, at an elevation of 282 feet below sea level. (*Source:* U.S. Geological Survey)

5. The record high temperature in Nevada is 118 degrees Fahrenheit above zero. (*Source:* National Climatic Data Center)

6. The Minnesota Viking football team lost 15 yards on a play.

7. The average depth of the Atlantic Ocean is 11,730 feet below the surface of the ocean. (*Source: 2003 World Almanac*)

8. The Dow Jones stock market average fell 317 points in one day.

9. In fiscal year 2001, US Airways posted a net loss of $1683 million. (*Source:* US Airways)

10. For the fourth quarter in fiscal year 2002, Apple Computer, Inc., reported a net loss of $45 million. (*Source:* Apple Computer, Inc.)

11. Two divers are exploring the bottom of a trench in the Pacific Ocean. Joe is at 135 feet below the surface of the ocean and Sara is at 157 feet below the surface. Represent each quantity by an integer and determine who is deeper in the water.

12. The temperature on one January day in Chicago was 10° below 0° Celsius. Represent this quantity by an integer and tell whether this temperature is cooler or warmer than 5° below 0° Celsius.

13. In 2002, the number of music CD singles shipped to retailers reflected an 81 percent loss from the previous year. Write an integer to represent the percent loss in CD singles shipped. (*Source:* Recording Industry Association of America)

14. In 2002, the number of music cassettes shipped to retailers reflected a 49 percent loss from the previous year. Write an integer to represent the percent loss of cassettes shipped. (*Source:* Recording Industry Association of America)

B *Graph each integer in the list on the same number line. See Example 2.*

15. 1, 2, 4, 6

<---+---+---+---+---+---+---+---+---+---+---+---+---+---+--->
−7 −6 −5 −4 −3 −2 −1 0 1 2 3 4 5 6 7

16. 3, 5, 2, 0

<---+---+---+---+---+---+---+---+---+---+---+---+---+---+--->
−7 −6 −5 −4 −3 −2 −1 0 1 2 3 4 5 6 7

17. 1, −1, −2, −4, −7

<---+---+---+---+---+---+---+---+---+---+---+---+---+---+--->
−7 −6 −5 −4 −3 −2 −1 0 1 2 3 4 5 6 7

18. 2, −2, −4, 6

<---+---+---+---+---+---+---+---+---+---+---+---+---+---+--->
−7 −6 −5 −4 −3 −2 −1 0 1 2 3 4 5 6 7

19. 0, 2, 5, 7

<---+---+---+---+---+---+---+---+---+---+---+---+---+---+--->
−7 −6 −5 −4 −3 −2 −1 0 1 2 3 4 5 6 7

20. 0, 3, 6, 10

<---+---+---+---+---+---+---+---+---+---+---+---+---+---+---+---+---+--->
−7 −6 −5 −4 −3 −2 −1 0 1 2 3 4 5 6 7 8 9 10

21. 0, −2, −7, −5

<---+---+---+---+---+---+---+---+---+---+---+---+---+---+--->
−7 −6 −5 −4 −3 −2 −1 0 1 2 3 4 5 6 7

22. 0, −7, 3, −6

<---+---+---+---+---+---+---+---+---+---+---+---+---+---+--->
−7 −6 −5 −4 −3 −2 −1 0 1 2 3 4 5 6 7

C *Insert* $<$ *or* $>$ *between each pair of integers to make a true statement. See Example 3.*

23. −4   0

24. −8   0

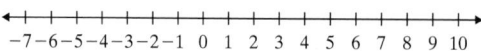 **25.** −7   −5

26. −12   −10

27. −30   −35

28. −27   −29

29. −26   26

30. 13   −13

D **E** *Simplify. See Example 4.*

31. $|5|$

32. $|7|$

33. $|-8|$

34. $|-19|$

35. $|0|$

36. $|100|$

37. $|-5|$

38. $|-10|$

Find the opposite of each integer. See Example 5.

39. 5

40. 8

41. −4

42. −6

43. 23

44. 123

45. −10

46. −23

Simplify. See Example 6.

47. $|-7|$

48. $|-11|$

49. $-|20|$

50. $-|43|$

51. $-|-3|$

52. $-|-18|$

53. $-(-8)$

54. $-(-7)$

55. $|-14|$

56. $-(-14)$

57. $-(-29)$

58. $-|-29|$

Evaluate. See Example 7.

59. $|-x|$ if $x = -8$ **60.** $-|x|$ if $x = -8$ **61.** $-|-x|$ if $x = 3$ **62.** $-|-x|$ if $x = 7$

63. $|x|$ if $x = -23$ **64.** $|x|$ if $x = 23$ **65.** $-|x|$ if $x = 4$ **66.** $|-x|$ if $x = 1$

Insert $<, >,$ or $=$ between each pair of numbers to make a true statement.

67. -3 -5 **68.** -17 -6 **69.** $|-9|$ $|-14|$

70. $|-8|$ $|-4|$ **71.** $|-33|$ $-(-33)$ **72.** $-|17|$ $-(-17)$

73. $-|-10|$ $-(-10)$ **74.** $|-24|$ $-(-24)$ **75.** 0 -9

76. -45 0 **77.** $|0|$ $|-9|$ **78.** $|-45|$ $|0|$

79. $-|-2|$ $-|-10|$ **80.** $-|-8|$ $-|-4|$

81. $-(-12)$ $-(-18)$ **82.** -22 $-(-38)$

Write the given integers in order from least to greatest.

83. $2^2, -|3|, -(-5), -|-8|$ **84.** $|10|, 2^3, -|-5|, -(-4)$ **85.** $|-1|, -|-6|, -(-6), -|1|$

86. $1^4, -(-3), -|7|, |-20|$ **87.** $-(-2), 5^2, -10, -|-9|, |-12|$ **88.** $3^3, -|-11|, -(-10), -4, -|2|$

F *The bar graph shows elevation of selected lakes. Use this graph For Exercises 89–92. (Source: U.S. Geological Survey). See Example 8.*

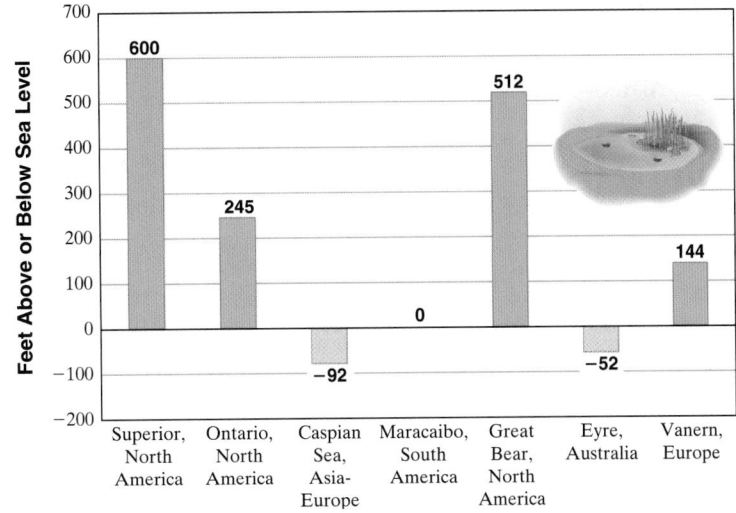

89. Which lake has an elevation at sea level?

90. Which lake shown has the lowest elevation?

91. Which lake shown has the second lowest elevation?

92. Which lake shown has the highest elevation?

Use the bar graph from Example 8 to answer exercises 93 through 96.

93. Which planet has an average temperature closest to 0°F?

94. Which planet has a negative average temperature closest to 0°F?

95. Which planet has an average temperature closest to −200°F?

96. Which planet has an average temperature closest to −300°F?

Review and Preview

Add. See Section 1.3.

97. $0 + 13$

98. $9 + 0$

99. $15 + 4^2 + 4$

100. $20 + 3^2 + 6$

101. $47 + 236 + 77$

102. $362 + 37 + 90$

 Combining Concepts

Choose all numbers for x from each given list that make each statement true.

103. $|x| > 8$
 a. 0 **b.** −5 **c.** 8 **d.** −12

104. $|x| > 4$
 a. 0 **b.** 4 **c.** −1 **d.** −100

105. Evaluate: $-(-|-5|)$

106. Evaluate: $-(-|-(-7)|)$

Answer true or false for Exercises 107 through 111.

107. If $a > b$, then a must be a positive number.

108. The absolute value of a number is *always* a positive number.

109. A positive number is always greater than a negative number.

110. Zero is always less than a positive number.

111. The number $-a$ is always a negative number. (*Hint:* Read "−" as "the opposite of.")

112. Given the number line $\xleftarrow{\bullet\ \bullet\ \ +\ +\ +}\rightarrow$, is it true that $b < a$? $\quad$ a b −1 0 1

113. Write in your own words how to find the absolute value of an integer.

114. Explain how to determine which of two integers is larger.

2.2   Adding Integers

Ⓐ   Adding Integers

Adding integers can be visualized using a number line. A positive number can be represented on the number line by an arrow of appropriate length pointing to the right, and a negative number by an arrow of appropriate length pointing to the left.

Both arrows represent 2 or +2. They both point to the right and they are both 2 units long.

Both arrows represent −3. They both point to the left and they are both 3 units long.

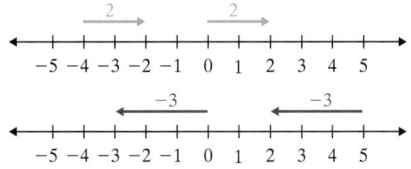

EXAMPLE 1   Add using a number line: $5 + (-2)$

Solution: To add integers on a number line, such as $5 + (-2)$, we start at 0 on the number line and draw an arrow representing 5. From the tip of this arrow, we draw another arrow representing −2. The tip of the second arrow ends at their sum, 3.

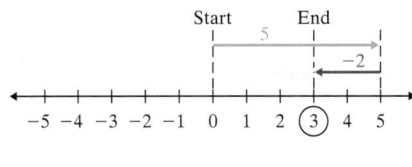

$$5 + (-2) = 3$$

EXAMPLE 2   Add using a number line: $-1 + (-4)$

Start at 0 and draw an arrow representing −1. From the tip of this arrow, we draw another arrow representing −4. The tip of the second arrow ends at their sum, −5.

Solution:

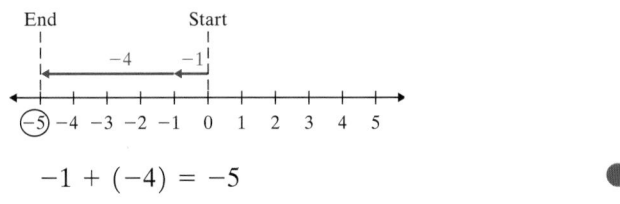

$$-1 + (-4) = -5$$

EXAMPLE 3   Add using a number line: $-7 + 3$

Solution:

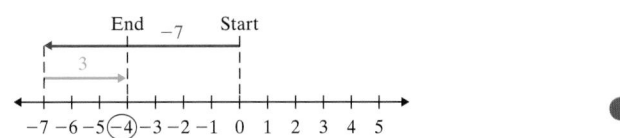

Using a number line each time we add two numbers can be time consuming. Instead, we can notice patterns in the previous examples and write rules for adding signed numbers.

Rules for adding signed numbers depend on whether we are adding numbers with the same sign or different signs. When adding two numbers with the same sign, as in Example 2, notice that the sign of the sum is the same as the sign of the addends.

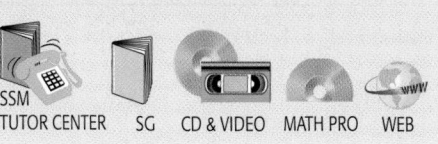

OBJECTIVES

Ⓐ   Add integers.

Ⓑ   Evaluate an algebraic expression by adding.

Ⓒ   Solve problems by adding integers.

Practice Problem 1

Add using a number line: $5 + (-1)$

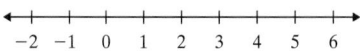

Practice Problem 2

Add using a number line: $-6 + (-2)$

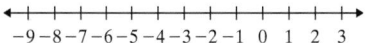

Practice Problem 3

Add using a number line: $-8 + 3$

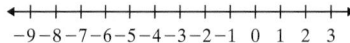

Answers

1.

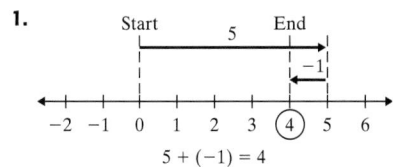

$$5 + (-1) = 4$$

2.

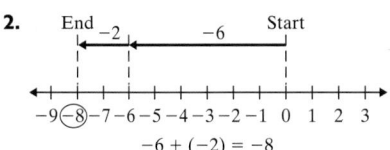

$$-6 + (-2) = -8$$

3.

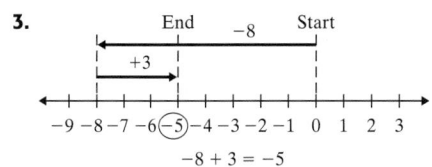

$$-8 + 3 = -5$$

Adding Two Numbers with the Same Sign

Step 1. Add their absolute values.

Step 2. Use their common sign as the sign of the sum.

Practice Problem 4

Add: $(-3) + (-9)$

EXAMPLE 4 Add: $-2 + (-21)$

Solution:

Step 1. $|-2| = 2, |-21| = 21$, and $2 + 21 = 23$.

Step 2. Their common sign is negative, so the sum is negative:

$$-2 + (-21) = -23$$

Practice Problems 5–6

Add.

5. $-12 + (-3)$
6. $9 + 5$

EXAMPLES Add.

5. $-5 + (-1) = -6$
6. $2 + 6 = 8$

When adding two numbers with different signs, as in Examples 1 and 3, the sign of the result may be positive, negative, or the result may be 0.

Adding Two Numbers with Different Signs

Step 1. Find the larger absolute value minus the smaller absolute value.

Step 2. Use the sign of the number with the larger absolute value as the sign of the sum.

Practice Problem 7

Add: $-3 + 9$

EXAMPLE 7 Add: $-2 + 5$

Solution:

Step 1. $|-2| = 2, |5| = 5$, and $5 - 2 = 3$.

Step 2. 5 has the larger absolute value and its sign is an understood $+$:

$$-2 + 5 = +3 \text{ or } 3$$

Practice Problem 8

Add: $2 + (-8)$

EXAMPLE 8 Add: $3 + (-7)$

Solution:

Step 1. $|3| = 3, |-7| = 7$, and $7 - 3 = 4$.

Step 2. -7 has the larger absolute value and its sign is $-$:

$$3 + (-7) = -4$$

Practice Problems 9–11

Add.

9. $-46 + 20$
10. $8 + (-6)$
11. $-2 + 0$

EXAMPLES Add.

9. $-18 + 10 = -8$
10. $12 + (-8) = 4$
11. $0 + (-5) = -5$ The sum of 0 and any number is the number.

Recall that numbers such as 7 and -7 are called opposites. In general, the sum of a number and its opposite is always 0.

$$7 + (-7) = 0 \qquad -26 + 26 = 0 \qquad 1008 + (-1008) = 0$$

opposites   opposites   opposites

Answers

4. -12 **5.** -15 **6.** 14 **7.** 6
8. -6 **9.** -26 **10.** 2 **11.** -2

If a is a number, then

   $-a$ is its opposite. Also,

$$\left.\begin{array}{l} a + (-a) = 0 \\ -a + a = 0 \end{array}\right\}$$ The sum of a number and its opposite is 0.

EXAMPLES   Add.

12. $-21 + 21 = 0$
13. $36 + (-36) = 0$ ●

Try the Concept Check in the margin.

In the following examples, we add three or more integers. Remember that by the associative and commutative properties for addition, we may add numbers in any order that we wish. In Examples 14 and 15, let's add the numbers from left to right.

EXAMPLE 14   Add: $(-3) + 4 + (-11)$

Solution: $(-3) + 4 + (-11) = 1 + (-11)$
$$= -10$$ ●

EXAMPLE 15   Add: $1 + (-10) + (-8) + 9$

Solution: $1 + (-10) + (-8) + 9 = -9 + (-8) + 9$
$$= -17 + 9$$
$$= -8$$

The sum will be the same if we add the numbers in any order. To see this, let's add the positive numbers together and then the negative numbers together first.

$$1 + 9 = 10$$ Add the positive numbers.

$$(-10) + (-8) = -18$$ Add the negative numbers.

$$10 + (-18) = -8$$ Add these results.

The sum is -8.   ●

Helpful Hint

Don't forget that addition is commutative and associative. In other words, numbers may be added in any order.

B   Evaluating Algebraic Expressions

We can continue our work with algebraic expressions by evaluating expressions given integer replacement values.

EXAMPLE 16   Evaluate $x + y$ for $x = 3$ and $y = -5$.

Solution: Replace x with 3 and y with -5 in $x + y$.

$$x + y = 3 + (-5)$$
$$= -2$$ ●

Practice Problems 12–13

Add.
12. $15 + (-15)$
13. $-80 + 80$

Concept Check

What is wrong with the following calculation?

$6 + (-22) = 16$

Practice Problem 14

Add: $8 + (-3) + (-13)$

Practice Problem 15

Add: $5 + (-3) + 12 + (-14)$

Practice Problem 16

Evaluate $x + y$ for $x = -4$ and $y = 1$.

Answers

12. 0   **13.** 0   **14.** -8 **15.** 0   **16.** -3

Concept Check: $6 + (-22) = -16$

Practice Problem 17

Evaluate $x + y$ for $x = -11$ and $y = -6$.

Practice Problem 18

If the temperature was $-8°$ Fahrenheit at 6 a.m., and it rose 4 degrees by 7 a.m. and then rose another 7 degrees in the hour from 7 a.m. to 8 a.m., what was the temperature at 8 a.m.?

EXAMPLE 17   Evaluate $x + y$ for $x = -2$ and $y = -10$.

Solution: $x + y = (-2) + (-10)$ Replace x with -2 and y with -10.

$$= -12$$

C Solving Problems by Adding Integers

Next, we practice solving problems that require adding integers.

EXAMPLE 18   Calculating Temperature

On January 6, the temperature in Caribou, Maine, at 8 a.m. was $-12°$ Fahrenheit. By 9 a.m., the temperature had risen 4 degrees, and by 10 a.m. it had risen 6 degrees from the 9 a.m. temperature. What was the temperature at 10 a.m.?

Solution:

In words:

temperature at 10 a.m.	=	8 a.m. temperature	+	rise of 4°	+	rise of 6°

Translate:

$$\text{temperature at 10 a.m.} = -12 + (+4) + (+6)$$

$$= -8 + (+6)$$

$$= -2$$

The temperature was $-2°$F at 10 a.m.

CALCULATOR EXPLORATIONS

Entering Negative Numbers

To enter a negative number on a calculator, find the key marked $\boxed{+/-}$. (Some calculators have a key marked $\boxed{\text{CHS}}$ and some calculators have a special key $\boxed{(-)}$ for entering a negative sign.) To enter the number -2, for example, press the keys $\boxed{2}\ \boxed{+/-}$. The display will read $\boxed{\quad -2}$.

To find $-32 + (-131)$, press the keys

$$\boxed{32}\ \boxed{+/-}\ \boxed{+}\ \boxed{131}\ \boxed{+/-}\ \boxed{=}\ \text{or}$$

$$\boxed{(-)}\ \boxed{32}\ \boxed{+}\ \boxed{(-)}\ \boxed{131}\ \boxed{\text{ENTER}}$$

The display will read $\boxed{\quad -163}$. Thus $-32 + (-131) = -163$.

Use a calculator to perform each indicated operation.

1. $-256 + 97$ **2.** $811 + (-1058)$

3. $6(15) + (-46)$ **4.** $-129 + 10(48)$

5. $-108,650 + (-786,205)$

6. $-196,662 + (-129,856)$

Answers

17. -17 **18.** 3°F

Name _____ Section _____ Date _____

Mental Math

Add.

1. $5 + 0$ **2.** $(-2) + 0$ **3.** $0 + (-35)$ **4.** $0 + 3$

5. $-12 + 12$ **6.** $48 + (-48)$ **7.** $28 + (-28)$ **8.** $-9 + 9$

EXERCISE SET 2.2

A   *Add using a number line. See Examples 1 and 2.*

1. $-1 + (-6)$

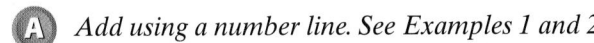

2. $9 + (-4)$

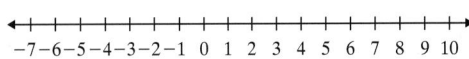

3. $-4 + 7$

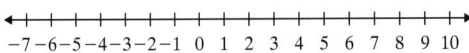

4. $10 + (-3)$

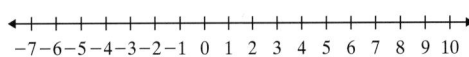

5. $-13 + 7$

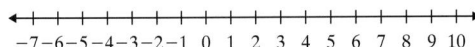

6. $-6 + (-5)$

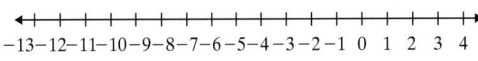

Add. See Examples 3 through 13.

7. $23 + 12$ **8.** $15 + 42$ **9.** $-6 + (-2)$ **10.** $-5 + (-4)$

11. $-43 + 43$ **12.** $-62 + 62$ 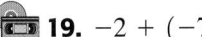 **13.** $6 + (-2)$ **14.** $8 + (-3)$

15. $-6 + 8$ **16.** $-8 + 12$ **17.** $3 + (-5)$ **18.** $5 + (-9)$

19. $-2 + (-7)$ **20.** $-6 + (-1)$ **21.** $-12 + (-12)$ **22.** $-23 + (-23)$

23. $-25 + (-32)$ **24.** $-45 + (-90)$ **25.** $-123 + (-100)$ **26.** $-500 + (-230)$

27. $-7 + 7$ **28.** $-10 + 10$ **29.** $12 + (-5)$ **30.** $24 + (-10)$

31. $-6 + 3$

32. $-8 + 2$

33. $-12 + 3$

34. $-15 + 5$

35. $56 + (-26)$

36. $89 + (-37)$

37. $-37 + 57$

38. $-25 + 65$

39. $-42 + 93$

40. $-64 + 164$

41. $34 + (-67)$

42. $42 + (-83)$

43. $124 + (-144)$

44. $325 + (-375)$

45. $-82 + (-43)$

46. $-56 + (-33)$

Add. See Examples 14 and 15.

47. $-4 + 2 + (-5)$

48. $-1 + 5 + (-8)$

49. $-52 + (-77) + (-117)$

50. $-103 + (-32) + (-27)$

51. $12 + (-4) + (-4) + 12$

52. $18 + (-9) + 5 + (-2)$

53. $(-10) + 14 + 25 + (-16)$

54. $34 + (-12) + (-11) + 213$

Add.

55. $-8 + (-14) + (-11)$

56. $-10 + (-6) + (-1)$

57. $-26 + 5$

58. $-35 + (-12)$

59. $5 + (-1) + 17$

60. $3 + (-23) + 6$

61. $-14 + (-31)$

62. $-100 + 70$

63. $13 + 14 + (-18)$

64. $(-45) + 22 + 20$

65. $-87 + 87$

66. $-87 + 0$

67. $-3 + (-8) + 12 + (-1)$

68. $-16 + 6 + (-14) + (-20)$

69. $0 + (-103)$

70. $94 + (-94)$

B *Evaluate $x + y$ for the given replacement values. See Examples 16 and 17.*

71. $x = -2$ and $y = 3$

72. $x = -7$ and $y = 11$

73. $x = -20$ and $y = -50$

74. $x = -1$ and $y = -29$

75. $x = 3$ and $y = -30$

76. $x = 13$ and $y = -17$

The bar graph below shows the yearly net income for Gateway, Inc. Net income is one indication of a company's health. It measures revenue (money taken in) minus cost (money spent). Use this graph for Exercises 77 through 80. (Source: Gateway, Inc.)

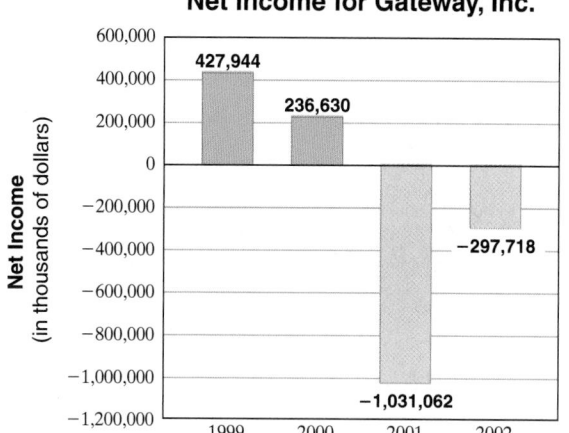

Net Income for Gateway, Inc.

77. What was the net income (in dollars) for Gateway, Inc. in 2000?

78. What was the net income (in dollars) for Gateway, Inc. in 2001?

79. Find the total net income for years 2001 and 2002.

80. Find the total net income for all the years shown.

81. The temperature at 4 p.m. on February 2 was −10° Celsius. By 11 p.m. the temperature had risen 12 degrees. Find the temperature at 11 p.m.

82. Scores in golf can be positive or negative integers. For example, a score of 3 *over* par can be represented by +3 and a score of 5 *under* par can be represented by −5. If Fred Couples had scores of 3 over par, 6 under par, and 7 under par for three games of golf, what was his total score?

83. A small business company reports the following net incomes. Find their sum.

Year	Net income (in dollars)
2000	$75,083
2001	−$10,412
2002	−$1,786
2003	$96,398

84. Suppose a deep-sea diver dives from the surface to 248 meters below the surface and then swims up 6 meters, down 17 meters, down another 24 meters, and then up 23 meters. Use positive and negative numbers to represent this situation. Then find the diver's depth after these movements.

In some card games, it is possible to have positive and negative scores. The table shows the scores for two teams playing a series of four card games. Use this table to answer Exercises 85 and 86.

	Game 1	Game 2	Game 3	Game 4
Team 1	−2	−13	20	2
Team 2	5	11	−7	−3

85. Find each team's total score after four games. If the winner is the team with the greater score, find the winning team.

86. Find each team's total score after three games. If the winner is the team with the greater score, which team was winning after three games?

87. The all-time record low temperature for Wyoming is $-66°F$, which was recorded on February 13, 1905. Kansas's all-time record low temperature is $26°F$ higher than Wyoming's record low. What is Kansas's record low temperature? (*Source:* National Climatic Data Center)

88. The all-time record low temperature for New York is $-52°F$, which occurred on February 13, 1905. In Mississippi's, the lowest temperature ever recorded is $33°F$ higher than New York's all-time low temperature. What is the all-time record low temperature for Mississippi? (*Source:* National Climatic Data Center)

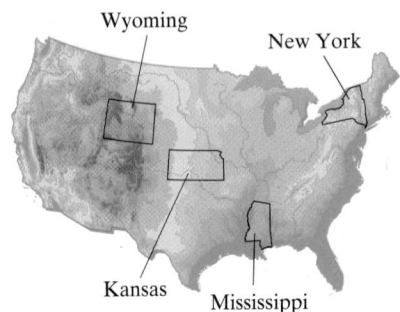

89. The difference between a country's exports and imports is called the country's *trade balance*. The United States had a trade balance of $-\$230$ billion in 1998, $-\$370$ billion in 2000, and $-\$346$ billion in 2001. What was the total U.S. trade balance for these years? (*Source:* U.S. Department of Commerce)

90. The U.S. trade balance for natural gas was -3538 billion cubic feet in 2000 and -3604 billion cubic feet in 2001. What was the total U.S. trade balance for natural gas for these years? (*Source:* U.S. Energy Information Administration)

Review and Preview

Subtract. See Section 1.4.

91. $44 - 0$
92. $91 - 0$
93. $52 - 52$
94. $103 - 103$
95. $87 - 59$

96. $32 - 18$
97. $-8602 + (-1056)$
98. $-7959 + 2335$
99. $-|-56| + |-56|$
100. $-|86| + (-|17|)$

Combining Concepts

For Exercises 101–104, determine whether each statement is true or false.

101. The sum of two negative numbers is always a negative number.

102. The sum of two positive numbers is always a positive number.

103. The sum of a positive number and a negative number is always a negative number.

104. The sum of zero and a negative number is always a negative number.

105. In your own words, explain how to add two negative numbers.

106. In your own words, explain how to add a positive number and a negative number.

2.3   Subtracting Integers

In Section 2.1, we discussed the opposite of an integer.

The opposite of 3 is -3.
The opposite of -6 is 6.

In this section, we use opposites to subtract integers.

OBJECTIVES

(A) Subtract integers.
(B) Add and subtract integers.
(C) Evaluate an algebraic expression by subtracting.
(D) Solve problems by subtracting integers.

SSM
TUTOR CENTER   SG   CD & VIDEO   MATH PRO   WEB

(A) Subtracting Integers

To subtract integers, we will write the subtraction problem as an addition problem. To see how to do this, study the examples below.

$$10 - 4 = 6$$
$$10 + (-4) = 6$$

Since both expressions simplify to 6, this means that

$$10 - 4 = 10 + (-4) = 6$$

Also,

$$3 - 2 = 3 + (-2) = 1$$
$$15 - 1 = 15 + (-1) = 14$$

Thus, to subtract two numbers, we add the first number to the opposite of the second number. (The opposite of a number is also known as its **additive inverse**.)

Subtracting Two Numbers

If a and b are numbers, then $a - b = a + (-b)$.

EXAMPLES   Subtract.

	subtraction	=	first number	+	opposite of the second number	
1.	$8 - 5$	=	8	+	(-5)	$= 3$
2.	$-4 - 10$	=	-4	+	(-10)	$= -14$
3.	$6 - (-5)$	=	6	+	5	$= 11$
4.	$-11 - (-7)$	=	-11	+	7	$= -4$

EXAMPLES   Subtract.

5. $-10 - 5 = -10 + (-5) = -15$

6. $8 - 15 = 8 + (-15) = -7$

7. $-4 - (-5) = -4 + 5 = 1$

Practice Problems 1–4

Subtract.

1. $12 - 7$ 2. $-6 - 4$
3. $11 - (-14)$ 4. $-9 - (-1)$

Practice Problems 5–7

Subtract.

5. $5 - 9$ 6. $-12 - 4$
7. $-2 - (-7)$

Answers

1. 5   **2.** -10 **3.** 25   **4.** -8
5. -4 **6.** -16 **7.** 5

Concept Check

What is wrong with the following calculation?

$-9 - (-6) = -15$

Practice Problem 8

Subtract 5 from -10.

Practice Problem 9

Simplify: $-4 - 3 - 7 - (-5)$

Practice Problem 10

Simplify: $3 + (-5) - 6 - (-4)$

Practice Problem 11

Evaluate $x - y$ for $x = -2$ and $y = 14$.

> **Helpful Hint:**
>
> To visualize subtraction, try the following:
>   The difference between $5°F$ and $-2°F$ can be found by subtracting. That is,
>
> $$5 - (-2) = 5 + 2 = 7$$
>
> Can you visually see from the thermometer on the right that there is actually 7 degrees between $5°F$ and $-2°F$?

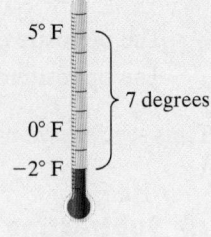

Try the Concept Check in the margin.

EXAMPLE 8   Subtract 7 from -3.

Solution: To subtract 7 *from* -3, we find
$$-3 - 7 = -3 + (-7) = -10$$

B Adding and Subtracting Integers

If a problem involves adding or subtracting more than two integers, we rewrite differences as sums and add. Recall that by associative and commutative properties, we may add numbers in any order. In Examples 9 and 10, we will add from left to right.

EXAMPLE 9   Simplify: $7 - 8 - (-5) - 1$

Solution:
$$
\begin{aligned}
7 - 8 - (-5) - 1 &= \underbrace{7 + (-8)} + 5 + (-1) \\
&= \underbrace{-1 + 5} + (-1) \\
&= \underbrace{4 + (-1)} \\
&= 3
\end{aligned}
$$

EXAMPLE 10   Simplify: $7 + (-12) - 3 - (-8)$

Solution:
$$
\begin{aligned}
7 + (-12) - 3 - (-8) &= \underbrace{7 + (-12)} + (-3) + 8 \\
&= -5 + (-3) + 8 \\
&= -8 + 8 \\
&= 0
\end{aligned}
$$

C Evaluating Expressions

Now let's practice evaluating expressions when the replacement values are integers.

EXAMPLE 11   Evaluate $x - y$ for $x = -3$ and $y = 9$.

Solution: Replace x with -3 and y with 9 in $x - y$.

$$
\begin{array}{ccc}
x & - & y \\
\downarrow & \downarrow & \downarrow
\end{array}
$$
$$
\begin{aligned}
&= (-3) - 9 \\
&= (-3) + (-9) \\
&= -12
\end{aligned}
$$

EXAMPLE 12 Evaluate $a - b$ for $a = 8$ and $b = -6$.

Solution: Watch your signs carefully!

$$
\begin{array}{ccc}
a & - & b \\
\downarrow & \downarrow & \downarrow \\
= 8 & - & (-6) \quad \text{Replace } a \text{ with 8 and } b \text{ with } -6. \\
= 8 & + & 6 \\
= 14
\end{array}
$$

Practice Problem 12

Evaluate $y - z$ for $y = -3$ and $z = -4$.

> **Helpful Hint**
>
> Watch carefully when replacing variables in the expression $x - y$. Make sure that all symbols are inserted and accounted for.

D Solving Problems by Subtracting Integers

Solving problems often requires subtraction of integers.

EXAMPLE 13 Finding a Change in Elevation

The highest point in the United States is the top of Mount McKinley, in Denali County, Alaska, at a height of 20,320 feet above sea level. The lowest point is Death Valley, California, which is 282 feet below sea level. How much higher is Mount McKinley than Death Valley? (*Source:* U.S. Geological Survey)

Practice Problem 13

The highest point in Asia is the top of Mount Everest, at a height of 29,028 feet above sea level. The lowest point is the Dead Sea, which is 1312 feet below sea level. How much higher is Mount Everest than the Dead Sea? (*Source:* National Geographic Society)

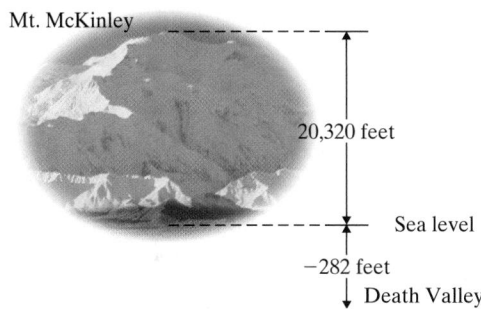

Mt. McKinley

20,320 feet

Sea level

−282 feet
Death Valley

Solution:

In words:

how much higher Mt. McKinley is	=	height of Mt. McKinley	minus	height of Death Valley
↓		↓	↓	↓

Translate:

$$
\begin{array}{ccccc}
\text{how much higher} & = & 20{,}320 & - & (-282) \\
\text{Mt. McKinley is} & & & & \\
& = & 20{,}320 + 282 & = & 20{,}602
\end{array}
$$

Mt. McKinley is 20,602 feet higher than Death Valley.

Answers

12. 1   **13.** 30,340 feet

Focus on History

MAGIC SQUARES

A magic square is a set of numbers arranged in a square table so that the sum of the numbers in each column, row, and diagonal is the same. For instance, in the magic square below, the sum of each column, row, and diagonal is 15. Notice that no number is used more than once in the magic square.

2	9	4
7	5	3
6	1	8

The properties of magic squares have been known for a very long time and once were thought to be good luck charms. The ancient Egyptians and Greeks understood their patterns. A magic square even made it into a famous work of art. The engraving titled *Melencolia I*, created by German artist Albrecht Dürer in 1514, features the following four-by-four magic square on the building behind the central figure.

16	3	2	13
5	10	11	8
9	6	7	12
4	15	14	1

CRITICAL THINKING

1. Verify that what is shown in the Dürer engraving is, in fact, a magic square. What is the common sum of the columns, rows, and diagonals?

2. Negative numbers can also be used in magic squares. Complete the following magic square:

		−2
	−1	
0		−4

3. Use the numbers −16, −12, −8, −4, 0, 4, 8, 12, and 16 to form a magic square:

EXERCISE SET 2.3

Ⓐ *Perform each indicated subtraction. See Examples 1 through 8.*

1. $5 - 5$ **2.** $-6 - (-6)$ **3.** $8 - 3$ **4.** $5 - 2$ **5.** $3 - 8$

6. $2 - 5$ **7.** $7 - (-7)$ **8.** $12 - (-12)$ **9.** $-5 - (-8)$ **10.** $-25 - (-25)$

11. $-14 - 4$ **12.** $-2 - 42$ **13.** $2 - 16$ **14.** $8 - 9$ **15.** $-10 - (-10)$

16. $-5 - (-5)$ **17.** $-15 - (-15)$ **18.** $-24 - (-24)$ **19.** $3 - 7$ **20.** $4 - 12$

21. $30 - 45$ **22.** $29 - 56$ **23.** $-4 - 10$ **24.** $-5 - 8$ **25.** $-230 - 0$

26. $-15 - 0$ **27.** $4 - (-6)$ **28.** $6 - (-9)$ **29.** $-7 - (-3)$ **30.** $-12 - (-5)$

31. $-16 - (-23)$ **32.** $-45 - (-16)$ **33.** Subtract 18 from -20.   **34.** Subtract 10 from -22.

35. Find the difference of -20 and -3.   **36.** Find the difference of -8 and -13.

37. Subtract -11 from 2.   **38.** Subtract -50 from -50.

Ⓑ *Simplify. See Examples 9 and 10.*

39. $7 - 3 - 2$ **40.** $8 - 4 - 1$ **41.** $12 - 5 - 7$ **42.** $30 - 7 - 12$

43. $-5 - 8 - (-12)$ **44.** $-10 - 6 - (-9)$ **45.** $-10 + (-5) - 12$ **46.** $-15 + (-8) - 4$

47. $12 - (-34) + (-6)$ **48.** $23 - (-17) + (-9)$ **49.** $-(-6) - 12 + (-16)$ **50.** $-(-9) - 7 + (-23)$

51. $-9 - (-12) + (-7) - 4$ **52.** $-6 - (-8) + (-12) - 7$

53. $-3 + 4 - (-23) - 10$ **54.** $5 + (-18) - (-21) - 2$

Ⓒ *Evaluate $x - y$ for the given replacement values. See Examples 11 and 12.*

55. $x = -3$ and $y = 5$ **56.** $x = -7$ and $y = 1$ **57.** $x = 6$ and $y = -30$ **58.** $x = 9$ and $y = -2$

59. $x = -4$ and $y = -4$ **60.** $x = -8$ and $y = -10$ **61.** $x = 1$ and $y = -18$ **62.** $x = 14$ and $y = -12$

Recall that the bar graph from Section 2.1 below shows the average temperature in Fahrenheit of known planets. Notice that a negative temperature is illustrated by a bar below the horizontal line representing 0°F, and a positive temperature is illustrated by a bar above the horizontal line representing 0°F.

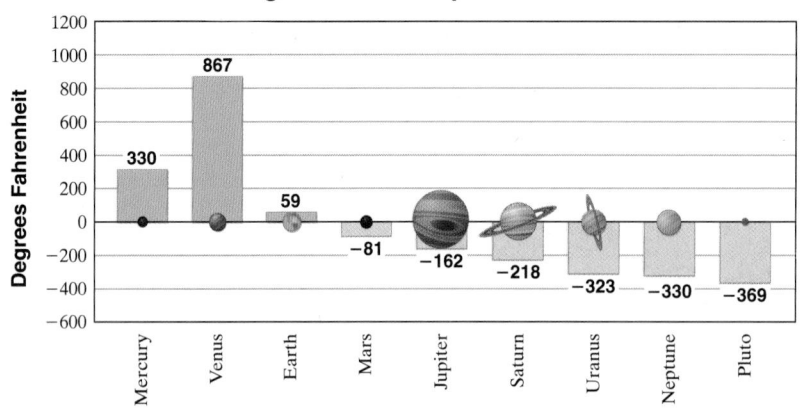

Average Surface Temperature of Planets*

*(For some planets, the temperature given is the temperature where the atmosphere pressure equals 1 Earth atmosphere; Source: *The World Almanac*, 2003)

63. Find the difference in temperature between Earth and Pluto.

64. Find the difference in temperature between Venus and Mars.

65. Find the difference in temperature between the two plants with the lowest temperature.

66. Find the difference in temperature between Jupiter and Saturn.

D *Solve. See Example 13.*

67. The coldest temperature ever recorded on Earth was −129°F in Antarctica. The warmest temperature ever recorded was 136°F in the Sahara Desert. How many degrees warmer is 136°F than −129°F? (*Source: Questions Kids Ask*, Grolier Limited, 1991, and *The World Almanac, 2003*)

68. The coldest temperature ever recorded in the United States was −80°F in Alaska. The warmest temperature ever recorded was 134°F in California. How many degrees warmer is 134°F than −80°F? (*Source*: *The World Almanac*, 2003)

69. Aaron Aiken has $125 in his checking account. He writes a check for $117, makes a deposit of $45, and then writes another check for $69. Find the balance in his account. (Write the amount as an integer.)

70. In the card game canasta, it is possible to have a negative score. If Juan Santanilla's score is 15, what is his new score if he loses 20 points?

71. The temperature on a February morning is −6° Celsius at 6 a.m. If the temperature drops 3 degrees by 7 a.m., rises 4 degrees between 7 a.m. and 8 a.m., and then drops 7 degrees between 8 a.m. and 9 a.m., find the temperature at 9 a.m.

72. Mauna Kea in Hawaii has an elevation of 13,796 feet above sea level. The Mid-America Trench in the Pacific Ocean has an elevation of 21,857 feet below sea level. Find the difference in elevation between those two points. (*Source:* National Geographic Society and Defense Mapping Agency)

Some places on Earth lie below sea level, which is the average level of the surface of the oceans. Use this diagram to answer Exercises 73 through 76. (Source: Fantastic Book of Comparisons, Russell Ash, 1999)

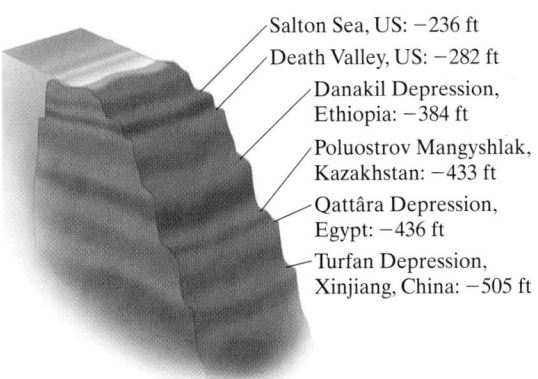

Salton Sea, US: −236 ft
Death Valley, US: −282 ft
Danakil Depression, Ethiopia: −384 ft
Poluostrov Mangyshlak, Kazakhstan: −433 ft
Qattâra Depression, Egypt: −436 ft
Turfan Depression, Xinjiang, China: −505 ft

73. Find the difference in elevation between Death Valley and Quattâra Depression.

74. Find the difference in elevation between Danakil and Turfan Depressions.

75. Find the difference in elevation between the two lowest elevations shown.

76. Find the difference in elevation between the highest elevation shown and the lowest elevation shown.

The bar graph shows heights of selected lakes. For Exercises 77–80, find the difference in elevation for the lakes listed. (Source: U.S. Geological Survey)

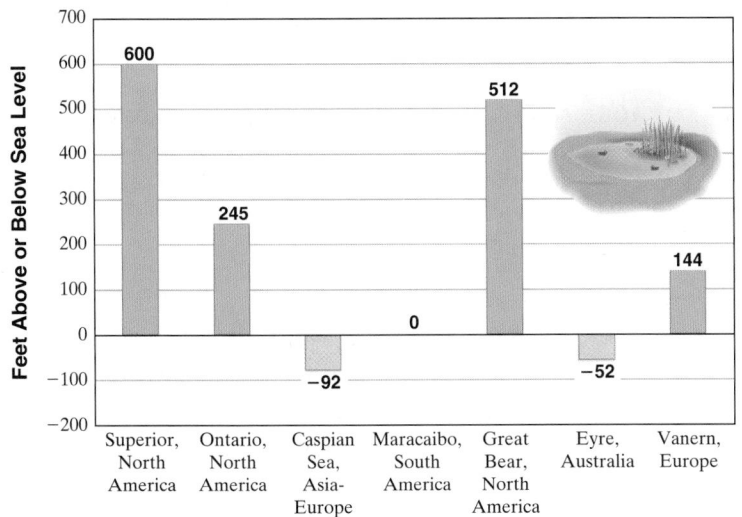

77. Lake Superior and Lake Eyre

78. Great Bear Lake and Caspian Sea

79. Lake Maracaibo and Lake Vanern

80. Lake Eyre and Caspian Sea

81. Recall that the difference between a country's exports and imports is called the country's *trade balance*. In 2001, the United States had $731 billion in exports and $1141 billion in imports. What was the U.S. trade balance in 2001? (*Source:* U.S. Department of Commerce)

82. In 2000, the United States exported 380 million barrels of petroleum products and imported 788 million barrels of petroleum products. What was the U.S. trade balance for petroleum products in 1997? (*Source:* U.S. Energy Information Administration)

Review and Preview

Multiply. See Section 1.6.

83. $436 \cdot 0$ **84.** $0 \cdot 86$ **85.** $436 \cdot 1$ **86.** $1 \cdot 704$ **87.** $\begin{array}{r} 23 \\ \times\ 46 \end{array}$ **88.** $\begin{array}{r} 51 \\ \times\ 89 \end{array}$

Simplify. See Section 1.8.

89. $5^2 - 6 \cdot 2 + 8$ **90.** $80 - 7 \cdot 10 + 8^2$

Combining Concepts

Evaluate each expression for the given replacement values.

91. $x - y - z$ for $x = -4$, $y = 3$, and $z = 15$

92. $x - y - z$ for $x = -14$, $y = 8$, and $z = -6$

93. $a + b - c$ for $a = -16$, $b = 14$, and $c = -22$

94. $a + b - c$ for $a = -1$, $b = -1$, and $c = 100$

Simplify. (Evaluate absolute values first.)

 **95.** $|-3| - |-7|$

96. $|-12| - |-5|$

97. $|-6| - |6|$

98. $|-23| - |-42|$

99. $-8067 - 1129$

100. $76{,}804 - 96{,}971$

For Exercises 101 and 102, determine whether each statement is true or false.

101. $|-8 - 3| = 8 - 3$

102. $|-2 - (-6)| = |-2| - |-6|$

103. In your own words, explain how to subtract one integer from another.

104. A student explains to you that the first step to simplify $8 + 12 \cdot 5 - 100$ is to add 8 and 12. Is the student correct? Explain why or why not.

Internet Excursions

 Go To: http://www.prenhall.com/martin-gay_prealgebra   What's Related

The World Wide Web address listed here will direct you to the Web site of the CNN Financial Network, or a related site. You will be able to collect data on prices of stocks traded on the New York Stock Exchange and answer the questions below.

105. Pick stocks for any four companies. You can use the Web site to look up the stock market ticker symbol for the companies. Then get a stock quote for each company. Complete the table below. For each company in the table, show that the sum of the previous closing price and the change in price give the last price.

106. For the same four companies you used in Exercise 105, show how to use the last price and the change in price to calculate the previous closing price.

Date of stock quotes:_____ Time of stock quotes: _____

Company Name	Ticker Symbol	Previous Close	Change	Last

Name _____ Section _____ Date _____

Integrated Review–Integers

Represent each quantity by an integer.

1. The peak of Mount Everest in Asia is 29,028 feet above sea level. (*Source:* U.S. Geological Survey)

2. The Marianas Trench in the Pacific Ocean is 35,840 feet below sea level. (*Source:* The World Almanac)

3. The deepest hole ever drilled in the Earth's crust is in Russia and its depth is over 7 miles below sea level. (*Source: Fantastic Book of Comparisons*, 1999)

4. Graph the signed numbers on the given number line. −4, 0, −1, 3

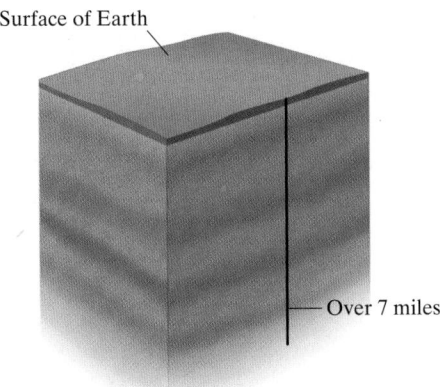

Surface of Earth

Over 7 miles

Insert < or > between each pair of numbers to make a true statement.

5. 0   −3   6. −15   −5   7. −1   1   8. −2   −7

Simplify.

9. |−1| 10. −|−4| 11. |−8| 12. −(−5)

Find the opposite of each number.

13. 6   14. −3   15. 89   16. 0

Add or subtract as indicated.

17. −7 + 12   18. −9 + (−11)   19. 25 + (−35)

1. _____
2. _____
3. _____
4. see number line
5. _____
6. _____
7. _____
8. _____
9. _____
10. _____
11. _____
12. _____
13. _____
14. _____
15. _____
16. _____
17. _____
18. _____
19. _____

20. _____

21. _____

22. _____

23. _____

24. _____

25. _____

26. _____

27. _____

28. _____

29. _____

30. _____

31. _____

32. _____

33. _____

34. _____

20. $1 - 3$ 　　　　　　**21.** $26 - (-26)$ 　　　　**22.** $-2 - 1$

23. $-18 - (-102)$ 　　**24.** $-8 + (-6) + 20$ 　　**25.** $-11 - 7 - (-19)$

26. $-4 + (-8) - 16 - (-9)$ 　**27.** Subtract 14 from 26. 　**28.** Subtract -8 from -12.

Choose all numbers for x from each given list that make each statement true.

29. $|x| > 0$ 　　　　　　　　　　　　**30.** $|x| > -5$
　　a. 0 　**b.** 18 　**c.** -3 　**d.** -21 　　　　**a.** 0 　**b.** 3 　**c.** -1 　**d.** -1000

Evaluate the expressions below for $x = -1$ and $y = 11$

31. $x + y$ 　　　　　**32.** $x - y$

33. $y - x$ 　　　　　**34.** $y + x$

2.4   Multiplying and Dividing Integers

Multiplying and dividing integers is similar to multiplying and dividing whole numbers. One difference is that we need to determine whether the result is a positive number or a negative number.

(A) Multiplying Integers

Consider the following pattern of products.

First factor decreases by 1 each time.

$$3 \cdot 2 = 6$$
$$2 \cdot 2 = 4 \qquad \text{Product decreases by 2 each time.}$$
$$1 \cdot 2 = 2$$
$$0 \cdot 2 = 0$$

This pattern can be continued, as follows.

$$-1 \cdot 2 = -2$$
$$-2 \cdot 2 = -4$$
$$-3 \cdot 2 = -6$$

This suggests that the product of a negative number and a positive number is a negative number.

What is the sign of the product of two negative numbers? To find out, we form another pattern of products. Again, we decrease the first factor by 1 each time, but this time the second factor is negative.

$$2 \cdot (-3) = -6$$
$$1 \cdot (-3) = -3 \qquad \text{Product increases by 3 each time.}$$
$$0 \cdot (-3) = 0$$

This pattern continues as:

$$-1 \cdot (-3) = 3$$
$$-2 \cdot (-3) = 6$$
$$-3 \cdot (-3) = 9$$

This suggests that the product of two negative numbers is a positive number. Thus we can determine the sign of a product when we know the signs of the factors.

Multiplying Numbers

The product of two numbers having the same sign is a positive number.

The product of two numbers having different signs is a negative number.

Product of Like Signs

$$(+)(+) = +$$
$$(-)(-) = +$$

Product of Different Signs

$$(-)(+) = -$$
$$(+)(-) = -$$

EXAMPLES   Multiply.

1. $-7 \cdot 3 = -21$
2. $-2(-5) = 10$
3. $0 \cdot (-4) = 0$
4. $10(-8) = -80$

OBJECTIVES

(A) Multiply integers.
(B) Divide integers.
(C) Evaluate an algebraic expression by multiplying or dividing.
(D) Solve problems by multiplying or dividing integers.

SSM
TUTOR CENTER   SG   CD & VIDEO   MATH PRO   WEB

Practice Problems 1–4

Multiply.

1. $-2 \cdot 6$
2. $-4(-3)$
3. $0 \cdot (-10)$
4. $5(-15)$

Answers

1. -12 **2.** 12 **3.** 0 **4.** -75

Recall that by the associative and commutative properties for multiplication, we may multiply numbers in any order that we wish. In Example 5, we multiply from left to right.

Practice Problems

Multiply.

5. $7(-2)(-4)$
6. $(-5)(-6)(-1)$
7. $(-2)(-5)(-6)(-1)$

EXAMPLES Multiply.

5. $7(-6)(-2) = -42(-2)$
$$= 84$$

6. $(-2)(-3)(-4) = 6(-4)$
$$= -24$$

7. $(-1)(-2)(-3)(-4) = -1(-24)$ We have -24 from Example 6.
$$= 24$$

Concept Check

What is the sign of the product of five negative numbers? Explain.

Try the Concept Check in the margin.

Recall from our study of exponents that $2^3 = 2 \cdot 2 \cdot 2 = 8$. We can now work with bases that are negative numbers. For example,

$$(-2)^3 = (-2)(-2)(-2) = -8$$

EXAMPLE 8 Evaluate: $(-5)^2$

Practice Problem 8

Evaluate $(-3)^4$.

Solution: Remember that $(-5)^2$ means 2 factors of -5.

$$(-5)^2 = (-5)(-5) = 25$$

Helpful Hint:

Have you noticed a pattern when multiplying signed numbers?

If we let $(-)$ represent a negative number and $(+)$ represent a positive number, then

$$(-)(-) = (+)$$

The product of an even number of negative numbers is a positive result.

$(-)(-)(-) = (-)$ The product of an odd number of negative numbers is a negative result.

$(-)(-)(-)(-) = (+)$

$(-)(-)(-)(-)(-) = (-)$

Notice in Example 8 the parentheses around -5 in $(-5)^2$. With these parentheses, -5 is the base that is squared. Without parentheses, such as -5^2, only the 5 is squared.

Practice Problem 9.

Evaluate: -9^2

EXAMPLE 9 Evaluate: -7^2

Solution: Remember that without parentheses, only the 7 is squared.

$$-7^2 = -(7 \cdot 7) = -49$$

Answers

5. 56 **6.** -30 **7.** 60 **8.** 81 **9.** -81

Concept Check: Negative

Make sure you understand the difference between Examples 8 and 9.

⌐→ parentheses, so −5 is squared

$$(-5)^2 = (-5)(-5) = 25$$

⌐→ no parentheses, so only the 7 is squared

$$-7^2 = -(7 \cdot 7) = -49$$

B Dividing Integers

Division of integers is related to multiplication of integers. The sign rules for division can be discovered by writing a related multiplication problem. For example,

$$\frac{6}{2} = 3 \quad \text{because} \quad 3 \cdot 2 = 6$$

$$\frac{-6}{2} = -3 \quad \text{because} \quad -3 \cdot 2 = -6$$

$$\frac{6}{-2} = -3 \quad \text{because} \quad -3 \cdot (-2) = 6$$

$$\frac{-6}{-2} = 3 \quad \text{because} \quad 3 \cdot (-2) = -6$$

Helpful Hint

Just as for whole numbers, division can be checked by multiplication.

Dividing Numbers

The quotient of two numbers having the same sign is a positive number.	The quotient of two numbers having different signs in a negative number.
Quotient of Like Signs	**Quotient of Different Signs**
$\dfrac{(+)}{(+)} = + \qquad \dfrac{(-)}{(-)} = +$	$\dfrac{(+)}{(-)} = - \qquad \dfrac{(-)}{(+)} = -$

EXAMPLES   Divide.

10. $\dfrac{-12}{6} = -2$

11. $-20 \div (-4) = 5$

12. $\dfrac{48}{-3} = -16$

Try the Concept Check in the margin.

EXAMPLES   Divide, if possible.

13. $\dfrac{0}{-5} = 0$ because $0 \cdot -5 = 0$

14. $\dfrac{-7}{0}$ is undefined because there is no number that gives a product of -7 when multiplied by 0.

Practice Problems 10-12

Divide.

10. $\dfrac{28}{-7}$ **11.** $-18 \div (-2)$ **12.** $\dfrac{-60}{10}$

Concept Check

What is wrong with the following calculation?

$$\frac{-27}{-9} = -3$$

Practice Problems 13–14

Divide, if possible.

13. $\dfrac{-1}{0}$ **14.** $\dfrac{0}{-2}$

Answers

10. -4 **11.** 9   **12.** -6 **13.** undefined   **14.** 0

Concept Check: $\dfrac{-27}{-9} = 3$

Ⓒ Evaluating Expressions

Next, we practice evaluating expressions given integer replacement values.

Practice Problem 15

Evaluate xy for $x = 5$ and $y = -9$.

EXAMPLE 15 Evaluate xy for $x = -2$ and $y = 7$.

Solution: Recall that xy means $x \cdot y$.
  Replace x with -2 and y with 7 in xy.

$$xy = -2 \cdot 7$$
$$= -14$$

●

Practice Problem 16

Evaluate $\dfrac{x}{y}$ for $x = -9$ and $y = -3$.

EXAMPLE 16 Evaluate $\dfrac{x}{y}$ for $x = -24$ and $y = 6$.

Solution: $\dfrac{x}{y} = \dfrac{-24}{6}$ Replace x with -24 and y with 6.

$$= -4$$

●

Ⓓ Solving Problems by Multiplying or Dividing Integers

Many real-life problems involve multiplication and division of integers.

Practice Problem 17

A card player had a score of -12 for each of four games. Find her total score.

EXAMPLE 17 Calculating Total Golf Score

A professional golfer finished seven strokes under par (-7) for each of three days of a tournament. What was his total score for the tournament?

Solution:

In words: | golfer's total score | = | number of days | · | score each day |

Translate: golfer's total score = 3 · (-7)

$$= -21$$

The golfer's total score is -21, or 21 strokes under par.   ●

Answers

15. -45 **16.** 3 **17.** -48

EXERCISE SET 2.4

Ⓐ *Multiply. See Examples 1 through 4.*

1. $-2(-3)$

2. $5(-3)$

3. $-4(9)$

4. $-7(-2)$

5. $8(-8)$

6. $-9(9)$

7. $0(-14)$

8. $-6(0)$

Multiply. See Example 5 through 7.

9. $6(-4)(2)$

10. $-2(3)(-7)$

11. $-1(-2)(-4)$

12. $8(-3)(3)$

13. $-4(4)(-5)$

14. $-2(-5)(-4)$

15. $10(-5)(0)$

16. $2(-1)(3)(-2)$

17. $-5(3)(-1)(-1)$

18. $3(0)(-4)(-8)$

Evaluate. See Examples 8 and 9.

19. -2^2

20. -2^4

21. $(-3)^3$

22. $(-1)^4$

23. -5^2

24. -4^3

25. $(-2)^3$

26. $(-3)^2$

Ⓑ *Find each quotient. See Examples 10 through 14.*

27. $-24 \div 6$

28. $90 \div (-9)$

29. $\dfrac{-30}{6}$

30. $\dfrac{56}{-8}$

31. $\dfrac{-88}{-11}$

32. $\dfrac{-32}{4}$

33. $\dfrac{0}{14}$

34. $\dfrac{-13}{0}$

35. $\dfrac{2}{0}$

36. $\dfrac{0}{-5}$

37. $\dfrac{39}{-3}$

38. $\dfrac{-24}{-12}$

Ⓐ Ⓑ *Multiply or divide as indicated.*

39. $-12(0)$

40. $0(-100)$

41. $-4(3)$

42. $-6 \cdot 2$

43. $-9 \cdot 6$

44. $-12(13)$

45. $-7(-6)$

46. $-9(-5)$

47. $-3(-4)(-2)$

48. $-7(-5)(-3)$

49. $(-4)^2$

50. $(-5)^2$

51. $-\dfrac{10}{5}$

52. $-\dfrac{25}{5}$

53. $-\dfrac{56}{8}$

54. $-\dfrac{49}{7}$

55. $-12 \div 3$ **56.** $-15 \div 3$ **57.** $4(-4)(-3)$ **58.** $6(-5)(-2)$

59. $-30(6)(-2)(-3)$ **60.** $-20 \cdot 5 \cdot (-5) \cdot (-3)$ **61.** $3 \cdot (-2) \cdot 0$ **62.** $-5(4)(0)$

63. $\dfrac{100}{-20}$ **64.** $\dfrac{45}{-9}$ **65.** $240 \div (-40)$ **66.** $480 \div (-8)$

**67.** $\dfrac{-12}{-4}$ **68.** $\dfrac{-36}{-3}$ **69.** -1^4 **70.** -2^3

71. $(-3)^5$ **72.** $(-9)^2$ **73.** $-2(3)(5)(-6)$ **74.** $-1(2)(7)(-3)$

75. $(-1)^{32}$ **76.** $(-1)^{33}$ **77.** $-2(-2)(-5)$ **78.** $-2(-2)(-3)(-2)$

79. $-42 \cdot 23$ **80.** $-56 \cdot 43$ **81.** $25 \cdot (-82)$ **82.** $70 \cdot (-23)$

C *Evaluate ab for the given replacement values. See Example 15.*

83. $a = -4$ and $b = 7$ **84.** $a = 5$ and $b = -1$ **85.** $a = 3$ and $b = -2$

86. $a = -9$ and $b = -6$ **87.** $a = -5$ and $b = -5$ **88.** $a = -8$ and $b = 8$

Evaluate $\dfrac{x}{y}$ for the given replacement values. See Example 16.

89. $x = 5$ and $y = -5$ **90.** $x = 9$ and $y = -3$ **91.** $x = -12$ and $y = 0$

92. $x = -10$ and $y = -10$ **93.** $x = -36$ and $y = -6$ **94.** $x = 0$ and $y = -5$

Evaluate xy and also $\dfrac{x}{y}$ for the given replacement values.

95. $x = -4$ and $y = -2$ **96.** $x = 20$ and $y = -5$ **97.** $x = 0$ and $y = -6$ **98.** $x = -3$ and $y = 0$

D *The graph shows melting points in degrees Celsius of selected elements. Use this graph to answer Exercises 99 through 102. See Example 17.*

99. The melting point of nitrogen is 3 times the melting point of radon. Find the melting point of nitrogen.

100. The melting point of rubidium is -1 times the melting point of mercury. Find the melting point of rubidium.

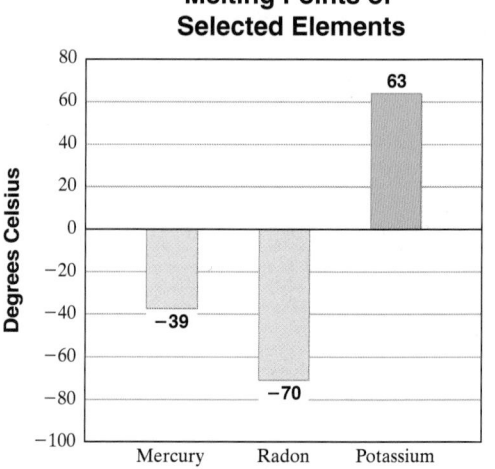

Melting Points of Selected Elements

101. The melting point of argon is −3 times the melting point of potassium. Find the melting point of argon.

102. The melting point of strontium is −11 times the melting point of radon. Find the melting point of strontium.

Solve. See Example 17

103. A football team lost 4 yards on each of three consecutive plays. Represent the total loss as a product of integers, and find the total loss.

104. Joe Norstrom lost $400 on each of seven consecutive days in the stock market. Represent his total loss as a product of integers, and find his total loss.

105. A deep-sea diver must move up or down in the water in short steps to keep from getting a physical condition called the bends. Suppose the diver moves down from the surface in five steps of 20 feet each. Represent his movement as a product of integers, and find his final depth.

106. A weather forecaster predicts that the temperature will drop 5 degrees each hour for the next six hours. Represent this drop as a product of integers, and find the total drop in temperature.

107. For fiscal year 2001, Kmart reported a net loss of $2420 million. If this continues, what would Kmart's income be after three years (*Source*: Kmart Corporation)

108. In 2001, the earnings for a share of Kmart stock were approximately $5 per share. If an investor owned 15 shares of Kmart stock in 2001, what were his total earnings for these shares? (*Source*: Kmart corporation)

109. In 1987, there were only 27 California Condors in the entire world. In 2002, there were 192 California Condors. (*Source:* California Department of Fish and Game)
  a. Find the change in the number of California Condors from 1987 to 2002.
  b. Find the average change per year in the California Condor population over this period.

110. In 1997, music cassettes with a total value of $2227 million were produced by American music manufacturers. In 2001, this value had dropped to $467 million. (*Source:* Recording Industry Association of America)
  a. Find the change in the value of music cassettes produced from 1997 to 2001.
  b. Find the average change per year in the value of music cassettes produced over this period.

Recall that the average of a list of numbers is the sum of the numbers divided by how many numbers are in the list. Use this for Exercises 111 and 112, (We will review this again in Section 2.5.)

111. During the 2002 PGA Championship in Chaska, MN, Fred Funk had scores of −4, −2, 1 and 1 in four rounds of golf. Find his average score per tournament round. (*Source*: Professional Golf Association)

112. During the 2002 Senior PGA Championship, Fuzzy Zoeller had scores of −1, +1, and 0 for three rounds of golf. Find this average score per tournament round. (*Source*: Professional Golf Association)

Review and Preview

Perform each indicated operation. See Section 1.8.

113. $(3 \cdot 5)^2$

114. $(12 - 3)^2(18 - 10)$

115. $90 + 12^2 - 5^3$

116. $3 \cdot (7 - 4) + 2 \cdot 5^2$

117. $12 \div 4 - 2 + 7$

118. $12 \div (4 - 2) + 7$

 Combining Concepts

Let a and b be positive numbers. Answer true or false for each statement.

119. $a(-b)$ is a negative number.

120. $(-a)(-b)$ is a negative number.

121. $(-a)(-a)$ is a positive number.

122. $(-a)(-a)(-a)$ is a positive number.

Let a and b be positive numbers. Determine the sign of each expression.

123. $(-a)^{30}$

124. $(-b)^{29}$

125. $(-a)^5(-b)^4$

126. $(-a)^7(-b)^5$

Without actually finding the product, write the list of numbers in Exercises 127 and 128 in order from least to greatest. For help, see a helpful hint box in this section.

127. $(-2)^{12}, (-2)^{17}, (-5)^{12}, (-5)^{17}$

128. $(-1)^{50}, (-1)^{55}, 0^{15}, (-7)^{20}, (-7)^{23}$

129. In 2001, there were 2190 commercial country music radio stations in the United States. By 2002, that number had declined to 2131. (*Source:* M Street Corporation)
   a. Find the change in the number of country music radio stations from 2001 to 2002.
   b. If this change continues, what will be the total change in the number of country music stations four years after 2002?
   c. Based on your answer to part (b), how many country music radio stations will there be in 2006?

130. In 2000, Saturn sold a total of 260,899 cars. By 2001, the number of Saturns sold had increased to 275,425 (*Source:* Ward's Auto Info Bank)
   a. Find the change in the number of Saturns sold from 2000 to 2001.
   b. If this change continues, what will be the total change in the number of Saturns sold five years after 2001?
   c. Based on your answer to part (b), how many Saturns would have been sold in 2006?

131. In your own words, explain how to multiply two integers.

132. In your own words, explain how to divide two integers.

133. $87^2 - (-12)^5$

134. $\dfrac{(-60 - 21)^3 - 9^4}{(-3)^6}$

MODELING ADDING AND SUBTRACTING INTEGERS

Just as we used physical models to represent fractions and operations on whole numbers, we can use a physical model to help us add and subtract integers. In this model, we use objects called counters to represent numbers. Black counters • represent positive numbers, and red counters • represent negative numbers. The key to this model is remembering that taking a • and • together creates a neutral or zero pair. Once a neutral pair has been formed, it can be removed from or added to the model without changing the overall value.

ADDING INTEGERS

$5 + (-8)$

Begin with 5 black counters and add to that 8 red counters.

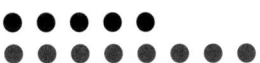

Next, form and remove neutral pairs.

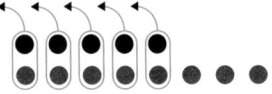

Now there are only 3 red counters left, so $5 + (-8) = -3$

SUBTRACTING INTEGERS

$6 - (-2)$
Begin with 6 black counters.

Subtracting a −2 indicates taking away 2 red counters. However, there are no red counters in the model at this point. Add enough neutral pairs to the model to obtain 2 red counters.

Now take away 2 red counters.

Because there are 8 black counters remaining, $6 - (-2) = 8$.

$-4 + (-3)$

Begin with 4 red counters and add to that 3 red counters.

Because this group does not contain a mixture of red and black counters, there is no need to form neutral pairs. Simply find the total number of red counters and remember that red counters represent negative numbers. So,

$$-4 + (-3) = -7$$

$-3 - 9$

Begin with 3 red counters.

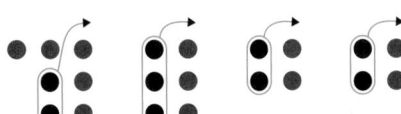

Subtracting 9 indicates taking away 9 black counters, However, there are no black counters in the model at this point. Add enough neutral pairs to the model to obtain 9 black counters.

Now take away 9 black counters.

Because there are 12 red counters remaining, $-3 - 9 = -12$.

CRITICAL THINKING

Use the counter model to perform each indicated operation.

1. $3 + 7$ **2.** $(-5) + 6$ **3.** $(-8) + (-4)$ **4.** $9 + (-11)$ **5.** $(-2) + 2$

6. $8 - 4$ **7.** $(-3) - (-1)$ **8.** $2 - (-5)$ **9.** $(-6) - 4$ **10.** $5 - 7$

STUDY SKILLS REMINDER

How are your homework assignments going?

It is so important in mathematics to keep up with homework. Why? Many concepts build on each other. Often, your understanding of a day's lecture in mathematics depends on an understanding of the previous day's material.

Remember that completing your homework assignment involves a lot more than attempting a few of the problems assigned.

To complete a homework assignment, remember these four things:

1. Attempt all of it. **3.** Correct it.

2. Check it. **4.** If needed, ask questions about it.

Focus on **Business and Career**

NET INCOME AND NET LOSS

For most businesses, a financial goal is to "make money." But what does this mean from a mathematical point of view? To find out, we must first discuss some common business terms.

■ **Revenue** is the amount of money a business takes in. A company's annual revenue is the amount of money it collects during its fiscal, or business, year. For most companies, the largest source of revenue is from the sales of their products or services. For instance, a grocery store's annual revenue is the amount of money it collects during the year from selling groceries to customers. Large companies may also have revenues from interest or rentals.

■ **Expenses** are the costs of doing business. For instance, a large part of a grocery store's expenses includes the cost of the food items it buys from wholesalers to resell to customers. Other expenses include salaries, mortgage payments, equipment, taxes, advertising, and so on.

■ **Net income/loss** is the difference between a company's annual revenues and expenses. If the company's revenues are larger than its expenses, the difference is a positive number, and the company posts a net income for the year. Posting a net income can be interpreted as "making money." If the company's revenues are smaller than its expenses, the difference is a negative number, and the company posts a net loss for the year. Posting a net loss can be interpreted as "losing money."

Net income can also be thought of as profit. A negative profit is a net loss.

GROUP ACTIVITY

Search for corporate annual reports or articles in financial newspapers and magazines that report a company's net income or net loss. Describe what the income or loss means for the company. What are some of the factors that contributed to the net income or loss?

2.5   Order of Operations

OBJECTIVES

A Simplify expressions by using the order of operations.

B Evaluate an algebraic expression.

C Find the average of a list of numbers.

SSM
TUTOR CENTER   SG   CD & VIDEO   MATH PRO   WEB

A **Simplifying Expressions**

We first discussed the order of operations in Chapter 1. In this section, you are given an opportunity to practice using the order of operations when expressions contain integers. The rules for order of operations from Section 1.8 are repeated here.

If there are no other grouping symbols such as fraction bars or absolute values, perform operations in the following order.

Order of Operations

1. Do all operations within grouping symbols such as parentheses or brackets. (Start with the innermost set.)

2. Evaluate any expressions with exponents.

3. Multiply or divide in order from left to right.

4. Add or subtract in order from left to right.

EXAMPLES   Find the value of each expression.

1. $(-3)^2 = (-3)(-3) = 9$ The base of the exponent is -3.

2. $-3^2 = -(3)(3) = -9$ The base of the exponent is 3.

3. $2 \cdot 5^2 = 2 \cdot (5 \cdot 5) = 2 \cdot 25 = 50$ The base of the exponent is 5.

Helpful Hint

When simplifying expressions with exponents, remember that parentheses make an important difference.

$(-3)^2$ and -3^2 **do not** mean the same thing.

$(-3)^2$ means $(-3)(-3) = 9$.

-3^2 means the opposite of $3 \cdot 3$, or -9.

Only with parentheses around it is the -3 squared.

Spend time studying Examples 1 through 3 and the Helpful Hint above. It is important to be able to determine the base of an exponent. After reading these examples, if you have trouble completing the Practice Problems, review this material again.

EXAMPLE 4   Simplify: $\dfrac{-6(2)}{-3}$

Solution: The fraction bar serves as a grouping symbol. First we multiply -6 and 2. Then we divide.

$$\frac{-6(2)}{-3} = \frac{-12}{-3}$$
$$= 4$$

EXAMPLE 5   Simplify: $\dfrac{12 - 16}{-1 + 3}$

Solution: We simplify above and below the fraction bar separately. Then we divide.

$$\frac{12 - 16}{-1 + 3} = \frac{-4}{2}$$
$$= -2$$

Practice Problems 1–3

Find the value of each expression.

1. $(-2)^4$

2. -2^4

3. $3 \cdot 6^2$

Practice Problem 4

Simplify: $\dfrac{25}{5(-1)}$

Practice Problem 5

Simplify: $\dfrac{-18 + 6}{-3 - 1}$

Answers

1. 16   **2.** -16 **3.** 108   **4.** -5 **5.** 3

Practice Problem 6

Simplify: $-20 + 2 \cdot 7 + 4$

EXAMPLE 6 Simplify: $3 + 4 \cdot 5 - 27$

Solution: Follow the order of operations.

$$3 + \overbrace{4 \cdot 5} - 27 = 3 + 20 - 27 \qquad \text{Multiply.}$$
$$= 23 - 27 \qquad \text{Add or subtract from left to right.}$$
$$= -4 \qquad \text{Subtract.}$$

Practice Problem 7

Simplify: $-2^3 + (-4)^2 + 1^5$

EXAMPLE 7 Simplify: $-4^2 + (-3)^2 - 1^3$

Solution: Follow the order of operations.

$$-4^2 + (-3)^2 - 1^3 = -16 + 9 - 1 \qquad \text{Simplify expressions with exponents.}$$
$$= -7 - 1 \qquad \text{Add or subtract from left to right.}$$
$$= -8$$

Practice Problem 8

Simplify: $2(2 - 8) + (-12) - 3$

EXAMPLE 8 Simplify: $3(4 - 7) + (-2) - 5$

Solution: Follow the order of operations.

$$3(4 - 7) + (-2) - 5 = 3(-3) + (-2) - 5 \qquad \text{Simplify inside parentheses.}$$
$$= -9 + (-2) - 5 \qquad \text{Multiply.}$$
$$= -11 - 5 \qquad \text{Add or subtract from left to right.}$$
$$= -16$$

Practice Problem 9

Simplify: $(-5) \cdot |-4| + (-3) + 2^3$

EXAMPLE 9 Simplify: $(-3) \cdot |-5| - (-2) + 4^2$

Solution: Follow the order of operations.

$$(-3) \cdot |-5| - (-2) + 4^2 = (-3) \cdot 5 - (-2) + 4^2 \qquad \text{Write } |-5| \text{ as 5.}$$
$$= (-3) \cdot 5 - (-2) + 16 \qquad \text{Write } 4^2 \text{ as 16.}$$
$$= -15 - (-2) + 16 \qquad \text{Multiply.}$$
$$= -13 + 16 \qquad \text{Add or subtract from left to right.}$$
$$= 3$$

Practice Problem 10

Simplify: $4(-6) \div [3(5 - 7)^2]$

EXAMPLE 10 Simplify: $-2[-3 + 2(-1 + 6)] - 5$

Solution: By the order of operations, we begin with the innermost set of parentheses.

$$-2[-3 + 2(-1 + 6)] - 5 = -2[-3 + 2(5)] - 5 \qquad \text{Write } -1 + 6 \text{ as 5.}$$
$$= -2[-3 + 10] - 5 \qquad \text{Multiply.}$$
$$= -2(7) - 5 \qquad \text{Add.}$$
$$= -14 - 5 \qquad \text{Multiply.}$$
$$= -19 \qquad \text{Subtract.}$$

Concept Check

True or false? Explain your answer. The result of

$$-4 \cdot (3 - 7) - 8 \cdot (9 - 6)$$

is positive because there are four negative signs.

Try the Concept Check in the margin.

B **Evaluating Expressions**

Now we practice evaluating expressions.

Practice Problem 11

Evaluate x^2 and $-x^2$ for $x = -12$.

EXAMPLE 11 Evaluate x^2 and $-x^2$ for $x = -11$.

Solution: $x^2 = (-11)^2 = (-11)(-11) = 121$

$$-x^2 = -(-11)^2 = -(-11)(-11) = -121$$

Answers

6. -2 **7.** 9 **8.** -27 **9.** -15
10. -2 **11.** 144; -144

Concept Check: False;
$-4 \cdot (3 - 7) - 8 \cdot (9 - 6) = -8$

EXAMPLE 12   Evaluate $6z^2$ for $z = 2$ and $z = -2$.

Solution: $6z^2 = 6(2)^2 = 6(4) = 24$

$6z^2 = 6(-2)^2 = 6(4) = 24$ ●

EXAMPLE 13   Evaluate $x + 2y - z$ for $x = 3$ and $y = -5$ and $z = -4$.

Solution: Replace x with 3, y with -5, z with -4 and simplify.

$x + 2y - z = 3 + 2(-5) - (-4)$ Let $x = 3$, $y = -5$, and $z = -4$.

$= 3 + (-10) + 4$ Replace $2(-5)$ with its product, -10.

$= -3$ Add.   ●

EXAMPLE 14   Evaluate $7 - x^2$ for $x = -4$.

Solution: Replace x with -4 and simplify carefully!

$7 - x^2 = 7 - (-4)^2$

$\downarrow \qquad \downarrow$

$= 7 - 16$ $(-4)^2 = (-4)(-4) = 16$

$= -9$ Subtract.   ●

C Finding Averages

Recall from Chapter 1 that the average of a list of numbers is

$$\text{average} = \frac{\text{sum of numbers}}{\textit{number} \text{ of numbers}}$$

EXAMPLE 15   The graph shows the monthly normal temperatures for Barrow, Alaska. Use this graph to find the average of the temperatures for months January through May.

Monthly Normal Temperatures for Barrow, Alaska

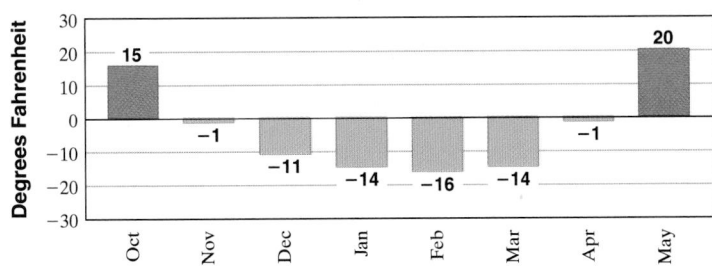

Solution: By reading the graph, we have

$$\text{average} = \frac{-14 + (-16) + (-14) + (-1) + 20}{5}$$ There are 5 months from January through May.

$$= \frac{-25}{5}$$

$$= -5$$

The average for the temperatures is $-5°F$.   ●

CALCULATOR EXPLORATIONS

Simplifying an Expression Containing a Fraction Bar

Recall that even though most calculators follow the order of operations, parentheses must sometimes be inserted. For example, to simplify $\dfrac{-8 + 6}{-2}$ on a calculator, enter parentheses about the expression above the fraction bar so that it is simplified separately.

To simplify $\dfrac{-8 + 6}{-2}$, press the keys

$(\boxed{8} \boxed{+/-} \boxed{+} \boxed{6}) \boxed{\div} \boxed{2} \boxed{+/-} \boxed{=}$ or

$(\boxed{(-)} \boxed{8} \boxed{+} \boxed{6}) \boxed{\div} \boxed{(-)} \boxed{2} \boxed{\text{ENTER}}$

The display will read $\boxed{1}$.

Thus $\dfrac{-8 + 6}{-2} = 1$.

Use a calculator to simplify.

1. $\dfrac{-120 - 360}{-10}$

2. $\dfrac{4750}{-2 + (-17)}$

3. $\dfrac{-316 + (-458)}{28 + (-25)}$

4. $\dfrac{-234 + 86}{-18 + 16}$

STUDY SKILLS REMINDER

Are you prepared for a test on Chapter 2?

Below I have listed some *common trouble areas* for topics covered in Chapter 2. After studying for your test—but before taking your test—read these.

- Don't forget the difference between $-(-5)$ and $-|-5|$.

  $-(-5) = 5$ The opposite of -5 is 5.

  $-|-5| = -5$ The opposite of the absolute value of -5 is the opposite of 5, which is -5

- Remember how to simplify $(-7)^2$ and -7^2.

  $(-7)^2 = (-7)(-7) = 49$

  $-7^2 = -(7)(7) = -49$

- Don't forget the order of operations.

  $1 + 3(4 - 6) = 1 + 3(-2)$ Simplify inside parentheses.

  $= 1 + (-6)$ Multiply.

  $= -5$ Add.

  Remember: This is simply a checklist of common trouble spots. For a review of Chapter 2, see the Highlights and Chapter Review at the end of this chapter.

Mental Math

Identify the base and exponent of each expression. Do not simplify.

1. -3^2

2. $(-3)^2$

3. $4 \cdot 2^3$

4. $9 \cdot 5^6$

5. $(-7)^5$

6. -9^4

7. $5^7 \cdot 10$

8. $2^8 \cdot 11$

EXERCISE SET 2.5

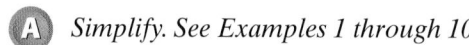 *Simplify. See Examples 1 through 10.*

1. $(-4)^3$

2. -2^4

3. -4^3

4. $(-2)^4$

5. $6 \cdot 2^2$

6. $5 \cdot 2^3$

7. $-1(-2) + 1$

8. $3 + (-8) \div 2$

9. $9 - 12 - 4$

10. $10 - 23 - 12$

11. $4 + 3(-6)$

12. $-8 + 4(3)$

13. $5(-9) + 2$

14. $7(-6) + 3$

15. $(-10) + 4 \div 2$

16. $(-12) + 6 \div 3$

17. $6 + 7 \cdot 3 - 40$

18. $5 + 9 \cdot 4 - 52$

19. $\dfrac{16 - 13}{-3}$

20. $\dfrac{20 - 15}{-1}$

21. $\dfrac{24}{10 + (-4)}$

22. $\dfrac{88}{-8 - 3}$

23. $5(-3) - (-12)$

24. $7(-4) - (-6)$

25. $(-19) - 12(3)$

26. $(-24) - 14(2)$

27. $-8 + 4^2$

28. $-12 + 3^3$

29. $[8 + (-4)]^2$

30. $[9 + (-2)]^3$

31. $8 \cdot 6 - 3 \cdot 5 + (-20)$

32. $7 \cdot 6 - 6 \cdot 5 + (-10)$

33. $16 - (-3)^4$

34. $20 - (-5)^2$

35. $|5 + 3| \cdot 2^3$

36. $|-3 + 7| \cdot 7^2$

37. $7 \cdot 8^2 + 4$

38. $10 \cdot 5^3 + 7$

39. $5^3 - (4 - 2^3)$

40. $8^2 - (5 - 2)^4$

41. $|3 - 12| \div 3$

42. $|12 - 19| \div 7$

43. $-(-2)^2$

44. $-(-2)^3$

45. $(5 - 9)^2 \div (4 - 2)^2$

46. $(2 - 7)^2 \div (4 - 3)^4$

47. $|8 - 24| \cdot (-2) \div (-2)$

48. $|3 - 15| \cdot (-4) \div (-16)$

49. $(-12 - 20) \div 16 - 25$

50. $(-20 - 5) \div 5 - 15$

51. $5(5 - 2) + (-5)^2 - 6$

52. $3 \cdot (8 - 3) + (-4) - 10$

53. $(2 - 7) \cdot (6 - 19)$ **54.** $(4 - 12) \cdot (8 - 17)$ **55.** $2 - 7 \cdot 6 - 19$ **56.** $4 - 12 \cdot 8 - 17$

57. $(-36 \div 6) - (4 \div 4)$ **58.** $(-4 \div 4) - (8 \div 8)$ **59.** $-5^2 - 6^2$ **60.** $-4^4 - 5^4$

61. $(-5)^2 - 6^2$ **62.** $(-4)^4 - (5)^4$ **63.** $(10 - 4^2)^2$ **64.** $(11 - 3^2)^3$

65. $2(8 - 10)^2 - 5(1 - 6)^2$ **66.** $-3(4 - 8)^2 + 5(14 - 16)^3$ **67.** $3(-10) \div [5(-3) - 7(-2)]$

68. $12 - [7 - (3 - 6)] + (2 - 3)^3$ **69.** $\dfrac{(-7)(-3) - (4)(3)}{3[7 \div (3 - 10)]}$ **70.** $\dfrac{10(-1) - (-2)(-3)}{2[-8 \div (-2 - 2)]}$

71. $-5[4 + 5(-3 + 5)] + 11$ **72.** $-2[1 + 3(7 - 12)] - 35$

B *Evaluate each expression for $x = -2$, $y = 4$, and $z = -1$. See Examples 11 through 14.*

73. $x + y + z$ **74.** $x - y - z$ **75.** $2x - 3y - 4z$ **76.** $5x - y + 4z$

77. $x^2 - y$ **78.** $x^2 + z$ **79.** $\dfrac{5y}{z}$ **80.** $\dfrac{4x}{y}$

Evaluate each expression for $x = -3$ and $z = -4$. See Examples 11 through 14.

81. x^2 **82.** z^2 **83.** $-z^2$ **84.** $-x^2$

85. $10 - x^2$ **86.** $3 - z^2$ **87.** $2x^3 - z$ **88.** $3z^2 - x$

C *Find the average of each list of numbers. See Example 15.*

89. $-10, 8, -4, 2, 7, -5, -12$ **90.** $-18, -8, -1, -1, 0, 4$

91. $-17, -26, -20, -13$ **92.** $-40, -20, -10, -15, -5$

Scores in golf can be 0 (also called par), a positive integer (also called above par), or a negative integer (also called below par). On the next page are scores of selected golfers from the 2003 Memorial Tournament in Dublin, Ohio. Use this graph for Exercises 93 through 98.

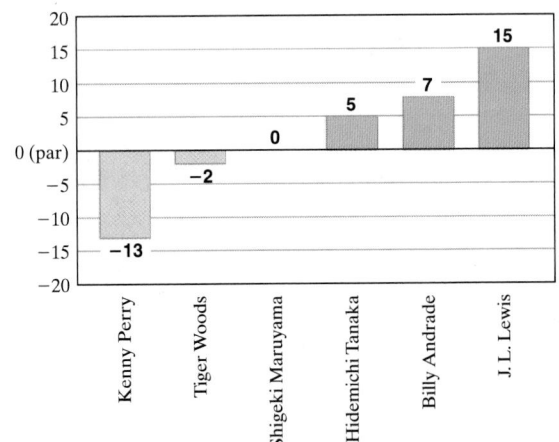

**Golf Scores of Selected Players
2003 Memorial Tournament**

93. Find the difference between the lowest score shown and the highest score shown.

94. Find the difference between the two lowest scores.

95. Find the average of the scores shown. (*Hint*: Here, the average is the sum of the scores divided by the number of players.)

96. Find the average of the scores for Perry, Woods, Maruyama, and Lewis.

97. Can the average for these scores be greater than the highest score, 15? Explain why or why not.

98. Can the average of all the scores shown be less than the lowest score, −13. Explain why or why not.

Review and Preview

Perform each indicated operation. See Sections 1.3, 1.4, 1.6, and 1.7.

99. $45 \cdot 90$

100. $90 \div 45$

101. $90 - 45$

102. $45 + 90$

Find the perimeter of each figure. See Section 1.2.

△ **103.** Square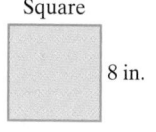
8 in.

△ **104.** Parallelogram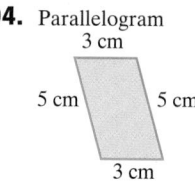
3 cm
5 cm   5 cm
3 cm

△ **105.** Rectangle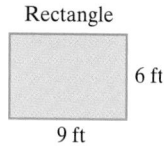
6 ft
9 ft

△ **106.**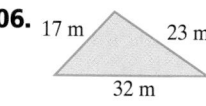
17 m   23 m
32 m

Combining Concepts

Insert parentheses where needed so that each expression evaluates to the given number.

107. $2 \cdot 7 - 5 \cdot 3$; evaluates to 12

108. $7 \cdot 3 - 4 \cdot 2$; evaluates to 34

109. $-6 \cdot 10 - 4$; evaluates to −36

110. $2 \cdot 8 \div 4 - 20$; evaluates to −36

Evaluate.

111. $(-12)^4$

112. $(-17)^5$

113. $x^3 - y^2$ for $x = 21$ and $y = -19$

114. $3x^2 + 2x - y$ for $x = -18$ and $y = 2868$

115. $(xy + z)^x$ for $x = 2$, $y = -5$, and $z = 7$

116. $5(ab + 3)^b$ for $a = -2$, $b = 3$

117. Discuss the effect parentheses have in an exponential expressions. For example, what is the difference between $(-6)^2$ and -6^2.

118. Discuss the effect parentheses have in an exponential expressions. For example, what is the difference between $(2 \cdot 4)^2$ and $2 \cdot 4^2$?

CHAPTER 2 ACTIVITY Investigating Positive and Negative Numbers

MATERIALS:

■ colored thumbtacks

■ coin

■ cardboard

■ six-sided die

■ tape

Work with a partner or a small group to try the following activity. The object is to have the largest absolute value at the end of the game.

1. Attach this page to a piece of cardboard with tape. Each person should choose a different colored thumbtack as his or her playing piece. Insert each thumbtack at the starting place 0 on the number line.

2. Each player takes a turn as follows: Roll the die and flip the coin. "Heads" on the coin makes the number that lands faceup on the die positive. "Tails" on the coin makes the number that lands faceup on the die negative. Record your number along with its sign (positive or negative) in the table below. Move your thumbtack on the number line according to your number.

3. Continue taking turns, until each person has taken five turns. Verify your final position on the number line by finding the total of the integers in the table. Do your total and final position agree?

	Positive or Negative Number
Turn 1	
Turn 2	
Turn 3	
Turn 4	
Turn 5	
Total	

4. The winner is the player having the final position with the largest absolute value. Find the absolute value of your final position. Create a table listing the absolute values of each person's final position. Who won? How could you tell who won just by looking at the number line?

5. Many board games include instructions for moving playing pieces forward or backward. Make a list of games that include such instructions. Then explain how these instructions for moving forward or backward are related to positive and negative numbers.

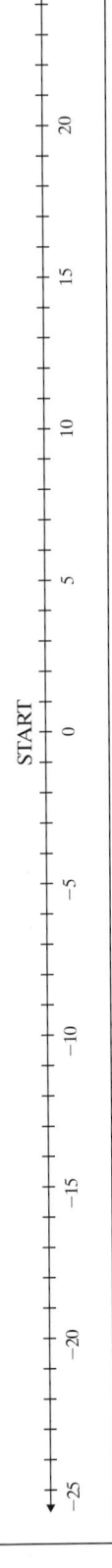

Chapter 2 Vocabulary Check

Fill in each blank with one of the words or phrases listed below.

opposites	integers	positive integers
signed	absolute value	negative integers

1. The _____ are integers less than 0.
2. Two numbers that are the same distance from 0 on the number line but are on opposite sides of 0 are called _____.
3. Together, positive numbers, negative numbers, and 0 are called _____ numbers.
4. The _____ are integers greater than 0.
5. The _____ of a number is that number's distance from 0 on the number line.
6. The _____ are $\ldots, -3, -2, -1, 0, 1, 2, 3, \ldots$.

CHAPTER 2 | Highlights

DEFINITIONS AND CONCEPTS	EXAMPLES

SECTION 2.1 INTRODUCTION TO INTEGERS

The **integers** are $\ldots, -3, -2, -1, 0, 1, 2, 3, \ldots$.	Integers: $-432, \quad -10, \quad 0, \quad 15$
The **absolute value** of a number is that number's distance from 0 on the number line. The symbol for absolute value is $\lvert \; \rvert$.	$\lvert -2 \rvert = 2$ 2 units, on number line $-3 \, (-2) \, -1 \; 0 \; 1 \; 2$
	$\lvert 2 \rvert = 2$ 2 units, on number line $-3 \; -2 \; -1 \; 0 \; 1 \; (2)$
Two numbers that are the same distance from 0 on the number line but are on opposite sides of 0 are called **opposites**.	5 and -5 are opposites. 5 units, 5 units on number line $(-5) \; -4 \; -3 \; -2 \; -1 \; 0 \; 1 \; 2 \; 3 \; 4 \; (5)$
If a is a number, then $-(-a) = a$.	$-(-11) = 11, \qquad -\lvert -3 \rvert = -3$

SECTION 2.2 ADDING INTEGERS

ADDING TWO NUMBERS WITH THE SAME SIGN	Add:
Step 1. Add their absolute values.	$-3 + (-2) = -5$
Step 2. Use their common sign as the sign of the sum.	$-7 + (-15) = -22$
ADDING TWO NUMBERS WITH DIFFERENT SIGNS	$-6 + 4 = -2$
Step 1. Find the larger absolute value minus the smaller absolute value.	$17 + (-12) = 5$
	$-32 + (-2) + 14 = -34 + 14$
Step 2. Use the sign of the number with the larger absolute value as the sign of the sum.	$= -20$

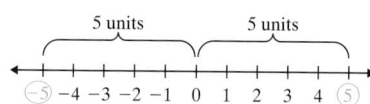

DEFINITIONS AND CONCEPTS	**EXAMPLES**

SUBTRACTING TWO NUMBERS

If a and b are numbers, then $a - b = a + (-b)$.

Subtract:

$$-35 - 4 = -35 + (-4) = -39$$

$$3 - 8 = 3 + (-8) = -5$$

$$-10 - (-12) = -10 + 12 = 2$$

$$7 - 20 - 18 - (-3) = 7 + (-20) + (-18) + (+3)$$

$$= -13 + (-18) + 3$$

$$= -31 + 3$$

$$= -28$$

MULTIPLYING NUMBERS

The product of two numbers having the same sign is a positive number.

The product of two numbers having unlike signs is a negative number.

Multiply:

$$(-7)(-6) = 42$$

$$9(-4) = -36$$

Evaluate:

$$(-3)^2 = (-3)(-3) = 9$$

DIVIDING NUMBERS

The quotient of two numbers having the same sign is a positive number.

The quotient of two numbers having unlike signs is a negative number.

Divide:

$$-100 \div (-10) = 10$$

$$\frac{14}{-2} = -7, \quad \frac{0}{-3} = 0, \quad \frac{22}{0} \text{ is undefined.}$$

ORDER OF OPERATIONS

1. Do all operations within grouping symbols such as parentheses or brackets.

2. Evaluate any expressions with exponents.

3. Multiply or divide in order from left to right.

4. Add or subtract in order from left to right.

Simplify:

$$3 + 2 \cdot (-5) = 3 + (-10) \quad \text{Multiply.}$$

$$= -7 \quad \text{Add.}$$

$$\frac{-2(5 - 7)}{-7 + |-3|} = \frac{-2(-2)}{-7 + 3}$$

$$= \frac{4}{-4}$$

$$= -1$$

$$-3 + 5[2^3 - (12 - 9)] = -3 + 5[2^3 - 3] \quad \text{Subtract.}$$

$$= -3 + 5[8 - 3] \quad \text{Evaluate exponent.}$$

$$= -3 + 5(5) \quad \text{Subtract.}$$

$$= -3 + 25 \quad \text{Multiply.}$$

$$= 22 \quad \text{Add.}$$

Name_____ Section_____ Date _____

Chapter 2 Review

The map below shows selected cities and their normal high and low temperatures. Use this map as indicated throughout the review to fill-in each missing temperature in the picture.

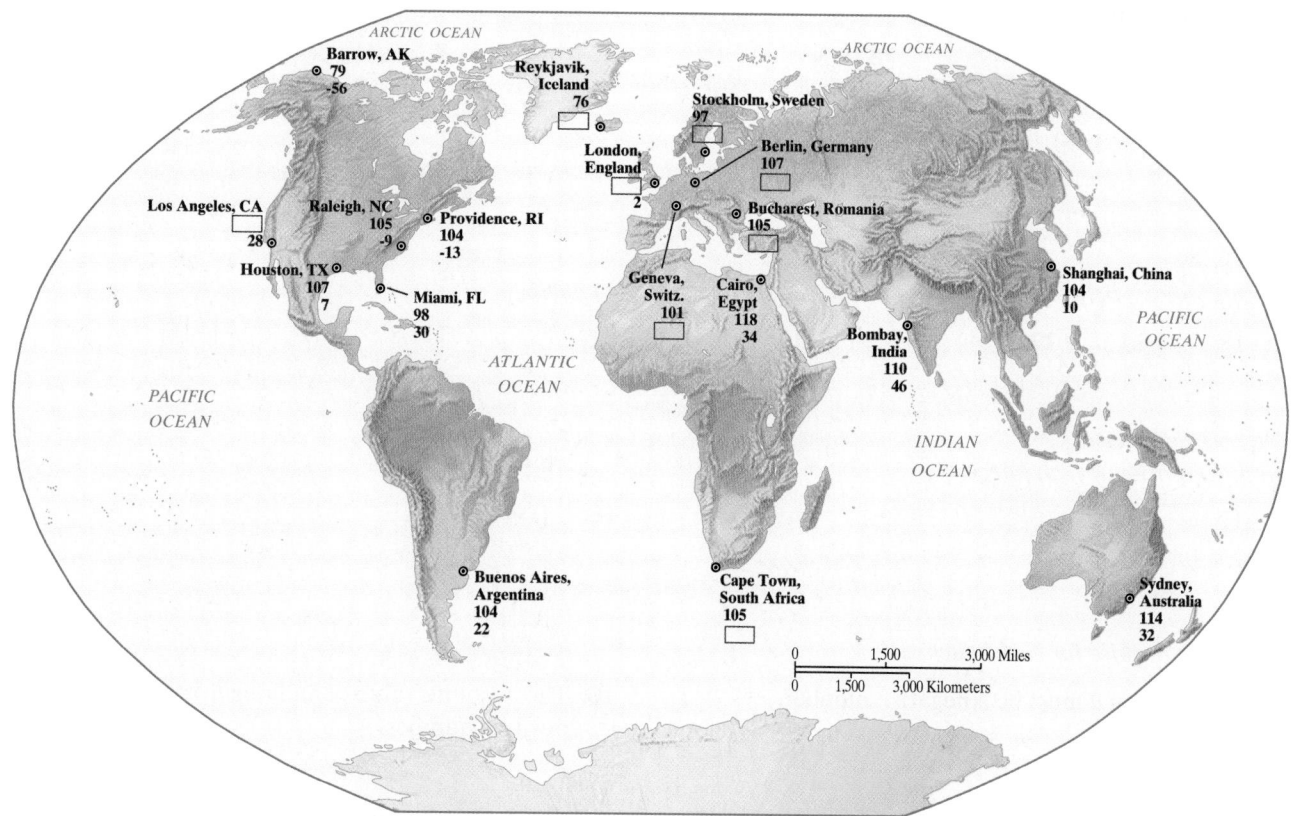

Source: World Almanac 2003; temperatures are for a 30-year period ending in 1990.

Extreme High and Low Temperatures for Selected Locations (in degrees Fahrenheit)					
	Max	Min		Max	Min
Berlin, Germany	107		Barrow, Alaska	79	−56
Raleigh, NC	105	−9	London, England		2
Houston, TX	107	7	Cairo, Egypt	118	34
Miami, FL	98	30	Sydney, Australia	114	32
Los Angeles, CA		28	Shanghai, China	104	10
Bucharest, Romania	105		Reykjavik, Iceland	76	
Geneva, Switzerland	101		Capetown, South Africa	105	
Providence, RI	104	−13	Buenos Aires, Argentina	104	22
Stockholm, Sweden	97		Bombay, India	110	46

(2.1) *Represent each quantity by an integer.*

1. A gold miner is working 1435 feet down in a mine.

2. A mountain peak is 7562 meters above sea level.

Graph each integer in the list on the same number line.

3. $-2, -5, 0, 5$

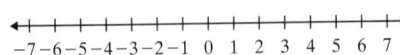

4. $-7, -1, 0, 7$

$$\longleftrightarrow \begin{array}{cccccccccccccccc} & & & & & & & & & & & & & & \\ -7 & -6 & -5 & -4 & -3 & -2 & -1 & 0 & 1 & 2 & 3 & 4 & 5 & 6 & 7 \end{array} \longrightarrow$$

Simplify.

5. $|-12|$

6. $|0|$

7. $-|6|$

8. $-(-9)$

9. $-|-9|$

10. $-(-2)$

Insert $<$ or $>$ between each pair of integers to make a true statement.

11. $-18 \quad -20$

12. $-5 \quad 5$

13. $|-123| \quad -|-198|$

14. $8 - |-12| \quad -|-16|$

Find the opposite of each integer.

15. -12

16. $-(-3)$

Answer true or false for each statement.

17. If $a < b$, then a must be a negative number.

18. The absolute value of an integer is always 0 or a positive number.

19. A negative number is always less than a positive number.

20. If a is a negative number, then $-a$ is a positive number.

(2.2) *Add.*

21. $5 + (-3)$

22. $18 + (-4)$

23. $-12 + 16$

24. $-23 + 40$

25. $-8 + (-15)$

26. $-5 + (-17)$

27. $-24 + 3$

28. $-89 + 19$

29. $15 + (-15)$

30. $-24 + 24$

31. $-43 + (-108)$

32. $-100 + (-506)$

33. During the 2002 Kellogg-Keebler Classic, the winner, Annika Sorenstam had scores of -9, -5, and -7 for 3 rounds of golf. Find her total score (*Source*: Ladies Professional Golf Association)

34. During the 2002 LPGA Corning Classic, Joanne Morley had scores of -1, 3, -5, and 3. Find her total score. (*Source*: Ladies Professional Golf Association)

35. The temperature at 5 a.m. on a day in January was $-15°C$. By 6 a.m. the temperature had fallen 5 degrees. Find the temperature at 6 a.m.

36. A diver starts out at 127 feet below the surface and then swims downward another 23 feet. Find the diver's current depth.

For Exercises 37 and 38, use the map at the beginning of this review.

37. The high temperature for London, England is 155 degrees greater than the low temperature for Barrow, Alaska. Find the high temperature for London.

38. The high temperature for Los Angeles, California is 125 degrees greater than the low temperature for Providence, Rhode Island. Find the high temperature for Los Angeles.

(2.3) *Subtract.*

39. $12 - 4$

40. $-12 - 4$

41. $8 - 19$

42. $-8 - 19$

43. $7 - (-13)$

44. $-6 - (-14)$

45. $16 - 16$

46. $-16 - 16$

47. $-12 - (-12)$

48. $|-5| - |-12|$

49. $-(-5) - 12 + (-3)$

50. $-8 + |-12| - 10 - |-3|$

Solve.

51. Josh Weidner has $142 in his checking account. He writes a check for $125, makes a deposit for $43, and then writes another check for $85. Represent the balance in his account by an integer.

52. If the elevation of Lake Superior is 600 feet above sea level and the elevation of the Caspian Sea is 92 feet below sea level, find the difference of the elevations.

For exercises 53 and 54, use the map at the beginning of this review.

53. The low temperature for Reykjavik is 35 degrees less than the low temperature for Sydney, Australia Find the low temperature for Reykjavik.

54. The low temperature for Berlin, Germany is 14 degrees less than the low temperature for Shanghai, China. Find the low temperature for Berlin.

Answer true or false for each statement.

55. $|-5| - |-6| = 5 - 6$

56. $|-5 - (-6)| = 5 + 6$

57. If $b > a$, then $b - a$ is a positive number.

58. If $b < a$, then $b - a$ is a negative number.

(2.4) *Multiply.*

59. $-3(-7)$

60. $-6(3)$

61. $-4(16)$

62. $-5(-12)$

63. $(-5)^2$

64. $(-1)^5$

65. $12(-3)(0)$

66. $-1(6)(2)(-2)$

Divide.

67. $-15 \div 3$

68. $\dfrac{-24}{-8}$

69. $\dfrac{0}{-3}$

70. $\dfrac{-46}{0}$

71. $\dfrac{100}{-5}$

72. $\dfrac{-72}{8}$

73. $\dfrac{-38}{-1}$

74. $\dfrac{45}{-9}$

75. A football team lost 5 yards on each of two consecutive plays. Represent the total loss by a product of integers, and find the product.

76. A race horse bettor loss $50 on each of four consecutive races. Represent the total loss by a product of integers, and find the product.

For exercises 77 through 80, use the map at the beginning of this review.

77. The low temperature for Bucharest, Romania is 2 times the low temperature for Raleigh, North Carolina. Find the low temperature for Bucharest.

78. The low temperature for Geneva, Switzerland is the same as the low temperature for Miami, Florida divided by -10. Find the low temperature for Geneva.

79. The low temperature for Capetown, South Africa is the same as the low temperature for Barrow, Alaska divided by -2. Find the low temperature for Capetown.

80. The low temperature for Stockholm, Sweden, is 2 times the low temperature for Providence, Rhode Island. Find the low temperature for Stockholm.

(2.5) *Simplify.*

81. $(-7)^2$

82. -7^2

83. -2^5

84. $(-2)^5$

85. $5 - 8 + 3$

86. $-3 + 12 + (-7) - 10$

87. $-10 + 3 \cdot (-2)$

88. $5 - 10 \cdot (-3)$

89. $16 \cdot (-2) + 4$

90. $3 \cdot (-12) - 8$

91. $5 + 6 \div (-3)$

92. $-6 + (-10) \div (-2)$

93. $16 + (-3) \cdot 12 \div 4$

94. $(-12) + 25 \cdot 1 \div (-5)$

95. $4^3 - (8 - 3)^2$

96. $4^3 - 90$

97. $-(-4) \cdot |-3| - 5$

98. $|5 - 1|^2 \cdot (-5)$

99. $\dfrac{(-4)(-3) - (-2)(-1)}{-10 + 5}$

100. $\dfrac{4(12 - 18)}{-10 \div (-2 - 3)}$

Find the average of each list of numbers.

101. $-18, 25, -30, 7, 0, -2$

102. $-45, -40, -30, -25$

Evaluate each expression for $x = -2$ and $y = 1$.

103. $2x - y$

104. $y^2 + x^2$

105. $\dfrac{3x}{6}$

106. $\dfrac{5y - x}{-y}$

107. x^2

108. $-x^2$

109. $7 - x^2$

110. $100 - x^3$

Chapter 2 Test

Remember to check your answers and use the Chapter Test Prep Video to view solutions.

Simplify each expression.

1. $-5 + 8$

2. $18 - 24$

3. $5 \cdot (-20)$

4. $(-16) \div (-4)$

5. $(-18) + (-12)$

6. $-7 - (-19)$

7. $(-5) \cdot (-13)$

8. $\dfrac{-25}{-5}$

9. $|-25| + (-13)$

10. $14 - |-20|$

11. $|5| \cdot |-10|$

12. $\dfrac{|-10|}{-|-5|}$

13. $(-8) + 9 \div (-3)$

14. $-7 + (-32) - 12 + 5$

15. $(-5)^3 - 24 \div (-3)$

16. $(5 - 9)^2 \cdot (8 - 2)^3$

17. $-(-7)^2 \div 7 \cdot (-4)$

18. $3 - (8 - 2)^3$

Answers

1. _____

2. _____

3. _____

4. _____

5. _____

6. _____

7. _____

8. _____

9. _____

10. _____

11. _____

12. _____

13. _____

14. _____

15. _____

16. _____

17. _____

18. _____

19. $-6 + (-15) \div (-3)$

20. $\dfrac{4}{2} - \dfrac{8^2}{16}$

21. $\dfrac{-3(-2) + 12}{-1(-4 - 5)}$

22. $\dfrac{|25 - 30|^2}{2(-6) + 7}$

23. $5(-8) - [6 - (2 - 4)] + (12 - 16)^2$

24. $-2^3 - 2^2$

Evaluate each expression for $x = 0$, $y = -3$, and $z = 2$.

25. $3x + y$

26. $|y| + |x| + |z|$

27. $\dfrac{3z}{2y}$

28. $2y^3$

29. $10 - y^2$

30. $7x + 3y - 4z$

31. Mary Dunstan, a diver, starts at sea level and then makes 4 successive descents of 22 feet. After the descents, what is her elevation?

32. Aaron Hawn has $129 in his checking account. He writes a check for $79, withdraws $40 from an ATM, and then deposits $35. Represent the new balance in his account by an integer.

33. Mt. Washington in New Hampshire has an elevation of 6288 feet above sea level. The Romanche Gap in the Atlantic Ocean has an elevation of 25,354 feet below sea level. Represent the difference in elevation between these two points by an integer. (*Source:* National Geographic Society and Defense Mapping Agency)

34. Lake Baykal in Siberian Russia is the deepest lake in the world with a maximum depth of 5315 feet. The elevation of the lake's surface is 1495 feet above sea level. What is the elevation (with respect to sea level) of the deepest point in the lake? (*Source:* U.S. Geological Survey)

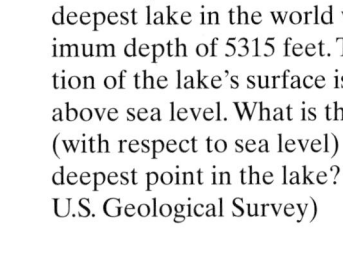

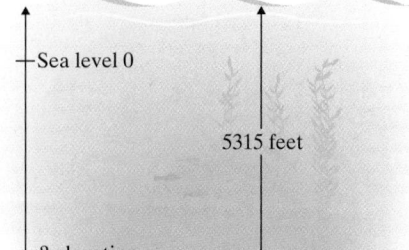

1495 feet above sea level

Sea level 0

5315 feet

? elevation

35. Find the average of $-12, -13, 0, 9$.

Chapter 2 Cumulative Review

Find the place value of the digit 4 in each whole number.

1. 48,761

2. 3408

3. 249

4. 694,298

5. 524,007,656

6. 267,401,818

7. Insert $<$ or $>$ to make a true statement.
   a. 5   50
   b. 101   0
   c. 29   27

8. Insert $<$ or $>$ to make a true statement
   a. 12   4
   b. 13   31
   c. 82   79

9. Add:
   $13 + 2 + 7 + 8 + 9$

10. Add: $11 + 3 + 9 + 16$

11. Subtract: $7826 - 505$
   Check by adding.

12. Subtract:
   $3285 - 272$
   Check by adding.

13. The radius of Venus is 6052 kilometers. The radius of Mercury is 3612 kilometers less than the radius of Venus. Find the radius of Mercury. (*Source:* National Space Science Data Center)

14. C.J. Dufour wants to buy a digital camera. She has $762 in her savings account. If the camera costs $237, how much money will she have in her account after buying the camera?

15. Round 568 to the nearest ten.

16. Round 568 to the nearest hundred.

17. Round each number to the nearest hundred to find an estimated difference.
   $$4725$$
   $$-2879$$

18. Round each number to the nearest thousand to find an estimated difference.
   $$8394$$
   $$-2913$$

19. Rewrite each using the distributive property.
   a. $3(4 + 5)$
   b. $10(6 + 8)$
   c. $2(7 + 3)$

20. Rewrite each using the distributive property.
   a. $5(2 + 12)$
   b. $9(3 + 6)$
   c. $4(8 + 1)$

21. Multiply: 631×125

22. Multiply: 299×104

23. Find each quotient. Check by multiplying.
   a. $42 \div 7$
   b. $\dfrac{81}{9}$
   c. $4\overline{)24}$

24. Find each quotient. Check by multiplying.
   a. $\dfrac{35}{5}$
   b. $64 \div 8$
   c. $4\overline{)48}$

Answers

1. _____
2. _____
3. _____
4. _____
5. _____
6. _____
7. a. _____
   b. _____
   c. _____
8. a. _____
   b. _____
   c. _____
9. _____
10. _____
11. _____
12. _____
13. _____
14. _____
15. _____
16. _____
17. _____
18. _____
19. a. _____
    b. _____
    c. _____
20. a. _____
    b. _____
    c. _____
21. _____
22. _____
23. a. _____
    b. _____
    c. _____
24. a. _____
    b. _____
    c. _____

25. Divide: $3705 \div 5$
Check by multiplying.

26. Divide: $3648 \div 8$
Check by multiplying.

27. How many boxes are needed to ship 56 pairs of Nikes to a shoe store in Texarkana if 9 pairs of shoe will fit in each shipping box?

28. Mrs. Mallory's first grade class is going to the zoo. She pays a total of $324 for 36 admission tickets. How much did each ticket cost?

Evaluate.

29. 8^2

30. 5^3

31. 7^1

32. 4^1

33. $5 \cdot 6^2$

34. $2^3 \cdot 7$

35. Simplify: $\dfrac{7 - 2 \cdot 3 + 3^2}{2^2 + 1}$

36. Simplify: $\dfrac{6^2 + 4 \cdot 4 + 2^3}{37 - 5^2}$

37. Evaluate $x + 7$ if x is 8.

38. Evaluate $5 + x$ if x is 9.

39. Simplify:
   a. $|-2|$
   b. $|5|$
   c. $|0|$

40. Simplify:
   a. $|4|$
   b. $|-7|$

25. _____

26. _____

27. _____

28. _____

29. _____

30. _____

31. _____

32. _____

33. _____

34. _____

35. _____

36. _____

37. _____

38. _____

39. a. _____

   b. _____

   c. _____

40. a. _____

   b. _____

41. Add: $-2 + 5$

42. Add: $8 + (-3)$

43. Evaluate $a - b$ for $a = 8$ and $b = -6$.

44. Evaluate $x - y$ for $x = -2$ and $y = -7$.

45. Multiply: $-7 \cdot 3$

46. Multiply: $5(-2)$

47. Multiply: $0 \cdot (-4)$

48. Multiply: $-6 \cdot 9$

49. Simplify: $3(4 - 7) + (-2) - 5$

50. Simplify: $4 - 8(7 - 3) - (-1)$

41. _____

42. _____

43. _____

44. _____

45. _____

46. _____

47. _____

48. _____

49. _____

50. _____

STUDY SKILLS REMINDER

Are you getting all the mathematics help that you need?

Remember that, in addition to your instructor, there are many places to get help with your mathematics course. For example, see which of the list below are available.

- This text has an accompanying video for every section in this text and also for the chapter tests.

- The back of this book contains answers to odd-numbered exercises and selected solutions.

- MathPro is available with this text. It is a tutorial software program with lessons corresponding to each section in the text.

- A student solutions manual is available that contains worked-out solutions to odd-numbered exercises as well as solutions to every exercise in the Chapter Pretests, Integrated Reviews, Chapter Reviews, Chapter Tests, and Cumulative Reviews.

- Don't forget to check with your instructor for other local resources available to you, such as a tutoring center.

Solving Equations and Problem Solving

Throughout this text, we have been making the transition from arithmetic to algebra. We have said that in algebra letters called variables represent numbers. Using variables is a very powerful method for solving problems that cannot be solved with arithmetic alone. This chapter introduces operations on algebraic expressions and solving variable equations.

Americans are in love with traveling the open roads. Throughout the country, roads and highways lead travelers on an exciting journey through every period of U.S. history. The *Santa Fe Trail Scenic and Historic Byway* in Colorado retraces the famous road taken by wagon trains of the Old West. The *Seward Highway* in Alaska is named for the U.S. Secretary of State who arranged the purchase of Alaska from Russia in 1867. The *Selma to Montgomery March Byway* in Alabama traces the historic route taken by Dr. Martin Luther King, Jr. in 1965. Whether by scenic byways or interstate highways, touring the country by automobile remains a favorite pastime of vacationers today. Before you venture out on the open road, check out Exercises 107–110 on page 198 and see how we can use equations to relate the distance, rate, and time of an automobile trip.

1. _____

2. _____

3. _____

4. _____

5. _____

6. _____

7. _____

8. _____

9. _____

10. _____

11. _____

12. _____

13. _____

14. _____

15. _____

16. _____

17. _____

18. _____

19. _____

20. _____

Name _____ Section _____ Date_____

Chapter 3 Pretest

Simplify.

1. $9x - 4 + 6x + 8$

2. $3(2x - 1) - (x - 8)$

Multiply.

3. $8(7b)$

4. $-5(2y - 7)$

△ **5.** Find the perimeter of the triangle.

△ **6.** Find the area of the rectangle.

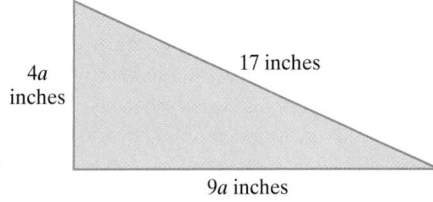

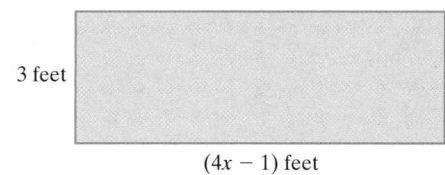

7. Determine whether -4 is a solution of the equation $2y - 3 = -5$.

Solve.

8. $-7 = x + 5$

9. $9n + 3 - 8n = 8 - 14$

10. $-6x = 42$

11. $2m - 9m = -77$

Translate each phrase to an algebraic expression. Let x be the unknown number.

12. Twice the difference of a number and 12

13. The product of a number and 3

Solve.

14. $18 - 3x = -9$

15. $8a - 5 = 2a + 7$

16. $4(x + 1) = 9x - 1$

17. $-2(3x + 4) - 10 = 0$

Translate each sentence into an equation. Use x to represent "a number."

18. The quotient of 54 and -6 is -9.

19. Five less than a number is 12.

Solve.

20. Sixty-eight is 5 more than 9 times a number. Find the number.

3.1   Simplifying Algebraic Expressions

OBJECTIVES

A Use properties of numbers to combine like terms.

B Use properties of numbers to multiply expressions.

C Simplify expressions by multiplying and then combining like terms.

D Find the perimeter and area of figures.

SSM
TUTOR CENTER   SG   CD & VIDEO   MATH PRO   WEB

Just as we can add, subtract, multiply, and divide numbers, we can add, subtract, multiply, and divide algebraic expressions. In previous sections we evaluated algebraic expressions like $x + 3$, $4x$, and $x + 2y$ for particular values of the variables. In this section, we explore working with variable expressions without evaluating them. We begin with a definition of a term.

A Combining Like Terms

The addends of an algebraic expression are called the **terms** of the expression.

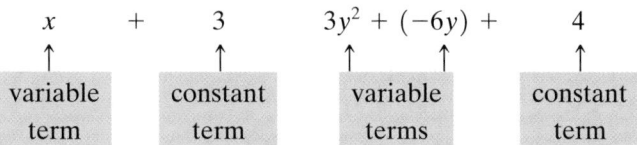

A term that is only a number has a special name. It is called a **constant term**, or simply a **constant**. A term that contains a variable is called a **variable term**.

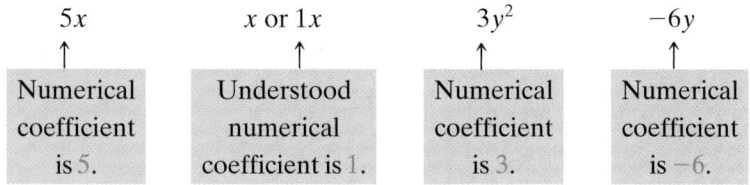

The number factor of a variable term is called the **numerical coefficient**. A numerical coefficient of 1 is usually not written.

$5x$	x or $1x$	$3y^2$	$-6y$
Numerical coefficient is 5.	Understood numerical coefficient is 1.	Numerical coefficient is 3.	Numerical coefficient is -6.

Terms that contain the same variables raised to the same exponents are called like terms.

Like Terms	Unlike Terms
$3x, \dfrac{1}{2}x$	$5x, x^2$
$-6y, 2y, y$	$7x, 7y$
$-5, 4$	$4y, 4$

Try the Concept Check in the margin.

A sum or difference of like terms can be simplified using the **distributive property**. Recall from Chapter 1 that the distributive property says that multiplication distributes over addition (and subtraction). For example,

$$(2 + 7)5 = 2 \cdot 5 + 7 \cdot 5$$

Using variables, we have

$$(a + b)c = a \cdot c + b \cdot c$$

If we write this from right to left, we can state the distributive property as follows.

Concept Check

True or false? The terms $-7xz^2$ and $3z^2x$ are like terms. Explain.

Answer

Concept Check: True.

Distributive Property

If a, b, and c are numbers, then

$$ac + bc = (a + b)c$$

Also,

$$ac - bc = (a - b)c$$

By the distributive property then,
$$7x + 2x = (7 + 2)x$$
$$= 9x$$

We have combined like terms and the expression $7x + 2x$ simplifies to $9x$.

Practice Problem 1

Combine like terms.

a. $8m - 11m$

b. $5a + a$

c. $-y^2 + 3y^2 + 7$

EXAMPLE 1 Simplify by combining like terms.

a. $3x + 2x$

b. $y - 7y$

c. $3x^2 + 5x^2 - 2$

Solution: Add or subtract like terms.

a. $3x + 2x = (3 + 2)x$
$$= 5x$$

b. $y - 7y = 1y - 7y$
$$= (1 - 7)y$$
$$= -6y$$

c. $3x^2 + 5x^2 - 2 = (3 + 5)x^2 - 2$
$$= 8x^2 - 2$$

The commutative and associative properties of addition and multiplication can also help us simplify expressions. We presented these properties in Sections 1.2 and 1.5 and state them again using variables.

Properties of Addition and Multiplication

If a, b, and c are numbers, then

$$a + b = b + a \qquad \text{Commutative property of addition}$$
$$a \cdot b = b \cdot a \qquad \text{Commutative property of multiplication}$$

That is, the **order** of adding or multiplying two numbers can be changed without changing their sum or product.

$$(a + b) + c = a + (b + c) \qquad \text{Associative property of addition}$$
$$(a \cdot b) \cdot c = a \cdot (b \cdot c) \qquad \text{Associative property of multiplication}$$

That is, the **grouping** of numbers in addition or multiplication can be changed without changing their sum or product.

Answers

1. a. $-3m$ **b.** $6a$ **c.** $2y^2 + 7$

Examples of these properties are

$$2 + 3 = 3 + 2$$ Commutative property of addition
$$7 \cdot 9 = 9 \cdot 7$$ Commutative property of multiplication
$$(1 + 8) + 10 = 1 + (8 + 10)$$ Associative property of addition
$$(4 \cdot 2) \cdot 3 = 4 \cdot (2 \cdot 3)$$ Associative property of addition

EXAMPLE 2   Simplify: $2y - 6 + 4y + 8$

Solution: We begin by writing subtraction as the opposite of addition.

$$
\begin{aligned}
2y - 6 + 4y + 8 &= 2y + (-6) + 4y + 8 && \text{Apply the commutative} \\
&= 2y + 4y + (-6) + 8 && \text{property of addition.} \\
&= (2 + 4)y + (-6) + 8 && \text{Apply the distributive property.} \\
&= 6y + 2 && \text{Simplify.}
\end{aligned}
$$

EXAMPLES   Simplify each expression by combining like terms.

3. $6x + 2x - 5 = 8x - 5$

4. $4x + 2 - 5x + 3 = 4x + 2 + (-5x) + 3$
$$
\begin{aligned}
&= 4x + (-5x) + 2 + 3 \\
&= -1x + 5 \quad \text{or} \quad -x + 5
\end{aligned}
$$

5. $2x - 5 + 3y + 4x - 10y + 11$
$$
\begin{aligned}
&= 2x + (-5) + 3y + 4x + (-10y) + 11 \\
&= 2x + 4x + 3y + (-10y) + (-5) + 11 \\
&= 6x - 7y + 6
\end{aligned}
$$

As we practice combining like terms, keep in mind that some of the steps may be performed mentally.

B   Multiplying Expressions

We can also use properties of numbers to multiply expressions such as $3(2x)$. By the associative property of multiplication, we can write the product $3(2x)$ as $(3 \cdot 2)x$, which simplifies to $6x$.

EXAMPLES   Multiply.

6. $\quad 5(3y) = (5 \cdot 3)y$ Apply the associative property of multiplication.
$$\qquad\quad = 15y \qquad \text{Multiply.}$$

7. $-2(4x) = (-2 \cdot 4)x$ Apply the associative property of multiplication.
$$\qquad\quad = -8x \qquad \text{Multiply.}$$

We can use the distributive property to combine like terms, which we have done, and also to multiply expressions such as $2(3 + x)$. By the distributive property, we have

$$
\begin{aligned}
2(3 + x) &= 2 \cdot 3 + 2 \cdot x && \text{Apply the distributive property.} \\
&= 6 + 2x && \text{Multiply.}
\end{aligned}
$$

Practice Problem 2

Simplify: $8m + 5 + m - 4$

Practice Problems 3–5

Simplify each expression by combining like terms.

3. $7y + 11y - 8$
4. $2y - 6 + y + 7$
5. $-9y + 2 - 4y - 8x + 12 - x$

Practice Problems 6–7

Multiply.

6. $7(8a)$
7. $-5(9x)$

Answers

2. $9m + 1$ **3.** $18y - 8$ **4.** $3y + 1$
5. $-13y - 9x + 14$ **6.** $56a$ **7.** $-45x$

Practice Problem 8

Use the distributive property to multiply: $7(y + 2)$

EXAMPLE 8   Use the distributive property to multiply: $6(x + 4)$

Solution: By the distributive property,

$$6(x + 4) = 6 \cdot x + 6 \cdot 4 \qquad \text{Apply the distributive property.}$$
$$= 6x + 24 \qquad \text{Multiply.}$$

Try the Concept Check in the margin.

Concept Check

What's wrong with the following?
$8(a - b) = 8a - b$

EXAMPLE 9   Multiply: $-3(5a + 2)$

Solution: By the distributive property,

$$-3(5a + 2) = -3(5a) + (-3)(2) \qquad \text{Apply the distributive property.}$$
$$= (-3 \cdot 5)a + (-6) \qquad \text{Use the associative property and multiply.}$$
$$= -15a - 6 \qquad \text{Multiply.}$$

Practice Problem 9

Multiply: $4(7a - 5)$

To simplify expressions containing parentheses, we first use the distributive property and multiply.

Practice Problem 10

Multiply: $6(5 - y)$

EXAMPLE 10   Multiply: $8(x - 4)$

Solution:

$$8(x - 4) = 8 \cdot x - 8 \cdot 4$$
$$= 8x - 32$$

C   **Simplifying Expressions**

Next we will **simplify** expressions by first using the distributive property to multiply and then **combining** any like terms.

Practice Problem 11

Simplify: $5(y - 3) - 8 + y$

EXAMPLE 11   Simplify: $2(3 + x) - 15$

Solution: First we use the distributive property to remove parentheses.

$$2(3 + x) - 15 = 2 \cdot 3 + 2 \cdot x + (-15) \qquad \text{Apply the distributive property.}$$
$$= 6 + 2x + (-15) \qquad \text{Multiply.}$$
$$= 2x + (-9) \quad \text{or} \quad 2x - 9 \qquad \text{Combine like terms.}$$

> **Helpful Hint**
>
> 2 is *not* distributed to the -15 since -15 is not within the parentheses.

Practice Problem 12

Simplify: $5(2x - 3) + 7(x - 1)$

Answers

8. $7y + 14$ **9.** $28a - 20$ **10.** $30 - 6y$
11. $6y - 23$ **12.** $17x - 22$

Concept Check: Did not distribute the 8;
$8(a - b) = 8a - 8b$

EXAMPLE 12   Simplify: $-2(x - 5) + 4(2x + 2)$

Solution: First we use the distributive property to remove parentheses.

$$-2(x - 5) + 4(2x + 2) = -2 \cdot x + (-2)(-5) + 4 \cdot 2x + 4 \cdot 2$$
$$= -2x + 10 + 8x + 8$$
$$= 6x + 18$$

EXAMPLE 13 Simplify: $-(x + 4) + 5x + 16$

Solution: The expression $-(x + 4)$ means $-1(x + 4)$.

$$-(x + 4) + 5x + 16 = -1(x + 4) + 5x + 16$$

$$= -1 \cdot x + (-1)(4) + 5x + 16 \quad \text{Apply the distributive property.}$$

$$= -x + (-4) + 5x + 16 \quad \text{Multiply.}$$

$$= 4x + 12 \quad \text{Combine like terms.}$$

⬤

Practice Problem 13

Simplify: $-(y + 1) + 3y - 12$

D Finding Perimeter and Area

△ **EXAMPLE 14** Find the perimeter of the triangle.

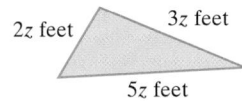
2z feet   3z feet
5z feet

Solution: Recall that the perimeter of a figure is the distance around the figure. To find the perimeter, then, we find the sum of the lengths of the sides.

$$\text{perimeter} = 2z + 3z + 5z$$
$$= 10z$$

> **Helpful Hint**
>
> Don't forget to insert proper units.

The perimeter is $10z$ feet.

⬤

Practice Problem 14

Find the perimeter of the square.

2x
centimeters

△ **EXAMPLE 15** Find the area of the rectangular deck.

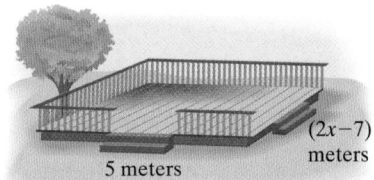
(2x−7) meters
5 meters

Solution: Recall how to find the area of a rectangle.

$$A = \text{length} \cdot \text{width}$$

$$= 5(2x - 7) \quad \text{Let length} = 5 \text{ and width} = (2x - 7).$$

$$= 10x - 35 \quad \text{Multiply.}$$

The area is $(10x - 35)$ *square* meters.

⬤

> **Helpful Hint:**
>
> Don't forget …
>
> | Area: | Perimeter: |
> | --- | --- |
> | • surface enclosed | • distance around |
> | • measured in square units | • measured in units |

Practice Problem 15

Find the area of the rectangular garden.

3 yards
(12y + 9) yards

Answers

13. $2y - 13$ **14.** $8x$ centimeters
15. $(36y + 27)$ square yards

Focus on **History**

THE EQUAL SIGN

The most important symbol in an equation is the equal sign, =. It indicates that a statement is, in fact, an equation and not just an algebraic expression. However, the equal sign has not always been around. In fact, the = symbol has been used only since 1557 when it first appeared in a book called *The Whetstone of Witte* by English physician and mathematician Robert Recorde. Recorde's original symbol used much longer lines, =====, and can be seen in this page from *The Whetstone of Witte*.

In his book, Recorde gave an interesting reason for his choice of symbol. He explained, "I will sette as I doe often in woorke use, a paire of parralles, or Gemowe lines of one lengthe, thus: =====, bicause noe 2, thynges, can be moare equalle."

Prior to the appearance of the equal sign, authors of mathematics texts often used words, such as "aequales" or "gleich" to indicate equality. These words continued to be used regularly until the use of the = symbol became widely accepted in the 1700s.

Name _____ Section _____ Date _____

Mental Math

Find the numerical coefficient of each variable term.

1. $5y$

2. $-2z$

3. z

4. $3xy^2$

5. $11a$

6. $-x$

EXERCISE SET 3.1

A *Simplify each expression by combining like terms. See Examples 1 through 5.*

 1. $3x + 5x$

2. $8y + 3y$

3. $5n - 9n$

4. $7z - 10z$

 5. $4c + c - 7c$

6. $5b - 8b - b$

7. $5x - 7x + x - 3x$

8. $8y + y - 2y - y$

9. $4a + 3a + 6a - 8$

10. $5b - 4b + b - 15$

B *Multiply. See Examples 6 and 7.*

11. $6(5x)$

12. $4(4x)$

13. $-2(11y)$

14. $-3(21z)$

15. $12(6a)$

16. $9(7b)$

Multiply. See Examples 8 through 10.

17. $2(y + 2)$

18. $3(x + 1)$

19. $5(a - 8)$

20. $4(y - 6)$

21. $-4(3x + 7)$

22. $-8(8y + 10)$

C *Simplify each expression. First use the distributive property to multiply and remove parentheses. See Examples 11 through 13.*

23. $2(x + 4) + 7$

24. $5(6 - y) - 2$

25. $-4(6n - 5) + 3n$

26. $-3(5 - 2b) - 4b$

27. $8 + 5(3c - 1)$

28. $10 + 4(6d - 2)$

29. $3 + 6(w + 2) + w$

30. $8z + 5(6 + z) + 20$

31. $2(3x + 1) + 5(x - 2)$

32. $3(5x - 2) + 2(3x + 1)$

33. $-(5x - 1) - 10$

34. $-(2y - 6) + 10$

35. $18y - 20y$

36. $x + 12x$

37. $z - 8z$

38. $12x - 8x$

39. $9d - 3c - d$

40. $8r + s - 7s$

41. $2y - 6 + 4y - 8$

42. $a + 4 - 7a - 5$

43. $5q + p - 6q - p$

44. $m - 8n + m + 8n$

45. $2(x + 1) + 20$

46. $5(x - 1) + 18$

47. $5(x - 7) - 8x$

48. $3(x + 2) - 11x$

49. $-5(z + 3) + 2z$

50. $-8(1 + v) + 6v$

51. $8 - x + 4x - 2 - 9x$

52. $5y - 4 + 9y - y + 15$

53. $-7(x + 5) + 5(2x + 1)$

54. $-2(x + 4) + 8(x - 1)$

55. $3r - 5r + 8 + r$

56. $6x - 4 + 2x - x + 3$

57. $-3(n - 1) - 4n$

58. $5(c + 2) + 7c$

59. $4(z - 3) + 5z - 2$

60. $8(m + 3) - 20 + m$

61. $6(2x - 1) - 12x$

62. $5(2a + 3) - 10a$

63. $-(4x - 5) + 5$

64. $-(7y - 2) + 6$

65. $-(4xy - 10) + 2(3xy + 5)$

66. $-(12ab - 10) + 5(3ab - 2)$

67. $3a + 4(a + 3)$

68. $b + 2(b - 5)$

69. $5y - 2(y - 1) + 3$

70. $3x - 4(x + 2) + 1$

D *Find the perimeter of each figure. See Example 14.*

 71.

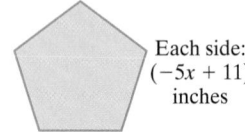

Each side:
$(-5x + 11)$
inches

72.

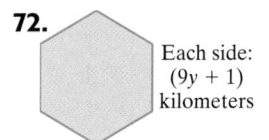

Each side:
$(9y + 1)$
kilometers

73.

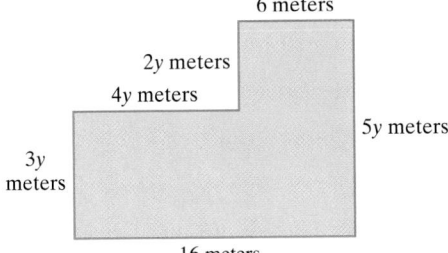
6 meters
2y meters
4y meters
5y meters
3y meters
16 meters

74.

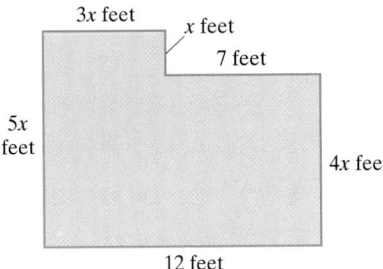
3x feet
x feet
7 feet
5x feet
4x feet
12 feet

75.

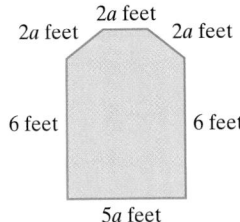
2a feet
2a feet
2a feet
6 feet
6 feet
5a feet

76.

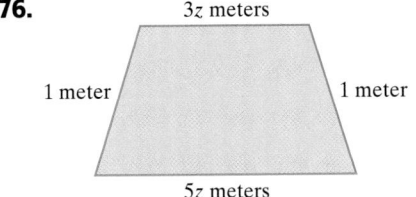
3z meters
1 meter
1 meter
5z meters

Find the area of each Square or rectangle. See Example 15.

77.

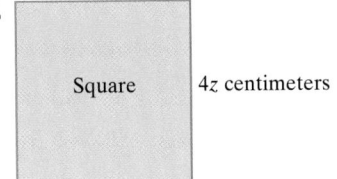
Square
4z centimeters

78.

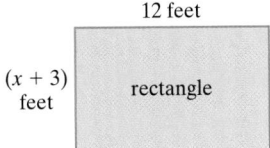
12 feet
(x + 3) feet
rectangle

79.

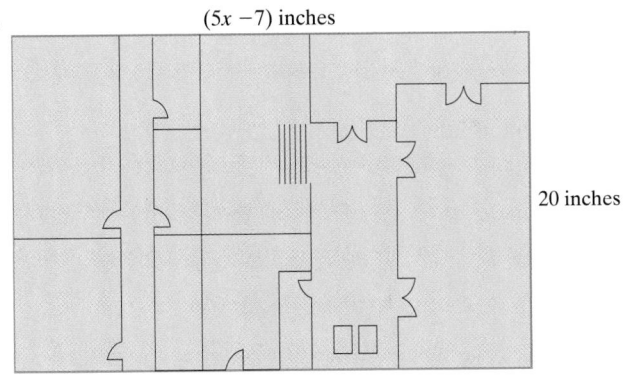
(5x − 7) inches
20 inches

80.

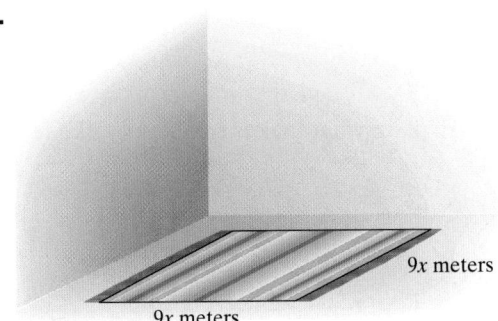
9x meters
9x meters

Review and Preview

Perform each indicated operation. See Sections 2.2 and 2.3.

81. $-13 + 10$

82. $-7 - (-4)$

83. $-4 - (-12)$

84. $-15 + 23$

85. $-4 + 4$

86. $8 + (-8)$

Combining Concepts

Simplify.

87. $9684q - 686 - 4860q + 12{,}960$

88. $76(268x + 592) - 2960$

89. If x is a whole number, which expression is the greatest: $2x$ or $5x$? Explain your answer.

90. If x is a whole number, which expression is the greatest: $-2x$ or $-5x$? Explain your answer.

Find the area of each figure.

△ **91.**

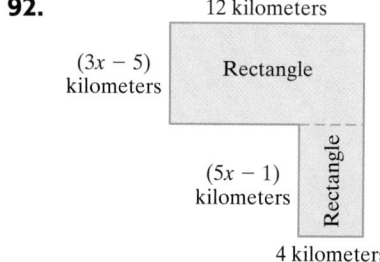

△ **92.**

93. Explain what makes two terms "like terms."

94. Explain how to combine like terms.

3.2 Solving Equations: The Addition Property

Frequently in this book, we have written statements like $7 + 4 = 11$ or area $=$ length $\cdot$ width. Each of these statements is called an **equation**. An equation is of the form

expression = expression

An equation can be labeled as

$$\underbrace{x + 7}_{\uparrow} = \overset{\downarrow}{10}$$
$$\text{left side} \quad \text{right side}$$

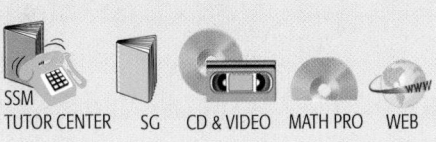

OBJECTIVES

- **A** Determine whether a given number is a solution of an equation.
- **B** Use the addition property of equality to solve equations.

SSM TUTOR CENTER   SG   CD & VIDEO   MATH PRO   WEB

A Determining Whether a Number is a Solution

When an equation contains a variable, deciding which values of the variable make an equation a true statement is called **solving** an equation for the variable. A **solution** of an equation is a value for the variable that makes an equation a true statement. For example, 2 is a solution of the equation $x + 5 = 7$, since replacing x with 2 results in the *true* statement $2 + 5 = 7$. Similarly, 3 is not a solution of $x + 5 = 7$, since replacing x with 3 results in the *false* statement $3 + 5 = 7$.

EXAMPLE 1   Determine whether 6 is a solution of the equation $4(x - 3) = 12$.

Solution: We replace x with 6 in the equation.

$$4(x - 3) = 12$$
$$\downarrow$$
$$4(6 - 3) \overset{?}{=} 12 \qquad \text{Replace } x \text{ with 6.}$$
$$4(3) \overset{?}{=} 12$$
$$12 = 12 \qquad \text{True.}$$

Since $12 = 12$ is a true statement, 6 *is* a solution of the equation. ●

EXAMPLE 2   Determine whether -1 is a solution of the equation $3y + 1 = 3$.

Solution:
$$3y + 1 = 3$$
$$3(-1) + 1 \overset{?}{=} 3$$
$$-3 + 1 \overset{?}{=} 3$$
$$-2 = 3 \qquad \text{False.}$$

Since $-2 = 3$ is a false statement, -1 *is not* a solution of the equation. ●

B Using the Addition Property to Solve Equations

To solve an equation, we will use properties of equality to write simpler equations, all equivalent to the original equation, until the final equation has the form

$$x = \textbf{number} \quad \text{or} \quad \textbf{number} = x$$

Practice Problem 1

Determine whether 4 is a solution of the equation $3(y - 6) = 6$.

Practice Problem 2

Determine whether -2 is a solution of the equation $-4x - 3 = 5$.

Answers

1. no   **2.** yes

Equivalent equations have the same solution. For example, $x + 3 = 5$ and $x = 2$ are equivalent equations because they both have the same solution, 2. Since we will be writing equivalent equations, the word "number" in the equations on the previous page represents the solution of the original equation. The first property of equality to help us write simpler equations is the **addition property of equality**.

Addition Property of Equality

Let a, b, and c represent numbers.
  If $a = b$, then

$$a + c = b + c \quad \text{and} \quad a - c = b - c$$

In other words, the same number may be added to or subtracted from both sides of an equation without changing the solution of the equation.

(Recall in Section 2.3 that we defined subtraction as addition of the first number and the opposite of the second number. Because of this, the Addition Property of Equality also allows us to subtract the same number from both sides.)

A good way to visualize a true equation is to picture a balanced scale. Since the scale is balanced, each side weighs the same amount. Similarly, in a true equation the expressions on each side have the same value. Picturing our balanced scale, if we add the same weight to each side, the scale remains balanced.

Practice Problem 3

Solve for y: $y - 5 = -3$

EXAMPLE 3   Solve for x: $x - 2 = 1$

Solution: To solve the equation for x, we need to rewrite the equation in the form $x =$ number. In other words, our goal is to get x alone on one side of the equation. To do so, we add 2 to both sides of the equation.

$$x - 2 = 1$$
$$x - 2 + 2 = 1 + 2 \qquad \text{Add 2 to both sides of the equation.}$$
$$x = 3 \qquad \text{Simplify.}$$

To check, we replace x with 3 in the *original* equation.

$$x - 2 = 1 \qquad \text{Original equation}$$
$$3 - 2 \overset{?}{=} 1 \qquad \text{Replace } x \text{ with 3.}$$
$$1 = 1 \qquad \text{True.}$$

Since $1 = 1$ is a true statement, 3 is the solution of the equation.   ●

Helpful Hint

Remember to check the solution in the *original* equation to see that it makes the equation a true statement.

Let's visualize how we used the addition property of equality to solve the equation in Example 3. Picture the original equation $x - 2 = 1$ as a balanced scale. The left side of the equation has the same value as the right side.

Answer

3. 2

If the same weight is added to each side of a scale, the scale remains balanced. Likewise, if the same number is added to each side of an equation, the left side continues to have the same value as the right side.

EXAMPLE 4   Solve: $-8 = x + 1$

Solution: To get x alone on one side of the equation, we subtract 1 from both sides of the equation.

$$-8 = x + 1$$
$$-8 - 1 = x + 1 - 1 \qquad \text{Subtract 1 from both sides.}$$
$$-8 + (-1) = x + 1 + (-1)$$
$$-9 = x \qquad \text{Simplify.}$$

Check: $-8 = x + 1$
$$-8 \stackrel{?}{=} -9 + 1 \qquad \text{Replace x with } -9$$
$$-8 = -8 \qquad \text{True.}$$

The solution is -9.

Practice Problem 4

Solve: $z + 9 = 1$

> Helpful Hint
>
> Remember that we can get the variable alone on either side of the equation. For example, the equations $-9 = x$ and $x = -9$ both have the solution of -9.

EXAMPLE 5   Solve: $7x = 6x + 4$

Solution: Subtract $6x$ from both sides so that variable terms will be on the same side of the equation.

$$7x = 6x + 4$$
$$7x - 6x = 6x + 4 - 6x \qquad \text{Subtract } 6x \text{ from both sides.}$$
$$1x = 4 \quad \text{or} \quad x = 4 \qquad \text{Simplify.}$$

Check to see that 4 is the solution.

Practice Problem 5

Solve: $10x = -2 + 9x$

If one or both sides of an equation can be simplified, do that first.

EXAMPLE 6   Solve: $y - 5 = -2 - 6$

Solution: First we simplify the right side of the equation.

$$y - 5 = -2 - 6$$
$$y - 5 = -8 \qquad \text{Combine like terms.}$$

Practice Problem 6

Solve: $x + 6 = 1 - 3$

Answers

4. -8 **5.** -2 **6.** -8

Concept Check

What number should be added to or subtracted from both sides of the equation in order to solve the equation $-3 = y + 2$?

Practice Problem 7

Solve: $-6y - 1 + 7y = 17$

Practice Problem 8

Solve: $13x = 4(3x - 1)$

Next we get y alone by adding 5 to both sides of the equation.

$$y - 5 + 5 = -8 + 5 \qquad \text{Add 5 to both sides.}$$
$$y = -3 \qquad \text{Simplify.}$$

Check to see that -3 is the solution.

Try the Concept Check in the margin.

EXAMPLE 7 Solve: $5x + 2 - 4x = 7 - 9$

Solution: First we simplify each side of the equation separately.

$$5x + 2 - 4x = 7 - 9$$
$$\underbrace{5x - 4x} + 2 = \underbrace{7 - 9}$$
$$1x + 2 = -2 \qquad \text{Combine like terms.}$$

To get x alone on the left side, we subtract 2 from both sides.

$$1x + 2 - 2 = -2 - 2 \qquad \text{Subtract 2 from both sides.}$$
$$1x = -4 \quad \text{or} \quad x = -4 \qquad \text{Simplify.}$$

Check to see that -4 is the solution.

EXAMPLE 8 Solve: $3(3x - 5) = 10x$

Solution: First we multiply on the left side to remove the parentheses.

$$3(3x - 5) = 10x$$
$$3 \cdot 3x - 3 \cdot 5 = 10x \qquad \text{Use the distributive property.}$$
$$9x - 15 = 10x$$

Now we subtract $9x$ from both sides.

$$9x - 15 - 9x = 10x - 9x \qquad \text{Subtract } 9x \text{ from both sides.}$$
$$-15 = 1x \quad \text{or} \quad x = -15 \qquad \text{Simplify.}$$

Answers

7. 18 **8.** -4

Concept Check: Subtract 2 from both sides.

EXERCISE SET 3.2

 A *Decide whether the given number is a solution of the given equation. See Examples 1 and 2.*

1. Is 10 a solution of $x - 8 = 2$?

2. Is 9 a solution of $y - 2 = 7$?

3. Is 6 a solution of $z + 8 = 14$?

4. Is 5 a solution of $n + 11 = 16$?

5. Is -5 a solution of $x + 12 = 7$?

6. Is -7 a solution of $a + 23 = 16$?

 7. Is 8 a solution of $7f = 64 - f$?

8. Is 3 a solution of $12 - k = 9$?

9. Is 0 a solution of $h - 8 = -8$?

10. Is -2 a solution of $3 + d = 0$?

11. Is 3 a solution of $4c + 2 - 3c = -1 + 6$?

12. Is 1 a solution of $2(b - 3) = 10$?

B *Solve and check the solution. See Examples 3 through 5.*

13. $a + 5 = 23$

14. $s - 7 = 15$

 15. $d - 9 = 17$

16. $f + 4 = -6$

17. $7 = y - 2$

18. $-10 = z - 15$

19. $-12 = x + 4$

20. $1 = y + 7$

21. $3x = 2x + 11$

22. $5y = 4y + 12$

23. $-4 + y = 2y$

24. $8x - 4 = 9x$

Solve and check the solution. See Examples 6 through 8.

25. $x - 3 = -1 + 4$

26. $x + 7 = 2 + 3$

27. $y + 1 = -3 + 4$

28. $y - 8 = -5 - 1$

29. $-7 + 10 = m - 5$

30. $1 - 8 = n + 2$

31. $-2 - 3 = -4 + x$

32. $7 - (-10) = x - 5$

33. $2(5x - 3) = 11x$

34. $6(3x + 1) = 19x$

35. $3y = 2(y + 12)$

36. $17x = 4(4x - 6)$

37. $-8x + 4 + 9x = -1 + 7$ **38.** $3x - 2x + 5 = 5$ **39.** $2 - 2 = 5x - 4x$

40. $11 - 15 = 6x - 4 - 5x$ **41.** $7x + 14 - 6x = -4 - 10$ **42.** $-10x + 11x + 5 = 9 - 5$

Solve each equation. See Examples 3 through 8.

43. $57 = y - 16$ **44.** $56 = -45 + x$ **45.** $67 = z + 67$ **46.** $66 = 8 + z$

47. $x + 5 = 4 - 3$ **48.** $x - 7 = -14 + 10$ **49.** $z - 23 = -88$ **50.** $x + 3 = -8$

51. $7a + 7 - 6a = 20$ **52.** $-8a - 11 + 9a = -5$ **53.** $-12 + x = -15$ **54.** $10 + x = 12$

55. $-8 - 9 = 3x + 5 - 2x$ **56.** $-7 + 10 = 4x - 6 - 3x$ **57.** $8(3x - 2) = 25x$ **58.** $9(4x + 5) = 37x$

59. $7x + 7 - 6x = 10$ **60.** $-3 + 5x - 4x = 13$ **61.** $50y = 7(7y + 4)$ **62.** $65y = 8(8y - 9)$

Review and Preview

The trumpeter swan is the largest waterfowl in the United States. Although it was thought to be nearly extinct at the beginning of the twentieth century, recent conservation efforts have been succeeding. Use the bar graph to answer Exercises 63–66. See Section 1.3.

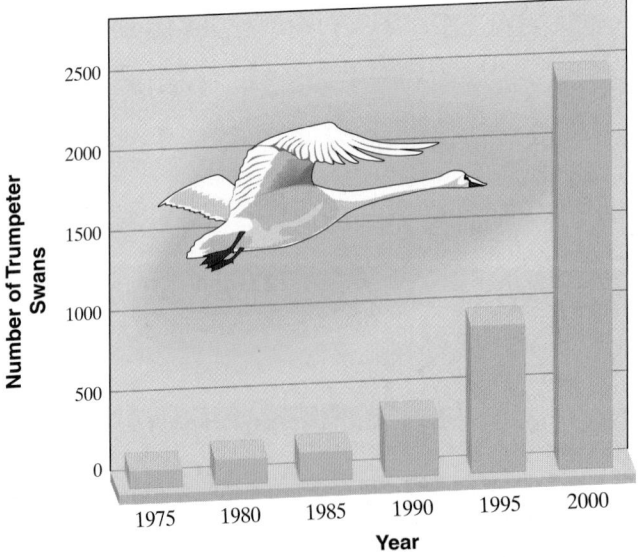

Great Lakes Trumpeter Swan Population

Source: *U.S. Fish and Wildlife Service*

63. Estimate the number of trumpeter swans in the Great Lakes region in 2000.

64. Estimate the number of trumpeter swans in the Great Lakes region in 1995.

65. How many more trumpeter swans in the Great Lakes region were there in 2000 than in 1985?

66. Describe any trends shown in this graph.

Simplify. See Section 1.7.

67. $\dfrac{8}{8}$

68. $\dfrac{11}{11}$

69. $\dfrac{-3}{-3}$

70. $\dfrac{-7}{-7}$

◆ Combining Concepts

71. In your own words, explain the addition property of equality.

72. Write an equation that can be solved using the addition property of equality.

73. Are the equations below equivalent? Why or why not?
$$x + 7 = 4 + (-9)$$
$$x + 7 = 5$$

74. Write 2 equivalent equations.

Solve.

75. $x - 76{,}862 = 86{,}102$

76. $-968 + 432 = 86y - 508 - 85y$

77. $5^3 = x + 4^4$

78. $7y - 10^4 = 8y$

79. $|-13| + 3^2 = 100y - |-20| - 99y$

80. $4(x - 11) + |90| - |-86| + 2^5 = 5x$

A football team's total offense T is found by adding the total passing yardage P to the total rushing yardage R:
$T = P + R.$

81. During the 2002 football season, the Green Bay Packers' total offense was 5560 yards. The Packers' rushing yardage for the season was 1933 yards. How many yards did the Packers gain by passing during the season? (*Source:* National Football League)

82. During the 2002 football season, the Denver Broncos' total offense was 6090 yards. The Broncos' passing yardage for the season was 3824 yards. How many yards did the Broncos gain by rushing during the season? (*Source:* National Football League)

In accounting, a company's annual net income, I, can be computed using the relation I = R − E, where R is the company's total revenues for a year and E is the company's total expenses for the year.

83. At the end of fiscal year 2002, Wal-Mart had a net income of $6671 million. During the year, Wal-Mart incurred a total of $213,141 million in expenses. What was Wal-Mart's total revenues for the year? (*Source:* Wal-Mart Stores, Inc.)

84. At the end of fiscal year 2001, Home Depot had a net income of $3044 million. During the year, Home Depot had a total of $50,559 million in expenses. What was Home Depot's total revenues for the year? (*Source:* Home Depot, Inc.)

 **Focus on Mathematical Connections**

MODELING EQUATION SOLVING WITH ADDITION AND SUBTRACTION

We can use positive counters ● and negative counters ● to help us model the equation-solving process. We also need to use an object that represents a variable. We use small slips of paper with the variable name written on them.

Recall that taking a ● and ● together creates a neutral or zero pair. After a neutral pair has been formed, it can be removed from or added to an equation model without changing the overall value. We also need to remember that we can add or remove the same number of positive or negative counters from both sides of an equation without changing the overall value.

We can represent the equation $x + 5 = 2$ as follows:

To get the variable by itself, we must remove 5 black counters from both sides of the model. Because there are only 2 counters on the right side, we must add 5 negative counters to both sides of the model! Then we can remove neutral pairs: 5 from the left side and 2 from the right side.

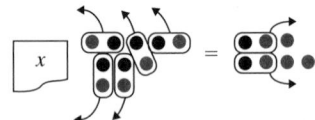

We are left with the following model, which represents the solution, $x = -3$.

Similarly, we can represent the equation $x - 4 = -6$ as follows:

To get the variable by itself, we must remove 4 red counters from both sides of the model

We are left with the following model, which represents the solution, $x = -2$.

CRITICAL THINKING

Use the counter model to solve each equation.

1. $x - 3 = -7$ **2.** $x - 1 = -9$

3. $x + 2 = 8$ **4.** $x + 4 = 5$

5. $x + 8 = 3$ **6.** $x - 5 = -1$

7. $x - 2 = 1$ **8.** $x - 5 = 10$

9. $x + 3 = -7$ **10.** $x + 8 = -2$

3.3 Solving Equations: The Multiplication Property

A Using the Multiplication Property to Solve Equations

Although the addition property of equality is a powerful tool for helping us solve equations, it cannot help us solve all types of equations. For example, it cannot help us solve an equation such as $2x = 6$. To solve this equation, we use a second property of equality called the **multiplication property of equality**.

Multiplication Property of Equality

Let a, b, and c represent numbers and let $c \neq 0$. If $a = b$, then

$$a \cdot c = b \cdot c \quad \text{and} \quad \frac{a}{c} = \frac{b}{c}$$

In other words, both sides of an equation may be multiplied or divided by the same nonzero number without changing the solution of the equation. (We will see in Chapter 4 how the Multiplication Property also allows us to divide both sides of an equation by the same nonzero number.)

Picturing again our balanced scale, if we multiply or divide the weight on each side by the same nonzero number, the scale (or equation) remains balanced.

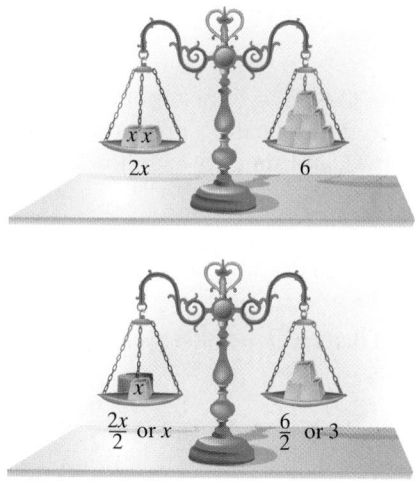

The multiplication property allows us to solve equations like

$$\frac{x}{5} = 2$$

Here, x is *divided* by 5. To get x alone, we use the multiplication property to *multiply* both sides by 5.

$$5 \cdot \frac{x}{5} = 5 \cdot 2$$

OBJECTIVES

A Use the multiplication property to solve equations.

B Translate word phrases into algebraic expressions.

SSM
TUTOR CENTER   SG   CD & VIDEO   MATH PRO   WEB

It can be shown that an expression such as $5 \cdot \dfrac{x}{5}$ is equivalent to $\dfrac{5}{5} \cdot x$, so we have that $5 \cdot \dfrac{x}{5} = 5 \cdot 2$ can be written as

$$\frac{5}{5} \cdot x = 5 \cdot 2$$

$$1 \cdot x = 10 \quad \text{or} \quad x = 10.$$

Practice Problem 1

Solve: $\dfrac{x}{-4} = 7$

EXAMPLE 1 Solve: $\dfrac{x}{3} = -2$

Solution: To get x alone, multiply both sides by 3.

$$\frac{x}{3} = -2$$

$$3 \cdot \frac{x}{3} = 3 \cdot (-2) \qquad \text{Multiply both sides by 3.}$$

$$\frac{3}{3} \cdot x = 3 \cdot (-2)$$

$$1x = -6 \quad \text{or} \quad x = -6 \qquad \text{Simplify.}$$

Check: Replace x with -6 in the original equation.

$$\frac{x}{3} = -2 \qquad \text{Original equation}$$

$$\frac{-6}{3} \stackrel{?}{=} -2 \qquad \text{Let } x = -6.$$

$$-2 = -2 \qquad \text{True.}$$

The solution is -6.

To solve an equation like $2x = 6$ for x, notice that 2 is *multiplied* by x. To get x alone, we use the multiplication property of equality to *divide* both sides of the equation by 2, and simplify as follows:

$$2x = 6$$

$$\frac{2 \cdot x}{2} = \frac{6}{2} \qquad \text{Divide both sides by 2.}$$

Then it can be shown that an expression such as $\dfrac{2 \cdot x}{2}$ is equivalent to $\dfrac{2}{2} \cdot x$, so

$$\frac{2 \cdot x}{2} = \frac{6}{2} \quad \text{can be written as} \quad \frac{2}{2} \cdot x = \frac{6}{2}$$

$$1 \cdot x = 3 \quad \text{or} \quad x = 3$$

Practice Problem 2

Solve: $3y = -18$

EXAMPLE 2 Solve: $-5x = 15$

Solution: To get x alone, divide both sides by -5.

$$-5x = 15 \qquad \text{Original equation}$$

$$\frac{-5x}{-5} = \frac{15}{-5} \qquad \text{Divide both sides by } -5.$$

$$\frac{-5}{-5} \cdot x = \frac{15}{-5}$$

$$1x = -3 \quad \text{or} \quad x = -3 \qquad \text{Simplify.}$$

Answers

1. -28 **2.** -6

To check, replace x with -3 in the original equation.

$$-5x = 15 \qquad \text{Original equation}$$
$$-5(-3) \stackrel{?}{=} 15 \qquad \text{Let } x = -3.$$
$$15 = 15 \qquad \text{True.}$$

The solution is -3.

EXAMPLE 3   Solve: $-8 = 2y$

Solution: To get y alone, divide both sides of the equation by 2.

$$-8 = 2y$$
$$\frac{-8}{2} = \frac{2y}{2} \qquad \text{Divide both sides by 2.}$$
$$\frac{-8}{2} = \frac{2}{2} \cdot y$$
$$-4 = 1y \quad \text{or} \quad y = -4$$

Check to see that -4 is the solution.

EXAMPLE 4   Solve: $-12x = -36$

Solution: Divide both sides of the equation by the coefficient of x, which is -12.

$$-12x = -36$$
$$\frac{-12x}{-12} = \frac{-36}{-12}$$
$$\frac{-12}{-12} \cdot x = \frac{-36}{-12}$$
$$x = 3$$

To check, replace x with 3 in the original equation.

$$-12x = -36$$
$$-12(3) \stackrel{?}{=} -36 \qquad \text{Let } x = 3.$$
$$-36 = -36 \qquad \text{True.}$$

Since $-36 = -36$ is a true statement, the solution is 3.

Try the Concept Check in the margin.

We often need to simplify one or both sides of an equation before applying the properties of equality to get the variable alone.

EXAMPLE 5   Solve: $3y - 7y = 12$

Solution: First combine like terms.

$$3y - 7y = 12$$
$$-4y = 12 \qquad \text{Combine like terms.}$$
$$\frac{-4y}{-4} = \frac{12}{-4} \qquad \text{Divide both sides by } -4.$$
$$y = -3 \qquad \text{Simplify.}$$

Practice Problem 3

Solve: $-16 = 8x$

Practice Problem 4

Solve: $-3y = -27$

Concept Check

Which operation is appropriate for solving each of the following equations, addition or division?

a. $12 = x - 3$
b. $12 = 3x$

Practice Problem 5

Solve: $10 = 2m - 4m$

Answers

3. -2 **4.** 9 **5.** -5

Concept Check:
a. Addition.
b. Division.

Check: Replace y with -3.

$$3y - 7y = 12$$

$$3(-3) - 7(-3) \stackrel{?}{=} 12$$
$$-9 + 21 \stackrel{?}{=} 12$$
$$12 = 12 \quad \text{True.}$$

The solution is -3.

Practice Problem 6

Solve: $-8 + 6 = \dfrac{a}{3}$

EXAMPLE 6   Solve: $\dfrac{z}{-4} = 11 - 5$

Solution: Simplify the right side of the equation first.

$$\frac{z}{-4} = 11 - 5$$

$$\frac{z}{-4} = 6$$

Next, to get z alone, multiply both sides by -4.

$$-4 \cdot \frac{z}{-4} = -4 \cdot 6 \qquad \text{Multiply both sides by } -4$$

$$\frac{-4}{-4} \cdot z = -4 \cdot 6$$

$$1z = -24 \quad \text{or} \quad z = -24$$

Check to see that -24 is the solution.

Practice Problem 7

Solve: $-4 - 10 = 4y - 5y$

EXAMPLE 7   Solve: $8x - 9x = 12 - 17$

Solution: First combine like terms on each side of the equation.

$$8x - 9x = 12 - 17$$
$$-x = -5$$

Recall that $-x$ means $-1x$ and divide both sides by -1.

$$\frac{-1x}{-1} = \frac{-5}{-1} \qquad \text{Divide both sides by } -1.$$

$$x = 5 \qquad \text{Simplify.}$$

Check to see that the solution is 5.

B   **Translating Word Phrases into Expressions**

A section later in this chapter contains a formal introduction to problem solving. To prepare for this section, let's review writing phrases as algebraic expressions using the following key words and phrases as a guide:

Addition	Subtraction	Multiplication	Division	Equal Sign
sum	difference	product	quotient	equals
plus	minus	times	divided by	gives
added to	subtracted from	multiply	into	is/was
more than	less than	twice	per	yields
increased by	decreased by	of		amounts to
total	less	double		is equal to

Answers

6. -6 **7.** 14

EXAMPLE 8   Write each phrase as an algebraic expression. Use x to represent "a number."

a. a number increased by -5

b. the product of -7 and a number

c. a number less 20

d. the quotient of -18 and a number

e. a number subtracted from -2

Solution:

a. In words: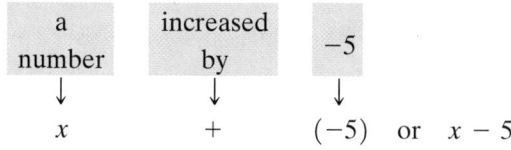

   Translate: x $+$ (-5) or $x - 5$

b. In words: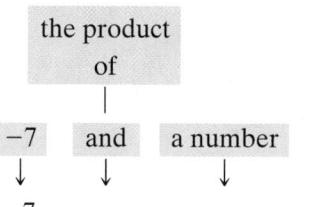

   Translate: -7 $\cdot$ x or $-7x$

c. In words: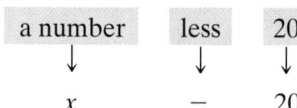

   Translate: x $-$ 20

d. In words: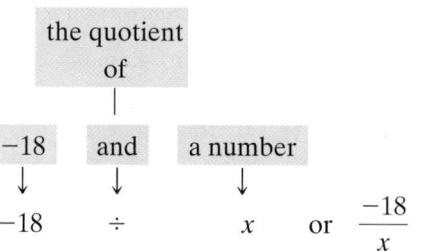

   Translate: -18 $\div$ x or $\dfrac{-18}{x}$ or $-\dfrac{18}{x}$

e. In words: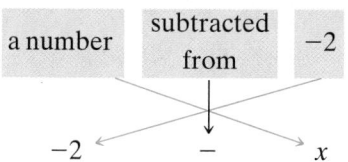

   Translate: -2 $-$ x

Helpful Hint

As we reviewed in Chapter 1, don't forget that order is important when subtracting. Notice the translation order of numbers and variables below.

Phrase	Translation
a number less 9	$x - 9$
a number subtracted from 9	$9 - x$

Practice Problem 9

Translate each phrase into an algebraic expression. Let x be the unknown number.

a. The product of 5 and a number, decreased by 25

b. Twice the sum of a number and 3

c. The quotient of 39 and twice a number

EXAMPLE 9 Write each phrase as an algebraic expression. Let x be the unknown number.

a. Twice a number, increased by -9

b. Three times the difference of a number and 11

c. The quotient of 5 times a number and 17

Solution:

a. In words:

Twice a number	increased by	-9
↓	↓	↓

Translate: $\quad 2x \qquad\qquad + \qquad\qquad (-9)$

b. In words:

	the difference of		
Three times	a number	and	11
↓	↓	↓	↓

Translate: $\quad 3 \qquad (x \qquad - \qquad 11)$

c. In words:

	The quotient of	
5 times a number	and	17
↓	↓	↓

$\qquad 5x \qquad\quad \div \qquad 17 \quad$ or $\quad \dfrac{5x}{17}$

STUDY SKILLS REMINDER

Are you making this class a priority?

Make sure your study habits are ones that will help you successfully complete this course. One useful habit is to make this class a priority by

- paying attention in class
- attending class
- arriving on time
- Completing assigned work on time

Answers

9. a. $5x - 25$ **b.** $2(x + 3)$ **c.** $39 \div 2x$ or $\dfrac{39}{2x}$

EXERCISE SET 3.3

A *Solve each equation. See Examples 1 through 4.*

1. $5x = 20$ **2.** $6y = 48$  **3.** $-3z = 12$ **4.** $-2x = 26$ **5.** $\dfrac{x}{7} = 1$

6. $\dfrac{y}{4} = 2$ **7.** $\dfrac{z}{-9} = 9$ **8.** $\dfrac{x}{-4} = 10$ **9.** $4y = 0$ **10.** $8x = -8$

11. $2z = -34$ **12.** $7y = 21$ **13.** $\dfrac{x}{-8} = -4$ **14.** $\dfrac{x}{-7} = -5$ **15.** $\dfrac{y}{-20} = 3$

16. $\dfrac{y}{-11} = 11$ **17.** $-3x = -15$ **18.** $-4z = -12$ **19.** $\dfrac{x}{-17} = 0$ **20.** $\dfrac{y}{-15} = 0$

Solve each equation. First combine any like terms on each side of the equation. See Examples 5 through 7.

21. $2w - 12w = 40$ **22.** $8y + y = 45$ **23.** $16 = 10t - 8t$ **24.** $100 = 15y + 5y$

 25. $2z = 12 - 14$ **26.** $-3x = 11 - 2$ **27.** $4 - 10 = \dfrac{z}{-3}$ **28.** $20 - 22 = \dfrac{z}{-4}$

29. $-3x - 3x = 50 - 2$ **30.** $5y - 9y = -14 + (-14)$ **31.** $\dfrac{x}{5} = -26 + 16$ **32.** $\dfrac{y}{3} = 32 - 52$

Solve each equation. See Examples 1 through 7.

33. $-10x = 10$ **34.** $2z = -14$ **35.** $5x = -35$ **36.** $-3y = -27$

37. $0 = \dfrac{x}{3}$ **38.** $0 = \dfrac{x}{7}$ **39.** $24 = t + 3t$ **40.** $35 = 3r + 2r$

41. $10z - 3z = -63$ **42.** $x + 3x = -48$ **43.** $3z - 10z = -63$ **44.** $6y - y = 20$

45. $12 = 13y - 10y$ **46.** $20 = y - 6y$ **47.** $-4x = 20 - (-4)$ **48.** $6x = 5 - 35$

49. $18 - 11 = \dfrac{x}{-5}$ **50.** $9 - 14 = \dfrac{x}{-12}$ **51.** $-20 - (-50) = \dfrac{x}{9}$ **52.** $-2 - 10 = \dfrac{z}{10}$

53. $10p - 11p = 25$ **54.** $5q - 6q = 21$ **55.** $6x - x = 4 - 14$ **56.** $3x + x = 12 - 4$

57. $10 = 7t - 12t$ **58.** $-30 = t + 9t$ **59.** $5 - 5 = 3x + 2x$ **60.** $-42 + 20 = -2x + 13x$

61. $4r - 9r = -20$ **62.** $11v + 3v = 28$ **63.** $\dfrac{x}{-4} = -1 - (-8)$ **64.** $\dfrac{y}{-6} = 6 - (-1)$

65. $3w - 12w = -27$ **66.** $5t + 3t = -72$ **67.** $-36 = 9u - 10u$ **68.** $-50 = 4y - 5y$

69. $23x - 25x = 7 - 9$ **70.** $8x - 6x = 12 - 22$

B   *Write each phrase as a variable expression. Use x to represent "a number." See Examples 8 and 9.*

71. The sum of -7 and a number

72. Negative eight plus a number

73. The product of -11 and a number

74. The quotient of -20 and a number, decreased by three

75. A number divided by -12

76. The quotient of six and a number, added to -80

77. Eleven subtracted from a number

78. A number subtracted from twelve

79. Negative ten decreased by 7 times a number

80. Twice a number decreased by thirty

81. The product of -13 and a number

82. Twice a number

83. The quotient of seventeen and a number, increased by -15

84. The difference of -9 times a number, and 1

85. Seven added to the product of 4 and a number

86. The product of 7 and a number, added to 100

87. Twice a number, decreased by 17

88. A number decreased by 14 and increased by 5 times the number

89. The product of −6 and the sum of a number and 15

90. Twice the sum of a number and −5

91. The quotient of 45 and the product of a number and −5

92. The quotient of ten times a number, and −4

Review and Preview

Evaluate each expression for x = −5. See Section 1.9.

93. $3x + 10$

94. $40x$

95. $\dfrac{x - 5}{2}$

96. $7x - 20$

97. $\dfrac{3x + 4}{x + 4}$

98. $\dfrac{2x - 4}{x - 2}$

Combining Concepts

99. Why does the multiplication property of equality not allow us to divide both sides of an equation by zero?

100. Is the equation $-x = 6$ solved for the variable? Explain why or why not.

Solve.

 101. $-25x = 900$

 102. $3y = -1259 - 3277$

 103. $\dfrac{y}{72} = -86 - (-1029)$

 104. $\dfrac{x}{-13} = 4^6 - 5^7$

105. $\dfrac{x}{-2} = 5^2 - |-10| - (-9)$

106. $\dfrac{y}{10} = (-8)^2 - |20| + (-2)^2$

For Exercises 107 through 110, the equation d = r · t describes the relationship between distance d in miles, rate r in miles per hour, and time t in hours.

107. The distance between New Orleans, Louisiana, and Kansas City, Missouri, by road, is approximately 780 miles. How long will it take to drive from New Orleans to Kansas City if the driver maintains a speed of 65 miles per hour? (*Source: 2003 World Almanac*)

108. The distance between Boston, Massachusetts, and Memphis, Tennessee, by road, is approximately 1320 miles. How long will it take to drive from Boston to Memphis if the driver maintains a speed of 60 miles per hour? (*Source: 2003 World Almanac*)

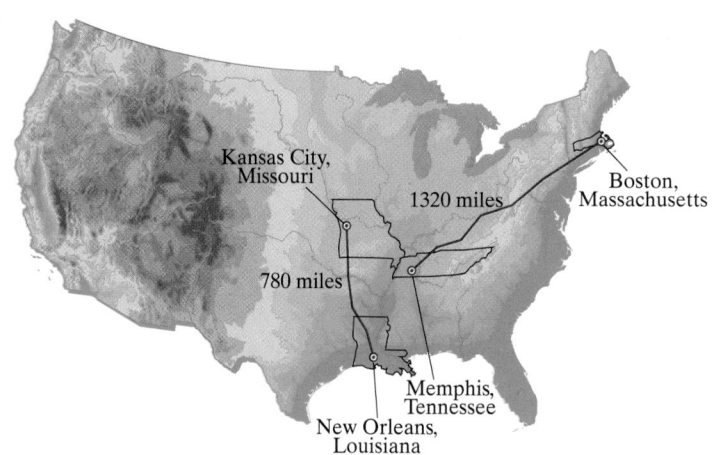

109. The distance between Toledo, Ohio, and Chicago, Illinois, by road, is approximately 232 miles. At what speed should a driver drive if he or she would like to make the trip in 4 hours? (*Source: 2003 World Almanac*)

110. The distance between St. Louis, Missouri, and Minneapolis, Minnesota, by road, is approximately 549 miles. If it took 9 hours to drive from St. Louis to Minneapolis, what was the driver's speed? (*Source: 2003 World Almanac*)

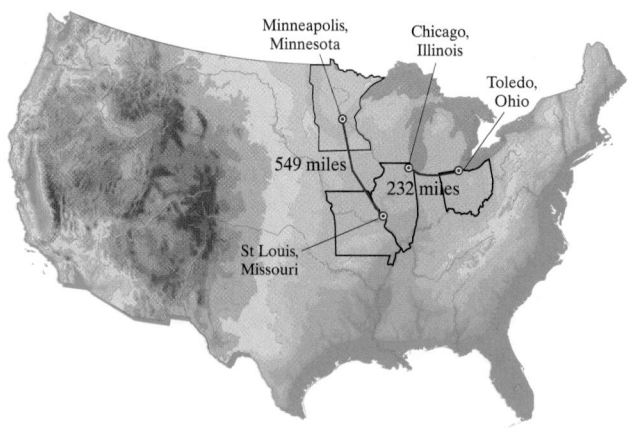

Name _____ Section _____ Date _____

Integrated Review—Expressions and Equations

Answers

1. _____

2. _____

3. _____

4. _____

5. _____

6. _____

7. _____

8. _____

9. _____

10. _____

11. _____

12. _____

13. _____

14. _____

15. _____

16. _____

17. _____

18. _____

19. _____

20. _____

21. _____

22. _____

Simplify each expression by combining like terms.

1. $7x + x$

2. $6y - 10y$

3. $2a + 5a - 9a - 2$

4. $6a - 12 - a - 14$

Multiply and simplify if possible.

5. $-2(4x + 7)$

6. $-3(2x - 10)$

7. $5(y + 2) - 20$

8. $12x + 3(x - 6) - 13$

9. Find the area of the rectangle.

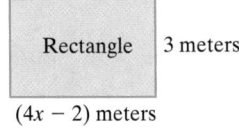

Rectangle | 3 meters
$(4x - 2)$ meters

10. Find the perimeter of the triangle.

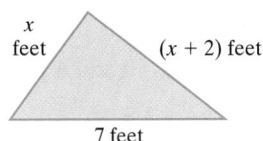

x feet $(x + 2)$ feet
7 feet

Solve and check.

11. $x + 7 = 20$

12. $-11 = x - 2$

13. $11x = 55$

14. $-7y = 0$

15. $\dfrac{x}{-3} = -13$

16. $\dfrac{z}{-5} = -11$

17. $12 = 11x - 14x$

18. $8y + 7y = -45$

19. $-3x = -15$

20. $-2m = -16$

21. $x - 12 = -45 + 23$

22. $6 - (-5) = x + 5$

199

23. _____

24. _____

25. _____

26. _____

27. _____

28. _____

29. _____

30. _____

31. _____

32. _____

33. _____

34. _____

35. _____

36. _____

23. $6(3x - 4) = 19x$

24. $25x = 6(4x - 9)$

25. $-36x - 10 + 37x = -12 - (-14)$

26. $-8 + (-14) = -80y + 20 + 81y$

27. $\dfrac{x}{13} = -1 - (-3)$

28. $\dfrac{y}{4} = 7 - 10$

29. $-8z - 2z = 26 - (-4)$

30. $-12 + (-13) = 5x - 10x$

Write each phrase as an algebraic expression. Use x to represent "a number."

31. The difference of a number and 10

32. The sum of -20 and a number

33. The product of 10 and a number

34. The quotient of 10 and a number

35. Five added to the product of -2 and a number

36. The product of -4 and the difference of a number and 1

3.4   Solving Linear Equations in One Variable

In this chapter, the equations we are solving are called **linear equations in one variable** or **first degree equations in one variable**. For example, an equation such as $5x - 2 = 6x$ is a linear equation in one variable. It is called linear or first degree because the exponent on each x is 1 and there is no variable below a fraction bar. It is an equation in one variable because it contains one variable, x.

OBJECTIVES

- **A** Solve linear equations using the addition and multiplication properties.
- **B** Solve linear equations containing parentheses.
- **C** Write sentences as equations.

SSM
TUTOR CENTER   SG   CD & VIDEO   MATH PRO   WEB

A Solving Equations Using the Addition and Multiplication Properties

We will now solve linear equations in one variable using more than one property of equality. To solve an equation such as $2x - 6 = 18$, we first get the variable term $2x$ alone on one side of the equation.

EXAMPLE 1   Solve: $2x - 6 = 18$

Solution: We start by adding 6 to both sides to get the variable term $2x$ alone.

$$2x - 6 = 18$$
$$2x - 6 + 6 = 18 + 6 \qquad \text{Add 6 to both sides.}$$
$$2x = 24 \qquad \text{Simplify.}$$

To finish solving, we divide both sides by 2.

$$\frac{2x}{2} = \frac{24}{2} \qquad \text{Divide both sides by 2.}$$
$$x = 12 \qquad \text{Simplify.}$$

Helpful Hint

Don't forget to check the proposed solution in the *original* equation.

Check: $2x - 6 = 18$
$$2(12) - 6 \stackrel{?}{=} 18 \qquad \text{Replace } x \text{ with 12 and simplify.}$$
$$24 - 6 \stackrel{?}{=} 18$$
$$18 = 18 \qquad \text{True.}$$

The solution is 12.

EXAMPLE 2   Solve: $20 - x = 21$

Solution: First we get the variable term alone on one side of the equation.

$$20 - x = 21$$
$$20 - x - 20 = 21 - 20 \qquad \text{Subtract 20 from both sides.}$$
$$-1x = 1 \qquad \text{Simplify. Recall that } -x \text{ means } -1x.$$
$$\frac{-1x}{-1} = \frac{1}{-1} \qquad \text{Divide both sides by } -1.$$
$$x = -1 \qquad \text{Simplify.}$$

Check: $20 - x = 21$
$$20 - (-1) \stackrel{?}{=} 21$$
$$21 = 21 \qquad \text{True.}$$

The solution is -1.

Practice Problem 1

Solve: $5y + 2 = 17$

Practice Problem 2

Solve: $45 = -10 - y$

Answers

1. 3   **2.** -55

Make sure you understand which property to use to solve an equation.

Addition┐	Understood multiplication
$x + 2 = 10$	$2x = 10$

To undo addition of 2, we subtract 2 from both sides.

To undo multiplication of 2, we divide both sides by 2.

$$x + 2 - 2 = 10 - 2$$
Use Addition Property of Equality.

$$x = 8$$

$$\frac{2x}{2} = \frac{10}{2}$$ Use Multiplication Property of Equality.

$$x = 5$$

Check: $x + 2 = 10$

$$8 + 2 = 10$$

$$10 = 10$$

Check: $2x = 10$

$$2 \cdot 5 = 10$$

$$10 = 10$$

If an equation contains variable terms on both sides, we use the addition property of equality to get all the variable terms on one side and all the constants or numbers on the other side.

Practice Problem 3

Solve: $7x + 12 = 3x - 4$

EXAMPLE 3 Solve: $3a - 6 = a + 4$

Solution:

$$3a - 6 = a + 4$$

$$3a - 6 + 6 = a + 4 + 6 \qquad \text{Add 6 to both sides.}$$

$$3a = a + 10 \qquad \text{Simplify.}$$

$$3a - a = a + 10 - a \qquad \text{Subtract } a \text{ from both sides.}$$

$$2a = 10 \qquad \text{Simplify.}$$

$$\frac{2a}{2} = \frac{10}{2} \qquad \text{Divide both sides by 2.}$$

$$a = 5 \qquad \text{Simplify.}$$

Check to see that the solution is 5.

Try the Concept Check in the margin.

Concept Check

In Example 3, the solution is 5. If you wanted to check this solution, what equation should you use?

B Solving Equations Containing Parentheses

If an equation contains parentheses, we must first use the distributive property to remove them.

Practice Problem 4

Solve: $6(a - 5) = 4(a + 1)$

EXAMPLE 4 Solve: $7(x - 2) = 9x - 6$

Solution: First we apply the distributive property.

$$7(x - 2) = 9x - 6$$

$$7x - 14 = 9x - 6 \qquad \text{Apply the distributive property.}$$

Next we move variable terms to one side of the equation and constants to the other side.

$$7x - 14 - 9x = 9x - 6 - 9x \qquad \text{Subtract } 9x \text{ from both sides.}$$

$$-2x - 14 = -6 \qquad \text{Simplify.}$$

$$-2x - 14 + 14 = -6 + 14 \qquad \text{Add 14 to both sides.}$$

$$-2x = 8 \qquad \text{Simplify.}$$

$$\frac{-2x}{-2} = \frac{8}{-2} \qquad \text{Divide both sides by } -2.$$

$$x = -4 \qquad \text{Simplify.}$$

Check to see that -4 is the solution.

Answers

3. -4 **4.** 17

Concept Check: $3a - 6 = a + 4$

You may want to use the following steps to solve equations.

Steps for Solving an Equation

Step 1. If parentheses are present, use the distributive property.

Step 2. Combine any like terms on each side of the equation.

Step 3. Use the addition property of equality to rewrite the equation so that variable terms are on one side of the equation and constant terms are on the other side.

Step 4. Use the multiplication property of equality to divide both sides by the numerical coefficient of the variable to solve.

Step 5. Check the solution in the *original equation*.

EXAMPLE 5 Solve: $3(2x - 6) + 6 = 0$

Solution: $3(2x - 6) + 6 = 0$

Step 1. $6x - 18 + 6 = 0$ — Apply the distributive property.

Step 2. $6x - 12 = 0$ — Combine like terms on the left side of the equation.

Step 3. $6x - 12 + 12 = 0 + 12$ — Add 12 to both sides.

$6x = 12$ — Simplify.

Step 4. $\dfrac{6x}{6} = \dfrac{12}{6}$ — Divide both sides by 6.

$x = 2$ — Simplify.

Check:

Step 5. $3(2x - 6) + 6 = 0$

$3(2 \cdot 2 - 6) + 6 \stackrel{?}{=} 0$

$3(4 - 6) + 6 \stackrel{?}{=} 0$

$3(-2) + 6 \stackrel{?}{=} 0$

$-6 + 6 \stackrel{?}{=} 0$

$0 = 0$ True.

The solution is 2.

Practice Problem 5

Solve: $4(x + 3) = 12$

C Writing Sentences as Equations

Next we practice translating sentences into equations. Below are key words and phrases that translate to an equal sign.

Key Words or Phrases	Examples	Symbols
equals	3 equals 2 plus 1	$3 = 2 + 1$
gives	the quotient of 10 and −5 gives −2	$\dfrac{10}{-5} = -2$
is/was/will be	x is 5	$x = 5$
yields	y plus 2 yields 13	$y + 2 = 13$
amounts to	twice x amounts to −30	$2x = -30$
is equal to	−24 is equal to 2 times −12	$-24 = 2(-12)$

Answer

5. 0

Practice Problem 6

Translate each sentence into an equation.

a. The difference of 110 and 80 is 30.

b. The product of 3 and the sum of -9 and 11 amounts to 6.

c. The quotient of twice 12 and -6 yields -4.

EXAMPLE 6 Translate each sentence into an equation.

a. The product of 7 and 6 is 42.

b. Twice the sum of 3 and 5 is equal to 16.

c. The quotient of -45 and 5 yields -9.

Solution:

a. In words:

| The product of 7 and 6 | is | 42 |

Translate: $7 \cdot 6 = 42$

b. In words:

| Twice | the sum of 3 and 5 | is equal to | 16 |

Translate: $2 \quad (3 + 5) \quad = \quad 16$

c. In words:

| The quotient of -45 and 5 | yields | -9 |

Translate: $\dfrac{-45}{5} = -9$

Answers

6. a. $110 - 80 = 30$ **b.** $3(-9 + 11) = 6$

c. $\dfrac{2(12)}{-6} = -4$

CALCULATOR EXPLORATIONS

Checking Equations

A calculator can be used to check possible solutions of equations. To do this, replace the variable by the possible solution and evaluate each side of the equation separately. For example, to see whether 7 is a solution of the equation $52x = 15x + 259$, replace x with 7 and use your calculator to evaluate each side separately.

Equation: $52x = 15x + 259$

$52 \cdot 7 \stackrel{?}{=} 15 \cdot 7 + 259$ Replace x with 7.

Evaluate left side: 52 × 7 = or ENTER . Display: 364 .

Evaluate right side: 15 × 7 + 259 = or ENTER . Display: 364 .

Since the left side equals the right side, 7 is a solution of the equation $52x = 15x + 259$.

Use a calculator to determine whether the numbers given are solutions of each equation.

1. $76(x - 25) = -988$; 12
2. $-47x + 862 = -783$; 35
3. $x + 562 = 3x + 900$; -170
4. $55(x + 10) = 75x + 910$; -18
5. $29x - 1034 = 61x - 362$; -21
6. $-38x + 205 = 25x + 120$; 25

EXERCISE SET 3.4

A *Solve each equation. See Examples 1 through 3.*

1. $2x - 6 = 0$

2. $3y - 12 = 0$

3. $\dfrac{n}{5} + 10 = 0$

4. $\dfrac{z}{4} + 8 = 0$

5. $6 - n = 10$

6. $7 - y = 9$

 **7.** $10x + 15 = 6x + 3$

8. $5x - 3 = 2x - 18$

9. $3x - 7 = 4x + 5$

10. $3x + 1 = 8x - 4$

B *Solve each equation. See Examples 4 and 5.*

11. $3(x - 1) = 12$

12. $2(x + 5) = -8$

13. $-2(y + 4) = 2$

14. $-1(y + 3) = 10$

15. $35 = 17 + 3(x - 2)$

16. $22 - 42 = 4(x - 1)$

A **B** *Solve each equation. See Examples 1 through 5.*

17. $8 - t = 3$

18. $6 - x = 4$

19. $0 = 4x - 4$

20. $0 = 5y + 5$

21. $\dfrac{x}{-2} - 8 = 0$

22. $\dfrac{w}{-8} - 40 = 0$

23. $7 = 4c - 1$

24. $9 = 2b - 5$

 **25.** $3r + 4 = 19$

26. $5m + 1 = 46$

27. $2x - 1 = -7$

28. $3t - 2 = -11$

29. $2 = 3z - 4$

30. $4 = 4p - 12$

31. $5x - 2 = -12$

32. $7y - 3 = -24$

33. $-7c + 1 = -20$

34. $-2b + 5 = -7$

35. $9 = \dfrac{x}{2} - 15$

36. $16 = \dfrac{y}{6} - 14$

37. $8m + 79 = -1$

38. $9a + 29 = -7$

39. $10 + 4v = -6$

40. $13 + 6b = 1$

41. $-5 = -13 - 8k$

42. $-7 = -17 - 10d$

43. $4x + 3 = 2x + 11$

44. $6y - 8 = 3y + 7$

45. $-2y - 10 = 5y + 18$

46. $7n + 5 = 12n - 10$

47. $-8n + 1 = -6n - 5$

48. $10w + 8 = w - 10$

49. $9 - 3x = 14 + 2x$

50. $4 - 7m = -3m$

51. $2(y - 3) = y - 6$

52. $3(z + 2) = 5z + 6$

53. $2t - 1 = 3(t + 7)$

54. $4 + 3c = 2(c + 2)$

 **55.** $3(5c - 1) - 2 = 13c + 3$

56. $4(3t + 4) - 20 = 3 + 5t$

57. $10 + 5(z - 2) = 4z + 1$

58. $14 + 4(w - 5) = 6 - 2w$

59. $7(6 + w) = 6(2 + w)$

60. $6(5 + c) = 5(c - 4)$

C *Write each sentence as an equation. See Example 6.*

61. The sum of -42 and 16 is -26.

62. The difference of -30 and 10 equals -40.

 63. The product of -5 and -29 gives 145.

64. The quotient of -16 and 2 yields -8.

65. Three times the difference of -14 and 2 amounts to -48.

66. Negative 2 times the sum of 3 and 12 is -30.

67. The quotient of 100 and twice 50 is equal to 1.

68. Seventeen subtracted from -12 equals -29.

The following bar graph shows the estimated number of U.S. federal individual income tax returns that will be filed electronically during the years shown. Electronically filed returns include Telefile and online returns. Use this graph to answer Exercises 69 through 72. See Section 1.3

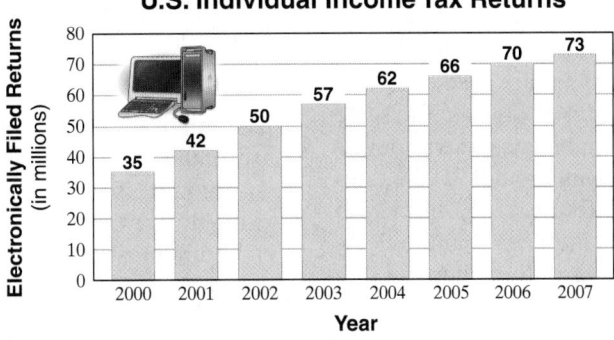

**Total Electronically Filed
U.S. Individual Income Tax Returns**

Source: IRS Compliance Research Division

69. What year shows the greatest increase in the number of electronically filed returns.

70. What year shows the smallest increase in the number of electronically filed returns.

71. By how much is the number of electronically filed returns expected to increase from 2000 to 2007?

72. Describe any trends shown in this graph.

Evaluate each expression for $x = 3$, $y = -1$, and $z = 0$. See Section 2.5.

73. $x^3 - 2xy$

74. $y^3 + 3xyz$

75. $y^5 - 4x^2$

76. $(-y)^3 + 3xyz$

77. $(2x - y)^2$

78. $(6z - 5y)^3$

79. $4xy + 6y^2$

80. $2x^2 - 7z$

Combining Concepts

Solve.

81. $(-8)^2 + 3x = 5x + 4^3$

82. $3^2 \cdot x = (-9)^3$

83. $2^3(x + 4) = 3^2(x + 4)$

84. $x + 45^2 = 54^2$

85. A classmate tries to solve $3x = 39$ by subtracting 3 from both sides of the equation. Will this step solve the equation for x? Why or why not?

86. A classmate tries to solve $2 + x = 20$ by dividing both sides by 2. Will this step solve the equation for x? Why or why not?

Focus on the **Real World**

SCALES

In this chapter, we have been using a scale to help illustrate true equations and the properties of equality. Scales of different types are used everyday to measure the mass or weight of objects. The scale shown is called an equal-arm balance. It was developed by the ancient Egyptians around 2500 B.C. to assist in everyday business. To use it, an object with an unknown mass is placed in one pan and objects with known masses are placed in the other pan one at a time, until the arm to which the pans are attached is perfectly level. When the pans are balanced in this way, the user knows that the object with the unknown mass has the same mass as the sum of the known masses in the other pan.

The ancient Romans invented a different kind of scale, called a steelyard, about 2000 years ago. The Romans' steelyard used a pan to hold the unknown mass on one end of the scale's arm and a known mass that slid along the other end of the arm. The position of the sliding mass over markings along the arm indicated the mass of the object in the pan. A variation of the steelyard can still be seen today in many doctors' offices.

Today, most of the scales we use on a daily basis measure weights and masses using springs, pendulums, or changes in electrical resistance. However, the most accurate scales in the world are based on the same balancing principles used in the Egyptians' equal-arm balance and the Romans' steelyard. These highly precise scales can measure accurately to a millionth of a gram.

GROUP ACTIVITY

Borrow an equal-arm balance or construct a simple one of your own. Use the scale to investigate and demonstrate the properties of equality.

3.5 Linear Equations in One Variable and Problem Solving

OBJECTIVES

Ⓐ Translate sentences into mathematical equations.

Ⓑ Use problem-solving steps to solve problems.

SSM
TUTOR CENTER   SG   CD & VIDEO   MATH PRO   WEB

Ⓐ Writing Sentences as Equations

Now that we have practiced solving equations for a variable, we can extend our problem-solving skills considerably. We begin by reviewing how to translate sentences into algebraic equations using the same key words and phrases table from section 3.3.

Addition	Subtraction	Multiplication	Division	Equal Sign
sum	difference	product	quotient	equals
plus	minus	times	divided by	gives
added to	subtracted from	multiply	into	is/was/will be
more than	less than	twice	per	yields
increased by	decreased by	of		amounts to
total	less	double		is equal to

EXAMPLE 1 Write each sentence as an equation. Use x to represent "a number."

a. Nine more than a number is 5.

b. Twice a number equals -10.

c. A number minus 6 amounts to 168.

d. Three times the sum of a number and 5 is -30.

e. The quotient of 8 and twice a number is equal to 2.

Solution:

a. In words: Nine | more than | a number | is | 5

Translate: $9 \quad + \quad x \quad = \quad 5$

b. In words: Twice a number | equals | -10

Translate: $2x \quad = \quad -10$

c. In words: A number | minus | 6 | amounts to | 168

Translate: $x \quad - \quad 6 \quad = \quad 168$

d. In words: Three times | the sum of a number and 5 | is | -30

Translate: $3 \quad (x+5) \quad = \quad -30$

e. In words: The quotient of

Translate: 8 | and | twice a number | is equal to | 2

$8 \quad \div \quad 2x \quad = \quad 2$

or $\dfrac{8}{2x} = 2$

Practice Problem 1

Write each sentence as an equation. Use x to represent "a number."

a. Five times a number is 20.

b. The sum of a number and -5 yields 14.

c. Ten subtracted from a number amounts to -23.

d. Five times a number added to 7 is equal to -8.

e. The quotient of 6 and the sum of a number and 4 gives 1.

Answers

1. **a.** $5x = 20$ **b.** $x + (-5) = 14$
c. $x - 10 = -23$ **d.** $7 + 5x = -8$
e. $\dfrac{6}{x+4} = 1$

B Use Problem-Solving Steps to Solve Problems

Our main purpose for studying arithmetic and algebra is to solve problems. In previous sections, we have prepared for problem solving by writing phrases as algebraic expressions and sentences as equations. We now draw upon this experience as we solve problems. The following problem-solving steps will be used throughout this text.

Problem-Solving Steps

1. UNDERSTAND the problem. During this step, become comfortable with the problem. Some ways of doing this are as follows:
   - Read and reread the problem.
   - Choose a variable to represent the unknown.
   - Construct a drawing, if possible.
   - Propose a solution and check it. Pay careful attention to how you check your proposed solution. This will help when writing an equation to model the problem.
2. TRANSLATE the problem into an equation.
3. SOLVE the equation.
4. INTERPRET the results. *Check* the proposed solution in the stated problem and *state* your conclusion.

Practice Problem 2

Translate "the sum of a number and 2 equals 6 added to three times the number" into an equation and solve.

EXAMPLE 2 Finding an Unknown Number

Twice a number, added to 3, is the same as the number minus 6. Find the number.

Solution:

1. UNDERSTAND the problem. To do so, we read and reread the problem. To help us understand the problem better, let's propose a solution and check it. Suppose that the unknown number is 10. From the stated problem, "twice 10, added to 3" is 23, but "10 minus 6" is 4. Since 23 is not equal to 4, the solution is not 10. The purpose of proposing a solution is not to guess correctly but to better understand the problem.

   Now, let's assign a variable to the unknown. We will let

   x = the unknown number.

2. TRANSLATE the problem into an equation.

In words:	Twice a number	added to 3	is the same as	the number minus 6
	↓	↓	↓	↓
Translate:	$2x$	$+\ 3$	$=$	$x - 6$

3. SOLVE the equation. To solve the equation, we first subtract x from both sides.

   $$2x + 3 = x - 6$$
   $$2x + 3 - x = x - 6 - x \quad \text{Subtract } x \text{ from both sides.}$$
   $$x + 3 = -6 \quad \text{Simplify.}$$
   $$x + 3 - 3 = -6 - 3 \quad \text{Subtract 3 from both sides.}$$
   $$x = -9 \quad \text{Simplify.}$$

Answer

2. -2

4. INTERPRET the results. First, *check* the proposed solution in the stated problem. Twice "−9" is −18 and −18 + 3 is −15. This is equal to the number minus 6 or "−9" −6 or −15. Then *state* your conclusion: The unknown number is −9.

Try the Concept Check in the margin.

EXAMPLE 3   Determining Voter Counts

In the 2000 Senate election in Wyoming, incumbent Craig Thomas received 100,280 *more* votes than his challenger. If a total of 204,358 votes were cast, find how many votes Craig Thomas received. (*Source: World Almanac*)

Solution:

1. UNDERSTAND the problem. We read and reread the problem. Then we assign a variable to an unknown. We use this variable to represent any other unknown quantities. We let

    x = the number of challenger votes

    Then

    $x + 110,280$ = the number of incumbent votes
    since Thomas received 110,280 more votes

2. TRANSLATE the problem into an equation.

    In words: | challenger votes | plus | incumbent votes | is | total votes |

    Translate: x + $x + 110,280$ = $204,358$

3. SOLVE the equation:

    $$x + x + 110,280 = 204,358$$
    $$2x + 110,280 = 204,358 \quad \text{Combine like terms.}$$
    $$2x + 110,280 - 110,280 = 204,358 - 110,280 \quad \text{Subtract 110,280 from both sides.}$$
    $$2x = 94,078 \quad \text{Simplify.}$$
    $$\frac{2x}{2} = \frac{94,078}{2} \quad \text{Divide both sides by 2.}$$
    $$x = 47,039 \quad \text{Simplify.}$$

4. INTERPRET the results. First *Check* the proposed solution in the stated problem. Since x represents the number of votes the challenger received, the challenger received 47,039 votes. The incumbent received $x + 110,280 = 47,039 + 110,280 = 157,319$ votes. To check, notice that the total number of challenger votes and incumbent votes is $47,039 + 157,319 = 204,358$ votes, the given total of votes cast. Also, 157,319 is 110,280 more votes than 47,039, so the solution checks. Then, *state* your conclusion: The incumbent, Craig Thomas, received 157,319 votes.

Try the Concept Check in the margin.

Concept Check

Suppose you have solved an equation involving perimeter to find the length of a rectangular table. Explain why you would want to recheck your math if you obtain the result of −5.

Practice Problem 3

At a recent United States/Japan summit meeting, 121 delegates attended. If the United States sent 19 more delegates than Japan, find how many the United States sent.

Concept Check

When solving the problem given in Example 3, why should you be skeptical if you obtain an answer of 15,714 votes for the incumbent?

Answers

3. 70 delegates

Concept Check: Length cannot be negative.

Concept Check: Answers may vary.

Practice Problem 4

A woman's $21,000 estate is to be divided so that her husband receives twice as much as her son. How much will each receive?

EXAMPLE 4 Calculating Separate Costs

Leo Leal sold a used computer system and software for $2100, receiving four times as much money for the computer system as for the software. Find the price of each.

Solution:

1. UNDERSTAND the problem. We read and reread the problem. Then we assign a variable to an unknown. We use this variable to represent any other unknown quantities. We let

    x = the software price

    $4x$ = the computer system price

2. TRANSLATE the problem into an equation.

In words:	Software price	plus	computer price	is	2100
	↓	↓	↓	↓	↓
Translate:	x	$+$	$4x$	$=$	2100

3. SOLVE the equation.

    $$x + 4x = 2100$$
    $$5x = 2100 \qquad \text{Combine like terms.}$$
    $$\frac{5x}{5} = \frac{2100}{5} \qquad \text{Divide both sides by 5.}$$
    $$x = 420 \qquad \text{Simplify.}$$

4. INTERPRET the results. *Check* the proposed solution in the stated problem. The software sold for $420. The computer system sold for $4x = 4(\$420) = \$1680$. Since $420 + $1680 = $2100, the total price, and $1680 is four times $420, the solution checks. State your conclusion: The software sold for $420 and the computer system sold for $1680.

Answers

4. husband: $14,000; son: $7000

EXERCISE SET 3.5

A *Write each sentence as an equation. Use x to represent "a number." See Example 1.*

1. A number added to −5 is −7.

2. Five subtracted from a number equals 10.

3. Three times a number yields 27.

4. The quotient of 8 and a number is −2.

5. A number subtracted from −20 amounts to 104.

6. Two added to twice a number gives −14.

7. Twice the sum of a number and −1 is equal to 50.

8. Three times the difference of 7 and a number amounts to −40.

B *Solve. See Example 2.*

9. Three times a number, added to 9, is 33. Find the number.

10. Twice a number, subtracted from 60, is 20. Find the number.

11. The product of 9 and a number gives 54. Find the number.

12. The product of 7 and a number is 14. Find the number.

13. The sum of 3, 4, and a number amounts to 16. Find the number.

14. A number less 5 is 11. Find the number.

15. Seventy-two is 8 times a number added to 24. Find the number.

16. The product of 11 and a number is 121. Find the number.

17. The difference of a number and 3 is the same as 45 less the number. Find the number.

18. Sixty-four less half of 64 is equal to the sum of some number and 20. Find the number.

19. Three times the difference of some number and 5 amounts to 9. Find the number.

20. Five times the sum of some number plus 2 is equal to 8 times the number, minus 11. Find the number.

21. Eight decreased by some number equals the quotient of 15 and 5. Find the number.

22. Ten less some number is twice the sum of that number and 5. Find the number.

23. Thirteen added to the product of 3 and some number amounts to 5 times the number added to 3.

24. The product of 4 and a number is the same as 30 less twice that same number.

25. Five times a number, less 40, is 8 more than the number.

26. Thirty less a number is equal to 3 times the sum of the number and 6.

27. The difference of a number and 3 is equal to the quotient of 10 and 5.

28. The product of a number and 3 is twice the sum of that number and 5.

Solve. See Examples 3 and 4.

29. In the 2000 presidential election, George W. Bush received 5 more electoral votes than Al Gore. If a total of 527 electoral votes were cast for the two candidates, find how many votes each candidate received. (*Source:* Voter News Service)

30. Based on the 2000 Census, California has 21 more electoral votes for president than Texas. If the total number of electoral votes for these two states is 89, find the number for each state. (*Source: The World Almanac* 2003)

31. Bamboo and Pacific Kelp, a kind of sea weed, are two fast-growing plants. Bamboo grows twice as fast as kelp. If in one day both can grow a total of 54 inches, find how many inches each plant can grow in one day.

32. Norway has had three times as many rulers as Liechtenstein. If the total rulers for both countries is 56, find the number of rulers for Norway and the number for Liechtenstein.

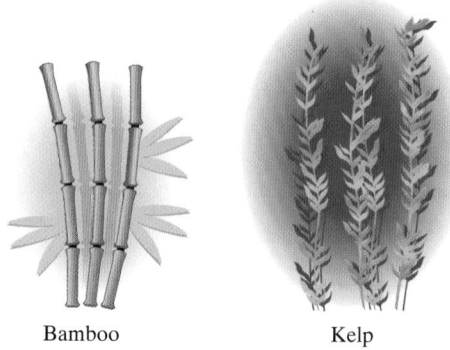

Bamboo   Kelp

Norway   Liechtenstein

33. The country with the most universities[*] is India, followed by the United States. If India has 2649 more universities than the U.S. and their combined total is 14,165, find the number of universities in India and the number in the U.S. [* Includes all further education establishments. (*Source:* The Top 10 of Everything, 2003)]

34. The average life expectancy for a man is 34 years longer than the life expectancy for a polar bear. If the total of these life expectancies is 110 years, find the life expectancy of each.

35. A Nintendo Gamecube and several games are sold for $600. The cost of the games is 3 times as much as the cost of the Gamecube. Find the cost of the Gamecube and the cost of the games.

36. Disneyland®, in Anaheim, California, has 12 more rides than its neighboring sister park, Disney's California Adventure, which opened in 2001. Together, the two parks have 36 rides. How many rides does each park have? (*Source:* The Walt Disney Company)

37. The two NCAA stadiums with the largest capacities are Michigan Stadium (Univ. of Michigan) and Neyland Stadium (Univ. of Tennessee). Michigan Stadium has a capacity of 4647 more than Neyland. If the combined capacity for the two stadiums is 210,355, find the capacity for each stadium. (*Source:* National Collegiate Athletic Association)

38. An NHRA top fuel dragster has a top speed of 95 mph faster than an Indy Racing League car. If the combined top speeds for these two cars is 565 mph, find the top speed of each car. (*Source:* USA Today)

39. California contains the largest state population of native Americans. This population is three times the native American population of Washington state. If the total of these two populations is 412 thousand find the native American population in each of these two states. (*Source*: U.S. Census Bureau)

40. In 2020, China is projected to be the country with the greatest number of visiting tourists. This number is twice the number of tourists projected for Spain. If the total number of tourists for these two countries is projected to be 210 million, find the number projected for each. (*Source: The State of the World Atlas* by Dan Smith)

41. During the 2002 Women's NCAA Division I basketball championship game, the Connecticut Huskies scored 12 more points than the Oklahoma Sooners. Together, both teams scored a total of 152 points. How many points did the 2002 Champion Connecticut Huskies score during the game? (*Source:* National Collegiate Athletic Association)

42. During the 2002 Men's NCAA Division I Basketball Championship game, the Indiana Hoosiers scored 12 fewer points than the Maryland Terrapins. Together, both teams scored a total of 116 points. How many points did the 2002 Champion Maryland Terrapins score during the game? (*Source:* National Collegiate Athletic Association)

43. A crow will eat five more ounces of food a day than a finch. If together they eat 13 ounces of food, find how many ounces of food the crow consumes and how many ounces of food the finch consumes.

44. A Toyota Camry is traveling twice as fast as a Dodge truck. If their combined speed is 105 miles per hour, find the speed of the car and find the speed of the truck.

Review and Preview

Round each whole number to the given place. See Section 1.5.

45. 586 to the nearest ten

46. 82 to the nearest ten

47. 1026 to the nearest hundred

48. 52,333 to the nearest thousand

49. 2986 to the nearest thousand

50. 101,552 to the nearest hundred

Combining Concepts

51. Solve Example 3 again, but this time let *x* be the number of incumbent votes. Did you get the same results? Explain why or why not.

In real estate, a house's selling price P is found by adding the real estate agent's commission C to the amount A that the seller of the house receives: P = A + C.

52. Brianna Morley's house sold for $230,000. Her real estate agent received a commission of $13,800. How much did Brianna receive? (*Hint:* Substitute the known values into the equation, then solve the equation for the remaining unknown.)

53. Duncan Bostic plans to use a real estate agent to sell his house. He hopes to sell the house for $165,000 and keep $156,750 of that. If everything goes as he has planned, how much will his real estate agent receive as a commission?

In retailing, the retail price P of an item can be computed using the equation $P = C + M$, where C is the wholesale cost of the item and M is the amount of markup.

54. The retail price of a computer system is $999 after a markup of $450. What is the wholesale cost of the computer system? (*Hint:* Substitute the known values into the equation, then solve the equation for the remaining unknown.)

55. Slidell Feed and Seed sells a bag of cat food for $12. If the store paid $7 for the cat food, what is the markup on the cat food?

Internet Excursions

 Go To: http://www.prenhall.com/martin-gay_prealgebra What's Related

Project Vote Smart is a nonpartisan, nonprofit organization that strives to provide the American public with unbiased information about more than 13,000 of their elected public officials as well as candidates running for public office. The Project Vote Smart goal is to help American voters make informed choices.

This World Wide Web address will provide you with access to Project Vote Smart's listing of links to official state election information, or a related site. Use this listing to connect to election result information for your state.

56. Use election results from your state's most recent governor's race to write a problem similar to Exercises 29 and 30. Exchange problems with a classmate to solve.

57. Using election results from another recent election in your state, write another problem like Exercises 29 and 30. Exchange with another student in your class to solve.

CHAPTER 3 ACTIVITY **Using Equations**

This activity may be completed by working in groups or individually.

A hospital nurse working second shift has been keeping track of the fluid intake and output on the chart of one of her patients, Mr. Ramirez. At the end of her shift, she totals the intakes and outputs and returns Mr. Ramirez's chart to the floor nurses' station.

Suppose you are one of the night nurses for this floor and have been assigned to Mr. Ramirez. Shortly after you come on duty, someone at the nurses' station spills coffee over several patients' charts, including the one for Mr. Ramirez. After blotting up the coffee on the chart, you notice that several entries on his chart have been smeared by the coffee spill.

MEDICAL CHART

Patient: <u>Juan Ramirez</u>   Room: <u>314</u>

Shift: <u>Second</u>   Nurse's Initials: <u>*SRJ*</u>   Date: <u>3/27</u>

Fluid Intake (cubic centimeters):

					Totals
Blood	500				500
Intravenous	250		**100**		500
Oral Fluid	300	150		100	900
Oral Meds	4 doses of		cc each		200
TOTAL					2100

Fluid Output (cubic centimeters):

				Totals
Urine	240	310		550
Emesis	120			120
Irrigation	50	25		75
TOTAL				815

---

1. On the Fluid Intake chart, is it possible to tell if Mr. Ramirez was given blood more than once? If so, how many times was he given blood and how can you tell?

2. On the Fluid Intake chart, is it possible to tell if Mr. Ramirez was given intravenous fluid more than once? Is it possible to tell exactly how many times Mr. Ramirez was given intravenous fluid? Explain.

3. For the Oral Fluid row of the Fluid Intake chart, notice that the third entry in that row is obliterated by the coffee stain. Let the variable x represent this third entry. Write an equation that describes the total amount of oral fluids Mr. Ramirez received during the

second shift. Then solve the equation for x to find the obliterated entry.

4. For the Oral Meds row of the Fluid Intake chart, notice that the second-shift nurse indicated that Mr. Ramirez was given four equal doses of oral medication. We cannot see the size of each dose, but the chart shows that he received a total of 200 cc. Write and solve an equation to find the size of each dose.

5. Using the Fluid Output chart, write and solve an equation to find the total urine output.

6. Using the Fluid Output chart and the result from Question 5, write and solve an equation to find the obliterated entry in the Urine output row.

Chapter 3 Vocabulary Check

Fill in each blank with one of the words or phrases listed below.

Variable	Simplified	numerical coefficient	equation
terms	combined	algebraic expression	solution
like	constant	evaluating the expression	

1. An algebraic expression is _____ when all like terms have been _____ .
2. Terms that are exactly the same, except that they may have different numerical coefficients, are called _____ terms.
3. A letter used to represent a number is called a _____ .
4. A combination of operations on variables and numbers is called an _____ .
5. The addends of an algebraic expression are called the _____ of the expression.
6. The number factor of a variable term is called the _____ .
7. Replacing a variable in an expression by a number and then finding the value of the expression is called _____ _____ for the variable.
8. A term that is a number only is called a _____ .
9. An _____ is of the form expression = expression.
10. A _____ of an equation is a value for the variable that makes an equation a true statement.

CHAPTER 3

Highlights

DEFINITIONS AND CONCEPTS	EXAMPLES
SECTION 3.1 SIMPLIFYING ALGEBRAIC EXPRESSIONS	

DEFINITIONS AND CONCEPTS	EXAMPLES
The addends of an algebraic expression are called the **terms** of the expression.	$5x^2 + (-4x) + (-2)$ 3 terms
The number factor of a variable term is called the **numerical coefficient**.	**Term** **Numerical Coefficient** $7x$ 7 $-6y$ -6 x or $1x$ 1
Terms that are exactly the same, except that they may have different numerical coefficients, are called **like terms**.	$5x + 11x = (5 + 11)x = 16x$ like terms
An algebraic expression is **simplified** when all like terms have been **combined**.	$y - 6y = (1 - 6)y = -5y$
Use the **distributive property** to multiply an algebraic expression by a term.	Simplify: $-4(x + 2) + 3(5x - 7)$ $= -4(x) + (-4)(2) + 3(5x) + 3(-7)$ $= -4x + (-8) + 15x + (-21)$ $= 11x + (-29)$ or $11x - 29$

219

DEFINITIONS AND CONCEPTS	**EXAMPLES**

SECTION 3.2 SOLVING EQUATIONS: THE ADDITION PROPERTY

ADDITION PROPERTY OF EQUALITY

Let a, b, and c represent numbers.

If $a = b$, then

$$a + c = b + c \quad \text{and} \quad a - c = b - c$$

In other words, the same number may be added to or subtracted from both sides of an equation without changing the solution of the equation.

Solve:

$$\begin{aligned} x + 8 &= 2 + (-1) \\ x + 8 &= 1 \quad &&\text{Combine like terms.} \\ x + 8 - 8 &= 1 - 8 \quad &&\text{Subtract 8 from both sides.} \\ x &= -7 \quad &&\text{Simplify.} \end{aligned}$$

The solution is -7.

SECTION 3.3 SOLVING EQUATIONS: THE MULTIPLICATION PROPERTY

MULTIPLICATION PROPERTY OF EQUALITY

Let a, b, and c represent numbers and let $c \neq 0$.

If $a = b$, then

$$a \cdot c = b \cdot c \quad \text{and} \quad \frac{a}{c} = \frac{b}{c}$$

In other words, both sides of an equation may be multiplied or divided by the same nonzero number without changing the solution of the equation.

Solve:

$$\begin{aligned} y - 7y &= 30 \\ -6y &= 30 \quad &&\text{Combine like terms.} \\ \frac{-6y}{-6} &= \frac{30}{-6} \quad &&\text{Divide both sides by } -6. \\ y &= -5 \quad &&\text{Simplify.} \end{aligned}$$

The solution is -5.

SECTION 3.4 SOLVING LINEAR EQUATIONS IN ONE VARIABLE

STEPS FOR SOLVING AN EQUATION

Step 1. If parentheses are present, use the distributive property to remove them.

Step 2. Combine any like terms on each side of the equation.

Step 3. Use the addition property of equality to rewrite the equation so that variable terms are on one side of the equation and constant terms are on the other side.

Step 4. Use the multiplication property of equality to divide both sides by the numerical coefficient of the variable to solve.

Step 5. Check the solution in the *original equation*.

Solve: $5(3x - 1) + 15 = -5$

Step 1. $\quad 15x - 5 + 15 = -5$ Apply the distributive property.

Step 2. $\quad 15x + 10 = -5$ Combine like terms.

Step 3. $\quad 15x + 10 - 10 = -5 - 10$ Subtract 10 from both sides.

$$15x = -15$$

Step 4. $\quad \dfrac{15x}{15} = \dfrac{-15}{15}$ Divide both sides by 15.

$$x = -1$$

Step 5. Check to see that -1 is the solution.

SECTION 3.5 LINEAR EQUATIONS IN ONE VARIABLE AND PROBLEM SOLVING

PROBLEM-SOLVING STEPS

The incubation period for a golden eagle is three times the incubation period for a hummingbird. If the total of their incubation periods is 60 days, find the incubation period for each bird. (*Source: Wildlife Fact File*, International Masters Publishers)

1. UNDERSTAND the problem. Some ways of doing this are as follows:
- Read and reread the problem.
- Construct a drawing, if possible.
- Assign a variable to an unknown in the problem.
- Propose a solution and check.

1. UNDERSTAND the problem. Then assign a variable. Let

$$x = \text{incubation period of a hummingbird}$$
$$3x = \text{incubation period of a golden eagle}$$

220

DEFINITIONS AND CONCEPTS	EXAMPLES

2. TRANSLATE the problem into an equation.

2. TRANSLATE.

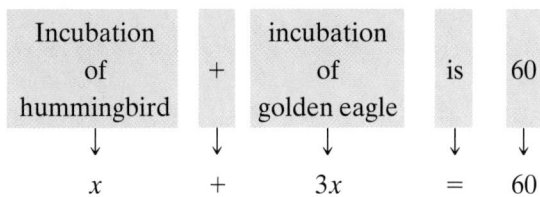

Incubation of hummingbird	+	incubation of golden eagle	is	60
↓	↓	↓	↓	↓
x	+	$3x$	=	60

3. SOLVE the equation.

3. SOLVE.

$$x + 3x = 60$$
$$4x = 60$$
$$\frac{4x}{4} = \frac{60}{4}$$
$$x = 15$$

4. INTERPRET the results. *Check* the proposed solution in the stated problem and *state* your conclusion.

4. INTERPRET the solution in the stated problem. The incubation period for the hummingbird is 15 days. The incubation period for the golden eagle is $3x = 3 \cdot 15 = 45$ days. Since 15 days + 45 *days* = 15 days + 45 days = 60 days and 45 is 3(15), the solution checks. State your conclusion: The incubation period for the hummingbird is 15 days. The incubation period for the golden eagle is 45 days.

As a convenient summary and for quick reference, these problem-solving steps are provided.

PROBLEM-SOLVING STEPS

1. UNDERSTAND the problem. During this step, become comfortable with the problem. Some ways of doing this are as follows:
- Read and reread the problem.
- Choose a variable to represent the unknown.
- Construct a drawing.
- Propose a solution and check it. Pay careful attention to how you check your proposed solution. This will help when writing an equation to model the problem.

2. TRANSLATE the problem into an equation.

3. SOLVE the equation.

4. INTERPRET the results. *Check* the proposed solution in the stated problem and *state* your conclusion.

STUDY SKILLS REMINDER

Are you organized?

Have you ever had trouble finding a completed assignment? When it's time to study for a test, are your notes neat and organized? Have you ever had trouble reading your own mathematics handwriting? (Be honest—I have.)

When any of these things happen, it's time to get organized. Here are a few suggestions:

Write your notes and complete your homework assignment in a notebook with pockets (spiral or ring binder.) Take class notes in this notebook, and then follow the notes with your completed homework assignment. When you receive graded papers or handouts, place them in the notebook pocket so that you will not lose them.

Remember to mark (possibly with an exclamation point) any note(s) that seem extra important to you. Also remember to mark (possibly with a question mark) any notes or homework that you are having trouble with. Don't forget to see your instructor or a math tutor to help you with the concepts or exercises that you are having trouble understanding.

Also, if you are having trouble reading your own handwriting, *slow down* and write your mathematics work clearly!

Chapter 3 Review

(3.1) *Simplify each expression by combining like terms.*

1. $3y + 7y - 15$ **2.** $2y - 10 - 8y$ **3.** $8a + a - 7 - 15a$ **4.** $y + 3 - 9y - 1$

Multiply.

5. $2(x + 5)$ **6.** $-3(y + 8)$

Simplify.

7. $7x + 3(x - 4) + x$ **8.** $-(3m + 2) - m - 10$

9. $3(5a - 2) - 20a + 10$ **10.** $6y + 3 + 2(3y - 6)$

Find the perimeter of each figure.

△ **11.**

2x yards

3 yards | Rectangle

△ **12.**

Square | 5y meters

Find the area of each figure.

△ **13.**

(2x − 1) yards

3 yards | Rectangle

△ **14.**

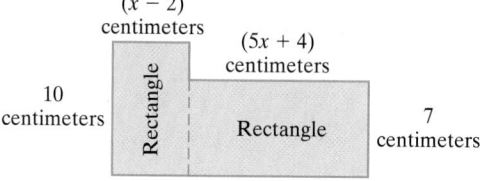

(x − 2) centimeters

(5x + 4) centimeters

10 centimeters | Rectangle | Rectangle | 7 centimeters

Simplify.

15. $85(7068x - 108) + 42x$ **16.** $-4268y + 120(63y - 32)$

(3.2)

17. Is 4 a solution of $5(2 - x) = -10$? **18.** Is 0 a solution of $6y + 2 = 23 + 4y$?

223

Solve each equation.

19. $z - 5 = -7$

20. $3x + 10 = 4x$

21. $n + 18 = 10 - (-2)$

22. $c - 5 = -13 + 7$

23. $7x + 5 - 6x = -20$

24. $17x = 2(8x - 4)$

(3.3) *Solve each equation.*

25. $3y = -21$

26. $-8x = 72$

27. $-5n = -5$

28. $-3a = -15$

29. $\dfrac{x}{-6} = 2$

30. $\dfrac{y}{-15} = -3$

31. $-5t + 32 + 4t = 32$

32. $3z + 72 - 2z = -56$

33. $\dfrac{z}{4} = -8 - (-6)$

34. $-1 + (-8) = \dfrac{x}{5}$

35. $6y - 7y = 100 - 105$

36. $19x - 16x = 45 - 60$

Translate each phrase into an algebraic expression. Let x represent "a number."

37. The product of -5 and a number

38. Three subtracted from a number

39. The sum of -5 and a number

40. The quotient of -2 and a number

41. Eleven added to twice a number

42. The product of -5 and a number, decreased by 50

43. The quotient of 70 and the sum of a number and 6

44. Twice the difference of a number and 13

(3.4) *Solve each equation.*

45. $3x - 4 = 11$

46. $6y + 1 = 73$

47. $14 - y = -3$

48. $7 - z = 0$

49. $4z - z = -6$

50. $t - 9t = -64$

51. $2x + 5 = 7x - 100$

52. $-6x - 4 = x + 66$

53. $2x + 7 = 6x - 1$

54. $5x - 18 = -4x$

55. $\dfrac{x}{9} + 3 = 0$

56. $\dfrac{z}{-2} - 11 = 0$

57. $5(n - 3) = 7 + 3n$

58. $7(2 + x) = 4x - 1$

Write each sentence as an equation.

59. The difference of 20 and -8 is 28.

60. Five times the sum of 2 and -6 yields -20.

61. The quotient of -75 and the sum of 5 and 20 is equal to -3.

62. Nineteen subtracted from -2 amounts to -21.

(3.5) *Write each sentence as an equation using x as the variable.*

63. Twice a number minus 8 is 40.

64. Twelve subtracted from the quotient of a number and 2 is 10.

65. The difference of a number and 3 is the quotient of the number and 4.

66. The product of some number and 6 is equal to the sum of the number and 2.

Solve.

67. Five times a number subtracted from 40 is the same as three times the number. Find the number.

68. The product of a number and 3 is twice the difference of that number and 8. Find the number.

69. In an election the incumbent received 14,000 votes of the 18,500 votes cast. Of the remaining votes, the Democratic candidate received 272 more than the Independent candidate. Find how many votes the Democratic candidate received.

70. Rajiv Puri has twice as many movies on DVDs as he has video tapes. Find the number of DVDs if he has a total of 126 movie recordings.

STUDY SKILLS REMINDER

Are you prepared for a test on Chapter 3?

Below I have listed some *common trouble areas* for topics covered in Chapter 3. After studying for your test, but before taking your test, read these.

- Be careful when evaluating expressions. For example, evaluate $3x - y$ when $x = -2$ and $y = -3$.

$$3x - y = 3(-2) - (-3) \quad \text{Let } x = -2 \text{ and } y = -3.$$
$$= -6 - (-3) \quad \text{Multiply.}$$
$$= -6 + 3$$
$$= -3 \quad \text{Add.}$$

- Remember the distributive property.

$$5(4x - 3) + 2 = 5 \cdot 4x - 5 \cdot 3 + 2 \quad \text{Use the distributive property.}$$
$$= 20x - 15 + 2 \quad \text{Multiply.}$$
$$= 20x - 13 \quad \text{Combine like terms.}$$

- Don't forget the steps for solving a linear equation.

$$2(3x - 2) + 16 = 6$$
$$6x - 4 + 16 = 6 \quad \text{Apply the distributive property.}$$
$$6x + 12 = 6 \quad \text{Combine like terms.}$$
$$6x + 12 - 12 = 6 - 12 \quad \text{Subtract 12 from both sides.}$$
$$6x = -6 \quad \text{Simplify.}$$
$$\frac{\overset{1}{\cancel{6}}x}{\cancel{6}_{1}} = \frac{-6}{6} \quad \text{Divide both sides by 6.}$$
$$x = -1 \quad \text{Simplify.}$$

Remember: This is simply a checklist of common trouble areas. For a review of Chapter 3 see the Highlights and Chapter Review at the end of this chapter.

Name_____ Section_____ Date _____

Chapter 3 Test   Remember to check your answers and use the Chapter Test Prep Video to view solutions.

1. Simplify $7x - 5 - 12x + 10$ by combining like terms.

2. Multiply: $-2(3y + 7)$

3. Simplify: $-(3z + 2) - 5z - 18$

△ **4.** Write an expression that represents the perimeter of the equilateral triangle. Simplify the expression. (A triangle with three sides of equal length.)

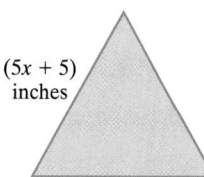

(5x + 5)
inches

5. Write an expression that represents the area of the rectangle. Simplify the expression.

4 meters

Rectangle | (3x − 1)
meters

6. Solve: $9x = -90$

7. Solve: $\dfrac{y}{-4} = 4$

Solve each equation.

8. $12 = y - 3y$

9. $\dfrac{x}{2} = -5 - (-2)$

10. $5 + 4z = 37$

11. $5x + 12 - 4x - 14 = 22$

12. $-4x + 7 = 15$

13. $2(x - 6) = 0$

14. $5x - 2 = x - 10$

15. $4(5x + 3) = 2(7x + 6)$

16. Translate the following phrases into mathematical expressions. If needed, use x to represent "a number."

  a. The sum of -23 and a number

  b. Three times a number, subtracted from -2

17. Translate each sentence into an equation. If needed, use x to represent "a number."

  a. The sum of twice 5 and -15 is -5.

  b. Six added to three times a number equals -30.

Answers

1. _____

2. _____

3. _____

4. _____

5. _____

6. _____

7. _____

8. _____

9. _____

10. _____

11. _____

12. _____

13. _____

14. _____

15. _____

16. a. _____

  b. _____

17. a. _____

  b. _____

Solve.

18. The difference of three times a number and five times the same number is 4. Find the number.

19. In a championship basketball game, Paula Zimmerman made twice as many free throws as Maria Kaminsky. If the total number of free throws made by both women was 12, find how many free throws Paula made.

20. In a 10-kilometer race, there are 112 more men entered than women. Find the number of female runners if the total number of runners in the race is 600.

Name_____ Section_____ Date _____

Chapter 3 Cumulative Review

1. Write 106,052,447 in words.

2. Write 276,004 in words.

△**3.** Find the perimeter of the polygon shown.

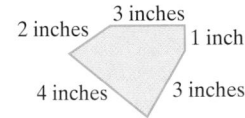

4. Find the perimeter of the rectangle shown.

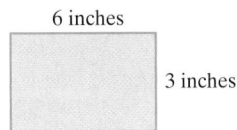

5. Subtract: $900 - 174$. Check by adding.

6. Subtract: $17,801 - 8216$ Check by adding.

7. Round 248,982 to the nearest hundred.

8. Round 844,497 to the nearest thousand.

9. Multiply: 25×8

10. Multiply: 395×74

11. Divide and check: $1872 \div 9$

12. Divide and check: $3956 \div 46$

13. Simplify: $2 \cdot 4 - 3 \div 3$

14. Simplify: $8 \cdot 4 + 9 \div 3$

15. Evaluate $x^2 + z - 3$ for $x = 5$ and $z = 4$.

16. Evaluate $2a^2 + 5 - c$ for $a = 2$ and $c = 3$.

17. Insert $<$ or $>$ between each pair of numbers to make a true statement.

    a. -7 7
    b. 0 -4
    c. -9 -11

18. Insert $<$ or $>$ to make a true statement.

    a. -14 0
    b. $-(-7)$ -8

19. Add using a number line: $5 + (-2)$

20. Add using a number line: $-3 + (-4)$

Add.

21. $-5 + (-1)$

22. $3 + (-7)$

Answers

1. _____

2. _____

3. _____

4. _____

5. _____

6. _____

7. _____

8. _____

9. _____

10. _____

11. _____

12. _____

13. _____

14. _____

15. _____

16. _____

17. a. _____

b. _____

c. _____

18. a. _____

b. _____

19. _____

20. _____

21. _____

22. _____

23. _____

24. _____

25. _____

26. _____

27. _____

28. _____

29. _____

30. _____

31. _____

32. _____

33. _____

34. _____

35. _____

36. _____

37. _____

38. _____

39. _____

40. _____

41. _____

42. _____

43. _____

44. _____

45. _____

46. _____

47. _____

48. _____

49. _____

50. _____

23. $2 + 6$

24. $21 + 15 + (-19)$

Subtract.

25. $-4 - 10$

26. $-2 - 3$

27. $6 - (-5)$

28. $19 - (-10)$

29. $-11 - (-7)$

30. $-16 - (-13)$

Divide.

31. $\dfrac{-12}{6}$

32. $\dfrac{-30}{-5}$

33. $-20 \div (-4)$

34. $26 \div (-2)$

35. $\dfrac{48}{-3}$

36. $\dfrac{-120}{12}$

Find the value of each expression.

37. $(-3)^2$

38. -2^5

39. -3^2

40. $(-5)^2$

41. Simplify:
$2y - 6 + 4y + 8$

42. Simplify:
$6x + 2 - 3x + 7$

43. Determine whether -1 is a solution of the equation $3y + 1 = 3$.

44. Determine whether 2 is a solution of $5x - 3 = 7$.

45. Solve: $-12x = -36$

46. Solve: $7y - 4y = 15$

47. Solve: $2x - 6 = 18$

48. Solve: $3a + 5 = -1$

49. Leo Leal sold a used computer system and software for $2100, receiving four times as much money for the computer system as for the software. Find the price of each.

50. Rose Daunis is thinking of a number. Two times the number, plus four is the same amount as three times the number minus seven. Find Rose's number.

Fractions

Fractions are numbers and, like whole numbers and integers, they can be added, subtracted, multiplied, and divided. Fractions are very useful and appear frequently in everyday language, in common phrases such as "half an hour," "quarter of a pound," and "third of a cup." This chapter reviews the concept of fractions and demonstrates how to add, subtract, multiply, and divide fractions. In this chapter we also solve linear equations containing fractions.

Veterinarians are doctors who diagnose, treat, and work to prevent animal diseases. According to the American Veterinary Medicine Association (AVMA), there are over 67,000 professionally active veterinarians in the United States today. Many of these work in private practice, caring for household pets or livestock. Some veterinarians work in the area of public health, mainly with the control of rabies. Research veterinarians work to establish and improve animal vaccines for diseases. There are currently 28 accredited colleges of veterinary medicine in the United States. Following graduation, the veterinarian must pass a licensing examination for the state in which he or she plans to practice. In Exercise 75 on page 270, we will see how veterinarians use fractions to compute the amount of flea medication to include in a dipping vat.

Name _____ Section _____ Date _____

Chapter **4** Pretest

1. Write a fraction to represent the shaded area of the figure.

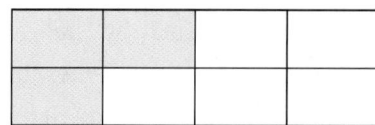

2. Write $\dfrac{5}{6}$ as an equivalent fraction whose denominator is 18.

3. Simplify: $\dfrac{0}{-4}$

4. Write the prime factorization of 140.

5. Simplify: $\dfrac{30}{54}$

6. There are 1000 milligrams in 1 gram. What fraction of a gram is 450 milligrams?

Perform each indicated operation and simplify.

7. $\dfrac{3}{4} \cdot \dfrac{24}{15}$

8. $\dfrac{5x}{7} \div 10x$

9. $\dfrac{8}{11} - \dfrac{3}{11}$

10. $-\dfrac{3}{10} + \dfrac{2}{10}$

11. Evaluate: $\left(-\dfrac{2}{3}\right)^3$

12. Evaluate $x \div y$ if $x = \dfrac{2}{9}$ and $y = -\dfrac{2}{3}$.

13. Add: $\dfrac{5}{9} + \dfrac{1}{12}$

14. Evaluate $x - y$ if $x = -\dfrac{3}{14}$ and $y = -\dfrac{2}{7}$.

Simplify.

15. $\dfrac{\frac{x}{3}}{\frac{7}{9}}$

16. $\left(\dfrac{2}{5}\right)^2 - 2$

17. Solve: $\dfrac{x}{4} + 3 = \dfrac{1}{8}$

18. Write $2\dfrac{3}{5}$ as an improper fraction.

Perform each indicated operation and simplify.

19. $3\dfrac{1}{5} \cdot 2\dfrac{3}{4}$

20. $5\dfrac{2}{3} + 4\dfrac{1}{6}$

4.1   Introduction to Fractions and Equivalent Fractions

A   Identifying Numerators and Denominators

Whole numbers are used to count whole things or units, such as cars, ball games, horses, dollars, and people. To refer to a part of a whole, fractions are used. For example, a whole baseball game is divided into nine parts called innings. If a player pitched five complete innings, the fraction $\frac{5}{9}$ can be used to show the part of a whole game he or she pitched. The 9 in the fraction $\frac{5}{9}$ is called the denominator, and it refers to the total number of equal parts (innings) in the whole game. The 5 in the fraction $\frac{5}{9}$ is called the numerator, and it tells how many of those equal parts (innings) the pitcher pitched.

$$\frac{5}{9} \quad \begin{array}{l} \leftarrow \text{ how many of the parts being considered} \\ \leftarrow \text{ number of equal parts in the whole} \end{array}$$

EXAMPLES   Identify the numerator and the denominator of each fraction.

1. $\frac{3}{7} \quad \begin{array}{l} \leftarrow \text{ numerator} \\ \leftarrow \text{ denominator} \end{array}$ **2.** $\frac{3}{7} \quad \begin{array}{l} \leftarrow \text{ numerator} \\ \leftarrow \text{ denominator} \end{array}$

 **Helpful Hint**

$\frac{3}{7} \leftarrow$ Remember that the bar in a fraction means division. Since division by 0 is undefined, a fraction with a denominator of 0 is undefined.

B   Writing Fractions to Represent Shaded Areas of Figures

One way to become familiar with the concept of fractions is to visualize fractions with shaded figures. We can then write a fraction to represent the shaded area of the figure.

EXAMPLES   Write a fraction to represent the shaded area of each figure.

3.

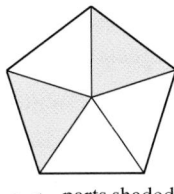

$\frac{2}{5} \quad \begin{array}{l} \leftarrow \text{ parts shaded} \\ \leftarrow \text{ equal parts} \end{array}$

The figure is divided into 5 equal parts and 2 of them are shaded.

Thus, $\frac{2}{5}$ of the figure is shaded.

4.

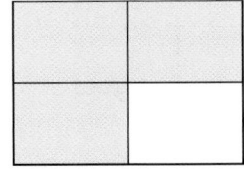

$\frac{3}{4} \quad \begin{array}{l} \leftarrow \text{ parts shaded} \\ \leftarrow \text{ equal parts} \end{array}$

The figure is divided into 4 equal parts and 3 of them are shaded.

Thus, $\frac{3}{4}$ of the figure is shaded.

OBJECTIVES

Ⓐ Identify the numerator and the denominator of a fraction.

Ⓑ Write a fraction to represent the shaded area of a figure.

Ⓒ Graph fractions on a number line.

Ⓓ Simplify fractions of the form $\frac{a}{a}$, $\frac{a}{1}$, and $\frac{0}{a}$.

Ⓔ Write equivalent fractions.

SSM   SG   CD & VIDEO   MATH PRO   WEB
TUTOR CENTER

Practice Problems 1–2

Identify the numerator and the denominator of each fraction.

1. $\frac{9}{2}$ 2. $\frac{10y}{17}$

Practice Problems 3–4

Write a fraction to represent the shaded area of each figure.

3.

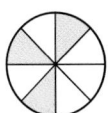

4.

Answers

1. $\frac{9}{2} \quad \begin{array}{l} \leftarrow \text{ numerator} \\ \leftarrow \text{ denominator} \end{array}$ **2.** $\frac{10y}{17} \quad \begin{array}{l} \leftarrow \text{ numerator} \\ \leftarrow \text{ denominator} \end{array}$

3. $\frac{3}{8}$ **4.** $\frac{1}{6}$

Practice Problems 5–6

Draw and shade a part of a diagram to represent each fraction.

5. $\frac{7}{11}$ of a diagram

6. $\frac{2}{3}$ of a diagram

Concept Check

Identify each as a proper fraction or improper fraction.

a. $\frac{6}{7}$ b. $\frac{13}{12}$ c. $\frac{2}{2}$ d. $\frac{99}{101}$

Answers

5. answers may vary; for example,

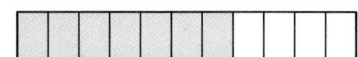

6. answers may vary; for example,

Concept Check:
a. proper   **b.** improper   **c.** improper   **d.** proper

EXAMPLES Draw and shade a part of a diagram to represent each fraction.

5. $\frac{5}{6}$ of a diagram

We can use a geometric figure such as a rectangle and divide it into 6 equal parts. Then we will shade 5 of the equal parts.

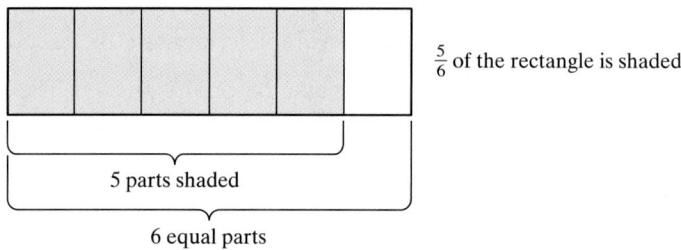

$\frac{5}{6}$ of the rectangle is shaded

5 parts shaded

6 equal parts

6. $\frac{3}{8}$ of a diagram

If you'd like, our diagram can consist of 8 triangles of the same size. We will shade 3 of the triangles.

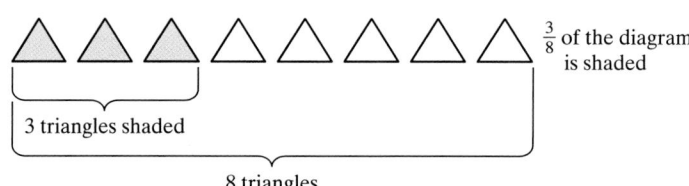

$\frac{3}{8}$ of the diagram is shaded

3 triangles shaded

8 triangles

A **proper fraction** is a fraction whose numerator is less than its denominator. Proper fractions have values that are less than 1. The shaded portion of the triangle's area is represented by $\frac{2}{3}$.

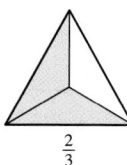

$\frac{2}{3}$

An **improper fraction** is a fraction whose numerator is greater than or equal to its denominator. Improper fractions have values that are greater than or equal to 1. The area of the shaded part of the group of circles is $\frac{9}{4}$. The shaded part of the rectangle's area is $\frac{6}{6}$.

Whole circle

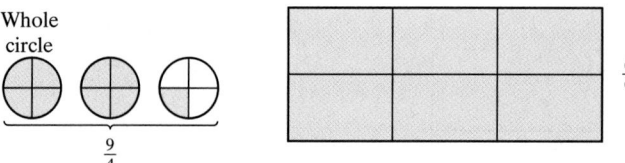

$\frac{9}{4}$

$\frac{6}{6}$

Try the Concept Check in the margin.

EXAMPLES   Represent the shaded part of each figure group with an improper fraction.

7.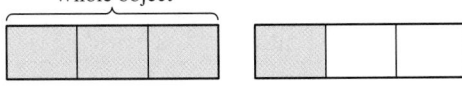
Whole object

improper fraction: $\frac{4}{3}$

8.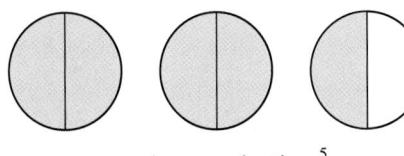

improper fraction: $\frac{5}{2}$

EXAMPLE 9   Writing Fractions from Real-Life Data

Of the nine planets in our solar system, two are closer to the sun than Earth is. What fraction of the planets are closer to the sun than Earth is?

Solution: The fraction closer to the sun is

$\frac{2}{9}$ ← number of planets that are closer
        ← number of planets in our solar system

Thus, $\frac{2}{9}$ of the planets in our solar system are closer to the sun than Earth is. ●

C   Graphing Fractions on a Number Line

Another way to visualize fractions is to graph them on a number line. To do this, think of 1 unit on the number line as a whole. To graph $\frac{2}{5}$, for example, divide the distance from 0 to 1 into 5 equal parts. Then start at 0 and count 2 parts to the right.

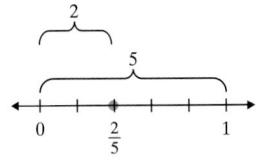

Practice Problems 7–8

Represent the shaded part of each figure group with an improper fraction.

7.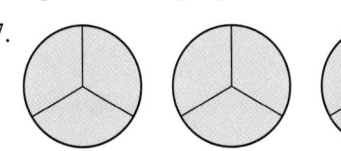

8.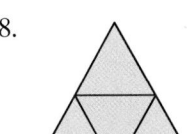

●

Practice Problem 9

Of the nine planets in our solar system, seven are farther from the sun than Venus is. What fraction of the planets are farther from the sun than Venus is?

Answers

7. $\frac{8}{3}$ **8.** $\frac{5}{4}$ **9.** $\frac{7}{9}$

Practice Problem 10

Graph each proper fraction on a number line.

a. $\dfrac{5}{7}$ b. $\dfrac{2}{3}$ c. $\dfrac{4}{6}$

Practice Problem 11

Graph each improper fraction on a number line.

a. $\dfrac{8}{3}$ b. $\dfrac{5}{4}$ c. $\dfrac{7}{7}$

EXAMPLE 10   Graph each proper fraction on a number line.

a. $\dfrac{3}{4}$ b. $\dfrac{1}{2}$ c. $\dfrac{3}{6}$

Solution: **a.** Divide the distance from 0 to 1 into 4 parts. Then start at 0 and count over 3 parts.

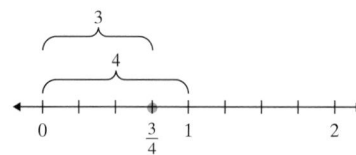

b.

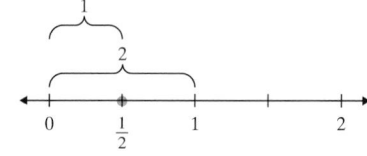

c.

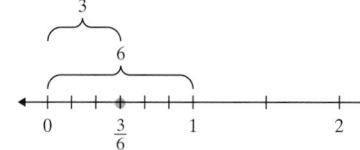

The fractions in Example 10 are all proper fractions. Notice that the value of each is less than 1. This is always true for proper fractions since the numerator of a proper fraction is less than the denominator.

EXAMPLE 11   Graph each improper fraction on a number line.

a. $\dfrac{7}{6}$ b. $\dfrac{9}{5}$ c. $\dfrac{6}{6}$ d. $\dfrac{3}{1}$

Solution:

a. 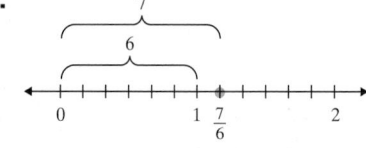   **b.**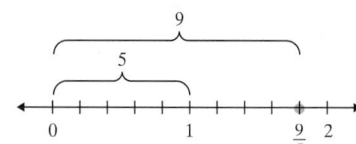

c.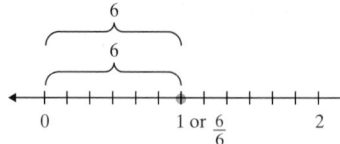

d. Each 1-unit distance has 1 equal part. Count over 3 parts.

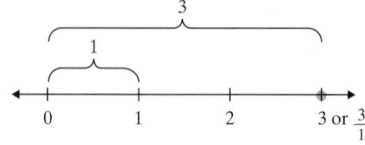

Answers

10. a.

0    $\dfrac{5}{7}$ 1

b.

0    $\dfrac{2}{3}$ 1

c.

0    $\dfrac{4}{6}$ 1

11. a.

0   1   2 $\dfrac{8}{3}$ 3

b.

0   1 $\dfrac{5}{4}$ 2

c.

0   1 or $\dfrac{7}{7}$

The fractions in Example 11 are all improper fractions. Notice that the value of each is greater than or equal to 1. This is always true since the numerator of an improper fraction is greater than or equal to the denominator.

Ⓓ Simplifying $\frac{a}{a}$, $\frac{a}{1}$, and $\frac{0}{a}$.

The graphs of $\frac{6}{6}$ and $\frac{3}{1}$ from Example 11 are shown next.

Notice that the graph of $\frac{6}{6}$ is the same as the graph of 1, and the graph of $\frac{3}{1}$ is the same as 3. This makes sense if we recall that the fraction bar indicates division. Let's review some division properties for 1 and 0.

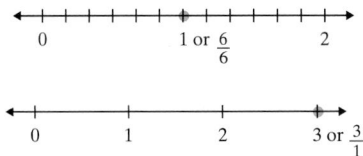

$\frac{6}{6} = 1$ because $1 \cdot 6 = 6$ $\frac{3}{1} = 3$ because $3 \cdot 1 = 3$

$\frac{0}{6} = 0$ because $0 \cdot 6 = 0$

$\frac{6}{0}$ *is undefined* because there is no number that when multiplied by 0 gives 6.

In general, we can say the following.

Let *n* be any integer.

$\frac{n}{n} = 1$ as long as *n* is not 0. $\frac{0}{n} = 0$ as long as *n* is not 0.

$\frac{n}{1} = n$ $\frac{n}{0}$ is undefined.

EXAMPLES Simplify.

12. $\frac{5}{5} = 1$ **13.** $\frac{-2}{-2} = 1$ **14.** $\frac{0}{-5} = 0$

15. $\frac{-5}{1} = -5$ **16.** $\frac{41}{1} = 41$ **17.** $\frac{19}{0}$ is undefined   ●

Practice Problems 12–17

Simplify.

12. $\frac{9}{9}$ **13.** $\frac{-6}{-6}$ **14.** $\frac{0}{-1}$

15. $\frac{4}{1}$ **16.** $\frac{-13}{0}$ **17.** $\frac{-13}{1}$

Answers

12. 1   **13.** 1   **14.** 0

15. 4   **16.** undefined   **17.** −13

E Writing Equivalent Fractions

When we graph $\frac{1}{3}$ and $\frac{2}{6}$ on a number line,

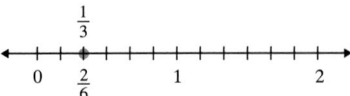

notice that both $\frac{1}{3}$ and $\frac{2}{6}$ correspond to the same point. These fractions are called **equivalent fractions**, and we write $\frac{1}{3} = \frac{2}{6}$.

Another way to visualize equivalent fractions is to use shaded figures.

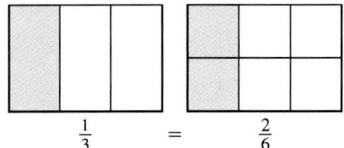

From the shaded figures, we see that equivalent fractions also represent the same portion of a whole.

Equivalent Fractions

Fractions that represent the same portion of a whole or the same point on a number line are called **equivalent fractions**.

To write equivalent fractions, we use the **fundamental property of fractions**. This property guarantees that, if we multiply or divide both the numerator and the denominator by the same nonzero number, the result is an equivalent fraction. For example, if we multiply both the numerator and denominator of $\frac{1}{3}$ by 2, the result is the equivalent fraction $\frac{2}{6}$.

$$\frac{1 \cdot 2}{3 \cdot 2} = \frac{2}{6}$$

Fundamental Property of Fractions

If a, b, and c are numbers, then

$$\frac{a}{b} = \frac{a \cdot c}{b \cdot c} \qquad \text{and also} \qquad \frac{a}{b} = \frac{a \div c}{b \div c}$$

as long as b and c are not 0. In other words, if the numerator and denominator of a fraction are multiplied or divided by the **same** nonzero number, the result is an equivalent fraction.

EXAMPLE 18 Write $\dfrac{2}{5}$ as an equivalent fraction whose denominator is 15.

Solution: Since $5 \cdot 3 = 15$, we use the fundamental property of fractions and multiply the numerator and denominator of $\dfrac{2}{5}$ by 3.

$$\frac{2}{5} = \frac{2 \cdot 3}{5 \cdot 3} = \frac{6}{15}$$

Then $\dfrac{2}{5}$ is equivalent to $\dfrac{6}{15}$. They both represent the same part of a whole. ●

EXAMPLE 19 Write $\dfrac{9x}{11}$ as an equivalent fraction whose denominator is 44.

Solution: Since $11 \cdot 4 = 44$, we multiply the numerator and denominator by 4.

$$\frac{9x}{11} = \frac{9x \cdot 4}{11 \cdot 4} = \frac{36x}{44}$$

Then $\dfrac{9x}{11}$ is equivalent to $\dfrac{36x}{44}$. ●

EXAMPLE 20 Write 3 as an equivalent fraction whose denominator is 7.

Solution: Recall that $3 = \dfrac{3}{1}$. Since $1 \cdot 7 = 7$, multiply the numerator and denominator by 7.

$$\frac{3}{1} = \frac{3 \cdot 7}{1 \cdot 7} = \frac{21}{7}$$ ●

In this chapter, we will perform operations on fractions containing variables. When the denominator of a fraction contains a variable, such as $\dfrac{8}{3x}$, we will assume that the variable does not represent 0. Recall that the denominator of a fraction cannot be 0.

EXAMPLE 21 Write $\dfrac{8}{3x}$ as an equivalent fraction whose denominator is 12x.

Solution: Since $3x \cdot 4 = 12x$, multiply the numerator and denominator by 4.

$$\frac{8}{3x} = \frac{8 \cdot 4}{3x \cdot 4} = \frac{32}{12x}$$ ●

Try the Concept Check in the margin.

Practice Problem 18

Write $\dfrac{1}{4}$ as an equivalent fraction whose denominator is 20.

Practice Problem 19

Write $\dfrac{3x}{7}$ as an equivalent fraction with a denominator of 42.

Practice Problem 20

Write 4 as an equivalent fraction whose denominator is 6.

Practice Problem 21

Write $\dfrac{9}{4x}$ as an equivalent fraction with a denominator of 20x.

Concept Check

What is the first step in writing $\dfrac{3}{10}$ as an equivalent fraction whose denominator is 100?

Answers

18. $\dfrac{5}{20}$ **19.** $\dfrac{18x}{42}$ **20.** $\dfrac{24}{6}$ **21.** $\dfrac{45}{20x}$

Concept Check: Answers may vary.

FOCUS ON **Mathematical Connections**

MODELING FRACTIONS

There are several different physical models for representing fractions.

SET MODEL

In this model, a fraction represents the portion of a set of objects that has a certain characteristic. For example, in the set of 10 shapes, 3 are hearts. That is, $\frac{3}{10}$ of the shapes are hearts.

AREA MODEL

In this model, a shape is divided into a number of equal-sized regions. A fraction can be represented by shading some of the regions. For example, both of the following models represent the fraction $\frac{7}{8}$.

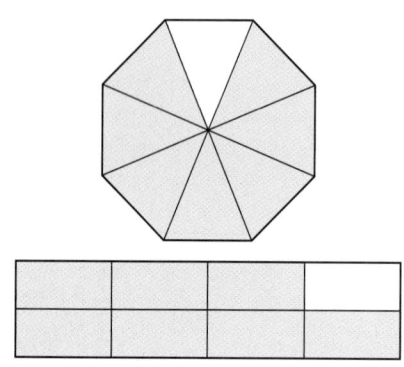

NUMBER LINE MODEL

For this model, draw a line and label a point 0 and a point to its right, 1. Now subdivide this distance from 0 to 1 depending on the denominator of the fraction that is to be represented. For example, the fraction $\frac{3}{4}$ is graphed by subdividing the portion of the number line between 0 and 1 into four equal lengths and placing a dot at the mark that represents $\frac{3}{4}$ of the distance between 0 and 1.

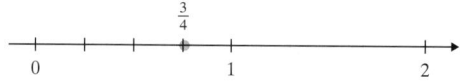

CRITICAL THINKING

1. a. Represent the fraction $\frac{5}{6}$ in two different ways using the set model.

b. Represent the fraction $\frac{5}{6}$ in two different ways using the area model.

c. Represent the fraction $\frac{5}{6}$ using the number line model.

2. Which model do you prefer? Why?

3. Which model do you think would be most useful for representing multiplication of fractions? Explain how this could be done.

Mental Math

A *Identify the numerator and the denominator of each fraction. See Examples 1 and 2.*

1. $\dfrac{1}{2}$

2. $\dfrac{1}{4}$

3. $\dfrac{10}{3}$

4. $\dfrac{53}{21}$

5. $\dfrac{3z}{7}$

6. $\dfrac{11x}{15}$

EXERCISE SET 4.1

B *Write a fraction to represent the shaded area of each figure. See Examples 3 and 4.*

 1.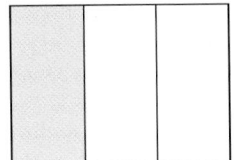

2.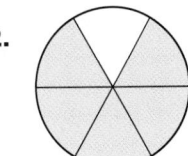

3.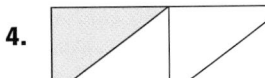

4.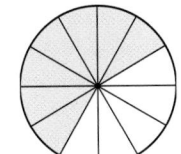

5.

6.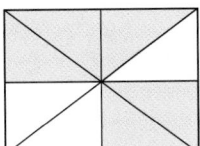

7.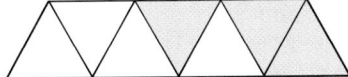

8.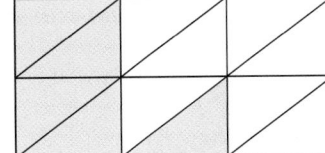

9.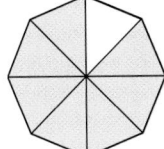

10.

Use the pizza illustration for Exercises 11 and 12.

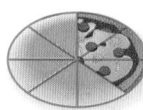

11. What fraction of the pizza is gone?

12. What fraction of the pizza remains?

Draw and shade a part of a diagram to represent each fraction. See Examples 5 and 6.

13. $\frac{5}{8}$ of a diagram

14. $\frac{1}{4}$ of a diagram

15. $\frac{1}{5}$ of a diagram

16. $\frac{3}{5}$ of a diagram

17. $\frac{6}{7}$ of a diagram

18. $\frac{7}{9}$ of a diagram

19. $\frac{4}{4}$ of a diagram

20. $\frac{6}{6}$ of a diagram

Represent the shaded area in each figure group with an improper fraction. See Examples 7 and 8.

21.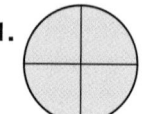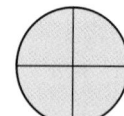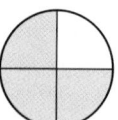

22.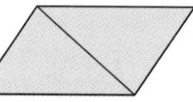

23.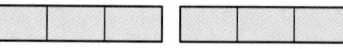

24.

25.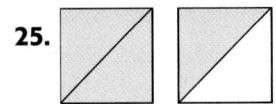

26.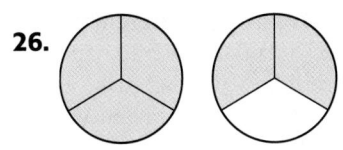

27.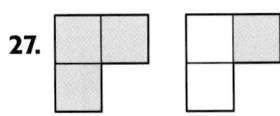

28.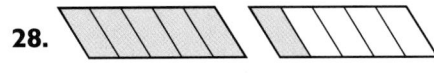

29.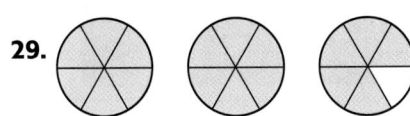

30.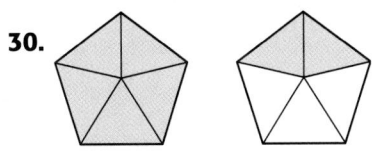

31.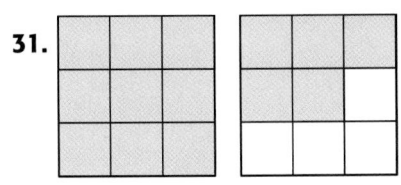

32.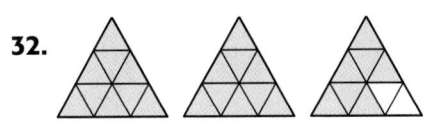

Write each fraction. See Example 9.

33. Of the 131 students at a small private school, 42 are freshmen. What fraction of the students are freshmen?

34. Of the 78 executives at a private accounting firm, 61 are men. What fraction of the executives are men?

35. From Exercise 33, how many students are *not* freshmen? What fraction of the students are *not* freshmen?

36. From Exercise 34, how many of the executives are women? What fraction of the executives are women?

37. According to a recent study, 4 out of 10 visits to U.S. hospital emergency rooms were for an injury. What fraction of emergency room visits are injury-related? (*Source*: National Center for Health Statistics)

38. The average American driver spends about 1 hour each day in a car or truck. What fraction of a day does an average American driver spend in a car or truck? (*Hint:* How many hours are in a day?) (*Source:* U.S. Department of Transportation—Federal Highway Administration)

39. As of 2003, the United States had 43 different presidents. A total of eight U.S. presidents were born in the state of Virginia, more than any other state. What fraction of U.S. presidents were born in Virginia? (*Source:* 2003 World Almanac)

Eight U.S. Presidents

40. Of the nine planets in our solar system, four have days that are longer than the 24-hour Earth day. What fraction of the planets have longer days than Earth has? (*Source:* National Space Science Data Center)

41. The hard drive in Aaron Hawn's Computer can hold 32 gigabytes of information. He has currently used 17 gigabytes. What fraction of his hard drive has he used?

42. There are 12 inches in a foot. What fractional part of a foot does 5 inches represent?

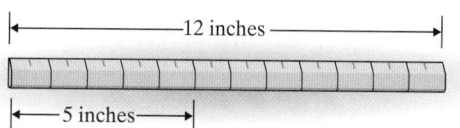

43. There are 31 days in the month of March. What fraction of the month does 11 days represent?

44. There are 60 minutes in an hour. What fraction of an hour does 37 minutes represent?

Mon.	Tue.	Wed.	Thu.	Fri.	Sat.	Sun.
					1	2
3	4	5	6	7	8	9
10	11	12	13	14	15	16
17	18	19	20	21	22	23
24	25	26	27	28	29	30
31						

45. In a prealgebra class containing 31 students, there are 18 freshmen, 10 sophomores, and 3 juniors. What fraction of the class is sophomores?

46. In a family with 11 children, there are 4 boys and 7 girls. What fraction of the children is girls?

47. Consumer fireworks are legal in 40 states in the United States.
   a. In what fraction of the states are consumer fireworks legal?
   b. In how many states are consumer fireworks illegal?
   c. In what fraction of the states are consumer fireworks illegal? (*Source:* United States Fireworks Safety Council)

48. Thirty-three states in the United States contain federal Indian reservations.
   a. What fraction of the states contain Indian reservations?
   b. How many states do not contain Indian reservations?
   c. What fraction of the states do not contain Indian reservations? (*Source:* Tiller Research, Inc., Albuquerque, NM)

C *Graph each fraction on a number line. See Examples 10 and 11.*

49. $\dfrac{1}{4}$

50. $\dfrac{1}{3}$

51. $\dfrac{4}{7}$

52. $\dfrac{5}{6}$

53. $\dfrac{8}{5}$

54. $\dfrac{9}{8}$

55. $\dfrac{7}{3}$

56. $\dfrac{15}{7}$

57. $\dfrac{3}{8}$

58. $\dfrac{8}{3}$

D *Simplify by dividing. See Examples 12 through 17.*

59. $\dfrac{12}{12}$

60. $\dfrac{-3}{-3}$

61. $\dfrac{-5}{1}$

62. $\dfrac{-10}{1}$

63. $\dfrac{0}{-2}$

64. $\dfrac{0}{-8}$

65. $\dfrac{-8}{-8}$

66. $\dfrac{-14}{-14}$

67. $\dfrac{-9}{0}$

68. $\dfrac{-7}{0}$

69. $\dfrac{3}{1}$

70. $\dfrac{5}{5}$

E *Write each fraction as an equivalent fraction with the given denominator. See Examples 18 through 21.*

71. $\frac{4}{7}$; denominator of 35

72. $\frac{3}{5}$; denominator of 20

73. $\frac{2}{3}$; denominator of 21

74. $\frac{1}{6}$; denominator of 24

75. $\frac{2y}{5}$; denominator of 25

76. $\frac{9a}{10}$; denominator of 70

77. $\frac{1}{2}$; denominator of 30

78. $\frac{1}{3}$; denominator of 30

79. $\frac{10}{7x}$; denominator of 21x

80. $\frac{5}{3b}$; denominator of 21b

81. 2; denominator of 5

82. 5; denominator of 8

Write each fraction as an equivalent fraction whose denominator is 12. See Examples 18 through 21.

83. $\frac{3}{4}$

84. $\frac{4}{6}$

85. $\frac{2y}{3}$

86. $\frac{2}{3}$

87. $\frac{1}{2}$

88. $\frac{3x}{2}$

Write each fraction as an equivalent fraction whose denominator is 36x. See Example 18 through 21.

89. $\frac{4}{3}$

90. $\frac{3}{4}$

91. $\frac{5}{9}$

92. $\frac{7}{6}$

93. 1

94. 2

The table shows the fraction of the population in each country that used cell phones in a recent year. Use this table to answer Exercises 95–98.

95. Complete the table by writing each fraction as an equivalent fraction with a denominator of 100.

96. Which of these countries has the largest fraction of cell phone users?

97. Which of these countries has the smallest fraction of cell phone users?

98. In which of these countries do over $\frac{3}{4}$ of the population use cell phones?

Country	Fraction of Population Using Cell Phones	Equivalent Fraction with a Denominator of 100
Denmark	$\frac{13}{25}$	
Finland	$\frac{39}{50}$	
Israel	$\frac{87}{100}$	$\frac{87}{100}$
Italy	$\frac{21}{25}$	
Japan	$\frac{59}{100}$	$\frac{59}{100}$
Norway	$\frac{83}{100}$	$\frac{83}{100}$
Singapore	$\frac{67}{100}$	$\frac{67}{100}$
South Korea	$\frac{6}{10}$	
Sweden	$\frac{79}{100}$	$\frac{79}{100}$
United States	$\frac{9}{20}$	

(*Source: International Telecommunication and World Almanac,* 2003)

Helpful Hint

Write $\frac{3}{4}$ as an equivalent fraction with a denominator of 100.

Review and Preview

Simplify. See Section 1.8.

99. 3^2 **100.** 4^3 **101.** 5^3 **102.** 3^4

103. 7^2 **104.** 5^4 **105.** $2^3 \cdot 3$ **106.** $4^2 \cdot 5$

Combining Concepts

107. Write $\frac{2}{9}$ as an equivalent fraction whose denominator is 2088.

108. Write $\frac{3}{11}$ as an equivalent fraction whose denominator is 6479.

109. In your own words, explain how to write equivalent fractions.

110. In your own words, explain why $\frac{0}{10} = 0$ and $\frac{10}{0}$ is undefined.

Write each fraction.

111. Habitat for Humanity is a nonprofit organization that helps provide affordable housing to families in need. Habitat for Humanity does its work of building and renovating houses through 1651 local affiliates in the United States and 634 international affiliates. What fraction of the total Habitat for Humanity affiliates are located in the United States? (*Source:* Habitat for Humanity International)

112. The United States Marine Corps (USMC) has five principal training centers in California, three in North Carolina, two in South Carolina, one in Arizona, one in Hawaii, and one in Virginia. What fraction of the total USMC principal training centers are located in California? (*Source:* U.S. Department of Defense)

113. The Wendy's Corporation owns restaurants with five different names, as shown on the bar graph. What fraction of restaurants owned by Wendy's corporation are named "Wendy's" restaurants? (*Source:* The Wendy's Corporation)

114. The Public Broadcasting Service (PBS) provides programming to the noncommercial public TV stations of the United States. The table shows a breakdown of the public television licensees by type. Each licensee operates one or more PBS member TV stations. What fraction of the public television licensees are universities or colleges? (*Source:* The Public Broadcast Service)

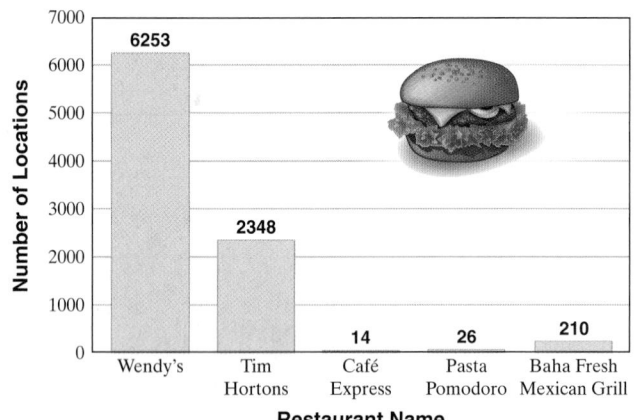

Wendy's Corporation Restaurant Ownership

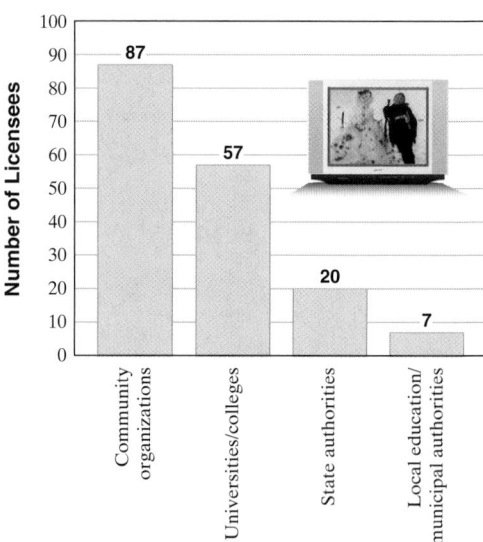

Public Television Licensees

4.2 Factors and Simplest Form

 **Writing a Number as a Product of Prime Numbers**

Of all the equivalent ways to write a particular fraction, one special way is called **simplest form** or **lowest terms**. To help us write a fraction in simplest form, we first practice writing a number as a product of prime numbers.

> A prime number is a natural number greater than 1 whose only factors are 1 and itself. The first few prime numbers are 2, 3, 5, 7, 11, 13, 17, 19, 23, 29, ...
>
> A **composite number** is a natural number greater than 1 that is not prime.

Helpful Hint

The natural number 1 is neither prime nor composite.

When a composite number is written as a product of prime numbers, this product is called the **prime factorization** of the number. For example, the prime factorization of 12 is $2 \cdot 2 \cdot 3$ because

$$12 = \underline{2 \cdot 2 \cdot 3}$$

This product is 12 and each number is a prime number.

Because multiplication is commutative, the order of the factors is not important. We can write the factorization $2 \cdot 2 \cdot 3$ as $2 \cdot 3 \cdot 2$ or $3 \cdot 2 \cdot 2$. Any of these is called the prime factorization of 12.

> Every whole number greater than 1 has exactly one prime factorization.

Recall from Section 1.5 that since $12 = 2 \cdot 2 \cdot 3$, the numbers 2 and 3 are called *factors* of 12. A **factor** is any number that divides a number evenly (with a remainder of 0).

One method for finding the prime factorization of a number is by using a factor tree, as shown in the next example.

EXAMPLE 1 Write the prime factorization of 45.

Solution: We can begin by writing 45 as the product of two numbers, say 5 and 9.

$$
\begin{array}{c}
45 \\
\diagup \quad \diagdown \\
5 \quad \cdot \quad 9
\end{array}
$$

The number 5 is prime but 9 is not, so we write 9 as $3 \cdot 3$.

$$
\begin{array}{c}
45 \\
\diagup \quad \diagdown \\
5 \quad \cdot \quad 9 \\
\diagup \quad \diagdown \\
5 \quad \cdot \quad 3 \quad \cdot \quad 3
\end{array}
\Bigg\} \text{A factor tree}
$$

Each factor is now a prime number, so the prime factorization of 45 is $3 \cdot 3 \cdot 5$ or $3^2 \cdot 5$.

OBJECTIVES

A Write a number as a product of prime numbers.

B Write a fraction in simplest form.

C Solve problems by writing fractions in simplest form.

  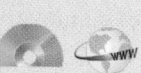

SSM
TUTOR CENTER   SG   CD & VIDEO   MATH PRO   WEB

Practice Problem 1

Write the prime factorization of 28.

Answer

1. $28 = 2 \cdot 2 \cdot 7$ or $2^2 \cdot 7$

Practice Problem 2

Write the prime factorization of 60.

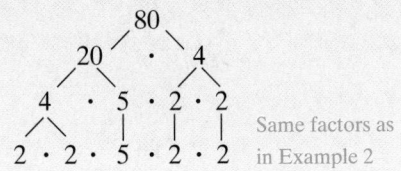
EXAMPLE 2   Write the prime factorization of 80.

Solution: Write 80 as a product of two numbers. Continue this process until all factors are prime.

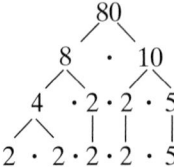

All factors are now prime, so the prime factorization of 80 is

$$2 \cdot 2 \cdot 2 \cdot 2 \cdot 5 \quad \text{or} \quad 2^4 \cdot 5.$$

There are a few quick **divisibility tests** to determine whether a number is divisible by the primes 2, 3, 5, or 10.

Divisibility Tests

A whole number is divisible by

- **2** if its last digit is 0, 2, 4, 6, or 8.

  132 is divisible by 2 since the last digit is a 2.

- **3** if the sum of its digits is divisible by 3.

  144 is divisible by 3 since $1 + 4 + 4 = 9$ is divisible by 3.

- **5** if its last digit is 0 or 5.

  1115 is divisible by 5 since the last digit is a 5.

- **10** if its last digit is 0.

  230 is divisible by 10 since the last digit is 0.

Helpful Hint

Here are a few other divisibility tests you may find interesting. A whole number is divisible by

- **4** if its last two digits are divisible by 4.

  1712 is divisible by 4.

- **6** if it is divisible by 2 and 3.

  9858 is divisible by 6.

- **9** if the sum of its digits is divisible by 9.

  5238 is divisible by 9 since $5 + 2 + 3 + 8 = 18$ is divisible by 9.

When finding the prime factorization of larger numbers, you may want to use the procedure shown in Example 3.

EXAMPLE 3   Write the prime factorization of 252.

Solution: For this method, we divide prime numbers into the given number. Since the ones digit of 252 is 2, we know that 252 is divisible by 2.

$$\begin{array}{r} 126 \\ 2\overline{)252} \end{array}$$

Practice Problem 3

Write the prime factorization of 297.

Answers

2. $60 = 2 \cdot 2 \cdot 3 \cdot 5$ or $2^2 \cdot 3 \cdot 5$

3. $297 = 3 \cdot 3 \cdot 3 \cdot 11$ or $3^3 \cdot 11$

126 is divisible by 2 also.

$$\begin{array}{r} 63 \\ 2\overline{)126} \\ 2\overline{)252} \end{array}$$

63 is not divisible by 2 but is divisible by 3. Divide 63 by 3 and continue in this same manner until the quotient is a prime number.

$$\begin{array}{r} 7 \\ 3\overline{)21} \\ 3\overline{)63} \\ 2\overline{)126} \\ 2\overline{)252} \end{array}$$

Helpful Hint:

The order of choosing prime numbers does not matter. For consistency, we use the order 2, 3, 5, 7, ...

The prime factorization of 252 is $2 \cdot 2 \cdot 3 \cdot 3 \cdot 7$ or $2^2 \cdot 3^2 \cdot 7$.   ●

Try the Concept Check in the margin.

B   Writing a Fraction in Simplest Form

We can use the prime factorization of a number to help us write a fraction in **simplest form** or **lowest terms**.

Simplest Form

A fraction is in **simplest form**, or **lowest terms**, when the numerator and the denominator have no common factors other than 1.

For example, the fraction $\dfrac{6}{10}$ *is not* in simplest form because 6 and 10 both have a factor of 2. That is, 2 is a common factor of 6 and 10.

We'll use the fundamental principle of fractions to divide the numerator and denominator by the common factor of 2.

$$\frac{6}{10} = \frac{6 \div 2}{10 \div 2} = \frac{3}{5}$$

The fraction $\dfrac{3}{5}$ is in lowest terms, since the numerator and the denominator have no common factors (other than 1).

In the future, we will write the prime factorization of the numerator and denominator to help us find common factors. Then we will divide out common factors.

Let's use the following notation to show dividing the numerator and the denominator by common factors.

$$\frac{6}{10} = \frac{2 \cdot 3}{2 \cdot 5} = \frac{\cancel{2} \cdot 3}{\cancel{2} \cdot 5} = \frac{3}{5}$$

Writing a Fraction in Simplest Form

To write a fraction in simplest form, write the prime factorization of the numerator and the denominator and then divide both by all common factors.

The process of writing a fraction in simplest form is called **simplifying** the fraction.

Concept Check

True or false? The prime factorization of 72 is $2 \cdot 4 \cdot 9$. Explain your reasoning.

Practice Problem 4

Simplify: $\dfrac{30}{45}$

EXAMPLE 4 Simplify: $\dfrac{12}{20}$

Solution: First write the prime factorization of the numerator and the denominator.

$$\frac{12}{20} = \frac{2 \cdot 2 \cdot 3}{2 \cdot 2 \cdot 5}$$

Next divide the numerator and the denominator by all common factors.

$$\frac{12}{20} = \frac{2 \cdot 2 \cdot 3}{2 \cdot 2 \cdot 5} = \frac{3}{5}$$

●

Practice Problem 5

Simplify: $\dfrac{39}{51}$

EXAMPLE 5 Simplify: $\dfrac{42}{66}$

Solution: $\dfrac{42}{66} = \dfrac{2 \cdot 3 \cdot 7}{2 \cdot 3 \cdot 11} = \dfrac{7}{11}$

●

Practice Problem 6

Simplify: $\dfrac{45}{105y}$

EXAMPLE 6 Simplify: $\dfrac{84x}{90}$

Solution: $\dfrac{84x}{90} = \dfrac{2 \cdot 2 \cdot 3 \cdot 7 \cdot x}{2 \cdot 3 \cdot 3 \cdot 5} = \dfrac{14x}{15}$

●

Practice Problem 7

Simplify: $\dfrac{9a}{50a}$

EXAMPLE 7 Simplify: $\dfrac{10y}{27y}$

Solution: $\dfrac{10y}{27y} = \dfrac{2 \cdot 5 \cdot y}{3 \cdot 3 \cdot 3 \cdot y} = \dfrac{10}{27}$

●

Practice Problem 8

Simplify: $\dfrac{38}{4}$

EXAMPLE 8 Simplify: $\dfrac{36}{28}$

Solution: $\dfrac{36}{28} = \dfrac{2 \cdot 2 \cdot 3 \cdot 3}{2 \cdot 2 \cdot 7} = \dfrac{9}{7}$

●

For the fraction $\dfrac{36}{28}$ in Example 8, you may immediately notice that a common factor of 36 and 28 is 4. If so, you may simply divide out that common factor instead of writing the prime factorizations.

$$\frac{36}{28} = \frac{4 \cdot 9}{4 \cdot 7} = \frac{9}{7}$$

The result is in simplest form. If it were not, we would repeat the same procedure until it was.

Answers

4. $\dfrac{2}{3}$ **5.** $\dfrac{13}{17}$ **6.** $\dfrac{3}{7y}$ **7.** $\dfrac{9}{50}$ **8.** $\dfrac{19}{2}$

EXAMPLE 9   Simplify: $\dfrac{6x^2}{60x^3}$

Solution: Notice that 6 and 60 have a common factor of 6. Let's factor x^2 and x^3 to see what variable common factors we have,

$$\frac{6x^2}{60x^3} = \frac{6 \cdot x \cdot x}{6 \cdot 10 \cdot x \cdot x \cdot x} = \frac{1}{10x}$$

●

> **Helpful Hint**
>
> When all the factors of the numerator or denominator are divided out, don't forget that 1 still remains in that numerator or denominator. If it helps, use the following notation.
>
> $$\frac{6x^2}{60x^3} = \frac{\overset{1}{\cancel{6}} \cdot \overset{1}{\cancel{x}} \cdot \overset{1}{\cancel{x}}}{\underset{1}{\cancel{6}} \cdot 10 \cdot \underset{1}{\cancel{x}} \cdot \underset{1}{\cancel{x}} \cdot x} = \frac{1}{10x}$$ $\leftarrow 1 \cdot 1 \cdot 1$ is 1.
> $\leftarrow 1 \cdot 10 \cdot 1 \cdot 1 \cdot x$ is $10x$.

Try the Concept Check in the margin.

C Solving Problems by Writing Fractions in Simplest Form

Many real-life problems can be solved by writing fractions. To make the answers more clear, these fractions should be written in simplest form.

EXAMPLE 10   Calculating Fraction of Parks in Washington State

In a recent year, there were 54 national parks in the United States. Three of these parks are located in the state of Washington. What fraction of the United States' national parks can be found in Washington state? Write the fraction in simplest form. (*Source:* National Park Service)

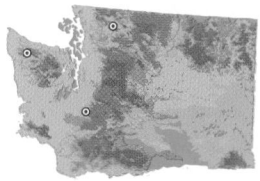

Solution: First we determine the fraction of parks found in Washington state.

$\dfrac{3}{54}$ $\leftarrow$ national parks in Washington
      $\leftarrow$ total natioal parks

Next we simplify the fraction.

$$\frac{3}{54} = \frac{3}{3 \cdot 18} = \frac{1}{18}$$

Thus, $\dfrac{1}{18}$ of the United States' national parks are in Washington state.   ●

Practice Problem 9

Simplify: $\dfrac{7a^3}{56a^2}$

Concept Check

Why is the following way to simplify the fraction $\dfrac{17}{75}$ incorrect?

$$\frac{\cancel{17}}{7\cancel{5}} = \frac{1}{5}$$

Practice Problem 10

Eighty pigs were used in a recent study of olestra, a calorie-free fat substitute. A group of 12 of these pigs were fed a diet high in fat. What fraction of the pigs were fed the high-fat diet in this study? Write your answer in simplest form. (*Source:* From a study conducted by the Procter & Gamble Company)

Answers

9. $\dfrac{a}{8}$ **10.** $\dfrac{3}{20}$

Concept Check: Answers may vary.

CALCULATOR EXPLORATIONS

Simplifying Fractions

SCIENTIFIC CALCULATOR

Many calculators have a fraction key, such as $\boxed{a\ b/c}$, that allows you to simplify a fraction on the calculator. For example, to simplify $\frac{324}{612}$, enter

$$\boxed{3}\ \boxed{2}\ \boxed{4}\ \boxed{a\ b/c}\ \boxed{6}\ \boxed{1}\ \boxed{2}\ \boxed{=}$$

The display will read

$$\boxed{\qquad 9\ |\ 17\qquad}$$

which represents $\frac{9}{17}$, the original fraction simplified.

GRAPHING CALCULATOR

Graphing calculators also allow you to simplify fractions. The fraction option on a graphing calculator may be found under the $\boxed{\text{MATH}}$ menu.

To simplify $\frac{324}{612}$, enter

$$\boxed{3}\ \boxed{2}\ \boxed{4}\ \boxed{\div}\ \boxed{6}\ \boxed{1}\ \boxed{2}\ \boxed{\text{MATH}}\ \boxed{\text{ENTER}}\ \boxed{\text{ENTER}}$$

The display will read

$$\boxed{324/612 \blacktriangleright \text{Frac } 9/17}$$

Helpful Hint

The Calculator Explorations boxes in this chapter provide only an introduction to fraction keys on calculators. Any time you use a calculator, there are both advantages and limitations to its use. Never rely solely on your calculator. It is very important that you understand how to perform all operations on fractions by hand in order to progress through later topics. For further information, talk to your instructor.

Use your calculator to simplify each fraction.

1. $\frac{128}{224}$

2. $\frac{231}{396}$

3. $\frac{340}{459}$

4. $\frac{999}{1350}$

5. $\frac{810}{432}$

6. $\frac{315}{225}$

7. $\frac{243}{54}$

8. $\frac{689}{455}$

Name _____ Section _____ Date _____

Mental Math

1. Is 2430 divisible by 2? By 3? By 5?

Write the prime factorization of each number.

2. 15   **3.** 10   **4.** 6   **5.** 21

6. 4   **7.** 9   **8.** 14

EXERCISE SET 4.2

Ⓐ *Write the prime factorization of each number. See Examples 1 through 3.*

1. 20   **2.** 12   **3.** 48   **4.** 75

5. 45   **6.** 64   **7.** 240   **8.** 128

Ⓑ *Simplify each fraction. See Examples 4 through 9.*

9. $\dfrac{3}{12}$ **10.** $\dfrac{5}{20}$ **11.** $\dfrac{7x}{35}$ **12.** $\dfrac{9}{48z}$

13. $\dfrac{14}{16}$ **14.** $\dfrac{18}{4}$ **15.** $\dfrac{24a}{30a}$ **16.** $\dfrac{70y}{80xy}$

17. $\dfrac{35}{42}$ **18.** $\dfrac{25}{55}$ **19.** $\dfrac{30x^2}{36x}$ **20.** $\dfrac{45b}{80b^2}$

21. $\dfrac{16}{24}$ **22.** $\dfrac{18}{45}$ **23.** $\dfrac{45xz}{60z}$ **24.** $\dfrac{22a}{99ab}$

25. $\dfrac{39ab}{26a^2}$

26. $\dfrac{42x^2}{24xy}$

27. $\dfrac{63}{72}$

28. $\dfrac{56}{64}$

29. $\dfrac{21}{49}$

30. $\dfrac{14}{35}$

31. $\dfrac{24y}{40}$

32. $\dfrac{36}{54x}$

33. $\dfrac{36z}{63z}$

34. $\dfrac{39b}{52b}$

35. $\dfrac{72x^3y^2}{90xy}$

36. $\dfrac{24a^2b}{54ab^3}$

37. $\dfrac{12}{15}$

38. $\dfrac{18}{24}$

39. $\dfrac{25x^2}{40x}$

40. $\dfrac{36y^2}{42y}$

41. $\dfrac{27xy}{90y}$

42. $\dfrac{60y}{150yz}$

43. $\dfrac{36a^3bc^2}{24ab^4c^2}$

44. $\dfrac{60x^2yz}{36x^3y^3z^3}$

45. $\dfrac{40xy}{64xyz}$

46. $\dfrac{28abc}{60ac}$

 Solve. Write each fraction in simplest form. See Example 10.

47. A work shift for an employee at McDonald's consists of 8 hours. What fraction of the employee's work shift is represented by 6 hours?

48. Two thousand golf caps were sold one year at the U.S. Open golf tournament. What fractional part of this total does 200 caps represent?

49. There are 5280 feet in a mile. What fraction of a mile is represented by 2640 feet?

50. There are 100 centimeters in 1 meter. What fraction of a meter is 20 centimeters?

51. As of the year 2000, a total of 414 individuals from around the world had flown in space. Of these, 261 were citizens of the United States. What fraction of individuals who have flown in space were Americans? (*Source:* Congressional Research Service)

52. Hallmark Cards employs 25,000 full-time employees worldwide. About 5600 employees work at the Hallmark headquarters in Kansas City, Missouri. What fraction of Hallmark employees work in Kansas City? (*Source:* Hallmark Cards,, Inc.)

53. There are 16,000 students at a local university. If 8800 are females, what fraction of the students are *male*?

54. Four out of 10 marbles are red. What fraction of marbles are *not red*?

55. Sixteen states in the United States have Ritz-Carlton hotels. (*Source:* Marriott International)
  a. What fraction of states can claim at least one Ritz-Carlton hotel?
  b. How many states do not have a Ritz-Carlton hotel?
  c. Write the fraction of states without a Ritz-Carlton hotel.

56. As of 2003, there were 56 national parks in the United States. Eight of these parks are located in Alaska. (*Source:* National Park Service)
  a. What fraction of the United States' national parks can be found in Alaska?
  b. How many of the United States' national parks are found outside Alaska?
  c. Write the fraction of national parks found in states other than Alaska.

57. The outer wall of the Pentagon is 24 inches wide. Ten inches is concrete, 8 inches is brick, and 6 inches is limestone. What fraction of the width is concrete? (*Source:* USA Today 1/28/2000)

 58. There are 35 students in a biology class. If 10 students made an A on the first test, what fraction of the students made an A?

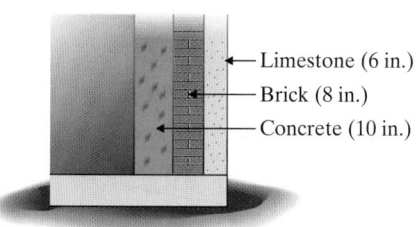

Limestone (6 in.)
Brick (8 in.)
Concrete (10 in.)

The following graph is called a circle graph or pie chart. Each sector (shaped like a piece of pie) shows the fraction of entering college freshmen who expect to major in each discipline shown. The whole circle represents the entire class of college freshmen. Use this graph to answer Exercises 59–62.

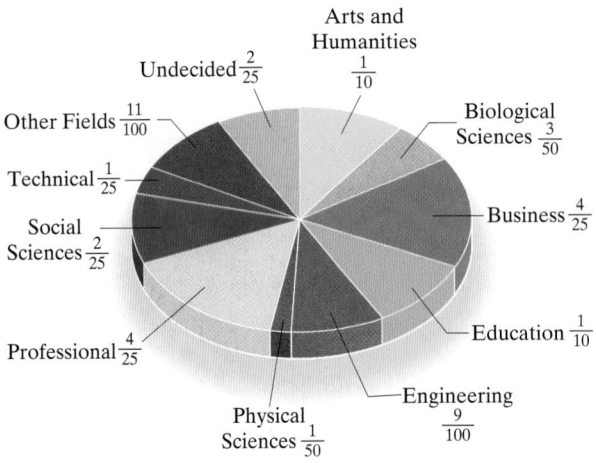

Arts and Humanities $\frac{1}{10}$

Undecided $\frac{2}{25}$

Other Fields $\frac{11}{100}$

Biological Sciences $\frac{3}{50}$

Technical $\frac{1}{25}$

Business $\frac{4}{25}$

Social Sciences $\frac{2}{25}$

Education $\frac{1}{10}$

Professional $\frac{4}{25}$

Engineering $\frac{9}{100}$

Physical Sciences $\frac{1}{50}$

Source: Higher Education Research Institute

59. What fraction of entering college freshmen plan to major in education?

60. What fraction of entering college freshmen plan to major in social sciences?

61. Why is the Professional sector the same size as the Business sector?

62. Why is the Physical Sciences sector smaller than the Biological Sciences sector?

Review and Preview

Evaluate each expression using the given replacement numbers. See Section 2.5.

63. $\dfrac{x^3}{9}$ when $x = -3$

64. $\dfrac{y^3}{5}$ when $y = -5$

65. $2y$ when $y = -7$

66. $-5a$ when $a = -4$

67. $3z - y$ when $z = 2$ and $y = 6$

68. $7a - b$ when $a = 1$ and $b = -5$

69. $a^2 + 2b + 3$ when $a = 4$ and $b = 5$

70. $yx - z^2$ when $y = 6$, $x = 6$, and $z = 6$

 **Combining Concepts**

71. Which fraction is closest to 0? Explain your answer.

    a. $\dfrac{1}{2}$ **b.** $\dfrac{1}{3}$

    c. $\dfrac{1}{4}$ **d.** $\dfrac{1}{5}$

72. Which fraction is closest to 1? Explain your answer.

    a. $\dfrac{1}{2}$ **b.** $\dfrac{2}{3}$

    c. $\dfrac{3}{4}$ **d.** $\dfrac{5}{6}$

Answer true or false for each statement.

73. $\dfrac{14}{42} = \dfrac{2 \cdot 7}{2 \cdot 3 \cdot 7} = \dfrac{0}{3}$

74. The fractions $\dfrac{8}{20}, \dfrac{4}{5},$ and $\dfrac{16}{60}$ are all equivalent.

75. A proper fraction cannot be equivalent to an improper fraction.

76. A fraction whose numerator and denominator are two different prime numbers cannot be simplified.

Simplify each fraction.

77. $\dfrac{372}{620}$

78. $\dfrac{9506}{12,222}$

There are generally considered to be eight basic blood types. The table shows the number of people with the various blood types in a typical group of 100 blood donors. Use this table to answer Exercises 79–83. Write each answer in simplest form.

79. What fraction of blood donors have A Rh-positive blood type?

80. What fraction of blood donors have O blood type?

81. What fraction of blood donors have AB blood type?

82. What fraction of blood donors have B blood type?

83. What fraction of blood donors have the negative Rh factor?

Distribution of Blood Types in Blood Donors	
Blood Type	Number of People
O Rh-positive	37
O Rh-negative	7
A Rh-positive	36
A Rh-negative	6
B Rh-positive	9
B Rh-negative	1
AB Rh-positive	3
AB Rh-negative	1

(*Source*: American Red Cross Biomedical Services)

Use the following numbers for Exercises 84–87.

| 8691 | 786 | 1235 | 2235 | 85 | 105 | 22 | 222 | 900 | 1470 |

84. List the numbers divisible by both 2 and 3.

85. List the numbers that are divisible by both 3 and 5.

86. The answers to Exercise 84, are also divisible by what number? Tell why.

87. The answers to Exercise 85 are also divisible by what number? Tell why.

FOCUS ON The Real World

BLOOD AND BLOOD DONATION

Blood is the workhorse of the body. It carries to the body's tissues everything they need, from nutrients to antibodies to heat. Blood also carries away waste products like carbon dioxide. Blood contains three types of cells—red blood cells, white blood cells, and platelets—suspended in clear, watery fluid called plasma. Blood is $\frac{11}{20}$ plasma, and plasma itself is $\frac{9}{10}$ water. In the average healthy adult human, blood accounts for $\frac{1}{11}$ of a person's body weight.

Roughly every 2 seconds someone in the United States needs blood. Although only $\frac{1}{20}$ of eligible donors donate blood, the American Red Cross is still able to collect nearly 6 million volunteer donations of blood each year. This volume makes Red Cross Biomedical Services the largest blood supplier for blood transfusions in the United States.

The modern Red Cross blood donation program has its roots in World War II. In 1940, Dr. Charles Drew headed the Red Cross-sponsored American blood collection efforts for bombing victims in Great Britain. During that time, Dr. Drew developed techniques for separating plasma from blood cells that allowed mass production of plasma. Dr. Drew had discovered that plasma could be preserved longer than whole blood. He also found that dried plasma could be stored longer than its liquid form. By 1941, Dr. Drew had become the first med-

ical director of the first American Red Cross Blood Bank in the United States. Plasma collected through this program saved the lives of many wounded civilians and Allied soldiers by reducing the high rate of death from shock.

GROUP ACTIVITY

Contact your local Red Cross Blood Service office. Find out how many people donated blood in your area in the past two months. Ask whether it is possible to get a breakdown of the blood donations by blood type. (For more on blood type, see Exercises 79 through 83 in Section 4.2.)

1. Research the population of the area served by your local Red Cross Blood Service office. Write the fraction of the local population who gave blood in the past two months.

2. Use the breakdown by blood type to write the fraction of donors giving each type of blood.

4.3 Multiplying and Dividing Fractions

(A) Multiplying Fractions

Let's use a diagram to discover how fractions are multiplied. For example, to multiply $\frac{1}{2}$ and $\frac{3}{4}$, we find $\frac{1}{2}$ of $\frac{3}{4}$. To do this, we begin with a diagram showing $\frac{3}{4}$ of rectangle's area shaded.

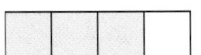 $\frac{3}{4}$ of the rectangle's area is shaded.

To find $\frac{1}{2}$ of $\frac{3}{4}$, we heavily shade — of the part that is already shaded.

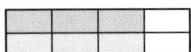

By counting smaller rectangles, we see that $\frac{3}{8}$ of the larger rectangle is now heavily shaded, so that $\frac{1}{2}$ of $\frac{3}{4}$ is $\frac{3}{8}$. This means that

$$\frac{1}{2} \cdot \frac{3}{4} = \frac{3}{8}$$ Notice that $\frac{1}{2} \cdot \frac{3}{4} = \frac{1 \cdot 3}{2 \cdot 4} = \frac{3}{8}$.

Notice that the numerator of the product is equal to the product of the numerators and that the denominator of the product is equal to the product of the denominators. This is how we multiply fractions.

Multiplying Two Fractions

If a, b, c, and d are numbers and b and d are not 0, then

$$\frac{a}{b} \cdot \frac{c}{d} = \frac{a \cdot c}{b \cdot d}$$

In other words, to multiply two fractions, multiply the numerators and multiply the denominators.

EXAMPLES Multiply:

1. $\dfrac{2}{3} \cdot \dfrac{5}{11} = \dfrac{2 \cdot 5}{3 \cdot 11} = \dfrac{10}{33}$ ← Product of numerators
 ← Product of denominators

2. $\dfrac{1}{4} \cdot \dfrac{1}{2} = \dfrac{1 \cdot 1}{4 \cdot 2} = \dfrac{1}{8}$

EXAMPLE 3   Multiply and simplify: $\dfrac{6}{7} \cdot \dfrac{14}{27}$

Solution: $\dfrac{6}{7} \cdot \dfrac{14}{27} = \dfrac{6 \cdot 14}{7 \cdot 27}$

Next, simplify by factoring into primes and dividing out common factors.

$$\frac{6 \cdot 14}{7 \cdot 27} = \frac{2 \cdot 3 \cdot 2 \cdot 7}{7 \cdot 3 \cdot 3 \cdot 3}$$

$$= \frac{4}{9}$$

OBJECTIVES

(A) Multiply fractions.

(B) Evaluate exponential expressions with fractional bases.

(C) Divide fractions.

(D) Multiply and divide given fractional replacement values.

(E) Solve applications that require multiplication or division of fractions.

SSM   TUTOR CENTER   SG   CD & VIDEO   MATH PRO   WEB

Practice Problems 1–2

Multiply:

1. $\dfrac{3}{8} \cdot \dfrac{5}{7}$ 2. $\dfrac{1}{3} \cdot \dfrac{1}{6}$

Practice Problem 3

Multiply and simplify: $\dfrac{6}{11} \cdot \dfrac{5}{8}$

Answers

1. $\dfrac{15}{56}$ 2. $\dfrac{1}{18}$ 3. $\dfrac{15}{44}$

Helpful Hint

In simplifying a product, it may be possible to identify common factors without actually writing the prime factorizations. For example,

$$\frac{10}{11} \cdot \frac{1}{20} = \frac{10 \cdot 1}{11 \cdot 20} = \frac{10 \cdot 1}{11 \cdot 10 \cdot 2} = \frac{1}{22}$$

Practice Problem 4

Multiply and simplify: $\dfrac{4}{15} \cdot \dfrac{3}{8}$

EXAMPLE 4 Multiply and simplify: $\dfrac{23}{32} \cdot \dfrac{4}{7}$

Solution: Notice that 4 and 32 have a common factor of 4.

$$\frac{23}{32} \cdot \frac{4}{7} = \frac{23 \cdot 4}{32 \cdot 7} = \frac{23 \cdot 4}{4 \cdot 8 \cdot 7} = \frac{23}{56}$$

After multiplying two fractions, *always* check to see whether the product can be simplified.

Practice Problem 5

Multiply: $\dfrac{1}{2} \cdot \left(-\dfrac{11}{28} \right)$

EXAMPLE 5 Multiply: $-\dfrac{1}{4} \cdot \dfrac{1}{2}$

Solution: Recall that the product of a negative number and a positive number is a negative number.

$$-\frac{1}{4} \cdot \frac{1}{2} = -\frac{1 \cdot 1}{4 \cdot 2} = -\frac{1}{8}$$

Practice Problem 6

Multiply: $\dfrac{9}{5} \cdot \dfrac{20}{12}$

EXAMPLE 6 Multiply: $\dfrac{13}{6} \cdot \dfrac{30}{26}$

Solution: $\dfrac{13}{6} \cdot \dfrac{30}{26} = \dfrac{13 \cdot 6 \cdot 5}{6 \cdot 2 \cdot 13} = \dfrac{5}{2}$

We multiply fractions in the same way if variables are involved.

Practice Problem 7

Multiply: $\dfrac{2}{3} \cdot \dfrac{3y}{2}$

EXAMPLE 7 Multiply: $\dfrac{3x}{4} \cdot \dfrac{8}{5x}$

Solution: $\dfrac{3x}{4} \cdot \dfrac{8}{5x} = \dfrac{3 \cdot x \cdot 8}{4 \cdot 5 \cdot x} = \dfrac{3 \cdot 4 \cdot 2}{4 \cdot 5} = \dfrac{6}{5}$

Helpful Hint

Recall that when the denominator of a fraction contains a variable, such as $\dfrac{8}{5x}$, we assume that the variable does not represent 0.

Practice Problem 8

Multiply: $\dfrac{a^3}{b^2} \cdot \dfrac{b}{a^2}$

EXAMPLE 8 Multiply: $\dfrac{x^2}{y} \cdot \dfrac{y^3}{x}$

Solution: $\dfrac{x^2}{y} \cdot \dfrac{y^3}{x} = \dfrac{x^2 \cdot y^3}{y \cdot x} = \dfrac{x \cdot x \cdot y \cdot y \cdot y}{y \cdot x} = \dfrac{x \cdot y \cdot y}{1} = xy^2$

Answers

4. $\dfrac{1}{10}$ **5.** $-\dfrac{11}{56}$ **6.** 3 **7.** y **8.** $\dfrac{a}{b}$

Ⓑ Evaluating Expressions with Fractional Bases

The base of an exponential expression can also be a fraction.

$$\left(\frac{1}{3}\right)^4 = \underbrace{\frac{1}{3}\cdot\frac{1}{3}\cdot\frac{1}{3}\cdot\frac{1}{3}}_{} = \frac{1\cdot1\cdot1\cdot1}{3\cdot3\cdot3\cdot3} = \frac{1}{81}$$

$\frac{1}{3}$ is a factor 4 times.

EXAMPLE 9 Evaluate.

a. $\left(\dfrac{2}{5}\right)^4$

b. $\left(-\dfrac{1}{4}\right)^2$

Solution:

a. $\left(\dfrac{2}{5}\right)^4 = \dfrac{2}{5}\cdot\dfrac{2}{5}\cdot\dfrac{2}{5}\cdot\dfrac{2}{5} = \dfrac{2\cdot2\cdot2\cdot2}{5\cdot5\cdot5\cdot5} = \dfrac{16}{625}$

b. $\left(-\dfrac{1}{4}\right)^2 = \left(-\dfrac{1}{4}\right)\cdot\left(-\dfrac{1}{4}\right) = \dfrac{1\cdot1}{4\cdot4} = \dfrac{1}{16}$

Practice Problem 9

Evaluate.

a. $\left(\dfrac{3}{4}\right)^3$ b. $\left(-\dfrac{4}{5}\right)^2$

Ⓒ Dividing Fractions

Before we can divide fractions, we need to know how to find the **reciprocal** of a fraction.

Reciprocal of a Fraction

Two numbers are **reciprocals** of each other if their product is 1. The reciprocal of the fraction $\dfrac{a}{b}$ is $\dfrac{b}{a}$ because $\dfrac{a}{b}\cdot\dfrac{b}{a} = \dfrac{a\cdot b}{b\cdot a} = 1.$

For example,

The reciprocal of $\dfrac{2}{5}$ is $\dfrac{5}{2}$ because $\dfrac{2}{5}\cdot\dfrac{5}{2} = \dfrac{10}{10} = 1.$

The reciprocal of 5 is $\dfrac{1}{5}$ because $5\cdot\dfrac{1}{5} = \dfrac{5}{1}\cdot\dfrac{1}{5} = \dfrac{5}{5} = 1.$

The reciprocal of $-\dfrac{7}{11}$ is $-\dfrac{11}{7}$ because $-\dfrac{7}{11}\cdot-\dfrac{11}{7} = \dfrac{77}{77} = 1.$

> **Helpful Hint**
>
> Every number has a reciprocal except 0. The number 0 has no reciprocal because there is no number such that $0\cdot a = 1.$

Division of fractions has the same meaning as division of whole numbers. For example,

$10 \div 5$ means: How many 5s are there in 10?

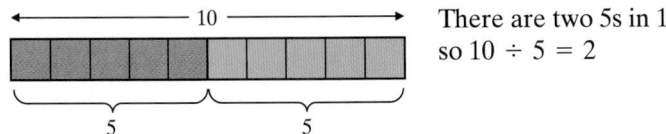

There are two 5s in 10, so $10 \div 5 = 2$

$\dfrac{3}{4} \div \dfrac{1}{8}$ means: How many $\dfrac{1}{8}$s are there in $\dfrac{3}{4}$?

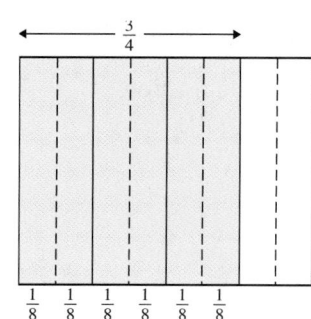

There are six $\frac{1}{8}$s in $\frac{3}{4}$, so

$$\frac{3}{4} \div \frac{1}{8} = 6$$

We use reciprocals to divide fractions.

Dividing Fractions

If b, c, and d are not 0, then $\dfrac{a}{b} \div \dfrac{c}{d} = \dfrac{a}{b} \cdot \dfrac{d}{c} = \dfrac{a \cdot d}{b \cdot c}$.

In other words, to divide fractions, multiply the first fraction by the reciprocal of the second fraction.

For example,

multiply by reciprocal

$$\frac{3}{4} \div \frac{1}{8} = \frac{3}{4} \cdot \frac{8}{1} = \frac{3 \cdot 8}{4 \cdot 1} = \frac{3 \cdot 2 \cdot 4}{4 \cdot 1} = \frac{6}{1} \quad \text{or} \quad 6$$

Practice Problems 10–11

Divide and simplify.

10. $\dfrac{3}{2} \div \dfrac{14}{5}$

11. $\dfrac{4}{9} \div \dfrac{1}{2}$

EXAMPLES Divide and simplify.

10. $\dfrac{7}{8} \div \dfrac{2}{9} = \dfrac{7}{8} \cdot \dfrac{9}{2} = \dfrac{7 \cdot 9}{8 \cdot 2} = \dfrac{63}{16}$

11. $\dfrac{2}{5} \div \dfrac{1}{2} = \dfrac{2}{5} \cdot \dfrac{2}{1} = \dfrac{2 \cdot 2}{5 \cdot 1} = \dfrac{4}{5}$

After dividing two fractions, *always* check to see whether the result can be simplified.

Helpful Hint

When dividing by a fraction, do not look for common factors to divide out until you rewrite the division as multiplication.

Do **not** try to divide out these two 2s.

$$\frac{1}{2} \div \frac{2}{3} = \frac{1}{2} \cdot \frac{3}{2} = \frac{3}{4}$$

Practice Problem 12

Divide: $\dfrac{10}{4} \div \dfrac{2}{9}$

EXAMPLE 12 Divide: $-\dfrac{5}{16} \div -\dfrac{3}{4}$

Solution: Recall that the quotient (or product) of two negative numbers is a positive number.

$$-\frac{5}{16} \div -\frac{3}{4} = -\frac{5}{16} \cdot -\frac{4}{3} = \frac{5 \cdot 4}{4 \cdot 4 \cdot 3} = \frac{5}{12}$$

Answers

10. $\dfrac{15}{28}$ 11. $\dfrac{8}{9}$ 12. $\dfrac{45}{4}$

EXAMPLE 13   Divide: $\dfrac{2x}{3} \div 3x^2$

Solution: $\dfrac{2x}{3} \div 3x^2 = \dfrac{2x}{3} \div \dfrac{3x^2}{1} = \dfrac{2x}{3} \cdot \dfrac{1}{3x^2} = \dfrac{2 \cdot x \cdot 1}{3 \cdot 3 \cdot x \cdot x} = \dfrac{2}{9x}$

Try the Concept Check in the margin.

EXAMPLE 14   Simplify: $\left(\dfrac{4}{7} \cdot \dfrac{3}{8}\right) \div -\dfrac{3}{4}$

Solution: Remember to perform the operations inside the ( ) first.

$$\left(\dfrac{4}{7} \cdot \dfrac{3}{8}\right) \div -\dfrac{3}{4} = \left(\dfrac{4 \cdot 3}{7 \cdot 2 \cdot 4}\right) \div -\dfrac{3}{4} = \dfrac{3}{14} \div -\dfrac{3}{4}$$

Now divide.

$$\dfrac{3}{14} \div -\dfrac{3}{4} = \dfrac{3}{14} \cdot -\dfrac{4}{3} = -\dfrac{3 \cdot 2 \cdot 2}{2 \cdot 7 \cdot 3} = -\dfrac{2}{7}$$

D   **Multiplying and Dividing with Fractional Replacement Values**

EXAMPLE 15   If $x = \dfrac{7}{8}$ and $y = -\dfrac{1}{3}$, evaluate (**a**) xy and (**b**) $x \div y$.

Solution: Replace x with $\dfrac{7}{8}$ and y with $-\dfrac{1}{3}$.

a. $xy = \dfrac{7}{8} \cdot -\dfrac{1}{3}$ **b.** $x \div y = \dfrac{7}{8} \div -\dfrac{1}{3}$

$= -\dfrac{7 \cdot 1}{8 \cdot 3}$ $= \dfrac{7}{8} \cdot -\dfrac{3}{1}$

$= -\dfrac{7}{24}$ $= -\dfrac{7 \cdot 3}{8 \cdot 1}$

$\qquad\qquad\qquad\qquad\qquad = -\dfrac{21}{8}$

EXAMPLE 16   Is $-\dfrac{2}{3}$ a solution of the equation $-\dfrac{1}{2}x = \dfrac{1}{3}$?

Solution: To check whether a number is a solution of an equation, recall that we replace the variable with the given number and see if a true statement results.

$-\dfrac{1}{2} \cdot x = \dfrac{1}{3}$ Recall that $-\dfrac{1}{2}x$ means $-\dfrac{1}{2} \cdot x$.

$-\dfrac{1}{2} \cdot -\dfrac{2}{3} = \dfrac{1}{3}$ Replace x with $-\dfrac{2}{3}$.

$\dfrac{1 \cdot 2}{2 \cdot 3} = \dfrac{1}{3}$ The product of two negative numbers is a positive number.

$\dfrac{1}{3} = \dfrac{1}{3}$ True.

Since we have a true statement, $-\dfrac{2}{3}$ is a solution.

Practice Problem 13

Divide: $\dfrac{3y}{4} \div 5y^3$

Concept Check

Which is the correct way to divide $\dfrac{3}{5}$ by $\dfrac{5}{12}$? Explain.

a. $\dfrac{3}{5} \div \dfrac{5}{12} = \dfrac{3}{5} \cdot \dfrac{12}{5}$

b. $\dfrac{3}{5} \div \dfrac{5}{12} = \dfrac{5}{3} \cdot \dfrac{5}{12}$

Practice Problem 14

Simplify: $\left(-\dfrac{2}{3} \cdot \dfrac{9}{14}\right) \div \dfrac{7}{15}$

Practice Problem 15

If $x = -\dfrac{3}{4}$ and $y = \dfrac{9}{2}$, evaluate (**a**) xy, and (**b**) $x \div y$.

Practice Problem 16

Is $-\dfrac{9}{8}$ a solution of the equation

$$2x = -\dfrac{9}{4}?$$

Answers

13. $\dfrac{3}{20y^2}$ 14. $-\dfrac{45}{49}$ 15. **a.** $-\dfrac{27}{8}$ **b.** $-\dfrac{1}{6}$

16. yes

Concept Check: a

E Solving Applications by Multiplying and Dividing Fractions

To solve real-life problems that involve multiplying and dividing fractions, we will use our four problem-solving steps from Chapter 3. In Example 17, a new key word that implies multiplication is used. That key word is "of."

EXAMPLE 17   Finding Number of Roller Coasters in an Amusement Park

Cedar Point is an amusement park located in Sandusky, Ohio. Its collection of 68 rides is the largest in the world. Of the rides, $\frac{7}{34}$ are roller coasters. How many roller coasters are in Cedar Point's collection of rides? (*Source:* Cedar Fair, L.P.)

Solution:

1. UNDERSTAND the problem. To do so, read and reread the problem. We are told that $\frac{7}{34}$ of Cedar Point's rides are roller coasters. The word "of" here means multiplication, since we are looking for the number of roller coasters.

   x = number of roller coasters

2. TRANSLATE.

In words:	Number of roller coasters	is	$\frac{7}{34}$	of	total rides at Cedar Point
Translate:	x	$=$	$\frac{7}{34}$	$\cdot$	68

3. SOLVE.

   $$x = \frac{7}{34} \cdot 68$$
   $$= \frac{7}{34} \cdot \frac{68}{1} = \frac{7 \cdot 68}{34 \cdot 1} = \frac{7 \cdot 34 \cdot 2}{34 \cdot 1} = \frac{14}{1} \text{ or } 14$$

4. INTERPRET. *Check* your work. *State* your conclusion: The number of roller coasters at Cedar Point is 14.

Practice Problem 17

About $\frac{1}{3}$ of all plant and animal species in the United States are at risk of becoming extinct. There are 20,439 known species of plants and animals in the United States. How many species are at risk of extinction? (*Source:* The Nature Conservancy)

Helpful Hint

To help visualize a fractional part of a whole number, look at the diagram below.

$\frac{1}{5}$ of 60 = ?

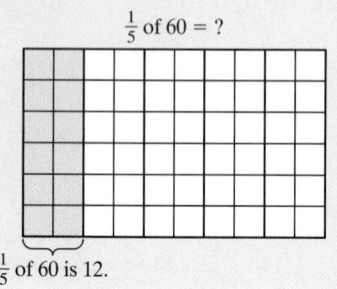

$\frac{1}{5}$ of 60 is 12.

Answer

17. 6813 species

Mental Math

Find each product.

1. $\dfrac{1}{3} \cdot \dfrac{2}{5}$

2. $\dfrac{2}{3} \cdot \dfrac{4}{7}$

3. $\dfrac{6}{5} \cdot \dfrac{1}{7}$

4. $\dfrac{7}{3} \cdot \dfrac{2}{3}$

5. $\dfrac{3}{1} \cdot \dfrac{3}{8}$

6. $\dfrac{2}{1} \cdot \dfrac{7}{11}$

EXERCISE SET 4.3

A *Multiply. Write the product in simplest form. See Examples 1 through 8.*

1. $\dfrac{7}{8} \cdot \dfrac{2}{3}$

2. $\dfrac{5}{9} \cdot \dfrac{7}{4}$

3. $-\dfrac{2}{7} \cdot \dfrac{5}{8}$

4. $\dfrac{5}{8} \cdot -\dfrac{1}{3}$

5. $-\dfrac{1}{2} \cdot -\dfrac{2}{15}$

6. $-\dfrac{3}{8} \cdot -\dfrac{5}{12}$

7. $\dfrac{18x}{20} \cdot \dfrac{36}{99}$

8. $\dfrac{5}{32} \cdot \dfrac{64y}{100}$

9. $3a^2 \cdot \dfrac{1}{4}$

10. $-\dfrac{2}{3} \cdot 6y^3$

11. $\dfrac{x^3}{y^3} \cdot \dfrac{y^2}{x}$

12. $\dfrac{a}{b^3} \cdot \dfrac{b}{a^3}$

B *Evaluate. See Example 9.*

13. $\left(\dfrac{1}{5}\right)^3$

14. $\left(-\dfrac{1}{2}\right)^4$

15. $\left(-\dfrac{2}{3}\right)^2$

16. $\left(\dfrac{8}{9}\right)^2$

17. $\left(-\dfrac{2}{3}\right)^3 \cdot \dfrac{1}{2}$

18. $\left(-\dfrac{3}{4}\right)^3 \cdot \dfrac{1}{3}$

C *Divide. Write all quotients in simplest form. See Examples 10 through 13.*

19. $\dfrac{2}{3} \div \dfrac{5}{6}$

20. $\dfrac{5}{8} \div \dfrac{2}{3}$

21. $-\dfrac{6}{15} \div \dfrac{12}{5}$

22. $-\dfrac{4}{15} \div -\dfrac{8}{3}$

23. $\dfrac{8}{9} \div \dfrac{x}{2}$

24. $\dfrac{10}{11} \div -\dfrac{4}{5}$

25. $\dfrac{11y}{20} \div \dfrac{3}{11}$

26. $\dfrac{9z}{20} \div \dfrac{2}{9}$

27. $-\dfrac{2}{3} \div 4$

28. $-\dfrac{5}{6} \div 10$

29. $\dfrac{1}{5x} \div \dfrac{5}{x^2}$

30. $\dfrac{3}{y^2} \div \dfrac{9}{y}$

A **B** **C** *Perform each indicated operation. See Examples 1 through 14.*

31. $\dfrac{2}{3} \cdot \dfrac{5}{9}$

32. $\dfrac{8}{15} \cdot \dfrac{5}{32}$

33. $\dfrac{3x}{7} \div \dfrac{5}{6x}$

34. $\dfrac{16}{27y} \div \dfrac{8}{15y}$

35. $-\dfrac{5}{28} \cdot \dfrac{35}{25}$

36. $\dfrac{24}{45} \cdot -\dfrac{5}{8}$

37. $-\dfrac{3}{5} \div -\dfrac{4}{5}$

38. $-\dfrac{11}{16} \div -\dfrac{13}{16}$

39. $\left(-\dfrac{3}{4}\right)^2$

40. $\left(-\dfrac{1}{2}\right)^5$

41. $\dfrac{x^2}{y} \cdot \dfrac{y^3}{x}$

42. $\dfrac{b}{a^2} \cdot \dfrac{a^3}{b^3}$

43. $7 \div \dfrac{2}{11}$

44. $-100 \div \dfrac{1}{2}$

45. $-3x \div \dfrac{x^2}{12}$

46. $7x \div \dfrac{14x}{3}$

47. $\left(\dfrac{2}{7} \div \dfrac{7}{2}\right) \cdot \dfrac{3}{4}$

48. $\dfrac{1}{2} \cdot \left(\dfrac{5}{6} \div \dfrac{1}{12}\right)$

49. $-\dfrac{19}{63y} \cdot 9y^2$

50. $16a^2 \cdot -\dfrac{31}{24a}$

51. $-\dfrac{2}{3} \cdot -\dfrac{6}{11}$

52. $-\dfrac{1}{5} \cdot -\dfrac{6}{7}$

53. $\dfrac{4}{8} \div \dfrac{3}{16}$

54. $\dfrac{9}{2} \div \dfrac{16}{15}$

55. $\dfrac{21x^2}{10y} \div \dfrac{14x}{25y}$

56. $\dfrac{17y^2}{24x} \div \dfrac{13y}{18x}$

57. $\left(1 \div \dfrac{3}{4}\right) \cdot \dfrac{2}{3}$

58. $\left(33 \div \dfrac{2}{11}\right) \cdot \dfrac{5}{9}$

59. $\dfrac{a^3}{2} \div 30a^3$

60. $15c^3 \div \dfrac{3c^2}{5}$

61. $\dfrac{ab^2}{c} \cdot \dfrac{c}{ab}$

62. $\dfrac{ac}{b} \cdot \dfrac{b^3}{a^2c}$

63. $\left(\dfrac{1}{2} \cdot \dfrac{2}{3}\right) \div \dfrac{5}{6}$

64. $\left(\dfrac{3}{4} \cdot \dfrac{8}{9}\right) \div \dfrac{2}{5}$

65. $-\dfrac{4}{7} \div \left(\dfrac{4}{5} \cdot \dfrac{3}{7}\right)$

66. $\dfrac{5}{8} \div \left(\dfrac{4}{7} \cdot -\dfrac{5}{16}\right)$

D *Given the following replacement values, evaluate (**a**) xy and (**b**) x ÷ y. See Example 15.*

67. $x = \dfrac{2}{5}$ and $y = \dfrac{5}{6}$

   a.

   b.

68. $x = \dfrac{8}{9}$ and $y = \dfrac{1}{4}$

   a.

   b.

69. $x = -\dfrac{4}{5}$ and $y = \dfrac{9}{11}$

   a.

   b.

70. $x = \dfrac{7}{6}$ and $y = -\dfrac{1}{2}$

   a.

   b.

Determine whether the given replacement values are solutions of the given equations. See Example 16.

71. Is $-\dfrac{5}{18}$ a solution to $3x = -\dfrac{5}{6}$?

72. Is $\dfrac{9}{11}$ a solution to $\dfrac{2}{3}y = \dfrac{6}{11}$?

73. Is $\dfrac{2}{5}$ a solution to $-\dfrac{1}{2}z = \dfrac{1}{10}$?

74. Is $\dfrac{3}{5}$ a solution to $5x = \dfrac{1}{3}$?

Solve. See Example 17.

75. A veterinarian's dipping vat holds 36 gallons of liquid. She usually fills it $\frac{5}{6}$ full of a medicated flea dip solution. Find how many gallons of solution are usually in the vat.

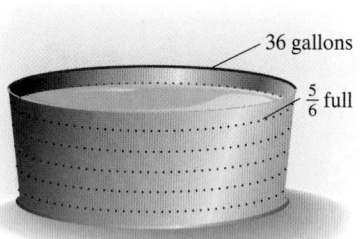

36 gallons

$\frac{5}{6}$ full

76. Each turn of a screw sinks it $\frac{3}{16}$ of an inch deeper into a piece of wood. Find how deep the screw is after 8 turns.

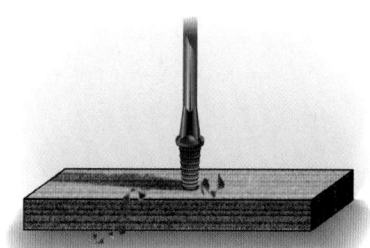

77. A special on a cruise to the Bahamas is advertised to be $\frac{2}{3}$ of the regular price. If the regular price is $2757, what is the sale price?

78. The Gonzales recently sold their house for $102,000, but $\frac{3}{50}$ of this amount goes to the real estate companies that helped them sell their house. How much money do the Gonzales pay to the real estate companies?

△ **79.** The radius of a circle is one-half of its diameter as shown. If the diameter of a circle is $\frac{3}{8}$ of an inch, what is its radius?

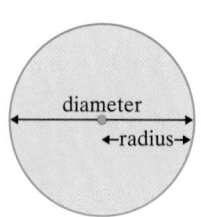

diameter

←radius→

80. A recipe calls for $\frac{1}{3}$ of a cup of flour. How much flour should be used if only $\frac{1}{2}$ of the recipe is being made?

81. The Oregon National Historic Trail is 2,170 miles long. It begins in Independence, Missouri, and ends in Oregon City, Oregon. Manfred Coulon has hiked $\frac{2}{5}$ of the trail before. How many miles has he hiked? (*Source:* National Park Service)

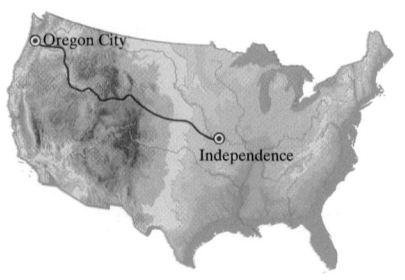

Oregon City

Independence

82. Movie theater owners received a total of $7660 million in movie admission tickets, about $\frac{7}{10}$ of this amount was for R-rated movies. Find the amount of money received from R-rated movies. (*Source:* Motion Picture Association of America)

83. As part of his research, famous tornado expert Dr. T. Fujita studied approximately 31,050 tornadoes that occurred in the United States between 1916 and 1985. He found that roughly $\frac{7}{10}$ of these tornadoes occurred during April, May, June, and July. How many of these tornadoes occurred during these four months? (*Source: U.S. Tornadoes Part 1*, T. Fujita, University of Chicago)

84. Campbell Soup Company ships its soup in boxes containing 24 cans. If each can weighs about $\frac{3}{4}$ pound, find how much the contents of a box weighs.

85. An estimate for the measure of an adult's wrist is $\frac{1}{4}$ of the waist size. If Jorge has a 34-inch waist, estimate the size of his wrist.

86. An estimate for an adult's waist measurement is found by dividing the neck size (in inches) by $\frac{1}{2}$. Jock's neck measures 18 inches. Estimate his waist measurement.

When setting a post for building a deck or fence, it is recommended that $\frac{1}{3}$ of the total length of the post be buried in the ground. Find the amount of post to be buried in the ground for the given post lengths.

87. a 9-foot post

88. a 12-foot post

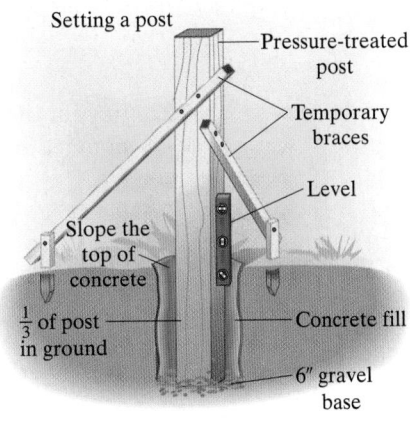

Setting a post
Pressure-treated post
Temporary braces
Level
Slope the top of concrete
$\frac{1}{3}$ of post in ground
Concrete fill
6″ gravel base

Source: *Southern Living Magazine*

89. A child needs a $\frac{1}{12}$-fluid ounce dose of a guaifenesin solution. How many doses are available in 2 fluid ounces?

90. How many $\frac{1}{3}$-ounce doses are available in a 5-ounce container of medicine.

Find the area of each rectangle. Recall that area = length · width, *or lw.*

△ **91.**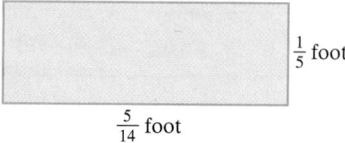
$\frac{1}{5}$ foot

$\frac{5}{14}$ foot

△ **92.**
$\frac{1}{2}$ mile

$\frac{3}{8}$ mile

Review and Preview

Write the prime factorization of each number. See Section 4.2.

93. 90

94. 42

95. 65

96. 72

97. 126

98. 112

 Combining Concepts

99. In a recent year, approximately $\frac{7}{10}$ of U.S. households that have on-line service made on-line purchases. If 10,300,000 U.S. households have on-line services, how many of these made on-line purchases? (*Source:* Inteco Corp.)

100. Approximately $\frac{3}{25}$ of the U.S. population lives in the state of California. If the U.S. population is approximately 281,422,000 find the population of California. (*Source:* U.S. Bureau of the Census, 2000)

101. In your own words, describe how to multiply fractions.

102. In your own words, describe how to divide fractions.

103. One-third of all native flowering plant species in the United States are at risk of becoming extinct. That translates into 5144 at-risk flowering plant species. Based on this data, how many flowering plant species are native to the United States overall? (*Source*: The Nature Conservancy) (*Hint*: How many $\frac{1}{3}$s are in 5144?)

104. The FedEx fleet of aircraft includes 264 Cessnas. These Cessnas make up $\frac{66}{149}$ of the FedEx fleet. What is the size of the entire FedEx fleet of aircraft? (*Source:* Federal Express Corp.)

105. $\frac{42}{25} \cdot \frac{125}{36} \div \frac{7}{6}$

106. $\left(\frac{8}{13} \cdot \frac{39}{16} \cdot \frac{8}{9}\right)^2$

4.4   Adding and Subtracting Like Fractions and Least Common Denominator

Fractions that have the same or a common denominator are called **like fractions**. Fractions that have different denominators are called **unlike fractions**.

Like Fractions	Unlike Fractions
$\dfrac{2}{5}$ and $\dfrac{3}{5}$	$\dfrac{2}{5}$ and $\dfrac{3}{4}$
$\dfrac{5}{21}, \dfrac{16}{21}$, and $\dfrac{7}{21}$	$-\dfrac{5}{7}$ and $\dfrac{5}{9}$
$-\dfrac{9}{15}$ and $\dfrac{13}{15}$	$\dfrac{3}{4}, \dfrac{9}{12}$, and $\dfrac{18}{24}$

A **Adding or Subtracting Like Fractions**

We can add like fractions on a number line just as we added whole numbers and integers on a number line.

To add $\dfrac{1}{5} + \dfrac{3}{5}$, start at 0 and draw an arrow $\dfrac{1}{5}$ of a unit long pointing to the right. From the tip of this arrow, draw an arrow $\dfrac{3}{5}$ of a unit long also pointing to the right. The tip of the second arrow ends at their sum, $\dfrac{4}{5}$.

Notice that the numerator of the sum is the sum of the numerators. Also, the denominator of the sum is the common denominator. This is how we add fractions. A similar method is used to subtract fractions.

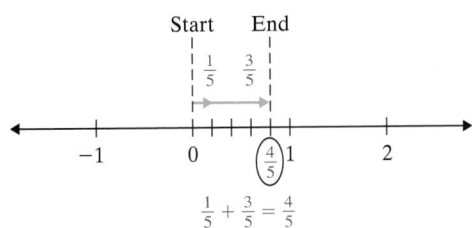

Adding or Subtracting Like Fractions (Fractions with the Same Denominator)

If a, b, and c, are numbers and b is not 0, then

$$\frac{a}{b} + \frac{c}{b} = \frac{a + c}{b} \qquad \text{and also} \qquad \frac{a}{b} - \frac{c}{b} = \frac{a - c}{b}$$

In other words, to add or subtract fractions with the same denominator, add or subtract their numerators and write the sum or difference over the **common** denominator.

For example,

$$\frac{1}{4} + \frac{2}{4} = \frac{1 + 2}{4} = \frac{3}{4} \qquad \text{Add the numerators.}$$
Keep the denominator.

$$\frac{4}{5} - \frac{2}{5} = \frac{4 - 2}{5} = \frac{2}{5} \qquad \text{Subtract the numerators.}$$
Keep the denominator.

 Helpful Hint

As usual, don't forget to write all answers in simplest form.

Practice Problems 1–3

Add and simplify.

1. $\dfrac{5}{9} + \dfrac{2}{9}$ 2. $\dfrac{5}{8} + \dfrac{1}{8}$

3. $\dfrac{10}{11} + \dfrac{1}{11} + \dfrac{7}{11}$

Concept Check

Find and correct the error in the following.

$$\dfrac{3}{7} + \dfrac{3}{7} = \dfrac{6}{14}$$

Practice Problems 4–5

Subtract and simplify.

4. $\dfrac{7}{12} - \dfrac{2}{12}$ 5. $\dfrac{9}{10} - \dfrac{1}{10}$

Practice Problem 6

Add: $-\dfrac{8}{5} + \dfrac{4}{5}$

Practice Problem 7

Subtract: $\dfrac{2}{5} - \dfrac{3y}{5}$

Answers

1. $\dfrac{7}{9}$ 2. $\dfrac{3}{4}$ 3. $\dfrac{18}{11}$ 4. $\dfrac{5}{12}$ 5. $\dfrac{4}{5}$ 6. $-\dfrac{4}{5}$

7. $\dfrac{2 - 3y}{5}$

Concept Check: $\dfrac{3}{7} + \dfrac{3}{7} = \dfrac{6}{7}$; keep the common denominator.

EXAMPLES Add and simplify.

1. $\dfrac{2}{7} + \dfrac{3}{7} = \dfrac{2+3}{7} = \dfrac{5}{7}$ Add the numerators.
Keep the common denominator.

2. $\dfrac{3}{16} + \dfrac{7}{16} = \dfrac{3+7}{16} = \dfrac{10}{16} = \dfrac{2 \cdot 5}{2 \cdot 8} = \dfrac{5}{8}$

3. $\dfrac{7}{8} + \dfrac{6}{8} + \dfrac{3}{8} = \dfrac{7+6+3}{8} = \dfrac{16}{8} = 2$

Try the Concept Check in the margin.

EXAMPLES Subtract and simplify.

4. $\dfrac{8}{9} - \dfrac{1}{9} = \dfrac{8-1}{9} = \dfrac{7}{9}$ ← Subtract the numerators
← Keep the common denominator.

5. $\dfrac{7}{8} - \dfrac{5}{8} = \dfrac{7-5}{8} = \dfrac{2}{8} = \dfrac{2}{2 \cdot 4} = \dfrac{1}{4}$

From our earlier work, we know that $\dfrac{-12}{6} = -2$ and that $\dfrac{12}{-6} = -2$. Also, $-\dfrac{12}{6} = -2$. Since all these fractions simplify to -2, we have that

$$\dfrac{-12}{6} = \dfrac{12}{-6} = -\dfrac{12}{6}$$

In general, the following is true:

$$\dfrac{-a}{b} = \dfrac{a}{-b} = -\dfrac{a}{b} \quad \text{as long as } b \text{ is not 0.}$$

For example, $\dfrac{-3}{4} = \dfrac{3}{-4} = -\dfrac{3}{4}$.

EXAMPLE 6 Add: $-\dfrac{11}{8} + \dfrac{6}{8}$

Solution: $-\dfrac{11}{8} + \dfrac{6}{8} = \dfrac{-11+6}{8}$

$$= \dfrac{-5}{8} \quad \text{or} \quad -\dfrac{5}{8}$$

EXAMPLE 7 Subtract: $\dfrac{3x}{4} - \dfrac{7}{4}$

Solution: $\dfrac{3x}{4} - \dfrac{7}{4} = \dfrac{3x-7}{4}$

Recall from Section 3.1 that the terms in the numerator are unlike terms and cannot be combined.

EXAMPLE 8 Subtract: $\dfrac{3}{7} - \dfrac{6}{7} - \dfrac{3}{7}$

Solution: $\dfrac{3}{7} - \dfrac{6}{7} - \dfrac{3}{7} = \dfrac{3 - 6 - 3}{7} = \dfrac{-6}{7}$ or $-\dfrac{6}{7}$

B **Adding and Subtracting Given Fractional Replacement Values**

EXAMPLE 9 Evaluate $y - x$ if $x = -\dfrac{3}{10}$ and $y = -\dfrac{8}{10}$.

Solution: Be very careful when replacing x and y with replacement values.

$$y - x = -\dfrac{8}{10} - \left(-\dfrac{3}{10}\right) \quad \text{Replace } x \text{ with } -\tfrac{3}{10} \text{ and } y \text{ with } -\tfrac{8}{10}.$$

$$= \dfrac{-8 - (-3)}{10}$$

$$= \dfrac{-5}{10} = \dfrac{-1 \cdot 5}{2 \cdot 5} = \dfrac{-1}{2} \text{ or } -\dfrac{1}{2}$$

C **Solving Equations Containing Fractions**

EXAMPLE 10 Solve: $x - \dfrac{1}{5} = \dfrac{3}{5}$

Solution: To solve, add $\dfrac{1}{5}$ to both sides of the equation.

$$x - \dfrac{1}{5} = \dfrac{3}{5}$$

$$x - \dfrac{1}{5} + \dfrac{1}{5} = \dfrac{3}{5} + \dfrac{1}{5}$$

$$x = \dfrac{4}{5}$$

To check, replace x with $\dfrac{4}{5}$ in the original equation.

$$x - \dfrac{1}{5} = \dfrac{3}{5}$$

$$\dfrac{4}{5} - \dfrac{1}{5} = \dfrac{3}{5} \quad \text{Replace } x \text{ with } \tfrac{4}{5}.$$

$$\dfrac{4 - 1}{5} = \dfrac{3}{5}$$

$$\dfrac{3}{5} = \dfrac{3}{5} \quad \text{True.}$$

The solution of $x - \dfrac{1}{5} = \dfrac{3}{5}$ is $\dfrac{4}{5}$.

Practice Problem 8

Subtract: $\dfrac{4}{11} - \dfrac{6}{11} - \dfrac{3}{11}$

Helpful Hint

Recall that

$\dfrac{-6}{7} = -\dfrac{6}{7}$ $\left(\text{Also, } \dfrac{6}{-7} = -\dfrac{6}{7}, \text{ if needed.}\right)$

Practice Problem 9

Evaluate $x + y$ if $x = -\dfrac{10}{12}$ and $y = \dfrac{5}{12}$.

Practice Problem 10

Solve: $\dfrac{7}{10} = x - \dfrac{1}{10}$

Answers

8. $-\dfrac{5}{11}$ **9.** $-\dfrac{5}{12}$ **10.** $\dfrac{4}{5}$

D Solving Problems by Adding and Subtracting Like Fractions

We can combine our skills in adding and subtracting like fractions with our four problem-solving steps from Chapter 3 to solve many kinds of real-life problems.

Practice Problem 11

If a piano student practices the piano $\frac{3}{4}$ of an hour in the morning and $\frac{1}{4}$ of an hour in the evening, how long did she practice that day?

EXAMPLE 11 Total Amount of an Ingredient in a Recipe

A recipe calls for $\frac{1}{3}$ of a cup of flour at the beginning and $\frac{2}{3}$ of a cup of flour later. How much total flour is needed to make that recipe?

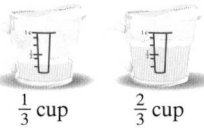

$\frac{1}{3}$ cup $\frac{2}{3}$ cup

Solution:

1. UNDERSTAND the problem. To do so, read and reread the problem. Since we are finding total flour, we will add. Let x = total flour.

2. TRANSLATE.

In words:	Total flour	is	flour at the beginning	added to	flour later
	↓	↓	↓	↓	↓
Translate:	x	=	$\frac{1}{3}$	+	$\frac{2}{3}$

3. SOLVE. $x = \frac{1}{3} + \frac{2}{3}$

$$= \frac{1+2}{3} = \frac{3}{3} = 1$$

4. INTERPRET. *Check* your work. *State* your conclusion: The total flour needed for the recipe is 1 cup. ●

EXAMPLE 12 Calculating Distance

The distance from home to the World Gym is $\frac{7}{8}$ of a mile and from home to the Post Office is $\frac{5}{8}$ of a mile. How much farther is it from home to the World Gym than from home to the Post Office?

Practice Problem 12

A walker attends a gym that has a $\frac{1}{8}$-mile track. If he walks 9 laps on the track on Monday and 3 laps on Wednesday, how much farther did he walk on Monday than on Wednesday?

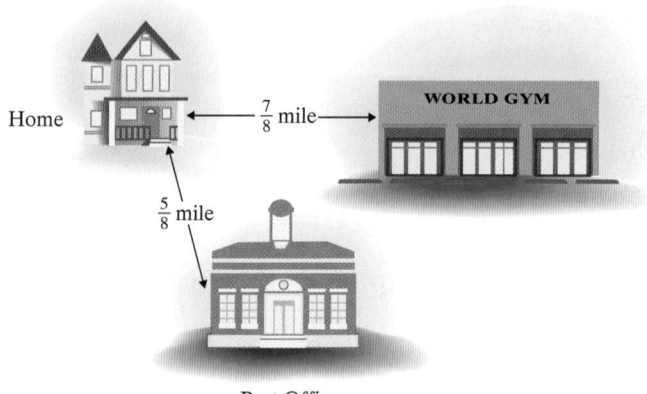

Home ← $\frac{7}{8}$ mile → WORLD GYM

$\frac{5}{8}$ mile

Post Office

Answers

11. 1 hour   **12.** $\frac{3}{4}$ mile

Solution:

1. UNDERSTAND. Read and reread the problem. The phrase "How much farther" tells us to subtract distances.

    Let x = distance farther.

2. TRANSLATE.

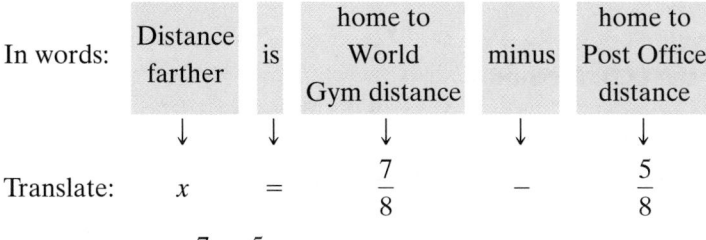

In words:	Distance farther	is	home to World Gym distance	minus	home to Post Office distance
	↓	↓	↓	↓	↓
Translate:	x	$=$	$\dfrac{7}{8}$	$-$	$\dfrac{5}{8}$

3. SOLVE: $x = \dfrac{7}{8} - \dfrac{5}{8}$

    $\qquad = \dfrac{7-5}{8} = \dfrac{2}{8} = \dfrac{2}{2\cdot 4} = \dfrac{1}{4}$

4. INTERPRET. *Check* your work. *State* your conclusion: The distance from home to the World Gym is $\dfrac{1}{4}$ mile farther than from home to the Post Office.   ●

Ⓔ Finding the Least Common Denominator

In the next section, we will add and subtract fractions that have different denominators. To add or subtract fractions that have unlike, or different, denominators, we first write them as equivalent fractions with a common denominator.

Although any common denominator can be used to add or subtract unlike fractions, we will use the **least common denominator (LCD)** or the **least common multiple (LCM)** of the denominators. Why? Since the LCD is the *smallest* of all common denominators, operations are usually less tedious with this number.

> The **least common denominator (LCD)** of a list of fractions is the smallest positive number divisible by all the denominators in the list. (The least common denominator is also the **least common multiple (LCM) of the denominators.**)

For example, the LCD of $\dfrac{1}{4}$ and $\dfrac{3}{10}$ is 20 because 20 is the smallest positive number divisible by both 4 and 10.

Finding the LCD: Method 1

One way to find the LCD is to see whether the larger denominator is divisible by the smaller denominator. If so, the larger number is the LCD. If not, then check consecutive multiples of the larger denominator until the LCD is found.

To find the LCD for $\dfrac{1}{4}$ and $\dfrac{3}{10}$, we check to see whether 10 is a multiple of 4. No, it is not, so we check consecutive multiples of 10.

$\quad 2 \cdot 10 = 20 \qquad$ 20 is divisible by 4, so LCD = 20.

Practice Problem 13

Find the LCD of $\dfrac{7}{12}$ and $\dfrac{2}{15}$.

Copyright 2004 Pearson Education, Inc.

Helpful Hint

The number 2 is a factor twice since that is the greatest number of times that 2 is a factor in the prime factorization of any one denominator.

Practice Problem 14

Find the LCD of $\dfrac{9}{14}$ and $\dfrac{11}{35}$.

EXAMPLE 13 Find the LCD of $\dfrac{3}{8}$ and $\dfrac{1}{6}$.

Solution: Is 8 divisible by 6? No, so we check multiples of 8.

$2 \cdot 8 = 16$ 16 is not divisible by 6.
$3 \cdot 8 = 24$ 24 is divisible by 6, so LCD $= 24$.   ●

Finding the LCD: Method 2

Another way to find the LCD is to first write each denominator as a product of primes. To find the LCD of $\dfrac{1}{4}$ and $\dfrac{3}{10}$, we write

$4 = 2 \cdot 2$
$10 = 2 \cdot 5$

If the LCD is divisible by 4, it must contain the factors $2 \cdot 2$. If the LCD is divisible by 10, it must contain the factors $2 \cdot 5$. Since 4 and 10 will divide into the LCD separately, the LCD needs to contain a factor the greatest number of times that the factor appears in any **one** prime factorization.

factors of 4

LCD $= 2 \cdot 2 \cdot 5 = 20$

factors of 10

EXAMPLE 14 Find the LCD of $\dfrac{5}{24}$ and $\dfrac{1}{18}$.

Solution: First write each denominator as a product of primes.

$24 = 2 \cdot 2 \cdot 2 \cdot 3$
$18 = 2 \cdot 3 \cdot 3$

Write each factor the greatest number of times that it appears in any **one** prime factorization.
The greatest number of times that 2 appears is **3** times:

$24 = 2 \cdot 2 \cdot 2 \cdot 3$

The greatest number of times that 3 appears is **2** times:

$18 = 2 \cdot 3 \cdot 3$

LCD $= 2 \cdot 2 \cdot 2 \cdot 3 \cdot 3 = 72$

Notice that 72 is the smallest positive number that is divisible by both 18 and 24.   ●

Answers

13. 70   **14.** 60

EXAMPLE 15 Find the LCD of $-\dfrac{2}{5}, \dfrac{1}{6},$ and $\dfrac{5}{12}$.

Solution: To help find the LCD, we write the prime factorization of each denominator and circle each different factor the greatest number of times that it appears in any one factor.

$$5 = \text{⑤}$$
$$6 = 2 \cdot \text{③}$$
$$12 = \text{②·②} \cdot 3$$
$$\text{LCD} = 2 \cdot 2 \cdot 3 \cdot 5 = 60$$

●

EXAMPLE 16 Find the LCD of $\dfrac{3}{5}, \dfrac{2}{x},$ and $\dfrac{7}{x^3}$.

Solution: $5 = \text{⑤}$
$$x = x$$
$$x^3 = \text{(x·x·x)}$$
$$\text{LCD} = 5 \cdot x \cdot x \cdot x = 5x^3$$

●

Try the Concept Check in the margin.

Practice Problem 15

Find the LCD of $\dfrac{7}{4}, \dfrac{7}{15},$ and $\dfrac{3}{10}$.

> **Helpful Hint**
>
> If you prefer working with exponents,
> Example 15: $5 = \text{⑤}$
> $$6 = 2 \cdot \text{③}$$
> $$12 = \text{②}^2 \cdot 3$$
> $$\text{LCD} = 2^2 \cdot 3 \cdot 5 = 60$$

Practice Problem 16

Find the LCD of $\dfrac{7}{y}$ and $\dfrac{6}{11}$.

Concept Check

True or false? The LCD of the fractions $\dfrac{1}{6}$ and $\dfrac{1}{8}$ is 48.

STUDY SKILLS REMINDER

Have you decided to successfully complete this course?

Ask yourself if one of your current goals is to successfully complete this course.

If it is not a goal of yours, ask yourself why? One common reason is fear of failure. Amazingly enough, fear of failure alone can be strong enough to keep many of us from doing our best in any endeavor. Another common reason is that you simply haven't taken the time to make successfully completing this course one of your goals.

If you are taking this mathematics course, then successfully completing this course probably should be one of your goals. To make it a goal, start by writing this goal in your mathematics notebook. Then read or reread Section 1.1 and make a commitment to try the suggestions in this section.

If successfully completing this course is already a goal of yours, also read or reread Section 1.1 and try some suggestions in that section so that you are actively working toward your goal.

Good luck and don't forget that a positive attitude will make a big difference.

Answers

15. 60   **16.** $11y$

Concept Check: False; it is 24.

FOCUS ON **Business and Career**

INTENDED MAJORS

Table 1 shows the breakdown of entering college freshmen who plan to major in a professional field by specific professional major. Table 2 shows a similar breakdown for technical majors.

TABLE 1	
Professional Majors	Fraction
Architecture or urban planning	$\frac{11}{155}$
Home economics	$\frac{1}{155}$
Health technology	$\frac{13}{155}$
Nursing	$\frac{7}{31}$
Pharmacy	$\frac{9}{155}$
Pre-dental, pre-medical, pre-veterinary	$\frac{39}{155}$
Therapy (occupational, physical, speech)	$\frac{36}{155}$
Other	$\frac{11}{155}$

(*Source:* Higher Education Research Institute)

TABLE 2	
Technical Majors	Fraction
Building trades	$\frac{1}{8}$
Data processing, computer programming	$\frac{3}{8}$
Drafting or design	$\frac{1}{8}$
Electronics	$\frac{3}{40}$
Mechanics	$\frac{7}{40}$
Other	$\frac{1}{8}$

(*Source:* Higher Education Research Institute)

CRITICAL THINKING

1. Interpret the meaning of the first line of Table 1.
2. Interpret the meaning of the third line of Table 2.
3. Which of the professional majors is most popular? Explain your reasoning.
4. Which of the technical majors is most popular? Explain your reasoning.

GROUP ACTIVITY

Research the number of majors offered through your discipline area (such as Arts and Humanities, Biological Sciences, Business, Education, Engineering, etc.) at your school. How many entering students plan to take each major? Make a table similar to Tables 1 and 2 to display your results. Show the fraction of entering students taking each major. Do additional research to find the number of students that graduate with each major. Make another table to display your results. How does the breakdown of majors (those intended by entering students and those actually completed by graduating) change?

Mental Math

State whether the fractions in each list are like or unlike fractions.

1. $\dfrac{7}{8}, \dfrac{7}{10}$

2. $\dfrac{2}{3}, \dfrac{2}{9}$

3. $\dfrac{9}{10}, \dfrac{1}{10}$

4. $\dfrac{8}{11}, \dfrac{2}{11}$

5. $\dfrac{2}{31}, \dfrac{30}{31}, \dfrac{19}{31}$

6. $\dfrac{3}{10}, \dfrac{3}{11}, \dfrac{3}{13}$

7. $\dfrac{5}{12}, \dfrac{7}{12}, \dfrac{12}{11}$

8. $\dfrac{1}{5}, \dfrac{2}{5}, \dfrac{4}{5}$

A *Add or subtract as indicated. See Examples 1 through 8.*

9. $\dfrac{3}{7} + \dfrac{2}{7}$

10. $\dfrac{5}{9} + \dfrac{2}{9}$

11. $\dfrac{10}{11} - \dfrac{4}{11}$

12. $\dfrac{9}{13} - \dfrac{5}{13}$

13. $\dfrac{5}{11} + \dfrac{2}{11}$

14. $\dfrac{4}{7} + \dfrac{2}{7}$

15. $\dfrac{9}{15} - \dfrac{1}{15}$

16. $\dfrac{3}{15} - \dfrac{1}{15}$

EXERCISE SET 4.4

Add or subtract as indicated. See Examples 1 through 8.

1. $-\dfrac{1}{2} + \dfrac{1}{2}$

2. $-\dfrac{3}{x} + \dfrac{1}{x}$

3. $\dfrac{2}{9x} + \dfrac{4}{9x}$

4. $\dfrac{3}{10y} + \dfrac{2}{10y}$

5. $-\dfrac{4}{13} + \dfrac{2}{13} + \dfrac{1}{13}$

6. $-\dfrac{5}{11} + \dfrac{1}{11} + \dfrac{2}{11}$

7. $\dfrac{7}{18} + \dfrac{3}{18} + \dfrac{2}{18}$

8. $\dfrac{2}{15} + \dfrac{4}{15} + \dfrac{9}{15}$

9. $\dfrac{1}{y} - \dfrac{4}{y}$

10. $\dfrac{4}{z} - \dfrac{7}{z}$

11. $\dfrac{7a}{4} - \dfrac{3}{4}$

12. $\dfrac{18b}{5} - \dfrac{3}{5}$

13. $\dfrac{1}{8} - \dfrac{7}{8}$

14. $\dfrac{1}{6} - \dfrac{5}{6}$

15. $\dfrac{20}{21} - \dfrac{10}{21} - \dfrac{17}{21}$

16. $\dfrac{27}{28} - \dfrac{5}{28} - \dfrac{28}{28}$

17. $\dfrac{9x}{15} + \dfrac{1x}{15}$

18. $\dfrac{2x}{15} - \dfrac{7}{15}$

19. $\dfrac{7x}{16} - \dfrac{15x}{16}$

20. $\dfrac{15b}{16} + \dfrac{7b}{16}$

21. $\dfrac{15}{16z} - \dfrac{3}{16z}$

22. $\dfrac{7}{16a} + \dfrac{15}{16a}$

23. $\dfrac{3}{10} - \dfrac{6}{10}$

24. $-\dfrac{6}{10} + \dfrac{3}{10}$

25. $\dfrac{15}{17} + \dfrac{5}{17} + \dfrac{14}{17}$

26. $\dfrac{1}{8} - \dfrac{15}{8} + \dfrac{2}{8}$

27. $\dfrac{9}{12} - \dfrac{7}{12} - \dfrac{10}{12}$

28. $\dfrac{9}{13} + \dfrac{10}{13} + \dfrac{7}{13}$

29. $\dfrac{x}{4} + \dfrac{3x}{4} - \dfrac{2x}{4} + \dfrac{x}{4}$

30. $\dfrac{9y}{8} + \dfrac{2y}{8} + \dfrac{5y}{8} - \dfrac{4y}{8}$

B *Evaluate each expression for the given replacement values. See Example 9.*

31. $x + y;\ x = \dfrac{3}{4},\ y = \dfrac{2}{4}$

32. $x - y;\ x = \dfrac{7}{8},\ y = \dfrac{9}{8}$

33. $x - y;\ x = -\dfrac{1}{5},\ y = -\dfrac{3}{5}$

34. $x + y;\ x = -\dfrac{1}{6},\ y = \dfrac{5}{6}$

35. $x - y + z;\ x = \dfrac{3}{12},\ y = \dfrac{5}{12},\ z = -\dfrac{7}{12}$

36. $x + y - z;\ x = \dfrac{2}{14},\ y = \dfrac{3}{14},\ z = \dfrac{8}{14}$

C *Solve and check. See Example 10.*

37. $x + \dfrac{1}{3} = -\dfrac{1}{3}$

38. $x + \dfrac{1}{9} = -\dfrac{7}{9}$

39. $y - \dfrac{3}{13} = -\dfrac{2}{13}$

40. $z - \dfrac{5}{14} = \dfrac{4}{14}$

41. $3x - \dfrac{1}{5} - 2x = \dfrac{1}{5} + \dfrac{2}{5}$

42. $5x + \dfrac{1}{11} - 4x = \dfrac{2}{11} - \dfrac{5}{11}$

D *Find the perimeter of each figure. Recall that the perimeter of a figure is the distance around a figure.*

△ **43.**
$\dfrac{4}{20}$ inch $\dfrac{7}{20}$ inch
$\dfrac{9}{20}$ inch

△ **44.**
Square $\dfrac{1}{6}$ centimeter

△ **45.**
$\dfrac{5}{12}$ meter Rectangle
$\dfrac{7}{12}$ meter

△ **46.**
$\dfrac{3}{13}$ foot
$\dfrac{2}{13}$ foot $\dfrac{6}{13}$ foot
$\dfrac{3}{13}$ foot
$\dfrac{4}{13}$ foot

Solve. Write each answer in simplest form. See Examples 11 and 12.

47. Nathan Payne worked in his yard for $\dfrac{5}{8}$ of an hour on Saturday and $\dfrac{7}{8}$ of an hour on Sunday. How long did Nathan work in his yard over the weekend?

48. A recipe for Heavenly Hash Cake calls for $\dfrac{3}{4}$ cup of flour and later $\dfrac{1}{4}$ cup of flour. How much flour is needed to make the recipe?

The chart shows the breakdown of all U.S. employees covered by health benefits in 2002 by type of health plan. Use this chart to answer Exercises 49–52.

49. Put the health plans in order from the smallest fraction of employees covered to the largest fraction of employees covered.

50. Find the fraction of employees that are *not* covered by a Health Maintenance Organization.

Type of Health Plan	Fraction of Employees with Health Benefits
Health Maintenance Organization	$\dfrac{29}{100}$
Point-of-Service	$\dfrac{14}{100}$
Preferred Provider Organization	$\dfrac{50}{100}$
Traditional fee-for-service	$\dfrac{7}{100}$

(*Source:* William M. Mercer, Inc.)

51. Find the fraction of the employees that are *not* covered by a Preferred Provider Organization.

52. Which type of health plan is the most popular?

53. As of 2002, the fraction of states in the United States with maximum interstate highway speed limits up to and including 70 mph was $\frac{39}{50}$. The fraction of states with 70 mph speed limits was $\frac{18}{50}$. What fraction of states had speed limits that were less than 70 mph? (*Source:* National Motorists Association)

54. When people take aspirin, $\frac{31}{50}$ of the time it is used to treat some type of pain. Approximately $\frac{7}{50}$ of all aspirin use is for treating headaches. What fraction of aspirin use is for treating pain other than headaches? (*Source:* Bayer Market Research)

E *Find the LCD of each list of fractions. See Examples 13 through 16.*

55. $\dfrac{1}{3}, \dfrac{3}{4}$

56. $\dfrac{1}{4}, \dfrac{5}{6}$

57. $-\dfrac{2}{9}, \dfrac{6}{15}$

58. $-\dfrac{7}{12}, \dfrac{3}{20}$

59. $\dfrac{5}{12}, \dfrac{5}{18}$

60. $\dfrac{7}{12}, \dfrac{7}{15}$

61. $-\dfrac{7}{24}, -\dfrac{5}{x}$

62. $-\dfrac{11}{y}, -\dfrac{13}{70}$

63. $\dfrac{2}{25}, \dfrac{3}{15}, \dfrac{5}{6}$

64. $\dfrac{3}{4}, \dfrac{1}{6}, \dfrac{13}{18}$

65. $\dfrac{23}{18}, \dfrac{1}{21}$

66. $\dfrac{45}{24}, \dfrac{2}{45}$

67. $-\dfrac{16}{15}, -\dfrac{11}{25}$

68. $-\dfrac{22}{21}, \dfrac{3}{14}$

69. $\dfrac{1}{8}, \dfrac{1}{24}$

70. $\dfrac{1}{15}, \dfrac{1}{90}$

71. $\dfrac{8}{25}, \dfrac{7}{10}$

72. $\dfrac{7}{8}, \dfrac{13}{12}$

73. $-\dfrac{1}{a}, -\dfrac{7}{12}$

74. $-\dfrac{1}{9}, -\dfrac{80}{b}$

75. $\dfrac{4}{3}, \dfrac{8}{21}, \dfrac{3}{56}$

76. $\dfrac{6}{70}, \dfrac{11}{80}, \dfrac{15}{90}$

77. $\dfrac{12}{11}, \dfrac{20}{33}, \dfrac{12}{121}$

78. $\dfrac{7}{10}, \dfrac{8}{15}, \dfrac{9}{100}$

Perform each indicated operation. See Sections 4.3 and 2.5.

79. $\dfrac{4}{5} \cdot \dfrac{3}{7}$

80. $\dfrac{5}{3} \cdot \dfrac{4}{9}$

81. $-2 + 10$

82. $-2(10)$

83. $\dfrac{2}{5} \div \dfrac{1}{2}$

84. $\dfrac{4}{7} \div \dfrac{1}{3}$

85. $-12 - 16$

86. $-18 - (-2)$

 Combining Concepts

Perform each indicated operation.

87. $\dfrac{4}{11} + \dfrac{5}{11} - \dfrac{3}{11} + \dfrac{2}{11}$

88. $\dfrac{9}{12} + \dfrac{1}{12} - \dfrac{3}{12} - \dfrac{5}{12}$

Solve. Write each answer in simplest form.

89. A person's marital status is either single (never married), married, widowed, or divorced. Of American men over the age of 65, $\dfrac{38}{50}$ are married and $\dfrac{7}{50}$ are widowed. What fraction of American men over age 65 are either single or divorced? (*Source:* Based on data from the U.S. Bureau of the Census)

90. Trey Nguyen jogged $\dfrac{3}{8}$ of a mile from home and then rested. Then he continued jogging for another $\dfrac{3}{8}$ of a mile until he discovered his watch had fallen off. He walked back along the same path for $\dfrac{4}{8}$ of a mile until he found his watch. Find how far he was from his *starting point*.

91. In your own words, explain how to add like fractions.

92. In your own words, explain how to subtract like fractions.

INDUCTIVE REASONING

Inductive reasoning is the process of drawing a general conclusion from just a few observations. Many times in mathematics, we observe similarities among numbers or calculations and notice a pattern emerging. Identifying a pattern in this way uses inductive reasoning.

For example, look at the following list of numbers. The dots at the end of the list indicate that the list continues indefinitely. What do you notice?

$$2, 4, 6, 8, 10, 12, \ldots$$

You probably noticed the pattern that each number in the list is 2 greater than the previous number and that all of the numbers are even numbers. If we were asked to guess the next number in the list, we could be confident that "14" would be a good response.

Let's try finding another pattern in a list of numbers.

$$10, 13, 18, 25, 34, 45, \ldots$$

What do you notice? It might be useful to find the difference between successive numbers in the list. Using the differences found below, we can see that each successive difference is 2 greater than the previous difference. We can guess that the next difference will be 13, so the next number in the list is probably $45 + 13 = 58$.

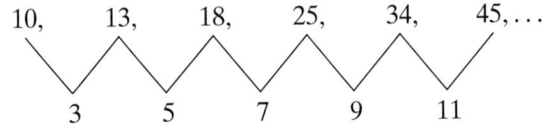

CRITICAL THINKING

Give the next two numbers in each list.

1. $5, 8, 11, 14, 17, 20, \ldots$

2. $100, 95, 90, 85, 80, \ldots$

3. $\dfrac{1}{2}, \dfrac{1}{3}, \dfrac{1}{4}, \dfrac{1}{5}, \ldots$

4. $5, 1, 6, 1, 1, 7, 1, 1, 1, 8, 1, 1, 1, \ldots$

5. $3, 4, 6, 9, 13, 18, \ldots$

6. $2, 4, 8, 16, 32, \ldots$

STUDY SKILLS REMINDER

What should you do the day of an exam?

On the day of an exam, try the following:

- Allow yourself plenty of time to arrive.

- Read the directions on the test carefully.

- Read each problem carefully as you take your test. Make sure that you answer the question asked.

- Watch your time and pace yourself so that you may attempt each problem on your test.

- If you have time, check your work and answers.

- Do not turn your test in early. If you have extra time, spend it double-checking your work.

Good luck!

4.5 Adding and Subtracting Unlike Fractions

Ⓐ Adding and Subtracting Unlike Fractions

In this section we add and subtract fractions with different denominators. To add or subtract these unlike fractions, first write the fractions as equivalent fractions with a common denominator and then add or subtract the like fractions. The common denominator we will use is the least common denominator (LCD).

For example, add the following unlike fractions: $\frac{3}{4} + \frac{1}{6}$. The LCD of the denominators 4 and 6 is 12. Write each fraction as an equivalent fraction with a denominator of 12.

$$\frac{3}{4} = \frac{3 \cdot 3}{4 \cdot 3} = \frac{9}{12} \quad \text{and} \quad \frac{1}{6} = \frac{1 \cdot 2}{6 \cdot 2} = \frac{2}{12}$$

Then

$$\frac{3}{4} + \frac{1}{6} = \frac{9}{12} + \frac{2}{12} = \frac{11}{12}$$

Adding or Subtracting Unlike Fractions

Step 1. Find the LCD of the denominators of the fractions.

Step 2. Write each fraction as an equivalent fraction whose denominator is the LCD.

Step 3. Add or subtract the like fractions.

Step 4. Write the sum or difference in simplest form.

EXAMPLE 1   Add: $\frac{2}{5} + \frac{4}{15}$

Solution: **Step 1.** The LCD of the denominators 5 and 15 is 15.

Step 2. $\frac{2}{5} = \frac{2 \cdot 3}{5 \cdot 3} = \frac{6}{15}, \qquad \frac{4}{15} = \frac{4}{15}$ ← This fraction already has a denominator of 15.

Step 3. Add. $\frac{2}{5} + \frac{4}{15} = \frac{6}{15} + \frac{4}{15} = \frac{10}{15}$

Step 4. Write in simplest form.

$$\frac{10}{15} = \frac{2 \cdot 5}{3 \cdot 5} = \frac{2}{3}$$

EXAMPLE 2   Subtract: $\frac{2}{3} - \frac{10}{11}$

Solution: **Step 1.** The LCD of the denominators 3 and 11 is 33.

Step 2. $\frac{2}{3} = \frac{2 \cdot 11}{3 \cdot 11} = \frac{22}{33} \qquad \text{and} \qquad \frac{10}{11} = \frac{10 \cdot 3}{11 \cdot 3} = \frac{30}{33}$

Step 3. Subtract. $\frac{2}{3} - \frac{10}{11} = \frac{22}{33} - \frac{30}{33}$

$$= \frac{22 - 30}{33}$$

$$= \frac{-8}{33} \quad \text{or} \quad -\frac{8}{33}$$

Step 4. $-\frac{8}{33}$ is in simplest form.

OBJECTIVES

Ⓐ Add or subtract unlike fractions.

Ⓑ Write fractions in order.

Ⓒ Evaluate expressions given fractional replacement values.

Ⓓ Solve equations containing fractions.

Ⓔ Solve problems by adding or subtracting unlike fractions.

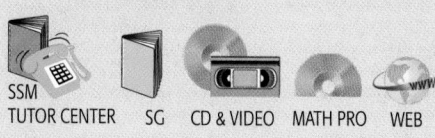

SSM
TUTOR CENTER   SG   CD & VIDEO   MATH PRO   WEB

Practice Problem 1

Add: $\frac{4}{7} + \frac{3}{14}$

Practice Problem 2

Subtract: $\frac{3}{7} - \frac{9}{10}$

Answers

1. $\frac{11}{14}$ **2.** $-\frac{33}{70}$

> **Helpful Hint**
>
> Remember that $-\dfrac{a}{b} = \dfrac{a}{-b} = \dfrac{-a}{b}$. For example, $-\dfrac{8}{33} = \dfrac{8}{-33} = \dfrac{-8}{33}$.

Practice Problem 3

Add: $-\dfrac{1}{5} + \dfrac{3}{20}$

EXAMPLE 3 Add: $-\dfrac{1}{6} + \dfrac{1}{2}$

Solution: The LCD of the denominators 6 and 2 is 6.

$$-\frac{1}{6} + \frac{1}{2} = \frac{-1}{6} + \frac{1 \cdot 3}{2 \cdot 3}$$

$$= \frac{-1}{6} + \frac{3}{6}$$

$$= \frac{2}{6}$$

Next, simplify $\dfrac{2}{6}$.

$$\frac{2}{6} = \frac{2}{2 \cdot 3} = \frac{1}{3}$$

When the fractions contain variables, we add and subtract the same way. ●

EXAMPLE 4 Subtract: $2 - \dfrac{x}{3}$

Solution: Recall that $2 = \dfrac{2}{1}$. The LCD of the denominators 1 and 3 is 3.

$$\frac{2}{1} - \frac{x}{3} = \frac{2 \cdot 3}{1 \cdot 3} - \frac{x}{3}$$

$$= \frac{6}{3} - \frac{x}{3}$$

$$= \frac{6 - x}{3}$$

The numerator $6 - x$ cannot be simplified further since 6 and $-x$ are unlike terms. ●

Practice Problem 4

Add: $5 + \dfrac{3y}{4}$

> **Helpful Hint**
>
> The expression $\dfrac{6 - x}{3}$ from Example 4 *does not simplify* to $2 - x$. The number 3 must be a factor of both terms in the numerator (not just 6) in order to factor it out.

Practice Problem 5

Find: $\dfrac{5}{8} - \dfrac{1}{3} - \dfrac{1}{12}$

EXAMPLE 5 Find: $-\dfrac{3}{4} - \dfrac{1}{14} + \dfrac{6}{7}$

Solution: The LCD of 4, 14, and 7 is 28.

$$-\frac{3}{4} - \frac{1}{14} + \frac{6}{7} = -\frac{3 \cdot 7}{4 \cdot 7} - \frac{1 \cdot 2}{14 \cdot 2} + \frac{6 \cdot 4}{7 \cdot 4}$$

$$= -\frac{21}{28} - \frac{2}{28} + \frac{24}{28}$$

$$= \frac{1}{28}$$

Answers

3. $-\dfrac{1}{20}$ **4.** $\dfrac{20 + 3y}{4}$ **5.** $\dfrac{5}{24}$

Try the Concept Check in the margin.

B Writing Fractions in Order

One important application of the least common denominator is to use the LCD to help order or compare fractions.

EXAMPLE 6.   Insert $<$ or $>$ to form a true sentence.

$$\frac{3}{4} \qquad \frac{9}{11}$$

Solution: The LCD for these fractions is 44. Let's write each fraction as an equivalent fraction with a denominator of 44.

$$\frac{3}{4} = \frac{3 \cdot 11}{4 \cdot 11} = \frac{33}{44} \qquad\qquad \frac{9}{11} = \frac{9 \cdot 4}{11 \cdot 4} = \frac{36}{44}$$

Since $33 < 36$, then $\dfrac{33}{44} < \dfrac{36}{44}$ or

$$\frac{3}{4} < \frac{9}{11}$$

EXAMPLE 7.   Insert $<$ or $>$ to form a true sentence.

$$-\frac{2}{7} \qquad -\frac{1}{3}$$

Solution: The LCD is 21.

$$-\frac{2}{7} = -\frac{2 \cdot 3}{7 \cdot 3} = -\frac{6}{21} \qquad -\frac{1}{3} = -\frac{1 \cdot 7}{3 \cdot 7} = -\frac{7}{21}$$

Since $-6 > -7$, then $-\dfrac{6}{21} > -\dfrac{7}{21}$ or

$$-\frac{2}{7} > -\frac{1}{3}$$

C Evaluating Expressions Given Fractional Replacement Values

EXAMPLE 8   Evaluate $x - y$ if $x = \dfrac{7}{18}$ and $y = \dfrac{2}{9}$.

Solution: Replace x with $\dfrac{7}{18}$ and y with $\dfrac{2}{9}$ in the expression $x - y$.

$$x - y = \frac{7}{18} - \frac{2}{9}$$

The LCD of the denominators 18 and 9 is 18. Then

$$\frac{7}{18} - \frac{2}{9} = \frac{7}{18} - \frac{2 \cdot 2}{9 \cdot 2}$$

$$= \frac{7}{18} - \frac{4}{18}$$

$$= \frac{3}{18} = \frac{1}{6} \qquad \text{Simplified.}$$

Concept Check

Find and correct the error in the following: $\dfrac{7}{12} - \dfrac{3}{4} = \dfrac{4}{8} = \dfrac{1}{2}$.

Practice Problem 6

Insert $<$ or $>$ to form a true sentence.

$$\frac{3}{8} \qquad \frac{7}{20}$$

Practice Problem 7

Insert $<$ or $>$ to form a true sentence.

$$-\frac{17}{20} \qquad -\frac{4}{5}$$

Practice Problem 8

Evaluate $x + y$ if $x = \dfrac{5}{11}$ and $y = \dfrac{4}{9}$.

Answers

6. $>$ **7.** $<$ **8.** $\dfrac{89}{99}$

Concept Check: $\dfrac{7}{12} - \dfrac{3}{4} = \dfrac{7}{12} - \dfrac{9}{12} = -\dfrac{2}{12} = -\dfrac{1}{6}$

Practice Problem 9

Solve: $y - \dfrac{2}{3} = \dfrac{5}{12}$

D Solving Equations Containing Fractions

EXAMPLE 9   Solve: $x - \dfrac{3}{4} = \dfrac{1}{20}$

Solution: To get x by itself, add $\dfrac{3}{4}$ to both sides.

$$x - \frac{3}{4} = \frac{1}{20}$$

$$x - \frac{3}{4} + \frac{3}{4} = \frac{1}{20} + \frac{3}{4} \qquad \text{Add } \frac{3}{4} \text{ to both sides.}$$

$$x = \frac{1}{20} + \frac{3 \cdot 5}{4 \cdot 5} \qquad \text{The LCD of 20 and 4 is 20.}$$

$$x = \frac{1}{20} + \frac{15}{20}$$

$$x = \frac{16}{20}$$

$$x = \frac{4 \cdot 4}{4 \cdot 5} = \frac{4}{5} \qquad \text{Write } \frac{16}{20} \text{ in simplest form.}$$

Check: To check, replace x with $\dfrac{4}{5}$ in the original equation.

$$x - \frac{3}{4} = \frac{1}{20}$$

$$\frac{4}{5} - \frac{3}{4} = \frac{1}{20} \qquad \text{Replace } x \text{ with } \frac{4}{5}.$$

$$\frac{4 \cdot 4}{5 \cdot 4} - \frac{3 \cdot 5}{4 \cdot 5} = \frac{1}{20} \qquad \text{The LCD of 5 and 4 is 20.}$$

$$\frac{16}{20} - \frac{15}{20} = \frac{1}{20}$$

$$\frac{1}{20} = \frac{1}{20} \qquad \text{True.}$$

Thus $\dfrac{4}{5}$ is the solution of $x - \dfrac{3}{4} = \dfrac{1}{20}$.

Practice Problem 10

To repair her sidewalk, a homeowner must pour small amounts of cement in three different locations. She needs $\dfrac{3}{5}$ of a cubic yard, $\dfrac{2}{10}$ of a cubic yard, and $\dfrac{2}{15}$ of a cubic yard for these locations. Find the total amount of concrete the homeowner needs. If she bought enough cement to mix 1 cubic yard, did she buy enough?

E Solving Problems by Adding or Subtracting Unlike Fractions

Very often, real-world problems involve adding or subtracting unlike fractions.

EXAMPLE 10   Finding Total Weight

A freight truck has $\dfrac{1}{4}$ ton of computers, $\dfrac{1}{3}$ ton of televisions, and $\dfrac{3}{8}$ ton of small appliances. Find the total weight of its load.

Solution: 1.   UNDERSTAND. Read and reread the problem. The phrase "total weight" tells us to add. Since the unknown is total weight, let $x = $ total weight.

Answers

9. $\dfrac{13}{12}$ **10.** $\dfrac{14}{15}$ cubic yard; yes, she bought enough.

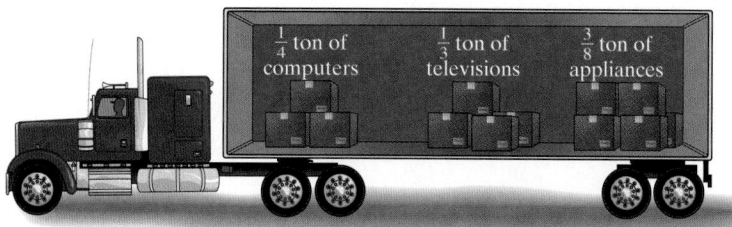

2. TRANSLATE.

In words:	Total weight	is	weight of computers	plus	weight of televisions	plus	weight of appliances
	↓	↓	↓	↓	↓	↓	↓
Translate:	x	$=$	$\frac{1}{4}$	$+$	$\frac{1}{3}$	$+$	$\frac{3}{8}$

3. SOLVE. The LCD is 24.

$$x = \frac{1}{4} + \frac{1}{3} + \frac{3}{8}$$

$$= \frac{1 \cdot 6}{4 \cdot 6} + \frac{1 \cdot 8}{3 \cdot 8} + \frac{3 \cdot 3}{8 \cdot 3}$$

$$= \frac{6}{24} + \frac{8}{24} + \frac{9}{24}$$

$$= \frac{23}{24}$$

4. INTERPRET. Check the solution. State your conclusion: The total weight of the truck's load is $\frac{23}{24}$ ton.

EXAMPLE 11   Calculating Flight Time

A flight from Tucson, Arizona, to Phoenix, Arizona, requires $\frac{5}{12}$ of an hour. If the plane has been flying $\frac{1}{4}$ of an hour, find how much time remains before landing.

Solution:

1. UNDERSTAND. Read and reread the problem. The phrase "how much time remains" tells us to subtract. Let x = time remaining.

Practice Problem 11

Find the difference in length of two boards if one board is $\frac{4}{5}$ of a foot long and the other is $\frac{2}{3}$ of a foot long.

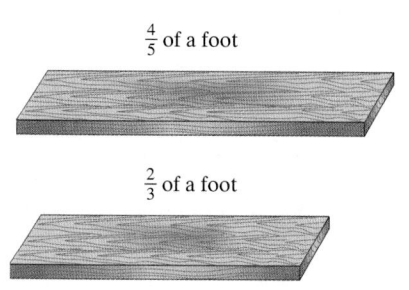

Answer

11. $\frac{2}{15}$ of a foot

2. TRANSLATE.

In words:	Time remaining	is	flight time from Tucson to Phoenix	minus	flight time already passed
Translate:	↓	↓	↓	↓	↓
	x	=	$\dfrac{5}{12}$	−	$\dfrac{1}{4}$

3. SOLVE. The LCD is 12.

$$
\begin{aligned}
x &= \frac{5}{12} - \frac{1}{4} \\
&= \frac{5}{12} - \frac{1 \cdot 3}{4 \cdot 3} \\
&= \frac{5}{12} - \frac{3}{12} \\
&= \frac{2}{12} = \frac{2}{2 \cdot 6} = \frac{1}{6}
\end{aligned}
$$

4. INTERPRET. *Check* the solution. *State* your conclusion: The flight time remaining is $\dfrac{1}{6}$ of an hour. ●

CALCULATOR EXPLORATIONS

Performing Operations on Fractions

SCIENTIFIC CALCULATOR

Many calculators have a fraction key, such as $\boxed{a\ b/c}$, that allows you to enter fractions, perform operations on fractions, and will give the result as a fraction. If your calculator has a fraction key, use it to calculate

$$\frac{3}{5} + \frac{4}{7}$$

Enter the keystrokes

$\boxed{3}\ \boxed{a\ b/c}\ \boxed{5}\ \boxed{+}\ \boxed{4}\ \boxed{a\ b/c}\ \boxed{7}\ \boxed{=}$

The display should read $\boxed{1\_6 \rfloor 35}$

which represents the mixed number $1\dfrac{6}{35}$. Until we discuss mixed numbers in Section 4.8, let's write the result as a fraction. To convert from mixed number notation to fractional notation, press

$\boxed{2^{nd}}\ \boxed{d/c}$

The display now reads $\boxed{41 \rfloor 35}$

which represents $\dfrac{41}{35}$, the sum in fractional notation.

GRAPHING CALCULATOR

Graphing calculators also allow you to perform operations on fractions and will give exact fractional results. The fraction option on a graphing calculator may be found under the $\boxed{\text{MATH}}$ menu. To perform the addition above, try the keystrokes.

$\boxed{3}\ \boxed{\div}\ \boxed{5}\ \boxed{+}\ \boxed{4}\ \boxed{\div}\ \boxed{7}\ \boxed{\text{MATH}}\ \boxed{\text{ENTER}}$
$\boxed{\text{ENTER}}$

The display should read

$\boxed{3/5 + 4/7 \blacktriangleright \text{Frac } 41/35}$

Use a calculator to add the following fractions. Give each sum as a fraction.

1. $\dfrac{1}{16} + \dfrac{2}{5}$ **2.** $\dfrac{3}{20} + \dfrac{2}{25}$ **3.** $\dfrac{4}{9} + \dfrac{7}{8}$

4. $\dfrac{9}{11} + \dfrac{5}{12}$ **5.** $\dfrac{10}{17} + \dfrac{12}{19}$ **6.** $\dfrac{14}{31} + \dfrac{15}{21}$

Mental Math

Find the LCD of each pair of fractions.

1. $\dfrac{1}{2}, \dfrac{2}{3}$

2. $\dfrac{1}{2}, \dfrac{3}{4}$

3. $\dfrac{1}{6}, \dfrac{5}{12}$

4. $\dfrac{2}{5}, \dfrac{7}{10}$

5. $\dfrac{4}{7}, \dfrac{1}{8}$

6. $\dfrac{23}{24}, \dfrac{1}{3}$

7. $\dfrac{11}{12}, \dfrac{3}{4}$

8. $\dfrac{2}{3}, \dfrac{3}{11}$

EXERCISE SET 4.5

A *Add or subtract as indicated. See Examples 1 through 4.*

1. $\dfrac{2}{3} + \dfrac{1}{6}$

2. $\dfrac{5}{6} + \dfrac{1}{12}$

3. $\dfrac{1}{2} - \dfrac{1}{3}$

4. $\dfrac{2}{3} - \dfrac{1}{4}$

5. $-\dfrac{2}{11} + \dfrac{2}{33}$

6. $-\dfrac{5}{9} + \dfrac{1}{3}$

7. $\dfrac{3x}{14} - \dfrac{3}{7}$

8. $\dfrac{2y}{5} - \dfrac{2}{15}$

9. $\dfrac{11}{35} + \dfrac{2}{7}$

10. $\dfrac{2}{5} + \dfrac{3}{25}$

11. $2y - \dfrac{5}{12}$

12. $5y - \dfrac{3}{20}$

13. $\dfrac{5}{12} - \dfrac{1}{9}$

14. $\dfrac{7}{12} - \dfrac{5}{18}$

15. $\dfrac{5}{7} + 1$

16. $-10 + \dfrac{7}{10}$

17. $\dfrac{5a}{11} + \dfrac{4a}{9}$

18. $\dfrac{7x}{18} + \dfrac{2x}{9}$

19. $\dfrac{2y}{3} - \dfrac{1}{6}$

20. $\dfrac{5}{6} - \dfrac{1}{12}$

21. $\dfrac{1}{2} + \dfrac{3}{x}$

22. $\dfrac{2}{5} + \dfrac{3}{x}$

23. $-\dfrac{2}{11} - \dfrac{2}{33}$

24. $-\dfrac{5}{9} - \dfrac{1}{3}$

25. $\dfrac{9}{14} - \dfrac{3}{7}$

26. $\dfrac{4}{5} - \dfrac{2}{15}$

27. $\dfrac{11y}{35} - \dfrac{2}{7}$

28. $\dfrac{2b}{5} - \dfrac{3}{25}$

29. $\dfrac{1}{9} - \dfrac{5}{12}$

30. $\dfrac{5}{18} - \dfrac{7}{12}$

31. $\dfrac{7}{15} - \dfrac{5}{12}$

32. $\dfrac{5}{8} - \dfrac{3}{20}$

33. $\dfrac{5}{7} - \dfrac{1}{8}$

34. $\dfrac{10}{13} - \dfrac{7}{10}$

35. $\dfrac{7}{8} + \dfrac{3}{16}$

36. $-\dfrac{7}{18} - \dfrac{2}{9}$

37. $\dfrac{5}{9} + \dfrac{3}{9}$

38. $\dfrac{4}{13} - \dfrac{1}{13}$

39. $\dfrac{5}{11} + \dfrac{y}{3}$

40. $\dfrac{5z}{13} + \dfrac{3}{26}$

41. $-\dfrac{5}{6} - \dfrac{3}{7}$

42. $\dfrac{1}{2} - \dfrac{3}{29}$

43. $\dfrac{7}{9} - \dfrac{1}{6}$

44. $\dfrac{9}{16} - \dfrac{3}{8}$

45. $\dfrac{2a}{3} + \dfrac{6a}{13}$

46. $\dfrac{3y}{4} + \dfrac{y}{7}$

47. $\dfrac{7}{30} - \dfrac{5}{12}$

48. $\dfrac{7}{30} - \dfrac{3}{20}$

49. $\dfrac{5}{9} + \dfrac{1}{y}$

50. $\dfrac{1}{12} - \dfrac{5}{x}$

51. $\dfrac{4}{5} + \dfrac{4}{9}$

52. $\dfrac{11}{12} - \dfrac{7}{24}$

53. $\dfrac{5}{9x} + \dfrac{1}{8}$

54. $\dfrac{3}{8} + \dfrac{5}{12x}$

Perform each indicated operation. See Example 5.

55. $-\dfrac{2}{5} + \dfrac{1}{3} - \dfrac{3}{10}$

56. $-\dfrac{1}{3} - \dfrac{1}{4} + \dfrac{2}{5}$

57. $\dfrac{x}{2} + \dfrac{x}{4} + \dfrac{2x}{16}$

58. $\dfrac{z}{4} + \dfrac{z}{8} + \dfrac{2z}{16}$

59. $\dfrac{6}{5} - \dfrac{3}{4} + \dfrac{1}{2}$

60. $\dfrac{6}{5} + \dfrac{3}{4} - \dfrac{1}{2}$

61. $-\dfrac{9}{12} + \dfrac{17}{24} - \dfrac{1}{6}$

62. $-\dfrac{5}{14} + \dfrac{3}{7} - \dfrac{1}{2}$

63. $\dfrac{3x}{8} + \dfrac{2x}{7} - \dfrac{5}{14}$

64. $\dfrac{9x}{10} - \dfrac{1}{2} + \dfrac{x}{5}$

B *Insert < or > to form a true sentence. See examples 6 and 7.*

65. $\dfrac{2}{7} \quad \dfrac{3}{10}$

66. $\dfrac{5}{9} \quad \dfrac{6}{11}$

67. $\dfrac{5}{6} \quad -\dfrac{13}{15}$

68. $-\dfrac{7}{8} \quad -\dfrac{5}{6}$

69. $-\dfrac{3}{4} \quad -\dfrac{11}{14}$

70. $-\dfrac{2}{9} \quad -\dfrac{3}{13}$

Evaluate each expression if $x = \dfrac{1}{3}$ *and* $y = \dfrac{3}{4}$. *See Example 8.*

71. $x + y$

72. $x - y$

73. xy

74. $x \div y$

75. $2y + x$

76. $2x + y$

D *Solve and check. See Example 9.*

77. $x - \dfrac{1}{12} = \dfrac{5}{6}$

78. $y - \dfrac{8}{9} = \dfrac{1}{3}$

79. $\dfrac{2}{5} + y = -\dfrac{3}{10}$

80. $\dfrac{1}{2} + a = -\dfrac{3}{8}$

81. $7z + \dfrac{1}{16} - 6z = \dfrac{3}{4}$

82. $9x - \dfrac{2}{7} - 8x = \dfrac{11}{14}$

83. $-\dfrac{2}{9} = x - \dfrac{5}{6}$

84. $-\dfrac{1}{4} = y - \dfrac{7}{10}$

E *Find the perimeter of each geometric figure. (Hint: Recall that perimeter means distance around.)*

 85.

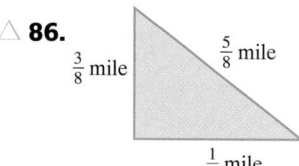

$\dfrac{4}{5}$ centimeter

$\dfrac{1}{3}$ centimeter | Parallelogram | $\dfrac{1}{3}$ centimeter

$\dfrac{4}{5}$ centimeter

△ **86.**

$\dfrac{3}{8}$ mile

$\dfrac{5}{8}$ mile

$\dfrac{1}{2}$ mile

△ **87.**

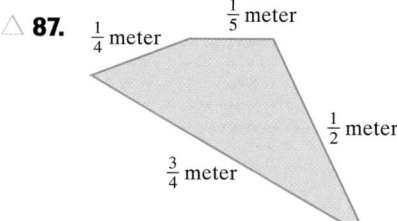

$\dfrac{1}{4}$ meter

$\dfrac{1}{5}$ meter

$\dfrac{1}{2}$ meter

$\dfrac{3}{4}$ meter

△ **88.**

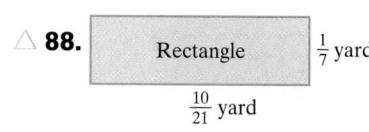

Rectangle | $\dfrac{1}{7}$ yard

$\dfrac{10}{21}$ yard

Solve. See Examples 10 and 11.

89. Killer bees have been known to chase people for up to $\frac{1}{4}$ of a mile, while domestic European honeybees will normally chase a person for no more than 100 feet, or $\frac{5}{264}$ of a mile. How much farther will a killer bee chase a person than a domestic honeybee? (*Source:* Coachella Valley Mosquito & Vector Control District)

90. The slowest mammal is the three-toed sloth from South America. The sloth has an average ground speed of $\frac{1}{10}$ mph. In the trees, it can accelerate to $\frac{17}{100}$ mph. How much faster can a sloth travel in the trees? (*Source: The Guiness Book of World Records*)

91. Given the following diagram, find L, its total length.

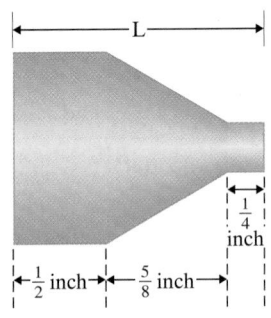

92. Given the following diagram, find W, its total width.

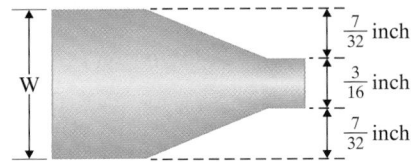

93. About $\frac{13}{20}$ of American students ages 10 to 17 name math, science, or art as their favorite subject in school. Art is the favorite subject for about $\frac{4}{25}$ of these students. For what fraction of students this age is math or science their favorite subject? (*Source:* Peter D. Hart Research Associates for the National Science Foundation)

94. Together, the United States' and Japan's postal services handle $\frac{49}{100}$ of the world's mail volume. Japan's postal service alone handles $\frac{3}{50}$ of the world's mail. What fraction of the world's mail is handled by the postal service of the United States? (*Source:* United States Postal Service)

The table gives the fraction of Americans who eat pasta at various intervals. Use this table to answer Exercises 95 and 96.

How Often Americans Eat Pasta	
Frequency	**Fraction**
3 times per week	$\frac{31}{100}$
1 or 2 times per week	$\frac{23}{50}$
1 or 2 times per month	$\frac{17}{100}$
Less often	$\frac{3}{50}$

(*Source:* Princeton Survey Research)

95. What fraction of Americans eat pasta 1, 2, or 3 times a week?

96. What fraction of Americans eat pasta 1 or 2 times a month or less often?

The circle graph shows the fraction of American adults who drive various distances in miles in an average week. Use this graph to answer Exercises 97 and 98.

97. What fraction of American adults drive less than 50 miles in an average week?

98. What fraction of American adults drive between 50 and 149 miles in an average week?

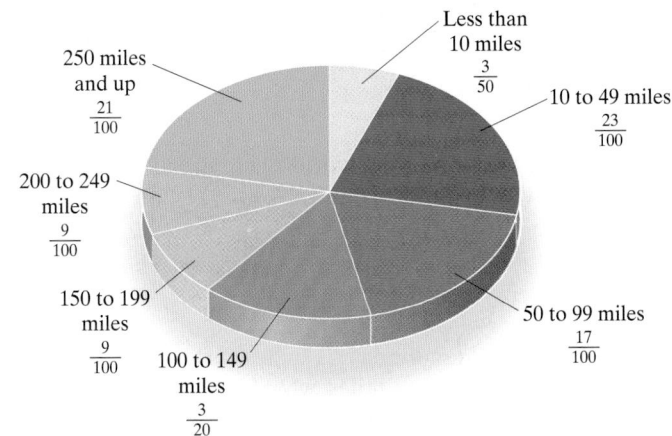

Miles Driven by American Adults in an Average Week

Source: Simmons

Review and Preview

Evaluate. See Section 4.3.

99. $\left(\dfrac{5}{6}\right)^2$

100. $\left(\dfrac{1}{2}\right)^2$

101. $\left(-\dfrac{5}{6}\right)^2$

102. $\left(-\dfrac{1}{2}\right)^2$

Round each number to the given place value. See Section 1.5.

103. 57,236 to the nearest hundred

104. 576 to the nearest hundred

105. 327 to the nearest ten

106. 2333 to the nearest ten

Combining Concepts

Perform each indicated operation.

107. $\dfrac{30}{55} + \dfrac{1000}{1760}$; the LCD of 55 and 1760 is 1760.

108. $\dfrac{19}{26} - \dfrac{968}{1352}$; the LCD of 26 and 1352 is 1352.

109. In your own words, describe how to add two fractions with different denominators.

110. Find the sum of the fractions in the circle graph on page 297. Did the sum surprise you? Why or why not?

Solve.

111. $x - \dfrac{5}{117} = \dfrac{71}{27}$

112. $-\dfrac{8}{81} = y - \dfrac{3}{45}$

The table shows the fraction of the world's land area occupied by each continent. Use this table to answer Exercises 113–117.

113. What fraction of the world's land area is accounted for by North and South America?

114. What fraction of the world's land area is accounted for by Asia and Europe?

115. If the total land area of Earth's surface is 57,900,000 square miles, what is the combined land area of the North and South American continents?

116. If the total land area of Earth's surface is 57,900,000 square miles, what is the combined land area of the European and Asian continents?

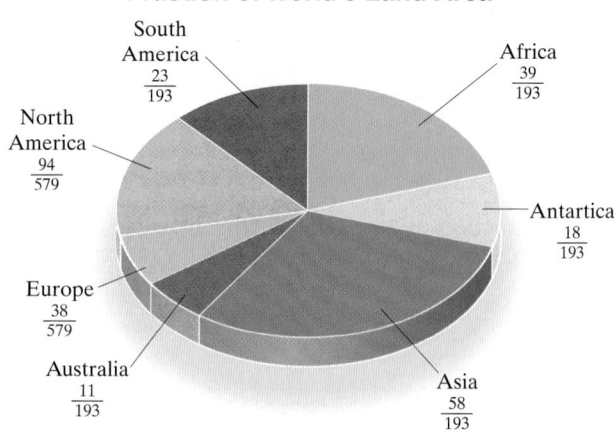

Fraction of World's Land Area

South America $\frac{23}{193}$

Africa $\frac{39}{193}$

North America $\frac{94}{579}$

Antartica $\frac{18}{193}$

Europe $\frac{38}{579}$

Australia $\frac{11}{193}$

Asia $\frac{58}{193}$

Source: *2000 World Almanac*

117. Antarctica is generally considered to be uninhabited. What fraction of the world's land area is accounted for by inhabited continents?

Solve.

118. In 2000, about $\frac{11}{67}$ of the total weight of mail delivered by the United States Postal Service was first-class mail. That same year, about $\frac{75}{134}$ of the total weight of mail delivered by the United States Postal Service was standard mail. Which of these two categories account for a greater portion of the mail handled by weight? (*Source:* U.S. Postal Service)

119. The National Park System (NPS) in the United States includes a wide variety of park types. National military parks account for $\frac{3}{128}$ of all NPS parks, and $\frac{1}{24}$ of NPS parks are classified as national preserves. Which category, national military park or national preserve, is bigger? (*Source:* National Park Service)

120. Approximately $\frac{7}{10}$ of U.S. adults have a savings account. About $\frac{11}{25}$ of U.S. adults have a non-interest bearing checking account. Which type of banking service, savings account or non-interest checking account, do adults in the United States use more? (*Source:* Scarborough Research/USData.com, Inc.)

121. About $\frac{127}{500}$ of U.S. adults rent one or two videos per month. Approximately $\frac{31}{200}$ of U.S. adults rent three or four videos per month. Which video rental category, 1–2 videos or 3–4 videos per month, is bigger? (*Source:* Telenation/Market Facts, Inc.)

Integrated Review–Summary of Fractions and Factors

Use a fraction to represent the shaded area of each figure or figure group.

1.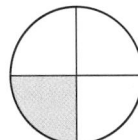

2.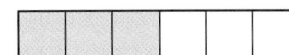

3.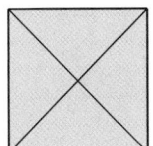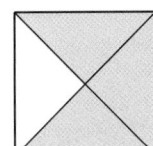

4. In a survey, 73 people out of 85 get less than 8 hours of sleep each night. What fraction of people in the survey get less than 8 hours of sleep?

Write the prime factorization of each number. Write any repeated factors using exponents.

5. 6

6. 70

7. 252

Write each fraction in simplest form.

8. $\dfrac{2}{14}$

9. $\dfrac{20}{24}$

10. $\dfrac{18}{38}$

11. $\dfrac{42}{110}$

12. $\dfrac{32}{64}$

13. $\dfrac{72}{80}$

14. Of the 50 United States, 2 states are not adjacent to any other states. What fraction of the states are not adjacent to other states? Write the fraction in simplest form.

The following summary will help you with this review of operations on fractions.

Operations on Fractions

Let a, b, c, and d be integers.

Addition: $\dfrac{a}{b} + \dfrac{c}{b} = \dfrac{a+c}{b}$

$(b \neq 0)$ ↑   ↑
     common denominator

Subtraction: $\dfrac{a}{b} - \dfrac{c}{b} = \dfrac{a-c}{b}$

$(b \neq 0)$ ↑   ↑
     common denominator

Multiplication: $\dfrac{a}{b} \cdot \dfrac{c}{d} = \dfrac{a \cdot c}{b \cdot d}$

$(b \neq 0, d \neq 0)$

Division: $\dfrac{a}{b} \div \dfrac{c}{d} = \dfrac{a}{b} \cdot \dfrac{d}{c} = \dfrac{a \cdot d}{b \cdot c}$

$(b \neq 0, d \neq 0, c \neq 0)$

Answers

1. _____

2. _____

3. _____

4. _____

5. _____

6. _____

7. _____

8. _____

9. _____

10. _____

11. _____

12. _____

13. _____

14. _____

Perform each indicated operation.

15. $\dfrac{1}{5} + \dfrac{3}{5}$

16. $\dfrac{1}{5} - \dfrac{3}{5}$

17. $\dfrac{1}{5} \cdot \dfrac{3}{5}$

18. $\dfrac{1}{5} \div \dfrac{3}{5}$

19. $\dfrac{2}{3} \div \dfrac{5}{6}$

20. $\dfrac{2}{3} \cdot \dfrac{5}{6}$

21. $\dfrac{2}{3} - \dfrac{5}{6}$

22. $\dfrac{2}{3} + \dfrac{5}{6}$

23. $-\dfrac{1}{7} \cdot -\dfrac{7}{18}$

24. $-\dfrac{4}{9} \cdot -\dfrac{3}{7}$

25. $-\dfrac{7}{8} \div 6$

26. $-\dfrac{9}{10} \div 5$

27. $\dfrac{7}{8} + \dfrac{1}{20}$

28. $\dfrac{5}{12} - \dfrac{1}{9}$

29. $\dfrac{9}{11} - \dfrac{2}{3}$

30. $\dfrac{2}{9} + \dfrac{1}{18}$

31. $\dfrac{2}{9} + \dfrac{1}{18} + \dfrac{1}{3}$

32. $\dfrac{3}{10} + \dfrac{1}{5} + \dfrac{6}{25}$

33. A contractor is using 18 acres of his land to sell $\dfrac{3}{4}$-acre lots. How many lots can he sell?

34. Suppose that the cross-section of a piece of pipe looks like the diagram shown. What is the inner diameter?

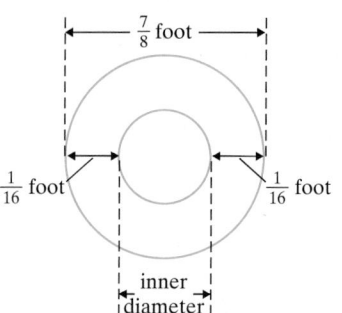

15. _____

16. _____

17. _____

18. _____

19. _____

20. _____

21. _____

22. _____

23. _____

24. _____

25. _____

26. _____

27. _____

28. _____

29. _____

30. _____

31. _____

32. _____

33. _____

34. _____

4.6 Complex Fractions and Review of Order of Operations

Ⓐ Simplifying Complex Fractions

Thus far, we have studied operations on fractions. We now practice simplifying fractions whose numerators or denominators themselves contain fractions. These fractions are called **complex fractions**.

Complex Fraction

A fraction whose numerator or denominator or both numerator and denominator contain fractions is called a **complex fraction**.

Examples of complex fractions are

$$\frac{\frac{x}{4}}{\frac{3}{2}} \qquad \frac{\frac{1}{2}+\frac{3}{8}}{\frac{3}{4}-\frac{1}{6}} \qquad \frac{\frac{y}{5}-2}{\frac{3}{10}}$$

Method 1 for Simplifying Complex Fractions

Two methods are presented to simplify complex fractions. The first method makes use of the fact that a fraction bar means division.

EXAMPLE 1 Simplify: $\dfrac{\frac{x}{4}}{\frac{3}{2}}$

Solution: Since a fraction bar means division, the complex fraction

$\dfrac{\frac{x}{4}}{\frac{3}{2}}$ can be written as $\dfrac{x}{4} \div \dfrac{3}{2}$. Then divide as usual to simplify.

$$\frac{x}{4} \div \frac{3}{2} = \frac{x}{4} \cdot \frac{2}{3} \qquad \text{Multiply by the reciprocal.}$$

$$= \frac{x \cdot 2}{2 \cdot 2 \cdot 3}$$

$$= \frac{x}{6}$$

EXAMPLE 2 Simplify: $\dfrac{\frac{1}{2}+\frac{3}{8}}{\frac{3}{4}-\frac{1}{6}}$

Solution: Recall the order of operations. Since the fraction bar is considered a grouping symbol, we simplify the numerator and the denominator of the complex fraction separately. Then we divide.

$$\frac{\frac{1}{2}+\frac{3}{8}}{\frac{3}{4}-\frac{1}{6}} = \frac{\frac{1 \cdot 4}{2 \cdot 4}+\frac{3}{8}}{\frac{3 \cdot 3}{4 \cdot 3}-\frac{1 \cdot 2}{6 \cdot 2}} = \frac{\frac{4}{8}+\frac{3}{8}}{\frac{9}{12}-\frac{2}{12}} = \frac{\frac{7}{8}}{\frac{7}{12}}$$

Practice Problem 1

Simplify: $\dfrac{\frac{7y}{10}}{\frac{1}{5}}$

Practice Problem 2

Simplify: $\dfrac{\frac{1}{2}+\frac{1}{6}}{\frac{3}{4}-\frac{2}{3}}$

Answers

1. $\dfrac{7y}{2}$ 2. $\dfrac{8}{1}$ or 8

Thus,

$$\frac{\dfrac{1}{2} + \dfrac{3}{8}}{\dfrac{3}{4} - \dfrac{1}{6}} = \frac{\dfrac{7}{8}}{\dfrac{7}{12}}$$

$$= \frac{7}{8} \div \frac{7}{12} \qquad \text{Rewrite the quotient using the } \div \text{ sign.}$$

$$= \frac{7}{8} \cdot \frac{12}{7} \qquad \text{Multiply by the reciprocal.}$$

$$= \frac{7 \cdot 3 \cdot 4}{2 \cdot 4 \cdot 7} \qquad \text{Multiply.}$$

$$= \frac{3}{2} \qquad \text{Simplify.} \qquad \bullet$$

Method 2 for Simplifying Complex Fractions

The second method for simplifying complex fractions is to multiply the numerator and the denominator of the complex fraction by the LCD of all the fractions in its numerator and its denominator. Since this LCD is divisible by all denominators, this has the effect of leaving sums and differences of integers in the numerator and the denominator. Let's use this second method to simplify the complex fraction in Example 2 again.

EXAMPLE 3 Simplify: $\dfrac{\dfrac{1}{2} + \dfrac{3}{8}}{\dfrac{3}{4} - \dfrac{1}{6}}$

Solution: The complex fraction contains fractions with denominators 2, 8, 4, and 6. The LCD is 24. By the fundamental property of fractions, we can multiply the numerator and the denominator of the complex fraction by 24. Notice below that by the distributive property, this means that we multiply each term in the numerator and denominator by 24.

$$\frac{\dfrac{1}{2} + \dfrac{3}{8}}{\dfrac{3}{4} - \dfrac{1}{6}} = \frac{24\left(\dfrac{1}{2} + \dfrac{3}{8}\right)}{24\left(\dfrac{3}{4} - \dfrac{1}{6}\right)}$$

$$= \frac{\left(24 \cdot \dfrac{1}{2}\right) + \left(24 \cdot \dfrac{3}{8}\right)}{\left(24 \cdot \dfrac{3}{4}\right) - \left(24 \cdot \dfrac{1}{6}\right)} \qquad \text{Apply the distributive property.}$$

$$= \frac{12 + 9}{18 - 4} \qquad \text{Multiply.}$$

$$= \frac{21}{14}$$

$$= \frac{7 \cdot 3}{7 \cdot 2} = \frac{3}{2} \qquad \text{Simplify.} \qquad \bullet$$

The simplified result is the same, of course, no matter which method is used.

Practice Problem 3

Use Method 2 to simplify: $\dfrac{\dfrac{1}{2} + \dfrac{1}{6}}{\dfrac{3}{4} - \dfrac{2}{3}}$

Answer

3. $\dfrac{8}{1}$ or 8

EXAMPLE 4   Simplify: $\dfrac{\dfrac{y}{5} - 2}{\dfrac{3}{10}}$

Practice Problem 4

Simplify: $\dfrac{\dfrac{3}{4}}{\dfrac{x}{5} - 1}$

Solution: Use the second method and multiply the numerator and the denominator of the complex fraction by the LCD of all fractions. Recall that $2 = \dfrac{2}{1}$. The LCD of the denominators 5, 1, and 10 is 10.

$$\frac{\dfrac{y}{5} - \dfrac{2}{1}}{\dfrac{3}{10}} = \frac{10\left(\dfrac{y}{5} - \dfrac{2}{1}\right)}{10\left(\dfrac{3}{10}\right)}$$ Multiply the numerator and denominator by 10.

$$= \frac{\left(10 \cdot \dfrac{y}{5}\right) - \left(10 \cdot \dfrac{2}{1}\right)}{10 \cdot \dfrac{3}{10}}$$ Apply the distributive property.

$$= \frac{2y - 20}{3}$$ Multiply.

> **Helpful Hint**
>
> Don't forget to multiply the numerator and the denominator of the complex fraction by the same number—the LCD.

B   Reviewing the Order of Operations

At this time, it is probably a good idea to review the order of operations on expressions containing fractions.

Order of Operations

1. Do all operations within grouping symbols such as parentheses or brackets.
2. Evaluate any expressions with exponents.
3. Multiply or divide in order from left to right.
4. Add or subtract in order from left to right.

EXAMPLE 5   Simplify: $\left(\dfrac{4}{5}\right)^2 - 1$

Solution: According to the order of operations, first evaluate $\left(\dfrac{4}{5}\right)^2$.

$$\left(\frac{4}{5}\right)^2 - 1 = \frac{16}{25} - 1$$ Write $\left(\dfrac{4}{5}\right)^2$ as $\dfrac{16}{25}$.

Next, combine the fractions. The LCD of 25 and 1 is 25.

$$\frac{16}{25} - 1 = \frac{16}{25} - \frac{25}{25}$$ Write 1 as $\dfrac{25}{25}$.

$$= \frac{-9}{25} \text{ or } -\frac{9}{25}$$ Subtract.

Practice Problem 5

Simplify: $\left(2 - \dfrac{2}{3}\right)^3$

Answers

4. $\dfrac{15}{4x - 20}$ **5.** $\dfrac{64}{27}$

Practice Problem 6

Simplify: $\left(-\dfrac{1}{2} + \dfrac{1}{5}\right)\left(\dfrac{7}{8} + \dfrac{1}{8}\right)$

EXAMPLE 6 Simplify: $\left(\dfrac{1}{4} + \dfrac{2}{3}\right)\left(\dfrac{11}{12} + \dfrac{1}{4}\right)$

Solution: First perform operations inside parentheses. Then multiply.

$$\left(\frac{1}{4} + \frac{2}{3}\right)\left(\frac{11}{12} + \frac{1}{4}\right) = \left(\frac{1 \cdot 3}{4 \cdot 3} + \frac{2 \cdot 4}{3 \cdot 4}\right)\left(\frac{11}{12} + \frac{1 \cdot 3}{4 \cdot 3}\right) \quad \text{Each LCD is 12.}$$

$$= \left(\frac{3}{12} + \frac{8}{12}\right)\left(\frac{11}{12} + \frac{3}{12}\right)$$

$$= \left(\frac{11}{12}\right)\left(\frac{14}{12}\right) \qquad \text{Add.}$$

$$= \frac{11 \cdot 2 \cdot 7}{2 \cdot 6 \cdot 12} \qquad \text{Multiply.}$$

$$= \frac{77}{72} \qquad \text{Simplify.}$$

Try the Concept Check in the margin.

Concept Check

What should be done first to simplify the expression $\dfrac{1}{5} \cdot \dfrac{5}{2} - \left(\dfrac{2}{3} + \dfrac{4}{5}\right)^2$?

EXAMPLE 7 Evaluate $2x + y^2$ if $x = -\dfrac{1}{2}$ and $y = \dfrac{1}{3}$.

Solution: Replace x and y with the given values and simplify.

$$2x + y^2 = 2\left(-\frac{1}{2}\right) + \left(\frac{1}{3}\right)^2 \quad \text{Replace } x \text{ with } -\frac{1}{2} \text{ and } y \text{ with } \frac{1}{3}.$$

$$= 2\left(-\frac{1}{2}\right) + \frac{1}{9} \qquad \text{Write } \frac{1}{3} \text{ as } \frac{1}{9}.$$

$$= -1 + \frac{1}{9} \qquad \text{Multiply.}$$

$$= -\frac{9}{9} + \frac{1}{9} \qquad \text{The LCD is 9.}$$

$$= -\frac{8}{9} \qquad \text{Add.}$$

Practice Problem 7

Evaluate $-\dfrac{3}{5} - xy$ if $x = \dfrac{3}{10}$ and $y = \dfrac{2}{3}$.

Answers

6. $-\dfrac{3}{10}$ **7.** $-\dfrac{4}{5}$

Concept Check: Add inside parentheses.

EXERCISE SET 4.6

A *Simplify each complex fraction. See Examples 1 through 4.*

1. $\dfrac{\frac{1}{8}}{\frac{3}{4}}$

2. $\dfrac{\frac{2}{3}}{\frac{2}{7}}$

3. $\dfrac{\frac{9}{10}}{\frac{21}{10}}$

4. $\dfrac{\frac{14}{5}}{\frac{7}{5}}$

5. $\dfrac{\frac{2x}{27}}{\frac{4}{9}}$

6. $\dfrac{\frac{3y}{11}}{\frac{1}{2}}$

7. $\dfrac{\frac{3}{4} + \frac{2}{5}}{\frac{1}{2} + \frac{3}{5}}$

8. $\dfrac{\frac{7}{6} + \frac{2}{3}}{\frac{3}{2} - \frac{8}{9}}$

9. $\dfrac{\frac{3x}{4}}{5 - \frac{1}{8}}$

10. $\dfrac{\frac{3}{10} + 2}{\frac{2}{5y}}$

B *Use the order of operations to simplify each expression. See Examples 5 and 6.*

11. $\dfrac{1}{5} + \dfrac{1}{3} \cdot \dfrac{1}{4}$

12. $\dfrac{1}{2} + \dfrac{1}{6} \cdot \dfrac{1}{3}$

13. $\dfrac{5}{6} \div \dfrac{1}{3} \cdot \dfrac{1}{4}$

14. $\dfrac{7}{8} \div \dfrac{1}{4} \cdot \dfrac{1}{7}$

15. $2^2 - \left(\dfrac{1}{3}\right)^2$

16. $3^2 - \left(\dfrac{1}{2}\right)^2$

17. $\left(\dfrac{2}{9} + \dfrac{4}{9}\right)\left(\dfrac{1}{3} - \dfrac{9}{10}\right)$

18. $\left(\dfrac{1}{5} - \dfrac{1}{10}\right)\left(\dfrac{1}{5} + \dfrac{1}{10}\right)$

19. $\left(\dfrac{7}{8} - \dfrac{1}{2}\right) \div \dfrac{3}{11}$

20. $\left(-\dfrac{2}{3} - \dfrac{7}{3}\right) \div \dfrac{4}{9}$

21. $2 \cdot \left(\dfrac{1}{4} + \dfrac{1}{5}\right) + 2$

22. $\dfrac{2}{5} \cdot \left(5 - \dfrac{1}{2}\right) - 1$

23. $\left(\dfrac{3}{4}\right)^2 \div \left(\dfrac{3}{4} - \dfrac{1}{12}\right)$

24. $\left(\dfrac{8}{9}\right)^2 \div \left(2 - \dfrac{2}{3}\right)$

25. $\left(\dfrac{2}{3} - \dfrac{5}{9}\right)^2$

26. $\left(1 - \dfrac{2}{5}\right)^3$

27. $\left(\dfrac{3}{4} + \dfrac{1}{8}\right)^2 - \left(\dfrac{1}{2} + \dfrac{1}{8}\right)$

28. $\left(\dfrac{1}{6} + \dfrac{1}{3}\right)^3 + \left(\dfrac{2}{5} \cdot \dfrac{3}{4}\right)^2$

Evaluate each expression if $x = -\dfrac{1}{3}$, $y = \dfrac{2}{5}$, *and* $z = \dfrac{5}{6}$. *See Example 7.*

29. $5y - z$

30. $2z - x$

31. $\dfrac{x}{z}$

32. $\dfrac{y + x}{z}$

33. $x^2 - yz$

34. $(1 + x)(1 + z)$

A **B** *Simplify the following. See Examples 1 through 6.*

35. $\dfrac{\dfrac{5}{24}}{\dfrac{1}{12}}$

36. $\dfrac{\dfrac{7}{10}}{\dfrac{14}{25}}$

37. $\left(\dfrac{3}{2}\right)^3 + \left(\dfrac{1}{2}\right)^3$

38. $\left(\dfrac{5}{21} \div \dfrac{1}{2}\right) + \left(\dfrac{1}{7} \cdot \dfrac{1}{3}\right)$

39. $\left(-\dfrac{1}{3}\right)^2 + \dfrac{1}{3}$

40. $\left(-\dfrac{3}{4}\right)^2 + \dfrac{3}{8}$

41. $\dfrac{2 + \dfrac{1}{6}}{1 - \dfrac{4}{3}}$

42. $\dfrac{3 - \dfrac{1}{2}}{4 + \dfrac{1}{5}}$

43. $\left(1 - \dfrac{2}{5}\right)^2$

44. $\left(-\dfrac{1}{2}\right)^2 - \left(\dfrac{3}{4}\right)^2$

45. $\left(\dfrac{3}{4} - 1\right)\left(\dfrac{1}{8} + \dfrac{1}{2}\right)$

46. $\left(\dfrac{1}{10} + \dfrac{3}{20}\right)\left(\dfrac{1}{5} - 1\right)$

47. $\left(-\dfrac{2}{9} - \dfrac{7}{9}\right)^4$

48. $\left(\dfrac{5}{9} - \dfrac{2}{3}\right)^2$

49. $\dfrac{\dfrac{1}{2} - \dfrac{3}{8}}{\dfrac{3}{4} + \dfrac{1}{2}}$

50. $\dfrac{\dfrac{7}{10} + \dfrac{1}{2}}{\dfrac{4}{5} + \dfrac{3}{4}}$

51. $\left(\dfrac{3}{4} \div \dfrac{6}{5}\right) - \left(\dfrac{3}{4} \cdot \dfrac{6}{5}\right)$

52. $\left(\dfrac{1}{2} \cdot \dfrac{2}{7}\right) - \left(\dfrac{1}{2} \div \dfrac{2}{7}\right)$

53. $\dfrac{\dfrac{x}{3} + 2}{5 + \dfrac{1}{3}}$

54. $\dfrac{1 - \dfrac{x}{4}}{2 + \dfrac{3}{8}}$

Review and Preview

Evaluate. See Section 1.8.

55. 2^3

56. 3^2

57. 5^2

58. 2^5

Multiply. See Section 2.4.

59. $\dfrac{1}{3}(3x)$

60. $\dfrac{1}{5}(5y)$

61. $\dfrac{2}{3}\left(\dfrac{3}{2}a\right)$

62. $-\dfrac{9}{10}\left(-\dfrac{10}{9}m\right)$

Combining Concepts

63. Calculate $\dfrac{2^3}{3}$ and $\left(\dfrac{2}{3}\right)^3$. Do both of these expressions simplify to the same number? Explain why or why not.

64. Calculate $\left(\dfrac{1}{2}\right)^2 \cdot \left(\dfrac{3}{4}\right)^2$ and $\left(\dfrac{1}{2} \cdot \dfrac{3}{4}\right)^2$. Do both of these expressions simplify to the same number? Explain why or why not.

Evaluate each expression if $x = \dfrac{3}{4}$ and $y = -\dfrac{4}{7}$.

65. $\dfrac{2 + x}{y}$

66. $4x + y$

67. $x^2 + 7y$

68. $\dfrac{\dfrac{9}{14}}{x + y}$

Recall that to find the average of two numbers, find their sum and divide by 2. For example, the average of $\frac{1}{2}$ and $\frac{3}{4}$ is $\dfrac{\frac{1}{2} + \frac{3}{4}}{2}$. Find the average of each pair of numbers.

69. $\dfrac{1}{2}, \dfrac{3}{4}$
70. $\dfrac{3}{5}, \dfrac{9}{10}$
71. $\dfrac{1}{4}, \dfrac{2}{14}$
72. $\dfrac{5}{6}, \dfrac{7}{9}$

73. Two positive numbers, a and b, are graphed below. Where should the graph of their average lie?

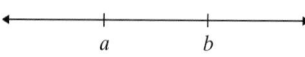

a b

Answer true or false for each statement.

74. It is possible for the average of two numbers to be greater than both numbers.

75. It is possible for the average of two numbers to be less than both numbers.

76. The sum of two negative fractions is always a negative number.

77. The sum of a negative fraction and a positive fraction is always a positive number.

78. It is possible for the sum of two fractions to be a whole number.

79. It is possible for the difference of two fractions to be a whole number.

80. What operation should be performed first to simplify
$$\frac{1}{5} \cdot \frac{5}{2} - \left(\frac{2}{3} + \frac{4}{5}\right)^2$$
Explain your answer.

81. A student is to evaluate $x - y$ when $x = \dfrac{1}{5}$ and $y = -\dfrac{1}{7}$. This student is asking you if he should evaluate $\dfrac{1}{5} - \dfrac{1}{7}$. What do you tell this student and why?

4.7   Solving Equations Containing Fractions

Ⓐ **Solving Equations Containing Fractions**

In Chapter 3, we solved linear equations in one variable. In this section, we practice this skill by solving linear equations containing fractions. To help us solve these equations, let's review the multiplication property of equality.

| **OBJECTIVES** |
| Ⓐ   Solve equations containing fractions. |
| Ⓑ   Review adding and subtracting fractions. |

Multiplication Property of Equality

Let a, b, and c be numbers, and $c \neq 0$.

If $a = b$, then $a \cdot c = b \cdot c$ and also $\dfrac{a}{c} = \dfrac{b}{c}$.

In other words, both sides of an equation may be multiplied or divided by the **same** nonzero number without changing the solution of the equation.

(We now know that we define division of a number as multiplication of its reciprocal. Because of this, the Multiplication Property of Equality also allows us to divide the same nonzero number from both sides.)

EXAMPLE 1   Solve: $\dfrac{1}{3}x = 7$

Solution: Recall that isolating x means that we want the coefficient of x to be 1. To do so, we use the multiplication property of equality and multiply both sides of the equation by the reciprocal of $\dfrac{1}{3}$, or 3. Since $\dfrac{1}{3} \cdot 3 = 1$, we will have isolated x.

$$\frac{1}{3}x = 7$$

$$3 \cdot \frac{1}{3}x = 3 \cdot 7 \qquad \text{Multiply both sides by 3.}$$

$$1 \cdot x = 21 \text{ or } x = 21 \qquad \text{Simplify.}$$

To check, replace x with 21 in the original equation.

$$\frac{1}{3}x = 7 \qquad \text{Original equation}$$

$$\frac{1}{3} \cdot 21 = 7 \qquad \text{Replace } x \text{ with 21.}$$

$$7 = 7 \qquad \text{True.}$$

Since $7 = 7$ is a true statement, 21 is the solution of $\dfrac{1}{3}x = 7$.

Practice Problem 1

Solve: $\dfrac{1}{5}y = 2$

Answer

1. 10

Practice Problem 2

Solve: $\dfrac{5}{7}b = 25$

Practice Problem 3

Solve: $-\dfrac{7}{10}x = \dfrac{2}{5}$

EXAMPLE 2 Solve: $\dfrac{3}{5}a = 9$

Solution: Multiply both sides by $\dfrac{5}{3}$, the reciprocal of $\dfrac{3}{5}$, so that the coefficient of a is 1.

$$\dfrac{3}{5}a = 9$$

$$\dfrac{5}{3} \cdot \dfrac{3}{5}a = \dfrac{5}{3} \cdot 9 \qquad \text{Multiply both sides by } \dfrac{5}{3}.$$

$$1a = \dfrac{5 \cdot 9}{3} \qquad \text{Multiply.}$$

$$a = 15 \qquad \text{Simplify.}$$

To check, replace a with 15 in the original equation.

$$\dfrac{3}{5}a = 9$$

$$\dfrac{3}{5} \cdot 15 = 9 \qquad \text{Replace } a \text{ with 15.}$$

$$\dfrac{3 \cdot 15}{5} = 9 \qquad \text{Multiply.}$$

$$9 = 9 \qquad \text{True.}$$

Since $9 = 9$ is true, 15 is the solution of $\dfrac{3}{5}a = 9$.

EXAMPLE 3 Solve: $\dfrac{3}{4}x = -\dfrac{1}{8}$

Solution: Multiply both sides of the equation by $\dfrac{4}{3}$, the reciprocal of $\dfrac{3}{4}$.

$$\dfrac{3}{4}x = -\dfrac{1}{8}$$

$$\dfrac{4}{3} \cdot \dfrac{3}{4}x = \dfrac{4}{3} \cdot -\dfrac{1}{8} \qquad \text{Multiply both sides by } \dfrac{4}{3}.$$

$$1x = -\dfrac{4 \cdot 1}{3 \cdot 8} \qquad \text{Multiply.}$$

$$x = -\dfrac{1}{6} \qquad \text{Simplify.}$$

To check, replace x with $-\dfrac{1}{6}$ in the original equation.

$$\dfrac{3}{4}x = -\dfrac{1}{8} \qquad \text{Original equation.}$$

$$\dfrac{3}{4} \cdot -\dfrac{1}{6} = -\dfrac{1}{8} \qquad \text{Replace } x \text{ with } -\dfrac{1}{6}.$$

$$-\dfrac{1}{8} = -\dfrac{1}{8} \qquad \text{True.}$$

Since we arrived at a true statement, $-\dfrac{1}{6}$ is the solution of $\dfrac{3}{4}x = -\dfrac{1}{8}$.

Answers

2. 35   **3.** $-\dfrac{4}{7}$

EXAMPLE 4   Solve: $3y = -\dfrac{2}{11}$

Solution: We can either divide both sides by 3 or multiply both sides by the reciprocal of 3.

$$3y = -\frac{2}{11}$$

$$\frac{1}{3} \cdot 3y = \frac{1}{3} \cdot -\frac{2}{11} \qquad \text{Multiply both sides by } \frac{1}{3}.$$

$$1y = -\frac{1 \cdot 2}{3 \cdot 11} \qquad \text{Multiply.}$$

$$y = -\frac{2}{33} \qquad \text{Simplify.}$$

Check to see that the solution is $-\dfrac{2}{33}$.   ●

Solving equations with fractions can be tedious. If an equation contains fractions, it is often helpful to first multiply both sides of the equation by the LCD of the fractions. This has the effect of eliminating the fractions in the equation, as shown in the next example.

EXAMPLE 5   Solve: $\dfrac{x}{6} + 1 = \dfrac{4}{3}$

Solution: First multiply both sides of the equation by the LCD of the fractions. The LCD of the denominators 6 and 3 is 6.

$$\frac{x}{6} + 1 = \frac{4}{3}$$

$$6\left(\frac{x}{6} + 1\right) = 6\left(\frac{4}{3}\right) \qquad \text{Multiply both sides by 6.}$$

$$\overset{1}{6}\left(\frac{x}{\underset{1}{6}}\right) + 6(1) = \overset{2}{6}\left(\frac{4}{\underset{1}{3}}\right) \qquad \text{Apply the distributive property.}$$

$$x + 6 = 8 \qquad \text{Simplify.}$$

$$x + 6 + (-6) = 8 + (-6) \qquad \text{Add } -6 \text{ to both sides.}$$

$$x = 2 \qquad \text{Simplify.}$$

To check, replace x with 2 in the original equation.

$$\frac{x}{6} + 1 = \frac{4}{3} \qquad \text{Original equation}$$

$$\frac{2}{6} + 1 = \frac{4}{3} \qquad \text{Replace } x \text{ with 2.}$$

$$\frac{1}{3} + \frac{3}{3} = \frac{4}{3} \qquad \text{Simplify } \frac{2}{6}. \text{ The LCD of 3 and 1 is 3.}$$

$$\frac{4}{3} = \frac{4}{3} \qquad \text{True.}$$

Since we arrived at a true statement, 2 is the solution of $\dfrac{x}{6} + 1 = \dfrac{4}{3}$.   ●

Practice Problem 4

Solve: $5x = -\dfrac{3}{4}$

Practice Problem 5

Solve: $\dfrac{y}{8} + \dfrac{3}{4} = 2$

Answers

4. $-\dfrac{3}{20}$ **5.** 10

Let's review the steps for solving equations in x. An extra step is now included to handle equations containing fractions.

Solving an Equation in x

Step 1. If fractions are present, multiply both sides of the equation by the LCD of the fractions.

Step 2. If parentheses are present, use the distributive property.

Step 3. Combine any like terms on each side of the equation.

Step 4. Use the addition property of equality to rewrite the equation so that variable terms are on one side of the equation and constant terms are on the other side.

Step 5. Divide both sides of the equation by the numerical coefficient of x to solve.

Step 6. Check the answer in the **original equation**.

Practice Problem 6

Solve: $\dfrac{x}{5} - x = \dfrac{1}{5}$

> **Helpful Hint**
>
> Don't forget to multiply *both* sides of the equation by the LCD.

EXAMPLE 6 Solve: $\dfrac{z}{5} - \dfrac{z}{3} = 6$

Solution:

$$\frac{z}{5} - \frac{z}{3} = 6$$

$$15\left(\frac{z}{5} - \frac{z}{3}\right) = 15(6) \quad \text{Multiply both sides by the LCD, 15.}$$

$$\overset{3}{\cancel{15}}\left(\frac{z}{\cancel{5}}\right) - \overset{5}{\cancel{15}}\left(\frac{z}{\cancel{3}}\right) = 15(6) \quad \text{Apply the distributive property.}$$

$$3z - 5z = 90 \quad \text{Simplify.}$$

$$-2z = 90 \quad \text{Combine like terms.}$$

$$\frac{-2z}{-2} = \frac{90}{-2} \quad \text{Divide both sides by } -2, \text{ the coefficient of } z.$$

$$z = -45 \quad \text{Simplify.}$$

To check, replace z with -45 in the **original equation** to see that a true statement results.

Practice Problem 7

Solve: $\dfrac{y}{2} = \dfrac{y}{5} + \dfrac{3}{2}$

EXAMPLE 7 Solve: $\dfrac{x}{2} = \dfrac{x}{3} + \dfrac{1}{2}$

Solution: First multiply both sides by the LCD, 6.

$$\frac{x}{2} = \frac{x}{3} + \frac{1}{2}$$

$$6\left(\frac{x}{2}\right) = 6\left(\frac{x}{3} + \frac{1}{2}\right) \quad \text{Multiply both sides by the LCD, 6.}$$

$$\overset{3}{\cancel{6}}\left(\frac{x}{\cancel{2}}\right) = \overset{2}{\cancel{6}}\left(\frac{x}{\cancel{3}}\right) + \overset{3}{\cancel{6}}\left(\frac{1}{\cancel{2}}\right) \quad \text{Apply the distributive property.}$$

$$3x = 2x + 3 \quad \text{Simplify.}$$

$$3x + (-2x) = 2x + 3 + (-2x) \quad \text{Add } (-2x) \text{ to both sides.}$$

$$x = 3 \quad \text{Simplify.}$$

To check, replace x with 3 in the original equation to see that a true statement results.

Answers

6. $-\dfrac{1}{4}$ **7.** 5

B   Review of Adding and Subtracting Fractions

Make sure you understand the difference between **solving an equation** containing fractions and **adding or subtracting two fractions**. To solve an equation containing fractions, we use the multiplication property of equality and multiply both sides by the LCD of the fractions, thus eliminating the fractions. This method does not apply to adding or subtracting fractions. The multiplication property of equality applies only to equations. To add or subtract unlike fractions, we write each fraction as an equivalent fraction using the LCD of the fractions as the denominator. See the next example for a review.

EXAMPLE 8   Add: $\dfrac{x}{3} + \dfrac{2}{5}$

Solution: This expression is not an equation. Here, we are adding two unlike fractions. To add unlike fractions, we need to find the LCD. The LCD of the denominators 3 and 5 is 15. Write each fraction as an equivalent fraction with a denominator of 15.

$$\frac{x}{3} + \frac{2}{5} = \frac{x \cdot 5}{3 \cdot 5} + \frac{2 \cdot 3}{5 \cdot 3}$$

$$= \frac{5x}{15} + \frac{6}{15}$$

$$= \frac{5x + 6}{15}$$

Try the Concept Check in the margin.

Practice Problem 8

Subtract: $\dfrac{9}{10} - \dfrac{y}{3}$

Concept Check

Which of the following are equations and which are expressions?

a. $\dfrac{1}{2} + 3x = 5$

b. $\dfrac{2}{3}x - \dfrac{x}{5}$

c. $\dfrac{x}{12} + \dfrac{5x}{24}$

d. $\dfrac{x}{5} = \dfrac{1}{10}$

Answer

8. $\dfrac{27 - 10y}{30}$

Concept Check: equations: a, d; expressions: b, c

FOCUS ON **History**

THE DEVELOPMENT OF FRACTIONS

Many ancient cultures were familiar with the use of fractions and used them in everyday life. Two examples are discussed here.

■ Although the ancient Egyptians were comfortable with the idea of fractions, their system of hieroglyphic numbers allowed them to only write fractions with 1 as the numerator, like $\frac{1}{3}$ or $\frac{1}{5}$. To write fractions with numerators other than 1, the Egyptians had to write a sum of fractions with 1 as the numerator. They preferred to never use the same denominator more than once in one of these sums of fractions. For example, instead of representing the fraction $\frac{2}{5}$ as the sum $\frac{1}{5} + \frac{1}{5}$ (which uses the denominator 5 more than once), they would prefer instead the sum $\frac{1}{3} + \frac{1}{15}$, which also is equal to $\frac{2}{5}$ but uses two different denominators. Similarly, $\frac{2}{13}$ would not be represented by $\frac{1}{13} + \frac{1}{13}$, as we might think would be easiest, but instead by $\frac{1}{8} + \frac{1}{52} + \frac{1}{104}$.

■ The ancient Babylonians also dealt with fractions but in a more direct way than did the Egyptians. For instance, they thought of the fraction $\frac{3}{8}$ as the product of 3 and the reciprocal of 8. They compiled detailed lists of reciprocals and could calculate with fractions very easily and quickly.

Although many ancient cultures used fractions, they didn't represent them the way we do today. Fractions did not evolve into anything similar to our present-day format until the seventh century A.D. Around that time, the Hindus became the first to write fractions with the numerator above the denominator. However, Hindu mathematicians, such as Brahmagupta in A.D. 628, wrote these fractions without the fraction bar we use today. Thus, the fraction $\frac{5}{8}$ would have been written by the Hindus as $\begin{smallmatrix}5\\8\end{smallmatrix}$.

It wasn't until between A.D. 1100 and 1200 that the next giant step in fraction notation was made. Up to this point, Arab mathematicians had copied Hindu notation for fractions. However, they then improved this notation by inserting a horizontal bar, called a *vinculum*, between the numerator and denominator to separate the two parts of the fraction.

Today we see fractions in the form $^4/_9$, or 4/9 just as often as in the $\frac{4}{9}$ form, developed by the Arabs.

The use of the diagonal fraction bar, called a *solidus*, was introduced with the appearance of the printing press. Typesetting fractions with a horizontal fraction bar was difficult and space-consuming because three lines of type were required for the fraction. Typesetting fractions with a diagonal fraction bar was far easier for printers because the entire fraction fit on a single line of type.

EXERCISE SET 4.7

A *Solve each equation. See Examples 1 through 4.*

1. $7x = 2$

2. $-5x = 4$

3. $\dfrac{1}{4}x = 3$

4. $\dfrac{1}{3}x = 6$

5. $\dfrac{2}{9}y = -6$

6. $\dfrac{4}{7}x = -8$

7. $-\dfrac{4}{9}z = -\dfrac{3}{2}$

8. $-\dfrac{11}{10}x = -\dfrac{2}{7}$

9. $7a = \dfrac{1}{3}$

10. $2z = -\dfrac{5}{12}$

11. $-3x = -\dfrac{6}{11}$

12. $-4z = -\dfrac{12}{25}$

Solve each equation. See Examples 5 through 7.

13. $\dfrac{x}{3} + 2 = \dfrac{7}{3}$

14. $\dfrac{x}{5} - 1 = \dfrac{7}{5}$

15. $\dfrac{x}{5} - \dfrac{5}{10} = 1$

16. $\dfrac{x}{3} - x = -6$

17. $\dfrac{1}{2} - \dfrac{3}{5} = \dfrac{x}{10}$

18. $\dfrac{2}{3} - \dfrac{1}{4} = \dfrac{x}{12}$

19. $\dfrac{x}{3} = \dfrac{x}{5} - 2$

20. $\dfrac{a}{2} = \dfrac{a}{7} + \dfrac{5}{2}$

B *Add or subtract as indicated. See Example 8.*

21. $\dfrac{x}{7} - \dfrac{4}{3}$

22. $-\dfrac{5}{9} + \dfrac{y}{8}$

23. $\dfrac{y}{2} + 5$

24. $2 + \dfrac{7x}{3}$

25. $\dfrac{3x}{10} + \dfrac{x}{6}$

26. $\dfrac{9x}{8} - \dfrac{5x}{6}$

A **B** *Solve. If no equation is given, perform the indicated operation.*

27. $\dfrac{3}{8}x = \dfrac{1}{2}$

28. $\dfrac{2}{5}y = \dfrac{3}{10}$

29. $\dfrac{2}{3} - \dfrac{x}{5} = \dfrac{4}{15}$

30. $\dfrac{4}{5} + \dfrac{x}{4} = \dfrac{21}{20}$

31. $\dfrac{9}{14}z = \dfrac{27}{20}$

32. $\dfrac{5}{16}a = \dfrac{5}{6}$

33. $-3m - 5m = \dfrac{4}{7}$

34. $30n - 34n = \dfrac{3}{20}$

35. $\dfrac{x}{4} + 1 = \dfrac{1}{4}$

39. $\dfrac{5}{9} - \dfrac{2}{3}$

40. $\dfrac{8}{11} - \dfrac{1}{2}$

41. $-\dfrac{3}{4}x = \dfrac{9}{2}$

36. $\dfrac{y}{7} - 2 = \dfrac{1}{7}$

37. $\dfrac{1}{5}y = 10$

38. $\dfrac{1}{4}x = -2$

42. $\dfrac{5}{7}y = -\dfrac{15}{49}$

43. $\dfrac{x}{2} - x = -2$

44. $\dfrac{y}{3} = -4 + y$

45. $-\dfrac{5}{8}y = \dfrac{3}{16} - \dfrac{9}{16}$

46. $-\dfrac{7}{9}x = -\dfrac{5}{18} - \dfrac{4}{18}$

47. $17x - 25x = \dfrac{1}{3}$

48. $27x - 30x = \dfrac{4}{9}$

49. $\dfrac{7}{6}x = \dfrac{1}{4} - \dfrac{2}{3}$

50. $\dfrac{5}{4}y = \dfrac{1}{2} - \dfrac{7}{10}$

51. $\dfrac{b}{4} = \dfrac{b}{12} + \dfrac{2}{3}$

52. $\dfrac{a}{6} = \dfrac{a}{3} + \dfrac{1}{2}$

53. $\dfrac{x}{3} + 2 = \dfrac{x}{2} + 8$

54. $\dfrac{y}{5} - 2 = \dfrac{y}{3} - 4$

Review and Preview

Perform each indicated operation. See Section 4.5.

55. $3 + \dfrac{1}{2}$ **56.** $2 + \dfrac{2}{3}$ **57.** $5 + \dfrac{9}{10}$ **58.** $1 + \dfrac{7}{8}$ **59.** $9 - \dfrac{5}{6}$ **60.** $4 - \dfrac{1}{5}$

Combining Concepts

61. Explain why the method for eliminating fractions in an **equation** does not apply to simplifying **expressions** containing fractions.

Solve.

62. $\dfrac{14}{11} + \dfrac{3x}{8} = \dfrac{x}{2}$

63. $\dfrac{19}{53} = \dfrac{353x}{1431} + \dfrac{23}{27}$

64. The area of the rectangle is $\dfrac{5}{12}$ square inches. Find its length, x.

x

$\dfrac{3}{4}$

4.8 Operations on Mixed Numbers

A Illustrating Mixed Numbers

Recall the graph of the improper fraction $\frac{9}{5}$.

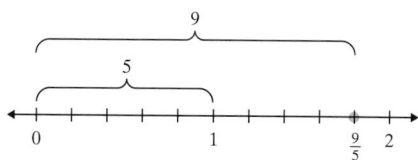

Notice that the graph of $\frac{9}{5}$ is $\frac{4}{5}$ past the number 1. Another way to name the improper fraction $\frac{9}{5}$ is by the mixed number $1\frac{4}{5}$ (read as "one and four fifths"). This **mixed number** contains a whole number and a proper fraction. The mixed number $1\frac{4}{5}$ means $1 + \frac{4}{5}$. Examples of mixed numbers are

$$2\frac{1}{4} = 2 + \frac{1}{4} \quad \text{and} \quad 5\frac{3}{8} = 5 + \frac{3}{8}$$

We can write a mixed number as an improper fraction by adding.

$$1\frac{4}{5} = 1 + \frac{4}{5} = \frac{1 \cdot 5}{1 \cdot 5} + \frac{4}{5} = \frac{5}{5} + \frac{4}{5} = \frac{9}{5}$$

We can also see that $1\frac{4}{5} = \frac{9}{5}$ by studying the following illustration.

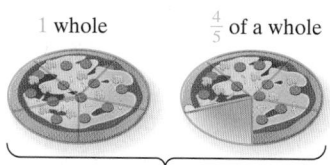

1 whole $\frac{4}{5}$ of a whole

$\frac{9}{5}$ Number of pieces of pizza
   Number of equal parts in a whole

EXAMPLES Represent the shaded area of each figure group with both an improper fraction and a mixed number.

1.

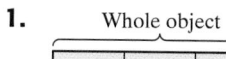
Whole object

improper fraction: $\frac{4}{3}$

mixed number: $1\frac{1}{3}$

2.

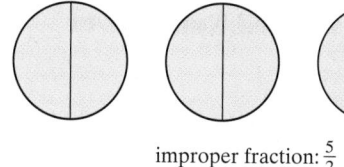

improper fraction: $\frac{5}{2}$

mixed number: $2\frac{1}{2}$

Practice Problems 1–2

Represent the shaded area of each figure group with both an improper fraction and a mixed number.

1.

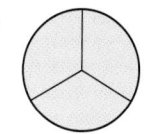 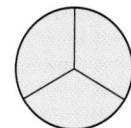 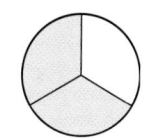

2.

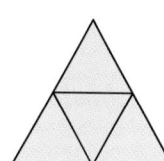 

Answers

1. $\frac{8}{3}$ or $2\frac{2}{3}$ **2.** $\frac{5}{4}$ or $1\frac{1}{4}$

Concept Check

Is $2\frac{1}{8}$ closer to the number 2 or the number 3? If you were to estimate $2\frac{1}{8}$ by a whole number, which number would you choose? Why?

Try the Concept Check in the margin.

B Writing Mixed Numbers as Improper Fractions

Notice from Examples 1 and 2 that mixed numbers and improper fractions were both used to represent the shaded area of the figure groups. For example,

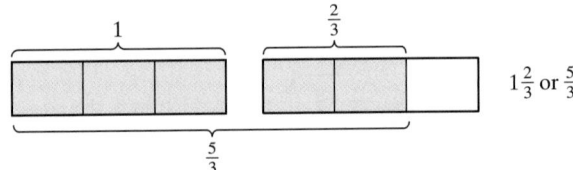

By studying these examples, we can see that a pattern occurs when writing mixed numbers as improper fractions. This pattern leads to the following method for writing a mixed number as an improper fraction.

Writing a Mixed Number as an Improper Fraction

To write a mixed number as an improper fraction,

Step 1. Multiply the whole number by the denominator of the fraction.

Step 2. Add the numerator of the fraction to the product from Step 1.

Step 3. Write the sum from Step 2 as the numerator of the improper fraction over the original denominator.

For example,

$$1\frac{2}{3} = \frac{3 \cdot 1 + 2}{3} = \frac{3 + 2}{3} = \frac{5}{3}$$

EXAMPLE 3 Write each mixed number as an improper fraction.

a. $4\frac{2}{9}$

b. $1\frac{8}{11}$

Solution:

a. $4\frac{2}{9} = \frac{9 \cdot 4 + 2}{9} = \frac{36 + 2}{9} = \frac{38}{9}$

b. $1\frac{8}{11} = \frac{11 \cdot 1 + 8}{11} = \frac{11 + 8}{11} = \frac{19}{11}$

C Writing Improper Fractions as Mixed Numbers or Whole Numbers

Just as there are times when an improper fraction is preferred, sometimes a mixed or a whole number better suits a situation. To write improper fractions as mixed or whole numbers, just remember that the fraction bar means division.

Practice Problem 3

Write each mixed number as an improper fraction.

a. $2\frac{5}{7}$

b. $5\frac{1}{3}$

c. $9\frac{3}{10}$

d. $1\frac{1}{5}$

Answers

3. a. $\frac{19}{7}$ **b.** $\frac{16}{3}$ **c.** $\frac{93}{10}$ **d.** $\frac{6}{5}$

Concept Check: 2; Answers may vary.

Writing an Improper Fraction as a Mixed Number or a Whole Number

To write an improper fraction as a mixed number or a whole number,

Step 1. Divide the denominator into the numerator.

Step 2. The whole-number part of the mixed number is the quotient. The fraction part of the mixed number is the remainder over the original denominator.

$$\text{quotient } \frac{\text{remainder}}{\text{original denominator}}$$

For example,

$$\frac{5}{3} = \begin{array}{r} 1 \\ 3\overline{)5} \\ \underline{3} \\ 2 \end{array} = 1\frac{2}{3} \begin{array}{l} \leftarrow \text{remainder} \\ \leftarrow \text{original denominator} \end{array}$$

$\uparrow$ quotient

EXAMPLE 4   Write each improper fraction as a mixed number or a whole number.

a. $\dfrac{30}{7}$ **b.** $\dfrac{16}{15}$ **c.** $\dfrac{84}{6}$

Solution:

a.
$$\begin{array}{r} 4 \\ 7\overline{)30} \\ \underline{-28} \\ 2 \end{array} \qquad \frac{30}{7} = 4\frac{2}{7}$$

b.
$$\begin{array}{r} 1 \\ 15\overline{)16} \\ \underline{-15} \\ 1 \end{array} \qquad \frac{16}{15} = 1\frac{1}{15}$$

c.
$$\begin{array}{r} 14 \\ 6\overline{)84} \\ \underline{-6} \\ 24 \\ \underline{-24} \\ 0 \end{array} \qquad \frac{84}{6} = 14 \quad \text{Since the remainder is 0, the result is the whole number 14.}$$

Helpful Hint

When the remainder is 0, the improper fraction is a whole number. For example, $\dfrac{92}{4} = 23$.

$$\begin{array}{r} 23 \\ 4\overline{)92} \\ \underline{-8} \\ 12 \\ \underline{-12} \\ 0 \end{array}$$

Practice Problem 4

Write each fraction as a mixed number or a whole number.

a. $\dfrac{8}{5}$ **b.** $\dfrac{17}{6}$

c. $\dfrac{48}{4}$ **d.** $\dfrac{35}{4}$

e. $\dfrac{51}{7}$ **f.** $\dfrac{21}{20}$

Answers

4. a. $1\frac{3}{5}$ **b.** $2\frac{5}{6}$ **c.** 12 **d.** $8\frac{3}{4}$ **e.** $7\frac{2}{7}$

f. $1\frac{1}{20}$

Concept Check

Which of the following are equivalent to 9?

a. $7\dfrac{6}{3}$ b. $8\dfrac{4}{4}$ c. $8\dfrac{9}{9}$

d. $\dfrac{18}{2}$ e. all of these

Practice Problem 5

Multiply: $2\dfrac{1}{2} \cdot \dfrac{8}{15}$

Practice Problems 6–7

Multiply.

6. $3\dfrac{1}{5} \cdot 2\dfrac{3}{4}$ 7. $\dfrac{2}{3} \cdot 18$

Concept Check

a. Find the error.

$$2\dfrac{1}{4} \cdot \dfrac{1}{2} = 2\dfrac{1 \cdot 1}{4 \cdot 2} = 2\dfrac{1}{8}$$

b. How could you estimate the product $3\dfrac{1}{7} \cdot 4\dfrac{5}{6}$?

Practice Problems 8–10

Divide.

8. $\dfrac{4}{9} \div 5$ 9. $\dfrac{8}{15} \div 3\dfrac{4}{5}$ 10. $3\dfrac{2}{5} \div 2\dfrac{2}{15}$

Answers

5. $\dfrac{4}{3}$ or $1\dfrac{1}{3}$ **6.** $\dfrac{44}{5}$ or $8\dfrac{4}{5}$ **7.** $\dfrac{12}{1}$ or 12   **8.** $\dfrac{4}{45}$

9. $\dfrac{8}{57}$ **10.** $\dfrac{51}{32}$ or $1\dfrac{19}{32}$

Concept Check: e

Concept Check:

a. $2\dfrac{1}{4} \cdot \dfrac{1}{2} = \dfrac{9}{2} \cdot \dfrac{1}{2} = \dfrac{9}{4} = 2\dfrac{1}{4}$

b. answers may vary

Try the Concept Check in the margin.

D Multiplying or Dividing with Mixed Numbers or Whole Numbers

To multiply or divide with mixed numbers or whole numbers, first write each number as an improper fraction. (Note: If an exercise contains a mixed number, we will write the solution as a mixed number, if possible.)

EXAMPLE 5   Multiply: $3\dfrac{1}{3} \cdot \dfrac{7}{8}$

Solution: The mixed number $3\dfrac{1}{3}$ can be written as the fraction $\dfrac{10}{3}$. Then,

$$3\dfrac{1}{3} \cdot \dfrac{7}{8} = \dfrac{10}{3} \cdot \dfrac{7}{8} = \dfrac{2 \cdot 5 \cdot 7}{3 \cdot 2 \cdot 4} = \dfrac{35}{12} \quad \text{or} \quad 2\dfrac{11}{12}$$

Don't forget that a whole number can be written as a fraction by writing the whole number over 1. For example,

$$20 = \dfrac{20}{1} \quad \text{and} \quad 7 = \dfrac{7}{1}$$

EXAMPLES   Multiply.

6. $1\dfrac{2}{3} \cdot 2\dfrac{1}{4} = \dfrac{5}{3} \cdot \dfrac{9}{4} = \dfrac{5 \cdot 9}{3 \cdot 4} = \dfrac{5 \cdot 3 \cdot 3}{3 \cdot 4} = \dfrac{15}{4}$ or $3\dfrac{3}{4}$

7. $\dfrac{3}{4} \cdot 20 = \dfrac{3}{4} \cdot \dfrac{20}{1} = \dfrac{3 \cdot 20}{4 \cdot 1} = \dfrac{3 \cdot 4 \cdot 5}{4 \cdot 1} = \dfrac{15}{1}$ or 15

Try the Concept Check in the margin.

EXAMPLES   Divide.

8. $\dfrac{3}{4} \div 5 = \dfrac{3}{4} \div \dfrac{5}{1} = \dfrac{3}{4} \cdot \dfrac{1}{5} = \dfrac{3 \cdot 1}{4 \cdot 5} = \dfrac{3}{20}$

9. $\dfrac{11}{18} \div 2\dfrac{5}{6} = \dfrac{11}{18} \div \dfrac{17}{6} = \dfrac{11}{18} \cdot \dfrac{6}{17} = \dfrac{11 \cdot 6}{18 \cdot 17}$

   $= \dfrac{11 \cdot 6}{6 \cdot 3 \cdot 17} = \dfrac{11}{51}$

10. $5\dfrac{2}{3} \div 2\dfrac{5}{9} = \dfrac{17}{3} \div \dfrac{23}{9} = \dfrac{17}{3} \cdot \dfrac{9}{23} = \dfrac{17 \cdot 9}{3 \cdot 23}$

   $= \dfrac{17 \cdot 3 \cdot 3}{3 \cdot 23} = \dfrac{51}{23}$ or $2\dfrac{5}{23}$

E Adding or Subtracting Mixed Numbers

We can add or subtract mixed numbers, too, by first writing each mixed number as an improper fraction. But it is often easier to add or subtract the whole-number parts and add or subtract the proper-fraction parts vertically, as shown in the next examples.

EXAMPLE 11 Add: $2\dfrac{1}{3} + 5\dfrac{3}{8}$

Solution: The LCD of 3 and 8 is 24.

$$2\dfrac{1\cdot 8}{3\cdot 8} = \quad 2\dfrac{8}{24}$$
$$+5\dfrac{3\cdot 3}{8\cdot 3} = +5\dfrac{9}{24}$$
$$7\dfrac{17}{24} \quad \leftarrow \text{Add the fractions.}$$
$$\underset{\uparrow}{\phantom{7\dfrac{17}{24}}}\text{Add the whole numbers.}$$

EXAMPLE 12 Add: $3\dfrac{4}{5} + 1\dfrac{4}{15}$

Solution: The LCD of 5 and 15 is 15.

$$3\dfrac{4}{5} = \quad 3\dfrac{12}{15}$$
$$+1\dfrac{4}{15} = +1\dfrac{4}{15} \quad \text{Add the fractions, then add the whole numbers.}$$
$$4\dfrac{16}{15} \quad \text{Notice that the fractional part is improper.}$$

Since $\dfrac{16}{15}$ is ——, we can write the sum as

$$4\dfrac{16}{15} = 4 + 1\dfrac{1}{15} = 5\dfrac{1}{15}$$

EXAMPLE 13 Add: $1\dfrac{4}{5} + 4 + 2\dfrac{1}{2}$

Solution: The LCD of 5 and 2 is 10.
$$1\dfrac{4}{5} = \quad 1\dfrac{8}{10}$$
$$4 = \quad 4$$
$$+2\dfrac{1}{2} = +2\dfrac{5}{10}$$
$$7\dfrac{13}{10} = 7 + 1\dfrac{3}{10} = 8\dfrac{3}{10}$$

Try the Concept Check in the margin.

EXAMPLE 14 Subtract: $9\dfrac{3}{7} - 5\dfrac{4}{21}$

Solution: The LCD of 7 and 21 is 21.

$$9\dfrac{3}{7} = \quad 9\dfrac{9}{21}$$
$$-5\dfrac{4}{21} = -5\dfrac{4}{21}$$
$$4\dfrac{5}{21} \quad \leftarrow \text{Subtract the fractions.}$$
$$\underset{\uparrow}{\phantom{4\dfrac{5}{21}}}\text{Subtract the whole numbers.}$$

Practice Problem 11

Add: $4\dfrac{2}{5} + 5\dfrac{3}{10}$

Practice Problem 12

Add: $2\dfrac{5}{14} + 5\dfrac{6}{7}$

Practice Problem 13

Add: $10 + 2\dfrac{1}{7} + 3\dfrac{1}{5}$

Concept Check

Explain how you could estimate the following sum: $5\dfrac{1}{9} + 14\dfrac{10}{11}$.

Practice Problem 14

Subtract: $29\dfrac{7}{8} - 13\dfrac{3}{16}$

Answers

11. $9\dfrac{7}{10}$ **12.** $8\dfrac{3}{14}$ **13.** $15\dfrac{12}{35}$ **14.** $16\dfrac{11}{16}$

Concept Check: Round each mixed number to the nearest whole number and add. $5\dfrac{1}{9}$ rounds to 5 and $14\dfrac{10}{11}$ rounds to 15, and the estimated sum is $5 + 15 = 20$.

When subtracting mixed numbers, borrowing may be needed, as shown in the next example.

Practice Problem 15

Subtract: $9\dfrac{7}{15} - 5\dfrac{4}{5}$

EXAMPLE 15 Subtract: $7\dfrac{3}{14} - 3\dfrac{6}{7}$

Solution: The LCD of 7 and 14 is 14.

$$7\dfrac{3}{14} = \quad 7\dfrac{3}{14}$$
$$-3\dfrac{6}{7} = -3\dfrac{12}{14}$$

Notice that we cannot subtract $\dfrac{12}{14}$ from $\dfrac{3}{14}$, so we borrow from the whole number 7.

Borrow 1 from 7.

$$7\dfrac{3}{14} = 6 + 1\dfrac{3}{14} = 6 + \dfrac{17}{14} \quad \text{or} \quad 6\dfrac{17}{14}$$

Now subtract.

$$7\dfrac{3}{14} = \quad 7\dfrac{3}{14} = \quad 6\dfrac{17}{14}$$
$$-3\dfrac{6}{7} = -3\dfrac{12}{14} = -3\dfrac{12}{14}$$
$$\phantom{-3\dfrac{6}{7} = -3\dfrac{12}{14} =} 3\dfrac{5}{14} \leftarrow \text{Subtract the fractions.}$$

↑ Subtract the whole numbers.

Concept Check

In the subtraction problem $5\dfrac{1}{4} - 3\dfrac{3}{4}$, $5\dfrac{1}{4}$ must be rewritten because $\dfrac{3}{4}$ cannot be subtracted from $\dfrac{1}{4}$. Why is it incorrect to rewrite $5\dfrac{1}{4}$ as $5\dfrac{5}{4}$?

Try the Concept Check in the margin.

EXAMPLE 16 Subtract: $12 - 8\dfrac{3}{7}$

Practice Problem 16

Subtract: $25 - 10\dfrac{2}{9}$

Solution:

$$12 \quad = \quad 11\dfrac{7}{7}$$

Borrow 1 from 12 and write it as $\dfrac{7}{7}$.

$$-8\dfrac{3}{7} = -8\dfrac{3}{7}$$
$$\phantom{-8\dfrac{3}{7} =} 3\dfrac{4}{7} \leftarrow \text{Subtract the fractions.}$$

↑ Subtract the whole numbers.

F **Solving Problems Containing Mixed Numbers**

Now that we know how to perform operations on mixed numbers, we can solve real-life problems.

Answers

15. $3\dfrac{2}{3}$ **16.** $14\dfrac{7}{9}$

Concept Check: Rewrite $5\dfrac{1}{4}$ as $4\dfrac{5}{4}$ by borrowing from the 5.

EXAMPLE 17   Finding Legal Lobster Size

Lobster fishermen must measure the upper body shells of the lobsters they catch. Lobsters that are too small are thrown back into the ocean. Each state has its own size standard for lobsters to help control the breeding stock. In 1988, Massachusetts increased its legal lobster size from $3\frac{3}{16}$ inches to $3\frac{7}{32}$ inches. How much of an increase was this?(*Source:* Peabody Essex Museum, Salem, MA)

Solution:

1. UNDERSTAND. Read and reread the problem carefully. The word "increase" found in the problem might make you think that we add to solve the problem. But the phrase "how much of an increase" tells us to subtract to find the increase. Let

   x = the increase

2. TRANSLATE.

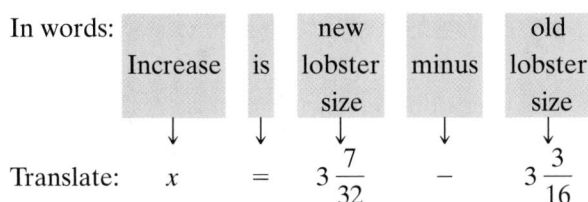

In words:	Increase	is	new lobster size	minus	old lobster size
	↓	↓	↓	↓	↓
Translate:	x	$=$	$3\frac{7}{32}$	$-$	$3\frac{3}{16}$

3. Solve. Since x is by itself on one side of the equation, the equation is solved for x. We just need to perform the subtraction.

$$3\frac{7}{32} = \quad 3\frac{7}{32}$$
$$-3\frac{3}{16} = -3\frac{6}{32}$$
$$\overline{\qquad\quad \frac{1}{32}}$$

Thus, $x = \dfrac{1}{32}$.

4. INTERPRET. *Check* your work. *State* your conclusion: The increase in lobster size is $\dfrac{1}{32}$ of an inch.

●

Practice Problem 17

The measurement around the trunk of a tree just below shoulder height is called its girth. The largest known American Beech tree in the United States has a girth of $23\frac{1}{4}$ feet. The largest known Sugar Maple tree in the United States has a girth of $19\frac{5}{12}$ feet. How much larger is the girth of the largest known American Beech tree than the girth of the largest known Sugar Maple tree? (*Source*: American Forests)

girth

Answer

17. $3\frac{5}{6}$ feet

A designer of women's clothing designs a woman's dress that requires $2\frac{1}{7}$ yards of material. How many dresses can be made from a 30-yard bolt of material?

EXAMPLE 18 Calculating Manufacturing Materials Needed

In a manufacturing process, a metal-cutting machine cuts strips $1\frac{3}{5}$ inches long from a piece of metal stock. How many such strips can be cut from a 48-inch piece of stock?

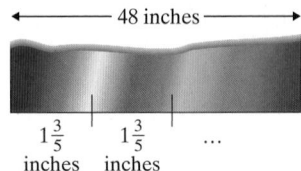

Solution:

1. UNDERSTAND the problem. To do so, read and reread the problem. Then draw a diagram.

   We want to know how many $1\frac{3}{5}$s there are in 48. Let x = number of strips

2. TRANSLATE.

   In words:

Number of strips	is	48	divided by	$1\frac{3}{5}$
↓	↓	↓	↓	↓

   Translate: x =   48   ÷ $1\frac{3}{5}$

3. SOLVE.

   $$x = 48 \div 1\frac{3}{5}$$

   $$= 48 \div \frac{8}{5} = \frac{48}{1} \cdot \frac{5}{8} = \frac{48 \cdot 5}{1 \cdot 8} = \frac{8 \cdot 6 \cdot 5}{1 \cdot 8} = \frac{30}{1} \quad \text{or} \quad 30$$

4. INTERPRET. *Check* your work. *State* your conclusion: Thirty strips can be cut from the 48-inch piece of stock. ●

Ⓒ Operating on Negative Mixed Numbers

To perform operations on negative mixed numbers, let's first practice writing these numbers as negative fractions and negative fractions as negative mixed numbers.

To understand negative mixed numbers, we simply need to know that, for example,

$$-3\frac{2}{5} \text{ means } -\left(3\frac{2}{5}\right)$$

Thus, to write a negative mixed number as a fraction, we do the following.

$$-3\frac{2}{5} = -\left(3\frac{2}{5}\right) = -\left(\frac{17}{5}\right) \quad \text{or} \quad -\frac{17}{5}$$

$$3\frac{2}{5} = \frac{5 \cdot 3 + 2}{5} = \frac{17}{5}$$

Answer

18. 14 dresses

EXAMPLES Write each as a fraction.

19. $-1\dfrac{7}{8} = -\dfrac{8 \cdot 1 + 7}{8} = -\dfrac{15}{8}$ Write $1\dfrac{7}{8}$ as an improper fraction and keep the negative sign.

20. $-23\dfrac{1}{2} = -\dfrac{2 \cdot 23 + 1}{2} = -\dfrac{47}{2}$ Write $23\dfrac{1}{2}$ as an improper fraction and keep the negative sign.

To write a negative fraction as a negative mixed number, we use a similar procedure. We simply disregard the negative sign, convert the improper fraction to a mixed number, then reinsert the negative sign.

EXAMPLES Write each as a mixed number.

21. $-\dfrac{22}{5} = -4\dfrac{2}{5}$

$\begin{array}{r} 4 \\ 5\overline{)22} \\ -20 \\ \hline 2 \end{array}$ $\dfrac{22}{5} = 4\dfrac{2}{5}$

22. $-\dfrac{9}{4} = -2\dfrac{1}{4}$

$\begin{array}{r} 2 \\ 4\overline{)9} \\ -8 \\ \hline 1 \end{array}$ $\dfrac{9}{4} = 2\dfrac{1}{4}$

We multiply or divide with negative mixed numbers the same way that we multiply or divide with positive mixed numbers. First, write each mixed number as a fraction.

EXAMPLES Perform the indicated operations.

23. $-4\dfrac{2}{5} \cdot 1\dfrac{3}{11} = -\dfrac{22}{5} \cdot \dfrac{14}{11} = -\dfrac{22 \cdot 14}{5 \cdot 11} = -\dfrac{2 \cdot 11 \cdot 14}{5 \cdot 11} = -\dfrac{28}{5}$ or $-5\dfrac{3}{5}$

24. $-2\dfrac{1}{3} \div \left(-2\dfrac{1}{2}\right) = -\dfrac{7}{3} \div \left(-\dfrac{5}{2}\right) = -\dfrac{7}{3} \cdot \left(-\dfrac{2}{5}\right) = \dfrac{7 \cdot 2}{3 \cdot 5} = \dfrac{14}{15}$

Helpful Hint.

Recall that
$$(-) \cdot (-) = +$$

To add or subtract with negative mixed numbers, we must be very careful! Problems arise because recall that

$$-3\dfrac{2}{5} \text{ means } -\left(3\dfrac{2}{5}\right)$$

This means that

$$-3\dfrac{2}{5} = -\left(3\dfrac{2}{5}\right) = -\left(3 + \dfrac{2}{5}\right) = -3 - \dfrac{2}{5}$$ This can sometimes be easily overlooked.

To avoid problems, we will add or subtract negative mixed numbers by rewriting as addition and recalling how to add signed numbers.

Practice Problem 25

Add: $7\dfrac{2}{3} + \left(-11\dfrac{3}{4}\right)$

EXAMPLE 25 Add: $6\dfrac{3}{5} + \left(-9\dfrac{7}{10}\right)$

Solution: Here we are adding two numbers with different signs. Recall that we then subtract the absolute values and keep the sign of the larger absolute value.

Since $-9\dfrac{7}{10}$ has the larger absolute value, the answer is negative.

First, subtract absolute values:

$$
\begin{aligned}
9\dfrac{7}{10} &= 9\dfrac{7}{10} \\
-6\dfrac{3\cdot 2}{5\cdot 2} &= -6\dfrac{6}{10} \\
\hline
&3\dfrac{1}{10}
\end{aligned}
$$

Thus,

$$6\dfrac{3}{5} + \left(-9\dfrac{7}{10}\right) = -3\dfrac{1}{10}$$
The result is negative since $-9\dfrac{7}{10}$ has the larger absolute value.

Practice Problem 26

Subtract: $-9\dfrac{2}{7} - 15\dfrac{11}{14}$

EXAMPLE 26 Subtract: $-11\dfrac{5}{6} - 20\dfrac{4}{9}$

Solution: Let's write as an equivalent addition: $-11\dfrac{5}{6} + \left(-20\dfrac{4}{9}\right)$. Here, we are adding two numbers with like signs. Recall that we add their absolute values and keep the common negative sign.

First, add absolute values:

$$
\begin{aligned}
11\dfrac{5\cdot 3}{6\cdot 3} &= 11\dfrac{15}{18} \\
+20\dfrac{4\cdot 2}{9\cdot 2} &= +20\dfrac{8}{18} \\
\hline
&31\dfrac{23}{18} \text{ or } 32\dfrac{5}{18}
\end{aligned}
$$

Since $\dfrac{23}{18} = 1\dfrac{5}{18}$

Thus,

$$-11\dfrac{5}{6} - 20\dfrac{4}{9} = -32\dfrac{5}{18}$$
Keep the common sign.

Answers

25. $-4\dfrac{1}{12}$ **26.** $-25\dfrac{1}{14}$

CHAPTER 4 ACTIVITY   Comparing Fractions

Make seven identical copies of the strip shown in Figure 1. Cut out each strip.

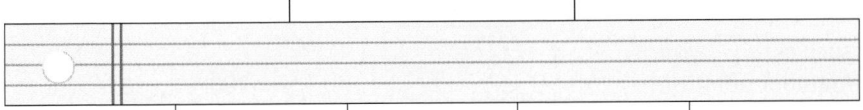

Figure 1

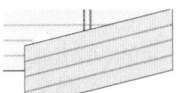

Figure 2

1. Fold and crease one of the strips (as shown in Figure 2) to represent two equal halves. Now unfold the strip so that it lies flat again. You have just created a fraction strip for the denominator 2.

2. Make similar fraction strips for the denominators 3, 4, 5, 6, 8, and 12 by folding. (Note: The markings along the top of Figure 1 will help you gauge equal thirds. The markings along the bottom of Figure 1 will help you gauge equal fifths.)

3. Shade one of the fraction strips to represent $\frac{1}{2}$ and another to represent $\frac{2}{5}$. Compare the fraction strips to decide which fraction is larger or whether the fractions are equal.

4. Shade one of the fraction strips to represent $\frac{3}{8}$ and another to represent $\frac{2}{3}$. Compare the fraction strips to decide which fraction is larger or whether the fractions are equal.

5. Shade one of the fraction strips to represent $\frac{3}{4}$ and another to represent $\frac{5}{6}$. Compare the fraction strips to decide which fraction is larger or whether the fractions are equal.

6. Shade one of the fraction strips to represent $\frac{2}{8}$ and another to represent $\frac{3}{12}$. (To reuse a fraction strip, just flip it over and use the blank side.) Compare the fraction strips to decide which fraction is larger or whether the fractions are equal.

7. Shade one of the fraction strips to represent $\frac{4}{5}$ and another to represent $\frac{7}{12}$. (To reuse a fraction strip just flip it over and use the blank side.) Compare the fraction strips to decide which fraction is larger or whether the fractions are equal.

CALCULATOR EXPLORATIONS

Converting Between Mixed-Number and Fraction Notation

If your calculator has a fraction key, such as $\boxed{a\ b/c}$, you can use it to convert between mixed-number notation and fraction notation.

To write $13\frac{7}{16}$ as an improper fraction, press

$\boxed{1}\ \boxed{3}\ \boxed{a\ b/c}\ \boxed{7}\ \boxed{a\ b/c}\ \boxed{1}\ \boxed{6}\ \boxed{2nd}\ \boxed{d/c}$

The display will read

$$\boxed{215\,|\,16}$$

which represents $\frac{215}{16}$. Thus $13\frac{7}{16} = \frac{215}{16}$.

To convert $\frac{190}{13}$ to a mixed number, press

$\boxed{1}\ \boxed{9}\ \boxed{0}\ \boxed{a\ b/c}\ \boxed{1}\ \boxed{3}\ \boxed{=}$

The display will read

$$\boxed{14\_8/13}$$

which represents $14\frac{8}{13}$. Thus $\frac{190}{13} = 14\frac{8}{13}$.

Write each mixed number as a fraction and each fraction as a mixed number.

1. $25\frac{5}{11}$

2. $67\frac{14}{15}$

3. $107\frac{31}{35}$

4. $186\frac{17}{21}$

5. $\frac{365}{14}$

6. $\frac{290}{13}$

7. $\frac{2769}{30}$

8. $\frac{3941}{17}$

EXERCISE SET 4.8

 Represent the shaded area in each figure group with (a) an improper fraction and (b) a mixed number. See Examples 1 and 2.

1.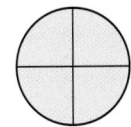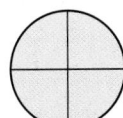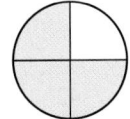

2.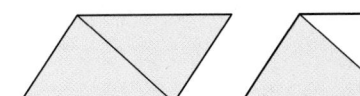

3.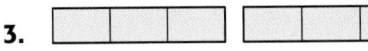

4.

5.

6.

7.

8.

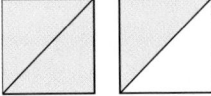 Write each mixed number as an improper fraction. See Example 3.

9. $2\frac{1}{3}$

10. $6\frac{3}{4}$

**11.** $3\frac{3}{8}$

12. $2\frac{5}{8}$

13. $11\frac{6}{7}$

14. $8\frac{9}{10}$

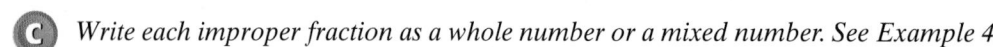

C *Write each improper fraction as a whole number or a mixed number. See Example 4.*

15. $\dfrac{13}{7}$

16. $\dfrac{42}{13}$

17. $\dfrac{47}{15}$

18. $\dfrac{65}{12}$

19. $\dfrac{37}{8}$

20. $\dfrac{42}{6}$

D *Multiply or divide. See Examples 5 through 10.*

21. $2\dfrac{2}{3} \cdot \dfrac{1}{7}$

22. $\dfrac{5}{9} \cdot 4\dfrac{1}{5}$

23. $8 \div 1\dfrac{5}{7}$

24. $5 \div 3\dfrac{3}{4}$

25. $3\dfrac{2}{3} \cdot 1\dfrac{1}{2}$

26. $2\dfrac{4}{5} \cdot 2\dfrac{5}{8}$

27. $2\dfrac{2}{3} \div \dfrac{1}{7}$

28. $\dfrac{5}{9} \div 4\dfrac{1}{5}$

E *Add. See Examples 11 through 13.*

29. $4\dfrac{7}{10} + 2\dfrac{1}{10}$

30. $7\dfrac{4}{9} + 3\dfrac{2}{9}$

31. $\begin{array}{r} 15\dfrac{4}{7} \\ +9\dfrac{11}{14} \\ \hline \end{array}$

32. $\begin{array}{r} 23\dfrac{3}{5} \\ +8\dfrac{8}{15} \\ \hline \end{array}$

33. $\begin{array}{r} 3\dfrac{5}{8} \\ 2\dfrac{1}{6} \\ +7\dfrac{3}{4} \\ \hline \end{array}$

34. $\begin{array}{r} 4\dfrac{1}{3} \\ 9\dfrac{2}{5} \\ +3\dfrac{1}{6} \\ \hline \end{array}$

Subtract. See Examples 14 through 16.

35. $\quad 4\frac{7}{10}$

$\quad\;\; -2\frac{1}{10}$

36. $\quad 7\frac{4}{9}$

$\quad\;\; -3\frac{2}{9}$

37. $\quad 10\frac{13}{14}$

$\quad\;\; -3\frac{4}{7}$

38. $\quad 12\frac{5}{12}$

$\quad\;\; -4\frac{1}{6}$

39. $\quad 9\frac{1}{5}$

$\quad\;\; -8\frac{6}{25}$

40. $\quad 6\frac{2}{13}$

$\quad\;\; -4\frac{7}{26}$

D **E** *Perform each indicated operation. See Examples 5 through 16.*

41. $\quad 2\frac{3}{4}$

$\quad\;\; +1\frac{1}{4}$

42. $\quad 5\frac{5}{8}$

$\quad\;\; +2\frac{3}{8}$

43. $\quad 15\frac{4}{7}$

$\quad\;\; -9\frac{11}{14}$

44. $\quad 23\frac{3}{5}$

$\quad\;\; -8\frac{8}{15}$

45. $3\frac{1}{9}\cdot 2$

46. $4\frac{1}{2}\cdot 3$

47. $1\frac{2}{3}\div 2\frac{1}{5}$

48. $5\frac{1}{5}\div 3\frac{1}{4}$

49. $22\frac{4}{9}+13\frac{5}{18}$

50. $15\frac{3}{25}-5\frac{2}{5}$

51. $5\frac{2}{3}-3\frac{1}{6}$

52. $5\frac{3}{8}-2\frac{13}{16}$

53. $\quad 15\frac{1}{5}$

$\quad\;\; 20\frac{3}{10}$

$\quad\;\; +37\frac{2}{15}$

54. $\quad 7\frac{3}{7}$

$\quad\;\; 15$

$\quad\;\; +20\frac{1}{2}$

55. $6\frac{4}{7}-5\frac{11}{14}$

56. $47\frac{5}{12}-23\frac{19}{24}$

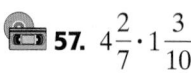 **57.** $4\frac{2}{7} \cdot 1\frac{3}{10}$ **58.** $6\frac{2}{3} \cdot 2\frac{3}{4}$

59.
$$6\frac{2}{11}$$
$$3$$
$$+4\frac{10}{33}$$
$$\overline{}$$

60.
$$3\frac{7}{16}$$
$$6\frac{1}{2}$$
$$+9\frac{3}{8}$$
$$\overline{}$$

    Solve. See Examples 17 and 18.

The first three total eclipses of the sun visible from the Atlantic Ocean in the 21st century are listed below. The duration of each eclipse is included in the table. Use this table to answer Exercises 61–64.

Total Solar Eclipses Visible from the Atlantic Ocean	
Date of Eclipse	**Duration (in minutes)**
June 21, 2001	$4\frac{14}{15}$
March 29, 2006	$4\frac{7}{60}$
November 3, 2013	$1\frac{2}{3}$

(*Source:* 2003 *World Almanac*)

61. What is the total duration for the three eclipses?

62. How much longer is the June 21, 2001, eclipse than the November 3, 2013, eclipse?

63. How much longer is the June 21, 2001, eclipse than the March 29, 2006, eclipse?

64. What is the total duration for the two eclipses occurring in odd-numbered years?

65. A sidewalk is built six bricks wide by laying each brick side by side. How many inches wide is the sidewalk if each brick measures $3\frac{1}{4}$ inches wide?

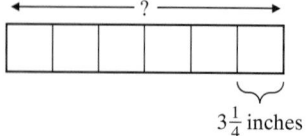

66. Vonshay Bartlet wants to build a bookcase with six equally spaced shelves. Disregarding the width of the shelves, he finds he has $5\frac{1}{2}$ feet of vertical space for shelves. How far apart should he space the shelves?

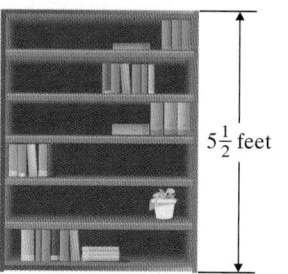

67. On March 12, 2003, the closing price for Kellogg Company stock was $28 1/5 per share. Brian Robinson spent $423 buying shares of Kellogg stock at the close of the stock market on that day. How many shares did he purchase? (Based on data from Commodity Systems, Inc.)

68. The nutrition label on a can of chili shows there are 26 grams of carbohydrates for each cup of chili. How many grams of carbohydrates are in a 2 1/2 cup can?

△ **69.** To prevent intruding birds, birdhouses built for Eastern Bluebirds should have an entrance hole measuring $1\frac{1}{2}$ inches in diameter. Entrance holes in bird houses for Mountain Bluebirds should measure $1\frac{9}{16}$ inches in diamater. How much wider should entrance holes for Mountain Bluebirds be than for Eastern Bluebirds? (*Source:* North American Bluebird Society)

70. If the total weight allowable without overweight charges is 50 pounds and the traveler's luggage weighs $60\frac{5}{8}$ pounds, on how many pounds will the traveler's overweight charges be based?

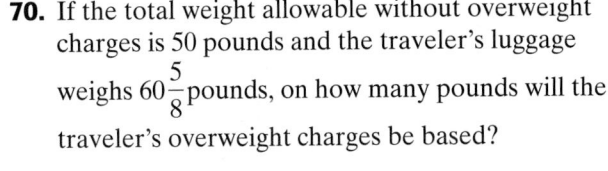

71. Charlotte Dowlin has $15\frac{2}{3}$ feet of plastic pipe. She cuts off a $2\frac{1}{2}$-foot length and then a $3\frac{1}{4}$-foot length. If she now needs a 10-foot piece of pipe, will the remaining piece do?

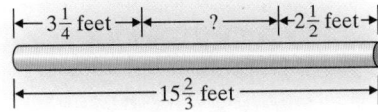

72. Shamalika Corning, a trim carpenter, cuts a board $3\frac{3}{8}$ feet long from one 6 feet long. How long is the remaining piece?

73. A small airplane used $58\frac{3}{4}$ gallons of fuel to fly a $7\frac{1}{2}$ hour trip. How many gallons of fuel were used for each hour?

74. Kesha Jonston is planning a Memorial Day barbeque. She has $27\frac{3}{4}$ pounds of hamburger. How many quarter-pound hamburgers can she make?

75. The Gauge Act of 1846 set the standard gauge for U.S. railroads at $56\frac{1}{2}$ inches. (See figure.) If the standard gauge in Spain is $65\frac{9}{10}$ inches, how much wider is Spain's standard gauge than the U.S. standard gauge? (*Source:* San Diego Railroad Museum)

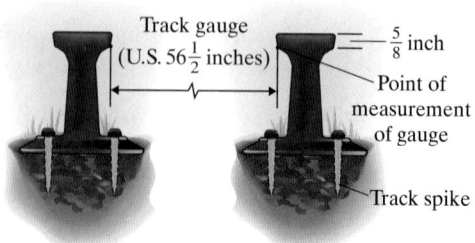

76. The standard railroad track gauge (see figure) in Spain is $65\frac{9}{10}$ inches, while in neighboring Portugal it is $65\frac{11}{20}$ inches. Which gauge is wider and by how much? (*Source:* San Diego Railroad Museum)

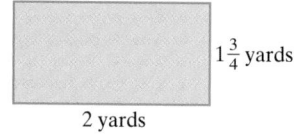 **77.** If Tucson's average rainfall is $11\frac{1}{4}$ inches and Yuma's is $3\frac{3}{5}$ inches, how much more rain, on the average, does Tucson get than Yuma?

78. A pair of crutches needs adjustment. One crutch is 43 inches and the other is $41\frac{5}{8}$ inches. Find how much the short crutch should be lengthened to make both crutches the same length.

For Exercises 79 and 80, find the area of each figure.

△ **79.**

$1\frac{3}{4}$ yards

2 yards

△ **80.**

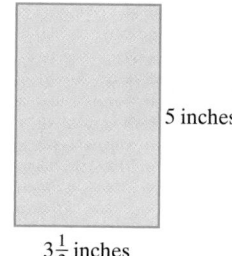

5 inches

$3\frac{1}{2}$ inches

△ **81.** A model for a proposed computer chip measures $\frac{3}{4}$ inch by $1\frac{1}{4}$ inches. Find its area.

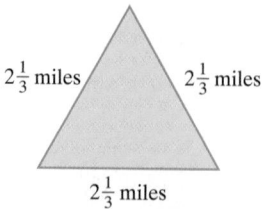

$\frac{3}{4}$ inch

$1\frac{1}{4}$ inches

△ **82.** The Saltalamachios are planning to build a deck that measures $4\frac{1}{2}$ yards by $6\frac{1}{3}$ yards. Find the area of their proposed deck.

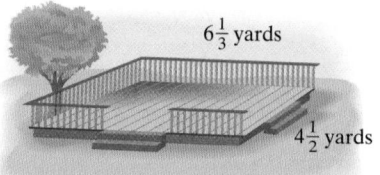

$6\frac{1}{3}$ yards

$4\frac{1}{2}$ yards

For Exercises 83 through 86, find the perimeter of each figure.

△ **83.**

$2\frac{1}{3}$ miles $2\frac{1}{3}$ miles

$2\frac{1}{3}$ miles

△ **84.**

$3\frac{1}{4}$ yards $3\frac{1}{4}$ yards

$3\frac{1}{4}$ yards $3\frac{1}{4}$ yards

$3\frac{1}{4}$ yards

△ **85.**

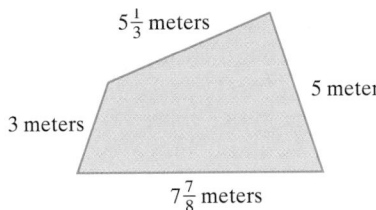

5$\frac{1}{3}$ meters

5 meters

3 meters

7$\frac{7}{8}$ meters

△ **86.**

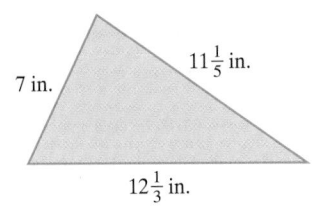

11$\frac{1}{5}$ in.

7 in.

12$\frac{1}{3}$ in.

87. The life expectancy of a circulating coin is 30 years. The life expectancy of a circulating dollar bill is only $\frac{1}{20}$ as long. Find the life expectancy of circulating paper money. (*Source:* The U.S. Mint)

88. A decorative wall in Ben and Joy Lander's garden is to be built using brick that is 2$\frac{3}{4}$ inches wide and a mortar joint that is $\frac{1}{2}$ inch wide. Use the diagram to find the height of the wall.

Height

Mortar joint

89. Located on an island in New York City's harbor, the Statue of Liberty is one of the largest statues in the world. The copper figure is 152$\frac{1}{6}$ feet tall from feet to tip of torch. The figure stands on a pedestal that is 154$\frac{1}{2}$ feet tall. What is the overall height of the Statue of Liberty from the base of the pedestal to the tip of the torch? (*Source: Microsoft Encarta Encyclopedia*)

90. The record for largest rainbow trout ever caught is 42$\frac{1}{8}$ pounds and was set in Alaska in 1970. The record for largest tiger trout ever caught is 20$\frac{13}{16}$ pounds and was set in Michigan in 1978. How much more did the record-setting rainbow trout weigh than the record-setting tiger trout? (*Source:* International Game Fish Association)

91. The record for rainfall during a 24-hour period in Alaska is $15\frac{1}{5}$ inches. This record was set in Angoon, Alaska, in October 1982. How much rain fell per hour on average? (*Source:* National Climatic Data Center)

92. The longest floating pontoon bridge in the United States is the Evergreen Point Bridge in Seattle, Washington. It is 2526 yards long. The second-longest pontoon bridge in the United States is the Hood Canal Bridge in Point Gamble, Washington which is $2173\frac{2}{3}$ yards long. How much longer is the Evergreen Point Bridge than the Hood Canal Bridge? (*Source:* Federal Highway Administration)

G *Perform the indicated operations. See Examples 19 through 26.*

93. $-4\frac{2}{5} \cdot 2\frac{3}{10}$

94. $-3\frac{5}{6} \div \left(-3\frac{2}{3}\right)$

95. $-5\frac{1}{8} - 19\frac{3}{4}$

96. $17\frac{5}{9} + \left(-14\frac{2}{3}\right)$

97. $-31\frac{2}{15} + 17\frac{3}{20}$

98. $-1\frac{5}{7} \cdot \left(-2\frac{1}{2}\right)$

99. $1\frac{3}{4} \div \left(-3\frac{1}{2}\right)$

100. $-31\frac{7}{8} - \left(-26\frac{5}{12}\right)$

101. $11\frac{7}{8} - 13\frac{5}{6}$

102. $-20\frac{2}{5} + \left(-30\frac{3}{10}\right)$

103. $-7\frac{3}{10} \div (-100)$

104. $-4\frac{1}{4} \div 2\frac{3}{8}$

105. $-3\frac{1}{6} \cdot \left(-2\frac{3}{4}\right)$

106. $42\frac{2}{9} - 50\frac{1}{3}$

107. $-21\frac{5}{12} + \left(-10\frac{3}{4}\right)$

108. $4\frac{1}{3} \cdot \left(-5\frac{1}{4}\right)$

Review and Preview

Simplify by combining like terms. See Section 1.9.

109. $2x - 5 + 7x - 8$

110. $11 + 8y - 5 - 11y$

111. $3(y - 2) - 6y$

112. $8(3 + 2y) - 20$

Simplify. See Section 1.8.

113. $2^3 + 3^2$

114. $8 \cdot 5 + 5 \div 5$

115. $\dfrac{7 - 3}{2^2}$

116. $4^2 - 2^4$

Combining Concepts

Solve.

117. Carmen's Candy Clutch is famous for its "Nut-stuff," a special blend of nuts and candy. A Supreme box of Nutstuff has $2\frac{1}{4}$ pounds of nuts and $3\frac{1}{2}$ pounds of candy. A Deluxe box has $1\frac{3}{8}$ pounds of nuts and $4\frac{1}{4}$ pounds of candy. Which box is heavier and by how much?

118. Willie Cassidie purchased three Supreme boxes and two Deluxe boxes of Nutstuff from Carmen's Candy Clutch. What is the total weight of his purchase? (See Exercise 117.)

 119. Explain in your own words why $9\frac{13}{9}$ is equal to $10\frac{4}{9}$.

120. In your own words, explain how to borrow when subtracting mixed numbers.

121. Find two mixed numbers whose sum is a whole number.

122. Find two mixed numbers whose difference is a whole number.

Internet Excursions

 Go To: http://www.prenhall.com/martin-gay_prealgebra What's Related

This World Wide Web site will provide access to a site called the Low Fat Vegetarian Archive, or a related site. It contains more than 2500 recipes for healthy low-fat and fat-free vegetarian dishes. Users can search the recipes by category, such as Mexican foods or desserts, or by ingredient.

123. Visit this site and find a recipe that is interesting to you. Be sure that the ingredient list contains at least two measures that are fractions (for example, $\frac{1}{3}$ cup). Print or copy the recipe. Then rewrite the list of ingredients to show the amounts that would be needed to make $\frac{1}{2}$ of the recipe.

124. Find a different recipe at this site that is interesting to you. Make sure that the recipe lists at least two fractional measures. Print or copy the recipe. Then rewrite the list of ingredients to show the amounts that would be needed to triple the recipe.

Chapter 4 Vocabulary Check

Fill in each blank with one of the words or phrases listed below.

prime factorization   simplest form   fraction   prime   equivalent   least common multiple   like

mixed number   improper   composite   reciprocals   numerator   proper   denominator

1. Fractions that have the same denominator are called _____ fractions.
2. The _____ is the smallest number that is a multiple of all numbers in a list of numbers.
3. _____ fractions represent the same portion of a whole.
4. A _____ has a whole number part and a fraction part.
5. A _____ number is a whole number greater than 1 that is not prime.
6. A _____ is a number of the form $\frac{a}{b}$ where a and b are integers and b is not 0. The number a is called the _____ and the number b is called the _____ .
7. An _____ fraction's numerator is greater than or equal to its denominator.
8. A _____ number is a whole number greater than 1 whose only divisions are 1 and itself.
9. A _____ fraction's numerator is less than its denominator.
10. Two numbers are _____ of each other if their product is 1.
11. A fraction is in _____ when the numerator and denominator have no common factors other than 1.
12. The _____ of a number is that number written as a product of prime numbers.

Highlights

CHAPTER 4

DEFINITIONS AND CONCEPTS	EXAMPLES

SECTION 4.1 INTRODUCTION TO FRACTIONS AND EQUIVALENT FRACTIONS

A **fraction** is a number of the form $\frac{a}{b}$, where a and b are integers and b is not 0. The number a is the **numerator** of the fraction, and the number b is the **denominator** of the fraction.

A fraction is called a **proper fraction** if its numerator is less than its denominator.

Proper Fractions: $\frac{1}{3}, \frac{2}{5}, \frac{7}{8}, \frac{100}{101}$

A fraction is called an **improper fraction** if its numerator is greater than or equal to its denominator.

Improper Fractions: $\frac{5}{4}, \frac{2}{2}, \frac{9}{7}, \frac{101}{100}$

Fractions that represent the same portion of a whole are called **equivalent fractions**.

FUNDAMENTAL PROPERTY OF FRACTIONS

If a, b, and c are numbers, then

$$\frac{a}{b} = \frac{a \cdot c}{b \cdot c} \quad \text{and also} \quad \frac{a}{b} = \frac{a \div c}{b \div c}$$

as long as b and c are not 0.

Write $\frac{3}{7}$ as an equivalent fraction with a denominator of 42.

$$\frac{3}{7} = \frac{3 \cdot 6}{7 \cdot 6} = \frac{18}{42}$$

DEFINITIONS AND CONCEPTS	**EXAMPLES**

SECTION 4.2 FACTORS AND SIMPLEST FORM

A **prime number** is a whole number greater than 1 whose only divisors are 1 and itself.

A **composite number** is a whole number greater than 1 that is not prime.

Every whole number greater than 1 has exactly one prime factorization.

A fraction is in **simplest form** or **lowest terms** when the numerator and the denominator have no common factors other than 1.

Prime Numbers: 2, 3, 5, 7, 11, 13, 17, ...
Composite Numbers: 4, 6, 8, 9, 10, 12, 14, 15, 16, ...
Write the prime factorization of 60.

$$60 = 6 \cdot 10$$
$$= 2 \cdot 3 \cdot 2 \cdot 5 \text{ or } 2^2 \cdot 3 \cdot 5$$

Write $\dfrac{30x}{36x}$ in simplest form.

$$\frac{30x}{36x} = \frac{2 \cdot 3 \cdot 5 \cdot x}{2 \cdot 2 \cdot 3 \cdot 3 \cdot x} = \frac{5}{6}$$

SECTION 4.3 MULTIPLYING AND DIVIDING FRACTIONS

To **multiply** two fractions,

$$\frac{a}{b} \cdot \frac{c}{d} = \frac{a \cdot c}{b \cdot d}, b \neq 0, d \neq 0$$

Two numbers are **reciprocals** of each other if their product is 1.

To **divide** two fractions,

$$\frac{a}{b} \div \frac{c}{d} = \frac{a}{b} \cdot \frac{d}{c} = \frac{a \cdot d}{b \cdot c}, b \neq 0, c \neq 0, d \neq 0$$

Multiply: $\dfrac{2x}{3} \cdot \dfrac{5}{7} = \dfrac{2x \cdot 5}{3 \cdot 7} = \dfrac{10x}{21}$

The reciprocal of $\dfrac{3}{5}$ is $\dfrac{5}{3}$.

Divide: $-\dfrac{3}{4} \div \dfrac{3}{8} = -\dfrac{3}{4} \cdot \dfrac{8}{3}$

$$= -\frac{3 \cdot 2 \cdot 4}{4 \cdot 3} = -2$$

SECTION 4.4 ADDING AND SUBTRACTING LIKE FRACTIONS AND LEAST COMMON DENOMINATOR

Fractions that have a common denominator are called **like fractions**.

To add or subtract like fractions,

$$\frac{a}{b} + \frac{c}{b} = \frac{a + c}{b}, \frac{a}{b} - \frac{c}{b} = \frac{a - c}{b}, b \neq 0$$

The **least common denominator (LCD)** of a list of fractions is the smallest positive number divisible by all the denominators in the list.

Like Fractions:

$$-\frac{1}{3} \text{ and } \frac{2}{3}; \quad \frac{5x}{7} \text{ and } \frac{6}{7}$$

$$\frac{10}{11} + \frac{3}{11} - \frac{8}{11} = \frac{10 + 3 - 8}{11} = \frac{5}{11}.$$

Find the LCD of $\dfrac{7}{30}$ and $\dfrac{1}{12}$.

$$30 = 2 \cdot 3 \cdot 5$$
$$12 = 2 \cdot 2 \cdot 3$$

Notice that the greatest number of times that 2 appears is 2 times.

$$LCD = 2 \cdot 2 \cdot 3 \cdot 5 = 60$$

SECTION 4.5 ADDING AND SUBTRACTING UNLIKE FRACTIONS

TO ADD OR SUBTRACT UNLIKE FRACTIONS

Step 1. Find the LCD.

Add: $\dfrac{3}{20} + \dfrac{2}{5}$

Step 1. The LCD is 20.

DEFINITIONS AND CONCEPTS	**EXAMPLES**

SECTION 4.5 ADDING AND SUBTRACTING UNLIKE FRACTIONS (*CONTINUED*)

Step 2. Write equivalent fractions with the LCD as the denominator.

Step 2. $\dfrac{2}{5} = \dfrac{2 \cdot 4}{5 \cdot 4} = \dfrac{8}{20}.$

Step 3. Add or subtract the like fractions.

Step 3. $\dfrac{3}{20} + \dfrac{2}{5} = \dfrac{3}{20} + \dfrac{8}{20} = \dfrac{11}{20}.$

Step 4. Write the result in simplest form.

Step 4. $\dfrac{11}{20}$ is in simplest form.

SECTION 4.6 COMPLEX FRACTIONS AND REVIEW OF ORDER OF OPERATIONS

A fraction whose numerator or denominator or both contain fractions is called a **complex fraction**.

Complex Fractions:

$$\dfrac{\dfrac{x}{4}}{\dfrac{7}{10}}, \quad \dfrac{\dfrac{y}{6} - 11}{\dfrac{4}{3}}$$

One method for simplifying complex fractions is to multiply the numerator and the denominator of the complex fraction by the LCD of all fractions in its numerator and its denominator.

$$\dfrac{\dfrac{y}{6} - 11}{\dfrac{4}{3}} = \dfrac{6\left(\dfrac{y}{6} - 11\right)}{6\left(\dfrac{4}{3}\right)} = \dfrac{6\left(\dfrac{y}{6}\right) - 6(11)}{6\left(\dfrac{4}{3}\right)}$$

$$= \dfrac{y - 66}{8}$$

SECTION 4.7 SOLVING EQUATIONS CONTAINING FRACTIONS

MULTIPLICATION AND DIVISION PROPERTIES OF EQUALITY
Let a, b, and c be numbers, and $c \neq 0$.

If $a = b$, then $a \cdot c = b \cdot c$ and also $\dfrac{a}{c} = \dfrac{b}{c}$.

TO SOLVE AN EQUATION IN *X*

Step 1. If fractions are present, multiply both sides of the equation by the LCD of the fractions.

Step 2. If parentheses are present, use the distributive property.

Step 3. Combine any like terms on each side of the equation.

Step 4. Use the addition property of equality to rewrite the equation so that variable terms are on one side of the equation and constant terms are on the other side.

Step 5. Divide both sides by the numerical coefficient of *x* to solve.

Step 6. Check the answer in the *original equation*.

Solve: $\dfrac{x}{15} + 2 = \dfrac{7}{3}$

$$15\left(\dfrac{x}{15} + 2\right) = 15\left(\dfrac{7}{3}\right) \quad \text{Multiply by the LCD 15.}$$

$$15\left(\dfrac{x}{15}\right) + 15 \cdot 2 = 15\left(\dfrac{7}{3}\right)$$

$$x + 30 = 35$$

$$x + 30 + (-30) = 35 + (-30)$$

$$x = 5$$

Check to see that 5 is the solution.

DEFINITIONS AND CONCEPTS	**EXAMPLES**

A **mixed number** is the sum of a whole number and a proper fraction.

Mixed Numbers: $1\frac{2}{5}$, $4\frac{7}{8}$

$$4\frac{7}{8} = \frac{8 \cdot 4 + 7}{8} = \frac{39}{8}$$

WRITING A MIXED NUMBER AS AN IMPROPER FRACTION

Step 1. Multiply the denominator of the fraction by the whole-number part. Add the numerator of the fraction to this product.

Step 2. Write this sum as the numerator of the improper fraction over the original denominator.

WRITING AN IMPROPER FRACTION AS A MIXED NUMBER OR WHOLE NUMBER

Step 1. Divide the denominator into the numerator.

$$\frac{29}{7} = 4\frac{1}{7}$$

$$\begin{array}{r} 4 \\ 7\overline{)29} \\ -28 \\ \hline 1 \end{array}$$

Step 2. The whole-number part of the quotient is the whole-number part of the mixed number and the $\dfrac{\text{remainder}}{\text{divisor}}$ is the fractional part.

To perform operations on mixed numbers, first write each mixed number as an improper fraction.

Multiply: $1\frac{3}{4} \cdot 2\frac{1}{5}$

$$\frac{7}{4} \cdot \frac{11}{5} = \frac{77}{20} = 3\frac{17}{20}$$

Add: $2\frac{1}{2} + 5\frac{7}{8}$

Addition and subtraction of mixed numbers can be performed using a vertical format.

$$2\frac{1}{2} = 2\frac{4}{8}$$
$$+5\frac{7}{8} = 5\frac{7}{8}$$
$$\overline{\phantom{+5\frac{7}{8}}}$$
$$7\frac{11}{8} = 7 + 1\frac{3}{8} = 8\frac{3}{8}$$

Chapter 4 Review

(4.1) *Represent the shaded area of each figure with a fraction.*

1.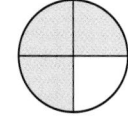

2.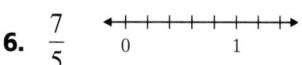

Graph each fraction on a number line.

3. $\dfrac{7}{9}$ ← 0 ───── 1 →

4. $\dfrac{4}{7}$ ← 0 ───── 1 →

5. $\dfrac{5}{4}$ ← 0 ── 1 →

6. $\dfrac{7}{5}$ ← 0 ── 1 →

7. The United States Army offers 242 job specialties to its soldiers. Certain specialties, such as infantry and field artillery, are closed to women. There are 207 army job specialties open to women and 35 army job specialties closed to women. What fraction of army job specialties are closed to women? (*Source:* U.S. Department of the Army)

8. In the United States Armed Forces today, roughly 43 out of every 50 personnel are men. What fraction of U.S. Armed Forces are men? (*Source:* U.S. Department of Defense)

Write each fraction as an equivalent fraction with the given denominator.

9. $\dfrac{2}{3} = \dfrac{?}{30}$

10. $\dfrac{5}{8} = \dfrac{?}{56}$

11. $\dfrac{7a}{6} = \dfrac{?}{42}$

12. $\dfrac{9b}{4} = \dfrac{?}{20}$

13. $\dfrac{4}{5x} = \dfrac{?}{50x}$

14. $\dfrac{5}{9y} = \dfrac{?}{18y}$

(4.2) *Write each fraction in simplest form.*

15. $\dfrac{12}{28}$

16. $\dfrac{15}{27}$

17. $\dfrac{25x}{75x^2}$

18. $\dfrac{36y^3}{72y}$

19. $\dfrac{29ab}{32abc}$

20. $\dfrac{18xyz}{23xy}$

21. $\dfrac{45x^2y}{27xy^3}$

22. $\dfrac{42ab^2c}{30abc^3}$

23. There are 12 inches in a foot. What fractional part of a foot does 8 inches represent?

12 inches
|← = 1 foot →|
|← 8 →|

24. Six out of 15 cars are white. What fraction of cars are *not* white?

(4.3) *Multiply.*

25. $\dfrac{3}{5} \cdot \dfrac{1}{2}$

26. $-\dfrac{6}{7} \cdot \dfrac{5}{12}$

27. $\dfrac{7}{8x} \cdot -\dfrac{2}{3}$

28. $\dfrac{6}{15} \cdot \dfrac{5y}{8}$

29. $-\dfrac{24x}{5} \cdot -\dfrac{15}{8x^3}$

30. $\dfrac{27y^3}{21} \cdot \dfrac{7}{18y^2}$

31. $\left(-\dfrac{1}{3}\right)^3$

32. $\left(-\dfrac{5}{12}\right)^2$

33. $\dfrac{x^3}{y} \cdot \dfrac{y^3}{x}$

34. $\dfrac{ac}{b} \cdot \dfrac{b^2}{a^3c}$

35. Evaluate xy if $x = \dfrac{2}{3}$ and $y = \dfrac{1}{5}$.

36. Evaluate ab if $a = -7$ and $b = \dfrac{9}{10}$.

Divide.

37. $-\dfrac{3}{4} \div \dfrac{3}{8}$

38. $\dfrac{21a}{4} \div \dfrac{7a}{5}$

39. $\dfrac{18x}{5} \div \dfrac{2}{5x}$

40. $-\dfrac{9}{2} \div -\dfrac{1}{3}$

41. $-\dfrac{5}{3} \div 2y$

42. $\dfrac{5x^2}{y} \div \dfrac{10x^3}{y^3}$

43. Evaluate $x \div y$ if $x = \dfrac{9}{7}$ and $y = \dfrac{3}{4}$.

44. Evaluate $a \div b$ if $a = -5$ and $b = \dfrac{2}{3}$.

Find the area of each figure.

△ **45.**

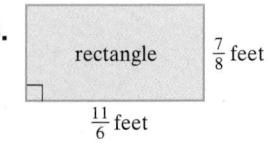

rectangle $\dfrac{7}{8}$ feet

$\dfrac{11}{6}$ feet

△ **46.**

square $\dfrac{2}{3}$ meter

47. $\dfrac{7}{11} + \dfrac{3}{11}$

48. $\dfrac{4}{9} + \dfrac{2}{9}$

49. $\dfrac{1}{12} - \dfrac{5}{12}$

50. $\dfrac{3}{y} - \dfrac{1}{y}$

51. $\dfrac{11x}{15} + \dfrac{x}{15}$

52. $\dfrac{4y}{21} - \dfrac{3}{21}$

53. $\dfrac{4}{15} + \dfrac{3}{15} - \dfrac{2}{15}$

54. $\dfrac{4}{15} - \dfrac{3}{15} - \dfrac{2}{15}$

Find the LCD of each list of fractions.

55. $\dfrac{2}{3}, \dfrac{5}{x}$

56. $\dfrac{3}{4}, \dfrac{3}{8}, \dfrac{7}{12}$

57. Determine whether $\dfrac{4}{5}$ is a solution of $z + \dfrac{1}{5} = 1$.

58. Determine whether $\dfrac{3}{4}$ is a solution of $x - \dfrac{2}{4} = \dfrac{1}{4}$.

Solve.

59. If a student studies math for $\dfrac{3}{8}$ of an hour and geography for $\dfrac{1}{8}$ of an hour, find how long she studies.

60. Beryl Goldstein mixed $\dfrac{5}{8}$ of a gallon of water with $\dfrac{1}{8}$ of a gallon of punch concentrate. Then she and her friends drank $\dfrac{3}{8}$ of a gallon of the punch. Find how much was left.

61. One evening Mark Alorenzo did $\dfrac{3}{8}$ of his homework before supper, another $\dfrac{2}{8}$ of it while his children did their homework, and $\dfrac{1}{8}$ of it after his children went to bed. Find what part of his homework he did that day.

△ **62.** The Simpsons will be fencing in their land. To do this, they need to find its perimeter. Find the perimeter of their land given that it is the shape of a rectangle.

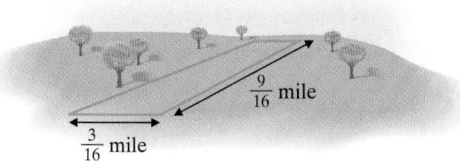

$\dfrac{9}{16}$ mile

$\dfrac{3}{16}$ mile

63. $\dfrac{7}{18} + \dfrac{2}{9}$

64. $\dfrac{4}{13} - \dfrac{1}{26}$

65. $-\dfrac{1}{3} + \dfrac{1}{4}$

66. $-\dfrac{2}{3} + \dfrac{1}{4}$

67. $\dfrac{5x}{11} + \dfrac{2}{55}$

68. $\dfrac{4}{15} + \dfrac{b}{5}$

69. $\dfrac{5y}{12} - \dfrac{2y}{9}$

70. $\dfrac{7x}{18} + \dfrac{2x}{9}$

71. $\dfrac{4}{9} + \dfrac{5}{y}$

72. $-\dfrac{9}{14} - \dfrac{3}{7}$

73. $\dfrac{4}{25} + \dfrac{23}{75} + \dfrac{7}{50}$

74. $\dfrac{2}{3} - \dfrac{2}{9} - \dfrac{1}{6}$

Solve each equation.

75. $a - \dfrac{2}{3} = \dfrac{1}{6}$

76. $9x + \dfrac{1}{5} - 8x = -\dfrac{7}{10}$

Find the perimeter of each figure.

△ **77.**

$\dfrac{2}{9}$ meter | Rectangle

$\dfrac{5}{6}$ meter

△ **78.**

$\dfrac{1}{5}$ foot $\quad$ $\dfrac{3}{5}$ foot

$\dfrac{7}{10}$ foot

79. Determine whether $\dfrac{8}{11}$ is a solution of $x + \dfrac{1}{3} = \dfrac{35}{11}$.

80. Determine whether $\dfrac{9}{2}$ is a solution of $\dfrac{1}{9}y - \dfrac{1}{4} = \dfrac{1}{4}$.

81. In a group of 100 blood donors, typically $\dfrac{9}{25}$ have type A Rh-positive blood and $\dfrac{3}{50}$ have type A Rh-negative blood. What fraction have type A blood?

82. Find the difference in length of two scarves if one scarf is $\dfrac{5}{12}$ of a yard long and the other is $\dfrac{2}{3}$ of a yard long.

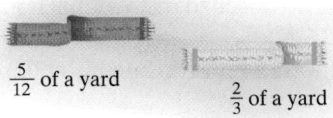

$\dfrac{5}{12}$ of a yard $\qquad$ $\dfrac{2}{3}$ of a yard

(4.6) *Simplify each complex fraction.*

83. $\dfrac{\frac{2x}{5}}{\frac{7}{10}}$

84. $\dfrac{\frac{3y}{7}}{\frac{11}{7}}$

85. $\dfrac{2 + \frac{3}{4}}{1 - \frac{1}{8}}$

86. $\dfrac{\dfrac{5}{6} + 2}{\dfrac{11}{3} - 1}$

87. $\dfrac{\dfrac{2}{5} - \dfrac{1}{2}}{\dfrac{3}{4} - \dfrac{7}{10}}$

88. $\dfrac{\dfrac{5}{6} - \dfrac{1}{4}}{\dfrac{-1}{12y}}$

Evaluate each expression if $x = \dfrac{1}{2}$, $y = -\dfrac{2}{3}$, *and* $z = \dfrac{4}{5}$.

89. $2x + y$

90. $\dfrac{x}{y + z}$

91. $\dfrac{x + y}{z}$

92. $x + y + z$

93. y^2

94. $x - z$

(4.7) *Solve.*

95. $-\dfrac{3}{5}x = 6$

96. $\dfrac{2}{9}y = -\dfrac{4}{3}$

97. $\dfrac{x}{7} - 3 = -\dfrac{6}{7}$

98. $\dfrac{y}{5} + 2 = \dfrac{11}{5}$

99. $\dfrac{1}{6} + \dfrac{x}{4} = \dfrac{17}{12}$

100. $\dfrac{x}{5} - \dfrac{5}{4} = \dfrac{x}{2} - \dfrac{1}{20}$

(4.8) *Write each improper fraction as a mixed number or a whole number.*

101. $\dfrac{15}{4}$

102. $\dfrac{39}{13}$

103. $\dfrac{7}{7}$

104. $\dfrac{125}{4}$

Write each mixed or whole number as an improper fraction.

105. $2\dfrac{1}{5}$

106. 5

107. $3\dfrac{8}{9}$

108. 3

Perform each indicated operation.

109. $31\frac{2}{7}$

$+14\frac{10}{21}$

110. $24\frac{4}{5}$

$+35\frac{1}{5}$

111. $69\frac{5}{22}$

$-36\frac{7}{11}$

112. $36\frac{3}{20}$

$-32\frac{5}{6}$

113. $29\frac{2}{9}$

$27\frac{7}{18}$

$+54\frac{2}{3}$

114. $7\frac{3}{8}$

$9\frac{5}{6}$

$+3\frac{1}{12}$

115. $1\frac{5}{8} \cdot \frac{2}{3}$

116. $3\frac{6}{11} \cdot \frac{5}{13}$

117. $4\frac{1}{6} \cdot 2\frac{2}{5}$

118. $5\frac{2}{3} \cdot 2\frac{1}{4}$

119. $6\frac{3}{4} \div 1\frac{2}{7}$

120. $5\frac{1}{2} \div 2\frac{1}{11}$

121. $\frac{7}{2} \div 1\frac{1}{2}$

122. $1\frac{3}{5} \div \frac{1}{4}$

123. Two packages of soup bones weigh $3\frac{3}{4}$ pounds and $2\frac{3}{5}$ pounds. Find their comined weight.

124. A ribbon $5\frac{1}{2}$ yards long is cut from a reel of ribbon with 50 yards on it. Find the length of the piece remaining on the reel.

125. The average annual snowfall at a certain ski resort is $62\frac{3}{10}$ inches. Last year it had $54\frac{1}{2}$ inches. Find how many inches below average last year's annual snowfall was.

△ **126.** Find the area of a rectangular sheet of gift wrap that is $2\frac{1}{4}$ feet by $3\frac{1}{3}$ feet.

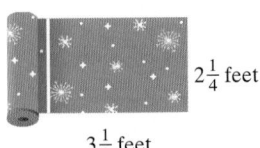

$2\frac{1}{4}$ feet

$3\frac{1}{3}$ feet

127. Find the perimeter of a sheet of shelf paper needed to fit exactly a square drawer $1\frac{1}{4}$ feet long on each side.

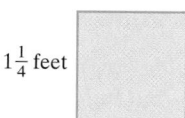

$1\frac{1}{4}$ feet

128. Find the area of the rectangle.

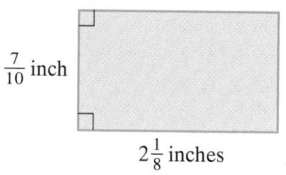

$\frac{7}{10}$ inch

$2\frac{1}{8}$ inches

129. A flower gardener has a square flower bed $\frac{2}{3}$ meter on a side and a rectangular one that is $\frac{7}{8}$ meter by $\frac{1}{3}$ meter. Find the total perimeter of his flower beds.

130. There are 58 calories in 1 ounce of turkey. Find how many calories there are in a $3\frac{1}{2}$-ounce serving of turkey.

131. There are $3\frac{1}{3}$ grams of fat in each ounce of lean hamburger. Find how many grams of fat are in a 4-ounce hamburger.

132. Herman Heltznutt walks 5 days a week for a total distance of $5\frac{1}{4}$ miles per week. If he walks the same distance each day, find the distance he walks each day.

Perform the indicated operations.

133. $-12\frac{1}{7} + \left(-15\frac{3}{14}\right)$

134. $-3\frac{1}{5} \div \left(-2\frac{7}{10}\right)$

135. $-2\frac{1}{4} \cdot 1\frac{3}{4}$

136. $23\frac{7}{8} - 24\frac{7}{10}$

Are you prepared for a test on Chapter 4?

Below I have listed some *common trouble areas* for topics covered in Chapter 4. After studying for your test—but before taking your test—read these.

Make sure you remember how to perform different operations on fractions!!! Try to add, subtract, multiply, then divide $\frac{3}{5}$ and $\frac{7}{15}$. Check your results below.

$$\frac{3}{5} + \frac{7}{15} = \frac{3 \cdot 3}{5 \cdot 3} + \frac{7}{15} = \frac{9}{15} + \frac{7}{15} = \frac{16}{15} = 1\frac{1}{15}$$

To add or subtract, you need common denominators.

$$\frac{3}{5} - \frac{7}{15} = \frac{3 \cdot 3}{5 \cdot 3} - \frac{7}{15} = \frac{9}{15} - \frac{7}{15} = \frac{2}{15}$$

$$\frac{3}{5} \cdot \frac{7}{15} = \frac{3 \cdot 7}{5 \cdot 15} = \frac{\overset{1}{\cancel{3}} \cdot 7}{5 \cdot \underset{1}{\cancel{3}} \cdot 5} = \frac{7}{25}$$

$$\frac{3}{5} \div \frac{7}{15} = \frac{3}{5} \cdot \frac{15}{7} = \frac{3 \cdot 3 \cdot \overset{1}{\cancel{5}}}{\underset{1}{\cancel{5}} \cdot 7} = \frac{9}{7} = 1\frac{2}{7}$$

To divide, multiply by the reciprocal.

Chapter 4 Test

Remember to check your answers and use the Chapter Test Prep Video to view solutions.

Write each mixed number as an improper fraction.

1. $7\dfrac{2}{3}$

2. $3\dfrac{6}{11}$

Write each improper fraction as a mixed number or a whole number.

3. $\dfrac{23}{5}$

4. $\dfrac{75}{4}$

Write each fraction in simplest form.

5. $\dfrac{54}{210}$

6. $-\dfrac{42}{70}$

Perform each indicated operation and write the answers in simplest form.

7. $\dfrac{4}{4} \div \dfrac{3}{4}$

8. $-\dfrac{4}{3} \cdot \dfrac{4}{4}$

9. $\dfrac{7x}{9} + \dfrac{x}{9}$

10. $\dfrac{1}{7} - \dfrac{3}{x}$

11. $\dfrac{xy^3}{z} \cdot \dfrac{z}{xy}$

12. $-\dfrac{2}{3} \cdot -\dfrac{8}{15}$

13. $\dfrac{9a}{10} + \dfrac{2}{5}$

14. $-\dfrac{8}{15y} - \dfrac{2}{15y}$

15. $8y^3 \div \dfrac{y}{3}$

Answers

1. _____

2. _____

3. _____

4. _____

5. _____

6. _____

7. _____

8. _____

9. _____

10. _____

11. _____

12. _____

13. _____

14. _____

15. _____

16. _____

17. _____

18. _____

19. _____

20. _____

21. _____

22. _____

23. _____

24. _____

25. _____

26. _____

27. _____

28. _____

29. _____

30. _____

16. $5\dfrac{1}{4} \div \dfrac{7}{12}$

17. $3\dfrac{7}{8}$
$7\dfrac{2}{5}$
$+2\dfrac{3}{4}$
$\overline{}$

18. $\dfrac{3a}{8} \cdot \dfrac{16}{6a^3}$

19. $-\dfrac{16}{3} \div -\dfrac{3}{12}$

20. $3\dfrac{1}{3} \cdot 6\dfrac{3}{4}$

21. $12 \div 3\dfrac{1}{3}$

22. $\left(\dfrac{14}{5} \cdot \dfrac{25}{21} \right) \div 10$

23. $\dfrac{11}{12} - \dfrac{3}{8} + \dfrac{5}{24}$

Simplify each complex fraction.

24. $\dfrac{\dfrac{5x}{7}}{\dfrac{20x^2}{21}}$

25. $\dfrac{5 + \dfrac{3}{7}}{2 - \dfrac{1}{2}}$

Solve.

26. $-\dfrac{3}{8}x = \dfrac{3}{4}$

27. $\dfrac{x}{5} + x = -\dfrac{24}{5}$

28. $\dfrac{2}{3} + \dfrac{x}{4} = \dfrac{5}{12} + \dfrac{x}{2}$

Evaluate each expression for the given replacement values.

29. $-5x;\ x = -\dfrac{1}{2}$

30. $x \div y;\ x = \dfrac{1}{2},\ y = 3\dfrac{7}{8}$

Solve.

31. A McDonald's Big Mac® sandwich has 560 calories. There are 280 calories from fat in a Big Mac. What fraction of a Big Mac's calories are from fat? (*Source:* McDonald's Corporation)

32. A carpenter cuts a piece $2\frac{3}{4}$ feet long from a cedar plank that is $6\frac{1}{2}$ feet long. How long is the remaining piece?

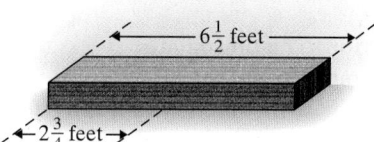

As shown in the circle graph, the market for backpacks is divided among five companies. For instance, Wilderness, Inc.'s backpack accounts for $\frac{1}{4}$ of all backpack sales. Use this graph to answer Questions 33 and 34.

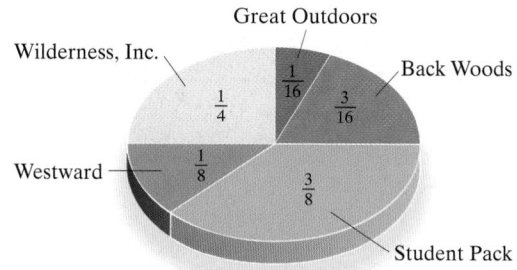

33. What fraction of backpack sales goes to Back Woods and Westward combined?

34. If a total of 500,000 backpacks are sold each year, how many backpacks does Wilderness, Inc. sell?

△ **35.** How many square yards of artificial turf are necessary to cover a football field, including the end zones and 10 yards beyond the side lines? (*Hint:* A football field measures $100 \times 53\frac{1}{3}$ yards and the end zones are 10 yards deep.)

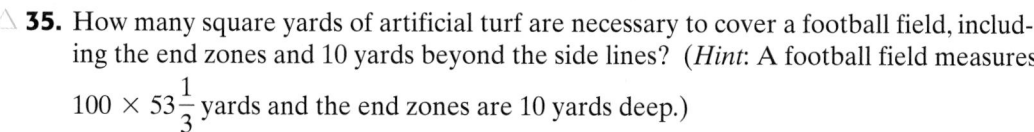

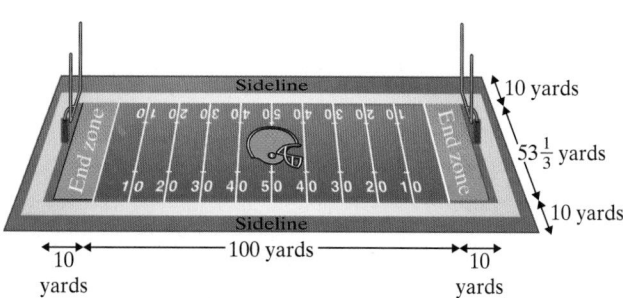

31. _____

32. _____

33. _____

34. _____

35. _____

353

△ **36.** Find the area of the figure.

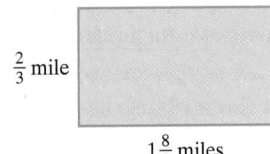

$\frac{2}{3}$ mile

$1\frac{8}{9}$ miles

37. During a 258-mile trip, a car used $10\frac{3}{4}$ gallons of gas. How many miles could we expect the car to travel on one gallon of gas?

38. Prior to an oil spill, the stock in an oil company sold for $120 per share. As a result of the liability that the company incurred from the spill, the price per share fell to $\frac{3}{4}$ of the price before the spill. What did the stock sell for after the spill?

Chapter 4 Cumulative Review

Write each number in words.

1. 126

2. 115

3. 3005

4. 6573

5. Add: 23 + 136

6. Add: 587 + 44

7. Subtract: 43 − 29. Check by adding.

8. Subtract: 995 − 62. Check by adding.

9. Round 278,362 to the nearest thousand.

10. Round 1436 to the nearest ten.

11. A certain computer disk can hold about 1510 thousand bytes of information. How many total bytes can 42 such disks hold?

12. On a trip across country, Daniel Daunis travels 435 miles per day. How many total miles does he travel in 3 days?

13. Divide and check: 56,717 ÷ 8

14. Divide and check: 4558 ÷ 12

1. _____

2. _____

3. _____

4. _____

5. _____

6. _____

7. _____

8. _____

9. _____

10. _____

11. _____

12. _____

13. _____

14. _____

15. _____

16. _____

17. _____

18. _____

19. _____

20. _____

21. _____

22. _____

23. _____

24. _____

25. _____

26. _____

27. _____

28. _____

29. _____

30. _____

Write using exponential notation.

15. $4 \cdot 4 \cdot 4$

16. $7 \cdot 7$

17. $6 \cdot 6 \cdot 6 \cdot 8 \cdot 8 \cdot 8 \cdot 8 \cdot 8$

18. $9 \cdot 9 \cdot 9 \cdot 9 \cdot 5 \cdot 5$

19. Evaluate $2(x - y)$ if $x = 8$ and $y = 4$.

20. Evaluate $8a + 3(b - 5)$ if $a = 5$ and $b = 9$.

21. Jack Mayfield, a miner for the Molly Kathleen Gold Mine, is presently 150 feet below the surface of the Earth. Represent this position using an integer.

22. The temperature on a cold day in Minneapolis, MN is 21° F below zero. Represent this temperature using an integer.

23. Add using a number line: $-7 + 3$

24. Add using a number line: $-3 + 8$

25. Simplify: $7 - 8 - (-5) - 1$

26. Simplify: $6 + (-8) - (-9) + 3$

27. Evaluate: $(-5)^2$

28. Evaluate: -2^4

29. Simplify: $\dfrac{12 - 16}{-1 + 3}$

30. Simplify: $(20 - 5^2)^2$

Multiply.

31. $5(3y)$

32. $12(3c)$

33. $-2(4x)$

34. $-7(14a)$

35. Solve: $-8 = x + 1$

36. Solve: $x - 3 = 5$

37. Solve: $8x - 9x = 12 - 17$

38. Solve: $2x + 5x = 0 - 7$

39. Translate each sentence into an equation.
   a. The product of 7 and 6 is 42.
   b. Twice the sum of 3 and 5 is equal to 16.
   c. The quotient of -45 and 5 yields -9.

40. Translate each sentence into an equation:
   a. The sum of 4 and 3 is 7.
   b. Four times the difference of 5 and 2 equals 12.
   c. The product of -4 and -6 yields 24.

41. Twice a number, added to 3, is the same as the number minus 6. Find the number.

42. A number is tripled, then added to 4. The result is equal to the original number minus 6. Find the number.

43. Write $\dfrac{9x}{11}$ as an equivalent fraction whose denominator is 44.

44. Write $\dfrac{2a}{3}$ as an equivalent fraction whose denominator is 15.

45. Write the prime factorization of 45.

46. Write the prime factorization of 92.

31. _____

32. _____

33. _____

34. _____

35. _____

36. _____

37. _____

38. _____

39. a. _____

b. _____

c. _____

40. a. _____

b. _____

c. _____

41. _____

42. _____

43. _____

44. _____

45. _____

46. _____

Multiply.

47. $\dfrac{2}{3} \cdot \dfrac{5}{11}$

48. $\dfrac{1}{7} \cdot \dfrac{2}{5}$

49. $\dfrac{1}{4} \cdot \dfrac{1}{2}$

50. $\dfrac{3}{5} \cdot \dfrac{1}{5}$

Decimals

Decimals are an important part of everyday life. For example, we use decimal numbers in our money system. One penny is 0.01 dollar, and one dime—or ten pennies, is 0.10 dollar. Among many other uses, decimals also express batting averages. A baseball player with a 0.333 batting average is a pretty good batter. Decimal numbers represent parts of a whole, just like fractions. In this chapter, we analyze the relationship between fractions and decimals.

The Nutrition Labeling and Education Act (NLEA) was signed into law on November 8, 1990. It requires food manufacturers to include nutrition information on their product labels. The NLEA provides specific guidelines concerning the use of terms such as "low fat," or "high fiber." Labels contain information about portion sizes, vitamins and minerals, and sodium, fat, and cholesterol content of foods. The result of this important legislation is to help consumers make more informed and healthier food choices. In Exercise 37 on page 412, we will see how comparing decimals can help consumers to determine if a product can truly be called "fat free."

Name _____ Section _____ Date_____

Chapter 5 Pretest

1. _____

2. _____

3. _____

4. _____

5. _____

6. _____

7. _____

8. _____

9. _____

10. _____

11. _____

12. _____

13. _____

14. _____

15. _____

16. _____

17. _____

18. _____

19. _____

20. _____

1. Write 0.27 as a fraction.`

2. Insert $<$, $>$, or $=$ to form a true statement. 0.205 ____ 0.213

3. Round 54.651 to the nearest tenth.

Perform each indicated operation.

4. $38.41 + 14.032 + 7.6$

5. $(-3.4)(-2.1)$

6. $(2.016)(100)$

7. $16.24 \div 0.4$

8. $\dfrac{891}{10,000}$

9. Evaluate $x - y$ for $x = 12.3$ and $y = 0.61$.

10. Simplify: $-9.8 - 6.2x - 7.9 + 1.4x$

11. Evaluate xy for $x = 4.2$ and $y = 0.03$.

△ **12.** Find the circumference of a circle whose radius is 6 inches. Then use the approximation 3.14 for π to approximate the circumference.

13. Perry Sitongia borrowed $576 from his father. He plans to pay it back over the next 20 months with equal payments. How much will each monthly payment be?

14. Simplify: $0.2(6.9 - 3.01)$

15. Write $\dfrac{3}{8}$ as a decimal.

16. Insert $<$, $>$, or $=$ to form a true statement. $\dfrac{9}{11}$ ____ 0.8182

17. Solve: $4(x + 0.22) = -3.4$

18. Find the square root: $\sqrt{\dfrac{36}{49}}$

19. Approximate $\sqrt{46}$ to the nearest hundredth.

△ **20.** Find the length of the hypotenuse of the given right triangle.

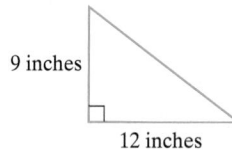

9 inches

12 inches

5.1   Introduction to Decimals

 **Decimal Notation and Writing Decimals in Words**

Like fractional notation, decimal notation is used to denote a part of a whole. Numbers written in decimal notation are called **decimal numbers**, or simply **decimals**. The decimal 17.758 has three parts.

$$1 \quad 7 \quad . \quad 7 \quad 5 \quad 8$$

Whole-number part   Decimal point   Decimal part

In Section 1.2, we introduced place value for whole numbers. Place names and place values for the whole-number part of a decimal number are exactly the same, as shown next. Place names and place values for the decimal part are also shown.

Place-Value Chart

hundreds	tens	ones		tenths	hundredths	thousandths	ten-thousandths	hundred-thousandths
	1	7	.	7	5	8		
100	10	1	decimal point	$\frac{1}{10}$	$\frac{1}{100}$	$\frac{1}{1000}$	$\frac{1}{10,000}$	$\frac{1}{100,000}$

Notice that the value of each place is $\frac{1}{10}$ of the value of the place to its left. For example,

$$1 \cdot \frac{1}{10} = \frac{1}{10} \quad \text{and} \quad \frac{1}{10} \cdot \frac{1}{10} = \frac{1}{100}$$

ones   tenths   tenths   hundredths

For the example above, 17.758, the digit 5 is in the hundredths place, so its value is 5 hundredths or $\frac{5}{100}$.

Writing or reading a decimal in words is similar to writing or reading a whole number. Use the following steps.

Writing (or Reading) a Decimal in Words

Step 1. Write the whole-number part in words.

Step 2. Write "and" for the decimal point.

Step 3. Write the decimal part in words as though it were a whole number, followed by the place value of the last digit.

EXAMPLE 1   Write each decimal in words.

a. 0.3   **b.** −5.82   **c.** 21.093

Solution:

a. Three tenths

b. Negative five and eighty-two hundredths

c. Twenty-one and ninety-three thousandths

OBJECTIVES

A Know the meaning of place value for a decimal number and write decimals in words.

B Write decimals in standard form.

C Write decimals as fractions.

D Compare decimals.

E Round decimals to a given place value.

SSM   TUTOR CENTER   SG   CD & VIDEO   MATH PRO   WEB

Practice Problem 1

Write each decimal in words.

a. 0.08   b. −500.025   c. 0.0329

Answers

1. a. eight hundredths   **b.** negative five hundred and twenty-five thousandths   **c.** three hundred twenty-nine ten-thousandths

Practice Problem 2

Write the decimal 97.28 in words.

Practice Problem 3

Write the decimal 72.1085 in words.

EXAMPLE 2 Write the decimal in the sentence in words: The Golden Jubilee Diamond is a 545.67 carat cut diamond. (*Source: The Guinness Book of Records*)

Solution: Five hundred forty-five and sixty-seven hundredths ●

EXAMPLE 3 Write the decimal in the sentence in words: The oldest known fragments of Earth's crust are Zircon crystals. They were discovered in Australia and are believed to be 4.276 billion years old. (*Source: The Guinness Book of Records*)

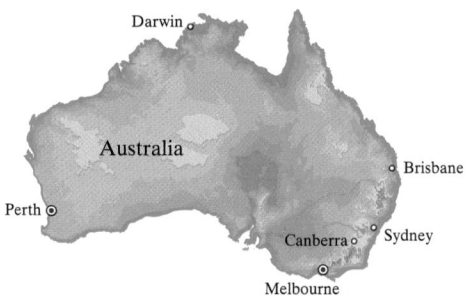

Solution: Four and two hundred seventy-six thousandths ●

Suppose that you are paying for a purchase of $368.42 at Circuit City by writing a check. Checks are usually written using the following format.

Answers

2. ninety-seven and twenty-eight hundredths
3. seventy-two and one thousand eighty-five ten-thousandths

EXAMPLE 4 Fill in the check to Camelot Music to pay for your purchase of $92.98.

Solution:

PAY TO THE ORDER OF *Camelot Music* $ 92.98

Your Preprinted Name
Your Preprinted Address   00-1110/1000   2236

Date *(Insert today's date)*

Ninety-two and ———————— 98/100 Dalars

Contains Security Features. Details on Back.

For _____ *(Your signature)* MP

⑆000000000⑆ 0000000000⑈ 0000

B Writing Decimals in Standard Form

A decimal written in words can be written in standard form by reversing the above procedure.

EXAMPLES Write each decimal in standard form.

5. Forty-eight and twenty-six hundredths
↓
48.26
↑———— hundredths place

6. Six and ninety-five thousandths
6.095
↑———— thousandths place

Helpful Hint

When writing a decimal from words to decimal notation, make sure the last digit is in the correct place by inserting 0s after the decimal point if necessary. For example,

Two and thirty-eight thousandths is 2.038
thousandths place

Practice Problem 4

Fill in the check to CLECO (Central Louisiana Electric Company) to pay for your monthy electric bill of $207.40.

Your Preprinted Name
Your Preprinted Address   00-1110/1000   2236
Date _____

PAY TO THE ORDER OF _____ $ _____

_____ Dollars

For _____

⑆000000000⑆ 0000000000⑈ 0000

Practice Problems 5–6

Write each decimal in standard form.

5. Three hundred and ninety-six hundredths

6. Thirty-nine and forty-two thousandths

Answers

4.

Your Preprinted Name
Your Preprinted Address   00-1110/1000   2237
Date *(Current date)*

PAY TO THE ORDER OF *CLECO* $ 207.40

Two hundred seven and ———— 40/100 Dollars

For _____ *(Your signature)*

⑆000000000⑆ 0000000000⑈ 0000

5. 300.96   **6.** 39.042

C Writing Decimals as Fractions

Once you master reading and writing decimals, writing a decimal as a fraction follows naturally.

Decimal	In Words	Fraction
0.7	seven tenths	$\dfrac{7}{10}$
0.51	fifty-one hundredths	$\dfrac{51}{100}$
0.009	nine thousandths	$\dfrac{9}{1000}$

Notice that the number of decimal places in a decimal number is the same as the number of zeros in the denominator of the equivalent fraction. We can use this fact to write decimals as fractions.

$$0.51 = \frac{51}{100} \qquad 0.009 = \frac{9}{1000}$$

2 decimal places   2 zeros   3 decimal places   3 zeros

Practice Problem 7

Write 0.037 as a fraction.

EXAMPLE 7 Write 0.43 as a fraction.

Solution: $0.43 = \dfrac{43}{100}$

2 decimal places   2 zeros

Practice Problem 8

Write 14.97 as a mixed number.

EXAMPLE 8 Write 5.6 as a mixed number.

Solution: $5.6 = 5\dfrac{6}{10} = 5\dfrac{3}{5}$ in simplest form

1 decimal place   1 zero

Practice Problems 9–11

Write each decimal as a fraction or mixed number. Write your answer in simplest form.

9. 0.12
10. 57.8
11. −209.986

EXAMPLES Write each decimal as a fraction or mixed number. Write your answer in simplest form.

9. $0.125 = \dfrac{125}{1000} = \dfrac{1}{8}$

10. $23.5 = 23\dfrac{5}{10} = 23\dfrac{1}{2}$

11. $-105.083 = -105\dfrac{83}{1000}$

Answers

7. $\dfrac{37}{1000}$ **8.** $14\dfrac{97}{100}$ **9.** $\dfrac{3}{25}$

10. $57\dfrac{4}{5}$ **11.** $-209\dfrac{493}{500}$

Ⓓ Comparing Decimals

One way to compare positive decimals is by comparing digits in corresponding places. To see why this works, let's compare 0.5 or $\frac{5}{10}$ and 0.8 or $\frac{8}{10}$. We know

$$\frac{5}{10} < \frac{8}{10} \text{ since } 5 < 8, \text{ so}$$

$$0.\underset{\downarrow}{5} < 0.\underset{\downarrow}{8} \text{ since } 5 < 8$$

This leads to the following.

Comparing Two Positive Decimals

Compare digits in the same places from left to right. When two digits are not equal, the number with the larger digit is the larger decimal. If necessary, insert 0s after the last digit to the right of the decimal point to continue comparing.

Compare hundredths place digits

28.253   28.263
  ↑   ↑
  5   <   6
so 28.253   <   28.263

Helpful Hint

For any decimal, writing 0s after the last digit to the right of the decimal point does not change the value of the number.
    7.6 = 7.60 = 7.600, and so on.
When a whole number is written as a decimal, the decimal point is placed to the right of the ones digit.
    25 = 25.0 = 25.00, and so on

EXAMPLE 12 Insert <, >, or = to form a true statement.

    0.378   0.368

Solution: 0.3̲ 78   0.3̲ 68   The tenths places are the same.

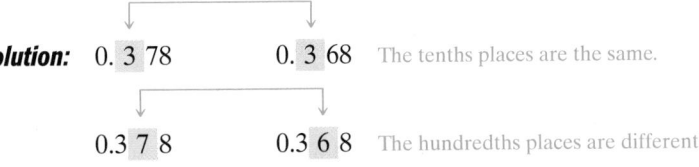

    0.3 7̲ 8   0.3 6̲ 8   The hundredths places are different.

    Since 7 > 6, then 0.378 > 0.368.

EXAMPLE 13 Insert <, >, or = to form a true statement.

    0.052   0.236

Solution: 0.0̲ 52 < 0.2̲ 36   0 is smaller than 2 in the tenths place.
              ↑   ↑

We can also use a number line to compare decimals. This is especially helpful when comparing negative decimals. Remember, the number whose graph is to the left is smaller and the number whose graph is to the right is larger.

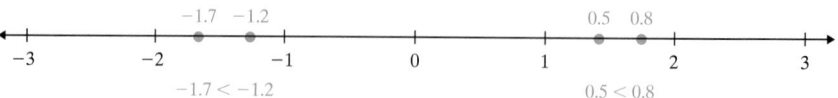

Helpful Hint:

If you have trouble comparing two negative decimals, try the following: Compare their absolute values. Then to correctly compare the negative decimals, reverse the direction of the inequality symbol.

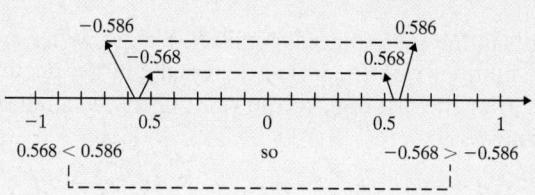

Practice Problem 14

Insert $<$, $>$, or $=$ to form a true statement.

$$-0.029 \quad -0.0209$$

EXAMPLE 14 Insert $<$, $>$, or $=$ to form a true statement.

$$-0.0101 \quad -0.00109$$

Solution: Since $0.0101 > 0.00109$, then $-0.0101 \; < \; -0.00109$

(E) Rounding Decimals

We **round the decimal part** of a decimal number in nearly the same way as we round whole numbers. The only difference is that we drop digits to the right of the rounding place, instead of replacing these digits with 0s. For example,

24.954 rounded to the nearest hundredth is 24.95.

Rounding Decimals to a Place Value to the Right of the Decimal Point

Step 1. Locate the digit to the right of the given place value.

Step 2. If this digit is 5 or greater, add 1 to the digit in the given place value and drop all digits to its right. If this digit is less than 5, drop all digits to the right of the given place.

Practice Problem 15

Round 123.7817 to the nearest thousandth.

EXAMPLE 15 Round 736.2359 to the nearest tenth.

Solution:

Step 1. We locate the digit to the right of the tenths place.

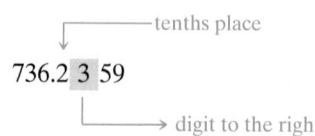

Answers

14. $-0.029 < -0.0209$ **15.** 123.782

Step 2. Since this digit to the right is less than 5, we drop it and all digits to its right.

Thus, 736.2359 rounded to the nearest tenth is 736.2.   ●

The same steps for rounding can be used when the decimal is negative.

EXAMPLE 16   Round −0.027 to the nearest hundredth.

Solution:

Step 1. Locate the digit to the right of the hundredths place.

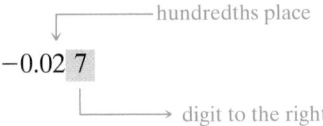

Step 2. Since this digit to the right is 5 or greater, we add 1 to the hundredths digit and drop all digits to its right.

Thus, −0.027 is −0.03 rounded to the nearest hundredth.   ●

The following number line illustrates this rounding.

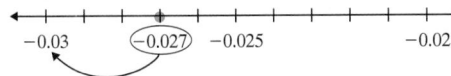

Rounding often occurs with money amounts. Since there are 100 cents in a dollar, each cent is $\frac{1}{100}$ of a dollar. This means that if we want to round to the nearest cent, we round to the nearest hundredth of a dollar.

EXAMPLE 17   Finding Gasoline Prices

The price of a gallon of gasoline in Aimsville is currently $1.3279 per gallon. Round this to the nearest cent.

Solution:

             ┌───── hundredths place
             ↓
     $1.32 **7** 9
           └────→ digit to the right

Since this digit to the right is 5 or greater, we add 1 to the hundredths digit and drop all digits to the right of the hundredths digit. Thus, the rounded gasoline price per gallon is $1.33.   ●

EXAMPLE 18   Determining State Taxable Income

A high school teacher's taxable income is $31,567.72. The tax tables in the teacher's state use amounts to the nearest dollar. Round the teacher's income to the nearest whole dollar.

Solution: Rounding to the nearest whole dollar means rounding to the ones place.

            ┌───── ones place
            ↓
  $31,567. **7** 2
          └────→ digit to the right

Since the digit to the right is 5 or greater, we add 1 to the digit in the ones place and drop all digits to the right of the ones place. The teacher's income rounded to the nearest dollar is $31,568.   ●

Practice Problem 16

Round −0.072 to the nearest hundredth.

Practice Problem 17

In Cititown, the price of a gallon of gasoline is $1.2789 per gallon. Round this to the nearest cent.

Practice Problem 18

Water bills in Gotham City are always rounded to the nearest dime. Lois's water bill was $24.43. Round her bill to the nearest dime (tenth).

Answers

16. −0.07   **17.** $1.28   **18.** $24.40

STUDY SKILLS REMINDER

Are you satisfied with your performance on a particular quiz or exam?

If not, analyze your quiz or exam like you would a good mystery novel. Look for common themes in your errors.

Were most of your errors a result of

- *Carelessness*? If your errors were careless, did you turn in your work before the allotted time expired? If so, resolve next time to use the entire time allotted. Any extra time can be spent checking your work.

- *Running out of time*? If so, make a point to better manage your time on your next exam. A few suggestions are to work any questions that you are unsure of last and to check your work after all questions have been answered.

- *Not understanding a concept*? If so, review that concept and correct your work. Remember next time to make sure that all concepts expected to be on a quiz or exam are understood before the exam.

Mental Math

Determine the place value for the 7 in each decimal or integer.

1. 70   **2.** 700   **3.** 0.7   **4.** 0.07

EXERCISE SET 5.1

A *Write each decimal in words. See Examples 1 through 3.*

1. 6.52   **2.** 7.59   **3.** 16.23   **4.** −47.65

5. −3.205   **6.** 7.495   **7.** 167.009   **8.** 233.056

Fill in each check for the described purchase. See Example 4.

9. Your monthly car loan of $321.42 to R.W. Financial.

10. Your part of the monthly apartment rent, which is $213.70. You pay this to Amanda Dupre.

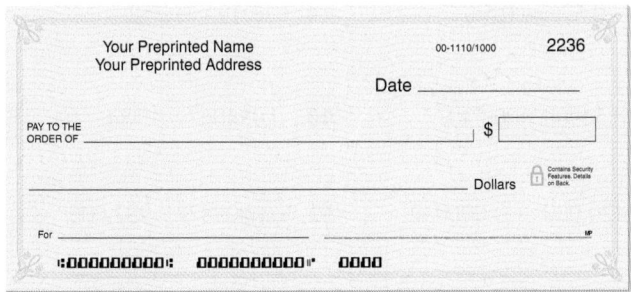

11. Your cell phone bill of $59.68 to Bell South.

12. Your grocery bill of $87.49 at Albertsons.

B *Write each decimal number in standard form. See Examples 5 and 6.*

13. Six and five tenths

14. Three and nine tenths

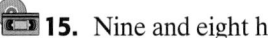**15.** Nine and eight hundredths

16. Twelve and six hundredths

17. Negative five and six hundred twenty-five thousandths

18. Negative four and three hundred ninety-nine thousandths

19. Sixty-four ten-thousandths

20. Thirty-eight ten-thousandths

21. The average annual rainfall for the state of Louisiana is sixty four and sixteen hundredths inches. (*Source:* National Climatic Data Center)

22. The United States Postal Service vehicle fleet averages nine and sixty-two hundredths miles per gallon of fuel. (*Source:* United States Postal Service)

23. In 2001, there was an average of five and four tenths on-the-job injuries for every 100 workers in the mining industry. (*Source:* Bureau of Labor Statistics)

24. The Olympic record for the 200 meter dash is nineteen and thirty two hundredths seconds, and was obtained by Michael Johnson at the 1996 Olympics in Atlanta, GA. (*Source: 2003 World Almanac*)

C *Write each decimal as a fraction or a mixed number. Write your answer in simplest form. See Examples 7 through 11.*

25. 0.3   **26.** 0.7   **27.** 0.27   **28.** 0.39   **29.** −5.47   **30.** −6.3

31. 0.048   **32.** 0.082   **33.** 7.07   **34.** 9.09   **35.** 15.802   **36.** 11.406

37. 0.3005   **38.** 0.2006   **39.** 487.32   **40.** 298.62

D *Insert <, >, or = between each pair of numbers to form a true statement. See Examples 12 through 14.*

41. 0.15   0.16   **42.** 0.12   0.15   **43.** −0.57   −0.54   **44.** −0.59   −0.52

45. 0.098   0.1   **46.** 0.0756   0.2   **47.** 0.54900   0.549   **48.** 0.98400   0.984

49. 167.908   167.980   **50.** 519.3405   519.3054   **51.** 420,000   0.000042   **52.** 0.000987   987,000

53. −1.0621   −1.07   **54.** −18.1   −18.01   **55.** −7.052   7.0052   **56.** 0.01   −0.1

57. −0.023   −0.024   **58.** −0.562   −0.652

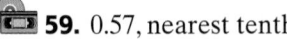 **E** *Round each decimal to the given place value. See Examples 15 through 18.*

59. 0.57, nearest tenth

60. 0.54, nearest tenth

61. 0.234, nearest hundredth

62. 0.452, nearest hundredth

63. 0.5942, nearest thousandth

64. 63.4523, nearest thousandth

65. 98,207.23, nearest ten

66. 68,934.543, nearest ten

67. 12.999, nearest tenth

68. 42.5799, nearest thousandth

69. −17.667, nearest hundredth

70. −0.766, nearest hundredth

71. −0.501, nearest tenth

72. −0.602, nearest tenth

Round each money amount to the nearest cent or dollar as indicated. See Examples 17 and 18.

73. $0.067, nearest cent

74. $0.025, nearest cent

75. $26.95, nearest dollar

76. $14,769.52, nearest dollar

77. $0.1992, nearest cent

78. $0.7633, nearest cent

79. Which number(s) rounds to 0.26?
0.26559 0.26499 0.25786 0.25186

80. Which number(s) rounds to 0.06?
0.066 0.0586 0.0506 0.0612

Round each number to the given place value. See Examples 17 through 18.

81. Brendan Schmelter bought the DVD "Who Framed Roger Rabbit" at Target for $15.99. Round this price to the nearest dollar.

82. The attendance at a Mets baseball game was reported to be 39,867 people. Round this number to the nearest thousand.

83. During the 2002 Boston Marathon, Margaret Okaya of Kenya was the women's winner, finishing with a time of 2.34528 hours. Round this time to the nearest hundredth (*Source: 2003 World Almanac*)

84. The population density of the state of Ohio is roughly 271.0961 people per square mile. Round this population density to the nearest tenth. (*Source:* U.S. Bureau of the Census)

85. The length of a day on Mars is 24.6229 hours. Round this figure to the nearest thousandth. (*Source:* National Space Science Data Center)

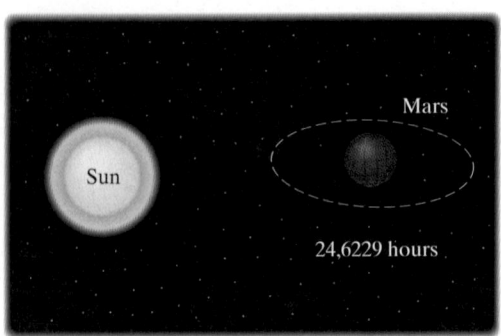

Mars

24,6229 hours

Sun

86. Venus makes a complete orbit around the sun every 224.695 days. Round this figure to the nearest whole day. (*Source:* National Space Science Data Center)

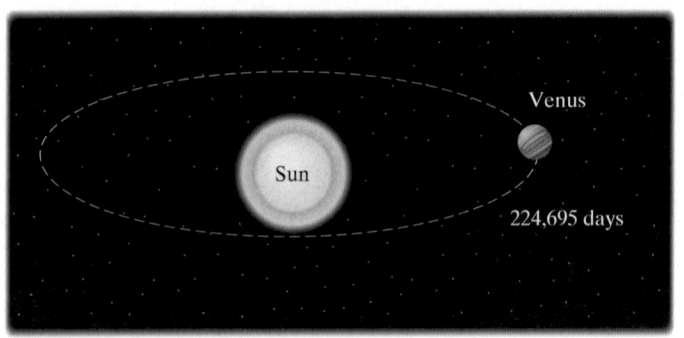

Venus

Sun

224,695 days

87. The official barefoot water-skiing record is 135.74 mph by Scott Pellaton.
Round this figure to the nearest mile per hour. (*Source: The Guinness Book of Records*)

88. Raptor is a roller coaster at Cedar Point, an amusement park in Sandusky, Ohio. It is one of the world's tallest, fastest, and steepest inverted roller coasters. A ride on Raptor lasts about 2.267 minutes. Round this figure to the nearest tenth. (*Source:* Cedar Fair, L.P.)

89. The leading NBA scorer for the 2002 regular season was Allen Iverson of the Philadelphia 76ers. The average number of points he scored per game was 31.3833. Round this figure to the nearest whole point. (*Source*: National Basketball Association)

90. The leading WNBA scorer for the 2002 regular season was Chamique Holdsclaw of the Washington Mystics. The average number of points she scored per game was 19.85. Round this figure to the nearest whole point. (*Source:* Women's National Basketball Association)

Write these numbers from smallest to largest.

91. 0.9
0.1038
0.10299
0.1037

92. 0.01
0.0839
0.09
0.1

Review and Preview

Perform each indicated operation. See Sections 1.3 and 1.4.

93. 3452 + 2314   **94.** 8945 + 4536   **95.** 94 − 23   **96.** 82 − 47   **97.** 482 − 239   **98.** 4002 − 3897

The table gives the leading bowling averages for the Professional Bowlers Association Tour for each of the years listed. Use this table to answer Exercises 99–101.

Leading PBA Averages By Year		
Year	Bowler	Average Score
1995	Mike Aulby	225.490
1996	Walter Ray Williams, Jr.	225.370
1997	Walter Ray Williams, Jr.	222.008
1998	Walter Ray Williams, Jr.	226.130
1999	Parker Bohn	228.040
2000	Chris Barnes	220.930
2001	Parker Bohn	221.546
2002	Walter Ray Williams, Jr.	224.940

(*Source:* Professional Bowlers Association)

99. What is the highest average score on the list? Which bowler achieved that average?

100. What is the lowest average score on the list? Which bowler achieved that average?

101. Make a list of the leading averages in order from greatest to least for the years shown in the table.

102. Write a 4-digit number that rounds to 26.3.

103. Write a 5-digit number that rounds to 1.7.

104. Explain how to identify the value of the 9 in the decimal 486.3297.

105. Write 0.0000203 in words.

How well do you know this textbook?

See if you can answer the questions below.

1. What does the icon  mean?

2. What does the icon ✏ mean?

3. What does the icon △ mean?

4. Where can you find a review for each chapter? Which answers to this review can be found in the back of your text?

5. Each chapter contains an overview of the chapter along with examples. What is this feature called?

6. Does this text contain any solutions to exercises? If so, where?

7. What help is available to you for each chapter test?

5.2   Adding and Subtracting Decimals

Adding or subtracting decimals is similar to adding or subtracting whole numbers. We add or subtract digits in corresponding place values from right to left, carrying or borrowing if necessary. To make sure that digits in corresponding place values are added or subtracted, we line up the decimal points vertically.

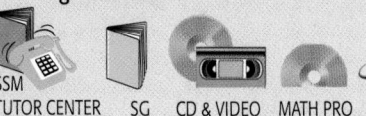

OBJECTIVES

A Add or subtract decimals.

B Evaluate expressions and check solutions with decimal replacement values.

C Simplify expressions containing decimals.

D Solve problems by adding or subtracting decimals.

SSM
TUTOR CENTER   SG   CD & VIDEO   MATH PRO   WEB

Adding or Subtracting Decimals

Step 1. Write the decimals so that the decimal points line up vertically.

Step 2. Add or subtract the same as for whole numbers.

Step 3. Place the decimal point in the sum or difference so that it lines up vertically with the decimal points in the problem.

EXAMPLE 1   Add: $23.85 + 1.604$

Solution: Line up the decimal points vertically and add as the same for whole numbers.

$$
\begin{array}{r}
\overset{1}{23.850} \\
+\ 1.604 \\
\hline
25.454
\end{array}
$$

One 0 is inserted.

Place the decimal point in the sum so that all decimal points line up.

Practice Problem 1

Add.

a. $15.52 + 2.371$

b. $20.06 + 17.612$

c. $0.125 + 122.8$

Helpful Hint

Recall that 0s may be inserted to the right of the decimal point after the last digit without changing the value of the decimal. This may be used to help line up place values when adding decimals.

$$
\begin{array}{r}
3.2 \\
15.567 \\
+\ 0.11 \\
\end{array}
\quad \text{becomes} \quad
\begin{array}{r}
3.\mathbf{200} \\
15.567 \\
+\ 0.11\mathbf{0} \\
\hline
18.877
\end{array}
$$

⟵ Two 0s are inserted.

⟵ One 0 is inserted.

EXAMPLE 2   Add: $763.7651 + 22.001 + 43.89$

Solution:

$$
\begin{array}{r}
763.7651 \\
22.0010 \\
+\ 43.8900 \\
\hline
829.6561
\end{array}
$$

⟵ One 0 is inserted.

⟵ Two 0s are inserted.

Add.

Practice Problem 2

Add.

a. $34.567 + 129.43 + 2.8903$

b. $11.21 + 46.013 + 362.526$

Answers

1. a. 17.891   **b.** 37.672   **c.** 122.925

2. a. 166.8873   **b.** 419.749

Helpful Hint

Don't forget that the decimal point in a whole number is after the last digit.

Practice Problem 3

Add: 27 + 0.00043

EXAMPLE 3 Add: 39 + 0.0021

Solution:

$$
\begin{array}{r}
39.0000 \\
+\ 0.0021 \\
\hline
39.0021
\end{array}
$$

For a whole number, place the decimal point to the right of the ones digit.

Concept Check

Find and correct the error made in the following addition of 3.05, 2.6, and 1.941.

$$
\begin{array}{r}
3.05 \\
2.6 \\
+1.941 \\
\hline
2.272
\end{array}
$$

Try the Concept Check in the margin.

EXAMPLE 4 Add: 3.62 + (−4.78)

Solution: Recall from Chapter 2 that to add two numbers with different signs we find the difference of the larger absolute value and the smaller absolute value. The sign of the answer is the same as the sign of the number with the larger absolute value.

$$
\begin{array}{r}
4.78 \\
-3.62 \\
\hline
1.16
\end{array}
$$

Subtract the absolute values.

Thus, 3.62 + (−4.78) = −1.16

The sign of the number with the larger absolute value −4.78 has the larger absolute value.

Practice Problem 4

Add: 8.1 + (−99.2)

Just as for whole numbers, borrowing may sometimes be needed when subtracting decimals.

Practice Problem 5

Subtract: 5.8 − 3.92

EXAMPLE 5 Subtract: 3.5 − 0.068

Solution:

$$
\begin{array}{r}
3.\overset{4}{\cancel{5}}\,\overset{9}{\cancel{0}}\,\overset{10}{\cancel{0}} \\
-0.068 \\
\hline
3.432
\end{array}
$$

← Insert two 0s.

Recall that we can check a subtraction problem by adding.

$$
\begin{array}{r}
3.432 \\
+0.068 \\
\hline
3.500
\end{array}
$$

Difference
Subtrahend
Minuend

Practice Problem 6

Subtract: 53 − 29.31

Answers

3. 27.00043   **4.** −91.1   **5.** 1.88   **6.** 23.69

Concept Check: Decimal points were not lined up before adding.

$$
\begin{array}{r}
3.050 \\
2.600 \\
+1.941 \\
\hline
7.591
\end{array}
$$

EXAMPLE 6 Subtract: 85 − 17.31

Solution:

$$
\begin{array}{r}
\overset{7}{\cancel{8}}\,\overset{14}{\cancel{8}}.\overset{9}{\cancel{0}}\,\overset{10}{\cancel{0}} \\
-1\ 7\ .\ 3\ 1 \\
\hline
6\ 7\ .\ 6\ 9
\end{array}
$$

Check:

$$
\begin{array}{r}
67.69 \\
+17.31 \\
\hline
85.00
\end{array}
$$

EXAMPLE 7   Subtract 3 from 6.98.

Solution:
$$\begin{array}{r} 6.98 \\ -3.00 \quad \text{Insert two 0s.} \\ \hline 3.98 \end{array}$$

Practice Problem 7

Subtract: 18 from 26.99.

EXAMPLE 8   Subtract: $-5.8 - 1.7$

Solution: Recall from Chapter 2 that to subtract 1.7 we add the opposite of 1.7, or -1.7. Thus

$$-5.8 - 1.7 = -5.8 + (-1.7) \quad \begin{array}{l}\text{To subtract, add the opposite} \\ \text{of 1.7 which is } -1.7.\end{array}$$

Add the absolute values.

$$= -7.5.$$

Use the common negative sign.

Practice Problem 8

Subtract: $-3.4 - 9.6$.

EXAMPLE 9   Subtract: $-2.56 - (-4.01)$

Solution: $-2.56 - (-4.01) = -2.56 + 4.01 \quad \begin{array}{l}\text{To subtract, add the opposite of} \\ -4.01, \text{which is } 4.01.\end{array}$

Subtract the absolute values.

$$= 1.45$$

The answer is (understood) positive since 4.01 has the larger absolute value.

Practice Problem 9

Subtract: $-1.05 - (-7.23)$.

B   **Using Decimals as Replacement Values**

Let's review evaluating expressions with given replacement values. This time the replacement values are decimals.

EXAMPLE 10   Evaluate $x - y$ for $x = 2.8$ and $y = 0.92$.

Solution: Replace x with 2.8 and y with 0.92 and simplify.

$$\begin{array}{rl} x - y = 2.8 - 0.92 & \quad 2.80 \\ = 1.88 & \quad \underline{-0.92} \\ & \quad 1.88 \end{array}$$

Practice Problem 10

Evaluate $y - z$ for $y = 11.6$ and $z = 10.8$.

EXAMPLE 11   Is 2.3 a solution of the equation $6.3 = x + 4$?

Solution: Replace x with 2.3 in the equation $6.3 = x + 4$ to see if the result is a true statement.

$$\begin{array}{rl} 6.3 = x + 4 & \\ 6.3 = 2.3 + 4 & \quad \text{Replace } x \text{ with 2.3.} \\ 6.3 = 6.3 & \quad \text{True.} \end{array}$$

Since $6.3 = 6.3$ is a true statement, 2.3 is a solution of $6.3 = x + 4$.

Practice Problem 11

Is 12.1 a solution of the equation $y - 4.3 = 7.8$?

Answers

7. 8.99   **8.** -13 **9.** 6.18   **10.** 0.8
11. yes

⒞ Simplify Expressions Containing Decimals

EXAMPLE 12 Simplify by combining like terms:

$$11.1x - 6.3 + 8.9x - 4.6$$

Solution:
$$11.1x - 6.3 + 8.9x - 4.6 = 11.1x + 8.9x + (-6.3) + (-4.6)$$
$$= 20x + (-10.9)$$
$$= 20x - 10.9 \qquad \bullet$$

Practice Problem 12

Simplify by combining like terms:
$$-4.3y + 7.8 - 20.1y + 14.6$$

⒟ Solving Problems by Adding or Subtracting Decimals

Decimals are very common in real-life problems.

EXAMPLE 13   Calculating the Cost of Owning an Automobile

Find the total monthly cost of owning and operating a certain automobile given the expenses shown.

Monthly car payment: $256.63
Monthly insurance cost: $ 47.52
Average gasoline bill per month: $ 95.33

Solution:

1. **UNDERSTAND.** Read and reread the problem. The phrase "total monthly cost" tells us to add.

   Note: If you would like to assign a variable to the unknown, let x = total monthly cost. Since the translated equation will already be solved for the unknown, variables will not be used here.

2. **TRANSLATE.**

   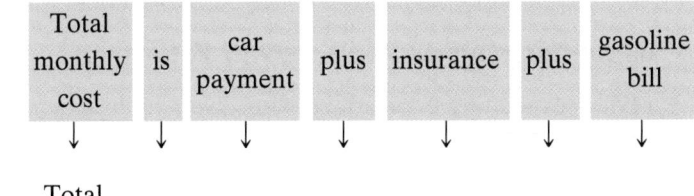

   Translate: Total
                 monthly = $256.63   + $47.52   + $95.33
                 cost

3. **SOLVE.**

   $$\begin{array}{r} {}^{1\,1\,1} \\ 256.63 \\ 47.52 \\ +\ 95.33 \\ \hline \$399.48 \end{array}$$

4. **INTERPRET.** *Check* your work. *State* your conclusion: The total monthly cost is $399.48.   ●

Practice Problem 13

Find the total monthly cost of owning and operating a certain automobile given the expenses shown.

Monthly car payment: $536.50
Monthly insurance cost: $52.70
Average gasoline bill
per month: $87.50

Answers

12. $-24.4y + 22.4$ **13.** $676.70

EXAMPLE 14 Comparing Average Heights

The bar graph shows the current average heights for adults in various countries. How much taller is the average height in Denmark than the average height in the United States?

Average Adult Height

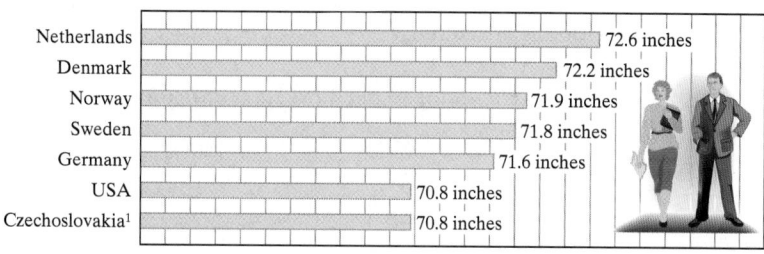

Netherlands	72.6 inches
Denmark	72.2 inches
Norway	71.9 inches
Sweden	71.8 inches
Germany	71.6 inches
USA	70.8 inches
Czechoslovakia[1]	70.8 inches

[1]Average for Czech Republic, Slovakia
Source: *USA Today*, 8/28/97

Solution:

1. UNDERSTAND. Read and reread the problem. Since we want to know "how much taller," we subtract.

2. TRANSLATE.

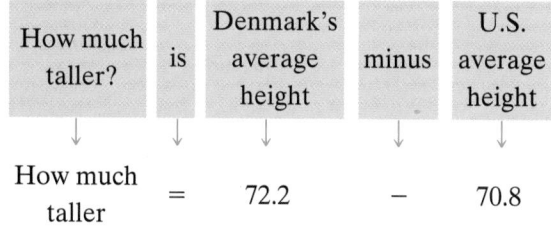

In words:

How much taller?	is	Denmark's average height	minus	U.S. average height

Translate:

$$\text{How much taller} = 72.2 - 70.8$$

3. SOLVE.

$$
\begin{array}{r}
7\overset{1}{2}.\overset{12}{2} \\
-\ 7\ 0\ .\ 8 \\
\hline
1\ .\ 4
\end{array}
$$

4. INTERPRET. *Check* your work. *State* your conclusion: The average height in Denmark is 1.4 inches more than the average height in the U.S. ●

Practice Problem 14

Use the bar graph for Example 14. How much taller is the average height in the Netherlands than the average height in Czechoslovakia?

Answer

14. 1.8 inches

CALCULATOR EXPLORATIONS

ENTERING DECIMAL NUMBERS

To enter a decimal number, find the key marked ⬚·⬚. To enter the number 2.56, for example, press the keys

| 2 | · | 5 | 6 |

The display will read ⬚ 2.56⬚.

OPERATIONS ON DECIMAL NUMBERS

Operations on decimal numbers are performed in the same way as operations on whole or signed numbers. For example, to find $8.625 - 4.29$, press the keys

| 8.625 | − | 4.29 | = | or | ENTER |

The display will read ⬚ 4.335⬚.

(Although entering 8.625, for example, requires pressing more than one key, we group numbers together for easier reading.)

Use a calculator to perform each indicated operation.

1. $315.782 + 12.96$

2. $29.68 + 85.902$

3. $6.249 - 1.0076$

4. $5.238 - 0.682$

5.
$$
\begin{array}{r}
12.555 \\
224.987 \\
5.2 \\
+622.65 \\
\hline
\end{array}
$$

6.
$$
\begin{array}{r}
47.006 \\
0.17 \\
313.259 \\
+139.088 \\
\hline
\end{array}
$$

Name _____ Section _____ Date _____

Mental Math

State each sum or difference.

1. $\begin{array}{r} 0.3 \\ +0.2 \\ \hline \end{array}$

2. $\begin{array}{r} 0.4 \\ +0.5 \\ \hline \end{array}$

3. $\begin{array}{r} 1.00 \\ +0.26 \\ \hline \end{array}$

4. $\begin{array}{r} 3.00 \\ +0.19 \\ \hline \end{array}$

5. $\begin{array}{r} 7.6 \\ +1.3 \\ \hline \end{array}$

6. $\begin{array}{r} 4.5 \\ +3.2 \\ \hline \end{array}$

7. $\begin{array}{r} 0.9 \\ -\ 0.3 \\ \hline \end{array}$

8. $\begin{array}{r} 0.6 \\ -\ 0.2 \\ \hline \end{array}$

EXERCISE SET 5.2

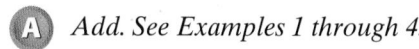
Ⓐ *Add. See Examples 1 through 4.*

1. $1.3 + 2.2$

2. $2.5 + 4.1$

3. $5.7 + 1.13$

4. $2.31 + 6.4$

5. $24.6 + 2.39 + 0.0678$

6. $32.4 + 1.58 + 0.0934$

7. $\begin{array}{r} 45.023 \\ 3.006 \\ +\ 8.403 \\ \hline \end{array}$

8. $\begin{array}{r} 65.0028 \\ 5.0903 \\ +\ 6.9003 \\ \hline \end{array}$

9. $-2.6 + (-5.97)$

10. $-18.2 + (-10.8)$

11. $15.78 + (-4.62)$

12. $6.91 + (-7.03)$

Subtract. See Examples 5 through 9.

13. $8.8 - 2.3$

14. $7.6 - 2.1$

15. $18 - 2.7$

16. $28 - 3.3$

17. $\begin{array}{r} 654.9 \\ -56.67 \\ \hline \end{array}$

18. $\begin{array}{r} 863.23 \\ -39.453 \\ \hline \end{array}$

19. Subtract 6.7 from 23

20. Subtract 9.2 from 45

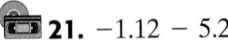**21.** $-1.12 - 5.2$ **22.** $-8.63 - 5.6$ **23.** $7.7 - 14.1$ **24.** $10.25 - 21.76$

25. $-2.6 - (-5.7)$ **26.** $-9.4 - (-10.4)$

Perform each indicated operation. See Examples 1 through 9.

27. $0.9 + 2.2$ **28.** $0.7 + 3.4$ **29.** $-5.9 - 4$ **30.** $-6.4 - 3.4$

31. $45.67 - 20$ **32.** $56.89 - 30$ **33.** $-6.06 + 0.44$ **34.** $-5.05 + 0.88$

35. $900.34 - 123.45$ **36.** $800.74 - 463.98$ **37.** $3490.23 + 8493.09$ **38.** $600.004 + 7983.0062$

39. $\begin{array}{r} 234.89 \\ +230.67 \\ \hline \end{array}$ **40.** $\begin{array}{r} 734.89 \\ +640.56 \\ \hline \end{array}$ **41.** $50.2 - 600$ **42.** $40.3 - 700$

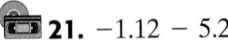**43.** Subtract 61.9 from 923.5   **44.** Subtract 45.8 from 845.93   **45.** $\begin{array}{r} 100.009 \\ 6.08 \\ + \ 9.034 \\ \hline \end{array}$

46. $\begin{array}{r} 200.89 \\ 7.49 \\ + \ 62.83 \\ \hline \end{array}$ **47.** $\begin{array}{r} 1000 \\ - \ 123.4 \\ \hline \end{array}$ **48.** $\begin{array}{r} 2000 \\ - \ 327.47 \\ \hline \end{array}$

49. $-0.003 + 0.091$ **50.** $-0.004 + 0.085$ **51.** $500 - 34.098$

52. $300 - 98.345$

53. $-102.4 - 78.04$

54. $-36.2 - 10.02$

55. $-2.9 - (-1.8)$

56. $-6.5 - (-3.3)$

B *Evaluate each expression for* $x = 3.6$, $y = 5$, *and* $z = 0.21$. *See Example 10.*

57. $x + z$

58. $y + x$

59. $x - z$

60. $y - z$

61. $y - x + z$

62. $x + y + z$

Determine whether the given values are solutions to the given equations. See Example 11.

63. Is 7 a solution to $x + 2.7 = 9.3$?

64. Is 3.7 a solution to $x + 5.9 = 8.6$?

65. Is -11.4 a solution to $27.4 + y = 16$?

66. Is -22.9 a solution to $45.9 + z = 23$?

67. Is 1 a solution to $2.3 + x = 5.3 - x$?

68. Is 0.9 a solution to $1.9 - x = x + 0.1$?

C *Simplify by combining like terms. See Example 12.*

69. $30.7x + 17.6 - 23.8x - 10.7$

70. $14.2z + 11.9 - 9.6z - 15.2$

71. $-8.61 + 4.23y - 2.36 - 0.76y$

72. $-8.96x - 2.31 - 4.08x + 9.68$

D *Solve. See Examples 13 and 14.*

73. Find the total monthly cost of owning and maintaining a car given the information shown.

Monthly car payment: $275.36
Monthly insurance cost: $83.00
Average cost of
   gasoline per month: $81.60
Average maintenance
   cost per month: $14.75

74. Find the total monthly cost of owning and maintaining a car given the information shown.

Monthly car payment: $306.42
Monthly insurance cost: $53.50
Average cost of
   gasoline per month: $123.00
Average maintenance
   cost per month: $23.50

75. Gasoline was $1.499 per gallon on one day and $1.559 per gallon the next day. By how much did the price change?

76. A pair of eyeglasses costs a total of $347.89. The frames of the glasses are $97.23. How much do the lenses of the eyeglasses cost?

77. Ann-Margaret Tober bought a book for $32.48. If she paid with two $20 bills, what was her change?

78. Tom Mackey bought a car part for $18.26. If he paid with two $10 bills, what was his change?

79. Americans' consumption of sugar is on the rise. During 1990, Americans consumed an average of 136.8 pounds of sugar in its various forms such as refined white sugar, honey, and corn sweeteners. By 2000, the average American was consuming 150.1 pounds of sugar products per year. How much more sugar was the average American consuming annually in 2000 than in 1990? (*Source:* Economic Research Service, U.S. Department of Agriculture)

80. In 2002, the average wage for U.S. production workers was $14.77 per hour. In 1997, this average wage was $12.28 per hour. How much of an increase is this? (*Source:* Bureau of Labor Statistics)

81. The average wind speed at the weather station on Mt. Washington in New Hampshire is 35.2 miles per hour. The highest speed ever recorded at the station is 321.0 miles per hour. How much faster is the highest speed than the average wind speed? (*Source:* National Climatic Data Center)

82. The average annual rainfall in Omaha, Nebraska, is 30.22 inches. The average annual rainfall in New Orleans, Louisiana, is 61.88 inches. On average, how much more rain does New Orleans receive annually than Omaha? (*Source:* National Climatic Data Center)

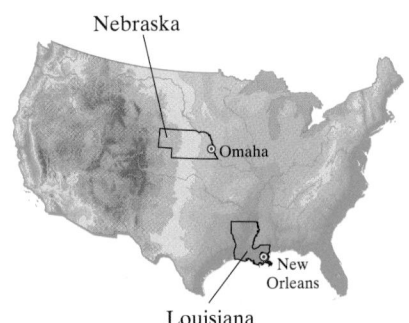

83. In October 1997, Andy Green set a new one-mile land speed record. This record was 129.567 miles per hour faster than a previous record of 633.468 set in 1983. What was Green's record-setting speed? (*Source:* United States Auto Club)

84. It costs $3.13 to send a 2-pound package locally via parcel post at a U.S. Post Office. To send the same package as Priority Mail, it costs $3.95. How much more does it cost to send a package as Priority Mail? (*Source*: USPS)

85. The three North America concert tours that have earned the most money are the Rolling Stones (1994) $121.2 million, Pink Floyd (1994). $103.5 million, and U2 (2001) $109.7 million What was the total amount of money earned from these three concerts? (*Source:* Pollstar, Fresno, CA)

86. In 1995, the average credit-card late fee was $12.53. In 2002, the average late fee had increased to $27.82. By how much did the average credit-card late fee increase from 1995 to 2002? (*Source:* Consumer Action)

87. The snowiest city in the United States is Blue Canyon, California, which receives an average of 111.6 more inches of snow than the second snowiest city. The second snowiest city in the United States is Marquette, Michigan. Marquette receives an average of 129.2 inches of snow annually. How much snow does Blue Canyon receive on average each year? (*Source:* National Climatic Data Center)

88. The driest city in the world is Aswan, Egypt, which receives an average of only 0.02 inches of rain per year. Yuma, Arizona, is the driest city in the United States. Yuma receives an average of 2.63 more inches of rain each year than Aswan. What is the average annual rainfall in Yuma? (*Source:* National Climatic Data Center)

89. A landscape architect is planning a border for a flower garden shaped like a triangle. The sides of the garden measure 12.4 feet, 29.34 feet, and 25.7 feet. Find the amount of border material needed.

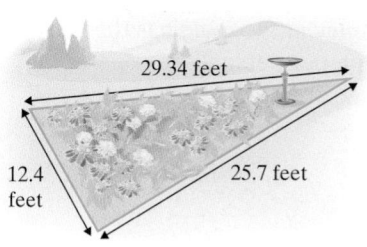

29.34 feet

12.4 feet

25.7 feet

90. A contractor purchased enough buy railing to completely enclose the newly built deck shown below. Find the amount of railing purchased.

10.6 feet

15.7 feet

The table shows the average speeds for the Daytona 500 winners for the years shown. Use this table to answer Exercises 91–92.

	Daytona 500 Winners	
Year	**Winner**	**Average Speed**
1959	Lee Petty	135.521
1969	Lee Roy Yarborough	160.875
1979	Richard Petty	143.977
1989	Darrell Waltrip	148.466
1999	Jeff Gordon	161.551

91. How much slower was the average Daytona 500 winning speed in 1989 than in 1969?

92. How much faster is the average Daytona 500 winning speed in 1999 than in 1969?

The Mercury space program had 6 manned flights. They are shown in the following table. Use this table to answer Exercises 93–94.

Year	**Mission**	**Duration (in minutes)**
1961	Redstone 3	15.467
1961	Redstone 4	15.167
1962	Atlas 6	295.383
1962	Atlas 7	296.083
1962	Atlas 8	553.183
1963	Atlas 9	2059.817

93. How many more minutes longer was the last Mercury mission than the first?

94. What is the difference in duration of the first two Mercury missions flown in 1961?

The bar graph shows the five top chocolate-consuming nations in the world. Use this table to answer Exercises 95–99.

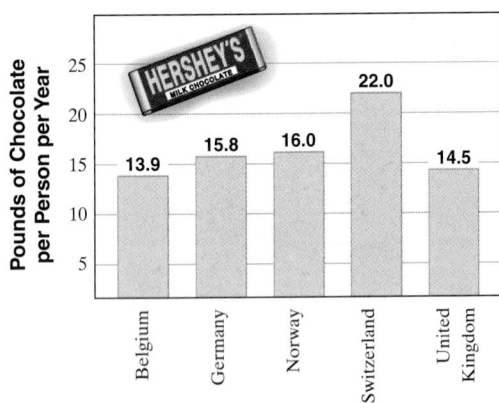

The World's Top Chocolate-Consuming Countries

Source: Hershey Foods Corporation

95. Which country in the table has the greatest chocolate consumption per person?

96. Which country in the table has the least chocolate consumption per person?

97. How much more is the greatest chocolate consumption than the least chocolate consumption shown in the table?

98. How much more chocolate does the average German consume than the average citizen of the United Kingdom?

99. Make a new chart listing the countries and their corresponding chocolate consumption in order from greatest to least.

Review and Preview

Multiply. See Section 4.3.

100. $\left(\dfrac{2}{3}\right)^2$

101. $\left(\dfrac{1}{5}\right)^3$

102. $-\dfrac{12}{7} \cdot \dfrac{14}{3}$

103. $\dfrac{25}{36} \cdot \dfrac{24}{40}$

Let's review the values of these popular U.S. coins in order to answer the following exercises.

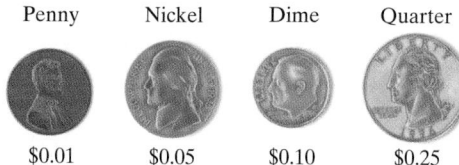

Penny	Nickel	Dime	Quarter
$0.01	$0.05	$0.10	$0.25

Write the value of each group of coins. To do so, it is usually easiest to start with the coin(s) of greatest value and end with the coin(s) of least value.

104.

105.

106. Name the different ways that coins can have a value of $0.17 given that you may use no more than 10 coins.

107. Name the different ways that coin(s) can have a value of $0.25 given that there are no pennies.

108. Laser beams can be used to measure the distance to the moon. One measurement showed the distance to the moon to be 256,435.235 miles. A later measurement showed that the distance is 256,436.012 miles. Find how much farther away the moon is in the second measurement compared to the first.

109. Explain how adding or subtracting decimals is similar to adding or subtracting whole numbers.

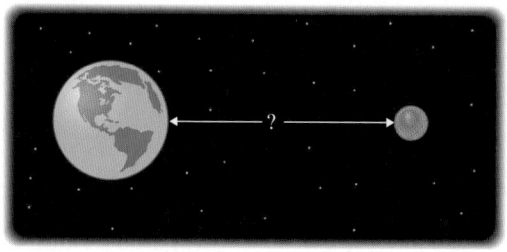

Combine like terms and simplify.

 110. $-8.689 + 4.286x - 14.295 - 12.966x + 30.861x$ **111.** $14.271 - 8.968x + 1.333 - 201.815x + 101.239x$

112. Can the sum of two negative decimals ever be a positive decimal? Why or why not?

Find the unknown length in each figure.

113.

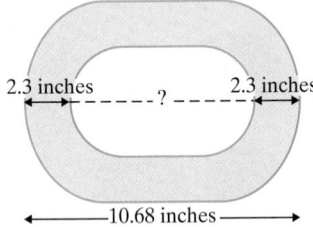

2.3 inches   ?   2.3 inches

—10.68 inches—

114.

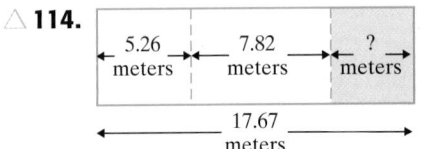

5.26 meters   7.82 meters   ? meters

17.67 meters

5.3   Multiplying Decimals and Circumference of a Circle

OBJECTIVES

Ⓐ Multiply decimals.

Ⓑ Multiply decimals by powers of 10.

Ⓒ Evaluate expressions and check solutions with decimal replacement values.

Ⓓ Find the circumference of a circle.

Ⓔ Solve problems by multiplying decimals.

SSM   TUTOR CENTER   SG   CD & VIDEO   MATH PRO   WEB

Ⓐ Multiplying Decimals

Multiplying decimals is similar to multiplying whole numbers. The only difference is that we place a decimal point in the product. To discover where a decimal point is placed in the product, let's multiply 0.6×0.03. We first write each decimal as an equivalent fraction and then multiply.

$$0.6 \times 0.03 = \frac{6}{10} \times \frac{3}{100} = \frac{18}{1000} = 0.018$$

1 decimal place   2 decimal places   3 decimal places

Now let's multiply 0.03×0.002.

$$0.03 \times 0.002 = \frac{3}{100} \times \frac{2}{1000} = \frac{6}{100,000} = 0.00006$$

2 decimal places   3 decimal places   5 decimal places

Instead of writing decimals as fractions each time we want to multiply, we notice a pattern from these examples and state a rule that we can use.

Multiplying Decimals

Step 1. Multiply the decimals as though they were whole numbers.

Step 2. The decimal point in the product is placed so the number of decimal places in the product is equal to the *sum* of the number of decimal places in the factors.

EXAMPLE 1   Multiply: 23.6×0.78

Solution:

$$
\begin{array}{r}
23.6 \quad \text{1 decimal place} \\
\times\, 0.78 \quad \text{2 decimal places} \\
\hline
1888 \\
16520 \\
\hline
18.408 \quad \text{3 decimal places}
\end{array}
$$

EXAMPLE 2   Multiply: 0.283×0.3

Solution:

$$
\begin{array}{r}
0.283 \quad \text{3 decimal places} \\
\times\quad 0.3 \quad \text{1 decimal place} \\
\hline
0.0849 \quad \text{4 decimal places}
\end{array}
$$

Insert one 0 since the product must have 4 decimal places.

EXAMPLE 3   Multiply: 0.0531×16

Solution:

$$
\begin{array}{r}
0.0531 \quad \text{4 decimal places} \\
\times\quad 16 \quad \text{0 decimal places} \\
\hline
3186 \\
05310 \\
\hline
0.8496 \quad \text{4 decimal places}
\end{array}
$$

Practice Problem 1

Multiply: 45.9×0.42

Practice Problem 2

Multiply: 0.112×0.6

Practice Problem 3

Multiply: 0.0721×48

Answers

1. 19.278   **2.** 0.0672   **3.** 3.4608

Concept Check

True or false? The number of decimal places in the product of 0.261 and 0.78 is 6. Explain.

Practice Problem 4

Multiply: $(5.4)(-1.3)$

Try the Concept Check in the margin.

EXAMPLE 4 Multiply: $(-2.6)(0.8)$

Solution: Recall that the product of a negative number and a positive number is a negative number.

$$(-2.6)(0.8) = -2.08$$ ●

B Multiplying Decimals by Powers of 10

There are some patterns that occur when we multiply a number by a power of 10, such as 10, 100, 1000, 10,000, and so on.

$$23.6951 \times 10 = 236.951 \qquad \text{Move the decimal point 1 } place \text{ to the } right.$$

1 zero

$$23.6951 \times 100 = 2369.51 \qquad \text{Move the decimal point 2 } places \text{ to the } right.$$

2 zeros

$$23.6951 \times 100,000 = 2,369,510. \qquad \text{Move the decimal point 5 } places \text{ to the } right \text{ (insert a 0).}$$

5 zeros

Notice that we move the decimal point the same number of places as there are zeros in the power of 10.

> **Multiplying Decimals by Powers of 10 such as 10, 100, 1000, 10,000, ...**
>
> Move the decimal point to the *right* the same number of places as there are *zeros* in the power of 10.

Practice Problems 5–7

Multiply.

5. 23.7×10
6. 203.004×100
7. 1.15×1000

EXAMPLES Multiply.

5. $7.68 \times 10 = 76.8$ $\qquad 7.68$

6. $23.702 \times 100 = 2370.2$ $\qquad 23.702$

7. $(-76.3)(1000) = -76,300$ $\qquad -76.300$ $\quad$ Recall that the product of a negative number and a positive number is a negative number. ●

There are also powers of 10 that are less than 1. The decimals 0.1, 0.01, 0.001, 0.0001, and so on, are examples of powers of 10 less than 1. Notice the pattern when we multiply by these powers of 10.

$$569.2 \times 0.1 = 56.92 \qquad \text{Move the decimal point 1 } place \text{ to the } left.$$

1 decimal place

$$569.2 \times 0.01 = 5.692 \qquad \text{Move the decimal point 2 } places \text{ to the } left.$$

2 decimal places

$$569.2 \times 0.0001 = 0.05692 \qquad \text{Move the decimal point 4 } places \text{ to the } left \text{ (insert one 0).}$$

4 decimal places

Answers

4. -7.02 **5.** 237 **6.** 20,300.4 **7.** 1150

Concept Check: False; 3 decimal places + 2 decimal places is 5 decimal places in the product.

Multiplying Decimals by Powers of 10 such as 0.1, 0.01, 0.001, 0.0001, . . .

Move the decimal point to the *left* the same number of places as there are *decimal places* in the power of 10.

EXAMPLES   Multiply.

8. $42.1 \times 0.1 = 4.21$ $\underset{\curvearrowleft}{42.1}$

9. $76{,}805 \times 0.01 = 768.05$ $76{,}805.$

10. $(-9.2)(-0.001) = 0.0092$ 0009.2 Recall that the product of a negative number and a negative number is a positive result.

Try the Concept Check in the margin.

Many times we see large numbers written, for example, in the form 281.4 million rather than in the longer standard notation. The next example shows how to interpret these numbers.

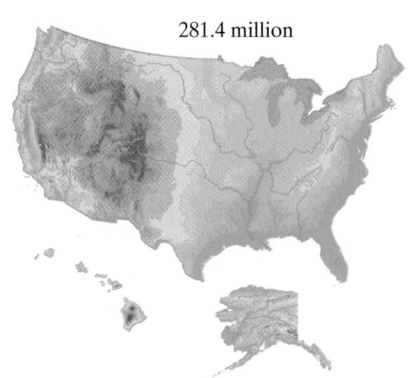

281.4 million

EXAMPLE 11   Population of the United States

The population of the United States is approximately 281.4 million. Write this number in standard notation. (*Source: The World Almanac*, 2003)

Solution: 281.4 million = 281.4 × 1 million

= 281.4 × 1,000,000 = 281,400,000

c   Using Decimals as Replacement Values

Now let's practice working with variables.

EXAMPLE 12   Evaluate xy for $x = 2.3$ and $y = 0.44$.

Solution: Recall that xy means $x \cdot y$.

$xy = (2.3)(0.44)$

$$\begin{array}{r} 2.3 \\ \times\ 0.44 \\ \hline 92 \\ 920 \\ \hline 1.012 \end{array}$$

= 1.012 ⟵

Practice Problems 8–10

Multiply.

8. 7.62×0.1

9. 1.9×0.01

10. 7682×0.001

Concept Check

True or false? 372.511 multiplied by 100 is 3.72511.

Practice Problem 11

There are 2158 thousand farms in the United States. Write this number in standard notation. (*Source: The World Almanac*, 2003)

Practice Problem 12

Evaluate $7y$ for $y = -0.028$.

Answers

8. 0.762   **9.** 0.019   **10.** 7.682   **11.** 2,158,000
12. −0.196

Concept Check: False.

Practice Problem 13

Is -5.5 a solution of the equation $-4x = 22$?

EXAMPLE 13 Is -9 a solution of the equation $3.7y = -3.33$?

Solution: Replace y with -9 in the equation $3.7y = -3.33$ to see if a true equation results.

$$3.7y = -3.33$$
$$3.7(-9) = -3.33 \quad \text{Replace } y \text{ with } -9.$$
$$-33.3 = -3.33 \quad \text{False.}$$

Since $-33.3 = -3.33$ is a false statement, -9 is **not** a solution of $3.7y = -3.33$. ⬤

D Finding the Circumference of a Circle

Recall that the distance around a polygon is called its perimeter. The distance around a circle is given a special name called the **circumference**, and this distance depends on the radius or the diameter of the circle.

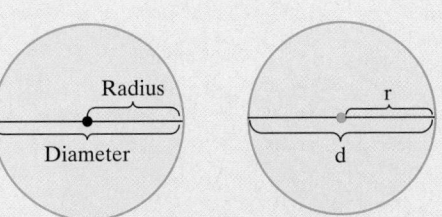

Circumference of a Circle

Circumference $= 2 \cdot \pi \cdot$ **r**adius or Circumference $= \pi \cdot$ **d**iameter

$$C = 2\pi r \qquad \text{or} \qquad C = \pi d$$

The symbol π is the Greek letter pi, pronounced "pie." It is a constant between 3 and 4. A decimal approximation for π is 3.14. Also, a fraction approximation for π is $\frac{22}{7}$.

Practice Problem 14

Find the circumference of a circle whose radius is 11 meters. Then use the approximation 3.14 for π to approximate this circumference.

EXAMPLE 14 Circumference of a Circle

Find the circumference of a circle whose radius is 5 inches. Then use the approximation 3.14 for π to approximate the circumference.

Solution: Let $r = 5$ in the formula $C = 2\pi r$.

$$C = 2\pi r$$
$$= 2\pi \cdot 5$$
$$= 10\pi$$

5 inches

Next, replace π with the approximation 3.14.

$$C = 10\pi$$
$$\text{(is approximately)} \longrightarrow \approx 10(3.14)$$
$$= 31.4$$

The **exact** circumference or distance around the circle is 10π inches, which is **approximately** 31.4 inches. ⬤

Answers

13. yes   **14.** 22π meters ≈ 69.08 meters

Solving Problems by Multiplying Decimals

The solutions to many real-life problems are found by multiplying decimals. We continue using our four problem-solving steps to solve such problems.

EXAMPLE 15   Finding Total Cost of Materials for a Job

A college student is hired to paint a billboard with paint costing $2.49 per quart. If the job requires 3 quarts of paint, what is the total cost of the paint?

Solution:

1. UNDERSTAND. Read and reread the problem. The phrase "total cost" might make us think addition, but since this is repeated addition, let's multiply.

2. TRANSLATE.

In words:	Total cost	is	cost per quart of paint	times	number of quarts
	↓	↓	↓	↓	↓
Translate:	Total cost	=	2.49	×	3

3. SOLVE.

$$\begin{array}{r} 2.49 \\ \times\ \ \ 3 \\ \hline 7.47 \end{array}$$

4. INTERPRET. *Check* your work. *State* your conclusion: The total cost of the paint is $7.47.

Practice Problem 15

Elaine Rehmann is fertilizing her garden. She uses 5.6 ounces of fertilizer per square yard. The garden measures 60.5 square yards. How much fertilizer does she need?

Answer

15. 338.8 ounces

FOCUS ON **Mathematical Connections**

MODELING MULTIPLICATION WITH DECIMAL NUMBERS

We can use an area model to represent multiplying decimal numbers. For instance, we can think of the 10 × 10 grid shown below on the left as representing 1, or a whole. We can show the product 0.5 × 0.2 by shading five tenths of the grid along one side of the square and shading two tenths of the grid along the other side.

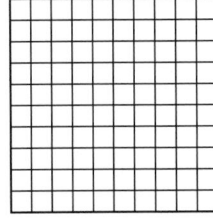

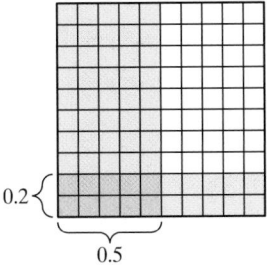

CRITICAL THINKING

1. What does the region where the shading overlaps represent? See whether you can figure it out by comparing it to the product of 0.5 × 0.2 found numerically. (Remember that one square within the grid represents one hundredth.)

2. How could you use this type of area model to represent the product of a decimal number (such as 0.6) and a whole number (such as 2)?

3. Use the model to show each of the following products:

   a. 0.4 × 0.4
   b. 0.9 × 0.1
   c. 0.8 × 2
   d. 0.7 × 1.5

FOCUS ON **The Real World**

McDonald's Menu

Today's worldwide chain of McDonald's fast-food restaurants started as a single drive-in restaurant run by the McDonald brothers, Dick and Mac, in San Bernardino, California. In the late 1940s, they decided to revamp their business, eliminating car hops in favor of counter-only service and reducing their 25-item menu to just 9 items. They focused on quick service, large volume, and low prices. The newly revamped business was hugely popular and attracted the attention of a milkshake machine salesman named Ray Kroc. Seeing the franchising potential of the McDonald brothers' fast-food formula, Ray Kroc became the exclusive McDonald's franchising agent for the entire United States in 1955. By 1956, there were 14 McDonald's restaurants, and that number had grown to 228 by 1960. Today, there are over 23,000 McDonald's restaurants worldwide—with a presence on every continent except Antarctica.

Ray Kroc's franchise McDonald's menu from 1955 is shown at the right. Local franchise owners have frequently been responsible for additions to the basic menu. The Filet-O-Fish sandwich was the first item to be added to the original menu in 1963 after it was developed by a franchise owner in Cincinnati, Ohio. The Big Mac sandwich, the invention of a Pittsburgh franchise owner, followed in 1968. And the Egg McMuffin was introduced nationally in 1973 after a Santa Barbara, California, franchisee searched for a breakfast product to offer at his restaurant. Other items that are standards on today's menu followed: the Quarter Pounder in 1972, Chicken McNuggets in 1983, and fresh tossed salads in 1987.

International McDonald's restaurants include local specialities on the menu. McDonald's restaurants in Norway include a grilled salmon sandwich called McLaks. McDonald's customers will find McSpaghetti on the menu in the Philippines, a Kiwi burger in New Zealand, and the Samurai Pork burger in Thailand.

Group Activity

Visit a McDonald's restaurant near you and record the present-day prices for each of the seven items on the original 1955 menu. (Unless otherwise noted, assume that sizes on the original menu are "small" for the sake of comparison with today's menu.) For each menu item, find the number of times greater that today's price is than the original price.

Original 1955 McDonald's Menu	
Hamburger	$0.15
Cheeseburger	$0.19
Fries	$0.10
Small soft drink	$0.10
Large soft drink	$0.15
Coffee	$0.10
Shake	$0.20

EXERCISE SET 5.3

A *Multiply. See Examples 1 through 4.*

1. $\begin{array}{r} 0.2 \\ \times\ 0.6 \\ \hline \end{array}$

2. $\begin{array}{r} 0.7 \\ \times\ 0.9 \\ \hline \end{array}$

3. $\begin{array}{r} 1.2 \\ \times\ 0.5 \\ \hline \end{array}$

4. $\begin{array}{r} 6.8 \\ \times\ 0.3 \\ \hline \end{array}$

5. $(-2.3)(7.65)$

6. $(4.7)(-9.02)$

7. $(-6.89)(-5.7)$

8. $(-6.45)(-2.8)$

B *Multiply. See Examples 5 through 10.*

9. 6.5×10

10. 7.2×10

11. 6.5×0.1

12. 7.2×0.1

13. $(-7.093)(1000)$

14. $(-1.123)(1000)$

15. $(-9.83)(-0.01)$

16. $(-4.72)(-0.01)$

A **B** *Multiply. See Examples 1 through 10.*

17. $\begin{array}{r} 5.62 \\ \times\ 7.7 \\ \hline \end{array}$

18. $\begin{array}{r} 8.03 \\ \times\ 5.5 \\ \hline \end{array}$

19. $\begin{array}{r} 1.0047 \\ \times\ 8.2 \\ \hline \end{array}$

20. $\begin{array}{r} 2.0005 \\ \times\ 5.5 \\ \hline \end{array}$

21. $(147.9)(100)$

22. $(345.2)(10)$

23. $(937.62)(-0.01)$

24. $(-0.001)(562.01)$

25. 49.02×0.023

26. 30.09×0.0032

27. $(-0.023)(6.28)$

28. $(0.071)(-5.19)$

Write each number in standard notation. See Example 11.

29. The storage silos at the main Hershey chocolate factory in Hershey, Pennsylvania, can hold enough cocoa beans to make 5.5 billion Hershey's milk chocolate bars. (*Source:* Hershey Foods Corporation)

30. Of the 105.5 million American households that have televisions, 78.07 million have more than one television set. Write both numbers in standard notation. (*Source*: Nielsen Media Research)

31. About 36.4 million American households own at least one dog. (*Source:* American Pet Products Manufacturers Association)

32. The world's most popular amusement park in 2001 was Tokyo Disneyland, which had an attendance of 17.708 million visitors. (*Source: Amusement Business*, August 2002)

33. Americans lose more than 1.6 million hours each day stuck in traffic. (*Source:* United States Department of Transportation—Federal Highway Administration)

34. In 2001, the top advertiser in the United States was General Motors, who spent $3.37 billion on advertising. (*Source:* Crain Communications, Inc.)

C *Evaluate each expression for $x = 3$, $y = -0.2$, and $z = 5.7$. See Example 12.*

35. xy

36. yz

37. $xz - y$

38. $-5y + z$

39. Evaluate $2LW + 2WH + 2HL$ for $L = 10$, $W = 2.1$ and $H = 0.6$. (Surface area of a rectangular solid)

40. Evaluate $6s^2$ for $s = 2.7$. (Surface area of a cube)

Determine whether the given value is a solution of each given equation. See Example 13.

41. Is 14.2 a solution of $0.6x = 4.92$?

42. Is 1414 a solution of $100z = 14.14$?

43. Is 0.08 a solution of $-3x = -2.4$?

44. Is 17.8 a solution of $2x = 3.56$?

45. Is -4 a solution of $3.5y = -14$?

46. Is -3.6 a solution of $0.7x = -2.52$?

D *Find the circumference of each circle. Then use the approximation 3.14 for π and approximate each circumference. See Example 14.*

△ **47.**

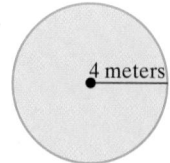
4 meters

△ **48.**

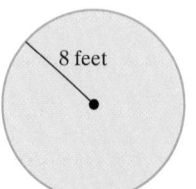
8 feet

49.

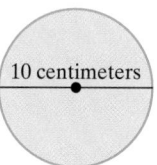
10 centimeters

50.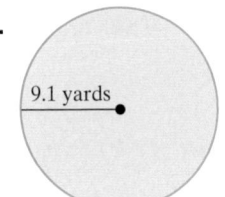

22 inches

51.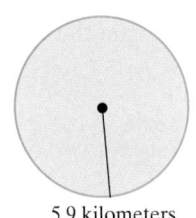

9.1 yards

52.

5.9 kilometers

53. In 1893, the first ride called a Ferris wheel was constructed by Washington Gale Ferris. Its diameter was 250 feet. Find its circumference. Give an exact answer and an approximation using 3.14 for π. (*Source: The Handy Science Answer Book*, Visible Ink Press, 1994)

Source: *The Handy Science Answer Book*, Visible Ink Press, 1994

54. The radius of Earth is approximately 3950 miles. Find the distance around Earth at the equator. Give an exact answer and an approximation using 3.14 for π. (*Hint*: Find the circumference of a circle with radius 3950 miles.)

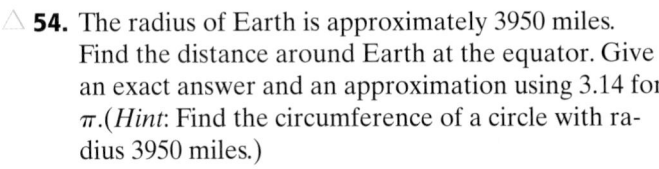

3950 miles

55. The London Eye, built for the Millennium celebration in London, resembles a gigantic ferris wheel with a diameter of 135 meters. If Adam Hawn rides the Eye for one revolution, find how far he travels. (*Hint*: Give an exact answer and an approximation using 3.14 for π. (*Source*: Londoneye.com)

56. The world's longest suspension bridge is the Akashi Kaikyo Bridge in Japan. This bridge has two circular caissons, which are under water foundations. If the diameter of a caisson is 80 meters, find its circumference. Give an exact answer and an approximation using 3.14 for π. (*Source: Scientific American*; How Things Work Today)

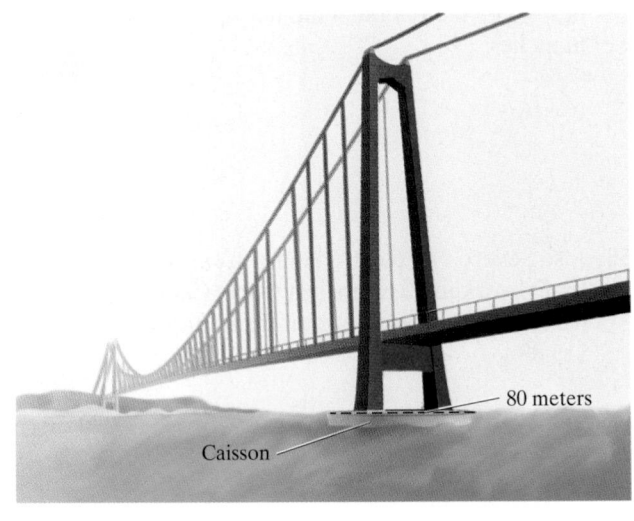

80 meters

Caisson

△ **57. a.** Approximate the circumference of each circle.

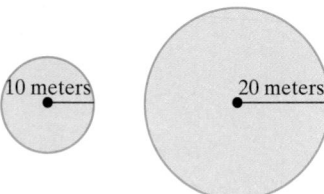

b. If the radius of a circle is doubled, is its corresponding circumference doubled?

△ **58. a.** Approximate the circumference of each circle.

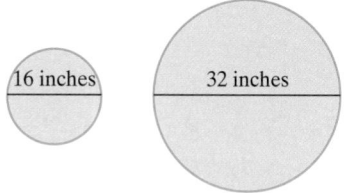

b. If the diameter of a circle is doubled, is its corresponding circumference doubled?

E *Solve. See Example 15.*

59. A 1-ounce serving of cream cheese contains 6.2 grams of saturated fat. How much saturated fat is in 4 ounces of cream cheese? (*Source: Home and Garden Bulletin No. 72*; U.S. Department of Agriculture)

60. A 3.5-ounce serving of lobster meat contains 0.1 gram of saturated fat. How much saturated fat is in 3 servings of lobster meat? (*Source:* The National Institute of Health)

61. In 2001, American farmers received an average of $2.80 per bushel of wheat. How much did a farmer receive for selling 1000 bushels of wheat? (*Source:* National Agricultural Statistics Service)

62. In 2001, American farmers received an average of $4.30 per bushel of soybeans. How much did a farmer receive for selling 10,000 bushels of soybeans? (*Source:* National Agricultural Statistics Service)

63. A meter is a unit of length in the metric system that is approximately equal to 39.37 inches. Sophia Wagner is 1.65 meters tall. Find her approximate height in inches.

64. The doorway to a room is 2.15 meters tall. Approximate this height in inches. (*Hint:* See Exercise 63.)

65. Jose Severos, an electrician for Central Power and Light, worked 40 hours last week. Calculate his pay before taxes for last week if his hourly wage is $13.88.

66. Maribel Chin, an assembly line worker, worked 20 hours last week. Her hourly rate is $8.52 per hour. Calculate her pay before taxes.

The table shows currency exchange rates for various countries on March 31, 2003. To find the amount of foreign currency equivalent to an amount of U.S. money, multiply the U.S. dollar amount by the exchange rate listed in the table. Use this table to answer Exercises 67–70.

Foreign Currency Exchange Rates	
Country	**Exchange Rate**
European Union Euro	0.9174
Australian Dollar	1.6543
British Pound	0.6333
Canadian Dollar	1.4695
Sri Lanka Rupee	96.97
Japanese Yen	118.0700

(*Source:* New York Federal Reserve Bank)

67. How many Japanese yen are equivalent to 675 U.S. dollars?

68. Suppose you wish to exchange 500 U.S. dollars into Sri Lanka Rupees. How much money, in Sri Lanka Rupees, would you receive?

69. The Goldthwaite family is taking a vacation to Canada. How much Canadian currency can they "buy" with 350 U.S. dollars?

70. A British tourist bought a $30 sweatshirt in New York City. What did he pay for the sweatshirt in British pounds?

Review and Preview

Divide. See Sections 1.7 and 4.3.

71. $130 \div 5$

72. $-495 \div -27$

73. $2016 \div 56$

74. $1863 \div 69$

75. $2920 \div 365$

76. $2916 \div 6$

77. $-\dfrac{24}{7} \div \dfrac{8}{21}$

78. $\dfrac{162}{25} \div \dfrac{9}{75}$

Combining Concepts

79. Find how far radio waves travel in 20.6 seconds. (Radio waves travel at a speed of $1.86 \times 100,000$ miles per second.)

80. If it takes radio waves approximately 8.3 minutes to travel from the sun to Earth, find approximately how far it is from the sun to Earth. (*Hint:* See Exercise 79.)

81. In your own words, explain how to find the number of decimal places in a product of decimal numbers.

82. In your own words, explain how to multiply by a power of 10.

83. Evaluate $4\pi r^2$ for r = 7.68. Round to the nearest hundredth. (Surface area of a sphere)

Internet Excursions

 **Go To:** http://www.prenhall.com/martin-gay_prealgebra What's Related

This World Wide Web address will provide you with access to a listing of current foreign currency exchange rates for the U.S. dollar, or a related site. The exchange rates in the "To United States Dollar" column are used to convert U.S. dollars to other currencies. The exchange rates in the "In United States Dollar" column are used to convert other currencies into U.S. dollars. Visit this site and use today's exchange rates to answer the following questions.

84. Convert $650 U.S. dollars to each given currancy.
  a. Chinese renminbi
  b. Italian lira
  c. Swiss francs

85. Determine the cost in U.S. dollars for each item.
  a. A silk scarf costing 1400 Japanese yen
  b. A paperback book costing 9.95 Canadian dollars
  c. A sweater costing 50 British pounds

5.4   Dividing Decimals

Ⓐ Dividing Decimals

Division of decimal numbers is similar to division of whole numbers. The only difference is the placement of a decimal point in the quotient. If the divisor is a whole number, see the example below, we divide as for whole numbers; then place the decimal point in the quotient directly above the decimal point in the dividend. Recall that division can be checked by multiplication.

EXAMPLE 1.   Divide: $32\overline{)8.32}$

Solution: We divide as usual. The decimal point in the quotient is directly above the decimal point in the dividend.

$$
\begin{array}{r}
0.26 \leftarrow \text{quotient} \\
\text{divisor} \rightarrow 32\overline{)8.32} \leftarrow \text{dividend} \\
-64 \\
\hline
192 \\
-192 \\
\hline
0
\end{array}
\qquad
\begin{array}{rl}
\textit{Check:} & 0.26 \quad \text{quotient} \\
& \times\ 32 \quad \text{divisor} \\
\hline
& 52 \\
& 7\ 80 \\
\hline
& 8.32 \quad \text{dividend}
\end{array}
$$

If the divisor is not a whole number, we need to move the decimal point to the right until the divisor is a whole number before we divide.

$$1.5\overline{)64.85}$$

divisor ⟶ ↑   ↑ ⟶ dividend

To understand how this works, let's rewrite

$$1.5\overline{)64.85} \quad \text{as} \quad \frac{64.85}{1.5}$$

and then multiply the numerator and the denominator by 10.

$$\frac{64.85}{1.5} = \frac{64.85 \times 10}{1.5 \times 10} = \frac{648.5}{15}$$

which can be written as $15\overline{)648.5}$. Notice that

$$1.5\overline{)64.85} \quad \text{is equivalent to} \quad 15.\overline{)648.5}$$

The decimal points in the dividend and the divisor were both moved one place to the right, and the divisor is now a whole number. This procedure is summarized next.

Dividing by a Decimal

Step 1. Move the decimal point in the divisor to the right until the divisor is a whole number.

Step 2. Move the decimal point in the dividend to the right the *same number of places* as the decimal point was moved in Step 1.

Step 3. Divide. Place the decimal point in the quotient directly over the moved decimal point in the dividend.

OBJECTIVES

Ⓐ Divide decimals.

Ⓑ Divide decimals by powers of 10.

Ⓒ Evaluate expressions and check solutions with decimal replacement values.

Ⓓ Solve problems by dividing decimals.

SSM
TUTOR CENTER   SG   CD & VIDEO   MATH PRO   WEB

Practice Problem 1

Divide: $48\overline{)34.08}$

Answer

1. 0.71

Practice Problem 2

Divide: $5.6\overline{)166.88}$

Practice Problem 3

Divide: $-2.808 \div (-104)$

Practice Problem 4

Divide: $23.4 \div 0.57$. Round the quotient to the nearest hundredth.

EXAMPLE 2   Divide: $2.3\overline{)10.764}$

Solution: Move the decimal points in the divisor and the dividend one place to the right so that the divisor is a whole number.

$$
2.3\overline{)10.764} \quad \text{becomes} \quad
\begin{array}{r}
4.68 \\
23.\overline{)107.64} \\
-92 \\
\hline
15\,6 \\
-13\,8 \\
\hline
1\,84 \\
-1\,84 \\
\hline
0
\end{array}
$$

To check, see that $4.68 \times 2.3 = 10.764$.   ●

EXAMPLE 3   Divide: $-5.98 \div 115$

Solution: Recall that a negative number divided by a positive number gives a negative quotient. The divisor is a whole number, so the decimal point in the divisor is not moved.

$$
\begin{array}{r}
0.052 \\
115\overline{)5.980} \\
-5\,75 \\
\hline
230 \\
-230 \\
\hline
0
\end{array}
$$
← Insert one 0.

Thus $-5.98 \div 115 = -0.052$.   ●

EXAMPLE 4   Divide: $17.5 \div 0.48$. Round the quotient to the nearest hundredth.

Solution: Move the decimal points in the divisor and the dividend 2 places.

hundredths place

$$
\begin{array}{r}
36.458 \approx 36.46 \\
48.\overline{)1750.000} \\
-144 \\
\hline
310 \\
-288 \\
\hline
220 \\
-192 \\
\hline
280 \\
-240 \\
\hline
400 \\
-\ 384 \\
\hline
16
\end{array}
$$

approximately

If rounding to the nearest hundredth, carry the division process out to one more decimal place, the thousandths place.

●

Answers

2. 29.8   **3.** 0.027   **4.** 41.05

Try the Concept Check in the margin.

There are patterns that occur when we divide decimals by powers of 10 such as 10, 100, 1000, and so on.

$$\frac{569.2}{10} = 56.92$$ Move the decimal point 1 *place* to the *left*.

— 1 zero

$$\frac{569.2}{10,000} = 0.05692$$ Move the decimal point 4 *places* to the *left*.

— 4 zeros

This pattern suggests the following rule.

Dividing Decimals by Powers of 10 such as 10, 100, 1000, . . .

Move the decimal point of the dividend to the *left* the same number of places as there are *zeros* in the power of 10.

Notice that this is the same pattern as *multiplying* by powers of 10 such as 0.1, 0,01, or 0.001. This is because dividing by a power of 10 such as 100 is the same as multiplying by its reciprocal $\frac{1}{100}$, or 0.01. For example,

$$\frac{569.2}{10} = 569.2 \times \frac{1}{10} = 569.2 \times 0.1 = 56.92$$

EXAMPLES Divide.

5. $\dfrac{786}{10,000} = 0.0786$ Move the decimal point 4 *places* to the *left*.

— 4 zeros

6. $\dfrac{0.12}{10} = 0.012$ Move the decimal point 1 *place* to the *left*.

—1 zero

Try the Concept Check in the margin.

C Using Decimals as Replacement Values

EXAMPLE 7   Evaluate $x \div y$ for $x = 2.5$ and $y = 0.05$.

Solution: Replace x with 2.5 and y with 0.05.

$$x \div y = 2.5 \div 0.05 \qquad 0.05\overline{)2.5} \quad \text{becomes} \quad 5\overline{)250}^{\,50}$$
$$= 50$$

Concept Check

If a quotient is to be rounded to the nearest thousandth, to what place should the division be carried out?

Practice Problems 5–6

Divide.

5. $\dfrac{28}{1000}$

6. $\dfrac{8.56}{100}$

Concept Check

Describe how to check the answer in Example 5.

Evaluate $x \div y$ for $x = 0.035$ and $y = 0.02$.

Answers

5. 0.028   **6.** 0.0856   **7.** 1.75

Concept Check: The ten-thousandths place.

Concept Check: Answers may vary.

Practice Problem 8

Is 39 a solution of the equation
$\frac{x}{100} = 3.9$?

Practice Problem 9

A bag of fertilizer covers 1250 square feet of lawn. Tim Parker's lawn measures 14,800 square feet. How many bags of fertilizer does he need? If he can buy whole bags of fertilizer only, how many whole bags does he need?

EXAMPLE 8 Is 720 a solution of the equation $\frac{y}{100} = 7.2$?

Solution: Replace y with 720 to see if a true statement results.

$$\frac{y}{100} = 7.2 \quad \text{Original equation}$$

$$\frac{720}{100} = 7.2 \quad \text{Replace } y \text{ with 720.}$$

$$7.2 = 7.2 \quad \text{True.}$$

Since $7.2 = 7.2$ is a true statement, 720 is a solution of the equation. ●

Ⓓ Solving Problems by Dividing Decimals

Many real-life problems involve dividing decimals.

EXAMPLE 9   Calculating Materials Needed for a Job

A gallon of paint covers a 250-square-foot area. If Betty Adkins wishes to paint a wall that measures 1450 square feet, how many gallons of paint does she need? If she can buy gallon containers of paint only, how many gallon containers does she need?

Solution:

1. UNDERSTAND. Read and reread the problem. To find the number of gallons, divide 1450 by 250.

2. TRANSLATE.

	Number of gallons	is	square feet	divided by	square feet per gallon
In words:	↓	↓	↓	↓	↓
Translate:	Number of gallons	=	1450	÷	250

3. SOLVE.

$$
\begin{array}{r}
5.8 \\
250\overline{)1450.0} \\
-1250 \\
\hline
200\ 0 \\
-200\ 0 \\
\hline
0
\end{array}
$$

4. INTERPRET. *Check* your work. *State* your conclusion: Betty needs 5.8 gallons of paint. If she can buy gallon containers of paint only, she needs 6 gallon containers of paint to complete the job. ●

Answers

8. no

9. 11.84 bags; 12 bags

Name _____ Section _____ Date _____

EXERCISE SET 5.4

 Divide. *See Examples 1 through 3.*

1. $0.47 \div 5$ **2.** $11.8 \div 2$ **3.** $-18 \div 0.06$ **4.** $20 \div (-0.04)$

 5. $4.756 \div 0.82$ **6.** $3.312 \div 0.92$ **7.** $-36.3 \div -5.5$ **8.** $-21.78 \div -2.2$

Divide, and round each quotient as indicated. See Example 4.

9. Divide 429.34 by 2.4 and round the quotient to the nearest hundred.

10. Divide 54.8 by 2.6 and round the quotient to the nearest ten.

 11. Divide 0.549 by 0.023 and round the quotient to the nearest hundredth.

12. Divide 0.0453 by 0.98 and round the quotient to the nearest thousandth.

13. Divide -45.23 by 0.4 and round the quotient to the nearest ten.

14. Divide -983.32 by 0.061 and round the quotient to the nearest thousand.

 Divide. *See Examples 5 and 6.*

15. $54.982 \div 100$ **16.** $342.54 \div 100$ **17.** $12.9 \div (-1000)$

18. $13.49 \div (-10,000)$ **19.** $87 \div 10$ **20.** $0.27 \div 10$

B *Divide. See Examples 1 through 6.*

21. $1.239 \div 3$

22. $0.54 \div 12$

23. Divide -4.2 by 0.6

24. Divide 3.6 by 0.9

25. $1.296 \div 0.27$

26. $-2.176 \div 0.34$

27. Divide 42 by 0.02

28. Divide 24 by 0.03

29. Divide -18 by -0.6

30. Divide 20 by 0.4

31. Divide 35 by 0.005

32. Divide -35 by -0.0007

33. $-1.104 \div 1.6$

34. $-2.156 \div 0.98$

35. $-2.4 \div (-100)$

36. $-86.79 \div (-1000)$

37. $\dfrac{4.615}{0.071}$

38. $\dfrac{23.8}{-0.035}$

39. Divide 0.00263 by 8.9 and round the quotient to the nearest ten-thousandth.

40. Divide 200 by 0.054 and round the quotient to the nearest thousand.

41. Divide 500 by 0.0043 and round the quotient to the nearest ten-thousand.

42. Divide 4.56 by 9.34 and round the quotient to the nearest hundredth.

Review and Preview

Round each decimal to the given place value. See Section 5.1.

71. 345.219, nearest hundredth

72. 902.155, nearest hundredth

73. −1000.994, nearest tenth

74. 234.1029, nearest thousandth

Use the order of operations to simplify each expression. See Section 1.8.

75. $2 + 3 \cdot 6$

76. $9 \cdot 2 + 8 \cdot 3$

77. $20 - 10 \div (-5)$

78. $(20 - 10) \div 5$

Combining Concepts

Recall from Section 1.7 that the average of a list of numbers is their total divided by how many numbers there are in the list. Use this to find the average of the numbers listed. If necessary, round to the nearest tenth.

79. 86, 78, 91, 85

80. 56, 75, 80

81. 9.6, 8.5, 4.1, 4.1, 9.8

82. 1.7, 3.8, 4.6, 8.7

△ **83.** The area of this rectangle is 38.7 square feet. If its width is 4.5 feet, find its length.

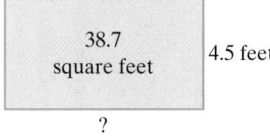

△ **84.** The perimeter of the square is 180.8 centimeters. Find the length of a side.

Perimeter is 180.8 centimeters

?

85. When dividing decimals, describe the process you use to place the decimal point in the quotient.

86. In your own words, describe how to quickly divide a number by a power of 10 such as 10, 100, 1000, etc.

To convert wind speeds in miles per hour to knots, divide by 1.15. Use this information and the Saffir-Simpson Hurricane Intensity chart below to answer Exercises 87–88. Round to the nearest tenth.

87. The chart gives wind speeds in miles per hour. What is the range of wind speeds for a Category 1 hurricane in knots?

88. What is the range of wind speeds for a Category 4 hurricane in knots?

Saffir-Simpson Hurricane Intensity Scale				
Category	**Wind Speed**	**Barometric Pressure [inches of mercury (Hg)]**	**Storm Surge**	**Damage Potential**
1 (Weak)	75–95 mph	≥28.94 in.	4–5 ft	Minimal damage to vegetation
2 (Moderate)	96–110 mph	28.50–28.93 in.	6–8 ft	Moderate damage to houses
3 (Strong)	111–130 mph	27.91–28.49 in.	9–12 ft	Extensive damage to small buildings
4 (Very Strong)	131–155 mph	27.17–27.90 in.	13–18 ft	Extreme structural damage
5 (Devastating)	>155 mph	<27.17 in.	>18 ft	Catastrophic building failures possible

Integrated Review–Operations on Decimals

Answers
1. _____
2. _____
3. _____
4. _____
5. _____
6. _____
7. _____
8. _____
9. _____
10. _____
11. _____
12. _____
13. _____
14. _____
15. _____
16. _____
17. _____
18. _____
19. _____
20. _____
21. _____
22. _____
23. _____
24. _____
25. _____
26. _____
27. _____

Perform each indicated operation.

1. $1.6 + 0.97$ **2.** $3.2 + 0.85$ **3.** $9.8 - 0.9$ **4.** $10.2 - 6.7$

5. $\begin{array}{r} 0.8 \\ \times\ 0.2 \\ \hline \end{array}$ **6.** $\begin{array}{r} 0.6 \\ \times\ 0.4 \\ \hline \end{array}$ **7.** $8\overline{)2.16}$ **8.** $6\overline{)3.12}$

9. $(9.6)(-0.5)$ **10.** $(-8.7)(-0.7)$ **11.** $\begin{array}{r} 123.6 \\ -\ 48.04 \\ \hline \end{array}$ **12.** $\begin{array}{r} 325.2 \\ -\ 36.08 \\ \hline \end{array}$

13. $-25 + 0.026$ **14.** $0.125 + (-44)$ **15.** $29.24 \div (-3.4)$ **16.** $-10.26 \div (-1.9)$

17. -2.8×100 **18.** 1.6×1000 **19.** $\begin{array}{r} 96.21 \\ 7.028 \\ +121.7 \\ \hline \end{array}$ **20.** $\begin{array}{r} 0.268 \\ 1.93 \\ +142.881 \\ \hline \end{array}$

21. $-25.76 \div -46$ **22.** $-27.09 \div 43$ **23.** $\begin{array}{r} 12.004 \\ \times\ \ \ \ 2.3 \\ \hline \end{array}$ **24.** $\begin{array}{r} 28.006 \\ \times\ \ \ \ 5.2 \\ \hline \end{array}$

25. Subtract 4.6 from 10   **26.** Subtract 18 from 0.26   **27.** $-268.19 - 146.25$

28. _____

29. _____

30. _____

31. _____

32. _____

33. _____

34. _____

35. _____

36. _____

37. _____

38. _____

28. $-860.18 - 434.85$ **29.** $\dfrac{2.958}{-0.087}$ **30.** $\dfrac{-1.708}{0.061}$

31. $160 - 43.19$ **32.** $120 - 101.21$ **33.** 15.62×10

34. $15.62 \div 10$ **35.** $15.62 + 10$ **36.** $15.62 - 10$

37. According to the federal Nutrition Labeling and Education Act of 1990, a food manufacturer may label a food product "Fat Free" if it contains less than 0.5 grams of fat per serving. A new type of cookie contains 0.38 grams of fat per cookie. Can the packaging for this cookie be labeled "Fat Free"? Explain.

38. The highest NBA scorer for the 2002–2003 regular season was Tracy McGrady of the Orlando Magic. He scored a total of 2407 points in 75 games. Find his average number of points per game to the nearest tenth of a point. (*Source:* National Basketball Association)

5.5   Estimating and Order of Operations

A   Estimating Operations on Decimals

Estimating sums, differences, products, and quotients of decimal numbers is an important skill whether you use a calculator or perform decimal operations by hand. When you can estimate results as well as calculate them, you can judge whether the calculations are reasonable. If they aren't, you know you've made an error somewhere along the way. To estimate, we round numbers.

EXAMPLE 1   Subtract: $78.62 − $16.85. Then estimate the difference to see whether the proposed result is reasonable by rounding each decimal to the nearest dollar and then subtracting.

Solution:

Given		Estimate
$78.62	rounds to	$79
−$16.85	rounds to	−$17
$61.77		$62

The estimated difference is $62, so $61.77 is reasonable.

Try the Concept Check in the margin.

EXAMPLE 2   Multiply: 28.06 × 1.95. Then estimate the product to see whether the proposed result is reasonable.

Solution:

Given	Estimate 1	Estimate 2
28.06	28	30
× 1.95	× 2	× 2
14030	56	60
252540		
280600		
54.7170		

The answer 54.7170 is reasonable.

As shown in Example 2, estimated results will vary depending on what estimates are used. Notice that estimating results is a good way to see whether the decimal point has been correctly placed.

> **Helpful Hint**
>
> When rounding to check a calculation, you may want to round the numbers to a place value of your choosing so that your estimates are easy to compute mentally.

Practice Problem 1

Subtract: $65.34 − $14.68. Then estimate the difference to see whether the proposed result is reasonable.

Concept Check

Should you estimate the sum 21.98 + 42.36 as 30 + 50 = 80? Why or why not?

Practice Problem 2

Multiply: 30.26 × 2.98. Then estimate the product to see whether the proposed solution is reasonable.

Answers

1. $50.66   **2.** 90.1748

Concept Check: No; each number is rounded incorrectly. The estimate is too high.

Practice Problem 3

Divide: 713.7 ÷ 91.5. Then estimate the quotient to see whether the proposed answer is reasonable.

Practice Problem 4

Using the figure below, estimate how much farther it is between Dodge City and Pratt than it is between Garden City and Dodge City by rounding each given distance to the nearest mile.

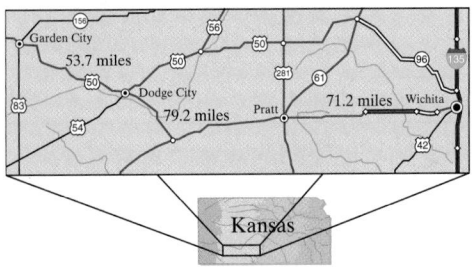

Practice Problem 5

Use the estimation to determine whether the calculaor result is reasonable or not.

Simplify: 51.793 ÷ 13

Calculator display: 19.612

EXAMPLE 3   Divide: 272.356 ÷ 28.4. Then estimate the quotient to see whether the proposed result is reasonable.

Solution: **Given**   **Estimate**

$$\begin{array}{r} 9.59 \\ 284.\overline{)2723.56} \\ -2556 \\ \hline 1675 \\ -1420 \\ \hline 2556 \\ -2556 \\ \hline 0 \end{array}$$

$$\begin{array}{r} 9 \\ 30\overline{)270} \end{array} \quad \text{or} \quad \begin{array}{r} 9 \\ 300\overline{)2700} \end{array}$$

The estimate is 9, so 9.59 is reasonable.   ●

EXAMPLE 4   Estimating Distance

Use the figure in the margin to estimate the distance in miles between Garden City, Kansas, and Wichita, Kansas, by rounding each given distance to the nearest ten.

Solution: **Calculated**
Distance   **Estimate**

53.7	rounds to	50
79.2	rounds to	80
+71.2	rounds to	+70
		200

The distance between Garden City and Wichita is approximately 200 miles. (The calculated distance is 204.1 miles.)   ●

EXAMPLE 5   Use estimation to determine whether the calculator result is reasonable or not. (For example, a result that is not reasonable can occur if proper keys are not pressed.)

Simplify: 82.064 ÷ 23   Calculator display: 35.68

Solution: Round each number to the nearest 10. Since 80 ÷ 20 = 4, the calculator display 35.68 is not reasonable.   ●

Ⓑ Simplifying Expressions Containing Decimals

In the remaining examples, we will review the order of operations by simplifying expressions that contain decimals.

Order of Operations

1. Do all operations within grouping symbols such as parentheses or brackets.

2. Evaluate any expressions with exponents.

3. Multiply or divide in order from left to right.

4. Add or subtract in order from left to right.

Answers

3. 7.8   **4.** 25 miles   **5.** not reasonable

EXAMPLE 6   Simplify: $-0.5(8.6 - 1.2)$

Solution: According to the order of operations, simplify inside the parentheses first.

$$-0.5(8.6 - 1.2) = -0.5(7.4) \quad \text{Subtract.}$$
$$= -3.7 \quad \text{Multiply. The product of a negative number and a positive number is a negative result.} \ \bullet$$

EXAMPLE 7   Simplify: $(-1.3)^2 + 2.4$

Solution: Recall the meaning of an exponent.

$$(-1.3)^2 = (-1.3)(-1.3) + 2.4 \quad \text{Use the definition of an exponent.}$$
$$= 1.69 + 2.4 \quad \text{Multiply. The product of two negative numbers is a positive number}$$
$$= 4.09 \quad \text{Add.} \ \bullet$$

EXAMPLE 8   Simplify: $\dfrac{0.7 + 1.84}{0.4}$

Solution: First simplify the numerator of the fraction. Then divide.

$$\frac{0.7 + 1.84}{0.4} = \frac{2.54}{0.4} \quad \text{Simplify the numerator.}$$
$$= 6.35 \quad \text{Divide.} \ \bullet$$

(C) Using Decimals as Replacement Values

EXAMPLE 9   Evaluate $-2x + 5$ for $x = 3.8$.

Solution: Replace x with 3.8 in the expression $-2x + 5$ and simplify.

$$-2x + 5 = -2(3.8) + 5 \quad \text{Replace } x \text{ with 3.8.}$$
$$= -7.6 + 5 \quad \text{Multiply.}$$
$$= -2.6 \quad \text{Add.} \ \bullet$$

EXAMPLE 10   Determine whether -3.3 is a solution of $2x - 1.2 = -7.8$.

Solution: Replace x with -3.3 in the equation $2x - 1.2 = -7.8$ to see if a true statement results.

$$2x - 1.2 = -7.8$$
$$2(-3.3) - 1.2 = -7.8 \quad \text{Replace } x \text{ with}$$
$$-6.6 - 1.2 = -7.8 \quad \text{Multiply.}$$
$$-7.8 = -7.8 \quad \text{True.}$$

Since $-7.8 = -7.8$ is a true statement, -3.3 is a solution of $2x - 1.2 = -7.8$. $\bullet$

Practice Problem 6

Simplify: $-8.6(3.2 - 1.8)$

Practice Problem 7

Simplify: $(0.7)^2 + 2.1$

Practice Problem 8

Simplify: $\dfrac{8.78 - 2.8}{20}$

Practice Problem 9

Evaluate $1.7y - 2$ for $y = 2.3$.

Practice Problem 10

Is -2.1 a solution of $3x + 7.5 = 1.2$?

Answers

6. -12.04 **7.** 2.59   **8.** 0.299   **9.** 1.91   **10.** yes

FOCUS ON **Business and Career**

TIME SHEETS

Many jobs require employees to fill out a weekly time sheet. A time sheet allows employees to show how much time they worked on various projects and to find how much time they worked overall. For ease of calculation, it is standard to round times on a time sheet to the nearest quarter hour. The following is an example of a time sheet.

Time Sheet for the week of: 5/9								
Employee: Darla Williams							**Employee Number: 630087**	
	Project Number	**Monday**	**Tuesday**	**Wednesday**	**Thursday**	**Friday**	**Saturday/Sunday**	**TOTALS**
1	315	5.5	1		3.25	1.5		
2	629	2.5		4	1.25	1.5	3	
3	471		3.25	2		1.75		
4	511		3.75	1.5	2.75			
5	512				0.75	2.5	3.25	
6								
7								
8								
9								
10								
	TOTALS							

CRITICAL THINKING

Find the column totals for columns Monday through Saturday/Sunday on the time sheet. What do these column totals represent?

1. Find the row totals for rows 1 through 5 on the time sheet. What do these row totals represent?

2. Find the sum of the row totals and the sum of the column totals. Are these sums the same? If so, explain why. Otherwise, explain why having different row sums and column sums should prompt you to recheck your math.

3. Darla Williams is a full-time employee who is expected to work 40 hours per week. Any time over 40 hours is considered overtime. Did Darla work any overtime this week? If so, how much?

4. Darla is paid an hourly rate of $16.60 per hour. For overtime, she receives "time and a quarter," or 1.25 times her normal hourly pay rate. How much will Darla be paid for this week of work?

5. Describe how Darla's time was distributed among the five projects on which she worked during this week.

Name _____ Section _____ Date _____

EXERCISE SET 5.5

A *Perform each indicated operation. Then estimate to see whether each proposed result is reasonable. See Examples 1 through 3.*

1. 4.9 − 2.1

2. 7.3 − 3.8

3. 6 × 483.11

4. 98.2 × 405.8

5. 62.16 ÷ 14.8

6. 186.75 ÷ 24.9

7. 69.2 + 32.1 + 48.57

8. 79.52 + 21.47 + 97.2

9. 34.92 − 12.03

10. 349.87 − 251.32

Solve. See Example 4.

11. Estimate the perimeter of the rectangle by rounding each measurement to the nearest whole inch.

5.9 inches

12.2 inches

12. Joe Armani wants to put plastic edging around his rectangular flower garden to help control the weeds. The length and width are 16.2 meters and 12.9 meters, respectively. Estimate how much edging he needs by rounding each measurement to the nearest whole meter.

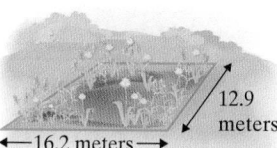

12.9 meters

←—16.2 meters—→

13. Estimate the perimeter of the triangle by rounding each measurement to the nearest whole foot.

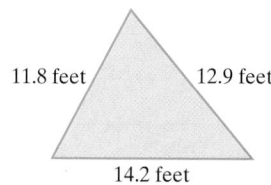

11.8 feet   12.9 feet

14.2 feet

14. The area of a triangle is area $= \frac{1}{2} \cdot$ base $\cdot$ height.

Approximate the area of a triangle whose base is 21.9 centimeters and whose height is 9.9 centimeters. Round each measurement to the nearest whole centimeter.

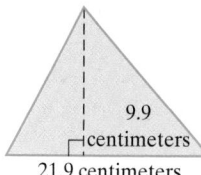

9.9 centimeters

21.9 centimeters

△ **15.** Use 3.14 for π to approximate the circumference of a circle whose radius is 7 meters.

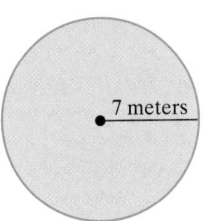

7 meters

17. Mike and Sandra Hallahan are taking a trip and plan to travel about 1550 miles. Their car gets 32.8 miles to the gallon. Approximate how many gallons of gasoline their car will use by rounding 32.8 to the nearest ten. Round the result to the nearest gallon.

19. Robert and Kathi Hawn purchased a new car. They pay $398.79 per month on their car loan for 5 years. Approximate how much they will pay on the loan altogether by rounding the monthly payment to the nearest hundred.

21. Estimate the total distance to the nearest mile between Grove City and Jerome by rounding each distance to the nearest mile.

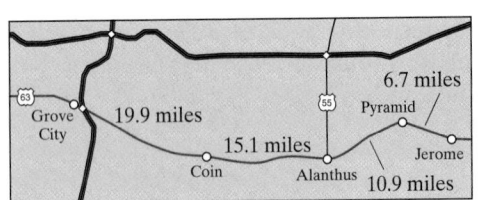

23. The all-time top six movies (that is, those that have earned the most money in the United States) along with the approximate amount of money they have earned are listed in the table. Estimate the total amount of money that these movies have earned by rounding each earning to the nearest hundred million. (*Source: Variety* magazine)

△ **16.** Use 3.14 for π to approximate the circumference of a circle whose radius is 6 centimeters.

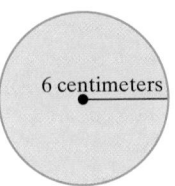

6 centimeters

18. Refer to Exercise 17. Suppose gasoline costs $1.659 per gallon. Approximate how much money Mike and Sandra need for gasoline for their 1550-mile trip by rounding $1.659 to the nearest cent.

20. Yoshikazu Sumo is purchasing the following groceries. He has only a ten dollar bill in his pocket. Estimate the cost of each item to see whether he has enough money. Bread, $1.49; milk, $2.09; carrots, $0.97; corn, $0.89; salt, $0.53; and butter, $2.89.

△ **22.** Estimate the perimeter of the figure shown by rounding each measurement to the nearest whole foot.

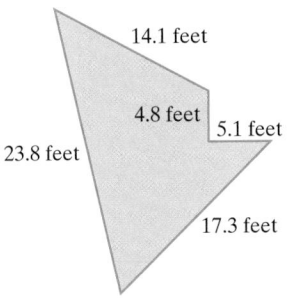

All-Time Top Six American Movies	
Movie	**Gross Domestic Earnings**
Titanic (1997)	$600.8 million
Star Wars (1977)	$461.0 million
E.T. (1982)	$435.0 million
The Phantom Menace (1999)	$431.1 million
Spider Man (2002)	$403.7 million
Jurassic Park (1993)	$357.1 million

24. San Marino is one of the smallest countries in the world. The current San Marino population density is 1197 people per square mile. The area of San Marino is about 20 square miles. Estimate the population of San Marino by rounding 1197 to the nearest ten. (*Source: 2003 World Almanac*)

25. Micronesia is a string of islands in the western Pacific Ocean. The current Micronesia has a population density is 501 people per square mile. The area of Micronesia is about 271 square miles. Estimate the population of Micronesia by rounding both 501 and 271 to the nearest ten. (*Source: 2003 World Almanac*)

26. In 2001, American manufacturers shipped approximately 17.3 million music CD singles to retailers. The value of these shipments was about $79.4 million. Estimate the value of an individual CD single by rounding $79.4 and 17.3 to the nearest ten. (*Source: Recording Industry Association of America*)

Choose the best estimate. See Example 5.

27. 8.62×41.7
 a. 36
 b. 32
 c. 360
 d. 3.6

28. $1.437 + 20.69$
 a. 34
 b. 22
 c. 3.4
 d. 2.2

29. $78.6 \div 97$
 a. 7.86
 b. 0.786
 c. 786
 d. 7860

30. $302.729 - 28.697$
 a. 270
 b. 20
 c. 27
 d. 300

Use estimation to determine whether each result is reasonable or not. See Example 5.

31. 102.62×41.8 Result: 428.9516

32. $174.835 \div 47.9$ Result: 3.65

33. $1025.68 - 125.42$ Result: 900.26

34. $562.781 + 2.96$ Result: 858.781

B *Simplify each expression. See Examples 6 through 8.*

35. $(0.4)^2 - 0.1$

36. $(-100)(2.3) - 30$

37. $\dfrac{1 + 0.8}{-0.6}$

38. $(-0.09)^2 + 1.16$

39. $1.4(2 - 1.8)$

40. $\dfrac{0.29 + 1.69}{3}$

41. $4.83 \div 2.1$

42. $62.1 \div 2.7$

43. $(-2.3)^2(0.3 + 0.7)$

44. $(8.2)(100) - (8.2)(10)$

45. $(3.1 + 0.7)(2.9 - 0.9)$

46. $\dfrac{0.707 - 3.19}{13}$

47. $\dfrac{(4.5)^2}{100}$

48. $0.9(5.6 - 6.5)$

49. $\dfrac{7 + 0.74}{-6}$

50. $(1.5)^2 + 0.5$

C *Evaluate each expression for x = 6, y = 0.3, and z = −2.4. See Example 9.*

51. z^2

52. y^2

53. $x - y$

54. $x - z$

55. $4y - z$

56. $\dfrac{x}{y}$

Determine whether the given value is a solution of the given equation. See Example 10.

57. Is -1.3 a solution to $7x + 2.1 = -7$?

58. Is 6.2 a solution to $5y - 8.6 = 23.9$?

59. Is -4.7 a solution to $x - 6.5 = 2x + 1.8$?

60. Is -0.9 a solution to $2x - 3.7 = x - 4.6$?

Review and Preview

Perform each indicated operation. See Sections 4.3, 4.5, and 4.8.

61. $\dfrac{3}{4} \cdot \dfrac{5}{12}$

62. $\dfrac{56}{23} \cdot \dfrac{46}{8}$

63. $-\dfrac{36}{56} \div \dfrac{30}{35}$

64. $\dfrac{5}{7} \div \left(-\dfrac{14}{10}\right)$

65. $\dfrac{5}{12} - \dfrac{1}{3}$

66. $\dfrac{12}{15} + \dfrac{5}{21}$

Combining Concepts

Simplify each expression. Then estimate to see whether your proposed result is reasonable.

67. $1.96(7.852 - 3.147)^2$

68. $(6.02)^2 + (2.06)^2$

69. Suppose that your approximation of 12,743 is 13,000 and your friend's approximation is 12,700. Who is correct? Explain your answer.

70. When simplifying $5 + 0.1(1.26 - 0.23)$, do you add, multiply, or subtract first? Explain your answer.

71. When simplifying $3(1.5)^2$, do you square first or multiply by 3 first? Explain your answer.

5.6 Fractions and Decimals

Ⓐ Writing Fractions as Decimals

To write a fraction as a decimal, we interpret the fraction bar as division and find the quotient.

> **Writing Fractions as Decimals**
>
> To write a fraction as a decimal, divide the numerator by the denominator.

EXAMPLE 1 Write $\frac{1}{4}$ as a decimal.

Solution: $\frac{1}{4} = 1 \div 4$

$$
\begin{array}{r}
0.25 \\
4\overline{)1.00} \\
\underline{-8} \\
20 \\
\underline{-20} \\
0
\end{array}
$$

Thus, $\frac{1}{4}$ written as a decimal is 0.25.

EXAMPLE 2 Write $-\frac{5}{8}$ as a decimal.

Solution: $-\frac{5}{8} = -(5 \div 8) = -0.625$

$$
\begin{array}{r}
0.625 \\
8\overline{)5.000} \\
\underline{-48} \\
20 \\
\underline{-16} \\
40 \\
\underline{-40} \\
0
\end{array}
$$

EXAMPLE 3 Write $\frac{2}{3}$ as a decimal.

Solution:
$$
\begin{array}{r}
0.666\ldots \\
3\overline{)2.000} \\
\underline{-18} \\
20 \\
\underline{-18} \\
20 \\
\underline{-18} \\
2
\end{array}
$$
This pattern will continue so that $\frac{2}{3} = 0.6666\ldots$.

We can place a bar over the digit 6 to indicate that it repeats.

$$\frac{2}{3} = 0.666\ldots = 0.\overline{6}$$

We can also write a decimal approximation for $\frac{2}{3}$. For example, $\frac{2}{3}$ rounded to the nearest hundredth is 0.67. This can be written as $\frac{2}{3} \approx 0.67$.

OBJECTIVES

Ⓐ Write fractions as decimals.

Ⓑ Compare fractions and decimals.

Ⓒ Solve area problems containing fractions and decimals.

SSM
TUTOR CENTER   SG   CD & VIDEO   MATH PRO   WEB

Practice Problem 1

a. Write $\frac{2}{5}$ as a decimal.

b. Write $\frac{9}{40}$ as a decimal.

Practice Problem 2

Write $-\frac{3}{8}$ as a decimal.

Practice Problem 3

a. Write $\frac{5}{6}$ as a decimal.

b. Write $\frac{2}{9}$ as a decimal.

Answers

1. a. 0.4 **b.** 0.225
2. −0.375
3. a. $0.8\overline{3} \approx 0.83$ **b.** $0.\overline{2} \approx 0.22$

Practice Problem 4

Write $\dfrac{4}{11}$ as a decimal. Give an exact answer and a 3 decimal place approximation.

EXAMPLE 4   Write   as a decimal. Give an exact answer and a 3 decimal place approximation.

Solution:

$$11\overline{)6.0000}$$

$$\begin{array}{r} 0.5454\ldots \\ 11\overline{)6.0000} \\ \underline{-55} \\ 50 \\ \underline{-44} \\ 60 \\ \underline{-55} \\ 50 \\ \underline{-44} \\ 6 \end{array}$$

This pattern will continue, so $\dfrac{6}{11} = 0.\overline{54}$.

Rounded to 3 decimal places, $\dfrac{6}{11} \approx 0.545$. ●

Practice Problem 5

Write $\dfrac{1}{9}$ as a decimal. Round to the nearest thousandth.

EXAMPLE 5   Write $\dfrac{22}{7}$ as a decimal. Round to the nearest hundredth. (The fraction $\dfrac{22}{7}$ is an approximation for π.)

Solution:

$$\begin{array}{r} 3.142 \approx 3.14 \\ 7\overline{)22.000} \\ \underline{-21} \\ 10 \\ \underline{-7} \\ 30 \\ \underline{-28} \\ 20 \\ \underline{-14} \\ 6 \end{array}$$

Carry the division out to the thousandths place.

The fraction $\dfrac{22}{7}$ in decimal form is approximately 3.14. ●

Concept Check

Suppose you are rewriting the fraction $\dfrac{9}{16}$ as a decimal. How do you know you have made a mistake if you come up with 1.735?

Try the Concept Check in the margin.

B   **Comparing Fractions and Decimals**

Now we can compare decimals and fractions by writing fractions as equivalent decimals.

Practice Problem 6

Insert $<$, $>$, or $=$ to form a true statement.

$$\dfrac{1}{5} \qquad 0.25$$

EXAMPLE 6   Insert $<$, $>$, or $=$ to form a true statement.

$$\dfrac{1}{8} \qquad 0.12$$

Answers

4. $0.\overline{36} \approx 0.364$ **5.** ≈ 0.111 **6.** $<$

Concept Check: Answers may vary.

Solution: First we write $\frac{1}{8}$ as an equivalent decimal. Then we compare decimal places.

$$\begin{array}{r} 0.125 \\ 8)\overline{1.000} \\ -8 \\ \hline 20 \\ -16 \\ \hline 40 \\ -40 \\ \hline 0 \end{array}$$

Original numbers	$\frac{1}{8}$	0.12
Decimals	0.125	0.120
Compare	0.125 > 0.12	
Thus,	$\frac{1}{8} > 0.12$	

EXAMPLE 7 Insert <, >, or = to form a true statement.

$$0.\overline{7} \qquad \frac{7}{9}$$

Solution: We write $\frac{7}{9}$ as a decimal and then compare.

$$\begin{array}{r} 0.77\ldots = 0.\overline{7} \\ 9)\overline{7.00} \\ -63 \\ \hline 70 \\ -63 \\ \hline 7 \end{array}$$

Original numbers	$0.\overline{7}$	$\frac{7}{9}$
Decimals	$0.\overline{7}$	$0.\overline{7}$
Compare	$0.\overline{7} = 0.\overline{7}$	
Thus,	$0.7 = \frac{7}{9}$	

EXAMPLE 8 Write the numbers in order from smallest to largest.

$$\frac{9}{20}, \quad \frac{4}{9}, \quad 0.456$$

Solution:

Original numbers	$\frac{9}{20}$	$\frac{4}{9}$	0.456
Decimals	0.450	0.444…	0.456
Compare in order	2nd	1st	3rd

The numbers written in order are

$$\frac{4}{9}, \quad \frac{9}{20}, \quad 0.456$$

Practice Problem 7

Insert <, >, or = to form a true statement.

a. $\frac{1}{2} \qquad 0.54$

b. $0.\overline{2} \qquad \frac{2}{9}$

c. $\frac{5}{7} \qquad 0.72$

Practice Problem 8

Write the numbers in order from smallest to largest.

a. $\frac{1}{3}, \quad 0.302, \quad \frac{3}{8}$

b. $1.26, \quad 1\frac{1}{4}, \quad 1\frac{2}{5}$

c. $0.4, \quad 0.41, \quad \frac{5}{7}$

Answers

7. a. <   **b.** =   **c.** <

8. a. $0.302, \frac{1}{3}, \frac{3}{8}$ **b.** $1\frac{1}{4}, 1.26, 1\frac{2}{5}$

c. $0.4, 0.41, \frac{5}{7}$

C Solving Area Problems Containing Fractions and Decimals

Sometimes real-life problems contain both fractions and decimals. In this section, we will solve such problems about area.

Practice Problem 9

Find the area of the triangle.

2.1 meters
7 meters

Answer

9. 7.35 square meters

EXAMPLE 9 Finding the Area of a Triangle

The area of a triangle is Area $= \frac{1}{2} \cdot$ base $\cdot$ height. Find the area of the triangle shown.

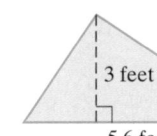

3 feet
5.6 feet

Solution:

$$
\begin{aligned}
\text{area} &= \frac{1}{2} \cdot \text{base} \cdot \text{height} \\
&= \frac{1}{2} \cdot 5.6 \cdot 3 \qquad \text{Let base} = 5.6 \text{ and height} = 3. \\
&= 0.5 \cdot 5.6 \cdot 3 \qquad \text{Write } \frac{1}{2} \text{ as the decimal } 0.5. \\
&= 8.4
\end{aligned}
$$

The area of the triangle is 8.4 square feet.

STUDY SKILLS REMINDER

Are you prepared for a test on Chapter 5?

Below I have listed some *common trouble areas* for topics covered in Chapter 5. After studying for your test—but before taking your test—read these.

- Don't forget the order of operations. To simplify $0.7 + 1.3(5 - 0.1)$, should you add, subtract, or multiply first? First, perform the subtraction within parentheses, then multiply, and finally add.

$$
\begin{aligned}
0.7 + 1.3(5 - 0.1) &= 0.7 + 1.3(4.9) \qquad \text{Subtract.} \\
&= 0.7 + 6.37 \qquad \text{Multiply.} \\
&= 7.07 \qquad \text{Add.}
\end{aligned}
$$

- If you are having trouble with ordering or operations on decimals, don't forget that you can insert 0s after the last digit to the right of the decimal point as needed.

Addition	Addition with zeros inserted	Subtraction	Subtraction with zeros inserted
			6 9 10
8.1	8.100	7	7.00
0.6	0.600	−0.28	−0.28
+23.003	+23.003		6.72
	31.703		

Place in order from smallest to largest: 0.108, 0.18, 0.0092.
If we insert zeros, we have 0.1080, 0.1800, 0.0092.
The decimals in order are 0.0092, 0.1080, 0.1800 or 0.0092, 0.108, 0.18.

EXERCISE SET 5.6

A *Write each fraction as a decimal. See Examples 1 through 5.*

1. $\dfrac{1}{5}$

2. $\dfrac{2}{5}$

3. $\dfrac{4}{8}$

4. $\dfrac{6}{8}$

 **5.** $\dfrac{3}{4}$

6. $\dfrac{6}{5}$

7. $-\dfrac{2}{25}$

8. $-\dfrac{3}{25}$

9. $\dfrac{3}{8}$

10. $\dfrac{1}{4}$

 **11.** $\dfrac{11}{12}$

12. $\dfrac{19}{25}$

13. $\dfrac{17}{40}$

14. $\dfrac{5}{12}$

15. $\dfrac{9}{20}$

16. $\dfrac{31}{40}$

17. $-\dfrac{1}{3}$

18. $-\dfrac{7}{9}$

19. $\dfrac{7}{16}$

20. $\dfrac{27}{16}$

21. $\dfrac{2}{9}$

22. $\dfrac{9}{11}$

23. $\dfrac{5}{3}$

24. $\dfrac{4}{11}$

Round each number as indicated.

25. Round your decimal answer to Exercise 17 to the nearest hundredth.

26. Round your decimal answer to Exercise 18 to the nearest hundredth.

27. Round your decimal answer to Exercise 19 to the nearest hundredth.

28. Round your decimal answer to Exercise 20 to the nearest hundredth.

29. Round your decimal answer to Exercise 21 to the nearest tenth.

30. Round your decimal answer to Exercise 22 to the nearest tenth.

31. Round your decimal answer to Exercise 23 to the nearest tenth.

32. Round your decimal answer to Exercise 24 to the nearest tenth.

Write each fraction as a decimal. When necessary, round all decimal answers to the nearest hundreth. See Examples 1 through 5.

33. In 2000, about $\frac{701}{1048}$ Americans used the Internet in some form. (*Source:* UCLA Center for Communications Policy)

34. About $\frac{21}{50}$ of all blood donors have type A blood. (*Source:* American Red Cross Biomedical Services)

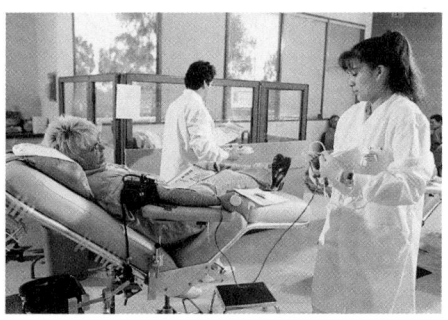

35. The National Athletic Trainers' Association is an organization dedicated to advancing the athletic training profession. Women make up approximately $\frac{11}{25}$ of its membership. (*Source:* National Athletic Trainers' Association)

36. In a recent year, approximately $\frac{29}{46}$ of all individuals who had flown in space were citizens of the United States. (*Source:* Congressional Research Service)

37. Of the U.S. mountains that are over 14,000 feet in elevation, $\frac{56}{91}$ are located in Colorado. (*Source:* U.S. Geological Survey)

38. In the United States, $\frac{1}{40}$ of all births are twins. (*Source:* Columbia Encyclopedia)

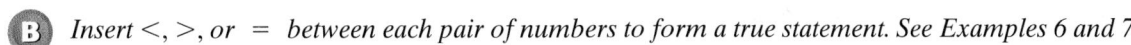

B *Insert* $<, >, or =$ *between each pair of numbers to form a true statement. See Examples 6 and 7.*

39. 0.562   0.569

40. 0.983   0.988

41. 0.823   0.813

42. 0.824   0.821

43. −0.0923   −0.0932

44. −0.00536   −0.00563

45. $\dfrac{2}{3}$ $\dfrac{5}{6}$

46. $\dfrac{1}{9}$ $\dfrac{2}{17}$

47. $\dfrac{5}{9}$ $\dfrac{51}{91}$

48. $\dfrac{7}{12}$ $\dfrac{6}{11}$

49. $\dfrac{4}{7}$ 0.14

50. $\dfrac{5}{9}$ 0.557

51. 1.38 $\dfrac{18}{13}$

52. 0.372 $\dfrac{22}{59}$

53. 7.123 $\dfrac{456}{64}$

54. 12.713 $\dfrac{89}{7}$

Write each list of numbers in order from smallest to largest. See Example 8.

55. 0.34, 0.35, 0.32

56. 0.47, 0.42, 0.40

57. 0.49, 0.491, 0.498

58. 0.72, 0.727, 0.728

59. $\dfrac{3}{4}$, 0.78, 0.73

60. $\dfrac{2}{5}$, 0.49, 0.42

61. $\dfrac{4}{7}$, 0.453, 0.412

62. $\dfrac{6}{9}$, 0.663, 0.668

63. 5.23, $\dfrac{42}{8}$, 5.34

64. 7.56, $\dfrac{67}{9}$, 7.562

65. $\dfrac{12}{5}$, 2.37, $\dfrac{17}{8}$

66. $\dfrac{29}{16}$, 1.75, $\dfrac{58}{32}$

Find the area of each triangle or rectangle. See Example 9.

67.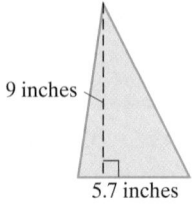

9 inches

5.7 inches

△ **68.**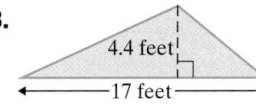

4.4 feet

17 feet

△ **69.**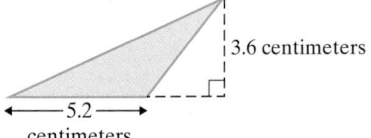

3.6 centimeters

5.2 centimeters

△ **70.**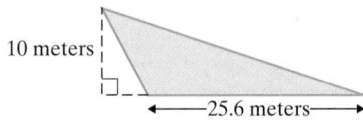

10 meters

25.6 meters

△ **71.**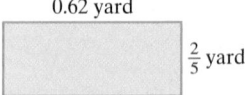

0.62 yard

$\frac{2}{5}$ yard

△ **72.**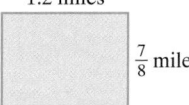

1.2 miles

$\frac{7}{8}$ mile

Review and Preview

Simplify. See Sections 1.8 and 4.3.

73. 2^3

74. 5^4

75. $6^2 \cdot 2$

76. $4 \cdot 3^4$

77. $\left(\dfrac{1}{3}\right)^4$

78. $\left(\dfrac{4}{5}\right)^3$

79. $\left(\dfrac{3}{5}\right)^2$

80. $\left(\dfrac{7}{2}\right)^2$

81. $\left(\dfrac{2}{5}\right)\left(\dfrac{5}{2}\right)^2$

82. $\left(\dfrac{2}{3}\right)\left(\dfrac{3}{2}\right)^2$

Combining Concepts

In 2002, there were 10,569 commercial radio stations in the United States. The most popular formats are listed in the table along with their counts. Use this graph to answer Exercises 83–86.

Top Commercial Radio Station Formats in 2002

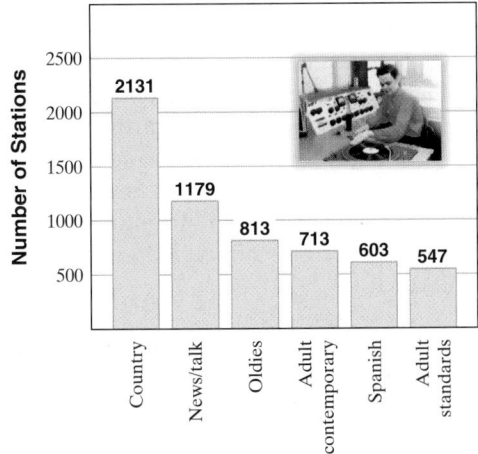

Format (Total stations: 10,569)

83. Write the fraction of radio stations with a country music format as a decimal. Round to the nearest thousandth.

84. Write the fraction of radio stations with a news/talk format as a decimal. Round to the nearest hundredth.

85. Estimate, by rounding each number in the table to the nearest hundred, the total number of stations with the top six formats in 2002.

86. Use your estimate from Exercise 85 to write the fraction of radio stations accounted for by the top six formats as a decimal. Round to the nearest hundredth.

87. Suppose that you are writing a paper on a word processor. You have been instructed to use $\frac{5}{8}$-inch margins. To change margins in the word processing software, you must enter margin widths in inches as a decimal. What number should you enter?

88. Suppose that you are using a word processor to format a table for a presentation. The first column of the table must be $3\frac{7}{16}$ inches wide. To adjust the width of the column in the word processing software, you must enter the column width in inches as a decimal. What number should you enter?

89. Describe how to determine the larger of two fractions.

Find the value of each expression. Give the result as a decimal.

90. $\frac{1}{5} - 2(7.8)$

91. $\frac{3}{4} - (9.6)(5)$

92. $8.25 - \left(\frac{1}{2}\right)^2$

93. $\left(\frac{1}{10}\right)^2 + (1.6)(2.1)$

94. $\frac{1}{4}(-9.6 - 5.2)$

95. $\frac{3}{8}(4.7 - 5.9)$

THE GOLDEN RECTANGLE IN ART

The golden rectangle is a rectangle whose length is approximately 1.6 times its width. The early Greeks thought that a rectangle with these dimensions was the most pleasing to the eye. Examples of the golden rectangle are found in many ancient, as well as modern, works of art. For example, the Parthenon in Athens, Greece, shows the golden rectangle in many aspects of its design. Modern-era artists, including Piet Mondrian (1872–1944) and Georges Seurat (1859–1891), also frequently used the proportions of a golden rectangle in their paintings.

To test whether a rectangle is a golden rectangle, divide the rectangle's length by its width. If the result is approximately 1.6, we can consider the rectangle to be a golden rectangle. For instance, consider Mondrian's *Composition with Gray and Light Brown*, which was painted on an 80.2 cm × 49.9 cm canvas. Because $\frac{80.2}{49.9} \approx 1.6$, the dimensions of the canvas form a golden rectangle. In what other ways are golden rectangles connected with this painting?

Examples of golden rectangles can be found in the designs of many everyday objects. Visual artists, from architects to product and package designers, use the golden rectangle shape in such things as the face of a building, the floor of a room, the front of a food package, the front cover of a book, and even the shape of a credit card.

GROUP ACTIVITY

Find an example of a golden rectangle in a building or an everyday object. Use a ruler to measure its dimensions and verify that the length is approximately 1.6 times the width.

5.7   Equations Containing Decimals

A   Solving Equations Containing Decimals

In this section, we continue our work with decimals and algebra by solving equations containing decimals. First, we review the steps given earlier for solving an equation.

Steps for Solving an Equation in *x*

Step 1. If fractions are present, multiply both sides of the equation by the LCD of the fractions.

Step 2. If parentheses are present, use the distributive property.

Step 3. Combine any like terms on each side of the equation.

Step 4. Use the addition property of equality to rewrite the equation so that variable terms are on one side of the equation and constant terms are on the other side.

Step 5. Divide both sides by the numerical coefficient of x to solve.

Step 6. Check the answer in the **original equation**.

EXAMPLE 1   Solve: $x - 1.5 = 8$

Solution: Steps 1 through 3 are not needed for this equation, so we begin with Step 4. To get *x* alone on one side of the equation, add 1.5 to both sides.

$$
\begin{aligned}
x - 1.5 &= 8 &&\text{Original equation} \\
x - 1.5 + 1.5 &= 8 + 1.5 &&\text{Add 1.5 to both sides.} \\
x &= 9.5 &&\text{Simplify.}
\end{aligned}
$$

Check: To check, replace *x* with 9.5 in the *original equation*.

$$
\begin{aligned}
x - 1.5 &= 8 &&\text{Original equation} \\
9.5 - 1.5 &= 8 &&\text{Replace } x \text{ with 9.5.} \\
8 &= 8 &&\text{True.}
\end{aligned}
$$

Since $8 = 8$ is a true statement, 9.5 is a solution of the equation.

Practice Problem 1

Solve: $z + 0.9 = 1.3$

EXAMPLE 2   Solve: $-2y = 6.7$

Solution: Steps 1 through 4 are not needed for this equation, so we begin with Step 5. To solve for *y*, divide both sides by the coefficient of *y*, which is -2.

$$
\begin{aligned}
-2y &= 6.7 &&\text{Original equation} \\
\frac{-2y}{-2} &= \frac{6.7}{-2} &&\text{Divide both sides by } -2. \\
y &= -3.35 &&\text{Simplify.}
\end{aligned}
$$

Check: To check, replace *y* with -3.35 in the original equation.

$$
\begin{aligned}
-2y &= 6.7 &&\text{Original equation} \\
-2(-3.35) &= 6.7 &&\text{Replace } y \text{ with } -3.35. \\
6.7 &= 6.7 &&\text{True.}
\end{aligned}
$$

Thus -3.35 is a solution of the equation $-2y = 6.7$.

Practice Problem 2

Solve: $0.17x = -0.34$

Answers

1. 0.4   **2.** -2

Practice Problem 3

Solve: $2.9 = 1.7 + 0.3x$

EXAMPLE 3 Solve: $1.2x + 5.8 = 8.2$

Solution: We begin with Step 4 and get the variable term alone by subtracting 5.8 from both sides.

$$1.2x + 5.8 = 8.2$$
$$1.2x + 5.8 - 5.8 = 8.2 - 5.8 \quad \text{Subtract 5.8 from both sides.}$$
$$1.2x = 2.4 \quad \text{Simplify.}$$
$$\frac{1.2x}{1.2} = \frac{2.4}{1.2} \quad \text{Divide both sides by 1.2.}$$
$$x = 2 \quad \text{Simplify.}$$

To check, replace x with 2 in the original equation. The solution is 2. ●

Practice Problem 4

Solve: $8x + 4.2 = 10x + 11.6$

EXAMPLE 4 Solve: $7x + 3.2 = 4x - 1.6$

Solution:

$$7x + 3.2 = 4x - 1.6$$
$$7x + 3.2 - 3.2 = 4x - 1.6 - 3.2 \quad \text{Subtract 3.2 from both sides.}$$
$$7x = 4x - 4.8 \quad \text{Simplify.}$$
$$7x - 4x = 4x - 4.8 - 4x \quad \text{Subtract } 4x \text{ from both sides.}$$
$$3x = -4.8 \quad \text{Simplify.}$$
$$\frac{\overset{1}{\cancel{3}}x}{\underset{1}{\cancel{3}}} = -\frac{4.8}{3} \quad \text{Divide both sides by 3.}$$
$$x = -1.6 \quad \text{Simplify.}$$

Check to see that -1.6 is the solution. ●

Practice Problem 5

Solve: $6.3 - 5x = 3(x + 2.9)$

EXAMPLE 5 Solve: $5(x - 0.36) = -x + 2.4$

Solution: First use the distributive property to distribute the factor 5.

$$5(x - 0.36) = -x + 2.4 \quad \text{Original equation}$$
$$5x - 1.8 = -x + 2.4 \quad \text{Apply the distributive property.}$$

Next, get x alone on the left side of the equation by adding 1.8 to both sides of the equation and then adding x to both sides of the equation.

$$5x - 1.8 + 1.8 = -x + 2.4 + 1.8 \quad \text{Add 1.8 to both sides.}$$
$$5x = -x + 4.2 \quad \text{Simplify.}$$
$$5x + x = -x + 4.2 + x \quad \text{Add } x \text{ to both sides.}$$
$$6x = 4.2 \quad \text{Simplify.}$$
$$\frac{6x}{6} = \frac{4.2}{6} \quad \text{Divide both sides by 6.}$$
$$x = 0.7 \quad \text{Simplify.}$$

To verify that 0.7 is the solution, replace x with 0.7 in the original equation. ●

Instead of solving equations with decimals, sometimes it may be easier to first rewrite the equation so it contains integers only. Recall that multiplying a decimal by a power of 10, such as 10, 100, or 1000, has the effect of moving the decimal point to the right. We can use the multiplication property of equality to multiply both sides of the equation through by an appropriate power of 10. The resulting equivalent equation will contain integers only.

Answers

3. 4 **4.** -3.7 **5.** -0.3

EXAMPLE 6   Solve: $0.5y + 2.3 = 1.65$

Solution: Multiply the equation through by 100. This will move the decimal point in each term two places to the right.

$$0.5y + 2.3 = 1.65 \qquad \text{Original equation}$$

$$100(0.5y + 2.3) = 100(1.65) \qquad \text{Multiply both sides by 100.}$$

$$100(0.5y) + 100(2.3) = 100(1.65) \qquad \text{Apply the distributive property.}$$

$$50y + 230 = 165 \qquad \text{Simplify.}$$

Now the equation contains integers only. Finish solving by subtracting 230 from both sides.

$$50y + 230 = 165$$

$$50y + 230 - 230 = 165 - 230 \qquad \text{Subtract 230 from both sides.}$$

$$50y = -65 \qquad \text{Simplify.}$$

$$\frac{50y}{50} = \frac{-65}{50} \qquad \text{Divide both sides by 50.}$$

$$y = -1.3 \qquad \text{Simplify.}$$

Check to see that -1.3 is the solution by replacing y with -1.3 in the original equation.

Try the Concept Check in the margin.

Practice Problem 6

Solve: $0.2y + 2.6 = 4$

Concept Check

By what number would you multiply both sides of the following equation to make calculations easier? Explain your choice.

$$1.7x + 3.655 = -14.2$$

Answer

6. 7

Concept Check: Multiply by 1000.

FOCUS ON Business and Career

IN-DEMAND OCCUPATIONS

According to U.S. Bureau of Labor Statistics projections, the careers listed below will have the largest job growth in the next decade:

Occupation	Employment (numbers in thousands)		
	1998	2008	Change
1. Systems analysts	617	1194	+577
2. Retail salespersons	4056	4620	+564
3. Cashiers	3198	3754	+556
4. General managers and top executives	3362	3913	+551
5. Truck drivers, light and heavy	2970	3463	+493
6. Office clerks, general	3021	3484	+463
7. Registered nurses	2079	2530	+451
8. Computer support specialists	429	869	+440
9. Personal care and home-health aides	746	1179	+433
10. Teacher assistants	1192	1567	+375

(*Source:* Bureau of Labor Statistics, U.S. Department of Labor)

What do all of these in-demand occupations have in common? They all require a knowledge of math! For some careers like systems analysts, salespersons, cashiers, and nurses, the ways math is used on the job may be obvious. For other occupations, the use of math may not be quite as obvious. However, tasks common to many jobs, such as filling in a time sheet or a mileage log, writing up an expense report, planning a budget, figuring a bill, ordering supplies, completing a packing list, and even making a work schedule, all require math.

CRITICAL THINKING

Suppose that your college placement office is planning to publish an occupational handbook on math in popular occupations. Choose one of the occupations from the list above that interests you. Research the occupation. Then write a brief entry for the occupational handbook that describes how a person in that career would use math in his or her job. Include an example if possible.

EXERCISE SET 5.7

A *Solve each equation. See Examples 1 and 2.*

1. $x + 1.2 = 7.1$

2. $y - 0.5 = 9$

3. $-5y = 2.15$

4. $-0.4x = 50$

5. $6.2 = y - 4$

6. $9.7 = x + 11.6$

7. $3.1x = -13.95$

8. $3y = -25.8$

Solve each equation. See Examples 3 through 5.

9. $-3.5x + 2.8 = -11.2$

10. $7.1 - 0.2x = 6.1$

11. $6x + 8.65 = 3x + 10$

12. $7x - 9.64 = 5x + 2.32$

 13. $2(x - 1.3) = 5.8$

14. $5(x + 2.3) = 19.5$

Solve each equation by first multiplying both sides through by an appropriate power of 10 so that the equation contains integers only. See Example 6.

15. $0.4x + 0.7 = -0.9$

16. $0.7x + 0.1 = 1.5$

17. $7x - 10.8 = x$

18. $3y = 7y + 24.4$

19. $2.1x + 5 - 1.6x = 10$

20. $1.5x + 2 - 1.2x = 12.2$

Solve.

21. $y - 3.6 = 4$

22. $x + 5.7 = 8.4$

23. $-0.02x = -1.2$

24. $-9y = -0.162$

25. $6.5 = 10x + 7.2$

26. $2x - 4.2 = 8.6$

27. $2.7x - 25 = 1.2x + 5$

28. $9y - 6.9 = 6y - 11.1$

29. $200x - 0.67 = 100x + 0.81$

30. $2.3 + 500x = 600x - 0.2$

31. $3(x + 2.71) = 2x$

32. $7(x + 8.6) = 6x$

33. $8x - 5 = 10x - 8$

34. $24y - 10 = 20y - 17$

35. $1.2 + 0.3x = 0.9$

36. $1.5 = 0.4x + 0.5$

37. $-0.9x + 2.65 = -0.5x + 5.45$

38. $-50x + 0.81 = -40x - 0.48$

39. $4x + 7.6 = 2(3x - 3.2)$

40. $4(2x - 1.6) = 5x - 6.4$

41. $0.7x + 13.8 = x - 2.16$

42. $y - 5 = 0.3y + 4.1$

Review and Preview

Simplify each expression by combining like terms. If parentheses are present, first use the distributive property. See Section 3.1.

43. $2x - 6 + 4x - 10$

44. $x - 4 - x - 4$

45. $3(x - 5) + 10$

46. $9x + 4(x + 20)$

47. $5y - 1.2 - 7y + 8$

48. $7.76 + 8z - 12z + 8.91$

Combining Concepts

49. Explain in your own words the property of equality that allows us to multiply an equation through by a power of 10.

50. Construct an equation whose solution is 1.4.

Solve.

51. $-5.25x = -40.33575$

52. $7.68y = -114.98496$

53. $1.95y + 6.834 = 7.65y - 19.8591$

54. $6.11x + 4.683 = 7.51x + 18.235$

5.8   Square Roots and the Pythagorean Theorem

A   Finding Square Roots

The square of a number is the number times itself. For example,

The square of 5 is 25 because 5^2 or $5 \cdot 5 = 25$.
The square of -5 is also 25 because $(-5)^2$ or $(-5)(-5) = 25$.

The reverse process of squaring is finding a **square root**. For example,

A square root of 25 is 5 because $^2 = 25$.
A square root of 25 is also because $(-5)^2 = 25$.

Every positive number has two square roots. We see above that the square roots of 25 are 5 and -5.

We use the symbol $\sqrt{}$, called a **radical sign**, to indicate the positive square root of a nonnegative number. For example,

$\sqrt{25} = 5$ because $5^2 = 25$ and 5 is positive.
$\sqrt{9} = 3$ because $3^2 = 9$ and 3 is positive.

Square Root of a Number

The square root, $\sqrt{}$, of a positive number a is the positive number b whose square is a. In symbols,

$$\sqrt{a} = b, \quad \text{if } b^2 = a$$

Also, $\sqrt{0} = 0$.

> **Helpful Hint**
>
> Remember that the radical sign $\sqrt{}$ is used to indicate the **positive square root** of a nonnegative number.

EXAMPLES   Find each square root.

1. $\sqrt{49} = 7$ because $7^2 = 49$.
2. $\sqrt{36} = 6$ because $6^2 = 36$.
3. $\sqrt{1} = 1$ because $1^2 = 1$.
4. $\sqrt{81} = 9$ because $9^2 = 81$.
5. Find $\sqrt{\dfrac{1}{36}} = \dfrac{1}{6}$ because $\left(\dfrac{1}{6}\right)^2$ or $\dfrac{1}{6} \cdot \dfrac{1}{6} = \dfrac{1}{36}$.
6. Find $\sqrt{\dfrac{4}{25}} = \dfrac{2}{5}$ because $\left(\dfrac{2}{5}\right)^2$ or $\dfrac{2}{5} \cdot \dfrac{2}{5} = \dfrac{4}{25}$.

B   Approximating Square Roots

Thus far, we have found square roots of perfect squares. Numbers like $\dfrac{1}{4}, 36, \dfrac{4}{25}$, and 1 are called **perfect squares** because their square root is a whole number or a fraction. A square root such as $\sqrt{5}$ cannot be written as a whole number or a fraction since 5 is not a perfect square.

Practice Problems

Find each square root.

1. $\sqrt{100}$
2. $\sqrt{64}$
3. $\sqrt{121}$
4. $\sqrt{0}$
5. $\sqrt{\dfrac{1}{4}}$
6. $\sqrt{\dfrac{9}{16}}$

Answers

1. 10   **2.** 8   **4.** 11   **3.** 0   **5.** $\dfrac{1}{2}$ **6.** $\dfrac{3}{4}$

Although $\sqrt{5}$ cannot be written as a whole number or a fraction, it can be approximated by estimating, by using a table (as in Appendix B), or by using a calculator.

Use Appendix B or a calculator to approximate the square root of 11 to the nearest thousandth.

EXAMPLE 7 Use Appendix B or a calculator to approximate the square root of 43 to the nearest thousandth.

Solution: $\sqrt{43} \approx 6.557$ ●

> **Helpful Hint**
>
> $\sqrt{43}$ is *approximately* 6.557. This means that if we multiply 6.557 by 6.557, the product is *close* to 43.
>
> $$6.557 \times 6.557 = 42.994249$$

Practice Problem 8

Approximate $\sqrt{29}$ to the nearest thousandth.

EXAMPLE 8 Approximate $\sqrt{32}$ to the nearest thousandth.

Solution: $\sqrt{32} \approx 5.657$ ●

C Using the Pythagorean Theorem

One important application of square roots has to do with right triangles. Recall that a **right triangle** is a triangle in which one of the angles is a right angle, or measures 90° (degrees). The **hypotenuse** of a right triangle is the side opposite the right angle. The **legs** of a right triangle are the other two sides. These are shown in the following figure. The right angle in the triangle is indicated by the small square drawn in that angle.

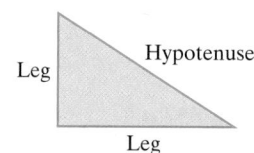

The following theorem is true for all right triangles.

> **Pythagorean Theorem**
>
> If a and b are the lengths of the legs of a right triangle and c is the length of the hypotenuse, then
>
> $$a^2 + b^2 = c^2$$
>
>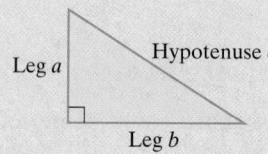
>
> In other words, $(\text{leg})^2 + (\text{other leg})^2 = (\text{hypotenuse})^2$.

Answers

7. ≈ 3.317 **8.** ≈ 5.385

△ **EXAMPLE 9**   Find the length of the hypotenuse of the given right triangle.

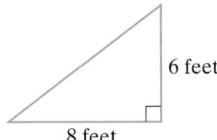

Solution: Let $a = 6$ and $b = 8$. According to the Pythagorean theorem,

$$a^2 + b^2 = c^2$$
$$6^2 + 8^2 = c^2 \quad \text{Let } a = 6 \text{ and } b = 8.$$
$$36 + 64 = c^2 \quad \text{Evaluate } 6^2 \text{ and } 8^2.$$
$$100 = c^2 \quad \text{Add.}$$

In the equation $c^2 = 100$, the solutions of c are the square roots of 100. This means that the solutions are 10 and -10. Since c represents a length, we are only interested in the positive square root of c^2.

$$c = \sqrt{100}$$
$$= 10$$

The hypotenuse is 10 feet long.   ●

△ **EXAMPLE 10**   Approximate the length of the hypotenuse of the given right triangle. Round the length to the nearest whole unit.

Solution: Let $a = 17$ and $b = 10$.

$$a^2 + b^2 = c^2$$
$$17^2 + 10^2 = c^2$$
$$289 + 100 = c^2$$
$$389 = c^2$$
$$\sqrt{389} = c \text{ or } c \approx 20 \quad \text{From Appendix B or a calculator.}$$

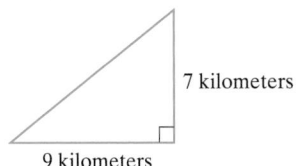

The hypotenuse is exactly $\sqrt{389}$ meters, which is approximately 20 meters.●

△ **EXAMPLE 11**   Find the length of the leg in the given right triangle. Give the exact length and a two-decimal-place approximation.

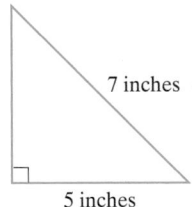

Solution: Notice that the hypotenuse measures 7 inches and that the length of one leg measures 5 inches. Thus, let $c = 7$ and a or b be 5. We will let $a = 5$.

$$a^2 + b^2 = c^2$$
$$5^2 + b^2 = 7^2 \quad \text{Let } a = 5 \text{ and } c = 7.$$
$$25 + b^2 = 49 \quad \text{Evaluate } 5^2 \text{ and } 7^2.$$
$$b^2 = 24 \quad \text{Subtract 25 from both sides.}$$
$$b = \sqrt{24} \approx 4.90$$

The length of the leg is exactly $\sqrt{24}$ inches and approximately 4.90 inches.   ●

Try the Concept Check in the margin.

Practice Problem 9

Find the length of the hypotenuse of the given right triangle.

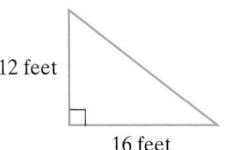

Practice Problem 10

Approximate the length of the hypotenuse of the given right triangle. Round to the nearest whole unit.

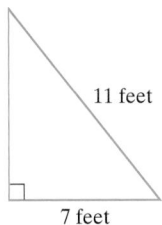

Practice Problem 11

Find the length of the leg in the given right triangle. Give the exact length and a two-decimal-place approximation.

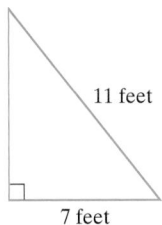

Concept Check

The following lists are the lengths of the sides of two triangles. Which set forms a right triangle?

a.   8, 15, 17

b.   24, 30, 40

Answers

9. 20 feet   **10.** 11 kilometers
11. $\sqrt{72}$ feet $\approx$ 8.49 feet

Concept Check: a

Practice Problem 12 △

A football field is a rectangle measuring 100 yards by 53 yards. Draw a diagram and find the length of the diagonal of a football field to the nearest yard.

Answer

12. ≈113 yards

EXAMPLE 12 Finding the Dimensions of a Park

An inner-city park is in the shape of a square that measures 300 feet on a side. A sidewalk is to be constructed along the diagonal of the park. Find the length of the sidewalk rounded to the nearest whole foot.

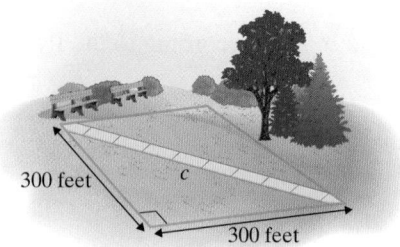

300 feet c 300 feet

Solution: The diagonal is the hypotenuse of a right triangle, which we label c.

$$a^2 + b^2 = c^2$$
$$300^2 + 300^2 = c^2 \qquad \text{Let } a = 300 \text{ and } b = 300.$$
$$90{,}000 + 90{,}000 = c^2 \qquad \text{Evaluate } (300)^2.$$
$$180{,}000 = c^2 \qquad \text{Add.}$$
$$\sqrt{180{,}000} = c \text{ or } c \approx 424$$

The length of the sidewalk is approximately 424 feet. ●

CALCULATOR EXPLORATIONS

Finding Square Roots

To simplify or approximate square roots using a calculator, locate the key marked $\boxed{\sqrt{}}$.

To simplify $\sqrt{64}$, for example, press the keys

$\boxed{64}\ \boxed{\sqrt{}}$ or $\boxed{\sqrt{}}\ \boxed{64}$

The display should read $\boxed{8}$. Then

$\sqrt{64} = 8$

To *approximate* $\sqrt{10}$, press the keys

$\boxed{10}\ \boxed{\sqrt{}}$ or $\boxed{\sqrt{}}\ \boxed{10}$

The display should read $\boxed{3.16227766}$. This is an *approximation* for $\sqrt{10}$. A three-decimal-place approximation is

$\sqrt{10} \approx 3.162$

Is this answer reasonable? Since 10 is between the perfect squares 9 and 16, $\sqrt{10}$ is between $\sqrt{9} = 3$ and $\sqrt{16} = 4$. Our answer is reasonable since 3.162 is between 3 and 4.

Simplify.

1. $\sqrt{1024}$ **2.** $\sqrt{676}$

Approximate each square root. Round each answer to the nearest thousandth.

3. $\sqrt{15}$ **4.** $\sqrt{19}$
5. $\sqrt{97}$ **6.** $\sqrt{56}$

Name _____ Section _____ Date _____

EXERCISE SET 5.8

A *Find each square root. See Examples 1 through 6.*

1. $\sqrt{4}$

2. $\sqrt{9}$

3. $\sqrt{625}$

4. $\sqrt{16}$

5. $\sqrt{\dfrac{1}{81}}$

6. $\sqrt{\dfrac{1}{64}}$

7. $\sqrt{\dfrac{144}{64}}$

8. $\sqrt{\dfrac{36}{81}}$

9. $\sqrt{256}$

10. $\sqrt{144}$

11. $\sqrt{\dfrac{9}{4}}$

12. $\sqrt{\dfrac{121}{169}}$

B *Use Appendix B or a calculator to approximate each square root. Round the square root to the nearest thousandth. See Examples 7 and 8.*

13. $\sqrt{3}$

14. $\sqrt{5}$

15. $\sqrt{15}$

16. $\sqrt{17}$

17. $\sqrt{14}$

18. $\sqrt{18}$

19. $\sqrt{47}$

20. $\sqrt{85}$

21. $\sqrt{8}$

22. $\sqrt{10}$

23. $\sqrt{26}$

24. $\sqrt{35}$

25. $\sqrt{71}$

26. $\sqrt{62}$

27. $\sqrt{7}$

28. $\sqrt{2}$

C *Find the unknown length in each right triangle. If necessary, approximate the length to the nearest thousandth. See Examples 9 through 12.*

△ **29.**

5 inches

12 inches

△ **30.**

24 feet

25 feet

△ **31.**

12 centimeters

10 centimeters

△ **32.**

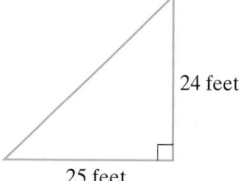

9 yards

3 yards

Square Roots and the Pythagorean Theorem   SECTION 5.8   **441**

Sketch each right triangle and find the length of the side not given. If necessary, approximate the length to the nearest thousandth. See Examples 9 through 12.

△ **33.** leg = 3, leg = 4

△ **34.** leg = 9, leg = 12

△ **35.** leg = 6, hypotenuse = 10

△ **36.** leg = 48, hypotenuse = 53

△ **37.** leg = 10, leg = 14

△ **38.** leg = 32, leg = 19

△ **39.** leg = 2, leg = 16

△ **40.** leg = 27, leg = 36

△ **41.** leg = 5, hypotenuse = 13

△ **42.** leg = 45, hypotenuse = 117

△ **43.** leg = 35, leg = 28

△ **44.** leg = 30, leg = 15

△ **45.** leg = 30, leg = 30

△ **46.** leg = 110, leg = 132

△ **47.** hypotenuse = 2, leg = 1

△ **48.** hypotenuse = 7, leg = 6

Solve. See Example 12.

△ **49.** A standard city block is a square with each side measuring 100 yards. Find the length of the diagonal of a city block to the nearest hundredth yard.

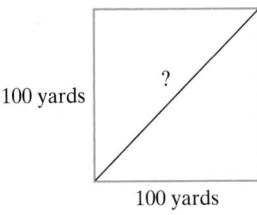

100 yards

?

100 yards

△ **50.** A section of land is a square with each side measuring 1 mile. Find the length of the diagonal of a section of land to the nearest thousandth mile.

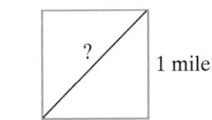

? 1 mile

△ **51.** Find the height of the tree. Round the height to one decimal place.

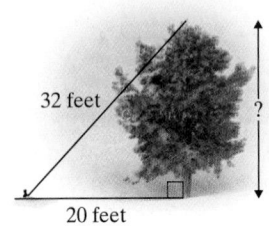

32 feet

?

20 feet

△ **52.** Find the height of the antenna. Round the height to one decimal place.

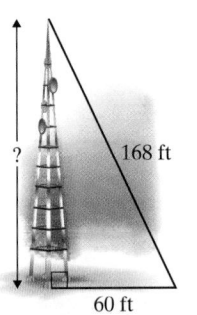

? 168 ft

60 ft

53. A football field is a rectangle that is 300 feet long by 160 feet wide. Find, to the nearest foot, the length of a straight-line run that started at one corner and went diagonally to end at the opposite corner.

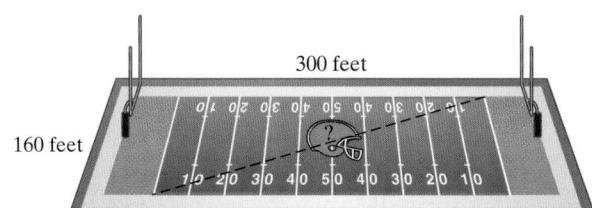

300 feet

160 feet

54. A baseball diamond is the shape of a square with 90-foot sides. Find the distance across the diamond from third base to first base to the nearest tenth of a foot.

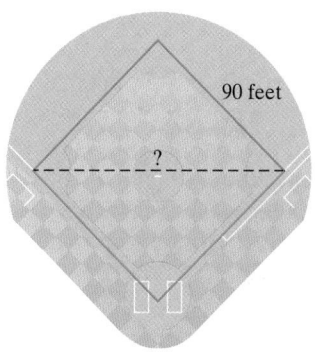

90 feet

?

Review and Preview

Write each fraction in simplest form. See Section 4.2.

55. $\dfrac{10}{12}$

56. $\dfrac{10}{15}$

57. $\dfrac{24}{60}$

58. $\dfrac{35}{75}$

59. $\dfrac{30}{72}$

60. $\dfrac{18}{30}$

Combining Concepts

Determine what two whole numbers each square root is between without using a calculator or table. Then use a calculator or table to check.

61. $\sqrt{38}$

62. $\sqrt{27}$

63. $\sqrt{101}$

64. $\sqrt{85}$

65. Without using a calculator, explain why you know that $\sqrt{11}$ is between 3 and 4.

66. Find the exact length of x. Then give a 2-decimal place approximation.

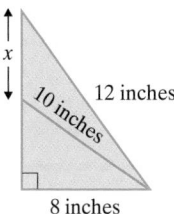

x

12 inches

10 inches

8 inches

CHAPTER 5 ACTIVITY Maintaining a Checking Account

This activity may be completed by working in groups or individually.

A checking account is a convenient way of handling money and paying bills. To open a checking account, the bank or savings and loan association requires a customer to make a deposit. Then the customer receives a checkbook that contains checks, deposit slips, and a register for recording checks written and deposits made. It is important to record all payments and deposits that affect the account. It is also important to keep the checkbook balance current by subtracting checks written (and service charges) and adding deposits made (or interest earned).

About once a month checking customers receive a statement from the bank listing all activity that the account has had in the last month. The statement lists a beginning balance, all checks and deposits, any service charges made against the account, and an ending balance. Because it may take several days for checks that a customer has written to clear the banking system, the check register may list checks that do not appear on the monthly bank statement. These checks are called **outstanding checks**. Deposits that are recorded in the check register but do not appear on the statement are called **deposits in transit**. Because of these differences, it is important to balance, or reconcile, the checkbook against the monthly statement. The steps for doing so are listed at the right.

BALANCING OR RECONCILING A CHECKBOOK

Step 1. Place a check mark in the checkbook register next to each check and deposit listed on the monthly bank statement. Any entries in the register without a check mark are outstanding checks or deposits in transit.

Step 2. Find the ending checkbook register balance and add to it any outstanding checks and any interest paid on the account.

Step 3. From the total in Step 2, subtract any deposits in transit and any service charges.

Step 4. Compare the amount found in Step 3 with the ending balance listed on the bank statement. If they are the same, the checkbook balances with the bank statement. Be sure to update the check register with service charges and interest.

Step 5. If the checkbook does not balance, recheck the balancing process. Next, make sure that the running checkbook register balance was calculated correctly. Finally, compare the checkbook register with the statement to make sure that each check was recorded for the correct amount.

For the checkbook register and monthly bank statement given:

a. update the checkbook register

b. list the outstanding checks and deposits in transit

c. balance the checkbook—be sure to update the register with any interest or service fees

Checkbook Register						
						Balance
#	**Date**	**Description**	**Payment**	√	**Deposit**	425.86
114	4/1	Market Basket	30.27			
115	4/3	May's Texaco	8.50			
	4/4	Cash at ATM	50.00			
116	4/6	UNO Bookstore	121.38			
	4/7	Deposit			100.00	
117	4/9	MasterCard	84.16			
118	4/10	Blockbuster	6.12			
119	4/12	Kroger	18.72			
120	4/14	Parking sticker	18.50			
	4/15	Direct deposit			294.36	
121	4/20	Rent	395.00			
122	4/25	Student fees	20.00			
	4/28	Deposit			75.00	

FIRST NATIONAL BANK		
Monthly Statement 4/30		
BEGINNING BALANCE:		425.86
Date	Number	Amount
CHECKS AND ATM WITHDRAWALS		
4/3	114	30.27
4/4	ATM	50.00
4/11	117	84.16
4/13	115	8.50
4/15	119	18.72
4/22	121	395.00
DEPOSITS		
4/7		100.00
4/15	Direct deposit	294.36
SERVICE CHARGES		
Low balance fee		7.50
INTEREST		
Credited 4/30 1.15		
ENDING BALANCE:		227.22

Chapter 5   Vocabulary Check

Fill in each blank with one of the words listed below.

vertically   decimal   and   sum

denominator   numerator

1. Like fractional notation, _____ notation is used to denote a part of a whole.

2. To write fractions as decimals, divide the _____ by the _____ .

3. To add or subtract decimals, write the decimals so that the decimal points line up _____ .

4. When writing decimals in words, write "___" for the decimal point.

5. When multiplying decimals, the decimal point in the product is placed so that the number of decimal places in the product is equal to the _____ of the number of decimal places in the factors.

CHAPTER 5

Highlights

DEFINITIONS AND CONCEPTS	**EXAMPLES**
SECTION 5.1 INTRODUCTION TO DECIMALS	

PLACE-VALUE CHART

hundreds	tens	ones		tenths	hundredths	thousandths	ten-thousandths	hundred-thousandths
	1	7	.	7	5	8		
100	10	1	decimal point	$\frac{1}{10}$	$\frac{1}{100}$	$\frac{1}{1000}$	$\frac{1}{10,000}$	$\frac{1}{100,000}$

WRITE 3.08 IN WORDS.

Three and eight hundredths

ROUNDING DECIMALS TO A PLACE VALUE TO THE RIGHT OF THE DECIMAL POINT

Step 1. Locate the digit to the right of the given place value.

Step 2. If this digit is 5 or greater, add 1 to the digit in the given place value and drop all digits to its right. If this digit is less than 5, drop all digits to the right of the given place value.

Round 86.1256 to the nearest hundredth.

       hundredths place

86.1256

       5 or greater

rounds to 86.13

SECTION 5.2 ADDING AND SUBTRACTING DECIMALS	

Adding or Subtracting Decimals

  Step 1. Write the decimals so that the decimal points line up vertically.

  Step 2. Add or subtract as for whole numbers.

  Step 3. Place the decimal point in the sum or difference so that it lines up vertically with the decimal points in the problem.

Subtract: 2.8 − 1.04

$$\begin{array}{r} {\scriptstyle 7\ 10} \\ 2.8\cancel{0} \\ -\ 1.04 \\ \hline 1.76 \end{array}$$

DEFINITIONS AND CONCEPTS	**EXAMPLES**

SECTION 5.3 MULTIPLYING DECIMALS AND CIRCUMFERENCE OF A CIRCLE

Multiplying Decimals

Step 1. Multiply the decimals as though they were whole numbers.

Step 2. The decimal point in the product is placed so that the number of decimal places in the product is equal to the *sum* of the number of decimal places in the factors.

The circumference of a circle is the distance around the circle.

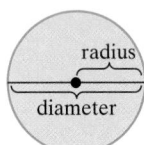

$C = \pi d$ or $C = 2\pi r$

where $\pi \approx 3.14$ or $\pi \approx \dfrac{22}{7}$.

or

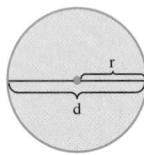

Multiply: 1.48×5.9

$$
\begin{array}{r}
1.48 \quad \leftarrow 2 \text{ decimal places.} \\
\times\, 5.9 \quad \leftarrow 1 \text{ decimal place.} \\
\hline
1332 \\
7400 \\
\hline
8.732 \quad \leftarrow 3 \text{ decimal places.}
\end{array}
$$

Find the exact circumference and a two-decimal-place approximation.

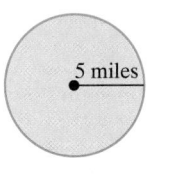

5 miles

$$
\begin{aligned}
C &= 2\pi r \\
&= 2\pi(5) \\
&= 10\pi \\
&\approx 10(3.14) \\
&= 31.4
\end{aligned}
$$

The circumference is exactly 10π miles and approximately 31.4 miles.

SECTION 5.4 DIVIDING DECIMALS

Dividing Decimals

Step 1. Move the decimal point in the divisor to the right until the divisor is a whole number.

Step 2. Move the decimal point in the dividend to the right the *same number of places* as the decimal point was moved in Step 1.

Step 3. Divide. The decimal point in the quotient is directly over the moved decimal point in the dividend.

Divide: $1.118 \div 2.6$

$$
\begin{array}{r}
0.43 \\
2.6\overline{)1.118} \\
-104 \\
\hline
78 \\
-78 \\
\hline
0
\end{array}
$$

SECTION 5.5 ESTIMATING AND ORDER OF OPERATIONS

Order of Operations

1. Do all operations within grouping symbols such as parentheses or brackets.

2. Evaluate any expressions with exponents.

3. Multiply or divide in order from left to right.

4. Add or subtract in order from left to right.

Simplify:

$$
\begin{aligned}
-1.9(12.8 - 4.1) &= -1.9(8.7) \quad \text{Subtract.} \\
&= -16.53 \quad \text{Multiply.}
\end{aligned}
$$

DEFINITIONS AND CONCEPTS	**EXAMPLES**

SECTION 5.6 FRACTIONS AND DECIMALS

To write fractions as decimals, divide the numerator by the denominator.

Write $\frac{3}{8}$ as a decimal.

$$
\begin{array}{r}
0.375 \\
8\overline{)3.000} \\
-24 \\
\hline
60 \\
-56 \\
\hline
40 \\
-\ 40 \\
\hline
0
\end{array}
$$

SECTION 5.7 EQUATIONS CONTAINING DECIMALS

Steps for Solving an Equation in x

Step 1. If fractions are present, multiply both sides of the equation by the LCD of the fractions.

Step 2. If parentheses are present, use the distributive property.

Step 3. Combine any like terms on each side of the equation.

Step 4. Use the addition property of equality to rewrite the equation so that variable terms are on one side of the equation and constant terms are on the other side.

Step 5. Divide both sides by the numerical coefficient of x to solve.

Step 6.   Check the answer in the *original equation*.

Solve:

$$
\begin{aligned}
3(x + 2.6) &= 10.92 \\
3x + 7.8 &= 10.92 & \text{Apply the distributive property.} \\
3x + 7.8 - 7.8 &= 10.92 - 7.8 & \text{Subtract 7.8 from both sides.} \\
3x &= 3.12 & \text{Simplify.} \\
\frac{3x}{3} &= \frac{3.12}{3} & \text{Divide both sides by 3.} \\
x &= 1.04 & \text{Simplify.}
\end{aligned}
$$

Check 1.04 in the original equation.

SECTION 5.8 SQUARE ROOTS AND THE PYTHAGOREAN THEOREM

SQUARE ROOT OF A NUMBER

The square root of a positive number a is the positive number b whose square is a. In symbols,

$$\sqrt{a} = b, \quad \text{if} \quad b^2 = a$$

Also, $\sqrt{0} = 0$.

$$\sqrt{9} = 3, \qquad \sqrt{100} = 10, \qquad \sqrt{1} = 1$$

PYTHAGOREAN THEOREM

If a and b are the lengths of the legs of a right triangle and c is the length of the hypotenuse, then

$$a^2 + b^2 = c^2$$

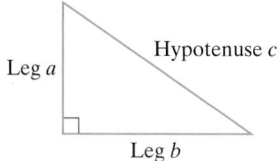

Leg a / Hypotenuse c / Leg b

Find c.

$a = 3$ c $b = 8$

$$
\begin{aligned}
a^2 + b^2 &= c^2 \\
3^2 + 8^2 &= c^2 & \text{Let } a = 3 \text{ and } b = 8. \\
9 + 64 &= c^2 & \text{Multiply.} \\
73 &= c^2 & \text{Simplify.} \\
\sqrt{73} &= c \text{ or } c \approx 8.5
\end{aligned}
$$

Chapter 5 Review

(5.1) *Determine the place value of the number 4 in each decimal.*

1. 23.45

2. 0.000345

Write each decimal in words.

3. −23.45

4. 0.00345

5. 109.23

6. 200.000032

Write each decimal in standard form.

7. Two and fifteen hundredths

8. Negative five hundred three and one hundred two thousandths

9. Sixteen thousand twenty-five and fourteen ten-thousandths

Write each decimal as a fraction or a mixed number.

10. 0.16   **11.** −12.023   **12.** 1.0045   **13.** 0.00231   **14.** 25.25

Insert <, >, or = between each pair of numbers to make a true statement.

15. 0.49   0.43

16. 0.973   0.9730

17. −402.00032   −402.000032

18. −0.230505   −0.23505

Round each decimal to the given place value.

19. 0.623, nearest tenth

20. 0.9384, nearest hundredth

21. −42.895, nearest hundredth

22. 16.34925, nearest thousandth

23. Every day in America an average of 13,490.5 people get married. Round this number to the nearest hundred.

24. A certain kind of chocolate candy bar contains 10.75 teaspoons of sugar. Convert this number to a mixed number.

(5.2) *Add.*

25. $2.4 + 7.1$

26. $3.9 + 1.2$

27. $-6.4 + (-0.88)$

28. $-19.02 + 6.98$

29. $200.49 + 16.82 + 103.002$

30. $0.00236 + 100.45 + 48.29$

Subtract.

31. $4.9 - 3.2$

32. $5.23 - 2.74$

33. $-892.1 - 432.4$

34. $0.064 - 10.2$

35. $100 - 34.98$

36. $200 - 0.00198$

37. The greatest one-day point loss on the Dow Jones Industrial Average (DJIA) occurred on September 17, 2001. On that day, the DJIA changed by -684.81 points and its value at the close of the stock market was 8920.70. What was the value of the DJIA when the stock market opened on September 17, 2001. (*Source:* Dow Jones & Co.)

38. Evaluate $x - y$ for $x = 1.2$ and $y = 6.9$.

Simplify by combining like terms.

39. $2.3x + 6.5 + 1.9x + 6.3$

40. $8.6y - 7.61 + 1.29y + 3.44$

(5.3) *Multiply.*

41. 7.2×10

42. 9.345×1000

43. -34.02×2.3

44. $-839.02 \times (-87.3)$

45. Find the exact circumference of the circle. Then use the approximation 3.14 for π and approximate the circumference.

7 meters

46. A kilometer is approximately 0.625 mile. It is 102 kilometers from Hays to Colby. Write 102 kilometers in miles to the nearest tenth mile.

(5.4) *Divide. Round the quotient to the nearest thousandth if necessary.*

47. $21 \div 0.3$

48. $-0.0063 \div 0.03$

49. $24.5 \div (-0.005)$

50. $54.98 \div 2.3$

51. $274 \div 34$

52. $-3165 \div (-20)$

53. There are approximately 3.28 feet in 1 meter. Find how many meters there are in 24 feet to the nearest tenth of a meter.

1 meter

~3.28 feet

54. George Strait pays $69.71 per month to pay back a loan of $3136.95. In how many months will the loan be paid off?

(5.5) *Perform each indicated operation. Then estimate to see whether your proposed result is reasonable.*

55. $2.4 + 6.7 + 9.1$

56. $15.9 + 34.1$

57. $340.03 - 240.98$

58. $100 - 45.9$

59. 6.02×5.91

60. 0.205×1.72

61. $62.13 \div 1.9$

62. $601.92 \div 19.8$

63. The table shows the world's top five consumers of primary energy (in quadrillion BTU's) for the year 2000. Estimate the total energy consumption of these countries by rounding each amount to the nearest whole.

United States	98.77
China	36.67
Russia	28.07
Japan	21.77
Germany	13.98

(*Source:* Energy Information Administration)

64. Julio Tomaso is going to fertilize his lawn, a rectangle measuring 77.3 feet by 115.9 feet. Approximate the area of the lawn by rounding each measurement to the nearest foot.

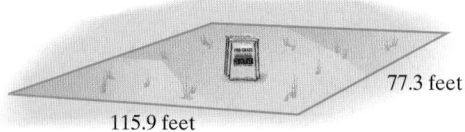

77.3 feet

115.9 feet

65. Estimate the cost of the items to see whether the groceries can be purchased with a $5 bill.

$1.89

BREAD

$1.07

3 cans for $0.99

Simplify each expression.

66. $(-7.6)(1.9) + 2.5$

67. $2.3^2 - 1.4$

68. $\dfrac{(-3.2)^2}{100}$

69. $(2.6 + 1.4)(4.5 - 3.6)$

(5.6) *Write each fraction as a decimal. Round to the nearest thousandth if necessary.*

70. $\dfrac{4}{5}$

71. $-\dfrac{12}{13}$

72. $-\dfrac{3}{7}$

73. $\dfrac{13}{60}$

74. $\dfrac{9}{80}$

75. $\dfrac{8935}{175}$

Insert $<, >,$ *or* $=$ *between each pair of numbers to form a true statement.*

76. $0.3920 \quad 0.392$

77. $\dfrac{4}{7} \quad \dfrac{5}{8}$

78. $0.293 \quad \dfrac{5}{17}$

79. $\dfrac{6}{11} \quad 0.55$

Write each list of numbers in order from smallest to largest.

80. $0.837, 0.839, 0.832$

81. $\dfrac{3}{7}, 0.42, 0.43$

82. $\dfrac{18}{11}, 1.63, \dfrac{19}{12}$

83. $\dfrac{6}{7}, \dfrac{8}{9}, \dfrac{3}{4}$

Find each area.

84.

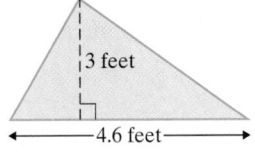

3 feet

4.6 feet

85.

2.1 inches

5.2 inches

(5.7) *Solve.*

86. $x + 3.9 = 4.2$

87. $70 = y + 22.81$

88. $2x = 17.2$

89. $-1.1y = 88$

90. $\dfrac{x}{4} = -0.12$

91. $6.8 = \dfrac{y}{5}$

92. $x + 0.78 = 1.2$

93. $0.56 = 2x$

94. $-1.3x - 9.4 = -0.4x + 8.6$

95. $3(x - 1.1) = 5x - 5.3$

(5.8) *Simplify.*

96. $\sqrt{64}$

97. $\sqrt{144}$

98. $\sqrt{36}$

99. $\sqrt{1}$

100. $\sqrt{\dfrac{4}{25}}$

101. $\sqrt{\dfrac{1}{100}}$

Find the unknown length in each given right triangle. If necessary, round to the nearest tenth.

△ **102.** leg = 12, leg = 5

△ **103.** leg = 20, leg = 21

△ **104.** leg = 9, hypotenuse = 14

△ **105.** leg = 124, hypotenuse = 155

△ **106.** leg = 66, leg = 56

△ **107.** Find the length to the nearest hundredth of the diagonal of a square that has a side of length 20 centimeters.

△ **108.** Find the height of the building rounded to the nearest tenth.

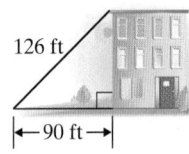

126 ft

90 ft

Chapter 5 Test   Remember to check your answers and use the Chapter Test Prep Video to view solutions.

Answers

1. _____

2. _____

3. _____

4. _____

5. _____

6. _____

7. _____

8. _____

9. _____

10. _____

11. _____

12. _____

13. _____

14. _____

15. _____

16. _____

17. _____

18. _____

19. _____

20. _____

21. _____

22. _____

23. _____

Write each decimal as indicated.

1. 45.092, in words

2. Three thousand and fifty-nine thousandths, in standard form

Perform each indicated operation. Round the result to the nearest thousandth if necessary.

3. $2.893 + 4.21 + 10.492$

4. $-47.92 - 3.28$

5. $9.83 - 30.25$

6. 10.2×4.01

7. $(-0.00843) \div (-0.23)$

Round each decimal to the indicated place value.

8. 34.8923, nearest tenth

9. 0.8623, nearest thousandth

Insert $<$, $>$, or $=$ between each pair of numbers to form a true statement.

10. 25.0909   25.9090

11. $\dfrac{4}{9}$ 0.445

Write each decimal as a fraction or a mixed number.

12. 0.345

13. -24.73

Write each fraction as a decimal. If necessary, round to the nearest thousandth.

14. $-\dfrac{13}{26}$

15. $\dfrac{16}{17}$

Simplify.

16. $(-0.6)^2 + 1.57$

17. $\dfrac{0.23 + 1.63}{-0.3}$

18. $2.4x - 3.6 - 1.9x - 9.8$

Find each square root and simplify. Round to the nearest thousandth if necessary.

19. $\sqrt{49}$

20. $\sqrt{157}$

21. $\sqrt{\dfrac{64}{100}}$

Solve.

22. $0.2x + 1.3 = 0.7$

23. $2(x + 5.7) = 6x - 3.4$

△ **24.** Approximate to the nearest hundredth of a centimeter the length of the missing side of a right triangle with legs of 4 centimeters each.

△ **25.** Find the area.

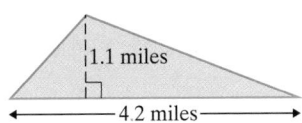

1.1 miles
4.2 miles

△ **26.** Vivian Thomas is going to put insecticide on her lawn to control grubworms. The lawn is a rectangle measuring 123.8 feet by 80 feet. The amount of insecticide required is 0.02 ounces per square foot. Find how much insecticide Vivian needs to purchase.

△ **27.** Find the exact circumference of the circle. Then use the approximation 3.14 for π and approximate the circumference.

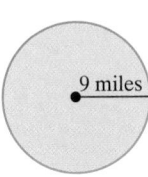

9 miles

28. A CD (compact disk) holds 700 megabytes of data. A DVD (digital video disc) holds 8740 megabytes of data. How many CDs does it take to hold as much data as a DVD? Round to the nearest whole CD.

29. Estimate the total distance from Bayette to Center City by rounding each distance to the nearest mile.

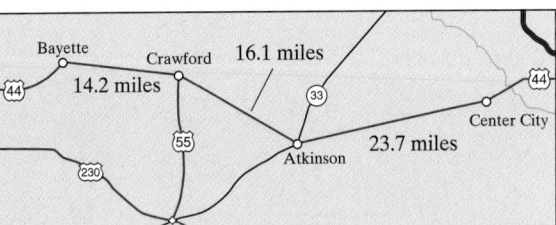

Bayette
Crawford   16.1 miles
44   14.2 miles
55
230   Atkinson   23.7 miles   Center City
33
44

Cumulative Review

1. Add: $34{,}285 + 149{,}761$

2. Add: $5{,}785 + 210{,}199$

3. Round each number to the nearest hundred to find an estimated sum.
$$294$$
$$625$$
$$1071$$
$$+\ 349$$

4. Round each number to the nearest hundred to find the estimated sum.
$$186$$
$$404$$
$$853$$
$$+1445$$

5. Divide: $6819 \div 17$

6. Divide: $2047 \div 14$

7. Evaluate $\dfrac{x - 5y}{y}$ for $x = 21$ and $y = 3$.

8. Evaluate $\dfrac{2a + 4}{c}$ for $a = 7$ and $c = 3$.

9. Find the opposite of each number.
   a. 11   **b.** -2 **c.** 0

10. Find the opposite of each number.
   a. -7 **b.** 4   **c.** -1

11. Add: $-2 + (-21)$

12. Add: $-7 + (-15)$

Find the value of each expression.

13. $2 \cdot 5^2$

14. $4 \cdot 2^3$

15. -3^2

16. $(-2)^5$

17. $(-3)^2$

18. -7^2

19. Simplify by combining like terms.
   a. $3x + 2x$
   b. $y - 7y$
   c. $3x^2 + 5x^2 - 2$

20. Simplify by combining like terms
   a. $2a - 5a$
   b. $-3z^2 - z^2$
   c. $4k + 2 + 7k$

21. Solve: $5x + 2 - 4x = 7 - 9$

22. Solve: $7x - 3 = 6x + 2$

23. Solve: $-5x = 15$

24. Solve: $-21 = -7x$

25. Solve: $7(x - 2) = 9x - 6$

26. Solve: $2(x + 4) = -x - 1$

27. _____

28. _____

29. _____

30. _____

31. _____

32. _____

33. _____

34. _____

35. _____

36. _____

37. _____

38. _____

39. _____

40. _____

41. _____

42. _____

43. _____

44. _____

45. _____

46. _____

47. _____

48. _____

49. _____

50. _____

27. In the 2000 Senate election in Wyoming, incumbent Craig Thomas received 110,280 _more_ votes than his challenger. If a total of 204,358 votes were cast, find how many votes Craig Thomas received. (_Source: World Almanac_)

28. Three times a number added to 84 is 153. Find the number.

Identify the numerator and the denominator of each fraction.

29. $\dfrac{3}{7}$

30. $\dfrac{2}{5a}$

31. $\dfrac{13x}{5}$

32. $\dfrac{7y}{3x}$

33. Simplify: $\dfrac{12}{20}$

34. Simplify: $\dfrac{64}{112}$

35. Multiply: $-\dfrac{1}{4} \cdot \dfrac{1}{2}$

36. Multiply: $\left(-\dfrac{1}{3}\right)\left(-\dfrac{1}{4}\right)$

Add and simplify.

37. $\dfrac{2}{7} + \dfrac{3}{7}$

38. $\dfrac{2}{5} + \dfrac{3}{5}$

39. $\dfrac{7}{8} + \dfrac{6}{8} + \dfrac{3}{8}$

40. $\dfrac{2}{3} + \dfrac{1}{3} + \dfrac{2}{3}$

41. $\dfrac{2}{5} + \dfrac{4}{15}$

42. $\dfrac{3}{4} + \dfrac{5}{12}$

43. Simplify: $\dfrac{\frac{x}{4}}{\frac{3}{2}}$

44. Simplify: $\dfrac{\frac{2}{3}}{\frac{4x}{9}}$

45. Solve: $\dfrac{3}{5}a = 9$

46. Solve: $\dfrac{2}{3}y = 12$

47. Add: $763.7651 + 22.001 + 43.89$

48. Add: $89.27 + 14.361 + 127.2318$

49. Multiply: 23.6×0.78

50. Multiply: 43.8×0.645

456

Ratio and Proportion

Having studied equations in Chapter 3 and quotients in fraction form in Chapter 4, we are ready to explore the useful notations of ratio and proportion. Ratio is another name for quotient and is usually written in fraction form. A proportion is an equation with 2 equal ratios.

The Olympic Games originated in 776 B.C. in Greece. There was only one event, a 192-meter foot race, but events such as wrestling, the pentathlon, boxing, and chariot racing were added later. After the Romans conquered Greece, the Olympic games were eventually banned in 393 A.D. The games resumed in modern times in Athens, Greece, in 1896. The first separate Winter Olympic Games were added in 1924. The Olympics are now held every two years, alternating between the winter and summer games. Each event awards three medals to the top finishers: gold, silver, and bronze. The award ceremony, which features the national anthem of the country of the gold medal winner, is one of the most moving ceremonies in modern sports. In Exercises 47 and 48 on page 463, we will see how ratios can be used to compare the success of different countries in the Olympic games.

Name _____ Section _____ Date_____

Chapter 6 Pretest

Write each ratio using fractional notation.

1. 15 to 19

2. $3\frac{4}{5}$ to $9\frac{1}{8}$

Write each ratio as a fraction in simplest form.

3. 5.1 to 7.9

4. $30 to $20

5. Write the rate as a fraction in simplest form: $6 for 3 pounds

6. Write as a unit rate: 292.5 miles every 15 gallons of gas

7. A store charges $3.27 for 3 boxes of cat food. What is the unit price in dollars per box?

Write each sentence as a proportion.

8. 18 credits is to 6 classes as 15 credits is to 5 classes.

9. 49 days is to 7 weeks as 35 days is to 5 weeks.

10. Is $\frac{3}{5} = \frac{21}{34}$ a true proportion?

11. Is $\frac{72}{9} = \frac{216}{27}$ a true proportion?

Solve.

12. $\frac{22}{x} = \frac{66}{12}$

13. $\frac{x}{8} = \frac{0.6}{2.4}$

14. $\frac{\frac{1}{2}}{8} = \frac{3}{y}$

15. $\frac{1\frac{1}{3}}{\frac{6}{5}} = \frac{n}{9}$

16. On Jill Tilbert's map, 1 inch represents 20 miles. Find the distance represented by $5\frac{3}{8}$ inches on her map.

△ **17.** Find the ratio of the corresponding sides of the given similar triangles.

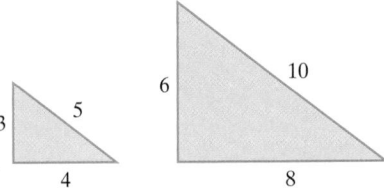

Given that the pairs of triangles are similar, find the length of the side labeled n.

△ **18.**

△ **19.**

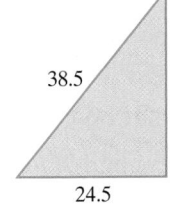

20. If a 40-foot tree casts a 22-foot shadow, find the length of the shadow cast by a 36-foot tree.

458

6.1 Ratios

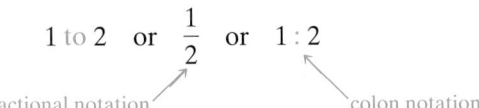 **Writing Ratios as Fractions**

A **ratio** is the quotient of two quantities. A ratio, in fact, is no different than a fraction, except that a ratio is sometimes written using notation other than fractional notation. For example, the ratio of 1 to 2 can be written as

$$1 \text{ to } 2 \quad \text{or} \quad \frac{1}{2} \quad \text{or} \quad 1 : 2$$

fractional notation   colon notation

These ratios are all read as "the ratio of 1 to 2."

Try the Concept Check in the margin.

In this section, we will write ratios using fractional notation.

> **Writing a Ratio as a Fraction**
>
> The order of the quantities is important when writing ratios. To write a ratio as a fraction, write the *first number* of the ratio as the *numerator* of the fraction and the *second number* as the *denominator*.

For example, the ratio of 6 to 11 is $\frac{6}{11}$, *not* $\frac{11}{6}$.

EXAMPLE 1 Write the ratio of 12 to 17 using fractional notation.

Solution: The ratio is $\frac{12}{17}$.

> **Helpful Hint**
>
> Don't forget that order is important when writing ratios.
> The ratio $\frac{17}{12}$ is *not* the same as the ratio $\frac{12}{17}$.

EXAMPLES Write each ratio using fractional notation. Do not simplify.

2. The ratio of 2.6 to 3.1 is $\frac{2.6}{3.1}$.

3. The ratio of $1\frac{1}{2}$ to $7\frac{3}{4}$ is $\dfrac{1\frac{1}{2}}{7\frac{3}{4}}$.

 **Simplifying Ratios**

To simplify a ratio, we just write the fraction in simplest form. Common factors can be divided out as well as common units.

EXAMPLE 4 Write the ratio of $15 to $10 as a fraction in simplest form.

Solution: $\dfrac{\$15}{\$10} = \dfrac{15}{10} = \dfrac{3 \cdot 5}{2 \cdot 5} = \dfrac{3}{2}$

The ratio of $15 to $10 as a fraction in simplest form is $\frac{3}{2}$.

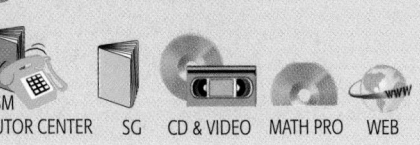

Concept Check

How should each ratio be read aloud?

a. $\dfrac{8}{5}$ b. $\dfrac{5}{8}$

Practice Problem 1

Write the ratio of 20 to 23 using fractional notation.

Practice Problems 2–3

Write each ratio using fractional notation.

2. The ratio of 10.3 to 15.1

3. The ratio of $3\frac{1}{3}$ to $12\frac{1}{5}$

Practice Problem 4

Write the ratio of $8 to $6 as a fraction in simplest form.

Answers

1. $\dfrac{20}{23}$ **2.** $\dfrac{10.3}{15.1}$ **3.** $\dfrac{3\frac{1}{3}}{12\frac{1}{5}}$ **4.** $\dfrac{4}{3}$

Concept Check:
a. Eight to five.   **b.** Five to eight.

In Example 4, although the fraction $\frac{3}{2}$ is equal to the mixed number $1\frac{1}{2}$, a ratio is a quotient of *two* quantities. For that reason, ratios are not written as mixed numbers.

If a ratio compares two decimal numbers, we will write the simplified ratio as a ratio of whole numbers.

EXAMPLE 5 Write the ratio of 2.5 to 3.15 as a fraction in simplest form.

Solution: The ratio is

$$\frac{2.5}{3.15}$$

Now let's clear the ratio of decimals.

$$\frac{2.5}{3.15} = \frac{2.5 \cdot 100}{3.15 \cdot 100} = \frac{250}{315} = \frac{50}{63} \qquad \text{Simplest form} \qquad \bullet$$

EXAMPLE 6 Writing a Ratio from a Circle Graph

The circle graph in the margin shows the parts of a car's total mileage that fall into particular categories. Write the ratio of medical miles to total miles as a fraction in simplest form.

Solution: $\dfrac{\text{medical miles}}{\text{total miles}} = \dfrac{150 \text{ miles}}{15{,}000 \text{ miles}} = \dfrac{150}{15{,}000} = \dfrac{150}{150 \cdot 100} = \dfrac{1}{100}$ $\bullet$

EXAMPLE 7 Given the rectangle shown:

a. Find the ratio of its width (shorter side) to its length (longer side).

b. Find the ratio of its length to its perimeter.

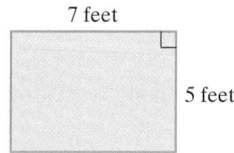

7 feet

5 feet

Solution:

a. The ratio of its width to its length is

$$\frac{\text{width}}{\text{length}} = \frac{5 \text{ feet}}{7 \text{ feet}} = \frac{5}{7}$$

b. Recall that the perimeter of the rectangle is the distance around the rectangle: $7 + 5 + 7 + 5 = 24$ feet. The ratio of its length to its perimeter is

$$\frac{\text{length}}{\text{perimeter}} = \frac{7 \text{ feet}}{24 \text{ feet}} = \frac{7}{24} \qquad \bullet$$

Try the Concept Check in the margin.

Practice Problem 5

Write the ratio of 1.68 to 4.8 as a fraction in simplest form.

Practice Problem 6

Use the circle graph below to write the ratio of work miles to total miles as a fraction in simplest form.

Total Yearly Mileage: 15,000

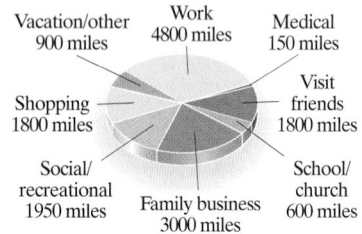

Vacation/other 900 miles
Work 4800 miles
Medical 150 miles
Shopping 1800 miles
Visit friends 1800 miles
Social/ recreational 1950 miles
Family business 3000 miles
School/ church 600 miles

△ **Practice Problem 7**

Given the triangle shown:

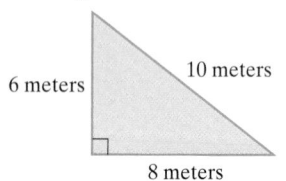

6 meters
10 meters
8 meters

a. Find the ratio of the length of the shortest side to the length of the longest side in simplest form.

b. Find the ratio of the length of the longest side to the perimeter of the triangle in simplest form.

Concept Check

Explain why the answer $\frac{7}{5}$ would be incorrect for part (a) of Example 7.

Answers

5. $\frac{7}{20}$ **6.** $\frac{8}{25}$ **7. a.** $\frac{3}{5}$ **b.** $\frac{5}{12}$

Concept Check: $\frac{7}{5}$ is the ratio of the rectangle's length to its width.

Name _____ Section _____ Date _____

EXERCISE SET 6.1

Ⓐ *Write each ratio using fractional notation. Do not simplify. See Examples 1 through 3.*

1. 11 to 14   **2.** 7 to 12   **3.** 23 to 10   **4.** 8 to 5   **5.** 151 to 201   **6.** 673 to 1000

7. 2.8 to 7.6   **8.** 3.9 to 4.2   **9.** 5 to $7\frac{1}{2}$ **10.** $5\frac{3}{4}$ to 3 **11.** $3\frac{3}{4}$ to $1\frac{2}{3}$ **12.** $2\frac{2}{5}$ to $6\frac{1}{2}$

Ⓑ *Write each ratio as a ratio of whole numbers using fractional notation. Write the fraction in simplest form. See Examples 4 and 5.*

13. 16 to 24   **14.** 25 to 150   **15.** 7.7 to 10   **16.** 8.1 to 10

17. 4.63 to 8.21   **18.** 9.61 to 7.62   **19.** 9 inches to 12 inches   **20.** 14 centimeters to 20 centimeters

21. 10 hours to 24 hours   **22.** 18 quarts to 30 quarts   **23.** $32 to $100   **24.** $46 to $102

25. 24 days to 14 days   **26.** 80 miles to 120 miles   **27.** 32,000 bytes to 46,000 bytes   **28.** 600 copies to 150 copies

29. 35¢ to $2   **30.** 9 feet to 2 yards

Find the ratio described in each exercise as a fraction in simplest form. For Exercises 31 and 32, use the circle graph by Practice Problem 6. See Examples 6 and 7.

31. Write the ratio of vacation/other miles to total miles as a fraction in simplest form.

32. Write the ratio of shopping miles to total miles as a fraction in simplest form.

△ **33.** Find the ratio of the length to the width of a regulation size basketball court.

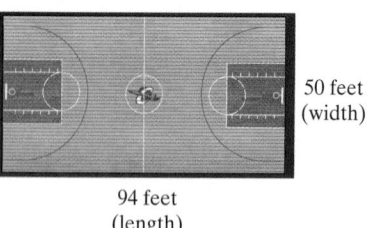

50 feet
(width)

94 feet
(length)

△ **34.** Find the ratio of the base to the height of the triangular mainsail.

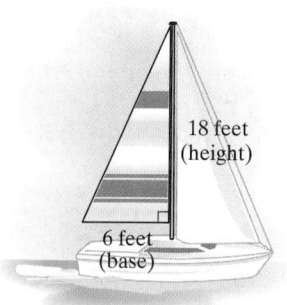

18 feet
(height)

6 feet
(base)

△ **35.** Find the ratio of the longest side to the perimeter of the right-triangular-shaped billboard.

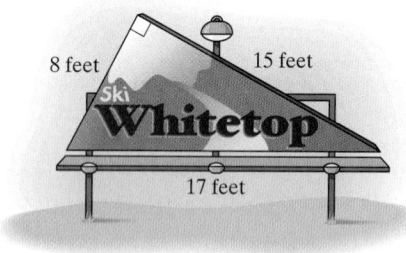

8 feet   15 feet

17 feet

△ **36.** Find the ratio of the width to the perimeter of the rectangular vegetable garden.

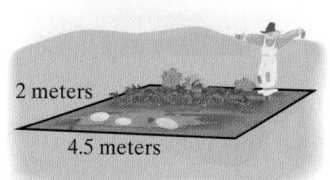

2 meters

4.5 meters

37. A large order of McDonald's french fries has 450 calories. Of this total, 200 calories are from fat. Find the ratio of calories from fat to total calories in a large order of McDonald's french fries. (*Source:* McDonald's Corporation)

38. A McDonald's Quarter Pounder® with Cheese contains 30 grams of fat. A McDonald's Grilled Chicken™ sandwich contains 20 grams of fat. Find the ratio of the amount of fat in a Quarter Pounder to the amount of fat in a Grilled Chicken sandwich. (*Source:* McDonald's Corporation)

At the Honey Island College Glee Club meeting one night, there were 125 women and 100 men present. Use this for Exercises 39 and 40.

 39. Find the ratio of women to men.

40. Find the ratio of men to the total number of people present.

A poll at State University revealed that 4500 students out of 6000 students are single, and the rest are married. Use this for Exercises 41 and 42.

41. Find the ratio of single students to married students.

42. Find the ratio of married students to the total student population.

Blood contains three types of cells: red blood cells, white blood cells, and platelets. For approximately every 600 red blood cells in healthy humans, there are 40 platelets and 1 white blood cell. (Source: American Red Cross Biomedical Services) Use this for Exercises 43 and 44.

43. Write the ratio of red blood cells to platelet cells.

44. Write the ratio of white blood cells to red blood cells.

45. When the first national minimum wage was enacted in 1938, minimum wage was $0.25 per hour. At this writing the minimum wage rate is currently $5.15. Find the ratio of this minimum wage to the 1938 minimum wage. (*Source:* U.S. Department of Labor)

46. Citizens of the United States eat an average of 25 pints of ice cream per year. Residents of the New England states eat an average of 39 pints of ice cream per year. Find the ratio of the average amount of ice cream eaten by New Englanders to the average amount eaten by U.S. citizens. (*Source:* International Dairy Foods Association)

47. At the 2002 Winter Olympic Games in Salt Lake City, Utah, a total of 234 medals were awarded. Italian athletes won a total of 12 medals. Find the ratio of medals won by Italy to the total medals awarded. (*Source: The World Almanac, 2003*)

48. At the 2002 Winter Olympic Games in Salt Lake City, Utah, a total of 234 medals were awarded. German athletes won a total of 35 medals. Find the ratio of medals won by Germany to the total medals awarded. (*Source: The World Almanac, 2003*)

49. Of the U.S. mountains that are over 14,000 feet in elevation, 57 are located in Colorado and 19 are located in Alaska. Find the ratio of the number of mountains over 14,000 feet found in Alaska to the number of mountains over 14,000 feet found in Colorado. (*Source:* U.S. Geological Survey)

50. Citizens of the United States eat an average of 25 pints of ice cream per year. Residents of the New England states eat an average of 39 pints of ice cream per year. Find the ratio of the average amount of ice cream eaten by U.S. citizens to the average amount eaten by New Englanders. (*Source:* International Dairy Foods Association)

Review and Preview

Divide. See Section 5.4.

51. $9\overline{)20.7}$　　　　**52.** $7\overline{)60.2}$　　　　**53.** $3.7\overline{)0.555}$　　　　**54.** $4.6\overline{)1.15}$

55. As of 2003, Target stores operate in 47 states. Find the ratio of states without Target stores to states with Target stores. (*Source:* Target Corporation)

56. A total of 32 states have 200 or more public libraries. Find the ratio of states with 200 or more public libraries to states with fewer than 200 public libraries. (*Source:* U.S. Department of Education)

57. Write the ratio $2\frac{1}{2}$ to $5\frac{3}{4}$ as a fraction in simplest form.

58. A panty hose manufacturing machine will be repaired if the ratio of defective panty hose to good panty hose is 1 to 20 or greater. A quality control engineer found 30 defective panty hose in a batch of 240. Determine whether the machine should be repaired.

59. A grocer will refuse a shipment of tomatoes if the ratio of bruised tomatoes to the total batch is at least 1 to 10. A sample is found to contain 3 bruised tomatoes and 33 good tomatoes. Determine whether the shipment should be refused.

60. Is the ratio $\frac{5}{7}$ the same as the ratio $\frac{7}{5}$? Explain your answer.

61. In 2003, 19 states have mandatory helmet laws. (*Source:* Bicycle Helmet Safety Institute)
  a. Find the ratio of states with mandatory helmet laws to total U.S. states.

  b. Find the ratio of states with mandatory helmet laws to states without mandatory helmet laws.

  c. Are your ratios for parts **a** and **b** the same? Explain why or why not.

Internet Excursions

 Go To: | http://www.prenhall.com/martin-gay_prealgebra | What's Related

A scale model of an object is one in which there is a particular ratio between each measurement of the model and the corresponding measurement of the object being modeled. For instance, model trains are scale models of actual trains. Model train enthusiasts know that model trains are available in several different scales. By going to the World Wide Web site listed above, you will be directed to the Lionel Glossary in the Lionel model trains Web site, or a related site, where you can look up information to help you answer the questions below.

62. The two most popular model train scales are the HO scale and the N scale. What ratio of measurements is used in HO scale? What ratio of measurements is used in N scale? Describe what each of these ratios means and give an example.

63. Other model train scales include the O scale and the S scale. What ratio of measurements is used in O scale? What ratio of measurements is used in S scale? Describe what each of these ratios means and give an example.

6.2 Rates

Ⓐ Writing Rates as Fractions

A special type of ratio is a rate. **Rates** are used to compare *different* kinds of quantities. For example, suppose that a recreational runner can run 3 miles in 33 minutes. This rate compares a measure of distance to a measure of time. If we write this rate as a fraction, we have

$$\frac{3 \text{ miles}}{33 \text{ minutes}} = \frac{1 \text{ mile}}{11 \text{ minutes}} \quad \text{In simplest form}$$

Helpful Hint

When comparing quantities with different units, write the units as part of the comparison. They do not divide out.

Same Units: $\dfrac{3 \text{ inches}}{12 \text{ inches}} = \dfrac{1}{4}$

Different Units: $\dfrac{2 \text{ miles}}{20 \text{ minutes}} = \dfrac{1 \text{ mile}}{10 \text{ minutes}}$ Units are still written.

EXAMPLE 1 Write the rate as a fraction in simplest form: 10 nails every 6 feet

Solution: $\dfrac{10 \text{ nails}}{6 \text{ feet}} = \dfrac{5 \text{ nails}}{3 \text{ feet}}$

EXAMPLES Write each rate as a fraction in simplest form.

2. $2160 for 12 weeks is

$$\frac{2160 \text{ dollars}}{12 \text{ weeks}} = \frac{180 \text{ dollars}}{1 \text{ week}}$$

3. 360 miles on 16 gallons of gasoline is

$$\frac{360 \text{ miles}}{16 \text{ gallons}} = \frac{45 \text{ miles}}{2 \text{ gallons}}$$

Try the Concept Check in the margin.

Ⓑ Finding Unit Rates

A **unit rate** is a rate with a denominator of 1. A familiar example of a unit rate is 55 mph, read as "55 **miles per hour**." This means 55 miles per 1 hour or

$\dfrac{55 \text{ miles}}{1 \text{ hour}}$ Denominator of 1

Helpful Hint

In this context, the word "per" translates to division

Writing a Rate as a Unit Rate

To write a rate as a unit rate, divide the numerator of the rate by the denominator.

Practice Problem 1

Write the rate as a fraction in simplest form: 12 commercials every 45 minutes

Practice Problems 2–3

Write each rate as a fraction in simplest form.

2. $1680 for 8 weeks

3. 236 miles on 12 gallons of gasoline

Concept Check

True or false: $\dfrac{16 \text{ gallons}}{4 \text{ gallons}}$ is a rate. Explain.

Answers

1. $\dfrac{4 \text{ commercials}}{15 \text{ minutes}}$ **2.** $\dfrac{210 \text{ dollars}}{1 \text{ week}}$ **3.** $\dfrac{59 \text{ miles}}{3 \text{ gallons}}$

Concept Check: False; a rate compares different kinds of quantities.

Practice Problem 4

Write as a unit rate: 3600 feet every 12 seconds

EXAMPLE 4 Write as a unit rate: $27,000 every 6 months

Solution: $\dfrac{27,000 \text{ dollars}}{6 \text{ months}}$ $6\overline{)27,000}$ with 4500

The unit rate is

$\dfrac{4500 \text{ dollars}}{1 \text{ month}}$ or   4500 dollars/month   Read as "4500 dollars per month." ●

Practice Problem 5

Write as a unit rate: 52 bushels of fruit from 8 trees

EXAMPLE 5 Write as a unit rate: 318.5 miles every 13 gallons of gas

Solution: $\dfrac{318.5 \text{ miles}}{13 \text{ gallons}}$ $13\overline{)318.5}$ with 24.5

The unit rate is

$\dfrac{24.5 \text{ miles}}{1 \text{ gallon}}$ or   24.5 miles/gallon

            or   24.5 mpg   Read as "24.5 miles per gallon." ●

Ⓒ Finding Unit Prices

Rates are used extensively in sports, business, medicine, and science applications. One of the most common uses of rates is in consumer economics. When a unit rate is "money per item" or "price per unit," it is also called a **unit price**.

Practice Problem 6

An automobile rental agency charges $170 for 5 days for a certain model car. What is the unit price in dollars per day?

EXAMPLE 6 Finding Unit Price

A store charges $3.36 for a 16-ounce jar of picante sauce. What is the unit price in dollars per ounce?

Solution: unit price $= \dfrac{\$3.36}{16 \text{ ounces}} = \dfrac{\$0.21}{1 \text{ ounce}}$ or $0.21 per ounce   ●

Practice Problem 7

Approximate each unit price to decide which is the better buy for a bag of nacho chips: 11 ounces for $2.32 or 16 ounces for $3.59.

EXAMPLE 7 Finding the Better Buy

Approximate each unit price to decide which is the better buy: $0.99 for 4 bars of soap or $1.19 for 5 bars of soap. Assume each bar is the same size.

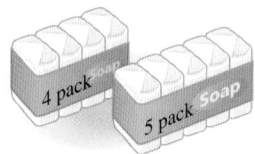

Solution:

unit price $= \dfrac{\$0.99}{4 \text{ bars}} \approx$ $0.25 per bar of soap $\begin{array}{r} 0.247 \approx 0.25 \\ 4\overline{)0.990} \end{array}$

                                      $\approx$ (is approximately)

unit price $= \dfrac{\$1.19}{5 \text{ bars}} \approx$ $0.24 per bar of soap $\begin{array}{r} 0.238 \approx 0.24 \\ 5\overline{)1.190} \end{array}$

Thus, the 5-bar package is the better buy. ●

Answers

4. $\dfrac{300 \text{ feet}}{1 \text{ second}}$ or 300 feet/second

5. $\dfrac{6.5 \text{ bushels}}{1 \text{ tree}}$ or 6.5 bushels/tree

6. $34 per day   **7.** 11-ounce bag

EXERCISE SET 6.2

(A) *Write each rate as a fraction in simplest form. See Examples 1 through 3.*

1. 5 shrubs every 15 feet

2. 14 lab tables for 28 students

3. 15 returns for 100 sales

4. 150 graduate students for 8 advisors

5. 8 phone lines for 36 employees

6. 6 laser printers for 28 computers

7. 18 gallons of pesticide for 4 acres of crops

8. 4 inches of rain in 18 hours

9. 6 flight attendants for 200 passengers

10. 240 pounds of grass seed for 9 lawns

11. 355 calories in a 10-fluid ounce chocolate milkshake (*Source: Home and Garden Bulletin No. 72*; U.S. Department of Agriculture)

12. 160 calories in an 8-fluid ounce serving of cream of tomato soup (*Source: Home and Garden Bulletin No. 72*; U.S. Department of Agriculture)

(B) *Write each rate as a unit rate. See Examples 4 and 5.*

13. 375 riders in 5 subway cars

14. 18 campaign yard signs in 6 blocks

15. 330 calories in a 3-ounce serving

16. 275 miles in 11 hours

17. A hummingbird moves its wings at a rate of 5400 wingbeats a minute write this rate in wingbeats per second.

18. A bat moves its wings at a rate of 1200 wingbeats a minute. Write this rate in wingbeats per second.

19. $1,000,000 lottery paid over 20 years

20. 400,000 library books for 8000 students

21. In 1999, the manned balloon *Breitling Orbiter* completed the first round-the-world flight. The balloon flew 26,600 miles in 20 days. (*Source: Fantastic Book of Comparisons*)

22. A bullet fired from a rifle moves at a rate of 1620 miles per 1 hour. Write this rate in miles per minute.

23. The state of Tennessee has 5,740,020 registered voters for 2 senators. (*Source:* U.S. Bureau of the Census)

25. 12,000 good assemblyline products to 40 defective products

27. 12 million tons of dust and dirt are trapped by the 25 million acres of lawns in the United States each year. (*Source:* Professional Lawn Care Association of America)

29. The National Zoo in Washington, D.C., has an annual budget of $28,600,000 for its approximately 500 different species. (*Source: World Almanac 2003*)

31. The top-grossing concert tour in North America is the 1994 Rolling Stones tour, which grossed $121,200,000 for 60 shows. (*Source:* Pollstar)

24. The state of Louisiana has about 4,465,280 residents for 64 parishes. *Note:* Louisiana is the only U.S. State with parishes instead of counties. (*Source:* U.S. Bureau of the Census)

26. 5,000,000 lottery tickets for 4 lottery winners

28. Approximately 65,000,000,000 checks are written each year by a total of approximately 260,000,000 Americans. (*Source:* Board of Governors of the Federal Reserve System)

30. On average, it costs $32,632 for 4 hours of flight in a B747-400 aircraft. (*Source:* Air Transport Association of America)

32. In Hawaii, there is $22,789,000 of funding for 50 public libraries. (*Source:* U.S. Department of Education)

33. Greer Krieger can assemble 250 computer boards in an 8-hour shift, while Lamont Williams can assemble 400 computer boards in a 12-hour shift.
   a. Find the unit rate of Greer.
   b. Find the unit rate of Lamont.
   c. Who can assemble computer boards faster, Greer or Lamont?

34. Jerry Stein laid 713 bricks in 46 minutes, while his associate, Bobby Burns, laid 396 bricks in 30 minutes.
   a. Find the unit rate of Jerry.
   b. Find the unit rate of Bobby.
   c. Who is the faster bricklayer?

C *Find each unit price. See Example 6.*

35. $57.50 for 5 compact disks

36. $0.87 for 3 apples

37. $1.19 for 7 bananas

38. $73.50 for 6 lawn chairs

Find each unit price and decide which is the better buy. Round to three decimal places. Assume that we are comparing different sizes of the same brand. See Example 7.

39. Crackers:
   8 ounces for $1.19
   12 ounces for $1.59

40. Pickles:
   18 ounces for $0.89
   32 ounces for $1.89

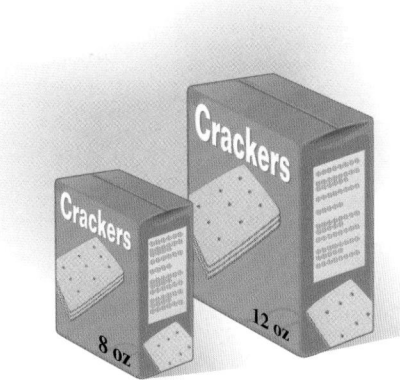

18 oz   32 oz

41. Frozen orange juice:
   6 ounces for $0.69
   16 ounces for $1.69

42. Eggs:
   A dozen for $0.69
   A flat $\left(2\frac{1}{2} \text{ dozen}\right)$ for $2.10

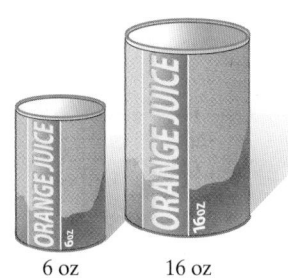

6 oz   16 oz

43. Soy sauce:
   8 ounces for $1.49
   12 ounces for $2.29

44. Shampoo:
   20 ounces for $1.89
   32 ounces for $3.19

20 oz   32 oz

45. Napkins:
   100 for $0.89
   180 for $1.49

100 napkins   180 napkins

46. Crackers:
   8 ounces for $0.99
   20 ounces for $2.39

Crackers

8 oz   20 oz

Review and Preview

Multiply or divide as indicated. See Sections 5.3 and 5.4.

47. $\begin{array}{r} 1.7 \\ \times\ 6 \\ \hline \end{array}$

48. $\begin{array}{r} 2.3 \\ \times\ 9 \\ \hline \end{array}$

49. $\begin{array}{r} 3.7 \\ \times\ 1.2 \\ \hline \end{array}$

50. $\begin{array}{r} 6.6 \\ \times\ 2.5 \\ \hline \end{array}$

51. $2.3\overline{)4.37}$

52. $3.5\overline{)22.75}$

◆ Combining Concepts

53. Fill in the table to calculate miles per gallon.

Beginning Odometer Reading	Ending Odometer Reading	Miles Driven	Gallons of Gas Used	Miles Per Gallon (round to the nearest tenth)
79,286	79,543		13.4	
79,543	79,895		15.8	
79,895	80,242		16.1	

Find each unit rate.

54. Sammy Joe Wingfield from Arlington, Texas, is the fastest bricklayer on record. On May 20, 1994, he laid 1048 bricks in 60 minutes. Find his unit rate of bricks per minute rounded to the nearest tenth of a brick. (*Source: The Guinness Book of Records*)

55. The longest stairway is the service stairway for the Niesenbahn Cable railway near Spiez, Switzerland. It has 11,674 steps and rises to a height of 7759 feet. Find the unit rate of steps per foot rounded to the nearest tenth of a step. (*Source: The Guinness Book of Records*)

56. In Kansas, the total enrollment in public schools was 470,610 in a recent year. At the same time, there were 32,680 public school teachers. Write a unit rate in students per teacher. Round to the nearest whole student. (*Source:* National Center for Education Statistics)

57. In Utah, the total number of students enrolled in public schools during a recent year was 481,687. At the same time, there were 22,008 public school teachers. Write a unit rate in students per teacher. Round to the nearest whole student. (*Source:* National Center for Education Statistics)

58. Suppose that the amount of a product decreases, say from an 80-ounce container to a 70-ounce container, but the price of the container remains the same. Does the unit price increase or decrease? Explain why.

59. In your own words, define "rate."

Focus on **Business and Career**

COST OF LIVING

In the working world, you may find it necessary to relocate to get the job you want. When considering job offers in different cities, you should keep in mind that the cost of living in one city is not necessarily the same as the cost of living in another. For example, an annual salary of $25,000 might be plenty to live on in one city but might barely cover housing expenses in another city.

A cost-of-living index helps us gauge how much more or less expensive it is to live in one city when compared to another. With it, we can find the salary level needed in one city to be equivalent to a given salary in another city using the following proportion:

$$\frac{\text{equivalent salary in City A}}{\text{index value for City A}} = \frac{\text{salary in City B}}{\text{index value for City B}}$$

Cost-of-Living Index	
City	**Index Value**
Atlanta, GA	99.5
Boston, MA	142.5
Chicago, IL	105.3
Cleveland, OH	104.4
Houston, TX	93.0
Los Angeles, CA	119.7
Miami, FL	107.7
New Orleans, LA	94.9
New York, NY	234.5
Oklahoma City, OK	90.0
Pittsburgh, PA	109.5

(*Source*: Virtual Relocation.com, Inc.)

For example, we can use the given cost-of-living index to find the salary that a person would need to earn in Chicago to be equivalent to his or her current annual salary of $25,000 in New Orleans. The index value is 105.3 for Chicago and 94.9 for New Orleans. We will let x represent the equivalent salary in Chicago,

$$\frac{\text{equivalent salary in Chicago}}{\text{index value for Chicago}} = \frac{\text{salary in New Orleans}}{\text{index value for New Orleans}}$$

$$\frac{x}{105.3} = \frac{25,000}{94.9}$$

$$94.9 \cdot x = 105.3(25,000)$$

$$94.9x = 2,632,500$$

$$\frac{94.9x}{94.9} = \frac{2,632,500}{94.9}$$

$$x \approx \$27,740$$

Thus, the person would need to earn $27,740 in Chicago to be equivalent to his or her current salary of $25,000 in New Orleans. Another way to look at this situation is that what this person can afford with $25,000 in New Orleans would require $27,740 in Chicago.

CRITICAL THINKING

1. In which city on the Cost-of-Living Index list is it most expensive to live? In which city is it least expensive to live? Explain your reasoning.

2. Suppose you currently live in Cleveland, Ohio, and earn an annual salary of $28,500. You are considering moving to Los Angeles, California. How much would you have to earn in Los Angeles to maintain the same standard of living you have in Cleveland?

3. Suppose you have just graduated from college and have been offered two comparable jobs in different cities. Job A is in Miami, Florida, and pays $32,000 per year. Job B is in Boston, Massachusetts, and pays $39,000 per year. Which job offer would you choose? Explain your reasoning.

6.3 Proportions

Ⓐ Writing Proportions

A **proportion** is a statement that two ratios or rates are equal.

> **Proportion**
>
> A proportion states that two ratios are equal. If $\dfrac{a}{b}$ and $\dfrac{c}{d}$ are two ratios, then
>
> $\dfrac{a}{b} = \dfrac{c}{d}$ is a proportion.

For example, $\dfrac{5}{6} = \dfrac{10}{12}$ is a proportion. When we want to emphasize this equation as a proportion, we read it as "5 is to 6 as 10 is to 12."

EXAMPLE 1   Write each sentence as a proportion.

a. 12 diamonds is to 15 rubies as 4 diamonds is to 5 rubies.

b. 5 hits is to 9 at bats as 20 hits is to 36 at bats.

Solution:

a. $\begin{array}{c}\text{diamonds} \rightarrow\\ \text{rubies} \rightarrow\end{array} \dfrac{12}{15} = \dfrac{4}{5} \begin{array}{c}\leftarrow \text{diamonds}\\ \leftarrow \text{rubies}\end{array}$

b. $\begin{array}{c}\text{hits} \rightarrow\\ \text{at bats} \rightarrow\end{array} \dfrac{5}{9} = \dfrac{20}{36} \begin{array}{c}\leftarrow \text{hits}\\ \leftarrow \text{at bats}\end{array}$

> **Helpful Hint**
>
> Notice in the preceding examples of proportions that numerators contain the same units and denominators contain the same units. In this text, proportions will be written in this manner.

Ⓑ Determining Whether Proportions are True

Like other mathematical statements, a proportion may be either true or false. A proportion is true if its ratios are equal. Since ratios are fractions, one way to determine whether a proportion is true is to write each fraction in simplest form and compare them. Another way to see if a proportion is true is by comparing cross products.

> **Using Cross Products to Determine Whether Proportions Are True or False**
>
> $$\dfrac{a}{b} = \dfrac{c}{d}$$
>
> cross product → $\overbrace{b \cdot c}$ $\overbrace{a \cdot d}$ ← cross product
>
> If cross products are *equal*, the proportion is *true*.
>
> If cross products are *not equal*, the proportion is *false*.

Practice Problem 1

Write each sentence as a proportion.

a. 24 right is to 6 wrong as 4 right is to 1 wrong.

b. 32 Cubs fans is to 18 Mets fans as 16 Cubs fans is to 9 Mets fans.

Answers

1. a. $\dfrac{24}{6} = \dfrac{4}{1}$ **b.** $\dfrac{32}{18} = \dfrac{16}{9}$

Practice Problem 2

Is $\dfrac{3}{6} = \dfrac{4}{8}$ a true proportion?

Practice Problem 3

Is $\dfrac{3.6}{6} = \dfrac{5.4}{8}$ a true proportion?

Practice Problem 4

Is $\dfrac{4\frac{1}{5}}{2\frac{1}{3}} = \dfrac{3\frac{3}{10}}{1\frac{5}{6}}$ a true proportion?

Concept Check

Using the numbers in the proportion $\dfrac{21}{27} = \dfrac{7}{9}$, write two other true proportions.

EXAMPLE 2 Is $\dfrac{2}{3} = \dfrac{4}{6}$ a true proportion?

Solution:

$$\dfrac{2}{3} = \dfrac{4}{6}$$

$3 \cdot 4 \qquad \overset{?}{=} \qquad 2 \cdot 6$

$12 \qquad = \qquad 12$ Equal, so proportion is true.

Since the cross products are equal, the proportion is true.

EXAMPLE 3 Is $\dfrac{4.1}{7} = \dfrac{2.9}{5}$ a true proportion?

Solution:

$$\dfrac{4.1}{7} = \dfrac{2.9}{5}$$

$7 \cdot 2.9 \qquad \overset{?}{=} \qquad 4.1 \cdot 5$

$20.3 \qquad \neq \qquad 20.5$ Not equal, so proportion is false.

Since the cross products are not equal, $\dfrac{4.1}{7} \neq \dfrac{2.9}{5}$. The proportion is false.

EXAMPLE 4 Is $\dfrac{1\frac{1}{6}}{10\frac{1}{2}} = \dfrac{\frac{1}{2}}{4\frac{1}{2}}$ a true proportion?

Solution: $\dfrac{1\frac{1}{6}}{10\frac{1}{2}} = \dfrac{\frac{1}{2}}{4\frac{1}{2}}$

$10\frac{1}{2} \cdot \frac{1}{2} \qquad \overset{?}{=} \qquad 1\frac{1}{6} \cdot 4\frac{1}{2}$

$\dfrac{21}{2} \cdot \dfrac{1}{2} \qquad \overset{?}{=} \qquad \dfrac{7}{6} \cdot \dfrac{9}{2}$

$\dfrac{21}{4} \qquad = \qquad \dfrac{21}{4}$ Equal, so proportion is true.

Since the cross products are equal, the proportion is true.

Try the Concept Check in the margin.

Ⓒ Finding Unknown Numbers in Proportions

When one number of a proportion is unknown, we can use cross products to find the unknown number. For example, to find the unknown number x in the proportion $\dfrac{2}{3} = \dfrac{x}{30}$, we use cross products.

EXAMPLE 5 Solve $\frac{2}{3} = \frac{x}{30}$ for x.

Solution: If the cross products are equal, then the proportion is true. We begin, then, by setting cross products equal to each other.

$$\frac{2}{3} = \frac{x}{30}$$

| $3 \cdot x$ | $=$ | $2 \cdot 30$ | Set cross products equal. |
| $3x$ | $=$ | 60 | Multiply. |

Recall that to find x, we divide both sides of the equation by 3.

$$\frac{3x}{3} = \frac{60}{3} \quad \text{Divide both sides by 3.}$$

$$x = 20 \quad \text{Simplify.}$$

To check, we replace x with 20 in the original proportion to see if the result is a true statement.

$$\frac{2}{3} = \frac{x}{30} \quad \text{Original proportion}$$

$$\frac{2}{3} = \frac{20}{30} \quad \text{Replace } x \text{ with 20.}$$

$$\frac{2}{3} = \frac{2}{3} \quad \text{True.}$$

Since $\frac{2}{3} = \frac{2}{3}$ is a true statement, 20 is the solution.

EXAMPLE 6 Solve $\frac{25}{x} = \frac{20}{4}$ for x and then check.

Solution:

$$\frac{25}{x} = \frac{20}{4}$$

$20x$	$=$	$25 \cdot 4$	Set cross products equal.
$20x$	$=$	100	Multiply.
$\dfrac{20x}{20}$	$=$	$\dfrac{100}{20}$	Divide both sides by 20.
x	$=$	5	Simplify.

Check: $\dfrac{25}{x} = \dfrac{20}{4}$ Original proportion

$$\frac{25}{5} = \frac{20}{4} \quad \text{Replace } x \text{ with 5.}$$

$$5 = 5 \quad \text{Simplify.}$$

Since 5 makes the original proportion true, 5 is the solution.

Practice Problem 5

Solve $\frac{2}{7} = \frac{x}{35}$ for x.

Practice Problem 6

Solve $\frac{2}{15} = \frac{y}{60}$ for y and then check.

Answers
5. 10   **6.** 8

Practice Problem 7

Solve for z:

$$\frac{\frac{7}{8}}{z} = \frac{\frac{2}{3}}{\frac{4}{7}}$$

EXAMPLE 7 Solve for y: $\dfrac{\frac{1}{2}}{\frac{4}{5}} = \dfrac{\frac{3}{4}}{y}$

Solution:

$$\frac{\frac{1}{2}}{\frac{4}{5}} = \frac{\frac{3}{4}}{y}$$

$\dfrac{3}{4} \cdot \dfrac{4}{5}$	$=$	$\dfrac{1}{2} \cdot y$	Set cross products equal.
$\dfrac{3}{5}$	$=$	$\dfrac{1}{2}y$	Multiply.
$2 \cdot \dfrac{3}{5}$	$=$	$2 \cdot \dfrac{1}{2}y$	Multiply both sides by 2.
$\dfrac{6}{5}$	$=$	y	Simplify.

Verify that $\dfrac{6}{5}$ is the solution.

Check: To check, replace y with $\dfrac{6}{5}$ in the original proportion and see if a true statement results. We will check by cross products.

$$\frac{\frac{1}{2}}{\frac{4}{5}} = \frac{\frac{3}{4}}{y}$$

$$\frac{\frac{1}{2}}{\frac{4}{5}} = \frac{3}{4}$$

$\dfrac{3}{4} \cdot \dfrac{4}{5}$	$\overset{?}{=}$	$\dfrac{1}{2} \cdot \dfrac{6}{5}$	Set cross products equal.
$\dfrac{3}{5}$	$=$	$\dfrac{3}{5}$	True.

True, so the solution is $\dfrac{6}{5}$.

Practice Problem 8

Solve for y: $\dfrac{y}{6} = \dfrac{0.7}{1.2}$

EXAMPLE 8 Solve for x: $\dfrac{x}{3} = \dfrac{0.8}{1.5}$

Solution:

$$\frac{x}{3} = \frac{0.8}{1.5}$$

$3(0.8)$	$=$	$1.5x$	Set cross products equal.
2.4	$=$	$1.5x$	Multiply.
$\dfrac{2.4}{1.5}$	$=$	$\dfrac{1.5x}{1.5}$	Divide both sides by 1.5.
1.6	$=$	x	Simplify.

Answers:

7. $\dfrac{3}{4}$ **8.** 3.5

Check: To check, replace x with 1.6 in the original proportion to see if the result is a true statement.

$$\frac{x}{3} = \frac{0.8}{1.5}$$

$$\frac{1.6}{3} = \frac{0.8}{1.5}$$

$$3(0.8) = (1.6)(1.5) \qquad \text{Set cross products equal.}$$
$$2.4 = 2.4 \qquad \text{True.}$$

True, so the solution is 1.6.

EXAMPLE 9   Solve for y: $\dfrac{14}{y} = \dfrac{12}{16}$

Solution:

$$\frac{14}{y} = \frac{12}{16}$$

$$12y = 224 \qquad \text{Set cross products equal.}$$
$$\frac{12y}{12} = \frac{224}{12} \qquad \text{Divide both sides by 12.}$$
$$y = \frac{56}{3} \qquad \text{Simplify.}$$

Check to see that the solution is $\dfrac{56}{3}$.

Try the Concept Check in the margin.

> **Helpful Hint**
>
> In Example 9, the fraction $\dfrac{12}{16}$ may be simplified to $\dfrac{3}{4}$ before solving the equation. The solution will remain the same.

EXAMPLE 10   Solve for x: $\dfrac{1.6}{1.1} = \dfrac{x}{0.3}$. Round the solution to the nearest hundredth.

Solution:

$$\frac{1.6}{1.1} = \frac{x}{0.3}$$

$$1.1x = 0.48 \qquad \text{Set cross products equal.}$$
$$\frac{1.1x}{1.1} = \frac{0.48}{1.1} \qquad \text{Divide both sides by 1.1.}$$
$$x \approx 0.44 \qquad \text{Round to the nearest hundredth.}$$

Practice Problem 9

Solve for z: $\dfrac{15}{z} = \dfrac{8}{10}$

Concept Check

True or false: the first step in solving the proportion $\dfrac{4}{z} = \dfrac{12}{15}$ yields the equation $4z = 180$.

Practice Problem 10

Solve for y: $\dfrac{3.4}{1.8} = \dfrac{y}{3}$. Round the solution to the nearest tenth.

Answers

9. $\dfrac{75}{4}$ or 18.75   **10.** 5.7

Concept Check: False.

FOCUS ON **Real Life**

BODY DIMENSIONS

Have you ever noticed how large a baby's head looks compared to the rest of its body? We tend to notice this because we are used to seeing adults with very different "proportions" or dimensions. Try the following group activity to see whether there are any patterns in adult body dimensions from person to person.

GROUP ACTIVITY

Have another group member help you measure the height of your head from under your chin to the top of your skull. (Remember that the top of your head is round. You may find it helpful to hold a flat surface on top of your head to more accurately locate and measure to the "top.") Use your head measurement to make a measuring stick from a piece of cardboard. You will now use this measuring stick to measure other parts of your body in units equal to the height of your head. With the help of another group member as needed, measure the following distances with your "head" measuring stick:

- Your overall height
- Your arm length from shoulder to elbow
- Your arm length from elbow to longest fingertip
- Your leg length from hip to ankle
- Your leg length from knee to ankle
- The distance from your chin to waistline

(Note: Each group member must make his or her own measurements in relation to the measurement of his or her own head height.)

1. For each of the above measurements, find the ratio of the measurement to the height measurement of your head.

2. As a group, compile a table summarizing the ratios for each group member by category. (You might list the six measurement categories along the side of the table and group members' names along the top of the table.)

3. Analyze the table. Do you see any patterns? Explain.

4. About how many "heads tall" would you say a normal adult is?

Name _____ Section _____ Date _____

Mental Math

State whether each proportion is true or false.

1. $\dfrac{2}{1} = \dfrac{6}{3}$ **2.** $\dfrac{3}{1} = \dfrac{15}{5}$ **3.** $\dfrac{1}{2} = \dfrac{3}{5}$ **4.** $\dfrac{2}{11} = \dfrac{1}{5}$ **5.** $\dfrac{2}{3} = \dfrac{4}{6}$ **6.** $\dfrac{3}{4} = \dfrac{6}{8}$

EXERCISE SET 6.3

Ⓐ *Write each sentence as a proportion. See Example 1.*

1. 10 diamonds is to 6 opals as 5 diamonds is to 3 opals. 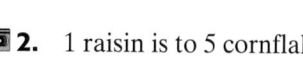 **2.** 1 raisin is to 5 cornflakes as 8 raisins is to 40 cornflakes.

3. 3 printers is to 12 computers as 1 printer is to 4 computers.

4. 4 hit songs is to 16 releases as 1 hit song is to 4 releases.

5. 6 eagles is to 58 sparrows as 3 eagles is to 29 sparrows.

6. 12 errors is to 8 pages as 1.5 errors is to 1 page.

7. $2\dfrac{1}{4}$ cups of flour is to 24 cookies as $6\dfrac{3}{4}$ cups of flour is to 72 cookies.

8. $1\dfrac{1}{2}$ cups milk is to 10 bagels as $\dfrac{3}{4}$ cup milk is to 5 bagels.

9. 22 vanilla wafers is to 1 cup of cookie crumbs as 55 vanilla wafers is to 2.5 cups of cookie crumbs. (*Source:* Based on data from *Family Circle* magazine)

10. 1 cup of instant rice is to 1.5 cups cooked rice as 1.5 cups of instant rice is to 2.25 cups of cooked rice. (*Source:* Based on data from *Family Circle* magazine)

Ⓑ *Determine whether each proportion is true or false. See Examples 2 through 4.*

11. $\dfrac{15}{9} = \dfrac{5}{3}$ **12.** $\dfrac{8}{6} = \dfrac{20}{15}$ **13.** $\dfrac{8}{6} = \dfrac{9}{7}$ **14.** $\dfrac{7}{12} = \dfrac{4}{7}$

15. $\dfrac{9}{36} = \dfrac{2}{8}$ **16.** $\dfrac{8}{24} = \dfrac{3}{9}$ **17.** $\dfrac{5}{8} = \dfrac{625}{1000}$ **18.** $\dfrac{30}{50} = \dfrac{600}{1000}$

19. $\dfrac{0.8}{0.3} = \dfrac{0.2}{0.6}$

20. $\dfrac{0.7}{0.4} = \dfrac{0.3}{0.1}$

21. $\dfrac{4.2}{8.4} = \dfrac{5}{10}$

22. $\dfrac{8}{10} = \dfrac{5.6}{0.7}$

23. $\dfrac{\frac{3}{4}}{\frac{4}{3}} = \dfrac{\frac{1}{2}}{\frac{8}{9}}$

24. $\dfrac{\frac{2}{5}}{\frac{2}{7}} = \dfrac{\frac{1}{10}}{\frac{1}{3}}$

25. $\dfrac{2\frac{2}{5}}{\frac{2}{3}} = \dfrac{\frac{10}{9}}{\frac{1}{4}}$

26. $\dfrac{5\frac{5}{8}}{\frac{5}{3}} = \dfrac{4\frac{1}{2}}{1\frac{1}{5}}$

C *Solve each proportion for the given variable. Round the solution where indicated. See Examples 5 through 10.*

27. $\dfrac{x}{5} = \dfrac{6}{10}$

28. $\dfrac{x}{3} = \dfrac{12}{9}$

29. $\dfrac{30}{10} = \dfrac{15}{y}$

30. $\dfrac{25}{100} = \dfrac{7}{y}$

31. $\dfrac{z}{8} = \dfrac{50}{100}$

32. $\dfrac{12}{18} = \dfrac{z}{21}$

33. $\dfrac{n}{6} = \dfrac{8}{15}$

34. $\dfrac{24}{n} = \dfrac{60}{96}$

35. $\dfrac{12}{10} = \dfrac{x}{16}$

36. $\dfrac{18}{54} = \dfrac{3}{x}$

37. $\dfrac{n}{\frac{6}{5}} = \dfrac{4\frac{1}{6}}{6\frac{2}{3}}$

38. $\dfrac{8}{\frac{1}{3}} = \dfrac{24}{n}$

39. $\dfrac{\frac{3}{4}}{12} = \dfrac{y}{48}$

40. $\dfrac{\frac{11}{4}}{\frac{25}{8}} = \dfrac{7\frac{3}{5}}{y}$

41. $\dfrac{\frac{2}{3}}{\frac{6}{9}} = \dfrac{12}{z}$

42. $\dfrac{z}{24} = \dfrac{\frac{5}{8}}{3}$

43. $\dfrac{n}{0.6} = \dfrac{0.05}{12}$

44. $\dfrac{0.2}{0.7} = \dfrac{8}{n}$

45. $\dfrac{3.5}{12.5} = \dfrac{7}{z}$

46. $\dfrac{7.8}{13} = \dfrac{z}{2.6}$

47. $\dfrac{3.2}{0.3} = \dfrac{x}{1.4}$
Round to the nearest tenth.

48. $\dfrac{1.8}{n} = \dfrac{2.5}{8.4}$
Round to the nearest tenth.

49. $\dfrac{z}{5.2} = \dfrac{0.08}{6}$
Round to the nearest hundredth.

50. $\dfrac{4.25}{6.03} = \dfrac{5}{y}$
Round to the nearest hundredth.

51. $\dfrac{9}{11} = \dfrac{x}{4}$
Round to the nearest tenth.

52. $\dfrac{24}{x} = \dfrac{7}{3}$
Round to the nearest thousandth.

53. $\dfrac{43}{17} = \dfrac{8}{z}$
Round to the nearest thousandth.

54. $\dfrac{n}{12} = \dfrac{18}{7}$
Round to the nearest hundredth.

Review and Preview

Insert $<$ or $>$ between each pair of numbers to form a true statement. See Sections 2.1 and 4.8.

55. -8 8 **56.** 7 -7 **57.** -2 -3 **58.** 0 -2

59. -5 0 **60.** $-5\dfrac{1}{3}$ $-6\dfrac{2}{3}$ **61.** $-1\dfrac{1}{2}$ $-2\dfrac{1}{2}$ **62.** -4 -1

63. -8.3 -7.9 **64.** -6.2 -6.02 **65.** 1.23 1.039 **66.** 3.068 3.15

◆ Combining Concepts

For each proportion, find the unknown number, n.

67. $\dfrac{n}{7} = \dfrac{0}{8}$ **68.** $\dfrac{0}{2} = \dfrac{n}{3.5}$ **69.** $\dfrac{n}{1150} = \dfrac{588}{483}$

70. $\dfrac{585}{n} = \dfrac{117}{474}$ **71.** $\dfrac{222}{1515} = \dfrac{37}{n}$ **72.** $\dfrac{1425}{1062} = \dfrac{n}{177}$

73. Explain the difference between a ratio and a proportion.

74. Explain how to find the unknown number in a proportion such as $\dfrac{n}{18} = \dfrac{12}{8}$.

Integrated Review—Ratio, Rate, and Proportion

Write each ratio as a ratio of whole numbers using fractional notation. Write the fraction in simplest form.

1. 18 to 20   **2.** 36 to 100   **3.** 8.6 to 10   **4.** 1.6 to 4.6

5. 8.65 to 6.95   **6.** 7.2 to 8.4   **7.** $3\frac{1}{2}$ to 13
(*Hint*: Recall that the fraction bar means division.)
    8. $1\frac{2}{3}$ to $2\frac{3}{4}$

9. 8 inches to 12 inches   **10.** 3 hours to 24 hours

Find the ratio described in each problem.

11. Find the ratio of the width to the length of the sign below.

12 inches

RESERVED PARKING

18 inches

12. At the end of 2002 Lockheed Martin Corporation had $26 hundred million in assets and $8 hundred million in debts. Find the ratio of assets to debt. (*Source*: Lockheed Martin Corporation)

Write each rate as a fraction in simplest form.

13. 5 offices for every 20 graduate assistants

14. 6 lights every 15 feet

15. 100 U.S. Senators for 50 states

16. 5 teachers for every 140 students

17. 20 inches every 2 feet

18. 40¢ every $3

1. _____
2. _____
3. _____
4. _____
5. _____
6. _____
7. _____
8. _____
9. _____
10. _____
11. _____
12. _____
13. _____
14. _____
15. _____
16. _____
17. _____
18. _____

19. _____

20. _____

21. _____

22. _____

23. _____

24. _____

25. _____

26. _____

27. _____

28. _____

29. _____

30. _____

31. _____

32. _____

19. 64 households with computers for every 100 households

20. 538 electoral votes for 50 states

Write each rate as a unit rate.

21. 165 miles in 3 hours

22. 560 feet in 4 seconds

23. 63 employees per 3 fax lines

24. 85 phone calls for 5 teenagers

25. 115 miles per 5 gallons

26. 112 teachers for 7 computers

27. 7524 books for 1254 college students

28. 2002 pounds for 13 adults

Write each unit price and decide which is the better buy.

29. Dog food:
8 pounds for $2.16
18 pounds for $4.99

30. Paper plates:
100 for $1.98
500 for $8.99

31. Microwave popcorn:
3 packs for $2.39
8 packs for $5.99

32. AA batteries:
4 for $3.69
10 for $9.89

6.4   Proportions and Problem Solving

O B J E C T I V E

Ⓐ   Solve problems by writing proportions.

SSM   SG   CD & VIDEO   MATH PRO   WEB
TUTOR CENTER

Ⓐ   Solving Problems by Writing Proportions

Writing proportions is a powerful tool for solving problems in almost every field, including business, chemistry, biology, health sciences, and engineering, as well as in daily life. Given a specified ratio (or rate) of two quantities, a proportion can be used to determine an unknown quantity.

EXAMPLE 1   Determining Distances from a Map

On a Chamber of Commerce map of Abita Springs, 5 miles corresponds to 2 inches. How many miles correspond to 7 inches?

Practice Problem 1

On an architect's blueprint, 1 inch corresponds to 12 feet. How long is a wall represented by a $3\frac{1}{2}$-inch line on the blueprint?

Solution:

1. UNDERSTAND. Read and reread the problem. You may want to draw a diagram.

$$\underset{\substack{\text{5 miles} \\ \text{2 inches}}}{\vdash}\underset{\substack{\text{5 miles} \\ \text{2 inches}}}{\quad}\underset{\substack{\text{5 miles} \\ \text{2 inches}}}{\quad}\underset{\substack{? \\ \text{between} \\ \text{0 and 5} \\ \text{miles} \\ \text{1 inch}}}{\quad}\dashv$$

= between 15 and 20 miles
= 7 inches

From the diagram we can see that our solution should be between 15 and 20 miles.

2. TRANSLATE. We will let n represent our unknown number. Since we are given that 5 miles corresponds to 2 inches, let's use this rate as the first fraction in our proportion. In this section, we will solve by writing proportions so that numerators have the same unit and denominators have the same unit. Since 5 miles corresponds to 2 inches as n miles corresponds to 7 inches, we have the proportion:

$$\text{miles} \rightarrow \frac{5}{2} = \frac{n}{7} \leftarrow \text{miles}$$
$$\text{inches} \rightarrow \qquad\qquad \leftarrow \text{inches}$$

3. SOLVE.

$$\frac{5}{2} = \frac{n}{7}$$

$2n$	$=$	35	Set cross products equal.
$\dfrac{2n}{2}$	$=$	$\dfrac{35}{2}$	Divide both sides by 2.
n	$=$	17.5	Simplify.

4. INTERPRET. *Check* your work. This result is reasonable since it is between 15 and 20 miles. *State* your conclusion: 7 inches corresponds to 17.5 miles.

Answer

1. 42 feet

Helpful Hint

We can also solve Example 1 by writing the proportion

$$\frac{2 \text{ inches}}{5 \text{ miles}} = \frac{7 \text{ inches}}{n \text{ miles}}$$

Although other proportions may be used to solve Example 1, we will solve by writing proportions so that the numerators have the same unit measures and the denominators have the same unit measures.

Practice Problem 2

An auto mechanic recommends that 3 ounces of isopropyl alcohol be mixed with a tankful of gas (14 gallons) to increase the octane of the gasoline for better engine performance. At this rate, how many gallons of gas can be treated with a 16-ounce bottle of alcohol?

EXAMPLE 2 Finding Medicine Dosage

The standard dose of an antibiotic is 4 cc (cubic centimeters) for every 25 pounds (lb) of body weight. At this rate, find the standard dose for a 140-lb woman.

Solution:

1. UNDERSTAND. Read and reread the problem. You may want to draw a diagram to estimate a reasonable solution.

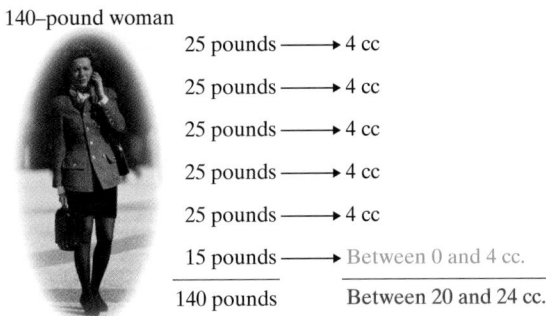

140–pound woman

25 pounds	⟶	4 cc
25 pounds	⟶	4 cc
25 pounds	⟶	4 cc
25 pounds	⟶	4 cc
25 pounds	⟶	4 cc
15 pounds	⟶	Between 0 and 4 cc.
140 pounds		Between 20 and 24 cc.

From the diagram, we can see that a reasonable solution is between 20 cc and 24 cc.

2. TRANSLATE. We will let n represent the unknown number. From the problem, we know that 4 cc is to 25 lb as n cc is to 140 lb, or

$$\begin{array}{c} cc \rightarrow \\ lb \rightarrow \end{array} \frac{4}{25} = \frac{n}{140} \begin{array}{c} \leftarrow cc \\ \leftarrow lb \end{array}$$

3. SOLVE.

$$\frac{4}{25} = \frac{n}{140}$$

$25n$	$=$	560	Set cross products equal.
$\dfrac{25n}{25}$	$=$	$\dfrac{560}{25}$	Divide both sides by 25.
n	$=$	22.4	Simplify.

4. INTERPRET. *Check* your work. This result is reasonable since it is between 20 and 24 cc. *State* your conclusion: The standard dose for a 140-lb woman is 22.4 cc. ●

Answer

2. $74\frac{2}{3}$ gallons

EXAMPLE 3   Calculating Supplies Needed to Fertilize a Lawn

A 50-pound bag of fertilizer covers 2400 square feet of lawn. How many bags of fertilizer are needed to cover a town square containing 15,360 square feet of lawn? Round the answer up to the nearest whole bag.

Solution:

1. UNDERSTAND. Read and reread the problem. Draw a picture.

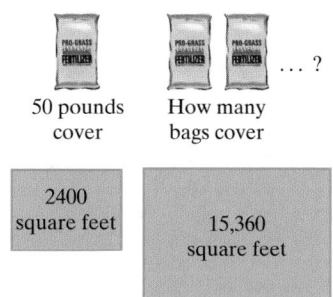

50 pounds cover   How many bags cover   … ?

2400 square feet   15,360 square feet

2. TRANSLATE. We'll let n represent the unknown number. From the problem, we know that 1 bag is to 2400 square feet as n bags is to 15,360 square feet.

$$\text{bags} \rightarrow \frac{1}{2400} = \frac{n}{15{,}360} \leftarrow \text{bags}$$
$$\text{square feet} \qquad\qquad\qquad \leftarrow \text{square feet}$$

3. SOLVE.

$$\frac{1}{2400} = \frac{n}{15{,}360}$$

$$2400n = 15{,}360 \qquad \text{Set cross products equal.}$$

$$\frac{2400n}{2400} = \frac{15{,}360}{2400} \qquad \text{Divide both sides by 2400.}$$

$$n = 6.4 \qquad \text{Simplify.}$$

4. INTERPRET. *Check* that replacing n with 6.4 makes the proportion true. Is the answer reasonable?

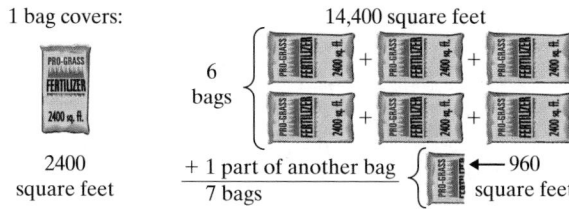

1 bag covers: 14,400 square feet

2400 square feet   6 bags

+ 1 part of another bag ← 960 square feet
7 bags

Yes. Since we must buy whole bags of fertilizer, 7 bags are needed. *State* your conclusion: To cover 15,360 square feet of lawn, 7 bags are needed.

Try the Concept Check in the margin.

Practice Problem 3

If a gallon of paint covers 400 square feet, how many gallons must be bought to paint a retaining wall 260 feet long and 4 feet high? Round the answer up to the nearest whole gallon.

Concept Check

You are told that 12 ounces of ground coffee will brew enough coffee to serve 20 people. How could you estimate how much ground coffee will be needed to serve 95 people?

Answer

3. 3 gallons

Concept Check: Find how much will be needed for 100 people (20 × 5) by multiplying 12 ounces by 5, which is 60 ounces.

Focus on **Business and Career**

CONSUMER PRICE INDEX

Do you remember when the regular price of a candy bar was 5¢, 10¢, or 25¢? It is certainly difficult to find a candy bar for that price these days. The reason is inflation: The tendency for the price of a given product to increase over time. Businesses and government agencies use the Consumer Price Index (CPI) to track inflation. The CPI measures the change in prices over time of basic consumer goods and services. It is calculated by the Bureau of Labor Statistics, part of the U.S. Department of Labor.

The CPI is very useful for comparing the prices of fixed items in various years. For instance, suppose an insurance company customer submits a claim for the theft of a fishing boat purchased in 1975. Because the customer's policy includes replacement cost coverage, the insurance company must calculate how much it would cost to replace the boat at today's prices. (Let's assume the theft took place in 2000.) The customer has a receipt for the boat showing that it cost $598 in 1975. The insurance company can use the following proportion to calculate the replacement cost:

$$\frac{\text{price in earlier year}}{\text{price in later year}} = \frac{\text{CPI value in earlier year}}{\text{CPI value in later year}}$$

Because the CPI value is 53.8 for 1975 and 172.2 for 2000, the insurance company would use the following proportion for this situation. (We will let n represent the unknown price in 2000.)

$$\frac{\text{price in 1975}}{\text{CPI value in 1975}} = \frac{\text{price in 2000}}{\text{CPI value in 2000}}$$

$$\frac{598}{53.8} = \frac{n}{172.2}$$

$$53.8 \cdot n = 598(172.2)$$

$$53.8 \cdot n = 102,975.6$$

$$\frac{53.8 \cdot n}{53.8} = \frac{102,975.6}{53.8}$$

$$n \approx 1914$$

The replacement cost of the fishing boat at 2000 prices is $1914.

CRITICAL THINKING

1. What trends do you see in the CPI values in the table? Do you think these trends make sense? Explain.

2. A piece of jewelry cost $800 in 1975. What is its 2000 replacement value?

3. In 1995, the cost of a loaf of bread was about $1.89. What would an equivalent loaf of bread cost in 1950?

4. Suppose a couple purchased a house for $22,000 in 1920. At what price could they have expected to sell the house in 1990?

5. An original Ford Model T cost about $850 in 1915. What is the equivalent cost of a Model T in 2001 dollars?

Consumer Price Index	
Year	**CPI**
1915	10.1
1920	20.0
1925	17.5
1930	16.7
1935	13.7
1940	14.0
1945	18.0
1950	24.1
1955	26.8
1960	29.6
1965	31.5
1970	38.8
1975	53.8
1980	82.4
1985	107.6
1990	130.7
1995	152.4
2000	172.2
2001	177.1

(*Source:* Bureau of Labor Statistics, U.S. Department of Labor)

EXERCISE SET 6.4

A *Solve. See Examples 1 through 3.*

The ratio of a quarterback's completed passes to attempted passes is 4 to 9.

1. If he attempted 27 passes, find how many passes he completed.

2. If he completed 20 passes, find how many passes he attempted.

It takes Sandra Hallahan 30 minutes to word process and spell check 4 pages.

 3. Find how long it takes her to word process and spell check 22 pages.

4. Find how many pages she can word process and spell check in 4.5 hours.

University Law School accepts 2 out of every 7 applicants.

5. If the school received 630 applications, find how many students were accepted.

6. If the school accepted 150 students, find how many applications were received.

On an architect's blueprint, 1 inch corresponds to 8 feet.

7. Find the length of a wall represented by a line $2\frac{7}{8}$ inches long on the blueprint.

8. If an exterior wall is 42 feet long, find how long the blueprint measurement should be.

A human factors expert recommends that there be at least 9 square feet of floor space in a college classroom for every student in the class.

9. Find the minimum floor space that 30 students require.

10. Due to a space crunch, a university converts a 21′ by 15′ conference room into a classroom. Find the maximum number of students the room can accommodate.

A Honda Civic averages 450 miles on a 12-gallon tank of gas.

11. If Dave Smythe runs out of gas in a Honda Civic and AAA comes to his rescue with $1\frac{1}{2}$ gallons of gas, determine how far he can go. Round to the nearest mile.

12. Find how many gallons of gas Denise Wolcott can expect to burn on a 2000-mile vacation trip in a Honda Civic. Round to the nearest gallon.

The scale on an Italian map states that 1 centimeter corresponds to 30 kilometers (units of length in the metric system).

13. Find how far apart Milan and Rome are if their corresponding points on the map are 15 centimeters apart.

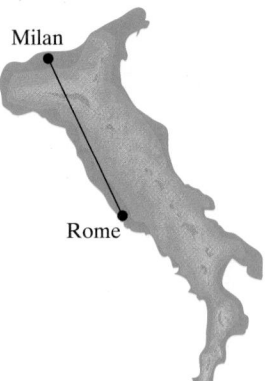

Milan

Rome

14. On the map, a small Italian village is located 0.4 centimeters from the Mediterranean Sea. Find the actual distance.

A drink called Sea Breeze punch is made by mixing 3 parts of grapefruit juice with 4 parts of cranberry juice.

15. Find how much grapefruit juice should be mixed with 32 ounces of cranberry juice.

16. For a party, 6 quarts of grapefruit juice have been purchased to make Sea Breeze punch. Find how much cranberry juice should be purchased.

A bag of Scott fertilizer covers 3000 square feet of lawn.

17. Find how many bags of fertilizer should be purchased to cover a rectangular lawn 260 feet by 180 feet.

18. Find how many bags of fertilizer should be purchased to cover a square lawn measuring 160 feet on each side.

Yearly homeowner property taxes are figured at a rate of $1.45 tax for every $100 of house value.

19. If Janet Blossom, a homeowner, pays $2349 in property taxes, find the value of her home.

20. Find the property taxes on a condominium valued at $72,000.

A Cubs baseball player makes 3 hits in every 8 times at bat.

21. If this Cubs player comes up to bat 40 times in a World Series playoff series, find how many hits he would be expected to make.

22. At this rate, if he made 12 hits, find how many times he batted.

A survey reveals that 2 out of 3 people prefer Coca Cola to Pepsi.

23. In a room of 40 people, how many people are likely to prefer Coca Cola? Round the answer to the nearest person.

24. In a college class of 36 students, find how many students are likely to prefer Pepsi.

An office uses 5 boxes of envelopes every 3 weeks.

25. Find how long a gross of envelope boxes is likely to last. (A gross of boxes is 144 boxes.) Round to the nearest week.

26. Find how many boxes should be purchased to last a month. Round to the nearest box.

27. The daily supply of oxygen for one person is provided by 625 square feet of lawn. A total of 3750 square feet of lawn would provide the daily supplies of oxygen for how many people? (*Source:* Professional Lawn Care Association of America)

28. In the United States, approximately 71 million of the 200 million cars and light trucks in service have driver side air bags. In a parking lot containing 800 cars and light trucks, how many would be expected to have driver side air bags? (*Source:* Insurance Institute for Highway Safety)

29. Mary Boyd would like to estimate the height of the Statue of Liberty in New York City's harbor. According to the *2003 World Almanac*, the length of the Statue of Liberty's right arm is 42 feet. Mary's right arm is 2 feet long and her height is $5\frac{1}{3}$ feet. Use this information to estimate the height of the Statue of Liberty. (The actual height of the Statue of Liberty, from heel to top of the head, is 111 feet, 1 inch. How close is the estimated height to the actual height of the statue?)

30. There are 72 milligrams of cholesterol in a 3.5 ounce serving of lobster. How much cholesterol is in 5 ounces of lobster? Round to the nearest tenth of a milligram. (*Source:* The National Institute of Health)

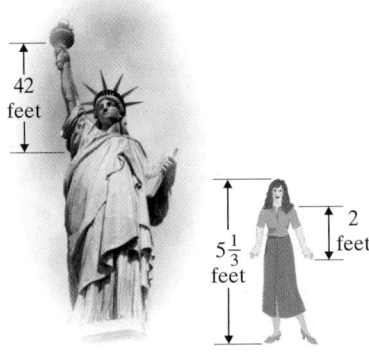

31. One pound of firmly packed brown sugar yields $2\frac{1}{4}$ cups. How many pounds of brown sugar will be required by a recipe that calls for 6 cups of firmly packed brown sugar? (*Source:* Based on data from *Family Circle* magazine)

32. Eleven out of every 25 greeting cards sold in the United States are Hallmark brand cards. If a consumer purchased 75 greeting cards in the past year, how many do you expect would have been Hallmark brand cards? (*Source:* Hallmark Cards, Inc.)

33. Most men are concerned about their health, but they continue to have poor eating habits. Two out of every 5 men blame their poor eating habits on too much fast food. In a room of 40 men, how many would you expect to blame their not eating well on fast food? (*Source:* Healthy Choice Mixed Grills survey)

34. Trump World Tower in New York City is 881 feet tall and contains 72 stories. The Empire State Building contains 102 stories. If the Empire State Building has the same number of feet per floor as the Trump World Tower, approximate its height rounded to the nearest foot. (*Source:* skyscrapers.com)

35. Medication is prescribed in 7 out of every 10 hospital emergency room visits that involve an injury. If a large urban hospital had 620 emergency room visits involving an injury in the past month, how many of these visits would you expect included a prescription for medication? (*Source:* National Center for Health Statistics)

36. Currently in the American population of people aged 65 years old and older, there are 145 women for every 100 men. In a nursing home with 280 male residents over the age of 65, how many female residents over the age of 65 would be expected? (*Source:* U.S. Bureau of the Census)

37. McDonald's four-piece Chicken McNuggets® has 190 calories. How many calories are in a nine-piece Chicken McNuggets? (*Source:* McDonald's Corporation)

38. A small order of McDonald's french fries weighs 68 grams and contains 10 grams of fat. McDonald's Super Size® french fries weighs 176 grams. How many grams of fat are in McDonald's SuperSize french fries? Round to the nearest tenth. (*Source:* McDonald's Corporation)

When making homemade ice cream in a hand-cranked freezer, the tub containing the ice cream mix is surrounded by a brine solution. To freeze the ice cream mix rapidly so that smooth and creamy ice cream results, the brine solution should combine crushed ice and rock salt in a ratio of 5 to 1. (Source: White Mountain Freezers, The Rival Company)

39. A small ice cream freezer requires 12 cups of crushed ice. How much rock salt should be mixed with the ice to create the necessary brine solution?

40. A large ice cream freezer requires $18\frac{3}{4}$ cups of crushed ice. How much rock salt will be needed?

41. The gas/oil ratio for a certain chainsaw is 50 to 1.
   a. How much oil (in gallons) should be mixed with 5 gallons of gasoline?

   b. If 1 gallon equals 128 fluid ounces, write the answer to part **a** in fluid ounces. Round to the nearest whole ounce.

42. The gas/oil ratio for a certain tractor mower is 20 to 1.
   a. How much oil (in gallons) should be mixed with 10 gallons of gas?

   b. If 1 gallon equals 4 quarts, write the answer to part **a** in quarts.

43. The adult daily dosage for a certain medicine is 150 mg (milligrams) of medicine for every 20 pounds of body weight.

    a. At this rate, find the daily dose for a man who weighs 275 pounds.

    b. If the man is to receive 500 mg of this medicine every 8 hours, is he receiving the proper dosage?

44. The adult daily dosage for a certain medicine is 80 mg (milligrams) for every 25 pounds of body weight.

    a. At this rate, find the daily dose for a woman who weighs 190 pounds.

    b. If she is to receive this medicine every 6 hours, find the amount to be given every 6 hours.

Review and Preview

Find the prime factorization of each number. See Section 4.2.

45. 15

46. 21

47. 20

48. 24

49. 200

50. 300

51. 32

52. 81

 Combining Concepts

As we have seen earlier, proportions are often used in medicine dosage calculations. The exercises below have to do with liquid drug preparations, where the weight of the drug is contained in a volume of solution. The description of mg and ml below will help. We will study metric units further in Chapter 9.

mg means milligrams (A paper clip weighs about a gram. A milligram is about the weight of $\frac{1}{1000}$ of a paper clip.)

ml means milliliter (A liter is about a quart, A milliliter is about the amount of liquid in $\frac{1}{1000}$ of a quart.)

One way to solve the applications below is to set up the proportion $\frac{mg}{ml} = \frac{mg}{ml}$.

A solution strength of 15 mg of medicine in 1 ml of solution is available.

53. If a patient needs 12 mg of medicine, how many ml do you administer?

54. If a patient need 33 mg of medicine, how many ml do you administer?

A solution strength of 8 mg of medicine in 1 ml of solution is available.

55. If a patient needs 10 mg of medicine, how many ml do you administer?

56. If a patient needs 6 mg of medicine, how many ml do you administer?

A board such as the one pictured below will balance if the following is true:

$$\frac{\text{first weight}}{\text{second distance}} = \frac{\text{second weight}}{\text{first distance}}$$

or, by using cross products,

$$\text{first weight} \cdot \text{first distance} = \text{second weight} \cdot \text{second distance}$$

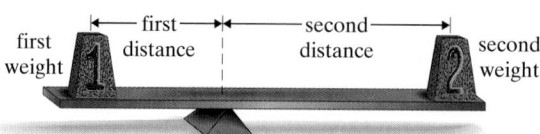

Use either equation to solve Exercises 57–58.

57. Find the distance n that will allow the board to balance.

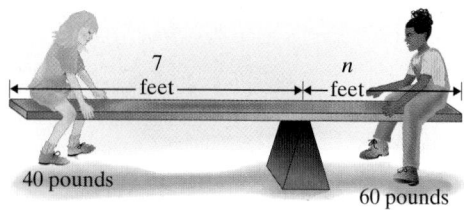

58. Find the length n needed to lift the weight below.

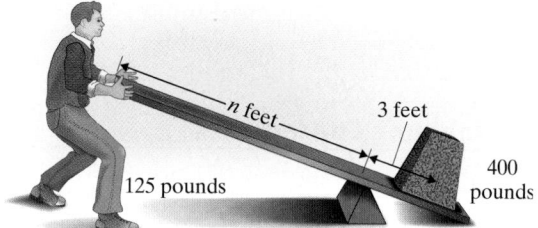

59. Describe a situation in which writing a proportion might solve a problem related to driving a car.

△ **6.5**   Congruent and Similar Triangles

A   Deciding Whether Two Triangles Are Congruent

Two triangles are **congruent** when they have the same shape and the same size. The triangles below are congruent. Angle A and D have the same measure, so they are called corresponding angles and they are marked the same way. Both have an arc with 1 tic mark. Angles C and F are corresponding angles. They have the same measure and are both marked by an arc with 2 tic marks. Finally, angles B and E are corresponding angles. They have the same measure and are both marked by an arc. In congruent triangles, the measures of corresponding angles are **equal** and the lengths of corresponding sides are **equal**.

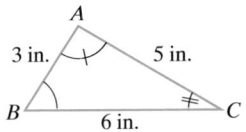 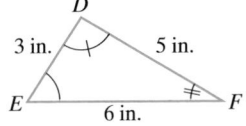

Now lets talk about corresponding sides. Sides AB and DE (denoted $\overline{AB}$ and $\overline{DE}$) are corresponding because they lie between corresponding angles. For the same reason, $\overline{AC}$ and $\overline{DF}$ are corresponding sides, and finally $\overline{BC}$ and $\overline{EF}$ are also corresponding sides.

Angles with equal measure: $\angle A$ and $\angle D$, $\angle B$ and $\angle E$, $\angle C$ and $\angle F$

(Sides with equal length:) $\overline{AB}$ and $\overline{DE}$, $\overline{BC}$ and $\overline{EF}$, $\overline{AC}$ and $\overline{DF}$

Any one of the following may be used to determine whether two triangles are congruent:

Congruent Triangles

Angle-Side-Angle (ASA)

If the measures of two angles of a triangle equal the measures of two angles of another triangle, and the lengths of the sides between each pair of angles are equal, the triangles are congruent.

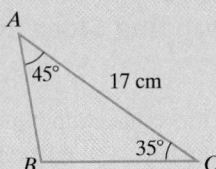 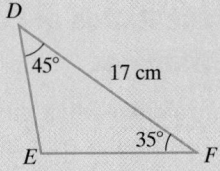

For example, these two triangles are congruent by Angle-Side-Angle.

Side-Side-Side (SSS)

If the lengths of the three sides of a triangle equal the lengths of the corresponding sides of another triangle, the triangles are congruent.

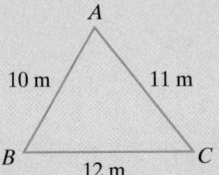 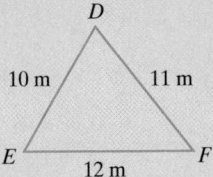

For example, these two triangles are congruent by Side-Side-Side.

Congruent Triangles, *continued*

Side-Angle-Side (SAS)

If the lengths of two sides of a triangle equal the lengths of corresponding sides of another triangle, and the measures of the angles between each pair of sides are equal, the triangle are congruent.

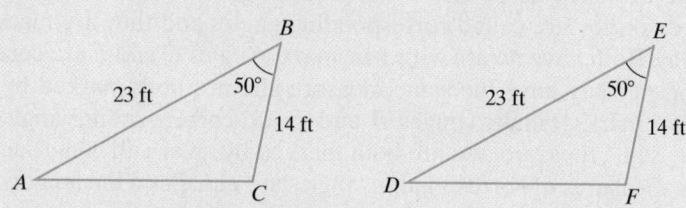

For example, these two triangles are congruent by Side-Angle-Side.

Practice Problem 1

Determine whether triangle *MNO* is congruent to triangle *RQS*.

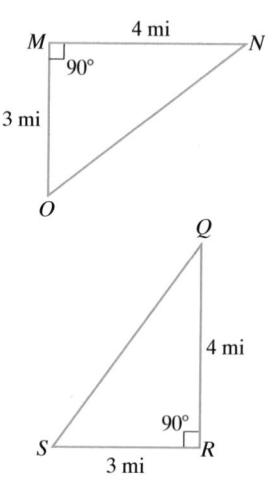

EXAMPLE 1   Determine whether triangle *ABC* is congruent to triangle *DEF*.

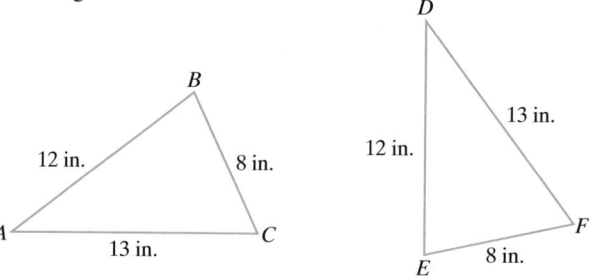

Solution: Since the lengths of all three sides of triangle *ABC* equal the lengths of all three sides of triangle *DEF*, the triangles are congruent.   ●

In Example 1, notice that as soon as we know that the two triangles are congruent, we know that all three corresponding angles also have equal measure.

Angles with equal measure: *B* and *E, C* and *F, A* and *D*

B   Finding the Ratios of Corresponding Sides in Similar Triangles

Two triangles are **similar** when they have the same shape but not necessarily the same size. In similar triangles, the measures of corresponding angles are **equal** and corresponding sides are **in proportion**. The following triangles are similar:

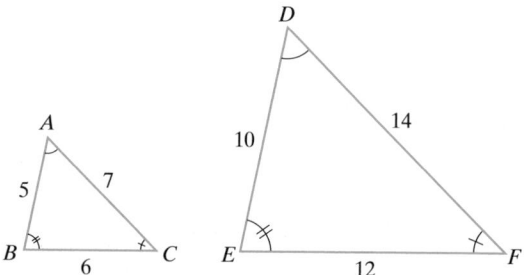

Since these triangles are similar, the measures of corresponding angles are equal.

Answer

1. congruent

Angles with equal measure: $\angle A$ and $\angle D$, $\angle B$ and $\angle E$, $\angle C$ and $\angle F$
Also, the lengths of corresponding sides are in proportion. This means that the ratios of the lengths of corresponding sides are equal.

Sides with lengths in proportion:

$$\frac{AB}{DE} = \frac{5}{10} = \frac{1}{2}, \frac{BC}{EF} = \frac{6}{12} = \frac{1}{2}, \frac{CA}{FD} = \frac{7}{14} = \frac{1}{2}$$

The ratio of corresponding sides is $\frac{1}{2}$.

EXAMPLE 2   Find the ratio of corresponding sides $\overline{AB}$ and $\overline{DE}$ for the similar triangles ABC and DEF.

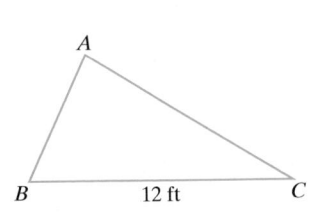
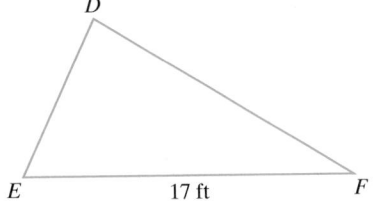

Solution: We are given the lengths of two corresponding sides. Their ratio is

$$\frac{12 \text{ feet}}{17 \text{ feet}} = \frac{12}{17}$$

●

ⓒ Finding Unknown Lengths of Sides in Similar Triangles

Because the ratios of lengths of corresponding sides are equal, we can use proportions to find unknown lengths in similar triangles.

Try the Concept Check in the margin.

EXAMPLE 3   Given that the triangles are similar, find the unknown length n.

Solution: Since the triangles are similar, corresponding sides are in proportion. Thus, the ratio of 2 to 3 is the same as the ratio of 10 to n. If we write this in symbols, we have

$$\frac{2}{3} = \frac{10}{n}$$

To find the unknown length n, we set cross products equal.

$$\frac{2}{3} = \frac{10}{n}$$

30	$=$	$2n$	Set cross products equal.
$\dfrac{30}{2}$	$=$	$\dfrac{2n}{2}$	Divide both sides by 2.
15	$=$	n	

The unknown length n is 15 inches.

●

Practice Problem 2

Find the ratio of corresponding sides $\overline{QR}$ and $\overline{XY}$ for the similar triangles QRS and XYZ.

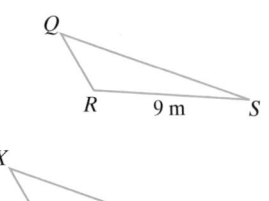

Practice Problem 3

Given that the triangles are similar, find the unknown length n.

Concept Check

The following two triangles are similar. Which vertices of the first triangle appear to correspond to which vertices of the second triangle?

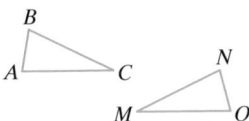

Answers

2. $\dfrac{9}{13}$ **3.** $n = 8$ ft

Concept Check: *A* corresponds to *O*; *B* corresponds to *N*; *C* corresponds to *M* (or *A* corresponds to *N*; *B* corresponds to *O*).

D Solving Problems Containing Similar Triangles

Many applications involve a diagram containing similar triangles. Surveyors, astronomers, and many other professionals use ratios of similar triangles frequently in their work.

EXAMPLE 4 Finding the Height of a Tree

Mel Rose is a 6-foot-tall park ranger who needs to know the height of a particular tree. He notices that when the shadow of the tree is 69 feet long, his own shadow is 9 feet long. Find the height of the tree.

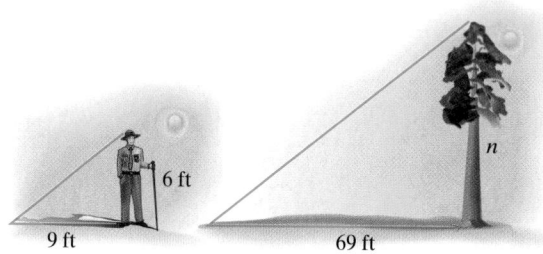

Solution:

1. UNDERSTAND. Read and reread the problem. Notice that the triangle formed by the sun's rays, Mel, and his shadow is similar to the triangle formed by the sun's rays, the tree, and its shadow.

2. TRANSLATE. Write a proportion from the similar triangles formed.

$$\text{Mel's height} \rightarrow \frac{6}{n} = \frac{9}{69} \leftarrow \text{length of Mel's shadow} \atop \text{height of tree} \rightarrow \qquad \leftarrow \text{length of tree's shadow}$$

$$\text{or} \quad \frac{6}{n} = \frac{3}{23} \qquad \text{Write in simplest form.}$$

3. SOLVE for n.

$$\frac{6}{n} = \frac{3}{23}$$

$$3n = 138 \qquad \text{Set cross products equal.}$$

$$\frac{3n}{3} = \frac{138}{3} \qquad \text{Divide both sides by 3.}$$

$$n = 46$$

4. INTERPRET. *Check* to see that replacing n with 46 in the proportion makes the proportion true. *State* your conclusion: The height of the tree is 46 feet. ●

Practice Problem 4

Tammy Shultz, a firefighter, needs to estimate the height of a burning building. She estimates the length of her shadow to be 8 feet long and the length of the building's shadow to be 200 feet long. Find the height of the building if she is 5 feet tall.

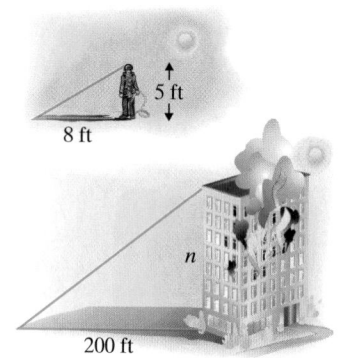

Answer

4. 125 feet

Name _____ Section _____ Date _____

EXERCISE SET 6.5

A *Determine whether each pair of triangles is congruent. If congruent, state the reason why, such as SSS, SAS, or ASA. See Example 1.*

1.

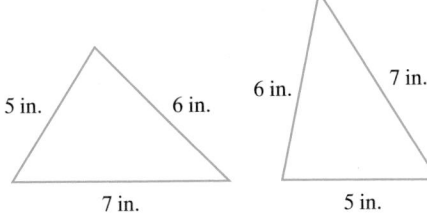

2.

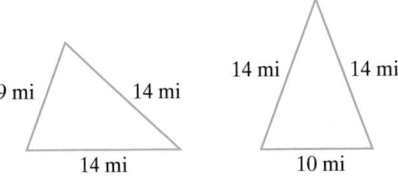

3.

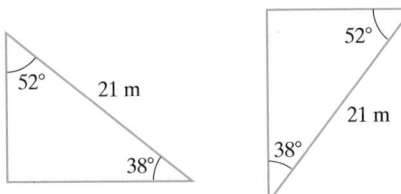

4.

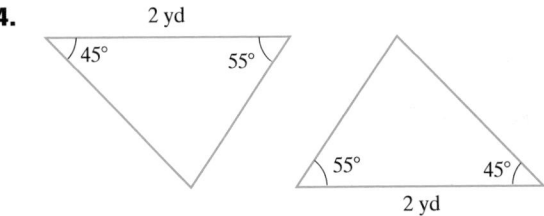

B *Find the ratio of the corresponding sides of the given similar triangles. See Example 2.*

5.

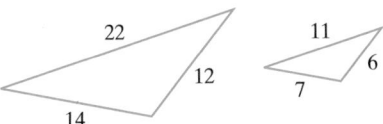

6.

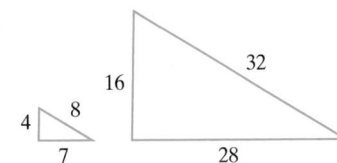

7.

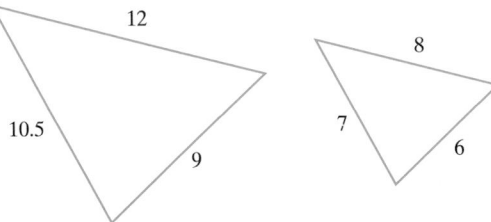

8.

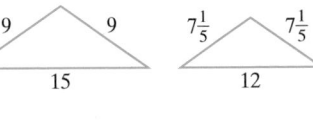

C *Given that the pairs of triangles are similar, find the length of the side labeled n. See Example 3.*

9.

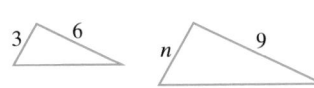

10.

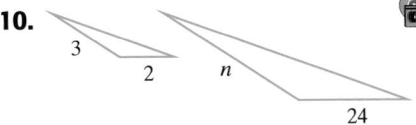

11.

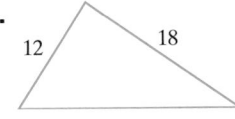

12.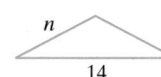

13.

14.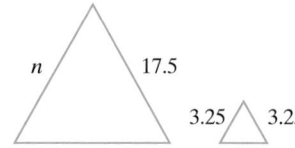

15.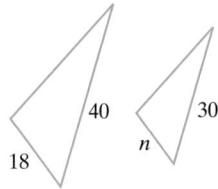

16.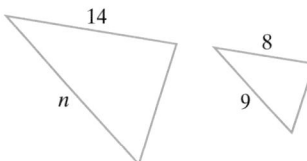

17.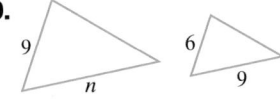

18.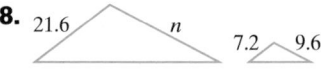

19.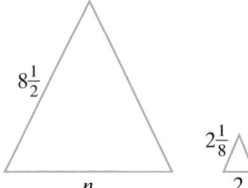

20.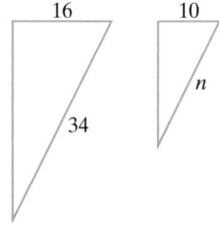

21.

22.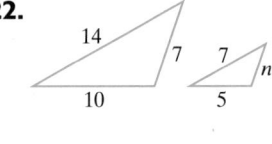

D *Solve. See Example 4.*

23. Given the following diagram, approximate the height of the First National Center in Oklahoma City, OK. (*Source: The World Almanac,* 2003)

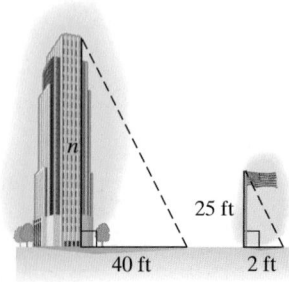

24. The tallest tree standing today is a redwood located near Ukiah, California. Given the following diagram, approximate its height. (*Source: Guinness World Record,* 2003)

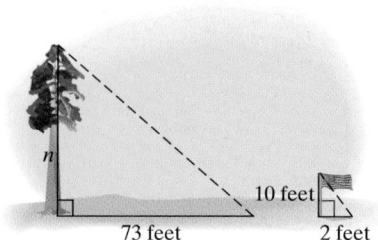

25. Samantha Black, a 5-foot-tall park ranger, needs to know the height of a tree. She notices that when the shadow of the tree is 48 feet long, her shadow is 4 feet long. Find the height of the tree.

 26. Lloyd White, a firefighter, needs to estimate the height of a burning building. He estimates the length of his shadow to be 9 feet long and the length of the building's shadow to be 75 feet long. Find the approximate height of the building if he is 6 feet tall.

27. If a 30-foot tree casts an 18-foot shadow, find the length of the shadow cast by a 24-foot tree.

28. If a 24-foot flagpole casts a 32-foot shadow, find the length of the shadow cast by a 44-foot antenna. Round to the nearest tenth.

Review and Preview

Find the average of each list of numbers. See Section 1.7.

29. 14, 17, 21, 18

30. 87, 84, 93

31. 76, 79, 88

32. 7, 8, 4, 6, 3, 8

Combining Concepts

Given that the pairs of triangles are similar, find the length of the side labeled n. Round your results to 1 decimal place.

 33.

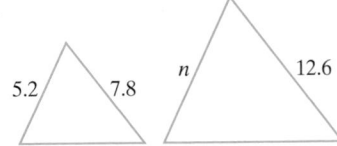

 34.

35. In your own words, describe any differences in similar triangles and congruent triangles.

36. The print on a particular page measures 7 inches by 9 inches where 7 inches is the width. A printing shop is to copy the page and reduce the print so that its length is 5 inches. What will its width be? Will the print now fit on a 3 inch by 5 inch index card?

37. Ben and Joyce Lander draw a rectangular deck on their house plans. Joyce measures the deck drawing on the plans to be 3 inches by $4\frac{1}{2}$ inches. If the scale on the drawing is $\frac{1}{4}$ in. = 1 foot, find the dimensions of the deck they want built.

CHAPTER 6 ACTIVITY **Investigating Scale Drawings**

MATERIALS:

- ruler

- tape measure

- grid paper (optional)

This activity may be completed by working in groups or individually.

Scale drawings are used by architects, engineers, interior designers, ship builders, and others. In a scale drawing, each unit measurement on the drawing represents a fixed length on the object being drawn. For instance, in an architect's scale drawing, 1 inch on the drawing may represent 10 feet on a building. The scale describes the relationship between the measurements. If the measurements have the same units, the scale can be expressed as a ratio. In this case, the ratio would be 1 : 120, representing 1 inch to 120 inches (or 10 feet).

Use a ruler and the scale drawing of an elementary school below to answer the following questions.

1. How wide are each of the front doors of the school?

2. How long is the front of the school?

3. How tall is the front of the school?

Now you will draw your own scale floor plan. First choose a room to draw—it can be your math classroom, your living room, your dormitory room, or any room that can be easily measured. Start by using a tape measure to measure the distances around the base of the walls in the room you are drawing.

4. Choose a scale for your floor plan.

5. Convert each measurement in the room you are drawing to the corresponding lengths needed for the scale drawing.

6. Complete your floor plan (you may find it helpful to use grid paper). Mark the locations of doors and windows on your floor plan. Be sure to indicate on the drawing the scale used in your floor plan.

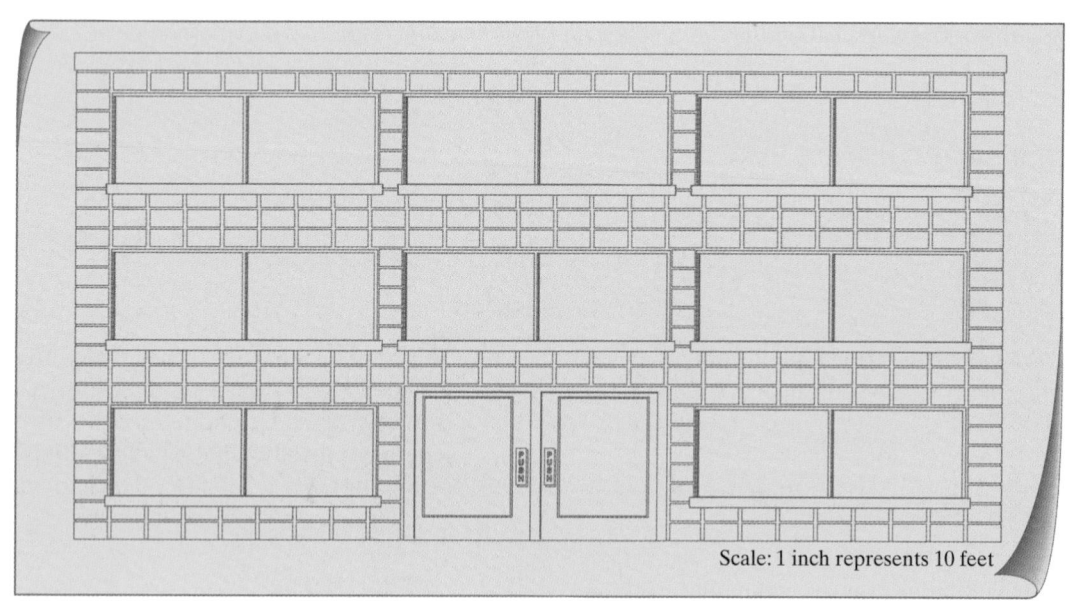

Scale: 1 inch represents 10 feet

Chapter 6   Vocabulary Check

Fill in each blank with one of the words or phrases listed below.

rate   unit rate   ratio   unit price   proportion   similar   congruent

1. A _____ is the quotient of two numbers. It can be written as a fraction, using a colon, or using the word *to*.
2. $\dfrac{x}{2} = \dfrac{7}{16}$ is an example of a _____.
3. A _____ is a rate with a denominator of 1.
4. A _____ is a "money per item" unit rate.
5. A _____ is used to compare different kinds of quantities.
6. _____ triangles have the same shape and the same size.
7. _____ triangles have exactly the same shape but not necessarily the same size.

C H A P T E R

Highlights

DEFINITIONS AND CONCEPTS	**EXAMPLES**
SECTION 6.1 RATIOS	
A **ratio** is the quotient of two quantities.	The ratio of 3 to 4 can be written as $\dfrac{3}{4}$ or 3:4 ↑ fractional notation ↑ colon notation
SECTION 6.2 RATES	
Rates are used to compare different kinds of quantities.	Write the rate 12 spikes every 8 inches as a fraction in simplest form. $\dfrac{12 \text{ spikes}}{8 \text{ inches}} = \dfrac{3 \text{ spikes}}{2 \text{ inches}}$
A **unit rate** is a rate with a denominator of 1.	Write as a unit rate: 117 miles on 5 gallons of gas $\dfrac{117 \text{ miles}}{5 \text{ gallons}} = \dfrac{23.4 \text{ miles}}{1 \text{ gallon}}$ or 23.4 miles per gallon or 23.4 mpg
A **unit price** is a "money per item" unit rate.	Write as a unit price: $5.88 for 42 ounces of detergent $\dfrac{\$5.88}{42 \text{ ounces}} = \dfrac{\$0.14}{1 \text{ ounce}} = \$0.14 \text{ per ounce}$

DEFINITIONS AND CONCEPTS	**EXAMPLES**

SECTION 6.3 PROPORTIONS

A **proportion** is a statement that two ratios or rates are equal.

$$\frac{a}{b} = \frac{c}{d}$$

cross product → $b \cdot c$ = $a \cdot d$ ← cross product

If cross products are equal, the proportion is true. If cross products are not equal, the proportion is false.

Find the value of x that makes the proportion true.

$$\frac{9}{45} = \frac{21}{x}$$

$45 \cdot 21$	=	$9 \cdot x$	Set cross products equal.
945	=	$9x$	Multiply.
$\dfrac{945}{9}$	=	$\dfrac{9x}{9}$	Divide both sides by 9.
105	=	x	Simplify.

SECTION 6.4 PROPORTIONS AND PROBLEM SOLVING

Given a specified ratio (or rate) of two quantities, a proportion can be used to determine an unknown quantity.

On a map, 50 miles corresponds to 3 inches. How many miles correspond to 10 inches?

1. UNDERSTAND. Read and reread the problem.

2. TRANSLATE. We let n represent the unknown number. We are given that 50 miles is to 3 inches as n miles is to 10 inches.

$$\begin{array}{c} \text{miles} \rightarrow \\ \text{inches} \rightarrow \end{array} \frac{50}{3} = \frac{n}{10} \begin{array}{c} \leftarrow \text{miles} \\ \leftarrow \text{inches} \end{array}$$

3. SOLVE.

$$\frac{50}{3} = \frac{n}{10}$$

$3n$	=	500	Set cross products equal.
$\dfrac{3n}{3}$	=	$\dfrac{500}{3}$	Divide both sides by 3.
n	=	$166\dfrac{2}{3}$	

4. INTERPRET. *Check* your work. *State* your conclusion: On the map, $166\dfrac{2}{3}$ miles corresponds to 10 inches.

Congruent triangles have the same shape and the same size. Corresponding angles are equal, and corresponding sides are equal.

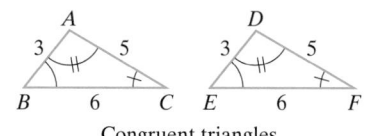

Congruent triangles

Similar triangles have exactly the same shape but not necessarily the same size. Corresponding angles are equal, and the ratios of the lengths of corresponding sides are equal.

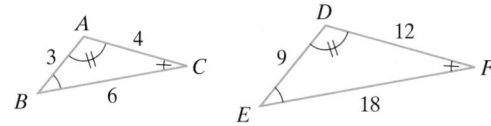

Similar triangles

$$\frac{AB}{DE} = \frac{3}{9} = \frac{1}{3}, \ \frac{BC}{EF} = \frac{6}{18} = \frac{1}{3},$$

$$\frac{AC}{DF} = \frac{4}{12} = \frac{1}{3}$$

STUDY SKILLS REMINDER

Tips for studying for an exam

To prepare for an exam, try the following study techniques.

■ Start the study process days before your exam.

■ Make sure that you are current and up to date on your assignments.

■ If there is a topic that you are unsure of, use one of the many resources that are available to you. For example,

See your instructor.

Visit a learning resource center on campus where math tutors are available.

Read the textbook material and examples on the topic.

View a videotape on the topic.

■ Reread your notes and carefully review the Chapter Highlights at the end of the chapter.

■ Work the review exercises at the end of the chapter and check your answers. Make sure that you correct any missed exercises. If you have trouble on a topic, use a resource listed above.

■ Find a quiet place to take the Chapter Test found at the end of the chapter. Do not use any resources when taking this sample test. This way you will have a clear indication of how prepared you are for your exam. Check your answers and make sure that you correct any missed exercises.

■ Get lots of rest the night before the exam. It's hard to show how well you know the material if your brain is foggy from lack of sleep.

Good luck and keep a positive attitude.

Chapter 6 Review

(6.1) *Write each ratio as a fraction in simplest form.*

1. 6000 people to 4800 people

2. 143 births to 121 births

3. $2\frac{1}{4}$ days to 10 days

4. 14 quarters to 5 quarters

5. 4 weeks to 15 weeks

6. 4 yards to 8 yards

7. $3\frac{1}{2}$ dollars to 7 dollars

8. 3.5 centimeters to 75 centimeters

(6.2) *Write each rate as a fraction in simplest form.*

9. 8 stillborn births to 1000 live births

10. 6 professors for 20 graduate research assistants

11. 15 word-processing pages printed in 6 minutes

12. 8 computers assembled in 6 hours

Find each unit rate.

13. 468 miles in 9 hours

14. 180 feet in 12 seconds

15. $0.93 for 3 pears

16. $6.96 for 4 diskettes

17. 260 kilometers in 4 hours

18. 8 gallons of pesticide for 6 acres of crops

19. $184 for books for 5 college courses

20. 52 bushels of fruit from 4 trees

Find each unit price and decide which is the better buy. Assume that we are comparing different sizes of the same brand.

21. Taco sauce:
8 ounces for $0.99
12 ounces for $1.69

22. Peanut butter:
18 ounces for $1.49
28 ounces for $2.39

23. 2% milk:
16 ounces for $0.59,
$\frac{1}{2}$ gallon for $1.69
1 gallon for $2.29 (1 gallon = 128 fluid ounces)

24. Soft drink:
12 ounces for $0.59
16 ounces for $0.79
32 ounces for $1.19

(6.3) *Write each sentence as a proportion.*

25. 20 men is to 14 women as 10 men is to 7 women.

26. 50 tries is to 4 successes as 25 tries is to 2 successes.

27. 16 sandwiches is to 8 players as 2 sandwiches is to 1 player.

28. 12 tires is to 3 cars as 4 tires is to 1 car.

Determine whether each proportion is true or false.

29. $\dfrac{21}{8} = \dfrac{14}{6}$

30. $\dfrac{3}{5} = \dfrac{60}{100}$

31. $\dfrac{3.1}{6.2} = \dfrac{0.8}{0.16}$

32. $\dfrac{3.75}{3} = \dfrac{7.5}{6}$

Solve each proportion for the given variable.

33. $\dfrac{x}{6} = \dfrac{15}{18}$

34. $\dfrac{y}{9} = \dfrac{5}{3}$

35. $\dfrac{4}{13} = \dfrac{10}{x}$

36. $\dfrac{8}{5} = \dfrac{9}{z}$

37. $\dfrac{16}{3} = \dfrac{y}{6}$

38. $\dfrac{x}{3} = \dfrac{9}{2}$

39. $\dfrac{x}{5} = \dfrac{27}{2\frac{1}{4}}$

40. $\dfrac{2\frac{1}{2}}{6} = \dfrac{3}{z}$

41. $\dfrac{x}{0.4} = \dfrac{4.7}{3}$ Round to the nearest hundredth.

42. $\dfrac{0.07}{0.3} = \dfrac{7.2}{n}$ Round to the nearest tenth.

(6.4) *Solve.*

The ratio of a quarterback's completed passes to attempted passes is 3 to 7.

43. If he attempts 32 passes, find how many passes he completed. Round to the nearest whole pass.

44. If he completed 15 passes, find how many passes he attempted.

One bag of pesticide covers 4000 square feet of crops.

45. Find how many bags of pesticide should be purchased to cover a rectangular garden 180 feet by 175 feet.

46. Find how many bags of pesticide should be purchased to cover a square garden 250 feet on each side.

An owner of a Ford Escort can drive 420 miles on 11 gallons of gas.

47. If Tom Aloiso ran out of gas in an Escort and AAA comes to his rescue with $1\frac{1}{2}$ gallons of gas, determine whether Tom can then drive to a gas station 65 miles away.

48. Find how many gallons of gas Tom can expect to burn on a 3000-mile trip. Round to the nearest gallon.

Yearly homeowner property taxes are figured at a rate of $1.15 tax for every $100 of house value.

49. If a homeowner pays $627.90 in property taxes, find the value of the house.

50. Find the property taxes on a townhouse valued at $89,000.

On an architect's blueprint, 1 inch corresponds to 12 feet.

51. Find the length of a wall represented by a $3\frac{3}{8}$-inch line on the blueprint.

52. If an exterior wall is 99 feet long, find how long the blueprint measurement should be.

(6.5) *Given that the pairs of triangles are similar, find the unknown length x.*

△ **53.**

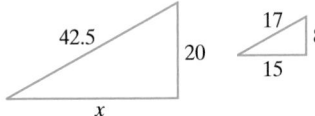

△ **54.**

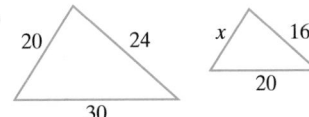

△ **55.**

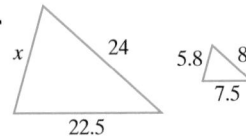

△ **56.**

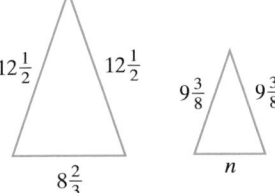

Solve.

57. A housepainter needs to estimate the height of a condominium. He estimates the length of his shadow to be 7 feet long and the length of the building's shadow to be 42 feet long. Find the height of the building if the housepainter is $5\frac{1}{2}$ feet tall.

△ **58.** Santa's elves are making a triangular sail for a toy sailboat. The toy sail is to be the same shape as a real sailboat's sail. Use the following diagram to find the unknown lengths *x* and *y*.

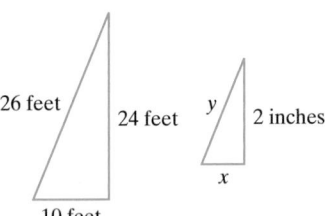

Determine whether each pair of triangles is congruent. If congruent, state the reason why, such as SSS, SAS, or ASA.

59.

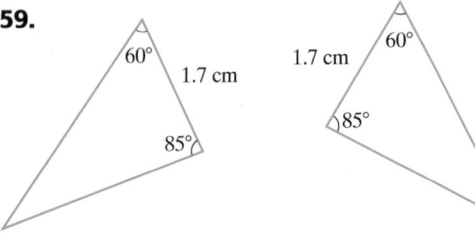

60.

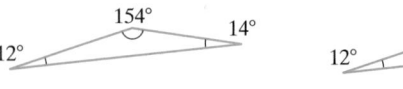 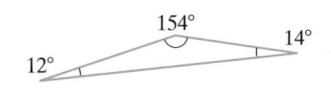

510

Chapter 6 Test   Remember to check your answers and use the Chapter Test Prep Video to view solutions.

Write each ratio as a fraction in simplest form.

1. 4500 trees to 6500 trees

2. $75 to $10

Write each rate as a fraction in simplest form.

3. 28 men to every 4 women

4. 9 inches of rain in 30 days

Find each unit rate.

5. 650 kilometers in 8 hours

6. 8 inches of rain in 12 hours

7. 140 students for 5 teachers

Find each unit price and decide which is the better buy.

8. Steak sauce:
   8 ounces for $1.19
   12 ounces for $1.89

9. Jelly:
   16 ounces for $1.49
   24 ounces for $2.39

Determine whether each proportion is true or false.

10. $\dfrac{28}{16} = \dfrac{14}{8}$

11. $\dfrac{3.6}{2.2} = \dfrac{1.9}{1.2}$

Solve each proportion for the given variable.

12. $\dfrac{n}{3} = \dfrac{15}{9}$

13. $\dfrac{8}{x} = \dfrac{11}{6}$

14. $\dfrac{\frac{4}{3}}{\frac{3}{7}} = \dfrac{y}{\frac{1}{4}}$

15. $\dfrac{1.5}{5} = \dfrac{2.4}{n}$

1. _____

2. _____

3. _____

4. _____

5. _____

6. _____

7. _____

8. _____

9. _____

10. _____

11. _____

12. _____

13. _____

14. _____

15. _____

16. _____

17. _____

18. _____

19. _____

20. _____

21. _____

22. _____

Solve.

16. On an architect's drawing, 2 inches corresponds to 9 feet. Find the length of a home represented by a line that is 11 inches long.

17. If a car can be driven 80 miles in 3 hours, how long will it take to travel 100 miles?

18. The standard dose of medicine for a dog is 10 grams for every 15 pounds of body weight. What is the standard dose for a dog that weighs 80 pounds?

19. Jerome Grant worked 6 hours and packed 86 cartons of books. At this rate, how many cartons can he pack in 8 hours?

20. Two of every 25 coffee drinkers prefers Maxwell House Master Blend coffee. In a group of 3200 coffee drinkers, how many prefer Maxwell House Master Blend? (_Source:_ Information Resources, Inc.)

△ **21.** Given that the following triangles are similar, find the unknown length _n_.

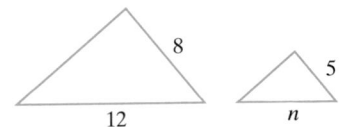

△ **22.** Tamara Watford, a surveyor, needs to estimate the height of a tower. She estimates the length of her shadow to be 4 feet long and the length of the tower's shadow to be 48 feet long. Find the height of the tower if she is $5\frac{3}{4}$ feet tall.

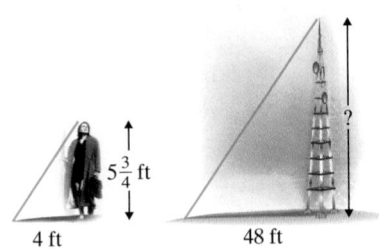

Cumulative Review

1. Divide: $51,600 \div 403$

2. Divide: $23,685 \div 194$

3. Write as an algebraic expression. Use x to represent "a number."
   a. 7 increased by a number
   b. 15 decreased by a number
   c. The product of 2 and a number
   d. The quotient of a number and 5
   e. 2 subtracted from a number

4. Write as an algebraic expression. Use y to represent "a number."
   a. a number increased by 3
   b. the quotient of 10 and a number
   c. a number subtracted from 4
   d. the product of 7 and a number

5. Simplify:
   a. $-(-4)$
   b. $-|-5|$
   c. $-|6|$

6. Simplify:
   a. $-|-3|$
   b. $-|2|$
   c. $-(-1)$

7. Subtract: $-10 - 5$

8. Subtract: $-7 - 3$

9. Simplify: $7 + (-12) - 3 - (-8)$

10. Simplify: $-7 + (-4) + 3 - (-11)$

Simplify each expression by combining like terms.

11. $6x + 2x - 5$

12. $-7x + 6 + 9x$

13. $2x - 5 + 3y + 4x - 10y + 11$

14. $3x + 2z - 4z + 2 + 7x - 5$

15. Determine whether 6 is a solution of the equation $4(x - 3) = 12$.

16. Determine whether 2 is a solution of the equation $-3(7 + 2x) = -33$

17. Solve: $-8 = 2y$

18. Solve: $-3k = 12$

19. Solve: $20 - x = 21$

20. Solve: $21 - 2x = -9x$

21. Write $\dfrac{2}{5}$ as an equivalent fraction whose denominator is 15.

22. Write $\dfrac{3}{7}$ as an equivalent fraction whose denominator is 56.

23. Write the prime factorization of 80.

24. Write the prime factorization of 44.

25. Multiply: $\dfrac{3x}{4} \cdot \dfrac{8}{5x}$

26. Multiply: $\dfrac{2a}{3} \cdot \dfrac{9}{4a}$

Answers

1. _____
2. _____
3. a. _____
   b. _____
   c. _____
   d. _____
   e. _____
4. a. _____
   b. _____
   c. _____
   d. _____
5. a. _____
   b. _____
   c. _____
6. a. _____
   b. _____
   c. _____
7. _____
8. _____
9. _____
10. _____
11. _____
12. _____
13. _____
14. _____
15. _____
16. _____
17. _____
18. _____
19. _____
20. _____
21. _____
22. _____
23. _____
24. _____
25. _____
26. _____

27. _____

28. _____

29. _____

30. _____

31. _____

32. _____

33. _____

34. _____

35. _____

36. _____

37. _____

38. _____

39. _____

40. _____

41. _____

42. _____

43. _____

44. _____

45. _____

46. _____

47. _____

48. _____

49. _____

50. _____

Subtract and simplify.

27. $\dfrac{8}{9} - \dfrac{1}{9}$

28. $\dfrac{3}{5} - \dfrac{1}{5}$

29. $\dfrac{7}{8} - \dfrac{5}{8}$

30. $\dfrac{7}{12} - \dfrac{5}{12}$

31. Subtract: $2 - \dfrac{x}{3}$

32. Subtract: $\dfrac{x}{10} - \dfrac{4}{5}$

33. Simplify: $\dfrac{\dfrac{y}{5} - 2}{\dfrac{3}{10}}$

34. Simplify: $\dfrac{\dfrac{3x}{2} + 5}{\dfrac{3}{8}}$

35. Solve for x: $\dfrac{1}{3}x = 7$

36. Solve for x: $\dfrac{2}{5}x = 8$

37. Multiply: $3\dfrac{1}{3} \cdot \dfrac{7}{8}$

38. Multiply: $2\dfrac{4}{5} \cdot \dfrac{1}{9}$

39. Write 0.43 as a fraction.

40. Write 0.65 as a fraction.

41. Divide: $2.3\overline{)10.764}$

42. Divide: $9.6\overline{)22.752}$

43. Simplify: $-0.5(8.6 - 1.2)$

44. Simplify: $0.7(5.9 - 2.3)$

45. Write the numbers in order from smallest to largest. $\dfrac{9}{20}, \dfrac{4}{9}, 0.456$

46. Write the numbers in order from smallest to largest. $\dfrac{4}{7}, \dfrac{1}{2}, 0.531$

47. Solve: $x - 1.5 = 8$

48. Solve: $3x + 2.5 = 13$

49. Solve for y: $\dfrac{\dfrac{1}{2}}{\dfrac{4}{5}} = \dfrac{\dfrac{3}{4}}{y}$

50. Solve for c: $\dfrac{2c}{5} = \dfrac{3}{10}$

Percent

This chapter is devoted to percent, a concept used virtually every day in ordinary and business life. Understanding percent and using it efficiently depends on understanding ratios, because a percent is a ratio whose denominator is 100. We present techniques to write percents as fractions and as decimals. We then solve problems relating to interest rates, sales tax, discounts, and other real-life situations by writing percent equations.

The Model T, developed by Henry Ford in 1908, was the world's first mass-produced automobile. It sold for $850. In 1913, Ford Motor Company introduced interchangeable parts and the moving assembly line, and the automobile industry as we know it today was born. For almost half a century, American automobile manufacturers produced the majority of the world's motor vehicles. However, in the 1970s and 1980s, the sales of foreign auto imports in the United States began to rise. Imports reached a high in 1987, when 31.3% of all auto sales in the United States were imports. However, American auto producers started to make a comeback, reducing the sales of imported vehicles to a low of 14.9% of all such sales in 1996. Sales of imports have rebounded since then, reaching 28.2% of all auto sales in the United States in the first half of 2001. In Exercises 103–106 on page 528 we see some of the ways percents are used by the automobile manufacturing industry.

Name _____ Section _____ Date_____

Chapter 7 Pretest

1. _____

2. _____

3. _____

4. _____

5. _____

6. _____

7. _____

8. _____

9. _____

10. _____

11. _____

12. _____

13. _____

14. _____

15. _____

16. _____

17. _____

18. _____

19. _____

20. _____

1. In a group of 100 people, 12 are female. What percent of the group is female?

2. Write 57% as a decimal.

3. Write 2.75 as a percent.

4. Write 7.5% as a fraction in simplest form.

5. Write $\frac{3}{20}$ as a percent.

Translate each question to an equation.

6. 18% of 50 is what number?

7. 4% of what number is 89?

Translate each question to a proportion.

8. 90% of what number is 82?

9. 48 is what percent of 112?

Solve.

10. What percent of 80 is 16?

11. What number is 1.5% of 220?

12. 32 is 16% of what number?

13. In a box of 250 lightbulbs, 1.2% were found to be defective. How many lightbulbs were defective?

14. The enrollment at a local high school decreased 2% over last year's enrollment of 4200. Find the decrease in enrollment and the current enrollment.

15. The sales tax on a $499 printer is $34.93. Find the sales tax rate.

16. Jerry Williams receives 2% commission on his sales of computer equipment. Last week his commission was $448. Find the amount of his sales last week.

17. A television that normally sells for $650 is on sale at 12% off. What is the discount and what is the sales price?

18. Find the simple interest after 3 years on $600 at an interest rate of 8%.

19. $5000 is invested at 6% compounded quarterly for 6 years. Find the total amount at the end of 6 years. Use Appendix F.

20. Find the monthly payment on a $700 loan for 2 years if the interest on the 2-year loan is $196.

7.1   Percents, Decimals, and Fractions

A   Understanding Percent

The word **percent** comes from the Latin phrase *per centum*, which means "**per 100**." For example, 53 percent (%) means 53 per 100. In the square below, 53 of the 100 squares are shaded. Thus 53% of the figure is shaded.

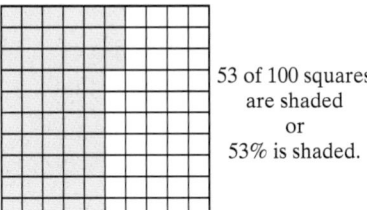

53 of 100 squares are shaded or 53% is shaded.

Since 53% means 53 per 100, 53% is the ratio of 53 to 100, or $\frac{53}{100}$.

$$53\% = \frac{53}{100}$$

OBJECTIVES

Ⓐ   Know the meaning of percent.
Ⓑ   Write percents as decimals.
Ⓒ   Write decimals as percents.
Ⓓ   Write percents as fractions.
Ⓔ   Write fractions as percents.
Ⓕ   Convert percents, decimals, and fractions.

SSM
TUTOR CENTER   SG   CD & VIDEO   MATH PRO   WEB

Percent

Percent means **per one hundred**. The "%" symbol is used to denote percent.

Also,

$$7\% = \frac{7}{100} \quad \text{7 parts per 100 parts}$$

$$73\% = \frac{73}{100} \quad \text{73 parts per 100 parts}$$

$$109\% = \frac{109}{100} \quad \text{109 parts per 100 parts}$$

Percent is used in a variety of everyday situations. For example:

The interest rate is 5.7%.

47.4% of U.S. homes have Internet access.

The store is having a 25% off sale.

78% of us trust our local fire department.

The enrollment in community colleges has increased 141% in the last 30 years.

EXAMPLE 1   In a survey of 100 people, 17 people drive blue cars. What percent of people drive blue cars?

Solution: Since 17 people out of 100 drive blue cars, the fraction is $\frac{17}{100}$. Then

$$\frac{17}{100} = 17\%$$

Practice Problem 1

Of 100 students in a club, 23 are freshmen. What percent of the students are freshmen?

Answer

1. 23%

29 out of 100 executives are in their forties. What percent of executives are in their forties?

EXAMPLE 2   46 out of every 100 college students live at home. What percent of students live at home? (*Source:* Independent Insurance Agents of America)

Solution: $\dfrac{46}{100} = 46\%$ ●

B   Writing Percents as Decimals

Since percent means "per hundred," we have that

$$1\% = \frac{1}{100} = 0.01$$

To write a percent as a decimal, we can first write the percent as a fraction.

$$53\% = \frac{53}{100}$$

Now we can write the fraction as a decimal as we did in Section 4.7.

$$\frac{53}{100} = 53(0.01) = 0.53 \text{ (53-hundredths)}$$

Notice that the result is

$$53\% = 53(0.01) = 0.53 \text{ Replace the percent symbol with 0.01. Then multiply.}$$

Practice Problem 3

Write 89% as a decimal.

Practice Problems 4–7

Write each percent as a decimal.

4.   2.7%

5.   150%

6.   0.69%

7.   500%

Concept Check

Why is it incorrect to write the percent 0.033% as 3.3 in decimal form?

Writing a Percent as a Decimal
Replace the percent symbol with its decimal equivalent, 0.01; then multiply.
$$43\% = 43(0.01) = 0.43$$

EXAMPLE 3   Write 23% as a decimal.

Solution: $23\% = 23(0.01)$ Replace the percent symbol with 0.01.
$= 0.23$ Multiply.   ●

EXAMPLES   Write each percent as a decimal.

4. $4.6\% = 4.6(0.01) = 0.046$ Replace the percent symbol with 0.01. Then multiply.

5. $190\% = 190(0.01) = 1.90$ or   1.9

6. $0.74\% = 0.74(0.01) = 0.0074$

7. $100\% = 100(0.01) = 1.00$ or   1   ●

Answers

2. 29%   **3.** 0.89   **4.** 0.027   **5.** 1.5   **6.** 0.0069
7. 5

Concept Check: To write a percent as a decimal, the decimal point should be moved two places to the left, not to the right. So the correct answer is 0.00033.

Try the Concept Check in the margin.

C Writing Decimals as Percents

To write a decimal as a percent, we use the result of Example 7 above. In this example, we found that $1 = 100\%$.

$$0.38 = 0.38(1) = 0.38(100\%) = 38\%$$

Notice that the result is

$$0.38 = 0.38(100\%) = 38.\% \text{ or } 38\% \quad \text{Multiply by 1 in the form of } 100\%.$$

Writing a Decimal as a Percent

Multiply by 1 in the form of 100%.

$$0.27 = 0.27(100\%) = 27.\% \text{ or } 27\%$$

EXAMPLE 8   Write 0.65 as a percent.

Solution: $0.65 = 0.65(100\%) = 65.\%$ Multiply by 100%.

$$= 65\%$$

Practice Problem 8

Write 0.19 as a percent.

EXAMPLES   Write each decimal as a percent.

9. $1.25 = 1.25(100\%) = 125.\%$ or 125%

10. $0.012 = 0.012(100\%) = 001.2\%$ or 1.2%

11. $0.6 = 0.6(100\%) = 060.\%$ or 60%

Practice Problems 9–11

Write each decimal as a percent.

9. 1.75   10. 0.044   11. 0.7

Helpful Hint

A zero was inserted as a placeholder.

Try the Concept Check in the margin.

Concept Check

Why is it incorrect to write the decimal 0.0345 as 34.5% in percent form?

Answers

8. 19%   **9.** 175%   **10.** 4.4%   **11.** 70%

Concept Check: To change a decimal to a percent, the decimal point should be moved *only* two places to the right. So the correct answer is 3.45%.

D Writing Percents as Fractions

When we write a percent as a fraction, we usually then write the fraction in simplest form. For example, recall from the previous section that

$$50\% = \frac{50}{100}$$

Then we write the fraction in simplest form:

$$\frac{50}{100} = \frac{\overset{1}{\cancel{50}}}{2 \cdot \cancel{50}} = \frac{1}{2}$$

Writing a Percent as a Fraction

Replace the percent symbol with its fraction equivalent, $\frac{1}{100}$; then multiply. Don't forget to simplify the fraction, if possible.

$$7\% = 7 \cdot \frac{1}{100} = \frac{7}{100}$$

EXAMPLES Write each percent as a fraction or mixed number in simplest form.

12. $40\% = 40 \cdot \dfrac{1}{100} = \dfrac{40}{100} = \dfrac{2 \cdot \overset{1}{\cancel{20}}}{5 \cdot \underset{1}{\cancel{20}}} = \dfrac{2}{5}$

13. $1.9\% = 1.9 \cdot \dfrac{1}{100} = \dfrac{1.9}{100}$. Next we multiply the numerator and denominator by 10.

$$\frac{1.9 \cdot 10}{100 \cdot 10} = \frac{19}{1000}$$

14. $125\% = 125 \cdot \dfrac{1}{100} = \dfrac{125}{100} = \dfrac{5 \cdot \overset{1}{\cancel{25}}}{4 \cdot \underset{1}{\cancel{25}}} = \dfrac{5}{4} \text{ or } 1\dfrac{1}{4}$

15. $33\frac{1}{3}\% = 33\dfrac{1}{3} \cdot \dfrac{1}{100} = \dfrac{100}{3} \cdot \dfrac{1}{100} = \dfrac{\overset{1}{\cancel{100}} \cdot 1}{3 \cdot \underset{1}{\cancel{100}}} = \dfrac{1}{3}$

Write as an improper fraction.

16. $100\% = 100 \cdot \dfrac{1}{100} = \dfrac{100}{100} = 1$

E Writing Fractions as Percents

Recall that to write a percent as a fraction, we drop the percent symbol and divide by 100. We reverse these steps to write a fraction as a percent.

Practice Problems 12–16

Write each percent as a fraction in simplest form.

12. 25%
13. 2.3%
14. 150%
15. $66\dfrac{2}{3}\%$
16. 8%

Answers

12. $\dfrac{1}{4}$ **13.** $\dfrac{23}{1000}$ **14.** $\dfrac{3}{2}$ **15.** $\dfrac{2}{3}$ **16.** $\dfrac{2}{25}$

Writing a Fraction as a Percent

Multiply by 1 in the form of 100%.

$$\frac{1}{8} = \frac{1}{8} \cdot 100\% = \frac{1}{8} \cdot \frac{100}{1}\% = \frac{100}{8}\% = 12\frac{1}{2}\% \quad \text{or} \quad 12.5\%$$

Helpful Hint

From Example 7, we know that

$$100\% = 1$$

Recall that when we multiply a number by 1, we are not changing the value of that number. Therefore, when we multiply a number by 100%, we are not changing its value but rather writing the number as an equivalent percent.

EXAMPLES   Write each fraction or mixed number as a percent.

17. $\dfrac{9}{20} = \dfrac{9}{20} \cdot 100\% = \dfrac{9}{20} \cdot \dfrac{100}{1}\% = \dfrac{900}{20}\% = 45\%$

18. $\dfrac{2}{3} = \dfrac{2}{3} \cdot 100\% = \dfrac{2}{3} \cdot \dfrac{100}{1}\% = \dfrac{200}{3}\% = 66\dfrac{2}{3}\%$

19. $1\dfrac{1}{2} = \dfrac{3}{2} \cdot 100\% = \dfrac{3}{2} \cdot \dfrac{100}{1}\% = \dfrac{300}{2}\% = 150\%$

Try the Concept Check in the margin.

EXAMPLE 20   Write $\dfrac{1}{12}$ as a percent. Round to the nearest hundredth percent.

Solution:

$$\frac{1}{12} = \frac{1}{12} \cdot 100\% = \frac{1}{12} \cdot \frac{100}{1} = \frac{100}{12}\% \approx 8.33\%$$

"approximately"

$$
\begin{array}{r}
8.333 \approx 8.33 \\
12\overline{)100.000} \\
-96 \\
\hline
40 \\
-36 \\
\hline
40 \\
-36 \\
\hline
40 \\
-36 \\
\hline
4
\end{array}
$$

Thus, $\dfrac{1}{12}$ is approximately 8.33%.

Practice Problems 17–19

Write each fraction or mixed number as a percent.

17. $\dfrac{1}{2}$ 18. $\dfrac{7}{40}$ 19. $2\dfrac{1}{4}$

Concept Check

Which digit in the percent 76.4582% represents

a.   A tenth percent?
b.   A thousandth percent?
c.   A hundredth percent?
d.   A whole percent?

Practice Problem 20

Write $\dfrac{3}{17}$ as a percent. Round to the nearest hundredth percent.

Answers

17. 50%   **18.** $17\dfrac{1}{2}\%$ **19.** 225%   **20.** 17.65%

Concept Check: **a.** 4,   **b.** 8,   **c.** 5,   **d.** 6

(F) Converting Percents, Decimals, and Fractions

Let's summarize what we have learned so far about percents, decimals, and fractions:

> **Summary of Converting Percents, Decimals, and Fractions**
>
> - *To write a percent as a decimal*, replace the % symbol with its decimal equivalent, 0.01; then multiply.
> - *To write a percent as a fraction*, replace the % symbol with its fraction equivalent, $\frac{1}{100}$; then multiply.
> - *To write a decimal or fraction as a percent*, multiply by 100%.

Practice Problem 21

A family decides to spend no more than 25% of its monthly income on rent. Write 25% as a decimal.

EXAMPLE 21 17.8% of automobile thefts in the continental United States occur in the Midwest. Write this percent as a decimal. (*Source:* The American Automobile Manufacturers Association)

Solution:

$$17.8\% = 17.8(0.01) = 0.178.$$

Thus, 17.8% written as a decimal is 0.178. ●

Practice Problem 22

Provincetown's budget for waste disposal increased by $1\frac{1}{4}$ times over the budget from last year. What percent increase is this?

EXAMPLE 22 An advertisement for a stereo system reads "$\frac{1}{4}$ off". What percent off is this?

Solution: Write $\frac{1}{4}$ as a percent.

$$\frac{1}{4} = \frac{1}{4} \cdot 100\% = \frac{1}{4} \cdot \frac{100}{1}\% = \frac{100}{4}\% = 25\%$$

Thus, "$\frac{1}{4}$ off" is the same as "25% off." ●

It is helpful to know a few basic percent conversions. Appendix D contains a handy reference of percent, decimal, and fraction equivalencies.

STUDY SKILLS REMINDER

Is your notebook still organized?

Is your notebook still organized? If it's not, it's not too late to start organizing it. Start writing your notes and completing your homework assignment in a notebook with pockets (spiral or ring binder). Take class notes in this notebook, and then follow the notes with your completed homework assignment. When you receive graded papers or handouts, place them in the notebook pocket so that you will not lose them.

Remember to mark (possibly with an explanation point) any note(s) that seem extra important to you. Also remember to mark (possibly with a question mark) any notes or homework that you are having trouble with. Don't forget to see your instructor or a math tutor to help you with the concepts or exercises that you are having trouble understanding.

Also—don't forget to write neatly and keep a positive attitude.

Answers

21. 0.25   **22.** 125%

Name _____ Section _____ Date _____

Mental Math

Write each fraction as a percent.

1. $\dfrac{13}{100}$

2. $\dfrac{92}{100}$

3. $\dfrac{87}{100}$

4. $\dfrac{71}{100}$

5. $\dfrac{1}{100}$

6. $\dfrac{2}{100}$

EXERCISE SET 7.1

A *Solve. See Examples 1 and 2.*

1. A basketball player made 81 out of 100 attempted free throws. What percent of free throws was made?

2. In a survey of 100 people, 54 preferred chocolate syrup on their ice cream. What percent preferred chocolate syrup?

Adults were asked what type of cookie was their favorite. The circle graph shows the results for every 100 people. Use this graph to answer Exercises 3–6. See Examples 1 and 2.

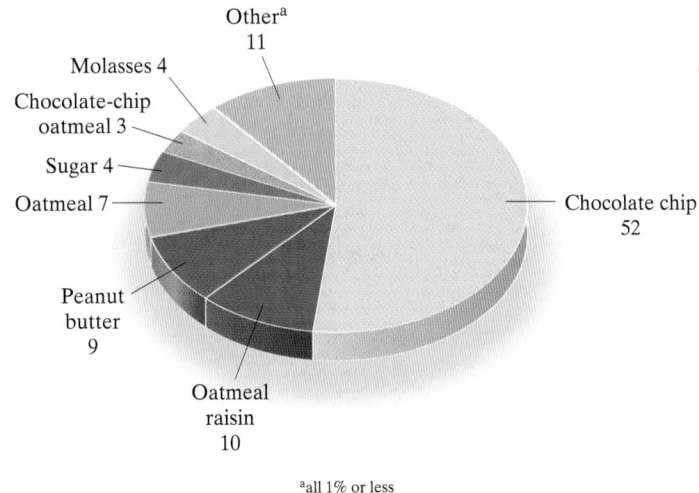

Other[a]
11

Molasses 4

Chocolate-chip
oatmeal 3

Sugar 4

Oatmeal 7

Peanut
butter
9

Oatmeal
raisin
10

Chocolate chip
52

[a]all 1% or less
Source: *USA Today*, 2/23/96

3. What percent preferred peanut butter cookies?

4. What percent preferred oatmeal raisin cookies?

5. What type of cookie was preferred by most adults? What percent preferred this type of cookie?

6. What two types of cookies were preferred by the same number of adults? What percent preferred each type?

7. 12 out of 100 adults have watched an entire infomercial. What percent is this? (*Source:* Aragon Consulting Group)

8. In 2002, 93 out of 100 elementary schools had computers. What percent is this? (*Source:* Quality Education Data, Inc.)

B *Write each percent as a decimal. See Examples 3 through 7.*

9. 48%   **10.** 64%   **11.** 6%   **12.** 9%

13. 100%   **14.** 136%   **15.** 61.3%   **16.** 52.7%

17. 2.8%   **18.** 1.7%   **19.** $64\frac{1}{4}$% (Hint: First write $\frac{1}{4}$ as a decimal)   **20.** $18\frac{3}{5}$%

21. 300%   **22.** 500%   **23.** 32.58%   **24.** 72.18%

Write each percent as a decimal. See Examples 3 through 7.

25. 73.7% of the workforce in the United States works 35 hours or more per week. (*Source:* U.S. Bureau of Labor)

26. In 2001, approximately 18.4% of new luxury cars sold were silver. (*Source:* Du Pont Automotive Products)

27. The United States is the largest consumer of energy in the world, using 25% of all energy produced worldwide each year. (*Source:* Energy Information Administration)

28. Some health insurance companies pay 80% of a person's medical costs.

29. In January, 2003, 11.1% of all Southwest Airlines' flights arrived late. (*Source:* Bureau of Transportation Statistics)

30. In 2002, 88% of all public high schools in the United States owned at least one computer. (*Source:* Quality Education Data, Inc.)

31. 46.2% of registered dental hygienists in the United States have earned an associate's degree or higher. (*Source:* The American Dental Hygienists' Association)

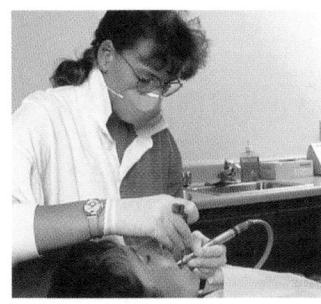

32. Video games made up 21.2% of the total toy market in the United States in 2000. (*Source:* The NPD Group Worldwide)

C *Write each decimal as a percent. See Examples 8 through 11.*

33. 3.1

34. 4.8

35. 29

36. 56

37. 0.003

38. 0.006

39. 0.22

40. 0.45

41. 0.056

42. 0.027

43. 0.3328

44. 0.1115

45. 3.00

46. 5.00

47. 0.7

48. 0.8

Write each decimal as a percent. See Examples 8 through 11.

49. The Munoz family saves 0.10 of their take-home pay.

50. The cost of an item for sale is 0.7 of the sales price.

51. In the state of California, the Hispanic population is 0.324 of the total population. (*Source:* 2003 *World Almanac*)

52. In a recent year, 0.522 of all mail delivered by the United States Postal Service was first-class mail. (*Source:* United States Postal Service)

53. People take aspirin for a variety of reasons. The most common use of aspirin is to prevent heart disease, accounting for 0.38 of all aspirin use. (*Source:* Bayer Market Research)

54. The highest state income tax rate in the state of Iowa is 0.0898 on taxable income over $51,660. (*Source: CCH State Tax Guide*)

D *Write each percent as a fraction or mixed number in simplest form. See Examples 12 through 16.*

55. 4%

56. 2%

57. 4.5%

58. 7.5%

59. 175%

60. 250%

61. 73%

62. 86%

63. 12.5%

64. 62.5%

65. 6.25%

66. 37.5%

67. $10\frac{1}{3}\%$

68. $7\frac{3}{4}\%$

69. $22\frac{3}{8}\%$

70. $15\frac{5}{8}\%$

E *Write each fraction or mixed number as a percent. See Examples 17 through 20.*

71. $\frac{3}{4}$

72. $\frac{1}{2}$

73. $\frac{7}{10}$

74. $\frac{3}{10}$

75. $\frac{2}{5}$

76. $\frac{4}{5}$

77. $\frac{59}{100}$

78. $\frac{73}{100}$

79. $\frac{17}{50}$

80. $\frac{47}{50}$

81. $\frac{3}{8}$

82. $\frac{5}{8}$

83. $\dfrac{5}{16}$　　　　**84.** $\dfrac{7}{16}$　　　　**85.** $\dfrac{2}{3}$　　　　**86.** $\dfrac{1}{3}$

87. $2\dfrac{1}{2}$　　　　**88.** $2\dfrac{1}{5}$　　　　**89.** $1\dfrac{9}{10}$　　　　**90.** $2\dfrac{7}{10}$

Write each fraction as a percent. Round to the nearest hundredth percent. See Example 20.

91. $\dfrac{7}{11}$　　　　**92.** $\dfrac{5}{12}$　　　**93.** $\dfrac{4}{15}$　　　　**94.** $\dfrac{10}{11}$

95. $\dfrac{1}{7}$　　　　**96.** $\dfrac{1}{9}$　　　　**97.** $\dfrac{11}{12}$　　　　**98.** $\dfrac{5}{6}$

(F) *Complete each table. See Examples 21 and 22.*

99.

Percent	Decimal	Fraction
35%		
		$\dfrac{1}{5}$
	0.5	
70%		
		$\dfrac{3}{8}$

see table

100.

Percent	Decimal	Fraction
	0.525	
		$\dfrac{3}{4}$
$66\dfrac{2}{3}\%$		
		$\dfrac{5}{6}$
100%		

see table

101.

Percent	Decimal	Fraction
40%		
	0.235	
		$\dfrac{4}{5}$
$33\dfrac{1}{3}\%$		
		$\dfrac{7}{8}$
7.5%		

see table

102.

Percent	Decimal	Fraction
50%		
		$\dfrac{2}{5}$
	0.25	
12.5%		
		$\dfrac{5}{8}$
		$\dfrac{7}{50}$

see table

Solve. See Examples 21 and 22.

103. Approximately 24.9% of new full-size cars are silver, making silver the most popular new vehicle color for that class. Write this percent as a fraction. (*Source:* American Automobile Manufacturers Association)

104. In 2001, 10.6% of all new cars sold in the United States were imported from Japan. Write this percent as a fraction. (*Source:* American Automobile Manufacturers Association)

105. In 1998, $\frac{17}{200}$ of all new cars sold in the United States were imported from Japan. Write this fraction as a percent. (*Source:* Ward's Communications)

106. In 1980, $\frac{53}{250}$ of all new cars sold in the United States were imported from Japan. Write this fraction as a percent. (*Source:* American Automobile Manufacturers Association)

107. In 2001, New Mexico had the highest percent of residents not covered by any type of health insurance at 20.7%. Write this percent as a fraction. (*Source:* U.S. Bureau of the Census)

108. In 2001, 70.2% of all households with televisions subscribed to a cable television service. Write this percent as a decimal. (*Source:* Nielsen Media Research)

109. For the 2001–2002 television season, the top-rated prime-time television program was *Friends*, which had an average audience share of approximately $\frac{1}{4}$ of all those watching television during that time slot. Write this fraction as a percent. (*Source:* Nielsen Media Research)

110. Approximately $\frac{37}{50}$ of all structure fires take place in personal residences. Write this fraction as a percent. (*Source:* National Fire Protection Association)

Write each percent in this circle graph as a fraction.

World Population by Continent

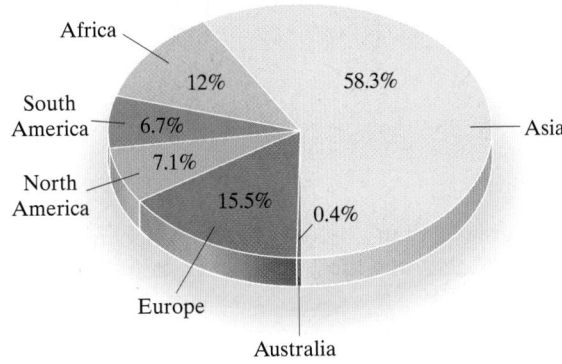

111. 0.4%   **112.** 58.3%   **113.** 12%

114. 6.7%   **115.** 7.1%   **116.** 15.5%

Review and Preview

Find the value of n. See Section 3.3.

117. $3n = 45$ **118.** $7n = 48$ **119.** $-8n = 80$

120. $-2n = 16$ **121.** $-6n = -72$ **122.** $5n = -35$

Combining Concepts

What percent of the figure is shaded?

123. 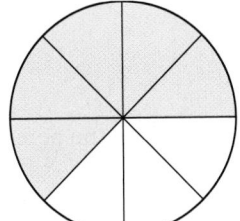   **124.** 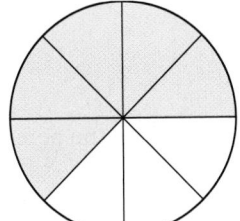   **125.** 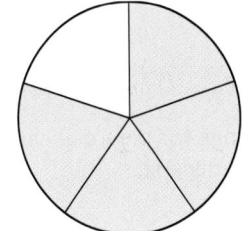   **126.**

Write each fraction as a decimal and then write each decimal as a percent. Round the decimal to three decimal places and the percent to the nearest tenth of a percent.

127. $\dfrac{850}{736}$

128. $\dfrac{506}{248}$

Fill in the blanks.

129. A fraction written as a percent is greater than 100% when the numerator is _____ than the denominator.
                                greater/less

130. A decimal written as a percent is less than 100% when the decimal is _____ than 1.
                           greater/less

131. In your own words, explain how to write a percent as a fraction.

132. In your own words, explain how to write a fraction as a decimal.

The bar graph shows the predicted fastest-growing occupations. Use this graph to answer Exercises 133–136.

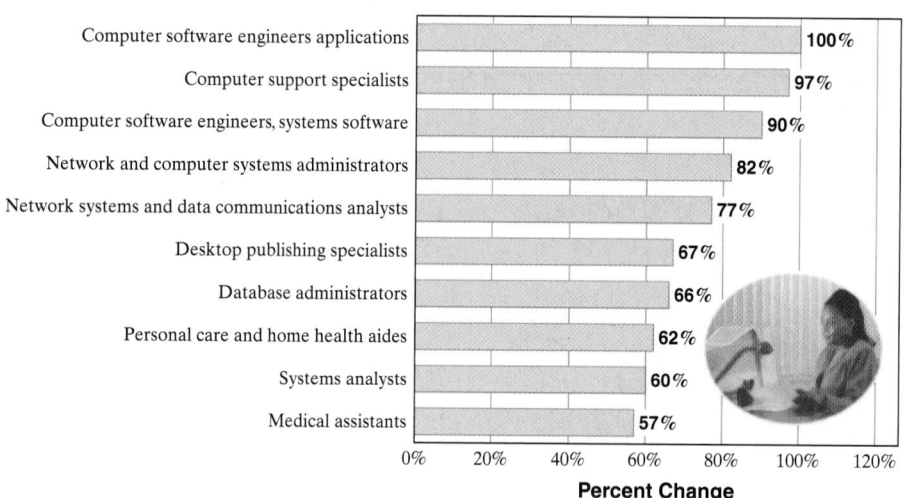

Fastest-Growing Occupations 2000–2010

Source: Bureau of Labor Statistics

133. What occupation is predicted to be the fastest growing?

134. What occupation is predicted to be the second fastest growing?

135. Write the percent change for personal care and home health aides as a decimal.

136. Write the percent change for systems analysts as a decimal.

7.2   Solving Percent Problems with Equations

Sections 7.2 and 7.3 introduce two methods for solving percent problems. It is not necessary that you study both sections. You may want to check with your instructor for further advice.

To solve percent problems in this section, we will translate the problems into mathematical statements, or equations.

Ⓐ   Writing Percent Problems as Equations

Recognizing key words in a percent problem is helpful in writing the problem as an equation. Three key words in the statement of a percent problem and their meanings are as follows:

of means **multiplication** (·)

is means **equals** (=)

what (or some equivalent) means **the unknown number**.

In our examples, we will let the letter x stand for the unknown number.

EXAMPLE 1   Translate to an equation:

5 is what percent of 20?

Solution: 5 is   what percent   of 20?

$$5 = \qquad x \qquad \cdot \quad 20$$

Practice Problem 1

Translate: 6 is what percent of 24?

> **Helpful Hint**
>
> Remember that an equation is simply a mathematical statement that contains an equal sign (=).
>
> $$5 = 20x$$
> ↑
> equal sign

EXAMPLE 2   Translate to an equation:

1.2 is 30% of what number?

Solution: 1.2 is 30% of   what number?

$$1.2 = 30\% \cdot \qquad x$$

EXAMPLE 3   Translate to an equation:

What number is 25% of 0.008?

Solution: What number   is 25% of 0.008?

$$x \qquad = 25\% \quad \cdot \quad 0.008$$

Practice Problem 2

Translate: 1.8 is 20% of what number?

Practice Problem 3

Translate: What number is 40% of 3.6?

Answers

1. $6 = x \cdot 24$ **2.** $1.8 = 20\% \cdot x$

3. $x = 40\% \cdot 3.6$

Practice Problems 4–6

Translate each question to an equation.

4. 42% of 50 is what number?
5. 15% of what number is 9?
6. What percent of 150 is 90?

Concept Check

In the equation $2x = 10$, what step is taken to solve the equation?

Practice Problem 7

What number is 20% of 85?

Answers

4. $42\% \cdot 50 = x$ **5.** $15\% \cdot x = 9$
6. $x \cdot 150 = 90$ **7.** 17

Concept Check: Divide both sides of the equation by 2.

EXAMPLES   Translate each question to an equation.

4.   38% of 200 is what number?
     ↓   ↓   ↓   ↓   ↓
     38% · 200 = x

5.   40% of what number is 80?
     ↓   ↓   ↓   ↓   ↓
     40% · x = 80

6.   What percent of 85 is 34?
          ↓   ↓   ↓ ↓
         x ·   85 = 34

Try the Concept Check in the margin.

B Solving Percent Problems

You may have noticed by now that each percent problem has contained three numbers—in our examples, two are known and one is unknown. Each of these numbers is given a special name.

          15%   of   60   is   9
           ↓   ↓   ↓   ↓   ↓

| 15% percent | · | 60 base | = | 9 amount |

We call this equation the **percent equation**.

> **Percent Equation**
>
> percent · base = amount

Once a percent problem has been written as a percent equation, we can use the equation to find the unknown number.

Solving Percent Equations for the Amount

EXAMPLE 7   What number is 35% of 60?
                      ↓   ↓   ↓   ↓   ↓

Solution: x =   35% · 60   Translate to an equation.

                         x =   0.35 · 60   Write 35% as 0.35.

                         x =   21   Multiply: 60
                                                            $\underline{\times 0.35}$
                                                              300
                                                            $\underline{1800}$
                                                            21.00

Then 21 is 35% of 40. Is this reasonable? To see, round 35% to 40%. Then 40% or 0.40(60) is 24. Our result is reasonable since 21 is close to 24.

Helpful Hint

When solving a percent equation, write the percent as a decimal or fraction.

EXAMPLE 8   85% of 300 is what number?

Solution:

$85\% \cdot 300 = x$ Translate to an equation.

$0.85 \cdot 300 = x$ Write 85% as 0.85.

$255 = x$ Multiply: $0.85 \cdot 300 = 255$.

Then 85% of 300 is 255. Is this result reasonable? To see, round 85% to 90%. Then 90% of 300 or $0.90(300) = 270$, which is close to 255.   ●

Practice Problem 8

90% of 150 is what number?

Solving Percent Equations for the Base

EXAMPLE 9   12% of what number is 0.6?

Solution:

$12\% \cdot x = 0.6$ Translate to an equation.

$0.12 \cdot x = 0.6$ Write 12% as 0.12.

$\dfrac{0.12 \cdot x}{0.12} = \dfrac{0.6}{0.12}$ Divide both sides by 0.12.

$x = 5$

$$0.12\overline{)0.60}$$
$$\underline{60}$$
$$0$$

Then 12% of 5 is 0.6. Is this reasonable? To see, round 12% to 10%. Then 10% of 0.6 or 0.10 (0.6) is 6, which is close to 5.   ●

Practice Problem 9

15% of what number is 1.2?

EXAMPLE 10   13   is $6\frac{1}{2}\%$ of what number?

Solution:

$13 = 6\frac{1}{2}\% \cdot x$ Translate to an equation.

$13 = 0.065 \cdot x$ $6\frac{1}{2}\% = 6.5\% = 0.065$.

$\dfrac{13}{0.065} = \dfrac{0.065 \cdot x}{0.065}$ Divide both sides by 0.065.

$200 = x$

$$0.065\overline{)13.000}$$
$$\underline{130}$$
$$0$$

Then 13 is $6\frac{1}{2}\%$ of 200. Check to see if this result is reasonable.   ●

Practice Problem 10

27 is $4\frac{1}{2}\%$ of what number?

Answers

8. 135   **9.** 8   **10.** 600

Solving Percent Equations for the Percent

Copyright 2004 Pearson Education, Inc.

Practice Problem 11

What percent of 80 is 8?

EXAMPLE 11 $\underbrace{\text{What percent}}\quad\underset{\downarrow}{\text{of}}\quad\underset{\downarrow}{12}\quad\underset{\downarrow}{\text{is}}\quad\underset{\downarrow}{9?}$

Solution:

$$x \cdot 12 = 9 \qquad \text{Translate to an equation.}$$

$$\frac{x \cdot 12}{12} = \frac{9}{12} \qquad \text{Divide both sides by 12.}$$

$$x = 0.75$$

Next, since we are looking for percent, we write 0.75 as a percent.

$$x = 75\%$$

Then 75% of 12 is 9. To check, see that $75\% \cdot 12 = 9$. ●

> **Helpful Hint**
>
> If your unknown in the percent equation is percent, don't forget to convert your answer to a percent.

Practice Problem 12

35 is what percent of 25?

EXAMPLE 12 $\underset{\downarrow}{78}\quad\underset{\downarrow}{\text{is}}\quad\underbrace{\text{what percent}}\quad\underset{\downarrow}{\text{of}}\ \underset{\downarrow}{65?}$

Solution:

$$78 = x \cdot 65 \qquad \text{Translate to an equation.}$$

$$\frac{78}{65} = \frac{x \cdot 65}{65} \qquad \text{Divide both sides by 65.}$$

$$1.2 = x$$

$$120\% = x \qquad \text{Write 1.2 as a percent.}$$

Then 78 is 120% of 65. Check this result. ●

Try the Concept Check in the margin.

Concept Check

Consider the problem

10 is 40% of what number?

Will the solution be greater than or less than 10? How do you know without solving the problem?

> **Helpful Hint**
>
> Use the following to see if your answers are reasonable.
>
> 100% of a number = the number
>
> $\left(\begin{array}{c}\text{a percent}\\\text{greater than}\\100\%\end{array}\right)$ of a number = $\begin{array}{c}\text{a number larger}\\\text{than the original number}\end{array}$
>
> $\left(\begin{array}{c}\text{a percent}\\\text{less than }100\%\end{array}\right)$ of a number = $\begin{array}{c}\text{a number less}\\\text{than the original number}\end{array}$

Answers

11. 10%   **12.** 140%

Concept Check: Greater. Answers may vary.

Mental Math

Identify the percent, the base, and the amount in each equation. Recall that percent · base = amount.

1. $42\% \cdot 50 = 21$

2. $30\% \cdot 65 = 19.5$

3. $107.5 = 125\% \cdot 86$

4. $99 = 110\% \cdot 90$

EXERCISE SET 7.2

A *Translate each question to an equation. Do not solve. See Examples 1 through 6.*

1. 15% of 72 is what number?

2. What number is 25% of 55?

3. 30% of what number is 80?

4. 0.5 is 20% of what number?

5. What percent of 90 is 20?

6. 8 is 50% of what number?

7. 1.9 is 40% of what number?

8. 72% of 63 is what number?

9. What number is 9% of 43?

10. 4.5 is what percent of 45?

B *Translate to an equation and solve. See Examples 7 and 8.*

 11. 10% of 35 is what number?

12. 25% of 60 is what number?

13. What number is 14% of 52?

14. What number is 30% of 17?

Solve. See Examples 9 and 10.

15. 30 is 5% of what number?

16. 25 is 25% of what number?

 17. 1.2 is 12% of what number?

18. 0.22 is 44% of what number?

Solve. See Examples 11 and 12.

19. 66 is what percent of 60?

20. 30 is what percent of 20?

21. 16 is what percent of 50?

22. 27 is what percent of 50?

Solve. See Examples 7 through 12.

23. 0.1 is 10% of what number?

24. 0.5 is 5% of what number?

25. 125% of 36 is what number?

26. 200% of 13.5 is what number?

27. 82.5 is $16\frac{1}{2}$% of what number?

28. 7.2 is $6\frac{1}{4}$% of what number?

 29. 2.58 is what percent of 50?

30. 264 is what percent of 33?

31. What number is 42% of 60?

32. What number is 36% of 80?

33. What percent of 150 is 67.5?

34. What percent of 105 is 88.2?

35. 120% of what number is 42?

36. 160% of what number is 40?

Review and Preview

Find the value of n in each proportion. See Section 6.3.

37. $\dfrac{27}{n} = \dfrac{9}{10}$ **38.** $\dfrac{35}{n} = \dfrac{7}{5}$ **39.** $\dfrac{n}{5} = \dfrac{8}{11}$ **40.** $\dfrac{n}{3} = \dfrac{6}{13}$

Write each sentence as a proportion.

41. 17 is to 12 as *n* is to 20.

42. 20 is to 25 as *n* is to 10.

43. 8 is to 9 as 14 is to *n*.

44. 5 is to 6 as 15 is to *n*.

Combining Concepts

Solve.

45. 1.5% of 45,775 is what number?

46. What percent of 75,528 is 27,945.36?

47. 22,113 is 180% of what number?

48. In your own words, explain how to solve a percent equation.

FOCUS ON **The Real World**

M&M's®

In the 1930s, many American stores did not stock chocolates during the summer because, without widespread air conditioning, the chocolate tended to melt and sales declined. In 1940, Forrest E. Mars, Sr., set out to develop a chocolate candy that could be sold year-round without the melting problem. So he formed a company in Newark, New Jersey, to make bite-sized melt-proof chocolate candies encased in a thin sugar shell. Thus M&M's® Plain Chocolate Candies were born!

The first Plain M&M's colors were brown, green, orange, red, violet, and yellow. Violet was replaced by tan in 1950. When the safety of a particular type of red food coloring was publicly questioned in 1976, red was completely eliminated from the M&M's color mix to avoid alarming consumers, even though the coloring in question was never used in M&Ms. After an 11-year hiatus, red returned to the color mix in 1987. In 1995, over 10 million Americans responded to a marketing campaign asking for help in choosing a new M&M's color. Given the choices of blue, pink, purple, or no change, 54% of the respondents chose blue, and it replaced tan. The new color mix of Plain M&M's became blue, brown, green, orange, red, and yellow

in the percentages shown in the circle graph. (Note: M&M continues to introduce different colors for a limited time. This may affect the outcome of the exercises below.)

M&M's Peanut Chocolate Candies debuted in 1954. At first, Peanut M&M's were all brown. However, in 1960, red, green, and yellow were added. Orange joined the mix in 1976, and, finally, blue was introduced in 1995. The current color mix for Peanut M&M's is shown in the circle graph.

**Color Mix for
M&M's® Peanut Chocolate Candies**

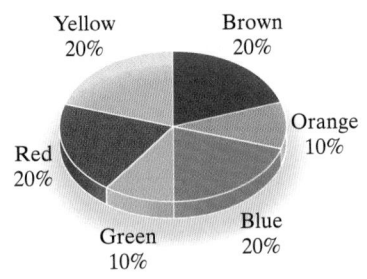

Source: Mars, Incorporated

GROUP ACTIVITY

1. Use a single-serving pouch of M&M's Plain Chocolate Candies and find the percentage of each color in the bag. How does it compare to the overall percentages officially reported?

2. Repeat Question 2 with a single-serving pouch of M&M's Peanut Chocolate Candies.

3. For each type of M&M's Candies, combine your group's color counts with those of the other groups in your class. Find the percentage of each color for the combined figures. How do these percentages compare to the official color mix?

**Color Mix for
M&M's® Plain Chocolate Candies**

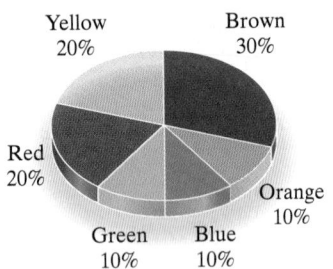

Source: Mars, Incorporated

7.3 Solving Percent Problems with Proportions

There is more than one method that can be used to solve percent problems. (See the note at the beginning of Section 7.2.) In the last section, we used the percent equation. In this section, we will use proportions.

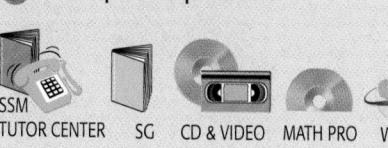

Ⓐ Write percent problems as proportions.

Ⓑ Solve percent problems.

SSM
TUTOR CENTER   SG   CD & VIDEO   MATH PRO   WEB

Ⓐ Writing Percent Problems as Proportions

To understand the proportion method, recall that 70% means the ratio of 70 to 100, or $\frac{70}{100}$.

$$\frac{7}{10} \text{ shaded}$$

$$70\% = \frac{70}{100} = \frac{7}{10}$$

$$70\% \text{ or } \tfrac{70}{100} \text{ shaded}$$

Since the ratio $\frac{70}{100}$ is equal to the ratio $\frac{7}{10}$, we have the proportion

$$\frac{7}{10} = \frac{70}{100}$$

We call this proportion the **percent proportion**. In general, we can name the parts of this proportion as follows.

Percent Proportion

$$\frac{\text{amount}}{\text{base}} = \frac{\text{percent}}{100} \quad \leftarrow \text{always 100}$$

or

$$\text{amount} \rightarrow \frac{a}{b} = \frac{p}{100} \quad \leftarrow \text{percent}$$
$$\text{base} \rightarrow$$

When we translate percent problems to proportions, the **percent** can be identified by looking for the symbol % or the word *percent*. The **base** usually follows the word *of*. The **amount** is the part compared to the whole.

Helpful Hint

This table may be useful when identifying the parts of a proportion.

Part of Proportion	How It's Identified
Percent	% or percent
Base	Appears after *of*
Amount	Part compared to whole

Practice Problem 1

Translate to a proportion: 15% of what number is 55?

Practice Problem 2

Translate to a proportion: 35 is what percent of 70?

Practice Problem 3

Translate to a proportion: What number is 25% of 68?

Practice Problem 4

Translate to a proportion: 520 is 65% of what number?

Answers

1. $\dfrac{55}{b} = \dfrac{15}{100}$ 2. $\dfrac{35}{70} = \dfrac{p}{100}$ 3. $\dfrac{a}{68} = \dfrac{25}{100}$

4. $\dfrac{520}{b} = \dfrac{65}{100}$

EXAMPLE 1 Translate to a proportion:

12% of what number is 47?

Solution: percent | base — It appears after the word *of*. | amount — It is the part compared to the whole.

$$\text{amount} \rightarrow \dfrac{47}{b} = \dfrac{12}{100} \leftarrow \text{percent}$$
$$\text{base} \rightarrow$$

EXAMPLE 2 Translate to a proportion:

101 is what percent of 200?

Solution: amount — It is the part compared to the whole. | percent | base — It appears after the word *of*.

$$\text{amount} \rightarrow \dfrac{101}{200} = \dfrac{p}{100} \leftarrow \text{percent}$$
$$\text{base} \rightarrow$$

EXAMPLE 3 Translate to a proportion:

What number is 90% of 45?

Solution: amount — It is the part compared to the whole. | percent | base — It appears after the word *of*.

$$\text{amount} \rightarrow \dfrac{a}{45} = \dfrac{90}{100} \leftarrow \text{percent}$$
$$\text{base} \rightarrow$$

EXAMPLE 4 Translate to a proportion:

238 is 40% of what number?

Solution: amount percent base

$$\dfrac{238}{b} = \dfrac{40}{100}$$

EXAMPLE 5   Translate to a proportion:

$$75 \quad \text{is} \quad \underline{\text{what percent}} \quad \text{of} \quad 30?$$

Solution: amount   percent   base

$$\frac{75}{30} = \frac{p}{100}$$

EXAMPLE 6   Translate to a proportion:

$$45\% \quad \text{of} \quad 105 \quad \text{is} \quad \underline{\text{what number?}}$$

Solution: percent   base   amount

$$\frac{a}{105} = \frac{45}{100}$$

Try the Concept Check in the margin.

B **Solving Percent Problems**

The proportions that we have written in this section contain three values that can change: the percent, the base, and the amount. If any two of these values are known, we can find the third (unknown value). To do this, we write a percent proportion and find the unknown value as we did in Section 6.3.

Try the Concept Check in the margin.

Solving Percent Proportions for the Amount

EXAMPLE 7 $\underline{\text{What number}} \quad \text{is} \quad 30\% \quad \text{of} \quad 9?$

Solution: amount   percent   base

$$\frac{a}{9} = \frac{30}{100}$$

To solve, we set cross products equal to each other.

$$\frac{a}{9} = \frac{30}{100}$$

$a \cdot 100 = 9 \cdot 30$ Set cross products equal.

$a \cdot 100 = 270$ Multiply.

$\dfrac{a \cdot 100}{100} = \dfrac{270}{100}$ Divide both sides by 100.

$a = 2.7$ Simplify.

Then 2.7 is 30% of 9.

Practice Problem 5

Translate to a proportion: 65 is what percent of 50?

Practice Problem 6

Translate to a proportion: 36% of 80 is what number?

Concept Check

When solving a percent problem with a proportion, describe how you can check the result.

Practice Problem 7

What number is 8% of 120?

Concept Check

Consider the problem

78 is what percent of 350?

Which part of the percent proportion is unknown?

a.   the amount
b.   the base
c.   the percent

Answers

5. $\dfrac{65}{50} = \dfrac{p}{100}$ **6.** $\dfrac{a}{80} = \dfrac{36}{100}$ **7.** 9.6

Concept Check: Put the result into the proportion and check that the proportion is true.

Concept Check: c

The proportion in Example 7 contained the ratio $\dfrac{30}{100}$. A ratio in a proportion may be simplified before solving the proportion. The unknown number in both

$$\frac{a}{9} = \frac{30}{100} \quad \text{and} \quad \frac{a}{9} = \frac{3}{10}$$

is 2.7

Practice Problem 8

75% of what number is 60?

Solving Percent Proportions for the Base

EXAMPLE 8 150% of what number is 30?

Solution: percent base amount

$$\frac{30}{b} = \frac{150}{100} \qquad \text{Write the proportion.}$$

$$\frac{30}{b} = \frac{3}{2} \qquad \text{Simplify } \tfrac{150}{100}.$$

$$30 \cdot 2 = b \cdot 3 \qquad \text{Set cross products equal.}$$

$$60 = b \cdot 3 \qquad \text{Multiply.}$$

$$\frac{60}{3} = \frac{b \cdot 3}{3} \qquad \text{Divide both sides by 3.}$$

$$20 = b \qquad \text{Simplify.}$$

Then 150% of 20 is 30.

Practice Problem 9

15 is 5% of what number?

EXAMPLE 9 20.8 is 40% of what number ?

Solution: amount percent base

$$\frac{20.8}{b} = \frac{40}{100} \quad \text{or} \quad \frac{20.8}{b} = \frac{2}{5} \qquad \begin{array}{l}\text{Write the proportion and} \\ \text{simplify } \tfrac{40}{100}.\end{array}$$

$$20.8 \cdot 5 = b \cdot 2 \qquad \text{Set cross products equal.}$$

$$104 = b \cdot 2 \qquad \text{Multiply.}$$

$$\frac{104}{2} = \frac{b \cdot 2}{2} \qquad \text{Divide both sides by 2.}$$

$$52 = b \qquad \text{Simplify.}$$

Then 20.8 is 40% of 52.

Answers

8. 80 **9.** 300

Solving Percent Proportions for the Percent

EXAMPLE 10   What percent   of   50   is   8?

Solution: percent   base   amount

$$\frac{8}{50} = \frac{p}{100} \quad \text{or} \quad \frac{4}{25} = \frac{p}{100} \quad \text{Write the proportion and simplify } \frac{8}{50}.$$

$$4 \cdot 100 = 25 \cdot p \quad \text{Set cross products equal.}$$

$$400 = 25 \cdot p \quad \text{Multiply.}$$

$$\frac{400}{25} = \frac{25 \cdot p}{25} \quad \text{Divide both sides by 25.}$$

$$16 = p \quad \text{Simplify.}$$

Then 16% of 50 is 8.

EXAMPLE 11   504   is   what percent   of   360?

Solution: amount   percent   base

$$\frac{504}{360} = \frac{p}{100}$$

Let's choose not to simplify the ratio $\frac{504}{360}$.

$$504 \cdot 100 = 360 \cdot p \quad \text{Set cross products equal.}$$

$$50,400 = 360 \cdot p \quad \text{Multiply.}$$

$$\frac{50,400}{360} = \frac{360 \cdot p}{360} \quad \text{Divide both sides by 360.}$$

$$140 = p \quad \text{Simplify.}$$

Notice by choosing not to simplify $\frac{504}{360}$, we had larger numbers in our equation. Either way, we find that 504 is 140% of 360.

You may have noticed the following while working examples.

Helpful Hint

Use the following to see whether your answers are reasonable.

$$100\% \text{ of a number} = \text{the number}$$

$$\left(\begin{array}{c} \text{a percent} \\ \text{greater than} \\ 100\% \end{array} \right) \text{of a number} = \begin{array}{c} \text{a number larger} \\ \text{than the original number} \end{array}$$

$$\left(\begin{array}{c} \text{a percent} \\ \text{less than } 100\% \end{array} \right) \text{of a number} = \begin{array}{c} \text{a number less} \\ \text{than the original number} \end{array}$$

Practice Problem 10

What percent of 40 is 5?

Helpful Hint

Recall from our percent proportion that this number already is a percent. Just keep the number the same and attach a % symbol.

Practice Problem 11

What percent of 160 is 336?

FOCUS ON The Real World

HOW MUCH CAN YOU AFFORD FOR A HOUSE?

When a home buyer takes out a mortgage to buy a house, the loan is generally repaid on a monthly basis with a monthly mortgage payment. (Some banks also offer biweekly payment programs.) An important consideration in choosing a house is the amount of the monthly payment. Usually, the amount that a home buyer can afford to make as a monthly payment will dictate the house purchase price that can be afforded.

The first step in deciding how much can be afforded for a house is finding out how much income the household has each month before taxes. The Mortgage Bankers Association of American (MBAA) suggests that the monthly mortgage payment be between 25% and 28% of the total monthly income. If other long-term debts exist (such as car or education loans and long-term credit card debt repayment), the MBAA further recommends that the total of housing costs and other monthly debt payments not exceed 36% of the total monthly income.

Once the size of the monthly payment that can be afforded has been found, a mortgage payment calculator can be used to work backward to estimate the mortgage amount that will give that desired monthly payment. For example, the Interest.com Web site includes a mortgage payment calculator at http://www.interest.com/calculators/monthly-payment.shtml. (Alternatively, visit www.interest.com and navigate to "Use our mortgage calculators." Look for the calculator to calculate the monthly payment for a particular mortgage loan.) With this mortgage payment calculator, the user can input the interest rate (as a percent), the term of the loan (in years), and total home loan amount (in dollars). This information is then used to calculate the associated monthly payment. To work backward with this mortgage payment calculator to find the total loan amount that can be afforded:

- Enter the interest rate that is likely for your loan and the term of the loan in which you are interested.
- Then make a guess (perhaps $100,000?) for the total home loan amount that can be afforded.
- Have the mortgage calculator calculate the monthly payment.
- If the monthly payment that is calculated is higher than the range that can be afforded, repeat the calculation using the same interest rate and loan term but a lower value for the total home loan amount.
- If the monthly payment that is calculated is lower than the range that can be afforded, repeat the calculation using the same interest rate and loan term but a higher value for the total home loan amount.
- Repeat these calculations methodically until a monthly payment is obtained that is in the range that can be afforded. The initial principal value that gave this monthly payment amount is an estimate of the mortgage amount that can be afforded to buy a home.

GROUP ACTIVITY

1. Research current interest rates on 30-year mortgages.

2. Use the method described above to find the size of mortgages that can be afforded by households with the following total monthly incomes before taxes. (Assume in each case that the household has no other debts.) Use a loan term of 30 years and a current interest rate on a 30-year mortgage.
   a. $1500   **b.** $2000   **c.** $2500
   d. $3000   **e.** $3500   **f.** $4000

3. Create a table of your results.

Mental Math

Identify the amount, the base, and the percent in each equation. Recall that $\dfrac{\text{amount}}{\text{base}} = \dfrac{\text{percent}}{100}$.

1. $\dfrac{12.6}{42} = \dfrac{30}{100}$

2. $\dfrac{201}{300} = \dfrac{67}{100}$

3. $\dfrac{20}{100} = \dfrac{102}{510}$

4. $\dfrac{40}{100} = \dfrac{248}{620}$

EXERCISE SET 7.3

A   *Translate each question to a proportion. Do not solve. See Examples 1 through 6.*

1. 32% of 65 is what number?

2. What number is 5% of 125?

3. 40% of what number is 75?

4. 1.2 is 47% of what number?

5. What percent of 200 is 70?

6. 520 is 85% of what number?

7. 2.3 is 58% of what number?

8. 92% of 30 is what number?

9. What number is 19% of 130?

10. 8.2 is what percent of 82?

B   *Solve. See Example 7.*

11. 10% of 55 is what number?

12. 25% of 84 is what number?

13. What number is 18% of 105?

14. What number is 40% of 29?

Solve. See Examples 8 and 9.

15. 60 is 15% of what number?

16. 75 is 75% of what number?

17. 7.8 is 78% of what number?

18. 1.1 is 44% of what number?

Solve. See Examples 10 and 11.

19. 105 is what percent of 84?

20. 77 is what percent of 44?

21. 14 is what percent of 50?

22. 37 is what percent of 50?

Solve. See Examples 7 through 11.

23. 2.9 is 10% of what number?

24. 6.2 is 5% of what number?

25. 2.4% of 80 is what number?

26. 6.5% of 120 is what number?

27. 160 is 16% of what number?

28. 30 is 6% of what number?

29. 348.6 is what percent of 166?

30. 262.4 is what percent of 82?

31. What number is 89% of 62?

32. What number is 53% of 130?

33. What percent of 8 is 3.6?

34. What percent of 5 is 1.6?

35. 140% of what number is 119?

36. 170% of what number is 221?

Review and Preview

Add or subtract as indicated. See Sections 4.4, 4.5, and 4.8.

37. $\dfrac{11}{16} + \dfrac{3}{16}$

38. $\dfrac{5}{8} - \dfrac{7}{12}$

39. $3\dfrac{1}{2} - \dfrac{11}{30}$

40. $2\dfrac{2}{3} + 4\dfrac{1}{2}$

Add or subtract as indicated. See Section 5.2.

41.
$$\begin{array}{r} 0.41 \\ +0.29 \\ \hline \end{array}$$

42.
$$\begin{array}{r} 10.78 \\ 4.3 \\ +0.21 \\ \hline \end{array}$$

43.
$$\begin{array}{r} 2.38 \\ -0.19 \\ \hline \end{array}$$

44.
$$\begin{array}{r} 16.37 \\ -\ 2.61 \\ \hline \end{array}$$

Combining Concepts

Solve. Round to the nearest tenth, if necessary.

45. What number is 22.3% of 53,862?

46. What percent of 110,736 is 88,542?

47. 8652 is 119% of what number?

48. In your own words, describe how to identify the percent, the base, and the amount in a percent problem.

Integrated Review—Percent and Percent Problems

Write each number as a percent.

1. 0.12

2. 0.68

3. $\frac{1}{4}$

4. $\frac{1}{2}$

5. 5.2

6. 7.8

7. $\frac{3}{50}$

8. $\frac{11}{25}$

9. $2\frac{1}{2}$

10. $3\frac{1}{4}$

11. 0.03

12. 0.05

Write each percent as a decimal.

13. 65%

14. 31%

15. 8%

16. 7%

17. 142%

18. 538%

19. 2.9%

20. 6.6%

Write each percent as a fraction or mixed number in simplest form.

21. 3%

22. 8%

23. 5.25%

24. 12.75%

25. 38%

26. 45%

27. $12\frac{1}{3}$%

28. $16\frac{2}{3}$%

Solve each percent problem.

29. 12% of 70 is what number?

30. 36 is 36% of what number?

31. 212.5 is 85% of what number?

32. 66 is what percent of 55?

1. _____
2. _____
3. _____
4. _____
5. _____
6. _____
7. _____
8. _____
9. _____
10. _____
11. _____
12. _____
13. _____
14. _____
15. _____
16. _____
17. _____
18. _____
19. _____
20. _____
21. _____
22. _____
23. _____
24. _____
25. _____
26. _____
27. _____
28. _____
29. _____
30. _____
31. _____
32. _____

33. _____

34. _____

35. _____

36. _____

37. _____

38. _____

39. _____

40. _____

33. 23.8 is what percent of 85?

34. 38% of 200 is what number?

35. What number is 25% of 44?

36. What percent of 99 is 128.7?

37. What percent of 250 is 215?

38. What number is 45% of 84?

39. 63 is 42% of what number?

40. 58.9 is 95% of what number?

7.4 Applications of Percent

(A) Solving Applications Involving Percent

The next few examples show just a few ways that percent occurs in real-life settings. (Each of these examples shows two ways of solving these problems. If you studied Section 7.2 only, see *Method 1*. If you studied Section 7.3 only, see *Method 2*.)

EXAMPLE 1.   Finding Increases in Nursing Schools

There is a world wide shortage of nurses that is projected to be 20% below requirements by 2020. Until 2003, there has also been a continual decline in enrollment in nursing schools.

In 2003, 2178 of the total 2593 nursing schools in the U.S. had an increase in applications or enrollment. What percent of nursing schools had an increase? Round to the nearest whole percent. (*Source*: CNN and *Nurse Week*)

Solution: *Method 1.* First, we state the problem in words.

In words: 2178   is   what percent   of   2593?

Translate: $2178 = x \cdot 2593$

Next, solve for x.

$$\frac{2178}{2593} = \frac{x \cdot 2593}{2593} \quad \text{Divide both sides by 2593.}$$

$$0.84 \approx x \qquad \text{Round to the nearest hundredth.}$$

$$84\% \approx x \qquad \text{Write as a percent.}$$

In 2003, about 84% of nursing schools had an increase in applications or enrollment.

Method 2.

In words: 2178   is   what percent of   2593?

    amount   percent   base

Translate: amount→ $\dfrac{2178}{2593} = \dfrac{p}{100}$ ←percent
            base→

Next, solve for p.

$$2178 \cdot 100 = 2593 \cdot p \quad \text{Set cross products equal.}$$

$$217{,}800 = 2593 \cdot p \quad \text{Multiply.}$$

$$\frac{217{,}800}{2593} = \frac{2593 \cdot p}{2593} \quad \text{Divide both sides by 2593.}$$

$$84 \approx p$$

In 2003, about 84% of nursing schools had an increase in applications or enrollment.

OBJECTIVES

(A) Solve applications involving percent.

(B) Find percent increase and percent decrease.

SSM
TUTOR CENTER   SG   CD & VIDEO   MATH PRO   WEB

Practice Problem 1

There are 106 nursing schools in Ohio. Of these schools, 61 offer RN (registered nurse) degrees. What percent of nursing schools in Ohio offer RN degrees? Round to the nearest whole percent.

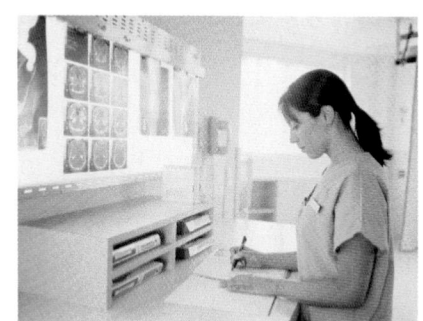

Answer

1. 58%

Practice Problem 2

The freshmen class of 775 students is 31% of all students at Euclid University. How many students go to Euclid University?

EXAMPLE 2 Finding Totals Using Percents

Mr. Buccaran, the principal at Slidell High School, counted 31 freshmen absent during a particular day. If this is 4% of the total number of freshmen, how many freshmen are there at Slidell High School?

Solution: *Method 1.* First we state the problem in words, then we translate.

In words: 31   is   4%   of   what number ?

Translate: 31   =   4%   · x

Next, we solve for x.

$31 = 0.04 \cdot x$ Write 4% as a decimal.

$\dfrac{31}{0.04} = \dfrac{0.04 \cdot x}{0.04}$ Divide both sides by 0.04.

$775 = x$ Simplify.

There are 775 freshmen at Slidell High School.

Method 2. First we state the problem in words, then we translate.

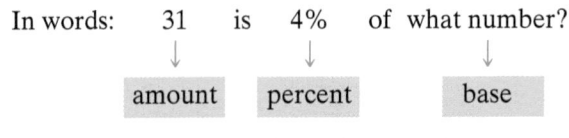

In words: 31   is   4%   of what number?
            amount percent   base

Translate: amount→$\dfrac{31}{b} = \dfrac{4}{100}$ ←percent
             base→

Next we solve for b.

$31 \cdot 100 = b \cdot 4$ Set cross products equal.

$3100 = b \cdot 4$ Multiply.

$\dfrac{3100}{4} = \dfrac{b \cdot 4}{4}$ Divide both sides by 4.

$775 = b$ Simplify.

There are 775 freshmen at Slidell High School.

Answer

2. 2500 students

EXAMPLE 3   **Finding Percents**

Standardized nutrition labeling like the one shown has been on foods since 1994. It is recommended that no more than 30% of your calorie intake be from fat. Find what percent of the total calories shown are fat.

Fruit snacks nutrition label

Solution: *Method 1.*

In words: 10   is   what percent   of   80?

Translate: 10   = x ·   80

Next we solve for x.

$$\frac{10}{80} = \frac{x \cdot 80}{80} \quad \text{Divide both sides by 80.}$$

$$0.125 = x \quad \text{Simplify.}$$

$$12.5\% = x \quad \text{Write 0.125 as a percent.}$$

This food contains 12.5% of its total calories as fat.

Method 2.

In words: 10 is what percent of 80?

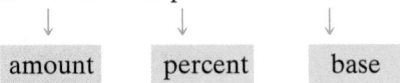

amount   percent   base

Translate: amount→ $\dfrac{10}{80} = \dfrac{p}{100}$ ←percent
base→

Next we solve for p.

$$10 \cdot 100 = 80 \cdot p \quad \text{Set cross products equal.}$$

$$1000 = 80 \cdot p \quad \text{Multiply.}$$

$$\frac{1000}{80} = \frac{80 \cdot p}{80} \quad \text{Divide both sides by 80.}$$

$$12.5 = p \quad \text{Simplify.}$$

This food contains 12.5% of its total calories as fat.

Practice Problem 3

The nutrition label below is from a can of cashews. Find what percent of total calories are from fat. Round to the nearest tenth of a percent.

Nutrition Facts

Serving Size $\frac{1}{4}$ cup (33g)
Servings Per Container About 9

Amount Per Serving

Calories 190 Calories from Fat 130

	% Daily Value
Total Fat 16g	**24%**
Saturated Fat 3g	**16%**
Cholesterol 0mg	**0%**
Sodium 135mg	**6%**
Total Carbohydrate 9g	**3%**
Dietary Fiber 1g	**5%**
Sugars 2g	
Protein 5g	

Vitamin A 0% • Vitamin C 0%
Calcium 0%   •   Iron 8%

Answer

3. 68.4%

Practice Problem 4

From 1980 to 2000, the number of U.S. registered vehicles on the road has increased by 10%. If the number of vehicles on the road in 1980 was 122 million, find the number of vehicles on the road in 2000. (*Source:* Federal Highway Administration)

B Finding Percent Increase and Percent Decrease

We often use percents to show how much an amount has increased or decreased.

EXAMPLE 4 Finding an Increase

From 1980 to 2000, the number of U.S. licensed drivers on the road increased by 35%. If the number of drivers on the road in 1980 was 145 million, find the number of drivers on the road in 2000. (*Source:* Federal Highway Administration)

Solution: *Method 1.* First we find the increase in drivers.

In words:	What number	is	35%	of	145?
Translate:	x	$=$	35%	$\cdot$	145

$x = 0.35 \cdot 145$ Write 35% as a decimal.

$x = 50.75$ Multiply.

The increase in drivers is 50.75 million. This means that the number of drivers in 2000 was

145 million + 50.75 million = 195.75 million drivers

Method 2. First we find the increase in drivers.

In words:	What number	is	35%	of	145?
	amount		percent		base

Translate: $\begin{array}{l} \text{amount} \rightarrow \\ \text{base} \quad\rightarrow \end{array} \dfrac{a}{145} = \dfrac{35}{100} \leftarrow\text{percent}$

$a \cdot 100 = 145 \cdot 35$ Set cross products equal.

$a \cdot 100 = 5075$ Multiply.

$\dfrac{a \cdot 100}{100} = \dfrac{5075}{100}$ Divide both sides by 100.

$a = 50.75$ Simplify.

The increase in drivers is 50.75 million. This means that the number of drivers in 2000 was

145 million + 50.75 million = 195.75 million drivers   ●

Suppose that the population of a town is 10,000 people and then it increases by 2000 people. The **percent increase** is

$\begin{array}{l} \text{amount of increase} \rightarrow \\ \text{original amount} \rightarrow \end{array} \dfrac{2000}{10{,}000} = 0.2 = 20\%$

Percent Increase

$$\text{percent increase} = \frac{\text{amount of increase}}{\text{original amount}}$$

Then write the quotient as a percent.

Answer

4. 134.2 million

EXAMPLE 5   Finding Percent Increase

The number of applications for a mathematics scholarship at Yale increased from 34 to 45 in one year. What is the percent increase? Round to the nearest whole percent.

Solution: First we find the amount of increase by subtracting the original number of applicants from the new number of applicants.

$$\text{amount of increase} = 45 - 34 = 11$$

The amount of increase is 11 applicants. To find the percent increase,

$$\text{percent increase} = \frac{\text{amount of increase}}{\text{original amount}} = \frac{11}{34} \approx 0.32 = 32\%$$

> **Helpful Hint**
>
> Make sure that this number in the denominator is the original number and not the new number.

The number of applications increased by about 32%.   ●

Try the Concept Check in the margin.

Suppose that your income was $300 a week and then it decreased by $30. The percent decrease is

$$\begin{array}{c}\text{amount of decrease} \rightarrow \\ \text{original amount} \rightarrow\end{array} \frac{\$30}{\$300} = 0.1 = 10\%$$

> **Percent Decrease**
>
> $$\text{percent decrease} = \frac{\text{amount of decrease}}{\text{original amount}}$$
>
> Then write the quotient as a percent.

Try the Concept Check in the margin.

EXAMPLE 6   Finding Percent Decrease

In response to a decrease in sales, a company with 1500 employees reduces the number of employees to 1230. What is the percent decrease?

Solution: First we find the amount of decrease by subtracting 1230 from 1500.

$$\text{amount of decrease} = 1500 - 1230 = 270$$

The amount of decrease is 270. To find the percent decrease,

$$\frac{\text{percent}}{\text{decrease}} = \frac{\text{amount of decrease}}{\text{original amount}} = \frac{270}{1500} = 0.18 = 18\%$$

The number of employees decreased by 18%.   ●

FOCUS ON The Real World

MORTGAGES

Buying a house may be one of the most expensive purchases we make. The amount borrowed from a lending institution for real estate is called a **mortgage**. The lending institutions that normally make mortgage loans include banks, savings and loan associations, credit unions, and mortgage companies.

There are basically three items that define a mortgage loan: the principal, the loan term, and the interest rate. The principal is the dollar amount being borrowed, or financed, by the home buyers. The loan term is the length of the loan, or how long it will take to pay off the loan. The interest rate, normally expressed as a percent, governs how much must be paid for the privilege of borrowing the money.

Mortgages come in all shapes and sizes. Loan terms can range anywhere from 10 years to 15, 20, 25, 30, or even 40 years. The interest rates on shorter loans are generally lower than the interest rates on longer loans. For instance, the interest rate on a 15-year loan might be 7.25% while the interest rate on a 30-year loan is 7.5%. Interest rates also tend to be lower on loans with a smaller principal as compared with a larger principal. For example, many banks offer a jumbo mortgage loan that applies only to principals over a certain limit, generally over $227,150. The interest rates on jumbo mortgages are higher (often by about 0.25%) than on other mortgage programs.

Most lending institutions require a home buyer to make a **down payment** in cash on a home. The size of the down payment usually depends on the buyer's circumstances and the specific mortgage program chosen, but down payments generally range from 3% to 20% of the home's value. A typical down payment on a house is 10% of the purchase price. After a down payment has been chosen, the mortgage amount can be calculated by subtracting the amount of the down payment from the purchase price:

$$\text{mortgage} = \text{purchase price} - \text{down payment}$$

Besides the down payment, there are a number of initial costs related to the mortgage that must be paid. These costs are called **closing costs**. Two expensive items on the list of closing costs are the **loan origination fee** and **loan discount points**. Both of these items are generally given in "points," where each point is equal to 1% of the mortgage amount. For example, "3 points" means 3% of the mortgage amount. The loan origination fee is the fee charged by the lender to cover the costs of preparing all the loan documents. Loan discount points are prepaid interest paid at closing. Home buyers generally can choose whether or not they will pay loan discount points. Doing so lowers the interest rate on the mortgage.

$$\text{loan origination fee} = \text{mortgage} \cdot \text{points}$$
$$\text{loan discount points} = \text{mortgage} \cdot \text{points}$$

CRITICAL THINKING

Suppose you are considering buying a house with a purchase price of $140,000.

1. Find the amount of the down payment if you plan to make a 10% down payment.

2. Find the mortgage amount.

3. Calculate the loan origination fee of 1 point.

4. To get a lower interest rate on your loan, suppose you choose to pay 2.5 loan discount points at closing. How much will you be paying in loan discount points?

Name _____ Section _____ Date _____

EXERCISE SET 7.4

Ⓐ *Solve. See Examples 1 through 3. If necessary, round percents to the nearest tenth and all other answers to the nearest whole.*

1. An inspector found 24 defective bolts during an inspection. If this is 1.5% of the total number of bolts inspected, how many bolts were inspected?

2. A daycare worker found 28 children absent one day during an epidemic of chicken pox. If this was 35% of the total number of children attending the daycare, how many children attend this daycare?

3. An owner of a repair service company estimates that for every 40 hours a repairperson is on the job, he can only bill for 75% of the hours. The remaining hours, the repairperson is idle or driving to or from a job. Determine the number of hours per 40-hour week the owner can bill for a repairperson.

4. The Hodder family paid 20% of the purchase price of a $75,000 home as a down payment. Determine the amount of the down payment.

5. Vera Faciane earns $2000 per month and budgets $300 per month for food. What percent of her monthly income is spent on food?

6. Last year, Mai Toberlan bought a share of stock for $83. She was paid a dividend of $4.15. Determine what percent of the stock price is the dividend.

7. A manufacturer of electronic components expects 1.04% of its product to be defective. Determine the number of defective components expected in a batch of 28,350 components. Round to the nearest whole component.

8. 18% of Marvin Frank's wages are withheld for income tax. Find the amount withheld from his wages of $3680 per month. Round to the nearest cent.

9. Of the 535 members of the 108th U.S. Congress, 73 have attended, a community college. What percent of the members of the 108th Congress is this? (*Source:* American Association of Community Colleges)

10. 31.6% of all households in the United States own at least one pet dog. There are 11,250 households in Anytown. How many of these households would you expect own a dog? (*Source:* American Veterinary Medical Association)

11. There are about 98,400 female dental hygienists registered in the United States. If this represents about 98.3% of the nation's dental hygienists, find the approximate number of dental hygienists in the United States. (*Source:* The American Dental Hygienists' Association)

12. The Los Angeles County courts excused 775,130 prospective jurors from jury duty in a recent year. This represented 28% of all juror qualification affidavits sent out that year. How many juror qualification affidavits were sent out that year? (*Source:* Los Angeles Superior Court)

For each food described, find what percent of total calories is from fat. If necessary, round to the nearest tenth of a percent. See Example 3.

13.

Nutrition Facts		
Serving Size 18 crackers (29g)		
Servings Per Container About 9		
Amount Per Serving		
Calories 120 Calories from Fat 35		
		% Daily Value*
Total Fat 4g		**6%**
Saturated Fat 0.5g		**3%**
Polyunsaturated Fat 0g		
Monounsaturated Fat 1.5g		
Cholesterol 0mg		**0%**
Sodium 220mg		**9%**
Total Carbohydrate 21g		**7%**
Dietary Fiber 2g		**7%**
Sugars 3g		
Protein 2g		
Vitamin A 0% • Vitamin C 0%		
Calcium 2% • Iron 4%		
Phosphorus 10%		

14.

Nutrition Facts		
Serving Size 28 crackers (31g)		
Servings Per Container About 6		
Amount Per Serving		
Calories 130 Calories from Fat 35		
		% Daily Value*
Total Fat 4g		**6%**
Saturated Fat 2g		**10%**
Polyunsaturated Fat 1g		
Monounsaturated Fat 1g		
Cholesterol 0mg		**0%**
Sodium 470mg		**20%**
Total Carbohydrate 23g		**8%**
Dietary Fiber 1g		**4%**
Sugars 4g		
Protein 2g		
Vitamin A 0% • Vitamin C 0%		
Calcium 0% • Iron 2%		

B *Solve. Round dollar amounts to the nearest cent and all other amounts to the nearest tenth. See Examples 4 through 6.*

15. Ace Furniture Company currently produces 6200 chairs per month. If production increases 8%, find the increase and the new number of chairs produced each month.

16. The enrollment at a local college increased 5% over last year's enrollment of 7640. Find the increase in enrollment and the current enrollment.

17. By carefully planning their meals, a family was able to decrease their weekly grocery bill by 20%. Their weekly grocery bill used to be $170. What is their new weekly grocery bill?

18. The profit of Ramone Company last year was $175,000. This year's profit decreased by 11%. Find this year's profit.

19. A car manufacturer announced that next year the price of a certain model car would increase 4.5%. This year the price is $19,286. Find the increase and the new price.

20. A union contract calls for a 6.5% salary increase for all employees. Determine the increase and the new salary that a worker currently making $28,500 under this contract can expect.

21. The population of Americans aged 65 and older was 35 million in 2000. That population is projected to increase by 80% by 2025. Find the increase and the projected 2025 population. (*Source*: Bureau of the Census)

22. The from 2000 to 2010, the number of masters degrees awarded to women is projected to increase by 8.3%. The number of women who received masters degrees in 2000 was 265,000. Find the predicted number of women to be awarded masters degrees in 2010. (*Source*: U.S. National Center for Education Statistics)

Find the amount of increase and the percent increase. See Example 4.

	Original Amount	New Amount	Amount of Increase	Percent Increase
23.	40	50	____	____
24.	10	15	____	____
25.	85	187	____	____
26.	78	351	____	____

Find the amount of decrease and the percent decrease. See Example 5.

	Original Amount	New Amount	Amount of Decrease	Percent Decrease
27.	8	6	____	____
28.	25	20	____	____
29.	160	40	____	____
30.	200	162	____	____

Solve. Round percents to the nearest tenth, if necessary. See Examples 4 through 6.

31. There are 150 calories in a cup of whole milk and only 84 in a cup of skimmed milk. In switching to skimmed milk, find the percent decrease in number of calories.

32. In reaction to a slow economy, the number of employees at a soup company decreased from 530 to 477. What was the percent decrease in employees?

33. By changing his driving routines, Alan Miller increased his car's rate of miles per gallon from 19.5 to 23.7. Find the percent increase.

34. John Smith decided to decrease the number of calories in his diet from 3250 to 2100. Find the percent decrease.

35. The number of cable TV systems recently decreased from 10,845 to 10,700. Find the percent decrease.

36. Before taking a typing course, Geoffry Landers could type 32 words per minute. By the end of the course, he was able to type 76 words per minute. Find the percent increase.

37. In 1940, there were approximately 52.1 million sheep on farms in the United States. In 2002, this number had decreased to 6.7 million sheep. What was the percent decrease? (*Source:* National Agricultural Statistics Service)

38. In 1970, there were approximately 12.1 million milk cows on farms in the United States. In 2002, this number had decreased to 9.1 million milk cows. What was the percent decrease? (*Source:* National Agricultural Statistics Service)

39. In 1999, discarded electronics, including obsolete computer equipment, accounted for 75,000 tons of solid waste per year in Massachusetts. By 2006, discarded electronic waste is expected to increase to 300,000 tons of waste per year in the state. Find the percent increase. (*Source:* Massachusetts Department of Environmental Protection)

40. The average soft-drink size has increased from 13.1 oz to 19.9 oz over the past two decades. Find the percent increase. (*Source*: University of North Carolina at Chapel Hill, *Journal for American Medicine*)

13.1 oz   19.9 oz

41. In 1994, approximately 16,000 occupational therapy assistants and aides were employed in the United States. According to one survey, by 2005, this number is expected to increase to 29,000 assistants and aides. What is the percent increase? (*Source:* Bureau of Labor Statistics)

42. In 1994, approximately 206,000 medical assistants were employed in the United States. By 2005, this number is expected to increase to 327,000 medical assistants. What is the percent increase? (*Source:* Bureau of Labor Statistics)

43. In 1995, 272.6 million recorded music cassettes were shipped to retailers in the United States. By 2000, this number had decreased to 76.0 million cassettes. What was the percent decrease? (*Source:* Recording Industry Association of America)

44. In 1940, the average size of a U.S. private owned farm was 174 acres. By 2000, the average size of a U.S. private owned farm had increased to 434 acres. What was the percent increase? (*Source:* National Agricultural Statistics Service)

45. In 1994, approximately 16,000,000 Americans subscribed to cellular phone service. By 2000, this number had increased to about 110,000,000 American subscribers. What was the percent increase? (*Source:* Network World, Inc.)

46. In 1998, approximately 299,000 computer engineers were employed in the United States. By 2008, this number is expected to increase to 622,000 computer engineers. What is the percent increase? (*Source:* Bureau of Labor Statistics)

47. The population of Tokyo is expected to increase from 26,518 thousand in 1994 to 28,700 thousand in 2015. Find the percent increase. (*Source:* United Nations, Department for Economic and Social Information and Policy Analysis)

48. In 1970, there were 1754 deaths from boating accidents in the United States. By 2000, the number of deaths from boating accidents had decreased to 698. What was the percent decrease? (*Source:* U.S. Coast Guard)

Japan

Tokyo

Perform each indicated operation. See Sections 5.2 and 5.3.

49.
$$\begin{array}{r} 0.12 \\ \times\ 38 \\ \hline \end{array}$$

50.
$$\begin{array}{r} 42 \\ \times\ 0.7 \\ \hline \end{array}$$

51. $9.20 + 1.98$

52. $46 + 7.89$

53. $78 - 19.46$

54. $64.80 - 10.72$

 Combining Concepts

55. If a number is increased by 100%, how does the increased number compare with the original number? Explain your answer.

56. In your own words, explain what is wrong with the following statement. "Last year we had 80 students attend. This year we have a 50% increase or a total of 160 students attend."

Internet Excursions

 Go To: http://www.prenhall.com/martin-gay_prealgebra What's Related

This World Wide Web address will provide you with access to an American Savings Education Council worksheet (or a related site) for calculating a ballpark estimate of the savings needed for retirement. Print out two copies of the worksheet to use with Exercises 57–58.

57. Fill out a copy of the ballpark estimate worksheet for a person who currently is 25 years old, earns $32,000 a year, plans to retire at age 60, expects no traditional employer pension, expects $5000 per year in part-time income in retirement, and has $2000 in retirement savings. How much does this person need to save each year toward retirement?

58. Fill out a copy of the ballpark estimate worksheet for your own situation. How much do you need to save each year toward retirement?

7.5   Percent and Problem Solving: Sales Tax, Commission, and Discount

OBJECTIVES

Ⓐ   Calculate sales tax and total price.

Ⓑ   Calculate commissions.

Ⓒ   Calculate discount and sale price.

SSM   TUTOR CENTER   SG   CD & VIDEO   MATH PRO   WEB

Ⓐ   Calculating Sales Tax and Total Price

Percents are frequently used in the retail trade. For example, most states charge a tax on certain items when purchased. This tax is called a **sales tax**, and retail stores collect it for the state. Sales tax is almost always stated as a percent of the purchase price.

A 6% sales tax rate on a purchase of a $10.00 item gives a sales tax of

sales tax = 6% of $10 = 0.06 · $10.00 = $0.60

The total price to the customer would be

purchase price	plus	sales tax	
↓	↓	↓	
$10.00	+	$0.60 =	$10.60

This example suggests the following equations.

> **Sales Tax and Total Price**
>
> sales tax = tax rate · purchase price
>
> total price = purchase price + sales tax

In this section, we will round dollar amounts to the nearest cent.

EXAMPLE 1   Finding Sales Tax and Purchase Price

Find the sales tax and the total price on a purchase of an $85.50 trench coat in a city where the sales tax rate is 7.5%.

Practice Problem 1

If the sales tax rate is 6%, what is the sales tax and the total amount due on a $29.90 Goodgrip tire?

Answers

1. tax: $1.79; total: $31.69

Solution: The purchase price is $85.50 and the tax rate is 7.5%.

sales tax	=	tax rate	·	purchase price
↓		↓		↓

sales tax = 7.5% · $85.50
= 0.075 · $85.5 Write 7.5% as a decimal,
≈ $6.41 Round to the nearest cent.

Thus

total price	=	purchase price	+	sales tax
↓		↓		↓

total price = $85.50 + $6.41
= $91.91

The sales tax on $85.50 is $6.41 and the total price is $91.91.

Try the Concept Check in the margin.

Concept Check

The purchase price of a textbook is $50 and sales tax is 10%. If you are told by the cashier that the total price is $75, how can you tell that a mistake has been made?

Practice Problem 2

The sales tax on a $13,500 automobile is $1080.00. Find the sales tax rate.

EXAMPLE 2 Finding a Sales Tax Rate

The sales tax on a $300 printer is $22.50. Find the sales tax rate.

SALE
$300
+$22.50 sales tax

Solution: Let r be the unknown sales tax rate. Then

sales tax	=	tax rate	·	purchase price
↓		↓		↓

$22.50 = r \cdot 300$

$$\frac{22.50}{300} = \frac{r \cdot 300}{300} \quad \text{Divide both sides by 300.}$$

$0.075 = r$ Simplify.

$7.5\% = r$ Write 0.075 as a percent.

The sales tax rate is 7.5%.

B Calculating Commissions

A **wage** is payment for performing work. Hourly wage, commissions, and salary are some of the ways wages can be paid. Many people who work in sales are paid a commission. An employee who is paid a **commission** is paid a percent of his or her total sales.

Answer

2. 8%

Concept Check: Since 10% = 0.10, the sales tax is $50 · 0.10 = $5. The total price should have been $55.

Commission

commission = commission rate · sales

EXAMPLE 3   **Finding a Commission**

Sherry Souter, a real estate broker for Wealth Investments, sold a house for $114,000 last week. If her commission is 1.5% of the selling price of the home, find the amount of her commission.

Solution:

commission	=	commission rate	·	sales
↓		↓		↓
commission	=	1.5%	·	114,000
	=	0.015	·	114,000	Write 1.5% as 0.015.
	=	1710			Multiply.

Her commission on the house is $1710.

EXAMPLE 4   **Finding a Commission Rate**

A salesperson earned $1560 for selling $13,000 worth of television and stereo systems. Find the commission rate.

Solution: Let r stand for the unknown commission rate. Then

commission	=	commission rate	·	sales
↓		↓		↓
1560	=	r	·	13,000
$\dfrac{1560}{13,000}$	=	$\dfrac{r \cdot 13,000}{13,000}$			Divide both sides by 13,000.
0.12	=	r			Simplify.
12%	=	r			Write 0.12 as a percent.

The commission rate is 12%.

C   Calculating Discount and Sale Price

Suppose that an item that normally sells for $40 is on sale for 25% off. This means that the **original price** of $40 is reduced, or **discounted** by 25% of $40, or $10. The **discount rate** is 25%, the **amount of discount** is $10, and the **sale price** is $40 − $10 or $30.

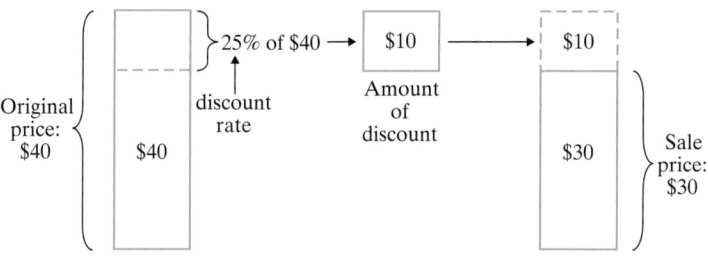

To calculate discounts and sale prices, we can use the following equations.

Discount and Sale Price

amount of discount = discount rate · original price

sale price = original price − amount of discount

Practice Problem 5

A Panasonic TV is advertised on sale for 15% off the regular price of $700. Find the discount and the sale price.

EXAMPLE 5   Finding a Discount and a Sale Price

A speaker that normally sells for $65.00 is on sale at 25% off. What is the discount and what is the sale price?

Solution: First we find the discount.

amount of discount	=	discount rate	·	original price
↓		↓		↓
amount of discount	=	25%	·	65
	=	0.25	·	65 Write 25% as 0.25.
	=	16.25 Multiply.		

The discount is $16.25. Next we find the sale price.

sale price	=	original price	−	discount
↓		↓		↓
sale price	=	$65	−	$16.25
	=	$48.75 Subtract.		

The sale price is $48.75.

Name _____ Section _____ Date _____

 A *Solve. See Examples 1 and 2.*

1. What is the sales tax on a suit priced at $150.00 if the sales tax rate is 5%?

2. If the sales tax rate is 6%, find the sales tax on a microwave oven priced at $188.

 3. The purchase price of a camcorder is $799. What is the total price if the sales tax rate is 7.5%?

4. A stereo system has a purchase price of $426. What is the total price if the sales tax rate is 8%?

5. A chair and ottoman have a purchase price of $600. If the sales tax on this purchase is $54, find the sales tax rate.

6. The sales tax on the purchase of a $2500 computer is $162.50. Find the sales tax rate.

7. A computer desk sells for $220. With a sales tax rate of 8.5%, find the total price.

8. A one-half carat diamond ring is priced at $800. The sales tax rate is 6.5%. Find the total price.

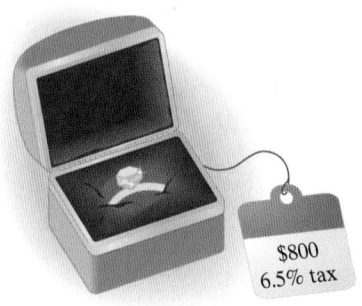

9. A gold and diamond bracelet sells for $1800. Find the total price if the sales tax rate is 6.5%.

10. The purchase price of a personal computer is $1890. If the sales tax rate is 8%, what is the total price?

11. The sales tax on the purchase of a used truck is $920. If the tax rate is 8%, find the purchase price of the truck.

12. The sales tax on the purchase of a desk is $27.50. If the tax rate is 5%, find the purchase price of the desk.

13. A cordless phone costs $90 and a battery recharger costs $15. What is the total price for purchasing these items if the sales tax rate is 7%?

$15

$90

14. Ms. Warner bought a blouse for $35, a skirt for $55, and a blazer for $95. Find the total price she paid, given a sales tax rate of 6.5%.

$55

$35

$95

15. The sales tax is $98.70 on a stereo sound system purchase of $1645. Find the sales tax rate.

16. The sales tax is $103.50 on a necklace purchase of $1150. Find the sales tax rate.

B *Solve. See Examples 3 and 4.*

17. Jane Moreschi, a sales representative for a large furniture warehouse, is paid a commission rate of 4%. Find her commission if she sold $1,236,856 worth of furniture last month.

18. Rosie Davis-Smith is a beauty consultant for a home cosmetic business. She is paid a commission rate of 4.8%. Find her commission if she sold $1638 in cosmetics last month.

 19. A salesperson earned a commission of $1380.40 for selling $9860.00 worth of paper products. Find the commission rate.

20. A salesperson earned a commission of $3575 for selling $32,500 worth of books to various book stores. Find the commission rate.

21. How much commission will Jack Pruet make on the sale of a $125,900 house if he receives 1.5% of the selling price?

22. Frankie Lopez sold $9638.00 of jewelry this week. Find her commission for the week if she is paid a commission rate of 5.6%.

23. A real estate agent earned a commission of $2565 for selling a house. If his rate is 3%, find the selling price of the house.

24. A salesperson earned $1750 for selling fertilizer. If her commission rate is 7%, find the selling price of the fertilizer.

C *Find the amount of discount and the sale price. See Example 5.*

	Original Price	Discount Rate	Amount of Discount	Sale Price
25.	$68.00	10%	_____	_____
26.	$47.00	20%	_____	_____
27.	$96.50	50%	_____	_____
28.	$110.60	40%	_____	_____
29.	$215.00	35%	_____	_____
30.	$370.00	25%	_____	_____
31.	$21,700.00	15%	_____	_____
32.	$17,800.00	12%	_____	_____

 33. A $300 fax machine is on sale at 15% off. Find the discount and the sale price.

34. A $2000 designer dress is on sale at 30% off. Find the discount and the sale price.

Find the missing amounts. See Example 5.

	Original Price	Discount Rate	Amount of Discount	Sale Price
35.	$75	20%		
36.	$40	15%		
37.	$120		$39.60	
38.	$87		$8.70	
39.		40%	$370	
40.		65%	$299	

Review and Preview

Multiply. See Sections 5.3 and 5.6.

41. $2000 \cdot 0.3 \cdot 2$

42. $500 \cdot 0.08 \cdot 3$

43. $400 \cdot 0.03 \cdot 11$

44. $1000 \cdot 0.05 \cdot 5$

45. $600 \cdot 0.04 \cdot \dfrac{2}{3}$

46. $6000 \cdot 0.06 \cdot \dfrac{3}{4}$

 Combining Concepts

47. A diamond necklace normally sells for $24,966. If it is first discounted by 15% and then taxed at 7.5%, find the total price.

48. A house recently sold for $562,560. The commission rate on the sale is 5.5%. If a real estate agent is to receive 60% of the commission, find the amount received by the agent.

49. Suppose that the original price of a shirt is $50. Which is better, a 60% discount or a discount of 30% followed by a discount of 35% of the reduced price. Explain your answer.

One very useful application of percent is mentally calculating a tip. Recall that to find 10% of a number, simply move the decimal point one place to the left. To find 20% of a number, just double 10% of the number. To find 15% of a number, find 10% and the add to that number half of the 10% amount. Mentally fill in the chart below. To do so, start by rounding the bill amount to the nearest dollar.

Tipping Chart

	Bill Amount	10%	15%	20%
50.	$40.21			
51.	$15.89			
52.	$72.17			
53.	$9.33			

FOCUS ON **Business and Career**

FASTEST-GROWING OCCUPATIONS

According to U.S. Bureau of Labor Statistics projections, the careers listed below are the top ten fastest-growing jobs, ranked by expected percent increase through the year 2010.

Occupation	Employment in 2000	Percent Increase from 2000 to 2010
Computer software engineers, applications	380,000	100%
Computer support specialists	506,000	97%
Computer software engineers, systems software	317,000	90%
Network and computer systems administrators	229,000	82%
Network systems and data communications analysts	119,000	77%
Desktop publishers	38,000	67%
Database Administrators	106,000	66%
Personal and home care aides	414,000	62%
Computer systems analysts	431,000	60%
Medical assistants	329,000	57%

(*Source*: November 2001, *Monthly Labor Review*, Bureau of Labor Statistics)

What do all of these fast-growing occupations have in common? They all require a knowledge of math! For some careers, such as desktop publishing specialists, medical assistants, and computer engineers, the ways math is used on the job may be obvious. For other occupations, the use of math may not be quite as apparent. However, tasks common to many jobs—filling in a time sheet, writing up an expense or mileage report, planning a budget, figuring a bill, ordering supplies, and even making a work schedule—all require math.

GROUP ACTIVITY

1. List the top five occupations by order of employment figures for 2000.

2. Using the 2000 employment figures and the percent increase from 2000 to 2010, find the expected 2010 employment figures for each occupation listed in the table.

3. List the top five occupations by order of employment figures for 2010. Did the order change at all from 2000? Explain.

4. How many of the occupations on the list are expected to increase by at least 80% the number of positions in 2010 than in 2000? Explain how you identified these occupations.

7.6 Percent and Problem Solving: Interest

A Calculating Simple Interest

Interest is money charged for using other people's money. When you borrow money, you pay interest. When you loan or invest money, you earn interest. The money borrowed, loaned, or invested is called the **principal amount**, or simply **principal**. Interest is normally stated in terms of a percent of the principal for a given period of time. The **interest rate** is the percent used in computing the interest. Unless stated otherwise, *the rate is understood to be per year*. When the interest is computed on the original principal, it is called **simple interest**. Simple interest is calculated using the following equation.

> **Simple Interest**
>
> simple interest = principal · rate · time
>
> or
>
> $I = P \cdot R \cdot T$
>
> where the rate is understood to be the rate per year and time is in years.

EXAMPLE 1   Finding Simple Interest

Find the simple interest after 2 years on $500 at an interest rate of 12%.

Solution: In this example, $P = \$500$, $R = 12\%$, and $T = 2$ years. Replace the variables with values in the formula $I = PRT$.

$I = P \cdot R \cdot T$
$I = \$500 \cdot 12\% \cdot 2$ Let $P = \$500$, $R = 12\%$, and $T = 2$.
$\quad = \$500 \cdot (0.12) \cdot 2$ Write 12% as a decimal.
$\quad = \$120$ Multiply.

The simple interest is $120.

Try the Concept Check in the margin.

If time is not given in years, we need to convert the given time to years.

EXAMPLE 2   Finding Simple Interest

Ivan Borski borrowed $2400 at 10% simple interest for 8 months to buy a used Chevy S-10. Find the simple interest he paid.

Solution: Since there are 12 months in a year, we first find what part of a year 8 months is.

$$8 \text{ months} = \frac{8}{12}\text{year} = \frac{2}{3}\text{year}$$

Now we find the simple interest.

$I = P \cdot R \cdot T$
$\quad = \$2400 \cdot (0.10) \cdot \dfrac{2}{3}$ Let $P = \$2400$, $R = 10\%$ or 0.10, and $T = \frac{2}{3}$.
$\quad = \$160$

The interest on Ivan's loan is $160.

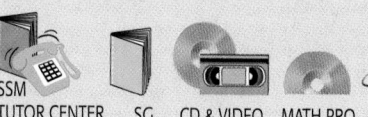
Practice Problem 1

Find the simple interest after 3 years on $750 at an interest rate of 8%.

Concept Check

If $2000 is borrowed at 12% simple interest for 14 months, is 2000 the variable T, P, or I in the simple interest formula

$I = P \cdot R \cdot T$

Practice Problem 2

Juanita Lopez borrowed $800 for 9 months at a simple interest rate of 20%. How much interest did she pay?

Answers
1. $180   **2.** $120

Concept Check: 2000 is P, principal.

Concept Check

Suppose in Example 2 you had obtained an answer of $16,000. How would you know that you had made a mistake in this problem?

Practice Problem 3

If $500 is borrowed at a simple interest rate of 12% for 6 months, find the total amount paid.

Concept Check

Which investment would earn more interest: an amount of money invested at 8% interest for 2 years or the same amount of money invested at 8% for 3 years? Explain.

Answer

3. $530

Concept Check: $16,000 is too much interest. Answers may vary.

Concept Check: 8% for 3 years. Since the interest rate is the same, the longer you keep the money invested, the more interest you earn.

Try the Concept Check in the margin.

When money is borrowed, the borrower pays the original amount borrowed, the principal, as well as the interest. When money is invested, the investor receives the original amount invested, or the principal, as well as the interest. In either case, the **total amount** is the sum of the principal and the interest.

> **Finding the Total Amount of a Loan or Investment**
>
> total amount (paid or received) = principal + interest

EXAMPLE 3   Finding the Total Amount of an Investment

An accountant invested $2000 at a simple interest rate of 10% for 2 years. What total amount of money will she have from her investment in 2 years?

Solution: First we find her interest.

$$I = P \cdot R \cdot T$$
$$= \$2000 \cdot (0.10) \cdot 2 \quad \text{Let } P = \$2000, R = 10\% \text{ or } 0.10, \text{ and } T = 2.$$
$$= \$400$$

The interest is $400.

Next we add the interest to the principal.

total amount	=	principal	+	interest
↓		↓		↓
total amount	=	$2000	+	$400
	=	$2400		

After 2 years, she will have a total amount of $2400.

Try the Concept Check in the margin.

B Calculating Compound Interest

Recall that simple interest depends on the original principal only. Another type of interest is compound interest. **Compound interest** is computed on not only the principal, but also on the interest already earned in previous compounding periods. Compound interest is used more often than simple interest in saving accounts.

Let's see how compound interest differs from simple interest. Suppose that $2000 is invested at 7% interest **compounded annually** for 3 years. This means that interest is added to the principal at the end of each year and next year's interest is computed on this new amount. In this section, we will round dollar amounts to the nearest cent.

Amount at Beginning of Year	Principal · Rate · Time = Interest	Amount at End of Year
1st year $2000	$2000 · 0.07 · 1 = $140	$2000 + $140 = $2140
2nd year $2140	$2140 · 0.07 · 1 = $149.80	$2140 + $149.80 = $2289.80
3rd year $2289.80	$2289.80 · 0.07 · 1 = $160.29	$2289.80 + $160.29 = $2450.09

The compound interest earned can be found by

total amount	−	original principal	=	compound interest
↓		↓		↓
$2450.09	−	$2000	=	$450.09

The simple interest earned would have been

principal	·	rate	·	time	=	interest
↓		↓		↓		↓
$2000	·	0.07	·	3	=	$420

Since compound interest earns "interest on interest," compound interest produces more interest than simple interest.

Computing compound interest using the method above can be tedious. We can use a **compound interest table** to compute interest more quickly. The compound interest table is found in Appendix F. This table gives the total compound interest and principal paid on $1 for given rates and number of years. Then we can use the following equation to find the total amount of interest and principal.

Finding Total Amounts with Compound Interest

total amount = original principal · compound interest factor
(from table)

EXAMPLE 4   Finding Total Amount Received on an Investment

$4000 is invested at 8% compounded semiannually for 10 years. Find the total amount at the end of 10 years.

Solution: Look in Appendix F. The compound interest factor for 10 years at 8% in the compounded semiannually section is 2.19112.

total amount	=	original principal	·	compound interest factor
↓		↓		↓
total amount	=	$4000	·	2.19112
	=	$8764.48		

Therefore, the total amount at the end of 10 years is $8764.48.   ●

EXAMPLE 5   Finding Compound Interest Earned

In Example 4, we found that the total amount for $4000 invested at 8% compounded semiannually for 10 years is $8764.48. Find the compound interest earned.

Solution:

interest earned	=	total amount	−	original principal
↓		↓		↓
interest earned	=	$8764.48	−	$4000
	=	$4764.48		

The compound interest earned is $4764.48.   ●

(c) Calculating a Monthly Payment

We conclude this section with a method to find the monthly payment on a loan.

Practice Problem 4

$5500 is invested at 7% compounded daily for 5 years. Find the total amount at the end of 5 years.

Practice Problem 5

If the total amount is $9933.14 when $5500 is invested, find the compound interest earned.

Answers
4. $7804.61   **5.** $4433.14

Finding the Monthly Payment of a Loan

$$\text{monthly payment} = \frac{\text{principal} + \text{interest}}{\text{total number of payments}}$$

Practice Problem 6

Find the monthly payment on a $3000 3-year loan if the interest on the loan is $1123.58.

EXAMPLE 6 Finding a Monthly Payment

Find the monthly payment on a $2000 loan for 2 years. The interest on the 2-year loan is $435.88.

Solution: First we determine the total number of monthly payments. The loan is for 2 years. Since there are 12 months per year, the number of payments is 2 · 12, or 24. Now we can calculate the monthly payment.

$$\text{monthly payment} = \frac{\text{principal} + \text{interest}}{\text{total number of payments}}$$

$$\text{monthly payment} = \frac{\$2000 + \$435.88}{24}$$

$$\approx \$101.50$$

The monthly payment is $101.50.

CALCULATOR EXPLORATIONS

Compound Interest Factor

A compound interest factor may be found by using your calculator and evaluating the formula

$$\textbf{compound interest factor} = \left(1 + \frac{r}{n}\right)^{nt}$$

where r is the interest rate, t is the time in years, and n is the number of times compounded per year. For example, we stated earlier that the compound interest factor for 10 years at 8% compounded semiannually is 2.19112. Let's find this factor by evaluating the compound interest factor formula when $r = 8\%$ or 0.08, $t = 10$, and $n = 2$ (compounded semiannually means 2 times per year at six month intervals). Thus,

$$\text{compound interest factor} = \left(1 + \frac{0.08}{2}\right)^{2 \cdot 10} \text{ or } \left(1 + \frac{0.08}{2}\right)^{20}$$

To evaluate, press the keys

$$(\quad 1 \quad + \quad 0.08 \quad \div \quad 2 \quad) \quad y^x \quad 20 \quad = \text{ or } \boxed{\text{ENTER}}$$

The display will read $\boxed{2.1911231}$. Rounded to five decimal places, this is 2.19112.

Find the compound interest factor. Use the table in Appendix F to check your answer.

1. 5 years, 9%, compounded quarterly

2. 15 years, 14%, compounded daily

3. 20 years, 11%, compounded annually

4. 1 year, 7%, compounded semiannually

5. Find the total amount after 4 years if $500 is invested at 6% compounded quarterly.

6. Find the total amount for 19 years if $2500 is invested at 5% compounded daily.

Answer

6. $114.54

EXERCISE SET 7.6

A *Find the simple interest. See Examples 1 and 2.*

	Principal	Rate	Time
1.	$200	8%	2 years
3.	$160	11.5%	4 years
5.	$5000	10%	$1\frac{1}{2}$ years
7.	$375	18%	6 months
9.	$2500	16%	21 months

	Principal	Rate	Time
2.	$800	9%	3 years
4.	$950	12.5%	5 years
6.	$1500	14%	$2\frac{1}{4}$ years
8.	$1000	10%	18 months
10.	$775	15%	8 months

Solve. See Examples 1 through 3.

11. A company borrows $62,500 for 2 years at a simple interest rate of 12.5% to buy an airplane. Find the total amount paid on the loan.

12. $65,000 is borrowed to buy a house. If the simple interest rate on the 30-year loan is 10.25%, find the total amount paid on the loan.

13. A money market fund advertises a simple interest rate of 9%. Find the total amount received on an investment of $5000 for 15 months.

14. The Real Service Company takes out a 270-day (9-month) short-term, simple interest loan of $4500 to finance the purchase of some new equipment. If the interest rate is 14%, find the total amount that the company pays back.

15. Marsha Waide borrows $8500 and agrees to pay it back in 4 years. If the simple interest rate is 12%, find the total amount she pays back.

16. Ms. Lapchinski gives her 18-year-old daughter a graduation gift of $2000. If this money is invested at 8% simple interest for 5 years, find the total amount after 5 years.

B *Find the total amount in each compound interest account. See Example 4.*

17. $6150 is compounded semiannually at a rate of 14% for 15 years.

18. $2060 is compounded annually at a rate of 15% for 10 years.

19. $1560 is compounded daily at a rate of 8% for 5 years.

20. $1450 is compounded quarterly at a rate of 10% for 15 years.

21. $10,000 is compounded semiannually at a rate of 9% for 20 years.

22. $3500 is compounded daily at a rate of 8% for 10 years.

Find the amount of compound interest earned. See Example 5.

23. $2675 is compounded annually at a rate of 9% for 1 year.

24. $6375 is compounded semiannually at a rate of 10% for 1 year.

25. $2000 is compounded annually at a rate of 8% for 5 years.

26. $2000 is compounded semiannually at a rate of 8% for 5 years.

27. $2000 is compounded quarterly at a rate of 8% for 5 years.

28. $2000 is compounded daily at a rate of 8% for 5 years.

Solve. See Example 6.

29. A college student borrows $1500 for 6 months to pay for a semester of school. If the interest is $61.88, find the monthly payment.

30. Jim Tillman borrows $1800 for 9 months. If the interest is $148.90, find his monthly payment.

 31. $20,000 is borrowed for 4 years. If the interest on the loan is $10,588.70, find the monthly payment.

32. $105,000 is borrowed for 15 years. If the interest on the loan is $181,125.00, find the monthly payment.

Review and Preview

Perform each indicated operation. See Sections 2.2 to 2.4.

33. $-5 + (-24)$

34. $7 - (-31)$

35. $(-5)(-30)$

36. $-30 \div (-5)$

37. $\dfrac{7 - 10}{3}$

38. $\dfrac{22 + (-4)}{-4 - 5}$

Combining Concepts

39. Explain how to look up the compound interest factor in the compound interest table.

40. Explain how to find the amount of interest on a compounded account.

41. Compare the following accounts: Account 1: $1000 is invested for 10 years at a simple interest rate of 6%. Account 2: $1000 is compounded semiannually at a rate of 6% for 10 years. Discuss how the interest is computed for each account. Determine which account earns more interest. Why?

This activity may be completed by working in groups or individually.

The measure of the chance of an event occurring is its probability. In this activity, you will investigate one way that probabilities can be estimated. You will need a cup and 30 thumbtacks.

1. Place the thumbtacks in the cup. Shake the cup and toss out the thumbtacks onto a flat surface. Count the number of tacks that land point up, and record this number in the table. Repeat for 60, 90, 120, and 150 tacks. Record your results in the second column of the table. (Hint: For 60 thumbtacks, count the number of tacks landing point up in two tosses of the 30 thumbtacks, etc.)

2. For each row of the table, find the fraction of tacks that landed point up. Express this fraction as a percent. Complete the third and fourth columns of the table.

3. Each of the percents you computed in Question 2 is an *estimate* of the probability that a single thumbtack will land point up when tossed. With an estimate like this, the larger the number of tack tosses used, the better the estimate of the actual probability. What do you suppose is the value of the actual probability? Explain your reasoning.

4. Combine your results for all of your tack tosses recorded in the table with the results of other students or groups in your class. Of this total number of tack tosses, compute the percent of tacks that landed point up. This is your best estimate of the probability that a tack will land point up when tossed.

5. If you tossed 200 thumbtacks, what percent would you expect to land point up? How many tacks would you expect to land point up? Use the percent (probability) you computed in Question 4 to make this calculation. What if you tossed 300 thumbtacks?

Number of Tacks	Number of Tacks Landing Point Up	Fraction of Point-Up Tacks	Percent of Point-Up Tacks
30			
60			
90			
120			
150			

Are you prepared for a test on Chapter 7?

Below I have listed some *common trouble areas* for topics covered in Chapter 7. After studying for your test—but before taking your test—read these.

- Can you convert from percents to fractions or decimals and from fractions or decimals to percents?

  Percent to decimal: $7.5\% = 7.5(0.01) = 0.075$

  Percent to fraction: $11\% = 11 \cdot \dfrac{1}{100} = \dfrac{11}{100}$

  Decimal to percent: $0.36 = 0.36(100\%) = 36\%$

  Fraction to percent: $\dfrac{6}{7} = \dfrac{6}{7} \cdot 100\% = \dfrac{6}{7} \cdot \dfrac{100}{1}\% = \dfrac{600}{7}\%$

  $\qquad\qquad\qquad = 85\dfrac{5}{7}\%$

- Do you remember how to find percent increase or percent decrease? The number of CDs increased from 40 to 48. Find the percent increase.

  $$\dfrac{\text{percent}}{\text{increase}} = \dfrac{\text{increase}}{\text{original number}} = \dfrac{8}{40} = 0.20 = 20\%$$

  Remember: This is simply a checklist of common trouble areas. For a review of Chapter 7, see the Highlights and Chapter Review at the end of the chapter.

Chapter 7 Vocabulary Check

Fill in each blank with one of the words or phrases listed below.

percent	of	amount	100%	compound interest
base	is	0.01	$\frac{1}{100}$	

1. In a mathematical statement, _____ usually means "multiplication."
2. In a mathematical statement, _____ means "equals."
3. _____ means "per hundred."
4. _____ is computed not only on the principal, but also on interest already earned in previous compounding periods.
5. In the percent proportion $\dfrac{}{\underline{}} = \dfrac{\text{percent}}{100}$
6. To write a decimal or fraction as a percent, multiply by _____ .
7. The decimal equivalent of the % symbol is _____ .
8. The fraction equivalent of the % symbol is _____ .

Chapter 7 Highlights

DEFINITIONS AND CONCEPTS	**EXAMPLES**

SECTION 7.1 PERCENTS, DECIMALS, AND FRACTIONS

Percent means per hundred. The % symbol denotes percent.

$$51\% = \frac{51}{100} \quad \text{51 per 100}$$

$$7\% = \frac{7}{100} \quad \text{7 per 100}$$

To write a percent as a decimal, replace the percent symbol with its decimal equivalent, 0.01; then multiply.

$$32\% = 32\,(0.01) = 0.32$$

To write a decimal as a percent, multiply by 1 in the form of 100%.

$$0.78 = 0.78\,(100\%)$$
$$= 7.8\%$$

To write a percent as a fraction, replace the percent symbol with its fraction equivalent, $\frac{1}{100}$; then multiply. Don't forget to simplify the fraction, if possible.

$$25\% = 25 \cdot \frac{1}{100} = \frac{25}{100} = \frac{\overset{1}{\cancel{25}}}{4 \cdot \cancel{25}} = \frac{1}{4}$$

To write a fraction as a percent, multiply the fraction by 100%.

$$\frac{1}{6} = \frac{1}{6} \cdot 100\% = \frac{1}{6} \cdot \frac{100}{1}\% = \frac{100}{6}\% = 16\frac{2}{3}\%$$

SECTION 7.2 SOLVING PERCENT PROBLEMS WITH EQUATIONS

Three key words in the statement of a percent problem are

of, which means multiplication ($\cdot$) **is**, which means equals ($=$)

what (or some equivalent word or phrase), which stands for the unknown (x)

Solve.

6	is	12%	of	what number?
↓	↓	↓	↓	↓
6	=	12%	$\cdot$	x

$$6 = 0.12 \cdot x \qquad \text{Write 12\% as a decimal.}$$

$$\frac{6}{0.12} = \frac{0.12 \cdot x}{0.12} \qquad \text{Divide both sides by 0.12.}$$

$$50 = x$$

Thus, 6 is 12% of 50.

SECTION 7.3 SOLVING PERCENT PROBLEMS WITH PROPORTIONS

PERCENT PROPORTION

$$\frac{\text{amount}}{\text{base}} = \frac{\text{percent}}{100} \leftarrow \text{always 100}$$

or

$$\text{amount} \rightarrow \frac{a}{b} = \frac{p}{100} \leftarrow \text{percent}$$
$$\text{base} \rightarrow$$

Solve.

20.4 is what percent of 85?

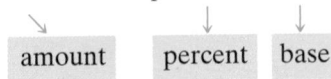

amount   percent   base

$$\begin{array}{l} \text{amount} \rightarrow \\ \text{base} \rightarrow \end{array} \frac{20.4}{85} = \frac{p}{100} \leftarrow \text{percent}$$

$$20.4 \cdot 100 = 85 \cdot p \qquad \text{Set cross products equal.}$$

$$2040 = 85 \cdot p \qquad \text{Multiply.}$$

$$\frac{2040}{85} = \frac{85 \cdot p}{85} \qquad \text{Divide both sides by 85.}$$

$$24 = p \qquad \text{Simplify.}$$

Thus, 20.4 is 24% of 85.

DEFINITIONS AND CONCEPTS	EXAMPLES

SECTION 7.4 APPLICATIONS OF PERCENT

PERCENT INCREASE

$$\text{percent increase} = \frac{\text{amount of increase}}{\text{original amount}}$$

PERCENT DECREASE

$$\text{percent decrease} = \frac{\text{amount of decrease}}{\text{original amount}}$$

A town with a population of 16,480 decreased to 13,870 over a 12-year period. Find the percent decrease. Round to the nearest whole percent.

$$\text{amount of decrease} = 16{,}480 - 13{,}870$$
$$= 2610$$

$$\text{percent decrease} = \frac{\text{amount of decrease}}{\text{original amount}}$$

$$= \frac{2610}{16{,}480} \approx 0.16$$

$$= 16\%$$

The town's population decreased by about 16%.

SECTION 7.5 PERCENT AND PROBLEM SOLVING: SALES TAX, COMMISSION, AND DISCOUNT

SALES TAX

$$\text{sales tax} = \text{sales tax rate} \cdot \text{purchase price}$$
$$\text{total price} = \text{purchase price} + \text{sales tax}$$

Find the sales tax and the total price of a purchase of $42.00 if the sales tax rate is 9%.

sales tax	=	sales tax rate	·	purchase price
↓		↓		↓
sales tax	=	9%	·	$42
	=	0.09 · $42		
	=	$3.78		

The total price is

total price	=	purchase price	+	sales tax
↓		↓		↓
total price	=	$42.00	+	$3.78
	=	$45.78		

The total price is $45.78.

COMMISSION

$$\text{commission} = \text{commission rate} \cdot \text{sales}$$

A salesperson earns a commission of 3%. Find the commission from sales of $12,500 worth of appliances.

commission	=	commission rate	·	sales
↓		↓		↓
commission	=	3%	·	$12,500
	=	0.03 · 12,500		
	=	$375		

The commission is $375.

DEFINITIONS AND CONCEPTS	EXAMPLES

DISCOUNT AND SALE PRICE

amount of discount = discount rate · original price

sale price = original price − amount of discount

A suit is priced at $320 and is on sale today for 25% off. What is the sale price?

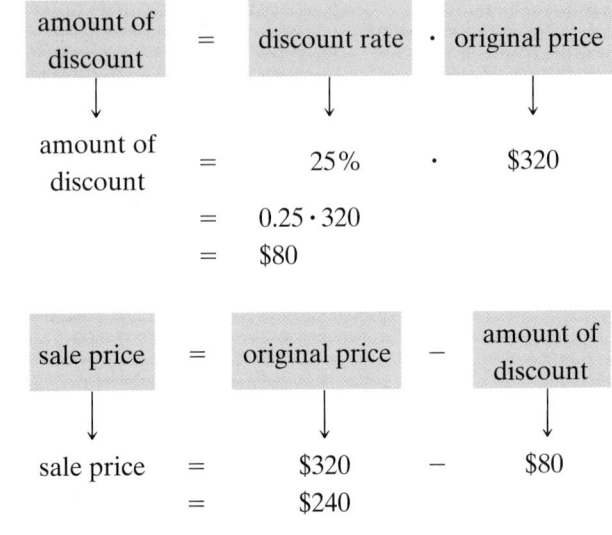

amount of discount	=	discount rate	· original price
amount of discount	=	25%	· $320
	=	0.25 · 320	
	=	$80	

sale price	=	original price	− amount of discount
sale price	=	$320	− $80
	=	$240	

The sale price is $240.

SECTION 7.6 PERCENT AND PROBLEM SOLVING: INTEREST

SIMPLE INTEREST

interest = principal · rate · time

where the rate is understood to be per year, unless told otherwise.

COMPOUND INTEREST

total amount = original principal · compound interest factor

Compound interest is computed not only on the principal, but also on interest already earned in previous compounding periods. (See Appendix F.)

Find the simple interest after 3 years on $800 at an interest rate of 5%.

interest	=	principal	·	rate	·	time
interest	=	$800	·	5%	·	3
	=	$800 · 0.05 · 3		Write 5% as 0.05.		
	=	$120		Multiply.		

The interest is $120.

$800 is invested at 5% compounded quarterly for 10 years. Find the total amount at the end of 10 years.

total amount	=	original principal	· compound interest factor
total amount	=	$800	· 1.64362
	≈	$1314.90	

Chapter 7 Review

(7.1) *Solve*

1. In a survey of 100 adults, 37 preferred pepperoni on their pizza. What percent preferred pepperoni?

2. A basketball player made 77 out of 100 attempted free throws. What percent of free throws was made?

Write each percent as a decimal.

3. 83%

4. 75%

5. 73.5%

6. 1.5%

7. 125%

8. 145%

9. 0.5%

10. 0.7%

11. 200%

12. 400%

13. 26.25%

14. 85.34%

Write each decimal as a percent.

15. 2.6

16. 0.055

17. 0.35

18. 1.02

19. 0.725

20. 0.25

21. 0.076

22. 0.085

23. 0.75

24. 0.65

25. 4.00

26. 9.00

Write each percent as a fraction or a mixed number in simplest form.

27. 1%

28. 10%

29. 25%

30. 8.5%

31. 10.2%

32. $16\frac{2}{3}\%$

33. $33\frac{1}{3}\%$

34. 110%

Write each fraction or mixed number as a percent.

35. $\dfrac{1}{5}$

36. $\dfrac{7}{10}$

37. $\dfrac{5}{6}$

38. $\dfrac{5}{8}$

39. $1\dfrac{2}{3}$

40. $1\dfrac{1}{4}$

41. $\dfrac{3}{5}$

42. $\dfrac{1}{16}$

43. More and more consumers are following the "90-10 rule," that is, eating healthy foods 90% of the time, but indulging in junk food the other 10% of the time. Write 90% and 10% as fractions. (*Source:* Grocery Manufacturers of America.)

44. In medium to large firms in the United States, 96% of full-time employees receive paid vacation benefits. Write 96% as a fraction. (*Source:* U.S. Bureau of Labor Statistics)

45. During a survey recently taken, $\dfrac{7}{10}$ of Americans said they would give up a gym membership if it meant they could buy a new home. Write $\dfrac{7}{10}$ as a percent. (*Source:* KRC Research for Century 21)

46. The number of U.S. coffee drinkers drinking cappuccino and other espresso drinks has grown 150% since the early 1990s. Write 150% as a decimal. (*Source:* National Coffee Association)

(7.2) *Translate each question into an equation and solve.*

47. 1250 is 1.25% of what number?

48. What number is $33\dfrac{1}{3}\%$ of 24,000?

49. 124.2 is what percent of 540?

50. 22.9 is 20% of what number?

51. What number is 40% of 7500?

52. 693 is what percent of 462?

(7.3) *Translate each question into a proportion and solve.*

53. 104.5 is 25% of what number?

54. 16.5 is 5.5% of what number?

55. What number is 36% of 180?

56. 63 is what percent of 35?

57. 93.5 is what percent of 85?

58. What number is 33% of 500?

(7.4) *Solve.*

59. In a survey of 2000 people, it was found that 1320 have a microwave oven. Find the percent of people who own microwaves.

60. Of the 12,360 freshmen entering County College, 2000 are enrolled in Prealgebra. Find the percent of entering freshmen who are enrolled in Prealgebra. Round to the nearest whole percent.

61. The current charge for dumping waste in a local landfill is $16 per cubic foot. To cover new environmental costs, the charge will increase to $33 per cubic foot. Find the percent increase.

62. The number of violent crimes in a city decreased from 675 to 534. Find the percent decrease. Round to the nearest tenth of a percent.

63. This year the fund drive for a charity collected $215,000. Next year, a 4% decrease is expected. Find how much is expected to be collected in next year's drive.

64. A local union negotiated a new contract that increases the hourly pay 15% over last year's pay. The old hourly rate was $11.50. Find the new hourly rate rounded to the nearest cent.

(7.5) *Solve.*

65. If the sales tax rate is 5.5%, what is the total amount charged for a $250 coat?

66. Find the sales tax paid on a $25.50 purchase if the sales tax rate is 4.5%.

67. Russ James is a sales representative for a chemical company and is paid a commission rate of 5% on all sales. Find his commission if he sold $100,000 worth of chemicals last month.

68. Carol Sell is a sales clerk in a clothing store. She receives a commission of 7.5% on all sales. Find her commission for the week if her sales for the week were $4005. Round to the nearest cent.

69. A $3000 mink coat is on sale for 30% off. Find the discount and the sale price.

70. A $90 calculator is on sale for 10% off. Find the discount and the sale price.

(7.6) *Solve.*

71. Find the simple interest due on $4000 loaned for 3 months at 12% interest.

72. Find the total amount due on an 8-month loan of $1200 at a simple interest rate of 15%.

73. Find the total amount in an account if $5500 is compounded annually at 12% for 15 years.

74. Find the total amount in an account if $6000 is compounded semiannually at 11% for 10 years.

75. Find the compound interest earned if $100 is compounded quarterly at 12% for 5 years.

76. Find the compound interest earned if $1000 is compounded quarterly at 18% for 20 years.

Chapter 7 Test   Remember to check your answers and use the Chapter Test Prep Video to view solutions.

Write each percent as a decimal.

1. 85%   **2.** 500%   **3.** 0.6%

Write each decimal as a percent.

4. 0.056   **5.** 6.1   **6.** 0.35

Write each percent as a fraction or a mixed number in simplest form.

7. 120%   **8.** 38.5%   **9.** 0.2%

Write each fraction or mixed number as a percent.

10. $\dfrac{11}{20}$ **11.** $\dfrac{3}{8}$ **12.** $1\dfrac{3}{4}$

13. Sales of bottled water have recently surged. Bottled water accounts for $\dfrac{1}{5}$ of the total noncarbonated beverage category in convenience stores. Write $\dfrac{1}{5}$ as a percent. (*Source:* Grocery Manufacturers of America)

14. In small firms in the United States, 64% of full-time employees receive medical insurance benefits. Write 64% as a fraction. (*Source:* U.S. Bureau of Labor Statistics)

Solve.

15. What number is 42% of 80?

16. 0.6% of what number is 7.5?

17. 567 is what percent of 756?

1. _____

2. _____

3. _____

4. _____

5. _____

6. _____

7. _____

8. _____

9. _____

10. _____

11. _____

12. _____

13. _____

14. _____

15. _____

16. _____

17. _____

Solve. If necessary, round percents to the nearest tenth, dollar amounts to the nearest cent, and all other numbers to the nearest whole.

18. _____

19. _____

20. _____

21. _____

22. _____

23. _____

24. _____

25. _____

26. _____

27. _____

28. _____

18. An alloy is 12% copper. How much copper is contained in 320 pounds of this alloy?

19. A farmer in Nebraska estimates that 20% of his potential crop, or $11,350, has been lost to a hard freeze. Find the total value of his potential crop.

20. If the local sales tax rate is 1.25%, find the total amount charged for a stereo system priced at $354.

21. A town's population increased from 25,200 to 26,460. Find the percent increase.

22. A $120 framed picture is on sale for 15% off. Find the discount and the sale price.

23. Randy Nguyen is paid a commission rate of 4% on all sales. Find Randy's commission if his sales were $9875.

24. A sales tax of $1.53 is added to an item's price of $152.99. Find the sales tax rate. Round to the nearest whole percent.

25. Find the simple interest earned on $2000 saved for $3\frac{1}{2}$ years at an interest rate of 9.25%.

26. $1365 is compounded annually at 8%. Find the total amount in the account after 5 years.

27. A couple borrowed $400 from a bank at 13.5% for 6 months for car repairs. Find the total amount due the bank at the end of the 6-month period.

28. The number of crimes reported in New York City was 125,587 in the first half of 2001 and 118,346 in the first half of 2002. Find the percent decrease in the number of crimes in New York City from 2001 to 2002. (*Source*: Federal Bureau of Investigation. Uniform Crime Reports, January–June 2002)

Cumulative Review

1. Multiply: 236×86

2. Multiply: 409×76

3. Subtract: 7 from -3

4. Subtract: -2 from 8.

5. Solve: $x - 2 = 1$

6. Solve: $x + 4 = 3$

7. Solve: $3(2x - 6) + 6 = 0$

8. Solve: $5(x - 2) = 3x$

9. Write 3 as an equivalent fraction whose denominator is 7.

10. Write 8 as an equivalent fraction whose denominator is 5.

11. Simplify: $\dfrac{84x}{90}$

12. Simplify: $\dfrac{10y}{32}$

13. Divide: $-\dfrac{5}{16} \div -\dfrac{3}{4}$

14. Divide: $\dfrac{-2}{5} \div \dfrac{7}{10}$

15. Evaluate $y - x$ if $x = -\dfrac{3}{10}$ and $y = -\dfrac{8}{10}$.

16. Evaluate $2x + 3y$ if $x = \dfrac{2}{5}$ and $y = \dfrac{-1}{5}$.

17. Find: $-\dfrac{3}{4} - \dfrac{1}{14} + \dfrac{6}{7}$

18. Find: $\dfrac{2}{9} + \dfrac{7}{15} - \dfrac{1}{3}$

19. Simplify: $\dfrac{\dfrac{1}{2} + \dfrac{3}{8}}{\dfrac{3}{4} - \dfrac{1}{6}}$

20. Simplify: $\dfrac{\dfrac{2}{3} + \dfrac{1}{6}}{\dfrac{3}{4} - \dfrac{3}{5}}$

21. Solve: $\dfrac{x}{2} = \dfrac{x}{3} + \dfrac{1}{2}$

22. Solve: $\dfrac{x}{2} + \dfrac{1}{5} = 3 - \dfrac{x}{5}$

23. Write each mixed number as an improper fraction.

    a. $4\dfrac{2}{9}$ **b.** $1\dfrac{8}{11}$

24. Write each mixed number as an improper fraction.

    a. $3\dfrac{2}{5}$ **b.** $6\dfrac{2}{7}$

Answers

1. _____
2. _____
3. _____
4. _____
5. _____
6. _____
7. _____
8. _____
9. _____
10. _____
11. _____
12. _____
13. _____
14. _____
15. _____
16. _____
17. _____
18. _____
19. _____
20. _____
21. _____
22. _____
23. a. _____

    b. _____

24. a. _____

    b. _____

Write each decimal as a fraction or mixed number. Write your answer in simplest form.

25. 0.125

26. 0.85

27. −105.083

28. 17.015

29. Subtract: 85 − 17.31

30. Subtract: 38 − 10.06

Multiply.

31. 7.68 × 10

32. 12.483 × 100

33. (−76.3)(1000)

34. −853.75 × 10

35. Is 720 a solution of the equation $\dfrac{y}{100} = 7.2$?

36. Is 470 a solution of the equation $\dfrac{x}{100} = 4.75$?

37. Find: $\sqrt{\dfrac{4}{25}}$

38. Find: $\sqrt{\dfrac{9}{16}}$

39. Write the ratio of 2.5 to 3.15 as a fraction in simplest form.

40. Write the ratio of 5.8 to 7.6 as a fraction in simplest form.

41. A store charges $3.36 for a 16-ounce jar of picante sauce. What is the unit price in dollars per ounce?

42. Flooring tiles cost $90 for a box with 40 tiles. Each tile is 1 square foot. Find the unit price in dollars per square foot.

43. Is $\dfrac{4.1}{7} = \dfrac{2.9}{5}$ a true proportion?

44. Is $\dfrac{6.3}{9} = \dfrac{3.5}{5}$ a true proportion?

45. On a Chamber of Commerce map of Abita Springs, 5 miles corresponds to 2 inches. How many miles correspond to 7 inches?

46. A student doing math homework can complete 7 problems in about 6 minutes. At this rate, how many problems can be complete in 30 minutes?

Write each percent as a fraction or mixed number in simplest form.

47. 1.9%

48. 2.3 %

49. $33\dfrac{1}{3}\%$

50. 108 %

25. _____
26. _____
27. _____
28. _____
29. _____
30. _____
31. _____
32. _____
33. _____
34. _____
35. _____
36. _____
37. _____
38. _____
39. _____
40. _____
41. _____
42. _____
43. _____
44. _____
45. _____
46. _____
47. _____
48. _____
49. _____
50. _____

590

Graphing and Introduction to Statistics

Numerical data is all around us—in newspaper and magazine articles and in television news reports. Frequently, numerical data is summarized in graphs and basic statistics. To understand this data, we must first know how to read graphs and interpret basic statistics. We learn to do just that in this chapter.

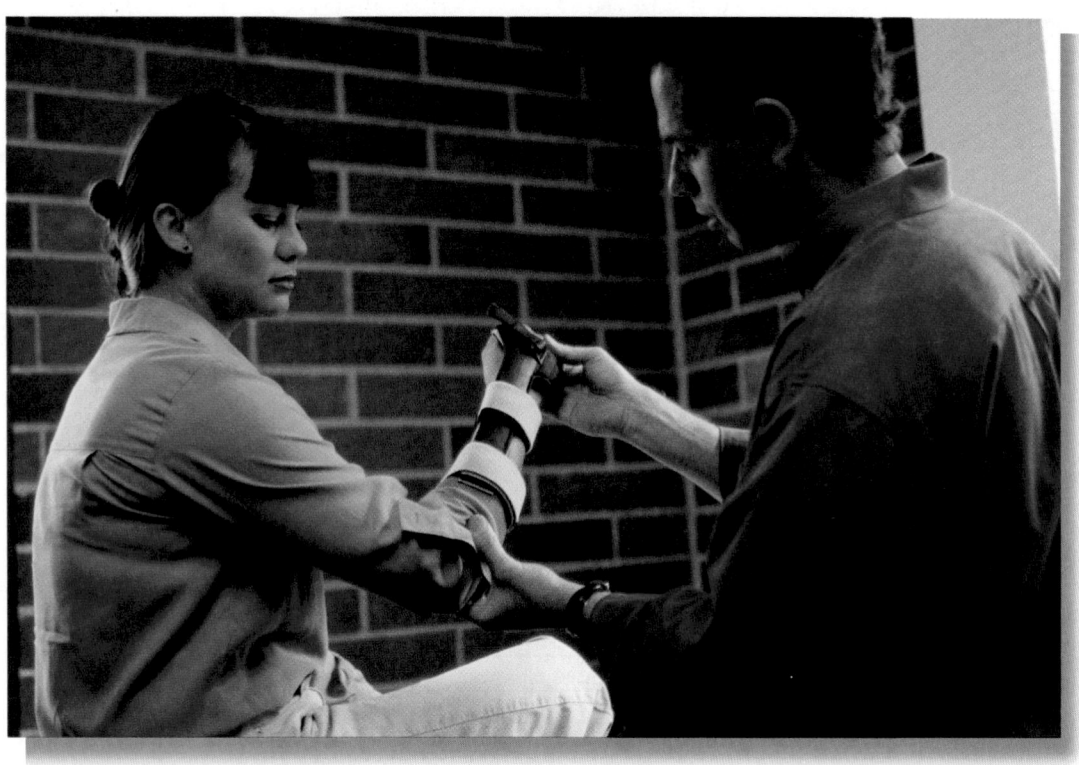

Carpal Tunnel Syndrome (CTS) is a disorder of the wrist and hand associated with pain and numbness in the fingers. It is caused by the compression of the median nerve inside the carpal tunnel, a small passage inside the wrist. Doctors can treat CTS by applying a splint to the wrist and prescribing an anti-inflammatory drug, such as ibuprofen. In some cases, however, surgery is required to enlarge the carpal tunnel and relieve the compression on the median nerve. Most people associate CTS with repetitive wrist motion, such as clicking a computer mouse. CTS can also occur as a result of fractures, arthritis, pregnancy, or diabetes. It is one of the most common work-related injuries. In Exercise 27 on page 615, we see how a circle graph can be used to describe the number of days of absence from work due to carpal tunnel syndrome.

Name _____ Section _____ Date_____

Chapter 8 Pretest

1. _____

2. _____

3. _____

4. _____

5. _____

6. _____

7. _____

8. _____

9. _____

10. _____

11. a. _____

    b. _____

    c. _____

12. see graph _____

13. _____

14. _____

15. _____

16. _____

The line graph below shows the number of burglaries in a town during the months of March through September.

Number of Burglaries

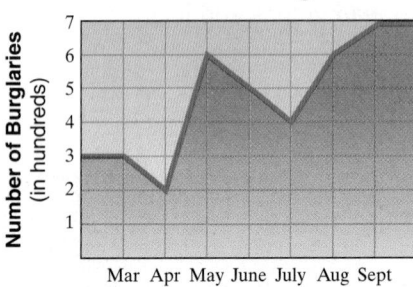

1. During which month, between March and September, did the fewest number of burglaries occur?

2. During which month, between March and September, were there 400 burglaries?

3. How many burglaries were there in September?

4. The following table shows a breakdown of an average day for Dawn Miller.

Attending college classes	4 hours
Studying	3 hours
Working	5 hours
Sleeping	8 hours
Driving	1 hour
Other	3 hours

Draw a circle graph showing this data.

Below is a list of scores from the final exam given in Mrs. Maxwell's basic college mathematics class. Use this list to complete the table below:

76 71 94 73 81 78 96 65
95 80 90 86 98 88 62 91

	Class Intervals (Scores)	Tally	Class Frequency (Number of Exams)
5.	60–69		
6.	70–79		
7.	80–89		
8.	90–99		

9. Use the table from Exercises 5 through 8 to draw a histogram.

10. Is $(-2, 7)$ a solution of the equation $3x + y = -1$?

11. Complete the ordered-pair solutions of the equation $y = 2x - 6$.
    a. $(0, \ )$ **b.** $(-3, \ )$ **c.** $(\ , 0)$

12. Graph $y = 4x - 1$

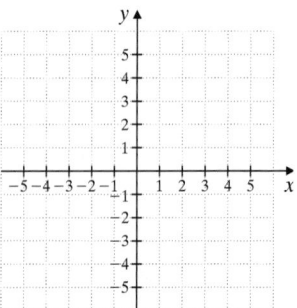

13. Find the mean, the median, and the mode for the given set of numbers. If necessary round the mean to one decimal place.
    28, 36, 81, 73, 28, 74, 31, 74, 64, 25, 74

A single die is tossed. Find the probability that the die lands showing each of the following.

14. a 4   **15.** a number greater than 3   **16.** a 3 or a 5

8.1 Reading Pictographs, Bar Graphs, and Line Graphs

Often data is presented visually in a graph. In this section, we practice reading several kinds of graphs including pictographs, bar graphs, and line graphs.

Ⓐ Reading Pictographs

A **pictograph** such as the one below is a graph in which pictures or symbols are used to represent some fixed amount. This type of graph contains a key that explains the meaning of the symbol used. An advantage of using a pictograph to display information is that comparisons can easily be made. A disadvantage of using a pictograph is that it is often hard to tell what fractional part of a symbol is shown. For example, in the pictograph below, Germany shows three complete symbols and a part of a fourth symbol. Notice that it's hard to read with any accuracy what fractional part of the 4th symbol is shown.

EXAMPLE 1   The following pictograph shows the approximate amount of nuclear energy generated by selected countries in the year 2001. Use this pictograph to answer the questions.

Nuclear Energy Generated by Selected Countries (2001)

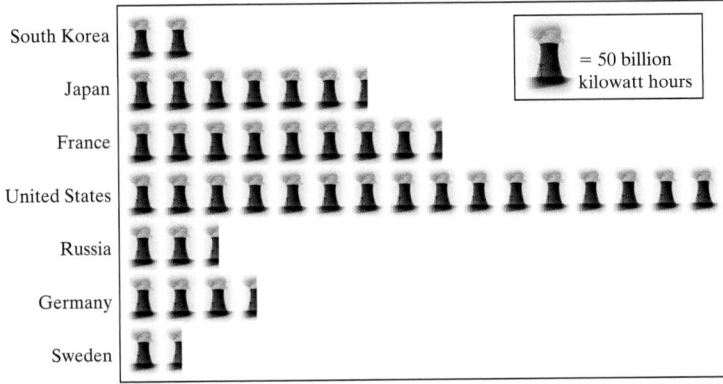

Source: Energy Information Administration

a. Approximate the amount of nuclear energy that is generated in South Korea.

b. Approximate how much more nuclear energy is generated in France than in South Korea.

Solution:

a. South Korea corresponds to 2 symbols, and each symbol represents 50 billion kilowatt hours of energy. This means that South Korea generates approximately $2 \cdot (50 \text{ billion})$ or 100 billion kilowatt hours of nuclear energy.

b. France shows about $6\frac{1}{2}$ more symbols than South Korea. This means that France generates $6\frac{1}{2} \cdot (50 \text{ billion})$ or 325 billion more kilowatt hours of nuclear energy than South Korea.   ●

Ⓑ Reading and Constructing Bar Graphs

Another way to present data graphically is with a **bar graph**. Bar graphs can appear with vertical bars or horizontal bars. Although we have studied bar

OBJECTIVES

Ⓐ Read pictographs.
Ⓑ Read and construct bar graphs.
Ⓒ Read and construct histograms.
Ⓓ Read line graphs.

SSM
TUTOR CENTER   SG   CD & VIDEO   MATH PRO   WEB

Practice Problem 1

Use the pictograph shown in Example 1 to answer the following questions:

a. Approximate the amount of nuclear energy that is generated in Sweden.

b. Approximate the total nuclear energy generated in Sweden and Russia.

Answers

1. a. 75 billion kilowatt hours.   **b.** 200 billion kilowatt hours

graphs in previous sections, we now practice reading the height of the bars contained in a bar graph. An advantage to using bar graphs is that a scale is usually included for greater accuracy. Care must be taken when reading bar graphs, as well as other types of graphs—they may be misleading, as shown later in this section.

Practice Problem 2

Use the bar graph in Example 2 to answer the following questions:

a. Approximate the number of endangered species that are insects.

b. Which category (or categories) shows the fewest endangered species?

EXAMPLE 2 The following bar graph shows the number of endangered species in 2003. Use this graph to answer the questions.

How Many U.S. Animal Species Are Endangered?

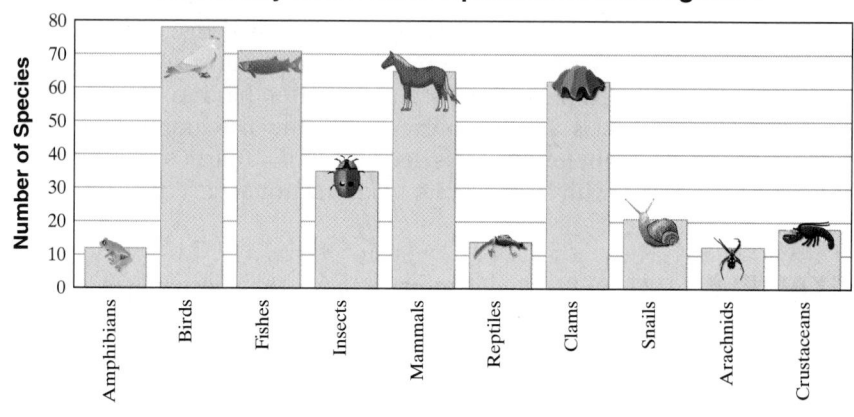

Source: U.S. Fish and Wildlife Service

a. Approximate the number of endangered species that are reptiles.

b. Which category has the most endangered species?

Solution:

a. To approximate the number of endangered species that are reptiles, we go to the top of the bar that represents reptiles. From the top of this bar, we move horizontally to the left until the scale is reached. We read the height of the bar on the scale as approximately 14. There are approximately 14 reptile species that are endangered, as shown in the next figure.

How Many U.S. Animal Species Are Endangered?

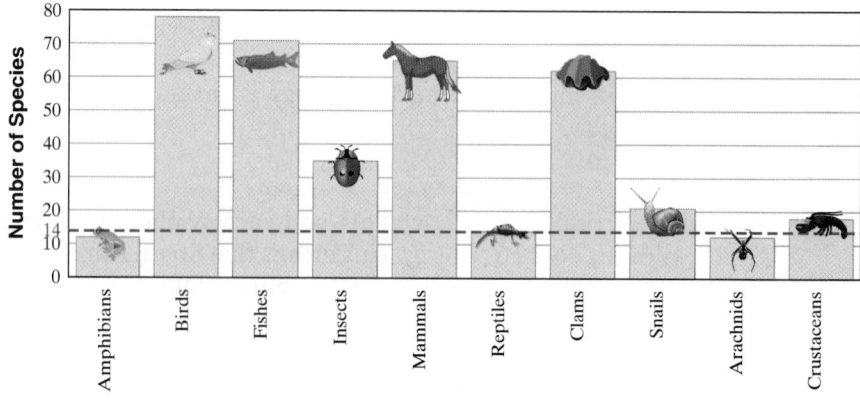

Source: U.S. Fish and Wildlife Service

b. The most endangered species is represented by the tallest bar. The tallest bar corresponds to birds.

As mentioned previously, graphs can be misleading. Both graphs below show the same information, but with different scales. Special care should be taken when forming conclusions from the appearance of a graph. (Recall that the ⚡ symbol on the vertical line with the scale means that some values have been omitted.)

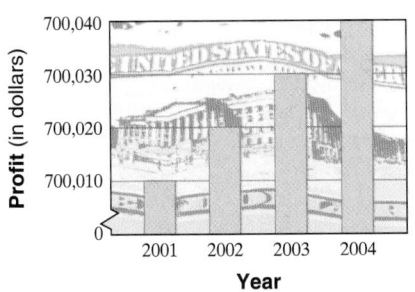

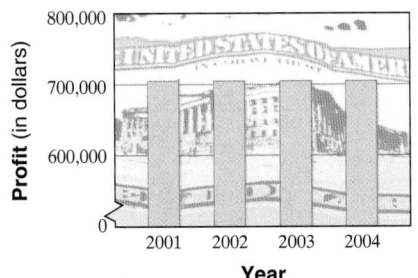

Are profits shown in the graphs above greatly increasing, or are they remaining about the same?

Next, we practice constructing a bar graph.

EXAMPLE 3   Draw a vertical bar graph using the information in the table below that gives the caffeine content of selected foods.

Average Caffeine Content of Selected Foods			
Food	**Milligrams**	**Food**	**Milligrams**
Brewed coffee (percolator, 8 ounces)	124	Instant coffee (8 ounces)	104
Brewed decaffeinated coffee (8 ounces)	3	Brewed tea (U.S. brands, 8 ounces)	64
Coca-Cola classic (8 ounces)	31	Mr. Pibb (8 ounces)	27
Dark chocolate (semi sweet, $1\frac{1}{2}$ ounces)	30	Milk chocolate (8 ounces)	9

(*Sources:* International Food Information Council and the Coca-Cola Company)

Solution: We draw and label a vertical line and a horizontal line as shown next to the left. These lines are also called axes. We place the different food categories along the horizontal axis. Along the vertical axis, we place a scale.

There are many choices of scales that would be appropriate. Notice that the milligrams range from a low of 3 to a high of 124. From this information, we use a scale that starts at 0 and then shows multiples of 20 so that the scale is not too cluttered. The scale stops at 140, the smallest multiple of 20 that will allow all milligrams to be graphed. It may also be helpful to draw horizontal lines along the scale markings to help draw the vertical bars at the correct height. The finished bar graph is shown on the top of the next page.

Practice Problem 3

Draw a vertical bar graph using the information in the table about electoral votes for selected states.

Electoral Votes for President by Selected States	
State	**Electoral Votes**
Texas	34
California	55
Florida	27
Nebraska	5
Indiana	11
Georgia	15

(*Source: World Almanac 2003*)

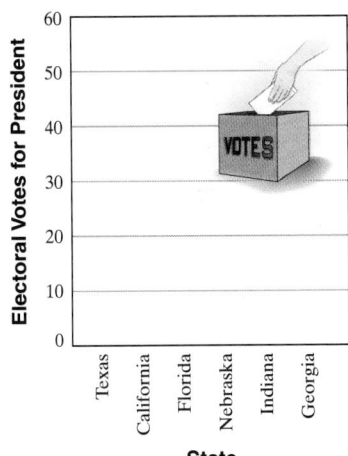

Answer

3.

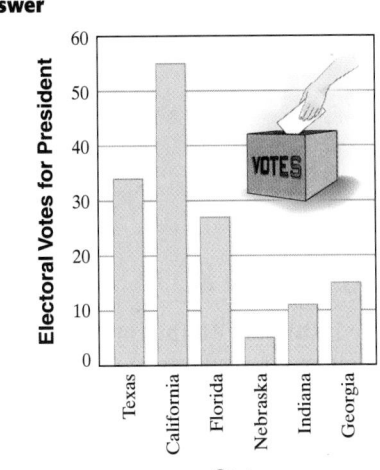

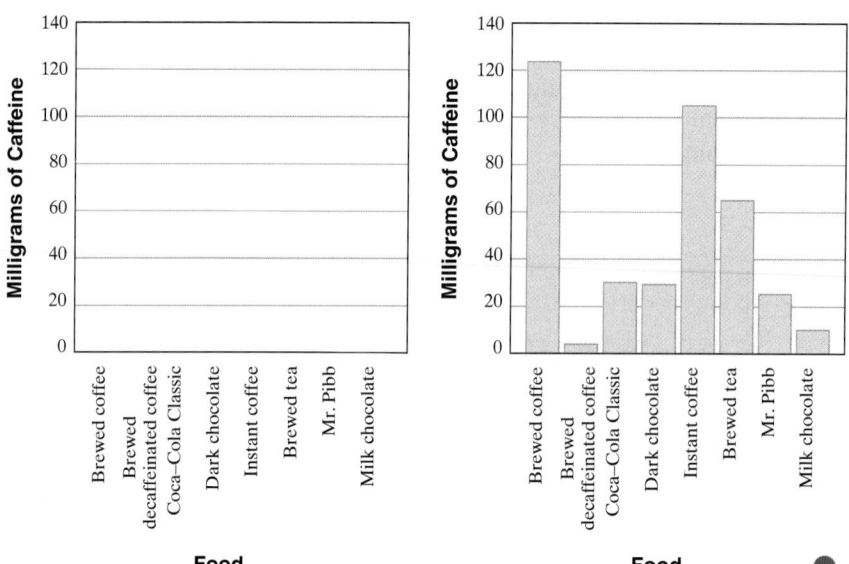

Helpful Hint

Can you see from the bar graph to the right that a scale too small may be too time-consuming to draw and too hard to read? Think of a scale that starts at 0 and shows multiples of 2. Think of the number of tick marks that would be needed.

C Reading and Constructing Histograms

Suppose that the test scores of 36 students are summarized in the table below:

Student Scores	Frequency (number of students)
40–49	1
50–59	3
60–69	2
70–79	10
80–89	12
90–99	8

The results in the table can be displayed in a histogram. A **histogram** is a special bar graph. The width of each bar represents a range of numbers called a **class interval**. The height of each bar corresponds to how many times a number in the class interval occurred and is called the **class frequency**. The bars in a histogram lie side by side with no space between them.

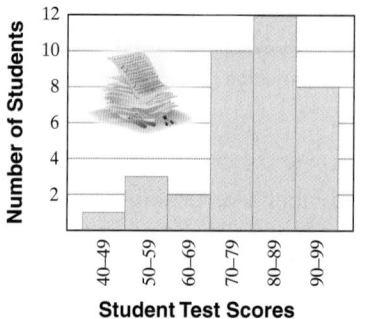

Practice Problem 4

Use the histogram on the right to determine how many students scored 70–79 on the test.

Answer

4. 10

EXAMPLE 4   Use the preceding histogram to determine how many students scored 50–59 on the test.

Solution: We find the bar representing 50–59. The height of this bar is 3, which means 3 students scored 50–59 on the test.

EXAMPLE 5   Use the preceding histogram to determine how many students scored 80 or above on the test.

Solution: We see that two different bars fit this description. There are 12 students who scored 80–89 and 8 students who scored 90–99. The sum of these two categories is 12 + 8 or 20 students. Thus, 20 students scored 80 or above on the test. ●

Now we will look at a way to construct histograms.

The daily high temperatures for 1 month in New Orleans, Louisiana, are recorded in the following list:

85°	90°	95°	89°	88°	94°
87°	90°	95°	92°	95°	94°
82°	92°	96°	91°	94°	92°
89°	89°	90°	93°	95°	91°
88°	90°	88°	86°	93°	89°

The data in this list have not been organized and can be hard to interpret. One way to organize the data is to place it in a **frequency distribution table**. We will do this in Example 6.

EXAMPLE 6   Complete the frequency distribution table for the preceding temperature data.

Solution: Go through the data and place a tally mark next to the class interval (in the second table column). Then count the tally marks and write each total in the third table column.

Class Intervals (Temperatures)	Tally	Class Frequency (Number of Days)
82°–84°	|	1
85°–87°	|||	3
88°–90°	⋕ ⋕ |	11
91°–93°	⋕ ||	7
94°–96°	⋕ |||	8

EXAMPLE 7   Construct a histogram from the frequency distribution table in Example 6.

Solution:

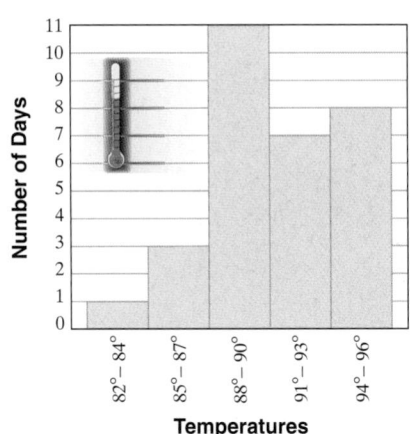

Temperatures

Practice Problem 5

Use the histogram for Example 4 to determine how many students scored less than 60 on the test.

Practice Problem 6

Complete the frequency distribution table for the data below. Each number represents a credit card owner's unpaid balance each month.

0	53	89	125
265	161	37	76
62	201	136	42

Class Intervals (Credit Card Balances)	Tally	Class Frequency (Number of Months)
$0–$49	_____	_____
$50–$99	_____	_____
$100–$149	_____	_____
$150–$199	_____	_____
$200–$249	_____	_____
$250–$299	_____	_____

Practice Problem 7

Construct a histogram from the frequency distribution table above.

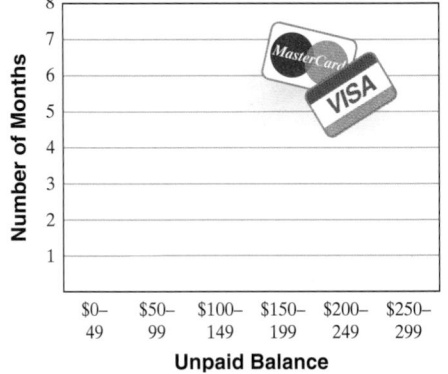

Unpaid Balance

Answers

5. 4

6.

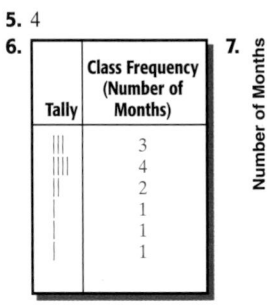

Tally	Class Frequency (Number of Months)
|||	3
||||	4
||	2
|	1
|	1
|	1

7.

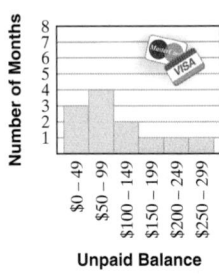

Unpaid Balance

Concept Check

Which of the following sets of data is better suited to representation by a histogram? Explain.

Grade on Final	# of Students	Section Number	Avg. Grade on Final
51–60	12	150	78
61–70	18	151	83
71–80	29	152	87
81–90	23	153	73
91–100	25		

Practice Problem 8

Use the temperature graph in Example 8 to answer the following questions:

a. During what month is the average daily high temperature the lowest?
b. During what month is the average daily high temperature 25°F?
c. During what months is the average daily high temperature greater than 70°F?

Try the Concept Check in the margin.

D Reading Line Graphs

Another common way to display information graphically is by using a **line graph**. An advantage of a line graph is that it can be used to visualize relationships between two quantities. A line graph can also be very useful in showing change over time.

EXAMPLE 8 The following line graph shows the average daily high temperature for each month for Omaha, Nebraska. Use this graph to answer the questions.

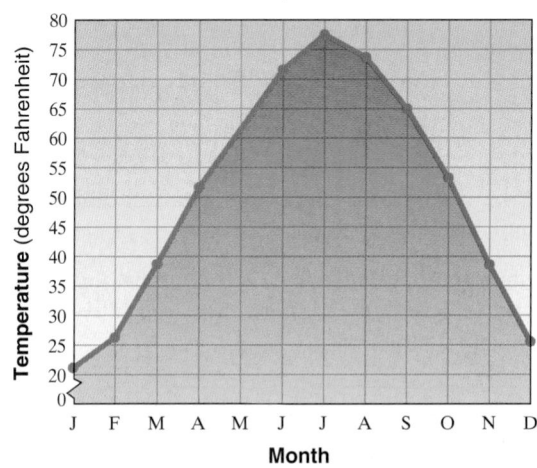

Average Daily High Temperature in Omaha, Nebraska

Source: National Climatic Data Center

a. During what month is the average daily high temperature the highest?
b. During what month is the average daily high temperature 65°F?
c. During what months is the average daily high temperature less than 30°F?

Solution:

a. The month with the highest temperature corresponds to the highest point on the graph. We follow the highest point downward to the horizontal month scale and see that this point corresponds to July.

Answers

8. a. January **b.** December
c. June, July, and August

Concept Check: first set of data

b. We find the 65°F mark on the vertical axis and move to the right until a darkened point on the graph is reached. From that point, we move downward to the Month axis and read the corresponding month. For the month of September, the average daily high temperature was 65°F.

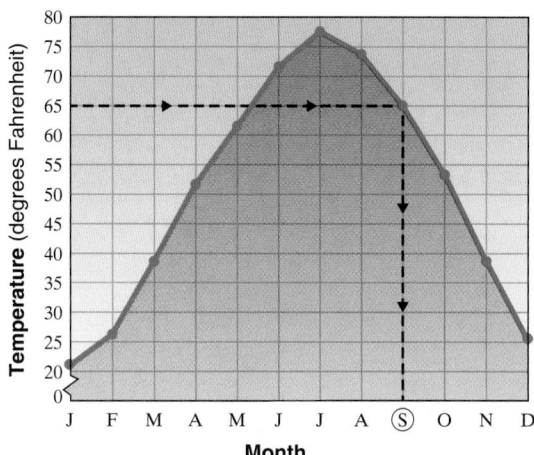

Average Daily High Temperature in Omaha, Nebraska

Source: National Climatic Data Center

c. To see what months the temperature is less than 30°F, we find what months correspond to darkened points that fall below the 30°F mark on the vertical axis. These months are January, February, and December.

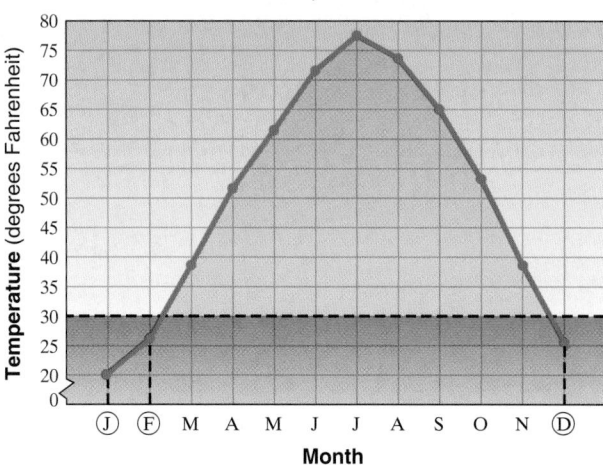

Average Daily High Temperature in Omaha, Nebraska

Source: National Climatic Data Center

Name _____ Section _____ Date _____

A *The following pictograph shows the annual automobile production by one plant for the years 1997–2003. Use this graph to answer Exercises 1 through 8. See Example 1.*

Automobile Production

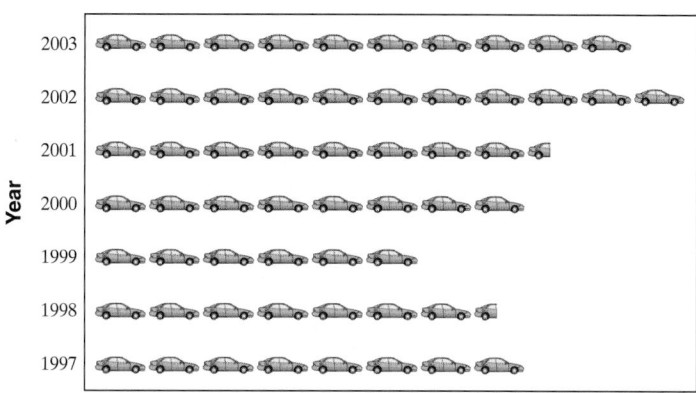

Each 🚗 represents 500 cars

1. In what year was the greatest number of cars manufactured?

2. In what year was the least number of cars manufactured?

3. Approximate the number of cars manufactured in the year 2000.

4. Approximate the number of cars manufactured in the year 2001.

5. In what year(s) did the production of cars decrease from the previous year?

6. In what year(s) did the production of cars increase from the previous year?

7. In what year(s) were 4000 cars manufactured?

8. In what year(s) were 5500 cars manufactured?

The following pictograph shows the average number of ounces of chicken consumed per person per week in the United States. Use this graph to answer Exercises 9 through 16. See Example 1.

Chicken Consumption

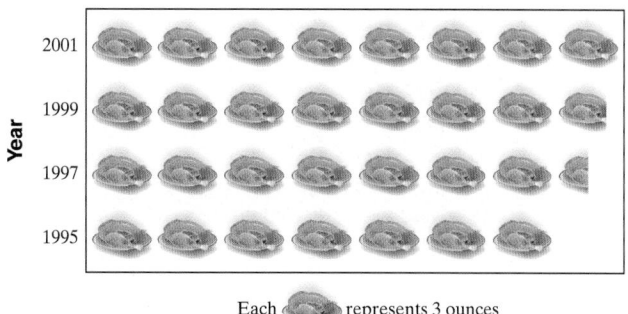

Each 🍗 represents 3 ounces

Source: National Agricultural Statistics Service

9. Approximate the number of ounces of chicken consumed per week in 1997.

10. Approximate the number of ounces of chicken consumed per week in 2001.

11. In what year(s) was the number of ounces of chicken consumed per week greater than 21 ounces?

12. In what year(s) was the number of ounces of chicken consumed per week 21 ounces or less?

13. What was the increase in average chicken consumption from 1995 to 2001?

14. What was the increase in average chicken consumption from 1997 to 2001?

15. Describe a trend in eating habits shown by this graph.

16. In 2001, did you consume less than, greater than, or about the same as the U.S. average number of ounces consumed per week?

B *The following bar graph shows the average number of people killed by tornadoes during the months of the year. Use this graph to answer Exercises 17 through 22. See Example 2.*

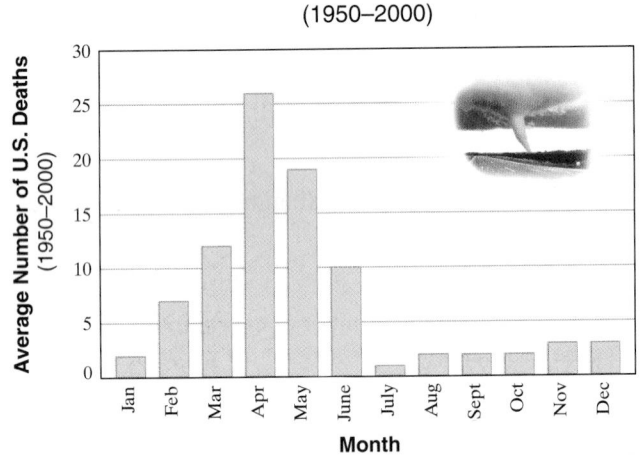

Tornado Deaths in U.S.
(1950–2000)

Source: Storm Prediction Center

17. In which month(s) did the most tornado-related deaths occur?

18. In which month(s) did the fewest tornado-related deaths occur?

19. Approximate the number of tornado-related deaths that occurred in May.

20. Approximate the number of tornado-related deaths that occurred in April.

21. In which month(s) did more than 5 deaths occur?

22. In which month(s) did more than 15 deaths occur?

The following horizontal bar graph shows the 2001 population of the world's largest cities (including their suburbs). Use this graph to answer Exercises 23 through 28. See Example 2.

World's Largest Cities (including suburbs)

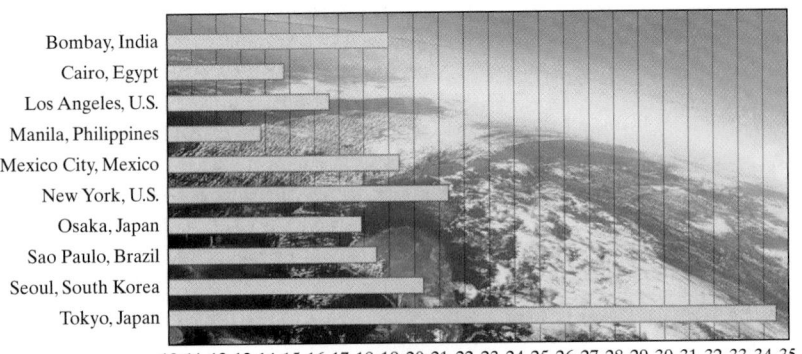

Source: Thomas Brinkhoff: *The Principal Agglomerations of the World*,
http://www.citypopulation.de, 13.05.2001

23. Name the city with the largest population and estimate its population.

24. Name the city whose population is between 16 and 17 million and estimate its population.

25. Name the city in the United States with the largest population and estimate its population.

26. Name the city with the smallest population and estimate its population.

27. How much larger is Tokyo than Sao Paulo?

28. How much larger is Bombay than Manila?

Use the information given to draw a vertical bar graph. Clearly label the bars. See Example 3.

29.

Fiber Content of Selected Foods	
Food	**Grams of Total Fiber**
Kidney beans ($\frac{1}{2}$c)	4.5
Oatmeal, cooked ($\frac{3}{4}$c)	3.0
Peanut butter, chunky (2 tbsp)	1.5
Popcorn (1 c)	1.0
Potato, baked with skin (1 med)	4.0
Whole wheat bread (1 slice)	2.5

(*Sources:* American Dietetic Association and National Center for Nutrition and Dietetics)

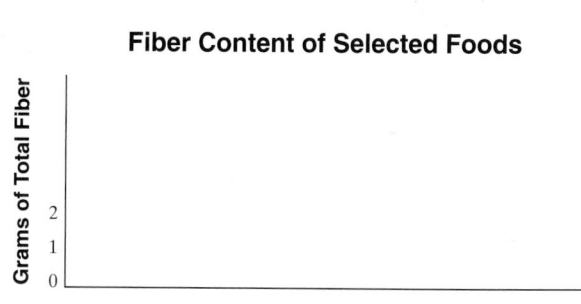

30.

U.S. Restaurant Industry Annual Food and Beverage Sales	
Year	**Sales in Billions of Dollars**
1970	43
1980	120
1990	239
2001	399

(*Source:* National Restaurant Association)

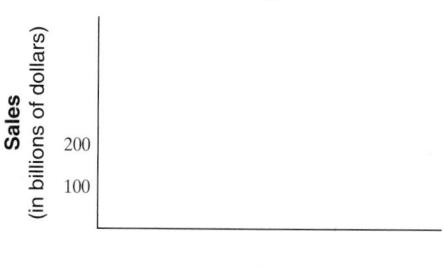

31.

Best-selling Albums of All Time (U.S. Sales)	
Album	**Estimated Sales (in millions)**
Pink Floyd: *The Wall* (1979)	23
Michael Jackson: *Thriller* (1982)	26
AC/DC: *Back in Black* (1980)	19
Billy Joel: *Greatest Hits Volumes I & II* (1985)	21
Eagles: *Their Greatest Hits* (1976)	27
Led Zeppelin: *Led Zeppelin IV* (1971)	22
Shania Twain: *Come on Over* (1997)	19

(*Source:* Recording Industry Association of America)

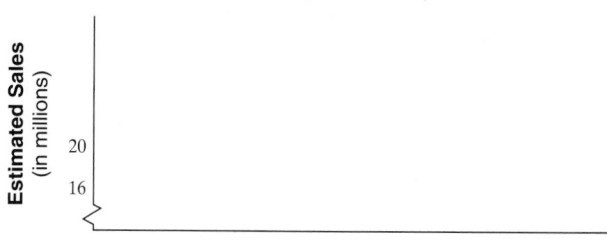

32.

Fuel Economy of the Top-Selling Vehicles in the United States for 2003[*]	
Vehicle (sales rank)	Highway Fuel Economy[*] (in miles per gallon)
Ford F-Series (1)	20
Chevrolet Silverado (2)	21
Dodge Ram (3)	21
Toyota Camry (4)	33
Honda Accord (5)	34
Ford Explorer (6)	22

[*]Maximum fuel economy available among all model trims, through June 2003
(*Sources:* U.S. Dept. of Energy, Reuters)

Fuel Economy of the Top-Selling Vehicles in the United States for 2003

Highway Fuel Economy (in miles per gallon)

22
20

Vehicle

c *The following histogram shows the number of miles that each adult, from a survey of 100 adults, drives per week. Use this histogram to answer Exercises 33 through 42. See Examples 4 and 5.*

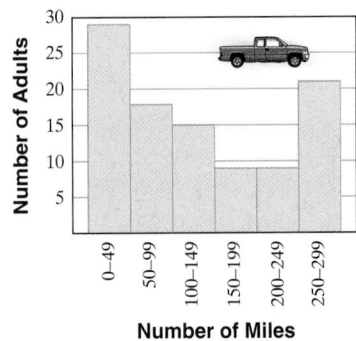

Number of Adults

30
25
20
15
10
5

0–49 50–99 100–149 150–199 200–249 250–299

Number of Miles

33. How many adults drive 100–149 miles per week?

34. How many adults drive 200–249 miles per week?

35. How many adults drive fewer than 150 miles per week?

36. How many adults drive 200 miles or more per week?

37. How many adults drive 100–199 miles per week?

38. How many adults drive 0–149 miles per week?

39. How many more adults drive 250–299 miles per week than 200–249 miles per week?

40. How many more adults drive 0–49 miles per week than 50–99 miles per week?

41. What is the ratio of adults who drive 150–199 miles per week to the total number of adults surveyed?

42. What is the ratio of adults who drive 50–99 miles per week to the total number of adults surveyed?

The following histogram shows the projected ages of householders for the year 2005. Use this histogram to answer Exercises 43 through 50. See Examples 4 and 5.

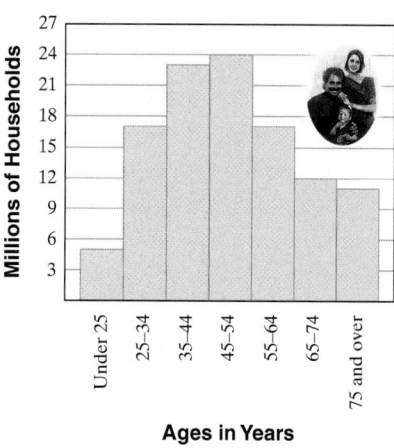

Source: U.S. Bureau of the Census, *Current Population Reports*

43. The most householders will be in what age range?

44. The least number of householders will be in what age range?

45. How many householders will be 55–64 years old?

46. How many householders will be 35–44 years old?

47. How many householders will be 44 years old or younger?

48. How many householders will be 55 years old or older?

49. Which bar represents the household you expect to be in during the year 2005?

50. How many more householders will be 45–54 years old than 55–64 years old?

The following list shows the golf scores for an amateur golfer. Use this list to complete the frequency distribution table to the right. See Example 6.

78	84	91	93	97
97	95	85	95	96
101	89	92	89	100

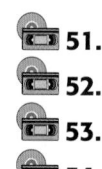

	Class Intervals (Scores)	Tally	Class Frequency (Number of Games)
51.	70–79	____	____
52.	80–89	____	____
53.	90–99	____	____
54.	100–109	____	____

Twenty-five people in a survey were asked to give their current checking account balances. Use the balances shown in the following list to complete the frequency distribution table to the right. See Examples.

$53	$105	$162	$443	$109
$468	$47	$259	$316	$228
$207	$357	$15	$301	$75
$86	$77	$512	$219	$100
$192	$288	$352	$166	$292

	Class Intervals (Account Balances)	Tally	Class Frequency (Number of People)
55.	$0–$99	____	____
56.	$100–$199	____	____
57.	$200–$299	____	____
58.	$300–$399	____	____
59.	$400–$499	____	____
60.	$500–$599	____	____

61. Use the table from Exercises 51 through 54 to construct a histogram. See Example 7.

Number of Games

2
1

Golf Scores

62. Use the table from Exercises 55 through 60 to construct a histogram. See Example 7.

Number of People

2
1

Account Balances

D *The following line graph shows the World Cup goals per game average during the years shown. Use this graph to answer Exercises 63 through 68. See Example 8.*

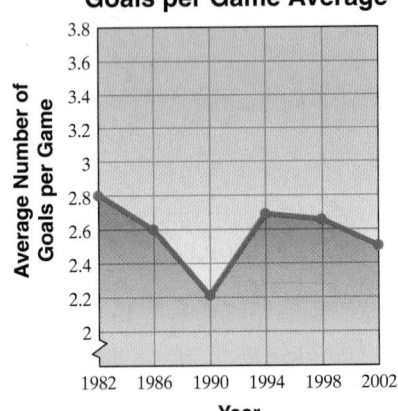

**World Cup
Goals per Game Average**

Average Number of
Goals per Game

3.8
3.6
3.4
3.2
3
2.8
2.6
2.4
2.2
2

1982 1986 1990 1994 1998 2002

Year

Source: *Soccer America Magazine*

63. Find the average number of goals per game in 1994.

64. Find the average number of goals per game in 2002.

65. During what year shown was the average number of goals per game the highest?

66. During what year shown was the average number of goals per game the lowest?

67. Between 1998 and 2000, did the average number of goals per game increase or decrease?

68. Between 1990 and 1994, did the average number of goals per game increase or decrease?

Review and Preview

Find each percent. See Sections 7.2 and 7.3.

69. 30% of 12

70. 45% of 120

71. 10% of 62

72. 95% of 50

Write each fraction as a percent. See Section 7.1.

73. $\frac{1}{4}$

74. $\frac{2}{5}$

75. $\frac{17}{50}$

76. $\frac{9}{10}$

Combining Concepts

The following double-line graph shows temperature highs and lows for a week. Use this graph to answer Exercises 77 through 82.

77. What was the high temperature reading on Thursday?

78. What was the low temperature reading on Thursday?

79. What day was the temperature the lowest? What was this low temperature?

80. What day of the week was the temperature the highest? What was this high temperature?

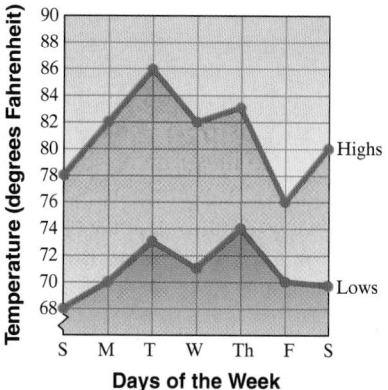

81. On what day of the week was the difference between the high temperature and the low temperature the greatest? What was this difference in temperature?

82. On what day of the week was the difference between the high temperature and the low temperature the least? What was this difference in temperature?

83. True or false? With a bar graph, the width of the bar is just as important as the height of the bar. Explain your answer.

Internet Excursions

 Go To: http://www.prenhall.com/martin-gay_prealgebra What's Related

The Bureau of Labor Statistics, within the U.S. Department of Labor, is the principal fact-finding agency for the Federal Government in the broad field of labor economics and statistics. The World Wide Web address listed here will provide you with access to the "U.S. Economy at a Glance" Web site of the Bureau of Labor Statistics, or a related site. You will find links to graphs of various data series.

84. Visit this Web site and view the graph of "Unemployment Rate." What type of graph is this? Use the graph to estimate when the highest unemployment rate occurred during the period of time covered by the graph, and estimate that unemployment rate.

85. Visit this Web site and view the graph of "Average Hourly Earnings." What type of graph is this? Describe any trends that you see in the graph.

FOCUS ON **the Real World**

MISLEADING GRAPHS

Graphs are very common in magazines and in newspapers such as *USA Today*. Graphs can be a convenient way to illustrate an idea because, as the old saying goes, "A picture is worth a thousand words." However, some graphs can be deceptive, which may or may not be intentional. It is important to know some of the ways that graphs can be misleading.

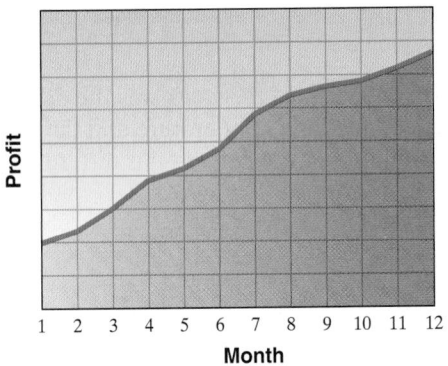

Beware of graphs like the one above. Notice that the graph shows a company's profit for various months. It appears that profit is growing quite rapidly. However, this impressive picture tells us little without knowing what units of profit are being graphed. Does the graph show profit in dollars or millions of dollars? An unethical company with profit increases of only a few pennies could use a graph like this one to make the profit increase seem much more substantial than it really is. A truthful graph describes the size of the units used along the vertical axis.

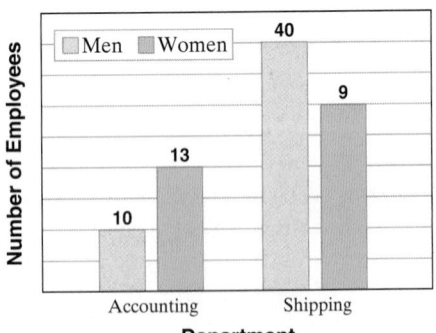

Another type of graph to watch for is one that misrepresents relationships. This can occur in both bar graphs and circle graphs. For example, the bar graph in the bottom of the left column shows the number of men and women employees in the accounting and shipping departments of a certain company. In the accounting department, the bar representing the number of women is twice as tall as the bar representing the number of men. However, the number of women (13) is not twice the number of men (10). This set of bars misrepresents the relationship between the number of men and women. Do you see how the relationship between the number of men and women in the shipping department is distorted by the heights of the bars used? A truthful graph will use bar heights or circle sectors that are proportional in size to the numbers they represent.

We have already seen that the impression a graph can give also depends on its vertical scale. Here is another example: The two graphs below represent exactly the same data. The only difference between the two graphs is the vertical scale—one shows enrollments from 246 to 260 students, and the other shows enrollments between 0 and 300 students. If you were trying to convince readers that algebra enrollment at UPH had changed drastically over the period 2000–2004, which graph would you use? Which graph do you think gives the more honest representation?

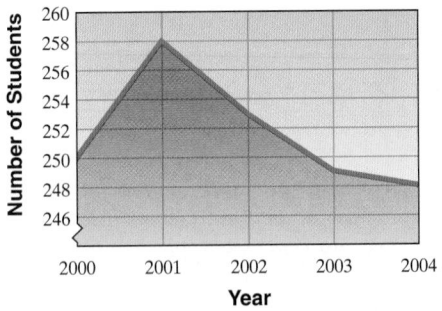

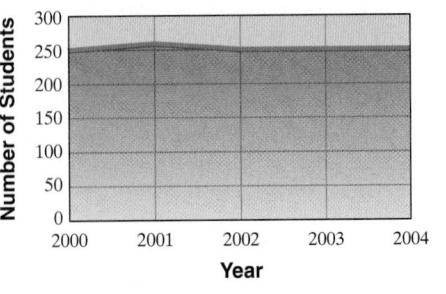

8.2   Reading Circle Graphs

Ⓐ   Reading Circle Graphs

In Section 7.1, the following graph was shown. This particular graph shows the favorite cookie for every 100 people.

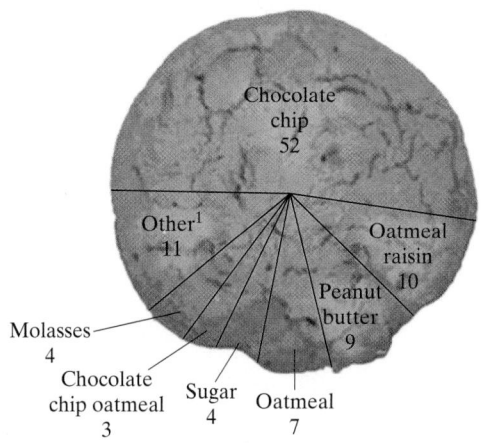

Chocolate chip
52

Other[1]
11

Oatmeal raisin
10

Molasses
4

Peanut butter
9

Chocolate chip oatmeal
3

Sugar
4

Oatmeal
7

[1]–all 1% or less

Each sector of the graph (shaped like a piece of pie) shows a category and the relative size of the category. In other words, the most popular cookie is the chocolate chip cookie because it is represented by the largest sector.

EXAMPLE 1   Find the ratio of people preferring chocolate chip cookies to the total number of people. Write the ratio as a fraction in simplest form.

Solution: The ratio is

$$\frac{52 \text{ people preferring chocolate chip}}{100 \text{ people}} = \frac{52}{100} = \frac{13}{25}$$ ●

A circle graph is often used to show percents in different categories, with the whole circle representing 100%. For example, in 2002 the population of the United States was about 288,400,000. The following circle graph shows the percent of Americans with various numbers of working computers at home. Notice that the percents in each category sum to 100%.

Number of Working Computers at Home

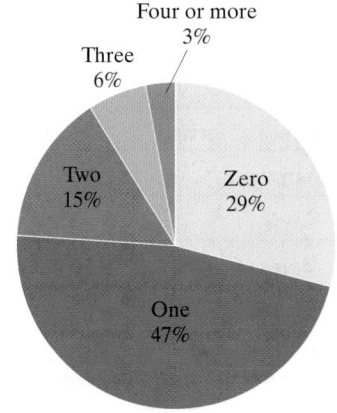

Four or more
3%

Three
6%

Two
15%

Zero
29%

One
47%

Source: UCLA Center for Communication Policy, 2003

Practice Problem 1

Find the ratio of people preferring oatmeal raisin cookies to total people. Write the ratio as a fraction in simplest form.

Answer

1. $\dfrac{1}{10}$

Practice Problem 2

Using the circle graph for Example 2, determine the percent of Americans that have two or more working computers at home.

Practice Problem 3

Using the circle graph for Example 2, find the number of Americans that have four or more working computers at home.

Concept Check

Can the following data be represented by a circle graph? Why or why not?

Responses to the question, "In which activities are you involved?"	
Intramural sports	60%
On-campus job	42%
Fraternity/sorority	27%
Academic clubs	21%
Music programs	14%

Practice Problem 4

Use the data shown to draw a circle graph.

Freshmen	30%
Sophomores	27%
Juniors	25%
Seniors	18%

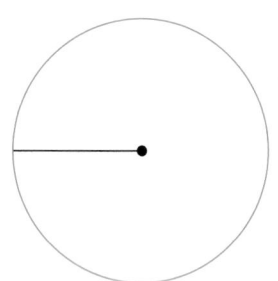

Answers

2. 24%   **3.** 8,652,000 Americans   **4.** see next page

Concept Check: no; the percents add up to more than 100%

EXAMPLE 2   Using the circle graph from the previous page, determine the percent of Americans that have one or more working computers at home.

Solution: To find this percent, we add the percents corresponding to one, two, three, and four or more working computers at home. The percent of Americans that have one or more working computers at home is

$$47\% + 15\% + 6\% + 3\% = 71\%$$ ●

EXAMPLE 3   Using the circle graph for Example 2, find the *number* of Americans that have no working computers at home.

Solution: Since the *percent* of Americans with no computer is 29%, we find the *number* of Americans by finding 29% of the population. To do this, we can use the percent equation.

amount	=	percent	·	base
amount	=	0.29	·	288,400,000

$$= 0.29(288,400,000)$$
$$= 83,636,000$$

Thus, 83,636,000 Americans have no working computer at home.   ●

Try the Concept Check in the margin.

B Drawing Circle Graphs

To draw a circle graph, we use the fact that a whole circle contains 360° (degrees).

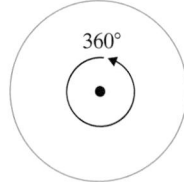

EXAMPLE 4   The following table shows the percent of U.S. armed forces personnel that are in each branch of service. (*Source:* U.S. Department of Defense)

Branch of Service	Percent
Army	33%
Navy	27%
Marine Corps	12%
Air Force	25%
Coast Guard	3%

Draw a circle graph showing this data.

Solution: First we find the number of degrees in each sector representing each branch of service. Remember that the whole circle contains 360°. (We will round degrees to the nearest whole.)

Sector	Degrees in Each Sector
Army	33% × 360° = 118.8° ≈ 119°
Navy	27% × 360° = 97.2° ≈ 97°
Marine Corps	12% × 360° = 43.2° ≈ 43°
Air Force	25% × 360° = 90° = 90°
Coast Guard	3% × 360° = 10.8° ≈ 11°

Next we draw a circle and mark its center. Then we draw a line from the center of the circle to the circle itself (that is, a radius).

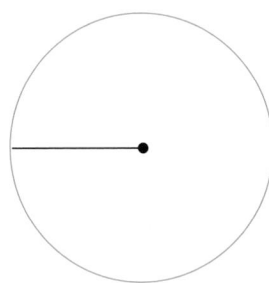

To construct the sectors, we will use a **protractor**. Recall that a protractor measures the number of degrees in an angle. We place the hole in the protractor over the center of the circle. Then we adjust the protractor so that 0° on the protractor is aligned with the line that we drew.

It makes no difference which sector we draw first. To construct the "Army" sector, we find 119° on the protractor and mark our circle. Then we remove the protractor and use this mark to draw a second line from the center to the circle itself.

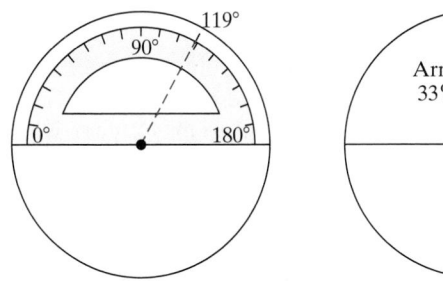

To construct the "Navy" sector, we follow the same procedure as above, except that we line up 0° with the second line we drew and mark the protractor at 97°.

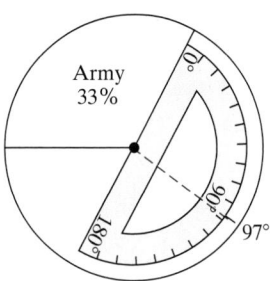
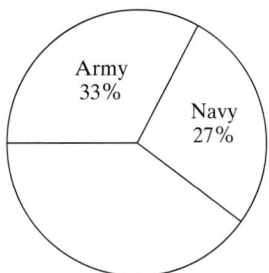

Helpful Hint

Check your calculations by finding the sum of the degrees.

119° + 97° + 43° + 90° + 11° = 360°.

The sum should be 360°. (It may vary only slightly because of rounding.)

Answer

4.

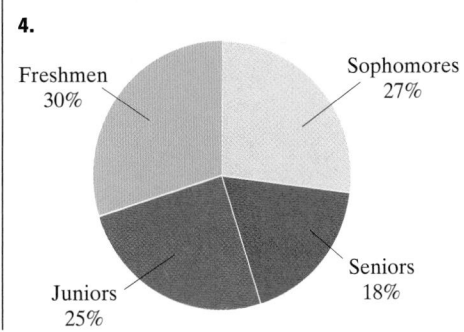

We continue in this manner until the circle graph is complete.

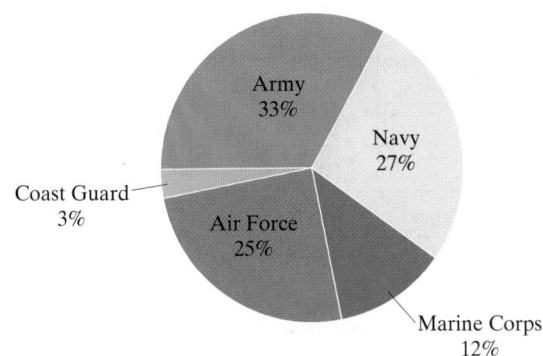

Concept Check

True or false? The larger a sector in a circle graph, the larger the percent of the total it represents. Explain your answer.

 Try the Concept Check in the margin.

STUDY SKILLS REMINDER

Are you satisfied with your performance on a particular quiz or exam?

If not, don't forget to analyze your quiz or exam and look for common errors.

Were most of your errors a result of

■ *Carelessness*? If your errors were careless, did you turn in your work before the allotted time expired? If so, resolve to use the entire time allotted next time. Any extra time can be spent checking your work.

■ *Running out of time*? If so, make a point to better manage your time on your next exam. A few suggestions are to work any questions that you are unsure of last and to check your work after all questions have been answered.

■ *Not understanding a concept*? If so, review that concept and correct your work. Remember next time to make sure that all concepts on a quiz or exam are understood before the exam.

■ *Test conditions*? When studying for your test, are you placing yourself in conditions similar to test conditions? In other words, once you feel that you know the material, use a few sheets of blank paper and take a sample test. (A sample test can be one provided by your instructor or you may use the Chapter Test found at the end of each chapter.)

Concept Check: true

Name _____ Section _____ Date _____

EXERCISE SET 8.2

A *The following circle graph is a result of surveying 700 college students. They were asked where they live while attending college. Use this graph to answer Exercises 1 through 6. Write all ratios as fractions in simplest form. See Example 1.*

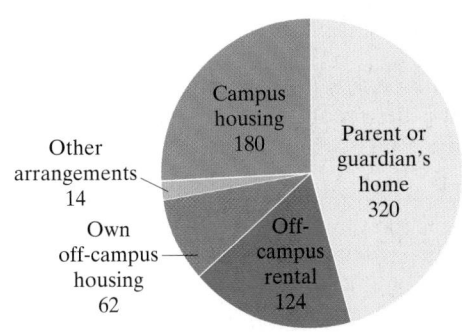

 1. Where do most of these college students live?

2. Besides the category "Other Arrangements," where do the fewest of these college students live?

3. Find the ratio of students living in campus housing to total students.

4. Find the ratio of students living in off-campus rentals to total students.

5. Find the ratio of students living in campus housing to students living in a parent or guardian's home.

6. Find the ratio of students living in off-campus rentals to students living in a parent or guardian's home.

The following circle graph shows the percent of Earth's land in each of the continents. Use this graph for Exercises 7 through 14. See Example 2.

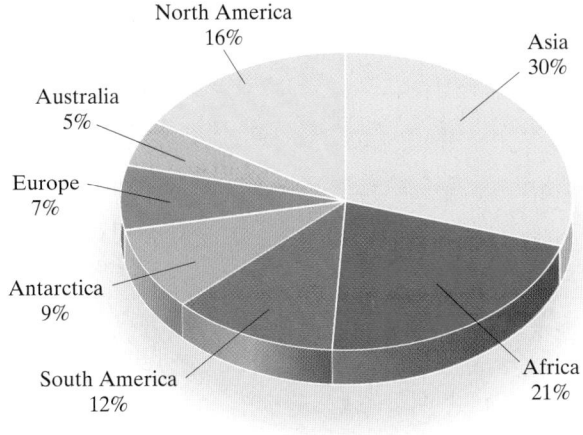

Source: National Geographic Society

7. Which continent is the largest?

8. Which continent is the smallest?

9. What percent of the land on Earth is accounted for by Asia and Europe together?

10. What percent of the land on Earth is accounted for by North and South America?

The total amount of land on Earth is approximately 57,000,000 square miles. Use the graph to find the area of the continents given in Exercises 11 through 14. See Example 3.

11. Asia

12. South America

13. Australia

14. Europe

The following circle graph shows the percent of the types of books available at Midway Memorial Library. Use this graph for Exercises 15 through 24. See Example 2.

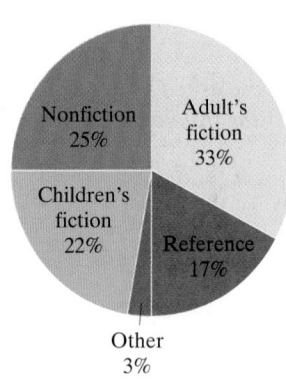

15. What percent of books are classified as some type of fiction?

16. What percent of books are nonfiction or reference?

17. What is the second-largest category of books?

18. What is the third-largest category of books?

If this library has 125,600 books, find how many books are in each category given in Exercises 19 through 24. See Example 3.

19. Nonfiction

20. Reference

21. Children's fiction

22. Adult's fiction

23. Reference or other

24. Nonfiction or other

B   *Fill in the table. Round to the nearest degree. Then draw a circle graph to represent the information given in each table. (Remember: The total of the Degrees in Sector column should equal 360° or very close to 360° because of rounding.) See Example 4.*

25.

2001 U.S. Car Sales by Vehicle Origin		
Country of Origin	**Percent**	**Degrees in Sector**
United States	75%	
Japan	11%	
Germany	6%	
Other Countries	8%	

(*Source:* Ward's AutoInfoBank)

26.

Size of Kellogg's Business Segments after Acquiring Keebler		
Business Segment	**Percent of Annual Sales**	**Degrees in Sector**
U.S. cereal	27%	
U.S. convenience foods	43%	
International	30%	

(*Source:* Kellogg Company's Annual Report)

27.

Absence from Work Due to Carpal Tunnel Syndrome		
Time	**Percent**	**Degrees in Sector**
Under 3 days	7%	
3–10 days	18%	
11–20 days	14%	
21 or more days	61%	

(*Note:* Carpal tunnel syndrome is a nerve disorder causing pain mostly in the wrist and hand.)
(*Source:* Bureau of Labor Statistics)

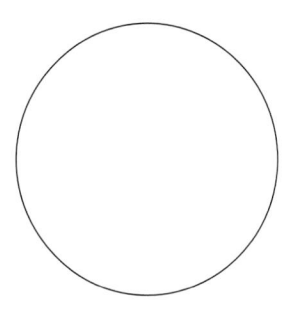

28.

Most Important Reason for Picking a Favorite Restaurant Among Wendy's, Burger King, and McDonald's		
Reason	**Percent**	**Degrees in Sector**
Convenience	13%	
Taste of food	70%	
Price	9%	
Atmosphere	4%	
Others	4%	

(*Source:* TeleNation)

Review and Preview

Write the prime factorization of each number. See Section 4.2.

29. 20   **30.** 25   **31.** 40   **32.** 16   **33.** 85   **34.** 105

Combining Concepts

The following circle graph shows the relative sizes of the great oceans. These oceans together make up 264,489,800 square kilometers of Earth's surface. Find the square kilometers for each ocean.

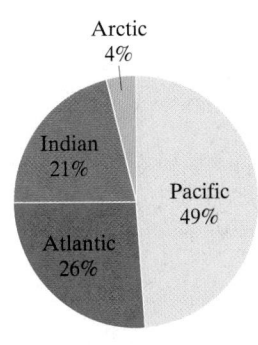

Arctic 4%
Indian 21%
Pacific 49%
Atlantic 26%

Source: *Philip's World Atlas*

35. Before actually calculating, determine which ocean is the largest. How can you answer this question by looking at the circle graph?

36. Before calculating, determine which ocean is the smallest. How can you answer this question by looking at the circle graph?

37. Pacific Ocean   **38.** Atlantic Ocean   **39.** Indian Ocean   **40.** Arctic Ocean

The following circle graph summarizes the results of a survey of 2800 Internet users who make purchases online. Use this graph for Exercises 41 through 46. Round to the nearest whole.

Online Spending per Month

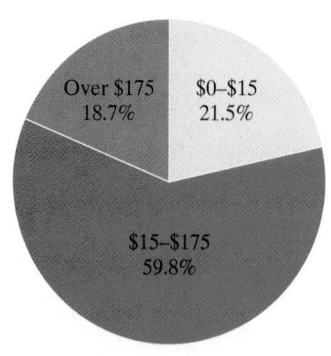

Over $175
18.7%

$0–$15
21.5%

$15–$175
59.8%

Source: UCLA Center for Communication Policy

41. How many of the survey respondents said that they spend $0–$15 online each month?

42. How many of the survey respondents said that they spend $15–$175 online each month?

43. How many of the survey respondents said that they spend at least $15 or over $175 online each month?

44. Find the ratio of *percent* of respondents who spend $0–$15 online to *percent* of those who spend $15–$175. Write the ratio as a fraction with integers in the numerator and denominator.

45. Find the ratio of *number* of respondents who spend $0–$15 online to *number* of respondents who spend $15–$175 online. Write the ratio as a fraction with integers in the numerator and denominator.

46. Compare the ratios in Exercises 44 and 45. How do they compare and why?

47. Can the following data be represented by a circle graph? Why or why not?

Responses to the Question "In which activities are you involved?"	
Intramural sports	60%
On-campus job	42%
Fraternity/sorority	27%
Academic clubs	21%
Music programs	14%

8.3 The Rectangular Coordinate System

Ⓐ Plotting Points

In the last section, we saw how bar and line graphs can be used to show relationships between items listed on the horizontal and vertical axes. We can use this same horizontal and vertical axis idea to describe the location of points in a plane.

The system that we use to describe the location of points in a plane is called the **rectangular coordinate system**. It consists of two number lines, one horizontal and one vertical, intersecting at the point 0 on each number line. This point of intersection is called the **origin**. We call the horizontal number line the **x-axis** and the vertical number line the **y-axis**. Notice that the axes divide the plane into four regions, called **quadrants**. They are numbered as shown.

OBJECTIVES

Ⓐ Plot points on a rectangular coordinate system.

Ⓑ Determine whether ordered pairs are solutions of equations.

Ⓒ Complete ordered-pair solutions of equations.

SSM   TUTOR CENTER   SG   CD & VIDEO   MATH PRO   WEB

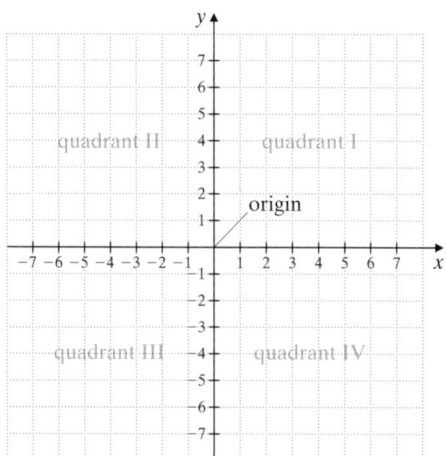

Every point in the rectangular coordinate system corresponds to an **ordered pair of numbers**, such as (3, 4). The first number, 3, of an ordered pair is associated with the x-axis and is called the **x-coordinate** or **x-value**. The second number, 4, is associated with the y-axis and is called the **y-coordinate** or **y-value**. To find the **single point** on the rectangular coordinate system corresponding to the ordered pair (3, 4), start at the origin. Move 3 units in the positive direction along the x-axis. From there, move 4 units in the positive direction parallel to the y-axis as shown to the right. This process of locating a point on the rectangular coordinate system is called **plotting the point**. Since the origin is located at 0 on the x-axis and 0 on the y-axis, the origin corresponds to the ordered pair (0, 0).

In general, to plot the ordered pair (x, y), start at the origin. Next,

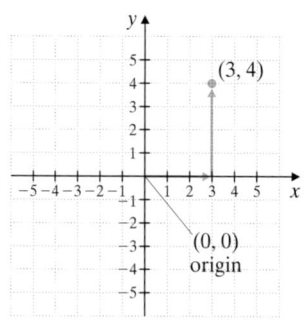

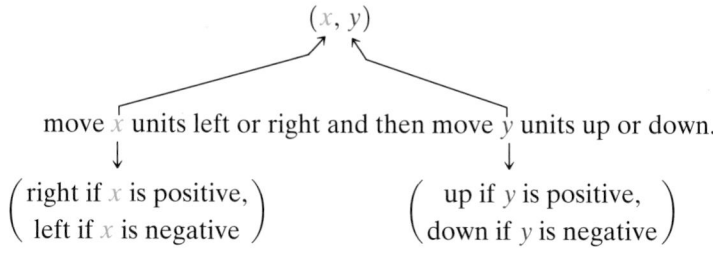

Since the first number, or x-coordinate, of an ordered pair is associated with the x-axis, it tells how many units to move left or right. Similarly, the second number, or y-coordinate, tells how many units to move up or down.

To plot $(-1, 5)$, start at the origin. Move 1 unit left (because the x-coordinate is negative) and then 5 units up and draw a dot at that point. This dot is the graph of the point that corresponds to the ordered pair $(-1, 5)$.

To plot $(0, -3)$, start at the origin, move 0 units (left or right), then 3 units down, then draw a dot.

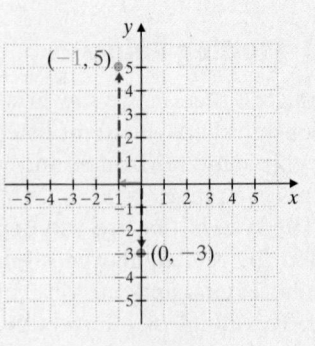

Below are some more plotted points with their corresponding ordered pairs. The rectangular coordinates system is also called the Cartesian coordinate system. This name comes from French scientist, philosopher, and mathematician René Descartes (1596–1650), who introduced the system.

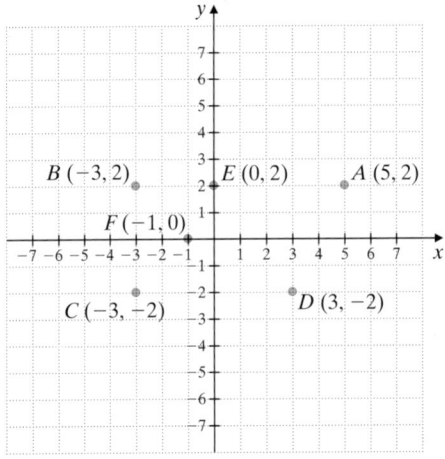

EXAMPLE 1 Plot each point corresponding to the ordered pairs on the same set of axes.

$$(5, 4), (-2, 3), (-1, -2), (6, -3), (0, 2), (-4, 0), \left(3, \frac{1}{2}\right)$$

Solution:

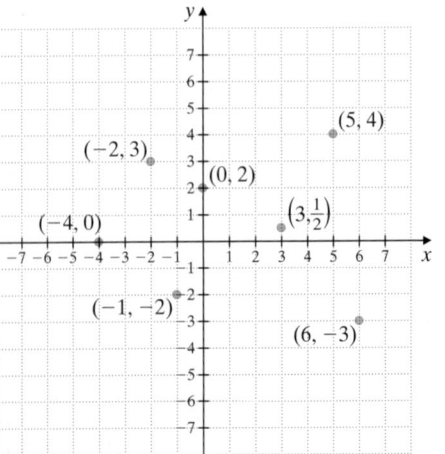

Helpful Hint

Remember that **each point** in the rectangular coordinate system corresponds to exactly **one ordered pair** and that **each ordered pair** corresponds to exactly **one point**.

Practice Problem 1

Plot each point corresponding to the ordered pairs on the same set of axes.
$(1, 3), (-3, 2)\ (-6, -5)\ (2, -2)\ (5, 0)$
$(0, -3)$

Answer

1.

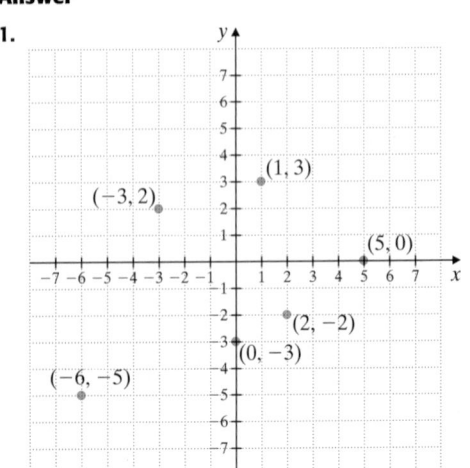

EXAMPLE 2   Find the ordered pair corresponding to each point plotted on the rectangular coordinate system.

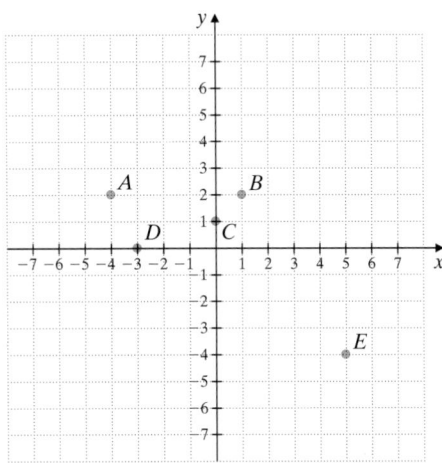

Practice Problem 2

Find the ordered pair corresponding to each point plotted on the rectangular coordinate system.

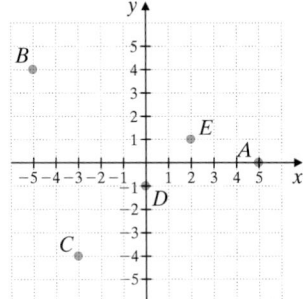

Solution:

Point A has coordinates $(-4, 2)$.
Point B has coordinates $(1, 2)$.
Point C has coordinates $(0, 1)$.
Point D has coordinates $(-3, 0)$.
Point E has coordinates $(5, -4)$

Have you noticed a pattern from the preceding examples? **If an ordered pair has a y-coordinate of 0, its graph lies on the x-axis. If an ordered pair has an x-coordinate of 0, its graph lies on the y-axis.**

Helpful Hint

Order is the key word in ordered pair. The first value always corresponds to the x-value and the second value always corresponds to the y-value. For example, $(-2, 4)$ does not describe the same point as $(4, -2)$.

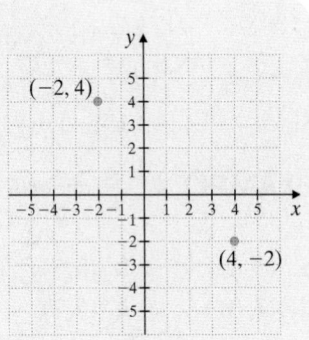

B   Determining Whether an Ordered Pair Is a Solution

Let's see how we can use ordered pairs of numbers to record solutions of equations containing two variables.

Recall that an equation with one variable, such as $3 + x = 5$, has one solution—in this case, 2 because 2 is the only value that can be substituted for x so that the resulting equation $3 + 2 = 5$ is a true statement. We can graph this solution on a number line.

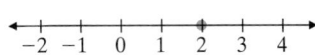

Answers

2. $A(5, 0)$; $B(-5, 4)$; $C(-3, -4)$; $D(0, -1)$; $E(2, 1)$

An equation with two variables such as $x + y = 7$ has many solutions. Each solution is a pair of numbers, one for each variable that makes the equation a true statement. For example, $x = 4$ and $y = 3$ is a solution of the equation $x + y = 7$ because, when x is replaced with 4 and y is replaced with 3, $x + y = 7$ becomes $4 + 3 = 7$, which is a true statement. We can write the solution $x = 4$ and $y = 3$ as the ordered pair $(4, 3)$. Study the chart below for more solutions.

Replacement Values	$x + y = 7$	True or False	Ordered-Pair Solution
$x = 4, y = 3$	$4 + 3 = 7$	True	$(4, 3)$
$x = 0, y = 7$	$0 + 7 = 7$	True	$(0, 7)$
$x = 7, y = 1$	$7 + 1 = 7$	False	No
$x = 11, y = -4$	$11 + (-4) = 7$	True	$(11, -4)$

In general, we say that an ordered pair of numbers is a solution of an equation if the equation is a true statement when the variables are replaced by the coordinates of the ordered pair.

Unlike equations with one variable, equations with two variables, such as $x + y = 7$, have so many solutions we cannot simply list them. Instead, to "see" all the solutions of a two-variable equation, we draw a graph of its solutions.

First, let's practice deciding whether an ordered pair of numbers is a solution of an equation in two variables.

EXAMPLE 3   Is $(-1, 6)$ a solution of the equation $2x + y = 4$?

Solution: Replace x with -1 and y with 6 in the equation $2x + y = 4$ to see if the result is a true statement.

$$2x + y = 4 \qquad \text{Original equation.}$$
$$2(-1) + 6 = 4 \qquad \text{Replace } x \text{ with } -1 \text{ and } y \text{ with 6.}$$
$$-2 + 6 = 4 \qquad \text{Multiply.}$$
$$4 = 4 \qquad \text{True.}$$

Since $4 = 4$ is true, $(-1, 6)$ is a solution of the equation $2x + y = 4$.   ●

EXAMPLE 4   Each ordered pair listed is a solution of the equation $x + y = 7$. Plot them on the same set of axes.

a. $(3, 4)$ **b.** $(0, 7)$ **c.** $(-1, 8)$

d. $(7, 0)$ **e.** $(5, 2)$ **f.** $(4, 3)$

Solution:

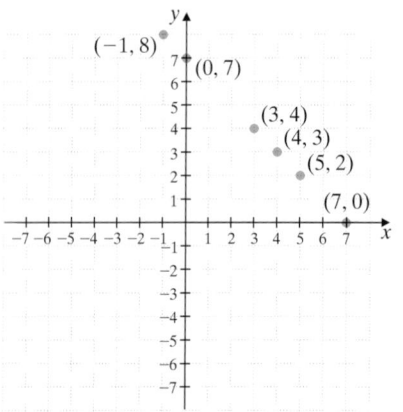

Practice Problem 3

Is $(0, -4)$ a solution of the equation $x + 3y = -12$?

Practice Problem 4

Each ordered pair listed is a solution of the equation $x - y = 5$. Plot them on the same set of axes.

a. $(6, 1)$ b. $(5, 0)$
c. $(0, -5)$ d. $(7, 2)$
e. $(-1, -6)$ f. $(2, -3)$

Answers

3. yes

4.

Notice that the points in Example 4 all seem to lie on the same line. We will discuss this more in the next section.

An equation such as $x + y = 7$ is called a **linear or first-degree equation in two variables**. It is called a **first-degree equation** because the exponent on x and y is an understood 1. The equation is called an equation **in two variables** because it contains two different variables, x and y.

© Completing Ordered-Pair Solutions

In the next section, we will graph linear equations in two variables. Before that, we need to practice finding ordered-pair solutions of these equations. If one coordinate of an ordered-pair solution is known, the other coordinate can be determined. To find the unknown coordinate, replace the appropriate variable with the known coordinate in the equation. Doing so results in an equation with one variable that we can solve.

EXAMPLE 5   Complete each ordered-pair solution of the equation $y = 2x$.

a. (3,)   **b.** (, 0)   **c.** (−2,)

Solution:

a. In the ordered pair (3,), the x-value is 3. To find the corresponding y-value, let $x = 3$ in the equation $y = 2x$ and calculate the value of y.

$y = 2x$ Original equation.

$y = 2(3)$ Replace x with 3.

$y = 6$ Multiply.

Check: To check, replace x with 3 and y with 6 in the original equation to see that a true statement results.

$y = 2x$ Original equation.

$6 \stackrel{?}{=} 2(3)$ Let $x = 3$ and $y = 6$.

$6 = 6$ True.

The ordered-pair solution is $(3, 6)$.

b. In the ordered pair (, 0), the y-value is 0. To find the corresponding x-value, replace y with 0 in the equation and solve for x.

$y = 2x$ Original equation.

$0 = 2x$ Replace y with 0.

$\dfrac{0}{2} = \dfrac{2x}{2}$ Divide both sides by 2.

$0 = x$ Simplify.

The ordered-pair solution is $(0, 0)$, the origin.

c. In the ordered pair (−2,), the x-value is −2. To find the corresponding y-value, replace x with −2 in the equation and calculate the value for y.

$y = 2x$ Original equation.

$y = 2(-2)$ Replace x with −2.

$y = -4$ Multiply.

The ordered-pair solution is $(-2, -4)$.

Practice Problem 5

Complete each ordered-pair solution of the equation $x + y = 10$.

a. (5,)   b. (0,)   c. (, −2)

Answers

5. a. $(5, 5)$ **b.** $(0, 10)$ **c.** $(12, -2)$

Practice Problem 6

Complete each ordered-pair solution of the equation $y = 5x + 2$.

a. $(0, \ )$ b. $(\ , -3)$

EXAMPLE 6 Complete each ordered-pair solution of the equation $y = 3x - 4$.

a. $(2, \ )$ b. $(\ , -16)$

Solution:

a. Replace x with 2 in the equation and calculate the value for y.

$y = 3x - 4$ Original equation.

$y = 3 \cdot 2 - 4$ Replace x with 2.

$y = 2$ Simplify.

The ordered-pair solution is $(2, 2)$.

b. Replace y with -16 in the equation and solve for x.

$y = 3x - 4$ Original equation.

$-16 = 3x - 4$ Replace y with -16.

$-16 + 4 = 3x - 4 + 4$ Add 4 to both sides.

$-12 = 3x$ Simplify.

$\dfrac{-12}{3} = \dfrac{3x}{3}$ Divide both sides by 3.

$-4 = x$ Simplify.

The ordered-pair solution is $(-4, -16)$.

Answers

6. a. $(0, 2)$ **b.** $(-1, -3)$

Name _____ Section _____ Date _____

EXERCISE SET 8.3

A *Plot the points corresponding to the ordered pairs on the same set of axes. See Example 1.*

1. $(1, 3), (-2, 4), (0, 7), (-5, 0), (-6, -3), (5, -5)$
see graph

2. $(5, 2), (3, -4), (-1, -1), (0, -6), (4, 0), (-2, 4)$
see graph

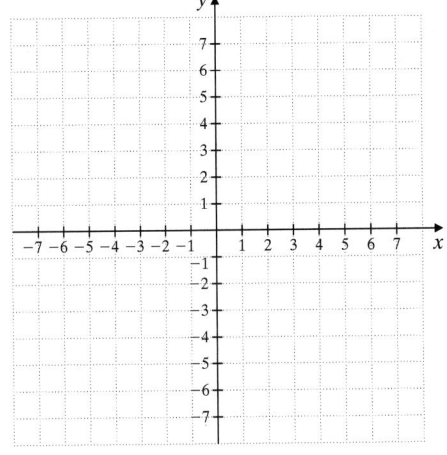

3. $\left(2\frac{1}{2}, 3\right), (0, -3), (-4, -6), \left(-1, 5\frac{1}{2}\right), (1, 0), (3, -5)$
see graph

4. $\left(5, \frac{1}{2}\right), \left(-3\frac{1}{2}, 0\right), (-1, 4), (4, -1), (0, 2), (-5, -5)$
see graph

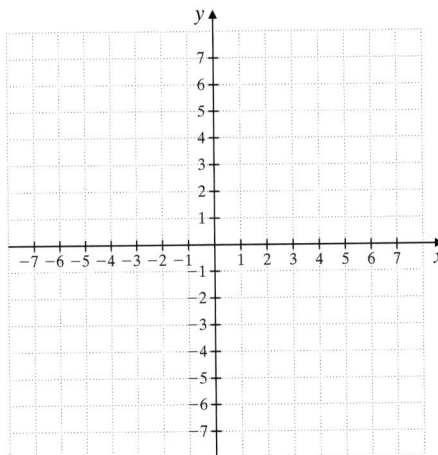

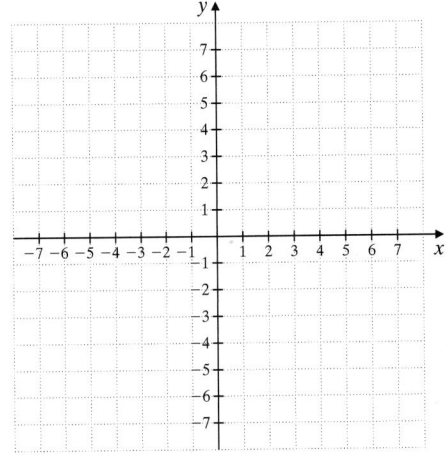

Find the x- and y-coordinates of each labeled point. See Example 2.

5.

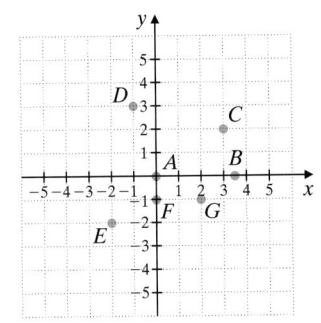

6.

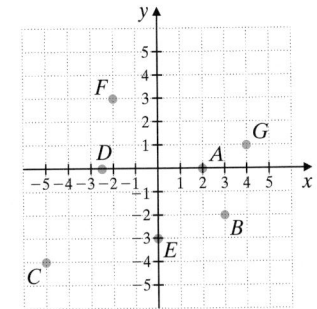

7. $(0, 0)$; $y = -10x$

8. $(1, 7)$; $x = 7y$

9. $(1, 2)$; $x - y = 3$

10. $(-1, 9)$; $x + y = 8$

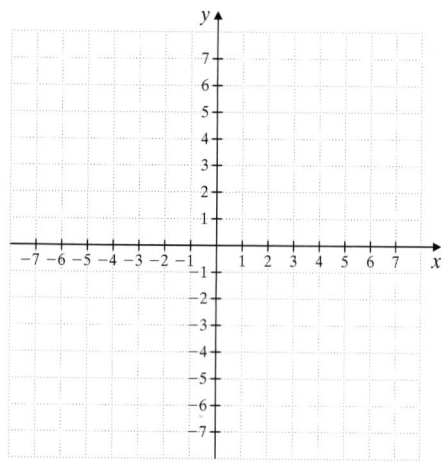 **11.** $(-2, -3)$; $y = 2x + 1$

12. $(1, 1)$; $y = -x$

13. $(2, -8)$; $y = -4x$

14. $(9, 1)$; $x = 9y$

15. $(5, 0)$; $3y + 2x = 10$

16. $(1, 1)$; $-5y + 4x = -1$

17. $(3, 1)$; $x - 5y = -1$

18. $(0, 2)$; $x - 7y = -15$

Plot the three ordered-pair solutions of the given equation. See Example 4.

19. $2x + y = 5$; $(1, 3), (0, 5), (3, -1)$
see graph

20. $x + y = 5$; $(-1, 6), (0, 5), (4, 1)$
see graph

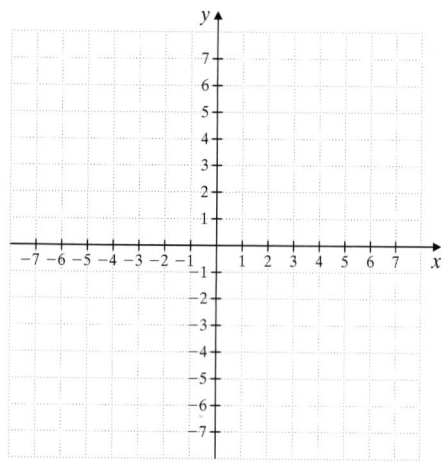

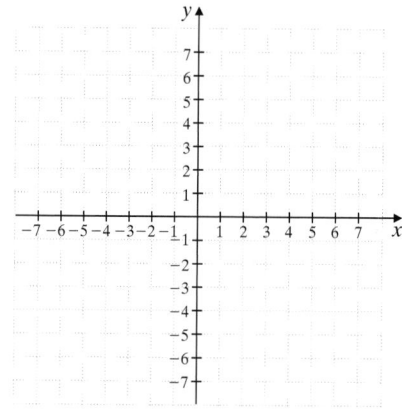

21. $x = 5y$; $(5, 1), (0, 0), (-5, -1)$
see graph

22. $y = -3x$; $(1, -3), (2, -6), (-1, 3)$
see graph

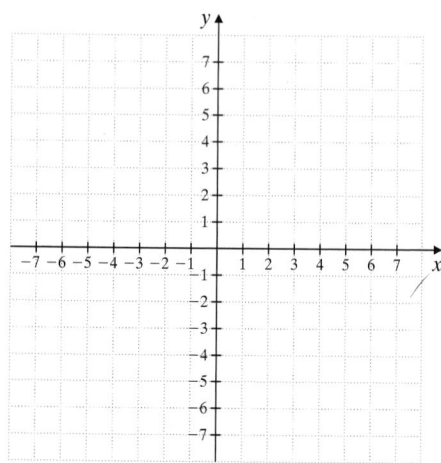

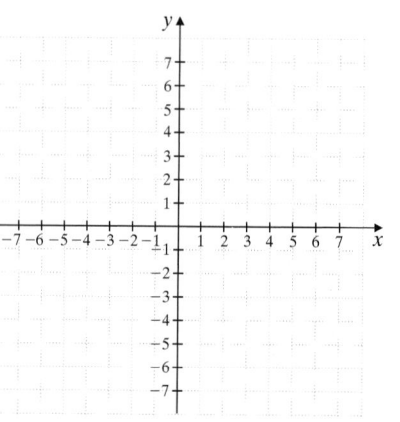

23. $x - y = 7$; $(8, 1)$, $(2, -5)$, $(0, -7)$
    see graph

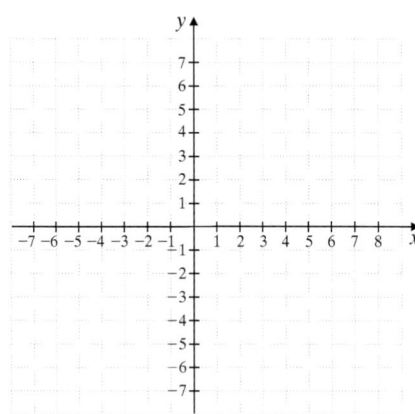

24. $y = 3x + 1$; $(0, 1)$, $(1, 4)$, $(-1, -2)$
    see graph

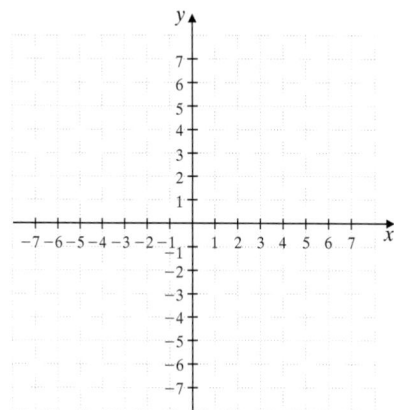

C *Complete each ordered-pair solution of the given equations. See Examples 5 and 6.*

25. $y = 8x$; $(1, \quad)$, $(0, \quad)$, $(\quad, -16)$

26. $x = -7y$; $(\quad, 2)$, $(14, \quad)$, $(\quad, -1)$

27. $x + y = 14$; $(2, \quad)$, $(\quad, -8)$, $(0, \quad)$

28. $x - y = 8$; $(0, \quad)$, $(\quad, 0)$, $(5, \quad)$

29. $y = x + 5$; $(1, \quad)$, $(\quad, 7)$, $(3, \quad)$

30. $x = y$; $(-8, \quad)$, $(\quad, 3)$, $(100, \quad)$

31. $y = 3x - 5$; $(1, \quad)$, $(2, \quad)$, $(3, \quad)$

32. $x = -12y$; $(\quad, -1)$, $(\quad, 1)$, $(36, \quad)$

33. $y = -x$; $(\quad, 0)$, $(2, \quad)$, $(\quad, 2)$

34. $y = 5x + 1$; $(0, \quad)$, $(-1, \quad)$, $(2, \quad)$

35. $x + y = -2$; $(-2, \quad)$, $(1, \quad)$, $(\quad, 5)$

36. $x - y = -3$; $(\quad, 0)$, $(0, \quad)$, $(4, \quad)$

Review and Preview

Perform each indicated operation on decimals. See Sections 5.2 to 5.4.

37. $5.6 - 3.9$

38. $5 + 2.54 + 8.7$

39. $5.6 \cdot 3.9$

40. $0.56 \div 0.8$

41. $(0.236)(-100)$

42. $44.72 \div 100$

Combining Concepts

Recall that the axes divide the plane into four quadrants as shown. If a and b are both positive numbers, determine whether each statement is true or false.

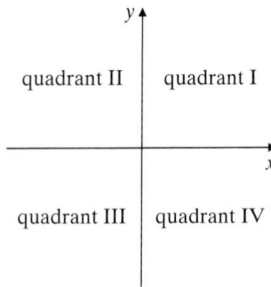

43. (a, b) lies in quadrant I.

44. $(-a, -b)$ lies in quadrant IV.

45. $(0, b)$ lies on the *y*-axis.

46. $(a, 0)$ lies on the *x*-axis.

47. $(0, -b)$ lies on the *x*-axis.

48. $(-a, 0)$ lies on the *y*-axis.

49. $(-a, b)$ lies in quadrant III.

50. $(a, -b)$ lies in quadrant IV.

51. a. Is the ordered pair $(4, -3)$ plotted to the left or right of the *y*-axis? Explain.

   b. Is the ordered pair $(6, -2)$ plotted above or below the *x*-axis? Explain.

52. In your own words, describe how to plot a point.

Plot the points $A(4, 3)$, $B(-2, 3)$, $C(-2, -1)$ and $D(4, -1)$.

53. Draw a segment connecting A and B, B and C, C and D, and D and A. Name the figure drawn.

54. Find the length and the width of the figure.

55. Find the perimeter of the figure.

56. Find the area of the figure.

Integrated Review–Reading Graphs

The following pictograph shows the average number of pounds of beef and veal consumed per person per year in the United States. Use this graph to answer Exercises 1 through 4.

Beef and Veal Consumption

Each 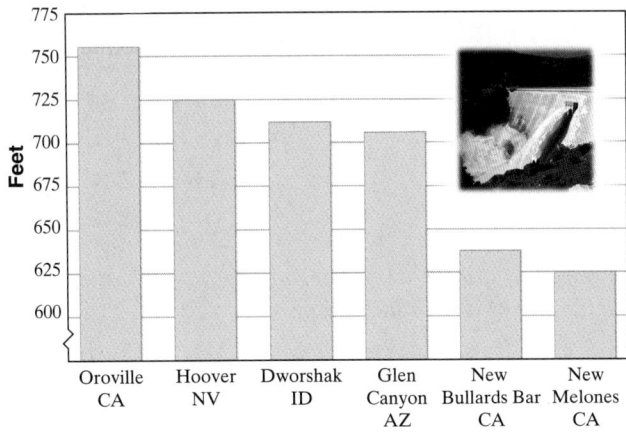 represents 10 pounds

Source: U.S. Department of Agriculture

1. Approximate the number of pounds of beef and veal consumed per person in 1995.

2. Approximate the number of pounds of beef and veal consumed per person in 1980.

3. In what year(s) was the number of pounds consumed the greatest?

4. In what year(s) was the number of pounds consumed the least?

The following bar graph shows the highest U.S. dams. Use this graph to answer Exercises 5 through 8.

Highest U.S. Dams

Source: Committee on Register of Dams

5. Name the U. S. dam with the greatest height and estimate its height.

6. Name the U.S. dam whose height is between 625 and 650 feet and estimate its height.

7. Estimate how much higher the Hoover Dam is than the Glen Canyon Dam.

8. How many U.S. dams have heights over 700 feet?

The following line graph shows the daily high temperatures for 1 week in Annapolis, Maryland. Use this graph to answer Exercises 9 through 12.

Days of the Week

9. Name the day(s) of the week with the highest high temperature and give that temperature.

10. Name the day(s) of the week with the lowest high temperature and give that temperature.

11. On what days of the week was the high temperature less than 90° Fahrenheit?

12. On what days of the week was the temperature greater than 90° Fahrenheit?

The following circle graph shows the type of beverage milk consumed in the United States. Use this graph for Exercises 13 through 16. If a store in Kerrville, Texas, sells 200 quart containers of milk per week, estimate how many quart containers are sold in each category below.

Types of Milk Consumed

Source: U.S. Department of Agriculture

13. Whole milk

14. Skim milk

15. Buttermilk

16. Flavored reduced fat and skim milk

14. _____

15. _____

16. _____

17. see table

18. see table

19. see table

20. see table

21. see table

The following list shows weekly quiz scores for a student in basic college mathematics. Use this list to complete the frequency distribution table.

50	80	71	83	86
67	89	93	88	97
75	80	78	93	99
	53	90		

	Class Intervals (Scores)	Tally	Class Frequency (Number of Quizzes)
17.	50–59	_____	_____
18.	60–69	_____	_____
19.	70–79	_____	_____
20.	80–89	_____	_____
21.	90–99	_____	_____

22. Use the table from Exercises 17 through 21 to construct a histogram.

Quiz Scores

23. Plot the points corresponding to the ordered pairs on the same set of axes.
$(0, 2), (-1, 4)(2, 1)(-3, 0)(-3, -5)$
$(4, -1)$

24. Determine whether the ordered pair $(1, 3)$ is a solution of the linear equation $x = 3y$.

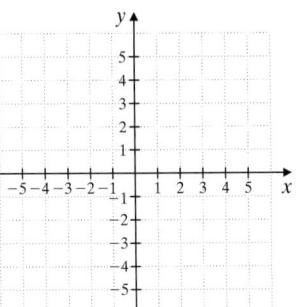

25. Determine whether the ordered pair $(-2, -4)$ is a solution of the linear equation $x + y = -6$.

26. Complete each ordered pair solution of the equation $x - y = 6$.
$(0, \ ), (\ , 0), (2, \ )$

FOCUS ON Mathematical Connections

SCATTER DIAGRAMS

In Section 8.1, we learned about presenting data visually in a graph, such as a pictograph, bar graph, or line graph. Now we learn about another type of graph, based on ordered pairs, that can be used to present data.

Data that can be represented as an ordered pair is called paired data. Many types of data collected from the real world are paired data. For instance, the total amount of rainfall a location receives each year can be written as an ordered pair of the form (year, total rainfall in inches) and is paired data. The graph of paired data as points in the rectangular coordinate system is called a **scatter diagram**, or scatter plot. Scatter diagrams can be used to look for patterns and trends in paired data.

For example, the data shown in the table is paired data that can be written as a set of ordered pairs of the form (year, number of restaurants in thousands), such as (1998, 25) and (1999, 26). A scatter diagram of the paired data is shown below. Notice that the horizontal axis is labeled "Year" to describe the x-coordinates in the ordered pairs. The vertical axis is labeled "Number of Restaurants (in thousands)" to describe the y-coordinates in the ordered pairs.

Number of McDonald's Restaurants	
Year	Number of Restaurants (in thousands)
1998	25
1999	26
2000	29
2001	30
2002	31

(*Source:* McDonald's Corporation)

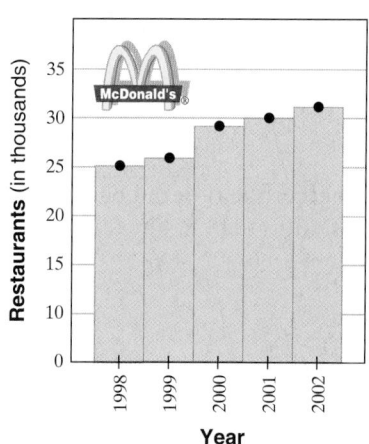

Number of McDonald's Restaurants Worldwide

CRITICAL THINKING

The table gives the annual revenue of McDonald's restaurants in operation worldwide each year. Use the table to answer the following questions.

1. Write this paired data as a set of ordered pairs of the form (year, revenue of restaurants in millions).

2. Create a scatter diagram of the paired data.

3. What trend in the paired data does the scatter diagram show?

GROUP ACTIVITY

4. Find or collect your own paired data and present it in a scatter diagram. What does the graph show?

Annual Revenue of McDonald's Restaurants Worldwide (in millions)	
Year	Annual Revenue of Restaurants (in millions)
1998	$36
1999	$38
2000	$40
2001	$41
2002	$42

(*Source:* McDonald's Corporation)

8.4 Graphing Linear Equations

Now that we know how to plot points in a rectangular coordinate system and how to find ordered-pair solutions, we are ready to graph linear equations in two variables. First, we give a formal definition.

Linear Equation in Two Variables

A *linear equation in two variables* is an equation that can be written in the form

$$ax + by = c \qquad \text{Examples:} \begin{cases} 2x + 3y = 6 \\ 2x = 5 \\ y = 3 \end{cases}$$

where a, b, and c are numbers, and a and b are not both 0.

In the last section, we discovered that a linear equation in two variables has many solutions. For the linear equation $x + y = 7$, we listed, for example, the solutions $(3, 4)$, $(0, 7)$, $(-1, 8)$, $(7, 0)$, $(5, 2)$, and $(4, 3)$. Are these all the solutions? No. There are infinitely many solutions of the equation $x + y = 7$ since there are infinitely many pairs of numbers whose sum is 7. Every linear equation in two variables has infinitely many ordered-pair solutions. Since it is impossible to list every solution, we graph the solutions instead.

Ⓐ Graphing Linear Equations by Plotting Points

Fortunately, the pattern described by the solutions of a linear equation makes "seeing" the solutions possible by graphing. This is so because **all the solutions of a linear equation in two variables correspond to points on a single straight line.** If we plot just a few of these points and draw the straight line connecting them, we have a complete graph of all the solutions.

To graph the equation $x + y = 7$, then, we plot a few ordered-pair solutions, say $(3, 4)$, $(0, 7)$, and $(-1, 8)$.

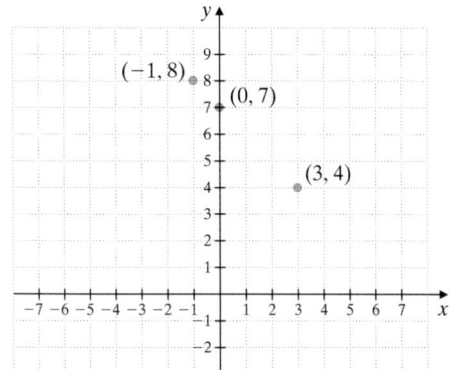

Now we connect these three points by drawing a straight line through them. The arrows at both ends of the line indicate that the line goes on forever in both directions.

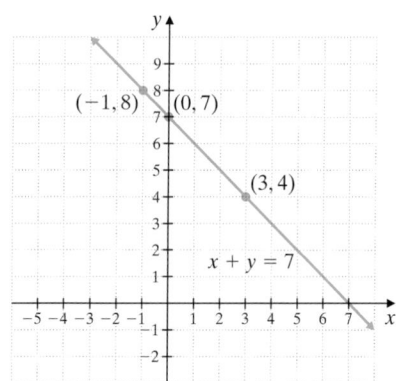

Every point on this line corresponds to an ordered-pair solution of the equation $x + y = 7$. Also, every ordered-pair solution of the equation $x + y = 7$ corresponds to a point on this line. In other words, this line is the graph of the equation $x + y = 7$. Although a line can be drawn using just two points, we will graph a third solution to check our work.

Practice Problem 1

Graph the equation $x - y = 1$ by plotting the following points that satisfy the equation and drawing a line through the points.

$$(3, 2), (0, -1), (1, 0)$$

EXAMPLE 1 Graph the equation $y = 3x$ by plotting the following points that satisfy the equation and drawing a line through the points.

$$(1, 3), (0, 0), (-1, -3)$$

Solution: Plot the points and draw a line through them. The line is the graph of the linear equation $y = 3x$.

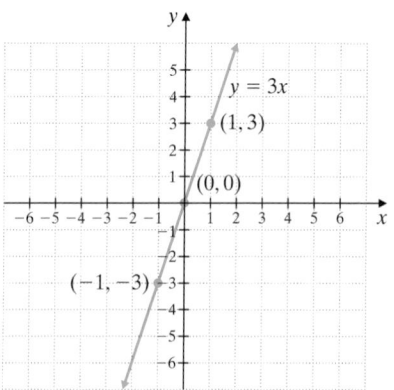

To graph a linear equation in two variables, find three ordered-pair solutions, graph the solutions, and draw the line through the plotted points. To find an ordered-pair solution of an equation, choose either an x-value or y-value of the ordered pair and complete the ordered pair as we did in the previous section.

Answer

1.

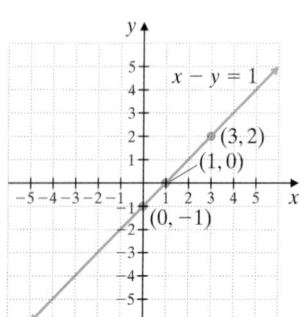

EXAMPLE 2 Graph: $x - y = 6$

Solution: Find any three ordered-pair solutions. For each solution, we choose a value for x or y, and replace x or y by its chosen value in the equation. Then we solve the resulting equation for the other variable.

For example, let $x = 3$.   Also, let $y = 0$.

$x - y = 6$	Original equation.	$x - y = 6$	Original equation.
$3 - y = 6$	Let $x = 3$.	$x - 0 = 6$	Let $y = 0$.
$-y = 6 - 3$	Subtract 3 from both sides.	$x = 6$	Simplify.
$-y = 3$	Simplify.		

The ordered pair is (6, 0).

$$\frac{-y}{-1} = \frac{3}{-1}$$ Divide both sides by -1.

$$y = -3$$ Simplify.

The ordered pair is $(3, -3)$.

If we let $x = 7$, and solve the resulting equation for y, we will see that $y = 1$. **A third ordered-pair solution is, then, $(7, 1)$.**

Plot the three ordered-pair solutions on a rectangular coordinate system and then draw a line through the three points. This line is the graph of $x - y = 6$.

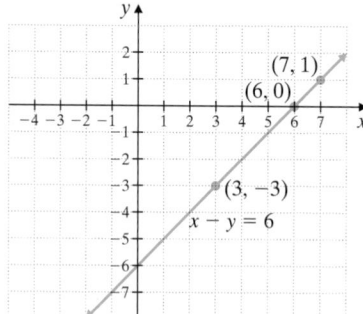

Helpful Hint

All three points should fall on the same straight line. If not, check your ordered-pair solutions for a mistake, since the graph of every linear equation is a line.

Try the Concept Check in the margin.

EXAMPLE 3 Graph: $y = 5x + 1$

Solution: For this equation, we will choose three x-values and find the corresponding y-values. The table shown will be used to list our ordered-pair solutions. We choose $x = 0$, $x = 1$, and $x = 2$.

If $x = 0$, then   If $x = 1$, then   If $x = 2$, then
  $y = 5x + 1$ becomes $y = 5x + 1$ becomes $y = 5x + 1$ becomes
  $y = 5 \cdot 0 + 1$ or $y = 5 \cdot 1 + 1$ or $y = 5 \cdot 2 + 1$ or
  $y = 1$ $y = 6$ $y = 11$

Practice Problem 2

Graph: $y - x = 6$

Concept Check

In Example 2, is the point $(9, -3)$ on the graph of $x - y = 6$?

Practice Problem 3

Graph: $y = -3x + 2$

Answers

2.

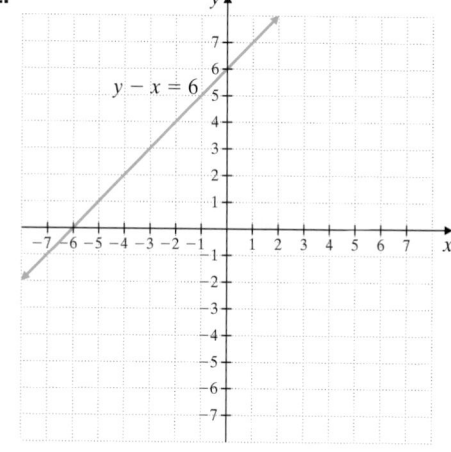

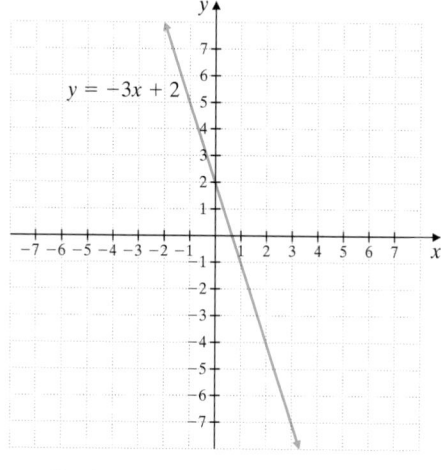

Concept Check: no

Now we complete the table of values, plot the ordered-pair solutions, and graph the equation $y = 5x + 1$.

x	y
0	
1	
2	

x	y
0	1
1	6
2	11

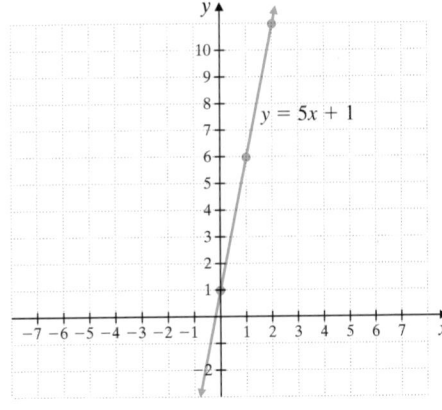

Next, we will graph a few special linear equations.

Practice Problem 4

Graph: $y = -1$

EXAMPLE 4   Graph: $y = 4$

Solution: The equation $y = 4$ can be written as $0x + y = 4$. When the equation is written in this form, notice that no matter what value we choose for x, y is always 4.

$$0 \cdot x + y = 4$$
$$0 \cdot (\text{any number}) + y = 4$$
$$0 + y = 4$$
$$y = 4$$

Fill in a table listing ordered-pair solutions of $y = 4$. Choose any three x-values. The y-values must be 4. Plot the ordered-pair solutions and graph $y = 4$.

x	y
0	4
1	4
−2	4

The graph is a horizontal line that crosses the y-axis at 4.

Answer

4.

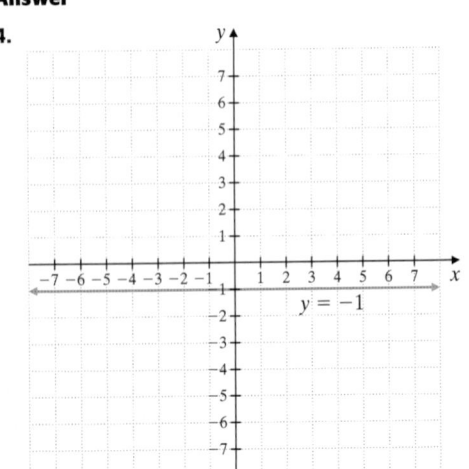

Horizontal Lines

If *a* is a number, then the graph of **y = a** is a *horizontal line* that crosses the *y*-axis at *a*. For example, the graph of $y = 2$ is a horizontal line that crosses the *y*-axis at 2.

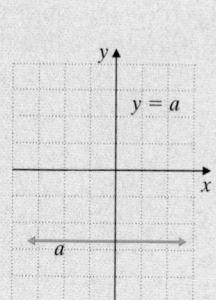

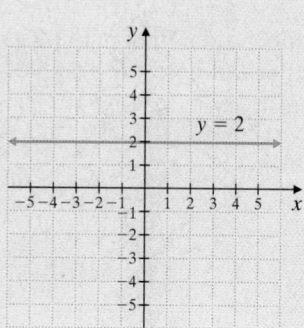

EXAMPLE 5   Graph: $x = -2$

Solution: The equation $x = -2$ can be written as $x + 0y = -2$. No matter what *y*-value we choose, *x* is always -2. Fill in a table listing ordered-pair solutions of $x = -2$. Choose any three *y*-values. The *x*-values must be -2. Plot the ordered-pair solutions and graph $x = -2$.

x	y
−2	3
−2	0
−2	−2

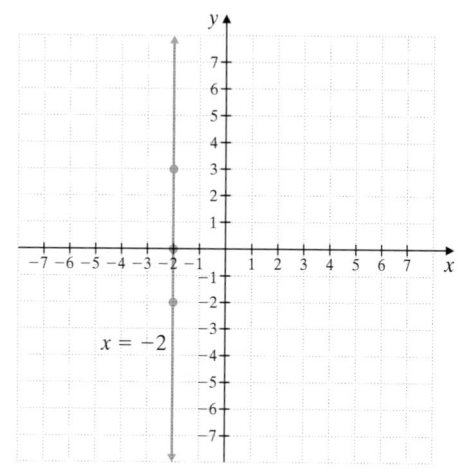

The graph is a vertical line that crosses the *x*-axis at -2.

Practice Problem 5

Graph: $x = 5$

Answer

5.

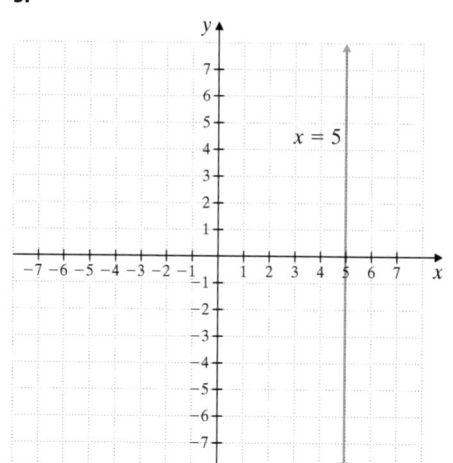

Concept Check

Determine whether the following equations represent vertical lines, horizontal lines, or neither.

a. $y = -6$

b. $x + y = 7$

c. $x = \dfrac{1}{2}$

d. $x = \dfrac{1}{2}y$

Concept Check: **a.** Horizontal line. **b.** Neither.
c. Vertical line. **d.** Neither.

Try the Concept Check in the margin.

Vertical Lines

If a is a number, then the graph of $\boldsymbol{x = a}$ is a *vertical line* that crosses the x-axis at a. For example, the graph of $x = -3$ is a vertical line that crosses the x-axis at -3.

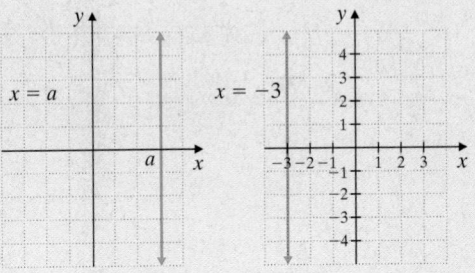

Name _____ Section _____ Date _____

EXERCISE SET 8.4

A *Graph each equation. See Examples 1 through 3.*

1. $x + y = 6$

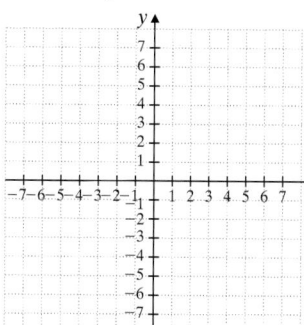

2. $x + y = 7$

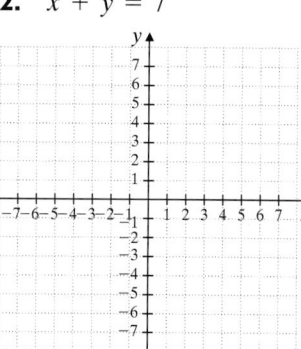

3. $x - y = -2$

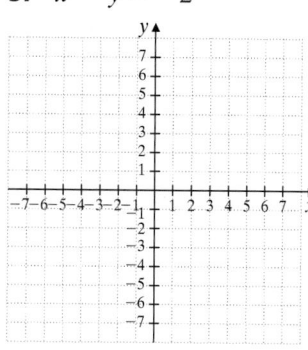

4. $y - x = 6$

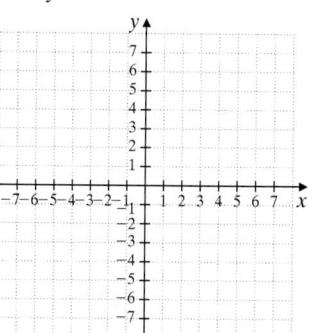

5. $y = 4x$

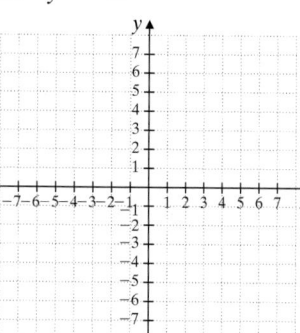

6. $x = 2y$

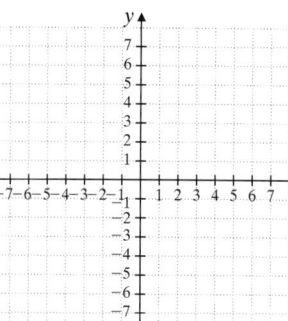

7. $y = 2x - 1$

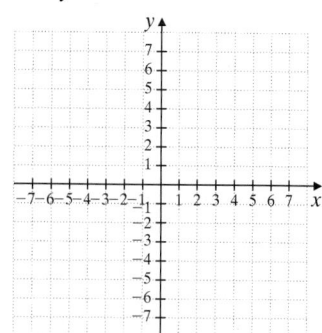

8. $y = x + 5$

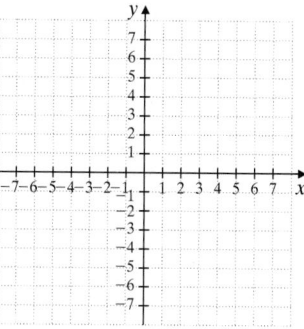

Graph each equation. See Examples 4 and 5.

9. $x = 5$

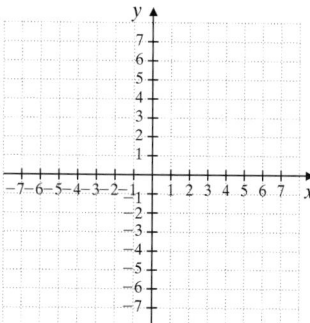

10. $y = 1$

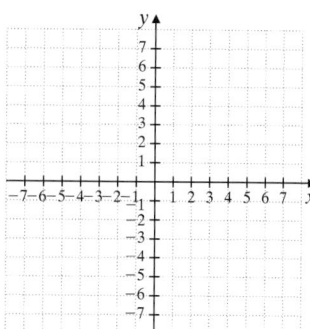

11. $y = -3$

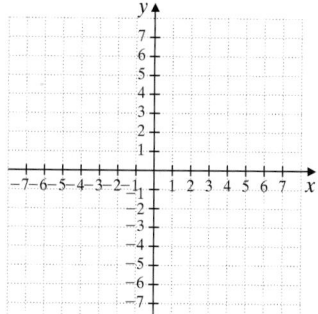

12. $x = -7$

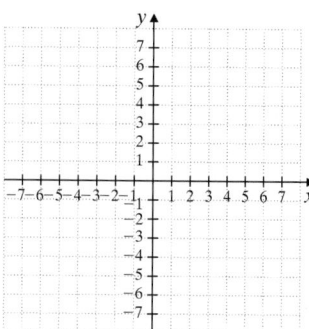

13. $x = 0$

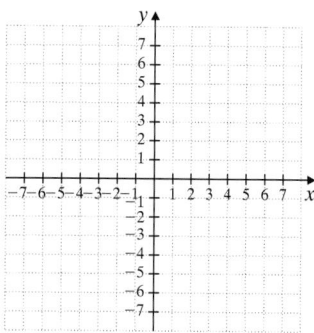

14. $y = 0$

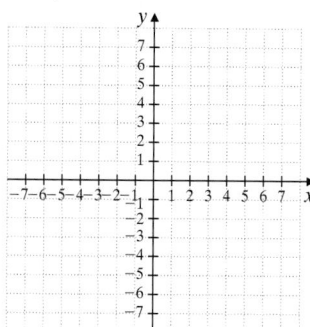

Graph each equation. See Examples 1 through 5.

15. $y = -2x$
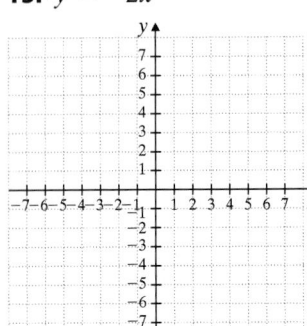

16. $x = y$
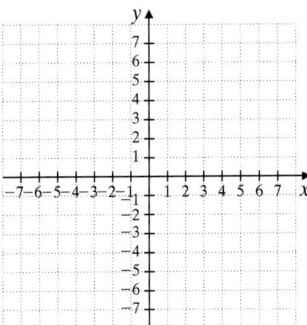

17. $y = -2$
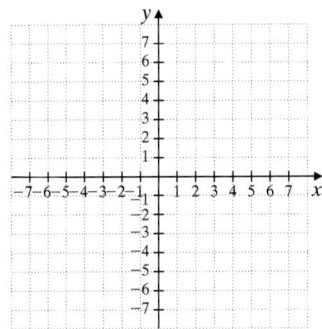

18. $x = 1$
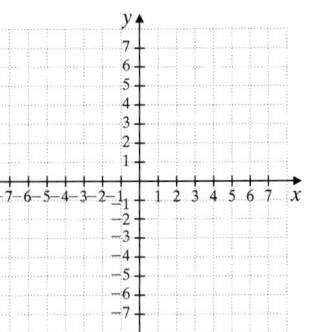

19. $x + 2y = 12$
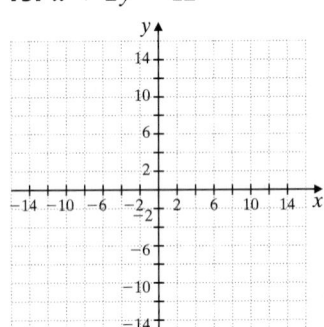

20. $3x - y = 3$
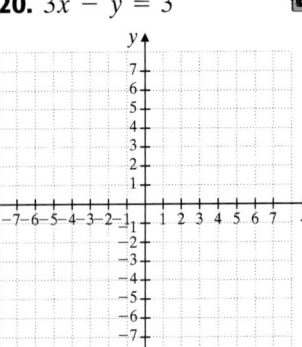

21. $x = 6$
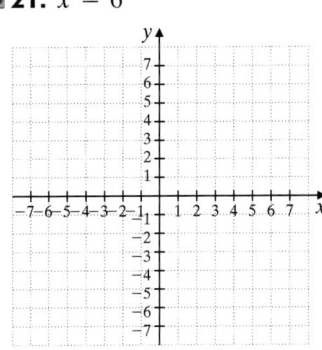

22. $y = x + 7$
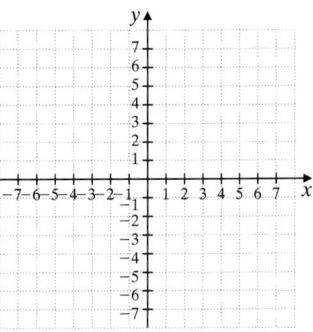

23. $y = x - 3$
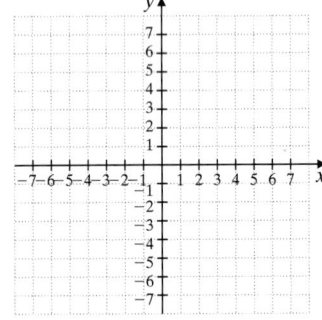

24. $x + y = -1$
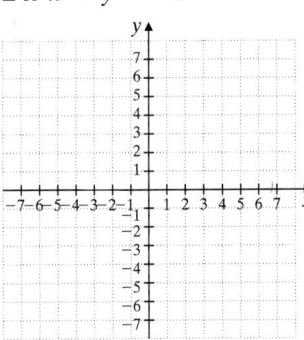

25. $x = y - 4$
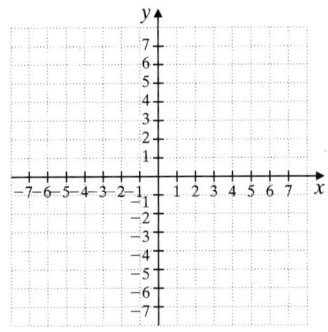

26. $x + y = 5$
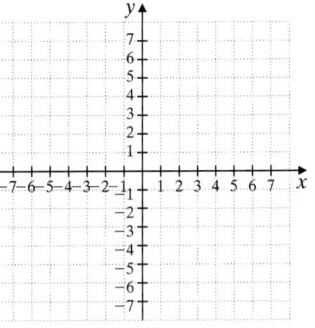

27. $x + 3 = 0$
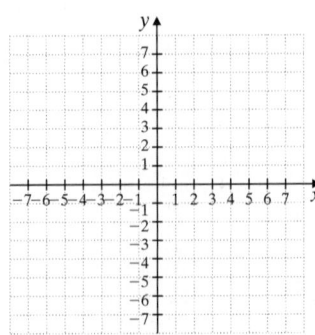

28. $y = -4$
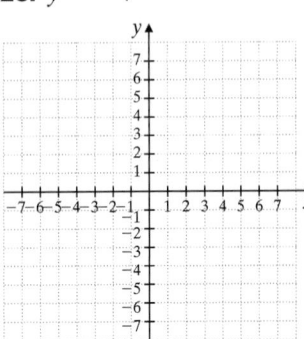

29. $x = 4y$
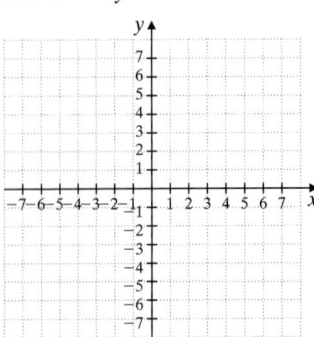

30. $y = -6x$
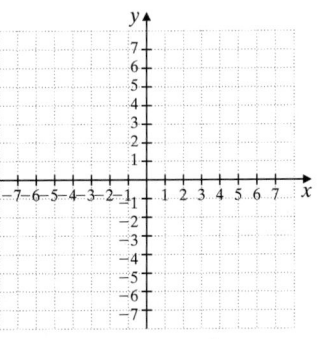

31. $y = \frac{1}{3}x$

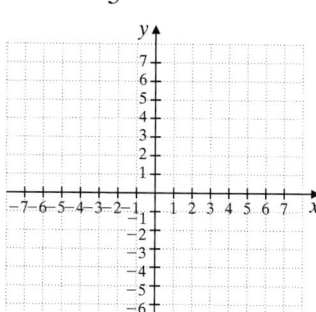

32. $y = -\frac{1}{3}x$

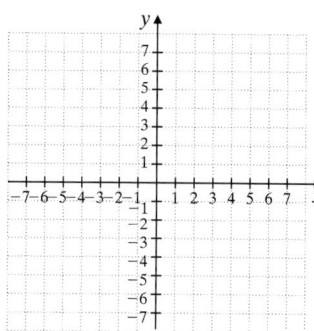

33. $y = 4x + 2$

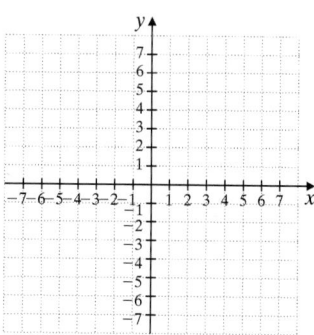

34. $y = -2x + 3$

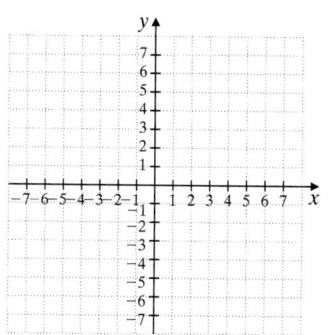

 **35.** $2x + 3y = 6$

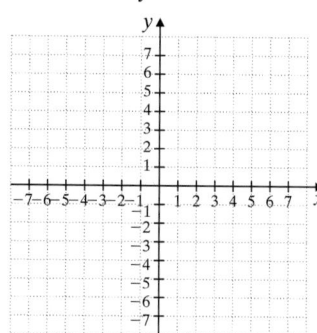

36. $5x - 2y = 10$

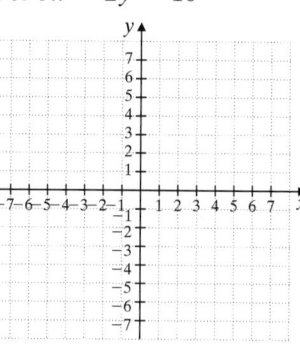

37. $x = -3.5$

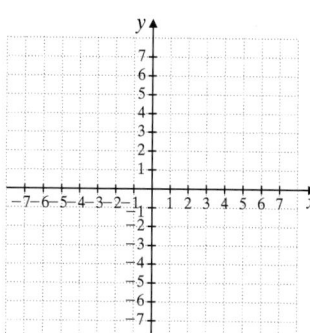

38. $y = 5.5$

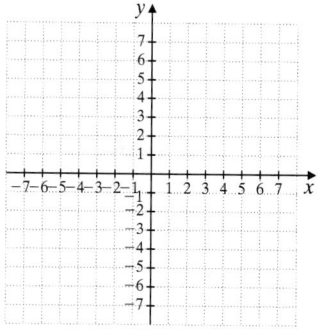

39. $3x - 4y = 24$

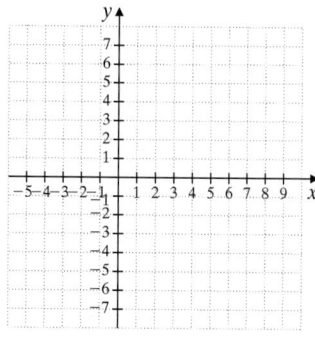

40. $4x + 2y = 16$

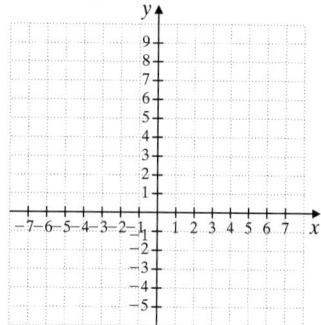

Review and Preview

Find the average of each list of numbers. See Section 1.7. (Recall that the average of a list of numbers is the sum of the numbers divided by the number of numbers.)

41. 86, 94

42. 75, 87

43. 12, 28, 20

44. 19, 10, 22

45. 30, 22, 23, 33

46. 39, 25, 31, 37

47. Fill in the table and use it to graph $y = |x|$.
see graph

x	y
−3	
−2	
−1	
0	
1	
2	
3	

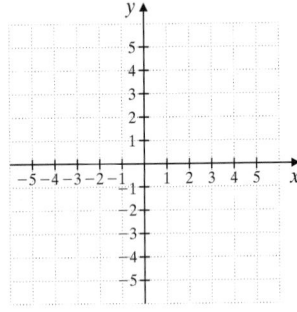

48. Graph: $15x - 18y = 270$

To do so, complete the table and find at least 2 additional ordered-pair solutions.
see graph

x	y
0	
	0

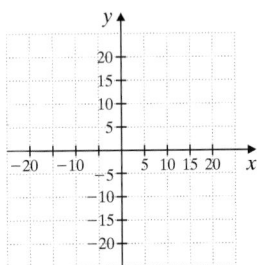

49. In your own words, explain how to graph a linear equation in two variables.

50. Suppose that a classmate tries to graph a linear equation in two variables and plots three ordered-pair solutions. If the three points do not all lie on the same line, what should the student do next?

8.5 Mean, Median, and Mode

(A) Finding the Mean

Sometimes we want to summarize data by displaying them in a graph, but sometimes it is also desirable to be able to describe a set of data, or a set of numbers, by a single "middle" number. Three such **measures of central tendency** are the **mean**, the **median**, and the **mode**.

The most common measure of central tendency is the mean (sometimes called the "arithmetic mean" or the "average"). Recall that we first introduced finding the average of a list of numbers in Section 1.7.

> The **mean (average)** of a set of number items is the sum of the items divided by the number of items.

OBJECTIVES

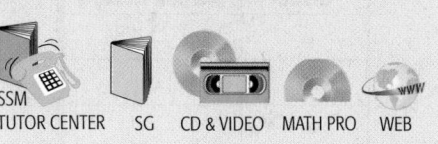

(A) Find the mean of a list of numbers.
(B) Find the median of a list of numbers.
(C) Find the mode of a list of numbers.

SSM   TUTOR CENTER   SG   CD & VIDEO   MATH PRO   WEB

EXAMPLE 1 Seven students in a psychology class conducted an experiment on mazes. Each student was given a pencil and asked to successfully complete the same maze. The timed results are below.

Student	Ann	Thanh	Carlos	Jesse	Melinda	Ramzi	Dayni
Time (seconds)	13.2	11.8	10.7	16.2	15.9	13.8	18.5

a. Who completed the maze in the shortest time? Who completed the maze in the longest time?

b. Find the mean time.

c. How many students took longer than the mean time? How many students took shorter than the mean time?

Solution:

a. Carlos completed the maze in 10.7 seconds, the shortest time. Dayni completed the maze in 18.5 seconds, the longest time.

b. To find the mean (or average), we find the sum of the number items and divide by 7, the number of items.

$$\text{mean} = \frac{13.2 + 11.8 + 10.7 + 16.2 + 15.9 + 13.8 + 18.5}{7}$$

$$= \frac{100.1}{7} = 14.3$$

c. Three students, Jesse, Melinda, and Dayni, had times longer than the mean time. Four students, Ann, Thanh, Carlos, and Ramzi, had times shorter than the mean time. ●

Try the Concept Check in the margin.

Often in college, the calculation of a **grade point average** (GPA) is a **weighted mean** and is calculated as shown in Example 2.

Practice Problem 1

Find the mean of the following test scores: 77, 85, 86, 91, and 88.

Concept Check

Find the mean of the following set of data:

5, 10, 10, 10, 10, 15

Answer

1. 85.4

Concept Check: 10

Practice Problem 2

Find the grade point average if the following grades were earned in one semester.

Grade	Credit Hours
A	2
C	4
B	5
D	2
A	2

EXAMPLE 2 The following grades were earned by a student during one semester. Find the student's grade point average.

Course	Grade	Credit Hours
College mathematics	A	3
Biology	B	3
English	A	3
PE	C	1
Social studies	D	2

Solution: To calculate the grade point average, we need to know the point values for the different possible grades. The point values of grades commonly used in colleges and universities are given below:

A: 4, B: 3, C: 2, D: 1, F: 0

Now, to find the grade point average, we multiply the number of credit hours for each course by the point value of each grade. The grade point average is the sum of these products divided by the sum of the credit hours.

Course	Grade	Point Value of Grade	Credit Hours	Point Value · Credit Hours
College mathematics	A	4	3	12
Biology	B	3	3	9
English	A	4	3	12
PE	C	2	1	2
Social studies	D	1	2	2
		Totals:	12	37

$$\text{grade point average} = \frac{37}{12} \approx 3.08 \text{ rounded to two decimal places}$$

The student earned a grade point average of 3.08.

B Finding the Median

You may have noticed that a very low number or a very high number can affect the mean of a list of numbers. Because of this, you may sometimes want to use another measure of central tendency. A second measure of central tendency is called the **median**. The median of a list of numbers is not affected by a low or high number in the list.

> The **median** of an *ordered set* of numbers is the middle number. If the number of items is even, the median is the mean (average) of the two middle numbers.

Helpful Hint:

In order to compute the median, the numbers must first be placed in order.

EXAMPLE 3 Find the median of the following list of numbers:

25, 54, 56, 57, 60, 71, 98

Solution: Because this list is in numerical order, the median is the middle number, 57.

25, 54, 56, 57, 60, 71, 98
          ↑
    middle number ●

EXAMPLE 4 Find the median of the following list of scores:

67, 91, 75, 86, 55, 91

Solution: First we list the scores in numerical order and then find the middle number.

55, 67, 75, 86, 91, 91
         ⏝
         ↑
    middle numbers

Since there is an even number of scores, there are two middle numbers. The median is the mean of the two middle numbers.

$$\text{median} = \frac{75 + 86}{2} = 80.5$$

The median is 80.5. ●

Ⓒ Finding the Mode

The last common measure of central tendency is called the **mode**.

> The **mode** of a set of numbers is the number that occurs most often. (It is possible for a set of numbers to have more than one mode or to have no mode.)

EXAMPLE 5 Find the mode of the list of numbers:

11, 14, 14, 16, 31, 56, 65, 77, 77, 78, 79

Solution: There are two numbers that occur the most often. They are 14 and 77. This list of numbers has two modes, 14 and 77. ●

EXAMPLE 6 Find the median and the mode of the following list of numbers. These numbers were high temperatures for 14 consecutive days in a city in Montana.

76, 80, 85, 86, 89, 87, 82, 77, 76, 79, 82, 89, 89, 92

Practice Problem 3

Find the median of the list of numbers:
7, 9, 13, 23, 24, 35, 38, 41, 43

Practice Problem 4

Find the median of the list of scores:
43, 89, 78, 65, 95, 95, 88, 71

Practice Problem 5

Find the mode of the list of numbers:
9, 10, 10, 13, 15, 15, 15, 17, 18, 18, 20

Practice Problem 6

Find the median and the mode of the list of numbers:
26, 31, 15, 15, 26, 30, 16, 18, 15, 35

Answers

3. 24 **4.** 83 **5.** 15 **6.** median: 22; mode: 15

Solution: First we write the numbers in numerical order.

$$76, 76, 77, 79, 80, 82, 82, 85, 86, 87, 89, 89, 89, 92$$

Since there is an even number of items, the median is the mean of the two middle numbers.

$$\text{median} = \frac{82 + 85}{2} = 83.5$$

The mode is 89, since 89 occurs most often. ●

Concept Check

True or false? Every set of numbers *must* have a mean, median, and mode. Explain your answer.

Try the Concept Check in the margin.

> **Helpful Hint**
>
> Don't forget that it is possible for a list of numbers to have no mode. For example, the list
>
> $$2, 4, 5, 6, 8, 9$$
>
> has no mode. There is no number or numbers that occur more often than the others.

Concept Check: false; a set of numbers may have no mode

Name _____ Section _____ Date _____

Mental Math

State the mean for each list of numbers.

1. 3, 5

2. 10, 20

3. 1, 3, 5

4. 7, 7, 7

EXERCISE SET 8.5

Ⓐ Ⓑ Ⓒ *For each set of numbers, find the mean, the median, and the mode. If necessary, round the mean to one decimal place. See Examples 1 and 3 through 6.*

1. 21, 28, 16, 42, 38

2. 42, 35, 36, 40, 50

 3. 7.6, 8.2, 8.2, 9.6, 5.7, 9.1

4. 4.9, 7.1, 6.8, 6.8, 5.3, 4.9

5. 0.2, 0.3, 0.5, 0.6, 0.6, 0.9, 0.2, 0.7, 1.1

6. 0.6, 0.6, 0.8, 0.4, 0.5, 0.3, 0.7, 0.8, 0.1

7. 231, 543, 601, 293, 588, 109, 334, 268

8. 451, 356, 478, 776, 892, 500, 467, 780

The eight tallest buildings in the world are listed in the following table. Use this table to answer Exercises 9 through 14. If necessary, round results to one decimal place. See Examples 1 and 3 through 6.

9. Find the mean height of the five tallest buildings.

10. Find the median height of the five tallest buildings.

11. Find the median height of the eight tallest buildings.

12. Find the mean height of the eight tallest buildings.

Building	Height (in feet)
Petronas Tower 1, Kuala Lumpur	1483
Petronas Tower 2, Kuala Lumpur	1483
Sears Tower, Chicago	1450
Jin Mao Building, Shanghai	1381
Citic Plaza, Guangzhou	1283
Shun Hing Square, Shenzhen	1260
Empire State Building, New York	1250
Central Plaza, Hong Kong	1227

(*Source:* Council on Tall Buildings and Urban Habitat)

13. Given the building heights, explain how you know, without calculating, that the answer to Exercise 10 is more than the answer to Exercise 11.

14. Given the building heights, explain how you know, without calculating, that the answer to Exercise 12 is less than the answer to Exercise 9.

For Exercises 15 through 18, the grades are given for a student for a particular semester. Find the grade point average. If necessary, round the grade point average to the nearest hundredth. See Example 2.

15.

Grade	Credit Hours
B	3
C	3
A	4
C	4

16.

Grade	Credit Hours
D	1
F	1
C	4
B	5

17.

Grade	Credit Hours
A	3
A	3
B	4
B	1
B	2

18.

Grade	Credit Hours
B	2
B	2
A	3
C	3
B	3

During an experiment, the following times (in seconds) were recorded:

$$7.8, 6.9, 7.5, 4.7, 6.9, 7.0$$

19. Find the mean. Round to the nearest tenth.

20. Find the median.

21. Find the mode.

In a mathematics class, the following test scores were recorded for a student:

$$86, 95, 91, 74, 77, 85$$

22. Find the mean. Round to the nearest hundredth.

23. Find the median.

24. Find the mode.

The following pulse rates were recorded for a group of 15 students:

$$78, 80, 66, 68, 71, 64, 82, 71, 70, 65, 70, 75, 77, 86, 72$$

25. Find the mean.

26. Find the median.

27. Find the mode.

28. How many rates were higher than the mean?

29. How many rates were lower than the mean?

Review and Preview

Write each fraction in simplest form. See Section 4.2.

30. $\dfrac{12}{20}$

31. $\dfrac{6}{18}$

32. $\dfrac{4}{36}$

33. $\dfrac{18}{30}$

34. $\dfrac{35}{100}$

35. $\dfrac{55}{75}$

Combining Concepts

Find the missing numbers in each set of numbers.

36. $16, 18, \_, \_, \_.$ The mode is 21. The median is 20.

37. $\_, \_, \_, 40, \_.$ The mode is 35. The median is 37. The mean is 38.

38. Write a list of numbers for which you feel the median would be a better measure of central tendency than the mean.

39. Without making any computations, decide whether the median of the following list of numbers will be a whole number. Explain your reasoning.

$$36, 77, 29, 58, 43$$

646 CHAPTER 8 Graphing and Introduction to Statistics

8.6   Counting and Introduction to Probability

A   Using a Tree Diagram

In our daily conversations, we often talk about the likelihood or the probability of a given result occurring. For example,

The *chance* of thundershowers is 70 percent.

What are the *odds* that the Saints will go to the Super Bowl?

What is the *probability* that you will finish cleaning your room today?

Each of these chance happenings—thundershowers, the Saints playing in the Super Bowl, and cleaning your room today—is called an **experiment**. The possible results of an experiment are called **outcomes**. For example, flipping a coin is an experiment and the possible outcomes are heads (H) or tails (T) and are equally likely to happen.

One way to picture the outcomes of an experiment is to draw a tree diagram. Each outcome is shown on a separate branch. For example, the outcomes of tossing a coin are

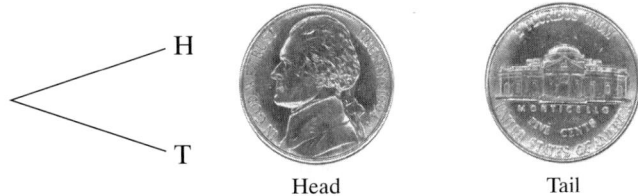

Head   Tail

EXAMPLE 1   Draw a tree diagram for tossing a coin twice. Then use the diagram to find the number of possible outcomes.

Solution: There are 4 possible outcomes when tossing a coin twice.

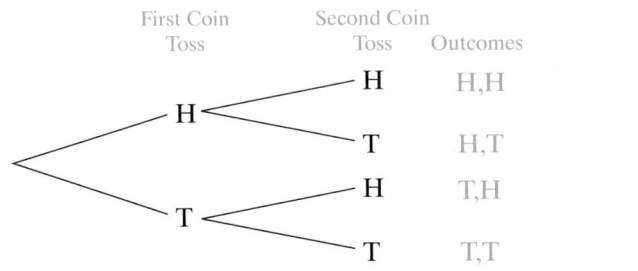

EXAMPLE 2   Draw a tree diagram for an experiment consisting of rolling a die and then tossing a coin. Then use the diagram to find the number of possible outcomes.

Die

Practice Problem 1

Draw a tree diagram for tossing a coin three times. Then use the diagram to find the number of possible outcomes.

Practice Problem 2

Draw a tree diagram for an experiment consisting of tossing a coin and then rolling a die. Then use the diagram to find the number of possible outcomes.

Answer

1.

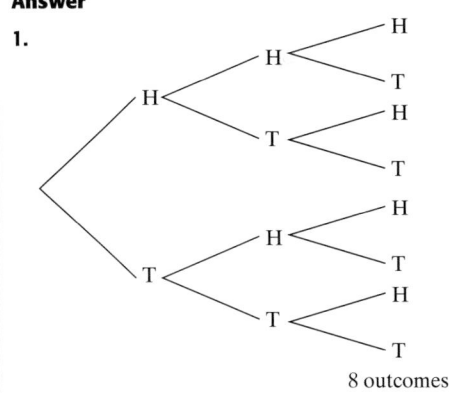

8 outcomes

2.

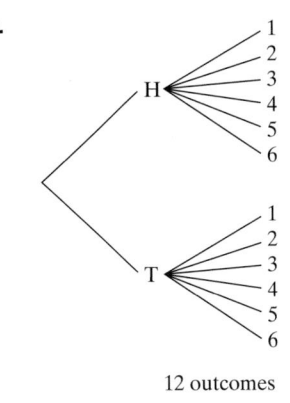

12 outcomes

Solution: Recall that a die has six sides and that each side represents a number, 1 through 6. Each side is as likely to be rolled as another side.

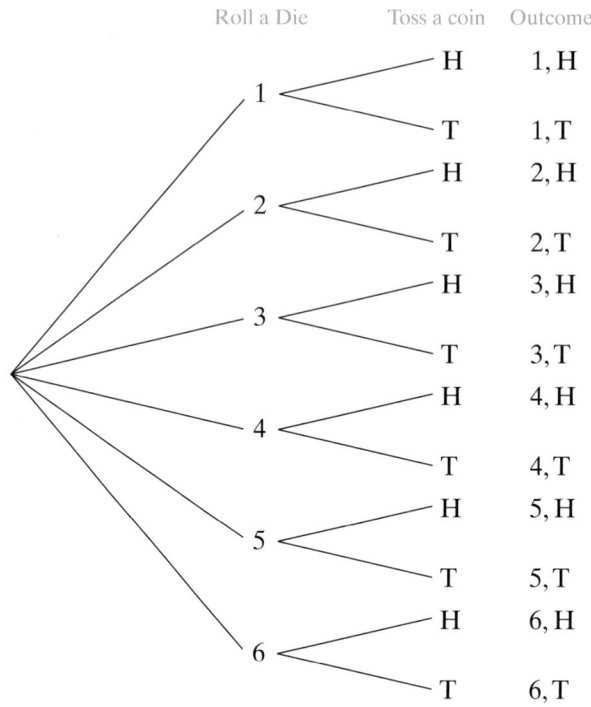

| Roll a Die | Toss a coin | Outcomes |

There are 12 possible outcomes for rolling a die and then tossing a coin. ●

Any number of outcomes considered together is called an **event**. For example, when tossing a coin twice, (H, H) is an event. The event is tossing heads first and tossing heads second. Another event would be tossing tails first and then heads (T, H), and so on.

Ⓑ Finding the Probability of an Event

As we mentioned earlier, the probability of an event is a measure of the chance or likelihood of it occurring. For example, if a coin is tossed, what is the probability that heads occurs? Since one of two equally likely possible outcomes is heads, the probability is $\frac{1}{2}$.

The Probability of an Event

$$\text{probability of an event} = \frac{\text{number of ways that the event can occur}}{\text{number of possible outcomes}}$$

Helpful Hint

Note from the definition of probability that the probability of an event is always between 0 and 1, inclusive (i.e., including 0 and 1). A probability of 0 means that an event won't occur, and a probability of 1 means that an event is certain to occur.

EXAMPLE 3   If a coin is tossed twice, find the probability of tossing heads and then heads (H, H).

Solution: 1 way the event can occur

$$\underbrace{(H, H)}, (H, T), (T, H), (T, T)$$
4 possible outcomes

probability $= \dfrac{1}{4}$ Number of ways the event can occur
Number of possible outcomes

The probability of tossing heads and then heads is $\dfrac{1}{4}$.   ●

Practice Problem 3

If a coin is tossed three times, find the probability of tossing heads, then tails, then tails (H, T, T).

EXAMPLE 4   If a die is rolled one time, find the probability of rolling a 3 or a 4.

Solution: Recall that there are 6 possible outcomes when rolling a die.

2 ways that the event can occur

possible outcomes: $\underbrace{1, \quad 2, \quad 3, \quad 4, \quad 5, \quad 6}$
6 possible outcomes

probability of a 3 or a 4 $= \dfrac{2}{6}$ Number of ways the event can occur
Number of possible outcomes

$= \dfrac{1}{3}$ Simplest form   ●

Practice Problem 4

If a die is rolled one time, find the probability of rolling a 1 or a 2.

Concept Check

Suppose you have calculated a probability of $\dfrac{11}{9}$. How do you know that you have made an error in your calculation?

Try The Concept Check in the margin.

EXAMPLE 5   Find the probability of choosing a red marble from a box containing 1 red, 1 yellow, and 2 blue marbles.

Solution: 1 way the event can occur

yellow   blue   blue   red
4 possible outcomes

Practice Problem 5

Use the diagram from Example 5 and find the probability of choosing a blue marble from the box.

Answers
3. $\dfrac{1}{8}$ **4.** $\dfrac{1}{3}$ **5.** $\dfrac{1}{2}$

Concept Check: The number of ways an event can occur can't be larger than the number of possible outcomes.

probability $= \dfrac{1}{4}$ ●

FOCUS ON **Mathematical Connections**

RANGE

In addition to measures of central tendency, *measures of dispersion* can also help to describe or summarize a set of data. These measures indicate how "spread out" the numbers are in a set of data. One measure of dispersion is the **range** of a set of data. The range is computed as the difference between the largest and the smallest number in the data set.

Take, for example, the data set 39, 27, 30, 66, 63, and 57. The smallest number in the set is 27 and the largest number in the set is 66. The difference between these two numbers is $66 - 27 = 39$. Thus, the range of the data is 39.

What can the range tell us about a set of data? Suppose we have two different sets of data. The median (middle number) of the first set is 18 and the median of the second set is also 18. So far, the data sets seem to be similar. But if we are then told that the range of the first set is 2 and the range of the second set is 28, we know that the two sets of data are likely to be quite different. In the first set, the numbers are clustered very closely to the median, 18. In the second set, the numbers are probably spread out much farther, covering a wider slice of values. In fact, the first set are ages of the students in a high school math class and the second set are ages of the students in a college math class (see below). We would expect the age makeup of these two classes to be quite different, and comparing the ranges of the ages confirms this.

Ages in High School Math Class

17   17   17   17   17   17   17   18   18   18   18   18   18   18   18   18   18   18   18   18   19

Median: 18   Range: 2

Ages in College Math Class:

18   18   18   18   18   18   18   18   18   18   18   18   19   19   20   21   23   27   33   41   46

Median: 18   Range: 28

GROUP ACTIVITY

Collect class grade information for two different tests or quizzes in this class or another class; both sets of grades should be for the same class. Find the mean, median, mode, and range for each set of grades. Compare statistics for each set of grades. What can you conclude?

Name _____ Section _____ Date _____

Mental Math

If a coin is tossed once, find the probability of each event.

1. The coin lands heads up.

2. The coin lands tails up.

If the spinner shown is spun once, find the probability of each event.

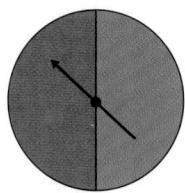

3. The spinner stops on red.

4. The spinner stops on blue.

EXERCISE SET 8.6

A *Draw a tree diagram for each experiment. Then use the diagram to find the number of possible outcomes. See Examples 1 and 2.*

1. Choosing a vowel, (a, e, i, o, u) and then a number (1, 2, or 3).

2. Choosing a number (1 or 2) and then a vowel (a, e, i, o, u).

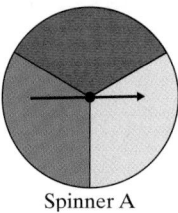

Spinner A

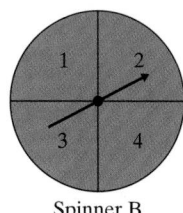

Spinner B

3. Spinning Spinner A once.

4. Spinning Spinner B once.

5. Spinning Spinner B twice.

6. Spinning Spinner A twice.

7. Spinning Spinner A and then Spinner B.

8. Spinning Spinner B and then Spinner A.

9. Tossing a coin and then spinning Spinner B.

10. Tossing a coin and then spinning Spinner A.

B *If a single die is tossed once, find the probability of each event. See Examples 3 through 5.*

11. A 5

12. A 7

13. A 1 or a 4

14. A 2 or a 3

15. An even number

16. An odd number

Suppose the spinner shown is spun once. Find the probability of each event. See Examples 3 through 5.

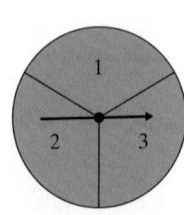

17. The result of the spin is 2.

18. The result of the spin is 3.

19. The result of the spin is an odd number.

20. The result of the spin is an even number.

If a single choice is made from the bag of marbles shown, find the probability of each event. See Examples 3 through 5.

21. A red marble is chosen.

22. A blue marble is chosen.

23. A yellow marble is chosen.

24. A green marble is chosen.

A new drug is being tested that is supposed to lower blood pressure. This drug was given to 200 people and the results are below.

Lower Blood Pressure	Higher Blood Pressure	Blood Pressure Not Changed
152	38	10

25. If a person is testing this drug, what is the probability that their blood pressure will be higher?

26. If a person is testing this drug, what is the probability that their blood pressure will be lower?

27. If a person is testing this drug, what is the probability that their blood pressure will not change?

28. What is the sum of the answers to exercises 25, 26, and 27? In your own words, explain why.

Review and Preview

Perform each indicated operation. See Sections 4.3, 4.4, and 4.5.

29. $\dfrac{1}{2} + \dfrac{1}{3}$

30. $\dfrac{7}{10} - \dfrac{2}{5}$

31. $\dfrac{1}{2} \cdot \dfrac{1}{3}$

32. $\dfrac{7}{10} \div \dfrac{2}{5}$

33. $5 \div \dfrac{3}{4}$

34. $\dfrac{3}{5} \cdot 10$

Combining Concepts

Recall that a deck of cards contains 52 cards. These cards consist of four suits (hearts, spades, clubs, and diamonds) of each of the following: 2, 3, 4, 5, 6, 7, 8, 9, 10, jack, queen, king, and ace. If a card is chosen from a deck of cards, find the probability of each event.

35. The king of hearts

36. The 10 of spades

37. A king

38. A 10

39. A heart

40. A club

Two dice are tossed. Find the probability of each sum of the dice. (Hint: Draw a tree diagram of the possibilities of two tosses of a die, and then find the sum of the numbers on each branch.)

41. A sum of 4

42. A sum of 11

43. A sum of 13

44. A sum of 2

45. In your own words, explain why the probability of an event cannot be greater than 1.

46. In your own words, explain when the probability of an event is 0.

FOCUS ON **Business and Career**

SURVEYS

How often have you read an article in a newspaper or a magazine that included results from a survey or a poll? Surveys have become very popular ways for businesses and organizations to get feedback on a variety of topics. A political organization may hire a polling group to estimate the public's response to a political candidate. A food company may send out surveys to customers to gather market research on a new food product. A health club may ask its patrons to fill out a brief comment card to report their views on the services offered at the club. Surveys are useful for collecting information needed to make a decision: Should we do a media blitz to increase public awareness of our candidate? Do we need to change the recipe for our new food product to make it more appealing to a wider audience? Should we add a new service to attract new customers and retain our current clientele?

After data have been collected from a survey, they must be summarized to be useful in decision making. Survey results can be summarized in any of the types of graphs presented in this chapter. Survey results can also be summarized by reporting average responses or with a combination of basic statistics and graphs.

GROUP ACTIVITY

1. Conduct a survey of 30 students in one of your classes. Ask each student to report his or her age.

2. Find the difference between the ages of the youngest and oldest survey respondents (this difference between the largest and smallest value is called the "range"). Divide the range into five or six equal age categories. Tally the number of your respondents that fall into each category. Make a histogram of your results. What does this graph tell you about the ages of your survey respondents?

3. Find the average age of your survey respondents.

4. Find the median age of your survey respondents.

5. Find the mode of your survey respondents.

6. Compare the mean, median, and mode of your age data. Are these measures similar? Which is the largest? Which is the smallest? If there is a noticeable difference between any of these measures, can you explain why?

CHAPTER 8 ACTIVITY **Conducting a Survey**

This activity may be completed by working in groups or individually.

How often have you read an article in a newspaper or in a magazine that included results from a survey or poll? Surveys seem to have become very popular ways of getting feedback on anything from a political candidate, to a new product, to services offered by a health club. In this activity, you will conduct a survey and analyze the results.

1. Conduct a survey of 30 students in one of your classes. Ask each student to report his or her age.

2. Classify each age according to the following categories: under 20, 20 to 24, 25 to 29, 30 to 39, 40 to 49, and 50 or over. Tally the number of your survey respondents that fall into each category. Make a bar graph of your results. What does this graph tell you about the ages of your survey respondents?

3. Find the average age of your survey respondents.

4. Find the median age of your survey respondents.

5. Find the mode of the ages of your survey respondents.

6. Compare the mean, median, and mode of your age data. Are these measures similar? Which is largest? Which is smallest? If there is a noticeable difference between any of these measures, can your explain why?

FOCUS ON Mathematical Connections

STEM-AND-LEAF DISPLAYS

Stem-and-leaf displays are another way to organize data. After data are logically organized, it can be much easier to draw conclusions from them.

Suppose we have collected the following set of data. It could represent the test scores for an algebra class or the pulse rates of a group of small children.

$$
\begin{array}{cccccccccc}
90 & 73 & 93 & 99 & 79 & 95 & 69 & 78 & 93 & 80 \\
89 & 85 & 97 & 78 & 75 & 79 & 72 & 76 & 97 & 88 \\
83 & 98 & 72 & 94 & 92 & 79 & 70 & 98 & 85 & 99
\end{array}
$$

In a stem-and-leaf display, the last digit of each number forms the *leaf*, and the remaining digits to the left form the *stem*. For the first number in the list, 90, 9 is the stem and 0 is the leaf. To make the stem-and-leaf display, we write all of the stems in numerical order in a column. Then we write each leaf on the horizontal line next to its stem, aligning leaves in vertical columns. In this case, because the data range from 69 to 99, we use the stems 6, 7, 8, and 9. Each line of the display represents an interval of data; for instance, the line corresponding to the stem **7** represents all data that fall in the interval **70** to **79**, inclusive. After the data have been divided into stems and leaves on the display as shown in the table on the left, we simply rearrange the leaves on each line to appear in numerical order, as shown in the table on the right.

Stem	Leaf	Stem	Leaf
6	9	6	9
7	39885926290 ⟶	7	02235688999
8	095835	8	035589
9	039537784289	9	023345778899

Now that the data have been organized into a stem-and-leaf display, it is easy to answer questions about the data such as: What are the least and greatest values in the set of data? Which data interval contains the most items from the data set? How many values fall between 74 and 84? Which data value occurs most frequently in the data set? What patterns or trends do you see in the data?

CRITICAL THINKING

Make a stem-and-leaf display of the weekend emergency room admission data shown on the right. Then answer the following questions:

1. What is the difference between the least number and the greatest number of weekend ER admissions?

2. How many weekends had between 125 and 165 ER admissions?

3. What number of weekend ER admissions occurred most frequently?

4. Which interval contains the most weekend ER admissions?

Number of Emergency Room Admissions on Weekends				
198	168	117	185	159
160	177	169	112	175
170	188	137	117	145
198	169	154	163	192
167	179	155	133	121
162	188	124	145	146
128	181	198	149	140
122	162	161	180	177

Chapter 8   Vocabulary Check

Fill in each blank with one of the words or phrases listed below.

outcomes   bar   experiment   mean   tree diagram
pictograph   line   circle   median   probability
histogram   class frequency mode   class interval

1. A _____ graph presents data using vertical or horizontal bars.

2. The _____ of a set of number items is
   $$\frac{\text{sum of items}}{\text{number of items}}.$$

3. The possible results of an experiment are the
   _____ .

4. A _____ is a graph in which pictures or symbols are used to visually present data.

5. The _____ of a set of numbers is the number that occurs most often.

6. A _____ graph displays information with a line that connects data points.

7. The _____ of an ordered set of numbers is the middle number.

8. A _____ is one way to picture and count outcomes.

9. An _____ is an activity being considered, such as tossing a coin or rolling a die.

10. In a _____ graph, each section (shaped like a piece of pie) shows a category and the relative size of the category.

11. The _____ of an event is
    $$\frac{\text{number of ways that the event can occur}}{\text{number of possible outcomes}}.$$

12. A _____ is a special bar graph in which the width of each bar represents a _____ and the height of each bar represents the _____ .

CHAPTER 8

Highlights

DEFINITIONS AND CONCEPTS	EXAMPLES

SECTION 8.1 READING PICTOGRAPHS, BAR GRAPHS, AND LINE GRAPHS

A **pictograph** is a graph in which pictures or symbols are used to visually present data.

A **line graph** displays information with a line that connects data points.

A **bar graph** presents data using vertical or horizontal bars.

The bar graph on the right shows the number of acres of wheat harvested in 1996 for leading states.

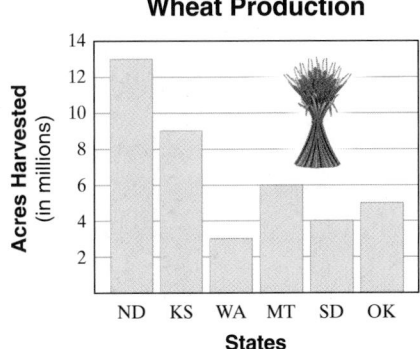

Wheat Production

Source: U.S. Department of Agriculture

1. Approximately how many acres of wheat were harvested in Kansas?

   9,000,000 acres

2. About how many more acres of wheat were harvested in North Dakota than South Dakota?

   $$\begin{array}{r} 13 \text{ million} \\ -\ 4 \text{ million} \\ \hline 9 \text{ million} \quad \text{or} \quad 9{,}000{,}000 \text{ acres} \end{array}$$

DEFINITIONS AND CONCEPTS	EXAMPLES

SECTION 8.1 READING PICTOGRAPHS, BAR GRAPHS, AND LINE GRAPHS (*continued*)

A **histogram** is a special bar graph in which the width of each bar represents a **class interval** and the height of each bar represents the **class frequency**. The histogram on the right shows student quiz scores.

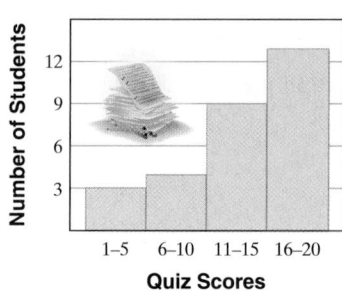

Quiz Scores

1. How many students received a score of 6–10?

   4 students

2. How many students received a score of 11–20?

   9 + 13 = 22 students

SECTION 8.2 READING CIRCLE GRAPHS

In a **circle graph**, each section (shaped like a piece of pie) shows a category and the relative size of the category.

The circle graph on the right classifies tornadoes by wind speed.

Tornado Wind Speeds

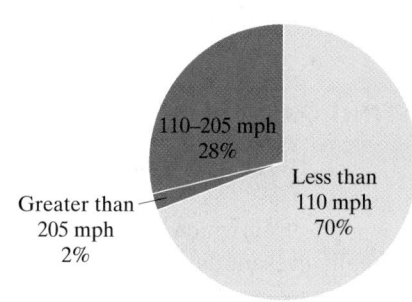

Source: National Oceanic and Atmospheric Administration

1. What percent of tornadoes have wind speeds of 110 mph or greater?

   28% + 2% = 30%

2. If there were 1235 tornadoes in the United States in 1995, how many of these might we expect to have had wind speeds less than 110 mph? Find 70% of 1235.

   70%(1235) = 0.70(1235) = 864.5 ≈ 865

   Around 865 tornadoes would be expected to have had wind speeds of less than 110 mph.

DEFINITIONS AND CONCEPTS	**EXAMPLES**

SECTION 8.3 THE RECTANGULAR COORDINATE SYSTEM

The **rectangular coordinate system** consists of two number lines intersecting at the point 0 on each number line. The horizontal number line is called the **x-axis** and the vertical number line is called the **y-axis**.

Every point in the rectangular coordinate system corresponds to an **ordered pair of numbers** such as

$$(2, -2)$$

x-coordinate
or x-value

y-coordinate
or y-value

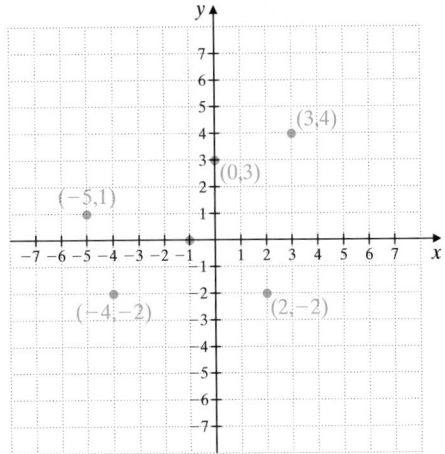

An ordered pair is a **solution** of an equation if the equation is a true statement when the variables are replaced by the coordinates of the ordered pair.

Is $(2, -1)$ a solution of $5x - y = 11$?

$$5x - y = 11$$

$5(2) - (-1) = 11$ Replace x with 2 and y with −1.

$10 + 1 = 11$ Multiply.

$11 = 11$ True.

Yes, $(2, -1)$ is a solution of $5x - y = 11$.

SECTION 8.4 GRAPHING LINEAR EQUATIONS

LINEAR EQUATIONS IN TWO VARIABLES

A linear equation in two variables is an equation that can be written in the form

$$ax + by = c$$

where a, b, and c are numbers, and a and b are not both 0.

To graph a linear equation in two variables, find three ordered-pair solutions and draw the line through the plotted points.

Graph: $y = 4x$

Let $x = 1$.   Let $x = -1$.   Let $x = 0$.

$y = 4(1)$ $y = 4(-1)$ $y = 4(0)$

$y = 4$ $y = -4$ $y = 0$

x	y
1	4
−1	−4
0	0

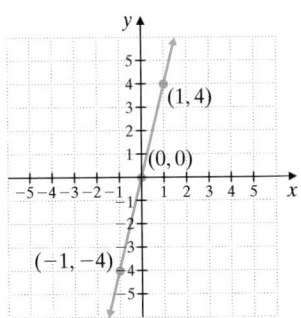

The graph of $y = a$ is a **horizontal line** that crosses the y-axis at a.

The graph of $x = a$ is a **vertical line** that crosses the x-axis at a.

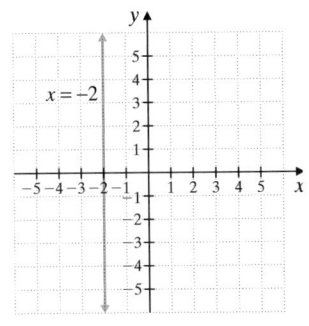

DEFINITIONS AND CONCEPTS	**EXAMPLES**

SECTION 8.5 MEAN, MEDIAN, AND MODE

The **mean** (or **average**) of a set of number items is

$$\text{mean} = \frac{\text{sum of items}}{\text{number of items}}$$

The **median** of an ordered set of numbers is the middle number. If the number of items is even, the median is the mean of the two middle numbers.

The **mode** of a set of numbers is the number that occurs most often. (A set of numbers may have no mode or more than one mode.)

Find the mean, median, and mode of the following set of numbers: 33, 35, 35, 43, 68, 68

$$\text{mean} = \frac{33 + 35 + 35 + 43 + 68 + 68}{6} = 47$$

The median is the mean of the two middle numbers:

$$\text{median} = \frac{35 + 43}{2} = 39$$

There are two modes because there are two numbers that occur twice:

35 and 68

SECTION 8.6 COUNTING AND INTRODUCTION TO PROBABILITY

An **experiment** is an activity being considered, such as tossing a coin or rolling a die. The possible results of an experiment are the **outcomes**. A **tree diagram** is one way to picture and count outcomes.

Draw a tree diagram for tossing a coin and then choosing a number from 1 to 4.

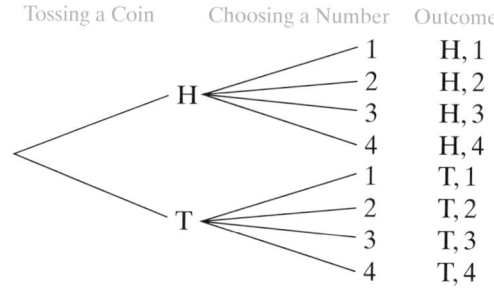

Tossing a Coin	Choosing a Number	Outcomes
H	1	H, 1
	2	H, 2
	3	H, 3
	4	H, 4
T	1	T, 1
	2	T, 2
	3	T, 3
	4	T, 4

Any number of outcomes considered together is called an **event**. The **probability** of an event is a measure of the chance or likelihood of it occurring.

$$\text{probability of an event} = \frac{\text{number of ways that the event can occur}}{\text{number of possible outcomes}}$$

Find the probability of tossing a coin twice and tails occurring each time.

1 way the event can occur

$$\text{(H,H), (H,T), (T,H), (T,T)}$$

4 possible outcomes

$$\text{probability} = \frac{1}{4}$$

Chapter 8 Review

(8.1) *The following pictograph shows the number of new homes constructed, by state. Use this graph to answer Exercises 1 through 6.*

**Housing Starts by Region
of United States**

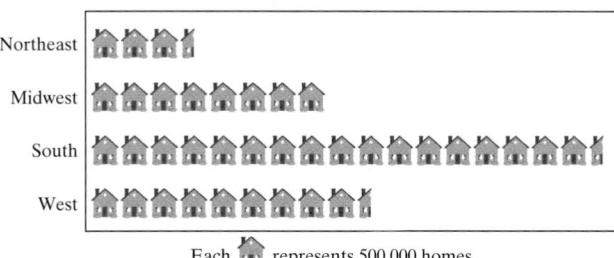

Each 🏠 represents 500,000 homes

Source: U.S. Census Bureau

1. How many housing starts were there in the Midwest in 2000?

2. How many housing starts were there in the Northeast in 2000?

3. Which region had the most housing starts?

4. Which region had the fewest housing starts?

5. Which region(s) had 4,000,000 or more housing starts?

6. Which region(s) had fewer than 4,000,000 housing starts?

The following bar graph shows the percent of persons age 25 or over who completed four or more years of college. Use this graph to answer Exercises 7 through 10.

**Four or More Years of College
by Persons Age 25 or Over**

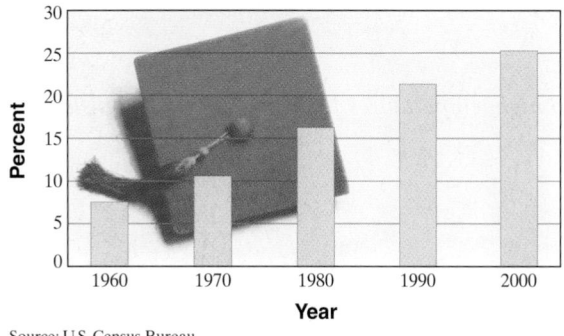

Source: U.S. Census Bureau

7. Approximate the percent of persons who completed four or more years of college in 1960.

8. What year shown had the greatest percent of persons completing four or more years of college?

9. What years shown had 15% or more of persons completing four or more years of college?

10. Describe any patterns you notice in this graph.

The following line graph shows the average price of a 30-second television advertisement during the Super Bowl for the years shown. Use this graph to answer Exercises 11 through 15.

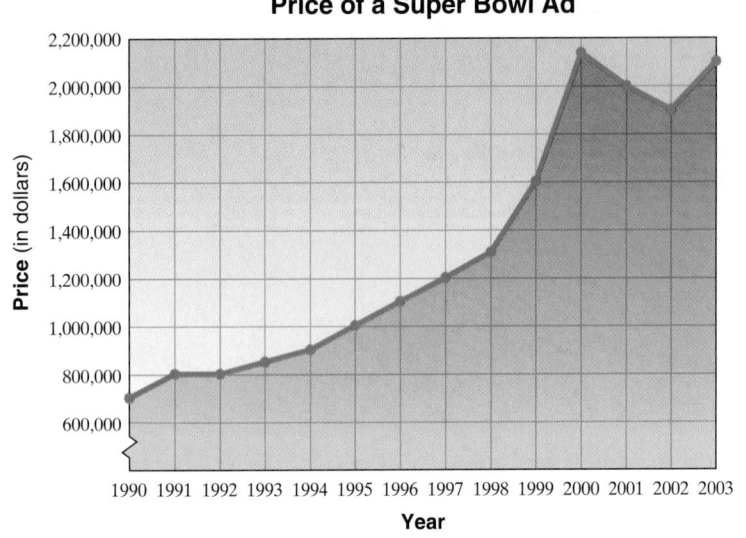

Price of a Super Bowl Ad

Sources: Nielsen Media Research and *Advertising Age* research

11. Approximate the price of a Super Bowl ad in 2003.

12. Approximate the price of a Super Bowl ad in 1997.

13. Between which two years did the price of a Super Bowl ad *not* increase?

14. Between which two years did the price of a Super Bowl ad increase the most?

15. During which years was the price of a Super Bowl ad *less than* $1,000,000?

The following histogram shows the hours worked per week by the employees of Southern Star Furniture. Use this histogram to answer Exercises 16 through 19.

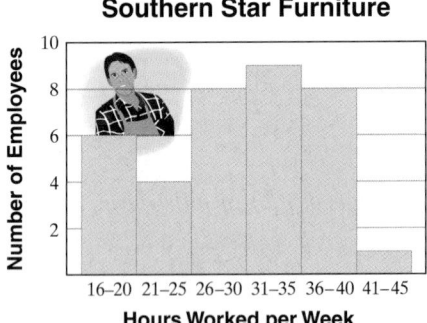

Southern Star Furniture

16. How many employees work 21–25 hours per week?

17. How many employees work 41–45 hours per week?

18. How many employees work 36 hours or more per week?

19. How many employees work 30 hours or less per week?

Following is a list of monthly record high temperatures for New Orleans, Louisiana. Use this list to complete the frequency distribution table below.

83	96	101	92
85	100	92	102
89	101	87	84

	Class Intervals (Temperatures)	**Tally**	**Class Frequency (Number of Months)**
20.	80°–89°	_____	_____
21.	90°–99°	_____	_____
22.	100°–109°	_____	_____

23. Use the table from Exercises 20, 21, and 22 to draw a histogram.

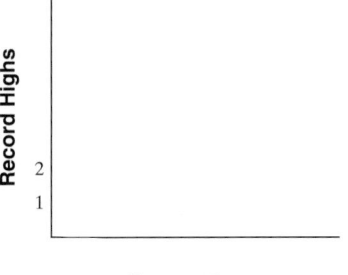

Record Highs

Temperatures

(8.2) *The following circle graph shows a family's $4000 monthly budget. Use this graph to answer Exercises 24 through 30. Write all ratios as fractions in simplest form.*

24. What is the largest budget item?

25. What is the smallest budget item?

26. How much money is budgeted for the mortgage payment and utilities?

27. How much money is budgeted for savings and contributions?

28. Find the ratio of the mortage payment to the total monthly budget.

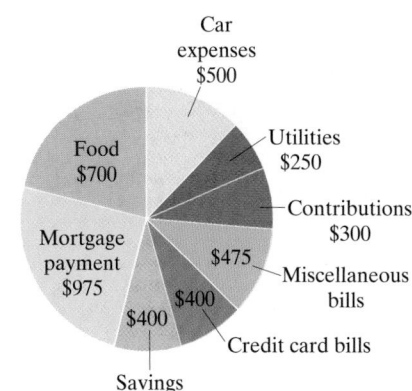

29. Find the ratio of food to the total monthly budget.

30. Find the ratio of car expenses to food.

The following circle graph shows the percent of states with various rural interstate highway speed limits in 2000. Use this graph to determine the number of states with each speed limit in Exercises 31 through 34.

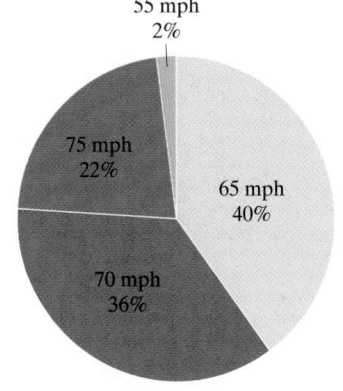

Percent of States with Rural Interstate Highway Speed Limits

Source: Insurance Institute for Highway Safety

31. How many states have a rural interstate highway speed limit of 65 mph?

32. How many states have a rural interstate highway speed limit of 75 mph?

33. How many states have a rural interstate highway speed limit of 55 mph?

34. How many states have a rural interstate highway speed limit of 70 mph or 75 mph?

(8.3) *Complete and graph the ordered-pair solutions of each given equation.*

35. $x = -7y;$ $(0, \quad), (\quad, -1), (-7, \quad)$

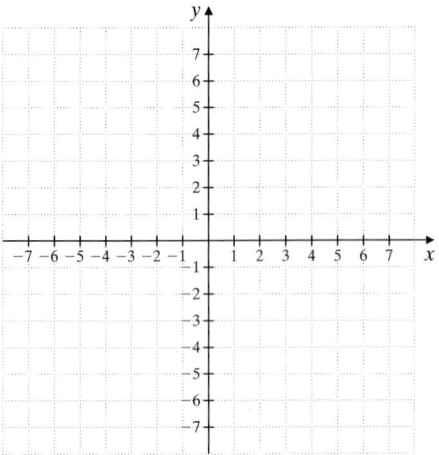

36. $y = 3x - 2;$ $(0, \quad), (1, \quad), (-2, \quad)$

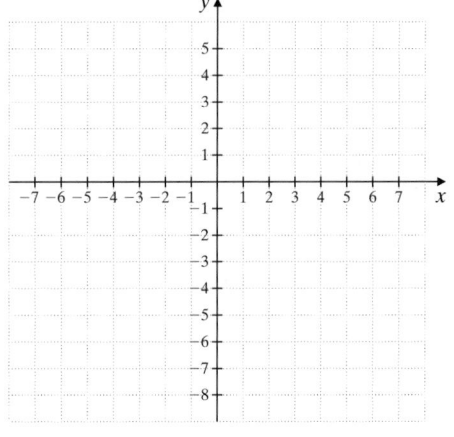

37. $x + y = -9;$ $(-1, \quad), (\quad, 0), (-5, \quad)$

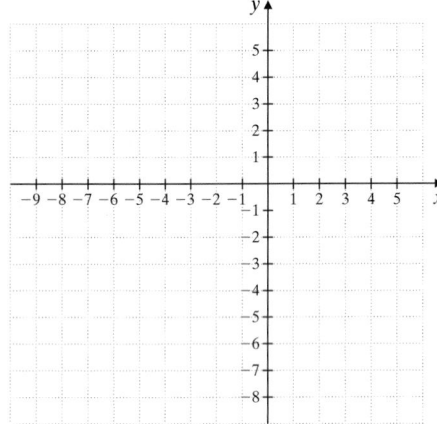

38. $x - y = 3;$ $(4, \quad), (0, \quad), (\quad, 3)$

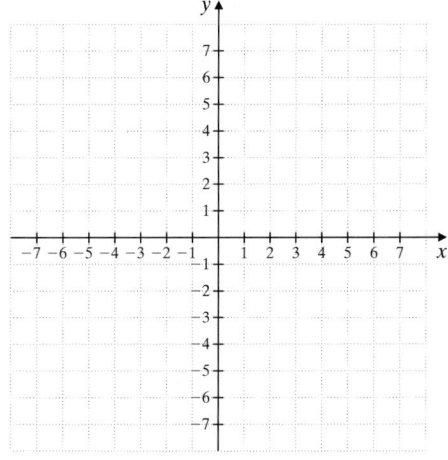

39. $y = 3x;$ $(1, \quad), (-2, \quad), (\quad, 0)$

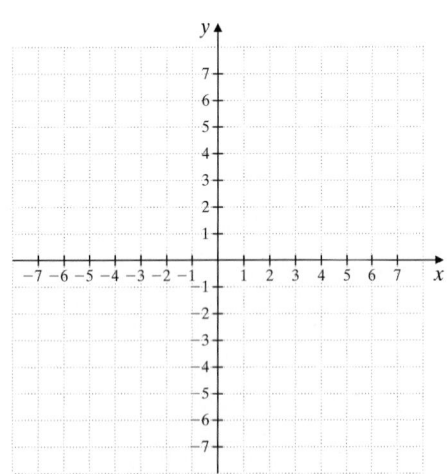

40. $x = y + 6;$ $(1, \quad), (6, \quad), (-1, \quad)$

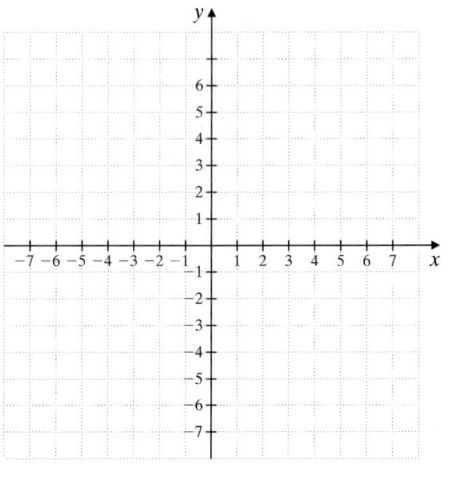

664

(8.4) *Graph each linear equation.*

41. $x = -6$

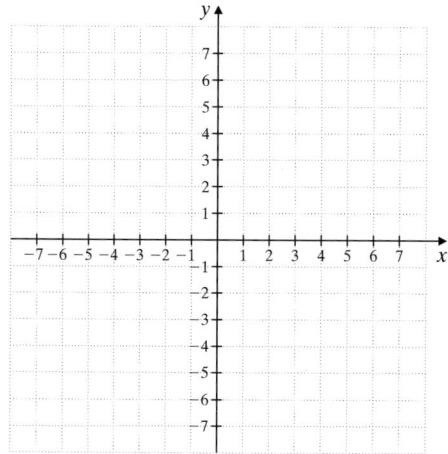

42. $y = 0$

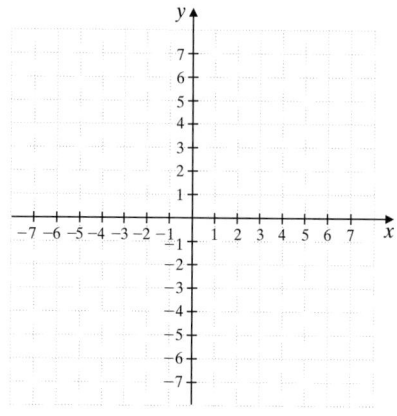

43. $x + y = 11$

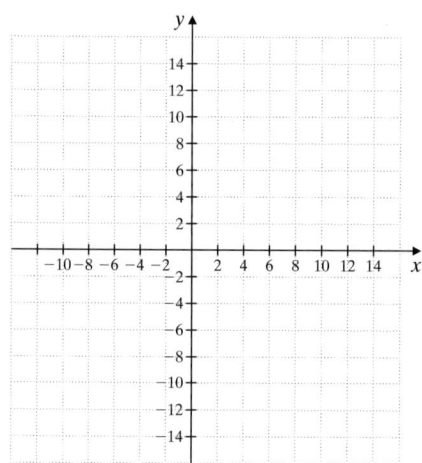

44. $x - y = 11$

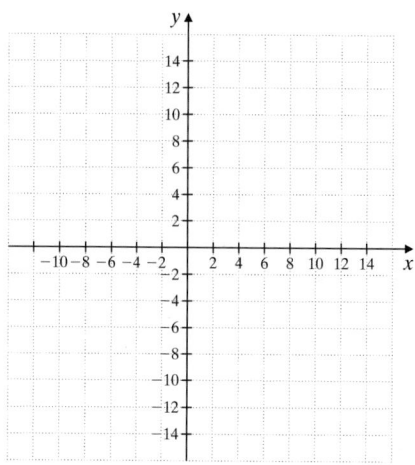

45. $y = 4x - 2$

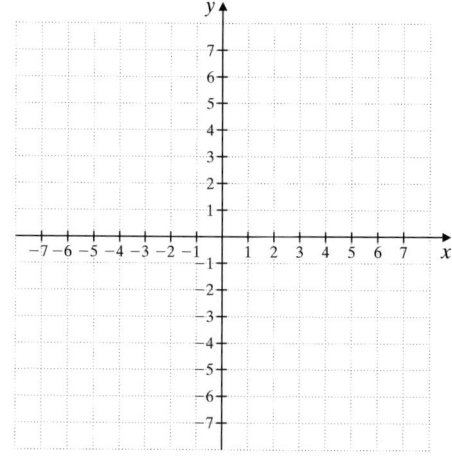

46. $y = 5x$

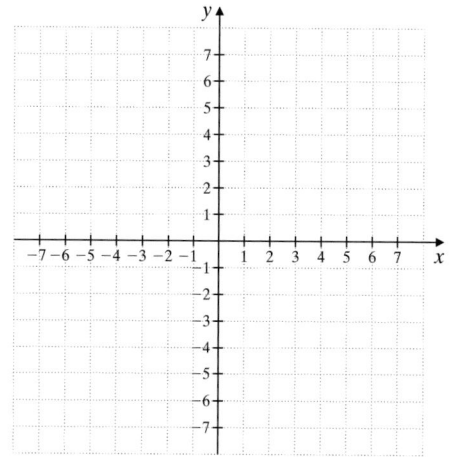

47. $x = -2y$

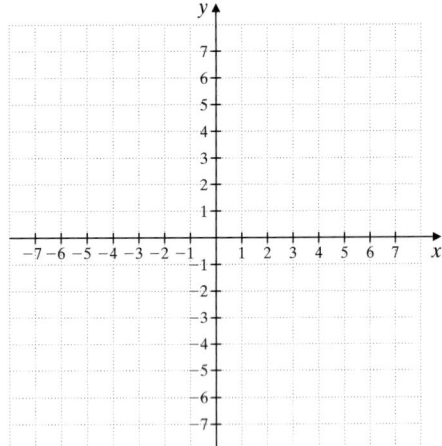

48. $x + y = -1$

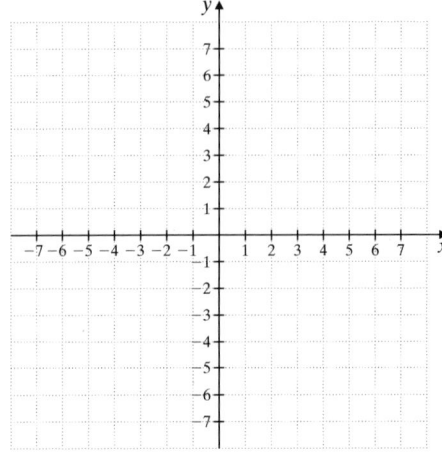

49. $2x - 3y = 12$

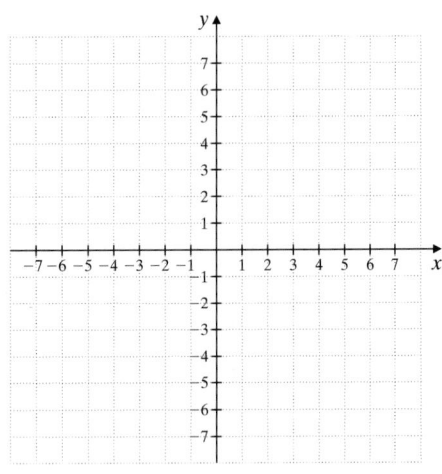

50. $x = \frac{1}{2}y$

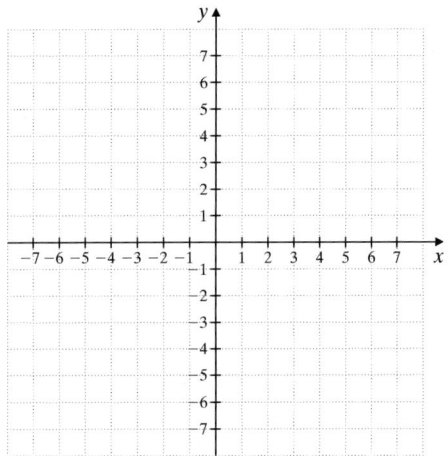

(8.5) *Find the mean, median, and any mode(s) for each list of numbers.*

51. $13, 23, 33, 14, 6$

52. $45, 21, 60, 86, 64$

53. $\$14{,}000, \$20{,}000, \$12{,}000, \$20{,}000, \$36{,}000, \$45{,}000$

54. $560, 620, 123, 400, 410, 300, 400, 780, 430, 450$

For Exercises 55 and 56, the grades are given for a student for a particular semester. Find each grade point average. If necessary, round the grade point average to the nearest hundredth.

55.

Grade	Credit Hours
A	3
A	3
C	2
B	3
C	1

56.

Grade	Credit Hours
B	3
B	4
C	2
D	2
B	3

(8.6) *Draw a tree diagram for each experiment. Then use the diagram to determine the number of outcomes.*

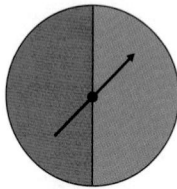

Spinner 1   Spinner 2

57. Tossing a coin and then spinning Spinner 1

58. Spinning Spinner 2 and then tossing a coin

59. Spinning Spinner 1 twice

60. Spinning Spinner 2 twice

61. Spinning Spinner 1 and then Spinner 2

Find the probability of each event.

Die

62. Rolling a 4 on a die

63. Rolling a 3 on a die

64. Spinning a 4 on Spinner 1

65. Spinning a 3 on Spinner 1

66. Spinning either a 1, 3, or 5 on Spinner 1

67. Spinning either a 2 or a 4 on Spinner 1

STUDY SKILLS REMINDER

Tips for studying for an exam

To prepare for an exam, try the following study techniques.

- Start the study process days before your exam.

- Make sure that you are current and up-to-date on your assignments.

- If there is a topic that you are unsure of, use one of the many resources that are available to you. For example,

    See your instructor.

    Visit a learning resource center on campus where math tutors are available.

    Read the textbook material and examples on the topic.

    View a videotape on the topic.

- Reread your notes and carefully review the Chapter Highlights at the end of the chapter.

- Work the review exercises at the end of the chapter and check your answers. Make sure that you correct any missed exercises. If you have trouble on a topic, use a resource listed above.

- Find a quiet place to take the Chapter Test found at the end of the chapter. Do not use any resources when taking this sample test. This way you will have a clear indication of how prepared you are for your exam. Check your answers and make sure that you correct any missed exercises.

- Get lots of rest the night before the exam. It's hard to show how well you know the material if your brain is foggy from lack of sleep.

Good luck, and keep a positive attitude.

Name_____ Section_____ Date _____

Chapter 8 Test   Remember to check your answers and use the Chapter Test Prep Video to view solutions.

The following pictograph shows the money collected each week from a wrapping paper fundraiser. Use this graph to answer Exercises 1 through 3.

Weekly Wrapping Paper Sales

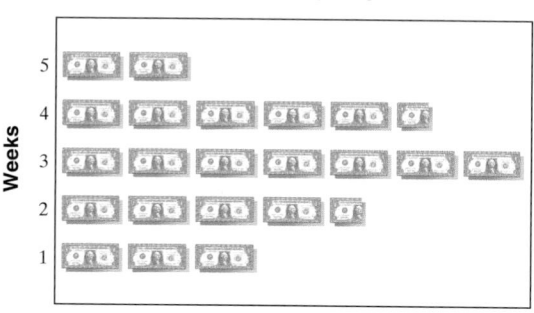

Each ▭ represents $50

1. How much money was collected during the second week?

2. During which week was the most money collected? How much money was collected during that week?

3. What was the total money collected for the fundraiser?

The bar graph shows the normal monthly precipitation in centimeters for Chicago, Illinois. Use this graph to answer Exercises 4–6.

Chicago Rainfall

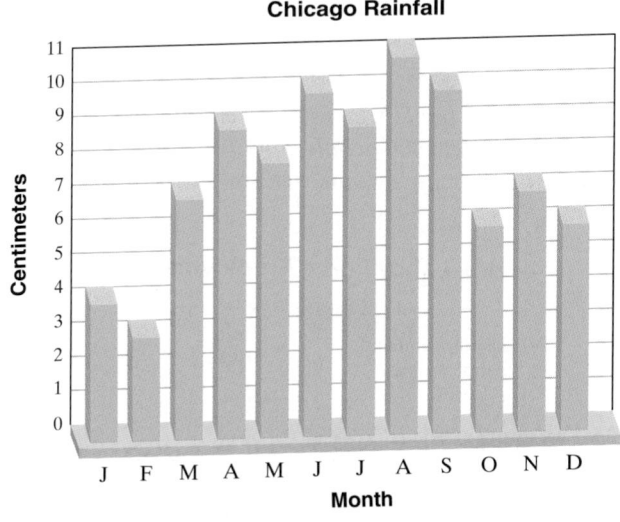

Source: U.S. National Oceanic and Atmospheric Administration, *Climatography of the United States*, No. 81

4. During which month(s) does Chicago normally have greater than 9 centimeters of rainfall?

5. During which month does Chicago normally have the least amount of rainfall? How much rain falls during that month?

6. During which month(s) does 7 centimeters of rain normally fall?

7. Use the information in the table to draw a bar graph. Clearly label each bar.

Countries with the Highest Newspaper Cirulations	
Country	**Average Daily Circulation (in millions)**
Japan	72
US	56
China	50
India	31
Germany	24
Russia	24
UK	19

(_Source:_ World Association of Newspapers

The result of a survey of 200 people is shown in the following circle graph. Each person was asked to tell his or her favorite type of music. Use this graph to answer Exercises 8 and 9.

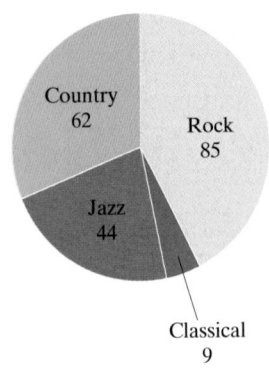

8. Find the ratio of those who prefer rock music to the total number surveyed.

9. Find the ratio of those who prefer country music to those who prefer jazz.

The following circle graph shows the U.S. labor force employment by industry for a recent year. This graph is based on 132,000,000 people employed by these industries in the United States. Use the graph to find how many people were employed by the industries given in Exercises 10 and 11.

U.S. Labor Force Employment by Industry

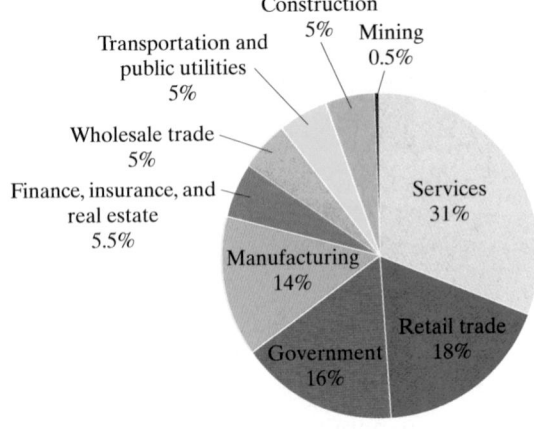

Source: Bureau of Labor Statistics

10. Services

11. Government

A professor measures the heights of the students in her class. The results are shown in the following histogram. Use this histogram to answer Exercises 12 and 13.

Student Heights

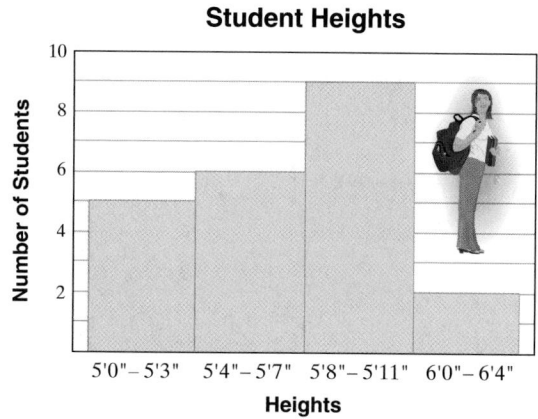

12. How many students are 5′8″–5′11″ tall?

13. How many students are 5′7″ or shorter?

14. The history test scores of 25 students are shown below. Use these scores to complete the frequency distribution table.

70	86	81	65	92
43	72	85	69	97
82	51	75	50	68
88	83	85	77	99
77	63	59	84	90

15. Use the results of Exercise 14 to draw a histogram.

Scores

Class Intervals (Scores)	Tally	Class Frequency (Number of Students)
40–49	_____	_____
50–59	_____	_____
60–69	_____	_____
70–79	_____	_____
80–89	_____	_____
90–99	_____	_____

Find the coordinates of each point in the graph below.

16. *A*   **17.** *B*   **18.** *C*   **19.** *D*

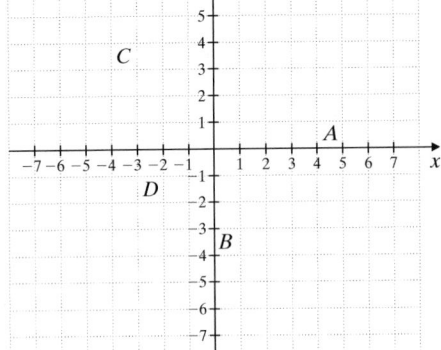

Complete and graph the ordered-pair solutions of each given equation.

20. $x = -6y$; $(0, \quad), (\quad, 1), (12, \quad)$ **21.** $y = 7x - 4$; $(2, \quad), (-1, \quad), (0, \quad)$

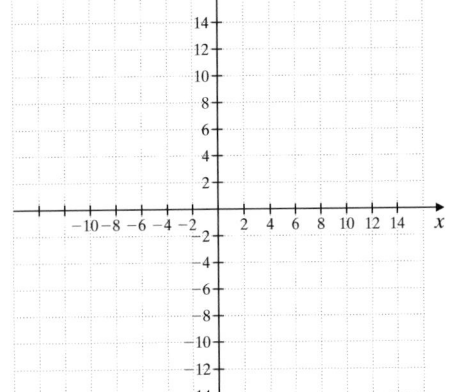

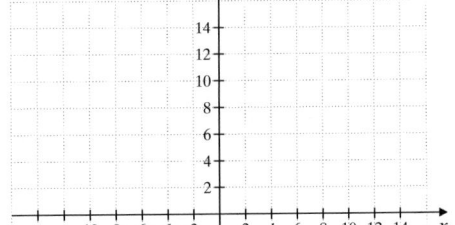

Graph each linear equation.

22. $y + x = -4$ **23.** $y = -4$

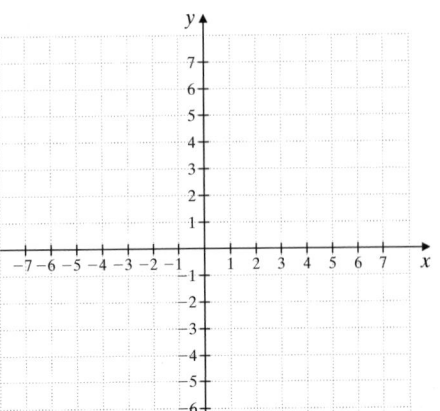

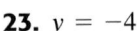

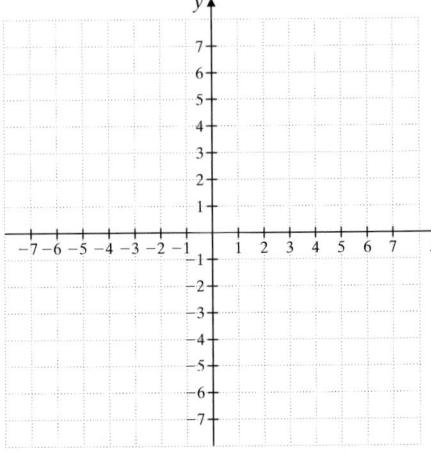

24. $y = 3x - 5$

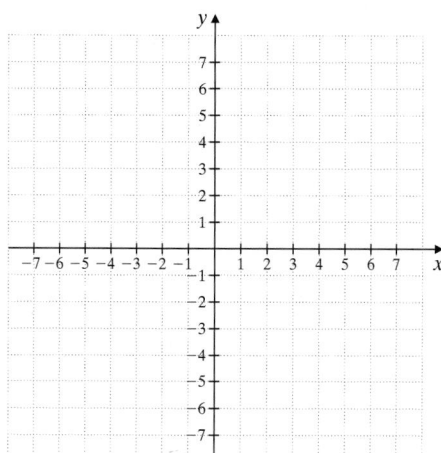

25. $x = 5$

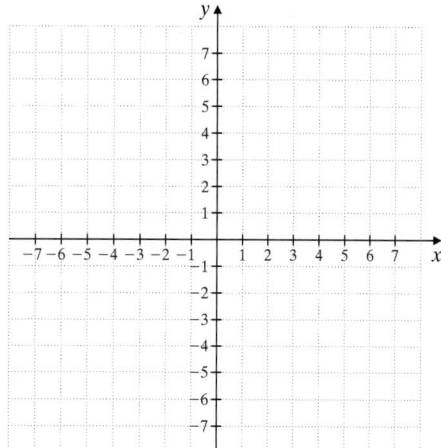

26. $y = -\dfrac{1}{2}x$

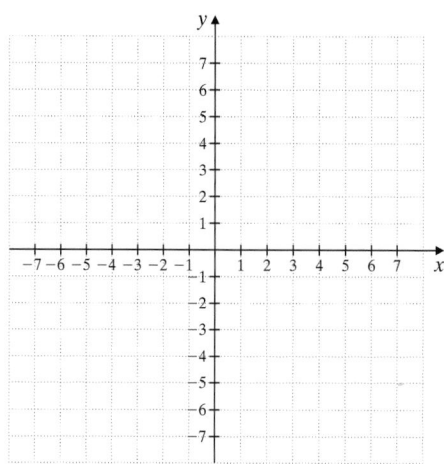

27. $3x - 2y = 12$

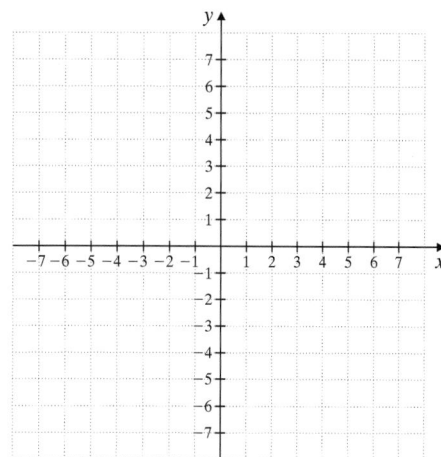

Find the mean, median, and mode of each list of numbers.

28. 26, 32, 42, 43, 49

29. 8, 10, 16, 16, 14, 12, 12, 13

24. see graph

25. see graph

26. see graph

27. see graph

28. _____

29. _____

673

Find the grade point average. If necessary, round to the nearest hundredth.

30.

Grade	Credit Hours
A	3
B	3
C	3
B	4
A	1

31. Draw a tree diagram for the experiment of spinning the spinner twice.

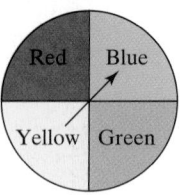

32. Draw a tree diagram for the experiment of tossing a coin twice.

Suppose that the numbers 1 through 10 are each written on the same size sheet of paper and placed in a bag. You then select one sheet of paper from the bag.

33. What is the probability of choosing a 6 from the bag?

34. What is the probability of choosing a 3 or a 4 from the bag?

Cumulative Review

1. Simplify: $4^3 + [3^2 - (10 \div 2)] - 7 \cdot 3$ **2.** $7^2 - [5^3 + (6 \div 3)] + 4 \cdot 2$

3. Evaluate $x - y$ for $x = -3$ and $y = 9$.   **4.** Evaluate $x - y$ for $x = 7$ and $y = -2$.

5. Solve: $3y - 7y = 12$ **6.** Solve: $2x - 6x = 24$

7. Solve: $\dfrac{x}{6} + 1 = \dfrac{4}{3}$ **8.** Solve a: $\dfrac{7}{2} + \dfrac{a}{4} = 1$

9. Add: $2\dfrac{1}{3} + 5\dfrac{3}{8}$ **10.** Add: $3\dfrac{2}{5} + 4\dfrac{3}{4}$

11. Write 5.6 as a mixed number.   **12.** Write 2.8 as a mixed number.

13. Subtract: $3.5 - 0.068$ **14.** Subtract: $7.4 - 0.073$.

15. Multiply: 0.283×0.3 **16.** Multiply: 0.147×0.2.

17. Divide: $-5.98 \div 115$ **18.** Divide: $27.88 \div 205$.

19. Simplify: $(-1.3)^2 + 2.4$ **20.** Simplify: $(-2.7)^2$

21. Write $\dfrac{1}{4}$ as a decimal.   **22.** Write $\dfrac{3}{8}$ as a decimal.

23. Solve: $5(x - 0.36) = -x + 2.4$ **24.** Solve: $4(0.35 - x) = x - 7$

25. Approximate $\sqrt{32}$ to the nearest thousandth.   **26.** Approximate $\sqrt{60}$ to the nearest thousandth.

Answers

1. _____
2. _____

3. _____
4. _____

5. _____
6. _____

7. _____
8. _____

9. _____

10. _____

11. _____

12. _____

13. _____

14. _____

15. _____

16. _____

17. _____

18. _____

19. _____

20. _____

21. _____

22. _____

23. _____

24. _____

25. _____

26. _____

27. _____

28. _____

29. _____

30. _____

31. _____

32. _____

33. _____

34. _____

35. _____

36. _____

37. _____

38. _____

39. _____

40. _____

41. _____

42. _____

43. _____

44. _____

45. _____

46. _____

47. _____

48. _____

49. _____

50. _____

27. Write the ratio of 12 to 17 using fractional notation.

28. Write the ratio of 7 to 15 using fractional notation.

29. Write as a unit rate: 318.5 miles every 13 gallons of gas.

30. Write as a unit rate: $1.59 for 3 ounces.

31. Solve $\dfrac{25}{x} = \dfrac{20}{4}$ for x and then check.

32. Solve $\dfrac{8}{5} = \dfrac{x}{10}$ for x and then check.

△ **33.** Find the ratio of corresponding sides $\overline{AB}$ and $\overline{DE}$ for the similar triangles ABC and DEF.

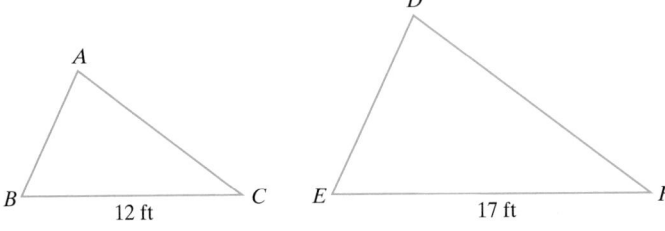

△ **34.** Find the ratio of corresponding sides for the similar triangles GHJ and KLM.

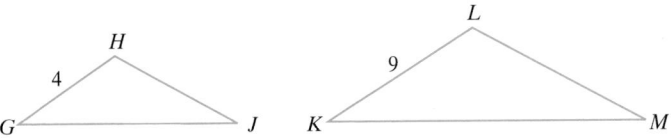

Write each percent as a decimal.

35. 4.6%

36. 32%

37. 0.74%

38. 2.7%

39. What number is 35% of 60?

40. What number is 40% of 36?

41. 20.8 is 40% of what number?

42. 9.5 is 25% of what number?

43. The sales tax on a $300 printer is $22.50. Find the sales tax rate.

44. The sales tax on a $2.00 yo-yo is $0.13. Find the sales tax rate.

45. Find the monthly payment on a $2000 loan for 2 years. The interest on the 2-year loan is $435.88.

46. Linda Bonnett borrows $1600 for 1 year. If the interest is $128.60, find the monthly payment.

47. Find the median of the list of scores: 67, 91, 75, 86, 55, 91

48. Find the median of the list of numbers. 43, 46, 47, 50, 52, 83

49. If a die is rolled one time, find the probability of rolling a 3 or a 4.

50. If a die is rolled once, find the probability of rolling an even number.

Geometry and Measurement

The word "geometry" is formed from the Greek words *geo*, meaning Earth, and *metron*, meaning measure. Geometry literally means to measure the Earth. In this chapter, we learn about various geometric figures and their properties such as perimeter, area, and volume. Knowledge of geometry can help us solve practical problems in real-life situations. For instance, knowing certain measures of a circular swimming pool allows us to calculate how much water it can hold.

The Statue of Liberty, built in the late 1800s, was the inspiration of Frenchman Edouard-Rene Lefebvre de Laboulaye. His idea was to present a monument to the American people, from the people of France, that celebrated their common ideals of freedom and liberty. French designer and sculptor Frederic Auguste Bartholdi oversaw the building of the statue and chose its location in New York Harbor. The statue was dedicated on October 28, 1886, under its official title *Liberty Enlightening the World*. The poem "The New Colossus," by Emma Lazarus, was placed on the statue in 1903. Its verse, "Give me your tired, your poor, your huddled masses yearning to breathe free," identifies the statue as a powerful symbol of welcome to immigrants from all over the world. In Exercises 39 and 40 on page 700, we will explore some measurements of parts of the Statue of Liberty.

Name _____ Section _____ Date_____

Chapter 9 Pretest

1. _____

Classify each angle as acute, right, obtuse, or straight.

△ **1.**
A

2. _____

△ **2.**
B

△ **3.** Find the supplement of a 92° angle.

3. _____

△ **4.** Find the measures of *x*, *y*, and *z*.

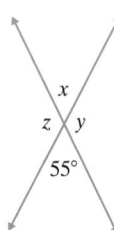
x
z *y*
55°

4. _____

5. _____

△ **5.** Find the measure of ∠*x*.

6. _____

6. Convert 11 feet to yards.

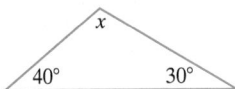
x
40°   30°

7. Convert 6,250,000 cm to kilometers.

7. _____

△ **8.** Find the perimeter of a rectangle with a length of 24 inches and a width of 6 inches.

8. _____

△ **9.** Find the circumference of the given circle. Use $\pi \approx 3.14$.

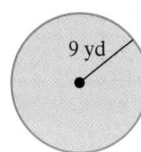
9 yd

9. _____

10. _____

△ **10.** Find the area of the given triangle.

11. _____

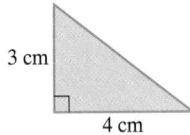
3 cm
4 cm

12. _____

11. Divide 15 lb 8 oz by 2.

12. Subtract 2 qt from 9 gal 1 qt.

13. _____

13. Convert 25 L to milliters.

14. _____

14. Convert 118°F to Celsius. If necessary, round to the nearest tenth of a degree.

△ **9.1 Lines and Angles**

Ⓐ Identify Lines, Line Segments, Rays, and Angles

Let's begin with a review of two important concepts—plane and space.

A **plane** is a flat surface that extends indefinitely in all directions. Surfaces like portions of a plane are a classroom floor or a blackboard or whiteboard.

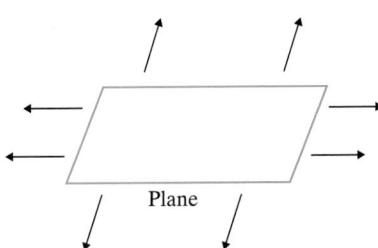

Plane

Space extends in all directions indefinitely. Examples of objects in space are houses, grains of salt, bushes, your *Prealgebra* textbook, and you.

The most basic concept of geometry is the idea of a point in space. A **point** has no length, no width, and no height, but it does have location. We represent a point by a dot, and we label points with letters.

Point *P*

A **line** is a set of points extending indefinitely in two directions. A line has no width or height, but it does have length. We name a line by any two of its points. The line below to the left is named $\overleftrightarrow{AB}$ or $\overleftrightarrow{BA}$. A **line segment** is a piece of a line with two endpoints. The line segment below with endpoints *A* and *B* is named $\overline{AB}$ or $\overline{BA}$.

Line *AB* or $\overleftrightarrow{AB}$ Line segment *AB* or $\overline{AB}$

A **ray** is a part of a line with one endpoint. A ray extends indefinitely in one direction. To name a ray, name its endpoint first, then any other point on the ray. Below is ray *AB*, or $\overrightarrow{AB}$. An **angle** is made up of two rays that share the same endpoint. The common endpoint is called the **vertex**.

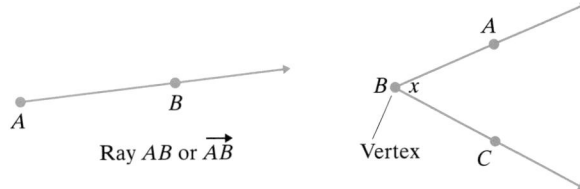

Ray *AB* or $\overrightarrow{AB}$ Vertex

The angle in the figure above can be named

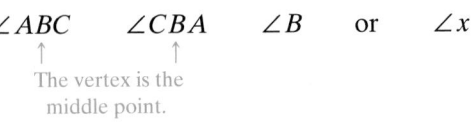

$$\angle ABC \qquad \angle CBA \qquad \angle B \qquad or \qquad \angle x$$

The vertex is the middle point.

Helpful Hint

Use the vertex alone to name an angle only when there is no confusion as to what angle is being named.

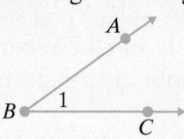

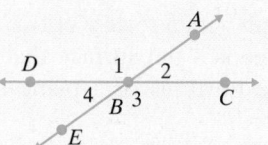

You may use ∠B to name the angle above. There is no confusion. ∠B means ∠1.

Do not use ∠B to name the angle above There is confusion. Does ∠B mean ∠1, ∠2, ∠3, or ∠4?

Rays *BA* and *BC* are **sides** of the angle.

Practice Problem 1

Identify each figure as a line, a ray, a line segment, or an angle. Then name the figure using the given points.

a. 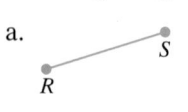   b.

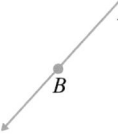

c.   d.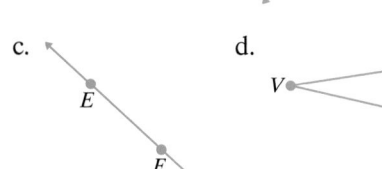

EXAMPLE 1 Identify each figure as a line, a ray, a line segment, or an angle. Then name the figure using the given points.

a. 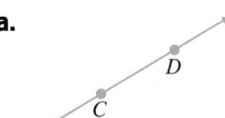   b.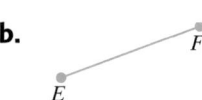

c. 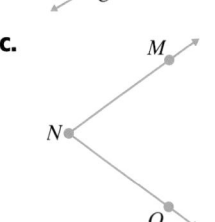   d.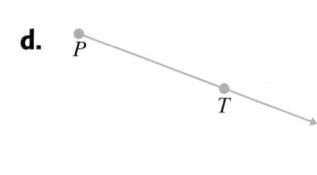

Solution: Figure (a) extends indefinitely in two directions. It is line *CD* or $\overleftrightarrow{CD}$. It can also be named line *DC* or $\overleftrightarrow{DC}$.

Figure (b) has two endpoints. It is line segment *EF* or $\overline{EF}$. It can also be named line segment *FE* or $\overline{FE}$.

Figure (c) has two rays with a common endpoint. It is an angle that can be named ∠*MNO*, ∠*ONM*, or ∠*N*.

Figure (d) is part of a line with one endpoint. It is ray *PT* or $\overrightarrow{PT}$. ●

EXAMPLE 2 List other ways to name ∠*y*.

Solution: Two other ways to name ∠*y* are ∠*QTR* and ∠*RTQ*. We may *not* use the vertex alone to name this angle because three different angles have *T* as their vertex.

Practice Problem 2

Use the figure in Example 2 to list other ways to name ∠*z*.

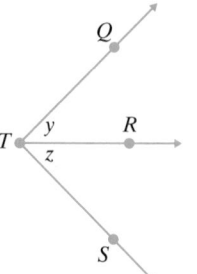

Answers

1. a. line segment; line segment *RS* or $\overline{RS}$ as well as *SR* or $\overline{SR}$.

  b. ray; ray *AB* or $\overrightarrow{AB}$

  c. line; line *EF* or $\overleftrightarrow{EF}$ as well as *FE* or $\overleftrightarrow{FE}$

  d. angle; ∠*TVH* or ∠*HVT* or ∠*V*

2. ∠*RTS*, ∠*STR*

Ⓑ Classifying Angles as Acute, Right, Obtuse, or Straight

An angle can be measured in **degrees**. The symbol for degrees is a small, raised circle, °. There are 360° in a full revolution, or full circle.

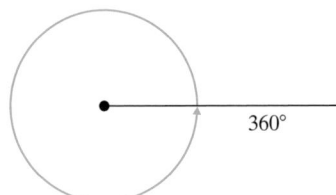

$\frac{1}{2}$ of a revolution measures $\frac{1}{2}(360°) = 180°$. An angle that measures $180°$ is called a **straight angle**.

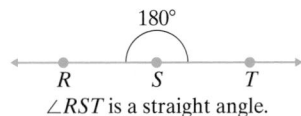

$\angle RST$ is a straight angle.

$\frac{1}{4}$ of a revolution measures $\frac{1}{4}(360°) = 90°$. An angle that measures $90°$ is called a **right angle**. The symbol ∟ is used to denote a right angle.

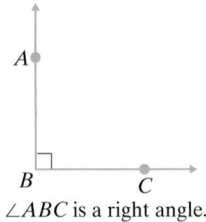

$\angle ABC$ is a right angle.

An angle whose measure is between $0°$ and $90°$ is called an **acute angle**.

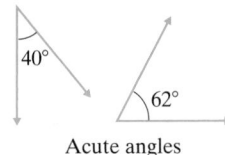

Acute angles

An angle whose measure is between $90°$ and $180°$ is called an **obtuse angle**.

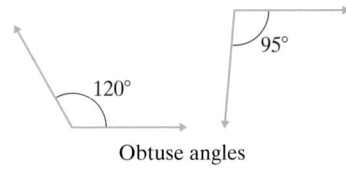

Obtuse angles

EXAMPLE 3   Classify each angle as acute, right, obtuse, or straight.

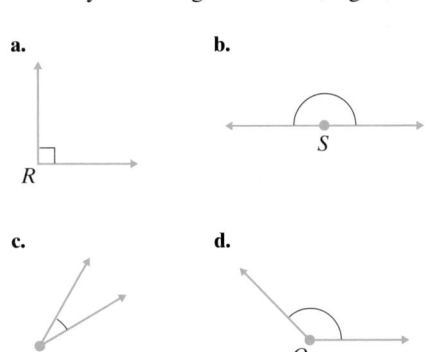

Practice Problem 3

Classify each angle as acute, right, obtuse, or straight.

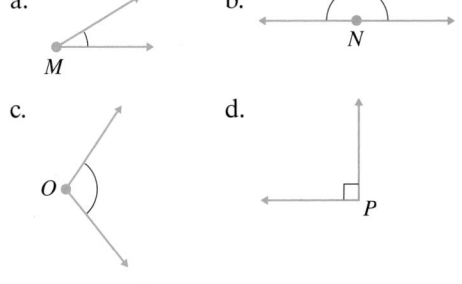

Answer

3. a. acute   **b.** straight   **c.** obtuse   **d.** right

Solution:

a. $\angle R$ is a right angle, denoted by ⌐ .

b. $\angle S$ is a straight angle.

c. $\angle T$ is an acute angle. It measures between 0° and 90°.

d. $\angle Q$ is an obtuse angle. It measures between 90° and 180°.

C Identifying Complementary and Supplementary Angles

Two angles that have a sum of 90° are called **complementary angles**. We say that each angle is the **complement** of the other.

$\angle R$ and $\angle S$ are complementary angles because

60° + 30° = 90°

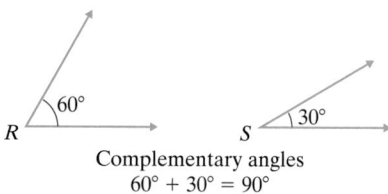

Complementary angles
60° + 30° = 90°

Two angles that have a sum of 180° are called **supplementary angles**. We say that each angle is the **supplement** of the other.

$\angle M$ and $\angle N$ are supplementary angles because

125° + 55° = 180°

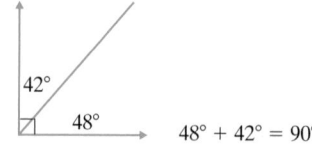

Supplementary angles
125° + 55° = 180°

Practice Problem 4

Find the complement of a 36° angle.

EXAMPLE 4 Find the complement of a 48° angle.

Solution: The complement of an angle that measures 48° is an angle that measures 90° − 48° = 42°.

Check:

42°

48°   48° + 42° = 90°

Practice Problem 5

Find the supplement of an 88° angle.

Concept Check

True or false? The supplement of a 48° angle is 42°. Explain.

EXAMPLE 5 Find the supplement of a 107° angle.

Solution: The supplement of an angle that measures 107° is an angle that measures 180° − 107° = 73°.

Check:

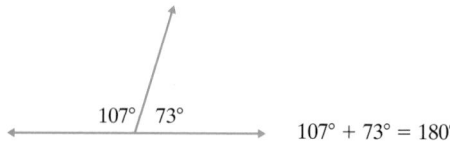

107° 73°   107° + 73° = 180°

Answers

4. 54° **5.** 92°

Concept Check: false; the complement of a 48° angle is 42°; the supplement of a 48° angle is 132°

Try the Concept Check in the margin.

Ⓓ **Finding Measures of Angles**

Measures of angles can be added or subtracted to find measures of related angles.

EXAMPLE 6   Find the measure of $\angle x$.

Solution: The measure of $\angle x = 87° - 52° = 35°$

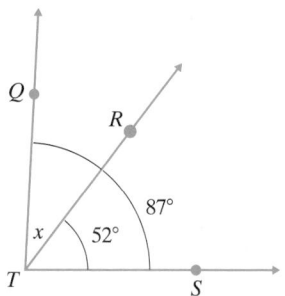

Two lines in a plane can be either parallel or intersecting. **Parallel lines** never meet. **Intersecting lines** meet at a point. The symbol ∥ is used to indicate "is parallel to." For example, in the figure $p\|q$.

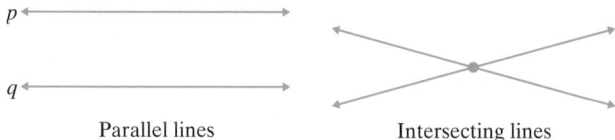

Parallel lines   Intersecting lines

Some intersecting lines are perpendicular. Two lines are **perpendicular** if they form right angles when they intersect. The symbol ⊥ is used to denote "is perpendicular to." For example, in the figure below, $n \perp m$.

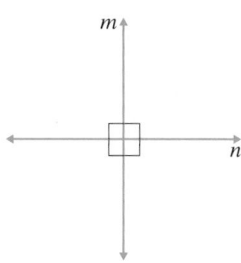

Perpendicular lines

When two lines intersect, four angles are formed. Two of these angles that are opposite each other are called **vertical angles**. Vertical angles have the same measure.

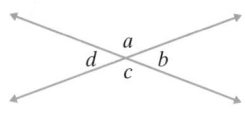

Vertical angles:
$\angle a$ and $\angle c$
$\angle d$ and $\angle b$

Practice Problem 6

Find the measure of $\angle y$.

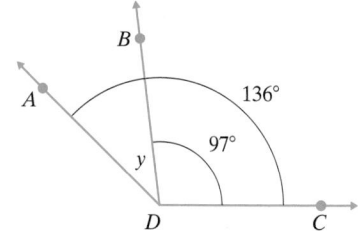

Answer

6. $39°$

Practice Problem 7

Find the measure of ∠ABY and ∠ZBA.

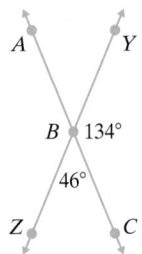

EXAMPLE 7 Find the measures of ∠FDG and ∠GDC.

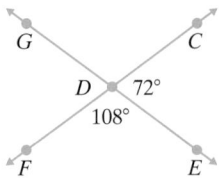

Solution: ∠FDG and ∠CDE are vertical angles, so they have the same measure.

measure of ∠FDG = 72°

∠GDC and ∠FDE are vertical angles, so they have the same measure.

measure of ∠GDC = 108°

Notice in Example 7 that ∠CDE and ∠EDF share a common side, ray DE. When this happens, the angles are called *adjacent angles*.

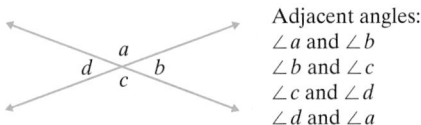

Adjacent angles:
∠a and ∠b
∠b and ∠c
∠c and ∠d
∠d and ∠a

Also notice in Example 7 that adjacent angles ∠CDE and ∠EDF are also supplementary angles since the sum of their measures is 180°.

72° + 108° = 180°

This is true in general. Adjacent angles formed by intersecting lines are supplementary.

Practice Problem 8

Find the measure of ∠a, ∠b, and ∠c.

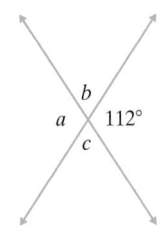

EXAMPLE 8 Find the measure of ∠x, ∠y, and ∠z if the measure of ∠t is 42°.

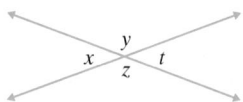

Solution: Since ∠t and ∠x are vertical angles, they have the same measure, so ∠x measures 42°.

Since ∠t and ∠y are adjacent angles, their measures have a sum of 180°, so ∠y measures 180° − 42° = 138°.

Since ∠y and ∠z are vertical angles, they have the same measure. So ∠z measures 138°.

A line that intersects two or more lines at different points is called a **transversal**. Line *l* is a transversal that intersects lines *m* and *n*.

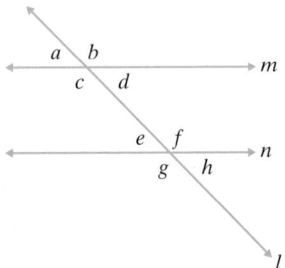

Answers

7. ∠ABY measures 46°.
∠ZBA measures 134°.

8. ∠a = 112°; ∠b = 68°; ∠c = 68°

The eight angles formed have special names. To understand these names, let's first learn what angles are called interior and what angles are called exterior by the illustration below.

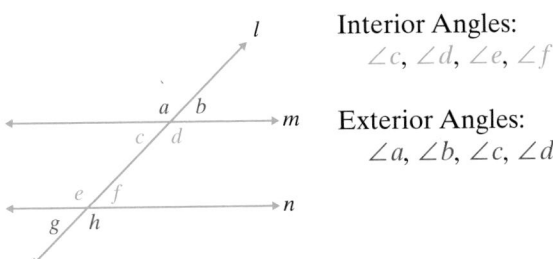

Interior Angles:
∠c, ∠d, ∠e, ∠f

Exterior Angles:
∠a, ∠b, ∠c, ∠d

Now that you know which angles are called interior angles, below we illustrate which pairs of these angles we call alternate interior angles.

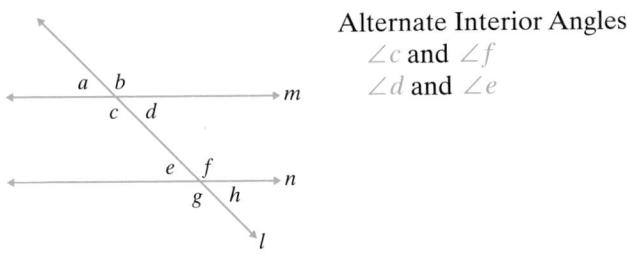

Alternate Interior Angles:
∠c and ∠f
∠d and ∠e

Another special name we give certain pairs of angles is corresponding angles. If lines *m* and *n* are cut by transveral *l*, then corresponding angles have the same exact position: one with part of line *m* as a side and one with part of line *n* as a side.

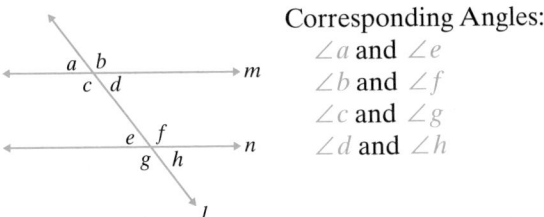

Corresponding Angles:
∠a and ∠e
∠b and ∠f
∠c and ∠g
∠d and ∠h

Note: There are other special names for angles formed by two lines cut by a transversal, such as alternate exterior angles, but we will not review them here.

When two lines cut by a transversal are *parallel*, the following are true:

Parallel Lines Cut by a Transversal

If two parallel lines are cut by a transversal, then the measures of **corresponding angles are equal** and **alternate interior angles are equal**.

Practice Problem 9

Given that $m\|n$ and that the measure of $\angle w = 40°$, find the measures of $\angle x$, $\angle y$, and $\angle z$.

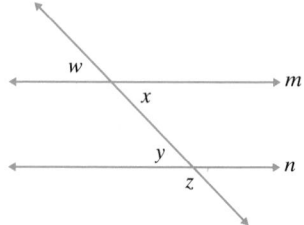

Answer

9. $\angle x = 40°$; $\angle y = 40°$; $\angle z = 140°$

EXAMPLE 9 Given that $m\|n$ and that the measure of $\angle w$ is $100°$, find the measures of $\angle x$, $\angle y$, and $\angle z$.

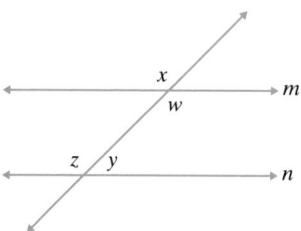

Solution:

The measure of $\angle x = 100°$. $\angle x$ and $\angle w$ are vertical angles.

The measure of $\angle z = 100°$. $\angle x$ and $\angle z$ are corresponding angles.

The measure of $\angle y = 180° - 100° = 80°$. $\angle z$ and $\angle y$ are supplementary angles. ●

STUDY SKILLS REMINDER

How are you doing?

If you haven't done so yet, take a few moments and think about how you are doing in this course. Are you working toward your goal of successfully completing this course? Is your performance on homework, quizzes, and tests satisfactory? If not, you might want to see your instructor to see if he/she has any suggestions on how you can improve your performance. Let me once again remind you that, in addition to your instructor, there are many places to get help with your mathematics course. A few suggestions are below.

- This text has an accompanying video lesson for every text section.

- The back of this book contains answers to odd-numbered exercises and selected solutions.

- MathPro is available with this text. It is a tutorial software program with lessons corresponding to each text section.

- There is a student solutions manual available that contains worked-out solutions to odd-numbered exercises as well as solutions to every exercise in the Chapter Pretests, Integrated Reviews, Chapter Reviews, Chapter Tests, and Cumulative Reviews.

- Don't forget to check with your instructor for other local resources available to you, such as a tutoring center.

EXERCISE SET 9.1

A *Identify each figure as a line, a ray, a line segment, or an angle. Then name the figure using the given points. See Examples 1 and 2.*

△ **1.**
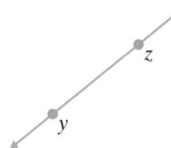

△ **2.**
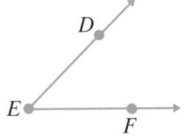

△ **3.**

△ **4.**
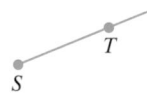

△ **5.**
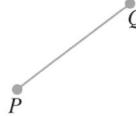

△ **6.**
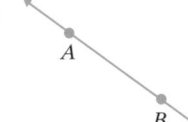

△ **7.**
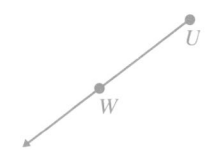

△ **8.**
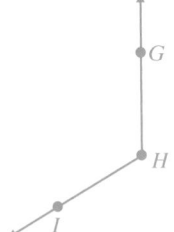

B *Find the measure of each angle in the following figure:*

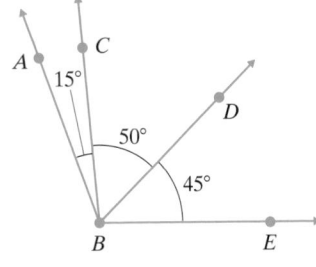

 **9.** ∠*ABC*

△ **10.** ∠*EBD*

 11. ∠*CBD*
△

△ **12.** ∠*CBA*

△ **13.** ∠*DBA*

△ **14.** ∠*EBC*

△ **15.** ∠*CBE*

△ **16.** ∠*ABE*

Fill in each blank. See Example 3.

△ **17.** A right angle has a measure of _____.

△ **18.** A straight angle has a measure of _____.

△ **19.** An acute angle measures between _____ and _____.

△ **20.** An obtuse angle measures between _____ and _____.

Classify each angle as acute, right, obtuse, or straight. See Example 3.

 21.
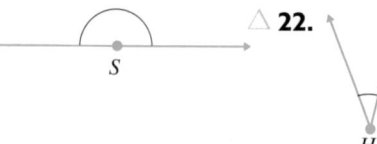 **22.**
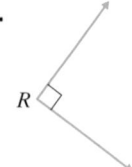 **23.**
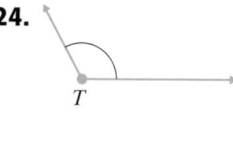 **24.**

S

H

R

T

25.
 Q

26.
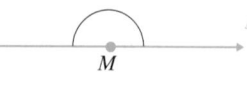 M

27. N
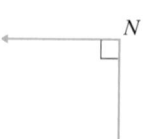

28. P

C *Find each complementary or supplementary angle as indicated. See Examples 4 and 5.*

29. Find the complement of a 17° angle.

30. Find the complement of an 87° angle.

31. Find the supplement of a 17° angle.

32. Find the supplement of an 87° angle.

33. Find the complement of a 48° angle.

34. Find the complement of a 22° angle.

35. Find the supplement of a 125° angle.

36. Find the supplement of a 155° angle.

37. Identify the pairs of complementary angles.

38. Identify the pairs of complementary angles.

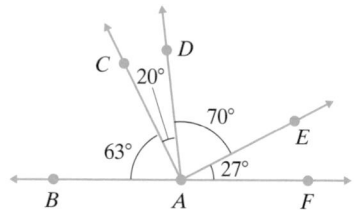

39. Identify the pairs of supplementary angles.

40. Identify the pairs of supplementary angles.

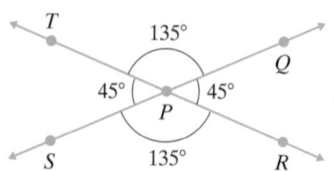

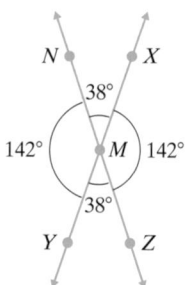

△ **41.** The angle between the two walls of the Vietnam Veterans Memorial in Washington, D.C., is 125.2°. Find the supplement of this angle. (*Source:* National Park Service)

△ **42.** The faces of Khafre's Pyramid at Giza, Egypt, are inclined at an angle of 53.13°. Find the complement of this angle. (*Source:* PBS *NOVA* Online)

D *Find the measure of ∠x in each figure. See Example 6.*

△ **43.**

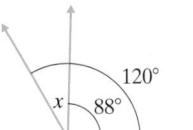

△ **44.**

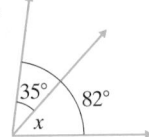

△ **45.**

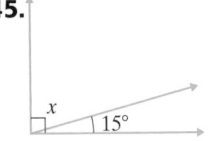

△ **46.**

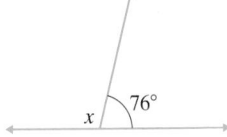

Find the measures of angles x, y, and z in each figure. See Examples 7, 8, and 9.

△ **47.**

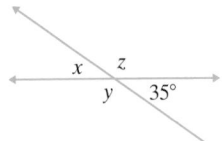

△ **48.**

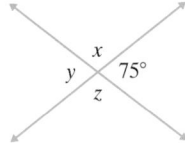

△ **49.**

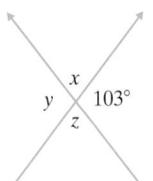

△ **50.**

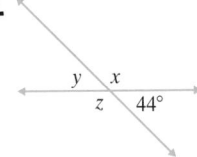

△ **51.** m∥n

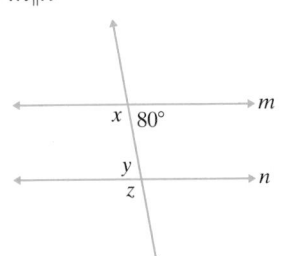

△ **52.** m∥n

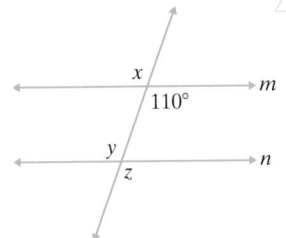

△ **53.** m∥n

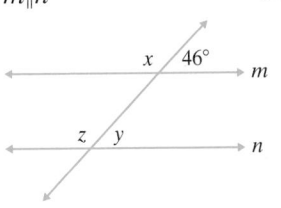

△ **54.** m∥n

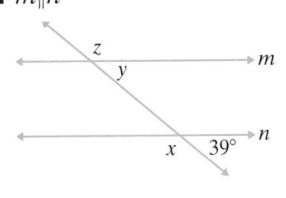

Review and Preview

Perform each indicated operation. See Sections 4.3, 4.5, and 4.8.

55. $\dfrac{7}{8} + \dfrac{1}{4}$

56. $\dfrac{7}{8} - \dfrac{1}{4}$

57. $\dfrac{7}{8} \cdot \dfrac{1}{4}$

58. $\dfrac{7}{8} \div \dfrac{1}{4}$

59. $3\dfrac{1}{3} - 2\dfrac{1}{2}$

60. $3\dfrac{1}{3} + 2\dfrac{1}{2}$

61. $3\dfrac{1}{3} \div 2\dfrac{1}{2}$

62. $3\dfrac{1}{3} \cdot 2\dfrac{1}{2}$

◆ Combining Concepts

△ **63.** If lines *m* and *n* are parallel, find the measures of angles *a* through *e*.

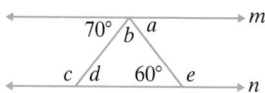

64. In your own words, describe how to find the complement and the supplement of a given angle.
△

△ **65.** Find two complementary angles with the same measure.

△ **66.** Can two supplementary angles both be acute?

67. Is the figure below possible? Why or why not?
△

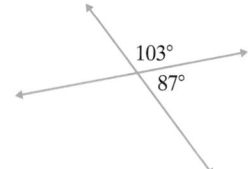

△ **68.** Below is a rectangle. List which segments, if extended, would be parallel lines.

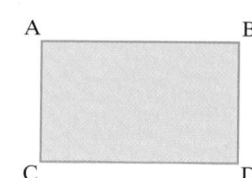

9.2   Linear Measurement

A   Defining and Converting U.S. System Units of Length

In the United States, two systems of measurement are commonly used. They are the **United States (U.S.), or English, measurement system** and the **metric system**. The U.S. measurement system is familiar to most Americans. Units such as feet, miles, ounces, and gallons are used. However, the metric system is also commonly used in fields such as medicine, sports, international marketing, and certain physical sciences. We are accustomed to buying 2-liter bottles of soft drinks, watching televised coverage of the 100-meter dash at the Olympic Games, or taking a 200-milligram dose of pain reliever.

The U.S. system of measurement uses the **inch**, **foot**, **yard**, and **mile** to measure **length**. The following is a summary of equivalencies between units of length:

U.S. Units of Length	Unit Fractions
12 inches (in.) = 1 foot (ft)	$\dfrac{12 \text{ in.}}{1 \text{ ft}} = \dfrac{1 \text{ ft}}{12 \text{ in.}} = 1$
3 feet = 1 yard (yd)	$\dfrac{3 \text{ ft}}{1 \text{ yd}} = \dfrac{1 \text{ yd}}{3 \text{ ft}} = 1$
5280 feet = 1 mile (mi)	$\dfrac{5280 \text{ ft}}{1 \text{ mi}} = \dfrac{1 \text{ mi}}{5280 \text{ ft}} = 1$

To convert from one unit of length to another, **unit fractions** may be used. A unit fraction is a fraction that equals 1. For example, since 12 in. = 1 ft, we have the unit fractions

$$\frac{12 \text{ in.}}{1 \text{ ft}} = \frac{1 \text{ ft}}{12 \text{ in.}} = 1$$

For example, to convert 48 inches to feet, we *multiply by a unit fraction that relates feet to inches*. The unit fraction should be written so that *the units we are converting to*, feet, *are in the numerator and the original units*, inches, *are in the denominator*. We do this so that like units will divide out, as shown next:

$$
\begin{aligned}
48 \text{ in.} &= \frac{48 \text{ in.}}{1} \cdot \frac{1 \text{ ft}}{12 \text{ in.}} \qquad \leftarrow \text{ Units to convert to} \\
&\qquad\qquad\qquad\quad \leftarrow \text{ Original units} \\
&= \frac{48 \cdot 1 \text{ ft}}{1 \cdot 12} \\
&= \frac{48 \text{ ft}}{12} \\
&= 4 \text{ ft}
\end{aligned}
$$

Therefore, 48 inches equals 4 feet, as seen in the diagram:

```
  12 in.   12 in.   12 in.   12 in.
 ┌──────┬──────┬──────┬──────┐
 │      │      │      │      │  48 in. = 4 ft
 └──────┴──────┴──────┴──────┘
  1 ft    1 ft    1 ft    1 ft
```

OBJECTIVES

A   Define U.S. units of length and convert from one unit to another.

B   Use mixed units of length.

C   Perform arithmetic operations on U.S. units of length.

D   Define metric units of length and convert from one unit to another.

E   Perform arithmetic operations on metric units of length.

SSM   TUTOR CENTER   SG   CD & VIDEO   MATH PRO   WEB

Helpful Hint:

When converting from one unit to another, select a unit fraction with the properties below:

$$\frac{\text{units you are converting to}}{\text{original units}}$$

By using this unit fraction, the original units will divide out, as wanted.

Practice Problem 1

Convert 5 feet to inches.

EXAMPLE 1 Convert 8 feet to inches.

Solution: We multiply 8 feet by a unit fraction that compares 12 inches to 1 foot. The unit fraction should be $\dfrac{\text{units to convert to}}{\text{original units}}$ or $\dfrac{12 \text{ inches}}{1 \text{ foot}}$.

$$8 \text{ ft} = \frac{8 \cancel{\text{ft}}}{1} \cdot \overbrace{\frac{12 \text{ in.}}{1 \cancel{\text{ft}}}}^{\text{Unit fraction}}$$

$$= 8 \cdot 12 \text{ in.}$$

$$= 96 \text{ in.} \qquad \text{Multiply.}$$

Thus, 8 ft = 96 in., as shown in the diagram:

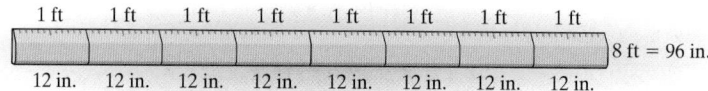

Practice Problem 2

Convert 7 yards to feet.

EXAMPLE 2 Convert 7 feet to yards.

Solution: We multiply by a unit fraction that compares 1 yard to 3 feet.

$$7 \text{ ft} = \frac{7 \cancel{\text{ft}}}{1} \cdot \frac{1 \text{ yd}}{3 \cancel{\text{ft}}} \qquad \leftarrow \text{Units to convert to}$$
$$\qquad\qquad\qquad \leftarrow \text{Original units}$$

$$= \frac{7 \text{ yd}}{3}$$

$$= 2\frac{1}{3} \text{ yd} \qquad \text{Divide.}$$

Thus, 7 ft = $2\dfrac{1}{3}$ yd.

B Using Mixed U.S. System Units of Length

Sometimes it is more meaningful to express a measurement of length with mixed units, such as 1 ft and 5 in. We usually condense this and write 1 ft 5 in.

In Example 2, we found that 7 feet was the same as $2\dfrac{1}{3}$ yards. The measurement can also be written as a mixture of yards and feet. That is,

$$7 \text{ ft} = \underline{\quad} \text{ yd} \underline{\quad} \text{ ft}$$

Because 3 ft = 1 yd, we divide 3 into 7 to see how many whole yards are in 7 feet. The quotient is the number of yards, and the remainder is the number of feet.

$$\begin{array}{r} 2 \text{ yd } 1 \text{ ft} \\ 3\overline{)7} \\ \underline{-6} \\ 1 \end{array}$$

Thus, 7 ft = 2 yd 1 ft, as seen in the diagram:

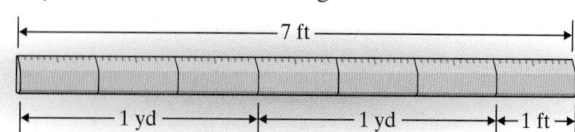

EXAMPLE 3   Convert: 134 in. = ____ ft ____ in.

Solution: Because 12 in. = 1 ft, we divide 12 into 134. The quotient is the number of feet. The remainder is the number of inches. To see why we divide 12 into 134, notice that

$$134 \text{ in.} = \frac{134 \text{ in.}}{1} \cdot \frac{1 \text{ ft}}{12 \text{ in.}} = \frac{134}{12} \text{ ft}$$

$$\begin{array}{r} 11 \text{ ft } 2 \text{ in.} \\ 12\overline{)134} \\ -12 \\ \hline 14 \\ -12 \\ \hline 2 \end{array}$$

Thus, 134 in. = 11 ft 2 in.

EXAMPLE 4   Convert 3 feet 7 inches to inches.

Solution: First, we will convert 3 feet to inches. Then we add 7 inches.

$$3 \text{ ft} = \frac{3 \text{ ft}}{1} \cdot \frac{12 \text{ in.}}{1 \text{ ft}} = 36 \text{ in.}$$

Then

$$3 \text{ ft } 7 \text{ in.} = 36 \text{ in.} + 7 \text{ in.} = 43 \text{ in.}$$

C   Performing Operations on U.S. System Units of Length

Finding sums or differences of measurements often involves converting units, as shown in the next example. Just remember that, as usual, only like units can be added or subtracted.

EXAMPLE 5   Add 3 ft 2 in. and 5 ft 11 in.

Solution: To add, we line up the similar units.

$$\begin{array}{r} 3 \text{ ft } 2 \text{ in.} \\ +5 \text{ ft } 11 \text{ in.} \\ \hline 8 \text{ ft } 13 \text{ in.} \end{array}$$

Since 13 inches is the same as 1 ft 1 in., we have

$$8 \text{ ft } 13 \text{ in.} = 8 \text{ ft } + 1 \text{ ft } 1 \text{ in.}$$
$$= 9 \text{ ft } 1 \text{ in.}$$

Try the Concept Check in the margin.

EXAMPLE 6   Multiply 8 ft 9 in. by 3.

Solution: By the distributive property, we multiply 8 ft by 3 and 9 in. by 3.

$$\begin{array}{r} 8 \text{ ft } 9 \text{ in.} \\ \times 3 \\ \hline 24 \text{ ft } 27 \text{ in.} \end{array}$$

Practice Problem 3

Convert: 68 in. = ____ ft ____ in.

Practice Problem 4

Convert 5 yards 2 feet to feet.

Practice Problem 5

Add 4 ft 8 in. to 8 ft 11 in.

Concept Check

How could you estimate the following sum?

$$\begin{array}{r} 7 \text{ yd } 4 \text{ in.} \\ +3 \text{ yd } 27 \text{ in.} \end{array}$$

Practice Problem 6

Multiply 4 ft 7 in. by 4.

Answers

3. 5 ft 8 in.   **4.** 17 ft   **5.** 13 ft 7 in.   **6.** 18 ft 4 in.

Concept Check: round each to the nearest yard:
7 yd + 4 yd = 11 yd

Since 27 in. is the same as 2 ft 3 in., we simplify the product as

24 ft 27 in. = 24 ft + 2 ft 3 in.

= 26 ft 3 in.

Practice Problem 7

Divide 18 ft 6 in. by 2.

EXAMPLE 7   Divide 24 yd 6 in. by 3.

Solution: We divide each of the units by 3.

$$
\begin{array}{r}
8 \text{ yd } 2 \text{ in.} \\
3\overline{)24 \text{ yd } 6 \text{ in.}} \\
-24 \text{ yd} \\
\hline
6 \text{ in.} \\
-6 \text{ in.} \\
\hline
0
\end{array}
$$

The quotient is 8 yd 2 in.

To check, see that 8 yd 2 in. multiplied by 3 is 24 yd 6 in.

Practice Problem 8

A carpenter cuts 1 ft 9 in. from a board of length 5 ft 8 in. Find the remaining length of the board.

EXAMPLE 8   Finding the Length of a Piece of Rope

A rope of length 6 yd 1 ft has 2 yd 2 ft cut from one end. Find the length of the remaining rope.

Solution: Subtract 2 yd 2 ft from 6 yd 1 ft.

beginning length	→	6 yd 1 ft
− amount cut	→	−2 yd 2 ft
remaining length		

We cannot subtract 2 ft from 1 ft, so we borrow 1 yd from the 6 yd. One yard is converted to 3 ft and combined with the 1 ft already there.

Borrow 1 yd = 3 ft   The problem now reads:

5 yd + 1 yd 3 ft

$$
\begin{array}{r}
\cancel{6 \text{ yd}} \ 1 \text{ ft} \\
- \ 2 \text{ yd } 2 \text{ ft} \\
\end{array}
\qquad
\begin{array}{r}
5 \text{ yd } 4 \text{ ft} \\
- \ 2 \text{ yd } 2 \text{ ft} \\
\hline
3 \text{ yd } 2 \text{ ft}
\end{array}
$$

The remaining rope is 3 yd 2 ft long.

Ⓓ Defining and Converting Metric System Units of Length

The basic unit of length in the metric system is the **meter**. A meter is slightly longer than a yard. It is approximately 39.37 inches long. Recall that a yard is 36 inches long.

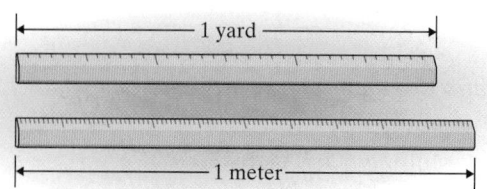

All units of length in the metric system are based on the meter. The following is a summary of the prefixes used in the metric system. Also shown are equivalencies between units of length. Like the decimal system, the metric system uses powers of 10 to define units.

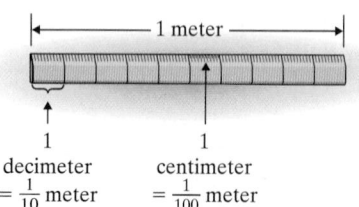

1
decimeter
= $\frac{1}{10}$ meter

1
centimeter
= $\frac{1}{100}$ meter

Answers

7. 9 ft 3 in.   **8.** 3 ft 11 in.

Prefix	Meaning	Metric Unit of Length	
kilo	1000	1 **kilo**meter (km) = 1000 meters (m)	
hecto	100	1 **hecto**meter (hm) = 100 m	
deka	10	1 **deka**meter (dam) = 10 m	
		1 meter (m) = 1 m	
deci	1/10	1 **deci**meter (dm) = 1/10 m	or 0.1 m
centi	1/100	1 **centi**meter (cm) = 1/100 m	or 0.01 m
milli	1/1000	1 **milli**meter (mm) = 1/1000 m	or 0.001 m

These same prefixes are used in the metric system for mass and capacity. The most commonly used measurements of length in the metric system are the **meter**, **millimeter**, **centimeter**, and **kilometer**.

Being comfortable with the metric units of length means gaining a "feeling" for metric lengths, just as you have a "feeling" for the length of an inch, a foot, and a mile. To help you accomplish this, study the following examples:

A millimeter is about the thickness of a large paper clip.

A centimeter is about the width of a large paper clip.

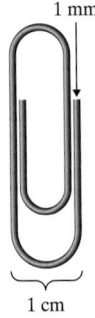

1 mm

1 cm

A meter is slightly longer than a yard.

A kilometer is about two-thirds of a mile.

$2\frac{1}{2}$ centimeters is about 1 inch.

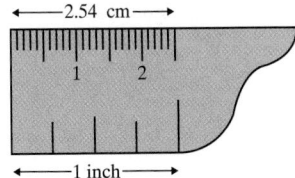

2.54 cm

1

2

1 inch

The length of this book is approximately 27.5 centimeters.

The width of this book is approximately 21.5 centimeters.

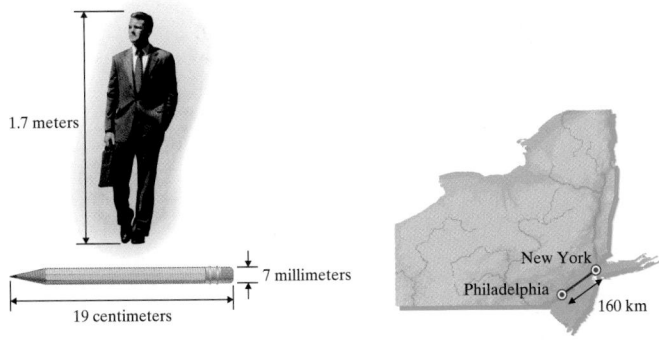

1.7 meters

7 millimeters

19 centimeters

New York

Philadelphia

160 km

As with the U.S. system of measurement, unit fractions may be used to convert from one unit of length to another. The metric system does, however, have a distinct advantage over the U.S. system of measurement: The ease of converting from one unit of length to another. Since all units of length are powers of 10 of the meter, converting from one unit of length to another is as simple as moving the decimal point. Listing units of length in order from largest to smallest helps to keep track of how many places to move the decimal point when converting.

For example, let's convert 1200 meters to kilometers. To convert from meters to kilometers, we move along the chart below 3 units to the left, from meters to kilometers. This means that we move the decimal point 3 places to the left.

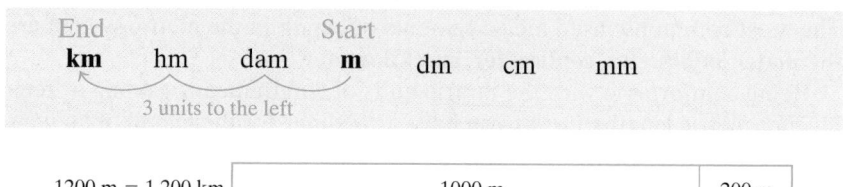

The same conversion can be made using unit fractions.

$$1200 \text{ m} = \frac{1200 \ \cancel{\text{m}}}{1} \cdot \overbrace{\frac{1 \text{ km.}}{1000 \ \cancel{\text{m}}}}^{\text{Unit fraction}} = \frac{1200 \text{ km}}{1000} = 1.2 \text{ km}$$

EXAMPLE 9   Convert 2.3 m to centimeters.

Solution: First we will convert by using a unit fraction.

$$2.3 \text{ m} = \frac{2.3 \ \cancel{\text{m}}}{1} \cdot \overbrace{\frac{1 \text{ km.}}{1000 \ \cancel{\text{m}}}}^{\text{Unit fraction}} = 230 \text{ cm}$$

Now we will convert by listing the units of length in a chart and moving from meters to centimeters.

km   hm   dam   **m**   dm   **cm**   mm

Start   End

2 units to the right

2.30 m = 230. cm or 230 cm

2 places to the right

With either method, we get 230 cm.

EXAMPLE 10   Convert 450,000 mm to meters.

Solution: We list the units of length in a chart and move from millimeters to meters.

km   hm   dam   **m**   dm   cm   **mm**

End   Start

3 units to the left

450,000 mm = 450.000 m   or   450 m

3 places to the left

Try the Concept Check in the margin.

Practice Problem 9

Convert 3.5 m to kilometers.

Practice Problem 10

Convert 2.5 m to millimeters.

Concept Check

What is wrong with the following conversion of 150 cm to meters?

150.00 cm = 15,000 m

Answers

9. 0.0035 km   **10.** 2500 mm

Concept Check: decimal should be moved to the left: 1.5 m

Performing Operations on Metric System Units of Length

To add, subtract, multiply, or divide with metric measurements of length, we write all numbers using the same unit of length and then add, subtract, multiply, or divide as with decimals.

EXAMPLE 11   Subtract 430 m from 1.3 km.

Solution: First we convert both measurements to kilometers or both to meters.

$$430 \text{ m} = 0.43 \text{ km} \quad\quad \text{or} \quad\quad 1.3 \text{ km} = 1300 \text{ m}$$

$$
\begin{array}{r}
1.30 \text{ km} \\
- \; 0.43 \text{ km} \\
\hline
0.87 \text{ km}
\end{array}
\quad\quad
\begin{array}{r}
1300 \text{ m} \\
- \; 430 \text{ m} \\
\hline
870 \text{ m}
\end{array}
$$

The difference is 0.87 km or 870 m. ●

EXAMPLE 12   Multiply 5.7 mm by 4.

Solution: Here we simply multiply the two numbers. Note that the unit of measurement remains the same.

$$
\begin{array}{r}
5.7 \text{ mm} \\
\times \quad 4 \\
\hline
22.8 \text{ mm}
\end{array}
$$

EXAMPLE 13   Finding a Person's Height

Fritz Martinson was 1.2 meters tall on his last birthday. Since then, he has grown 14 centimeters. Find his current height in meters.

Solution:
$$
\begin{array}{lll}
\text{original height} & \rightarrow & 1.20 \text{ m} \\
+ \text{ height grown} & \rightarrow & + \; 0.14 \text{ m} \quad \text{(Since 14 cm} = 0.14 \text{ m)} \\
\hline
\text{current height} & & 1.34 \text{ m}
\end{array}
$$

Fritz is now 1.34 meters tall. ●

Practice Problem 11

Subtract 640 m from 2.1 km.

Practice Problem 12

Multiply 18.3 hm by 5.

Practice Problem 13

Doris Blackwell is knitting a scarf that is currently 0.8 meter long. If she knits an additional 45 centimeters, how long will the scarf be?

Answers

11. 1.46 km or 1460 m   **12.** 91.5 hm
13. 125 cm or 1.25 m

FOCUS ON **Mathematical Connections**

Copyright 2004 Pearson Education, Inc.

POLYGONS

A **plane figure** is a figure that lies on a plane. Plane figures, like planes, have length and width but no thickness or depth.

A **polygon** is a closed plane figure that basically consists of three or more line segments that meet at their endpoints. A polygon is named according to the number of its sides. The table summarizes information about some basic polygons.

Polygons			
Number of Sides	Name	Sum of Interior Angles	Example
3	Triangle	180°	
4	Quadrilateral	360°	
5	Pentagon	540°	
6	Hexagon	720°	
7	Heptagon		
8	Octagon		
9	Nonagon		
10	Decagon		

CRITICAL THINKING

Study the sums of interior angles shown in the table for the triangle, quadrilateral, pentagon, and hexagon. What pattern do you notice. Use this pattern to complete the Sum of Interior Angles column in the table.

Name _____ Section _____ Date _____

Mental Math

Convert as indicated.

1. 12 inches to feet

2. 6 feet to yards

3. 24 inches to feet

4. 36 inches to feet

5. 36 inches to yards

6. 2 yards to inches

Determine whether the measurement in each statement is reasonable.

7. The screen of a home television set has a 30-meter diagonal.

8. A window measures 1 meter by 0.5 meter.

9. A drinking glass is made of glass 2 millimeters thick.

10. A paper clip is 4 kilometers long.

11. The distance across the Colorado River is 50 kilometers.

12. A model's hair is 30 centimeters long.

EXERCISE SET 9.2

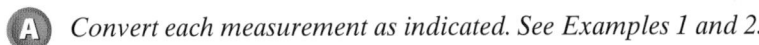

Ⓐ *Convert each measurement as indicated. See Examples 1 and 2.*

1. 60 in. to feet

2. 84 in. to feet

3. 12 yd to feet

4. 18 yd to feet

5. 42,240 ft to miles

6. 36,960 ft to miles

7. 102 in. to feet

8. 150 in. to feet

9. 10 ft to yards

10. 25 ft to yards

11. 6.4 mi to feet

12. 3.8 mi to feet

Ⓑ *Convert each measurement as indicated. See Examples 3 and 4.*

13. 40 ft = ____ yd ____ ft

14. 100 ft = ____ yd ____ ft

15. 41 in. = ____ ft ____ in.

16. 75 in. = ____ ft ____ in.

17. 10,000 ft = ____ mi ____ ft

18. 25,000 ft = ____ mi ____ ft

19. 5 ft 2 in. = _____ in.

20. 4 ft 11 in. = _____ in.

21. 5 yd 2 ft = _____ ft

22. 7 yd 1 ft = _____ ft

23. 2 yd 1 ft = _____ in.

24. 1 yd 2 ft = _____ in.

C *Perform each indicated operation. Simplify the result if possible. See Examples 5 through 8.*

25. 5 ft 8 in. + 6 ft 7 in.

26. 9 ft 10 in. + 8 ft 4 in.

27. 12 yd 2 ft + 9 yd 2 ft

28. 16 yd 2 ft + 8 yd 1 ft

29. 24 ft 8 in. − 16 ft 3 in.

30. 15 ft 5 in. − 8 ft 2 in.

31. 16 ft 3 in. − 10 ft 9 in.

32. 14 ft 8 in. − 3 ft 11 in.

33. 6 ft 8 in. ÷ 2

34. 26 ft 10 in. ÷ 2

35. 12 yd 2 ft × 4

36. 15 yd 1 ft × 8

Solve. Remember to insert units when writing your answers. See Example 8.

37. The National Zoo maintains a small patch of bamboo, which it grows as a food supply for its pandas. Two weeks ago, the bamboo was 6 ft 10 in. tall. Since then, the bamboo has grown 3 ft 8 in. taller. How tall is the bamboo now?

38. While exploring in the Marianas Trench, a submarine probe was lowered to a point 1 mile 1400 feet below the ocean's surface. Later it was lowered an additional 1 mile 4000 feet below this point. How far is the probe below the surface of the Pacific?

39. The length of one of the Statue of Liberty's hands is 16 ft 5 in. One of the Statue's eyes is 2 ft 6 in. across. How much longer is a hand than the width of an eye? (*Source:* National Park Service)

40. The width of the Statue of Liberty's head from ear to ear is 10 ft. The height of the Statue's head from chin to cranium is 17 ft 3 in. How much taller is the Statue's head than its width? (*Source:* National Park Service)

41. The Amana Corporation stacks up its microwave ovens in a distribution warehouse. Each stack is 1 ft 9 in. wide. How far from the wall would 9 of these stacks extend?

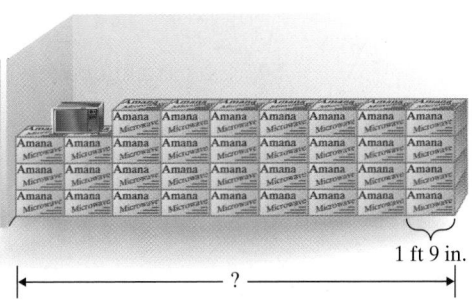

1 ft 9 in.

?

42. The highway commission is installing concrete sound barriers along a highway. Each barrier is 1 yd 2 ft long. How far will 25 barriers in a row reach?

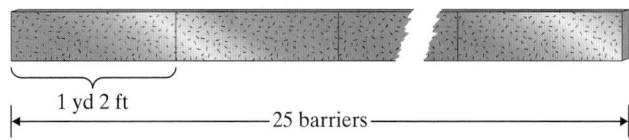

1 yd 2 ft

25 barriers

43. A carpenter needs to cut a board into thirds. If the board is 9 ft 3 in. long originally, how long will each cut piece be?

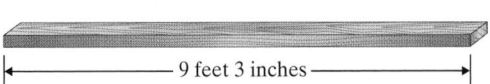

9 feet 3 inches

44. A wall is erected exactly halfway between two buildings that are 192 ft 8 in. apart. If the wall is 8 in. wide, how far is it from the wall to either of the buildings?

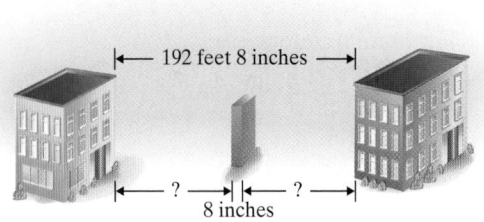

192 feet 8 inches

? 8 inches ?

△ **45.** Evelyn Pittman plans to erect a fence around her garden to keep the rabbits out. If the garden is a rectangle 24 ft 9 in. long by 18 ft 6 in. wide, what is the length of the fencing material she must purchase?

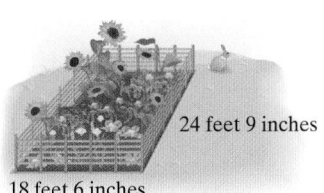

24 feet 9 inches

18 feet 6 inches

△ **46.** Ronnie Hall needs new gutters for the front and *both sides* of his home. The front of the house is 50 ft. 8 in., and each side is 22 ft 9 in. wide. What length of gutter must he buy?

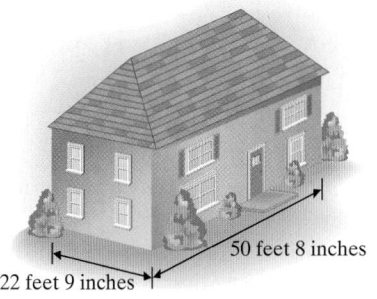

50 feet 8 inches

22 feet 9 inches

47. The world's longest Coca-Cola truck is in Sweden and is 79 feet long. How many *yards* long are 4 of these trucks? (*Source: Coca-Cola Today*)

△ **48.** The world's largest Coca-Cola sign is in Arica, Chile. It is in the shape of a rectangle whose length is 400 feet and whose width is 131 feet. Find the area of the sign. (*Source: Coca-Cola Today*) (*Hint*: Recall that area of a rectangle is the product of length times width.)

Convert as indicated. See Examples 9 and 10.

49. 40 m to centimeters   **50.** 18 m to centimeters   **51.** 40 mm to centimeters

52. 18 mm to centimeters   **53.** 300 m to kilometers   **54.** 400 m to kilometers

55. 1400 mm to meters   **56.** 6400 mm to meters   **57.** 1500 cm to meters

58. 6400 cm to meters   **59.** 8.3 cm to millimeters   **60.** 4.6 cm to millimeters

61. 20.1 mm to decimeters   **62.** 140.2 mm to decimeters

63. 0.04 m to millimeters   **64.** 0.2 m to millimeters

E *Perform each indicated operation. See Examples 11 and 12.*

65. 8.6 m + 0.34 m   **66.** 14.1 cm + 3.96 cm   **67.** 2.9 m + 40 mm   **68.** 30 cm + 8.9 m

69. 24.8 mm − 1.19 cm   **70.** 45.3 m − 2.16 dam   **71.** 15 km − 2360 m   **72.** 14 cm − 15 mm

73. 18.3 m × 3   **74.** 14.1 m × 4   **75.** 6.2 km ÷ 4   **76.** 9.6 m ÷ 5

Solve. Remember to insert units when writing your answers. See Example 13.

77. A 3.4-m rope is attached to a 5.8-m rope. However, when the ropes are tied, 8 cm of length are lost to form the knot. What is the length of the tied ropes?

78. A 2.15-m-long sash cord has become frayed at both ends so that 1 cm is trimmed from each end. How long is the remaining cord?

79. The ice on Doc Miller's pond is 5.33 cm thick. For safe skating, Doc insists that it must be 80 mm thick. How much thicker must the ice be before Doc goes skating?

80. The sediment on the bottom of the Towamencin Creek is normally 14 cm thick, but the recent flood washed away 22 mm of sediment. How thick is it now?

81. An art class is learning how to make kites. The two sticks used for each kite have lengths of 1 m and 65 cm. What total length of wood must be ordered for the sticks if 25 kites are to be built?

82. The total pages of a hardbound economics text are 3.1 cm thick. The front and back covers are each 2 mm thick. How high would a stack of 10 of these texts be?

83. A logging firm needs to cut a 67 m long redwood log into 20 equal pieces before loading it onto a truck for shipment. How long will each piece be?

84. An 18.3 m tall flagpole is mounted on a 65 cm high pedestal. How far is the top of the flagpole from the ground?

85. At one time it was believed that the fort of Basasi, on the Indian-Tibetan border, was the highest located structure at an elevation of 5.988 km above sea level. However, a settlement has been located that is 21 m higher than the fort. What is the elevation of this settlement?

86. The average American male at age 35 is 1.75 m tall. The average 65-year-old male is 48 mm shorter. How tall is the average 65-year-old male?

△ **87.** A floor tile is 22.86 cm wide. How many tiles in a row are needed to cross a room 3.429 m wide?

△ **88.** A standard postcard is 1.6 times longer than it is wide. If it is 9.9 cm wide, what is its length?

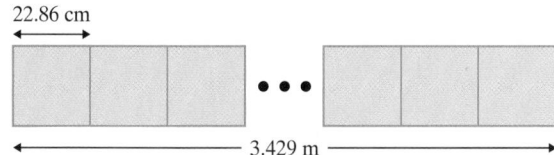

Review and Preview

Write each decimal or fraction as a percent. See Sections 6.1 and 6.2.

89. 0.21

90. 0.86

91. $\dfrac{13}{100}$

92. $\dfrac{47}{100}$

93. $\dfrac{1}{4}$

94. $\dfrac{3}{20}$

95. To convert from meters to centimeters, the decimal point is moved two places to the right. Explain how this relates to the fact that the prefix *centi* means $\frac{1}{100}$.

96. Explain why conversions in the metric system are easier to make than conversions in the U.S. system of measurement.

97. Anoa Longway plans to use 26.3 meters of leftover fencing material to enclose a square garden plot for her daughter. How long will each side of the garden be?

98. A marathon is a running race over a distance of 26 mi 385 yd. If a runner runs five marathons in a year, what is the total distance he or she runs? (*Source: Microsoft Encarta Encyclopedia*)

9.3 Perimeter

Ⓐ Using Formulas to Find Perimeters

Recall from Section 1.3 that the perimeter of a polygon is the distance around the polygon. This means that the perimeter of a polygon is the sum of the lengths of its sides.

EXAMPLE 1   Find the perimeter of the rectangle below.

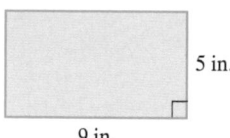

5 in.

9 in.

Solution: perimeter = 9 inches + 9 inches + 5 inches + 5 inches

= 28 inches

Notice that the perimeter of the rectangle in Example 1 can be written as 2 · (9 inches) + 2 · (5 inches).

length   width

In general, we can say that the perimeter of a rectangle is always

2 · length + 2 · width

As we have seen above and in previous sections, the perimeter of some special figures such as rectangles form patterns. These patterns are given as **formulas**. The formula for the perimeter of a rectangle is shown next:

Perimeter of a Rectangle

perimeter = 2 · length + 2 · width

In symbols, this can be written as

$P = 2l + 2w$

length

width   width

length

EXAMPLE 2   Find the perimeter of a rectangle with a length of 11 inches and a width of 3 inches.

11 in.

3 in.

Solution: We use the formula for perimeter and replace the letters by their known measures.

$P = 2l + 2w$

$= 2 \cdot 11 \text{ in.} + 2 \cdot 3 \text{ in.}$ Replace *l* with 11 in. and *w* with 3 in.

$= 22 \text{ in.} + 6 \text{ in.}$

$= 28 \text{ in.}$

The perimeter is 28 inches.

Practice Problem 1

Find the perimeter of the rectangular lot shown below:

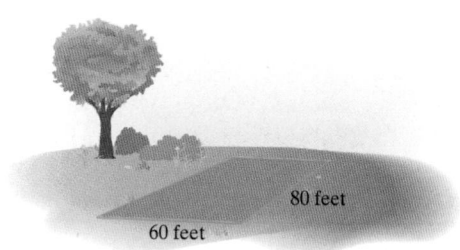

80 feet

60 feet

Practice Problem 2

Find the perimeter of a rectangle with a length of 22 centimeters and a width of 10 centimeters.

Answers

1. 280 ft   **2.** 64 cm

Recall that a square is a special rectangle with all four sides the same length. The formula for the perimeter of a square is shown next:

Perimeter of a Square

> Perimeter = side + side + side + side
> = 4 · side
>
> In symbols,
>
> $P = 4s$

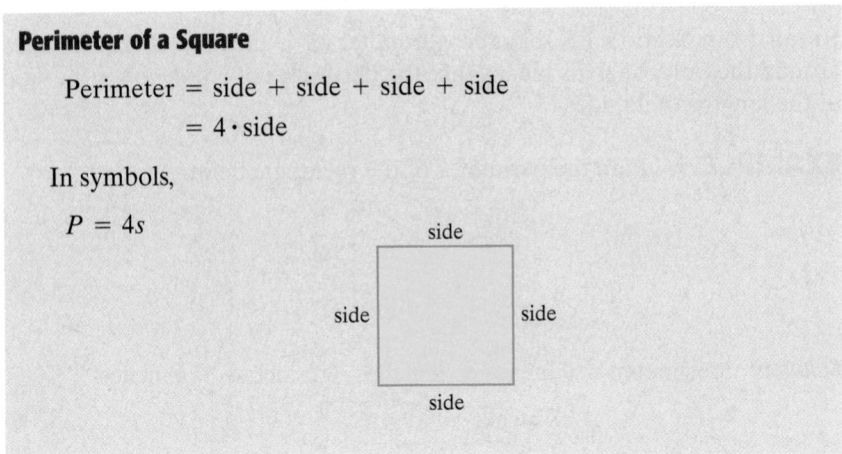

Practice Problem 3

How much fencing is needed to enclose a square field 50 yards on a side?

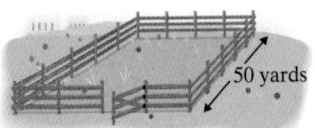

50 yards

EXAMPLE 3   Finding the Perimeter of a Tabletop

Find the perimeter of a square tabletop if each side is 5 feet long.

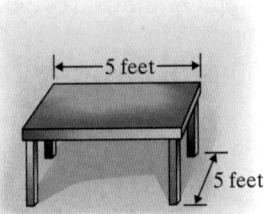

5 feet

5 feet

Solution: The formula for the perimeter of a square is $P = 4s$. We use this formula and replace s by 5 feet.

> $P = 4s$
> $= 4 \cdot 5$ feet
> $= 20$ feet

The perimeter of the square tabletop is 20 feet.   ●

The formula for the perimeter of a triangle with sides of lengths $a, b,$ and c is given next:

Perimeter of a Triangle

> Perimeter = side a + side b + side c
>
> In symbols,
>
> $P = a + b + c$

side a

side b

side c

Answer

3. 200 yd

EXAMPLE 4   Find the perimeter of a triangle when the sides are 3 inches, 7 inches, and 6 inches.

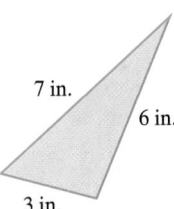

Solution: The formula is $P = a + b + c$, where $a, b,$ and c are the lengths of the sides. Thus,

$P = a + b + c$

$\quad = 3 \text{ in.} + 7 \text{ in.} + 6 \text{ in.}$

$\quad = 16 \text{ in.}$

The perimeter of the triangle is 16 inches.

Recall that to find the perimeter of other polygons, we find the sum of the lengths of their sides.

EXAMPLE 5   Find the perimeter of the quadrilateral shown below:

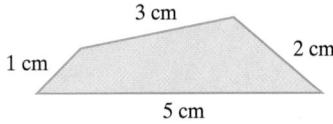

Solution: To find the perimeter, we find the sum of the lengths of its sides.

perimeter $= 3 \text{ cm} + 2 \text{ cm} + 5 \text{ cm} + 1 \text{ cm} = 11 \text{ cm}$

The perimeter is 11 centimeters.

EXAMPLE 6   Finding the Perimeter of a Room

Find the perimeter of the room shown below:

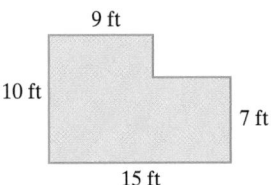

Solution: To find the perimeter of the room, we first need to find the lengths of all sides of the room.

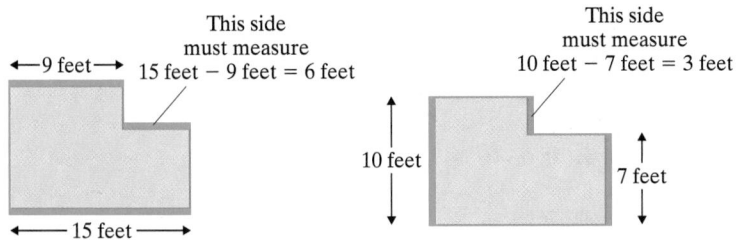

Practice Problem 4

Find the perimeter of a triangle when the sides are 5 centimeters, 9 centimeters, and 7 centimeters in length.

Practice Problem 5

Find the perimeter of the trapezoid shown.

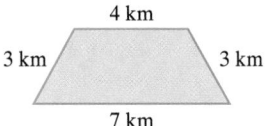

Practice Problem 6

Find the perimeter of the room shown.

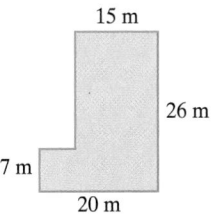

Answers

4. 21 cm   **5.** 17 km   **6.** 92 m

Now that we know the measures of all sides of the room, we can add the measures to find the perimeter.

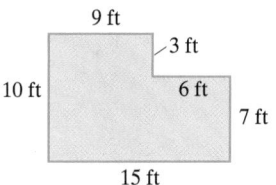

$$\text{perimeter} = 10 \text{ ft} + 9 \text{ ft} + 3 \text{ ft} + 6 \text{ ft} + 7 \text{ ft} + 15 \text{ ft}$$
$$= 50 \text{ ft}$$

The perimeter of the room is 50 feet. ●

Practice Problem 7

A rectangular lot measures 60 feet by 120 feet. Find the cost to install fencing around the lot if the cost of fencing is $1.90 per foot.

EXAMPLE 7   Calculating the Cost of Baseboard

A rectangular room measures 10 feet by 12 feet. Find the cost to hang a wallpaper border on the walls close to the ceiling if the cost of the wallpaper border is $1.09 per foot.

Solution: First we find the perimeter of the room.

$$P = 2l + 2w$$
$$= 2 \cdot 12 \text{ feet} + 2 \cdot 10 \text{ feet} \quad \text{Replace } l \text{ with 12 feet and } w \text{ with 10 feet.}$$
$$= 24 \text{ feet} + 20 \text{ feet}$$
$$= 44 \text{ feet}$$

The cost of the wallpaper is

$$\text{cost} = 1.09 \cdot 44 = 47.96$$

The cost of the wallpaper is $47.96 ●

B   Using Formulas to Find Circumferences

Recall from Section 5.3 that the distance around a circle is called the **circumference**. This distance depends on the radius or the diameter of the circle.

The formulas for circumference are shown next:

Circumference of a Circle

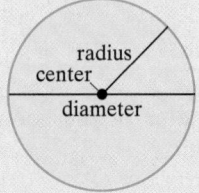

Circumference = $2 \cdot \pi \cdot$ radius or Circumference = $\pi \cdot$ diameter

In symbols,

$$C = 2\pi r \quad \text{or} \quad C = \pi d,$$

where $\pi \approx 3.14$ or $\pi \approx \dfrac{22}{7}$.

Answer

7. $684

To better understand circumference and π (pi), try the following experiment. Take any can and measure its circumference and its diameter.

The can in the figure above has a circumference of 23.5 centimeters and a diameter of 7.5 centimeters. Now divide the circumference by the diameter.

$$\frac{\text{circumference}}{\text{diameter}} = \frac{23.5 \text{ cm}}{7.5 \text{ cm}} \approx 3.1$$

Try this with other sizes of cylinders and circles—you should always get a number close to 3.1. The exact ratio of circumference to diameter is π. (Recall that $\pi \approx 3.14$ or $\approx \frac{22}{7}$).

EXAMPLE 8   Installing a Border in a Circular Spa

Mary Catherine Dooley plans to install a border of new tiling around the circumference of her circular spa. If her spa has a diameter of 14 feet, find its circumference.

Solution: Because we are given the diameter, we use the formula $C = \pi d$.

$C = \pi d$

$\quad = \pi \cdot 14 \text{ ft}$ Replace d with 14 feet.

$\quad = 14\pi \text{ ft}$

The circumference of the spa is *exactly* 14π feet. By replacing π with the *approximation* 3.14, we find that the circumference is *approximately* 14 feet $\cdot$ 3.14 = 43.96 feet.

Try the Concept Check in the margin.

Practice Problem 8

An irrigation device waters a circular region with a diameter of 20 yards. What is the circumference of the watered region? Use $\pi \approx 3.14$.

20 yd

Concept Check

The distance around which figure is greater: a square with side length 5 inches or a circle with radius 3 inches?

Answer

8. 20π yd ≈ 62.8 yd

Concept Check: a square with length 5 in.

FOCUS ON **The Real World**

SPEEDS

A speed measures how far something travels in a given unit of time. You already learned in Section 6.2 that the speed 55 miles per hour is a unit rate that can be written as $\dfrac{55 \text{ miles}}{1 \text{ hour}}$. Just as there are different units of measurement for length or distance, there are different units of measurement for speed as well. It is also possible to perform unit conversions on speeds. Before we learn about converting speeds, we will review units of time. The following is a summary of equivalencies between various units of time.

Units of Time	Unit Fractions
60 seconds (s) = 1 minute (min)	$\dfrac{60 \text{ s}}{1 \text{ min}} = \dfrac{1 \text{ min}}{60 \text{ s}} = 1$
60 minutes = 1 hour (h)	$\dfrac{60 \text{ min}}{1 \text{ h}} = \dfrac{1 \text{ h}}{60 \text{ min}} = 1$
3600 seconds = 1 hour	$\dfrac{3600 \text{ s}}{1 \text{ h}} = \dfrac{1 \text{ h}}{3600 \text{ s}}$

Here are some common speeds.

Speeds

Miles per hour (mph)
Miles per minute (mi/min)
Miles per second (mi/s)
Feet per second (ft/s)
Feet per minute (ft/min)
Kilometers per hour (kmph or km/h)
Kilometers per second (kmps or km/s)
Meters per second (m/s)
Knots

> **Helpful Hint**
>
> A **knot** is 1 nautical mile per hour and is a measure of speed used for ships.
>
> 1 nautical mile (nmi) ≈ 1.15 miles (mi)
>
> 1 nautical mile (nmi) ≈ 6076.12 feet (ft)

To convert from one speed to another, unit fractions may be used. To convert from mph to ft/s, first write the original speed as a unit rate. Then multiply by a unit fraction that relates miles to feet and by a unit fraction that relates hours to seconds. The unit fractions should be written so that the given units in the original rate will divide out. For example, to convert 55 mph to ft/s:

$$55 \text{ mph} = \frac{55 \text{ miles}}{1 \text{ hour}} = \frac{55 \text{ \sout{miles}}}{1 \text{ \sout{hour}}} \cdot \frac{5280 \text{ ft}}{1 \text{ \sout{mile}}} \cdot \frac{1 \text{ \sout{hour}}}{3600 \text{ s}}$$

$$= \frac{55 \cdot 5280 \text{ ft}}{3600 \text{ s}}$$

$$= \frac{290{,}400 \text{ ft}}{3600 \text{ s}}$$

$$= 80\frac{2}{3} \text{ ft/s}$$

GROUP ACTIVITY

1. Research the current world land speed record. Convert the speed from mph to feet per second.

2. Research the current world water speed record. Convert from mph to knots.

3. Research and then describe the Beaufort Wind Scale, its origins, and how it is used. Give the scale keyed to both miles per hour and knots. Why would both measures be useful?

Name _____ Section _____ Date _____

EXERCISE SET 9.3

A *Find the perimeter of each figure. See Examples 1 through 6.*

1.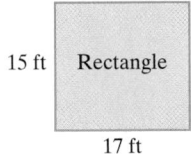

15 ft · Rectangle

17 ft

2.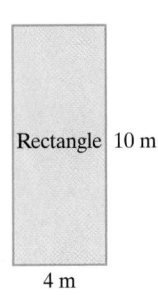

Rectangle 10 m

4 m

3.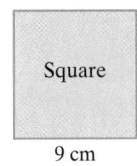

Square

9 cm

4.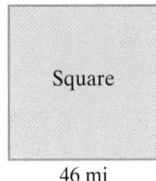

Square

46 mi

5.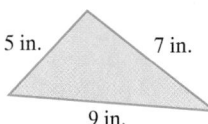

5 in.   7 in.

9 in.

6.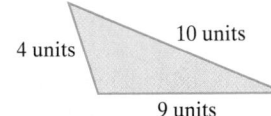

4 units   10 units

9 units

7.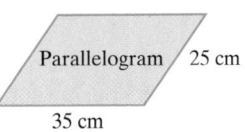

Parallelogram 25 cm

35 cm

8.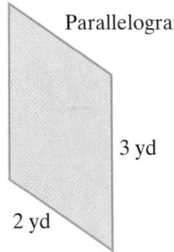

Parallelogram

3 yd

2 yd

9.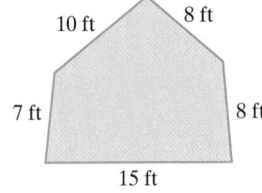

10 ft   8 ft

7 ft   8 ft

15 ft

10.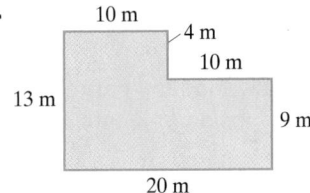

10 m   4 m

10 m

13 m   9 m

20 m

11.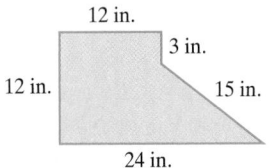

12 in.   3 in.

12 in.   15 in.

24 in.

12.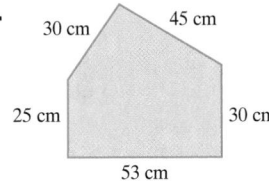

30 cm   45 cm

25 cm   30 cm

53 cm

Solve. See Examples 1-7.

13. A polygon has sides of length 5 feet, 3 feet, 2 feet, 7 feet, and 4 feet. Find its perimeter.

14. A triangle has sides of length 8 inches, 12 inches, and 10 inches. Find its perimeter.

15. Baseboard is to be installed in a square room that measures 15 feet on one side. Find how much baseboard is needed. *(Note:* Ignore the baseboard not needed because of doors.)

16. Find how much fencing is needed to enclose a rectangular rose garden 85 feet by 15 feet.

17. If a football field is 53 yards wide and 120 yards long, what is the perimeter?

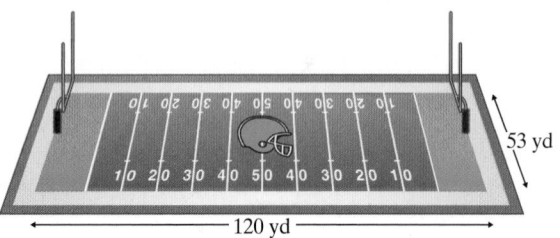

18. A stop sign has eight equal sides of length 12 inches. Find its perimeter.

19. A metal strip is being installed around a workbench that is 8 feet long and 3 feet wide. Find how much stripping is needed.

20. Find how much fencing is needed to enclose a rectangular garden 70 feet by 21 feet.

21. If the stripping in Exercise 19 costs $3 per foot, find the total cost of the stripping.

22. If the fencing in Exercise 20 costs $2 per foot, find the total cost of the fencing.

23. A regular hexagon has a side length of 6 inches. Find its perimeter.

24. A regular pentagon has a side length of 14 meters. Find its perimeter.

25. Find the perimeter of the top of a square compact disc case if the length of one side is 7 inches.

26. Find the perimeter of a square ceramic tile with a side of length 5 inches.

27. A rectangular room measures 6 feet by 8 feet. Find the cost of installing a strip of wallpaper around the top of the room if the wallpaper costs $0.86 per foot.

28. A rectangular house measures 75 feet by 60 feet. Find the cost of installing gutters around the house if the cost is $2.36 per foot.

Find the perimeter of each figure. See Example 6.

29.

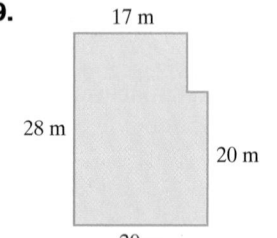

30.

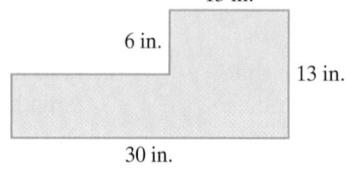

31.

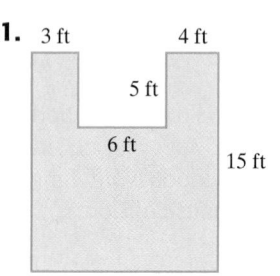

32.

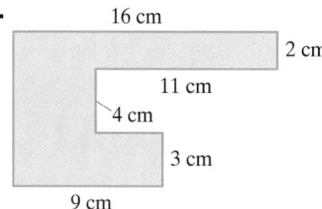

16 cm
2 cm
11 cm
4 cm
3 cm
9 cm

33.

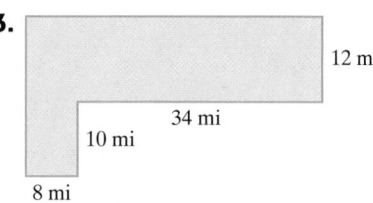

12 mi
34 mi
10 mi
8 mi

34.

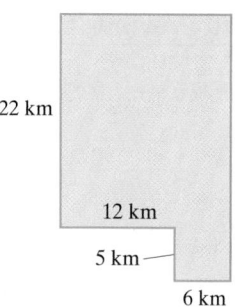

22 km
12 km
5 km
6 km

B *Find the circumference of each circle. Give the exact circumference and then an approximation. Use $\pi \approx 3.14$. See Example 8.*

35.

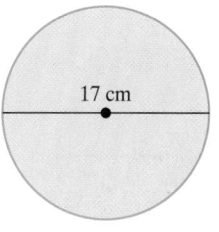

17 cm

36.

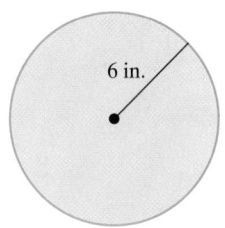

6 in.

37.

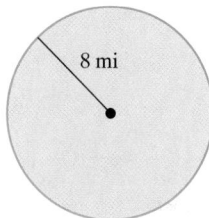

8 mi

38.

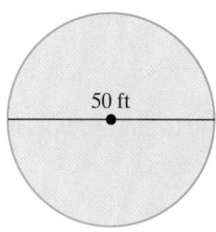

50 ft

39.

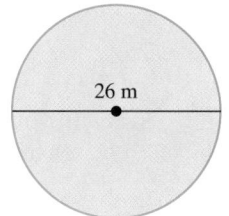

26 m

40.

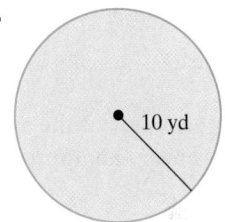

10 yd

41. A circular fountain has a radius of 5 feet. Approximate the distance around the fountain. Use $\frac{22}{7}$ for π.

42. A circular walkway has a radius of 40 meters. Approximate the distance around the walkway. Use 3.14 for π.

43. Meteor Crater, near Winslow, Arizona, is about 4000 feet in diameter. Approximate the distance around the crater. Use 3.14 for π. (*Source: The Handy Science Answer Book*)

44. The largest pearl, the *Pearl of Lao-tze*, has a diameter of $5\frac{1}{2}$ inches. Approximate the distance around the pearl. Use $\frac{22}{7}$ for π. (*Source: The Guinness Book of Records*)

Review and Preview

Simplify. See Section 1.9.

45. $5 + 6 \cdot 3$

46. $25 - 3 \cdot 7$

47. $(20 - 16) \div 4$

48. $6 \cdot (8 + 2)$

49. $(18 + 8) - (12 + 4)$

50. $72 \div (2 \cdot 6)$

51. $(72 \div 2) \cdot 6$

52. $4^1 \cdot (2^3 - 8)$

 Combining Concepts

53. a. Find the circumference of each circle. Approximate the circumference by using 3.14 for π.

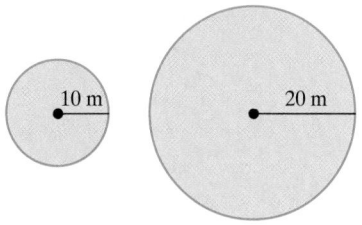

b. If the radius of a circle is doubled, is its corresponding circumference doubled?

54. a. Find the circumference of each circle. Approximate the circumference by using 3.14 for π.

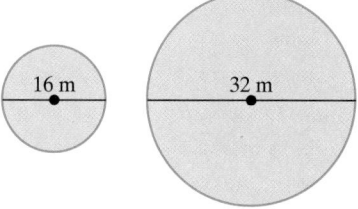

b. If the diameter of a circle is doubled, is its corresponding circumference doubled?

55. Find the perimeter of the skating rink. Give the exact answer and an approximate answer using 3.14 for π.

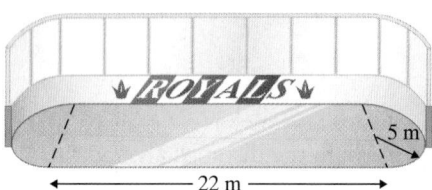

56. In your own words, explain how to find the perimeter of any polygon.

57. The perimeter of this rectangle is 30 feet. Find its width.

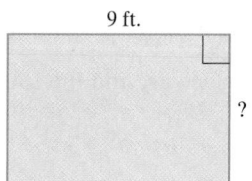

Find the perimeter. Round your results to the nearest tenth.

58.

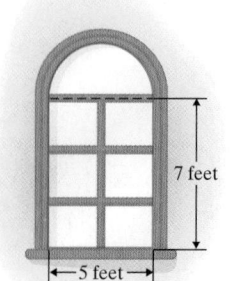

59.

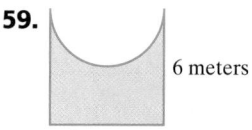

9.4   Area and Volume

Ⓐ   Finding Area

Recall that area measures the number of square units that cover the surface of a region. Thus far, we know how to find the area of a rectangle and a square. These formulas, as well as formulas for finding the areas of other common geometic figures, are given next.

Area Formulas of Common Geometric Figures

Geometric Figure   **Area Formula**

RECTANGLE   Area of a rectangle:

                                      area = length · width

                                      $A = lw$

SQUARE   Area of a square:

                                      area = side · side

                                      $A = s \cdot s = s^2$

TRIANGLE   Area of a triangle:

                                      area $= \dfrac{1}{2} \cdot$ **base · height**

                                      $A = \dfrac{1}{2}bh$

PARALLELOGRAM   Area of a parallelogram:

                                      area = base · height

                                      $A = bh$

TRAPEZOID   Area of a trapezoid:

                                      area $= \dfrac{1}{2} \cdot$ (**one base + other base**) **· height**

                                      $A = \dfrac{1}{2}(b + B)h$

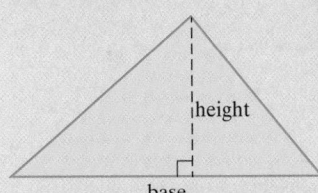

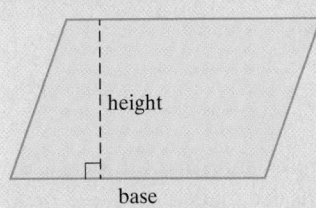

Use these formulas for the following examples.

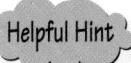

Helpful Hint

Area is always measured in square units.

Practice Problem 1

Find the area of the triangle.

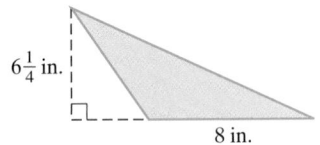

$6\frac{1}{4}$ in.

8 in.

EXAMPLE 1 Find the area of the triangle.

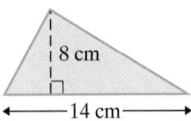

8 cm

←——14 cm——→

Solution: $A = \frac{1}{2}bh$

$= \frac{1}{2} \cdot 14 \text{ cm} \cdot 8 \text{ cm}$ Replace b, base, with 14 cm and h, height, with 8 cm.

$= \frac{\overset{1}{\cancel{2}} \cdot 7 \cdot 8}{\underset{1}{\cancel{2}}}$ square centimeters   Write 14 as $2 \cdot 7$.

$= 56$ square centimeters

The area is 56 square centimeters.

Practice Problem 2

Find the area of the trapezoid.

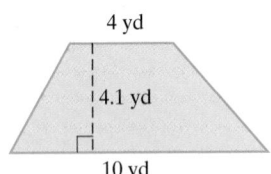

4 yd

4.1 yd

10 yd

EXAMPLE 2 Find the area of the parallelogram.

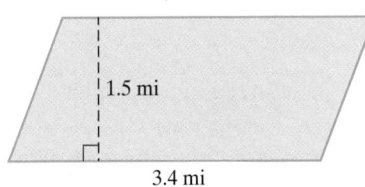

1.5 mi

3.4 mi

Solution: $A = bh$

$= 3.4 \text{ miles} \cdot 1.5 \text{ miles}$ Replace b, base, with 3.4 miles and h, height, with 1.5 miles.

$= 5.1$ square miles

The area is 5.1 square miles.

Helpful Hint

When finding the area of figures, check to make sure that all measurements are in the same units before calculations are made.

Practice Problem 3

Find the area of the figure.

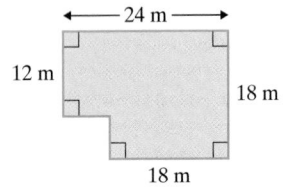

←—— 24 m ——→

12 m

18 m

18 m

EXAMPLE 3 Find the area of the figure.

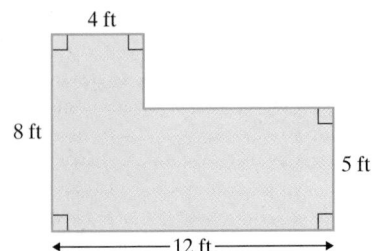

4 ft

8 ft

5 ft

←——— 12 ft ———→

Answers

1. 25 sq. in.   **2.** 28.7 sq. yd   **3.** 396 sq. m

Solution: Split the figure into two rectangles. To find the area of the figure, we find the sum of the areas of the two rectangles.

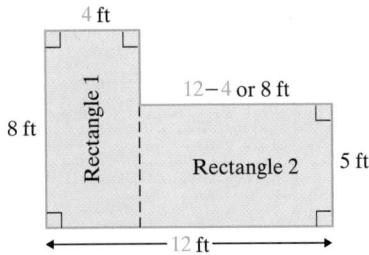

area of Rectangle 1 $= lw$

$\qquad = 8 \text{ feet} \cdot 4 \text{ feet}$

$\qquad = 32 \text{ square feet}$

Notice that the length of Rectangle 2 is 12 feet −4 feet or 8 feet.

area of Rectangle 2 $= lw$

$\qquad = 8 \text{ feet} \cdot 5 \text{ feet}$

$\qquad = 40 \text{ square feet}$

area of the figure $=$ area of Rectangle 1 $+$ area of Rectangle 2

$\qquad = 32 \text{ square feet} + 40 \text{ square feet}$

$\qquad = 72 \text{ square feet}$ ●

Helpful Hint

The figure in Example 3 could also be split into two rectangles as shown.

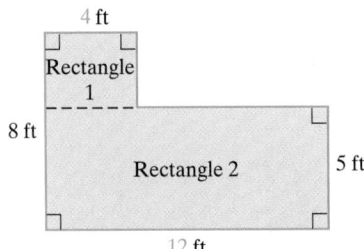

To better understand the formula for area of a circle, try the following. Cut a circle into many pieces, as shown.

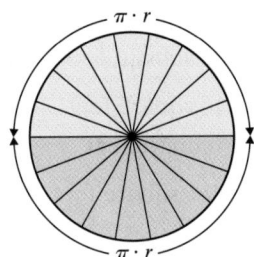

The circumference of a circle is $2\pi r$. This means that the circumference of half a circle is half of $2\pi r$, or πr.

Then unfold the two halves of the circle and place them together, as shown.

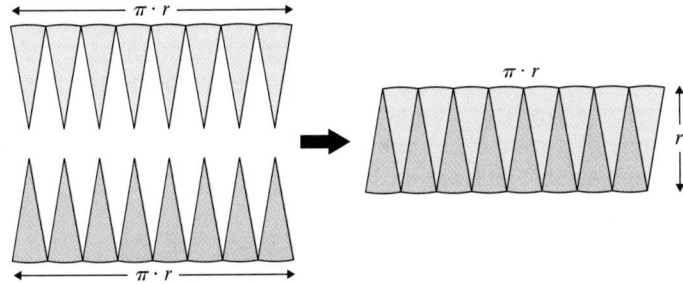

The figure on the right is almost a parallelogram with a base of πr and a height of r. The area is

$$A = \boxed{\text{base}} \quad \cdot \quad \boxed{\text{height}}$$

$$ = (\pi r) \quad \cdot \quad r$$

$$ = \pi r^2$$

This is the formula for area of a circle.

Area Formula of a Circle

Circle

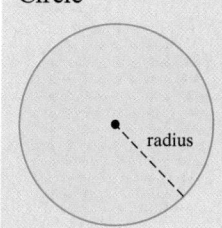

Area of a circle

area $= \pi \cdot (\text{radius})^2$

$A = \pi r^2$

(A fraction approximation for π is $\frac{22}{7}$.)

(A decimal approximation for π is 3.14.)

Practice Problem 4

Find the area of the given circle. Find the exact area and an approximation. Use 3.14 as an approximation for π.

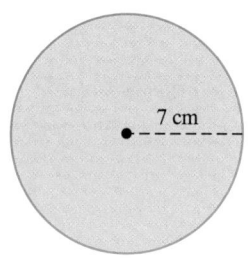

7 cm

Concept Check

Use estimation to decide which figure would have a larger area: a circle of diameter 10 in. or a square 10 in. long on each side.

EXAMPLE 4 Find the area of a circle with a radius of 3 feet. Find the exact area and an approximation. Use 3.14 as an approximation for π.

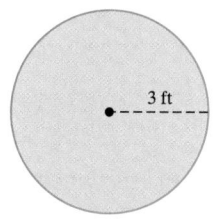

3 ft

Solution: We let $r = 3$ feet and use the formula

$A = \pi r^2$

$= \pi \cdot (3 \text{ feet})^2$ Replace r with 3 feet.

$= 9 \cdot \pi$ square feet   Replace $(3 \text{ feet})^2$ with 9 sq ft.

To approximate this area, we substitute 3.14 for π.

$9 \cdot \pi$ square feet $\approx 9 \cdot 3.14$ square feet

$= 28.26$ square feet

The *exact* area of the circle is 9π square feet, which is *approximately* 28.26 square feet. ●

Try the Concept Check in the margin.

B Finding Volume

Volume measures the number of cubic units that fill the space of a solid. The volume of a box or can, for example, is the amount of space inside. Volume can be used to describe the amount of juice in a pitcher or the amount of concrete needed to pour a foundation for a house.

The volume of a solid is the number of **cubic units** in the solid. A cubic centimeter and a cubic inch are illustrated.

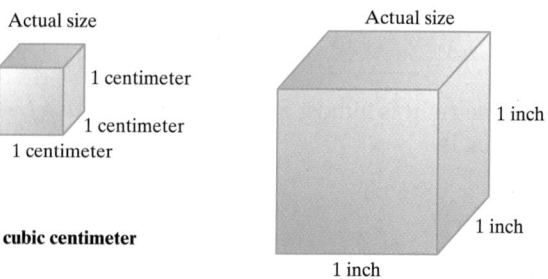

Formulas for finding the volumes of some common solids are given next.

Answers

4. 49π sq cm ≈ 153.86 sq cm

Concept Check: A square 10 in. long on each side would have a larger area.

Volume Formulas of Common Solids

Solid	Volume Formulas

RECTANGULAR SOLID

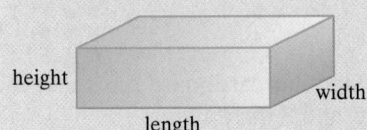

Volume of a rectangular solid:

volume = length · width · height

$$V = lwh$$

CUBE

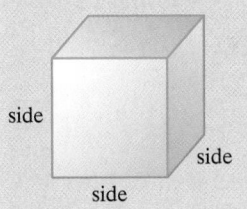

Volume of a cube:

volume = side · side · side

$$V = s \cdot s \cdot s = s^3$$

SPHERE

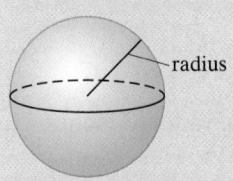

Volume of a sphere:

volume $= \dfrac{4}{3} \cdot \pi \cdot ($ radius $)^3$

$$V = \frac{4}{3}\pi r^3$$

CIRCULAR CYLINDER

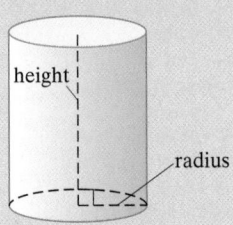

Volume of a circular cylinder:

volume $= \pi \cdot ($ radius $)^2 \cdot$ height

$$V = \pi r^2 h$$

CONE

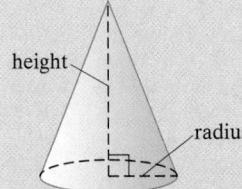

Volume of a cone:

volume $= \dfrac{1}{3} \cdot \pi \cdot ($ radius $)^2 \cdot$ height

$$V = \frac{1}{3}\pi r^2 h$$

SQUARE-BASED PYRAMID

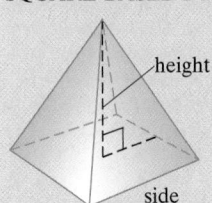

Volume of a square-based pyramid:

volume $= \dfrac{1}{3} \cdot ($ side $)^2 \cdot$ height

$$V = \frac{1}{3}s^2 h$$

Practice Problem 5

Draw a diagram and find the volume of a rectangular box that is 5 ft long, 2 ft wide, and 4 ft deep.

EXAMPLE 5 Find the volume of a rectangular cardboard box that is 12 inches long, 6 inches wide, and 3 inches high.

3 inches

6 inches

12 inches

Fragile

Solution: $V = lwh$

$V = 12 \text{ inches} \cdot 6 \text{ inches} \cdot 3 \text{ inches} = 216 \text{ cubic inches}$

The volume of the rectangular box is 216 cubic inches.

Try the Concept Check in the margin.

Concept Check

Juan Lopez is calculating the volume of the rectangular solid shown. Find the error in his calculation.

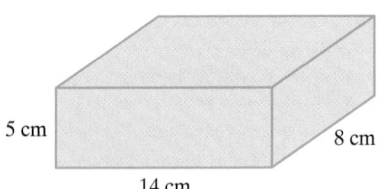

5 cm

8 cm

14 cm

volume $= l + w + h$

$= 14 + 8 + 5$

$= 27 \text{ cubic cm}$

EXAMPLE 6 Approximate the volume of a ball of radius 3 inches. Use the approximation $\dfrac{22}{7}$ for π.

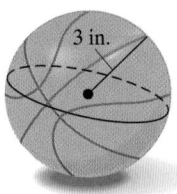

3 in.

Solution: $V = \dfrac{4}{3}\pi r^3$

$\approx \dfrac{4}{3} \cdot \dfrac{22}{7} \cdot (3 \text{ inches})^3$

$= \dfrac{4}{3} \cdot \dfrac{22}{7} \cdot 27 \text{ cubic inches}$

$= \dfrac{4 \cdot 22 \cdot \overset{1}{3} \cdot 9}{\underset{1}{3} \cdot 7} \text{ cubic inches}$ Multiply and write 27 as $3 \cdot 9$.

$= \dfrac{792}{7} \text{ cubic inches or }\quad 113\dfrac{1}{7} \text{ cubic inches}$

The volume is *approximately* $113\dfrac{1}{7}$ cubic inches.

Practice Problem 6

Draw a diagram and approximate the volume of a ball of radius $\dfrac{1}{2}$ cm. Use $\dfrac{22}{7}$ for π.

Answers

5. 40 cu ft   **6.** $\dfrac{11}{21}$ cu cm

Concept Check: volume $= l \cdot w \cdot h$

$= 14 \cdot 8 \cdot 5$

$= 560 \text{ cu cm}$

EXAMPLE 7   Approximate the volume of a can that has a $3\frac{1}{2}$-inch radius and a height of 6 inches. Use $\frac{22}{7}$ for π.

$3\frac{1}{2}$ in.

6 in.

Solution: Using the formula for volume of a circular cylinder, we have

$$V = \pi r^2 h$$

$$3\frac{1}{2} = \frac{7}{2}$$

$$= \pi \cdot \left(\frac{7}{2} \text{ inches}\right)^2 \cdot 6 \text{ inches}$$

or approximately

$$\approx \frac{22}{7} \cdot \frac{49}{4} \cdot 6 \text{ cubic inches}$$

$$= 231 \text{ cubic inches}$$

The volume is approximately 231 cubic inches.

EXAMPLE 8   Approximate the volume of a cone that has a height of 14 centimeters and radius of 3 centimeters. Use $\frac{22}{7}$ for π.

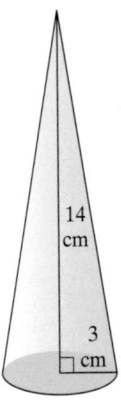

14 cm

3 cm

Solution: Using the formula for volume of a cone, we have

$$V = \frac{1}{3}\pi r^2 h$$

$$= \frac{1}{3} \cdot \pi \cdot (3 \text{ centimeters})^2 \cdot 14 \text{ centimeters} \quad \text{Replace } r \text{ with 3 cm and } h \text{ with 14 cm.}$$

$$= 42\pi \text{ cubic centimeters}$$

or approximately

$$\approx 42 \cdot \frac{22}{7} \text{ cubic centimeters}$$

$$= 132 \text{ cubic centimeters}$$

The volume is approximately 132 cubic centimeters.

Practice Problem 7

Approximate the volume of a cylinder of radius 5 in. and height 7 in. Use 3.14 for π.

Practice Problem 8

Find the volume of a square-based pyramid that has a 3-meter side and height of 5.1 meters.

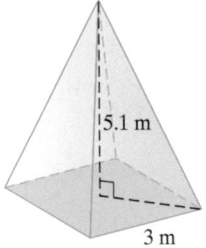

5.1 m

3 m

Answers

7. 549.5 cu in.   **8.** 15.3 cu m

Name _____ Section _____ Date _____

EXERCISE SET 9.4

A *Find the area of each geometric figure. If the figure is a circle, give an exact area and then use the given* **approximation** *for* π *to approximate the area. See Examples 1 through 4.*

 1.

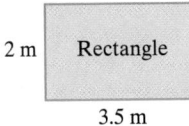

2 m | Rectangle

3.5 m

2.

2.75 ft | Rectangle

7 ft

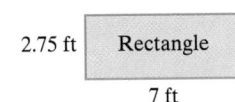 **3.**

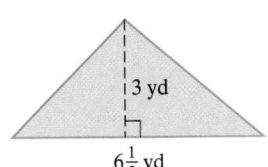

3 yd

$6\frac{1}{2}$ yd

4.

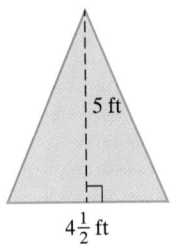

5 ft

$4\frac{1}{2}$ ft

5.

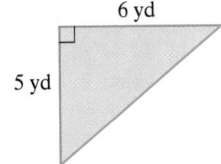

6 yd

5 yd

6.

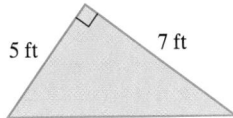

5 ft   7 ft

 7.   Use 3.14 for π.

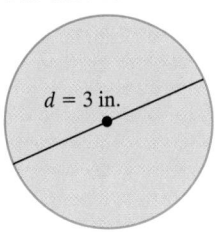

$d = 3$ in.

8.   Use $\dfrac{22}{7}$ for π.

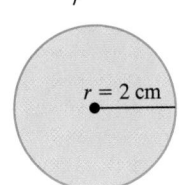

$r = 2$ cm

9.

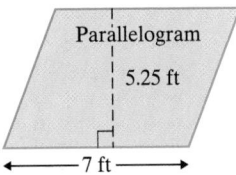

Parallelogram

5.25 ft

7 ft

10.

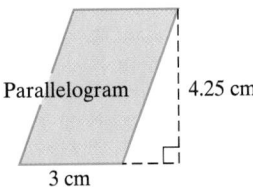

Parallelogram   4.25 cm

3 cm

11.

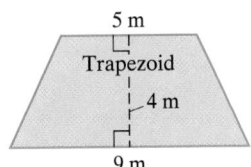

5 m

Trapezoid

4 m

9 m

12.

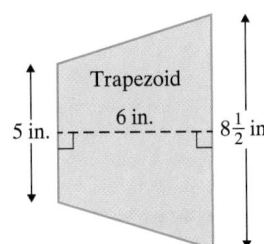

Trapezoid

6 in.

5 in. $8\frac{1}{2}$ in.

13.

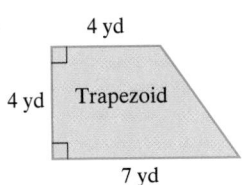

4 yd

4 yd | Trapezoid

7 yd

14.

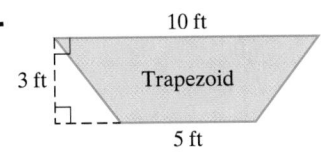

10 ft

3 ft   Trapezoid

5 ft

15.

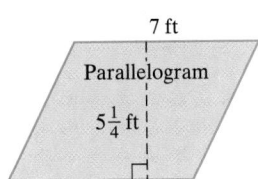

7 ft

Parallelogram

$5\frac{1}{4}$ ft

16.

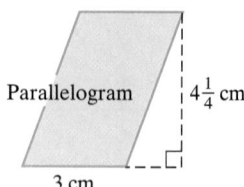

Parallelogram $4\frac{1}{4}$ cm

3 cm

17.

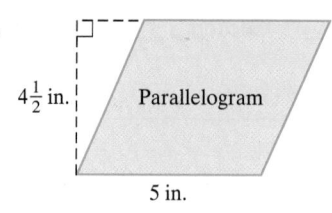

$4\frac{1}{2}$ in. Parallelogram

5 in.

18.

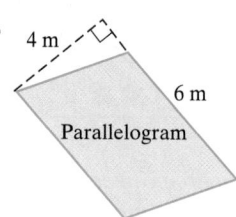

4 m

6 m

Parallelogram

19.

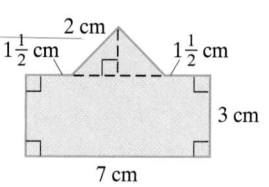

2 cm

$1\frac{1}{2}$ cm $1\frac{1}{2}$ cm

3 cm

7 cm

20.

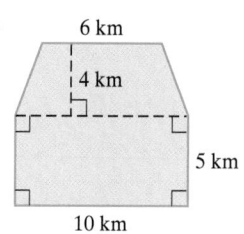

6 km

4 km

5 km

10 km

21.

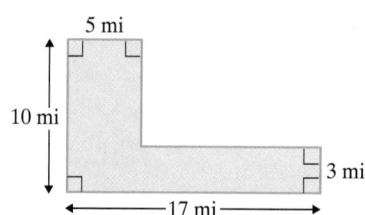

5 mi

10 mi

3 mi

17 mi

22.

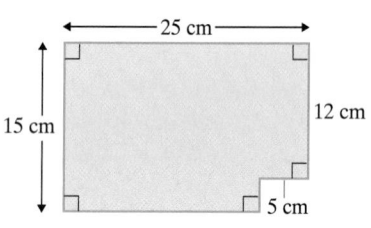

25 cm

15 cm

12 cm

5 cm

23.

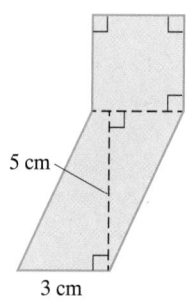

5 cm

3 cm

24.

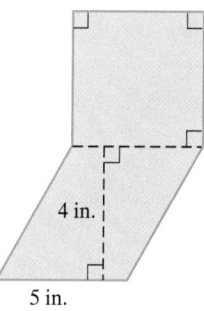

4 in.

5 in.

25. Use $\frac{22}{7}$ for π.

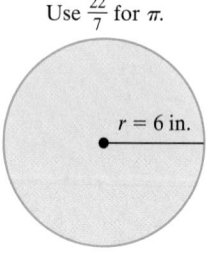

$r = 6$ in.

26. Use 3.14 for π.

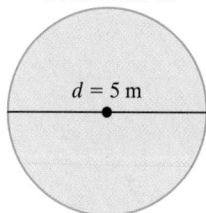

$d = 5$ m

B *Find the volume of each solid. Use* $\dfrac{22}{7}$ *as an approximation for* π. *See Examples 5 through 8.*

27.

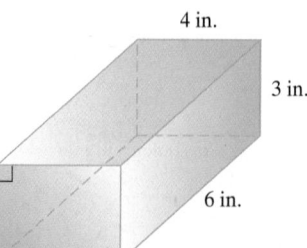

4 in.

3 in.

6 in.

28.

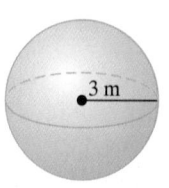

3 m

29.

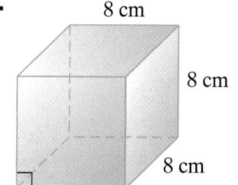

8 cm

8 cm

8 cm

30.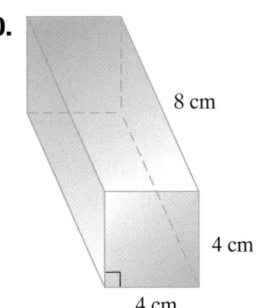

8 cm

4 cm

4 cm

31.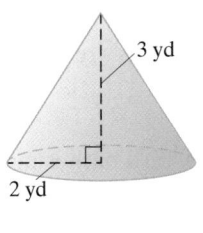

3 yd

2 yd

32.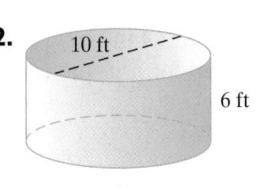

10 ft

6 ft

33.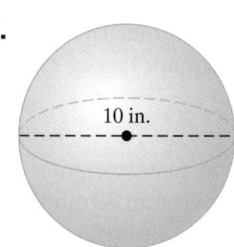

10 in.

34.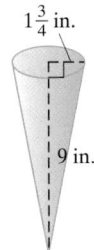

$1\frac{3}{4}$ in.

9 in.

35.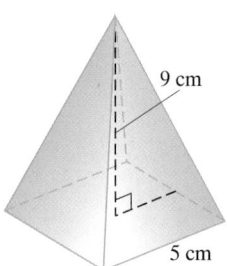

9 cm

5 cm

36.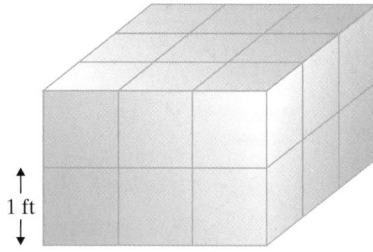

1 ft

A **B** *Solve. See Examples 1 through 8.*

37. Find the volume of a cube with edges of $1\frac{1}{3}$ inches.

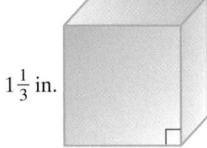

$1\frac{1}{3}$ in.

38. A water storage tank is in the shape of a cone with the pointed end down. If the radius is 14 ft and the depth of the tank is 15 ft, approximate the volume of the tank in cubic feet. Use $\frac{22}{7}$ for π.

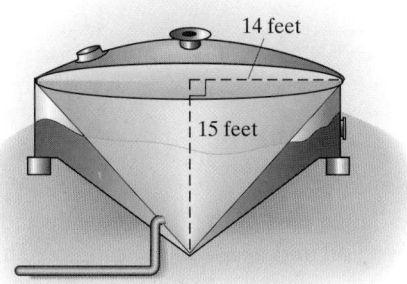

14 feet

15 feet

39. Find the volume of a rectangular box 2 ft by 1.4 ft by 3 ft.

40. Find the volume of a box in the shape of a cube that is 5 ft on each side.

41. The world's largest flag measures 505 feet by 225 feet. It's the U.S. "Super flag" owned by "Ski" Demski of Long Beach, California. Find its area. (*Source: Guinness World Records*)

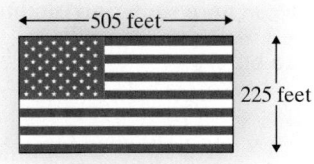

←—505 feet—→

225 feet

42. The longest illuminated sign is in Ramat Gan, Israel, and measures 197 feet by 66 feet. Find its area. (*Source: The Guinness Book of Records*)

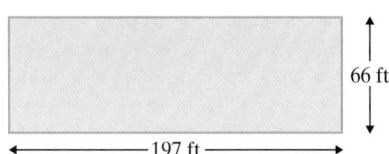

66 ft

←———197 ft———→

43. A drapery panel measures 6 ft by 7 ft. Find how many square feet of material are needed for *four* panels.

44. A page in this book measures 27.5 cm by 21.8 cm. Find its area.

45. A paperweight is in the shape of a square-based pyramid 20 cm tall. If an edge of the base is 12 cm, find the volume of the paperweight.

46. A bird bath is made in the shape of a hemisphere (half-sphere). If its radius is 10 in., approximate the volume. Use $\frac{22}{7}$ for π.

10 inches

47. Find how many square feet of land are in the following plot:

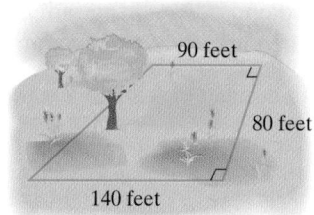

90 feet

80 feet

140 feet

48. For Gerald Gomez to determine how much grass seed he needs to buy, he must know the size of his yard. Use the drawing to determine how many square feet are in his yard.

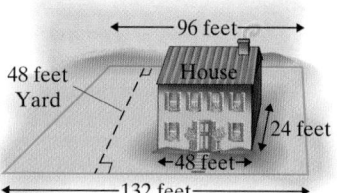

96 feet

48 feet Yard

House

24 feet

48 feet

132 feet

49. The shaded part of the roof shown is in the shape of a trapezoid and needs to be shingled. The number of packages of shingles to buy depends on the area. Use the dimensions given to find the area of the shaded part of the roof to the nearest whole square foot.

36 feet

$12\frac{1}{2}$ feet

25 feet

50. The end of the building shaded in the drawing is to be bricked. The number of bricks to buy depends on the area.

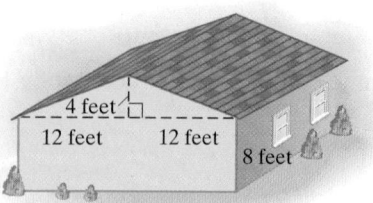

4 feet

12 feet   12 feet

8 feet

a. Find the area.

b. If the side area of each brick (including mortar room) is $\frac{1}{6}$ square foot, find the number of bricks needed to buy.

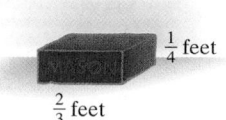

$\frac{1}{4}$ feet

$\frac{2}{3}$ feet

51. Find the exact volume of a sphere with a radius of 7 inches.

52. A tank is in the shape of a cylinder 8 feet tall and 3 feet in radius. Find the exact volume of the tank.

53. Find the volume of a rectangular block of ice 2 feet by $2\frac{1}{2}$ feet by $1\frac{1}{2}$ feet.

54. Find the capacity (volume in cubic feet) of a rectangular ice chest with inside measurements of 3 feet by $1\frac{1}{2}$ feet by $1\frac{3}{4}$ feet.

55. An ice cream cone with a 4-centimeter diameter and 3-centimeter depth is filled exactly level with the top of the cone. Approximate how much ice cream (in cubic centimeters) is in the cone. Use $\frac{22}{7}$ for π.

56. A child's toy is in the shape of a square-based pyramid 10 inches tall. If an edge of the base is 7 inches, find the volume of the toy.

57. Ball lightning is a rare form of lightning in which a moving white or colored luminous sphere is seen. It can last from a few seconds to a few minutes and travels at about walking pace. An average sphere size is 6 inches in diameter. Find the exact volume of a sphere with this diameter and then approximate the volume using 3.14 for π.

58. A monkey ball tree produces large green fruit in the shape of spheres. These fruits are approximately 4 inches (or 10 centimeters) in diameter and have a coarse surface. Find the exact volume of a sphere with diameter 4 inches and then approximate the volume using 3.14 for π. (Round to the nearest tenth.)

59. Find the volume of a pyramid with a square base 5 inches on a side and a height of 1.3 inches.

60. Approximate to the nearest hundredth the volume of a sphere with a radius of 2 centimeters. Use 3.14 for π.

Review and Preview

Evaluate. See Section 1.9.

61. 5^2

62. 7^2

63. 3^2

64. 20^2

65. $1^2 + 2^2$

66. $5^2 + 3^2$

67. $4^2 + 2^2$

68. $1^2 + 6^2$

Given the following situations, tell whether you are more likely to be concerned with area or perimeter.

69. ordering fencing to fence a yard

70. ordering grass seed to plant in a yard

71. buying carpet to install in a room

72. buying gutters to install on a house

73. ordering paint to paint a wall

74. ordering baseboards to install in a room

75. buying a wallpaper border to go on the walls around a room

76. buying fertilizer for your yard

77. A pizza restaurant recently advertised two specials. The first special was a 12-inch pizza for $10. The second special was two 8-inch pizzas for $9. Determine the better buy. (*Hint*: First compare the areas of the two specials and then find a price per square inch for both specials.)

78. Find the approximate area of the state of Utah.

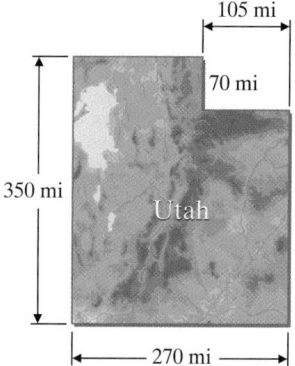

79. Find the area of a rectangle that measures 2 *feet* by 8 *inches*. Give the area in square feet and in square inches.

80. In your own words, explain why perimeter is measured in units and area is measured in square units. (*Hint*: See Section 1.6 for an introduction on the meaning of area.)

Find the area of each figure. Round results to the nearest tenth.

81.

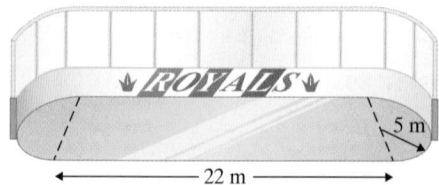

82.

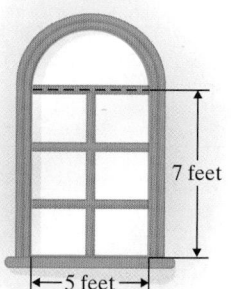

83. The Great Pyramid of Khufu at Giza is the largest of the ancient Egyptian pyramids. Its original height was 146.5 meters. The length of each side of its square base was originally 230 meters. Find the volume of the Great Pyramid of Khufu as it was originally built. Round to the nearest whole cubic meter. (*Source*: PBS *NOVA* Online)

84. The second-largest pyramid at Giza is Khafre's Pyramid. Its original height was 471 feet. The length of each side of its square base was originally 704 feet. Find the volume of Khafre's Pyramid as it was originally built. (*Source*: PBS *NOVA* Online)

85. Menkaure's Pyramid, the smallest of the three Great Pyramids at Giza, was originally 65.5 meters tall. Each of the sides of its square base was originally 344 meters long. What was the volume of Menkaure's Pyramid as it was originally built? Round to the nearest whole cubic meter. (*Source*: PBS *NOVA* Online)

86. Due to factors such as weathering and loss of outer stones, the Great Pyramid of Khufu now stands only 137 meters tall. Its square base is now only 227 meters on a side. Find the current volume of the Great Pyramid of Khufu to the nearest whole cubic meter. How much has its volume decreased since it was built? See Exercise 83 for comparison. (*Source*: PBS *NOVA* Online)

87. The centerpiece of the New England Aquarium in Boston is its Giant Ocean Tank. This exhibit is a four-story cylindrical saltwater tank containing sharks, sea turtles, stingrays, and tropical fish. The radius of the tank is 16.3 feet and its height is 32 feet (assuming that a story is 8 feet). What is the volume of the Giant Ocean Tank? Use $\pi \approx 3.14$ and round to the nearest tenth of a cubic foot. (*Source*: New England Aquarium)

88. Except for service dogs for guests with disabilities, Walt Disney World does not allow pets in its parks or hotels. However, the resort does make pet-boarding services available to guests. The pet-care kennels at Walt Disney World offer three different sizes of indoor kennels. Of these, the smaller two kennels measure
a. 2'1″ × 1'8″ × 1'7″ and
b. 1'1″ × 2' × 8″
What is the volume of each kennel rounded to the nearest cubic foot? Which is larger? (*Source*: Walt Disney World Resort)

89. Can you compute the volume of a rectangle? Why or why not?

90. Do two rectangles with the same perimeter have the same area? To see, find the perimeter and the area of each rectangle.

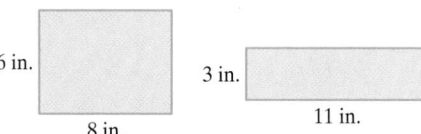

Internet Excursions

The World Wide Web address listed above will direct you to the official Web site of the U.S. Department of Defense and the Pentagon in Washington, D.C., or a related site. You will find information that helps you complete the questions below.

91. Visit this Web site and locate information on the length of each outer wall of the Pentagon. Then calculate the outer perimeter of the Pentagon.

92. Notice that each of the five outer walls of the Pentagon is in the shape of a rectangle. Visit this Web site and locate information on the height of the building. Then use this height information along with the length information found in Exercise 91 to calculate the area of each outer wall of the Pentagon.

Integrated Review–Geometry Concepts

△ **1.** Find the supplement and the complement of a 27° angle.

Find the measures of angles x, y, and z in each figure.

△ **2.**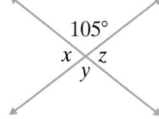

△ **3.** $m \| n$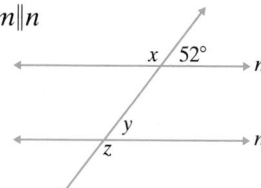

△ **4.** Recall that the sum of the measures of the angles of a triangle is 180°. Find the measure of $\angle x$.

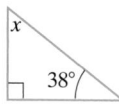

Convert each measurement of length as indicated.

5. 36 in. = ____ ft

6. 10,560 ft = ____ mi

7. 20 ft = ____ yd

8. 6 yd = ____ ft

9. 2.1 mi = ____ ft

10. 3.2 ft = ____ in.

11. 30 m = ____ cm

12. 24 mm = ____ cm

13. 2000 mm = ____ m

14. 18 m = ____ cm

15. 7.2 cm = ____ mm

16. 600 m = ____ km

1. _____

2. _____

3. _____

4. _____

5. _____

6. _____

7. _____

8. _____

9. _____

10. _____

11. _____

12. _____

13. _____

14. _____

15. _____

16. _____

Find the perimeter (or circumference) and area of each figure. For the circle, give an exact circumference and area, then use $\pi \approx 3.14$ to approximate each. Don't forget to attach correct units.

△ **17.**

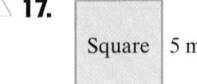

△ **18.**

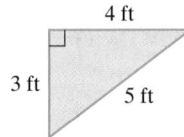

△ **19.**

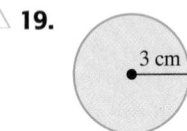

△ **20.**

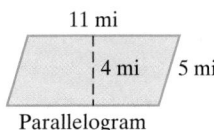

△ **21.** The smallest cathedral is in Highlandville, Missouri, and its floor is a rectangle that measures 14 feet by 17 feet. Find its perimeter and its area. (_Source: The Guinness Book of Records_)

Find the volume of each solid. Don't forget to attach correct units.

△ **22.** A cube with edges of 4 inches each

△ **23.** A rectangular box 2 feet by 3 feet by 5.1 feet

△ **24.** A pyramid with a square base 10 centimeters on a side and a height of 12 centimeters

△ **25.** A sphere with a diameter of 16 miles; give the exact volume and then use $\pi \approx \dfrac{22}{7}$ to approximate the volume.

9.5 Weight and Mass

Ⓐ Defining and Converting U.S. System Units of Weight

Whenever we talk about how heavy an object is, we are concerned with the object's **weight**. We discuss weight when we refer to a 12-ounce box of Rice Krunchies, an overweight 19-pound tabby cat, or a barge hauling 24 tons of garbage.

The most common units of weight in the U.S. measurement system are the **ounce**, the **pound**, and the **ton**. The following is a summary of equivalencies between units of weight:

U.S. Units of Weight	Unit Fractions
16 ounces (oz) = 1 pound (lb)	$\dfrac{16\text{ oz}}{1\text{ lb}} = \dfrac{1\text{ lb}}{16\text{ oz}} = 1$
2000 pounds = 1 ton	$\dfrac{2000\text{ lb}}{1\text{ ton}} = \dfrac{1\text{ ton}}{2000\text{ lb}} = 1$

Try the Concept Check in the margin.

Unit fractions that equal 1 are used to convert between units of weight in the U.S. system. When converting using unit fractions, recall that the numerator of a unit fraction should contain the units we are converting to and the denominator should contain the original units.

To convert 40 ounces to pounds, multiply by $\dfrac{1\text{ lb}}{16\text{ oz}}$ ← Units to convert to
  ← Original units

$$40\text{ oz} = \frac{40\ \cancel{\text{oz}}}{1} \cdot \overbrace{\frac{1\text{ lb}}{16\ \cancel{\text{oz}}}}^{\text{Unit fraction}}$$

$$= \frac{40\text{ lb}}{16} \quad \text{Multiply.}$$

$$= \frac{5}{2}\text{lb} \quad \text{or} \quad 2\frac{1}{2}\text{lb, as a mixed number}$$

EXAMPLE 1   Convert 9000 pounds to tons.

Solution: We multiply 9000 lb by the unit fraction

$\dfrac{1\text{ ton}}{2000\text{ lb}}$ ← Units to convert to
   ← Original units

Concept Check

If you were describing the weight of a semitrailer, which type of unit would you use: ounce, pound, or ton? Why?

Practice Problem 1

Convert 4500 pounds to tons.

Answer

1. $2\frac{1}{4}$ tons

Concept Check: ton

$$9000 \text{ lb} = \frac{9000 \text{ lb}}{1} \cdot \frac{1 \text{ ton}}{2000 \text{ lb}} = \frac{9000 \text{ tons}}{2000} = \frac{9}{2} \text{ tons or } 4\frac{1}{2} \text{ tons}$$

| 2000 lb | 2000 lb | 2000 lb | 2000 lb | 1000 lb | 9000 lb |
| 1 ton | 1 ton | 1 ton | 1 ton | $\frac{1}{2}$ ton | $= 4\frac{1}{2}$ tons |

●

Practice Problem 2

Convert 56 ounces to pounds.

EXAMPLE 2   Convert 3 pounds to ounces.

Solution: We multiply by the unit fraction $\dfrac{16 \text{ oz}}{1 \text{ lb}}$ to convert from pounds to ounces.

$$3 \text{ lb} = \frac{3 \text{ lb}}{1} \cdot \frac{16 \text{ oz}}{1 \text{ lb}} = 3 \cdot 16 \text{ oz} = 48 \text{ oz}$$

| 1 lb | 1 lb | 1 lb | 3 lb |
| 16 oz | 16 oz | 16 oz | = 48 oz |

●

As with length, it is sometimes useful to simplify a measurement of weight by writing it in terms of mixed units.

$$33 \text{ ounces} = \underline{\quad} \text{ lb} \underline{\quad} \text{ oz}$$

Because 16 oz = 1 lb, divide 16 into 33 to see how many pounds are in 33 ounces. The quotient is the number of pounds and the remainder is the number of ounces. To see why we divide 16 into 33, notice that

$$33 \text{ oz} = 33 \text{ oz} \cdot \frac{1 \text{ lb}}{16 \text{ oz}} = \frac{33 \text{ lb}}{16}$$

$$\begin{array}{r} 2 \text{ lb } 1 \text{ oz} \\ 16 \overline{)33} \\ -32 \\ \hline 1 \end{array}$$

Thus 33 ounces is the same as 2 lb 1 oz.

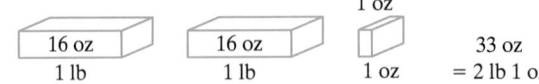

| 16 oz | 16 oz | 1 oz | 33 oz |
| 1 lb | 1 lb | 1 oz | = 2 lb 1 oz |

B Performing Operations on U.S. System Units of Weight

Performing arithmetic operations on units of weight works the same way as performing arithmetic operations on units of length.

Practice Problem 3

Subtract 5 tons 1200 lb from 8 tons 100 lb.

EXAMPLE 3   Subtract 3 tons 1350 lb from 8 tons 1000 lb.

Solution: To subtract, we line up similar units.

$$\begin{array}{r} 8 \text{ tons } 1000 \text{ lb} \\ -3 \text{ tons } 1350 \text{ lb} \end{array}$$

Since we cannot subtract 1350 lb from 1000 lb, we borrow 1 ton from the 8 tons. To do so, we write 1 ton as 2000 lb and combine it with the 1000 lb.

Answers

1. $3\frac{1}{2}$ lb   **2.** 2 tons 900 lb

7 tons + (1 ton) 2000 lb

	becomes	
8 tons 1000 lb		7 tons 3000 lb
−3 tons 1350 lb		−3 tons 1350 lb
		4 tons 1650 lb

To check, see that the sum of 4 tons 1650 lb and 3 tons 1350 lb is 8 tons 1000 lb.

EXAMPLE 4   Multiply 5 lb 9 oz by 6.

Solution: We multiply 5 lb by 6 and 9 oz by 6.

        5 lb 9 oz
    ×   6
    30 lb 54 oz

To write 54 oz as mixed units, we divide by 16 (1 lb = 16 oz).

        3 lb 6 oz
    16)54
      −48
        6

Thus,

30 lb 54 oz = 30 lb + 3 lb 6 oz = 33 lb 6 oz

Practice Problem 4

Multiply 4 lb 11 oz by 8.

EXAMPLE 5   Divide 9 lb 6 oz by 2.

Solution: We divide each of the units by 2.

        4 lb   11 oz
    2)9 lb   6 oz
     −8
      1 lb = 16 oz
            22 oz   Divide 2 into 22 oz to get 11 oz.

To check, multiply 4 pounds 11 ounces by 2. The result is 9 pounds 6 ounces.

Practice Problem 5

Divide 5 lb 8 oz by 4.

EXAMPLE 6   Finding the Weight of a Child

Bryan weighed 8 lb 8 oz at birth. By the time he was 1 year old, he had gained 11 lb 14 oz. Find his weight at age 1 year.

Solution:

birth weight	→	8 lb 8 oz
+ weight gained	→	+ 11 lb 14 oz
total weight	→	19 lb 22 oz

Since 22 oz equals 1 lb 6 oz.,

19 lb 22 oz = 19 lb + 1 lb 6 oz
            = 20 lb 6 oz

Bryan weighed 20 lb 6 oz on his first birthday.

Practice Problem 6

A 5-lb 14-oz batch of cookies is packed into a 6-oz container before it is mailed. Find the total weight.

Answers
4. 37 lb 8 oz   **5.** 1 lb 6 oz   **6.** 6 lb 4 oz

(c) Defining and Converting Metric System Units of Mass

In scientific and technical areas, a careful distinction is made between **weight** and **mass**. **Weight** is really a measure of the pull of gravity. The farther from Earth an object gets, the less it weighs. However, **mass** is a measure of the amount of substance in the object and does not change. Astronauts orbiting Earth weigh much less than they weigh on Earth, but they have the same mass in orbit as they do on Earth. Here on Earth weight and mass are the same, so either term may be used.

The basic unit of mass in the metric system is the **gram**. It is defined as the mass of water contained in a cube 1 centimeter (cm) on each side.

1 cm
1 cm
1 cm

The following examples may help you get a feeling for metric masses.

A tablet contains 200 milligrams of ibuprofen.

A large paper clip weighs approximately 1 gram.

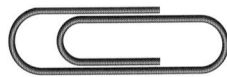

A box of crackers weighs 453 grams.

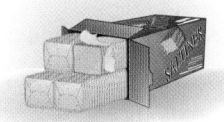

A kilogram is slightly over 2 pounds. An adult woman may weigh 60 kilograms.

The prefixes for units of mass in the metric system are the same as for units of length, as shown in the following table:

Prefix	Meaning	Metric Unit of Mass
kilo	1000	1 kilogram (kg) = 1000 grams (g)
hecto	100	1 hectogram (hg) = 100 g
deka	10	1 dekagram (dag) = 10 g
		1 gram (g) = 1 g
deci	1/10	1 decigram (dg) = 1/10 g or 0.1 g
centi	1/100	1 centigram (cg) = 1/100 g or 0.01 g
milli	1/1000	1 milligram (mg) = 1/1000 g or 0.001 g

Try the Concept Check in the margin.

The **milligram**, the **gram**, and the **kilogram** are the three most commonly used units of mass in the metric system.

As with lengths, all units of mass are powers of 10 of the gram, so converting from one unit of mass to another involves only moving the decimal point. To convert from one unit of mass to another in the metric system, list the units of mass in order from largest to smallest.

Let's convert 4300 milligrams to grams. To convert from milligrams to grams, we move along the table 3 units to the left.

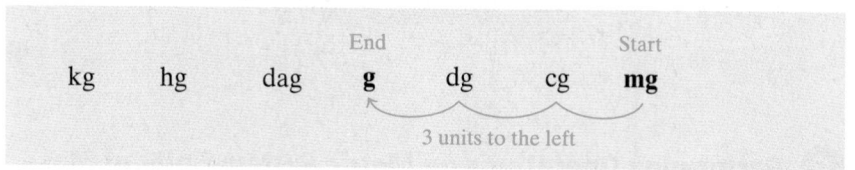

This means that we move the decimal point 3 places to the left to convert from milligrams to grams.

$$4300 \text{ mg} = 4.3 \text{ g}$$

The same conversion can be done with unit fractions.

$$4300 \text{ mg} = \frac{4300 \text{ mg}}{1} \cdot \frac{0.001 \text{ g}}{1 \text{ mg}}$$

$$= 4300 \cdot 0.001 \text{ g}$$

$$= 4.3 \text{ g}$$

To multiply by 0.001, move the decimal point 3 places to the left.

To see that this is reasonable, study the diagram:

1000 mg	1000 mg	1000 mg	1000 mg	300 mg	4300 mg
1 g	1 g	1 g	1 g	0.3 g	= 4.3 g

Thus 4300 mg = 4.3 g

EXAMPLE 7 Convert 3.2 kg to grams.

Solution: First we convert by using a unit fraction.

$$3.2 \text{ kg} = 3.2 \text{ kg} \cdot \frac{\overbrace{1000 \text{ g}}^{\text{Unit fraction}}}{1 \text{ kg}} = 3200 \text{ g}$$

Now let's list the units of mass in a chart and move from kilograms to grams.

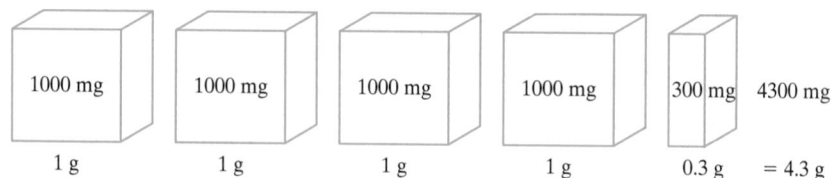

$$3.200 \text{ kg} = 3200. \text{ g} \text{ or } 3200 \text{ g}$$

3 places to the right

1 kg	1 kg	1 kg	0.2 kg	3.2 kg
1000 g	1000 g	1000 g	200 g	= 3200 g

Concept Check

True or false? A decigram is larger than a dekagram. Explain.

Practice Problem 7

Convert 3.41 g to milligrams.

Answer

7. 3410 mg

Concept Check: false

Practice Problem 8

Convert 56.2 cg to grams.

EXAMPLE 8 Convert 2.35 cg to grams.

Solution: We list the units of mass in a chart and move from centigrams to grams.

End   Start

kg   hg   dag   g   dg   cg   mg

2 units to the left

02.35 cg = 0.0235 g

2 places to the left

D   Performing Operations on Metric System Units of Mass

Arithmetic operations can be performed with metric units of mass just as we performed operations with metric units of length. We convert each number to the same unit of mass and add, subtract, multiply, or divide as with decimals.

Practice Problem 9

Subtract 3.1 dg from 2.5 g.

EXAMPLE 9 Subtract 5.4 dg from 1.6 g.

Solution: We convert both numbers to decigrams or to grams before subtracting.

5.4 dg = 0.54 g   1.6 g = 16 dg

     1.60 g   or   16.0 dg
    −0.54 g   −5.4 dg
     1.06 g   10.6 dg

The difference is 1.06 g or 10.6 dg.

Practice Problem 10

Multiply 12.6 kg by 4.

EXAMPLE 10 Multiply 15.4 kg by 5.

Solution: We multiply the two numbers together.

    15.4 kg
  ×   5
    77.0 kg

The result is 77.0 kg.

Practice Problem 11

Twenty-four bags of cement weigh a total of 550 kg. Find the average weight of 1 bag, rounded to the nearest kilogram.

EXAMPLE 11   Calculating Allowable Weight in an Elevator

An elevator has a weight limit of 1400 kg. A sign posted in the elevator indicates that the maximum capacity of the elevator is 17 persons. What is the average allowable weight for each passenger, rounded to the nearest kilogram?

Answers

8. 0.562 g   **9.** 2.19 g or 21.9 dg   **10.** 50.4 kg
11. 23 kg

Solution: To solve, notice that the total weight of 1400 kilograms ÷ 17 = average weight

$$
\begin{array}{r}
82.3 \text{ kg} \approx 82 \text{ kg} \\
17\overline{)1400.0 \text{ kg}} \\
-136 \phantom{.0 \text{ kg}} \\
\hline
40 \\
-34 \\
\hline
60 \\
-51 \\
\hline
9
\end{array}
$$

Each passenger can weigh an average of 82 kg. (Recall that a kilogram is slightly over 2 pounds, so 82 kilograms is over 164 pounds.)   ●

FOCUS ON **The Real World**

THE COST OF ROAD SIGNS

There are nearly 4 million miles of streets and roads in the United States. With streets, roads, and highways comes the need for traffic control, guidance, warning, and regulation. Road signs perform many of these tasks. Just in our routine travels, we see a wide variety of road signs every day. Think how many road signs must exist on the 4 million miles of roads in the United States. Have you ever wondered how much signs like these cost?

The cost of a road sign generally depends on the type of sign. Costs for several types of signs and signposts are listed in the table. Examples of various types of signs are shown below.

Road Sign Costs	
Type of Sign	**Cost**
Regulatory, warning, marker	$15–$18 per square foot
Large guide	$20–$25 per square foot
Type of Post	**Cost**
U-channel	$125–$200 each
Square tube	$10–$15 per foot
Steel breakaway posts	$15–$25 per foot

The cost of a sign is based on its area. For diamond, square, or rectangular signs, the area is found by multiplying the length (in feet) times the width (in feet). Then the area is multiplied by the cost per square foot. For signs with irregular shapes, costs are generally figured *as if* the sign were a rectangle, multiplying the height and width at the tallest and widest parts of the sign.

GROUP ACTIVITY

Locate four different kinds of road signs on or near your campus. Measure the dimensions of each sign and the height of the post on which it is mounted. Using the cost data given in the table, find the minimum and maximum costs of each sign, including its post. Summarize your results in a table, and include a sketch of each sign.

Name _____ Section _____ Date _____

Mental Math

Convert.

1. 16 ounces to pounds   **2.** 32 ounces to pounds   **3.** 1 ton to pounds   **4.** 4 tons to pounds

5. 1 pound to ounces   **6.** 6 pounds to ounces   **7.** 2000 pounds to tons   **8.** 4000 pounds to tons

Determine whether the measurement in each statement is reasonable.

9. The doctor prescribed a pill containing 2 kg of medication.

10. A full-grown cat weighs approximately 15 g.

11. A bag of flour weighs 4.5 kg.

12. A staple weighs 15 mg.

13. A professor weighs less than 150 g.

14. A car weighs 2000 mg.

EXERCISE SET 9.5

(A) *Convert as indicated. See Examples 1 and 2.*

1. 2 pounds to ounces   **2.** 3 pounds to ounces   **3.** 5 tons to pounds   **4.** 3 tons to pounds

5. 12,000 pounds to tons   **6.** 32,000 pounds to tons 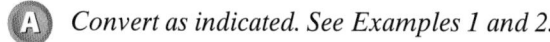 **7.** 60 ounces to pounds   **8.** 90 ounces to pounds

9. 3500 pounds to tons   **10.** 9000 pounds to tons   **11.** 16.25 pounds to ounces   **12.** 14.5 pounds to ounces

13. 4.9 tons to pounds   **14.** 8.3 tons to pounds   **15.** $4\frac{3}{4}$ pounds to ounces   **16.** $9\frac{1}{8}$ pounds to ounces

17. 2950 pounds to the nearest tenth of a ton

18. 51 ounces to the nearest tenth of a pound

Perform each indicated operation. See Examples 3 through 5.

19. 34 lb 12 oz + 18 lb 14 oz

20. 6 lb 10 oz + 10 lb 8 oz

21. 6 tons 1540 lb + 2 tons 850 lb

22. 2 tons 1575 lb + 1 ton 480 lb

23. 5 tons 1050 lb − 2 tons 875 lb

24. 4 tons 850 lb − 1 ton 260 lb

25. 12 lb 4 oz − 3 lb 9 oz

26. 45 lb 6 oz − 26 lb 10 oz

27. 5 lb 3 oz × 6

28. 2 lb 5 oz × 5

29. 6 tons 1500 lb ÷ 5

30. 5 tons 400 lb ÷ 4

Solve. Remember to insert units when writing your answers. See Example 6.

31. Doris Johnson has two open containers of Uncle Ben's rice. If she combines 1 lb 10 oz from one container with 3 lb 14 oz from the other container, how much total rice does she have?

32. Dru Mizel maintains the records of the amount of coal delivered to his department in the steel mill. In January, 3 tons 1500 lb were delivered. In February, 2 tons 1200 lb were delivered. Find the total amount delivered in these two months.

33. Carla Hamtini was amazed when she grew a 28 lb 10 oz zucchini in her garden, but later she learned that the heaviest zucchini ever grown weighed 64 lb 8 oz in Llanharry, Wales, by B. Lavery in 1990. How far below the record weight was Carla's zucchini? (*Source: The Guinness Book of Records*)

34. The heaviest baby born in good health weighed an incredible 22 lb 8 oz. He was born in Italy in September, 1955. How much heavier is this than a 7 lb 12 oz baby? (*Source: The Guinness Book of Records*)

35. The Shop 'n Bag supermarket chain ships hamburger meat by placing 10 packages of hamburger in a box, with each package weighing 3 lb 4 oz. How much will 4 boxes of hamburger weigh?

36. The Quaker Oats Company ships its 1-lb 2-oz boxes of oatmeal in cartons containing 12 boxes of oatmeal. How much will 3 such cartons weigh?

37. A carton of Del Monte Pineapple weighs 55 lb 4 oz, but 2 lb 8 oz of this weight is due to packaging. Subtract the weight of the packaging to find the actual weight of the pineapple in 4 cartons.

38. The Hormel Corporation ships cartons of canned ham weighing 43 lb 2 oz each. Of this weight, 3 lb 4 oz is due to packaging. Find the actual weight of the ham found in 3 cartons.

39. One bag of Pepperidge Farm Bordeaux cookies weighs $6\frac{3}{4}$ ounces. How many pounds will a dozen bags weigh?

40. One can of Payless Red Beets weighs $8\frac{1}{2}$ ounces. How much will eight cans weigh?

C *Convert as indicated. See Examples 7 and 8.*

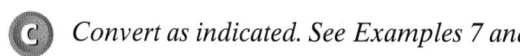 **41.** 500 g to kilograms

42. 650 g to kilograms

43. 4 g to milligrams

44. 9 g to milligrams

45. 25 kg to grams

46. 18 kg to grams

47. 48 mg to grams

48. 112 mg to grams

49. 6.3 g to kilograms

50. 4.9 g to kilograms

51. 15.14 g to milligrams

52. 16.23 g to milligrams

53. 4.01 kg to grams

54. 3.16 kg to grams

D *Perform each indicated operation. See Examples 9 and 10.*

55. 3.8 mg + 9.7 mg

56. 41.6 g + 9.8 g

57. 205 mg + 5.61 g

58. 2.1 g + 153 mg

59. 9 g − 7150 mg

60. 4 kg − 2410 g

61. 1.61 kg − 250 g

62. 6.13 g − 418 mg

63. 5.2 kg × 2.6

64. 4.8 kg × 9.3

65. 17 kg ÷ 8

66. 8.25 g ÷ 6

Solve. Remember to insert units when writing your answers. See Example 11.

67. A can of 7-Up weighs 336 grams. Find the weight in kilograms of 24 cans.

68. Guy Green normally weighs 73 kg, but he lost 2800 grams after being sick with the flu. Find Guy's new weight.

69. Sudafed is a decongestant that comes in two strengths. Regular strength contains 60 mg of medication. Extra strength contains 0.09 g of medication. How much extra medication is in the extra-strength tablet?

70. A small can of Planters sunflower seeds weighs 177 g. If each can contains 6 servings, find the weight of one serving.

71. Amanda Caucutt's doctor recommends that Amanda limit her daily intake of sodium to 0.6 gram. A one-ounce serving of Cheerios with $\frac{1}{2}$ cup of fortified skim milk contains 350 mg of sodium. How much more sodium can Amanda have after she eats a bowl of Cheerios for breakfast, assuming she intends to follow the doctor's orders?

72. A large bottle of Hire's Root Beer weighs 1900 grams. If a carton contains 6 large bottles of root beer, find the weight in kilograms of 5 cartons.

73. Three milligrams of preservatives are added to a 0.5-kg box of dried fruit. How many milligrams of preservatives are in 3 cartons of dried fruit if each carton contains 16 boxes?

74. One box of Swiss Miss Cocoa Mix weighs 0.385 kg, but 39 grams of this weight is the packaging. Find the actual weight of the cocoa in 8 boxes.

75. A carton of 12 boxes of Quaker Oats Oatmeal weighs 6.432 kg. Each box includes 26 grams of packaging material. What is the actual weight of the oatmeal in the carton?

76. The supermarket prepares hamburger in 85-gram market packages. When Leo Gonzalas gets home, he divides the package in half before refrigerating the meat. How much will each package weigh?

77. A package of Trailway's Gorp, a high-energy hiking trail mix, contains 0.3 kg of nuts, 0.15 kg of chocolate bits, and 400 grams of raisins. Find the total weight of the package.

78. The manufacturer of Anacin wants to reduce the caffeine content of its aspirin by $\frac{1}{4}$. Currently, each regular tablet contains 32 mg of caffeine. How much caffeine should be removed from each tablet?

79. A regular-size bag of Lay's potato chips weighs 198 grams. Find the weight of a dozen bags, rounded to the nearest hundredth of a kilogram.

80. A cat weighs a hefty 9 kg. The vet has recommended that this cat lose 1500 grams. How much should the cat weigh?

Review and Preview

Write each fraction as a decimal. See Section 4.7.

81. $\frac{1}{4}$ **82.** $\frac{1}{20}$ **83.** $\frac{4}{25}$ **84.** $\frac{3}{5}$ **85.** $\frac{7}{8}$ **86.** $\frac{3}{16}$

Combining Concepts

87. Why is the decimal point moved to the right when grams are converted to milligrams?

88. To change 8 pounds to ounces, multiply by 16. Why is this the correct procedure?

9.6 Capacity

Ⓐ Defining and Converting U.S. System Units of Capacity

Units of **capacity** are generally used to measure liquids. The number of gallons of gasoline needed to fill a gas tank in a car, the number of cups of water needed in a bread recipe, and the number of quarts of milk sold each day at a supermarket are all examples of using units of capacity. The following summary shows equivalencies between units of capacity:

OBJECTIVES

Ⓐ Define U.S. units of capacity and convert from one unit to another.

Ⓑ Perform arithmetic operations on U.S. units of capacity.

Ⓒ Define metric units of capacity and convert from one unit to another.

Ⓓ Perform arithmetic operations on metric units of capacity.

SSM TUTOR CENTER   SG   CD & VIDEO   MATH PRO   WEB

> **U.S. Units of Capacity**
>
> 8 fluid ounces (fl oz) = 1 cup (c)
> 2 cups = 1 pint (pt)
> 2 pints = 1 quart (qt)
> 4 quarts = 1 gallon (gal)

Just as with units of length and weight, we can form unit fractions to convert between different units of capacity. For instance,

$$\frac{2\text{ c}}{1\text{ pt}} = \frac{1\text{ pt}}{2\text{ c}} = 1 \quad \text{and} \quad \frac{2\text{ pt}}{1\text{ qt}} = \frac{1\text{ qt}}{2\text{ pt}} = 1$$

EXAMPLE 1 Convert 9 quarts to gallons.

Solution: We multiply by the unit fraction $\frac{1\text{ gal}}{4\text{ qt}}$.

$$9\text{ qt} = \frac{9\text{ qt}}{1} \cdot \frac{1\text{ gal}}{4\text{ qt}}$$
$$= \frac{9\text{ gal}}{4}$$
$$= 2\frac{1}{4}\text{ gal}$$

Thus, 9 quarts is the same as $2\frac{1}{4}$ gallons, as shown in the diagram:

Practice Problem 1

Convert 43 pints to quarts.

1 gallon   1 gallon $\frac{1}{4}$ gallon = $2\frac{1}{4}$ gal

9 quarts

EXAMPLE 2 Convert 14 cups to quarts.

Solution: Our equivalency table contains no direct conversion from cups to quarts. However, from this table we know that

1 qt = 2 pt = 4 c

so 1 qt = 4 c. Now we have the unit fraction $\frac{1\text{ qt}}{4\text{ c}}$. Thus,

$$14\text{ c} = \frac{14\text{ c}}{1} \cdot \frac{1\text{ qt}}{4\text{ c}} = \frac{7}{2}\text{qt} \quad \text{or} \quad 3\frac{1}{2}\text{qt}$$

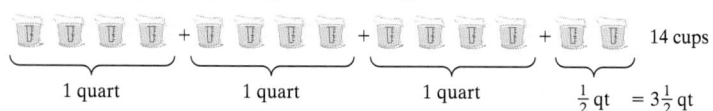

1 quart   1 quart   1 quart $\frac{1}{2}$ qt = $3\frac{1}{2}$ qt

14 cups

Practice Problem 2

Convert 26 quarts to cups.

Answers

1. $21\frac{1}{2}$ qt **2.** 104 c

Note: Another way to solve this example is to use two unit fractions.

$$14 \text{ c} = \frac{\overset{7}{\cancel{14} \text{ } \cancel{c}}}{1} \cdot \frac{1 \text{ pt}}{\underset{1}{\cancel{2} \text{ } \cancel{c}}} \cdot \frac{1 \text{ qt}}{2 \text{ } \cancel{pt}} = \frac{7}{2} \text{qt} \quad \text{or} \quad 3\frac{1}{2}\text{qt}$$

Concept Check

If 50 cups are converted to quarts, will the equivalent number of quarts be less than or greater than 50? Explain.

Practice Problem 3

Subtract 2 qt from 1 gal 1 qt.

Try the Concept Check in the margin.

B **Performing Operations on U.S. System Units of Capacity**

As is true of units of length and weight, units of capacity can be added, subtracted, multiplied, and divided.

EXAMPLE 3 Subtract 3 qt from 4 gal 2 qt.

Solution: To subtract, we line up similar units.

$$\begin{array}{r} 4 \text{ gal } 2 \text{ qt} \\ - \quad \quad 3 \text{ qt} \\ \hline \end{array}$$

We cannot subtract 3 qt from 2 qt. We need to borrow 1 gallon from the 4 gallons, convert it to 4 quarts, and then combine it with the 2 quarts.

Borrow 1 gal = 4 qt

3 gal + (1 gal) 4 qt

$$\begin{array}{r} 4 \text{ gal } 2 \text{ qt} \\ - \quad \quad 3 \text{ qt} \\ \hline \end{array} \quad \text{or} \quad \begin{array}{r} 3 \text{ gal } 6 \text{ qt} \\ - \quad \quad 3 \text{ qt} \\ \hline 3 \text{ gal } 3 \text{ qt} \end{array}$$

To check, see that the sum of 3 gal 3 qt and 3 qt is 4 gal 2 qt.

Practice Problem 4

Multiply 2 gal 3 qt by 2.

EXAMPLE 4 Multiply 3 qt 1 pt by 3.

Solution: We multiply each of the units of capacity by 3.

$$\begin{array}{r} 3 \text{ qt } 1 \text{ pt} \\ \times \quad \quad 3 \\ \hline 9 \text{ qt } 3 \text{ pt} \end{array}$$

Since 3 pints is the same as 1 quart and 1 pint, we have

9 qt 3 pt = 9 qt + 1 qt 1 pt = 10 qt 1 pt

The 10 quarts can be changed to gallons by dividing by 4, since there are 4 quarts in a gallon. To see why we divide, notice that

$$10 \text{ qt} = \frac{10 \text{ } \cancel{qt}}{1} \cdot \frac{1 \text{ gal}}{4 \text{ } \cancel{qt}} = \frac{10}{4}\text{gal}$$

$$\begin{array}{r} 2 \text{ gal } 2 \text{ qt} \\ 4 \overline{) 10 \text{ qt}} \\ \underline{-8} \\ 2 \end{array}$$

Then the product is 10 qt 1 pt or 2 gal 2 qt 1 pt.

Practice Problem 5

Divide 6 gal 1 qt by 2.

EXAMPLE 5 Divide 3 gal 2 qt by 2.

Solution: We divide each unit of capacity by 2.

$$\begin{array}{r} 1 \text{ gal } \quad 3 \text{ qt} \\ 2 \overline{) 3 \text{ gal} \quad 2 \text{ qt}} \\ \underline{-2} \\ 1 \text{ gal} = 4 \text{ qt} \\ 6 \text{ qt} \quad 6 \text{ qt} \div 2 = 3 \text{ qt} \end{array}$$

Answers

3. 3 qt **4.** 5 gal 2 qt **5.** 3 gal 1 qt

Concept Check: less than 50

EXAMPLE 6 **Finding the Amount of Water in an Aquarium**

An aquarium contains 6 gal 3 qt of water. If 2 gal 2 qt of water is added, what is the total amount of water in the aquarium?

Solution:

	beginning water	→	6 gal 3 qt
	+ water added	→	+ 2 gal 2 qt
	total water	→	8 gal 5 qt

Since 5 qt = 1 gal 1 qt, we have

$$\underbrace{8 \text{ gal}} \quad \underbrace{5 \text{ qt}}$$

= 8 gal + 1 gal 1 qt

= 9 gal 1 qt

The total amount of water is 9 gal 1 qt. ●

(c) **Defining and Converting Metric System Units of Capacity**

Thus far, we know that the basic unit of length in the metric system is the meter and that the basic unit of mass in the metric system is the gram. What is the basic unit of capacity? The **liter**. By definition, a **liter** is the capacity or volume of a cube measuring 10 centimeters on each side.

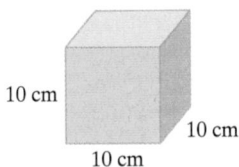

10 cm
10 cm
10 cm

The following examples may help you get a feeling for metric capacities:

One liter of liquid is slightly more than one quart.

1 quart 1 liter

Many soft drinks are packaged in 2-liter bottles.

The metric system was designed to be a consistent system. Once again, the prefixes for metric units of capacity are the same as for metric units of length and mass, as summarized in the following table:

Prefix	Meaning		Metric Unit of Capacity
kilo	1000	1 kiloliter	(kl) = 1000 liters (L)
hecto	100	1 hectoliter	(hl) = 100 L
deka	10	1 dekaliter	(dal) = 10 L
		1 liter (L) = 1 L	
deci	1/10	1 deciliter	(dl) = 1/10 L or 0.1 L
centi	1/100	1 centiliter	(cl) = 1/100 L or 0.01 L
milli	1/1000	1 milliliter	(ml) = 1/1000 L or 0.001 L

The **milliliter** and the **liter** are the two most commonly used metric units of capacity.

Converting from one unit of capacity to another involves multiplying by powers of 10 or moving the decimal point to the left or to the right. Listing units of capacity in order from largest to smallest helps to keep track of how many places to move the decimal point when converting.

Let's convert 2.6 liters to milliliters. To convert from liters to milliliters, we move along the chart 3 units to the right.

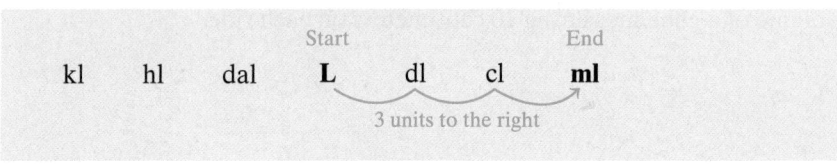

This means that we move the decimal point 3 places to the right to convert from liters to milliliters.

$$2.600 \text{ L} = 2600. \text{ ml or } 2600 \text{ ml}$$

This same conversion can be done with unit fractions.

$$2.6 \text{ L} = \frac{2.6 \text{ L}}{1} \cdot \frac{1000 \text{ ml}}{1 \text{ L}}$$
$$= 2.6 \cdot 1000 \text{ ml}$$
$$= 2600 \text{ ml} \qquad \text{To multiply by 1000, move the decimal point 3 places to the right.}$$

To visualize the result, study the diagram below:

Thus 2.6 L = 2600 ml.

Practice Problem 7

Convert 2100 ml to liters.

EXAMPLE 7 Convert 3210 ml to liters.

Solution: Let's use the unit fraction method first.

$$3210 \text{ ml} = 3210 \text{ ml} \cdot \frac{\overset{\text{Unit fraction}}{1 \text{ L}}}{1000 \text{ ml}} = 3.21 \text{ L}$$

Now let's list the unit measures in a chart and move from milliliters to liters.

Answer

7. 2.1 L

End   Start

kl   hl   dal   L   dl   cl   ml

3 units to the left

3210 ml = 3.210 L

3 places to the left

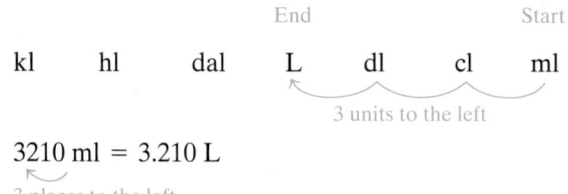

1000 ml   1000 ml   1000 ml

210 ml

3210 ml

1 L   1 L   1 L   0.210 L   = 3.210 L

EXAMPLE 8 Convert 0.185 dl to milliliters.

Solution: We list the unit measures in a chart and move from deciliters to milliliters.

kl   hl   dal   L   Start   End
                               dl   cl   ml

2 units to the right

0.185 dl = 18.5 ml

2 places to the right

D **Performing Operations on Metric System Units of Capacity**

As was true for length and weight, arithmetic operations involving metric units of capacity can also be performed. Make sure the metric units of capacity are the same before adding, subtracting, multiplying, or dividing.

EXAMPLE 9 Add 2400 ml to 8.9 L.

Solution: We must convert both to liters or both to milliliters before adding the capacities together.

$$2400 \text{ ml} = 2.4 \text{ L}$$ or $$8.9 \text{ L} = 8900 \text{ ml}$$
$$\phantom{2400 \text{ ml} = }2.4 \text{ L}$$ $$\phantom{8.9 \text{ L} = }2400 \text{ ml}$$
$$\underline{+8.9 \text{ L}}$$ $$\underline{+8900 \text{ ml}}$$
$$11.3 \text{ L}$$ $$11{,}300 \text{ ml}$$

The total is 11.3 L or 11,300 ml. They both represent the same capacity.

Try the Concept Check in the margin.

EXAMPLE 10 Divide 18.08 ml by 16.

Solution:

$$
\begin{array}{r}
1.13 \text{ ml} \\
16\overline{)18.08 \text{ ml}} \\
-16\phantom{.08 \text{ ml}} \\
\hline
20\phantom{8 \text{ ml}} \\
-16\phantom{8 \text{ ml}} \\
\hline
48\phantom{ \text{ml}} \\
-48\phantom{ \text{ml}} \\
\hline
0\phantom{ \text{ml}}
\end{array}
$$

The solution is 1.13 ml.

Practice Problem 11

If 28.6 L of water can be pumped every minute, how much water can be pumped in 85 minutes?

Answer

11. 2431 L

EXAMPLE 11 Finding the Amount of Medication a Person Has Received

A patient hooked up to an IV unit in the hospital is to receive 12.5 ml of medication every hour. How much medication does the patient receive in 3.5 hours?

Solution: We multiply 12.5 ml by 3.5.

$$
\begin{array}{rcl}
\text{medication per hour} & \rightarrow & 12.5 \text{ ml} \\
\times \quad\quad\quad \text{hours} & \rightarrow & \times\ 3.5 \\
\hline
\text{total medication} & & 625 \\
& & \underline{375} \\
& & 43.75 \text{ ml}
\end{array}
$$

The patient receives 43.75 ml of medication.

FOCUS ON **The Real World**

COMPUTER STORAGE CAPACITY

Have you ever heard someone describe a computer as having a 4 gig hard drive and wondered what exactly that means? In this chapter, we focused on measuring basic physical characteristics such as length, weight/mass, and capacity. However, another type of measurement that we frequently encounter in the real world deals with computer storage capacity; that is, how much information can be stored on a computer device such as a floppy diskette, data cartridge, CD-ROM, hard drive, or computer memory (such as RAM).

To understand computer storage capacities, we must first understand the building blocks of information in the world of computers. The smallest piece of information handled by a computer is called a **bit** (abbreviated "b"). Bit is short for "binary digit" and consists of either a 0 or a 1. A collection of 8 bits is called a **byte** (abbreviated "B"). A single byte represents a standard character such as a letter of the alphabet, a number, or a punctuation mark. The following table shows the relationships among various terms used to describe computer storage capacities:

1 nibble = 4 bits

1 byte (B) = 8 bits

1 kilobyte (KB) = 1024 bytes = 2^{10} bytes

1 megabyte (MB) = 1024 kilobytes = 1024^2 bytes = 2^{20} bytes

1 gigabyte (GB) = 1024 megabytes = 1024^3 bytes = 2^{30} bytes

1 terabyte (TB) = 1024 gigabytes = 1024^4 bytes = 2^{40} bytes

Note: Sometimes a megabyte is referred to as a "meg" and a gigabyte as a "gig."

GROUP ACTIVITY

Using advertisements for computer systems, find five different examples of uses of any of the computer storage capacity terms defined above. Then convert each example to both bytes and bits. Make a table to organize your results. Be sure to include an explanation of the original use of each example.

Mental Math

Convert as indicated.

1. 2 c to pints

2. 4 c to pints

3. 4 qt to gallons

4. 8 qt to gallons

5. 2 pt to quarts

6. 6 pt to quarts

7. 8 fl oz to cups

8. 24 fl oz to cups

9. 1 pt to cups

10. 3 pt to cups

11. 1 gal to quarts

12. 2 gal to quarts

Determine whether the measurement in each statement is reasonable.

13. Clair took a dose of 2 L of cough medicine to cure her cough.

14. John drank 250 ml of milk for lunch.

15. Jeannie likes to relax in a tub filled with 3000 ml of hot water.

16. Sarah pumped 20 L of gasoline into her car yesterday.

EXERCISE SET 9.6

A *Convert each measurement as indicated. See Examples 1 and 2.*

1. 32 fluid ounces to cups

2. 16 quarts to gallons

3. 8 quarts to pints

4. 9 pints to quarts

5. 10 quarts to gallons

6. 15 cups to pints

7. 80 fluid ounces to pints

8. 18 pints to gallons

9. 2 quarts to cups

10. 3 pints to fluid ounces

11. 120 fluid ounces to quarts

12. 20 cups to gallons

13. 6 gallons to fluid ounces

14. 5 quarts to cups

15. $4\frac{1}{2}$ pints to cups

16. $6\frac{1}{2}$ gallons to quarts

17. $2\frac{3}{4}$ gallons to pints

18. $3\frac{1}{4}$ quarts to cups

B *Perform each indicated operation. See Examples 3 through 5.*

19. 4 gal 3 qt + 5 gal 2 qt

20. 2 gal 3 qt + 8 gal 3 qt

21. 1 c 5 fl oz + 2 c 7 fl oz

22. 2 c 3 fl oz + 2 c 6 fl oz

23. 3 gal − 1 gal 3 qt

24. 2 pt − 1 pt 1 c

25. 3 gal 1 qt − 1 qt 1 pt

26. 3 qt 1 c − 1 c 4 fl oz

27. 1 pt 1 c × 3

28. 1 qt 1 pt × 2

29. 8 gal 2 qt × 2

30. 6 gal 1 pt × 2

31. 9 gal 2 qt ÷ 2

32. 5 gal 6 fl oz ÷ 2

Solve. Remember to insert units when writing your answers. See Example 6.

33. A can of Hawaiian punch holds $1\frac{1}{2}$ quarts of liquid. How many fluid ounces is this?

34. Weight Watchers Double Fudge bars contain 21 fluid ounces of ice cream. How many cups of ice cream is this?

35. Many diet experts advise individuals to drink 64 ounces of water each day. How many quarts of water is this?

36. A recipe for walnut fudge cake calls for $1\frac{1}{4}$ cups of water. How many fluid ounces is this?

37. Can 5 pt 1 c of fruit punch and 2 pt 1 c of ginger ale be poured into a 1-gal container without it overflowing?

38. Three cups of prepared Jell-O are poured into 6 dessert dishes. How many fluid ounces of Jell-O are in each dish?

39. A garden tool engine requires a 30 to 1 gas to oil mixture. This means that $\frac{1}{30}$ of a gallon of oil should be mixed with 1 gallon of gas. Convert $\frac{1}{30}$ gallon to ounces. Round to the nearest tenth.

40. Henning's Supermarket sells homemade soup in 1 qt 1 pt containers. How much soup is contained in three such containers?

41. A case of Pepsi Cola holds 24 cans, each of which contains 12 fluid ounces of Pepsi Cola. How many *quarts* are there in a case of Pepsi?

42. Manuela's Service Station has a drum that holds 40 gallons of oil. If 6 gallons and 3 quarts have been used, how much oil remains?

Convert as indicated. See Examples 7 and 8.

 43. 5 L to milliliters

44. 8 L to milliliters

45. 4500 ml to liters

46. 3100 ml to liters

47. 410 L to kiloliters

48. 250 L to kiloliters

49. 64 ml to liters

50. 39 ml to liters

 51. 0.16 kl to liters

52. 0.48 kl to liters

53. 3.6 L to milliliters

54. 1.9 L to milliliters

55. 0.16 L to kiloliters

56. 0.127 L to kiloliters

D *Perform each indicated operation. See Examples 9 and 10.*

57. 2.9 L + 19.6 L

58. 18.5 L + 4.6 L

59. 2700 ml + 1.8 L

60. 4.6 L + 1600 ml

61. 8.6 L − 190 ml

62. 4.8 L − 283 ml

63. 11,400 ml − 0.8 L

64. 6850 ml − 0.3 L

65. 480 ml × 8

66. 290 ml × 6

67. 81.2 L ÷ 0.5

68. 5.4 L ÷ 3.6

Solve. Remember to insert units when writing your answers. See Example 11.

69. Mike Schaferkotter drank 410 ml of Mountain Dew from a 2-liter bottle. How much Mountain Dew remains in the bottle?

70. A Volvo has a 54.5-L gas tank. Only 3.8 liters of gasoline still remain in the tank. How much is needed to fill it?

71. A woman added 354 ml of Prestone dry gasoline to the 18.6 L of gasoline in her car's tank. Find the total amount of gasoline in the tank.

72. Chris Peckaitis wishes to share a 2-L bottle of Coca Cola equally with 7 of his friends. How much will each person get?

73. A salesman paid $14 to fill his car with 44.3 liters of gasoline. Find the price per liter of gasoline to the nearest tenth of a cent.

74. A student carelessly misread the scale on a cylinder in the chemistry lab and added 40 cl of water to a mixture instead of 40 ml. Find the excess amount of water.

75. A large bottle of Langers Apple Juice contains 1.89 L of beverage. A smaller bottle of apple juice contains only 946 ml. How much more is in the larger bottle?

76. In a lab experiment, a student added 400 ml of salt water to 1.65 L of water. Later 320 ml of the solution was drained off. How much of the solution still remained?

Review and Preview

Write each decimal as a fraction. See Section 5.1.

77. 0.7   **78.** 0.9   **79.** 0.03   **80.** 0.007   **81.** 0.006   **82.** 0.08

Combining Concepts

83. Explain how to borrow in order to subtract 1 gal 2 qt from 3 gal 1 qt.

A cubic centimeter (cc) is the amount of space that a volume of 1 mL occupies. Because of this, we will say that 1 cc = 1 mL.

A common syringe is one with a capacity of 3 cc. Use the diagram below and give the measurement indicated by each arrow. The first measurement is done for you.

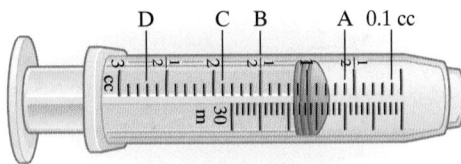

84. A   **85.** B   **86.** C   **87.** D

In order to measure small dosages, such as for insulin, u-100 syringes are used. For these syringes, 1 cc has been divided into 100 equal units (u). Use the diagram below and give the measurement indicated by each arrow in units (u) and then cubic centimeters. Use 100 u = 1 cc and round to the nearest hundredth.

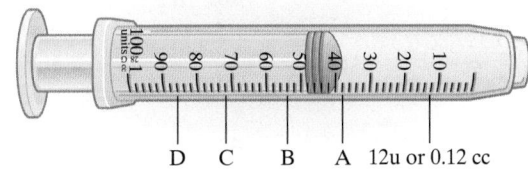

88. A   **89.** B

90. C   **91.** D

Internet Excursions

 Go To: http://www.prenhall.com/martin-gay_prealgebra 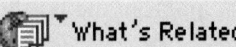 What's Related

By going to the World Wide Web address listed above, you will be directed to a site called A Dictionary of Units, or a related site, that will help you answer the questions below.

92. There are more units of capacity in use than the ones mentioned in this section. For instance, there are special measures for the capacity of dry items such as grains or fruits. Visit this Web site and locate the "U.S. System of Measurements" area. List the measure equivalencies for dry capacity. Then convert 38 pecks of apples to bushels.

93. There are also special apothecaries' measures for capacity of liquid drugs and medicines. Although metric measures are commonly used in the pharmacy industry today, awareness of these measures is still useful. Within the "U.S. System of Measurements" area of this Web site, find and list the apothecaries' measures equivalencies. Then convert 2 pints to fl drams.

9.7   Conversions between the U.S. and Metric Systems

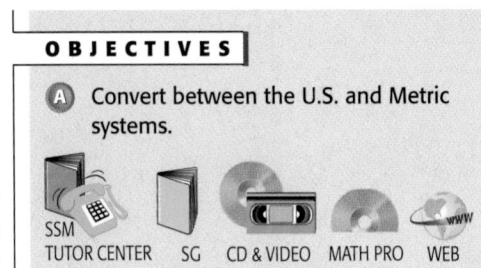
Ⓐ Converting Between the U.S. and Metric Systems

The metric system probably had its beginnings in France in the 1600s, but it was the Metric Act of 1866 that made the use of this system legal (but not mandatory) in the United States. Other laws have followed that allow for a slow, but deliberate, transfer to the modernized metric system. In April, 2001, for example, the U.S. Stock Exchanges completed their change to decimal trading instead of fractions. By the end of 2009, all products sold in Europe (with some exceptions) will be required to have only metric units on their labels. (*Source*: U.S. Metric Association and National Institute of Standards and Technology)

You may be surprised at the number of everyday items we use that are already manufactured in metric units. We easily recognize 1L and 2L soda bottles, but what about the following?

  Pencil leads (0.5 mm or 0.7 mm)
  Camera film (35 mm)
  Sporting events (5 km or 10 km races)
  Medicines (500 mg capsules)
  Labels on retail goods (dual-labeled since 1994)

Since the United States has not completely converted to the metric system, we need to practice converting from one system to the other. Below is a table of mostly approximate conversions.

Length:

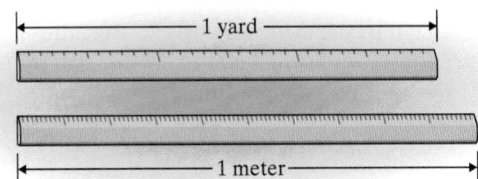

metric	U.S. System
1 m	≈ 1.09 yd
1 m	≈ 3.28 ft
1 km	≈ 0.62 mi
2.54 cm	= 1 in.
0.30 m	≈ 1 ft
1.61 km	≈ 1 mi

Capacity:

metric	U.S. System
1 L	≈ 1.06 qt
1 L	≈ 0.26 gal
3.79 L	≈ 1 gal
0.95 L	≈ 1 qt
29.57 ml	≈ 1 fl oz

Weight (mass):

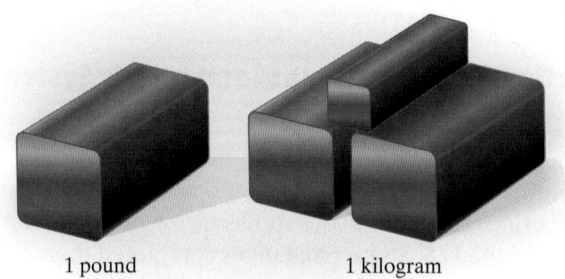

1 pound   1 kilogram

metric U.S. System
  1 kg ≈ 2.20 lb
   1 g ≈ 0.04 oz
0.45 kg ≈ 1 lb
28.35 g ≈ 1 oz

There are many ways to perform these metric to U.S. Conversions. We will do so by using unit fractions.

EXAMPLE 1. Compact Disks

Compact disks are 12 centimeters in diameter. Convert this length to inches. Round the result to two decimal places.(*Source*: usByte.com)

Solution: From our length conversion table, we know that 2.54 cm = 1 in. This fact gives us two unit fractions: $\dfrac{2.54 \text{ cm}}{1 \text{ in.}}$ and $\dfrac{1 \text{ in.}}{2.54 \text{ cm}}$. We use the unit fraction with cm in the denominator so that these units divide out.

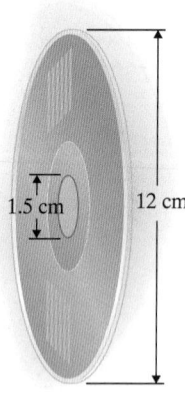

1.5 cm   12 cm

$$12 \text{ cm} = \frac{12 \text{ cm}}{1} \cdot \overbrace{\frac{1 \text{ in.}}{2.54 \text{ cm}}}^{\text{Unit fraction}} \quad \leftarrow \quad \text{Units to convert to}$$
$$\phantom{12 \text{ cm}} \qquad\qquad\qquad \leftarrow \quad \text{Original units}$$

$$= \frac{12 \text{ in.}}{2.54}$$

$$\approx 4.72 \text{ in.} \qquad \text{Divide.}$$

Thus, the diameter of a compact disk is exactly 12 cm or approximately 4.72 inches. For a dimension this size, you can use a ruler to check. Another method is to approximate. Our result, 4.72 in., is close to 5 inches. Since 1 in. is about 2.5 cm, then 5 in. is about 5(2.5 cm) = 12.5 cm, which is close to 12 cm.

Practice Problem 1

The center hole of a compact disk is 15 millimeters 1.5 centimeters in diameter. Convert this length to inches. Round the result to 2 decimal places.

Answer

1. 0.59 in.

EXAMPLE 2   Liver

The liver is your largest internal organ. It weighs about 3.5 pounds in a grown man. Convert this weight to kilograms. Round to the nearest tenth. (*Source*: *Some Body!* by Dr. Pete Rowan)

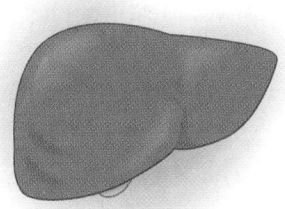

Solution: $3.5 \text{ lb} \approx \dfrac{3.5 \; \cancel{\text{lb}}}{1} \cdot \overset{\text{Unit fraction}}{\dfrac{0.45 \text{ kg.}}{1 \; \cancel{\text{lb}}}}$

$= 3.5(0.45 \text{ kg})$

$\approx 1.6 \text{ kg}$

Thus 3.5 pounds are approximately 1.6 kilograms. From the table of conversions, we know that 1 kg ≈ 2.2 lb. So that 0.5 kg ≈ 1.1 lb and adding, we have 1.5 kg ≈ 3.3 lb. Our result is reasonable.   ●

EXAMPLE 3   Postage Stamp

Australia converted to the metric system in 1973. In that year, four postage stamps were issued to publicize this conversion. One such stamp is shown below. Let's check the mathematics on the stamp by converting 7 fluid ounces to milliliters. Round to the nearest hundred.

Solution: $7 \text{ fl oz} \approx \dfrac{7 \; \cancel{\text{fl oz}}}{1} \cdot \overset{\text{Unit fraction}}{\dfrac{29.57 \text{ ml}}{1 \; \cancel{\text{fl oz}}}}$

$= 206.99 \text{ ml}$

Rounded to the nearest hundred, 7 fl oz ≈ 200 ml.   ●

Practice Problem 2

A full-grown human heart weighs about $\frac{1}{2}$ pound or 8 ounces. Convert this weight to grams. If necessary, round your result to the nearest tenth of a gram.

Practice Problem 3

Convert 237 ml to ounces. Round to the nearest whole ounce.

Answers

2. 226.8 g   **3.** 8 fl oz

STUDY SKILLS REMINDER

Are you preparing for a test on Chapter 9?

Below I have listed some common trouble areas for topics covered in Chapter 9. After studying for your test—but before taking your test—read these.

■ Don't forget the difference between complementary and supplementary angles.

Complementary angles have a sum of 90°.

Supplementary angles have a sum of 180°.

The complement of a 15° angle measures 90° − 15° = 75°.

The supplement of a 15° angle measures 180° − 15° = 165°.

■ Remember:

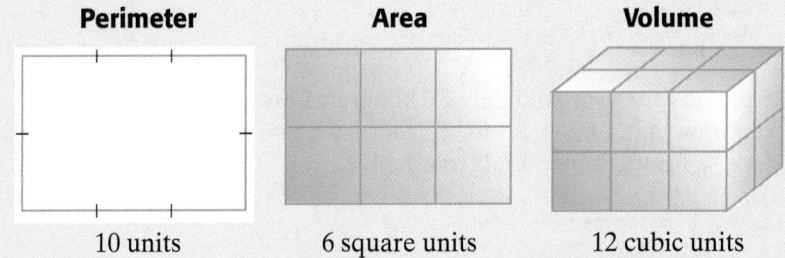

Perimeter	**Area**	**Volume**
10 units	6 square units	12 cubic units

Remember: This is simply a checklist of common trouble areas. For a review of Chapter 9, see the Highlights and Chapter Review at the end of this chapter.

Name _____ Section _____ Date _____

EXERCISE SET 9.7

A *Convert as indicated. If necessary, round answers to two decimal places. See Examples 1 through 3. Because approximations are used, your answers may vary slightly from the answers given in the back of the book.*

1. 578 milliliters to fluid ounces

2. 5 liters to quarts

3. 86 inches to centimeters

4. 86 miles to kilometers

5. 1000 grams to ounces

6. 100 kilograms to pounds

7. 93 kilometers to miles

8. 9.8 meters to feet

9. 14.5 liters to gallons

10. 150 milliliters to fluid ounces

11. 30 pounds to kilograms

12. 15 ounces to grams

Solve. If necessary, round answers to two decimal places. See Examples 1 through 3.

13. The balance beam for female gymnasts is 10 centimeters wide. Convert this width to inches.

14. In men's gymnastics, the rings are 250 centimeters from the floor. Convert this height to inches, then to feet.

15. The speed limit is 70 miles per hour. Convert this to kilometers per hour.

16. The speed limit is 40 kilometers per hour. Convert this to miles per hour.

17. Ibuprofen comes in 200 milligram tablets. Convert this to ounces.

18. Vitamin C tablets come in 500 milligram caplets. Convert this to ounces.

19. A stone is a unit in the British customary system. Use the conversion: 14 pounds = 1 stone to check the equivalencies in this 1973 Australian stamp. Is 100 kilograms approximately 15 stone 10 pounds?

20. Convert 5 feet 11 inches to centimeters and check the conversion on this 1973 Australian stamp. Is it correct?

21. You find two soda sizes at the store. One is 12 fluid ounces and the other is 380 milliliters. Which is larger?

22. A punch recipe calls for 2 gallons of pineapple juice. You have 8 liters of pineapple juice. Do you have enough for the recipe?

23. A $3\frac{1}{2}$-inch diskette is not really $3\frac{1}{2}$ inches. To find its actual width, convert this measurement to centimeters, then to millimeters. Round the result to the nearest ten.

24. The average two-year-old is 84 centimeters tall. Convert this to feet and inches.

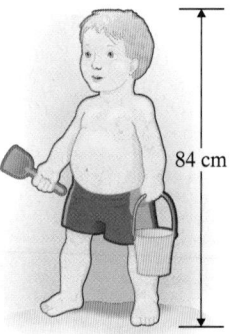

25. For an average adult, the weight of a right lung is greater than the weight of a left lung. If the right lung weighs 1.5 pounds and the left lung weighs 1.25 pounds, find the difference in grams. (*Source: Some Body!*)

26. The skin of an average adult weighs 9 pounds and is the heaviest organ. Find the weight in grams. (*Source: Some Body!*)

27. A fast sneeze has been clocked at about 167 kilometers per hour. Convert this to miles per hour. Round to the nearest whole hour.

28. The Boeing 747 has a cruising speed of about 980 kilometers per hour. Convert this to miles per hour. Round to the nearest whole hour.

29. The General Sherman giant sequoia tree has a diameter of about 8 meters at its base. Convert this to feet. (*Source*: *Fantastic Book of Comparisions*)

30. The largest crater on the near side of the moon is Billy Crater. It has a diameter of 303 kilometers, Convert this to miles. (*Source*: *Fantastic Book of Comparisions*)

31. The total length of the track on a CD is about 4.5 kilometers. Convert this to miles. Round to the nearest whole mile.

32. The distance between Mackinaw City, Michigan and Cheyenne, Wyoming is 2079 kilometers. Convert this to miles. Round to the nearest whole mile.

33. A doctor orders a dosage of 5 ml of medicine every 4 hours for 1 week. How many fluid ounces of medicine should be purchased? Round up to the next whole fluid ounce.

34. A doctor orders a dosage of 12 ml of medicine every 6 hours for 10 days. How many fluid ounces of medicine should be purchased? Round up to the next whole fluid ounce.

Without actually converting, choose the most reasonable answer.

35. A twin mattress has a width of about _____.

A. 1 m   B. 100 m   C. 10 m   D. 1000 m

36. A pie plate has a diameter of about _____.

A. 22 m   B. 22 km   C. 22 cm   D. 22 g

37. A liter has _____ capacity than a quart.

A. less   B. greater   C. the same

38. A foot is _____ a meter.

A. shorter than   B. longer than   C. the same length as

39. A kilogram weighs _____ a pound.

A. the same as   B. less than   C. greater than

40. A football field is 100 yards, which is about _____.

A. 9 m   B. 90 m   C. 900 m   D. 9000 m

41. An $8\frac{1}{2}$ ounce glass of water has a capacity of about _____.

A. 250 L   B. 25 L   C. 2.5 L   D. 250 ml

42. A 5-gallon gasoline can has a capacity of about _____.

A. 19 L   B. 1.9 L   C. 19 ml   D. 1.9 ml

43. The weight of an average man is about _____.

A. 700 kg   B. 7 kg   C. 0.7 kg   D. 70 kg

44. The weight of a pill is about _____.

A. 200 kg   B. 20 kg   C. 2 kg   D. 200 mg

Review and Preview

Perform the indicated operations. See Sections 2.2 through 2.5.

45. $-6 \cdot 4 + 5 \div (-1)$

46. $-10 \div (-2) + 9(-8)$

47. $\dfrac{-10 + 8}{-10 - 8}$

48. $\dfrac{-14 + (-1)}{-5(-3)}$

49. $3 + 5(17 - 19) - 8$

50. $1 + 4(9 - 19) + 5$

51. $3[(-1 + 5) \cdot (6 - 8)]$

52. $-5[9 - (18 - 8)]$

Combining Concepts

Body surface area (BSA) is often used to calculate dosages for some drugs. BSA is calculated in square meters using a person's weight and height.

$$\text{BSA} = \sqrt{\frac{(\text{weight in kg}) \times (\text{height in cm})}{3600}}$$

Calculate the BSA for each person. Round to the nearest hundredth.

53. An adult whose height is 182 cm and weight is 90 kg.

54. An adult whose height is 157 cm and weight is 63 kg.

55. A child whose height is 40 in. and weight is 50 kg. (Hint: Don't forget to first convert inches to centimeters)

56. A child whose height is 26 in. and weight is 13 kg.

57. An adult whose height is 60 in. and weight is 150 lb.

58. An adult whose height is 69 in. and weight is 172 lb.

59. Suppose the adult from Exercise 53 is to receive a drug that has a recommended dosage range of 10-12 mg per sq meter. Find the dosage range for the adult.

60. Suppose the Child from Exercise 56 is to receive a drug that has a recommended dosage of 30 mg per sq meter. Find the dosage for the Child.

61. A handball court is a rectangle that measures 20 meters by 40 meters. Find its area in square meters and square feet.

62. A backpack measures 16 inches by 13 inches by 5 inches. Find the volume of a box with these dimensions. Find the volume in cubic inches and cubic centimeters. Round the cubic centimeters to the nearest whole cubic centimeter.

9.8 Temperature

When Gabriel Fahrenheit and Anders Celsius independently established units for temperature scales, each based his unit on the heat of water the moment it boils compared to the moment it freezes. One degree Celsius is $\frac{1}{100}$ of the difference in heat. One degree Fahrenheit is $\frac{1}{180}$ of the difference in heat. Celsius arbitrarily labeled the temperature at the freezing point at 0°C, making the boiling point 100°C; Fahrenheit labeled the freezing point 32°F, making the boiling point 212°F. Water boils at 212°F or 100°C.

By comparing the two scales in the figure, we see that a 20°C day is as warm as a 68°F day. Similarly, a sweltering 104°F day in the Mojave desert corresponds to a 40°C day.

Try the Concept Check in the margin.

A   Converting Degrees Celsius to Degrees Fahrenheit

To convert from Celsius temperatures to Fahrenheit temperatures, see the box below. In this box, we use the symbol F to represent degrees Fahrenheit and the symbol C to represent degrees Celsius.

Converting Celsius to Fahrenheit

$$F = \frac{9}{5}C + 32 \quad \text{or} \quad F = 1.8C + 32$$

(To convert to Fahrenheit temperature, multiply the Celsius temperature by $\frac{9}{5}$ or 1.8, and then add 32.)

Concept Check

Which of the following statements is correct? Explain.

a. 6°C is below the freezing point of water.

b. 6°F is below the freezing point of water.

Answer

Concept Check:
a. false   **b.** true

Practice Problem 1

Convert 50°C to degrees Fahrenheit.

EXAMPLE 1   Convert 15°C to degrees Fahrenheit.

Solution: $F = \dfrac{9}{5}C + 32$

$= \dfrac{9}{5} \cdot 15 + 32$ Replace C with 15.

$= 27 + 32$ Simplify.

$= 59$ Add.

Thus, 15°C is equivalent to 59°F.

Practice Problem 2

Convert 18°C to degrees Fahrenheit.

EXAMPLE 2   Convert 29°C to degrees Fahrenheit.

Solution: $F = 1.8\,C + 32$

$= 1.8 \cdot 29 + 32$ Replace C with 29.

$= 52.2 + 32$ Multiply 1.8 by 29.

$= 84.2$ Add.

Therefore, 29°C is the same as 84.2°F.

B   Converting Degrees Fahrenheit to Degrees Celsius

To convert from Fahrenheit temperatures to Celsius temperatures, see the box below. The symbol C represents degrees Celsius and the symbol F represents degrees Fahrenheit.

Converting Fahrenheit to Celsius

$$C = \dfrac{5}{9}(F - 32)$$

(To convert to Celsius temperature, subtract 32 from the Fahrenheit temperature, and then multiply by $\dfrac{5}{9}$.)

Practice Problem 3

Convert 86°F to degrees Celsius.

EXAMPLE 3   Convert 59°F to degrees Celsius.

Solution: We evaluate the formula $C = \dfrac{5}{9}(F - 32)$ when F is 59.

$C = \dfrac{5}{9}(F - 32)$

$= \dfrac{5}{9} \cdot (59 - 32)$ Replace F with 59.

$= \dfrac{5}{9} \cdot (27)$ Subtract inside parentheses.

$= 15$ Multiply.

Therefore, 59°F is the same temperature as 15°C.

Answers

1. 122°F   **2.** 64.4°F   **3.** 30°C

EXAMPLE 4 Convert 114°F to degrees Celsius. If necessary, round to the nearest tenth of a degree.

Solution: $C = \dfrac{5}{9}(F - 32)$

$\qquad = \dfrac{5}{9}(114 - 32)$ Replace F with 114.

$\qquad = \dfrac{5}{9} \cdot (82)$ Subtract inside parentheses.

$\qquad \approx 45.6$ Multiply.

Therefore, 114°F is approximately 45.6°C.

EXAMPLE 5 Body Temperature

Normal body temperature is 98.6°F. What is this temperature in degrees Celsius?

Solution: We evaluate the formula $C = \dfrac{5}{9}(F - 32)$ when F is 98.6.

$C = \dfrac{5}{9}(F - 32)$

$\qquad = \dfrac{5}{9}(98.6 - 32)$ Replace F with 98.6.

$\qquad = \dfrac{5}{9} \cdot (66.6)$ Subtract inside parentheses.

$\qquad = 37$ Multiply.

Therefore, normal body temperature is 37°C.

Try the Concept Check in the margin.

Practice Problem 4

Convert 113°F to degrees Celsius. If necessary, round to the nearest tenth of a degree.

Practice Problem 5

During a bout with the flu, Albert's temperature reaches 102.8°F. What is his temperature measured in degrees Celsius? Round to the nearest tenth of a degree.

Concept Check

Clarissa must convert 40°F to degrees Celsius. What is wrong with her work shown below?

$\qquad F = 1.8 \cdot C + 32$
$\qquad F = 1.8 \cdot 40 + 32$
$\qquad F = 72 + 32$
$\qquad F = 104$

Answers

4. 45°C **5.** 39.3°C

Concept Check: She used the conversion for Celsius to Fahrenheit instead of Fahrenheit to Celsius.

FOCUS ON **History**

THE DEVELOPMENT OF UNITS OF MEASURE

The earliest units of measure were based on the human body. The ancient Egyptians, Babylonians, Hebrews, and Mesopotamians used a unit of length called the **cubit**, which represents the distance between a human elbow and fingertips. For instance, in the book of Genesis in the Bible, Noah's ark is described as having a length of "three hundred cubits, its width fifty cubits, and its height thirty cubits." Other commonly used measures found in documents of these ancient cultures include the digit, hand, span, and foot. These measures are shown at the right.

Several thousand years later, the English system of measurement also consisted of a mixed bag of body- and nature-related units of measure. A rod was the combined length of the left feet of 16 men. An inch was the distance spanned by three grains of barley. A foot was the length of the foot of the king currently in power. Around 1100 A.D. King Henry I of England decreed that a yard was the distance between the king's nose and the thumb of his outstretched arm.

Although the English system became somewhat standardized by the 13th century, the problem with it and the ancient systems of measurement was that the relationship between the various units was not necessarily easy to remember. This made it difficult to convert from one type of unit to another.

The metric system grew out of the French Revolution during the 1790s. As a reaction against the lack of consistency and utility of existing systems of measurement, the French Academy of Sciences set out to develop an easy-to-use, internationally standardized system of measurements. Originally the basic unit of the metric system, the meter, was to be one ten-millionth of the distance between the North Pole and the Equator on a line through Paris. However, it was soon discovered that this distance was nearly impossible to measure accurately. The meter has been redefined several times since the end of the French Revolution, most recently in 1983 as the distance that light travels in a vacuum during $\dfrac{1}{299,792,458}$ of a second.

Although the definition of the meter has evolved over time, the original idea of developing an easy-to-use system of measurements has endured. The metric system uses a standard set of prefixes for all basic unit types (length, mass, capacity, etc.) and all conversions within a unit type are based on powers of 10.

CRITICAL THINKING

Develop your own unit of measure for (a) length and (b) area. Explain how you chose your units and for what types of relative sizes of measurements they would be most useful. Discuss the advantages and disadvantages of your units of measurement. Then use your units to measure the width of your classroom desk and the area of the front cover of this textbook.

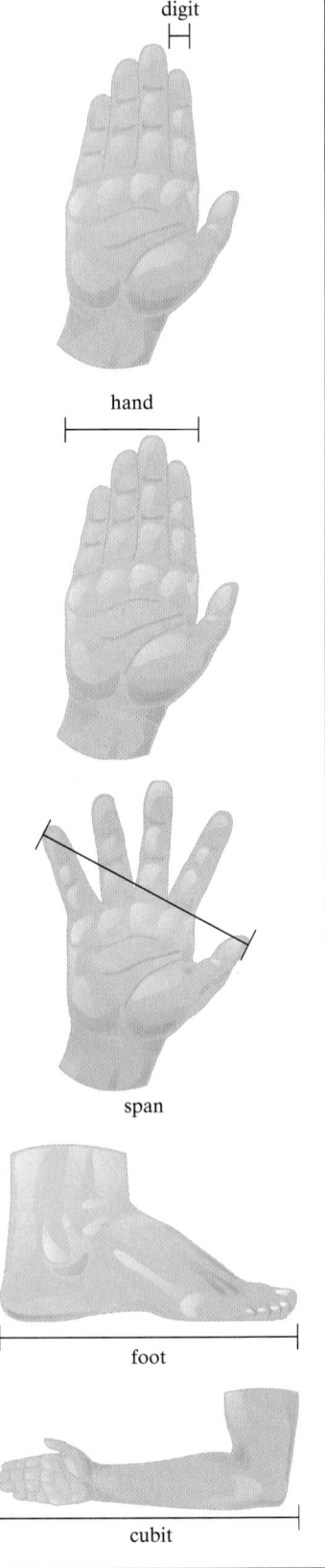

digit

hand

span

foot

cubit

Mental Math

Determine whether the measurement in each statement is reasonable.

1. A 72°F room feels comfortable.

2. Water heated to 110°F will boil.

3. Josiah has a fever if a thermometer shows his temperature to be 40°F.

4. An air temperature of 20°F on a Vermont ski slope can be expected in the winter.

5. When the temperature is 30°C outside, an overcoat is needed.

6. An air-conditioned room at 60°C feels quite chilly.

7. Barbara has a fever when a thermometer records her temperature at 40°C.

8. Water cooled to 32°C will freeze.

EXERCISE SET 9.8

(A) (B) *Convert as indicated. When necessary, round to the nearest tenth of a degree. See Examples 1 through 5.*

1. 41°F to degrees Celsius

2. 68°F to degrees Celsius

3. 104°F to degrees Celsius

4. 86°F to degrees Celsius

5. 60°C to degrees Fahrenheit

6. 80°C to degrees Fahrenheit

7. 115°C to degrees Fahrenheit

8. 35°C to degrees Fahrenheit

9. 62°F to degrees Celsius

10. 182°F to degrees Celsius

11. 142.1°F to degrees Celsius

12. 43.4°F to degrees Celsius

13. 92°C to degrees Fahrenheit

14. 75°C to degrees Fahrenheit

15. 16.3°C to degrees Fahrenheit

16. 48.6°C to degrees Fahrenheit

17. The hottest temperature ever recorded in New Mexico was 122°F. Convert this temperature to degrees Celsius. (*Source*: National Climatic Data Center)

18. The hottest temperature ever recorded in Rhode Island was 104°F. Convert this temperature to degrees Celsius. (*Source*: National Climatic Data Center)

19. A weather forecaster in Caracas predicts a high temperature of 27°C. Find this measurement in degrees Fahrenheit.

20. While driving to work, Alan Olda notices a temperature of 18°C flash on the local bank's temperature display. Find the corresponding temperature in degrees Fahrenheit.

21. At Mack Trucks' headquarters, the room temperature is to be set at 70°F, but the thermostat is calibrated in degrees Celsius. Find the temperature to be set.

22. The computer room at Merck, Sharp, and Dohm is normally cooled to 66°F. Find the corresponding temperature in degrees Celsius.

23. Najib Tan is running a fever of 100.2°F. Find his temperature as it would be shown on a Celsius thermometer.

24. William Saylor generally has a temperature of 98.2°F. Find what this temperature would be on a Celsius thermometer.

25. In a European cookbook, a recipe requires the ingredients for caramels to be heated to 118°C, but the cook has access only to a Fahrenheit thermometer. Find the temperature in degrees Fahrenheit that should be used to make the caramels.

26. The ingredients for divinity should be heated to 127°C, but the candy thermometer that Myung Kim has is calibrated to degrees Fahrenheit. Find how hot he should heat the ingredients.

27. Mark Tabbey's recipe for Yorkshire pudding calls for a 500°F oven. Find the temperature setting he should use with an oven having Celsius controls.

28. The temperature of Earth's core is estimated to be 4000°C. Find the corresponding temperature in degrees Fahrenheit.

29. The surface temperature of Venus can reach 864°F. Find this temperature in degrees Celsius.

Review and Preview

Find the perimeter of each figure. See Section 1.3.

△ **30.**
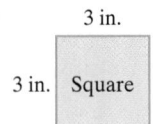
3 in.
3 in. Square

△ **31.**
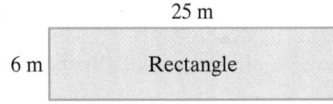
25 m
6 m Rectangle

△ **32.**
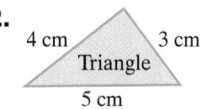
4 cm   3 cm
Triangle
5 cm

△ **33.**
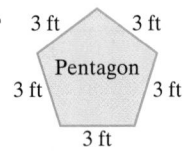
3 ft   3 ft
Pentagon
3 ft   3 ft
3 ft

△ **34.**
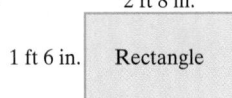
2 ft 8 in.
1 ft 6 in. Rectangle

△ **35.**
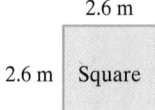
2.6 m
2.6 m Square

Combining Concepts

36. On February 17, 1995, in the Tokamak Fusion Test Reactor at Princeton University, the highest temperature produced in a laboratory was achieved. This temperature was 918,000,000°F. Convert this temperature to degrees Celsius. Round your answer to the nearest ten million of a degree. (*Source: Guinness Book of Records*)

37. The hottest-burning substance known is carbon subnitride. Its flame at one atmospheric pressure reaches 9010°F. Convert this temperature to degrees Celsius. (*Source: Guinness Book of Records*)

38. In your own words, describe how to convert from degrees Celsius to degrees Fahrenheit.

MATERIALS:

- ruler
- string
- calculator

This activity may be completed by working in groups or individually.

Investigate the route you would take from Santa Rosa, New Mexico, to San Antonio, New Mexico. Use the map in the figure to answer the following questions. You may find that using string to match the roads on the map is useful when measuring distances.

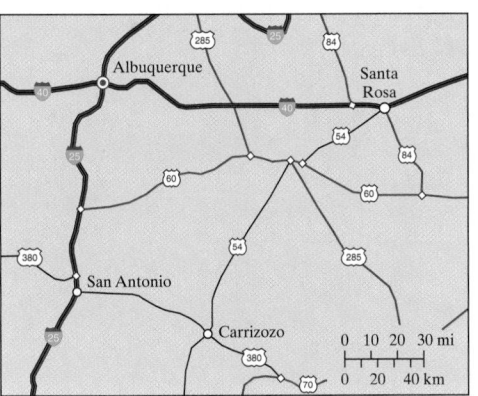

1. How many miles is it from Santa Rosa to San Antonio via Interstate 40 and Interstate 25? Convert this distance to kilometers.

2. How many miles is it from Santa Rosa to San Antonio via U.S. 54 and U.S. 380? Convert this distance to kilometers.

3. Assume that the speed limit on Interstates 40 and 25 is 65 miles per hour. How long would the trip take if you took this route and traveled 65 miles per hour the entire trip?

4. At what average speed would you have to travel on the U.S. routes to make the trip from Santa Rosa to San Antonio in the same amount of time that it would take on the interstate routes? Do you think this speed is reasonable on this route? Explain your reasoning.

5. Discuss in general the factors that might affect your decision among the different routes.

6. Explain which route you would choose in this case and why.

How are your home work assignments going?

By now, you should have good homework habits. If not, it's never too late to begin. Why is it so important in mathematics to keep up with homework? You probably now know the answer to that question. You have probably realized by now that many concepts in mathematics build on each other. Your understanding of one chapter in mathematics usually depends on you understanding of the previous chapter's material.

Don't forget that completing your homework assignment involves a lot more than attempting a few of the problems assigned.

To complete a homework assignment, remember these four things:

1. Attempt all of it.
2. Check it.
3. Correct it.
4. If needed, ask questions about it.

Chapter 9 Vocabulary Check

Fill in each blank with one of the words or phrases listed below.

transversal	line segment	obtuse	straight	adjacent	right	volume	area
acute	perimeter	vertical	supplementary	ray	angle	line	complementary
vertex	mass	unit fractions	gram	weight	meter	liter	

1. _____ is a measure of the pull of gravity.

2. _____ is a measure of the amount of substance in an object. This measure does not change.

3. The basic unit of length in the metric system is the _____.

4. To convert from one unit of length to another, _____ may be used.

5. A _____ is the basic unit of mass in the metric system.

6. The _____ is the basic unit of capacity in the metric system.

7. A _____ is a piece of a line with two endpoints.

8. Two angles that have a sum of 90° are called _____ angles.

9. A _____ is a set of points extending indefinitely in two directions.

10. The _____ of a polygon is the distance around the polygon.

11. An _____ is made up of two rays that share the same endpoint. The common endpoint is called the _____.

12. _____ measures the amount of surface of a region.

13. A _____ is a part of a line with one endpoint. A ray extends indefinitely in one direction.

14. A line that intersects two or more lines at different points is called a _____.

15. An angle that measures 180° is called a _____ angle.

16. The measure of the space of a solid is called its _____.

17. When two lines intersect, four angles are formed. Two of these angles that are opposite each other are called _____ angles.

18. Two of the angles from #17 that share a common side are called _____ angles.

19. An angle whose measure is between 90° and 180° is called an _____ angle.

20. An angle that measures 90° is called a _____ angle.

21. An angle whose measure is between 0° and 90° is called an _____ angle.

22. Two angles that have a sum of 180° are called _____ angles.

| **DEFINITIONS AND CONCEPTS** | **EXAMPLES** |

SECTION 9.1 LINES AND ANGLES

A **line** is a set of points extending indefinitely in two directions. A line has no width or height, but it does have length. We name a line by any two of its points.
A **line segment** is a piece of a line with two endpoints.

A **ray** is a part of a line with one endpoint. A ray extends indefinitely in one direction.

An **angle** is made up of two rays that share the same endpoint. The common endpoint is called the **vertex**

Line AB or $\overleftrightarrow{AB}$

Line segment AB or $\overline{AB}$

Ray AB or $\overrightarrow{AB}$

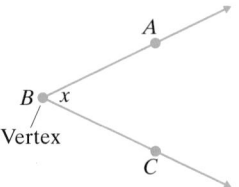

SECTION 9.2 LINEAR MEASUREMENT

To convert from one unit of length to another, unit **fractions** may be used.
The unit fraction should be in the form

$$\frac{\text{units converting to}}{\text{original units}}$$

LENGTH: U.S. SYSTEM OF MEASUREMENT

$$12 \text{ inches (in.)} = 1 \text{ foot (ft)}$$
$$3 \text{ feet} = 1 \text{ yard (yd)}$$
$$5280 \text{ feet} = 1 \text{ mile (mi)}$$

The basic unit of length in the metric system is the **meter**. A meter is slightly longer than a yard.

LENGTH: METRIC SYSTEM OF MEASUREMENT

$$\frac{12 \text{ inches}}{1 \text{ foot}}, \quad \frac{1 \text{ foot}}{12 \text{ inches}}, \quad \frac{3 \text{ feet}}{1 \text{ yard}}$$

Convert 6 feet to inches.

$$6 \text{ feet} = \frac{6 \text{ feet}}{1} \cdot \frac{12 \text{ inches}}{1 \text{ foot}} \quad \leftarrow \text{ units converting to}$$
$$\leftarrow \text{ original units}$$
$$= 6 \cdot 12 \text{ inches}$$
$$= 72 \text{ inches}$$

Convert 3650 centimeters to meters.

$$3650 \text{ cm} = \frac{3650 \text{ cm}}{1} \cdot \frac{0.01 \text{ m}}{1 \text{ cm}} = 36.5 \text{ m}$$

or

| km | hm | dam | m | dm | cm | mm |

End → Start above m/dm

2 units to the left

$$36.50 \text{ cm} = 36.5 \text{ m}$$

2 places to the left

Prefix	Meaning	Metric Unit of Length
kilo	1000	1 **kilo**meter (km) = 1000 meters (m)
hecto	100	1 **hecto**meter (hm) = 100 m
deka	10	1 **deka**meter (dam) = 10 m
		1 meter (m) = 1 m
deci	1/10	1 **deci**meter (dm) = 1/10 m or 0.1 m
centi	1/100	1 **centi**meter (cm) = 1/100 m or 0.01 m
milli	1/1000	1 **milli**meter (mm) = 1/1000 m or 0.001 m

DEFINITIONS AND CONCEPTS	EXAMPLES

SECTION 9.3 PERIMETER

PERIMETER FORMULAS

Rectangle: $P = 2l + 2w$

Square: $P = 4s$

Triangle: $P = a + b + c$

Circumference of a Circle: $C = 2\pi r$ or $C = \pi d$

where $\pi \approx 3.14$ or $\pi \approx \dfrac{22}{7}$

Find the perimeter of the rectangle.

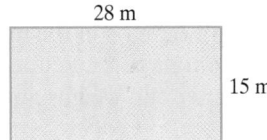

28 m

15 m

$$P = 2l + 2w$$
$$= 2 \cdot 28 \text{ meters} + 2 \cdot 15 \text{ meters}$$
$$= 56 \text{ meters} + 30 \text{ meters}$$
$$= 86 \text{ meters}$$

The perimeter is 86 meters.

SECTION 9.4 AREA AND VOLUME

AREA FORMULAS

Rectangle: $A = lw$

Square: $A = s^2$

Triangle: $A = \dfrac{1}{2}bh$

Parallelogram: $A = bh$

Trapezoid: $A = \dfrac{1}{2}(b + B)h$

Circle: $A = \pi r^2$

VOLUME FORMULAS

Rectangular Solid: $V = lwh$

Cube: $V = s^3$

Sphere: $V = \dfrac{4}{3}\pi r^3$

Right Circular Cylinder: $V = \pi r^2 h$

Cone: $V = \dfrac{1}{3}\pi r^2 h$

Square-Based Pyramid: $V = \dfrac{1}{3}s^2 h$

Find the area of the square.

8 cm

$$A = s^2$$
$$= (8 \text{ centimeters})^2$$
$$= 64 \text{ square centimeters}$$

The area of the square is 64 square centimeters.

Find the volume of the sphere. Use $\dfrac{22}{7}$ for π.

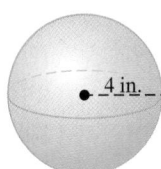

4 in.

$$V = \dfrac{4}{3}\pi r^3$$
$$\approx \dfrac{4}{3} \cdot \dfrac{22}{7} \cdot (4 \text{ inches})^3$$
$$= \dfrac{4 \cdot 22 \cdot 64}{3 \cdot 7} \text{ cubic inches}$$
$$= \dfrac{5632}{21} \quad \text{or} \quad 268\dfrac{4}{21} \text{ cubic inches}$$

DEFINITIONS AND CONCEPTS	EXAMPLES

SECTION 9.5 WEIGHT AND MASS

Weight is really a measure of the pull of gravity.

Mass is a measure of the amount of substance in the object and does not change.

WEIGHT: U.S. SYSTEM OF MEASUREMENT

16 ounces (oz) = 1 pound (lb)

2000 pounds = 1 ton

A **gram** is the basic unit of mass in the metric system. It is the mass of water contained in a cube 1 centimeter on each side. A large paper clip weighs about 1 gram.

Convert 5 pounds to ounces.

$$5 \text{ lb} = \frac{5 \text{ lb}}{1} \cdot \frac{16 \text{ oz}}{1 \text{ lb}} = 80 \text{ oz}$$

Convert 260 grams to kilograms.

$$260 \text{ g} = \frac{260 \text{ g}}{1} \cdot \frac{1 \text{ kg}}{1000 \text{ g}} = 0.26 \text{ kg}$$

End   Start

kg   hg   dag   g   dg   cg   mg

3 units to the left

260 g = 0.260 kg

3 places to the left

MASS: METRIC SYSTEM OF MEASUREMENT

Prefix	Meaning	Metric Unit of Mass
kilo	1000	1 kilogram (kg) = 1000 grams (g)
hecto	100	1 hectogram (hg) = 100 g
deka	10	1 dekagram (dag) = 10 g
1 gram (g) = 1 g		
deci	1/10	1 decigram (dg) = 1/10 g or 0.1 g
centi	1/100	1 centigram (cg) = 1/100 g or 0.01 g
milli	1/1000	1 milligram (mg) = 1/1000 g or 0.001 g

SECTION 9.6 CAPACITY

CAPACITY: U.S. SYSTEM OF MEASUREMENT

8 fluid ounces (fl oz) = 1 cup (c)

2 cups = 1 pint (pt)

2 pints = 1 quart (qt)

4 quarts = 1 gallon (gal)

The **liter** is the basic unit of capacity in the metric system. It is capacity or volume of a cube measuring 10 centimeters on each side. A liter of liquid is slightly more than 1 quart.

Convert 5 pints to gallons.

1 gal = 4 qt = 8 pt

$$5 \text{ pt} = \frac{5 \text{ pt}}{1} \cdot \frac{1 \text{ gal}}{8 \text{ pt}} = \frac{5}{8} \text{ gal}$$

Convert 1.5 liters to milliliters.

$$1.5 \text{ L} = \frac{1.5 \text{ L}}{1} \cdot \frac{1000 \text{ ml}}{1 \text{ L}} = 1500 \text{ ml}$$

or

Start   End

kl   hl   dal   L   dl   cl   ml

3 units to the right

1.500 L = 1500 mL

3 places to the right

(continued)

SECTION 9.6 CAPACITY (CONTINUED)

CAPACITY: METRIC SYSTEM OF MEASUREMENT

Prefix	Meaning	Metric Unit of Capacity
kilo	1000	1 kiloliter (kl) = 1000 liters (L)
hecto	100	1 hectoliter (hl) = 100 L
deka	10	1 dekaliter (dal) = 10 L
		1 liter (L) = 1 L
deci	1/10	1 deciliter (dl) = 1/10 L or 0.1 L
centi	1/100	1 centiliter (cl) = 1/100 L or 0.01 L
milli	1/1000	1 milliliter (ml) = 1/1000 L or 0.001 L

SECTION 9.7 CONVERSIONS BETWEEN THE U.S. AND METRIC SYSTEMS

To convert between systems, use approximate unit fractions.

Convert 7 feet to meters.

$$7 \text{ ft} \approx \frac{7 \text{ ft}}{1} \cdot \frac{0.30 \text{ m}}{1 \text{ ft}} = 2.1 \text{ m}$$

Convert 8 liters to quarts.

$$8 \text{ L} \approx \frac{8 \text{ L}}{1} \cdot \frac{1.06 \text{ qt}}{1 \text{ L}} = 8.48 \text{ qt}$$

Convert 363 grams to ounces.

$$363 \text{ g} \approx \frac{363 \text{ g}}{1} \cdot \frac{0.04 \text{ oz}}{1 \text{ g}} = 14.52 \text{ oz}$$

SECTION 9.8 TEMPERATURE

CELSIUS TO FAHRENHEIT

$$F = \frac{9}{5}C + 32 \quad \text{or} \quad F = 1.8C + 32$$

FAHRENHEIT TO CELSIUS

$$C = \frac{5}{9}(F - 32)$$

Convert 35°C to degrees Fahrenheit.

$$F = \frac{9}{5} \cdot 35 + 32 = 63 + 32 = 95$$

$$35°C = 95°F$$

Convert 50°F to degrees Celsius.

$$C = \frac{5}{9} \cdot (50 - 32) = \frac{5}{9} \cdot (18) = 10$$

$$50°F = 10°C$$

STUDY SKILLS REMINDER

Do you remember what to do the day of an exam?

On the day of an exam, don't forget to try the following:

- Allow yourself plenty of time to arrive.

- Read the directions on the test carefully.

- Read each problem carefully as you take your test. Make sure that you answer the question asked.

- Watch your time and pace yourself so that you may attempt each problem on your test.

- If you have time, check your work and answers.

- Do not turn your test in early. If you have extra time, spend it double-checking your work.

Good luck!

Chapter 9 Review

(9.1) *Classify each angle as acute, right, obtuse, or straight.*

1.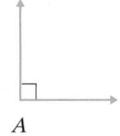

A

2.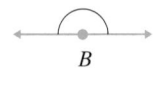

B

3.

C

4.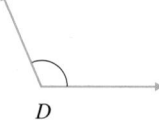

D

5. Find the complement of a 25° angle.

6. Find the supplement of a 105° angle.

7. Find the supplement of a 72° angle.

8. Find the complement of a 1° angle.

Find the measure of x in each figure.

9.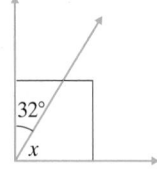

32°

x

10.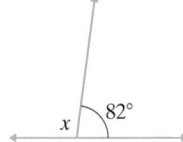

x 82°

11.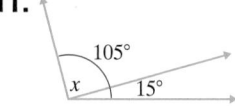

105°

x   15°

12.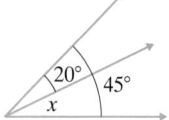

20° 45°

x

13. Identify the pairs of supplementary angles.

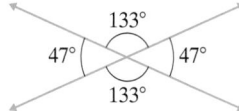

133°

47°   47°

133°

14. Identify the pairs of complementary angles.

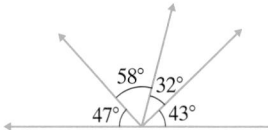

58° 32°

47° 43°

Find the measures of angles x, y, and z in each figure.

15.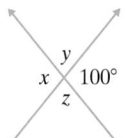

y

x   100°

z

16.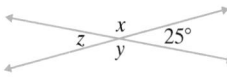

x

z   25°

y

17. *m∥n*

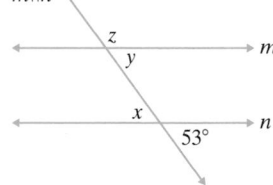

z

y   m

x

53°   n

18. *m∥n*

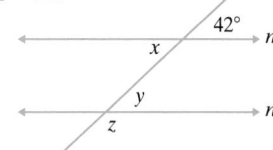

42°

x   m

y

z   n

9.2 *Convert.*

19. 108 in. to feet   **20.** 72 ft to yards   **21.** 2.5 mi to feet   **22.** 6.25 ft to inches

23. 52 ft = ____ yd ____ ft   **24.** 46 in. = ____ ft ____ in.   **25.** 42 m to centimeters

26. 82 cm to millimeters   **27.** 12.18 mm to meters   **28.** 2.31 m to kilometers

Perform each indicated operation.

29. 4 yd 2 ft + 16 yd 2 ft   **30.** 12 ft 1 in. − 4 ft 8 in.   **31.** 8 ft 3 in. × 5

32. 7 ft 4 in. ÷ 2   **33.** 8 cm + 15 mm   **34.** 4 m + 126 cm

35. 9.3 km − 183 m   **36.** 4100 mm − 3 m

Solve.

37. A bolt of cloth contains 333 yd 1 ft of cotton ticking. Find the amount of material that remains after 163 yd 2 ft is removed from the bolt.

38. The local ambulance corps plans to award 20 framed certificates of valor to some of its outstanding members. If each frame requires 6 ft 4 in. of framing material, how much material is needed for all the frames?

39. The trip from Philadelphia to Washington, DC, is 217 km. Four friends agree to share the driving equally. How far must each drive on this round-trip vacation?

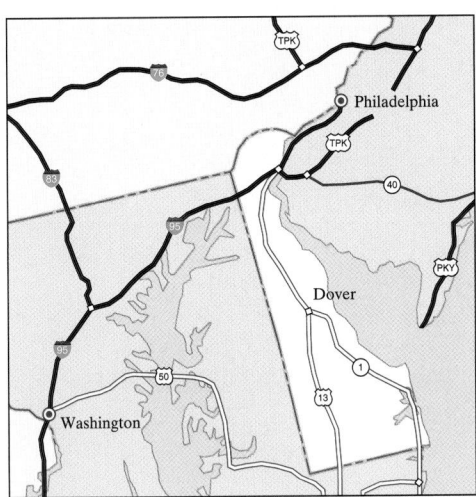

△ **40.** The college has ordered that NO SMOKING signs be placed above the doorway of each classroom. Each sign is 0.8 m long and 30 cm wide. Find the area of each sign. (*Hint*: Recall that the area of a rectangle = length · width.)

30 cm

0.8 m

△ **(9.3)** *Find the perimeter of each figure.*

41.

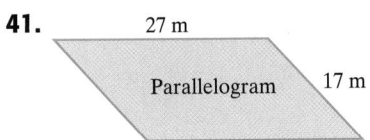

27 m

Parallelogram   17 m

42.

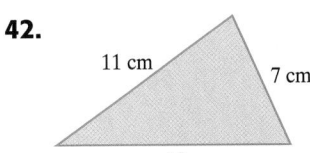

11 cm   7 cm

12 cm

43.

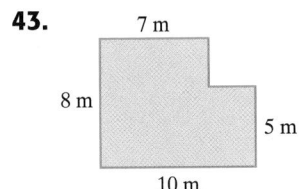

7 m

8 m

5 m

10 m

44.

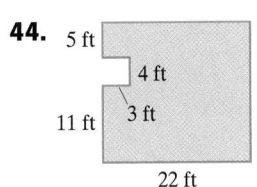

5 ft

4 ft

11 ft   3 ft

22 ft

Solve.

45. Find the perimeter of a rectangular sign that measures 6 feet by 10 feet.

46. Find the perimeter of a town square that measures 110 feet on a side.

Find the circumference of each circle. Use π ≈ 3.14.

47.

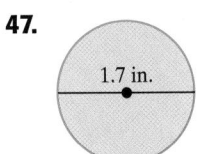

1.7 in.

48.

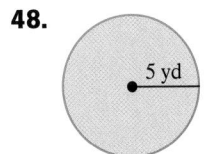

5 yd

(9.4) *Find the area of each figure. For the circles, find the exact area and then use $\pi \approx 3.14$ to approximate the area.*

49.
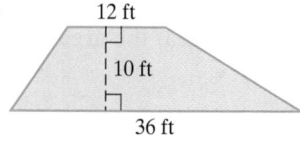
12 ft
10 ft
36 ft

50.
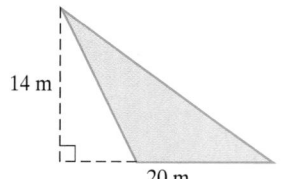
14 m
20 m

51.
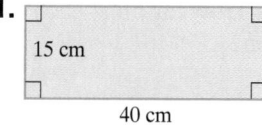
15 cm
40 cm

52.
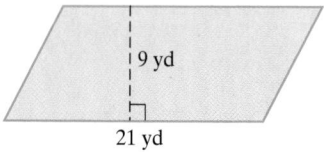
9 yd
21 yd

53.
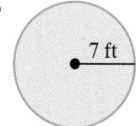
7 ft

54.
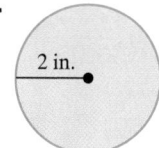
2 in.

55.
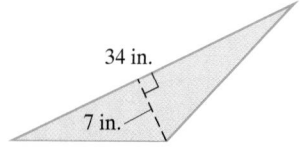
34 in.
7 in.

56.
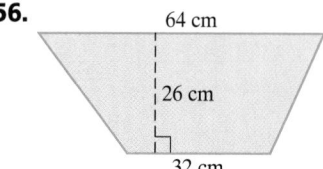
64 cm
26 cm
32 cm

57.
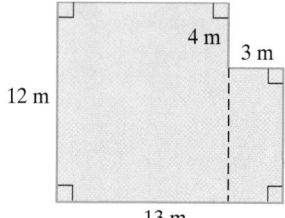
4 m
3 m
12 m
13 m

58. The amount of sealer necessary to seal a driveway depends on the area. Find the area of a rectangular driveway 36 feet by 12 feet.

59. Find how much carpet is needed to cover the floor of the room shown.

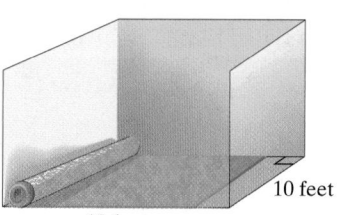
10 feet
13 feet

Find the volume of each solid. For Exercises 62 and 63, use $\pi \approx \dfrac{22}{7}$.

60.

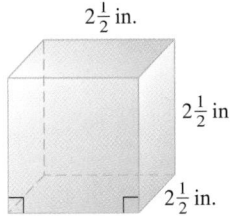

$2\frac{1}{2}$ in.

$2\frac{1}{2}$ in.

$2\frac{1}{2}$ in.

61.

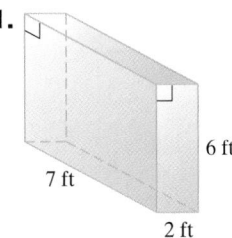

6 ft

7 ft

2 ft

62.

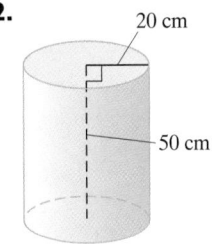

20 cm

50 cm

63.

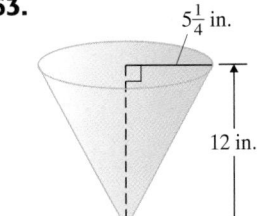

$5\frac{1}{4}$ in.

12 in.

64. Find the volume of a pyramid with a square base 2 feet on a side and a height of 2 feet.

65. Approximate the volume of a tin can 8 inches high and 3.5 inches for the radius. Use $\pi \approx 3.14$

66. A chest has 3 drawers. If each drawer has inside measurements of $2\frac{1}{2}$ feet by $1\frac{1}{2}$ feet by $\frac{2}{3}$ feet, find the total volume of the 3 drawers.

67. A cylindrical canister for a shop vacuum is 2 feet tall and 1 foot for the *diameter*. Find its exact volume.

68. Find the volume of air in a rectangular room 15 feet by 12 feet with a 7-foot ceiling.

69. A mover has two boxes left for packing. Both are cubical, one 3 feet on a side and the other 1.2 feet on a side. Find their combined volume.

(9.5) *Convert.*

70. 66 oz to pounds

71. 2.3 tons to pounds

72. 52 oz = ____ lb ____ oz

73. 8200 lb = ____ tons ____ lb

74. 1400 mg to grams

75. 40 kg to grams

76. 2.1 hg to dekagrams

77. 0.03 mg to decigrams

Perform each indicated operation.

78. 6 lb 5 oz − 2 lb 12 oz

79. 5 tons 1600 lb + 4 tons 1200 lb

80. 6 tons 2250 lb ÷ 3

81. 8 lb 6 oz × 4

82. 1300 mg + 3.6 g

83. 4.8 kg + 4200 g

84. 9.3 g − 1200 mg

85. 6.3 kg × 8

Solve the following.

86. Donshay Berry ordered 1 lb 12 oz of soft-center candies and 2 lb 8 oz of chewy-center candies for his party. Find the total weight of the candy ordered.

87. Four local townships jointly purchase 38 tons 300 lb of cinders to spread on their roads during an ice storm. Determine the weight of the cinders each township receives if they share the purchase equally.

88. Linda Holden ordered 8.3 kg of whole wheat flour from the health food store, but she received 450 g less. How much flour did she actually receive?

89. Eight friends spent a weekend in the Poconos tapping maple trees and preparing 9.3 kg of maple syrup. Find the weight each friend receives if they share the syrup equally.

(9.6) *Convert.*

90. 16 pt to quarts

91. 40 fl oz to cups

92. 6.75 gal to quarts

93. 8.5 pt to cups

94. 9 pt = ____ qt ____ pt

95. 15 qt = ____ gal ____ qt

96. 3.8 L to milliliters

97. 4.2 ml to deciliters

98. 14 hl to kiloliters

99. 30.6 L to centiliters

Perform each indicated operation.

100. 1 qt 1 pt + 3 qt 1 pt

101. 3 gal 2 qt 1 pt × 2

102. 0.946 L − 210 ml

103. 6.1 L + 9400 ml

Solve.

104. Carlos Perez prepares 4 gal 2 qt of iced tea for a block party. During the first 30 minutes of the party, 1 gal 3 qt of the tea is consumed. How much iced tea remains?

105. A recipe for soup stock calls for 1 c 4 fl oz of beef broth. How much should be used if the recipe is cut in half?

106. Each bottle of Kiwi liquid shoe polish holds 85 ml of the polish. Find the number of liters of shoe polish contained in 8 boxes if each box contains 16 bottles.

107. Ivan Miller wants to pour three separate containers of saline solution into a single vat with a capacity of 10 liters. Will 6 liters of solution in the first container combined with 1300 milliliters in the second container and 2.6 liters in the third container fit into the larger vat?

(9.7) *Convert as indicated. If necessary, round to two decimal places. Because approximations are used, your answers may vary slightly from the answers given in the back of the book.*

108. 7 meters to feet.

109. 11.5 yards to meters.

110. 17.5 liters to gallons

111. 7.8 liters to quarts

112. 15 ounces to grams

113. 23 pounds to kilograms

114. A 100-meter dash is being held today. How many yards is this?

115. If a person weighs 82 kilograms, how many pounds is this?

116. How many quarts are contained in a 3-liter bottle of cola?

117. A compact disk is 1.2 mm thick. Find the height (in inches) of 50 disks.

(9.8) *Convert. Round to the nearest tenth of a degree, if necessary.*

118. 245°C to degrees Fahrenheit

119. 160°C to degrees Fahrenheit

120. 42°C to degrees Fahrenheit

121. 86°C to degrees Fahrenheit

122. 93.2°F to degrees Celsius

123. 51.8°F to degrees Celsius

124. 41.3°F to degrees Celsius

125. 80°F to degrees Celsius

Solve. Round to the nearest tenth of a degree, if necessary.

126. A sharp dip in the jet stream caused the temperature in New Orleans to drop to 35°F. Find the corresponding temperature in degrees Celsius.

127. A recipe for meat loaf calls for a 165°C oven. Find the setting used if the oven has a Fahrenheit thermometer.

Chapter 9 Test
Remember to check your answers and use the Chapter Test Prep Video to view solutions.

| Answers |
| 1. _____ |

△ **1.** Find the complement of a 78°angle.   △ **2.** Find the supplement of a 124° angle.

△ **3.** Find the measure of ∠x.

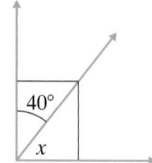

Find the measure of x, y, and z in each figure.

△ **4.**

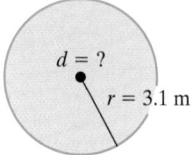

△ **5.**

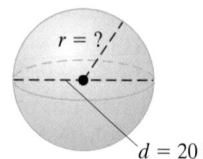

Find the unknown diameter or radius as indicated.

△ **6.**

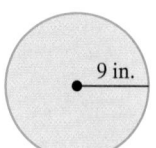

△ **7.**

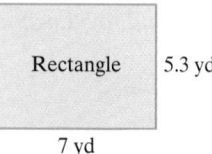

Find the perimeter (or circumference) and area of each figure. For the circle, give the exact value and then use $\pi \approx 3.14$.

△ **8.**

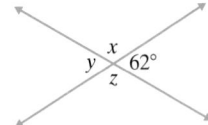

9 in.

△ **9.**

Rectangle   5.3 yd

7 yd

△ **10.**

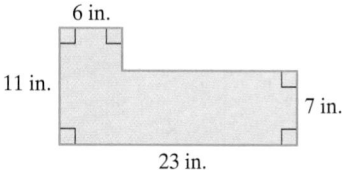

6 in.

11 in.   7 in.

23 in.

Answers

1. _____

2. _____

3. _____

4. _____

5. _____

6. _____

7. _____

8. _____

9. _____

10. _____

Find the volume of each solid. For the cylinder, use $\pi \approx \frac{22}{7}$.

△ **11.**

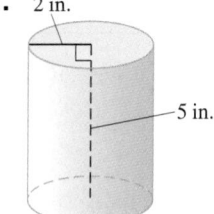

2 in.

5 in.

△ **12.**

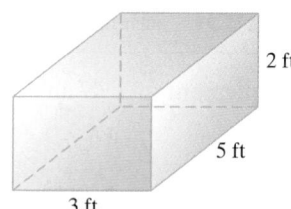

2 ft

5 ft

3 ft

Solve.

△ **13.** Find the perimeter of a square frame with a side length of 4 inches.

4 in.

△ **14.** How much soil is needed to fill a rectangular hole 3 feet by 3 feet by 2 feet?

△ **15.** Vivian Thomas is going to put insecticide on her lawn to control grubworms. The lawn is a rectangle measuring 123.8 feet by 80 feet. The amount of insecticide required is 0.02 ounces per square foot. Find how much insecticide Vivian needs to purchase.

Convert. If necessary, round to 3 decimal places.

16. 280 in. to feet and inches

17. $2\frac{1}{2}$ gal to quarts

18. 30 oz to pounds

19. 2.8 tons to pounds

20. 2.4 km to meters

21. 3.6 cm to millimeters

22. 4.3 dg to grams

23. 0.83 L to milliliters

24. 7 kg to pounds

25. 8.5 in. to centimeters

Convert. Round to the nearest tenth of a degree, if necessary.

26. 84°F to degrees Celsius

27. 12.6°C to degrees Fahrenheit

28. The sugar maples in front of Bette MacMillan's house are 8.4 meters tall. Because they interfere with the phone lines, the telephone company plans to remove the top third of the trees. How tall will the maples be after they are cut back?

29. A total of 15 gal 1 qt of oil has been removed from a 20-gallon drum. How much oil still remains in the container?

30. The engineer in charge of the bridge construction said that the span of the bridge would be 88 m. But the actual construction required it to be 340 cm longer. Find the span of the bridge.

31. If 2 ft 9 in. of material is used to manufacture one scarf, how much material is needed for 6 scarves?

20. _____

21. _____

22. _____

23. _____

24. _____

25. _____

26. _____

27. _____

28. _____

29. _____

30. _____

31. _____

32. _____

33. _____

34. _____

35. _____

32. The largest ice cream sundae ever made in the U.S. was assembled in Anaheim, California, in 1985. This giant sundae used 4667 gallons of ice cream. How many pints of ice cream were used?

33. A piece of candy weighs 5 grams. How many ounces is this?

34. A 5-gallon container holds how many liters?

35. A 5-kilometer race is being held today. How many miles is this?

Cumulative Review

1. Solve: $3a - 6 = a + 4$

2. Solve: $2x + 1 = 3x - 5$.

3. Evaluate:

    a. $\left(\dfrac{2}{5}\right)^4$ **b.** $\left(-\dfrac{1}{4}\right)^2$

4. Evaluate:

    a. $\left(-\dfrac{1}{3}\right)^3$ **b.** $\left(\dfrac{3}{7}\right)^2$.

5. Add: $1\dfrac{4}{5} + 4 + 2\dfrac{1}{2}$

6. Add: $2\dfrac{1}{3} + 4\dfrac{2}{5} + 3$.

7. Simplify by combining like terms:

    $11.1x - 6.3 + 8.9x - 4.6$

8. Simplify by combining like terms:

    $2.5y + 3.7 - 1.3y - 1.9$

9. Simplify: $\dfrac{0.7 + 1.84}{0.4}$

10. Simplify: $\dfrac{0.12 + 0.96}{0.5}$

11. Insert $<$, $>$, or $=$ to form a true statement.

    $0.\overline{7}$ $\dfrac{7}{9}$

12. Insert $<$, $>$, or $=$ to form a true statement.

    0.43 $\dfrac{2}{5}$

13. Solve: $0.5y + 2.3 = 1.65$

14. Solve: $0.4x - 9.3 = 2.7$

△ **15.** An inner city park is in the shape of a square that measures 300 feet on a side. Find the length of the diagonal of the park rounded to the nearest whole foot.

△ **16.** A rectangular field is 200 feet by 125 feet. Find the length of the diagonal of the field, rounded to the nearest whole foot.

Answers

1. _____

2. _____

3. a. _____

   b. _____

4. a. _____

   b. _____

5. _____

6. _____

7. _____

8. _____

9. _____

10. _____

11. _____

12. _____

13. _____

14. _____

15. _____

16. _____

△ **17.** Given the rectangle shown:

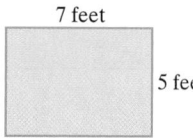

7 feet

5 feet

   a. Find the ratio of its width to its length.

   b. Find the ratio of its length to its perimeter.

19. Write the rate as a fraction in simplest form: 10 nails every 6 feet.

21. Solve: $\dfrac{1.6}{1.1} = \dfrac{x}{0.3}$

Round the solution to the nearest hundredth.

23. The standard dose of an antibiotic is 4 cc (cubic centimeters) for every 25 pounds (lb) of body weight. At this rate, find the standard dose for a 140 lb woman.

25. In a survey of 100 people, 17 people drive blue cars. What percent of people drive blue cars?

27. 13 is $6\frac{1}{2}\%$ of what number?

29. What number is 30% of 9?

△ **18.** A square is 9 inches by 9 inches.

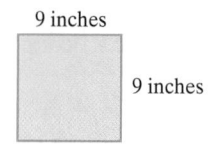

9 inches

9 inches

   a. Find the ratio of a side to its perimeter.

   b. Find the ratio of its perimeter to its area.

20. Write the rate as a fraction in simplest form: 8 chaperones for 40 students

22. Solve: $\dfrac{2.4}{3.5} = \dfrac{0.7}{x}$

Round the solution to the nearest hundredth.

24. A recipe that makes 2 pie crusts calls for 3 cups of flour. How much flour is needed to make 5 pie crusts?

26. Of 100 shoppers surveyed at a mall, 17 paid for their purchases using only cash. What percent of shoppers used only cash to pay for their purchases?

28. 54 is $4\frac{1}{2}\%$ of what number?

30. What number is 42% of 30?

31. The number of applications for a mathematics scholarship at Yale increased from 34 to 45 in one year. What is the percent increase? Round to the nearest whole percent.

32. The price of a gallon of paint rose from $15 to $19. Find the percent increase, rounded to the nearest whole percent.

33. Find the sales tax and the total price on a purchase of an $85.50 trench coat in a city where the sales tax rate is 7.5%.

34. A sofa has a purchase price of $375. If the sales tax rate is 8%, find the amount of sales tax and the total cost of the sofa.

35. Find the ordered pair corresponding to each point plotted on the rectangular coordinate system.

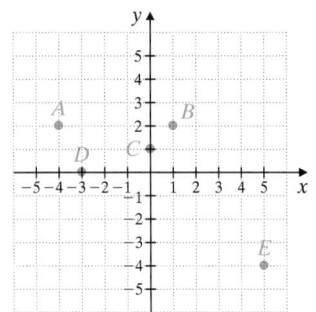

36. Find the ordered pair corresponding to each point plotted on the rectangular coordinate system.

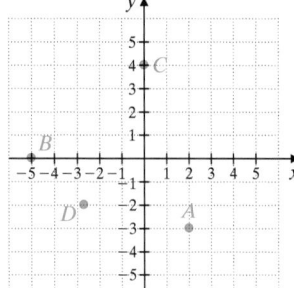

37. Graph $y = 4$.

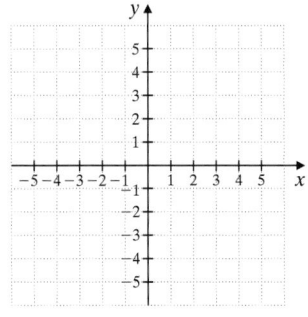

38. Graph $y = -2$.

39. Find the median of the list of numbers: 25, 54, 56, 57, 60, 71, 98

40. Find the median in the list of scores. 60, 95, 89, 72, 83

31. _____

32. _____

33. _____

34. _____

35. _____

36. _____

37. see graph _____

38. see graph _____

39. _____

40. _____

41. _____

42. _____

43. _____

44. _____

45. _____

46. _____

47. _____

48. _____

49. _____

50. _____

41. Find the probability of choosing a red marble from a box containing 1 red, 1 yellow, and 2 blue marbles.

42. Find the probability of choosing a nickel at random in a coin purse that contains 2 pennies, 2 nickels, and 3 quarters.

43. Find the complement of a 48° angle.

44. Find the supplement of a 137° angle.

45. Convert 8 feet to inches.

46. Convert 7 yards to feet.

47. Subtract 3 tons 1350 lb from 8 tons 1000 lb.

48. Add 8 lb 15 oz to 9 lb 3 oz.

49. Convert 59°F to degrees Celsius.

50. Convert 86°F to degrees Celsius.

Polynomials

CHAPTER

Recall that an exponent is a shorthand way of representing repeated multiplication. In this chapter, we learn more about exponents and a special type of expression containing exponents, called a *polynomial*. Studying polynomials is a major part of algebra. Polynomials are also useful for modeling many real-world situations. This chapter serves as an introduction to polynomials and some operations that can be performed on them.

The Royal Gorge Bridge is the world's highest suspension bridge. Located on U.S. Highway 50 near Cañon City, Colorado, the bridge spans the Arkansas River, which lies 1053 feet below the bridge. The Royal Gorge Bridge was built in 1929 at a cost of $350,000. Today it would cost over $10,000,000 to replace the bridge. It is 1260 feet long, 18 feet wide, and can support more than 2,000,000 pounds. In Exercises 55 and 56 on page 803 we will see how a polynomial can be used to describe the height of an object above the Arkansas River after being dropped from the Royal Gorge Bridge.

Chapter 10 Pretest

1. _____

2. _____

3. _____

4. _____

5. _____

6. _____

7. _____

8. _____

9. _____

10. _____

11. _____

12. _____

13. _____

14. _____

15. _____

16. _____

17. _____

18. _____

19. _____

20. _____

Perform each indicated operation.

1. $(7y^2 - 15y + 9) + (-8y + 17)$

2. $(9b - 5) - (-8b + 7)$

3. Subtract $(-3z^4 + 6z^2 + 2z)$ from $(2z^4 - z^3 + 5z)$.

4. Find the value of the polynomial $6t^3 - 5t + 18$ when $t = -2$.

Multiply.

5. $9m^6 \cdot 4m^{12}$

6. $4x \cdot 5x \cdot 6x$

7. $(t^{18})^3$

8. $(3n^2)^4$

9. $(3a^2bc^3)^5(8ab^4c)^2$

10. $2d(9d^4 - 5d^2 + 11)$

11. $(x + 6)(x + 3)$

12. $(y - 4)(2y + 5)$

13. $(3x - 2)^2$

14. $(n + 10)(n - 10)$

15. $(4a + 1)(2a^2 - a + 7)$

Find the greatest common factor of each list of terms.

16. 18 and 45   **17.** y^9, y^3, y^8 **18.** $6m^5, 14m, 18m^4$

Factor.

19. $10y^2 + 6y - 14$

20. $8n^6 - 12n^5 + 24n^3$

10.1 Adding and Subtracting Polynomials

Before we add and subtract polynomials, let's first review some definitions presented in Section 3.1. Recall that the *addends* of an algebraic expression are the *terms* of the expression.

Expression

$3x + 5$ ⟶ 2 terms

$7y^2 + (-6y) + 4$ ⟶ 3 terms

Also, recall that *like terms* can be added or subtracted by using the distributive property. For example,

$$7x + 3x = (7 + 3)x = 10x$$

Ⓐ Adding Polynomials

Some terms are also **monomials**. A term is a monomial if the term contains only whole-number exponents and no variable in the denominator.

Monomials	Not Monomials	
$3x^2$	$\dfrac{2}{y}$	Variable in denominator
$-\dfrac{1}{2}a^2bc^3$	$-2x^{-5}$	Not a whole number exponent
7		

A monomial or a sum and/or difference of monomials is called a **polynomial**.

Polynomial

A **polynomial** is a monomial or a sum and/or difference of monomials.

Examples of Polynomials

$$5x^3 - 6x^2 + 2x + 10, \quad -1.2y^3 + 0.7y, \quad z, \quad \frac{1}{3}r - \frac{1}{2}, \quad 0$$

Some polynomials are given special names depending on their number of terms.

Types of Polynomials

A **monomial** is a polynomial with exactly one term.
A **binomial** is a polynomial with exactly two terms.
A **trinomial** is a polynomial with exactly three terms.

The next page contains examples of monomials, binomials, and trinomials. Each of these examples is also a polynomial.

Polynomials			
Monomials	**Binomials**	**Trinomials**	**More Than Three Terms**
z	$x + 2$	$x^2 - 2x + 1$	$5x^3 - 6x^2 + 2x - 10$
4	$\dfrac{1}{3}r - \dfrac{1}{2}$	$y^5 + 3y^2 - 1.7$	$t^7 - t^5 + t^3 - t + 1$
$0.2x^2$	$-1.2y^3 + 0.7y$	$-a^3 + 2a^2 - 5a$	$z^8 - z^4 + 3z^2 - 2z$
$\uparrow$	$\uparrow$	$\uparrow$	

To add polynomials, we use the commutative and associative properties to rearrange and group like terms. Then, we combine like terms.

Adding Polynomials

To add polynomials, combine like terms.

EXAMPLE 1 Add: $(3x - 1) + (-6x + 2)$

Solution:

$$
\begin{aligned}
(3x - 1) + (-6x + 2) &= (3x - 6x) + (-1 + 2) && \text{Group like terms.}\\
&= (-3x) + (1) && \text{Combine like terms.}\\
&= -3x + 1
\end{aligned}
$$

EXAMPLE 2 Add: $(9y^2 - 6y) + (7y^2 + 10y + 2)$

Solution:

$$
\begin{aligned}
(9y^2 - 6y) + (7y^2 + 10y + 2) &= 9y^2 + 7y^2 - 6y + 10y + 2 && \text{Group like terms.}\\
&= 16y^2 + 4y + 2
\end{aligned}
$$

EXAMPLE 3 Find the sum of $(-y^2 + 2y + 1.7)$ and $(12y^2 - 6y - 3.6)$.

Solution: Recall that "sum" means addition.

$$
\begin{aligned}
&(-y^2 + 2y + 1.7) + (12y^2 - 6y - 3.6)\\
&= \underbrace{-y^2 + 12y^2}\ +\ \underbrace{2y - 6y}\ +\ \underbrace{1.7 - 3.6} && \text{Group like terms.}\\
&= 11y^2 - 4y - 1.9 && \text{Combine like terms.}
\end{aligned}
$$

Polynomials can also be added vertically. To do this, line up like terms underneath one another. Let's vertically add the polynomials in Example 3.

EXAMPLE 4 Find the sum of $(-y^2 + 2y + 1.7)$ and $(12y^2 - 6y - 3.6)$. Use a vertical format.

Solution: Line up like terms underneath one another.

$$
\begin{array}{r}
-y^2 + 2y + 1.7\\
\underline{+\,12y^2 - 6y - 3.6}\\
11y^2 - 4y - 1.9
\end{array}
$$

Notice that we are finding the same sum in Example 4 as in Example 3. Of course, the results are the same.

Practice Problem 1

Add: $(2y + 7) + (9y - 14)$

Practice Problem 2

Add: $(5x^2 + 4x - 3) + (x^2 - 6x)$

Practice Problem 3

Find the sum of $(7z^2 - 4.2z + 11)$ and $(-9z^2 - 1.9z + 4)$.

Practice Problem 4

Add the polynomials in Practice Problem 3 vertically.

Answers

1. $11y - 7$ **2.** $6x^2 - 2x - 3$
3. $-2z^2 - 6.1z + 15$ **4.** same as 3.

B **Subtracting Polynomials**

To subtract one polynomial from another, recall how we subtract numbers. To subtract a number, we add its opposite: $a - b = a + (-b)$.

For example,

$$7 - 10 = 7 + (-10)$$
$$= -3$$

To subtract a polynomial, we also add its opposite. Just as the opposite of 3 is -3, the opposite of $(2x^2 - 5x + 1)$ is $-(2x^2 - 5x + 1)$. Let's practice simplifying the opposite of a polynomial.

EXAMPLE 5 Simplify: $-(2x^2 - 5x + 1)$

Solution: Rewrite $-(2x^2 - 5x + 1)$ as $-1(2x^2 - 5x + 1)$ and use the distributive property.

$$-(2x^2 - 5x + 1) = -1(2x^2 - 5x + 1)$$
$$= -1(2x^2) + (-1)(-5x) + (-1)(1)$$
$$= -2x^2 + 5x - 1$$

Notice the result of Example 5.

$$-(2x^2 - 5x + 1) = -2x^2 + 5x - 1$$

This means that **the opposite of a polynomial can be found by changing the signs of the terms of the polynomial**. This leads to the following.

> **Subtracting Polynomials**
>
> To subtract polynomials, change the signs of the terms of the polynomial being subtracted, then add.

EXAMPLE 6 Subtract: $(5a + 7) - (2a - 10)$

Solution:

$$(5a + 7) - (2a - 10) = (5a + 7) + (-2a + 10) \quad \text{Add the opposite of } 2a - 10.$$
$$= 5a - 2a + 7 + 10 \quad \text{Group like terms.}$$
$$= 3a + 17$$

EXAMPLE 7 Subtract: $(8x^2 - 4x + 1) - (10x^2 + 4)$

Solution:

$$(8x^2 - 4x + 1) - (10x^2 + 4) = (8x^2 - 4x + 1) + (-10x^2 - 4) \quad \text{Add the opposite of } 10x^2 + 4.$$
$$= 8x^2 - 10x^2 - 4x + 1 - 4 \quad \text{Group like terms.}$$
$$= -2x^2 - 4x - 3$$

EXAMPLE 8 Subtract $(-6z^2 - 2z + 13)$ from $(4z^2 - 20z)$.

Solution: Be careful when arranging the polynomials in this example.

$$(4z^2 - 20z) - (-6z^2 - 2z + 13) = (4z^2 - 20z) + (6z^2 + 2z - 13)$$
$$= 4z^2 + 6z^2 - 20z + 2z - 13 \quad \text{Group like terms.}$$
$$= 10z^2 - 18z - 13$$

Practice Problem 5

Simplify: $-(7y^2 + 4y - 6)$

Practice Problem 6

Subtract: $(3b - 2) - (7b + 23)$

Practice Problem 7

Subtract:
$(11x^2 + 7x + 2) - (15x^2 + 4x)$

Practice Problem 8

Subtract $(3x^2 - 12x)$ from $(-4x^2 + 20x + 17)$.

Answers:
5. $-7y^2 - 4y + 6$ **6.** $-4b - 25$
7. $-4x^2 + 3x + 2$ **8.** $-7x^2 + 32x + 17$

Concept Check

Find and explain the error in the following subtraction.

$$(3x^2 + 4) - (x^2 - 3x)$$
$$= (3x^2 + 4) + (-x^2 - 3x)$$
$$= 3x^2 - x^2 - 3x + 4$$
$$= 2x^2 - 3x + 4$$

Practice Problem 9

Subtract $(3x^2 - 12x)$ from $(-4x^2 + 20x + 17)$. Use a vertical format.

Practice Problem 10

Find the value of the polynomial $2y^3 + y^2 - 6$ when $y = 3$.

Practice Problem 11

An object is dropped from the top of a 530-foot cliff. Its height in feet at time t seconds is given by the polynomial $-16t^2 + 530$. Find the height of the object when $t = 1$ second and when $t = 4$ seconds.

Try the Concept Check in the margin.

Just as with adding polynomials, we can subtract polynomials using a vertical format. Let's subtract the polynomials in Example 8 using a vertical format.

EXAMPLE 9 Subtract $(-6z^2 - 2z + 13)$ from $(4z^2 - 20z)$. Use a vertical format.

Solution: Line up like terms underneath one another.

$$\begin{array}{r} 4z^2 - 20z \\ -(-6z^2 - 2z + 13) \\ \hline \end{array} \qquad \begin{array}{r} 4z^2 - 20z \\ +6z^2 + 2z - 13 \\ \hline 10z^2 - 18z - 13 \end{array}$$ ●

C Evaluating Polynomials

Polynomials have different values depending on the replacement values for the variables.

EXAMPLE 10 Find the value of the polynomial $3t^3 - 2t + 5$ when $t = 1$.

Solution: Replace t with 1 and simplify.

$$3t^3 - 2t + 5 = 3(1)^3 - 2(1) + 5 \qquad \text{Let } t = 1.$$
$$= 3(1) - 2 + 5 \qquad (1)^3 = 1.$$
$$= 3 - 2 + 5$$
$$= 6$$

The value of $3t^3 - 2t + 5$ when $t = 1$ is 6.   ●

Many real-world applications are modeled by polynomials.

EXAMPLE 11 Finding the Height of an Object

An object is dropped from the top of an 800-foot-tall building. Its height at time t seconds is given by the polynomial $-16t^2 + 800$. Find the height of the object when $t = 1$ second and when $t = 3$ seconds.

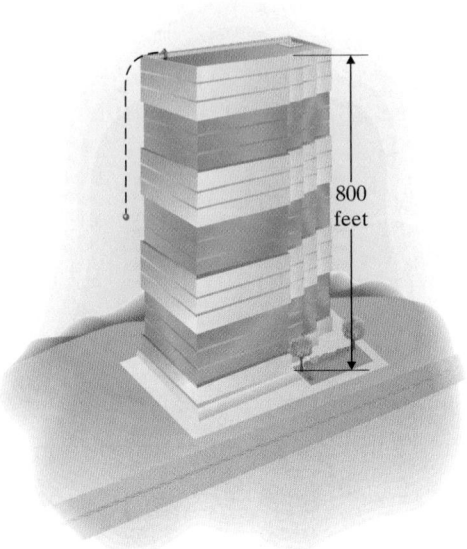

800 feet

Answers

9. $-7x^2 + 32x + 17$ **10.** 57
11. 514 feet; 274 feet

Concept Check:

$$(3x^2 + 4) - (x^2 - 3x)$$
$$= (3x^2 + 4) + (-x^2 + 3x)$$
$$= 3x^2 - x^2 + 3x + 4$$
$$= 2x^2 + 3x + 4$$

Solution: To find each height, we evaluate the polynomial when $t = 1$ and when $t = 3$.

$$-16t^2 + 800 = -16(1)^2 + 800$$
$$= -16 + 800$$
$$= 784$$

The height of the object at 1 second is 784 feet.

$$-16t^2 + 800 = -16(3)^2 + 800$$
$$= -16(9) + 800$$
$$= -144 + 800$$
$$= 656$$

Helpful Hint

Don't forget to insert units, if appropriate.

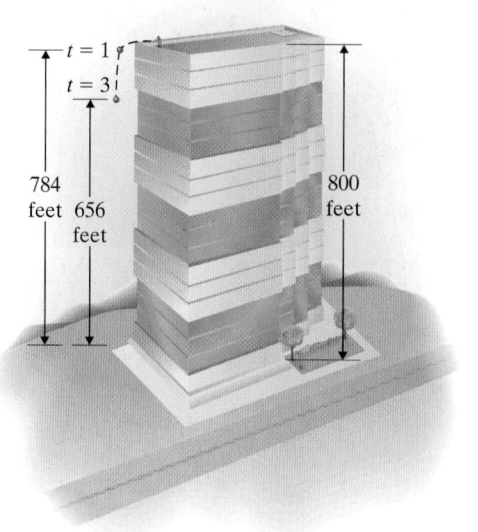

The height of the object at 3 seconds is 656 feet.

FOCUS ON **History**

EXPONENTIAL NOTATION

The French mathematician and philosopher René Descartes (1596–1650) is generally credited with devising the system of exponents that we use in math today. His book *La Géométrie* was the first to show successive powers of an unknown quantity x as x, xx, x^3, x^4, x^5, and so on. No one knows why Descartes preferred to write xx instead of x^2. However, the use of xx for the square of the quantity x continued to be popular. Those who used the notation defended it by saying that xx takes up no more space when written than x^2 does.

Before Descartes popularized the use of exponents to indicate powers, other less convenient methods were used. Some mathematicians preferred to write out the Latin words *quadratus* and *cubus* whenever they wanted to indicate that a quantity was to be raised to the second power or the third power. Other mathematicians used the abbreviations of *quadratus* and *cubus*, Q and C, to indicate second and third powers of a quantity.

Name _____ Section _____ Date _____

EXERCISE SET 10.1

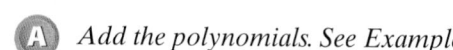 **A** *Add the polynomials. See Examples 1 through 4.*

 1. $(2x + 3) + (-7x - 27)$ **2.** $(9y - 16) + (-43y + 16)$

3. $(-4z^2 - 6z + 1) + (-5z^2 + 4z + 5)$ **4.** $(17a^2 - 6a + 3) + (16a^2 - 6a - 10)$

5. $(12y - 20) + (9y^2 + 13y - 20)$ **6.** $(5x^2 - 6) + (-3x^2 + 17x - 2)$

7. $(4.3a^4 + 5) + (-8.6a^4 - 2a^2 + 4)$ **8.** $(-12.7z^3 - 14z) + (-8.9z^3 + 12z + 2)$

B *Subtract the polynomials. See Examples 5 through 9.*

9. $(5a - 6) - (a + 2)$ **10.** $(12b + 7) - (-b - 5)$

11. $(3x^2 - 2x + 1) - (5x^2 - 6x)$ **12.** $(-9z^2 + 6z + 2) - (3z^2 + 1)$

13. $(10y^2 - 7) - (20y^3 - 2y^2 - 3)$ **14.** $(11x^3 + 15x - 9) - (-x^3 + 10x^2 - 9)$

15. Subtract $(3x - 4)$ from $(2x + 12)$.   **16.** Subtract $(6a + 1)$ from $(-7a + 7)$.

17. Subtract $(5y^2 + 4y - 6)$ from $(13y^2 - 6y - 14)$.   **18.** Subtract $(16x^2 - x + 1)$ from $(12x^2 - 3x - 12)$.

A **B** *Perform each indicated operation.*

19. $(25x - 5) + (-20x - 7)$ **20.** $(14x + 2) + (-7x - 1)$

21. $(4y + 4) - (3y + 8)$

22. $(6z - 3) - (8z + 5)$

23. $(9x^2 - 6) + (-5x^2 + x - 10)$

24. $(12a^2 - 4a - 4) + (-5a - 5)$

25. $(10x + 4.5) + (-x - 8.6)$

26. $(20x - 0.8) + (x + 1.2)$

27. $(12a - 5) - (-3a + 2)$

28. $(8t + 9) - (-2t + 6)$

 **29.** $(21y - 4.6) - (36y - 8.2)$

30. $(8.6x + 4) - (9.7x - 93)$

31. $(18t^2 - 4t + 2) - (-t^2 + 7t - 1)$

32. $(35x^2 + x - 5) - (17x^2 - x + 5)$

33. $(b^3 - 2b^2 + 10b + 11) + (b^2 - 3b - 12)$

34. $(-2z^3 + 5z^2 - 13z + 6) + (3z^2 - 7z - 6)$

35. Add $(6x^2 - 7)$ and $(-11x^2 - 11x + 20)$.

36. Add $(-2x^2 + 3x)$ and $(9x^2 - x + 14)$.

37. Subtract $\left(3z - \dfrac{3}{7} \right)$ from $\left(3z + \dfrac{6}{7} \right)$.

38. Subtract $\left(8y^2 - \dfrac{7}{10}y \right)$ from $\left(-5y^2 + \dfrac{3}{10}y \right)$.

C *Find the value of each polynomial when* $x = 2$. *See Examples 10 and 11.*

39. $-3x + 7$

40. $-5x - 7$

41. $x^2 - 6x + 3$

42. $5x^2 + 4x - 100$

43. $\dfrac{3x^2}{2} - 14$

44. $\dfrac{7x^3}{14} - x + 5$

Find the value of each polynomial when $x = 5$. *See Examples 10 and 11.*

45. $2x + 10$

46. $-5x - 6$

47. x^2

48. x^3

 **49.** $2x^2 + 4x - 20$

50. $4x^2 - 5x + 10$

Solve. See Example 11.

The distance in feet traveled by a free-falling object in t seconds is given by the polynomial

$$16t^2$$

Use this polynomial for Exercises 51 and 52.

51. Find the distance traveled by an object that falls for 6 seconds.

52. It takes 8 seconds for a hard hat to fall from the top of a building. How high is the building?

Office Supplies, Inc. manufactures office products. They determine that the total cost for manufacturing x file cabinets is given by the polynomial

$$3000 + 20x$$

Use this polynomial for Exercises 53 and 54.

53. Find the total cost to manufacture 10 file cabinets.

54. Find the total cost to manufacture 100 file cabinets.

An object is dropped from the deck of the Royal Gorge Bridge, which stretches across Royal Gorge at a height of 1053 feet above the Arkansas River. The height of the object above the river after t seconds is given by the polynomial

$$1053 - 16t^2$$

Use this polynomial for Exercises 55 and 56. (*Source:* Royal Gorge Bridge Co.)

55. How far above the river is an object that has been falling for 3 seconds?

56. How far above the river is an object that has been falling for 8 seconds?

Review and Preview

Evaluate. See Sections 1.8 and 2.4.

57. 3^4

58. $(-2)^5$

59. $(-5)^2$

60. 4^3

Write using exponential notation. See Section 1.8.

61. $x \cdot x \cdot x$

62. $y \cdot y \cdot y \cdot y \cdot y$

63. $2 \cdot 2 \cdot a \cdot a \cdot a \cdot a$

64. $5 \cdot 5 \cdot 5 \cdot b \cdot b$

Combining Concepts

Find the perimeter of each figure.

△ **65.**

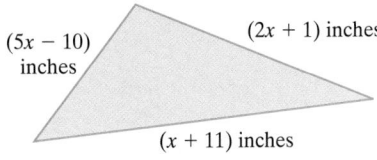

$(5x - 10)$ inches

$(2x + 1)$ inches

$(x + 11)$ inches

△ **66.**

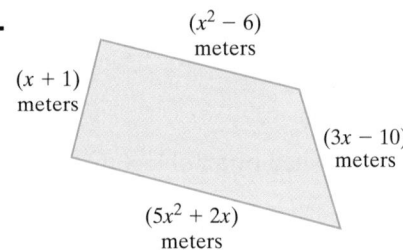

$(x^2 - 6)$ meters

$(x + 1)$ meters

$(3x - 10)$ meters

$(5x^2 + 2x)$ meters

Given the lengths in the figure below, we find the unknown length by subtracting. Use the information to find the unknown lengths in Exercises 67 and 68.

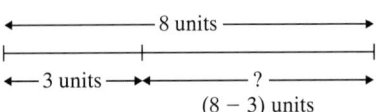

8 units

3 units

?

$(8 - 3)$ units

67.

$(7x - 10)$ units

$(3x + 5)$ units

? units

68.

$(x^2 - 7x + 6)$ units

$(x^2 + 2)$ units

? units

Fill in the blanks.

69. $(3x^2 + \underline{\quad} x - \underline{\quad}) + (\underline{\quad} x^2 - 6x + 2) = 5x^2 + 14x - 4$

70. $(\underline{\quad} y^2 + 4y - 3) + (8y^2 - \underline{\quad} y + \underline{\quad}) = 9y^2 + 2y + 7$

71. Find the value of $7a^4 - 6a^2 + 2a - 1$ when $a = 1.2$.

72. Find the value of $3b^3 + 4b^2 - 100$ when $b = -2.5$.

73. For Exercises 55 and 56, the polynomial $1053 - 16t^2$ was used to give the height of an object above the river after t seconds. Find the height when $t = 8$ seconds and $t = 9$ seconds. Explain what happened and why.

10.2 Multiplication Properties of Exponents

OBJECTIVES

Ⓐ Use the product rule for exponents.

Ⓑ Use the power rule for exponents.

Ⓒ Use the power of a product rule for exponents.

SSM
TUTOR CENTER   SG   CD & VIDEO   MATH PRO   WEB

Ⓐ Using the Product Rule

Recall from Section 1.8 that an exponent has the same meaning whether the base is a number or a variable. For example,

$$5^3 = \underbrace{5 \cdot 5 \cdot 5}_{3 \text{ factors of } 5} \quad \text{and} \quad x^3 = \underbrace{x \cdot x \cdot x}_{3 \text{ factors of } x}$$

We can use this definition of an exponent to discover properties that will help us to simplify products and powers of exponential expressions.

For example, let's use the definition of an exponent to find the product of x^3 and x^4.

$$\begin{aligned} x^3 \cdot x^4 &= (x \cdot x \cdot x)(x \cdot x \cdot x \cdot x) \\ &= \underbrace{x \cdot x \cdot x \cdot x \cdot x \cdot x \cdot x}_{7 \text{ factors of } x} \\ &= x^7 \end{aligned}$$

Notice that the result is the same if we add the exponents.

$$x^3 \cdot x^4 = x^{3+4} = x^7$$

This suggests the following product rule or property for exponents.

Product Property for Exponents

If m and n are positive integers and a is a real number, then

$$a^m \cdot a^n = a^{m+n}$$

In other words, to multiply two exponential expressions with the same base, keep the base and add the exponents.

EXAMPLE 1 Multiply: $y^7 \cdot y^2$

Solution:

$$\begin{aligned} y^7 \cdot y^2 &= y^{7+2} \quad \text{Use the product property for exponents.} \\ &= y^9 \quad \text{Simplify.} \end{aligned}$$

EXAMPLE 2 Multiply: $3x^5 \cdot 6x^3$

Solution:

$$\begin{aligned} 3x^5 \cdot 6x^3 &= (3 \cdot 6)(x^5 \cdot x^3) \quad \text{Apply the commutative and associative properties.} \\ &= 18x^{5+3} \quad \text{Use the product property for exponents.} \\ &= 18x^8 \quad \text{Simplify.} \end{aligned}$$

EXAMPLE 3 Multiply: $(-2a^4b^{10})(9a^5b^3)$

Solution: Use properties of multiplication to group numbers and like variables together.

$$\begin{aligned} (-2a^4b^{10})(9a^5b^3) &= (-2 \cdot 9)(a^4 \cdot a^5)(b^{10} \cdot b^3) \\ &= -18a^{4+5}b^{10+3} \\ &= -18a^9b^{13} \end{aligned}$$

Practice Problem 1

Multiply: $z^4 \cdot z^8$

Practice Problem 2

Multiply: $7y^5 \cdot 4y^9$

Practice Problem 3

Multiply: $(-7r^6s^2)(-3r^2s^5)$

Answers

1. z^{12} **2.** $28y^{14}$ **3.** $21r^8s^7$

Practice Problem 4

Multiply: $9y^4 \cdot 3y^2 \cdot y$. (Recall that $y = y^1$.)

> **Helpful Hint**
>
> Don't forget that if an exponent is not written, it is assumed to be 1.

EXAMPLE 4 Multiply: $2x^3 \cdot 3x \cdot 5x^6$

Solution: First notice the factor $3x$. Since there is one factor of x in $3x$, it can also be written as $3x^1$.

$$2x^3 \cdot 3x^1 \cdot 5x^6 = (2 \cdot 3 \cdot 5)(x^3 \cdot x^1 \cdot x^6)$$
$$= 30x^{10}$$

> **Helpful Hint**
>
> These examples will remind you of the difference between adding and multiplying terms.
>
> **Addition**
>
> $$5x^3 + 3x^3 = (5 + 3)x^3 = 8x^3$$
> $$7x + 4x^2 = 7x + 4x^2$$
>
> **Multiplication**
>
> $$(5x^3)(3x^3) = 5 \cdot 3 \cdot x^3 \cdot x^3 = 15x^{3+3} = 15x^6$$
> $$(7x)(4x^2) = 7 \cdot 4 \cdot x \cdot x^2 = 28x^{1+2} = 28x^3$$

B Using the Power Rule

Next suppose that we want to simplify an exponential expression raised to a power. To see how we simplify $(x^2)^3$, we again use the definition of an exponent.

$$(x^2)^3 = \underbrace{(x^2) \cdot (x^2) \cdot (x^2)}_{3 \text{ factors of } x^2} \qquad \text{Apply the definition of an exponent.}$$

$$= x^{2+2+2} \qquad \text{Use the product property for exponents.}$$
$$= x^6 \qquad \text{Simplify.}$$

Notice the result is exactly the same if we multiply the exponents.

$$(x^2)^3 = x^{2 \cdot 3} = x^6$$

This suggests the following power rule or property for exponents.

> **Power Property for Exponents**
>
> If m and n are positive integers and a is a real number, then
>
> $$(a^m)^n = a^{m \cdot n}$$

In other words, to raise a power to a power, keep the base and multiply the exponents.

Answer

4. $27y^7$

Helpful Hint

Take a moment to make sure that you understand when to apply the product rule and when to apply the power rule.

Product Property → Add Exponents	Power Property → Multiply Exponents
$x^5 \cdot x^7 = x^{5+7} = x^{12}$	$(x^5)^7 = x^{5 \cdot 7} = x^{35}$
$y^6 \cdot y^2 = y^{6+2} = y^8$	$(y^6)^2 = y^{6 \cdot 2} = y^{12}$

EXAMPLE 5 Simplify: $(y^8)^2$

Solution:

$$(y^8)^2 = y^{8 \cdot 2} \quad \text{Use the power property.}$$
$$= y^{16}$$

●

EXAMPLE 6 Simplify: $(a^3)^4 \cdot (a^2)^9$

Solution:

$$(a^3)^4 \cdot (a^2)^9 = a^{12} \cdot a^{18} \quad \text{Use the power property.}$$
$$= a^{12+18} \quad \text{Use the product property.}$$
$$= a^{30} \quad \text{Simplify.}$$

●

C Using the Power of a Product Rule

Next, let's simplify the power of a product.

$$(xy)^3 = xy \cdot xy \cdot xy \quad \text{Apply the definition of an exponent.}$$
$$= (x \cdot x \cdot x)(y \cdot y \cdot y) \quad \text{Group like bases.}$$
$$= x^3 y^3 \quad \text{Simplify.}$$

Notice that the power of a product can be written as the product of powers. This leads to the following power of a product rule or property.

Power of a Product Property for Exponents

If n is a positive integer and a and b are real numbers, then

$$(ab)^n = a^n b^n$$

In other words, to raise a product to a power, raise each factor to the power.

Try the Concept Check in the margin.

EXAMPLE 7 Simplify: $(5t)^3$

Solution:

$$(5t)^3 = 5^3 t^3 \quad \text{Apply the power of a product property.}$$
$$= 125t^3 \quad \text{Write } 5^3 \text{ as 125.}$$

●

Practice Problem 5

Simplify: $(z^3)^{10}$

Practice Problem 6

Simplify: $(z^4)^5 \cdot (z^3)^7$

Concept Check

Which property is needed to simplify $(x^6)^3$? Explain.

a. Product Property for Exponents
b. Power Property for Exponents
c. Power of a Product Property for Exponents

Practice Problem 7

Simplify: $(3b)^4$

Answers

5. z^{30} **6.** z^{41} **7.** $81b^4$

Concept Check: b

Practice Problem 8

Simplify: $(4x^2y^6)^3$

Practice Problem 9

Simplify: $(2x^2y^4)^4(3x^6y^9)^2$

EXAMPLE 8 Simplify: $(2a^5b^3)^3$

Solution:

$$(2a^5b^3)^3 = 2^3(a^5)^3(b^3)^3 \qquad \text{Apply the power of a product property.}$$
$$= 8a^{15}b^9 \qquad \text{Apply the power property.}$$

EXAMPLE 9 Simplify: $(3y^4z^2)^4(2y^3z^5)^5$

Solution:

$$(3y^4z^2)^4(2y^3z^5)^5 = 3^4(y^4)^4(z^2)^4 \cdot 2^5(y^3)^5(z^5)^5 \qquad \text{Apply the power of a product property.}$$
$$= 81y^{16}z^8 \cdot 32y^{15}z^{25} \qquad \text{Apply the power property.}$$
$$= (81 \cdot 32)(y^{16} \cdot y^{15})(z^8 \cdot z^{25}) \qquad \text{Group like bases.}$$
$$= 2592y^{31}z^{33} \qquad \text{Apply the product property.}$$

Answers

8. $64x^6y^{18}$ **9.** $144x^{20}y^{34}$

EXERCISE SET 10.2

(A) *Multiply. See Examples 1 through 4.*

1. $x^5 \cdot x^9$

2. $y^4 \cdot y^7$

3. $a^6 \cdot a$

4. $b \cdot b^8$

5. $3z^3 \cdot 5z^2$

6. $8r^2 \cdot 2r^{15}$

7. $-4x \cdot 10x$

8. $-9y \cdot 3y$

9. $(-5x^2y^3)(-5x^4y)$

10. $(-2xy^4)(-6x^3y^7)$

11. $(7ab)(4a^4b^5)$

12. $(3a^3b^6)(12a^2b^9)$

13. $2x \cdot 3x \cdot 7x$

14. $4y \cdot 3y \cdot 5y$

15. $a \cdot 4a^{11} \cdot 3a^5$

16. $b \cdot 7b^{10} \cdot 5b^8$

(B) **(C)** *Simplify. See Examples 5 through 9.*

17. $(x^5)^3$

18. $(y^4)^7$

19. $(z^2)^{10}$

20. $(a^6)^9$

21. $(b^7)^6(b^2)^{10}$

22. $(x^2)^9 \cdot (x^5)^3$

23. $(3a)^4$

24. $(2y)^5$

25. $(a^{11}b^8)^3$

26. $(x^7y^4)^8$

27. $(11x^3y^6)^2$

28. $(9a^4b^3)^2$

29. $(-3y)(2y^7)^3$

30. $(-2x)(5x^2)^4$

31. $(4xy)^3(2x^3y^5)^2$

32. $(2xy)^4(3x^4y^3)^3$

Review and Preview

Multiply. See Section 3.1.

33. $7(x - 3)$

34. $4(y + 2)$

35. $-2(3a + 2b)$

36. $-3(8r + 3s)$

37. $9(x + 2y - 3)$

38. $5(a + 7b - 3)$

Find the area of each figure.

△ **39.**

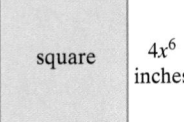

square $4x^6$ inches

△ **40.**

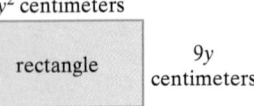

$9y^2$ centimeters
rectangle
$9y$ centimeters

△ **41.**

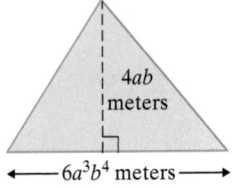

$4ab$ meters
$6a^3b^4$ meters

△ **42.**

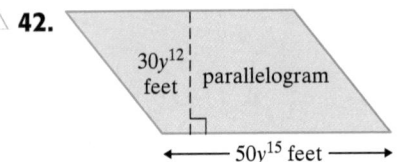

$30y^{12}$ feet | parallelogram
$50y^{15}$ feet

(*Hint:* Area $=$ base $\cdot$ height)

Multiply and simplify.

43. $(14a^7b^6)^3(9a^6b^3)^4$

44. $(5x^{14}y^6)^7(3x^{20}y^{19})^5$

45. $(8.1x^{10})^5$

46. $(4.6a^{14})^4$

47. In your own words, explain why $x^2 \cdot x^3 = x^5$ and $(x^2)^3 = x^6$.

48. $(x^{90}y^{72})^3$

49. $(a^{20}b^{10}c^5)^5(a^9b^{12})^3$

Name _____ Section _____ Date _____

Integrated Review—Operations on Polynomials

Add or subtract the polynomials as indicated.

1. $(7x + 1) + (-3x - 2)$

2. $(14y - 6) + (19y - 2)$

3. $(7x + 1) - (-3x - 2)$

4. $(14y - 6) - (19y - 2)$

5. $(a^3 + 1) + (2a^3 + 5a - 9)$

6. $(1.2y^2 - 3.6y) + (0.6y^2 + 1.2y - 5.6)$

7. $(3.5x^2 - 0.5x) - (5.3x^2 - 2.9x + 1.7)$

8. $(2a^3 - 6a^2 + 11) - (6a^3 + 6a^2 + 11)$

9. Subtract $(2x - 6)$ from $(8x + 1)$.

10. Subtract $(3x^2 - x + 2)$ from $(5x^2 + 2x - 10)$.

Find the value of each polynomial when $x = 3$.

11. $2x - 7$

12. $x^2 + 5x + 2$

Multiply.

13. $x^7 \cdot x^{11}$

14. $x^6 \cdot x^6$

15. $y^3 \cdot y$

16. $a \cdot a^{10}$

17. $\left(x^7\right)^{11}$

18. $\left(x^6\right)^6$

19. $\left(x^3\right)^4 \cdot \left(x^5\right)^6$

20. $\left(y^2\right)^9 \cdot \left(y^3\right)^3$

21. $(2x)^4$

Answers

1. _____

2. _____

3. _____

4. _____

5. _____

6. _____

7. _____

8. _____

9. _____

10. _____

11. _____

12. _____

13. _____

14. _____

15. _____

16. _____

17. _____

18. _____

19. _____

20. _____

21. _____

22. $(3y)^3$

23. $(-6xy^2)(2xy^5)$

24. $(-4a^2b^3)(-3ab)$

25. $(x^9y^5)^4$

26. $(a^{10}b^{12})^2$

27. $(10x^2y)^2(3y)$

28. $(8y^3z)^2(2z^5)$

29. $(2a^5b)^4(3a^9b^4)^2$

30. $(5x^4y^6)^3(x^2y^2)^5$

10.3 Multiplying Polynomials

(A) Multiplying a Monomial and a Polynomial

Recall that a polynomial that consists of one term is called a **monomial**. For example, $5x$ is a monomial. To multiply a monomial and any polynomial, we use the distributive property

$$a(b + c) = a \cdot b + a \cdot c$$

and apply properties of exponents.

EXAMPLE 1 Multiply: $5x(3x^2 + 2)$

Solution:

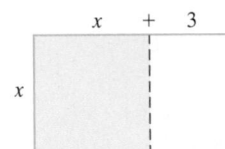

$$5x(3x^2 + 2) = 5x \cdot 3x^2 + 5x \cdot 2 \qquad \text{Apply the distributive property.}$$
$$= 15x^3 + 10x$$

EXAMPLE 2 Multiply: $2z(4z^2 + 6z - 9)$

Solution:

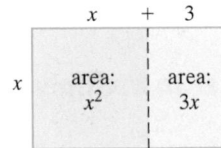

$$2z(4z^2 + 6z - 9) = 2z \cdot 4z^2 + 2z \cdot 6z + 2z(-9)$$
$$= 8z^3 + 12z^2 - 18z$$

To visualize multiplication by a monomial, let's look at two ways we can represent the area of the same rectangle.

The width of the rectangle is x and its length is $x + 3$. One way to calculate the area of the rectangle is

area = width · length
$$= x(x + 3)$$

Another way to calculate the area of the rectangle is to find the sum of the areas of the smaller figures.

area $= x^2 + 3x$

Since the areas must be equal, we have that

$$x(x + 3) = x^2 + 3x \qquad \text{As expected by the distributive property}$$

(B) Multiplying Binomials

Recall from Section 10.1 that a polynomial that consists of exactly two terms is called a **binomial**. To multiply two binomials, we use a version of the distributive property:

$$(b + c)a = b \cdot a + c \cdot a$$

SSM   TUTOR CENTER   SG   CD & VIDEO   MATH PRO   WEB

OBJECTIVES

(A) Multiply a monomial and any polynomial.

(B) Multiply two binomials.

(C) Squaring a binomial.

(D) Use the FOIL order to multiply binomials.

(E) Multiply any two polynomials.

Practice Problem 1

Multiply: $3y(7y^2 + 5)$

Practice Problem 2

Multiply: $5r(8r^2 - r + 11)$

Answers

1. $21y^3 + 15y$ **2.** $40r^3 - 5r^2 + 55r$

Copyright 2004 Pearson Education, Inc.

Practice Problem 3

Multiply: $(b + 7)(b + 5)$

Practice Problem 4

Multiply: $(5x - 1)(5x + 4)$

Practice Problem 5

Multiply: $(6y - 1)^2$

EXAMPLE 3   Multiply: $(x + 2)(x + 3)$

Solution:

$$
\begin{aligned}
(x + 2)(x + 3) &= x(x + 3) + 2(x + 3) & &\text{Apply the distributive property.}\\
&= x \cdot x + x \cdot 3 + 2 \cdot x + 2 \cdot 3 & &\text{Apply the distributive property.}\\
&= x^2 + 3x + 2x + 6 & &\text{Multiply.}\\
&= x^2 + 5x + 6 & &\text{Combine like terms.}
\end{aligned}
$$

EXAMPLE 4   Multiply: $(4y + 9)(3y - 2)$

Solution:

$$
\begin{aligned}
(4y + 9)(3y - 2) &= 4y(3y - 2) + 9(3y - 2) & &\text{Apply the distributive property.}\\
&= 4y \cdot 3y + 4y(-2) + 9 \cdot 3y + 9(-2) & &\text{Apply the distributive property.}\\
&= 12y^2 - 8y + 27y - 18 & &\text{Multiply.}\\
&= 12y^2 + 19y - 18 & &\text{Combine like terms}
\end{aligned}
$$

C Squaring a Binomial

Raising a binomial to the power of 2 is also called squaring a binomial. To square a binomial, we use the definition of an exponent, and then multiply.

EXAMPLE 5   Multiply: $(2x + 1)^2$

Solution:

$$
\begin{aligned}
(2x + 1)^2 &= (2x + 1)(2x + 1) & &\text{Apply the definition of an exponent.}\\
&= 2x(2x + 1) + 1(2x + 1) & &\text{Apply the distributive property.}\\
&= 2x \cdot 2x + 2x \cdot 1 + 1 \cdot 2x + 1 \cdot 1 & &\text{Apply the distributive property.}\\
&= 4x^2 + 2x + 2x + 1 & &\text{Multiply.}\\
&= 4x^2 + 4x + 1 & &\text{Combine like terms.}
\end{aligned}
$$

D Using the FOIL Order to Multiply Binomials

Recall from Example 3 that

$$
\begin{aligned}
(x + 2)(x + 3) &= x \cdot x + x \cdot 3 + 2 \cdot x + 2 \cdot 3\\
&= x^2 + 5x + 6
\end{aligned}
$$

One way to remember these products $x \cdot x$, $x \cdot 3$, $2 \cdot x$, and $2 \cdot 3$ is to use a special order for multiplying binomials, called the FOIL order. Of course, the product is the same no matter what order or method you choose to use.

FOIL stands for the products of the First terms, Outer terms, Inner terms, then Last terms. For example,

Answers

3. $b^2 + 12b + 35$ **4.** $25x^2 + 15x - 4$
5. $36y^2 - 12y + 1$

$$(x + 2)(x + 3) = x \cdot x + x \cdot 3 + 2 \cdot x + 2 \cdot 3 = x^2 + 3x + 2x + 6$$
$$= x^2 + 5x + 6$$

Helpful Hint

The product is the same no matter what order or method you choose to use.

EXAMPLES Use the FOIL order to multiply.

6. $(3x - 6)(2x + 1)$
$$= 3x \cdot 2x + 3x \cdot 1 + (-6)(2x) + (-6)(1)$$
$$= 6x^2 + 3x - 12x - 6 \quad \text{Multiply.}$$
$$= 6x^2 - 9x - 6 \quad \text{Combine like terms.}$$

7. $(3x - 5)^2 = (3x - 5)(3x - 5)$
$$\quad\quad\quad\quad F \quad\quad O \quad\quad\quad I \quad\quad\quad\quad L$$
$$= 3x \cdot 3x + 3x(-5) + (-5)(3x) + (-5)(-5)$$
$$= 9x^2 - 15x - 15x + 25 \quad \text{Multiply.}$$
$$= 9x^2 - 30x + 25 \quad \text{Combine like terms.}$$

Practice Problems

Use the FOIL order to multiply.

6. $(10x - 7)(x + 3)$
7. $(3x + 2)^2$

Helpful Hint

Remember that the FOIL order can only be used to multiply **two binomials**.

E Multiplying Polynomials

Recall from Section 10.1 that a polynomial that consists of exactly three terms is called a **trinomial**. Next, we multiply a binomial by a trinomial.

EXAMPLE 8 Multiply: $(3a + 2)(a^2 - 6a + 3)$

Solution: Use the distributive property to multiply $3a$ by the trinomial $(a^2 - 6a + 3)$ and then 2 by the trinomial.

$$(3a + 2)(a^2 - 6a + 3) = 3a(a^2 - 6a + 3) + 2(a^2 - 6a + 3) \quad \begin{array}{l}\text{Apply the}\\ \text{distributive}\\ \text{property.}\end{array}$$
$$= 3a \cdot a^2 + 3a(-6a) + 3a \cdot 3 +$$
$$2 \cdot a^2 + 2(-6a) + 2 \cdot 3 \quad \begin{array}{l}\text{Apply the}\\ \text{distributive property.}\end{array}$$
$$= 3a^3 - 18a^2 + 9a + 2a^2 - 12a + 6 \quad \text{Multiply.}$$
$$= 3a^3 - 16a^2 - 3a + 6 \quad \text{Combine like terms.}$$

In general, we have the following.

Practice Problem 8

Multiply: $(2x + 5)(x^2 + 4x - 1)$

To Multiply Two Polynomials

Multiply each term of the first polynomial by each term of the second polynomial, and then combine like terms.

A convenient method of multiplying polynomials is to use a vertical format similar to multiplying real numbers.

Answers

6. $10x^2 + 23x - 21$ **7.** $9x^2 + 12x + 4$
8. $2x^3 + 13x^2 + 18x - 5$

Concept Check

True or false? When a trinomial is multiplied by a trinomial, the result will have at most nine terms. Explain.

Practice Problem 9

Multiply $(x^2 + 4x - 1)$ and $(2x + 5)$ vertically.

Try the Concept Check in the margin.

EXAMPLE 9   Find the product of $(a^2 - 6a + 3)$ and $(3a + 2)$ vertically.

Solution:

$$
\begin{array}{r}
a^2 - 6a + 3 \\
\times \qquad 3a + 2 \\
\hline
2a^2 - 12a + 6 \\
3a^3 - 18a^2 + 9a \\
\hline
3a^3 - 16a^2 - 3a + 6
\end{array}
$$

Multiply $a^2 - 6a + 3$ by 2.

Multiply $a^2 - 6a + 3$ by $3a$. Line up like terms.

Combine like terms

Notice that this example is the same as Example 8 and of course the products are the same.   ●

Answers

9. $2x^3 + 13x^2 + 18x - 5$

Concept Check: True.

Name _____ Section _____ Date _____

EXERCISE SET 10.3

A *Multiply. See Examples 1 and 2.*

 1. $3x(9x^2 - 3)$

2. $4y(10y^3 + 2y)$

3. $-5a(4a^2 - 6a + 1)$

4. $-2b(3b^2 - 2b + 5)$

5. $7x^2(6x^2 - 5x + 7)$

6. $6z^2(-3z^2 - z + 4)$

B **C** **D** *Multiply. See Examples 3 through 7.*

7. $(x + 3)(x + 10)$

8. $(y + 5)(y + 9)$

9. $(2x - 6)(x + 4)$

10. $(7z + 1)(z - 6)$

11. $(6a + 4)^2$

12. $(8b - 3)^2$

E *Multiply. See Examples 8 and 9.*

13. $(a + 6)(a^2 - 6a + 3)$

14. $(y + 4)(y^2 + 8y - 2)$

15. $(4x - 5)(2x^2 + 3x - 10)$

16. $(9z - 2)(2z^2 + z + 1)$

17. $(x^3 + 2x + x^2)(3x + 1 + x^2)$

18. $(y^2 - 2y + 5)(y^3 + 2 + y)$

A **B** **C** **D** **E** *Multiply.*

19. $10r(-3r + 2)$

20. $5x(4x^2 + 5)$

21. $-2y^2(3y + y^2 - 6)$

22. $3z^3(4z^4 - 2z + z^3)$

23. $(x + 2)(x + 12)$

24. $(y + 7)(y - 7)$

25. $(2a + 3)(2a - 3)$

26. $(6s + 1)(3s - 1)$

27. $(x + 5)^2$

28. $(x + 3)^2$

29. $\left(b + \dfrac{3}{5}\right)\left(b + \dfrac{4}{5}\right)$

30. $\left(a - \dfrac{7}{10}\right)\left(a + \dfrac{3}{10}\right)$

31. $(6x + 1)(x^2 + 4x + 1)$

32. $(9y - 1)(y^2 + 3y - 5)$

33. $(7x + 5)^2$

34. $(5x + 9)^2$

35. $(2x - 1)^2$

36. $(4a - 3)^2$

37. $(2x^2 - 3)(4x^3 + 2x - 3)$ **38.** $(3y^2 + 2)(5y^2 - y + 2)$ **39.** $(x^3 + x^2 + x)(x^2 + x + 1)$

40. $(a^4 + a^2 + 1)(a^4 + a^2 - 1)$ **41.** $(2z^2 - z + 1)(5z^2 + z - 2)$ **42.** $(2b^2 - 4b + 3)(b^2 - b + 2)$

Review and Preview

Write each number as a product of prime numbers. See Section 4.2.

43. 50 **44.** 48 **45.** 72

46. 36 **47.** 200 **48.** 300

 Combining Concepts

Find the area of each figure.

 49.

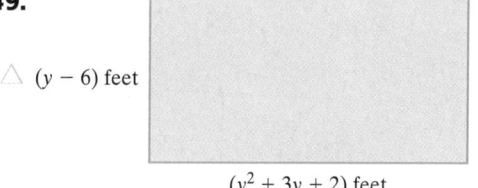

$(y - 6)$ feet

$(y^2 + 3y + 2)$ feet

△ **50.**

Square | $(2x + 11)$ centimeters

Find the area of the shaded figure. To do so, subtract the area of the smaller square from the area of the larger geometric figure.

△ **51.**

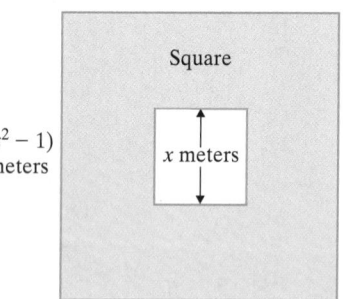

Square

$(x^2 - 1)$ meters

x meters

△ **52.**

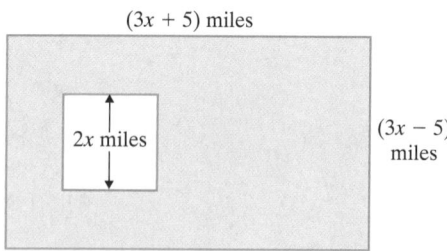

$(3x + 5)$ miles

$2x$ miles

$(3x - 5)$ miles

53. Suppose that a classmate asked you why $(2x + 1)^2$ is not $4x^2 + 1$. Write down your response to this classmate.

10.4   Introduction to Factoring Polynomials

Recall that when an integer is written as the product of two or more integers, each of these integers is called a *factor* of the product. This is true of polynomials also. When a polynomial is written as the product of two or more other polynomials, each of these polynomials is called a factor of the product.

factor · factor	=	product
$-2 \cdot 4$	=	-8
$x^3 \cdot x^7$	=	x^{10}
$5(x + 2)$	=	$5x + 10$

The process of writing a polynomial as a product is called **factoring**. Notice that factoring is the reverse process of multiplying.

$$\overbrace{5x + 10}^{\text{factoring}} = \underbrace{5(x + 2)}_{\text{multiplying}}$$

Ⓐ Finding the GCF of a List of Integers

Before we factor polynomials, let's practice finding the greatest common factor of a list of integers. The **greatest common factor (GCF)** of a list of integers is the largest integer that is a factor of all the integers in the list. For example,

the GCF of 30 and 18 is 6

because 6 is the largest integer that is a factor of both 30 and 18.

If the GCF cannot be found by inspection, the following steps can be used.

> **To Find the GCF of a List of Integers**
>
> **Step 1.** Write each number as a product of prime numbers.
>
> **Step 2.** Identify the common prime factors.
>
> **Step 3.** The product of all common prime factors found in *Step 2* is the greatest common factor. If there are no common prime factors, the greatest common factor is 1.

Concept Check

Which of the following is the prime factorization of 36?

a. $4 \cdot 9$
b. $2 \cdot 2 \cdot 3 \cdot 3$
c. $6 \cdot 6$

Practice Problem 1

Find the GCF of 42 and 28.

Practice Problem 2

Find the GCF of z^7, z^8, and z.

Practice Problem 3

Find the GCF of $6a^4$, $3a^5$, and $15a^2$.

Try the Concept Check in the margin.

Recall from Section 4.2 that a prime number is a whole number other than 1, whose only factors are 1 and itself.

EXAMPLE 1 Find the GCF of 12 and 20.

Solution:

Step 1. Write each number as a product of primes.

$$12 = 2 \cdot 2 \cdot 3$$
$$20 = 2 \cdot 2 \cdot 5$$

Step 2. $12 = \boxed{2} \cdot \boxed{2} \cdot 3$
$20 = \boxed{2} \cdot \boxed{2} \cdot 5$

$\downarrow \quad \downarrow$

$2 \cdot 2$ Identify the common factors.

Step 3. The GCF is $2 \cdot 2 = 4$

B Finding the GCF of a List of Terms

How do we find the GCF of a list of variables raised to powers? For example, what is the GCF of y^3, y^5, and y^{10}? Notice that each variable term contains a factor of y^3 and no higher power of y is a factor of each term.

$$y^3 = y^3$$
$$y^5 = y^3 \cdot y^2 \qquad \text{Recall the product property for exponents.}$$
$$y^{10} = y^3 \cdot y^7$$

The GCF of y^3, y^5, and y^{10} is y^3. From this example, we can see that **the GCF of a list of variables raised to powers is the variable raised to the smallest exponent in the list.**

EXAMPLE 2 Find the GCF of x^{11}, x^4, and x^6.

Solution: The GCF is x^4 since 4 is the smallest exponent to which x is raised.

In general, **the GCF of a list of terms is the product of all common factors.**

EXAMPLE 3 Find the GCF of $4x^3$, $12x$, and $10x^5$.

Solution:

The GCF of 4, 12, and 10 is 2.
The GCF of x^3, x^1, and x^5 is x^1.
Thus, the GCF of $4x^3$, $12x$, and $10x^5$ is $2x^1$ or $2x$.

Helpful Hint:

If you ever have trouble finding the GCF, remember that you can always use the method below.

Example 3:

$$4x^3 = 2 \cdot 2 \cdot x \cdot x \cdot x$$
$$12x = 2 \cdot 2 \cdot 3 \cdot x$$
$$10x^5 = 2 \cdot 5 \cdot x \cdot x \cdot x \cdot x \cdot x$$
$$\text{GCF} = 2 \cdot x \quad \text{or} \quad 2x$$

Answers

1. 14 **2.** z **3.** $3a^2$

Concept Check: b

(C) Factoring Out the GCF

Next, we practice factoring a polynomial by factoring the GCF from its terms. To do so, we write each term of the polynomial as a product of the GCF and another factor, and then apply the distributive property.

EXAMPLE 4   Factor: $7x^3 + 14x^2$

Solution: The GCF of $7x^3$ and $14x^2$ is $7x^2$.

$$7x^3 + 14x^2 = 7x^2 \cdot x^1 + 7x^2 \cdot 2$$
$$= 7x^2(x + 2) \qquad \text{Apply the distributive property.}$$

Notice in Example 4 that we factored $7x^3 + 14x^2$ by writing it as the product $7x^2(x + 2)$. Also notice that to check factoring, we multiply

$$7x^2(x + 2) = 7x^2 \cdot x + 7x^2 \cdot 2$$
$$= 7x^3 + 14x^2$$

which is the original binomial.

EXAMPLE 5   Factor: $6x^2 - 24x + 6$

Solution: The GCF of the terms is 6.

$$6x^2 - 24x + 6 = 6 \cdot x^2 - 6 \cdot 4x + 6 \cdot 1$$
$$= 6(x^2 - 4x + 1)$$

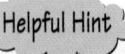 **Helpful Hint**

A common mistake in the example above is to forget to write down the term of 1. Remember to mentally check by multiplying.

$$6(x^2 - 4x) = 6x^2 - 24x \qquad \text{not the original trinomial}$$

EXAMPLE 6   Factor: $-2a + 20b - 4b^2$

Solution:

$$-2a + 20b - 4b^2 = 2 \cdot -a + 2 \cdot 10b - 2 \cdot 2b^2$$
$$= 2(-a + 10b - 2b^2)$$

When the coefficient of the first term is a negative number, we often factor out a negative common factor.

$$-2a + 20b - 4b^2 = (-2)(a) + (-2)(-10b) + (-2)(2b^2)$$
$$= -2(a - 10b + 2b^2)$$

Both $2(-a + 10b - 2b^2)$ and $-2(a - 10b + 2b^2)$ are factorizations of $-2a + 20b - 4b^2$.

Try the Concept Check in the margin.

 **Practice Problem 4**

Factor: $10y^7 + 5y^9$

Practice Problem 5

Factor: $4z^2 - 12z + 2$

Helpful Hint

Don't forget to include the term 1.

Practice Problem 6

Factor: $-3y^2 - 9y + 15x^2$

Concept Check

Check both factorizations given in Example 6.

Answers

4. $5y^7(2 + y^2)$ **5.** $2(2z^2 - 6z + 1)$
6. $-3(y^2 + 3y - 5x^2)$ or $3(-y^2 - 3y + 5x^2)$

Concept Check: Answers may vary.

STUDY SKILLS REMINDER

How are your homework assignments going?

By now, you should have good homework habits. If not, it's never too late to begin. Why is it so important in mathematics to keep up with homework? You probably now know the answer to that question. You have probably realized by now that many concepts in mathematics build on each other. Your understanding of one chapter in mathematics usually depends on your understanding of the previous chapter's material.

Don't forget that completing your homework assignment involves a lot more than attempting a few of the problems assigned.

To complete a homework assignment, remember these four things:

1. Attempt all of it.
2. Check it.
3. Correct it.
4. If needed, ask questions about it.

Name _____ Section _____ Date _____

EXERCISE SET 10.4

A *Find the greatest common factor of each list of numbers. See Example 1.*

1. 48 and 15

2. 36 and 20

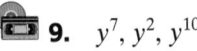 **3.** 60 and 72

4. 96 and 45

5. 12, 20, and 36

6. 18, 24, and 60

7. 8, 32, and 100

8. 30, 50, and 200

B *Find the greatest common factor of each list of terms. See Examples 2 and 3.*

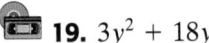 **9.** y^7, y^2, y^{10}

10. x^3, x, x^5

11. a^5, a^5, a^5

12. b^6, b^6, b^4

13. x^3y^2, xy^2, x^4y^2

14. a^5b^3, a^5b^2, a^5b

15. $3x^4, 5x^7, 10x$

16. $9z^6, z^5, 2z^3$

17. $2z^3, 14z^5, 18z^3$

18. $6y^7, 9y^6, 15y^5$

C *Factor. Check by multiplying. See Examples 4 through 6.*

19. $3y^2 + 18y$

20. $2x^2 + 18x$

21. $10a^6 - 5a^8$

22. $21y^5 + y^{10}$

23. $4x^3 + 12x^2 + 20x$

24. $9b^3 - 54b^2 + 9b$

25. $z^7 - 6z^5$

26. $y^{10} + 4y^5$

27. $-35 + 14y - 7y^2$

28. $-20x + 4x^2 - 2$

29. $12a^5 - 36a^6$

30. $25z^3 - 20z^2$

Review and Preview

Solve. See Sections 7.1–7.3.

31. Find 30% of 120.

32. Find 45% of 265.

33. Write 80% as a fraction in simplified form.

34. Write 65% as a fraction in simplified form.

35. Write $\dfrac{3}{8}$ as a percent.

36. Write $\dfrac{3}{4}$ as a percent.

 Combining Concepts

△ **37.** The area of the larger rectangle below is $x(x + 2)$. Find another expression for the area by writing the sum of the areas of the smaller rectangles.

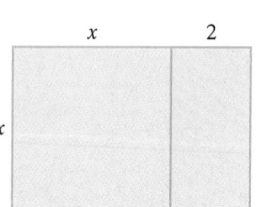

△ **38.** Write an expression for the area of the largest rectangle in two different ways.

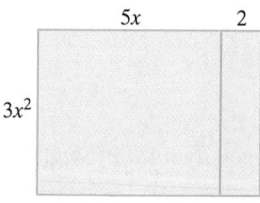

39. In your own words, define the greatest common factor of a list of numbers.

40. Suppose that a classmate asks you why $4x^2 + 6x + 2$ does not factor as $2(2x^2 + 3x)$. Write down your response to this classmate.

Internet Excursions

 Go To: 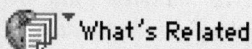 http://www.prenhall.com/martin-gay_prealgebra What's Related

Go to http://www.prenhall.com/martin-gay

This World Wide Web address will provide you with access to a Webmath calculator for Factoring Polynomials by Finding the Greatest Common Factor (GCF), or a related site. The instructions for using this calculator are included on the Web page. Use this calculator to check your work for the following exercises.

41. Exercise 24

42. Exercise 26

43. Exercise 28

44. Exercise 30

CHAPTER 10 ACTIVITY Business Analysis

This activity may be completed by working in groups or individually.

Suppose you own a business that manufactures baskets. You need to decide how many baskets to make. The more baskets you make, the lower the price you will have to charge to sell them all. Natu-rally, each basket you make costs you money be-cause you must buy the materials to make each bas-ket. The following table summarizes some factors you must consider in deciding how many baskets to make, along with algebraic representations of those factors.

Number of baskets	Description		Algebraic Expression
	Unknown		x
Total manufacturing expenses	This is the total amount that it will cost to manufacture all the baskets. It will cost $100 to buy special equipment to manufacture the baskets in addition to basket materials costing $0.50 per basket		$100 + 0.50x$
Price charged per basket	For each additional basket produced, the price that must be charged per basket decreases from $40 by an additional $0.05		$40 - 0.05x$

1. *Revenue* is the amount of money collected from selling the baskets. Revenue can be found by multiplying the price charged per basket by the number of baskets sold. Use the algebraic ex-pressions given in the table above to find a poly-nomial that represents the revenue from sales of baskets. Then write this polynomial in the Poly-nomial column next to "Revenue" in the table to the right.

2. *Profit* is the amount of money you make from selling the baskets after deducting the expenses for making the baskets. Profit can be found by subtracting total manufacturing expenses from revenue. Find a polynomial that represents the profit from the sales of baskets. Then write this polynomial in the Polynomial column next to "Profit" in the table to the right.

3. Complete the following table by evaluating each polynomial for each of the numbers of baskets given in the table.

	Polynomial	Number of Baskets, x				
		200	300	400	500	600
Revenue						
Total manufacturing expenses	$100 + 0.50x$					
Profit						

4. Study the table. Which number of baskets will give you the largest profit from making and sell-ing baskets?

Chapter 10 Vocabulary

Fill in each blank with one of the words or phrases listed below.

trinomial	monomial	greatest common factor	binomial
exponent	factoring	polynomials	FOIL

1. _____ is the process of writing an expression as a product.
2. The _____ of a list of terms is the product of all common factors.
3. The _____ method may be used when multiplying two binomials.
4. A polynomial with exactly 3 terms is called a _____ .
5. A polynomial with exactly 2 terms is called a _____ .
6. A polynomial with exactly 1 term is called a _____ .
7. Monomials, binomials, and trinomials are all examples of _____ .
8. In $5x^3$, the 3 is called an _____ .

CHAPTER 10 Highlights

DEFINITIONS AND CONCEPTS	EXAMPLES
SECTION 10.1 ADDING AND SUBTRACTING POLYNOMIALS	

DEFINITIONS AND CONCEPTS	EXAMPLES
A **monomial** is a polynomial with one term.	Monomial: $-2x^2y^3$
A **binomial** is a polynomial with two terms.	Binomial: $5x - y$
A **trinomial** is a polynomial with three terms.	Trinomial: $7z^3 + 0.5z + 1$
A **polynomial** is a monomial or a sum or difference of monomials.	Polynomials $$5x^2 - 6x + 2, \quad -\frac{9}{10}y, \quad 7$$
To add polynomials, combine like terms.	Add: $(7z^2 - 6z + 2) + (5z^2 - 4z + 5)$ $(7z^2 - 6z + 2) + (5z^2 - 4z + 5)$ $= \underbrace{7z^2 + 5z^2}\ \underbrace{-6z - 4z}\ \underbrace{+2 + 5}$ Group like terms. $= 12z^2 - 10z + 7$ Combine like terms.
To subtract polynomials, change the signs of the terms being subtracted, then add.	Subtract: $(20x - 6) - (30x - 6)$ $(20x - 6) - (30x - 6)$ $= (20x - 6) + (-30x + 6)$ $= \underbrace{20x - 30x}\ \underbrace{-6 + 6}$ Group like terms. $= -10x$ Combine like terms.

DEFINITIONS AND CONCEPTS	**EXAMPLES**

SECTION 10.2 MULTIPLICATION PROPERTIES OF EXPONENTS

Product property for exponents

$$a^m \cdot a^n = a^{m+n}$$

$$x^3 \cdot x^{11} = x^{3+11} = x^{14}$$

Power property for exponents

$$\left(a^m\right)^n = a^{m \cdot n}$$

$$\left(y^5\right)^3 = y^{5 \cdot 3} = y^{15}$$

Power of a product property for exponents

$$(ab)^n = a^n b^n$$

$$\left(2z^5\right)^4 = 2^4\left(z^5\right)^4 = 16z^{20}$$

SECTION 10.3 MULTIPLYING POLYNOMIALS

To multiply two polynomials, multiply each term of the first polynomial by each term of the second polynomial, and then combine like terms.

$$(x + 2)(x^2 + 5x - 1)$$

$$= x(x^2 + 5x - 1) + 2(x^2 + 5x - 1)$$
$$= x \cdot x^2 + x \cdot 5x + x(-1) + 2 \cdot x^2 + 2 \cdot 5x + 2(-1)$$
$$= x^3 + 5x^2 - x + 2x^2 + 10x - 2$$
$$= x^3 + 7x^2 + 9x - 2$$

SECTION 10.4 INTRODUCTION TO FACTORING POLYNOMIALS

TO FIND THE GREATEST COMMON FACTOR OF A LIST OF INTEGERS

Step 1. Write each number as a product of prime numbers.

Step 2. Identify the common prime factors.

Step 3. The product of all common prime factors found in *Step 2* is the greatest common factor. If there are no common prime factors, the greatest common factor is 1.

Find the GCF of 18 and 30.

$$18 = 2 \cdot 3 \cdot 5$$
$$30 = 2 \cdot 3 \cdot 5$$

The GCF is $2 \cdot 3$ or 6.

The **GCF of a list of variables** raised to powers is the variable raised to the smallest exponent in the list.

The GCF of x^6, x^8, and x^3 is x^3.

The **GCF of a list of terms** is the product of all common factors.

Find the GCF of $6y^3$, $12y$, and $4y^7$.

The GCF of 6, 12, and 4 is 2.

The GCF of y^3, y, and y^7 is y.

The GCF of $6y^3$, $12y$, and $4y^7$ is $2y$.

To factor the GCF from the terms of a polynomial, write each term as a product of the GCF and another factor, then apply the distributive property.

Factor $4y^6 + 6y^5$.

The GCF of $4y^6$ and $6y^5$ is $2y^5$.

$$4y^6 + 6y^5 = 2y^5 \cdot 2y + 2y^5 \cdot 3$$
$$= 2y^5(2y + 3)$$

Chapter 10 Review

(10.1) *Perform each indicated operation.*

1. $(2b + 7) + (8b - 10)$

2. $(7s - 6) + (14s - 9)$

3. $(3x + 0.2) - (4x - 2.6)$

4. $(10y - 6) - (11y + 6)$

5. $(4z^2 + 6z - 1) + (5z - 5)$

6. $(17a^3 + 11a^2 + a) + (14a^2 - a)$

7. $\left(9y^2 - y + \dfrac{1}{2}\right) - \left(20y^2 - \dfrac{1}{4}\right)$

8. Subtract $(x - 2)$ from $(x^2 - 6x + 1)$.

Find the value of each polynomial when $x = 3$.

9. $5x^2$

10. $2 - 7x$

△ **11.** Find the perimeter of the given rectangle.

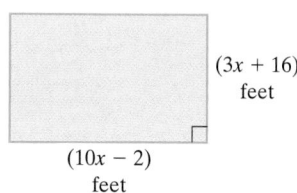

$(3x + 16)$
feet

$(10x - 2)$
feet

(10.2) *Multiply and simplify.*

12. $x^{10} \cdot x^{14}$

13. $y \cdot y^6$

14. $4z^2 \cdot 6z^5$

15. $(-3x^2y)(5xy^4)$

16. $(a^5)^7$

17. $(x^2)^4 \cdot (x^{10})^2$

18. $(9b)^2$

19. $(a^4b^2c)^5$

20. $(7x)(2x^5)^3$

21. $(3x^6y^5)^3(2x^6y^5)^2$

△ **22.** Find the area of the square.

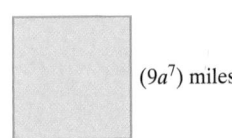

$(9a^7)$ miles

(10.3) *Multiply.*

23. $2a(5a^2 - 6)$

24. $-3y^2(y^2 - 2y + 1)$

25. $(x + 2)(x + 6)$

26. $(3x - 1)(5x - 9)$

27. $(y - 5)^2$

28. $(7a + 1)^2$

29. $(x + 1)(x^2 - 2x + 3)$

30. $(4y^2 - 3)(2y^2 + y + 1)$

31. $(3z^2 + 2z + 1)(z^2 + z + 1)$

△ **32.** Find the area of the given rectangle.

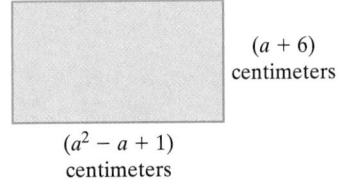

$(a + 6)$
centimeters

$(a^2 - a + 1)$
centimeters

(10.4) *Find the greatest common factor (GCF) of each list.*

33. 20 and 35

34. 12 and 32

35. 24, 30, and 60

36. 10, 20, and 25

37. x^3, x^2, x^{10}

38. y^{10}, y^7, y^7

39. xy^2, xy, x^3y^3

40. a^5b^4, a^6b^3, a^7b^2

41. $5a^3, 10a, 20a^4$

42. $12y^2z, 20y^2z, 24y^5z$

Factor out the GCF.

43. $2x^2 + 12x$

44. $6a^2 - 12a$

45. $6y^4 - y^6$

46. $7x^2 - 14x + 7$

47. $5a^7 - a^4 + a^3$

48. $10y^6 - 10y$

Tips for studying for an exam

To prepare for an exam, try the following study techniques.

- Start the study process days before your exam.

- Make sure that you are current and up-to-date on your assignments.

- If there is a topic that you are unsure of, use one of the many resources that are available to you. For example,

  See your instructor.

  Visit a learning resource center on campus where math tutors are available.

  Read the textbook material and examples on the topic.

  View a videotape on the topic.

- Reread your notes and carefully review the Chapter Highlights at the end of the chapter.

- Work the review exercises at the end of the chapter and check your answers. Make sure that you correct any missed exercises. If you have trouble on a topic, use a resource listed above.

- Find a quiet place to take the Chapter Test found at the end of the chapter. Do not use any resources when taking this sample test. This way you will have a clear indication of how prepared you are for your exam. Check your answers and make sure that you correct any missed exercises.

- Get lots of rest the night before the exam. It's hard to show how well you know the material if your brain is foggy from lack of sleep.

Good luck and keep a positive attitude.

Chapter 10 Test

Remember to check your answers and use the Chapter Test Prep Video to view solutions.

Add or subtract as indicated.

1. $(11x - 3) + (4x - 1)$

2. $(11x - 3) - (4x - 1)$

3. $(1.3y^2 + 5y) + (2.1y^2 - 3y - 3)$

4. Subtract $(8a^2 + a)$ from $(6a^2 + 2a + 1)$.

5. Find the value of $x^2 - 6x + 1$ when $x = 8$.

Multiply and simplify.

6. $y^3 \cdot y^{11}$

7. $(y^3)^{11}$

8. $(2x^2)^4$

9. $(6a^3)(-2a^7)$

10. $(p^6)^7(p^2)^6$

11. $(3a^4b)^2(2ba^4)^3$

12. $5x(2x^2 + 1.3)$

13. $-2y(y^3 + 6y^2 - 4)$

14. $(x - 3)(x + 2)$

15. $(5x + 2)^2$

16. $(a + 2)(a^2 - 2a + 4)$

△ **17.** Find the area and the perimeter of the parallelogram. (*Hint:* $A = b \cdot h$.)

$(x + 7)$ in. $2x$ in.

$(5x - 2)$ in.

Answers
1. _____
2. _____
3. _____
4. _____
5. _____
6. _____
7. _____
8. _____
9. _____
10. _____
11. _____
12. _____
13. _____
14. _____
15. _____
16. _____
17. _____

18.

19.

20.

21.

22.

23.

Find the greatest common factor of each list.

18. 45 and 60

19. $6y^3, 9y^5, 18y^4$

Factor out the GCF.

20. $3y^2 - 15y$

21. $10a^2 + 12a$

22. $6x^2 - 12x - 30$

23. $7x^6 - 6x^4 + x^3$

Cumulative Review

△ **1.** The state of Colorado is in the shape of a rectangle whose length is about 380 miles and whose width is about 280 miles. Find its area.

2. In a pecan orchard, there are 21 trees in each row and 7 rows of trees. How many pecan trees are there?

3. Add: $1 + (-10) + (-8) + 9$

4. Add: $-2 + (-7) + 3 + (-4)$.

Subtract.

5. $8 - 15$

6. $4 - 7$

7. $-4 - (-5)$

8. $3 - (-2)$

9. Solve: $7x = 6x + 4$

10. Solve: $4x = -2 + 3x$

11. Write $\dfrac{8}{3x}$ as an equivalent fraction whose denominator is $12x$.

12. Write $\dfrac{3}{2c}$ as an equivalent fraction with denominator $8c$.

13. Subtract: $12 - 8\dfrac{3}{7}$

14. Subtract: $15 - 4\dfrac{2}{5}$

15. Round 736.2359 to the nearest tenth.

16. Round 328.174 to the nearest tenth.

17. Add: $23.85 + 1.604$

18. Add: $12.762 + 4.29$

19. Is -9 a solution of the equation $3.7y = -3.33$?

20. Is 6 a solution of the equation $2.8x = 16.8$?

Answers
1.
2.
3.
4.
5.
6.
7.
8.
9.
10.
11.
12.
13.
14.
15.
16.
17.
18.
19.
20.

Divide.

21. $\dfrac{786}{10{,}000}$

22. $\dfrac{818}{1000}$

23. $\dfrac{0.12}{10}$

24. $\dfrac{5.03}{100}$

25. Evaluate $-2x + 5$ for $x = 3.8$.

26. Evaluate $6x - 1$ for $x = -2.1$.

27. Write $\dfrac{22}{7}$ as a decimal. Round to the nearest hundreth.

28. Write $\dfrac{37}{19}$ as a decimal. Round to the nearest thousandth.

29. Find: $\sqrt{\dfrac{1}{36}}$

30. Find: $\sqrt{\dfrac{4}{25}}$

31. Mel Rose is a 6-foot-tall park ranger who needs to know the height of a particular tree. He notices that when the shadow of the tree is 69 feet long, his own shadow is 9 feet long. Find the height of the tree.

32. Phoebe, a very intelligent dog, wants to estimate the height of a fire hydrant. She notices that when her shadow is 2 feet long, the shadow of the hydrant is 6 feet long. Find the height of the hydrant if Phoebe is 1 foot tall.

33. Translate to an equation: 1.2 is 30% of what number?

34. Translate to an equation: 9 is 45% of what number?

35. What percent of 50 is 8?

36. What percent of 16 is 4?

37. Mr. Buccaran, the principal at Slidell High School, counted 31 freshmen absent during a particular day. If this is 4% of the total number of freshmen, how many freshmen are there at Slidell High School?

38. 2% of the apples in a shipment are rotten. If there are 29 rotten apples, how many apples are in the shipment?

39. Ivan Borski borrowed $2400 at 10% simple interest for 8 months to buy a used Chevy S-10. Find the simple interest he paid.

40. Find the amount of simple interest earned on a $1000 CD for 10 months at an interest rate of 3%.

41. Using the circle graph shown, determine the percent of Americans that have one or more working computers at home.

42. Using the circle graph in Exercise 41, find the percent of Americans that have fewer than 3 working computers at home.

Number of Working Computers at Home

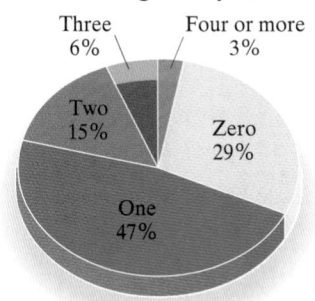

Source: UCLA Center for Communication Policy, 2003

△ **43.** Find the perimeter of a rectangle with a length of 11 inches and a width of 3 inches.

△ **44.** Find the perimeter of a triangular yard whose sides are 6 feet, 8 feet, and 11 feet.

△ **45.** Find the area of the parallelogram.

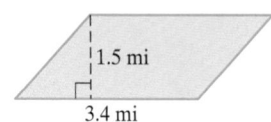

△ **46.** Find the area of the triangle.

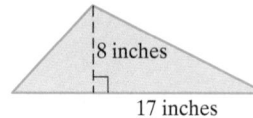

47. Multiply 5 lb 9 oz by 6.

48. Multiply 5 tons 700 lb by 3.

49. Convert 3210 ml to L.

50. Convert 4321 cl to L.

39. _____

40. _____

41. _____

42. _____

43. _____

44. _____

45. _____

46. _____

47. _____

48. _____

49. _____

50. _____

APPENDIX A

Percents, Decimals, and Fractions

Percent, Decimal, and Fraction Equivalent		
Percent	**Decimal**	**Fraction**
1%	0.01	$\frac{1}{100}$
5%	0.05	$\frac{1}{20}$
10%	0.1	$\frac{1}{10}$
12.5% or $12\frac{1}{2}$%	0.125	$\frac{1}{8}$
$16.\overline{6}$% or $16\frac{2}{3}$%	$0.1\overline{6}$	$\frac{1}{6}$
20%	0.2	$\frac{1}{5}$
25%	0.25	$\frac{1}{4}$
30%	0.3	$\frac{3}{10}$
$33.\overline{3}$% or $33\frac{1}{3}$%	$0.\overline{3}$	$\frac{1}{3}$
37.5% or $37\frac{1}{2}$%	0.375	$\frac{3}{8}$
40%	0.4	$\frac{2}{5}$
50%	0.5	$\frac{1}{2}$
60%	0.6	$\frac{3}{5}$
62.5% or $62\frac{1}{2}$%	0.625	$\frac{5}{8}$
$66.\overline{6}$% or $66\frac{2}{3}$%	$0.\overline{6}$	$\frac{2}{3}$
70%	0.7	$\frac{7}{10}$
75%	0.75	$\frac{3}{4}$
80%	0.8	$\frac{4}{5}$
$83.\overline{3}$% or $83\frac{1}{3}$%	$08.\overline{3}$	$\frac{5}{6}$
87.5% or $87\frac{1}{2}$%	0.875	$\frac{7}{8}$
90%	0.9	$\frac{9}{10}$
100%	1.0	1
110%	1.1	$1\frac{1}{10}$
125%	1.25	$1\frac{1}{4}$
$133.\overline{3}$% or $133\frac{1}{3}$%	$1.\overline{3}$	$1\frac{1}{3}$
150%	1.5	$1\frac{1}{2}$
$166.\overline{6}$% or $166\frac{2}{3}$%	$1.\overline{6}$	$1\frac{2}{3}$
175%	1.75	$1\frac{3}{4}$
200%	2.0	2

Table of Squares and Square Roots

\multicolumn{6}{c	}{Squares and Square Roots}
n	n^2	$\sqrt{n}$	n	n^2	$\sqrt{n}$
1	1	1.000	51	2601	7.141
2	4	1.414	52	2704	7.211
3	9	1.732	53	2809	7.280
4	16	2.000	54	2916	7.348
5	25	2.236	55	3025	7.416
6	36	2.449	56	3136	7.483
7	49	2.646	57	3249	7.550
8	64	2.828	58	3364	7.616
9	81	3.000	59	3481	7.681
10	100	3.162	60	3600	7.746
11	121	3.317	61	3721	7.810
12	144	3.464	62	3844	7.874
13	169	3.606	63	3969	7.937
14	196	3.742	64	4096	8.000
15	225	3.873	65	4225	8.062
16	256	4.000	66	4356	8.124
17	289	4.123	67	4489	8.185
18	324	4.243	68	4624	8.246
19	361	4.359	69	4761	8.307
20	400	4.472	70	4900	8.367
21	441	4.583	71	5041	8.426
22	484	4.690	72	5184	8.485
23	529	4.796	73	5329	8.544
24	576	4.899	74	5476	8.602
25	625	5.000	75	5625	8.660
26	676	5.099	76	5776	8.718
27	729	5.196	77	5929	8.775
28	784	5.292	78	6084	8.832
29	841	5.385	79	6241	8.888
30	900	5.477	80	6400	8.944
31	961	5.568	81	6561	9.000
32	1024	5.657	82	6724	9.055
33	1089	5.745	83	6889	9.110
34	1156	5.831	84	7056	9.165
35	1225	5.916	85	7225	9.220
36	1296	6.000	86	7396	9.274
37	1369	6.083	87	7569	9.327
38	1444	6.164	88	7744	9.381
39	1521	6.245	89	7921	9.434
40	1600	6.325	90	8100	9.487
41	1681	6.403	91	8281	9.539
42	1764	6.481	92	8464	9.592
43	1849	6.557	93	8649	9.644
44	1936	6.633	94	8836	9.695
45	2025	6.708	95	9025	9.747
46	2116	6.782	96	9216	9.798
47	2209	6.856	97	9409	9.849
48	2304	6.928	98	9604	9.899
49	2401	7.000	99	9801	9.950
50	2500	7.071	100	10,000	10.000

APPENDIX C

Compound Interest Table

Compounded Annually

	5%	6%	7%	8%	9%	10%	11%	12%	13%	14%	15%	16%	17%	18%
1 year	1.05000	1.06000	1.07000	1.08000	1.09000	1.10000	1.11000	1.12000	1.13000	1.14000	1.15000	1.16000	1.17000	1.18000
5 years	1.27628	1.33823	1.40255	1.46933	1.53862	1.61051	1.68506	1.76234	1.84244	1.92541	2.01136	2.10034	2.19245	2.28776
10 years	1.62889	1.79085	1.96715	2.15892	2.36736	2.59374	2.83942	3.10585	3.39457	3.70722	4.04556	4.41144	4.80683	5.23384
15 years	2.07893	2.39656	2.75903	3.17217	3.64248	4.17725	4.78459	5.47357	6.25427	7.13794	8.13706	9.26552	10.53872	11.97375
20 years	2.65330	3.20714	3.86968	4.66096	5.60441	6.72750	8.06231	9.64629	11.52309	13.74349	16.36654	19.46076	23.10560	27.39303

Compounded Semiannually

	5%	6%	7%	8%	9%	10%	11%	12%	13%	14%	15%	16%	17%	18%
1 year	1.05063	1.06090	1.07123	1.08160	1.09203	1.10250	1.11303	1.12360	1.13423	1.14490	1.15563	1.16640	1.17723	1.18810
5 years	1.28008	1.34392	1.41060	1.48024	1.55297	1.62889	1.70814	1.79085	1.87714	1.96715	2.06103	2.15892	2.26098	2.36736
10 years	1.63862	1.80611	1.98979	2.19112	2.41171	2.65330	2.91776	3.20714	3.52365	3.86968	4.24785	4.66096	5.11205	5.60441
15 years	2.09757	2.42726	2.80679	3.24340	3.74532	4.32194	4.98395	5.74349	6.61437	7.61226	8.75496	10.06266	11.55825	13.26768
20 years	2.68506	3.26204	3.95926	4.80102	5.81636	7.03999	8.51331	10.28572	12.41607	14.97446	18.04424	21.72452	26.13302	31.40942

Compounded Quarterly

	5%	6%	7%	8%	9%	10%	11%	12%	13%	14%	15%	16%	17%	18%
1 year	1.05095	1.06136	1.07186	1.08243	1.09308	1.10381	1.11462	1.12551	1.13648	1.14752	1.15865	1.16986	1.18115	1.19252
5 years	1.28204	1.34686	1.41478	1.48595	1.56051	1.63862	1.72043	1.80611	1.89584	1.98979	2.08815	2.19112	2.29891	2.41171
10 years	1.64362	1.81402	2.00160	2.20804	2.43519	2.68506	2.95987	3.26204	3.59420	3.95926	4.36038	4.80102	5.28497	5.81636
15 years	2.10718	2.44322	2.83182	3.28103	3.80013	4.39979	5.09225	5.89160	6.81402	7.87809	9.10513	10.51963	12.14965	14.02741
20 years	2.70148	3.29066	4.00639	4.87544	5.93015	7.20957	8.76085	10.64089	12.91828	15.67574	19.01290	23.04980	27.93091	33.83010

Compounded Daily

	5%	6%	7%	8%	9%	10%	11%	12%	13%	14%	15%	16%	17%	18%
1 year	1.05127	1.06183	1.07250	1.08328	1.09416	1.10516	1.11626	1.12747	1.13880	1.15024	1.16180	1.17347	1.18526	1.19716
5 years	1.28400	1.34983	1.41902	1.49176	1.56823	1.64861	1.73311	1.82194	1.91532	2.01348	2.11667	2.22515	2.33918	2.45906
10 years	1.64866	1.82203	2.01362	2.22535	2.45933	2.71791	3.00367	3.31946	3.66845	4.05411	4.48031	4.95130	5.47178	6.04696
15 years	2.11689	2.45942	2.85736	3.31968	3.85678	4.48077	5.20569	6.04786	7.02625	8.16288	9.48335	11.01738	12.79950	14.86983
20 years	2.71810	3.31979	4.05466	4.95216	6.04831	7.38703	9.02202	11.01883	13.45751	16.43582	20.07316	24.51533	29.94039	36.56577

Review of Geometric Figures

Plane Figures Have Length and Width But No Thickness or Depth.		
Name	**Description**	**Figure**
Polygon	Union of three or more coplanar line segments that intersect with each other only at each end point, with each end point shared by two segments.	
Triangle	Polygon with three sides (sum of measures of three angles is 180°).	
Scalene Triangle	Triangle with no sides of equal length.	
Isosceles Triangle	Triangle with two sides of equal length.	
Equilateral Triangle	Triangle with all sides of equal length.	
Right Triangle	Triangle that contains a right angle.	
Quadrilateral	Polygon with four sides (sum of measures of four angles is 360°).	
Trapezoid	Quadrilateral with exactly one pair of opposite sides parallel.	
Isosceles Trapezoid	Trapezoid with legs of equal length.	
Parallelogram	Quadrilateral with both pairs of opposite sides parallel.	

(continued)

Name	Description	Figure
Rhombus	Parallelogram with all sides of equal length.	
Rectangle	Parallelogram with four right angles.	
Square	Rectangle with all sides of equal length.	
Circle	All points in a plane the same distance from a fixed point called the **center**.	radius, center, diameter

Solid Figures Have Length, Width, and Height or Depth.		
Name	Description	Figure
Rectangular Solid	A solid with six sides, all of which are rectangles.	
Cube	A rectangular solid whose six sides are squares.	
Sphere	All points the same distance from a fixed point called the **center**.	radius, center
Right Circular Cylinder	A cylinder consisting of two circular bases that are perpendicular to its altitude.	
Right Circular Cone	A cone with a circular base that is perpendicular to its altitude.	

Review of Volume and Surface Area

A convex solid is a set of points, S, not all in one plane, such that for any two points A and B in S, all points between A and B are also in S. In this appendix, we will find the volume and surface area of special types of solids called polyhedrons. A solid formed by the intersection of a finite number of planes is called a **polyhedron.** The box below is an example of a polyhedron.

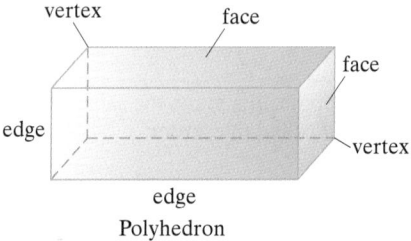

Polyhedron

Each of the plane regions of the polyhedron is called a **face** of the polyhedron. If the intersection of two faces is a line segment, this line segment is an **edge** of the polyhedron. The intersections of the edges are the **vertices** of the polyhedron.

    Volume is a measure of the space of a solid. The volume of a box or can, for example, is the amount of space inside. Volume can be used to describe the amount of juice in a pitcher or the amount of concrete needed to pour a foundation for a house.

    The volume of a solid is the number of **cubic units** in the solid. A cubic centimeter and a cubic inch are illustrated.

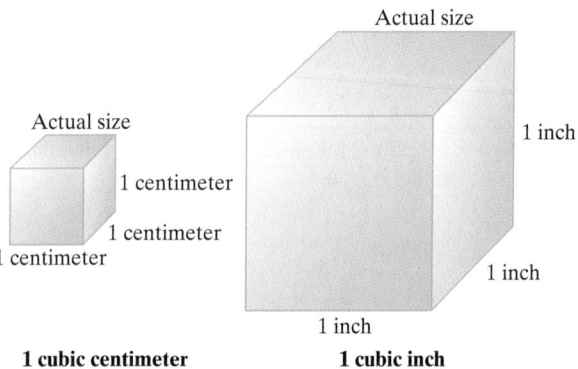

1 cubic centimeter   **1 cubic inch**

    The **surface area** of a polyhedron is the sum of the areas of the faces of the polyhedron. For example, each face of the cube to the left above has an area of 1 square centimeter. Since there are 6 faces of the cube, the sum of the areas of the faces is 6 square centimeters. Surface area can be used to describe the amount of material needed to cover a solid. Surface area is measured in square units.

Formulas for finding the volumes, V, and surface areas, SA, of some common solids are given next.

Volume and Surface Area Formulas of Common Solids	
Solid	**Formulas**
RECTANGULAR SOLID	$V = lwh$ $SA = 2lh + 2wh + 2lw$ where h = height, w = width, l = length
CUBE	$V = s^3$ $SA = 6s^2$ where s = side
SPHERE	$V = \dfrac{4}{3}\pi r^3$ $SA = 4\pi r^2$ where r = radius
CIRCULAR CYLINDER	$V = \pi r^2 h$ $SA = 2\pi rh + 2\pi r^2$ where h = height, r = radius
CONE	$V = \dfrac{1}{3}\pi r^2 h$ $SA = \pi r \sqrt{r^2 + h^2} + \pi r^2$ where h = height, r = radius
SQUARE-BASED PYRAMID	$V = \dfrac{1}{3}s^2 h$ $SA = B + \dfrac{1}{2}pl$ where B = area of base; p = perimeter of base, h = height, s = side, l = slant height

A8

Volume is measured in cubic units. Surface area is measured in square units.

EXAMPLE 1

Find the volume and surface area of a rectangular box that is 12 inches long, 6 inches wide, and 3 inches high.

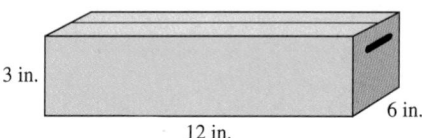

3 in.

6 in.

12 in.

Solution: Let $h = 3$ in., $l = 12$ in., and $w = 6$ in.

$$V = lwh$$

$$V = 12 \text{ inches} \cdot 6 \text{ inches} \cdot 3 \text{ inches} = 216 \text{ cubic inches}$$

The volume of the rectangular box is 216 cubic inches.

$$SA = 2lh + 2wh + 2lw$$

$$= 2(12 \text{ in.})(3 \text{ in.}) + 2(6 \text{ in.})(3 \text{ in.}) + 2(12 \text{ in.})(6 \text{ in.})$$

$$= 72 \text{ sq in.} + 36 \text{ sq in.} + 144 \text{ sq in.}$$

$$= 252 \text{ sq in.}$$

The surface area of rectangular box is 252 square inches. ●

EXAMPLE 2

Find the volume and surface area of a ball that has a radius of 2 inches. Give the exact volume and surface area and then use the approximation $\frac{22}{7}$ for π.

Solution:

$$V = \frac{4}{3}\pi r^3 \qquad \text{Formula for volume of a sphere.}$$

$$V = \frac{4}{3} \cdot \pi (2 \text{ in.})^3 \qquad \text{Let } r = 2 \text{ inches.}$$

$$= \frac{32}{3}\pi \text{ cu in.} \qquad \text{Simplify.}$$

$$\approx \frac{32}{3} \cdot \frac{22}{7} \text{ cu in.} \qquad \text{Approximate } \pi \text{ with } \frac{22}{7}.$$

$$= \frac{704}{21} \text{ or } 33\frac{11}{21} \text{ cu in.}$$

The volume of the sphere is exactly $\frac{32}{3}\pi$ cubic inches or approximately $33\frac{11}{21}$ cubic inches.

$$SA = 4\pi r^2 \qquad \text{Formula for surface area.}$$

$$SA = 4 \cdot \pi (2\,\text{in.})^2 \qquad \text{Let } r = 2 \text{ inches.}$$

$$= 16\pi \text{ sq in.} \qquad \text{Simplify.}$$

$$\approx 16 \cdot \frac{22}{7} \text{ sq in.} \qquad \text{Approximate } \pi \text{ with } \frac{22}{7}.$$

$$= \frac{352}{7} \text{ or } 50\frac{2}{7}\text{ sq in.}$$

The surface area of the sphere is exactly 16π square inches or approximately $50\frac{2}{7}$ square inches.

APPENDIX E EXERCISE SET

Find the volume and surface area of each solid. See Examples 1 and 2. For formulas that contain π, give an exact answer and then approximate using $\frac{22}{7}$ for π.

1.
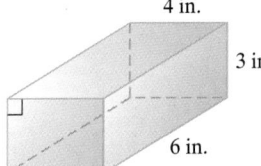
4 in.
3 in.
6 in.

2.
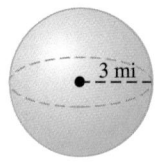
3 mi

3.
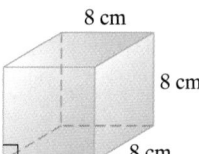
8 cm
8 cm
8 cm

4.
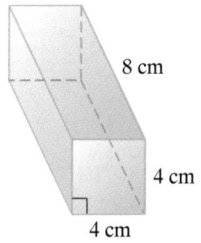
8 cm
4 cm
4 cm

5. (For surface area, use 3.14 for π.)
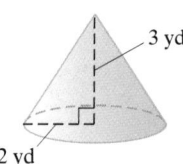
3 yd
2 yd

6.
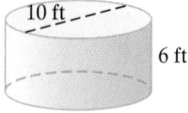
10 ft
6 ft

7.
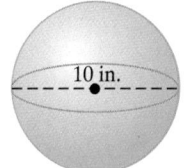
10 in.

8. Find the volume only.
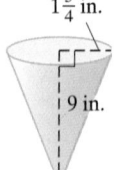
$1\frac{3}{4}$ in.
9 in.

9.
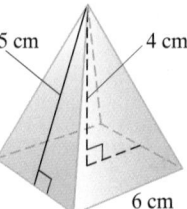
5 cm
4 cm
6 cm

10.
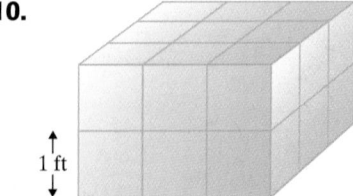
1 ft

A10

Solve.

11. Find the volume of a cube with edges of $1\frac{1}{3}$ inches.

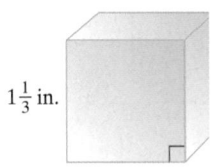

$1\frac{1}{3}$ in.

12. A water storage tank is in the shape of a cone with the pointed end down. If the radius is 14 ft and the depth of the tank is 15 ft, approximate the volume of the tank in cubic feet. Use $\frac{22}{7}$ for π.

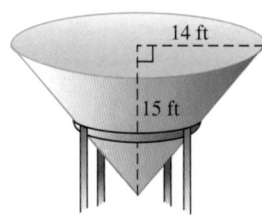

14 ft

15 ft

13. Find the surface area of a rectangular box 2 ft by 1.4 ft by 3 ft.

14. Find the surface area of a box in the shape of a cube that is 5 ft on each side.

15. Find the volume of a pyramid with a square base 5 in. on a side and a height of 1.3 in.

16. Approximate to the nearest hundredth the volume of a sphere with a radius of 2 cm. Use 3.14 for π.

17. A paperweight is in the shape of a square-based pyramid 20 cm tall. If an edge of the base is 12 cm, find the volume of the paperweight.

18. A bird bath is made in the shape of a hemisphere (half-sphere). If its radius is 10 in., approximate the volume. Use $\frac{22}{7}$ for π.

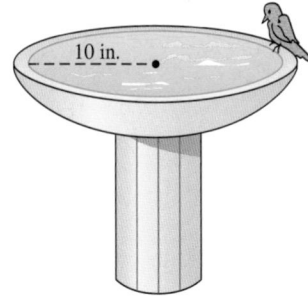

10 in.

19. Find the exact surface area of a sphere with a radius of 7 in.

20. A tank is in the shape of a cylinder 8 ft tall and 3 ft in radius. Find the exact surface area of the tank.

21. Find the volume of a rectangular block of ice 2 ft by $2\frac{1}{2}$ ft by $1\frac{1}{2}$ ft.

22. Find the capacity (volume in cubic feet) of a rectangular ice chest with inside measurements of 3 ft by $1\frac{1}{2}$ ft by $1\frac{3}{4}$ ft.

23. An ice cream cone with a 4-cm diameter and 3-cm depth is filled exactly level with the top of the cone. Approximate how much ice cream (in cubic centimeters) is in the cone. Use $\frac{22}{7}$ for π.

24. A child's toy is in the shape of a square-based pyramid 10 in. tall. If an edge of the base is 7 in., find the volume of the toy.

APPENDIX F

Energy: U.S. and Metric Systems of Measurement

Many people think of energy as a concept that involves movement or activity. However, **energy** is defined as "the capacity to do work." Often energy is stored, awaiting use at some later point in time.

 Defining and Using the U.S. System Units of Energy

In the U.S. system of measurement, energy is commonly measured in foot-pounds. One **foot-pound (ft-lb)** is the amount of energy needed to lift a 1-pound object a distance of 1 foot. To determine the amount of energy necessary to move a 50-pound weight a distance of 100-feet, we simply multiply these numbers. That is,

$$50 \text{ pounds} \cdot 100 \text{ feet} = 5000 \text{ ft-lb of energy}$$

EXAMPLE 1   Finding the Amount of Energy Needed to Move a Carton

An employee for the Jif Peanut Butter company must lift a carton of peanut butter jars 16 feet to the top of the warehouse. In the carton are 24 jars, each of which weighs 1.125 pounds. How much energy is required to lift the carton?

Solution: First we determine the weight of the carton.

$$\begin{aligned} \text{weight of carton} &= \text{weight of a jar} \cdot \text{number of jars} \\ &= 1.125 \text{ pounds} \cdot 24 \\ &= 27 \text{ pounds} \end{aligned}$$

Thus, the carton weighs 27 pounds.
To find the energy needed to lift the 27-pound carton, we multiply the weight times the distance.

$$\text{energy} = 27 \text{ pounds} \cdot 16 \text{ feet} = 432 \text{ ft-lb}$$

Thus, 432 ft-lb of energy are required to lift the carton.   ●

Another form of energy is heat. In the U.S. system of measurement, heat is measured in **British Thermal Units (BTU)**. A BTU is the amount of heat required to raise the temperature of 1 pound of water 1 degree Fahrenheit. To relate British Thermal Units to foot-pounds, we need to know that

$$1 \text{ BTU} = 778 \text{ ft-lb}$$

EXAMPLE 2   Converting BTU to Foot-Pounds

The Raywall Company produces several different furnace models. Their FC-4 model requires 13,652 BTU every hour to operate. Convert the required energy to foot-pounds.

Solution:

To convert BTU to foot-pounds, we multiply by the unit fraction $\dfrac{778 \text{ ft-lb}}{1 \text{ BTU}}$.

$$13{,}652 \text{ BTU} = 13{,}652 \; \cancel{\text{BTU}} \cdot \overbrace{\frac{778 \text{ ft-lb}}{1 \; \cancel{\text{BTU}}}}^{\text{Unit fraction}}$$

$$= 10{,}621{,}256 \text{ ft-lb}$$

Thus, 13,652 BTU is equivalent to 10,621,256 ft-lb.   ●

70 calories

B Defining and Using the Metric System Units of Energy

In the metric system, heat is measured in calories. A **calorie (cal)** is the amount of heat required to raise the temperature of 1 kilogram of water 1 degree Celsius.

The fact that an apple contains 70 calories means that 70 calories of heat energy are stored in our bodies whenever we eat an apple. This energy is stored in fat tissue and is burned (or "oxidized") by our bodies when we require energy to do work. We need 20 calories each hour just to stand still. This means that 20 calories of heat energy will be burned by our bodies each hour that we spend standing.

EXAMPLE 3   Finding the Number of Calories Needed

It takes 20 calories for Jim to stand for 1 hour. How many calories does he use when standing for 3 hours at a crowded party?

Solution: We multiply the number of calories used in 1 hour by the number of hours spent standing.

$$\text{total calories} = 20 \cdot 3 = 60 \text{ calories}$$

Therefore, Jim uses 60 calories to stand for 3 hours at the party.

EXAMPLE 4   Finding the Number of Calories Needed

It takes 115 calories for Kathy to walk slowly for 1 hour. How many calories does she use walking slowly for 1 hour each day for 6 days?

Solution: We multiply the total number of calories used in 1 hour each day by the number of days.

$$\text{total calories} = 115 \cdot 6 = 690 \text{ calories}$$

Therefore, Kathy uses 690 calories walking slowly for 1 hour each day for 6 days.

EXAMPLE 5   Finding the Number of Calories Needed

It requires 100 calories to play a game of cards for an hour. If Jason plays poker for 1.5 hours each day for 5 days, how many calories are required?

Solution: We first determine the number of calories Jason uses each day to play poker.

$$\text{calories used each day} = 100(1.5) = 150 \text{ calories}$$

Then we multiply the number of calories used each day by the number of days.

$$\text{calories used for 5 days} = 150 \cdot 5 = 750 \text{ calories}$$

Thus, Jason uses 750 calories to play poker for 1.5 hours each day for 5 days.

Mental Math

Solve.

1. How many foot-pounds of energy are needed to lift a 6-pound object 5 feet?

2. How many foot-pounds of energy are needed to lift a 10-pound object 4 feet?

3. How many foot-pounds of energy are needed to lift a 3-pound object 20 feet?

4. How many foot-pounds of energy are needed to lift a 5-pound object 9 feet?

5. If 30 calories are burned by the body in 1 hour, how many calories are burned in 3 hours?

6. If 15 calories are burned by the body in 1 hour, how many calories are burned in 2 hours?

7. If 20 calories are burned by the body in 1 hour, how many calories are burned in $\frac{1}{4}$ of an hour?

8. If 50 calories are burned by the body in 1 hour, how many calories are burned in $\frac{1}{2}$ of an hour?

APPENDIX F EXERCISE SET

Ⓐ *Solve. See Examples 1 and 2.*

 1. How much energy is required to lift a 3-pound math textbook 380 feet up a hill?

2. How much energy is required to lift a 20-pound sack of potatoes 55 feet?

3. How much energy is required to lift a 168-pound person 22 feet?

4. How much energy is needed to lift a 2250-pound car a distance of 45 feet?

5. How many foot-pounds of energy are needed to lift 2.5 tons of topsoil 85 feet from the pile delivered by the nursery to the garden?

6. How many foot-pounds of energy are needed to lift 4.25 tons of coal 16 feet into a new coal bin?

7. Convert 30 BTU to foot-pounds.

8. Convert 50 BTU to foot-pounds.

9. Convert 1000 BTU to foot-pounds.

10. Convert 10,000 BTU to foot-pounds.

11. A 20,000 BTU air conditioner requires how many foot-pounds of energy to operate?

12. A 24,000 BTU air conditioner requires how many foot-pounds of energy to operate?

13. The Raywall model FC-10 heater uses 34,130 BTU each hour to operate. How many foot-pounds of energy does it use each hour?

14. The Raywall model FC-12 heater uses 40,956 BTU each hour to operate. How many foot-pounds of energy does it use each hour?

15. 8,000,000 ft-lb is equivalent to how many BTU, rounded to the nearest whole number?

16. 450,000 ft-lb is equivalent to how many BTU, rounded to the nearest whole number?

17. While walking slowly, Janie Gaines burns 115 calories each hour. How many calories does she burn if she walks slowly for an hour every day of the week?

18. Dancing burns 270 calories per hour. How many calories are needed to go dancing an hour a night for 3 nights?

19. Approximately 300 calories are burned each hour skipping rope. How many calories are required to skip rope $\frac{1}{2}$ of an hour each day for 5 days?

20. Ebony Jordan burns 360 calories per hour while riding her stationary bike. How many calories does she burn when she rides her bicycle $\frac{1}{4}$ of an hour each day for 6 days?

21. Julius Davenport goes through a rigorous exercise routine each day. He burns calories at a rate of 720 calories per hour. How many calories does he need to exercise 20 minutes per day, 6 days a week?

22. A roller skater can easily use 325 calories per hour while skating. How many calories are needed to roller skate 75 minutes per day for 3 days?

23. A casual stroll burns 165 calories per hour. How long will it take to stroll off the 425 calories contained in a hamburger, to the nearest tenth of an hour?

24. Even when asleep, the body burns 15 calories per hour. How long must a person sleep to burn off the calories in an 80-calorie orange, to the nearest tenth of an hour?

25. One pound of body weight is lost whenever 3500 calories are burned off. If walking briskly burns 200 calories for each mile walked, how far must Sheila Osby walk to lose 1 pound?

26. Bicycling can burn as much as 500 calories per hour. How long must a person ride a bicycle at this rate to use up the 3500 calories needed to lose 1 pound?

27. A 123.9-pound pile of prepacked canned goods must be lifted 9 inches to permit a door to close. How much energy is needed to do the job?

28. A 14.3-pound framed picture must be lifted 6 feet 3 inches. How much energy is required to move the picture?

29. 6400 ft-lb of energy were needed to lift an anvil 25 feet. Find the weight of the anvil.

30. 825 ft-lb of energy were needed to lift 40 pounds of apples to a new container. How far were the apples moved?

Answers to Selected Exercises

Chapter 1 THE WHOLE NUMBERS

CHAPTER 1 PRETEST

1. hundreds; 1.2A **2.** twenty-three thousand, four hundred ninety; 1.2B **3.** 87; 1.3A **4.** 3717; 1.6B **5.** 626; 1.4A
6. 32; 1.4B **7.** 136 pages; 1.4C **8.** 9050; 1.5A **9.** 3100; 1.5B **10.** $9 \cdot 3 + 9 \cdot 11$; 1.6A **11.** 25 in.; 1.3B **12.** 184 sq yd; 1.6C
13. 576 seats; 1.6D **14.** 243; 1.7A **15.** 446 R 9; 1.7B **16.** 39; 1.7D **17.** $9880; 1.7C **18.** 9^7; 1.8A **19.** 2401; 1.8B
20. 39; 1.8C **21.** 3; 1.9A **22.** $2x + 6$; 1.9B

EXERCISE SET 1.2

1. tens **3.** thousands **5.** hundred-thousands **7.** millions **9.** five thousand, four hundred twenty **11.** twenty-six
thousand, nine hundred ninety **13.** one million, six hundred twenty thousand **15.** fifty-three million, five hundred twenty
thousand, one hundred seventy **17.** sixty-three thousand, nine hundred sixty **19.** one thousand, four hundred eighty-three
21. twelve thousand, six hundred sixty-two **23.** thirteen million, six hundred thousand **25.** three thousand, eight hundred
ninety-five **27.** 6508 **29.** 29,900 **31.** 6,504,019 **33.** 3,000,014 **35.** 220 **37.** 755 **39.** 73,500,000 **41.** 1815
43. 1262 **45.** 45,000 **47.** $400 + 6$ **49.** $5000 + 200 + 90$ **51.** $60,000 + 2000 + 400 + 7$ **53.** $30,000 + 600 + 80$
55. $30,000,000 + 9,000,000 + 600,000 + 80,000$ **57.** $1000 + 6$ **59.** < **61.** > **63.** > **65.** > **67.** four thousand
69. $4000 + 100 + 40 + 5$ **71.** Nile **73.** Labrador retriever; one hundred sixty-five thousand, nine hundred seventy
75. Golden retriever **77.** 7632 **79.** Answers may vary. **81.** Canton

CALCULATOR EXPLORATIONS

1. 134 **3.** 340 **5.** 2834

MENTAL MATH

1. 12 **3.** 9000 **5.** 1620

EXERCISE SET 1.3

1. 36 **3.** 92 **5.** 49 **7.** 5399 **9.** 117 **11.** 71 **13.** 117 **15.** 25 **17.** 62 **19.** 212 **21.** 94 **23.** 910 **25.** 8273
27. 11,926 **29.** 1884 **31.** 16,717 **33.** 1110 **35.** 8999 **37.** 35,901 **39.** 612,389 **41.** 29 in. **43.** 25 ft **45.** 24 in.
47. 8 yd **49.** 6684 ft **51.** 340 ft **53.** 263,700 motorcycles **55.** 28,000 people **57.** 8545 stores **59.** 2425 ft
61. 13,255 mi **63.** 7485 **65.** 10,413 **67.** Texas **69.** 678 stores **71.** 1495 stores **73.** Answers may vary.
75. 166,510,192 **77.** The computation is incorrect; answers may vary.

CALCULATOR EXPLORATIONS

1. 770 **3.** 109 **5.** 8978

MENTAL MATH

1. 7 **3.** 5 **5.** 0 **7.** 400 **9.** 500

EXERCISE SET 1.4

1. 44 **3.** 60 **5.** 265 **7.** 254 **9.** 545 **11.** 600 **13.** 25 **15.** 45 **17.** 146 **19.** 288 **21.** 168 **23.** 6 **25.** 447
27. 5723 **29.** 504 **31.** 89 **33.** 79 **35.** 39,914 **37.** 32,711 **39.** 5041 **41.** 31,213 **43.** 4 **45.** 20 **47.** 7
49. 72 **51.** 88 **53.** 264 pages **55.** 4 million sq km **57.** 6065 ft **59.** $175 **61.** 358 mi **63.** $239 **65.** 173 points
67. 37,035 boxers **69.** Jo; 271 votes **71.** 3,044,452 people **73.** 5920 sq ft **75.** Lake Pontchartrain Bridge; 2175 ft
77. Atlanta Hartsfield International **79.** 5842 thousand **81.** General Motors Corp. **83.** $1,389,100,000 **85.** $21,784,800,000
87. $5269 - 2385 = 2884$ **89.** Answers may vary.

EXERCISE SET 1.5

1. 630 **3.** 640 **5.** 790 **7.** 400 **9.** 1100 **11.** 43,000 **13.** 248,700 **15.** 36,000 **17.** 100,000 **19.** 60,000,000
21. 5280; 5300; 5000 **23.** 9440; 9400; 9000 **25.** 14,880; 14,900; 15,000 **27.** 11,000 **29.** 38,000 points **31.** 120,000,000 bushels
33. 18,000 women **35.** 130 **37.** 380 **39.** 5500 **41.** 300 **43.** 8500 **45.** correct **47.** incorrect **49.** correct
51. correct **53.** $3100 **55.** 80 mi **57.** 6000 ft **59.** 1,400,000 people **61.** 14,000,000 votes **63.** 154,000 children
65. 1100 mi **67.** $3,400,000,000 **69.** $2,000,000,000 **71.** 8550 **73.** 1,549,999 **75.** 21,900 mi **77.** Answers may vary.

CALCULATOR EXPLORATIONS

1. 3456 **3.** 15,322 **5.** 272,291

MENTAL MATH

1. 24 **3.** 0 **5.** 0 **7.** 87

EXERCISE SET 1.6

1. $4 \cdot 3 + 4 \cdot 9$ **3.** $2 \cdot 4 + 2 \cdot 6$ **5.** $10 \cdot 11 + 10 \cdot 7$ **7.** 252 **9.** 1872 **11.** 1362 **13.** 5310 **15.** 4172 **17.** 10,857
19. 11,326 **21.** 24,800 **23.** 0 **25.** 5900 **27.** 59,232 **29.** 142,506 **31.** 1,821,204 **33.** 456,135 **35.** 64,790
37. 199,548 **39.** 240,000 **41.** 300,000 **43.** 26,100 **45.** 3600 **47.** 63 sq m **49.** 390 sq ft **51.** 375 cal **53.** $1890
55. 192 cans **57.** 9900 sq ft **59.** 56,000 sq ft **61.** 5828 pixels **63.** 2000 characters **65.** 1280 cal **67.** 71,343 mi
69. 506 windows **71.** 21,700,000 qt **73.** $6,335,000,000 **75.** 50 students **77.** apple and orange **79.** 2, 9
81. Answers may vary. **83.** 1938 points

CALCULATOR EXPLORATIONS

1. 53 **3.** 62 **5.** 261 **7.** 0

MENTAL MATH

1. 5 **3.** 9 **5.** 0 **7.** 9 **9.** 1 **11.** 5 **13.** undefined **15.** 7 **17.** 0 **19.** 8

EXERCISE SET 1.7

1. 12 **3.** 37 **5.** 338 **7.** 16 R 2 **9.** 563 R 1 **11.** 37 R 1 **13.** 265 R 1 **15.** 49 **17.** 13 **19.** 97 R 40 **21.** 206
23. 506 **25.** 202 R 7 **27.** 45 **29.** 98 R 100 **31.** 202 R 15 **33.** 202 **35.** 58 students **37.** $252,000 **39.** 415 bushels
41. 89 bridges **43.** Yes, she needs 176 ft; she has 9 ft left over **45.** 24 touchdowns **47.** 1760 yd **49.** 26 **51.** 498 **53.** 79
55. 16° **57.** $2,957,500,000 **59.** increase; answers may vary **61.** No; answers may vary.

INTEGRATED REVIEW

1. 148 **2.** 6555 **3.** 1620 **4.** 562 **5.** 79 **6.** undefined **7.** 9 **8.** 1 **9.** 0 **10.** 0 **11.** 0 **12.** 3 **13.** 2433
14. 12,895 **15.** 213 R 3 **16.** 79,317 **17.** 27 **18.** 9 **19.** 138 **20.** 276 **21.** 1099 R 2 **22.** 111 R 1 **23.** 663 R 6
24. 1076 R 60 **25.** 1024 **26.** 9899 **27.** 30,603 **28.** 47,500 **29.** 0 **30.** undefined **31.** 0 **32.** 0 **33.** 86 **34.** 22
35. 8630; 8600; 9000 **36.** 1550; 1600; 2000 **37.** 10,900; 10,900; 11,000 **38.** 432,200; 432,200; 432,000
39. perimeter: 20 ft; area: 25 sq ft **40.** perimeter: 42 in.; area: 98 sq in. **41.** 26 mi **42.** 26 m **43.** 3987 mi **44.** 43 muscles
45. 206 cases with 12 cans left over; yes **46.** $17,820

CALCULATOR EXPLORATIONS

1. 729 **3.** 1024 **5.** 2048 **7.** 2526 **9.** 4295 **11.** 8

EXERCISE SET 1.8

1. 3^4 **3.** 7^8 **5.** 12^3 **7.** $6^2 \cdot 5^3$ **9.** $9^3 \cdot 8$ **11.** $3 \cdot 2^5$ **13.** $3 \cdot 2^2 \cdot 5^3$ **15.** 25 **17.** 125 **19.** 64 **21.** 1024 **23.** 7
25. 243 **27.** 256 **29.** 64 **31.** 81 **33.** 729 **35.** 100 **37.** 10,000 **39.** 10 **41.** 1920 **43.** 729 **45.** 21 **47.** 8
49. 29 **51.** 4 **53.** 17 **55.** 46 **57.** 28 **59.** 10 **61.** 7 **63.** 4 **65.** 14 **67.** 72 **69.** 2 **71.** 35 **73.** 4
75. undefined **77.** 52 **79.** 44 **81.** 12 **83.** 13 **85.** 400 sq mi **87.** 64 sq cm **89.** 10,000 sq m **91.** $(2 + 3) \cdot 6 - 2$
93. $24 \div (3 \cdot 2) + 2 \cdot 5$ **95.** 1260 ft **97.** 6,384,814

EXERCISE SET 1.9

1. 9 **3.** 26 **5.** 6 **7.** 3 **9.** 117 **11.** 94 **13.** 5 **15.** 626 **17.** 20 **19.** 4 **21.** 4 **23.** 0 **25.** 33 **27.** 121
29. 121 **31.** 100 **33.** 60 **35.** 4 **37.** 16, 64, 144, 256 **39.** $x + 5$ **41.** $x + 8$ **43.** $20 - x$ **45.** $512x$ **47.** $\dfrac{x}{2}$
49. $5x + (17 + x)$ **51.** $5x$ **53.** $11 - x$ **55.** $x - 5$ **57.** $6 \div x$ or $\dfrac{6}{x}$ **59.** $50 - 8x$ **61.** 274,657 **63.** 777
65. $5x$ **67.** As t gets larger, $16t^2$ gets larger.

CHAPTER 1 REVIEW

1. hundreds **2.** ten millions **3.** five thousand, four hundred eighty **4.** forty-six million, two hundred thousand, one
hundred twenty **5.** $6000 + 200 + 70 + 9$ **6.** $400,000,000 + 3,000,000 + 200,000 + 20,000 + 5000$ **7.** 59,800
8. 6,304,000,000 **9.** 1,630,553 **10.** 2,968,528 **11.** 531,341 **12.** 76,704 **13.** 13 **14.** 17 **15.** 3 **16.** 10 **17.** 38

18. 68   **19.** 56   **20.** 40   **21.** 110   **22.** 120   **23.** 950   **24.** 1250   **25.** 1711   **26.** 9867   **27.** 8032 mi
28. $197,699   **29.** 276 ft   **30.** 66 km   **31.** 33   **32.** 43   **33.** 14   **34.** 33   **35.** 362   **36.** 65   **37.** 304   **38.** 476
39. 2114   **40.** 321   **41.** 397 pages   **42.** $25,626   **43.** May   **44.** August   **45.** July, August, September   **46.** April,
May   **47.** 90   **48.** 50   **49.** 470   **50.** 500   **51.** 4800   **52.** 58,000   **53.** 50,000,000   **54.** 800,000   **55.** 7400
56. 4100   **57.** 68,000,000   **58.** 680,000   **59.** 42   **60.** 24   **61.** 0   **62.** 0   **63.** 1410   **64.** 2898   **65.** 800
66. 900   **67.** 3696   **68.** 1694   **69.** 0   **70.** 0   **71.** 16,994   **72.** 8954   **73.** 113,634   **74.** 44,763   **75.** 411,426
76. 636,314   **77.** 1500   **78.** 240,000   **79.** 1,040,000   **80.** 7,020,000   **81.** 24 g   **82.** $4,897,341   **83.** 60 sq mi
84. 500 sq cm   **85.** 3   **86.** 4   **87.** 6   **88.** 5   **89.** 5 R 2   **90.** 4 R 2   **91.** undefined   **92.** 0   **93.** 1   **94.** 10
95. undefined   **96.** 0   **97.** 15   **98.** 19 R 7   **99.** 24 R 2   **100.** 56   **101.** 1 R 17   **102.** 35 R 15   **103.** 500
104. 21 R 6   **105.** 506   **106.** 16   **107.** 199 R 8   **108.** 200   **109.** 458 ft   **110.** 51   **111.** 7^4 **112.** $6^2 \cdot 3^3$
113. $4 \cdot 2^3 \cdot 3^2$ **114.** $5^2 \cdot 7^3 \cdot 2^2$ **115.** 49   **116.** 64   **117.** 1125   **118.** 19,600   **119.** 13   **120.** 10   **121.** 3   **122.** 1
123. 32   **124.** 33   **125.** 49 sq m   **126.** 9 sq in.   **127.** 5   **128.** 17   **129.** undefined   **130.** 0   **131.** 121   **132.** 2
133. 4   **134.** 20   **135.** $x - 5$ **136.** $x + 7$ **137.** $10 \div x$ **138.** $5x$ **139.** 0, 8, 32, 72

Chapter 1 Test

1. 141   **2.** 113   **3.** 14,880   **4.** 766 R 42   **5.** 200   **6.** 48   **7.** 98   **8.** 0   **9.** undefined   **10.** 33   **11.** 21   **12.** 36
13. 52,000   **14.** 13,700   **15.** 1600   **16.** $17   **17.** $119   **18.** $126   **19.** 7 billion tablets   **20.** 170,000 cards   **21.** 30
22. 1   **23. a.** $17x$ **b.** $20 - 2x$ **24.** 20 cm, 25 sq cm   **25.** 60 yd, 200 sq yd   **26.** $403,706,375   **27.** $26,190,162
28. $86,148,484

Chapter 2   INTEGERS

Chapter 2 Pretest

1. -22; 2.1 A   **2.** ; 2.1B   **3.** $>$; 2.1 C   **4.** 8; 2.1D   **5.** 12; 2.1E   **6.** -11; 2.2A   **7.** -17; 2.2B
8. 3° F; 2.2C   **9.** 5; 2.3A   **10.** 8; 2.3B   **11.** -15; 2.3C   **12.** -4; 2.3D
13. -104; 2.4A   **14.** 9; 2.4B   **15.** 48; 2.4C   **16.** -20; 2.4C   **17.** -60; 2.4D   **18.** -25; 2.5A   **19.** 2; 2.5A   **20.** -8; 2.5A
21. -34; 2.5B   **22.** -6; 2.5C

Exercise zSet 2.1

1. -1445 **3.** $+14,433$ **5.** $+118$ **7.** $-11,730$ **9.** -1683 million   **11.** -135; -157; Sara   **13.** -81
15.   **17.**   **19.**
21.   **23.** $<$ **25.** $<$ **27.** $>$ **29.** $<$ **31.** 5   **33.** 8   **35.** 0   **37.** 5
39. -5 **41.** 4   **43.** -23 **45.** 10   **47.** 7   **49.** -20 **51.** -3 **53.** 8
55. 14   **57.** 29   **59.** 8   **61.** -3 **63.** 23   **65.** -4 **67.** $>$ **69.** $<$ **71.** $=$ **73.** $<$ **75.** $>$ **77.** $<$
79. $>$ **81.** $<$ **83.** $-|-8|, -|3|, 2^2, -(-5)$ **85.** $-|-6|, -|1|, -1, -(-6)$ **87.** $-10, -|-9|, -(-2), |-12|, 5^2$
89. Maracaibo Lake   **91.** Eyre Lake   **93.** Earth   **95.** Saturn   **97.** 13   **99.** 35   **101.** 360   **103.** d   **105.** 5
107. false   **109.** true   **111.** false   **113.** Answers may vary.

Calculator Explorations

1. -159 **3.** 44   **5.** $-894,855$

Mental Math

1. 5   **3.** -35 **5.** 0   **7.** 0

Exercise Set 2.2

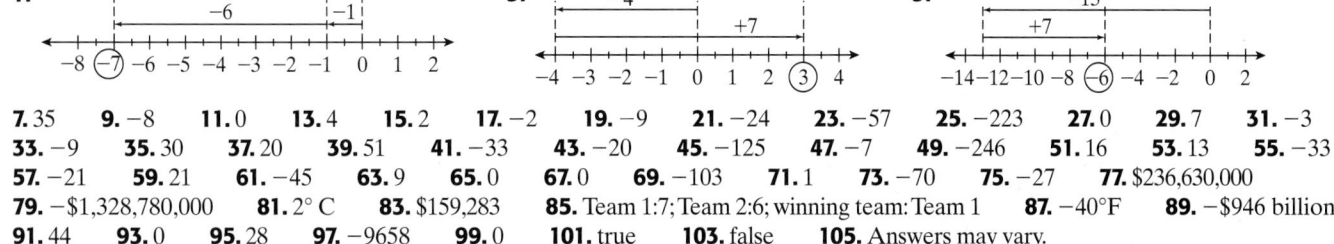

1.   **3.**   **5.**

7. 35   **9.** -8 **11.** 0   **13.** 4   **15.** 2   **17.** -2 **19.** -9 **21.** -24 **23.** -57 **25.** -223 **27.** 0   **29.** 7   **31.** -3
33. -9 **35.** 30   **37.** 20   **39.** 51   **41.** -33 **43.** -20 **45.** -125 **47.** -7 **49.** -246 **51.** 16   **53.** 13   **55.** -33
57. -21 **59.** 21   **61.** -45 **63.** 9   **65.** 0   **67.** 0   **69.** -103 **71.** 1   **73.** -70 **75.** -27 **77.** $236,630,000
79. $-$1,328,780,000   **81.** 2° C   **83.** $159,283   **85.** Team 1:7; Team 2:6; winning team: Team 1   **87.** $-40°F$ **89.** $-$946 billion
91. 44   **93.** 0   **95.** 28   **97.** -9658 **99.** 0   **101.** true   **103.** false   **105.** Answers may vary.

Exercise Set 2.3

1. 0   **3.** 5   **5.** −5   **7.** 14   **9.** 3   **11.** −18   **13.** −14   **15.** 0   **17.** 0   **19.** −4   **21.** −15   **23.** −14   **25.** −230
27. 10   **29.** −4   **31.** 7   **33.** −38   **35.** −17   **37.** 13   **39.** 2   **41.** 0   **43.** −1   **45.** −27   **47.** 40   **49.** −22
51. −8   **53.** 14   **55.** −8   **57.** 36   **59.** 0   **61.** 19   **63.** 428 degrees   **65.** 39 degrees   **67.** 265° F   **69.** −$16
71. −12°C   **73.** 154 ft   **75.** 69 ft   **77.** 652 ft   **79.** 144 ft   **81.** −$410 billion   **83.** 0   **85.** 436   **87.** 1058   **89.** 21
91. −22   **93.** 20   **95.** −4   **97.** 0   **99.** −9196   **101.** false   **103.** Answers may vary.   **105.** Answers may vary.

Integrated Review

1. +29,028   **2.** −35,840   **3.** −7   **4.** [number line from −4 to 4]   **5.** >   **6.** <   **7.** <   **8.** >   **9.** 1   **10.** −4
11. 8   **12.** 5   **13.** −6   **14.** 3   **15.** −89
16. 0   **17.** 5   **18.** −20   **19.** −10   **20.** −2   **21.** 52   **22.** −3   **23.** 84   **24.** 6   **25.** 1   **26.** −19   **27.** 12   **28.** −4
29. b, c, d   **30.** a, b, c, d   **31.** 10   **32.** −12   **33.** 12   **34.** 10

Exercise Set 2.4

1. 6   **3.** −36   **5.** −64   **7.** 0   **9.** −48   **11.** −8   **13.** 80   **15.** 0   **17.** −15   **19.** −4   **21.** −27   **23.** −25   **25.** −8
27. −4   **29.** −5   **31.** 8   **33.** 0   **35.** undefined   **37.** −13   **39.** 0   **41.** −12   **43.** −54   **45.** 42   **47.** −24   **49.** 16
51. −2   **53.** −7   **55.** −4   **57.** 48   **59.** −1080   **61.** 0   **63.** −5   **65.** −6   **67.** 3   **69.** −1   **71.** −243   **73.** 180
75. 1   **77.** −20   **79.** −966   **81.** −2050   **83.** −28   **85.** −6   **87.** 25   **89.** −1   **91.** undefined   **93.** 6   **95.** 8; 2
97. 0; 0   **99.** −210°C   **101.** −189°C   **103.** $(-4)(3) = -12$; a loss of 12 yd   **105.** $(-20)(5) = -100$; a depth of 100 ft
107. −$7260 million   **109. a.** 165 condors   **b.** 11 condors per year.   **111.** −1   **113.** 225   **115.** 109   **117.** 8   **119.** true
121. true   **123.** positive   **125.** negative   **127.** $(-5)^{17}, (-2)^{17}, (-2)^{12}, (-5)^{12}$ **129. a.** −59 radio stations
b. −236 radio stations   **c.** 1895 radio stations   **131.** Answers may vary.   **133.** 256,401

Calculator Explorations

1. 48   **3.** −258

Mental Math

1. base: 3; exponent: 2   **3.** base: 2; exponent: 3   **5.** base: −7; exponent: 5   **7.** base: 5; exponent: 7

Exercise Set 2.5

1. −64   **3.** −64   **5.** 24   **7.** 3   **9.** −7   **11.** −14   **13.** −43   **15.** −8   **17.** −13   **19.** −1   **21.** 4   **23.** −3
25. −55   **27.** −8   **29.** 16   **31.** 13   **33.** −65   **35.** 64   **37.** 452   **39.** 129   **41.** 3   **43.** −4   **45.** 4   **47.** 16
49. −27   **51.** 34   **53.** 65   **55.** −59   **57.** −7   **59.** −61   **61.** −11   **63.** 36   **65.** −117   **67.** 30   **69.** −3
71. −59   **73.** 1   **75.** −12   **77.** 0   **79.** −20   **81.** 9   **83.** −16   **85.** 1   **87.** −50   **89.** −2   **91.** −19
93. 28 points   **95.** 2 points   **97.** No; answers may vary.   **99.** 4050   **101.** 45   **103.** 32 in.   **105.** 30 ft
107. $2 \cdot (7 - 5) \cdot 3$ **109.** $-6 \cdot (10 - 4)$ **111.** 20,736   **113.** 8900   **115.** 9   **117.** Answers may vary.

Chapter 2 Review

1. −1435   **2.** +7562   **3.** [number line from −7 to 5]   **4.** [number line from −7 to 7]
5. 12   **6.** 0   **7.** −6   **8.** 9   **9.** −9   **10.** 2   **11.** >   **12.** <   **13.** >   **14.** >   **15.** 12   **16.** −3   **17.** false
18. true   **19.** true   **20.** true   **21.** 2   **22.** 14   **23.** 4   **24.** 17   **25.** −23   **26.** −22   **27.** −21   **28.** −70
29. 0   **30.** 0   **31.** −151   **32.** −606   **33.** −21   **34.** 0   **35.** −20°C   **36.** 150 ft below the surface   **37.** 99° F
38. 112° F   **39.** 8   **40.** −16   **41.** −11   **42.** −27   **43.** 20   **44.** 8   **45.** 0   **46.** −32   **47.** 0   **48.** −7
49. −10   **50.** −9   **51.** −25   **52.** 692 ft   **53.** −3°F   **54.** −4°F   **55.** true   **56.** false   **57.** true   **58.** true
59. 21   **60.** −18   **61.** −64   **62.** 60   **63.** 25   **64.** −1   **65.** 0   **66.** 24   **67.** −5   **68.** 3   **69.** 0
70. undefined   **71.** −20   **72.** −9   **73.** 38   **74.** −5   **75.** $(-5)(2) = -10$ **76.** $(-50)(4) = -200$ **77.** −18°F
78. −3°F   **79.** 28° F   **80.** −26°F   **81.** 49   **82.** −49   **83.** −32   **84.** −32   **85.** 0   **86.** −8   **87.** −16   **88.** 35
89. −28   **90.** −44   **91.** 3   **92.** −1   **93.** 7   **94.** −17   **95.** 39   **96.** −26   **97.** 7   **98.** −80   **99.** −2
100. −12   **101.** −3   **102.** −35   **103.** −5   **104.** 5   **105.** −1   **106.** −7   **107.** 4   **108.** −4   **109.** 3   **110.** 108

Chapter 2 Test

1. 3   **2.** −6   **3.** −100   **4.** 4   **5.** −30   **6.** 12   **7.** 65   **8.** 5   **9.** 12   **10.** −6   **11.** 50   **12.** −2   **13.** −11
14. −46   **15.** −117   **16.** 3456   **17.** 28   **18.** −213   **19.** −1   **20.** −2   **21.** 2   **22.** −5   **23.** −32   **24.** −12
25. −3   **26.** 5   **27.** −1   **28.** −54   **29.** 1   **30.** −17   **31.** 88 ft below sea level   **32.** 45   **33.** 31,642
34. 3820 ft below sea level   **35.** −4

CUMULATIVE REVIEW

1. ten-thousands; Sec. 1.2, Ex. 1   **2.** hundreds   **3.** tens; Sec. 1.2, Ex. 2   **4.** thousands   **5.** millions; Sec. 1.2, Ex. 3
6. hundred-thousands   **7. a.** <   **b.** >   **c.** >; Sec. 1.2, Ex. 12   **8. a.** >   **b.** <   **c.** >   **9.** 39; Sec. 1.3, Ex. 3
10. 39   **11.** 7321; Sec. 1.4, Ex. 2   **12.** 3013   **13.** 2440 km; Sec. 1.4, Ex. 5   **14.** $525   **15.** 570; Sec. 1.5, Ex. 1   **16.** 600
17. 1800; Sec. 1.5, Ex. 5   **18.** 5000   **19. a.** $3 \cdot 4 + 3 \cdot 5$ **b.** $10 \cdot 6 + 10 \cdot 8$ **c.** $2 \cdot 7 + 2 \cdot 3$; Sec. 1.6, Ex. 2
20. a. $5 \cdot 2 + 5 \cdot 12$ **b.** $9 \cdot 3 + 9 \cdot 6$ **b.** $4 \cdot 8 + 4 \cdot 1$ **21.** 78,875; Sec. 1.6, Ex. 5   **22.** 31,096   **23. a.** 6   **b.** 9   **c.** 6;
Sec. 1.7, Ex. 1   **24. a.** 7   **b.** 8   **c.** 12   **25.** 741; Sec. 1.7, Ex. 4   **26.** 456   **27.** 7 boxes; Sec. 1.7, Ex. 11   **28.** $9
29. 64; Sec. 1.8, Ex. 5   **30.** 125   **31.** 7; Sec. 1.8, Ex. 6   **32.** 4   **33.** 180; Sec. 1.8, Ex. 8   **34.** 56   **35.** 2; Sec. 1.8, Ex. 13
36. 5   **37.** 15; Sec. 1.9, Ex. 1   **38.** 14   **39. a.** 2   **b.** 5   **c.** 0; Sec. 2.1, Ex. 4   **40. a.** 4   **b.** 7   **41.** 3; Sec. 2.2, Ex. 7
42. 5   **43.** 14; Sec. 2.3, Ex. 12   **44.** 5   **45.** −21; Sec. 2.4, Ex. 1   **46.** −10   **47.** 0; Sec. 2.4, Ex. 3   **48.** −54   **49.** −16;
Sec. 2.5, Ex. 8   **50.** −27

Chapter 3   SOLVING EQUATIONS AND PROBLEM SOLVING

CHAPTER 3 PRETEST

1. $15x + 4$; 3.1A   **2.** $5x + 5$; 3.1C   **3.** $56b$; 3.1B   **4.** $-10y + 35$; 3.1B   **5.** $(13a + 17)$ in.; 3.1D   **6.** $(12x - 3)$ sq ft; 3.1D
7. not a solution; 3.2A   **8.** −12; 3.2B   **9.** −9; 3.2B   **10.** −7; 3.3A   **11.** 11; 3.3A   **12.** $2(x - 12)$; 3.3B   **13.** $3x$; 3.3B

14. 9; 3.4A   **15.** 2; 3.4A   **16.** 1; 3.4B   **17.** −3; 3.4B   **18.** $\dfrac{54}{-6} = -9$; 3.4C   **19.** $x - 5 = 12$; 3.5A   **20.** 7; 3.5B

MENTAL MATH

1. 5   **3.** 1   **5.** 11

EXERCISE SET 3.1

1. $8x$ **3.** $-4n$ **5.** $-2c$ **7.** $-4x$ **9.** $13a - 8$ **11.** $30x$ **13.** $-22y$ **15.** $72a$ **17.** $2y + 4$ **19.** $5a - 40$
21. $-12x - 28$ **23.** $2x + 15$ **25.** $-21n + 20$ **27.** $15c + 3$ **29.** $7w + 15$ **31.** $11x - 8$ **33.** $-5x - 9$ **35.** $-2y$
37. $-7z$ **39.** $8d - 3c$ **41.** $6y - 14$ **43.** $-q$ **45.** $2x + 22$ **47.** $-3x - 35$ **49.** $-3z - 15$ **51.** $-6x + 6$
53. $3x - 30$ **55.** $-r + 8$ **57.** $-7n + 3$ **59.** $9z - 14$ **61.** −6   **63.** $-4x + 10$ **65.** $2xy + 20$ **67.** $7a + 12$
69. $3y + 5$ **71.** $(-25x + 55)$ in.   **73.** $(14y + 22)$ m   **75.** $(11a + 12)$ ft   **77.** $16z^2$ sq cm   **79.** $(100x - 140)$ sq in.
81. −3   **83.** 8   **85.** 0   **87.** $4824q + 12{,}274$ **89.** Answers may vary.   **91.** $(20x + 16)$ sq mi   **93.** Answers may vary.

EXERCISE SET 3.2

1. yes   **3.** yes   **5.** yes   **7.** yes   **9.** yes   **11.** yes   **13.** 18   **15.** 26   **17.** 9   **19.** −16   **21.** 11   **23.** −4   **25.** 6
27. 0   **29.** 8   **31.** −1   **33.** −6   **35.** 24   **37.** 2   **39.** 0   **41.** −28   **43.** 73   **45.** 0   **47.** −4   **49.** −65   **51.** 13
53. −3   **55.** −22   **57.** −16   **59.** 3   **61.** 28   **63.** 2430   **65.** about 2200   **67.** 1   **69.** 1   **71.** Answers may vary.
73. No; answers may vary.   **75.** 162,964   **77.** −131   **79.** 42   **81.** 3627 yd   **83.** $219,812 million

EXERCISE SET 3.3

1. 4   **3.** −4   **5.** 7   **7.** −81   **9.** 0   **11.** −17   **13.** 32   **15.** −60   **17.** 5   **19.** 0   **21.** −4   **23.** 8   **25.** −1
27. 18   **29.** −8   **31.** −50   **33.** −1   **35.** −7   **37.** 0   **39.** 6   **41.** −9   **43.** 9   **45.** 4   **47.** −6   **49.** −35   **51.** 270
53. −25   **55.** −2   **57.** −2   **59.** 0   **61.** 4   **63.** −28   **65.** 3   **67.** 36   **69.** 1   **71.** $-7 + x$ **73.** $-11x$

75. $\dfrac{x}{-12}$ or $-\dfrac{x}{12}$ **77.** $x - 11$ **79.** $-10 - 7x$ **81.** $-13x$ **83.** $\dfrac{17}{x} + (-15)$ **85.** $4x + 7$ **87.** $2x - 17$

89. $-6(x + 15)$ **91.** $\dfrac{45}{-5x}$ **93.** −5   **95.** −5   **97.** 11   **99.** Answers may vary.   **101.** −36   **103.** 67,896   **105.** −48

107. 12 hr   **109.** 58 mph

INTEGRATED REVIEW

1. $8x$ **2.** $-4y$ **3.** $-2a - 2$ **4.** $5a - 26$ **5.** $-8x - 14$ **6.** $-6x + 30$ **7.** $5y - 10$ **8.** $15x - 31$
9. $(12x - 6)$ sq m   **10.** $(2x + 9)$ ft   **11.** 13   **12.** −9   **13.** 5   **14.** 0   **15.** 39   **16.** 55   **17.** −4   **18.** −3   **19.** 5
20. 8   **21.** −10   **22.** 6   **23.** −24   **24.** −54   **25.** 12   **26.** −42   **27.** 26   **28.** −12   **29.** −3   **30.** 5

31. $x - 10$ **32.** $-20 + x$ **33.** $10x$ **34.** $\dfrac{10}{x}$ **35.** $-2x + 5$ **36.** $-4(x - 1)$

CALCULATOR EXPLORATIONS

1. yes **3.** no **5.** yes

EXERCISE SET 3.4

1. 3 **3.** -50 **5.** -4 **7.** -3 **9.** -12 **11.** 5 **13.** -5 **15.** 8 **17.** 5 **19.** 1 **21.** -16 **23.** 2 **25.** 5
27. -3 **29.** 2 **31.** -2 **33.** 3 **35.** 48 **37.** -10 **39.** -4 **41.** -1 **43.** 4 **45.** -4 **47.** 3 **49.** -1
51. 0 **53.** -22 **55.** 4 **57.** 1 **59.** -30 **61.** $-42 + 16 = -26$ **63.** $-5(-29) = 145$ **65.** $3(-14 - 2) = -48$
67. $\dfrac{100}{2(50)} = 1$ **69.** 2002 **71.** 38 million returns **73.** 33 **75.** -37 **77.** 49 **79.** -6 **81.** 0 **83.** -4
85. No; answers may vary.

EXERCISE SET 3.5

1. $-5 + x = -7$ **3.** $3x = 27$ **5.** $-20 - x = 104$ **7.** $2[x + (-1)] = 50$ **9.** 8 **11.** 6 **13.** 9 **15.** 6 **17.** 24
19. 8 **21.** 5 **23.** 5 **25.** 12 **27.** 5 **29.** Bush: 266 votes; Gore: 261 votes **31.** bamboo: 36 in.; kelp: 18 in.
33. India: 8407; U.S.: 5758 **35.** Gamecube: $150; games: $450 **37.** Michigan Stadium, 107,501; Neyland Stadium, 102,854
39. California, 309 thousand; Washington, 103 thousand **41.** 82 points **43.** crow: 9 oz; finch: 4 oz **45.** 590 **47.** 1000
49. 3000 **51.** Answers may vary. **53.** $8250 **55.** $5 **57.** Answers may vary.

CHAPTER 3 REVIEW

1. $10y - 15$ **2.** $-6y - 10$ **3.** $-6a - 7$ **4.** $-8y + 2$ **5.** $2x + 10$ **6.** $-3y - 24$ **7.** $11x - 12$ **8.** $-4m - 12$
9. $-5a + 4$ **10.** $12y - 9$ **11.** $(4x + 6)$ yd **12.** $20y$ m **13.** $(6x - 3)$ sq yd **14.** $(45x + 8)$ sq cm
15. $600,822x - 9180$ **16.** $3292y - 3840$ **17.** yes **18.** no **19.** -2 **20.** 10 **21.** -6 **22.** -1 **23.** -25
24. -8 **25.** -7 **26.** -9 **27.** 1 **28.** 5 **29.** -12 **30.** 45 **31.** 0 **32.** -128 **33.** -8 **34.** -45 **35.** 5
36. -5 **37.** $-5x$ **38.** $x - 3$ **39.** $-5 + x$ **40.** $\dfrac{-2}{x}$ **41.** $2x + 11$ **42.** $-5x - 50$ **43.** $\dfrac{70}{x + 6}$ **44.** $2(x - 13)$
45. 5 **46.** 12 **47.** 17 **48.** 7 **49.** -2 **50.** 8 **51.** 21 **52.** -10 **53.** 2 **54.** 2 **55.** -27 **56.** -22
57. 11 **58.** -5 **59.** $20 - (-8) = 28$ **60.** $5[2 + (-6)] = -20$ **61.** $\dfrac{-75}{5 + 20} = -3$ **62.** $-2 - 19 = -21$
63. $2x - 8 = 40$ **64.** $\dfrac{x}{2} - 12 = 10$ **65.** $x - 3 = \dfrac{x}{4}$ **66.** $6x = x + 2$ **67.** 5 **68.** -16 **69.** 2386 votes **70.** 84 DVDs

CHAPTER 3 TEST

1. $-5x + 5$ **2.** $-6y - 14$ **3.** $-8z - 20$ **4.** $(15x + 15)$ in. **5.** $(12x - 4)$ sq m **6.** -10 **7.** -16 **8.** -6 **9.** -6
10. 8 **11.** 24 **12.** -2 **13.** 6 **14.** -2 **15.** 0 **16. a.** $-23 + x$ **b.** $-2 - 3x$ **17. a.** $2 \cdot 5 + (-15) = -5$
b. $3x + 6 = -30$ **18.** -2 **19.** 8 free throws **20.** 244 women

CUMULATIVE REVIEW

1. one hundred six million, fifty-two thousand, four hundred forty-seven; Sec. 1.2, Ex. 6 **2.** two hundred seventy-six thousand, four
3. 13 in.; Sec. 1.3, Ex. 5 **4.** 18 in. **5.** 726; Sec. 1.4, Ex. 4 **6.** 9585 **7.** 249,000; Sec. 1.5, Ex. 3 **8.** 844,000
9. 200; Sec. 1.6, Ex. 3 **10.** 29,230 **11.** 208; Sec. 1.7, Ex. 5 **12.** 86 **13.** 7; Sec. 1.8, Ex. 9 **14.** 35 **15.** 26; Sec. 1.9, Ex. 4
16. 10 **17. a.** $<$ **b.** $>$ **c.** $>$; Sec. 2.1, Ex. 3 **18. a.** $<$ **b.** $>$ **19.** 3; Sec. 2.2, Ex. 1 **20.** -7 **21.** -6; Sec. 2.2, Ex. 5
22. -4 **23.** 8; Sec. 2.2, Ex. 6 **24.** 17 **25.** -14; Sec. 2.3, Ex. 2 **26.** -5 **27.** 11; Sec. 2.3, Ex. 3 **28.** 29
29. -4; Sec. 2.3, Ex. 4 **30.** -3 **31.** -2; Sec. 2.4, Ex. 10 **32.** 6 **33.** 5; Sec. 2.4, Ex. 11 **34.** -13 **35.** -16; Sec. 2.4, Ex. 12
36. -10 **37.** 9; Sec. 2.5, Ex. 1 **38.** -32 **39.** -9; Sec. 2.5, Ex. 2 **40.** 25 **41.** $6y + 2$; Sec. 3.1, Ex. 2 **42.** $3x + 9$
43. not a solution; Sec. 3.2, Ex. 2 **44.** solution **45.** 3; Sec. 3.3, Ex. 4 **46.** 5 **47.** 12; Sec. 3.4, Ex. 1 **48.** -2
49. software, $420; computer system, $1680; Sec. 3.5, Ex. 4 **50.** 11

Chapter 4 FRACTIONS

CHAPTER 4 PRETEST

1. $\dfrac{3}{8}$; 4.1 B **2.** $\dfrac{15}{18}$; 4.1 E **3.** 0; 4.1 D **4.** $2 \cdot 2 \cdot 5 \cdot 7$ or $2^2 \cdot 5 \cdot 7$; 4.2A **5.** $\dfrac{5}{9}$; 4.2B **6.** $\dfrac{9}{20}$ of a gram; 4.2C **7.** $\dfrac{6}{5}$; 4.3 A
8. $\dfrac{1}{14}$; 4.3 C **9.** $\dfrac{5}{11}$; 4.4 A **10.** $-\dfrac{1}{10}$; 4.4A **11.** $-\dfrac{8}{27}$; 4.3B **12.** $-\dfrac{1}{3}$; 4.3D **13.** $\dfrac{23}{36}$; 4.5A **14.** $\dfrac{1}{14}$; 4.5C
15. $\dfrac{3x}{7}$; 4.6A **16.** $-\dfrac{46}{25}$; 4.6B **17.** $-\dfrac{23}{2}$; 4.7A **18.** $\dfrac{13}{5}$; 4.8B **19.** $\dfrac{44}{5}$ or $8\dfrac{4}{5}$; 4.8D **20.** $9\dfrac{5}{6}$; 4.8E

MENTAL MATH

1. numerator $= 1$; denominator $= 2$ **3.** numerator $= 10$; denominator $= 3$ **5.** numerator $= 3z$; denominator $= 7$

EXERCISE SET 4.1

1. $\frac{1}{3}$ **3.** $\frac{4}{7}$ **5.** $\frac{7}{12}$ **7.** $\frac{3}{7}$ **9.** $\frac{7}{8}$ **11.** $\frac{5}{8}$ **13.** 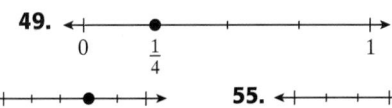   **15.**

17. ⚪⚪⚪⚪⚪⚪◯

19.   **21.** $\frac{11}{4}$ **23.** $\frac{11}{3}$

25. $\frac{3}{2}$ **27.** $\frac{4}{3}$ **29.** $\frac{17}{6}$ **31.** $\frac{14}{9}$

33. $\frac{42}{131}$ of the students   **35.** 89; $\frac{89}{131}$ of the students   **37.** $\frac{4}{10}$ of the visits   **39.** $\frac{8}{43}$ of the presidents

41. $\frac{17}{32}$ of the hard drive   **43.** $\frac{11}{31}$ of the month   **45.** $\frac{10}{31}$ of the class   **47. a.** $\frac{40}{50}$ of the states   **b.** 10 states

c. $\frac{10}{50}$ of the states   **49.**   **51.**

53.   **55.**   **57.**

59. 1   **61.** -5 **63.** 0   **65.** 1   **67.** undefined   **69.** 3   **71.** $\frac{20}{35}$ **73.** $\frac{14}{21}$ **75.** $\frac{10y}{25}$ **77.** $\frac{15}{30}$ **79.** $\frac{30}{21x}$ **81.** $\frac{10}{5}$

83. $\frac{9}{12}$ **85.** $\frac{8y}{12}$ **87.** $\frac{6}{12}$ **89.** $\frac{48x}{36x}$ **91.** $\frac{20x}{36x}$ **93.** $\frac{36x}{36x}$ **95.** $\frac{52}{100},\frac{78}{100},\frac{87}{100},\frac{84}{100},\frac{59}{100},\frac{83}{100},\frac{67}{100},\frac{60}{100},\frac{79}{100},\frac{45}{100}$

97. United States   **99.** 9   **101.** 125   **103.** 49   **105.** 24   **107.** $\frac{464}{2088}$ **109.** Answers may vary.

111. $\frac{1651}{2285}$ of the affiliates   **113.** $\frac{6253}{8851}$ of the restaurants

CALCULATOR EXPLORATIONS

1. $\frac{4}{7}$ **3.** $\frac{20}{27}$ **5.** $\frac{15}{8}$ **7.** $\frac{9}{2}$

MENTAL MATH

1. yes, yes, yes   **3.** $2 \cdot 5$ **5.** $3 \cdot 7$ **7.** 3^2

EXERCISE SET 4.2

1. $2^2 \cdot 5$ **3.** $2^4 \cdot 3$ **5.** $3^2 \cdot 5$ **7.** $2^4 \cdot 3 \cdot 5$ **9.** $\frac{1}{4}$ **11.** $\frac{x}{5}$ **13.** $\frac{7}{8}$ **15.** $\frac{4}{5}$ **17.** $\frac{5}{6}$ **19.** $\frac{5x}{6}$ **21.** $\frac{2}{3}$ **23.** $\frac{3x}{4}$

25. $\frac{3b}{2a}$ **27.** $\frac{7}{8}$ **29.** $\frac{3}{7}$ **31.** $\frac{3y}{5}$ **33.** $\frac{4}{7}$ **35.** $\frac{4x^2y}{5}$ **37.** $\frac{4}{5}$ **39.** $\frac{5x}{8}$ **41.** $\frac{3x}{10}$ **43.** $\frac{3a^2}{2b^3}$ **45.** $\frac{5}{8z}$

47. $\frac{3}{4}$ of a shift   **49.** $\frac{1}{2}$ mi   **51.** $\frac{29}{46}$ of individuals   **53.** $\frac{9}{20}$ of the students   **55. a.** $\frac{8}{25}$ of the states   **b.** 34 states

c. $\frac{17}{25}$ of the states   **57.** $\frac{5}{12}$ of the width   **59.** $\frac{1}{10}$ **61.** Answers may vary.   **63.** -3 **65.** -14 **67.** 0   **69.** 29

71. d   **73.** false   **75.** true   **77.** $\frac{3}{5}$ **79.** $\frac{9}{25}$ of the donors   **81.** $\frac{1}{25}$ of the donors   **83.** $\frac{3}{20}$ of the donors

85. $2235, 105, 900, 1470$ **87.** 15; Answers may vary.

MENTAL MATH

1. $\frac{2}{15}$ **3.** $\frac{6}{35}$ **5.** $\frac{9}{8}$

EXERCISE SET 4.3

1. $\frac{7}{12}$ **3.** $-\frac{5}{28}$ **5.** $\frac{1}{15}$ **7.** $\frac{18x}{55}$ **9.** $\frac{3a^2}{4}$ **11.** $\frac{x^2}{y}$ **13.** $\frac{1}{125}$ **15.** $\frac{4}{9}$ **17.** $-\frac{4}{27}$ **19.** $\frac{4}{5}$ **21.** $-\frac{1}{6}$ **23.** $\frac{16}{9x}$

25. $\dfrac{121y}{60}$ **27.** $-\dfrac{1}{6}$ **29.** $\dfrac{x}{25}$ **31.** $\dfrac{10}{27}$ **33.** $\dfrac{18x^2}{35}$ **35.** $-\dfrac{1}{4}$ **37.** $\dfrac{3}{4}$ **39.** $\dfrac{9}{16}$ **41.** xy^2 **43.** $\dfrac{77}{2}$ **45.** $-\dfrac{36}{x}$

47. $\dfrac{3}{49}$ **49.** $-\dfrac{19y}{7}$ **51.** $\dfrac{4}{11}$ **53.** $\dfrac{8}{3}$ **55.** $\dfrac{15x}{4}$ **57.** $\dfrac{8}{9}$ **59.** $\dfrac{1}{60}$ **61.** b **63.** $\dfrac{2}{5}$ **65.** $-\dfrac{5}{3}$ **67. a.** $\dfrac{1}{3}$ **b.** $\dfrac{12}{25}$

69. a. $-\dfrac{36}{55}$ **b.** $-\dfrac{44}{45}$ **71.** yes **73.** no **75.** 30 gal **77.** \$1838 **79.** $\dfrac{3}{16}$ in. **81.** 868 mi **83.** 21,735 tornados

85. $\dfrac{17}{2}$ in. **87.** 3 ft **89.** 24 **91.** $\dfrac{1}{14}$ sq ft **93.** $2 \cdot 3 \cdot 3 \cdot 5$ or $2 \cdot 3^2 \cdot 5$ **95.** $5 \cdot 13$ **97.** $2 \cdot 3 \cdot 3 \cdot 7$ or $2 \cdot 3^2 \cdot 7$

99. 7,210,000 households **101.** Answers may vary. **103.** 15,432 species **105.** 5

MENTAL MATH

1. unlike **3.** like **5.** like **7.** unlike **9.** $\dfrac{5}{7}$ **11.** $\dfrac{6}{11}$ **13.** $\dfrac{7}{11}$ **15.** $\dfrac{8}{15}$

EXERCISE SET 4.4

1. 0 **3.** $\dfrac{2}{3x}$ **5.** $-\dfrac{1}{13}$ **7.** $\dfrac{2}{3}$ **9.** $-\dfrac{3}{y}$ **11.** $\dfrac{7a-3}{4}$ **13.** $-\dfrac{3}{4}$ **15.** $-\dfrac{1}{3}$ **17.** $\dfrac{2x}{3}$ **19.** $-\dfrac{x}{2}$ **21.** $\dfrac{3}{4z}$ **23.** $-\dfrac{3}{10}$

25. 2 **27.** $-\dfrac{2}{3}$ **29.** $\dfrac{3x}{4}$ **31.** $\dfrac{5}{4}$ **33.** $\dfrac{2}{5}$ **35.** $-\dfrac{3}{4}$ **37.** $-\dfrac{2}{3}$ **39.** $\dfrac{1}{13}$ **41.** $\dfrac{4}{5}$ **43.** 1 in. **45.** 2 m **47.** $\dfrac{3}{2}$ hr

49. Traditional fee-for-service, Point-of-Service, Health Maintenance Organization, Preferred Provider Organization

51. $\dfrac{50}{100} = \dfrac{1}{2}$ of the employees **53.** $\dfrac{21}{50}$ of the states **55.** 12 **57.** 45 **59.** 36 **61.** $24x$ **63.** 150 **65.** 126 **67.** 75

69. 24 **71.** 50 **73.** $12a$ **75.** 168 **77.** 363 **79.** $\dfrac{12}{35}$ **81.** 8 **83.** $\dfrac{4}{5}$ **85.** -28 **87.** $\dfrac{8}{11}$ **89.** $\dfrac{1}{10}$ of men

91. Answers may vary.

CALCULATOR EXPLORATIONS

1. $\dfrac{37}{80}$ **3.** $\dfrac{95}{72}$ **5.** $\dfrac{394}{323}$

MENTAL MATH

1. 6 **3.** 12 **5.** 56 **7.** 12

EXERCISE SET 4.5

1. $\dfrac{5}{6}$ **3.** $\dfrac{1}{6}$ **5.** $-\dfrac{4}{33}$ **7.** $\dfrac{3x-6}{14}$ **9.** $\dfrac{3}{5}$ **11.** $\dfrac{24y-5}{12}$ **13.** $\dfrac{11}{36}$ **15.** $\dfrac{12}{7}$ **17.** $\dfrac{89a}{99}$ **19.** $\dfrac{4y-1}{6}$ **21.** $\dfrac{x+6}{2x}$

23. $-\dfrac{8}{33}$ **25.** $\dfrac{3}{14}$ **27.** $\dfrac{11y-10}{35}$ **29.** $-\dfrac{11}{36}$ **31.** $\dfrac{1}{20}$ **33.** $\dfrac{33}{56}$ **35.** $\dfrac{17}{16}$ **37.** $\dfrac{8}{9}$ **39.** $\dfrac{15+11y}{33}$ **41.** $-\dfrac{53}{42}$

43. $\dfrac{11}{18}$ **45.** $\dfrac{44a}{39}$ **47.** $-\dfrac{11}{60}$ **49.** $\dfrac{5y+9}{9y}$ **51.** $\dfrac{56}{45}$ **53.** $\dfrac{40+9x}{72x}$ **55.** $-\dfrac{11}{30}$ **57.** $\dfrac{7x}{8}$ **59.** $\dfrac{19}{20}$ **61.** $-\dfrac{5}{24}$

63. $\dfrac{37x-20}{56}$ **65.** < **67.** > **69.** > **71.** $\dfrac{13}{12}$ **73.** $\dfrac{1}{4}$ **75.** $\dfrac{11}{6}$ **77.** $\dfrac{11}{12}$ **79.** $-\dfrac{7}{10}$ **81.** $\dfrac{11}{16}$ **83.** $\dfrac{11}{18}$

85. $\dfrac{34}{15}$ cm **87.** $\dfrac{17}{10}$ m **89.** $\dfrac{61}{264}$ mi **91.** $\dfrac{11}{8}$ in. **93.** $\dfrac{49}{100}$ of students **95.** $\dfrac{77}{100}$ of Americans **97.** $\dfrac{29}{100}$ of adults

99. $\dfrac{25}{36}$ **101.** $\dfrac{25}{36}$ **103.** 57,200 **105.** 330 **107.** $\dfrac{49}{44}$ **109.** Answers may vary. **111.** $\dfrac{938}{351}$ **113.** $\dfrac{163}{579}$ of the land area

115. 16,300,000 sq mi **117.** $\dfrac{175}{193}$ of land area **119.** national preserves **121.** 1-2 videos per mo

INTEGRATED REVIEW

1. $\dfrac{1}{4}$ **2.** $\dfrac{3}{6}$ or $\dfrac{1}{2}$ **3.** $\dfrac{7}{4}$ **4.** $\dfrac{73}{85}$ **5.** $2 \cdot 3$ **6.** $2 \cdot 5 \cdot 7$ **7.** $2^2 \cdot 3^2 \cdot 7$ **8.** $\dfrac{1}{7}$ **9.** $\dfrac{5}{6}$ **10.** $\dfrac{9}{19}$ **11.** $\dfrac{21}{55}$ **12.** $\dfrac{1}{2}$

13. $\dfrac{9}{10}$ **14.** $\dfrac{1}{25}$ **15.** $\dfrac{4}{5}$ **16.** $-\dfrac{2}{5}$ **17.** $\dfrac{3}{25}$ **18.** $\dfrac{1}{3}$ **19.** $\dfrac{4}{5}$ **20.** $\dfrac{5}{9}$ **21.** $-\dfrac{1}{6}$ **22.** $\dfrac{3}{2}$ **23.** $\dfrac{1}{18}$ **24.** $\dfrac{4}{21}$

25. $-\dfrac{7}{48}$ **26.** $-\dfrac{9}{50}$ **27.** $\dfrac{37}{40}$ **28.** $\dfrac{11}{36}$ **29.** $\dfrac{5}{33}$ **30.** $\dfrac{5}{18}$ **31.** $\dfrac{11}{18}$ **32.** $\dfrac{37}{50}$ **33.** 24 lots **34.** $\dfrac{3}{4}$ ft

EXERCISE SET 4.6

1. $\frac{1}{6}$ **3.** $\frac{3}{7}$ **5.** $\frac{x}{6}$ **7.** $\frac{23}{22}$ **9.** $\frac{2x}{13}$ **11.** $\frac{17}{60}$ **13.** $\frac{5}{8}$ **15.** $\frac{35}{9}$ **17.** $-\frac{17}{45}$ **19.** $\frac{11}{8}$ **21.** $\frac{29}{10}$ **23.** $\frac{27}{32}$ **25.** $\frac{1}{81}$

27. $\frac{9}{64}$ **29.** $\frac{7}{6}$ **31.** $-\frac{2}{5}$ **33.** $-\frac{2}{9}$ **35.** $\frac{5}{2}$ **37.** $\frac{7}{2}$ **39.** $\frac{4}{9}$ **41.** $-\frac{13}{2}$ **43.** $\frac{9}{25}$ **45.** $-\frac{5}{32}$ **47.** 1   **49.** $\frac{1}{10}$

51. $-\frac{11}{40}$ **53.** $\frac{x+6}{16}$ **55.** 8   **57.** 25   **59.** $1x$ or x **61.** $1a$ or a **63.** No; answers may vary.   **65.** $-\frac{77}{16}$ **67.** $-\frac{55}{16}$

69. $\frac{5}{8}$ **71.** $\frac{11}{56}$ **73.** halfway between a and b **75.** false   **77.** false   **79.** true   **81.** No; answers may vary.

EXERCISE SET 4.7

1. $\frac{2}{7}$ **3.** 12   **5.** -27 **7.** $\frac{27}{8}$ **9.** $\frac{1}{21}$ **11.** $\frac{2}{11}$ **13.** 1   **15.** $\frac{15}{2}$ **17.** -1 **19.** -15 **21.** $\frac{3x-28}{21}$ **23.** $\frac{y+10}{2}$

25. $\frac{7x}{15}$ **27.** $\frac{4}{3}$ **29.** 2   **31.** $\frac{21}{10}$ **33.** $-\frac{1}{14}$ **35.** -3 **37.** 50   **39.** $-\frac{1}{9}$ **41.** -6 **43.** 4   **45.** $\frac{3}{5}$ **47.** $-\frac{1}{24}$

49. $-\frac{5}{14}$ **51.** 4   **53.** -36 **55.** $\frac{7}{2}$ **57.** $\frac{59}{10}$ **59.** $\frac{49}{6}$ **61.** Answers may vary.   **63.** -2

CALCULATOR EXPLORATIONS

1. $\frac{280}{11}$ **3.** $\frac{3776}{35}$ **5.** $26\frac{1}{14}$ **7.** $92\frac{3}{10}$

EXERCISE SET 4.8

1. a. $\frac{11}{4}$ **b.** $2\frac{3}{4}$ **3. a.** $\frac{11}{3}$ **b.** $3\frac{2}{3}$ **5. a.** $\frac{3}{2}$ **b.** $1\frac{1}{2}$ **7. a.** $\frac{4}{3}$ **b.** $1\frac{1}{3}$ **9.** $\frac{7}{3}$ **11.** $\frac{27}{8}$ **13.** $\frac{83}{7}$ **15.** $1\frac{6}{7}$

17. $3\frac{2}{15}$ **19.** $4\frac{5}{8}$ **21.** $\frac{8}{21}$ **23.** $4\frac{2}{3}$ **25.** $5\frac{1}{2}$ **27.** $18\frac{2}{3}$ **29.** $6\frac{4}{5}$ **31.** $25\frac{5}{14}$ **33.** $13\frac{13}{24}$ **35.** $2\frac{3}{5}$ **37.** $7\frac{5}{14}$

39. $\frac{24}{25}$ **41.** 4   **43.** $5\frac{11}{14}$ **45.** $6\frac{2}{9}$ **47.** $\frac{25}{33}$ **49.** $35\frac{13}{18}$ **51.** $2\frac{1}{2}$ **53.** $72\frac{19}{30}$ **55.** $\frac{11}{14}$ **57.** $5\frac{4}{7}$ **59.** $13\frac{16}{33}$

61. $10\frac{43}{60}$ min   **63.** $\frac{49}{60}$ min   **65.** $\frac{39}{2}$ or $19\frac{1}{2}$ in.   **67.** 15 shares   **69.** $\frac{1}{16}$ in.   **71.** No, she will be $\frac{1}{12}$ of a foot short.

73. $7\frac{5}{6}$ gal   **75.** $9\frac{2}{5}$ in.   **77.** $7\frac{13}{20}$ in.   **79.** $\frac{7}{2}$ or $3\frac{1}{2}$ sq yd   **81.** $\frac{15}{16}$ sq in.   **83.** 7 mi   **85.** $21\frac{5}{24}$ m   **87.** $1\frac{1}{2}$ yr

89. $306\frac{2}{3}$ ft   **91.** $\frac{19}{30}$ in.   **93.** $-10\frac{3}{25}$ **95.** $-24\frac{7}{8}$ **97.** $-13\frac{59}{60}$ **99.** $-\frac{1}{2}$ **101.** $-1\frac{23}{24}$ **103.** $\frac{73}{1000}$ **105.** $8\frac{17}{24}$

107. $-32\frac{1}{6}$ **109.** $9x-13$ **111.** $-3y-6$ **113.** 17   **115.** 1   **117.** Supreme is heavier by $\frac{1}{8}$ lb   **119.** Answers may vary.

121. Answers may vary.   **123.** Answers may vary.

CHAPTER 4 REVIEW

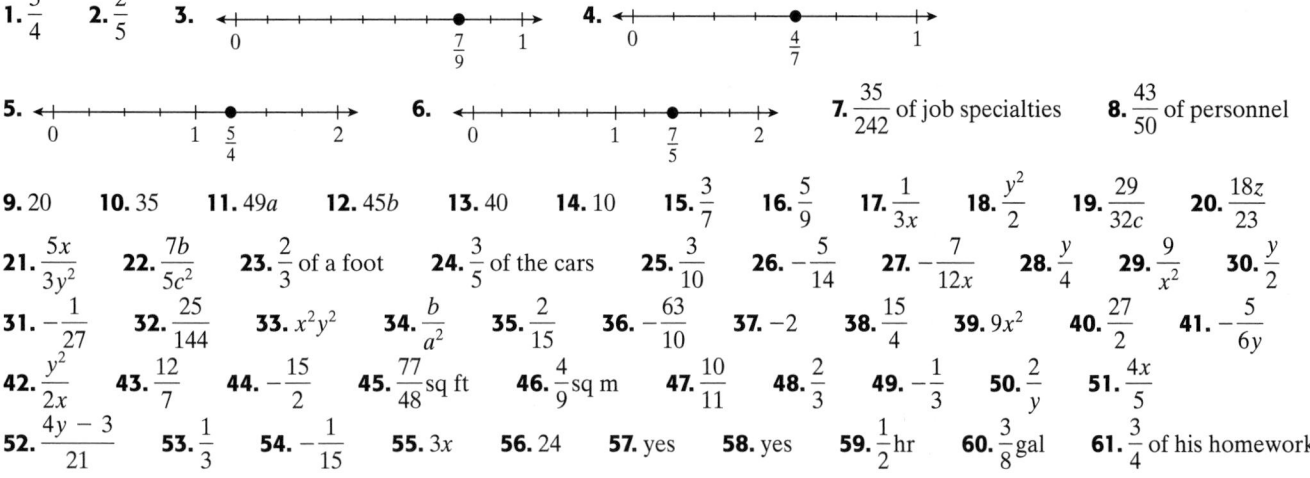

1. $\frac{3}{4}$ **2.** $\frac{2}{5}$ **3.**   **4.**

5.   **6.**   **7.** $\frac{35}{242}$ of job specialties   **8.** $\frac{43}{50}$ of personnel

9. 20   **10.** 35   **11.** $49a$ **12.** $45b$ **13.** 40   **14.** 10   **15.** $\frac{3}{7}$ **16.** $\frac{5}{9}$ **17.** $\frac{1}{3x}$ **18.** $\frac{y^2}{2}$ **19.** $\frac{29}{32c}$ **20.** $\frac{18z}{23}$

21. $\frac{5x}{3y^2}$ **22.** $\frac{7b}{5c^2}$ **23.** $\frac{2}{3}$ of a foot   **24.** $\frac{3}{5}$ of the cars   **25.** $\frac{3}{10}$ **26.** $-\frac{5}{14}$ **27.** $-\frac{7}{12x}$ **28.** $\frac{y}{4}$ **29.** $\frac{9}{x^2}$ **30.** $\frac{y}{2}$

31. $-\frac{1}{27}$ **32.** $\frac{25}{144}$ **33.** x^2y^2 **34.** $\frac{b}{a^2}$ **35.** $\frac{2}{15}$ **36.** $-\frac{63}{10}$ **37.** -2 **38.** $\frac{15}{4}$ **39.** $9x^2$ **40.** $\frac{27}{2}$ **41.** $-\frac{5}{6y}$

42. $\frac{y^2}{2x}$ **43.** $\frac{12}{7}$ **44.** $-\frac{15}{2}$ **45.** $\frac{77}{48}$ sq ft   **46.** $\frac{4}{9}$ sq m   **47.** $\frac{10}{11}$ **48.** $\frac{2}{3}$ **49.** $-\frac{1}{3}$ **50.** $\frac{2}{y}$ **51.** $\frac{4x}{5}$

52. $\frac{4y-3}{21}$ **53.** $\frac{1}{3}$ **54.** $-\frac{1}{15}$ **55.** $3x$ **56.** 24   **57.** yes   **58.** yes   **59.** $\frac{1}{2}$ hr   **60.** $\frac{3}{8}$ gal   **61.** $\frac{3}{4}$ of his homework

62. $\frac{3}{2}$ mi **63.** $\frac{11}{18}$ **64.** $\frac{7}{26}$ **65.** $-\frac{1}{12}$ **66.** $-\frac{5}{12}$ **67.** $\frac{25x+2}{55}$ **68.** $\frac{4+3b}{15}$ **69.** $\frac{7y}{36}$ **70.** $\frac{11x}{18}$ **71.** $\frac{4y+45}{9y}$

72. $-\frac{15}{14}$ **73.** $\frac{91}{150}$ **74.** $\frac{5}{18}$ **75.** $\frac{5}{6}$ **76.** $-\frac{9}{10}$ **77.** $\frac{19}{9}$ m **78.** $\frac{3}{2}$ ft **79.** no **80.** yes **81.** $\frac{21}{50}$ of the donors

82. $\frac{1}{4}$ yd **83.** $\frac{4x}{7}$ **84.** $\frac{3y}{11}$ **85.** $\frac{22}{7}$ **86.** $\frac{17}{16}$ **87.** -2 **88.** $-7y$ **89.** $\frac{1}{3}$ **90.** $\frac{15}{4}$ **91.** $-\frac{5}{24}$ **92.** $\frac{19}{30}$ **93.** $\frac{4}{9}$

94. $-\frac{3}{10}$ **95.** -10 **96.** -6 **97.** 15 **98.** 1 **99.** 5 **100.** -4 **101.** $3\frac{3}{4}$ **102.** 3 **103.** 1 **104.** $31\frac{1}{4}$

105. $\frac{11}{5}$ **106.** $\frac{5}{1}$ **107.** $\frac{35}{9}$ **108.** $\frac{3}{1}$ **109.** $45\frac{16}{21}$ **110.** 60 **111.** $32\frac{13}{22}$ **112.** $3\frac{19}{60}$ **113.** $111\frac{5}{18}$ **114.** $20\frac{7}{24}$

115. $1\frac{1}{12}$ **116.** $1\frac{4}{11}$ **117.** 10 **118.** $12\frac{3}{4}$ **119.** $5\frac{1}{4}$ **120.** $2\frac{29}{46}$ **121.** $2\frac{1}{3}$ **122.** $6\frac{2}{5}$ **123.** $6\frac{7}{20}$ lb **124.** $44\frac{1}{2}$ yd

125. $7\frac{4}{5}$ in. **126.** $7\frac{1}{2}$ sq ft **127.** 5 ft **128.** $\frac{119}{80}$ or $1\frac{39}{80}$ sq in. **129.** $5\frac{1}{12}$ m **130.** 203 calories **131.** $13\frac{1}{3}$ g

132. $\frac{21}{20}$ or $1\frac{1}{20}$ mi **133.** $-27\frac{5}{14}$ **134.** $1\frac{5}{27}$ **135.** $-3\frac{15}{16}$ **136.** $-\frac{33}{40}$

CHAPTER 4 TEST

1. $\frac{23}{3}$ **2.** $\frac{39}{11}$ **3.** $4\frac{3}{5}$ **4.** $18\frac{3}{4}$ **5.** $\frac{9}{35}$ **6.** $-\frac{3}{5}$ **7.** $\frac{4}{3}$ **8.** $-\frac{4}{3}$ **9.** $\frac{8x}{9}$ **10.** $\frac{x-21}{7x}$ **11.** y^2 **12.** $\frac{16}{45}$

13. $\frac{9a+4}{10}$ **14.** $-\frac{2}{3y}$ **15.** $24y^2$ **16.** 9 **17.** $14\frac{1}{40}$ **18.** $\frac{1}{a^2}$ **19.** $\frac{64}{3}$ **20.** $22\frac{1}{2}$ **21.** $3\frac{3}{5}$ **22.** $\frac{1}{3}$ **23.** $\frac{3}{4}$

24. $\frac{3}{4x}$ **25.** $\frac{76}{21}$ **26.** -2 **27.** -4 **28.** 1 **29.** $\frac{5}{2}$ **30.** $\frac{4}{31}$ **31.** $\frac{1}{2}$ of the calories **32.** $3\frac{3}{4}$ ft **33.** $\frac{5}{16}$ of the sales

34. 125,000 backpacks **35.** 8800 sq yd **36.** $\frac{34}{27}$ or $1\frac{7}{27}$ sq mi **37.** 24 mi **38.** $90 per share

CUMULATIVE REVIEW

1. one hundred twenty-six; Sec. 1.2, Ex. 4 **2.** one hundred fifteen **3.** three thousand five; Sec. 1.2, Ex. 5 **4.** six thousand five hundred seventy-three **5.** 159; Sec. 1.3, Ex. 1 **6.** 631 **7.** 14; Sec. 1.4, Ex. 3 **8.** 933 **9.** 278,000; Sec. 1.5, Ex. 2 **10.** 1440 **11.** 63,420 thousand bytes; Sec. 1.6, Ex. 7 **12.** 1305 mi **13.** 7089 R 5; Sec. 1.7, Ex. 7 **14.** 379 R 10 **15.** 4^3; Sec. 1.8, Ex. 1 **16.** 7^2 **17.** $6^3 \cdot 8^5$; Sec. 1.8, Ex. 4 **18.** $9^4 \cdot 5^2$ **19.** 8; Sec. 1.9, Ex. 2 **20.** 52 **21.** -150; Sec. 2.1, Ex. 1 **22.** -21 **23.** -4; Sec. 2.2, Ex. 3 **24.** 5 **25.** 3; Sec. 2.3, Ex. 9 **26.** 10 **27.** 25; Sec. 2.4, Ex. 8 **28.** -16 **29.** -2; Sec. 2.5, Ex. 5 **30.** 25 **31.** $15y$; Sec. 3.1, Ex. 6 **32.** $36c$ **33.** $-8x$; Sec. 3.1, Ex. 7 **34.** $-98a$ **35.** -9; Sec. 3.2, Ex. 4 **36.** 8

37. 5; Sec. 3.3, Ex. 7 **38.** -1 **39. a.** $7 \cdot 6 = 42$ **b.** $2(3+5) = 16$ **c.** $\frac{-45}{5} = -9$; Sec. 3.4, Ex. 6 **40. a.** $4+3 = 7$

b. $4(5-2) = 12$ **c.** $(-4)(-6) = 24$ **41.** -9; Sec. 3.5, Ex. 2 **42.** -5 **43.** $\frac{36x}{44}$; Sec. 4.1, Ex. 19 **44.** $\frac{10a}{15}$

45. $3 \cdot 3 \cdot 5$ or $3^2 \cdot 5$; Sec. 4.2, Ex. 1 **46.** $2 \cdot 2 \cdot 23$ or $2^2 \cdot 23$ **47.** $\frac{10}{33}$; Sec. 4.3, Ex. 1 **48.** $\frac{2}{35}$ **49.** $\frac{1}{8}$; Sec. 4.3, Ex. 2 **50.** $\frac{3}{25}$

Chapter 5 DECIMALS

CHAPTER 5 PRETEST

1. $\frac{27}{100}$; 5.1C **2.** $<$; 5.1D **3.** 54.7; 5.1E **4.** 60.042; 5.2A **5.** 7.14; 5.3A **6.** 201.6; 5.3B **7.** 40.6; 5.4A **8.** 0.0891; 5.4B **9.** 11.69; 5.2B **10.** $-17.7 - 4.8x$; 5.2C **11.** 0.126; 5.3C **12.** 12π in. ≈ 37.68 in.; 5.3D **13.** $28.80; 5.4D **14.** 0.778; 5.5B **15.** 0.375; 5.6A **16.** $<$; 5.6B **17.** -1.07; 5.7A **18.** $\frac{6}{7}$; 5.8A **19.** 6.78; 5.8B **20.** 15 in.; 5.8C

MENTAL MATH

1. tens **3.** tenths

EXERCISE SET 5.1

1. six and fifty-two hundredths **3.** sixteen and twenty-three hundredths **5.** negative three and two hundred five thousandths

7. one hundred sixty-seven and nine thousandths

9.

Preprinted Name	Current date
Preprinted Address	DATE

PAY TO THE
ORDER OF _R.W. Financial_ $ | 321.42 |

Three hundred twenty-one and 42/100   DOLLARS

FOR _____   _Signature_

11.

Preprinted Name	Current date
Preprinted Address	DATE

PAY TO THE
ORDER OF _Bell South_ $ | 59.68 |

Fifty-nine and 68/100   DOLLARS

FOR _____   _Signature_

13. 6.5   **15.** 9.08   **17.** -5.625 **19.** 0.0064   **21.** 64.16   **23.** 5.4   **25.** $\frac{3}{10}$ **27.** $\frac{27}{100}$ **29.** $-5\frac{47}{100}$ **31.** $\frac{6}{125}$
33. $7\frac{7}{100}$ **35.** $15\frac{401}{500}$ **37.** $\frac{601}{2000}$ **39.** $487\frac{8}{25}$ **41.** $<$ **43.** $<$ **45.** $<$ **47.** $=$ **49.** $<$ **51.** $>$ **53.** $>$
55. $<$ **57.** $>$ **59.** 0.6   **61.** 0.23   **63.** 0.594   **65.** 98,210   **67.** 13.0   **69.** -17.67 **71.** -0.5 **73.** $0.07
75. $27   **77.** $0.20   **79.** 0.26499; 0.25786   **81.** $16   **83.** 2.35 hr   **85.** 24.623 hr   **87.** 136 mph   **89.** 31 points
91. 0.10299, 0.1037, 0.1038, 0.9   **93.** 5766   **95.** 71   **97.** 243   **99.** 228.040; Parker Bohn   **101.** 228.040; 226.130; 225.490;
225.370; 224.940; 222.008; 221.546; 220.930   **103.** Answers may vary.   **105.** two hundred three ten-millionths

CALCULATOR EXPLORATIONS

1. 328.742   **3.** 5.2414   **5.** 865.392

MENTAL MATH

1. 0.5   **3.** 1.26   **5.** 8.9   **7.** 0.6

EXERCISE SET 5.2

1. 3.5   **3.** 6.83   **5.** 27.0578   **7.** 56.432   **9.** -8.57 **11.** 11.16   **13.** 6.5   **15.** 15.3   **17.** 598.23   **19.** 16.3
21. -6.32 **23.** -6.4 **25.** 3.1   **27.** 3.1   **29.** -9.9 **31.** 25.67   **33.** -5.62 **35.** 776.89   **37.** 11,983.32   **39.** 465.56
41. -549.8 **43.** 861.6   **45.** 115.123   **47.** 876.6   **49.** 0.088   **51.** 465.902   **53.** -180.44 **55.** -1.1 **57.** 3.81
59. 3.39   **61.** 1.61   **63.** no   **65.** yes   **67.** no   **69.** $6.9x + 6.9$ **71.** $-10.97 + 3.47y$ **73.** $454.71   **75.** $0.06
77. $7.52   **79.** 13.3 lb   **81.** 285.8 mph   **83.** 763.035 mph   **85.** $334.4 million   **87.** 240.8 in.   **89.** 67.44 ft   **91.** 12.409 mph
93. 2044.35 min   **95.** Switzerland   **97.** 8.1 lb   **99.**   **101.** $\frac{1}{125}$ **103.** $\frac{5}{12}$ **105.** $1.02
107. 5 nickels; 2 dimes and 1 nickel; 1 dime and 3 nickels;
1 quarter
109. Answers may vary.
111. $-109.544x + 15.604$
113. 6.08 in.

Country	Pounds of Chocolate Per Person
Switzerland	22.0
Norway	16.0
German	15.8
United Kingdom	14.5
Belgium	13.9

EXERCISE SET 5.3

1. 0.12   **3.** 0.6   **5.** -17.595 **7.** 39.273   **9.** 65   **11.** 0.65   **13.** -7093 **15.** 0.0983   **17.** 43.274   **19.** 8.23854
21. 14,790   **23.** -9.3762 **25.** 1.12746   **27.** -0.14444 **29.** 5,500,000,000 chocolate bars   **31.** 36,400,000 households
33. 1,600,000 hr   **35.** -0.6 **37.** 17.3   **39.** 56.52   **41.** no   **43.** no   **45.** yes   **47.** 8π m $\approx$ 24.12 m
49. 10π cm $\approx$ 31.4 cm   **51.** 18.2π yd $\approx$ 57.148 yd   **53.** 250π ft $\approx$ 785 ft   **55.** 135π m $\approx$ 423.9 m   **57. a.** 62.8 m; 125.6 m
b. yes   **59.** 24.8 g   **61.** $2800   **63.** 64.9605 in.   **65.** $555.20   **67.** 79,697.25 yen   **69.** 514 Canadian dollars   **71.** 26
73. 36   **75.** 8   **77.** -9 **79.** 3,831,600 mi   **81.** Answers may vary.   **83.** 740.82   **85.** Answers may vary.

EXERCISE SET 5.4

1. 0.094   **3.** -300 **5.** 5.8   **7.** 6.6   **9.** 200   **11.** 23.87   **13.** -110 **15.** 0.54982   **17.** -0.0129 **19.** 8.7
21. 0.413   **23.** -7 **25.** 4.8   **27.** 2100   **29.** 30   **31.** 7000   **33.** -0.69 **35.** 0.024   **37.** 65   **39.** 0.0003
41. 120,000   **43.** -5.65 **45.** -7.0625 **47.** 1.13   **49.** yes   **51.** no   **53.** yes   **55.** 24 mo   **57.** $2734.51
59. 202.1 lb   **61.** 5.1 m   **63.** 11.4 boxes   **65.** 132.5 mph   **67.** 24 tsp   **69.** 8 days   **71.** 345.22   **73.** -1001.0
75. 20   **77.** 22   **79.** 85   **81.** 7.2   **83.** 8.6 ft   **85.** Answers may vary.   **87.** $65.2 - 82.6$ knots

INTEGRATED REVIEW

1. 2.57 **2.** 4.05 **3.** 8.9 **4.** 3.5 **5.** 0.16 **6.** 0.24 **7.** 0.27 **8.** 0.52 **9.** −4.8 **10.** 6.09 **11.** 75.56
12. 289.12 **13.** −24.974 **14.** −43.875 **15.** −8.6 **16.** 5.4 **17.** −280 **18.** 1600 **19.** 224.938 **20.** 145.079
21. 0.56 **22.** −0.63 **23.** 27.6092 **24.** 145.6312 **25.** 5.4 **26.** −17.74 **27.** −414.44 **28.** −1295.03 **29.** −34
30. −28 **31.** 116.81 **32.** 18.79 **33.** 156.2 **34.** 1.562 **35.** 25.62 **36.** 5.62 **37.** Yes; answers may vary.
38. 32.1 points

EXERCISE SET 5.5

1. 2.8 **3.** 2898.66 **5.** 4.2 **7.** 149.87 **9.** 22.89 **11.** 36 in. **13.** 39 ft **15.** 43.96 m **17.** 52 gal **19.** $24,000
21. 53 mi **23.** $2700 million **25.** 135,000 people **27.** c **29.** b **31.** not reasonable **33.** reasonable **35.** 0.06
37. −3 **39.** 0.28 **41.** 2.3 **43.** 5.29 **45.** 7.6 **47.** 0.2025 **49.** −1.29 **51.** 5.76 **53.** 5.7 **55.** 3.6 **57.** yes
59. no **61.** $\frac{5}{16}$ **63.** $-\frac{3}{4}$ **65.** $\frac{1}{12}$ **67.** 43.388569 **69.** Answers may vary. **71.** Square; answers may vary.

EXERCISE SET 5.6

1. 0.2 **3.** 0.5 **5.** 0.75 **7.** −0.08 **9.** 0.375 **11.** $0.91\overline{6}$ **13.** 0.425 **15.** 0.45 **17.** $0.\overline{3}$ **19.** 0.4375 **21.** $0.\overline{2}$
23. $1.\overline{6}$ **25.** 0.33 **27.** 0.44 **29.** 0.2 **31.** 1.7 **33.** 0.67 **35.** 0.44 **37.** 0.62 **39.** < **41.** > **43.** >
45. < **47.** < **49.** > **51.** < **53.** < **55.** 0.32, 0.34, 0.35 **57.** 0.49, 0.491, 0.498 **59.** 0.73, $\frac{3}{4}$, 0.78
61. 0.412, 0.453, $\frac{4}{7}$ **63.** 5.23, $\frac{42}{8}$, 5.34 **65.** $\frac{17}{8}$, 2.37, $\frac{12}{5}$ **67.** 25.65 sq in. **69.** 9.36 sq cm **71.** 0.248 sq yd **73.** 8
75. 72 **77.** $\frac{1}{81}$ **79.** $\frac{9}{25}$ **81.** $\frac{5}{2}$ **83.** 0.202 **85.** 5900 stations **87.** 0.625 **89.** Answers may vary. **91.** −47.25
93. 3.37 **95.** −0.45

EXERCISE SET 5.7

1. 5.9 **3.** −0.43 **5.** 10.2 **7.** −4.5 **9.** 4 **11.** 0.45 **13.** 4.2 **15.** −4 **17.** 1.8 **19.** 10 **21.** 7.6 **23.** 60
25. −0.07 **27.** 20 **29.** 0.0148 **31.** −8.13 **33.** 1.5 **35.** −1 **37.** −7 **39.** 7 **41.** 53.2 **43.** $6x - 16$
45. $3x - 5$ **47.** $-2y + 6.8$ **49.** Answers may vary. **51.** 7.683 **53.** 4.683

CALCULATOR EXPLORATIONS

1. 32 **3.** 3.873 **5.** 9.849

EXERCISE SET 5.8

1. 2 **3.** 25 **5.** $\frac{1}{9}$ **7.** $\frac{12}{8} = \frac{3}{2}$ **9.** 16 **11.** $\frac{3}{2}$ **13.** 1.732 **15.** 3.873 **17.** 3.742 **19.** 6.856 **21.** 2.828
23. 5.099 **25.** 8.426 **27.** 2.646 **29.** 13 in. **31.** 6.633 cm **33.** 5 **35.** 8 **37.** 17.205 **39.** 16.125 **41.** 12
43. 44.822 **45.** 42.426 **47.** 1.732 **49.** 141.42 yd **51.** 25.0 ft **53.** 340 ft **55.** $\frac{5}{6}$ **57.** $\frac{2}{5}$ **59.** $\frac{5}{12}$ **61.** 6, 7
63. 10, 11 **65.** Answers may vary.

CHAPTER 5 REVIEW

1. tenths **2.** hundred-thousandths **3.** negative twenty-three and forty-five hundredths **4.** three hundred forty-five
hundred-thousandths **5.** one hundred nine and twenty-three hundredths **6.** two hundred and thirty-two millionths **7.** 2.15
8. −503.102 **9.** 16,025.0014 **10.** $\frac{4}{25}$ **11.** $-12\frac{23}{1000}$ **12.** $1\frac{9}{2000}$ **13.** $\frac{231}{100,000}$ **14.** $25\frac{1}{4}$ **15.** > **16.** =
17. < **18.** > **19.** 0.6 **20.** 0.94 **21.** −42.90 **22.** 16.349 **23.** 13,500 people **24.** $10\frac{3}{4}$ tsp **25.** 9.5 **26.** 5.1
27. −7.28 **28.** −12.04 **29.** 320.312 **30.** 148.74236 **31.** 1.7 **32.** 2.49 **33.** −1324.5 **34.** −10.136 **35.** 65.02
36. 199.99802 **37.** 9605.51 points **38.** −5.7 **39.** $4.2x + 12.8$ **40.** $9.89y - 4.17$ **41.** 72 **42.** 9345 **43.** −78.246
44. 73,246.446 **45.** 14π m $\approx$ 43.96 m **46.** 63.8 mi **47.** 70 **48.** −0.21 **49.** −4900 **50.** 23.904 **51.** 8.059
52. 158.25 **53.** 7.3 m **54.** 45 mo **55.** 18.2 **56.** 50 **57.** 99.05 **58.** 54.1 **59.** 35.5782 **60.** 0.3526 **61.** 32.7
62. 30.4 **63.** 200 quadrillion Btu **64.** 8932 sq ft **65.** Yes, $5 is enough. **66.** −11.94 **67.** 3.89 **68.** 0.1024
69. 3.6 **70.** 0.8 **71.** −0.923 **72.** −0.429 **73.** 0.217 **74.** 0.113 **75.** 51.057 **76.** = **77.** < **78.** <
79. < **80.** 0.832, 0.837, 0.839 **81.** 0.42, $\frac{3}{7}$, 0.43 **82.** $\frac{19}{12}$, 1.63, $\frac{18}{11}$ **83.** $\frac{3}{4}$, $\frac{6}{7}$, $\frac{8}{9}$ **84.** 6.9 sq ft **85.** 5.46 sq in.
86. 0.3 **87.** 47.19 **88.** 8.6 **89.** −80 **90.** −0.48 **91.** 34 **92.** 0.42 **93.** 0.28 **94.** −20 **95.** 1 **96.** 8

97. 12   **98.** 6   **99.** 1   **100.** $\frac{2}{5}$ **101.** $\frac{1}{10}$ **102.** 13   **103.** 29   **104.** 10.7   **105.** 93   **106.** 86.6   **107.** 28.28 cm
108. 88.2 ft

CHAPTER 5 TEST

1. forty-five and ninety-two thousandths   **2.** 3000.059   **3.** 17.595   **4.** -51.20 **5.** -20.42 **6.** 40.902   **7.** 0.037
8. 34.9   **9.** 0.862   **10.** $<$ **11.** $<$ **12.** $\frac{69}{200}$ **13.** $-24\frac{73}{100}$ **14.** -0.5 **15.** 0.941   **16.** 1.93   **17.** -6.2
18. $0.5x - 13.4$ **19.** 7   **20.** 12.530   **21.** $\frac{8}{10} = \frac{4}{5}$ **22.** -3 **23.** 3.7   **24.** 5.66 cm   **25.** 2.31 sq mi   **26.** 198.08 oz
27. 18π mi $\approx$ 56.52 mi   **28.** 12 CDs   **29.** 54 mi

CUMULATIVE REVIEW

1. 184,046; Sec. 1.3, Ex. 2   **2.** 215,984   **3.** 2300; Sec. 1.5, Ex. 4   **4.** 2900   **5.** 401 R 2; Sec. 1.7, Ex. 8   **6.** 146 R 3
7. 2; Sec. 1.9, Ex. 3   **8.** 6   **9. a.** -11 **b.** 2   **c.** 0; Sec. 2.1, Ex. 5   **10. a.** 7   **b.** -4 **c.** 1   **11.** -23; Sec. 2.2, Ex. 4
12. -22 **13.** 50; Sec. 2.5, Ex. 3   **14.** 32   **15.** -9; Sec. 2.5, Ex. 4   **16.** -32 **17.** 9; Sec. 2.5, Ex. 1   **18.** -9 **19. a.** $5x$
b. $-6y$ **c.** $8x^2 - 2$; Sec. 3.1, Ex. 1   **20. a.** $-3a$ **b.** $-4z^2$ **c.** $11k + 2$ **21.** -4; Sec. 3.2, Ex. 7   **22.** 5   **23.** -3; Sec.
3.3, Ex. 2   **24.** 3   **25.** -4; Sec. 3.4, Ex. 4   **26.** -3 **27.** 47,039 votes; Sec. 3.5, Ex. 3   **28.** 23   **29.** numerator: 3;
denominator: 7; Sec. 4.1, Ex. 1   **30.** numerator; 2: denominator; $5a$ **31.** numerator: $13x$; denominator: 5; Sec. 4.1, Ex. 2
32. numerator; $7y$: denominator; $3x$ **33.** $\frac{3}{5}$; Sec. 4.2, Ex. 4   **34.** $\frac{4}{7}$ **35.** $-\frac{1}{8}$; Sec. 4.3, Ex. 5   **36.** $\frac{1}{12}$ **37.** $\frac{5}{7}$; Sec. 4.4, Ex. 1
38. 1   **39.** 2; Sec. 4.4, Ex. 3   **40.** $1\frac{2}{3}$ **41.** $\frac{2}{3}$; Sec. 4.5, Ex. 1   **42.** $1\frac{1}{6}$ **43.** $\frac{x}{6}$; Sec. 4.6, Ex. 1   **44.** $\frac{3}{2x}$
45. 15; Sec. 4.7, Ex. 2   **46.** 18   **47.** 829.6561; Sec. 5.2, Ex. 2   **48.** 230.8628   **49.** 18.408; Sec. 5.3, Ex. 1   **50.** 28.251

Chapter 6   RATIO AND PROPORTION

CHAPTER 6 PRETEST

1. $\frac{15}{19}$; 6.1A   **2.** $\frac{3\frac{4}{5}}{9\frac{1}{8}}$; 6.1A   **3.** $\frac{51}{79}$; 6.1B   **4.** $\frac{3}{2}$; 6.1B   **5.** $\frac{\$2}{1\ \text{lb}}$; 6.2A   **6.** 19.5 mi/gal; 6.2B   **7.** \$1.09 per box; 6.2C
8. $\frac{18}{6} = \frac{15}{5}$; 6.3A   **9.** $\frac{49}{7} = \frac{35}{5}$; 6.3A   **10.** no; 6.3B   **11.** yes; 6.3B   **12.** 4; 6.3C   **13.** 2; 6.3C   **14.** 48; 6.3C
15. 10; 6.3C   **16.** $107\frac{1}{2}$ mi; 6.4A   **17.** $\frac{1}{2}$; 6.5B   **18.** 6; 6.5C   **19.** 5.5; 6.5C   **20.** 19.8 ft; 6.5D

EXERCISE SET 6.1

1. $\frac{11}{14}$ **3.** $\frac{23}{10}$ **5.** $\frac{151}{201}$ **7.** $\frac{2.8}{7.6}$ **9.** $\frac{5}{7\frac{1}{2}}$ **11.** $\frac{3\frac{3}{4}}{1\frac{2}{3}}$ **13.** $\frac{2}{3}$ **15.** $\frac{77}{100}$ **17.** $\frac{463}{821}$ **19.** $\frac{3}{4}$ **21.** $\frac{5}{12}$ **23.** $\frac{8}{25}$ **25.** $\frac{12}{7}$
27. $\frac{16}{23}$ **29.** $\frac{7}{40}$ **31.** $\frac{3}{50}$ **33.** $\frac{47}{25}$ **35.** $\frac{17}{40}$ **37.** $\frac{4}{9}$ **39.** $\frac{5}{4}$ **41.** $\frac{3}{1}$ **43.** $\frac{15}{1}$ **45.** $\frac{103}{5}$ **47.** $\frac{2}{39}$ **49.** $\frac{1}{3}$
51. 2.3   **53.** 0.15   **55.** $\frac{3}{47}$ **57.** $\frac{10}{23}$ **59.** No, the shipment should not be refused.   **61. a.** $\frac{19}{50}$ **b.** $\frac{19}{31}$
c. No; answers may vary.   **63.** $\frac{1}{48}$; $\frac{1}{64}$; answers may vary.

EXERCISE SET 6.2

1. $\frac{1\ \text{shrub}}{3\ \text{ft}}$ **3.** $\frac{3\ \text{returns}}{20\ \text{sales}}$ **5.** $\frac{2\ \text{phone lines}}{9\ \text{employees}}$ **7.** $\frac{9\ \text{gal}}{2\ \text{acres}}$ **9.** $\frac{3\ \text{flight attendants}}{100\ \text{passengers}}$ **11.** $\frac{71\ \text{cal}}{2\ \text{fl. oz}}$ **13.** 75 riders/car
15. 110 cal/oz   **17.** 90 wingbeats/sec   **19.** \$50,000/yr   **21.** 1330 mi/day   **23.** 2,870,010 voters/senator
25. 300 good/defective   **27.** 0.48 tons/acre   **29.** \$57,200/species   **31.** \$2,020,000/show   **33. a.** 31.25 computer boards/hr
b. $\approx$33.3 computer boards/hr   **c.** Lamont   **35.** \$11.50/compact disk   **37.** \$0.17/banana   **39.** 8 oz: \$0.149/oz; 12 oz:
\$0.133/oz; 12-oz size   **41.** 6 oz: \$0.115/oz; 16 oz: \$0.106/oz; 16-oz size   **43.** 8 oz: \$0.186/oz; 12 oz: \$0.191/oz; 8-oz size
45. 100: \$0.009/napkin; 180: \$0.008/napkin; 180 napkins   **47.** 10.2   **49.** 4.44   **51.** 1.9   **53.** miles driven: 257, 352, 347; miles
per gallon: 19.2, 22.3, 21.6   **55.** 1.5 steps/ft   **57.** 22 students/teacher   **59.** Answers may vary.

MENTAL MATH

1. true   **3.** false   **5.** true

EXERCISE SET 6.3

1. $\dfrac{10 \text{ diamonds}}{6 \text{ opals}} = \dfrac{5 \text{ diamonds}}{3 \text{ opals}}$ **3.** $\dfrac{3 \text{ printers}}{12 \text{ computers}} = \dfrac{1 \text{ printer}}{4 \text{ computers}}$ **5.** $\dfrac{6 \text{ eagles}}{58 \text{ sparrows}} = \dfrac{3 \text{ eagles}}{29 \text{ sparrows}}$

7. $\dfrac{2\frac{1}{4}\text{c flour}}{24 \text{ cookies}} = \dfrac{6\frac{3}{4}\text{c flour}}{72 \text{ cookies}}$ **9.** $\dfrac{22 \text{ vanilla wafers}}{1 \text{ c cookie crumbs}} = \dfrac{55 \text{ vanilla wafers}}{2.5 \text{ c cookie crumbs}}$ **11.** true   **13.** false   **15.** true   **17.** true

19. false   **21.** true   **23.** true   **25.** false   **27.** 3   **29.** 5   **31.** 4   **33.** 3.2   **35.** 19.2   **37.** $\dfrac{3}{4}$ **39.** 3   **41.** 12
43. 0.0025   **45.** 25   **47.** 14.9   **49.** 0.07   **51.** 3.3   **53.** 3.163   **55.** <   **57.** >   **59.** <
61. >   **63.** <   **65.** >   **67.** 0   **69.** 1400   **71.** 252.5   **73.** Answers may vary.

INTEGRATED REVIEW

1. $\dfrac{9}{10}$ **2.** $\dfrac{9}{25}$ **3.** $\dfrac{43}{50}$ **4.** $\dfrac{8}{23}$ **5.** $\dfrac{173}{139}$ **6.** $\dfrac{6}{7}$ **7.** $\dfrac{7}{26}$ **8.** $\dfrac{20}{33}$ **9.** $\dfrac{2}{3}$ **10.** $\dfrac{1}{8}$ **11.** $\dfrac{2}{3}$ **12.** $\dfrac{13}{4}$

13. $\dfrac{1 \text{ office}}{4 \text{ graduate assistants}}$ **14.** $\dfrac{2 \text{ lights}}{5 \text{ ft}}$ **15.** $\dfrac{2 \text{ Senators}}{1 \text{ state}}$ **16.** $\dfrac{1 \text{ teacher}}{28 \text{ students}}$ **17.** $\dfrac{10 \text{ in.}}{1 \text{ ft}}$ **18.** $\dfrac{40\text{¢}}{\$3}$

19. $\dfrac{16 \text{ households with computers}}{25 \text{ households}}$ **20.** $\dfrac{269 \text{ electoral votes}}{25 \text{ votes}}$ **21.** 55 mph   **22.** 140 ft/sec   **23.** 21 employees/fax line

24. 17 phone calls/teenager   **25.** 23 mpg   **26.** 16 teachers/computer   **27.** 6 books/student   **28.** 154 lb/adult

29. 8 lb: $0.27/lb; 18 lb: $0.28/lb; 8-lb size   **30.** 100: $0.020/plate; 500: $0.018/plate; 500 paper plates

31. 3 packs: $0.80/pack; 8 packs: $0.75/pack; 8 packs   **32.** 4: $0.92/battery; 10: $0.99/battery; 4 batteries

EXERCISE SET 6.4

1. 12 passes   **3.** 165 min   **5.** 180 students   **7.** 23 ft   **9.** 270 sq ft   **11.** 56 mi   **13.** 450 km   **15.** 24 oz   **17.** 16 bags

19. $162,000   **21.** 15 hits   **23.** 27 people   **25.** 86 weeks   **27.** 6 people   **29.** 112 ft; 11-in. difference   **31.** $2\frac{2}{3}$ lb

33. 16 men   **35.** 434 emergency room visits   **37.** 427.5 cal   **39.** 2.4 c   **41. a.** 0.1 gal   **b.** 13 fl oz   **43. a.** 2062.5 mg

b. no   **45.** $3 \cdot 5$ **47.** $2^2 \cdot 5$ **49.** $2^3 \cdot 5^2$ **51.** 2^5 **53.** 0.8 ml   **55.** 1.25 ml   **57.** $4\frac{2}{3}$ ft   **59.** Answers may vary.

EXERCISE SET 6.5

1. congruent; SSS   **3.** congruent; ASA   **5.** $\dfrac{2}{1}$ **7.** $\dfrac{3}{2}$ **9.** 4.5   **11.** 6   **13.** 5   **15.** 13.5   **17.** 17.5   **19.** 8

21. 21.25   **23.** 500 ft   **25.** 60 ft   **27.** 14.4 ft   **29.** 17.5   **31.** 81   **33.** 8.4   **35.** Answers may vary.   **37.** 12 ft by 18 ft

CHAPTER 6 REVIEW

1. $\dfrac{5}{4}$ **2.** $\dfrac{13}{11}$ **3.** $\dfrac{9}{40}$ **4.** $\dfrac{14}{5}$ **5.** $\dfrac{4}{15}$ **6.** $\dfrac{1}{2}$ **7.** $\dfrac{1}{2}$ **8.** $\dfrac{7}{150}$ **9.** $\dfrac{1 \text{ stillborn birth}}{125 \text{ live births}}$ **10.** $\dfrac{3 \text{ professors}}{10 \text{ assistants}}$ **11.** $\dfrac{5 \text{ pages}}{2 \text{ min}}$

12. $\dfrac{4 \text{ computers}}{3 \text{ hr}}$ **13.** 52 mph   **14.** 15 ft/sec   **15.** $0.31/pear   **16.** $1.74/diskette   **17.** 65 km/hr   **18.** $1\frac{1}{3}$ gal/acre

19. $36.80/course   **20.** 13 bushels/tree   **21.** 8 oz: $0.124 per oz; 12 oz: $0.141 per oz; 8-oz size   **22.** 18 oz: $0.83 per oz; 28 oz:

$0.85 per oz; 18-oz size   **23.** 16 oz: $0.037 per oz; $\frac{1}{2}$ gal: $0.026 per oz; 1 gal: $0.018 per oz; 1-gal size

24. 12 oz: $0.0492 per oz; 16 oz: $0.0494 per oz; 32 oz: $0.0372 per oz; 32-oz size   **25.** $\dfrac{20 \text{ men}}{14 \text{ women}} = \dfrac{10 \text{ men}}{7 \text{ women}}$

26. $\dfrac{50 \text{ tries}}{4 \text{ successes}} = \dfrac{25 \text{ tries}}{2 \text{ successes}}$ **27.** $\dfrac{16 \text{ sandwiches}}{8 \text{ players}} = \dfrac{2 \text{ sandwiches}}{1 \text{ player}}$ **28.** $\dfrac{12 \text{ tires}}{3 \text{ cars}} = \dfrac{4 \text{ tires}}{1 \text{ car}}$ **29.** false   **30.** true

31. false   **32.** true   **33.** 5   **34.** 15   **35.** 32.5   **36.** 5.625   **37.** 32   **38.** 13.5   **39.** 60   **40.** $7\frac{1}{5}$ **41.** 0.63

42. 30.9   **43.** 14 passes   **44.** 35 passes   **45.** 8 bags   **46.** 16 bags   **47.** no   **48.** 79 gal   **49.** $54,600   **50.** $1023.50

51. $40\frac{1}{2}$ ft   **52.** $8\frac{1}{4}$ in.   **53.** 37.5   **54.** $13\frac{1}{3}$ **55.** 17.4   **56.** $6\frac{1}{2}$ **57.** 33 ft   **58.** $x = \dfrac{5}{6}$ in., $y = 2\frac{1}{6}$ in.

59. congruent; ASA   **60.** not congruent

Chapter 6 Test

1. $\dfrac{9}{13}$ **2.** $\dfrac{15}{2}$ **3.** $\dfrac{7 \text{ men}}{1 \text{ woman}}$ **4.** $\dfrac{3 \text{ in.}}{10 \text{ days}}$ **5.** 81.25 km/hr   **6.** $\dfrac{2}{3}$ in./hr   **7.** 28 students/teacher
8. 8 oz: $0.149 per oz; 12 oz: $0.158 per oz; 8-oz size   **9.** 16 oz: $0.093 per oz; 24 oz: $0.100 per oz; 16-oz size   **10.** true
11. false   **12.** 5   **13.** $4\dfrac{4}{11}$ **14.** $\dfrac{7}{3}$ **15.** 8   **16.** $49\dfrac{1}{2}$ ft   **17.** $3\dfrac{3}{4}$ hr   **18.** $53\dfrac{1}{3}$ g   **19.** $114\dfrac{2}{3}$ cartons
20. 256 coffee drinkers   **21.** 7.5   **22.** 69 ft

Cumulative Review

1. 128 R 16; Sec. 1.7, Ex. 9   **2.** 122 R 17   **3. a.** $7 + x$ **b.** $15 - x$ **c.** $2x$ **d.** $x \div 5$ or $\dfrac{x}{5}$ **e.** $x - 2$; Sec. 1.9, Ex. 6
4. a. $y + 3$ **b.** $\dfrac{10}{y}$ **c.** $4 - y$ **d.** $7y$ **5. a.** 4   **b.** -5 **c.** -6; Sec. 2.1, Ex. 6   **6. a.** -3 **b.** -2 **c.** 1
7. -15; Sec. 2.3, Ex. 5   **8.** -10 **9.** 0; Sec. 2.3, Ex. 10   **10.** 3   **11.** $8x - 5$; Sec. 3.1, Ex. 3   **12.** $2x + 6$
13. $6x - 7y + 6$; Sec. 3.1, Ex. 5   **14.** $10x - 2z - 3$ **15.** yes; Sec. 3.2, Ex. 1   **16.** yes   **17.** -4; Sec. 3.3, Ex. 3   **18.** -4
19. -1; Sec. 3.4, Ex. 2   **20.** -3 **21.** $\dfrac{6}{15}$; Sec. 4.1, Ex. 18   **22.** $\dfrac{24}{56}$ **23.** $2^4 \cdot 5$; Sec. 4.2, Ex. 2   **24.** $2^2 \cdot 11$
25. $\dfrac{6}{5}$; Sec. 4.3, Ex. 7   **26.** $\dfrac{3}{2}$ **27.** $\dfrac{7}{9}$; Sec. 4.4, Ex. 4   **28.** $\dfrac{2}{5}$ **29.** $\dfrac{1}{4}$; Sec. 4.4, Ex. 5   **30.** $\dfrac{1}{6}$ **31.** $\dfrac{6 - x}{3}$; Sec. 4.5, Ex. 4
32. $\dfrac{x - 8}{10}$ **33.** $\dfrac{2y - 20}{3}$; Sec. 4.6, Ex. 4   **34.** $\dfrac{12x + 40}{3}$ **35.** 21; Sec. 4.7, Ex. 1   **36.** 20   **37.** $\dfrac{35}{12}$ or $2\dfrac{11}{12}$; Sec. 4.8, Ex. 5
38. $\dfrac{14}{45}$ **39.** $\dfrac{43}{100}$; Sec. 5.1, Ex. 7   **40.** $\dfrac{13}{20}$ **41.** 4.68; Sec. 5.4, Ex. 2   **42.** 2.37   **43.** -3.7; Sec. 5.5, Ex. 6   **44.** 2.52
45. $\dfrac{4}{9}, \dfrac{9}{20}$, 0.456; Sec. 5.6, Ex. 8   **46.** $\dfrac{1}{2}$, 0.531, $\dfrac{4}{7}$ **47.** 9.5; Sec. 5.7, Ex. 1   **48.** 3.5   **49.** $\dfrac{6}{5}$; Sec. 6.3, Ex. 7   **50.** $\dfrac{3}{4}$

Chapter 7   Percent

Chapter 7 Pretest

1. 12%; 7.1A   **2.** 0.57; 7.1B   **3.** 275%; 7.1C   **4.** $\dfrac{3}{40}$; 7.1D   **5.** 15%; 7.1E   **6.** $18\% \cdot 50 = x$; 7.2A   **7.** $4\% \cdot x = 89$; 7.2A
8. $\dfrac{82}{b} = \dfrac{90}{100}$; 7.3A   **9.** $\dfrac{48}{112} = \dfrac{p}{100}$; 7.3A   **10.** 20%; 7.2B, 7.3B   **11.** 3.3; 7.2B, 7.3B   **12.** 200; 7.2B, 7.3B   **13.** 3 lightbulbs;
7.4A   **14.** decrease of 84 students; current enrollment: 4116 students; 7.4B   **15.** 7%; 7.5A   **16.** $22,400; 7.5B
17. discount: $78, sales price: $572; 7.5C   **18.** $144; 7.6A   **19.** $7147.50; 7.6B   **20.** $37.33; 7.6C

Mental Math

1. 13%   **3.** 87%   **5.** 1%

Exercise Set 7.1

1. 81%   **3.** 9%   **5.** chocolate chip: 52%   **7.** 12%   **9.** 0.48   **11.** 0.06   **13.** 1   **15.** 0.613   **17.** 0.028   **19.** 0.6425
21. 3   **23.** 0.3258   **25.** 0.737   **27.** 0.25   **29.** 0.111   **31.** 0.462   **33.** 310%   **35.** 2900%   **37.** 0.3%   **39.** 22%
41. 5.6%   **43.** 33.28%   **45.** 300%   **47.** 70%   **49.** 10%   **51.** 32.4%   **53.** 38%   **55.** $\dfrac{1}{25}$ **57.** $\dfrac{9}{200}$ **59.** $1\dfrac{3}{4}$
61. $\dfrac{73}{100}$ **63.** $\dfrac{1}{8}$ **65.** $\dfrac{1}{16}$ **67.** $\dfrac{31}{300}$ **69.** $\dfrac{179}{800}$ **71.** 75%   **73.** 70%   **75.** 40%   **77.** 59%   **79.** 34%   **81.** $37\dfrac{1}{2}\%$
83. $31\dfrac{1}{4}\%$ **85.** $66\dfrac{2}{3}\%$ **87.** 250%   **89.** 190%   **91.** 63.64%   **93.** 26.67%   **95.** 14.29%   **97.** 91.67%
99. 0.35, $\dfrac{7}{20}$; 20%, 0.2; 50%, $\dfrac{1}{2}$; 0.7, $\dfrac{7}{10}$; 37.5%, 0.375   **101.** 0.4, $\dfrac{2}{5}$; $23\dfrac{1}{2}\%$, $\dfrac{47}{200}$; 80%, 0.8; $0.\overline{3}$, $\dfrac{1}{3}$; 87.5%, 0.875; 0.075, $\dfrac{3}{40}$
103. $\dfrac{249}{1000}$ **105.** 8.5%   **107.** $\dfrac{207}{1000}$ **109.** 25%   **111.** $\dfrac{1}{250}$ **113.** $\dfrac{3}{25}$ **115.** $\dfrac{71}{1000}$ **117.** 15   **119.** -10
121. 12   **123.** 75%   **125.** 80%   **127.** 1.155; 115.5%   **129.** greater   **131.** Answers may vary.
133. Computer software engineers, applications   **135.** 0.62

MENTAL MATH

1. percent: 42; base: 50; amount: 21   **3.** percent: 125; base: 86; amount: 107.5

EXERCISE SET 7.2

1. $15\% \cdot 72 = x$ **3.** $30\% \cdot x = 80$ **5.** $x \cdot 90 = 20$ **7.** $1.9 = 40\% \cdot x$ **9.** $x = 9\% \cdot 43$ **11.** 3.5   **13.** 7.28   **15.** 600
17. 10   **19.** 110%   **21.** 32%   **23.** 1   **25.** 45   **27.** 500   **29.** 5.16%   **31.** 25.2   **33.** 45%   **35.** 35   **37.** 30

39. $3\frac{7}{11}$ **41.** $\frac{17}{12} = \frac{n}{20}$ **43.** $\frac{8}{9} = \frac{14}{n}$ **45.** 686.625   **47.** 12,285

MENTAL MATH

1. amount: 12.6; base: 42; percent 30   **3.** amount: 102; base: 510; percent: 20

EXERCISE SET 7.3

1. $\frac{a}{65} = \frac{32}{100}$ **3.** $\frac{75}{b} = \frac{40}{100}$ **5.** $\frac{70}{200} = \frac{p}{100}$ **7.** $\frac{2.3}{b} = \frac{58}{100}$ **9.** $\frac{a}{130} = \frac{19}{100}$ **11.** 5.5   **13.** 18.9   **15.** 400   **17.** 10
19. 125%   **21.** 28%   **23.** 29   **25.** 1.92   **27.** 1000   **29.** 210%   **31.** 55.18   **33.** 45%   **35.** 85   **37.** $\frac{7}{8}$ **39.** $3\frac{2}{15}$
41. 0.7   **43.** 2.19   **45.** 12,011.2   **47.** 7270.6

INTEGRATED REVIEW

1. 12%   **2.** 68%   **3.** 25%   **4.** 50%   **5.** 520%   **6.** 780%   **7.** 6%   **8.** 44%   **9.** 250%   **10.** 325%   **11.** 3%
12. 5%   **13.** 0.65   **14.** 0.31   **15.** 0.08   **16.** 0.07   **17.** 1.42   **18.** 5.38   **19.** 0.029   **20.** 0.066   **21.** $\frac{3}{100}$
22. $\frac{2}{25}$ **23.** $\frac{21}{400}$ **24.** $\frac{51}{400}$ **25.** $\frac{19}{50}$ **26.** $\frac{9}{20}$ **27.** $\frac{37}{300}$ **28.** $\frac{1}{6}$ **29.** 8.4   **30.** 100   **31.** 250   **32.** 120%
33. 28%   **34.** 76   **35.** 11   **36.** 130%   **37.** 86%   **38.** 37.8   **39.** 150   **40.** 62

EXERCISE SET 7.4

1. 1600 bolts   **3.** 30 hr   **5.** 15%   **7.** 295 components   **9.** 13.6%   **11.** 100,102 dental hygienists   **13.** 29.2%
15. 496 chairs; 6696 chairs   **17.** $136   **19.** $867.87; $20,153.87   **21.** 28 million; 63 million   **23.** 10; 25%   **25.** 102; 120%
27. 2; 25%   **29.** 120; 75%   **31.** 44%   **33.** 21.5%   **35.** 1.3%   **37.** 87.1%   **39.** 300%   **41.** 81.3%   **43.** 72.1%
45. 587.5%   **47.** 8.2%   **49.** 4.56   **51.** 11.18   **53.** 58.54   **55.** The increased number is double the original number.
57. Answers may vary.

EXERCISE SET 7.5

1. $7.50   **3.** $858.93   **5.** 9%   **7.** $238.70   **9.** $1917   **11.** $11,500   **13.** $112.35   **15.** 6%   **17.** $49,474.24
19. 14%   **21.** $1888.50   **23.** $85,500   **25.** $6.80; $61.20   **27.** $48.25; $48.25   **29.** $75.25; $139.75   **31.** $3255.00;
$18,445.00   **33.** $45; $255   **35.** $15; $60   **37.** 33%; $80.40   **39.** $925; $555   **41.** 1200   **43.** 132   **45.** 16
47. $22,812.68   **49.** A discount of 60% is better.   **51.** $1.60; $2.40; $3.20   **53.** $0.90; $1.35; $1.80

CALCULATOR EXPLORATIONS

1. 1.56051   **3.** 8.06231   **5.** $634.50

EXERCISE SET 7.6

1. $32   **3.** $73.60   **5.** $750   **7.** $33.75   **9.** $700   **11.** $78,125   **13.** $5562.50   **15.** $12,580   **17.** $46,815.40
19. $2327.15   **21.** $58,163.60   **23.** $240.75   **25.** $938.66   **27.** $971.90   **29.** $260.31   **31.** $637.26   **33.** −29
35. 150   **37.** −1   **39.** Answers may vary.   **41.** Answers may vary.

CHAPTER 7 REVIEW

1. 37%   **2.** 77%   **3.** 0.83   **4.** 0.75   **5.** 0.735   **6.** 0.015   **7.** 1.25   **8.** 1.45   **9.** 0.005   **10.** 0.007   **11.** 2.00 or 2
12. 4.00 or 4   **13.** 0.2625   **14.** 0.8534   **15.** 260%   **16.** 5.5%   **17.** 35%   **18.** 102%   **19.** 72.5%   **20.** 25%
21. 7.6%   **22.** 8.5%   **23.** 75%   **24.** 65%   **25.** 400%   **26.** 900%   **27.** $\frac{1}{100}$ **28.** $\frac{1}{10}$ **29.** $\frac{1}{4}$ **30.** $\frac{17}{200}$
31. $\frac{51}{500}$ **32.** $\frac{1}{6}$ **33.** $\frac{1}{3}$ **34.** $1\frac{1}{10}$ **35.** 20%   **36.** 70%   **37.** $83\frac{1}{3}\%$ **38.** 62.5%   **39.** $166\frac{2}{3}\%$ **40.** 125%

41. 60%   **42.** 6.25%   **43.** $\dfrac{9}{10}, \dfrac{1}{10}$ **44.** $\dfrac{24}{25}$ **45.** 70%   **46.** 1.5   **47.** 100,000   **48.** 8000   **49.** 23%   **50.** 114.5

51. 3000   **52.** 150%   **53.** 418   **54.** 300   **55.** 64.8   **56.** 180%   **57.** 110%   **58.** 165   **59.** 66%   **60.** 16%

61. 106.25%   **62.** 20.9%   **63.** $206,400   **64.** $13.23   **65.** $263.75   **66.** $1.15   **67.** $5000   **68.** $300.38

69. discount: $900; sale price: $2100   **70.** discount: $9; sale price: $81   **71.** $120   **72.** $1320   **73.** $30,104.64

74. $17,506.56   **75.** $80.61   **76.** $32,830.10

Chapter 7 Test

1. 0.85   **2.** 5   **3.** 0.006   **4.** 5.6%   **5.** 610%   **6.** 35%   **7.** $1\dfrac{1}{5}$ **8.** $\dfrac{77}{200}$ **9.** $\dfrac{1}{500}$ **10.** 55%   **11.** 37.5%

12. 175%   **13.** 20%   **14.** $\dfrac{16}{25}$ **15.** 33.6   **16.** 1250   **17.** 75%   **18.** 38.4 lb   **19.** $56,750   **20.** $358.43   **21.** 5%

22. discount: $18; sale price: $102   **23.** $395   **24.** 1%   **25.** $647.50   **26.** $2005.64   **27.** $427   **28.** 5.8%

Cumulative Review

1. 20,296; Sec. 1.6, Ex. 4   **2.** 31,084   **3.** −10; Sec. 2.3, Ex. 8   **4.** 10   **5.** 3; Sec. 3.2, Ex. 3   **6.** −1   **7.** 2; Sec. 3.4, Ex. 5

8. 5   **9.** $\dfrac{21}{7}$; Sec. 4.1, Ex. 20   **10.** $\dfrac{40}{5}$ **11.** $\dfrac{14x}{15}$; Sec. 4.2, Ex. 6   **12.** $\dfrac{5y}{16}$ **13.** $\dfrac{5}{12}$; Sec. 4.3, Ex. 12   **14.** $-\dfrac{4}{7}$

15. $-\dfrac{1}{2}$; Sec. 4.4, Ex. 9   **16.** $\dfrac{1}{5}$ **17.** $\dfrac{1}{28}$; Sec. 4.5, Ex. 5   **18.** $\dfrac{16}{45}$ **19.** $\dfrac{3}{2}$; Sec. 4.6, Ex. 2   **20.** $\dfrac{50}{9}$ **21.** 3; Sec. 4.7, Ex. 7

22. 4   **23. a.** $\dfrac{38}{9}$ **b.** $\dfrac{19}{11}$; Sec. 4.8, Ex. 3   **24. a.** $\dfrac{17}{5}$ **b.** $\dfrac{44}{7}$ **25.** $\dfrac{1}{8}$; Sec. 5.1, Ex. 9   **26.** $\dfrac{17}{20}$

27. $-105\dfrac{83}{1000}$; Sec. 5.1, Ex. 11   **28.** $17\dfrac{3}{200}$ **29.** 67.69; Sec. 5.2, Ex. 6   **30.** 27.94   **31.** 76.8; Sec. 5.3, Ex. 5   **32.** 1248.3

33. −76,300; Sec. 5.3, Ex. 7   **34.** −8537.5   **35.** yes; Sec. 5.4, Ex. 8   **36.** no   **37.** $\dfrac{2}{5}$; Sec. 5.8, Ex. 6   **38.** $\dfrac{3}{4}$

39. $\dfrac{50}{63}$; Sec. 6.1, Ex. 5   **40.** $\dfrac{29}{38}$ **41.** $0.21/oz; Sec. 6.2, Ex. 6   **42.** $2.25 per sq ft   **43.** no; Sec. 6.3, Ex. 3   **44.** yes

45. 17.5 miles; Sec. 6.4, Ex. 1   **46.** 35   **47.** $\dfrac{19}{1000}$; Sec. 7.1, Ex. 13   **48.** $\dfrac{23}{1000}$ **49.** $\dfrac{1}{3}$; Sec. 7.1, Ex. 15   **50.** $1\dfrac{2}{25}$

Chapter 8   Graphing and Introduction to Statistics

Chapter 8 Pretest

1. April; 8.1D   **2.** July; 8.1D   **3.** 700; 8.1D   **4.** ;8.2B   **5.** ‖; 2; 8.1C   **6.** ‖‖‖; 4; 8.1C
7. ‖‖‖; 4; 8.1C   **8.** ‖‖‖ ‖; 6; 8.1C
9.   ; 8.1C   **10.** no; 8.3B   **11. a.** $(0, -6)$
b. $(-3, -12)$
c. $(3, 0)$; 8.3C
12. $y = 4x - 1$; 8.4A

13. mean: 53.5; median: 64; mode: 74; 8.5 ABC
14. $\dfrac{1}{6}$; 8.6B   **15** $\dfrac{1}{2}$; 8.6B   **16.** $\dfrac{1}{3}$; 8.6B

Exercise Set 8.1

1. 2002   **3.** 4000 cars   **5.** 1998, 1999, 2003   **7.** 1997, 2000   **9.** 22.5 oz   **11.** 1997, 1999, 2001   **13.** 3 oz/wk

15. consumption of chicken is increasing   **17.** April   **19.** 19 deaths   **21.** February, March, April, May, June

23. Tokyo; 34.5 million or 34,500,000   **25.** New York; 21.4 million or 21,400,000   **27.** 16 million or 16,000,000

29.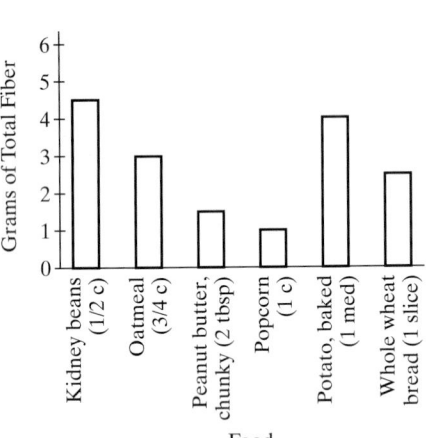

Fiber Content of Selected Foods

31.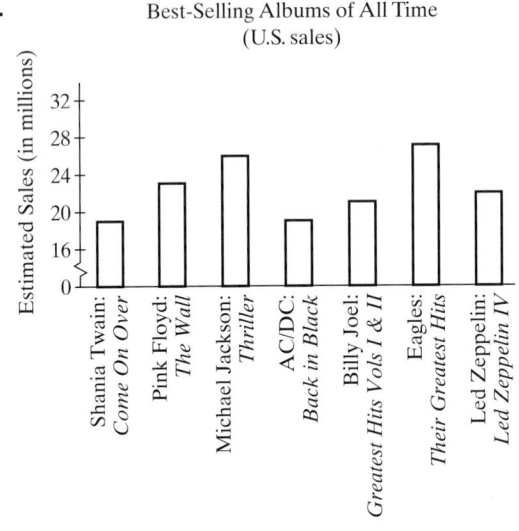

Best-Selling Albums of All Time
(U.S. sales)

33. 15 adults

35. 61 adults

37. 24 adults

39. 12 adults

41. $\dfrac{9}{100}$

43. 45–54

45. 17 million householders

47. 45 million householders

49. Answers may vary.

51. |; 1   **53.** ⊞|||; 8   **55.** ⊞|; 6   **57.** ⊞|; 6   **59.** ||; 2   **61.** 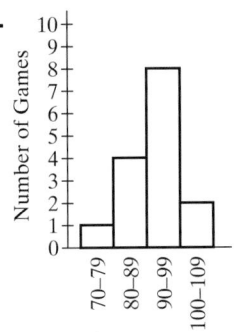   **63.** 2.7 goals   **65.** 1982

Golf Scores

67. decrease   **69.** 3.6   **71.** 6.2   **73.** 25%   **75.** 34%

77. 83°F   **79.** Sunday; 68°F   **81.** Tuesday; 13°F

83. Answers may vary.   **85.** Answers may vary.

Exercise Set 8.2

1. parent or guardian's home   **3.** $\dfrac{9}{35}$ **5.** $\dfrac{9}{16}$ **7.** Asia   **9.** 37%   **11.** 17,100,000 sq mi   **13.** 2,850,000 sq mi

15. 55%   **17.** nonfiction   **19.** 31,400 books   **21.** 27,632 books   **23.** 25,120 books

25. 270°; 40°; 22°; 29°   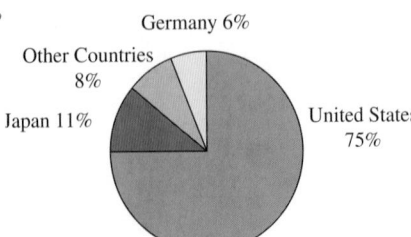   **27.** 25°; 65°; 50°; 220°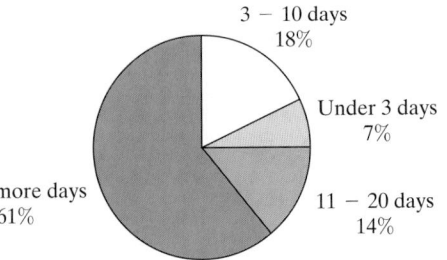

29. $2^2 \cdot 5$ **31.** $2^3 \cdot 5$ **33.** $5 \cdot 17$ **35.** Pacific; answers may vary.   **37.** 129,600,002 sq km   **39.** 55,542,858 sq km

41. 602 respondents   **43.** 2198 respondents   **45.** $\dfrac{301}{837}$ **47.** No; answers may vary.

EXERCISE SET 8.3

1.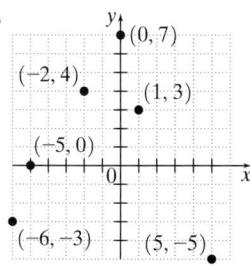

3.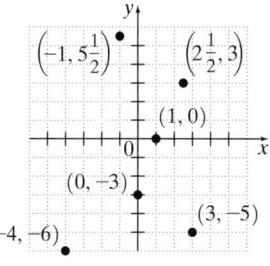

5. $A(0, 0)$; $B\left(3\frac{1}{2}, 0\right)$; $C(3, 2)$; $D(-1, 3)$; $E(-2, -2)$; $F(0, -1)$; $G(2, -1)$

7. yes   **9.** no   **11.** yes   **13.** yes   **15.** yes   **17.** no

19.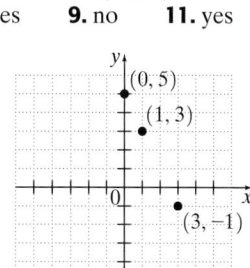

21.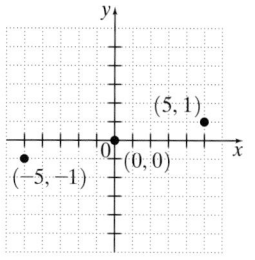

23.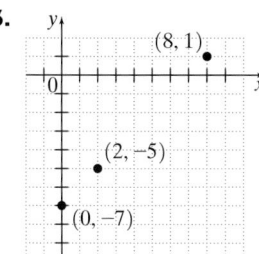

25. $(1, 8), (0, 0), (-2, -16)$ **27.** $(2, 12), (22, -8), (0, 14)$ **29.** $(1, 6), (2, 7), (3, 8)$

31. $(1, -2), (2, 1), (3, 4)$ **33.** $(0, 0), (2, -2), (-2, 2)$ **35.** $(-2, 0), (1, -3), (-7, 5)$

37. 1.7   **39.** 21.84   **41.** -23.6 **43.** true   **45.** true   **47.** false   **49.** false

51. a. right   **b.** below   **53.** rectangle   **55.** 20 units

INTEGRATED REVIEW

1. 69 lb   **2.** 78 lb   **3.** 1985   **4.** 1995 and 2000   **5.** Oroville Dam; 755 ft   **6.** New Bullards Bar Dam; 635 ft   **7.** 15 ft
8. 4 dams   **9.** Thursday and Saturday; 100° F   **10.** Monday; 82° F   **11.** Sunday, Monday, and Tuesday   **12.** Wednesday,
Thursday, Friday, and Saturday   **13.** 70 qt containers   **14.** 52 qt containers   **15.** 2 qt containers   **16.** 6 qt containers
17. ||; 2   **18.** |; 1   **19.** ||||; 3   **20.** ||||| |; 6   **21.** ||||; 5   **22.**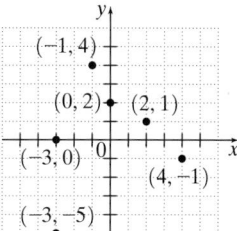
24. no   **25.** yes   **26.** $(0, -6), (6, 0), (2, -4)$

23.

EXERCISE SET 8.4

1.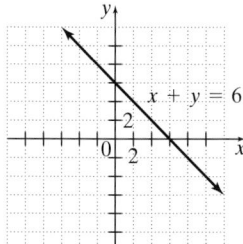

3.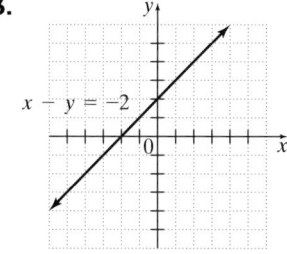

5.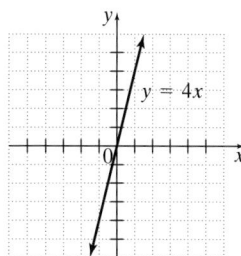

7.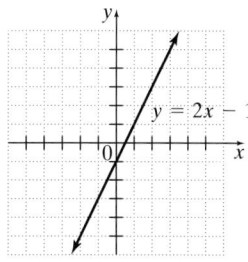

9.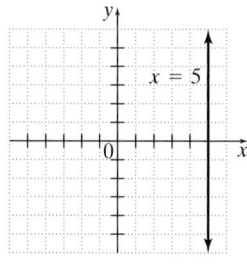

11.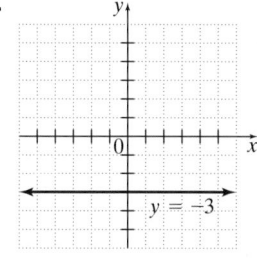

13.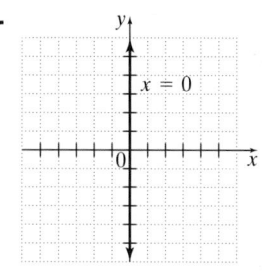

15.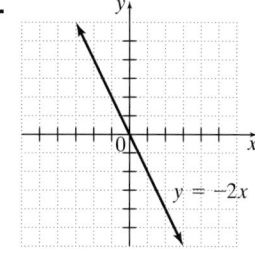

17.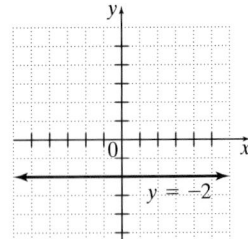

19.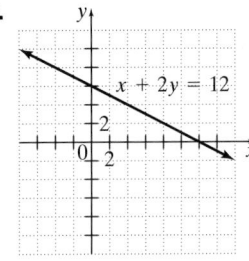

21.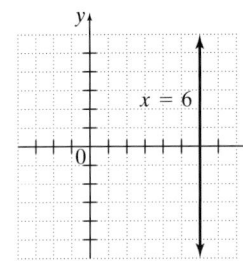

23.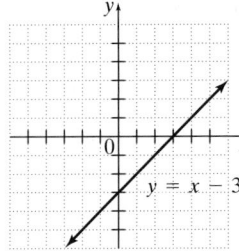

25.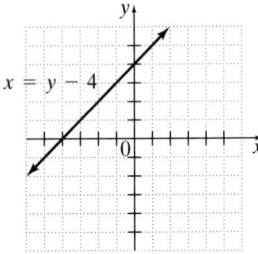

27.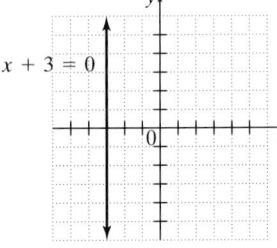

29.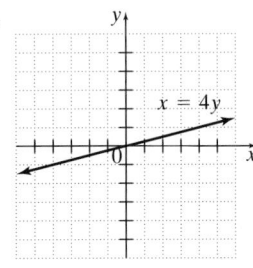

31.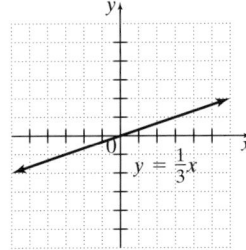

33.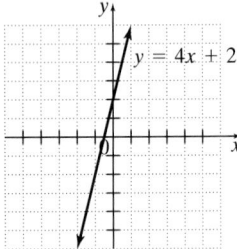

35.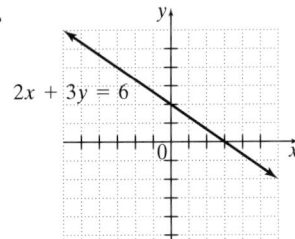

37.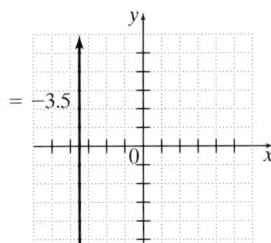

39.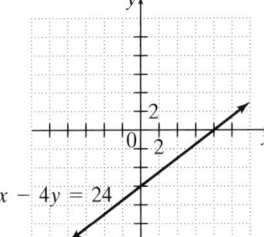

41. 90   **43.** 20   **45.** 27   **47.** 3, 2, 1, 0, 1, 2, 3

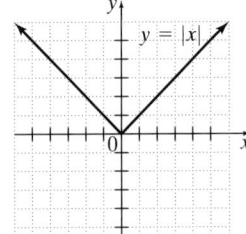

49. Answers may vary.

MENTAL MATH

1. 4   **3.** 3

EXERCISE SET 8.5

1. mean: 29; median: 28; no mode   **3.** mean: 8.1; median: 8.2; mode: 8.2   **5.** mean: 0.6; median: 0.6; mode: 0.2 and 0.6

7. mean: 370.9; median: 313.5; no mode   **9.** 1416 ft   **11.** 1332 ft   **13.** Answers may vary.   **15.** 2.79   **17.** 3.46   **19.** 6.8

21. 6.9   **23.** 85.5   **25.** 73   **27.** 70 and 71   **29.** 9 rates   **31.** $\frac{1}{3}$ **33.** $\frac{3}{5}$ **35.** $\frac{11}{15}$ **37.** 35, 35, 37, 43

39. Yes; answers may vary.

MENTAL MATH

1. $\frac{1}{2}$ **3.** $\frac{1}{2}$

EXERCISE SET 8.6

1.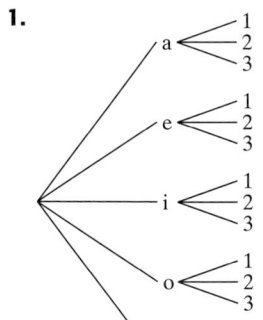

15 outcomes

3.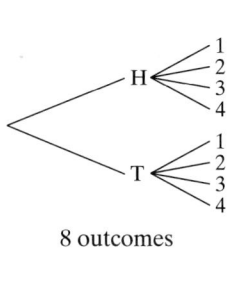

3 outcomes

5.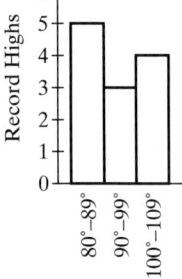

16 outcomes

7.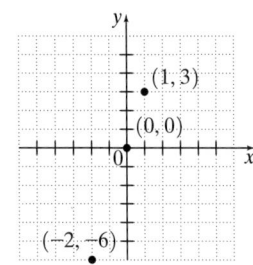

12 outcomes

9.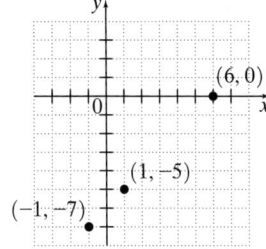

8 outcomes

11. $\dfrac{1}{6}$ **13.** $\dfrac{1}{3}$ **15.** $\dfrac{1}{2}$ **17.** $\dfrac{1}{3}$ **19.** $\dfrac{2}{3}$ **21.** $\dfrac{1}{7}$ **23.** $\dfrac{2}{7}$ **25.** $\dfrac{19}{100}$ **27.** $\dfrac{1}{20}$ **29.** $\dfrac{5}{6}$ **31.** $\dfrac{1}{6}$ **33.** $\dfrac{20}{3}$ or $6\dfrac{2}{3}$

35. $\dfrac{1}{52}$ **37.** $\dfrac{1}{13}$ **39.** $\dfrac{1}{4}$ **41.** $\dfrac{1}{12}$ **43.** 0 **45.** Answers may vary.

CHAPTER 80 REVIEW

1. 4,000,000 **2.** 1,750,000 **3.** South **4.** Northeast **5.** Midwest, South, and West **6.** Northeast **7.** 7.5% **8.** 2000

9. 1980, 1990, 2000 **10.** Answers may vary. **11.** $2,100,000 **12.** $1,200,000 **13.** 1991 and 1992, 2000 and 2001, 2001 and 2002

14. 1999 and 2000 **15.** 1990, 1991, 1992, 1993, 1994 **16.** 4 employees **17.** 1 employee **18.** 9 employees

19. 18 employees **20.** ||||; 5 **21.** |||; 3 **22.** ||||; 4 **23.**

24. mortgage payment **25.** utilities

26. $1225 **27.** $700 **28.** $\dfrac{39}{160}$ **29.** $\dfrac{7}{40}$ **30.** $\dfrac{5}{7}$

31. 20 states **32.** 11 states **33.** 1 state **34.** 29 states

35. $(0, 0), (7, -1), (-7, 1)$ **36.** $(0, -2), (1, 1), (-2, -8)$

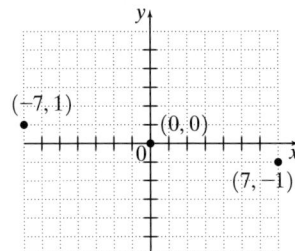

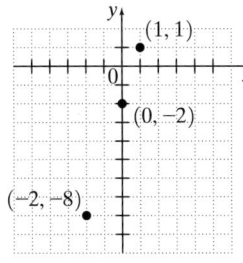

37. $(-1, -8), (-9, 0), (-5, -4)$ **38.** $(4, 1), (0, -3), (6, 3)$ **39.** $(1, 3), (-2, -6), (0, 0)$ **40.** $(1, -5), (6, 0), (-1, -7)$

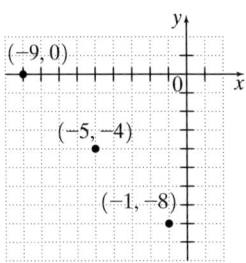

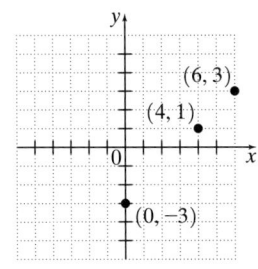

41.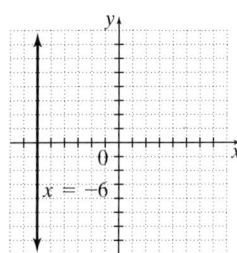

42.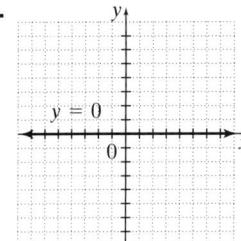

43.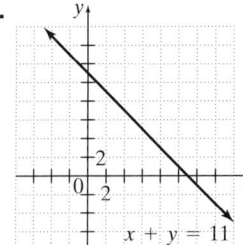

44.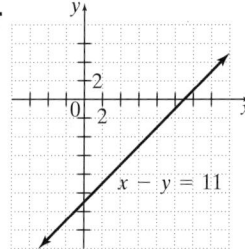

45.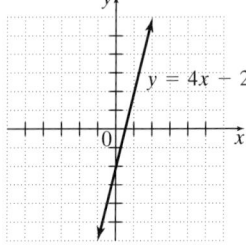

46.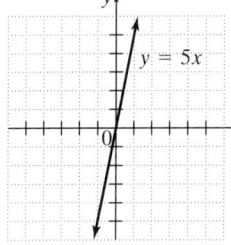

47.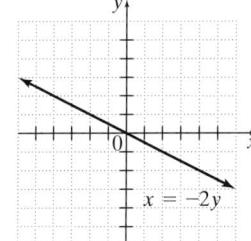

48.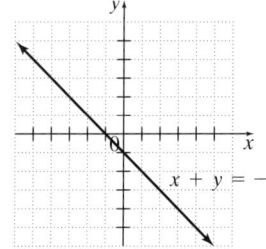

49.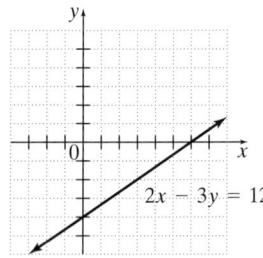

50.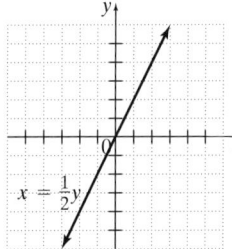

51. mean: 17.8; median: 14; no mode

52. mean: 55.2; median: 60; no mode

53. mean: $24,500; median: $20,000; mode: $20,000

54. mean: 447.3; median: 420; mode: 400

55. 3.25   **56.** 2.57

57.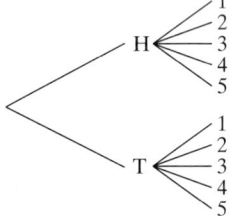

10 outcomes

58.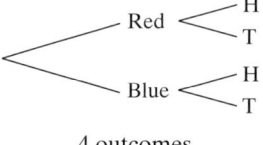

4 outcomes

59.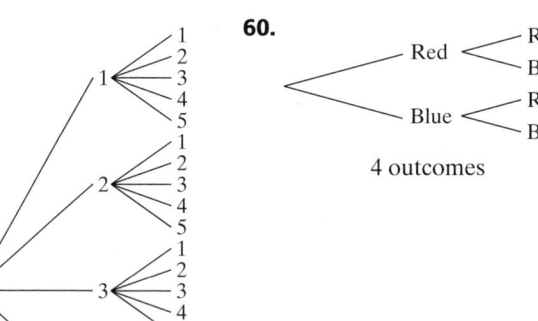

25 outcomes

60.

Red — Red
Red — Blue
Blue — Red
Blue — Blue

4 outcomes

61.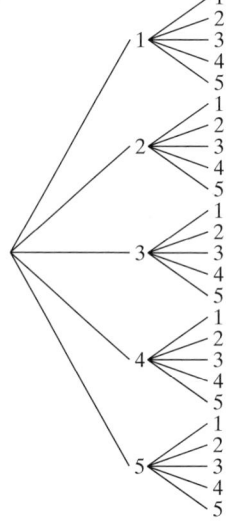

10 outcomes

62. $\frac{1}{6}$ **63.** $\frac{1}{6}$ **64.** $\frac{1}{5}$

65. $\frac{1}{5}$ **66.** $\frac{3}{5}$ **67.** $\frac{2}{5}$

CHAPTER 8 TEST

1. $225   **2.** 3rd week; $350   **3.** $1100   **4.** June, August, September   **5.** February, 3 cm   **6.** March and November

7.

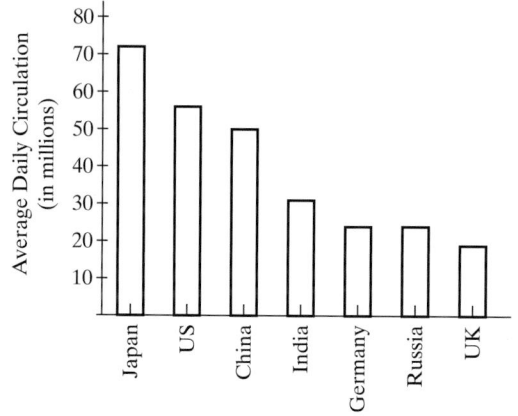

Countries with the Highest Newspaper Circulation

8. $\dfrac{17}{40}$ **9.** $\dfrac{31}{22}$ **10.** 40,920,000 people   **11.** 21,120,000 people

12. 9 students   **13.** 11 students

14.

Class Intervals (scores)	Tally	Class Frequency (Number of Students)
40–49	I	1
50–59	III	3
60–69	IIII	4
70–79	IIII	5
80–89	IIII III	8
90–99	IIII	4

15.

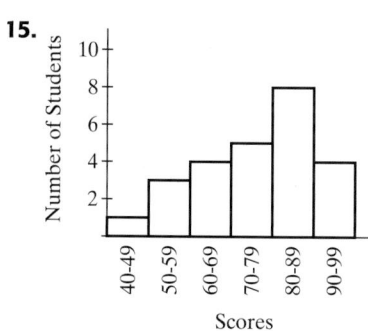

16. $(4, 0)$ **17.** $(0, -3)$ **18.** $(-3, 4)$ **19.** $(-2, -1)$

20. $(0, 0), (-6, 1), (12, -2)$ **21.** $(2, 10), (-1, -11), (0, -4)$

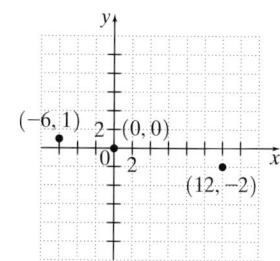   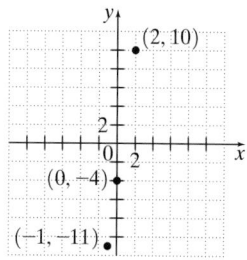

22. 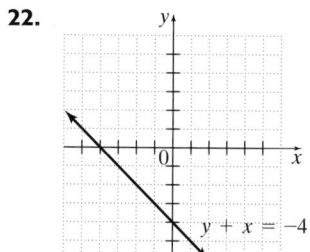   **23.** 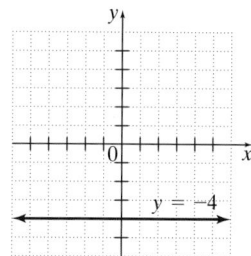   **24.** 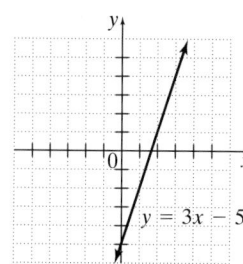   **25.**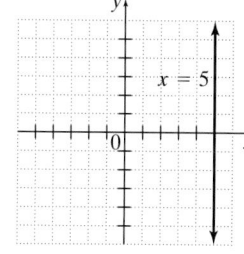

26. 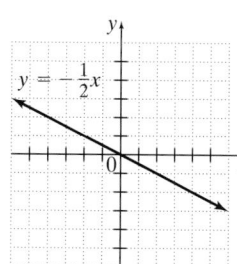   **27.**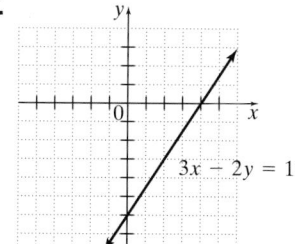

28. mean: 38.4; median: 42; no mode

29. mean: 12.625; median: 12.5; mode: 16 and 12

30. 3.07

31. 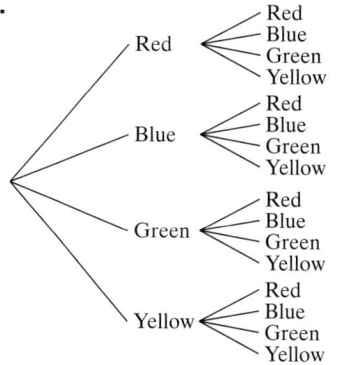   **32.**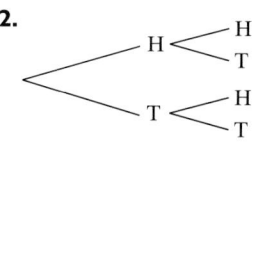

33. $\dfrac{1}{10}$ **34.** $\dfrac{1}{5}$

CUMULATIVE REVIEW

1. 47; Sec. 1.8, Ex. 12   **2.** −70   **3.** −12; Sec. 2.3, Ex. 11   **4.** 9   **5.** −3; Sec. 3.3, Ex. 5   **6.** −6   **7.** 2; Sec. 4.7, Ex. 5

8. −10   **9.** $7\dfrac{17}{24}$; Sec. 4.8, Ex. 11   **10.** $8\dfrac{3}{20}$ **11.** $5\dfrac{3}{5}$; Sec. 5.1, Ex. 8   **12.** $2\dfrac{4}{5}$ **13.** 3.432; Sec. 5.2, Ex. 5   **14.** 7.327

15. 0.0849; Sec. 5.3, Ex. 2   **16.** 0.0294   **17.** −0.052; Sec. 5.4, Ex. 3   **18.** 0.136   **19.** 4.09; Sec. 5.5, Ex. 7   **20.** 7.29

21. 0.25; Sec. 5.6, Ex. 1   **22.** 0.375   **23.** 0.7; Sec. 5.7, Ex. 5   **24.** 1.68   **25.** 5.657; Sec. 5.8, Ex. 8   **26.** 7.746

27. $\frac{12}{17}$; Sec. 6.1, Ex. 1   **28.** $\frac{7}{15}$ **29.** 24.5 mi/gal; Sec. 6.2, Ex. 5   **30.** $0.53 per oz   **31.** 5; Sec. 6.3, Ex. 6   **32.** 16

33. $\frac{12}{17}$; Sec. 6.5, Ex. 2   **34.** $\frac{4}{9}$ **35.** 0.046; Sec. 7.1, Ex. 4   **36.** 0.32   **37.** 0.0074; Sec. 7.1, Ex. 6   **38.** 0.027

39. 21; Sec. 7.2, Ex. 7   **40.** 14.4   **41.** 52; Sec. 7.3, Ex. 9   **42.** 38   **43.** 7.5%; Sec. 7.5, Ex. 2   **44.** 6.5%

45. $101.50; Sec. 7.6, Ex. 6   **46.** $144.05   **47.** 80.5; Sec. 8.5, Ex. 4   **48.** 48.5   **49.** $\frac{1}{3}$; Sec. 8.6, Ex. 4   **50.** $\frac{1}{2}$

Chapter 9   GEOMETRY AND MEASUREMENT

CHAPTER 9 PRETEST

1. acute; 9.1B   **2.** straight; 9.1B   **3.** 88°; 9.1C   **4.** $x = 55°$, $y = 125°$, $z = 125°$; 9.1D   **5.** 110°; 9.1D   **6.** $3\frac{2}{3}$ yd; 9.2A

7. 62.5 km; 9.2D   **8.** 60 in.; 9.3A   **9.** 56.52 yd; 9.3B   **10.** 6 sq cm; 9.4A   **11.** 7 lb 12 oz; 9.5B   **12.** 8 gal 3 qt; 9.6B

13. 25,000 ml; 9.6C   **14.** 47.8°C; 9.8B

EXERCISE SET 9.1

1. line; line yz or $\overleftrightarrow{yz}$, line zy or $\overleftrightarrow{zy}$ **3.** line segment; line segment LM or $\overline{LM}$, line segment ML or $\overline{ML}$ **5.** line segment; line segment PQ or $\overline{PQ}$, line segment QP or $\overline{QP}$ **7.** ray; ray UW or $\overrightarrow{UW}$ **9.** 15°   **11.** 50°   **13.** 65°   **15.** 95°   **17.** 90°

19. 0°; 90°   **21.** straight   **23.** right   **25.** obtuse   **27.** right   **29.** 73°   **31.** 163°   **33.** 42°   **35.** 55°

37. $\angle MNP$ and $\angle RNO$; $\angle PNQ$ and $\angle QNR$ **39.** $\angle SPT$ and $\angle TPQ$; $\angle SPR$ and $\angle RPQ$; $\angle SPT$ and $\angle SPR$; $\angle TPQ$ and $\angle QPR$

41. 54.8°   **43.** 32°   **45.** 75°   **47.** $\angle x = 35°$; $\angle y = 145°$; $\angle z = 145°$ **49.** $\angle x = 77°$; $\angle y = 103°$; $\angle z = 77°$

51. $\angle x = 100°$; $\angle y = 80°$; $\angle z = 100°$ **53.** $\angle x = 134°$; $\angle y = 46°$; $\angle z = 134°$ **55.** $\frac{9}{8}$ or $1\frac{1}{8}$ **57.** $\frac{7}{32}$ **59.** $\frac{5}{6}$ **61.** $\frac{4}{3}$ or $1\frac{1}{3}$

63. $\angle a = 60°$; $\angle b = 50°$; $\angle c = 110°$; $\angle d = 70°$; $\angle e = 120°$ **65.** 45°   **67.** No; answers may vary.

MENTAL MATH

1. 1 ft   **3.** 2 ft   **5.** 1 yd   **7.** no   **9.** yes   **11.** no

EXERCISE SET 9.2

1. 5 ft   **3.** 36 ft   **5.** 8 mi   **7.** $8\frac{1}{2}$ ft   **9.** $3\frac{1}{3}$ yd   **11.** 33,792 ft   **13.** 13 yd 1 ft   **15.** 3 ft 5 in.   **17.** 1 mi 4720 ft

19. 62 in.   **21.** 17 ft   **23.** 84 in.   **25.** 12 ft 3 in.   **27.** 22 yd 1 ft   **29.** 8 ft 5 in.   **31.** 5 ft 6 in.   **33.** 3 ft 4 in.   **35.** 50 yd 2 ft

37. 10 ft 6 in.   **39.** 13 ft 11 in.   **41.** 15 ft 9 in.   **43.** 3 ft 1 in.   **45.** 86 ft 6 in.   **47.** $105\frac{1}{3}$ yd   **49.** 4000 cm   **51.** 4 cm

53. 0.3 km   **55.** 1.4 m   **57.** 15 m   **59.** 83 mm   **61.** 0.201 dm   **63.** 40 mm   **65.** 8.94 m   **67.** 2.94 m or 2940 mm

69. 1.29 cm or 12.9 mm   **71.** 12.640 km or 12,640 m   **73.** 54.9 m   **75.** 1.55 km   **77.** 9.12 m   **79.** 26.7 mm

81. 41.25 m or 4125 cm   **83.** 3.35 m   **85.** 6.009 km or 6009 m   **87.** 15 tiles   **89.** 21%   **91.** 13%   **93.** 25%

95. Answers may vary.   **97.** 6.575 m

EXERCISE SET 9.3

1. 64 ft   **3.** 36 cm   **5.** 21 in.   **7.** 120 cm   **9.** 48 ft   **11.** 66 in.   **13.** 21 ft   **15.** 60 ft   **17.** 346 yd   **19.** 22 ft

21. $66   **23.** 36 in.   **25.** 28 in.   **27.** $24.08   **29.** 96 m   **31.** 66 ft   **33.** 128 mi   **35.** 17π cm; 53.38 cm

37. 16π mi; 50.24 mi   **39.** 26π m; 81.64 m   **41.** $31\frac{3}{7}$ ft   **43.** 12,560 ft   **45.** 23   **47.** 1   **49.** 10   **51.** 216

53. a. 62.8 m; 125.6 m   **b.** yes   **55.** $(44 + 10\pi)$ m ≈ 75.4 m   **57.** 6 ft   **59.** 27.4 m

EXERCISE SET 9.4

1. 7 sq m   **3.** $9\frac{3}{4}$ sq yd   **5.** 15 sq yd   **7.** 2.25π sq in. ≈ 7.065 sq in.   **9.** 36.75 sq ft   **11.** 28 sq m   **13.** 22 sq yd

15. $36\frac{3}{4}$ sq ft   **17.** $22\frac{1}{2}$ sq in.   **19.** 25 sq cm   **21.** 86 sq mi   **23.** 24 sq cm   **25.** 36π sq in. $\approx 113\frac{1}{7}$ sq in.   **27.** 72 cu in.

29. 512 cu cm   **31.** $12\frac{4}{7}$ cu yd   **33.** $523\frac{17}{21}$ cu in.   **35.** 75 cu cm   **37.** $2\frac{10}{27}$ cu in.   **39.** 8.4 cu ft   **41.** 113,625 sq ft

43. 168 sq ft   **45.** 960 cu cm   **47.** 9200 sq ft   **49.** 381 sq ft   **51.** $\frac{1372}{3}\pi$ cu in. or $\left(457\frac{1}{3}\right)\pi$ cu in.   **53.** $7\frac{1}{2}$ cu ft

55. $12\frac{4}{7}$ cu cm   **57.** 36π cu in. ≈ 113.04 cu in.   **59.** $10\frac{5}{6}$ cu in.   **61.** 25   **63.** 9   **65.** 5   **67.** 20   **69.** perimeter

71. area   **73.** area   **75.** perimeter   **77.** 12-in. pizza   **79.** $1\frac{1}{3}$ sq ft; 192 sq in.   **81.** 298.5 sq m   **83.** 2,583,283 cu m

85. 2,583,669 cu m   **87.** 26,696.5 cu ft   **89.** Answers may vary.   **91.** 4605 ft

INTEGRATED REVIEW

1. 153°; 63°   **2.** $\angle x = 75°$; $\angle y = 105°$; $\angle z = 75°$ **3.** $\angle x = 128°$; $\angle y = 52°$; $\angle z = 128°$ **4.** $\angle x = 52°$ **5.** 3 ft   **6.** 2 mi

7. $6\frac{2}{3}$ yd   **8.** 18 ft   **9.** 11,088 ft   **10.** 38.4 in.   **11.** 3000 cm   **12.** 2.4 cm   **13.** 2 m   **14.** 1800 cm   **15.** 72 mm

16. 0.6 km   **17.** perimeter = 20 m; area = 25 sq m   **18.** perimeter = 12 ft; area = 6 sq ft

19. circumference = 6π cm ≈ 18.84 cm; area = 9π sq cm ≈ 28.26 sq cm   **20.** perimeter = 32 mi; area = 44 sq mi

21. perimeter = 62 ft; area = 238 sq ft   **22.** 64 cu in.   **23.** 30.6 cu ft   **24.** 400 cu cm   **25.** $\frac{2048}{3}\pi$ cu mi $\approx 2145\frac{11}{21}$ cu mi

MENTAL MATH

1. 1 lb   **3.** 2000 lb   **5.** 16 oz   **7.** 1 ton   **9.** no   **11.** yes   **13.** no

EXERCISE SET 9.5

1. 32 oz   **3.** 10,000 lb   **5.** 6 tons   **7.** $3\frac{3}{4}$ lb   **9.** $1\frac{3}{4}$ tons   **11.** 260 oz   **13.** 9800 lb   **15.** 76 oz   **17.** 1.5 tons

19. 53 lb 10 oz   **21.** 9 tons 390 lb   **23.** 3 tons 175 lb   **25.** 8 lb 11 oz   **27.** 31 lb 2 oz   **29.** 1 ton 700 lb   **31.** 5 lb 8 oz

33. 35 lb 14 oz   **35.** 130 lb   **37.** 211 lb   **39.** 5 lb 1 oz   **41.** 0.5 kg   **43.** 4000 mg   **45.** 25,000 g   **47.** 0.048 g

49. 0.0063 kg   **51.** 15,140 mg   **53.** 4010 g   **55.** 13.5 mg   **57.** 5.815 g or 5815 mg   **59.** 1850 mg or 1.850 g

61. 1360 g or 1.360 kg   **63.** 13.52 kg   **65.** 2.125 kg   **67.** 8.064 kg   **69.** 30 mg   **71.** 250 mg   **73.** 144 mg   **75.** 6.12 kg

77. 850 g or 0.85 kg   **79.** 2.38 kg   **81.** 0.25   **83.** 0.16   **85.** 0.875   **87.** Answers may vary.

MENTAL MATH

1. 1 pt   **3.** 1 gal   **5.** 1 qt   **7.** 1 c   **9.** 2 c   **11.** 4 qt   **13.** no   **15.** no

EXERCISE SET 9.6

1. 4 c   **3.** 16 pt   **5.** $2\frac{1}{2}$ gal   **7.** 5 pt   **9.** 8 c   **11.** $3\frac{3}{4}$ qt   **13.** 768 fl oz   **15.** 9 c   **17.** 22 pt   **19.** 10 gal 1 qt

21. 4 c 4 fl oz   **23.** 1 gal 1 qt   **25.** 2 gal 3 qt 1 pt   **27.** 2 qt 1 c   **29.** 17 gal   **31.** 4 gal 3 qt   **33.** 48 fl oz   **35.** 2 qt

37. yes   **39.** 4.3 oz   **41.** 9 qt   **43.** 5000 ml   **45.** 4.5 L   **47.** 0.41 kl   **49.** 0.064 L   **51.** 160 L   **53.** 3600 ml

55. 0.00016 kl   **57.** 22.5 L   **59.** 4.5 L or 4500 ml   **61.** 8410 ml or 8.410 L   **63.** 10,600 ml or 10.6 L   **65.** 3840 ml

67. 162.4 L   **69.** 1.59 L   **71.** 18.954 L   **73.** $0.316   **75.** 944 ml   **77.** $\frac{7}{10}$ **79.** $\frac{3}{100}$ **81.** $\frac{3}{500}$

83. Answers may vary.   **85.** 1.5 cc   **87.** 2.7 cc   **89.** 54u or 0.54cc   **91.** 86u or 0.86cc   **93.** 256 fl drams

EXERCISE SET 9.7

1. 19.55 fl oz   **3.** 218.44 cm   **5.** 40 oz   **7.** 57.66 mi   **9.** 3.77 gal   **11.** 13.64 kg   **13.** 3.94 in.   **15.** 112.7 km per hour

17. 0.008 oz   **19.** yes   **21.** 380 ml   **23.** 90 mm   **25.** 112.5 g   **27.** 104 mph   **29.** 26.24 ft   **31.** 3 mi   **33.** 8 oz

35. A   **37.** B   **39.** C   **41.** D   **43.** D   **45.** -29 **47.** $\frac{1}{9}$ **49.** -15 **51.** -24 **53.** 2.13 sq m   **55.** 1.18 sq m

57. 1.70 sq m   **59.** 21.3 mg $-$ 25.56 mg   **61.** 800 sq m or 8606.72 sq ft

MENTAL MATH

1. yes   **3.** no   **5.** no   **7.** yes

EXERCISE SET 9.8

1. 5°C   **3.** 40°C   **5.** 140°F   **7.** 239°F   **9.** 16.7°C   **11.** 61.2°C   **13.** 197.6°F   **15.** 61.3°F   **17.** 50°C   **19.** 80.6°F
21. 21.1°C   **23.** 37.9°C   **25.** 244.4°F   **27.** 260°C   **29.** 462.2°C   **31.** 62 m   **33.** 15 ft   **35.** 10.4 m   **37.** 4988°C

CHAPTER 9 REVIEW

1. right   **2.** straight   **3.** acute   **4.** obtuse   **5.** 65°   **6.** 75°   **7.** 108°   **8.** 89°   **9.** 58°   **10.** 98°   **11.** 90°

12. 25°   **13.** 133° and 47°   **14.** 43° and 47°; 58° and 32°   **15.** $\angle x = 100°$; $\angle y = 80°$, $\angle z = 80°$

16. $\angle x = 155°$; $\angle y = 155°$; $\angle z = 25°$ **17.** $\angle x = 53°$; $\angle y = 53°$; $\angle z = 127°$ **18.** $\angle x = 42°$; $\angle y = 42°$; $\angle z = 138°$

19. 9 ft   **20.** 24 yd   **21.** 13,200 ft   **22.** 75 in.   **23.** 17 yd 1 ft   **24.** 3 ft 10 in.   **25.** 4200 cm   **26.** 820 mm

27. 0.01218 m   **28.** 0.00231 km   **29.** 21 yd 1 ft   **30.** 7 ft 5 in.   **31.** 41 ft 3 in.   **32.** 3 ft 8 in.   **33.** 9.5 cm or 95 mm

34. 5.26 m or 526 cm   **35.** 9117 m or 9.117 km   **36.** 1.1 m or 1100 mm   **37.** 169 yd 2 ft   **38.** 126 ft 8 in.   **39.** 108.5 km

40. 0.24 sq m   **41.** 88 m   **42.** 30 cm   **43.** 36 m   **44.** 90 ft   **45.** 32 ft   **46.** 440 ft   **47.** 5.338 in.   **48.** 31.4 yd

49. 240 sq ft   **50.** 140 sq m   **51.** 600 sq cm   **52.** 189 sq yd   **53.** 49π sq ft $\approx$ 153.86 sq ft   **54.** 4π sq in. $\approx$ 12.56 sq in.

55. 119 sq in.   **56.** 1248 sq cm   **57.** 144 sq m   **58.** 432 sq ft   **59.** 130 sq ft   **60.** $15\frac{5}{8}$ cu in.   **61.** 84 cu ft

62. $62{,}857\frac{1}{7}$ cu cm   **63.** $346\frac{1}{2}$ cu in.   **64.** $2\frac{2}{3}$ cu ft   **65.** 307.72 cu in.   **66.** $7\frac{1}{2}$ cu ft   **67.** 0.5π cu ft   **68.** 1260 cu ft

69. 28.728 cu ft   **70.** 4.125 lb   **71.** 4600 lb   **72.** 3 lb 4 oz   **73.** 4 tons 200 lb   **74.** 1.4 g   **75.** 40,000 g   **76.** 21 dag

77. 0.0003 dg   **78.** 3 lb 9 oz   **79.** 10 tons 800 lb   **80.** 2 tons 750 lb   **81.** 33 lb 8 oz   **82.** 4.9 g or 4900 mg   **83.** 9 kg or 9000 g

84. 8.1 g or 8100 mg   **85.** 50.4 kg   **86.** 4 lb 4 oz   **87.** 9 tons 1075 lb   **88.** 7.85 kg   **89.** 1.1625 kg   **90.** 8 qt

91. 5 c   **92.** 27 qt   **93.** 17 c   **94.** 4 qt 1 pt   **95.** 3 gal 3 qt   **96.** 3800 ml   **97.** 0.042 dl   **98.** 1.4 kl   **99.** 3060 cl

100. 1 gal 1 qt   **101.** 7 gal 1 qt   **102.** 736 ml or 0.736 L   **103.** 15.5 L or 15,500 ml   **104.** 2 gal 3 qt   **105.** 6 fl oz

106. 10.88 L   **107.** yes, 9.9 L   **108.** 22.96 ft   **109.** 10.55 m   **110.** 4.62 gal   **111.** 8.27 qt   **112.** 425.26 g

113. 10.44 kg   **114.** 109 yd   **115.** 180.4 lb   **116.** 3.18 qt   **117.** 2.36 in.   **118.** 473° F   **119.** 320° F   **120.** 107.6° F

121. 186.8° F   **122.** 34° C   **123.** 11° C   **124.** 5.2° C   **125.** 26.7° C   **126.** 1.7° C   **127.** 329° F

CHAPTER 9 TEST

1. 12°   **2.** 56°   **3.** 50°   **4.** $\angle x = 118°$; $\angle y = 62°$; $\angle z = 118°$ **5.** $\angle x = 73°$; $\angle y = 73°$; $\angle z = 73°$ **6.** 6.2 m   **7.** 10 in.

8. circumference $= 18\pi \approx 56.52$ in.; area $= 81\pi \approx 254.34$ sq in.   **9.** perimeter $= 24.6$ yd; area $= 37.1$ sq yd

10. perimeter $= 68$ in.; area $= 185$ sq in.   **11.** $62\frac{6}{7}$ cu in.   **12.** 30 cu ft   **13.** 16 in.   **14.** 18 cu ft   **15.** 198.08 oz

16. 23 ft 4 in.   **17.** 10 qt   **18.** 1.875 lb   **19.** 5600 lb   **20.** 2400 m   **21.** 36 mm   **22.** 0.43 g   **23.** 830 ml   **24.** 15.4 lb

25. 21.59 cm   **26.** 28.9° C   **27.** 54.7° F   **28.** 5.6 m   **29.** 4 gal 3 qt   **30.** 91.4 m   **31.** 16 ft 6 in. or $16\frac{1}{2}$ ft   **32.** 37,336 pt

33. 0.2 oz   **34.** 18.95 L   **35.** 3.1 mi

CUMULATIVE REVIEW

1. 5; Sec. 3.4, Ex. 3   **2.** 6   **3. a.** $\frac{16}{625}$ **b.** $\frac{1}{16}$; Sec. 4.3, Ex. 9   **4. a.** $-\frac{1}{27}$ **b.** $\frac{9}{49}$ **5.** $8\frac{3}{10}$; Sec. 4.8, Ex. 13   **6.** $9\frac{11}{15}$

7. $20x - 10.9$; Sec. 5.2, Ex. 12   **8.** $1.2y + 1.8$ **9.** 6.35; Sec. 5.5, Ex. 7   **10.** 2.16   **11.** $=$; Sec. 5.6, Ex. 7   **12.** $>$

13. -1.3; Sec. 5.7, Ex. 6 **14.** 30 **15.** 424 ft; Sec. 5.8, Ex. 12 **16.** 236 ft **17. a.** $\dfrac{5}{7}$ **b.** $\dfrac{7}{24}$; Sec. 6.1, Ex. 7 **18. a.** $\dfrac{1}{4}$

b. $\dfrac{4}{9}$ **19.** $\dfrac{5 \text{ nails}}{3 \text{ feet}}$; Sec. 6.2, Ex. 1 **20.** $\dfrac{1 \text{ chaperone}}{5 \text{ students}}$ **21.** 0.44; Sec. 6.3, Ex. 10 **22.** 1.02 **23.** 22.4 cc; Sec. 6.4, Ex. 2

24. 7.5 cups **25.** 17%; Sec. 7.1, Ex. 1 **26.** 17% **27.** 200; Sec. 7.2, Ex. 10 **28.** 1200 **29.** 2.7; Sec. 7.3, Ex. 7 **30.** 12.6

31. 32%; Sec. 7.4, Ex. 5 **32.** 27% **33.** sales tax: $6.41, total price: $91.91; Sec. 7.5; Ex. 1 **34.** tax: $30; total: $405

35. $A(-4, 2)$, $B(1, 2)$, $C(0, 1)$, $D(-3, 0)$, $E(5, -4)$; Sec. 8.3, Ex. 2 **36.** $A(2, -3)$, $B(-5, 0)$, $C(0, 4)$, $D(-3, -2)$

37. 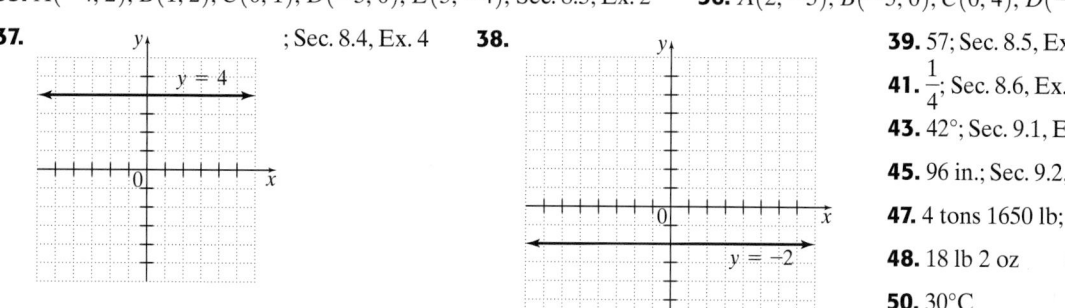 ; Sec. 8.4, Ex. 4 **38.**

39. 57; Sec. 8.5, Ex. 3 **40.** 83

41. $\dfrac{1}{4}$; Sec. 8.6, Ex. 5 **42.** $\dfrac{2}{7}$

43. 42°; Sec. 9.1, Ex. 4 **44.** 43°

45. 96 in.; Sec. 9.2, Ex. 1 **46.** 21 ft

47. 4 tons 1650 lb; Sec. 9.5, Ex. 3

48. 18 lb 2 oz **49.** 15°C; Sec. 9.8, Ex. 3

50. 30°C

Chapter 10 POLYNOMIALS

CHAPTER 10 PRETEST

1. $7y^2 - 23y + 26$; 10.1A **2.** $17b - 12$; 10.1B **3.** $5z^4 - z^3 - 6z^2 + 3z$; 10.1B **4.** -20; 10.1C **5.** $36m^{18}$; 10.2A
6. $120x^3$; 10.2A **7.** t^{54}; 10.2B **8.** $81n^8$; 10.2C **9.** $15{,}552a^{12}b^{13}c^{17}$; 10.2C **10.** $18d^5 - 10d^3 + 22d$; 10.3A
11. $x^2 + 9x + 18$; 10.3B **12.** $2y^2 - 3y - 20$; 10.3B **13.** $9x^2 - 12x + 4$; 10.3C **14.** $n^2 - 100$; 10.3C
15. $8a^3 - 2a^2 + 27a + 7$; 10.3E **16.** 9; 10.4A **17.** y^3; 10.4B **18.** $2m$; 10.4B **19.** $2(5y^2 + 3y - 7)$; 10.4C
20. $4n^3(2n^3 - 3n^2 + 6)$; 10.4C

EXERCISE SET 10.1

1. $-5x - 24$ **3.** $-9z^2 - 2z + 6$ **5.** $9y^2 + 25y - 40$ **7.** $-4.3a^4 - 2a^2 + 9$ **9.** $4a - 8$ **11.** $-2x^2 + 4x + 1$
13. $-20y^3 + 12y^2 - 4$ **15.** $-x + 16$ **17.** $8y^2 - 10y - 8$ **19.** $5x - 12$ **21.** $y - 4$ **23.** $4x^2 + x - 16$
25. $9x - 4.1$ **27.** $15a - 7$ **29.** $-15y + 3.6$ **31.** $19t^2 - 11t + 3$ **33.** $b^3 - b^2 + 7b - 1$ **35.** $-5x^2 - 11x + 13$
37. $\dfrac{9}{7}$ **39.** 1 **41.** -5 **43.** -8 **45.** 20 **47.** 25 **49.** 50 **51.** 576 ft **53.** $3200 **55.** 909 ft **57.** 81
59. 25 **61.** x^3 **63.** 2^2a^4 **65.** $(8x + 2)$ in. **67.** $(4x - 15)$ units **69.** 20; 6; 2 **71.** 7.2752
73. 29 ft; -243 ft; Answers may vary.

EXERCISE SET 10.2

1. x^{14} **3.** a^7 **5.** $15z^5$ **7.** $-40x^2$ **9.** $25x^6y^4$ **11.** $28a^5b^6$ **13.** $42x^3$ **15.** $12a^{17}$ **17.** x^{15} **19.** z^{20} **21.** b^{62}
23. $81a^4$ **25.** $a^{33}b^{24}$ **27.** $121x^6y^{12}$ **29.** $-24y^{22}$ **31.** $256x^9y^{13}$ **33.** $7x - 21$ **35.** $-6a - 4b$ **37.** $9x + 18y - 27$
39. $16x^{12}$ sq in. **41.** $12a^4b^5$ sq m **43.** $18{,}003{,}384a^{45}b^{30}$ **45.** $34{,}867.84401x^{50}$ **47.** Answers may vary. **49.** $a^{127}b^{86}c^{25}$

INTEGRATED REVIEW

1. $4x - 1$ **2.** $33y - 8$ **3.** $10x + 3$ **4.** $-5y - 4$ **5.** $3a^3 + 5a - 8$ **6.** $1.8y^2 - 2.4y - 5.6$ **7.** $-1.8x^2 + 2.4x - 1.7$
8. $-4a^3 - 12a^2$ **9.** $6x + 7$ **10.** $2x^2 + 3x - 12$ **11.** -1 **12.** 26 **13.** x^{18} **14.** x^{12} **15.** y^4 **16.** a^{11} **17.** x^{77}
18. x^{36} **19.** x^{42} **20.** y^{27} **21.** $16x^4$ **22.** $27y^3$ **23.** $-12x^2y^7$ **24.** $12a^3b^4$ **25.** $x^{36}y^{20}$ **26.** $a^{20}b^{24}$ **27.** $300x^4y^3$
28. $128y^6z^7$ **29.** $144a^{38}b^{12}$ **30.** $125x^{22}y^{28}$

EXERCISE SET 10.3

1. $27x^3 - 9x$ **3.** $-20a^3 + 30a^2 - 5a$ **5.** $42x^4 - 35x^3 + 49x^2$ **7.** $x^2 + 13x + 30$ **9.** $2x^2 + 2x - 24$

11. $36a^2 + 48a + 16$ **13.** $a^3 - 33a + 18$ **15.** $8x^3 + 2x^2 - 55x + 50$ **17.** $x^5 + 4x^4 + 6x^3 + 7x^2 + 2x$ **19.** $-30r^2 + 20r$

21. $-6y^3 - 2y^4 + 12y^2$ **23.** $x^2 + 14x + 24$ **25.** $4a^2 - 9$ **27.** $x^2 + 10x + 25$ **29.** $b^2 + \dfrac{7}{5}b + \dfrac{12}{25}$

31. $6x^3 + 25x^2 + 10x + 1$ **33.** $49x^2 + 70x + 25$ **35.** $4x^2 - 4x + 1$ **37.** $8x^5 - 8x^3 - 6x^2 - 6x + 9$

39. $x^5 + 2x^4 + 3x^3 + 2x^2 + x$ **41.** $10z^4 - 3z^3 + 3z - 2$ **43.** $2 \cdot 5^2$ **45.** $2^3 \cdot 3^2$ **47.** $2^3 \cdot 5^2$

49. $(y^3 - 3y^2 - 16y - 12)$ sq ft   **51.** $(x^4 - 3x^2 + 1)$ sq m   **53.** Answers may vary.

Exercise Set 10.4

1. 3   **3.** 12   **5.** 4   **7.** 4   **9.** y^2 **11.** a^5 **13.** xy^2 **15.** x   **17.** $2z^3$ **19.** $3y(y + 6)$ **21.** $5a^6(2 - a^2)$

23. $4x(x^2 + 3x + 5)$ **25.** $z^5(z^2 - 6)$ **27.** $-7(5 - 2y + y^2)$ or $7(-5 + 2y - y^2)$ **29.** $12a^5(1 - 3a)$ **31.** 36   **33.** $\dfrac{4}{5}$

35. 37.5%   **37.** $x^2 + 2x$ **39.** Answers may vary.

Chapter 10 Review

1. $10b - 3$ **2.** $21s - 15$ **3.** $-x + 2.8$ **4.** $-y - 12$ **5.** $4z^2 + 11z - 6$ **6.** $17a^3 + 25a^2$ **7.** $-11y^2 - y + \dfrac{3}{4}$

8. $x^2 - 7x + 3$ **9.** 45   **10.** -19 **11.** $(26x + 28)$ ft   **12.** x^{24} **13.** y^7 **14.** $24z^7$ **15.** $-15x^3y^5$ **16.** a^{35} **17.** x^{28}

18. $81b^2$ **19.** $a^{20}b^{10}c^5$ **20.** $56x^{16}$ **21.** $108x^{30}y^{25}$ **22.** $81a^{14}$ sq mi   **23.** $10a^3 - 12a$ **24.** $-3y^4 + 6y^3 - 3y^2$

25. $x^2 + 8x + 12$ **26.** $15x^2 - 32x + 9$ **27.** $y^2 - 10y + 25$ **28.** $49a^2 + 14a + 1$ **29.** $x^3 - x^2 + x + 3$

30. $8y^4 + 4y^3 - 2y^2 - 3y - 3$ **31.** $3z^4 + 5z^3 + 6z^2 + 3z + 1$ **32.** $(a^3 + 5a^2 - 5a + 6)$ sq cm   **33.** 5   **34.** 4   **35.** 6

36. 5   **37.** x^2 **38.** y^7 **39.** xy   **40.** a^5b^2 **41.** $5a$ **42.** $4y^2z$ **43.** $2x(x + 6)$ **44.** $6a(a - 2)$ **45.** $y^4(6 - y^2)$

46. $7(x^2 - 2x + 1)$ **47.** $a^3(5a^4 - a + 1)$ **48.** $10y(y^5 - 1)$

Chapter 10 Test

1. $15x - 4$ **2.** $7x - 2$ **3.** $3.4y^2 + 2y - 3$ **4.** $-2a^2 + a + 1$ **5.** 17   **6.** y^{14} **7.** y^{33} **8.** $16x^8$ **9.** $-12a^{10}$

10. p^{54} **11.** $72a^{20}b^5$ **12.** $10x^3 + 6.5x$ **13.** $-2y^4 - 12y^3 + 8y$ **14.** $x^2 - x - 6$ **15.** $25x^2 + 20x + 4$ **16.** $a^3 + 8$

17. perimeter: $(14x - 4)$ in.; area: $(5x^2 + 33x - 14)$ sq in.   **18.** 15   **19.** $3y^3$ **20.** $3y(y - 5)$ **21.** $2a(5a + 6)$

22. $6(x^2 - 2x - 5)$ **23.** $x^3(7x^3 - 6x + 1)$

Cumulative Review

1. 106,400 sq mi; Sec. 1.6, Ex. 6   **2.** 147 trees   **3.** -8; Sec. 2.2, Ex. 15   **4.** -10 **5.** -7; Sec. 2.3, Ex. 6   **6.** -3

7. 1; Sec. 2.3, Ex. 7   **8.** 5   **9.** 4; Sec. 3.2, Ex. 5   **10.** -2 **11.** $\dfrac{32}{12x}$; Sec. 4.1, Ex. 21   **12.** $\dfrac{12}{8c}$ **13.** $3\dfrac{4}{7}$; Sec. 4.8, Ex. 16

14. $10\dfrac{3}{5}$ **15.** 736.2; Sec. 5.1, Ex. 15   **16.** 328.2   **17.** 25.454; Sec. 5.2, Ex. 1   **18.** 17.052   **19.** no; Sec. 5.3, Ex. 13   **20.** yes

21. 0.0786; Sec. 5.4, Ex. 5   **22.** 0.818   **23.** 0.012; Sec. 5.4, Ex. 6   **24.** 0.0503   **25.** -2.6; Sec. 5.5, Ex. 9   **26.** -13.6

27. 3.14; Sec. 5.6, Ex. 5   **28.** 1.947   **29.** $\dfrac{1}{6}$; Sec. 5.8, Ex. 5   **30.** $\dfrac{2}{5}$ **31.** 46 ft; Sec.6.5, Ex. 4   **32.** 3 ft

33. $1.2 = 30\% \cdot x$; Sec. 7.2, Ex. 2   **34.** $9 = 45\% \cdot x$ **35.** 16%; Sec. 7.3, Ex. 10   **36.** 25%   **37.** 775 freshmen; Sec. 7.4, Ex. 1

38. 1450 apples   **39.** $160; Sec. 7.6, Ex. 2   **40.** $25   **41.** 71%; Sec. 8.2, Ex. 2   **42.** 91%   **43.** 28 in.; Sec. 9.3, Ex. 2

44. 25 ft   **45.** 5.1 sq mi; Sec. 9.4, Ex. 2   **46.** 68 sq in.   **47.** 33 lb 6 oz; Sec. 9.5, Ex. 4   **48.** 16 tons 100 lb

49. 3.21 L; Sec. 9.6, Ex. 7   **50.** 43.21 L

Solutions to Selected Exercises

Chapter 1

EXERCISE SET 1.2

1. The place value of the 5 in 352 is tens.
5. The place value of the 5 in 62,500,000 is hundred-thousands.
9. 5420 is written as five thousand, four hundred twenty.
13. 1,620,000 is written as one million, six hundred twenty thousand.
17. 63,960 is written as sixty-three thousand, nine hundred sixty.
21. 12,662 is written as twelve thousand, six hundred sixty-two.
25. 3895 is written as three thousand, eight hundred ninety-five.
29. Twenty-nine thousand, nine hundred in standard form is 29,900.
33. Three million, fourteen in standard form is 3,000,014.
37. Seven hundred fifty-five in standard form is 755.
41. One thousand, eight hundred fifteen in standard form is 1815.
45. Forty-five thousand in standard form is 45,000.
49. $5290 = 5000 + 200 + 90$
53. $30,680 = 30,000 + 600 + 80$
57. $1006 = 1000 + 6$
61. $9 > 0$
65. $22 > 0$
69. $4145 = 4000 + 100 + 40 + 5$
73. Labrador retrievers have the most registrations; 165,970 is written as one hundred sixty-five thousand, nine hundred seventy.
77. The largest number is 7632.
81. Based on the information and the diagram given, the Pro-Football Hall of Fame is located in Canton.

EXERCISE SET 1.3

1.
$$\begin{array}{r} 14 \\ + 22 \\ \hline 36 \end{array}$$

5.
$$\begin{array}{r} 12 \\ 13 \\ + 24 \\ \hline 49 \end{array}$$

9.
$$\begin{array}{r} \overset{1}{5}3 \\ + 64 \\ \hline 117 \end{array}$$

13.
$$\begin{array}{r} \overset{11}{3}8 \\ + 79 \\ \hline 117 \end{array}$$

17.
$$\begin{array}{r} \overset{2}{6} \\ 21 \\ 14 \\ 9 \\ + 12 \\ \hline 62 \end{array}$$

21.
$$\begin{array}{r} \overset{1}{6}2 \\ 18 \\ + 14 \\ \hline 94 \end{array}$$

25.
$$\begin{array}{r} \overset{111}{7}542 \\ 49 \\ + 682 \\ \hline 8273 \end{array}$$

29.
$$\begin{array}{r} \overset{12}{6}27 \\ 628 \\ + 629 \\ \hline 1884 \end{array}$$

33.
$$\begin{array}{r} \overset{111}{5}07 \\ 593 \\ + 10 \\ \hline 1110 \end{array}$$

37.
$$\begin{array}{r} \overset{1122}{4}9 \\ 628 \\ 5762 \\ + 29,462 \\ \hline 35,901 \end{array}$$

41. $8 + 3 + 5 + 7 + 5 + 1 = 8 + 1 + 3 + 7 + 5 + 5$
$$= 9 + 10 + 10$$
$$= 29$$
The perimeter is 29 inches.

45. Opposite sides of a rectangle have the same length.
$4 + 8 + 4 + 8 = 12 + 12 = 24$
The perimeter is 24 inches.

49.
$$\begin{array}{r} 3560 \\ + 3124 \\ \hline 6684 \end{array}$$
Mt. Mitchell is 6684 feet high.

53.
$$\begin{array}{r} \overset{1}{2}34,500 \\ + 29,200 \\ \hline 263,700 \end{array}$$
263,700 motorcycles were shipped in 2002.

57.
$$\begin{array}{r} \overset{11}{7}981 \\ + 564 \\ \hline 8545 \end{array}$$
There were 8545 stores in 2002.

61.
$$\begin{array}{r} \overset{11}{1}795 \\ + 11,460 \\ \hline 13,255 \end{array}$$
The total highway mileage in Alaska is 13,255 miles.

65.
$$\begin{array}{r} \overset{11}{9}821 \\ + 592 \\ \hline 10,413 \end{array}$$
There were 10,413 organ transplants that involved a kidney.

69.
$$\overset{1}{3}42$$
$$182$$
$$+\ 154$$
$$678$$

Texas, Florida, and California have the most Wal-Mart stores, with a total of 678 stores.

73. Answers may vary.

77. The computation is incorrect; answers may vary.

EXERCISE SET 1.4

1.
$$67$$
$$-\ 23$$
$$44$$

Check:
$$44$$
$$+\ 23$$
$$67$$

5.
$$389$$
$$-\ 124$$
$$265$$

Check:
$$265$$
$$+\ 124$$
$$389$$

9.
$$998$$
$$-\ 453$$
$$545$$

Check:
$$545$$
$$+\ 453$$
$$998$$

13.
$$62$$
$$-\ 37$$
$$25$$

Check:
$$\overset{1}{2}5$$
$$+\ 37$$
$$62$$

17.
$$938$$
$$-\ 792$$
$$146$$

Check:
$$\overset{1}{1}46$$
$$+\ 792$$
$$938$$

21.
$$600$$
$$-\ 432$$
$$168$$

Check:
$$\overset{1\ 1}{1}68$$
$$+\ 432$$
$$600$$

25.
$$923$$
$$-\ 476$$
$$447$$

Check:
$$\overset{1\ 1}{4}47$$
$$+\ 476$$
$$923$$

29.
$$533$$
$$-\ 29$$
$$504$$

Check:
$$\overset{1}{5}04$$
$$+\ 29$$
$$533$$

33.
$$1983$$
$$-\ 1904$$
$$79$$

Check:
$$\overset{1}{7}9$$
$$+\ 1904$$
$$1983$$

37.
$$50,000$$
$$-\ 17,289$$
$$32,711$$

Check:
$$\overset{1\ 1\ \ 1\ 1}{3}2,711$$
$$+\ 17,289$$
$$50,000$$

41.
$$51,111$$
$$-\ 19,898$$
$$31,213$$

Check:
$$\overset{1\ 1\ \ 1\ 1}{3}1,213$$
$$+\ 19,898$$
$$51,111$$

45.
$$41$$
$$-\ 21$$
$$20$$

Check:
$$20$$
$$+\ 21$$
$$41$$

49.
$$108$$
$$-\ 36$$
$$72$$

Check:
$$72$$
$$+\ 36$$
$$108$$

53.
$$503$$
$$-\ 239$$
$$264$$

Dyllis must read 264 pages.

57.
$$\begin{array}{r} 20{,}320 \\ -\ 14{,}255 \\ \hline 6{,}065 \end{array}$$
Mt. McKinley is 6065 feet higher than Long's Peak.

61.
$$\begin{array}{r} 645 \\ -\ 287 \\ \hline 358 \end{array}$$
The distance between Hays and Denver is 358 miles.

65.
$$\begin{array}{r} 2380 \\ -\ 2207 \\ \hline 173 \end{array}$$
Stackhouse scored 173 more points than Iverson.

69. The total number of votes cast for Jo was:
$$\begin{array}{r} \overset{1\,2\,1}{276} \\ 362 \\ 201 \\ +\ 179 \\ \hline 1018 \end{array}$$
The total number of votes cast for Trudy was:
$$\begin{array}{r} \overset{1\,1}{295} \\ 122 \\ 312 \\ +\ 18 \\ \hline 747 \end{array}$$
Since more votes were cast for Jo than for Trudy, Jo won the election.
$$\begin{array}{r} 1018 \\ -\ 747 \\ \hline 271 \end{array}$$
Jo won by 271 votes.

73.
$$\begin{array}{r} 100{,}000 \\ -\ 94{,}080 \\ \hline 5920 \end{array}$$
The Dole Plantation maze is 5920 square feet larger than the Ruurlo maze.

77. The busiest airport is the one corresponding to the highest bar, which is Atlanta Hartsfield International.

81. General Motors Corp. spent $3,374,400,000 on advertising, which is more than $3 billion.

85.
$$\begin{array}{r} 3{,}374{,}400{,}000 \\ 2{,}540{,}600{,}000 \\ 2{,}408{,}200{,}000 \\ 2{,}210{,}400{,}000 \\ 2{,}189{,}500{,}000 \\ 1{,}985{,}300{,}000 \\ 1{,}885{,}300{,}000 \\ 1{,}815{,}700{,}000 \\ 1{,}757{,}300{,}000 \\ +\ 1{,}618{,}100{,}000 \\ \hline 21{,}784{,}800{,}000 \end{array}$$
The top ten companies spent $21,784,800,000 on advertising.

89. Answers may vary.

EXERCISE SET 1.5

1. To round 632 to the nearest ten, observe that the digit in the ones place is 2. Since this digit is less than 5, we do not add 1 to the digit in the tens place. The number 632 rounded to the nearest ten is 630.

5. To round 792 to the nearest ten, observe that the digit in the ones place is 2. Since this digit is less than 5, we do not add 1 to the digit in the tens place. The number 792 rounded to the nearest ten is 790.

9. To round 1096 to the nearest ten, observe that the digit in the ones place is 6. Since this digit is at least 5, we need to add 1 to the digit in the tens place. The number 1096 rounded to the nearest ten is 1100.

13. To round 248,695 to the nearest hundred, observe that the digit in the tens place is 9. Since this digit is at least 5, we need to add 1 to the digit in the hundreds place. The number 248,695 rounded to the nearest hundred is 248,700.

17. To round 99,995 to the nearest ten, observe that the digit in the ones place is 5. Since this digit is at least 5, we need to add 1 to the digit in the tens place. The number 99,995 rounded to the nearest ten is 100,000.

21. Estimate 5281 to a given place value by rounding it to that place value. 5281 estimated to the tens place value is 5280, to the hundreds place value is 5300, and to the thousands place value is 5000.

25. Estimate 14,876 to a given place value by rounding it to that place value. 14,876 estimated to the tens place value is 14,880, to the hundreds place value is 14,900, and to the thousands place value is 15,000.

29. To round 38,387 to the nearest thousand, observe that the digit in the hundreds place is 3. Since this digit is less than 5, we do not add 1 to the digit in the thousands place. Therefore, 38,387 points rounded to the nearest thousand is 38,000 points.

33. To round 18,188 to the nearest thousand, observe that the digit in the hundreds place is 1. Since this digit is less than 5, we do not add 1 to the digit in the thousands place. Therefore, 18,188 women rounded to the nearest thousand is 18,000 women.

37.
$$\begin{array}{rll} 649 & \text{rounds to} & 650 \\ -\ 272 & \text{rounds to} & -\ 270 \\ \hline & & 380 \end{array}$$
The estimated difference is 380.

41.
$$\begin{array}{rll} 1774 & \text{rounds to} & 1800 \\ -\ 1492 & \text{rounds to} & -\ 1500 \\ \hline & & 300 \end{array}$$
The estimated difference is 300.

45. $362 + 419$ is approximately $360 + 420 = 780$.
The answer of 781 is correct.

49. $7806 + 5150$ is approximately $7800 + 5200 = 13{,}000$.
The answer of 12,956 is correct.

53.
$$\begin{array}{rll} 799 & \text{rounds to} & 800 \\ 1299 & \text{rounds to} & 1300 \\ +\ 999 & \text{rounds to} & +\ 1000 \\ \hline & & 3100 \end{array}$$
The total cost is approximately $3100.

57.

20,320	rounds to	20,000
− 14,410	rounds to	− 14,000
		6000

The difference in elevation is approximately 6000 feet.

61.

41,126,233	rounds to	41,000,000
− 27,174,898	rounds to	− 27,000,000
		14,000,000

Johnson won the election by approximately 14,000,000 votes.

65.

3274	rounds to	3300
− 2159	rounds to	− 2200
		1100

The difference in diameter is approximately 1100 miles.

69. 2,210,400,000 rounds to 2,000,000,000. PepsiCo spent approximately $2,000,000,000.

73. The largest possible number that rounds to 1,500,000 is 1,549,999.

77. Answers may vary.

EXERCISE SET 1.6

1. $4(3 + 9) = 4 \cdot 3 + 4 \cdot 9$

5. $10(11 + 7) = 10 \cdot 11 + 10 \cdot 7$

9.
$$\begin{array}{r} 624 \\ \times\ 3 \\ \hline 1872 \end{array}$$

13.
$$\begin{array}{r} 1062 \\ \times\ 5 \\ \hline 5310 \end{array}$$

17.
$$\begin{array}{r} 231 \\ \times\ 47 \\ \hline 1617 \\ 9240 \\ \hline 10,857 \end{array}$$

21.
$$\begin{array}{r} 620 \\ \times\ 40 \\ \hline 0 \\ 24,800 \\ \hline 24,800 \end{array}$$

25. $(590)(1)(10) = 5900$

29.
$$\begin{array}{r} 609 \\ \times\ 234 \\ \hline 2436 \\ 18,270 \\ 121,800 \\ \hline 142,506 \end{array}$$

33.
$$\begin{array}{r} 1941 \\ \times\ 235 \\ \hline 9705 \\ 58,230 \\ 388,200 \\ \hline 456,135 \end{array}$$

37.
$$\begin{array}{r} 964 \\ \times\ 207 \\ \hline 6748 \\ 0 \\ 192,800 \\ \hline 199,548 \end{array}$$

41.

604	rounds to	600
× 451	rounds to	× 500
		300,000

604×451 is approximately 300,000.

45.

36	rounds to	40
× 87	rounds to	× 90
		3600

36×87 is approximately 3600.

49. Area = length · width
= (30 feet)(13 feet)
= 390 square feet
The area is 390 square feet.

53.
$$\begin{array}{r} 54 \\ \times\ 35 \\ \hline 270 \\ 1620 \\ \hline 1890 \end{array}$$
The total cost for books is $1890.

57. Area = length · width
= (90 feet)(110 feet)
= 9900 square feet
The area is 9900 square feet.

61.
$$\begin{array}{r} 62 \\ \times\ 94 \\ \hline 248 \\ 5580 \\ \hline 5828 \end{array}$$
There are 5828 pixels on a screen.

65.
$$\begin{array}{r} 160 \\ \times\ 8 \\ \hline 1280 \end{array}$$
There are 1280 calories in 8 ounces of peanuts.

69. On 2 sides of the building there are $23 \times 7 = 161$ windows on each side, or $161 \times 2 = 322$ windows. On the other 2 sides of the building there are $23 \times 4 = 92$ windows on each side, or $92 \times 2 = 184$ windows. The total number of windows is $322 + 184 = 506$ windows.

73.

905,235	rounds to	905,000
× 6633	rounds to	× 7000
		6,335,000,000

The total cost of the Head Start program in 2001 was approximately $6,335,000,000.

77. The fruits that correspond to the most baskets are apple and orange, so apple and orange were the two most popular fruits.

81. Answers may vary.

EXERCISE SET 1.7

1.
$$\begin{array}{r} 12 \\ 9\overline{)108} \\ \underline{-9} \\ 18 \\ \underline{-18} \\ 0 \end{array}$$
Check: $12 \cdot 9 = 108$

5.
$$
\begin{array}{r}
338 \\
3\overline{)1014} \\
\underline{-9} \\
11 \\
\underline{-9} \\
24 \\
\underline{-24} \\
0
\end{array}
$$
Check: $338 \cdot 3 = 1014$

9.
$$
\begin{array}{r}
563 \ \text{R } 1 \\
2\overline{)1127} \\
\underline{-10} \\
12 \\
\underline{-12} \\
07 \\
\underline{-6} \\
1
\end{array}
$$
Check: $563 \cdot 2 + 1 = 1127$

13.
$$
\begin{array}{r}
265 \ \text{R } 1 \\
8\overline{)2121} \\
\underline{-16} \\
52 \\
\underline{-48} \\
41 \\
\underline{-40} \\
1
\end{array}
$$
Check: $265 \cdot 8 + 1 = 2121$

17.
$$
\begin{array}{r}
13 \\
55\overline{)715} \\
\underline{-55} \\
165 \\
\underline{-165} \\
0
\end{array}
$$
Check: $13 \cdot 55 = 715$

21.
$$
\begin{array}{r}
206 \\
18\overline{)3708} \\
\underline{-36} \\
10 \\
\underline{-0} \\
108 \\
\underline{-108} \\
0
\end{array}
$$
Check: $206 \cdot 18 = 3708$

25.
$$
\begin{array}{r}
202 \ \text{R } 7 \\
46\overline{)9299} \\
\underline{-92} \\
09 \\
\underline{-0} \\
99 \\
\underline{-92} \\
7
\end{array}
$$
Check: $202 \cdot 46 + 7 = 9299$

29.
$$
\begin{array}{r}
98 \ \text{R } 100 \\
103\overline{)10194} \\
\underline{-927} \\
924 \\
\underline{-824} \\
100
\end{array}
$$
Check: $98 \cdot 103 + 100 = 10{,}194$

33.
$$
\begin{array}{r}
202 \\
223\overline{)45046} \\
\underline{-446} \\
44 \\
\underline{-0} \\
446 \\
\underline{-446} \\
0
\end{array}
$$
Check: $202 \cdot 223 = 45{,}046$

37.
$$
\begin{array}{r}
252000 \\
21\overline{)5292000} \\
\underline{-42} \\
109 \\
\underline{-105} \\
42 \\
\underline{-42} \\
00 \\
\underline{-0} \\
00 \\
\underline{-0} \\
00 \\
\underline{-0} \\
0
\end{array}
$$
Each person received $252,000.

41.
$$
\begin{array}{r}
88 \ \text{R } 1 \\
3\overline{)265} \\
\underline{-24} \\
25 \\
\underline{-24} \\
1
\end{array}
$$
There are 88 bridges every 3 miles over 265 miles plus the first bridge for a total of 89 bridges.

45.
$$
\begin{array}{r}
24 \\
6\overline{)144} \\
\underline{-12} \\
24 \\
\underline{-24} \\
0
\end{array}
$$
He scored 24 touchdowns during 2002.

49.
$$
\begin{array}{r}
14 \\
22 \\
45 \\
18 \\
30 \\
+\ 27 \\
\hline
156
\end{array}
\qquad
\begin{array}{r}
26 \\
6\overline{)156} \\
\underline{-12} \\
36 \\
\underline{-36} \\
0
\end{array}
$$
Average $= \dfrac{156}{6} = 26$

53.

$$
\begin{array}{r}
86 \\
79 \\
81 \\
69 \\
+\ 80 \\
\hline
395
\end{array}
$$

$$
\begin{array}{r}
79 \\
5\overline{)395} \\
\underline{-35} \\
45 \\
\underline{-45} \\
0
\end{array}
$$

Average $= \dfrac{395}{5} = 79$

57. The top two companies are General Motors Corp. and Procter & Gamble Co.

$$
\begin{array}{r}
3{,}374{,}400{,}000 \\
+\ 2{,}540{,}600{,}000 \\
\hline
5{,}915{,}000{,}000
\end{array}
$$

$$
\begin{array}{r}
2957500000 \\
2\overline{)5915000000} \\
\underline{-4} \\
19 \\
\underline{-18} \\
11 \\
\underline{-10} \\
15 \\
\underline{-14} \\
10 \\
\underline{-10} \\
00 \\
\underline{-0} \\
00 \\
\underline{-0} \\
00 \\
\underline{-0} \\
00 \\
\underline{-0} \\
00 \\
\underline{-0} \\
0
\end{array}
$$

The average amount spent was $2,957,500,000.

61. No, the average is not 86; answers may vary.

Exercise Set 1.8

1. $3 \cdot 3 \cdot 3 \cdot 3 = 3^4$

5. $12 \cdot 12 \cdot 12 = 12^3$

9. $9 \cdot 9 \cdot 9 \cdot 8 = (9 \cdot 9 \cdot 9)8 = 9^3 \cdot 8$

13. $3 \cdot 2 \cdot 2 \cdot 5 \cdot 5 \cdot 5 = 3(2 \cdot 2)(5 \cdot 5 \cdot 5) = 3 \cdot 2^2 \cdot 5^3$

17. $5^3 = 5 \cdot 5 \cdot 5 = 125$

21. $2^{10} = 2 \cdot 2 \cdot 2 \cdot 2 \cdot 2 \cdot 2 \cdot 2 \cdot 2 \cdot 2 \cdot 2 = 1024$

25. $3^5 = 3 \cdot 3 \cdot 3 \cdot 3 \cdot 3 = 243$

29. $4^3 = 4 \cdot 4 \cdot 4 = 64$

33. $9^3 = 9 \cdot 9 \cdot 9 = 729$

37. $10^4 = 10 \cdot 10 \cdot 10 \cdot 10 = 10{,}000$

41. $1920^1 = 1920$

45. $15 + 3 \cdot 2 = 15 + 6 = 21$

49. $5 \cdot 9 - 16 = 45 - 16 = 29$

53. $14 + \dfrac{24}{8} = 14 + 3 = 17$

57. $0 \div 6 + 4 \cdot 7 = 0 + 28 = 28$

61. $(6 + 8) \div 2 = 14 \div 2 = 7$

65. $(3 + 5^2) \div 2 = (3 + 25) \div 2 = 28 \div 2 = 14$

69. $\dfrac{18 + 6}{2^4 - 4} = \dfrac{24}{16 - 4} = \dfrac{24}{12} = 2$

73. $\dfrac{7(9 - 6) + 3}{3^2 - 3} = \dfrac{7 \cdot 3 + 3}{9 - 3} = \dfrac{21 + 3}{6} = \dfrac{24}{6} = 4$

77.
$$
\begin{aligned}
3^4 - [35 - (12 - 6)] &= 3^4 - [35 - 6] \\
&= 81 - 29 \\
&= 52
\end{aligned}
$$

81.
$$
\begin{aligned}
8 \cdot [4 + (6 - 1) \cdot 2] - 50 \cdot 2 &= 8 \cdot [4 + 5 \cdot 2] - 50 \cdot 2 \\
&= 8 \cdot [4 + 10] - 100 \\
&= 8 \cdot 14 - 100 \\
&= 112 - 100 \\
&= 12
\end{aligned}
$$

85.
$$
\begin{aligned}
\text{Area of a square} &= (\text{side})^2 \\
&= (20 \text{ miles})^2 \\
&= 400 \text{ square miles}
\end{aligned}
$$

89.
$$
\begin{aligned}
\text{Area of base} &= (\text{side})^2 \\
&= (100 \text{ meters})^2 \\
&= 10{,}000 \text{ square meters}
\end{aligned}
$$

93.
$$
\begin{aligned}
24 \div (3 \cdot 2) + 2 \cdot 5 &= 24 \div 6 + 2 \cdot 5 \\
&= 4 + 10 \\
&= 14
\end{aligned}
$$

97.
$$
\begin{aligned}
(7 + 2^4)^5 - (3^5 - 2^4)^2 &= (7 + 16)^5 - (3^5 - 2^4)^2 \\
&= 23^5 - (243 - 16)^2 \\
&= 6{,}436{,}343 - 227^2 \\
&= 6{,}436{,}343 - 51{,}529 \\
&= 6{,}384{,}814
\end{aligned}
$$

Exercise Set 1.9

1.
$$
\begin{aligned}
3 + 2z &= 3 + 2(3) \\
&= 3 + 6 \\
&= 9
\end{aligned}
$$

5.
$$
\begin{aligned}
z - x + y &= 3 - 2 + 5 \\
&= 6
\end{aligned}
$$

9.
$$
\begin{aligned}
y^3 - 4x &= 5^3 - 4(2) \\
&= 125 - 4(2) \\
&= 125 - 8 \\
&= 117
\end{aligned}
$$

13.
$$
\begin{aligned}
8 - (y - x) &= 8 - (5 - 2) \\
&= 8 - 3 \\
&= 5
\end{aligned}
$$

17.
$$
\begin{aligned}
\dfrac{6xy}{z} &= \dfrac{6 \cdot 2 \cdot 5}{3} \\
&= \dfrac{60}{3} \\
&= 20
\end{aligned}
$$

21.
$$
\begin{aligned}
\dfrac{x + 2y}{z} &= \dfrac{2 + 2 \cdot 5}{3} \\
&= \dfrac{2 + 10}{3} \\
&= \dfrac{12}{3} \\
&= 4
\end{aligned}
$$

25.
$$
\begin{aligned}
2y^2 - 4y + 3 &= 2 \cdot 5^2 - 4 \cdot 5 + 3 \\
&= 2 \cdot 25 - 4 \cdot 5 + 3 \\
&= 50 - 20 + 3 \\
&= 33
\end{aligned}
$$

29. $(xy + 1)^2 = (2 \cdot 5 + 1)^2$
$= (10 + 1)^2$
$= (11)^2$
$= 121$

33. $xy(5 + z - x) = 2 \cdot 5(5 + 3 - 2)$
$= 2 \cdot 5(8 - 2)$
$= 2 \cdot 5(6)$
$= 10(6)$
$= 60$

37.

t	1	2	3	4
$16t^2$	$16(1)^2$	$16(2)^2$	$16(3)^2$	$16(4)^2$
	$16 \cdot 1$	$16 \cdot 4$	$16 \cdot 9$	$16 \cdot 16$
	16	64	144	256

41. $x + 8$

45. $512x$

49. $5x + (17 + x)$

53. $11 - x$

57. $6 \div x$ or $\dfrac{6}{x}$

61. $x^4 - y^2 = (23)^4 - (72)^2$
$= 279{,}841 - 5184$
$= 274{,}657$

65. Compare expressions:
$\dfrac{x}{3} = \left(\dfrac{1}{3}\right)x$
$\left(\dfrac{1}{3}\right)x < 2x < 5x$
$5x$ is the largest.

Chapter 1 Test

1. $59 + 82 = 141$

5. $2^3 \cdot 5^2 = 8 \cdot 25 = 200$

9. $62 \div 0$ is undefined.

13. 52,369 rounded to the nearest thousand is 52,000.

17.
$\begin{array}{r} 17 \\ \times\ 7 \\ \hline 119 \end{array}$
The total cost of the tickets is $119.

21. $5[(2)^3 - 2] = 5[8 - 2] = 5 \cdot 6 = 30$

25. Opposite sides of a rectangle have the same length. Find the perimeter by adding the lengths of the sides.
$20 + 10 + 20 + 10 = 60$.
The perimeter is 60 yards.
$\text{Area} = \text{length} \cdot \text{width}$
$= (20 \text{ yards})(10 \text{ yards})$
$= 200 \text{ square yards}$

Chapter 2

Exercise Set 2.1

1. If 0 represents ground level, then 1445 feet underground is -1445.

5. If 0 represents zero degrees Fahrenheit, then 118 degrees above zero is $+118$.

9. -1683 million

13. If 0 represents 0% a loss of 81% is -81.

17.
$-7\ -6\ -5\ -4\ -3\ -2\ -1\ \ 0\ \ 1$

21.
$-7\ -6\ -5\ -4\ -3\ -2\ -1\ \ 0\ \ 1$

25. $-7 < -5$

29. $-26 < 26$

33. $|-8| = 8$, because -8 is 8 units from 0.

37. $|-5| = 5$, because -5 is 5 units from 0.

41. The opposite of -4 is $-(-4) = 4$.

45. The opposite of -10 is $-(-10) = 10$.

49. $-|20| = -20$
The opposite of the absolute value of 20 is the opposite of 20.

53. $-(-8) = 8$
The opposite of negative 8 is 8.

57. $-(-29) = 29$
The opposite of negative 29 is 29.

61. $-|-3| = -3$

65. $-|4| = -4$

69. $|-9| ? |-14|$
$9 ? 14$
$9 < 14$

73. $-|-10| ? -(-10)$
$-10 ? 10$
$-10 < 10$

77. $|0| ? |-9|$
$0 ? 9$
$0 < 9$

81. $-(-12) ? -(-18)$
$12 < 18$

85. $|-1| = 1$, $-|-6| = -6$, $-(-6) = 6$, and $-|1| = -1$, so the integers in order from least to greatest are $-|-6|, -|1|, |-1|, -(-6)$.

89. Sea level corresponds to 0 feet above or below sea level on the graph. The lake with an elevation at sea level is the one that corresponds to the bar at 0. Maracaibo Lake has an elevation at sea level.

93. The planet with an average temperature closest to $0°$ F is the one that corresponds to the smallest bar. Earth has an average temperature closest to $0°$ F.

97. $0 + 13 = 13$

101.
$\begin{array}{r} 47 \\ 236 \\ +\ 77 \\ \hline 360 \end{array}$

105. $-(-|-5|) = -(-5) = 5$

109. True; consider the values on a number line.

113. Answers may vary.

Exercise Set 2.2

1.

5. $-13 + 7 = -6$

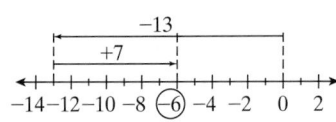

9. $|-6| + |-2| = 8$

Common sign is negative, so answer is -8.

13. $|6| - |-2| = 4$

$6 > 2$ so answer is $+4$.

17. $|-5| - |3| = 2$

$5 > 3$ so answer is -2.

21. $|-12| + |-12| = 24$

Common sign is negative, so answer is -24.

25. $|-123| + |-100| = 223$

Common sign is negative, so answer is -223.

29. $|12| - |-5| = 7$

$12 > 5$, so answer is $+7$.

33. $|-12| - |3| = 9$

$12 > 3$, so answer is -9.

37. $|57| - |-37| = 20$

$57 > 37$, so answer is $+20$.

41. $|-67| - |34| = 33$

$67 > 34$, so answer is -33.

45. $|-82| + |-43| = 125$

Common sign is negative, so -125.

49. $-52 + (-77) + (-117)$
$\qquad = -129 + (-117) \qquad$ add from left to right
$\qquad = -246 \qquad$ add from left to right

53. $(-10) + 14 + 25 + (-16)$
$\qquad = 4 + 25 + (-16) \qquad$ add from left to right
$\qquad = 29 + (-16) \qquad$ add from left to right
$\qquad = 13 \qquad$ add from left to right

57. $-26 + 5 = -21$

61. $-14 + (-31) = -45$

65. $-87 + 87 = 0$

69. $0 + (-103) = -103$

73. $x + y$
$\qquad = (-20) + (-50) \qquad$ substitute values
$\qquad = -70 \qquad$ add from left to right

77. The bar that corresponds to income for 2000 has a height of 236,630 (thousand dollars). Gateway, Inc.'s net income in 2000 was $236,630,000.

81. $-10 + 12 = 2$

The temperature at 11 p.m. was $2\,^{\circ}$C.

85. Team 1 Total $= -2 + (-13) + 20 + 2$
$\qquad\qquad\qquad = 7$
Team 2 Total $= 5 + 11 + (-7) + (-3)$
$\qquad\qquad\qquad = 6$
$7 > 6$ so Team 1 is the winning team.

89. $-230 + (-370) + (-346)$
$\qquad = -600 + (-346)$
$\qquad = -946$
The total U.S. trade balance was $-\$946$ billion.

93. $52 - 52 = 0$

97. $-8602 + (-1056) = -9658$

101. True; add any two negative numbers on a number line to verify.

105. Answers may vary.

Exercise Set 2.3

1. $5 - 5 = 5 + (-5) = 0$

5. $3 - 8 = 3 + (-8) = -5$

9. $-5 - (-8) = -5 + 8 = 3$

13. $2 - 16 = 2 + (-16) = -14$

17. $-15 - (-15) = -15 + 15 = 0$

21. $30 - 45 = 30 + (-45) = -15$

25. $-230 - 0 = -230$

29. $-7 - (-3) = -7 + 3 = -4$

33. $-20 - 18 = -20 + (-18) = -38$

37. $2 - (-11) = 2 + 11 = 13$

41. $12 - 5 - 7 = 12 + (-5) + (-7)$
$\qquad\qquad\qquad = 7 + (-7)$
$\qquad\qquad\qquad = 0$

45. $-10 + (-5) - 12 = -10 + (-5) + (-12)$
$\qquad\qquad\qquad\qquad = -15 + (-12)$
$\qquad\qquad\qquad\qquad = -27$

49. $-(-6) - 12 + (-16) = 6 + (-12) + (-16)$
$\qquad\qquad\qquad\qquad = -6 + (-16)$
$\qquad\qquad\qquad\qquad = -22$

53. $-3 + 4 - (-23) - 10 = -3 + 4 + 23 + (-10)$
$\qquad\qquad\qquad\qquad\qquad = 1 + 23 + (-10)$
$\qquad\qquad\qquad\qquad\qquad = 24 + (-10)$
$\qquad\qquad\qquad\qquad\qquad = 14$

57. $x - y = 6 - (-30) \qquad$ show substitution exactly
$\qquad\quad = 6 + (30) \qquad$ change subtraction to addition
$\qquad\quad = 36$

61. $x - y = 1 - (-18) \qquad$ show substitution exactly
$\qquad\quad = 1 + (18) \qquad$ change subtraction to addition
$\qquad\quad = 19$

65. The two planets with the lowest temperature correspond to the two longest bars on the graph below the horizontal line representing $0\,^{\circ}$F. Neptune's average temperature is $-330\,^{\circ}$F and Pluto's average temperature is $-369\,^{\circ}$F.
$-330 - (-369) = -330 + (-369)$
$\qquad\qquad\qquad = 39$
The difference in average temperature is $39\,^{\circ}$F.

69. Subtract check amounts from the checking account, and add the deposit.
$125 - 117 + 45 - 69$
$\quad = 125 + (-117) + 45 + (-69)$
$\quad = 8 + 45 + (-69)$
$\quad = 53 + (-69)$
$\quad = -16$
Aaron has overdrawn his checking account by $16.

73. $-282 - (-436) = -282 + 436$
$\qquad\qquad\qquad = 154$
The difference in elevation is 154 feet.

77. $600 - (-52) = 600 + 52 = 652$ feet

81. $731 - 1141 = 731 + (-1141) = -410$
The U.S. trade balance in 2001 was $-\$410$ billion

85. $436 \cdot 1 = 436$

89. $5^2 - 6 \cdot 2 + 8 = 25 - 12 + 8$
$\qquad\qquad\qquad = 13 + 8$
$\qquad\qquad\qquad = 21$

93. $a + b - c = -16 + 14 - (-22) \qquad$ show substitution exactly
$\qquad\qquad = -16 + 14 + (22) \qquad$ change subtraction to addition
$\qquad\qquad = -2 + 22 \qquad$ add left to right
$\qquad\qquad = 20$

97. $|-6| - |6| = 6 - 6 = 0$
101. False. $|-8 - 3| = |-8 + (-3)| = |-11| = 11$
105. Answers may vary.

Exercise Set 2.4

1. $-2(-3) = 6$
5. $8(-8) = -64$
9. $6(-4)(2) = -24(2) = -48$
13. $-4(4)(-5) = -16(-5) = 80$
17. $-5(3)(-1)(-1) = -15(-1)(-1)$
$$= 15(-1)$$
$$= -15$$
21. $(-3)^3 = (-3)(-3)(-3) = 9(-3) = -27$
25. $(-2)^3 = (-2)(-2)(-2) = 4(-2) = -8$
29. $\dfrac{-30}{6} = -5$
33. $\dfrac{0}{14} = 0$
37. $\dfrac{39}{-3} = -13$
41. $-4(3) = -12$
45. $-7(-6) = 42$
49. $(-4)^2 = (-4)(-4) = 16$
53. $-\dfrac{56}{8} = -7$
57. $4(-4)(-3) = -16(-3) = 48$
61. $3 \cdot (-2) \cdot 0 = (-6) \cdot 0 = 0$
65. $240 \div (-40) = -6$
69. $-1^4 = -(1 \cdot 1 \cdot 1 \cdot 1) = -1$
73. $-2(3)(5)(-6) = (-6)(5)(-6)$
$$= (-30)(-6)$$
$$= 180$$
77. $-2(-2)(-5) = 4(-5) = -20$
81.
$$25$$
$$\underline{\times\ 82}$$
$$50$$
$$\underline{2000}$$
$$2050$$
$$25 \cdot (-82) = -2050$$
85. $a \cdot b = (3) \cdot (-2) = -6$
89. $\dfrac{x}{y} = \dfrac{5}{-5}$ show substitution exactly
$$= -1$$
93. $\dfrac{x}{y} = \dfrac{-36}{-6}$ show substitution exactly
$$= 6$$
97. $x \cdot y = (0)(-6)$ show substitution exactly
$$= 0$$ multiplication property of zero
$\dfrac{x}{y} = \dfrac{0}{-6}$ show substitution exactly
$$= 0$$ zero division property
101. $(-3)(63) = -189$
The melting point of argon is $-189°$C.
105. $(-20)(5) = -100$
He is at a depth of 100 feet.
109. a. $192 - 27 = 165$
There were 165 more condors in 2002 than in 1987.

b. $\dfrac{165}{15} = 11$

Over the five years from 1987 to 2002 the average change in the number of condors per year was 11 condors per year.
113. $(3 \cdot 5)^2 = (3 \cdot 5)(3 \cdot 5) = (15)(15) = 225$
117. $12 \div 4 - 2 + 7 = 3 - 2 + 7 = 1 + 7 = 8$
121. True; a is positive so $-a$ is negative, and the product of two negatives is positive.
125. $(-a)^5$ is negative because a is positive and 5 is odd.
$(-b)^4$ is positive because 4 is even.
The product $(-a)^5(-b)^4$ is negative.
129. a. $2190 - 2131 = 59$
The number of country music radio stations in 2002 changed by -59 from the number in 2001.

b. $4(-59) = -236$
If this change continues, there will be 236 fewer radio stations four years after 2002.
c. Based on part b there will be $2131 - 236$ or 1895 radio stations in 2006.
133. $87^2 - (-12)^5 = 7569 - (-248,832)$
$$= 7569 + 248,832$$
$$= 256,401$$

Exercise Set 2.5

1. $(-4)^3 = (-4)(-4)(-4)$
$$= 16(-4)$$
$$= -64$$
5. $6 \cdot 2^2$
$$= 6 \cdot 4 \qquad \text{exponents first}$$
$$= 24$$
9. $9 - 12 - 4$
$$= 9 + (-12) + (-4) \quad \text{change to addition}$$
$$= -3 + (-4)$$
$$= -7$$
13. $5(-9) + 2$
$$= -45 + 2 \qquad \text{multiply before adding}$$
$$= -43$$
17. $6 + 7 \cdot 3 - 40$
$$= 6 + 21 - 40 \qquad \text{multiply before adding}$$
$$= 6 + 21 + (-40) \quad \text{change to addition}$$
$$= 27 + (-40)$$
$$= -13$$
21. $\dfrac{24}{10 + (-4)}$
$$= \dfrac{24}{6} \qquad \text{simplify the bottom of the fraction bar first}$$
$$= 4$$
25. $(-19) - 12(3)$
$$= (-19) - 36 \qquad \text{multiply before adding}$$
$$= (-19) + (-36) \quad \text{change to addition}$$
$$= -55$$
29. $\left(8 + (-4)\right)^2$
$$= (4)^2 \qquad \text{grouping symbols first}$$
$$= 16$$

33. $16 - (-3)^4$
$= 16 - 81$ exponents first
$= 16 + (-81)$ change to addition
$= -65$

37. $7 \cdot 8^2 + 4$
$= 7 \cdot 64 + 4$ exponents first
$= 448 + 4$ multiply before adding
$= 452$

41. $|3 - 12| \div 3$
$= |-9| \div 3$ grouping symbols first
$= 9 \div 3$
$= 3$

45. $(5 - 9)^2 \div (4 - 2)^2$
$= (-4)^2 \div (2)^2$ grouping symbols first
$= 16 \div 4$ exponents next
$= 4$

49. $(-12 - 20) \div 16 - 25$
$= (-32) \div 16 - 25$ grouping symbols first
$= -2 - 25$ divide before subtracting
$= -27$

53. $(2 - 7) \cdot (6 - 19)$
$= (-5) \cdot (-13)$ grouping symbols first
$= 65$

57. $(-36 \div 6) - (4 \div 4)$
$= -6 - 1$ divide before subtracting
$= -7$

61. $(-5)^2 - 6^2$
$= 25 - 36$ exponents first (watch the signs)
$= -11$

65. $2(8 - 10)^2 - 5(1 - 6)^2$
$= 2(-2)^2 - 5(-5)^2$
$= 2(-2)(-2) - 5(-5)(-5)$
$= 2(-2)(-2) + (-5)(-5)(-5)$
$= (-4)(-2) + (25)(-5)$
$= 8 + (-125)$
$= -117$

69.
$$\frac{(-7)(-3) - (4)(3)}{3[7 \div (3 - 10)]} = \frac{(-7)(-3) - (4)(3)}{3[7 \div (3 + (-10))]}$$
$$= \frac{21 + (-12)}{3[7 \div (-7)]}$$
$$= \frac{9}{3(-1)}$$
$$= \frac{9}{-3}$$
$$= -3$$

73. $x + y + z$
$= -2 + 4 + (-1)$ show substitution exactly
$= 2 + (-1)$
$= 1$

77. $x^2 - y$
$= (-2)^2 - 4$ show substitution exactly
$= (-2)(-2) - 4$
$= 4 - 4$
$= 0$

81. $x^2 = (-3)^2$ show substitution exactly
$= (-3)(-3)$
$= 9$

85. $10 - x^2 = 10 - (-3)^2$ show substitution exactly
$= 10 - (-3)(-3)$ multiplication first
$= 10 - 9$
$= 1$

89. average $= \dfrac{-10 + 8 + (-4) + 2 + 7 + (-5) + (-12)}{7}$
$= \dfrac{-14}{7}$
$= -2$

93. $|-13 - 15| = |-13 + (-15)|$
$= |-28|$
$= 28$
The difference between the lowest and highest scores is 28 points.

97. No; answers may vary.

101.
$$\begin{array}{r} 90 \\ -\ 45 \\ \hline 45 \end{array}$$

105. perimeter $= 2(9) + 2(6)$
$= 18 + 12$
$= 30$ feet

109. $-6 \cdot (10 - 4) = -6 \cdot 6 = -36$

113. $x^3 - y^2 = 21^3 - (-19)^2$ substitute
$= 21 \cdot 21 \cdot 21 - (-19)(-19)$
$= 21 \cdot 21 \cdot 21 + (19)(-19)$
$= 441 \cdot 21 + (-361)$
$= 9261 + (-361)$
$= 8900$

117. Answers may vary.

Chapter 2 Test

1. $-5 + 8 = 3$

5. $(-18) + (-12) = -30$

9. $|-25| + (-13) = 25 + (-13) = 12$

13. $(-8) + 9 \div (-3) = -8 + (-3) = -11$

17. $-(-7)^2 \div 7 \cdot (-4) = -49 \div 7 \cdot (-4)$
$= -7 \cdot (-4)$
$= 28$

21. $\dfrac{(-3)(-2) + 12}{-1(-4 - 5)} = \dfrac{6 + 12}{-1(-9)} = \dfrac{18}{9} = 2$

25. $3x + y = 3(0) + (-3)$ substitute exactly
$= 0 + (-3)$
$= 0 - 3$
$= -3$

29. $10 - y^2 = 10 - (-3)^2$ substitute exactly
$= 10 - (-3)(-3)$
$= 10 - 9$
$= 1$

33. $6288 - (-25,354) = 6288 + 25,354 = 31,642$
The difference in elevation is 31,642 feet. As an integer, the answer is 31,642.

Chapter 3

Exercise Set 3.1

1. $3x + 5x = (3 + 5)x = 8x$

5. $4c + c - 7c = (4 + 1 - 7)c = -2c$

9. $4a + 3a + 6a - 8 = (4 + 3 + 6)a - 8 = 13a - 8$

13. $-2(11y) = (-2 \cdot 11)y = -22y$

17. $2(y + 2) = 2 \cdot y + 2 \cdot 2 = 2y + 4$

21. $-4(3x + 7) = -4 \cdot 3x + (-4) \cdot 7$
$= -12x - 28$

25. $-4(6n - 5) + 3n = -4 \cdot 6n + (-4) \cdot (-5) + 3n$
$= -24n + 20 + 3n$
$= -21n + 20$

29. $3 + 6(w + 2) + w = 3 + 6 \cdot w + 6 \cdot 2 + w$
$= 3 + 6w + 12 + w$
$= 6w + w + 3 + 12$
$= 7w + 15$

33. $-(5x - 1) - 10 = -1(5x - 1) - 10$
$= -1 \cdot 5x + (-1) \cdot (-1) - 10$
$= -5x + 1 - 10$
$= -5x - 9$

37. $z - 8z = (1 - 8)z = -7z$

41. $2y - 6 + 4y - 8 = 2y + 4y - 6 - 8$
$= 6y - 14$

45. $2(x + 1) + 20 = 2 \cdot x + 2 \cdot 1 + 20$
$= 2x + 2 + 20$
$= 2x + 22$

49. $-5(z + 3) + 2z = -5 \cdot z + (-5) \cdot 3 + 2z$
$= -5z - 15 + 2z$
$= -5z + 2z - 15$
$= -3z - 15$

53. $-7(x + 5) + 5(2x + 1)$
$= -7 \cdot x + (-7) \cdot 5 + 5 \cdot 2x + 5 \cdot 1$
$= -7x - 35 + 10x + 5$
$= -7x + 10x - 35 + 5$
$= 3x - 30$

57. $-3(n - 1) - 4n = -3 \cdot n - (-3) \cdot 1 - 4n$
$= -3n + 3 - 4n$
$= -3n - 4n + 3$
$= -7n + 3$

61. $6(2x - 1) - 12x = 6 \cdot 2x - 6 \cdot 1 - 12x$
$= 12x - 6 - 12x$
$= 12x - 12x - 6$
$= -6$

65. $-(4xy - 10) + 2(3xy + 5)$
$= -1(4xy - 10) + 2 \cdot 3xy + 2 \cdot 5$
$= -1 \cdot 4xy - (-1) \cdot 10 + 6xy + 10$
$= -4xy + 10 + 6xy + 10$
$= -4xy + 6xy + 10 + 10$
$= 2xy + 20$

69. $5y - 2(y - 1) + 3 = 5y - 2y - 2(-1) + 3$
$= 3y + 2 + 3$
$= 3y + 5$

73. $5y + 16 + 3y + 4y + 2y + 6$
$= 5y + 3y + 4y + 2y + 16 + 6$
$= (14y + 22)$ meters

77. Area $= (\text{side})^2$
$= (4z)^2$
$= (4z)(4z)$
$= (4)(4)(z)(z)$
$= 16z^2$ square centimeters

81. $-13 + 10 = -3$

85. $-4 + 4 = 0$

89. Answers may vary.

93. Answers may vary.

Exercise Set 3.2

1. $x - 8 = 2$
$10 - 8 \overset{?}{=} 2$
$2 \overset{?}{=} 2$ True
Yes, 10 is a solution.

5. $x + 12 = 7$
$-5 + 12 \overset{?}{=} 7$
$7 \overset{?}{=} 7$ True
Yes, -5 is a solution.

9. $h - 8 = -8$
$0 - 8 \overset{?}{=} -8$
$-8 \overset{?}{=} -8$ True
Yes, 0 is a solution.

13. $a + 5 = 23$
$a + 5 - 5 = 23 - 5$
$a = 18$
Check:
$a + 5 = 23$
$18 + 5 \overset{?}{=} 23$
$23 \overset{?}{=} 23$ True
The solution is 18.

17. $7 = y - 2$
$7 + 2 = y - 2 + 2$
$9 = y$
Check:
$7 = y - 2$
$7 \overset{?}{=} 9 - 2$
$7 \overset{?}{=} 7$ True
The solution is 9.

21. $3x = 2x + 11$
$3x - 2x = 2x + 11 - 2x$
$x = 11$
Check:
$3x = 2x + 11$
$3(11) \overset{?}{=} 2(11) + 11$
$33 \overset{?}{=} 22 + 11$
$33 \overset{?}{=} 33$ True
The solution is 11.

25. $x - 3 = -1 + 4$
$x - 3 = 3$
$x - 3 + 3 = 3 + 3$
$x = 6$
Check:
$x - 3 = -1 + 4$
$6 - 3 \overset{?}{=} -1 + 4$
$3 \overset{?}{=} 3$ True
The solution is 6.

29. $-7 + 10 = m - 5$
$3 = m - 5$
$3 + 5 = m - 5 + 5$
$8 = m$
Check:
$-7 + 10 = m - 5$
$-7 + 10 \overset{?}{=} 8 - 5$
$3 \overset{?}{=} 3$ True
The solution is 8.

33.
$$2(5x - 3) = 11x$$
$$2 \cdot 5x - 2 \cdot 3 = 11x$$
$$10x - 6 = 11x$$
$$10x - 6 - 10x = 11x - 10x$$
$$-6 = x$$
Check:
$$2(5x - 3) = 11x$$
$$2[5 \cdot (-6) - 3] \stackrel{?}{=} 11 \cdot (-6)$$
$$2(-30 - 3) \stackrel{?}{=} -66$$
$$2(-33) \stackrel{?}{=} -66$$
$$-66 \stackrel{?}{=} -66 \text{ True}$$
The solution is -6.

37.
$$-8x + 4 + 9x = -1 + 7$$
$$x + 4 = 6$$
$$x + 4 - 4 = 6 - 4$$
$$x = 2$$
Check:
$$-8x + 4 + 9x = -1 + 7$$
$$-8(2) + 4 + 9(2) \stackrel{?}{=} -1 + 7$$
$$-16 + 4 + 18 \stackrel{?}{=} 6$$
$$6 \stackrel{?}{=} 6 \text{ True}$$
The solution is 2.

41.
$$7x + 14 - 6x = -4 - 10$$
$$x + 14 = -14$$
$$x + 14 - 14 = -14 - 14$$
$$x = -28$$
Check:
$$7x + 14 - 6x = -4 - 10$$
$$7(-28) + 14 - 6(-28) \stackrel{?}{=} -4 - 10$$
$$-196 + 14 + 168 \stackrel{?}{=} -14$$
$$-14 \stackrel{?}{=} -14 \text{ True}$$
The solution is -28.

45.
$$67 = z + 67$$
$$67 + (-67) = z + 67 + (-67)$$
$$0 = z$$

49.
$$z - 23 = -88$$
$$z - 23 + 23 = -88 + 23$$
$$z = -65$$

53.
$$-12 + x = -15$$
$$-12 + x + 12 = -15 + 12$$
$$x = -3$$

57.
$$8(3x - 2) = 25x$$
$$8 \cdot 3x - 8 \cdot 2 = 25x$$
$$24x - 16 = 25x$$
$$24x - 16 - 24x = 25x - 24x$$
$$-16 = x$$

61.
$$50y = 7(7y + 4)$$
$$50y = 7 \cdot 7y + 7 \cdot 4$$
$$50y = 49y + 28$$
$$50y - 49y = 49y + 28 - 49y$$
$$y = 28$$

65. In 2000 there were about 2400 trumpeter swans. In 1985 there were about 200 trumpeter swans. Thus, there were about $2400 - 200 = 2200$ more trumpeter swans in 2000 than in 1985.

69. $\dfrac{-3}{-3} = 1$

73. No; answers may vary.

77.
$$5^3 = x + 4^4$$
$$125 = x + 256$$
$$125 - 256 = x$$
$$-131 = x$$

81.
$$T = P + R$$
$$5560 = P + 1933$$
$$5560 - 1933 = P + 1933 - 1933$$
$$3627 = P$$
The Packers gained 3627 yards by passing.

Exercise Set 3.3

1.
$$5x = 20$$
$$\frac{5x}{5} = \frac{20}{5}$$
$$x = 4$$

5.
$$\frac{x}{7} = 1$$
$$7 \cdot \frac{x}{7} = 7 \cdot 1$$
$$x = 7$$

9.
$$4y = 0$$
$$\frac{4y}{4} = \frac{0}{4}$$
$$y = 0$$

13.
$$\frac{x}{-8} = -4$$
$$(-8) \cdot \frac{x}{-8} = (-8)(-4)$$
$$x = 32$$

17.
$$-3x = -15$$
$$\frac{-3x}{-3} = \frac{-15}{-3}$$
$$x = 5$$

21.
$$2w - 12w = 40$$
$$-10w = 40$$
$$\frac{-10w}{-10} = \frac{40}{-10}$$
$$w = -4$$

25.
$$2z = 12 - 14$$
$$2z = -2$$
$$\frac{2z}{2} = \frac{-2}{2}$$
$$z = -1$$

29.
$$-3x - 3x = 50 - 2$$
$$-6x = 48$$
$$\frac{-6x}{-6} = \frac{48}{-6}$$
$$x = -8$$

33.
$$-10x = 10$$
$$\frac{-10x}{-10} = \frac{10}{-10}$$
$$x = -1$$

37.
$$0 = \frac{x}{3}$$
$$3 \cdot 0 = 3 \cdot \frac{x}{3}$$
$$0 = x$$

41.
$$10z - 3z = -63$$
$$7z = -63$$
$$\frac{7z}{7} = \frac{-63}{7}$$
$$z = -9$$

45.
$$12 = 13y - 10y$$
$$12 = 3y$$
$$\frac{12}{3} = \frac{3y}{3}$$
$$4 = y$$

49. $18 - 11 = \dfrac{x}{-5}$

$\qquad\quad 7 = \dfrac{x}{-5}$

$\qquad -5 \cdot 7 = -5 \cdot \dfrac{x}{-5}$

$\qquad\quad -35 = x$

53. $10p - 11p = 25$

$\qquad\quad -1p = 25$

$\qquad\quad \dfrac{-1p}{-1} = \dfrac{25}{-1}$

$\qquad\qquad p = -25$

57. $10 = 7t - 12t$

$\qquad 10 = -5t$

$\qquad \dfrac{10}{-5} = \dfrac{-5t}{-5}$

$\qquad -2 = t$

61. $4r - 9r = -20$

$\qquad\quad -5r = -20$

$\qquad\quad \dfrac{-5r}{-5} = \dfrac{-20}{-5}$

$\qquad\qquad r = 4$

65. $3w - 12w = -27$

$\qquad\quad -9w = -27$

$\qquad\quad \dfrac{-9w}{-9} = \dfrac{-27}{-9}$

$\qquad\qquad w = 3$

69. $23x - 25x = 7 - 9$

$\qquad\quad -2x = -2$

$\qquad\quad \dfrac{-2x}{-2} = \dfrac{-2}{-2}$

$\qquad\qquad x = 1$

73. Let x represent a number. The product of -11 and a number is $-11x$.

77. Let x represent a number. Eleven subtracted from a number is $x - 11$.

81. Let x represent a number. The product of -13 and a number is $-13x$.

85. Let x represent a number. The product of 4 and a number is $4x$. Thus, seven added to the product of 4 and a number is $4x + 7$.

89. Let x represent a number. The sum of a number and 15 is $x + 15$. Thus, the product of -6 and the sum of a number and 15 is $-6(x + 15)$.

93. $3x + 10 = 3(-5) + 10 = -15 + 10 = -5$

97. $\dfrac{3x + 4}{x + 4} = \dfrac{3(-5) + 4}{-5 + 4} = \dfrac{-15 + 4}{-5 + 4} = \dfrac{-11}{-1} = 11$

101. $-25x = 900$

$\qquad \dfrac{-25x}{-25} = \dfrac{900}{-25}$

$\qquad\quad x = -36$

105. $\dfrac{x}{-2} = 5^2 - |-10| - (-9)$

$\qquad \dfrac{x}{-2} = 25 - 10 + 9$

$\qquad \dfrac{x}{-2} = 15 + 9$

$\qquad \dfrac{x}{-2} = 24$

$\qquad -2 \cdot \dfrac{x}{-2} = -2 \cdot 24$

$\qquad\qquad x = -48$

109. $d = r \cdot t$

$\qquad 232 = r(4)$

$\qquad \dfrac{232}{4} = \dfrac{4r}{4}$

$\qquad 58 = r$

The driver should drive 58 miles per hour.

EXERCISE SET 3.4

1. $2x - 6 = 0$

$\quad 2x - 6 + 6 = 0 + 6$

$\qquad\quad 2x = 6$

$\qquad\quad \dfrac{2x}{2} = \dfrac{6}{2}$

$\qquad\quad x = 3$

5. $6 - n = 10$

$\quad 6 - n - 6 = 10 - 6$

$\qquad\quad -n = 4$

$\qquad\quad \dfrac{-n}{-1} = \dfrac{4}{-1}$

$\qquad\quad n = -4$

9. $3x - 7 = 4x + 5$

$\quad 3x - 7 + 7 = 4x + 5 + 7$

$\qquad\quad 3x = 4x + 12$

$\quad 3x - 4x = 4x + 12 - 4x$

$\qquad\quad -x = 12$

$\qquad\quad \dfrac{-x}{-1} = \dfrac{12}{-1}$

$\qquad\quad x = -12$

13. $-2(y + 4) = 2$

$\qquad -2y - 8 = 2$

$\quad -2y - 8 + 8 = 2 + 8$

$\qquad\quad -2y = 10$

$\qquad\quad \dfrac{-2y}{-2} = \dfrac{10}{-2}$

$\qquad\quad y = -5$

17. $8 - t = 3$

$\quad 8 - t - 8 = 3 - 8$

$\qquad\quad -t = -5$

$\qquad\quad \dfrac{-t}{-1} = \dfrac{-5}{-1}$

$\qquad\quad t = 5$

21. $\dfrac{x}{-2} - 8 = 0$

$\quad \dfrac{x}{-2} - 8 + 8 = 0 + 8$

$\qquad\quad \dfrac{x}{-2} = 8$

$\quad -2 \cdot \dfrac{x}{-2} = -2 \cdot 8$

$\qquad\quad x = -16$

25. $3r + 4 = 19$

$\quad 3r + 4 - 4 = 19 - 4$

$\qquad\quad 3r = 15$

$\qquad\quad \dfrac{3r}{3} = \dfrac{15}{3}$

$\qquad\quad r = 5$

29. $2 = 3z - 4$

$\quad 2 + 4 = 3z - 4 + 4$

$\qquad\quad 6 = 3z$

$\qquad\quad \dfrac{6}{3} = \dfrac{3z}{3}$

$\qquad\quad 2 = z$

33.
$$-7c + 1 = -20$$
$$-7c + 1 - 1 = -20 - 1$$
$$-7c = -21$$
$$\frac{-7c}{-7} = \frac{-21}{-7}$$
$$c = 3$$

37.
$$8m + 79 = -1$$
$$8m + 79 - 79 = -1 - 79$$
$$8m = -80$$
$$\frac{8m}{8} = \frac{-80}{8}$$
$$m = -10$$

41.
$$-5 = -13 - 8k$$
$$-5 + 13 = -13 - 8k + 13$$
$$8 = -8k$$
$$\frac{8}{-8} = \frac{-8k}{-8}$$
$$-1 = k$$

45.
$$-2y - 10 = 5y + 18$$
$$-2y - 10 + 10 = 5y + 18 + 10$$
$$-2y = 5y + 28$$
$$-2y - 5y = 5y + 28 - 5y$$
$$-7y = 28$$
$$\frac{-7y}{-7} = \frac{28}{-7}$$
$$y = -4$$

49.
$$9 - 3x = 14 + 2x$$
$$9 - 3x - 9 = 14 + 2x - 9$$
$$-3x = 5 + 2x$$
$$-3x - 2x = 5 + 2x - 2x$$
$$-5x = 5$$
$$\frac{-5x}{-5} = \frac{5}{-5}$$
$$x = -1$$

53.
$$2t - 1 = 3(t + 7)$$
$$2t - 1 = 3t + 21$$
$$2t - 1 + 1 = 3t + 21 + 1$$
$$2t = 3t + 22$$
$$2t - 3t = 3t + 22 - 3t$$
$$-1t = 22$$
$$\frac{-t}{-1} = \frac{22}{-1}$$
$$t = -22$$

57.
$$10 + 5(z - 2) = 4z + 1$$
$$10 + 5z - 10 = 4z + 1$$
$$5z = 4z + 1$$
$$5z - 4z = 4z + 1 - 4z$$
$$z = 1$$

61. The sum of -42 and 16 is -26 translates to $-42 + 16 = -26$.

65. Three times the difference of -14 and 2 amounts to -48 translates to $3(-14 - 2) = -48$.

69. In 2002 there were 50–42 or 8 million more electronically filed income tax returns than in 2001.

73.
$$x^3 - 2xy = 3^3 - 2(3)(-1)$$
$$= 27 - (-6)$$
$$= 27 + 6$$
$$= 33$$

77.
$$(2x - y)^2 = [2(3) - (-1)]^2$$
$$= (6 + 1)^2$$
$$= (7)^2$$
$$= 49$$

81.
$$(-8)^2 + 3x = 5x + 4^3$$
$$64 + 3x = 5x + 64$$
$$64 + 3x - 5x = 5x + 64 - 5x$$
$$64 - 2x = 64$$
$$64 - 2x - 64 = 64 - 64$$
$$-2x = 0$$
$$\frac{-2x}{-2} = \frac{0}{-2}$$
$$x = 0$$

85. No; answers may vary.

EXERCISE SET 3.5

1. A number added to -5 is -7 translates to $-5 + x = -7$.

5. A number subtracted from -20 amounts to 104 translates to $-20 - x = 104$.

9.
$$3x + 9 = 33$$
$$3x = 24$$
$$x = 8$$

13.
$$3 + 4 + x = 16$$
$$7 + x = 16$$
$$x = 9$$

17.
$$x - 3 = 45 - x$$
$$2x = 48$$
$$x = 24$$

21.
$$8 - x = \frac{15}{5}$$
$$8 - x = 3$$
$$-x = -5$$
$$x = 5$$

25.
$$5x - 40 = x + 8$$
$$4x = 48$$
$$x = 12$$

29. Let x be the number of electoral votes for Al Gore. Then $x + 5$ is the number of electoral votes for George W. Bush.
$$x + x + 5 = 527$$
$$2x + 5 = 527$$
$$2x = 522$$
$$x = 261$$
Gore received 261 votes and Bush received $261 + 5 = 266$ votes.

33. Let x be the number of universities in the U.S. Then India has $x + 2649$ universities.
$$x + x + 2649 = 14{,}165$$
$$2x + 2649 = 14{,}165$$
$$2x = 11{,}516$$
$$x = 5758$$
There are 5758 universities in the U.S. and $5758 + 2649 = 8407$ universities in India.

37. Let x be the capacity of Neyland Stadium. Then $x + 4647$ is the capacity of Michigan Stadium.
$$x + x + 4647 = 210{,}355$$
$$2x + 4647 = 210{,}355$$
$$2x = 205{,}708$$
$$x = 102{,}854$$
Neyland Stadium has a capacity of 102,854 and Michigan Stadium has a capacity of $102{,}854 + 4647 = 107{,}501$.

41. Let x be the number of ounces of food a finch eats per day. Then $x + 5$ is the number of ounces of food a crow eats per day.

$$x + x + 5 = 13$$
$$2x + 5 = 13$$
$$2x = 8$$
$$x = 4$$

A finch eats 4 ounces of food per day and a crow eats $4 + 5 = 9$ ounces of food per day.

45. To round 586 to the nearest ten, observe that the digit in the ones place is 6. Since this digit is greater than 5, add 1 to the digit in the tens place. The number 586 rounded to the nearest ten is 590.

49. To round 2986 to the nearest thousand, observe that the digit in the hundreds place is 9. Since this digit is greater than 5, add 1 to the digit in the thousands place. The number 2986 rounded to the nearest thousand is 3000.

53.
$$P = A + C$$
$$165,000 = A + 156,750$$
$$165,000 - 156,750 = A + 156,750 - 156,750$$
$$8250 = A$$

The agent will receive a commission of $8250.

57. Answers may vary.

Chapter 3 Test

1. $7x - 5 - 12x + 10 = (7 - 12)x + (-5 + 10)$
$$= -5x + 5$$

5. $A = 4(3x - 1) = 12x - 4$
The area is $(12x - 4)$ square meters.

9.
$$\frac{x}{2} = -5 - (-2)$$
$$\frac{x}{2} = -3$$
$$2 \cdot \frac{x}{2} = 2(-3)$$
$$x = -6$$

13.
$$2(x - 6) = 0$$
$$2x - 12 = 0$$
$$2x = 12$$
$$x = 6$$

17. a. Twice 5 is $2 \cdot 5$ and the sum of twice 5 and -15 is $2 \cdot 5 + (-15)$. The statement translates to $2 \cdot 5 + (-15) = -5$.

b. Three times a number is $3x$ and six added to this is $3x + 6$. The statement translates to $3x + 6 = -30$.

Chapter 4

Exercise Set 4.1

1. 1 out of 3 equal parts is shaded: $\frac{1}{3}$.

5. 7 out of 12 equal parts are shaded: $\frac{7}{12}$.

9. 7 out of 8 equal parts are shaded: $\frac{7}{8}$.

13. Use a circle and divide it into 8 equal parts. Then shade 5 of the equal parts.

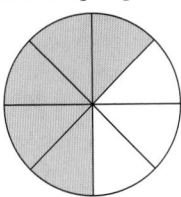

17. Draw 7 circles of the same size. Shade 6 of the circles.

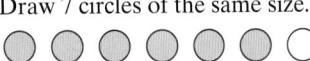

21. $\frac{11}{4}$

25. $\frac{3}{2}$

29. $\frac{17}{6}$

33. $\dfrac{\text{Freshmen} \rightarrow 42}{\text{Students} \rightarrow 131} = \dfrac{42}{131}$

Thus, $\frac{42}{131}$ of the students are freshmen.

37. $\dfrac{\text{Injury-related visits} \rightarrow 4}{\text{Total visits} \rightarrow 10} = \dfrac{4}{10}$

Thus, $\frac{4}{10}$ of the visits are injury–related.

41. $\dfrac{\text{Used gigabytes} \rightarrow 17}{\text{Total gigabytes} \rightarrow 32} = \dfrac{17}{32}$

Aaron has used $\frac{17}{32}$ of the hard drive.

45. $\dfrac{\text{Number of sophomores} \rightarrow 10}{\text{Number of students} \rightarrow 31} = \dfrac{10}{31}$

Thus, $\frac{10}{31}$ of the class is sophomores.

49. To graph $\frac{1}{4}$ on a number line, divide the distance from 0 to 1 into 4 equal parts. Then start at 0 and count over 1 part.

53. To graph $\frac{8}{5}$ on a number line, divide the distance from 0 to 1 into 5 equal parts and divide the distance from 1 to 2 into 5 equal parts. Then start at 0 and count over 8 parts.

57. To graph $\frac{3}{8}$ on a number line, divide the distance from 0 to 1 into 8 equal parts. Then start at 0 and count over 3 parts.

61. $\dfrac{-5}{1} = -5 \div 1 = -5$

65. $\dfrac{-8}{-8} = -8 \div (-8) = 1$

69. $\dfrac{3}{1} = 3 \div 1 = 3$

73. $\dfrac{2}{3} = \dfrac{?}{21}$

$\dfrac{2 \cdot 7}{3 \cdot 7} = \dfrac{14}{21}$

77. $\dfrac{1}{2} = \dfrac{?}{30}$

$\dfrac{1 \cdot 15}{2 \cdot 15} = \dfrac{15}{30}$

81. $2 = \dfrac{?}{5}$

$\dfrac{2}{1} = \dfrac{?}{5}$

$\dfrac{2 \cdot 5}{1 \cdot 5} = \dfrac{10}{5}$

85. $\dfrac{2y}{3} = \dfrac{?}{12}$

$\dfrac{2y \cdot 4}{3 \cdot 4} = \dfrac{8y}{12}$

89. $\dfrac{4}{3} = \dfrac{?}{36x}$

$\dfrac{4 \cdot 12x}{3 \cdot 12x} = \dfrac{48x}{36x}$

93. $1 = \dfrac{?}{36x}$

$\dfrac{1}{1} = \dfrac{?}{36x}$

$\dfrac{1 \cdot 36x}{1 \cdot 36x} = \dfrac{36x}{36x}$

97. The smallest numerator of all the fractions with a denominator of 100 is 45, which corresponds to the United States.

101. $5^3 = 5 \cdot 5 \cdot 5 = 125$

105. $2^3 \cdot 3 = 2 \cdot 2 \cdot 2 \cdot 3 = 24$

109. Answers may vary.

113. Total number of restaurants
$= 6253 + 2348 + 14 + 26 + 210$
$= 8851$
$\dfrac{6253}{8851}$ of the restaurants are "Wendy's" restaurants.

EXERCISE SET 4.2

1. $20 = 2 \cdot 10$
   ↓ ↓↘
   $2 \cdot 2 \cdot 5 = 2^2 \cdot 5$

5. $45 = 9 \cdot 5$
   ↓↘↘
   $3 \cdot 3 \cdot 5 = 3^2 \cdot 5$

9. $\dfrac{3}{12} = \dfrac{3}{3 \cdot 4} = \dfrac{1}{4}$

13. $\dfrac{14}{16} = \dfrac{2 \cdot 7}{2 \cdot 8} = \dfrac{7}{8}$

17. $\dfrac{35}{42} = \dfrac{7 \cdot 5}{7 \cdot 2 \cdot 3} = \dfrac{5}{6}$

21. $\dfrac{16}{24} = \dfrac{2 \cdot 2 \cdot 2 \cdot 2}{2 \cdot 2 \cdot 2 \cdot 3} = \dfrac{2}{3}$

25. $\dfrac{39ab}{26a^2} = \dfrac{3 \cdot 13 \cdot a \cdot b}{2 \cdot 13 \cdot a \cdot a} = \dfrac{3b}{2a}$

29. $\dfrac{21}{49} = \dfrac{3 \cdot 7}{7 \cdot 7} = \dfrac{3}{7}$

33. $\dfrac{36z}{63z} = \dfrac{2 \cdot 2 \cdot 3 \cdot 3 \cdot z}{3 \cdot 3 \cdot 7 \cdot z} = \dfrac{4}{7}$

37. $\dfrac{12}{15} = \dfrac{2 \cdot 2 \cdot 3}{3 \cdot 5} = \dfrac{4}{5}$

41. $\dfrac{27xy}{90y} = \dfrac{3 \cdot 3 \cdot 3 \cdot x \cdot y}{2 \cdot 3 \cdot 3 \cdot 5 \cdot y} = \dfrac{3x}{10}$

45. $\dfrac{40xy}{64xyz} = \dfrac{2 \cdot 2 \cdot 2 \cdot 5 \cdot x \cdot y}{2 \cdot 2 \cdot 2 \cdot 2 \cdot 2 \cdot 2 \cdot x \cdot y \cdot z} = \dfrac{5}{8z}$

49. $\dfrac{2640 \text{ feet}}{5280 \text{ feet}} = \dfrac{2 \cdot 2 \cdot 2 \cdot 2 \cdot 3 \cdot 5 \cdot 11}{2 \cdot 2 \cdot 2 \cdot 2 \cdot 2 \cdot 3 \cdot 5 \cdot 11} = \dfrac{1}{2}$

2640 feet represents $\dfrac{1}{2}$ mile.

53. $16{,}000 - 8800 = 7200$
   Number of male students → 7200
   Total number of students → 16,000
   $\dfrac{7200}{16{,}000} = \dfrac{9 \cdot 800}{20 \cdot 800} = \dfrac{9}{20}$
   $\dfrac{9}{20}$ of the students are male.

57. $\dfrac{10}{24} = \dfrac{2 \cdot 5}{2 \cdot 2 \cdot 2 \cdot 3} = \dfrac{5}{12}$
   $\dfrac{5}{12}$ of the width is concrete.

61. Answers may vary.

65. $2y = 2(-7) = -14$

69. $a^2 + 2b + 3 = 4^2 + 2(5) + 3 = 16 + 10 + 3 = 29$

73. False;
   $\dfrac{14}{42} = \dfrac{2 \cdot 7}{2 \cdot 3 \cdot 7} = \dfrac{1}{3}$

77. $\dfrac{372}{620} = \dfrac{2 \cdot 2 \cdot 3 \cdot 31}{2 \cdot 2 \cdot 5 \cdot 31} = \dfrac{3}{5}$

81. $3 + 1 = 4$
   $\dfrac{4}{100} = \dfrac{2 \cdot 2}{2 \cdot 2 \cdot 5 \cdot 5} = \dfrac{1}{5 \cdot 5} = \dfrac{1}{25}$
   $\dfrac{1}{25}$ of the donors have AB blood type (either Rh-positive or Rh-negative).

85. 2235, 105, 900, 1470 are divisible by 5 because each number ends with a 0 or 5. These numbers are also divisible by 3 because the sum of the digits for each number is divisible by 3.

EXERCISE SET 4.3

1. $\dfrac{7}{8} \cdot \dfrac{2}{3} = \dfrac{7 \cdot 2}{8 \cdot 3} = \dfrac{7 \cdot 2}{2 \cdot 2 \cdot 2 \cdot 3} = \dfrac{7}{2 \cdot 2 \cdot 3} = \dfrac{7}{12}$

5. $-\dfrac{1}{2} \cdot -\dfrac{2}{15} = \dfrac{1 \cdot 2}{2 \cdot 15} = \dfrac{1}{15}$

9. $3a^2 \cdot \dfrac{1}{4} = \dfrac{3 \cdot a^2 \cdot 1}{4} = \dfrac{3a^2}{4}$

13. $\left(\dfrac{1}{5}\right)^3 = \dfrac{1}{5} \cdot \dfrac{1}{5} \cdot \dfrac{1}{5} = \dfrac{1 \cdot 1 \cdot 1}{5 \cdot 5 \cdot 5} = \dfrac{1}{125}$

17. $\left(-\dfrac{2}{3}\right)^3 \cdot \dfrac{1}{2} = \left(-\dfrac{2}{3}\right) \cdot \left(-\dfrac{2}{3}\right) \cdot \left(-\dfrac{2}{3}\right) \cdot \left(\dfrac{1}{2}\right)$
   $= -\dfrac{2 \cdot 2 \cdot 2 \cdot 1}{3 \cdot 3 \cdot 3 \cdot 2}$
   $= -\dfrac{4}{27}$

21. $-\dfrac{6}{15} \div \dfrac{12}{5} = -\dfrac{6}{15} \cdot \dfrac{5}{12}$

$= -\dfrac{6 \cdot 5}{3 \cdot 5 \cdot 2 \cdot 6}$

$= -\dfrac{1}{6}$

25. $\dfrac{11y}{20} \div \dfrac{3}{11} = \dfrac{11y}{20} \cdot \dfrac{11}{3} = \dfrac{11 \cdot y \cdot 11}{2 \cdot 2 \cdot 5 \cdot 3} = \dfrac{121y}{60}$

29. $\dfrac{1}{5x} \div \dfrac{5}{x^2} = \dfrac{1}{5x} \cdot \dfrac{x^2}{5} = \dfrac{x \cdot x}{5 \cdot x \cdot 5} = \dfrac{x}{25}$

33. $\dfrac{3x}{7} \div \dfrac{5}{6x} = \dfrac{3x}{7} \cdot \dfrac{6x}{5}$

$= \dfrac{3 \cdot x \cdot 2 \cdot 3 \cdot x}{7 \cdot 5}$

$= \dfrac{18x^2}{35}$

37. $-\dfrac{3}{5} \div -\dfrac{4}{5} = \left(-\dfrac{3}{5}\right) \cdot \left(-\dfrac{5}{4}\right) = \dfrac{3 \cdot 5}{5 \cdot 2 \cdot 2} = \dfrac{3}{4}$

41. $\dfrac{x^2}{y} \cdot \dfrac{y^3}{x} = \dfrac{x \cdot x \cdot y \cdot y^2}{y \cdot x} = \dfrac{x \cdot y^2}{1} = xy^2$

45. $-3x \div \dfrac{x^2}{12} = -\dfrac{3x}{1} \cdot \dfrac{12}{x^2} = -\dfrac{3 \cdot x \cdot 12}{1 \cdot x \cdot x} = -\dfrac{36}{x}$

49. $-\dfrac{19}{63y} \cdot 9y^2 = -\dfrac{19}{63y} \cdot \dfrac{9y^2}{1} = -\dfrac{19 \cdot 9 \cdot y \cdot y}{7 \cdot 9 \cdot y \cdot 1} = -\dfrac{19y}{7}$

53. $\dfrac{4}{8} \div \dfrac{3}{16} = \dfrac{4}{8} \cdot \dfrac{16}{3} = \dfrac{4 \cdot 8 \cdot 2}{8 \cdot 3} = \dfrac{8}{3}$

57. $\left(1 \div \dfrac{3}{4}\right) \cdot \dfrac{2}{3} = \left(\dfrac{1}{1} \cdot \dfrac{4}{3}\right) \cdot \dfrac{2}{3} = \dfrac{1 \cdot 2 \cdot 2 \cdot 2}{1 \cdot 3 \cdot 3} = \dfrac{8}{9}$

61. $\dfrac{ab^2}{c} \cdot \dfrac{c}{ab} = \dfrac{a \cdot b \cdot b \cdot c}{c \cdot a \cdot b} = \dfrac{b}{1} = b$

65. $-\dfrac{4}{7} \div \left(\dfrac{4}{5} \cdot \dfrac{3}{7}\right) = -\dfrac{4}{7} \div \left(\dfrac{2 \cdot 2 \cdot 3}{5 \cdot 7}\right)$

$= -\dfrac{4}{7} \cdot \dfrac{5 \cdot 7}{2 \cdot 2 \cdot 3}$

$= -\dfrac{4 \cdot 5 \cdot 7}{7 \cdot 4 \cdot 3} = -\dfrac{5}{3}$

69. a. $xy = -\dfrac{4}{5} \cdot \dfrac{9}{11} = -\dfrac{2 \cdot 2 \cdot 3 \cdot 3}{5 \cdot 11} = -\dfrac{36}{55}$

b. $x \div y = -\dfrac{4}{5} \div \dfrac{9}{11}$

$= -\dfrac{4}{5} \cdot \dfrac{11}{9}$

$= -\dfrac{2 \cdot 2 \cdot 11}{5 \cdot 3 \cdot 3}$

$= -\dfrac{44}{45}$

73. $-\dfrac{1}{2}z = \dfrac{1}{10}$

$-\dfrac{1}{2} \cdot \dfrac{2}{5} = \dfrac{1}{10}$ replace z with $\dfrac{2}{5}$

$-\dfrac{1 \cdot 2}{2 \cdot 5} = \dfrac{1}{10}$

$-\dfrac{1}{5} = \dfrac{1}{10}$ False

No, it is not a solution.

77. $\dfrac{2}{3} \cdot 2757 = \dfrac{2}{3} \cdot \dfrac{2757}{1}$

$= \dfrac{2 \cdot 2757}{3 \cdot 1}$

$= \dfrac{2 \cdot 3 \cdot 919}{3 \cdot 1}$

$= \dfrac{1838}{1}$

$= 1838$

The sale price of the cruise is $1838.

81. $\dfrac{2}{5} \cdot 2170 = \dfrac{2}{5} \cdot \dfrac{2170}{1}$

$= \dfrac{2 \cdot 2170}{5 \cdot 1}$

$= \dfrac{2 \cdot 5 \cdot 434}{5 \cdot 1}$

$= \dfrac{868}{1}$

$= 868$

Manfred has hiked 868 miles.

85. Let x be the size of Jorge's wrist.

(wrist size) = (fraction of waist) $\cdot$ (waist size)

$x = \dfrac{1}{4} \cdot \dfrac{34}{1} = \dfrac{1 \cdot 2 \cdot 17}{2 \cdot 2 \cdot 1} = \dfrac{17}{2}$

Jorge's wrist is about $\dfrac{17}{2}$ inches.

89. $2 \div \dfrac{1}{12} = 2 \cdot \dfrac{12}{1}$

$= 24$

There are 24 doses available.

93. $90 = 9 \cdot 10 = 3 \cdot 3 \cdot 2 \cdot 5$ or $2 \cdot 3^2 \cdot 5$

97. $126 = 2 \cdot 63 = 2 \cdot 9 \cdot 7 = 2 \cdot 3 \cdot 3 \cdot 7$ or $2 \cdot 3^2 \cdot 7$

101. Answers may vary.

105. $\dfrac{42}{25} \cdot \dfrac{125}{36} \div \dfrac{7}{6}$

$= \dfrac{42}{25} \cdot \dfrac{125}{36} \cdot \dfrac{6}{7}$

$= \dfrac{42 \cdot 125 \cdot 6}{25 \cdot 36 \cdot 7}$

$= \dfrac{7 \cdot 6 \cdot 5 \cdot 25 \cdot 6}{25 \cdot 6 \cdot 6 \cdot 7} = \dfrac{5}{1} = 5$

EXERCISE SET 4.4

1. $-\dfrac{1}{2} + \dfrac{1}{2} = \dfrac{-1 + 1}{2} = \dfrac{0}{2} = 0$

5. $-\dfrac{4}{13} + \dfrac{2}{13} + \dfrac{1}{13} = \dfrac{-4 + 2 + 1}{13} = -\dfrac{1}{13}$

9. $\dfrac{1}{y} - \dfrac{4}{y} = \dfrac{1 - 4}{y} = \dfrac{-3}{y} = -\dfrac{3}{y}$

13. $\dfrac{1}{8} - \dfrac{7}{8} = \dfrac{1 - 7}{8} = \dfrac{-6}{8} = \dfrac{3 \cdot 2}{4 \cdot 2} = -\dfrac{3}{4}$

17. $\dfrac{9x}{15} + \dfrac{1x}{15} = \dfrac{9x + 1x}{15} = \dfrac{10x}{15} = \dfrac{5 \cdot 2x}{5 \cdot 3} = \dfrac{2x}{3}$

21. $\dfrac{15}{16z} - \dfrac{3}{16z} = \dfrac{15 - 3}{16z} = \dfrac{12}{16z} = \dfrac{4 \cdot 3}{4 \cdot 4 \cdot z} = \dfrac{3}{4z}$

25. $\dfrac{15}{17} + \dfrac{5}{17} + \dfrac{14}{17} = \dfrac{15 + 5 + 14}{17}$

$\qquad\qquad\qquad\quad = \dfrac{34}{17}$

$\qquad\qquad\qquad\quad = \dfrac{17 \cdot 2}{17}$

$\qquad\qquad\qquad\quad = \dfrac{2}{1}$

$\qquad\qquad\qquad\quad = 2$

29. $\dfrac{x}{4} + \dfrac{3x}{4} - \dfrac{2x}{4} + \dfrac{x}{4} = \dfrac{x + 3x - 2x + x}{4} = \dfrac{3x}{4}$

33. $x - y = -\dfrac{1}{5} - \left(-\dfrac{3}{5}\right) = \dfrac{-1 + 3}{5} = \dfrac{2}{5}$

37. $\qquad x + \dfrac{1}{3} = -\dfrac{1}{3}$

$\qquad x + \dfrac{1}{3} - \dfrac{1}{3} = -\dfrac{1}{3} - \dfrac{1}{3}$

$\qquad\qquad\quad x = \dfrac{-1 - 1}{3} = \dfrac{-2}{3} = -\dfrac{2}{3}$

Check:

$\qquad x + \dfrac{1}{3} = -\dfrac{1}{3}$

$\qquad -\dfrac{2}{3} + \dfrac{1}{3} \overset{?}{=} -\dfrac{1}{3}$

$\qquad\qquad -\dfrac{1}{3} \overset{?}{=} -\dfrac{1}{3}$ True

41. $\qquad 3x - \dfrac{1}{5} - 2x = \dfrac{1}{5} + \dfrac{2}{5}$

$\qquad (3x - 2x) - \dfrac{1}{5} = \dfrac{1 + 2}{5}$

$\qquad\qquad\quad x - \dfrac{1}{5} = \dfrac{3}{5}$

$\qquad x - \dfrac{1}{5} + \dfrac{1}{5} = \dfrac{3}{5} + \dfrac{1}{5}$

$\qquad\qquad\qquad x = \dfrac{4}{5}$

Check:

$\qquad 3x - \dfrac{1}{5} - 2x = \dfrac{1}{5} + \dfrac{2}{5}$

$\qquad 3 \cdot \dfrac{4}{5} - \dfrac{1}{5} - 2 \cdot \dfrac{4}{5} \overset{?}{=} \dfrac{1}{5} + \dfrac{2}{5}$

$\qquad \dfrac{3}{1} \cdot \dfrac{4}{5} - \dfrac{1}{5} - \dfrac{2}{1} \cdot \dfrac{4}{5} \overset{?}{=} \dfrac{3}{5}$

$\qquad \dfrac{3 \cdot 4}{1 \cdot 5} - \dfrac{1}{5} - \dfrac{2 \cdot 4}{1 \cdot 5} \overset{?}{=} \dfrac{3}{5}$

$\qquad \dfrac{12}{5} - \dfrac{1}{5} - \dfrac{8}{5} \overset{?}{=} \dfrac{3}{5}$

$\qquad \dfrac{12 - 1 - 8}{5} \overset{?}{=} \dfrac{3}{5}$

$\qquad\qquad \dfrac{3}{5} \overset{?}{=} \dfrac{3}{5}$ True

45. The perimeter is the distance around. A rectangle has 2 sets of equal sides. Add the lengths of the sides.

$\dfrac{5}{12} + \dfrac{7}{12} + \dfrac{5}{12} + \dfrac{7}{12} = \dfrac{5 + 7 + 5 + 7}{12}$

$\qquad\qquad\qquad\qquad\qquad = \dfrac{24}{12}$

$\qquad\qquad\qquad\qquad\qquad = \dfrac{2 \cdot 12}{1 \cdot 12}$

$\qquad\qquad\qquad\qquad\qquad = 2$

The perimeter is 2 meters.

49. The fraction of employees enrolled in each plan in order from smallest to largest are $\dfrac{7}{100}, \dfrac{14}{100}, \dfrac{29}{100}$, and $\dfrac{50}{100}$. Thus, the health plans in order from the smallest fraction of employees to the largest fraction of employees are Traditional fee-for-service, Point-of-Service, Health Maintenance Organization, and Preferred Provider Organization.

53. Subtract $\dfrac{18}{50}$ from $\dfrac{39}{50}$.

$\dfrac{39}{50} - \dfrac{18}{50} = \dfrac{39 - 18}{50} = \dfrac{21}{50}$

$\dfrac{21}{50}$ of the states had maximum speed limits that were less than 70 mph.

57. $\qquad 9 = 3 \cdot 3$

$\qquad 15 = 3 \cdot 5$

$\qquad \text{LCD} = 3 \cdot 3 \cdot 5 = 45$

61. $\qquad 24 = 2 \cdot 2 \cdot 2 \cdot 3$

$\qquad\quad x = x$

$\qquad \text{LCD} = 2 \cdot 2 \cdot 2 \cdot 3 \cdot x = 24x$

65. $\qquad 18 = 2 \cdot 3 \cdot 3$

$\qquad 21 = 3 \cdot 7$

$\qquad \text{LCD} = 2 \cdot 3 \cdot 3 \cdot 7 = 126$

69. $\qquad 8 = 2 \cdot 2 \cdot 2$

$\qquad 24 = 2 \cdot 2 \cdot 2 \cdot 3$

$\qquad \text{LCD} = 2 \cdot 2 \cdot 2 \cdot 3 = 24$

73. $\qquad a = a$

$\qquad 12 = 2 \cdot 2 \cdot 3$

$\qquad \text{LCD} = 2 \cdot 2 \cdot 3 \cdot a = 12a$

77. $\qquad 11 = 11$

$\qquad 33 = 3 \cdot 11$

$\qquad 121 = 11 \cdot 11$

$\qquad \text{LCD} = 3 \cdot 11 \cdot 11 = 363$

81. $-2 + 10 = 8$

85. $-12 - 16 = -12 + (-16) = -28$

89. Subtract the sum of $\dfrac{38}{50}$ and $\dfrac{7}{50}$ from the total of $\dfrac{50}{50}$.

$\dfrac{50}{50} - \left(\dfrac{38}{50} + \dfrac{7}{50}\right) = \dfrac{50}{50} - \dfrac{45}{50}$

$\qquad\qquad\qquad\qquad = \dfrac{50 - 45}{45}$

$\qquad\qquad\qquad\qquad = \dfrac{5}{50}$

$\qquad\qquad\qquad\qquad = \dfrac{1 \cdot 5}{10 \cdot 5}$

$\qquad\qquad\qquad\qquad = \dfrac{1}{10}$

$\dfrac{1}{10}$ of American men over age 65 are either single or divorced.

EXERCISE SET 4.5

1. The LCD of 3 and 6 is 6.

$\dfrac{2}{3} + \dfrac{1}{6} = \dfrac{2 \cdot 2}{3 \cdot 2} + \dfrac{1}{6} = \dfrac{4}{6} + \dfrac{1}{6} = \dfrac{5}{6}$

5. The LCD of 11 and 33 is 33.

$-\dfrac{2}{11} + \dfrac{2}{33} = -\dfrac{2 \cdot 3}{11 \cdot 3} + \dfrac{2}{33} = -\dfrac{6}{33} + \dfrac{2}{33} = -\dfrac{4}{33}$

9. The LCD of 35 and 7 is 35.

$$\frac{11}{35} + \frac{2}{7} = \frac{11}{35} + \frac{2 \cdot 5}{7 \cdot 5} = \frac{11}{35} + \frac{10}{35} = \frac{21}{35} = \frac{3 \cdot 7}{5 \cdot 7} = \frac{3}{5}$$

13. The LCD of 12 and 9 is 36.

$$\frac{5}{12} - \frac{1}{9} = \frac{5 \cdot 3}{12 \cdot 3} - \frac{1 \cdot 4}{9 \cdot 4} = \frac{15}{36} - \frac{4}{36} = \frac{11}{36}$$

17. The LCD of 11 and 9 is 99.

$$\frac{5a}{11} + \frac{4a}{9} = \frac{5a \cdot 9}{11 \cdot 9} + \frac{4a \cdot 11}{9 \cdot 11}$$

$$= \frac{45a}{99} + \frac{44a}{99}$$

$$= \frac{89a}{99}$$

21. The LCD of 2 and x is $2x$.

$$\frac{1}{2} + \frac{3}{x} = \frac{1 \cdot x}{2 \cdot x} + \frac{3 \cdot 2}{x \cdot 2} = \frac{x}{2x} + \frac{6}{2x} = \frac{x + 6}{2x}$$

25. The LCD of 14 and 7 is 14.

$$\frac{9}{14} - \frac{3}{7} = \frac{9}{14} - \frac{3 \cdot 2}{7 \cdot 2} = \frac{9}{14} - \frac{6}{14} = \frac{3}{14}$$

29. The LCD of 9 and 12 is 36.

$$\frac{1}{9} - \frac{5}{12} = \frac{1 \cdot 4}{9 \cdot 4} - \frac{5 \cdot 3}{12 \cdot 3} = \frac{4}{36} - \frac{15}{36} = -\frac{11}{36}$$

33. The LCD of 7 and 8 is 56.

$$\frac{5}{7} - \frac{1}{8} = \frac{5 \cdot 8}{7 \cdot 8} - \frac{1 \cdot 7}{8 \cdot 7} = \frac{40}{56} - \frac{7}{56} = \frac{33}{56}$$

37. $\dfrac{5}{9} + \dfrac{3}{9} = \dfrac{8}{9}$

41. The LCD of 6 and 7 is 42.

$$-\frac{5}{6} - \frac{3}{7} = -\frac{5 \cdot 7}{6 \cdot 7} - \frac{3 \cdot 6}{7 \cdot 6}$$

$$= -\frac{35}{42} - \frac{18}{42}$$

$$= -\frac{53}{42}$$

45. The LCD of 3 and 13 is 39.

$$\frac{2a}{3} + \frac{6a}{13} = \frac{2a \cdot 13}{3 \cdot 13} + \frac{6a \cdot 3}{13 \cdot 3}$$

$$= \frac{26a}{39} + \frac{18a}{39}$$

$$= \frac{44a}{39}$$

49. The LCD of 9 and y is $9y$.

$$\frac{5}{9} + \frac{1}{y} = \frac{5 \cdot y}{9 \cdot y} + \frac{1 \cdot 9}{y \cdot 9} = \frac{5y}{9y} + \frac{9}{9y} = \frac{5y + 9}{9y}$$

53. The LCD of $9x$ and 8 is $72x$.

$$\frac{5}{9x} + \frac{1}{8} = \frac{5 \cdot 8}{9x \cdot 8} + \frac{1 \cdot 9x}{8 \cdot 9x}$$

$$= \frac{40}{72x} + \frac{9x}{72x}$$

$$= \frac{40 + 9x}{72x}$$

57. The LCD of 2, 4, and 16 is 16.

$$\frac{x}{2} + \frac{x}{4} + \frac{2x}{16} = \frac{x \cdot 8}{2 \cdot 8} + \frac{x \cdot 4}{4 \cdot 4} + \frac{2x}{16}$$

$$= \frac{8x}{16} + \frac{4x}{16} + \frac{2x}{16}$$

$$= \frac{14x}{16} = \frac{2 \cdot 7x}{2 \cdot 8} = \frac{7x}{8}$$

61. The LCD of 12, 24, and 6 is 24.

$$-\frac{9}{12} + \frac{17}{24} - \frac{1}{6} = -\frac{9 \cdot 2}{12 \cdot 2} + \frac{17}{24} - \frac{1 \cdot 4}{6 \cdot 4}$$

$$= -\frac{18}{24} + \frac{17}{24} - \frac{4}{24}$$

$$= -\frac{5}{24}$$

65. The LCD of these fractions is 70.
Let's write each fraction as an equivalent fraction with a denominator of 70.

$$\frac{2}{7} = \frac{2 \cdot 10}{7 \cdot 10} = \frac{20}{70} \quad \text{and} \quad \frac{3}{10} = \frac{3 \cdot 7}{10 \cdot 7} = \frac{21}{70}$$

Since $20 < 21$, then $\dfrac{20}{70} < \dfrac{21}{70}$ or

$$\frac{2}{7} < \frac{3}{10}$$

69. The LCD of these fractions is 28.
Let's write each fraction as an equivalent fraction with a denominator of 28.

$$-\frac{3}{4} = -\frac{3 \cdot 7}{4 \cdot 7} = -\frac{21}{28} \quad \text{and} \quad -\frac{11}{14} = -\frac{11 \cdot 2}{14 \cdot 2} = -\frac{22}{28}$$

Since $-21 > -22$, then $-\dfrac{21}{28} > -\dfrac{22}{28}$ or

$$-\frac{3}{4} > -\frac{11}{14}$$

73. $xy = \dfrac{1}{3} \cdot \dfrac{3}{4} = \dfrac{1 \cdot 3}{3 \cdot 4} = \dfrac{1}{4}$

77. $\quad x - \dfrac{1}{12} = \dfrac{5}{6}$

$$x - \frac{1}{12} + \frac{1}{12} = \frac{5}{6} + \frac{1}{12}$$

$$x = \frac{5 \cdot 2}{6 \cdot 2} + \frac{1}{12}$$

$$x = \frac{10}{12} + \frac{1}{12}$$

$$x = \frac{11}{12}$$

Check:

$$x - \frac{1}{12} = \frac{5}{6}$$

$$\frac{11}{12} - \frac{1}{12} \stackrel{?}{=} \frac{5}{6}$$

$$\frac{10}{12} \stackrel{?}{=} \frac{5}{6}$$

$$\frac{2 \cdot 5}{2 \cdot 6} \stackrel{?}{=} \frac{5}{6}$$

$$\frac{5}{6} \stackrel{?}{=} \frac{5}{6} \text{ True}$$

81.

$$7z + \frac{1}{16} - 6z = \frac{3}{4}$$

$$(7z - 6z) + \frac{1}{16} = \frac{3}{4}$$

$$z + \frac{1}{16} = \frac{3}{4}$$

$$z + \frac{1}{16} - \frac{1}{16} = \frac{3}{4} - \frac{1}{16}$$

$$z = \frac{3 \cdot 4}{4 \cdot 4} - \frac{1}{16}$$

$$z = \frac{12}{16} - \frac{1}{16}$$

$$z = \frac{11}{16}$$

Check:

$$7z + \frac{1}{16} - 6z = \frac{3}{4}$$

$$7 \cdot \frac{11}{16} + \frac{1}{16} - 6 \cdot \frac{11}{16} \stackrel{?}{=} \frac{3}{4}$$

$$\frac{7}{1} \cdot \frac{11}{16} + \frac{1}{16} - \frac{6}{1} \cdot \frac{11}{16} \stackrel{?}{=} \frac{3}{4}$$

$$\frac{7 \cdot 11}{1 \cdot 16} + \frac{1}{16} - \frac{6 \cdot 11}{1 \cdot 16} \stackrel{?}{=} \frac{3}{4}$$

$$\frac{77}{16} + \frac{1}{16} - \frac{66}{16} \stackrel{?}{=} \frac{3}{4}$$

$$\frac{77 + 1 - 66}{16} \stackrel{?}{=} \frac{3}{4}$$

$$\frac{12}{16} \stackrel{?}{=} \frac{3}{4}$$

$$\frac{4 \cdot 3}{4 \cdot 4} \stackrel{?}{=} \frac{3}{4}$$

$$\frac{3}{4} \stackrel{?}{=} \frac{3}{4} \text{ True}$$

85. Add the lengths of the 4 sides. A parallelogram has two sets of equal sides.

$$\frac{1}{3} + \frac{4}{5} + \frac{1}{3} + \frac{4}{5} = \frac{1 \cdot 5}{3 \cdot 5} + \frac{4 \cdot 3}{5 \cdot 3} + \frac{1 \cdot 5}{3 \cdot 5} + \frac{4 \cdot 3}{5 \cdot 3}$$

$$= \frac{5}{15} + \frac{12}{15} + \frac{5}{15} + \frac{12}{15}$$

$$= \frac{34}{15}$$

The perimeter is $\frac{34}{15}$ centimeters.

89. Subtract $\frac{5}{264}$ from $\frac{1}{4}$.

$$\frac{1}{4} - \frac{5}{264} = \frac{1 \cdot 66}{4 \cdot 66} - \frac{5}{264} = \frac{66 - 5}{264} = \frac{61}{264}$$

A killer bee will chase a person $\frac{61}{264}$ mile farther.

93. Subtract the fraction of students for whom art is their favorite subject from the fraction of students for whom math, science, or art is their favorite subject.

$$\frac{13}{20} - \frac{4}{25} = \frac{13 \cdot 5}{20 \cdot 5} - \frac{4 \cdot 4}{25 \cdot 4}$$

$$= \frac{65}{100} - \frac{16}{100}$$

$$= \frac{49}{100}$$

$\frac{49}{100}$ of students name math or science as their favorite subject.

97. Add the fractions for less than 10 miles and 10 to 49 miles.

$$\frac{3}{50} + \frac{23}{100} = \frac{3 \cdot 2}{50 \cdot 2} + \frac{23}{100}$$

$$= \frac{6}{100} + \frac{23}{100}$$

$$= \frac{29}{100}$$

$\frac{29}{100}$ of adults drive less than 50 miles in an average week.

101.

$$\left(-\frac{5}{6}\right)^2 = \left(-\frac{5}{6}\right)\left(-\frac{5}{6}\right)$$

$$= \frac{5 \cdot 5}{6 \cdot 6}$$

$$= \frac{25}{36}$$

105. 327 rounded to the nearest ten is 330.

109. Answers may vary.

113. Find the sum of the fractions for North and South America.

$$\frac{94}{579} + \frac{23}{193} = \frac{94}{579} + \frac{23 \cdot 3}{193 \cdot 3}$$

$$= \frac{94}{579} + \frac{69}{579}$$

$$= \frac{163}{579}$$

$\frac{163}{579}$ of the world's land area is accounted for by North and South America.

117.

$$1 - \frac{18}{193} = \frac{193}{193} - \frac{18}{193}$$

$$= \frac{175}{193}$$

$\frac{175}{193}$ of the world's land area is inhabited continents.

121.

$$\frac{127}{500} = \frac{127 \cdot 2}{500 \cdot 2} = \frac{254}{1000}$$

$$\frac{31}{200} = \frac{31 \cdot 5}{200 \cdot 5} = \frac{155}{1000}$$

Since $\frac{254}{1000} > \frac{155}{1000}$, the rental category "1–2 videos per month" is bigger.

EXERCISE SET 4.6

1. $$\frac{\frac{1}{8}}{\frac{3}{4}} = \frac{1}{8} \div \frac{3}{4} = \frac{1}{8} \cdot \frac{4}{3} = \frac{1 \cdot 4}{2 \cdot 4 \cdot 3} = \frac{1}{6}$$

5. $$\frac{\frac{2x}{27}}{\frac{4}{9}} = \frac{2x}{27} \div \frac{4}{9} = \frac{2x}{27} \cdot \frac{9}{4} = \frac{2 \cdot x \cdot 3 \cdot 3}{3 \cdot 3 \cdot 3 \cdot 2 \cdot 2} = \frac{x}{6}$$

9. $$\frac{\frac{3x}{4}}{5 - \frac{1}{8}} = \frac{8 \cdot \left(\frac{3x}{4}\right)}{8 \cdot \left(\frac{5}{1} - \frac{1}{8}\right)}$$

$$= \frac{6x}{8 \cdot \left(\frac{5}{1}\right) - 8\left(\frac{1}{8}\right)}$$

$$= \frac{6x}{40 - 1}$$

$$= \frac{6x}{39}$$

$$= \frac{3 \cdot 2 \cdot x}{3 \cdot 13}$$

$$= \frac{2x}{13}$$

13. $\dfrac{5}{6} \div \dfrac{1}{3} \cdot \dfrac{1}{4} = \dfrac{5}{6} \cdot \dfrac{3}{1} \cdot \dfrac{1}{4}$

$$= \dfrac{5 \cdot 3 \cdot 1}{2 \cdot 3 \cdot 1 \cdot 2 \cdot 2}$$

$$= \dfrac{5}{8}$$

17. $\left(\dfrac{2}{9} + \dfrac{4}{9}\right)\left(\dfrac{1}{3} - \dfrac{9}{10}\right) = \left(\dfrac{6}{9}\right)\left(\dfrac{1 \cdot 10}{3 \cdot 10} - \dfrac{9 \cdot 3}{10 \cdot 3}\right)$

$$= \left(\dfrac{6}{9}\right)\left(\dfrac{10}{30} - \dfrac{27}{30}\right)$$

$$= \dfrac{6}{9} \cdot \left(-\dfrac{17}{30}\right)$$

$$= -\dfrac{6 \cdot 17}{9 \cdot 5 \cdot 6}$$

$$= -\dfrac{17}{45}$$

21. $2 \cdot \left(\dfrac{1}{4} + \dfrac{1}{5}\right) + 2 = 2 \cdot \left(\dfrac{1 \cdot 5}{4 \cdot 5} + \dfrac{1 \cdot 4}{5 \cdot 4}\right) + 2$

$$= 2 \cdot \left(\dfrac{5 + 4}{20}\right) + \dfrac{2}{1} \cdot \dfrac{20}{20}$$

$$= 2 \cdot \left(\dfrac{9}{20}\right) + \dfrac{2 \cdot 20}{20}$$

$$= \dfrac{18}{20} + \dfrac{40}{20}$$

$$= \dfrac{58}{20} = \dfrac{29}{10}$$

25. $\left(\dfrac{2}{3} - \dfrac{5}{9}\right)^2 = \left(\dfrac{2 \cdot 3}{3 \cdot 3} - \dfrac{5}{9}\right)^2$

$$= \left(\dfrac{6}{9} - \dfrac{5}{9}\right)^2$$

$$= \left(\dfrac{1}{9}\right)^2$$

$$= \left(\dfrac{1}{9}\right)\left(\dfrac{1}{9}\right)$$

$$= \dfrac{1 \cdot 1}{9 \cdot 9}$$

$$= \dfrac{1}{81}$$

29. $5y - z = 5\left(\dfrac{2}{5}\right) - \left(\dfrac{5}{6}\right)$

$$= 2 - \dfrac{5}{6}$$

$$= \dfrac{2 \cdot 6}{6} - \dfrac{5}{6}$$

$$= \dfrac{12}{6} - \dfrac{5}{6}$$

$$= \dfrac{7}{6}$$

33. $x^2 - yz = \left(-\dfrac{1}{3}\right)^2 - \left(\dfrac{2}{5}\right)\left(\dfrac{5}{6}\right)$

$$= \dfrac{1}{9} - \left(\dfrac{2}{5}\right)\left(\dfrac{5}{6}\right)$$

$$= \dfrac{1}{9} - \dfrac{2 \cdot 5}{5 \cdot 2 \cdot 3}$$

$$= \dfrac{1}{9} - \dfrac{1}{3}$$

$$= \dfrac{1}{9} - \dfrac{3}{9}$$

$$= -\dfrac{2}{9}$$

37. $\left(\dfrac{3}{2}\right)^3 + \left(\dfrac{1}{2}\right)^3 = \dfrac{27}{8} + \dfrac{1}{8} = \dfrac{28}{8} = \dfrac{7 \cdot 4}{2 \cdot 4} = \dfrac{7}{2}$

41. $\dfrac{2 + \frac{1}{6}}{1 - \frac{4}{3}} = \dfrac{6(2 + \frac{1}{6})}{6(1 - \frac{4}{3})}$

$$= \dfrac{6 \cdot 2 + 6 \cdot \frac{1}{6}}{6 \cdot 1 - 6 \cdot \frac{4}{3}}$$

$$= \dfrac{12 + 1}{6 - 8}$$

$$= \dfrac{13}{-2}$$

$$= -\dfrac{13}{2}$$

45. $\left(\dfrac{3}{4} - 1\right)\left(\dfrac{1}{8} + \dfrac{1}{2}\right) = \left(\dfrac{3}{4} - \dfrac{4}{4}\right)\left(\dfrac{1}{8} + \dfrac{4}{8}\right)$

$$= \left(-\dfrac{1}{4}\right)\left(\dfrac{5}{8}\right)$$

$$= -\dfrac{1 \cdot 5}{4 \cdot 8}$$

$$= -\dfrac{5}{32}$$

49. $\dfrac{\frac{1}{2} - \frac{3}{8}}{\frac{3}{4} + \frac{1}{2}} = \dfrac{8(\frac{1}{2} - \frac{3}{8})}{8(\frac{3}{4} + \frac{1}{2})}$

$$= \dfrac{8 \cdot \frac{1}{2} - 8 \cdot \frac{3}{8}}{8 \cdot \frac{3}{4} + 8 \cdot \frac{1}{2}}$$

$$= \dfrac{4 - 3}{6 + 4}$$

$$= \dfrac{1}{10}$$

53. $\dfrac{\frac{x}{3} + 2}{5 + \frac{1}{3}} = \dfrac{3(\frac{x}{3} + 2)}{3(5 + \frac{1}{3})}$

$$= \dfrac{3 \cdot \frac{x}{3} + 3 \cdot 2}{3 \cdot 5 + 3 \cdot \frac{1}{3}}$$

$$= \dfrac{x + 6}{15 + 1}$$

$$= \dfrac{x + 6}{16}$$

57. $5^2 = 5 \cdot 5 = 25$

61. $\dfrac{2}{3}\left(\dfrac{3}{2}a\right) = \dfrac{2}{3} \cdot \dfrac{3}{2} \cdot \dfrac{a}{1} = \dfrac{2 \cdot 3 \cdot a}{3 \cdot 2 \cdot 1} = \dfrac{a}{1} = a$

65.
$$\frac{2+x}{y} = \frac{2+\frac{3}{4}}{-\frac{4}{7}}$$
$$= \frac{\frac{2\cdot4}{1\cdot4}+\frac{3}{4}}{-\frac{4}{7}}$$
$$= \frac{\frac{8+3}{4}}{-\frac{4}{7}}$$
$$= \frac{\frac{11}{4}}{-\frac{4}{7}}$$
$$= \frac{11}{4}\cdot\left(-\frac{7}{4}\right)$$
$$= -\frac{77}{16}$$

69.
$$\frac{\frac{1}{2}+\frac{3}{4}}{2} = \frac{4\left(\frac{1}{2}+\frac{3}{4}\right)}{4\cdot2}$$
$$= \frac{4\cdot\frac{1}{2}+4\cdot\frac{3}{4}}{8}$$
$$= \frac{2+3}{8}$$
$$= \frac{5}{8}$$

73. The average of a and b should be halfway between a and b.

77. False, if the absolute value of the negative fraction is greater than the absolute value of the positive fraction.

81. No; answers may vary.

EXERCISE SET 4.7

1. $7x = 2$
$$\frac{7x}{7} = \frac{2}{7}$$
$$x = \frac{2}{7}$$

5. $\frac{2}{9}y = -6$
$$\frac{9}{2}\cdot\frac{2}{9}y = \frac{9}{2}\cdot(-6)$$
$$y = -27$$

9. $7a = \frac{1}{3}$
$$\frac{1}{7}\cdot7a = \frac{1}{7}\cdot\frac{1}{3}$$
$$a = \frac{1}{21}$$

13. Multiply both sides of the equation by 3.
$$\frac{x}{3} + 2 = \frac{7}{3}$$
$$3\left(\frac{x}{3}+2\right) = 3\cdot\frac{7}{3}$$
$$x + 6 = 7$$
$$x + 6 - 6 = 7 - 6$$
$$x = 1$$

17. Multiply both sides of the equation by the LCD of 2, 5, and 10: 10.
$$\frac{1}{2} - \frac{3}{5} = \frac{x}{10}$$
$$10\left(\frac{1}{2}-\frac{3}{5}\right) = 10\left(\frac{x}{10}\right)$$
$$5 - 6 = x$$
$$-1 = x$$

21.
$$\frac{x}{7} - \frac{4}{3} = \frac{x\cdot3}{7\cdot3} - \frac{4\cdot7}{3\cdot7}$$
$$= \frac{3x}{21} - \frac{28}{21}$$
$$= \frac{3x-28}{21}$$

25.
$$\frac{3x}{10} + \frac{x}{6} = \frac{3x\cdot3}{10\cdot3} + \frac{x\cdot5}{6\cdot5}$$
$$= \frac{9x}{30} + \frac{5x}{30}$$
$$= \frac{14x}{30}$$
$$= \frac{2\cdot7x}{2\cdot15}$$
$$= \frac{7x}{15}$$

29.
$$\frac{2}{3} - \frac{x}{5} = \frac{4}{15}$$
$$15\left(\frac{2}{3}-\frac{x}{5}\right) = 15\left(\frac{4}{15}\right)$$
$$10 - 3x = 4$$
$$10 - 10 - 3x = 4 - 10$$
$$\frac{-3x}{-3} = \frac{-6}{-3}$$
$$x = 2$$

33.
$$-3m - 5m = \frac{4}{7}$$
$$-8m = \frac{4}{7}$$
$$\left(-\frac{1}{8}\right)(-8m) = \left(-\frac{1}{8}\right)\left(\frac{4}{7}\right)$$
$$m = -\frac{1\cdot4}{2\cdot4\cdot7}$$
$$m = -\frac{1}{14}$$

37.
$$\frac{1}{5}y = 10$$
$$5\left(\frac{1}{5}y\right) = 5\cdot10$$
$$y = 50$$

41.
$$-\frac{3}{4}x = \frac{9}{2}$$
$$-\frac{4}{3}\cdot\left(-\frac{3}{4}\right)x = -\frac{4}{3}\cdot\left(\frac{9}{2}\right)$$
$$x = -\frac{2\cdot2\cdot3\cdot3}{3\cdot2}$$
$$x = -6$$

45.
$$-\frac{5}{8}y = \frac{3}{16} - \frac{9}{16}$$
$$-\frac{5}{8}y = -\frac{6}{16}$$
$$-\frac{5}{8}y = -\frac{2 \cdot 3}{2 \cdot 8}$$
$$-\frac{5}{8}y = -\frac{3}{8}$$
$$\left(-\frac{8}{5}\right)\left(-\frac{5}{8}y\right) = \left(-\frac{8}{5}\right)\left(-\frac{3}{8}\right)$$
$$y = \frac{8 \cdot 3}{5 \cdot 8}$$
$$y = \frac{3}{5}$$

49.
$$\frac{7}{6}x = \frac{1}{4} - \frac{2}{3}$$
$$12\left(\frac{7}{6}x\right) = 12\left(\frac{1}{4} - \frac{2}{3}\right)$$
$$14x = 3 - 8$$
$$14x = -5$$
$$\frac{14x}{14} = \frac{-5}{14}$$
$$x = -\frac{5}{14}$$

53.
$$\frac{x}{3} + 2 = \frac{x}{2} + 8$$
$$6\left(\frac{x}{3} + 2\right) = 6\left(\frac{x}{2} + 8\right)$$
$$2x + 12 = 3x + 48$$
$$2x + 12 - 2x = 3x + 48 - 2x$$
$$12 = x + 48$$
$$12 - 48 = x + 48 - 48$$
$$-36 = x$$

57.
$$5 + \frac{9}{10} = \frac{5}{1} + \frac{9}{10}$$
$$= \frac{5 \cdot 10}{1 \cdot 10} + \frac{9}{10}$$
$$= \frac{50}{10} + \frac{9}{10}$$
$$= \frac{50 + 9}{10}$$
$$= \frac{59}{10}$$

61.   Answers may vary.

EXERCISE SET 4.8

1.   Each part is $\frac{1}{4}$, and there are 11 parts shaded, or 2 wholes and 3 more parts.

   a. $\frac{11}{4}$ **b.** $2\frac{3}{4}$

5.   Each part is $\frac{1}{2}$, and there are 3 parts shaded or 1 whole part and 1 more part.

   a. $\frac{3}{2}$ **b.** $1\frac{1}{2}$

9. $2\frac{1}{3} = \frac{2 \cdot 3 + 1}{3} = \frac{7}{3}$

13. $11\frac{6}{7} = \frac{11 \cdot 7 + 6}{7} = \frac{83}{7}$

17.
$$\begin{array}{r} 3 \\ 15\overline{)47} \\ -45 \\ \hline 2 \end{array}$$
$$\frac{47}{15} = 3\frac{2}{15}$$

21. $2\frac{2}{3} \cdot \frac{1}{7} = \frac{8}{3} \cdot \frac{1}{7} = \frac{8 \cdot 1}{3 \cdot 7} = \frac{8}{21}$

25. $3\frac{2}{3} \cdot 1\frac{1}{2} = \frac{11}{3} \cdot \frac{3}{2} = \frac{11 \cdot 3}{3 \cdot 2} = \frac{11}{2} = 5\frac{1}{2}$

29. $4\frac{7}{10} + 2\frac{1}{10} = 6\frac{8}{10} = 6\frac{4}{5}$

33.
$$3\frac{5}{8} = \quad 3\frac{15}{24}$$
$$2\frac{1}{6} = \quad 2\frac{4}{24}$$
$$+ 7\frac{3}{4} = \quad + 7\frac{1}{24}$$
$$\overline{ 12\frac{37}{24} = \quad 13\frac{13}{24}}$$

37.
$$10\frac{13}{14} = \quad 10\frac{13}{14}$$
$$-3\frac{4}{7} = -3\frac{8}{14}$$
$$\overline{ 7\frac{5}{14}}$$

41.
$$2\frac{3}{4}$$
$$+ 1\frac{1}{4}$$
$$\overline{3\frac{4}{4} = 3 + 1 = 4}$$

45. $3\frac{1}{9} \cdot 2 = \frac{28}{9} \cdot \frac{2}{1} = \frac{56}{9} = 6\frac{2}{9}$

49. $22\frac{4}{9} + 13\frac{5}{18} = 22\frac{8}{18} + 13\frac{5}{18} = 35\frac{13}{18}$

53.
$$15\frac{1}{5} = \quad 15\frac{6}{30}$$
$$20\frac{3}{10} = \quad 20\frac{9}{30}$$
$$+ 37\frac{2}{15} = \quad + 37\frac{4}{30}$$
$$\overline{ 72\frac{19}{30}}$$

57. $4\frac{2}{7} \cdot 1\frac{3}{10} = \frac{30}{7} \cdot \frac{13}{10} = \frac{3 \cdot 10 \cdot 13}{7 \cdot 10} = \frac{39}{7} = 5\frac{4}{7}$

61.   The phrase "total duration" tells us to add. Find the sum of the three durations.
$$4\frac{14}{15} = \quad 4\frac{56}{60}$$
$$4\frac{7}{60} = \quad 4\frac{7}{60}$$
$$+ 1\frac{2}{3} = \quad + 1\frac{40}{60}$$
$$\overline{ 9\frac{103}{60} = \quad 10\frac{43}{60}}$$

The total duration of the three eclipses was $10\frac{43}{60}$ minutes.

65. $6 \cdot 3\frac{1}{4} = \frac{6}{1} \cdot \frac{13}{4}$

$\qquad = \frac{2 \cdot 3 \cdot 13}{1 \cdot 2 \cdot 2}$

$\qquad = \frac{39}{2} = 19\frac{1}{2}$

The sidewalk is $19\frac{1}{2}$ inches wide.

69. Subtract $1\frac{1}{2}$ inches from $1\frac{9}{16}$ inches.

$\begin{aligned} 1\frac{9}{16} &= 1\frac{9}{16} \\ -1\frac{1}{2} &= -1\frac{8}{16} \\ \hline &\quad\; \frac{1}{16} \end{aligned}$

The entrance holes for Mountain Bluebirds should be $\frac{1}{16}$ inch wider than the entrance for Eastern Bluebirds.

73. $58\frac{3}{4} \div 7\frac{1}{2} = \frac{235}{4} \div \frac{15}{2}$

$\qquad = \frac{235}{4} \cdot \frac{2}{15}$

$\qquad = \frac{5 \cdot 47 \cdot 2}{2 \cdot 2 \cdot 3 \cdot 5}$

$\qquad = \frac{47}{6} = 7\frac{5}{6}$

$7\frac{5}{6}$ gallons were used each hour.

77. $\begin{aligned} 11\frac{1}{4} &= 10\frac{25}{20} \\ -3\frac{3}{5} &= -3\frac{12}{20} \\ \hline &\quad 7\frac{13}{20} \end{aligned}$

Tucson gets $7\frac{13}{20}$ inches more rain than Yuma.

81. area $=$ length $\cdot$ width

$\qquad = \frac{3}{4} \cdot 1\frac{1}{4}$

$\qquad = \frac{3}{4} \cdot \frac{5}{4}$

$\qquad = \frac{15}{16}$

The area of the chip is $\frac{15}{16}$ square inch.

85. perimeter $= 3 + 5\frac{1}{3} + 5 + 7\frac{7}{8}$

$\qquad = 3 + 5\frac{8}{24} + 5 + 7\frac{21}{24}$

$\qquad = 20\frac{29}{24}$

$\qquad = 21\frac{5}{24}$

The perimeter is $21\frac{5}{24}$ meters.

89. $\begin{aligned} 152\frac{1}{6} &= \quad 152\frac{1}{6} \\ +154\frac{1}{2} &= +154\frac{3}{6} \\ \hline &\quad 306\frac{4}{6} \quad \text{or} \quad 306\frac{2}{3} \end{aligned}$

The overall height is $306\frac{2}{3}$ feet.

93. $-4\frac{2}{5} \cdot 2\frac{3}{10} = -\frac{22}{5} \cdot \frac{23}{10}$

$\qquad = -\frac{22 \cdot 23}{5 \cdot 10}$

$\qquad = -\frac{2 \cdot 11 \cdot 23}{5 \cdot 5 \cdot 2}$

$\qquad = -\frac{253}{25}$ or $-10\frac{3}{25}$

97. Subtract the absolute values of the numbers.

$\begin{aligned} 31\frac{2}{15} &= \quad 31\frac{8}{60} = \quad 30\frac{68}{60} \\ -17\frac{3}{20} &= -17\frac{9}{60} = -17\frac{9}{60} \\ \hline &\qquad\qquad\qquad\quad 13\frac{59}{60} \end{aligned}$

Thus, $-31\frac{2}{15} + 17\frac{3}{20} = -13\frac{59}{60}$ because $-31\frac{2}{15}$ has the larger absolute value.

101. Subtract the absolute values of the numbers.

$\begin{aligned} 13\frac{5}{6} &= \quad 13\frac{20}{24} = \quad 12\frac{44}{24} \\ -11\frac{7}{8} &= -11\frac{21}{24} = -11\frac{21}{24} \\ \hline &\qquad\qquad\qquad\quad 1\frac{23}{24} \end{aligned}$

Thus, $11\frac{7}{8} - 13\frac{5}{6} = -1\frac{23}{24}$ because $13\frac{5}{6}$ has the larger absolute value.

105. $-3\frac{1}{6} \cdot \left(-2\frac{3}{4}\right) = -\frac{19}{6} \cdot \left(-\frac{11}{4}\right)$

$\qquad = \frac{209}{24} = 8\frac{17}{24}$

109. $\begin{aligned} 2x - 5 + 7x - 8 &= 2x + 7x - 5 - 8 \\ &= 9x - 13 \end{aligned}$

113. $\begin{aligned} 2^3 + 3^2 &= 2 \cdot 2 \cdot 2 + 3 \cdot 3 \\ &= 8 + 9 \\ &= 17 \end{aligned}$

117. Supreme box:

$\begin{aligned} 2\frac{1}{4} &= \quad 2\frac{1}{4} \\ +3\frac{1}{2} &= +3\frac{2}{4} \\ \hline &\quad 5\frac{3}{4} \quad \text{or} \quad 5\frac{6}{8} \end{aligned}$

Deluxe box:

$$1\frac{3}{8} = 1\frac{3}{8}$$
$$+\ 4\frac{1}{4} = +\ 4\frac{2}{8}$$
$$5\frac{5}{8}$$

The supreme box is heavier by $5\frac{6}{8} - 5\frac{5}{8} = \frac{1}{8}$ pound.

121. Answers may vary.

CHAPTER 4 TEST

1. $7\frac{2}{3} = \frac{7 \cdot 3 + 2}{3} = \frac{23}{3}$

5. $\frac{54}{210} = \frac{2 \cdot 3 \cdot 3 \cdot 3}{2 \cdot 3 \cdot 5 \cdot 7} = \frac{3 \cdot 3}{5 \cdot 7} = \frac{9}{35}$

9. $\frac{7x}{9} + \frac{x}{9} = \frac{7x + x}{9} = \frac{8x}{9}$

13. $\frac{9a}{10} + \frac{2}{5} = \frac{9a}{10} + \frac{2 \cdot 2}{5 \cdot 2} = \frac{9a}{10} + \frac{4}{10} = \frac{9a + 4}{10}$

17.
$$3\frac{7}{8} = 3\frac{35}{40}$$
$$7\frac{2}{5} = 7\frac{16}{40}$$
$$+\ 2\frac{3}{4} = +\ 2\frac{30}{40}$$
$$12\frac{81}{40} = 14\frac{1}{40}$$

21. $12 \div 3\frac{1}{3} = 12 \div \frac{10}{3}$

$$= 12 \cdot \frac{3}{10}$$
$$= \frac{2 \cdot 6 \cdot 3}{2 \cdot 5}$$
$$= \frac{18}{5}$$
$$= 3\frac{3}{5}$$

25. $\dfrac{5 + \frac{3}{7}}{2 - \frac{1}{2}} = \dfrac{14\left(5 + \frac{3}{7}\right)}{14\left(2 - \frac{1}{2}\right)}$

$$= \frac{14 \cdot 5 + 14 \cdot \frac{3}{7}}{14 \cdot 2 - 14 \cdot \frac{1}{2}}$$
$$= \frac{70 + 6}{28 - 7}$$
$$= \frac{76}{21}$$

29. $-5x = -5\left(-\frac{1}{2}\right) = -\frac{5}{1} \cdot \left(-\frac{1}{2}\right) = \frac{5 \cdot 1}{1 \cdot 2} = \frac{5}{2}$

33. Find the sum of the fractions representing Back Woods and Westward.

$$\frac{3}{16} + \frac{1}{8} = \frac{3}{16} + \frac{1 \cdot 2}{8 \cdot 2}$$
$$= \frac{3}{16} + \frac{2}{16}$$
$$= \frac{5}{16}$$

$\frac{5}{16}$ of backpack sales go to Back Woods and Westward combined.

37. $258 \div 10\frac{3}{4} = \frac{258}{1} \div \frac{43}{4}$

$$= \frac{258}{1} \cdot \frac{4}{43}$$
$$= \frac{258 \cdot 4}{1 \cdot 43}$$
$$= \frac{43 \cdot 6 \cdot 4}{1 \cdot 43}$$
$$= \frac{24}{1}$$
$$= 24$$

The car will travel about 24 miles on one gallon of gas.

Chapter 5

EXERCISE SET 5.1

1. 6.52 is six and fifty-two hundredths.

5. -3.205 is negative three and two hundred five thousandths.

9.

Preprinted Name		Current date
Preprinted Address		DATE
PAY TO THE		
ORDER OF R.W. Financial		$ 321.42
Three hundred twenty-one and 42/100 DOLLARS		
FOR _____	Signature _____

13. Six and five tenths is 6.5.

17. Negative five and six hundred twenty-five thousandths is -5.625.

21. Sixty four and sixteen hundredths is 64.16.

25. $0.3 = \frac{3}{10}$

29. $-5.47 = -5\frac{47}{100}$

33. $7.07 = 7\frac{7}{100}$

37. $0.3005 = \frac{3005}{10{,}000} = \frac{601}{2000}$

41. $\begin{array}{cc} 0.15 & 0.16 \\ \uparrow & \uparrow \\ 5 & < \ 6 \end{array}$
so $0.15 < 0.16$

45. $\begin{array}{cc} 0.098 & 0.1 \\ \uparrow & \uparrow \\ 0 & < \ 1 \end{array}$
so $0.098 < 0.1$

49. $\begin{array}{cc} 167.908 & 167.980 \\ \uparrow & \uparrow \\ 0 & < \ 8 \end{array}$
so $167.908 < 169.980$

53. $\begin{array}{cc} 1.0621 & 1.07 \\ \uparrow & \uparrow \\ 6 & < \ 7 \end{array}$
so $1.0621 < 1.07$
Thus, $-1.0621 > -1.07$.

57. 0.023 0.024

↑ ↑

3 < 4

so 0.023 < 0.024

Thus, −0.023 > −0.024.

61. To round 0.234 to the nearest hundredth, observe that the digit in the thousandths place is 4. Since this digit is less than 5, we do not add 1 to the digit in the hundredths place. The number 0.234 rounded to the nearest hundredth is 0.23.

65. To round 98,207.23 to the nearest ten, observe that the digit in the ones place is 7. Since this digit is 5 or greater, add 1 to the digit in the tens place. The number 98,207.203 rounded to the nearest ten is 98,210.

69. To round −17.667 to the nearest hundredth, observe that the digit in the thousandths place is 7. Since this digit is 5 or greater, add 1 to the digit in the hundredths place. The number −17.667 rounded to the nearest hundredth is −17.67.

73. Cents are in the hundredths place, so to round 0.067 to the nearest hundredth observe that the digit in the thousandths place is 7. Since this digit is 5 or greater, add 1 to the digit in the hundredths place. The number 0.067 rounded to the nearest hundredth is 0.07. The amount is $0.07.

77. Cents are in the hundredths place, so to round 0.1992 to the nearest hundredth observe that the digit in the thousandths place is 9. Since this digit is 5 or greater, add 1 to the digit in the hundredths place. The number 0.1992 rounded to the nearest hundredth is 0.20. The amount is $0.20.

81. To round 15.99 to the nearest one, observe that the digit in the tenths place is 9. Since this digit is 5 or greater, add 1 to the digit in the ones place. The number 15.99 rounded to the nearest one is 16. The amount is $16.

85. To round 24.6229 to the nearest thousandth, observe that the digit in the ten-thousandths place is 9. Since this digit is 5 or greater, add 1 to the digit in the thousandths place. The number 24.6229 rounded to the nearest thousandth is 24.623. The length of a day on Mars is 24.623 hours.

89. To round 31.3833 to the nearest whole number, observe that the digit in the tenths place is 3. Since this digit is less than 5, we do not add 1 to the digit in the ones place. The number 31.3833 rounded to the nearest whole number is 31. Allen Iverson scored an average of 31 points per game.

93.   3452

     + 2314

     5766

97.   482

     − 239

     243

101. Comparing digits in the same place values from left to right, the averages in order from greatest to least are: 228.040; 226.130; 225.490; 225.370; 224.490; 222.008; 221.546; 220.930

105. two hundred three ten-millionths

EXERCISE SET 5.2

1.   1.3

     + 2.2

     3.5

5.   24.6000

     2.3900

     + 0.0678

     27.0578

9. −2.6 + (−5.97)

Add the absolute values.

     2.60

     + 5.97

     8.57

Attach the common sign.

−8.57

13.   8.8

     − 2.3

     6.5

17.   654.90

     − 56.67

     598.23

21. −1.12 − 5.2 = −1.12 + (−5.2)

Add the absolute values.

     1.12

     + 5.20

     6.32

Attach the common sign.

−6.32

25. −2.6 − (−5.7) = −2.6 + 5.7

Subtract the absolute values.

     5.7

     − 2.6

     3.1

Attach the sign of the larger absolute value.

3.1

29. −5.9 − 4 = −5.9 + (−4)

Add the absolute values.

     5.9

     + 4.0

     9.9

Attach the common sign.

−9.9

33. −6.06 + 0.44

Subtract the absolute values.

     6.06

     − 0.44

     5.62

Attach the sign of the larger absolute value.

−5.62

37.   3490.23

     + 8493.09

     11,983.32

41. 50.2 − 600 = 50.2 + (−600)

Subtract the absolute values.

     600.0

     − 50.2

     549.8

Attach the sign of the larger absolute value.

−549.8

45.   100.009
     6.080
  + 9.034
   115.123

49. $-0.003 + 0.091$
Subtract the absolute values.
   0.091
 − 0.003
   0.088
Attach the sign of the larger absolute value.
0.088

53. $-102.4 - 78.04 = -102.4 + (-78.04)$
Add the absolute values.
   102.40
 + 78.04
   180.44
Attach the common sign.
-180.44

57. $x + z = 3.6 + 0.21 = 3.81$

61. $y - x + z = 5 - 3.6 + 0.21$
$ = 5.00 - 3.60 + 0.21$
$ = 1.40 + 0.21$
$ = 1.61$

65. $27.4 + y = 16$
$27.4 + (-11.4) \overset{?}{=} 16$
$\phantom{27.4 + (-11.4) \overset{?}{=}} 16 = 16 \text{ True}$
Yes, it is a solution.

69. $30.7x + 17.6 - 23.8x - 10.7$
$= 30.7x + (-23.8x) + 17.6 + (-10.7)$
$= 6.9x + 6.9$

73.   The phrase "total monthly cost" tells us to add. Find the sum of the four expenses.
   275.36
    83.00
    81.60
 + 14.75
   454.71
The total monthly cost is $454.71.

77.   Subtract the cost of the book from what she paid: ($20 + $20 = $40).
   40.00
 − 32.48
    7.52
Check:
    7.52
 + 32.48
   40.00 or 40
Her change was $7.52.

81.   The phrase "how much faster" tells us to subtract. Subtract 35.2 from 321.0.
   321.0
 − 35.2
   285.8

Check:
   285.8
 + 35.2
   321.0
The highest speed is 285.8 miles per hour faster than the average speed.

85.   The phrase "total amount" tells us to add. Find the sum of the 3 concert's earnings.
   121.2
   103.5
 + 109.7
   334.4
The total amount of money these three concerts have earned is $334.4 million.

89.   Find the sum of the lengths of the three sides.
   12.40
   29.34
 + 25.70
   67.44
The architect needs 67.44 feet of border material.

93.   The first mission lasted 15.467 minutes. The last mission lasted 2059.817 minutes. Subtract 15.467 from 2059.817.
   2059.817
 −   15.467
   2044.350
The last mission lasted 2044.35 minutes longer than the first mission.

97.   Subtract 13.9 from 22.
   22.0
 − 13.9
    8.1
Check:
    8.1
 + 13.9
   22.0 or 22
The difference in consumption is 8.1 pounds per person.

101. $\left(\dfrac{1}{5}\right)^3 = \dfrac{1}{5} \cdot \dfrac{1}{5} \cdot \dfrac{1}{5} = \dfrac{1}{125}$

105.   There are 2 pennies, 4 nickels, 3 dimes, and 2 quarters.
$2(0.01) + 4(0.05) + 3(0.10) + 2(0.25)$
$ = 0.02 + 0.20 + 0.30 + 0.50$
$ = 1.02$
The value of the coins is $1.02.

109.   Answers may vary.

113. $10.68 - 2(2.3) = 10.68 - 4.6$
$ = 6.08$
The unknown length is 6.08 inches.

EXERCISE SET 5.3

1.   0.2   1 decimal place
  × 0.6   1 decimal place
    0.12   2 decimal places

5. The product, $(-2.3)(7.65)$, is negative.

$$
\begin{array}{rl}
7.65 & \text{2 decimal places} \\
\times\quad 2.3 & \text{1 decimal place} \\
\hline
2295 & \\
15300 & \\
\hline
-17.595 & \text{3 decimal places and include negative sign}
\end{array}
$$

9. 6.5×10 10 has 1 zero, so move decimal point 1 place to the right. $6.5 \times 10 = 65$

13. $(-7.093)(1000)$ 1000 has 3 zeros, so move decimal point 3 places to the right. $(-7.093)(1000) = -7093$

17.
$$
\begin{array}{rl}
5.62 & \text{2 decimal places} \\
\times\ 7.7 & \text{1 decimal place} \\
\hline
3934 & \\
39340 & \\
\hline
43.274 & \text{3 decimal places}
\end{array}
$$

21. $(147.9)(100)$ 100 has 2 zeros, so move decimal point 2 places to the right (a 0 is inserted).
$(147.9)(100) = 14{,}790$

25.
$$
\begin{array}{rl}
49.02 & \text{2 decimal places} \\
\times\ 0.023 & \text{3 decimal places} \\
\hline
14706 & \\
98040 & \\
\hline
1.12746 &
\end{array}
$$

29. 5.5 billion $= 5.5 \times 1{,}000{,}000{,}000$
$\qquad\qquad\quad = 5{,}500{,}000{,}000$

33. 1.6 million $= 1.6 \times 1{,}000{,}000$
$\qquad\qquad\quad = 1{,}600{,}000$

37. Recall that xz means $x \cdot z$.
$xz - y = (3)(5.7), -(-0.2)$
$\qquad\quad = 17.1 + 0.2$
$\qquad\quad = 17.3$

41. $\quad 0.6x = 4.92$
$0.6(14.2) = 4.92$
$\quad\ \ 8.52 = 4.92$ False
No, it is not a solution.

45. $\quad 3.5y = -14$
$3.5(-4) = -14$
$\quad -14.0 = -14$
$\qquad -14 = -14$ True
Yes, it is a solution.

49. Circumference $= \pi \cdot$ diameter
$C = \pi \cdot 10 = 10\pi$
$C \approx 10(3.14) = 31.4$
The circumference is 10π centimeters, which is approximately 31.4 centimeters.

53. Circumference $= \pi \cdot$ diameter
$C = \pi \cdot 250 = 250\pi$
$C \approx 250(3.14) = 785$
The circumference of the Ferris wheel is 250π feet which is approximately 785 feet.

57. **a.** Circumference $= 2\pi \cdot$ radius
$C = 2\pi r = 2\pi(10) = 20\pi$
$C \approx 20(3.14) = 62.8$
The circumference of the smaller circle is approximately 62.8 meters.
$C = 2\pi r = 2\pi(20) = 40\pi$
$C \approx 40(3.14) = 125.6$
The circumference of the larger circle is approximately 125.6 meters.

b. Yes, the circumference is doubled; for example, when the radius of 10 meters was doubled to 20 meters, the circumference of 62.8 was doubled:
$(62.8)(2) = 125.6.$

61. Multiply 2.80 by 1000. 1000 has 3 zeros, so move the decimal point 3 places to the right (a 0 is inserted).
$(2.80)(1000) = 2800$
The farmer received $2800.

65. Multiply Jose's hourly wage by the number of hours he worked.
$$
\begin{array}{rl}
13.88 & \text{2 decimal places} \\
\times\quad 40 & \\
\hline
555.20 & \text{2 decimal places}
\end{array}
$$
Jose earned $555.20.

69.
$$
\begin{array}{rl}
1.4695 & \text{4 decimal places} \\
\times\quad 350 & \\
\hline
734750 & \\
4408500 & \\
\hline
514.3250 & \text{4 decimal places}
\end{array}
$$
Rounding to the nearest dollar, they can buy 514 Canadian dollars.

73.
$$
\begin{array}{r}
36 \\
56\overline{)2016} \\
-168 \\
\hline
336 \\
-336 \\
\hline
0
\end{array}
$$

77. $-\dfrac{24}{7} \div \dfrac{8}{21} = -\dfrac{24}{7} \cdot \dfrac{21}{8}$
$\qquad\qquad = -\dfrac{3 \cdot 8 \cdot 3 \cdot 7}{7 \cdot 8}$
$\qquad\qquad = -9$

81. Answers may vary.

Exercise Set 5.4

1. $0.47 \div 5$
$$
\begin{array}{r}
0.094 \\
5\overline{)0.470} \\
-45 \\
\hline
20 \\
-20 \\
\hline
0
\end{array}
$$

5. $4.756 \div 0.82$

$0.82 \overline{)4.756}$ Move the decimal points 2 places.

$$
\begin{array}{r}
5.8 \\
82. \overline{)475.6} \\
-410 \\
\hline
65\,6 \\
-65\,6 \\
\hline
0
\end{array}
$$

9. $2.4 \overline{)429.34}$ Move the decimal points 1 place.

$$
\begin{array}{r}
178.8 \\
24. \overline{)4293.4} \\
-24 \\
\hline
189 \\
-168 \\
\hline
213 \\
-192 \\
\hline
214 \\
-192 \\
\hline
22
\end{array}
$$

178.8 rounds to 200.

13. $0.4 \overline{)45.23}$ Move the decimal points 1 place.

$$
\begin{array}{r}
113.0 \\
4. \overline{)452.3} \\
-4 \\
\hline
05 \\
-4 \\
\hline
12 \\
-12 \\
\hline
03 \\
-0 \\
\hline
3
\end{array}
$$

113.0 rounds to 110. The quotient is negative, so it is -110.

17. $12.9 \div (-1000)$

1000 has 3 zeros so move the decimal point 3 places to the left and attach the negative sign.

$12.9 \div (-1000) = -0.0129$

21. $1.239 \div 3$

$$
\begin{array}{r}
0.413 \\
3 \overline{)1.239} \\
-12 \\
\hline
03 \\
-3 \\
\hline
09 \\
-9 \\
\hline
0
\end{array}
$$

25. $1.296 \div 0.27$

$0.27 \overline{)1.296}$ Move the decimal points 2 places.

$$
\begin{array}{r}
4.8 \\
27. \overline{)129.6} \\
-108 \\
\hline
21\,6 \\
-21\,6 \\
\hline
0
\end{array}
$$

29. $-18 \div -0.6$

$0.6 \overline{)18.0}$ Move the decimal points 1 place.

$$
\begin{array}{r}
30. \\
6. \overline{)180.} \\
-18 \\
\hline
00 \\
-0 \\
\hline
0
\end{array}
$$

33. $-1.104 \div 1.6$

$$
\begin{array}{r}
0.69 \\
16 \overline{)11.04} \\
-96 \\
\hline
144 \\
-144 \\
\hline
0
\end{array}
$$

Thus $-1.104 \div 1.6 = -0.69$

37. $\dfrac{4.615}{0.071} = \dfrac{4615}{71}$

$$
\begin{array}{r}
65 \\
71 \overline{)4615} \\
-426 \\
\hline
355 \\
-355 \\
\hline
0
\end{array}
$$

41. $0.0043 \overline{)500.}$ Move the decimal point 4 places.

$$
\begin{array}{r}
116279. \\
43 \overline{)5000000} \\
-43 \\
\hline
70 \\
-43 \\
\hline
270 \\
-258 \\
\hline
120 \\
-86 \\
\hline
340 \\
-301 \\
\hline
390 \\
-387 \\
\hline
3
\end{array}
$$

116,279 rounded to nearest ten-thousand is 120,000.

45. $x \div y$

$5.65 \div -0.8$

$0.8 \overline{)5.65}$ Move the decimal points 1 place.

$$
\begin{array}{r}
7.0625 \\
8. \overline{)56.5000} \\
-56 \\
\hline
05 \\
-0 \\
\hline
50 \\
-48 \\
\hline
20 \\
-16 \\
\hline
40 \\
-40 \\
\hline
0
\end{array}
$$

The quotient is -7.0625.

49.
$$\frac{x}{4} = 3.04$$

$$\frac{12.16}{4} = 3.04$$

$$3.04 = 3.04 \quad \text{True}$$

Yes, it is a solution.

53.
$$\frac{z}{10} = 0.8$$

$$\frac{8}{10} = 0.8$$

$$0.8 = 0.8 \text{ True}$$

Yes, it is a solution.

57. There are 52 weeks per year and 40 hours per week. Therefore, there are $52 \times 40 = 2080$ hours per year.

```
              2 734.508
2080)5,687,777.000
    −4 160
     1 527 7
    −1 4560
       7177
      −6240
       9377
      −8320
      10570
     −104 00
       1700
        −0
      17000
     −16640
        360
```

His hourly wage was $2734.51

61. $39.37\overline{)200}$ Move the decimal points 2 places.

```
          5.08 ≈ 5.1
3937.)20,000.00
     −19685
       3150
        −0
      31500
     −31496
          4
```

There are about 5.1 meters in 200 inches.

65.
```
        132.5
24)3180.0
  −24
   78
  −72
   60
  −48
   120
  −120
     0
```

Their average speed was 132.5 miles per hour.

69. There are 24 teaspoons in 4 fluid ounces, and so there are 48 doses in the bottle. Find the number of hours the child will be on the medicine by multiplying the number of doses by the number of hours each dose lasts.

$48 \times 4 = 192$ hours

Then $192 \div 24 = 8$ days.

The medicine will last 8 days.

73. To round −1000.994 to the nearest tenth, observe that the digit in the hundredths place is 9. Since this digit is at least 5, we need to add 1 to the digit in the tenths place. The number −1000.994 rounded to the nearest tenth is −1001.0.

77. $20 − 10 \div (−5) = 20 − (−2) = 22$

81.
$$\text{average} = \frac{9.6 + 8.5 + 4.1 + 4.1 + 9.8}{5}$$

$$= \frac{36.1}{5}$$

$$= 7.22$$

The average, rounded to the nearest tenth, is 7.2

85. Answers may vary.

Exercise Set 5.5

1. $4.9 − 2.1 = 2.8$

$5 − 2 = 3$

3 is close to 2.8, so the answer is reasonable.

5. $62.16 \div 14.8 = 4.2$

$60 \div 15 = 4$

4 is close to 4.2, so the answer is reasonable.

9.
```
  34.92        35
− 12.03      − 12
  22.89        23
```
23 is close to 22.89, so the answer is reasonable.

13. $11.8 + 12.9 + 14.2 \approx 12 + 13 + 14 = 39$ ft

17.
```
       51.6
30)1550.0
  − 150
     50
    −30
    200
   −180
     20
```
Their car will use about 52 gallons.

21. $19.9 + 15.1 + 10.9 + 6.7 \approx 20 + 15 + 11 + 7$
$$= 53 \text{ miles}$$

The distance is about 53 miles.

25. 501 rounded to the nearest 10 is 500 and 271 rounded to the nearest 10 is 270.
```
    490        270
  × 270      × 500
      0          0
  34300         00
  98000     135000
132,300    135,000
```
The population is about 135,000 people.

29. $97 \approx 100$ so that $78.6 \div 97 \approx 78.6 \div 100 = 0.786$. Thus b is the best estimate.

33. $1025.68 \approx 1025$ and $125.42 \approx 125$ so that
$1025.68 - 125.42 \approx 1025 - 125 = 900$ Thus, 900.26 is
reasonable.

37. $\dfrac{1 + 0.8}{-0.6} = \dfrac{1.8}{-0.6} = \dfrac{18}{-6} = -3$

41. $4.83 \div 2.1 = 2.3$

45. $(3.1 + 0.7)(2.9 - 0.9) = (3.8)(2.0) = 7.6$

49. $\dfrac{7 + 0.74}{-6} = \dfrac{7.74}{-6} = -1.29$

53. $x - y$
$6 - (0.3)$
$= 5.7$

57. $\quad\quad 7x + 2.1 = -7$
$7(-1.3) + 2.1 = -7$
$\quad\; -9.1 + 2.1 = -7$
$\quad\quad\quad\quad\; -7 = -7 \quad$ True
Yes, it is a solution.

61. $\dfrac{3}{4} \cdot \dfrac{5}{12} = \dfrac{3 \cdot 5}{4 \cdot 3 \cdot 4} = \dfrac{5}{16}$

65. $\dfrac{5}{12} - \dfrac{1}{3} = \dfrac{5}{12} - \dfrac{4}{12} = \dfrac{1}{12}$

69. Answers may vary.

EXERCISE SET 5.6

1. $\dfrac{1}{5} = 0.2$

$$
\begin{array}{r}
0.2 \\
5\overline{)1.0} \\
-10 \\
\hline
0
\end{array}
$$

5. $\dfrac{3}{4} = 0.75$

$$
\begin{array}{r}
0.75 \\
4\overline{)3.00} \\
-28 \\
\hline
20 \\
-20 \\
\hline
0
\end{array}
$$

9. $\dfrac{3}{8} = 0.375$

$$
\begin{array}{r}
0.375 \\
8\overline{)3.000} \\
-24 \\
\hline
60 \\
-56 \\
\hline
40 \\
-40 \\
\hline
0
\end{array}
$$

13. $\dfrac{17}{40} = 0.425$

$$
\begin{array}{r}
0.425 \\
40\overline{)17.000} \\
-160 \\
\hline
100 \\
-80 \\
\hline
200 \\
-200 \\
\hline
0
\end{array}
$$

17. $-\dfrac{1}{3} = -0.\overline{3}$

$$
\begin{array}{r}
0.33\ldots \\
3\overline{)1.00} \\
-9 \\
\hline
10 \\
-9 \\
\hline
1
\end{array}
$$

21. $\dfrac{2}{9} = 0.\overline{2}$

$$
\begin{array}{r}
0.22\ldots \\
9\overline{)2.00} \\
-18 \\
\hline
20 \\
-18 \\
\hline
2
\end{array}
$$

25. $0.\overline{3} = 0.333\ldots \approx 0.33$

29. $0.\overline{2} = 0.222\ldots \approx 0.2$

33.
$$
\begin{array}{r}
0.668 \\
1048\overline{)701.000} \\
-6288 \\
\hline
7220 \\
-6288 \\
\hline
9320 \\
-8384 \\
\hline
936
\end{array}
$$
Approximately 0.67 of Americans used the Internet in some form.

37.
$$
\begin{array}{r}
0.615 \\
91\overline{)56.000} \\
-546 \\
\hline
140 \\
-91 \\
\hline
490 \\
-455 \\
\hline
35
\end{array}
$$
Approximately 0.62 of U.S. mountains over 14,000 feet are in Colorado.

41. $2 > 1$, so $0.823 > 0.813$

45. $\dfrac{2}{3} = \dfrac{4}{6}$ and $\dfrac{4}{6} < \dfrac{5}{6}$, so $\dfrac{2}{3} < \dfrac{5}{6}$.

49. $\dfrac{4}{7} \approx 0.5714$ and $0.5714 > 0.14$, so $\dfrac{4}{7} > 0.14$.

53. $\dfrac{456}{64} = 7.125$ and $7.123 < 7.125$, so $7.123 < \dfrac{456}{64}$.

57. $0.49 = 0.490$
$0.49, 0.491, 0.498$

61. $\dfrac{4}{7} \approx 0.571$

$0.412, 0.453, \dfrac{4}{7}$

65. $\dfrac{12}{5} = 2.4, \dfrac{17}{8} = 2.125$

$\dfrac{17}{8}, 2.37, \dfrac{12}{5}$

69. Area $= \dfrac{1}{2} \times$ base $\times$ height

$= \dfrac{1}{2} \times 5.2 \times 3.6$

$= 0.5 \times 5.2 \times 3.6$

$= 9.36$

The area is 9.36 square centimeters.

73. $2^3 = (2)(2)(2) = 8$

77. $\left(\dfrac{1}{3}\right)^4 = \dfrac{1}{3} \cdot \dfrac{1}{3} \cdot \dfrac{1}{3} \cdot \dfrac{1}{3} = \dfrac{1}{81}$

81. $\left(\dfrac{2}{5}\right)\left(\dfrac{5}{2}\right)^2 = \dfrac{2}{5} \cdot \dfrac{5}{2} \cdot \dfrac{5}{2} = \dfrac{5}{2}$

85.
2131	Estimate: 2100
1179	1200
813	800
713	700
603	600
$+\ 547$	$+\ 500$
	5900

There are about 5900 stations.

89. Answers may vary.

93. $\left(\dfrac{1}{10}\right)^2 + (1.6)(2.1) = \dfrac{1}{100} + (1.6)(2.1)$

$= 0.01 + 3.36$

$= 3.37$

EXERCISE SET 5.7

1. $x + 1.2 = 7.1$

$x + 1.2 - 1.2 = 7.1 - 1.2$

$x = 5.9$

5. $6.2 = y - 4$

$6.2 + 4 = y - 4 + 4$

$10.2 = y$

9. $-3.5x + 2.8 = -11.2$

$-3.5x + 2.8 - 2.8 = -11.2 - 2.8$

$-3.5x = -14$

$\dfrac{-3.5x}{-3.5} = \dfrac{-14}{-3.5}$

$x = 4$

13. $2(x - 1.3) = 5.8$

$2x - 2.6 = 5.8$

$2x - 2.6 + 2.6 = 5.8 + 2.6$

$2x = 8.4$

$\dfrac{2x}{2} = \dfrac{8.4}{2}$

$x = 4.2$

17. $7x - 10.8 = x$

$7x - 10.8 - 7x = x - 7x$

$-10.8 = -6x$

$\dfrac{-10.8}{-6} = \dfrac{-6x}{-6}$

$1.8 = x$

21. $y - 3.6 = 4$

$y - 3.6 + 3.6 = 4 + 3.6$

$y = 7.6$

25. $6.5 = 10x + 7.2$

$6.5 - 7.2 = 10x + 7.2 - 7.2$

$-0.7 = 10x$

$\dfrac{-0.7}{10} = \dfrac{10x}{10}$

$-0.07 = x$

29. $200x - 0.67 = 100x + 0.81$

$200x - 0.67 + 0.67 = 100x + 0.81 + 0.67$

$200x = 100x + 1.48$

$200x - 100x = 100x - 100x + 1.48$

$100x = 1.48$

$x = 0.0148$

33. $8x - 5 = 10x - 8$

$8x - 5 + 8 = 10x - 8 + 8$

$8x + 3 = 10x$

$8x + 3 - 8x = 10x - 8x$

$3 = 2x$

$\dfrac{3}{2} = \dfrac{2x}{2}$

$1.5 = x$

37. $-0.9x + 2.65 = -0.5x + 5.45$

Multiply each term by 100.

$-90x + 265 = -50x + 545$

$-90x + 265 + 90x = -50x + 545 + 90x$

$265 = 40x + 545$

$265 - 545 = 40x + 545 - 545$

$-280 = 40x$

$\dfrac{-280}{40} = \dfrac{40x}{40}$

$-7 = x$

41. $0.7x + 13.8 = x - 2.16$

Multiply each term by 100.

$70x + 1380 = 100x - 216$

$70x + 1380 + 216 = 100x - 216 + 216$

$70x + 1596 = 100x$

$70x + 1596 - 70x = 100x - 70x$

$1596 = 30x$

$\dfrac{1596}{30} = \dfrac{30x}{30}$

$53.2 = x$

45. $3(x - 5) + 10 = 3x - 15 + 10 = 3x - 5$

49. Answers may vary.

53. $1.95y + 6.834 = 7.65y - 19.8591$

$19.8591 + 6.834 = 7.65y - 1.95y$

$26.6931 = 5.7y$

$\dfrac{26.6931}{5.7} = \dfrac{5.7y}{5.7}$

$4.683 = y$

EXERCISE SET 5.8

1. $\sqrt{4} = 2$ because $2^2 = 4$.

5. $\sqrt{\dfrac{1}{81}} = \dfrac{1}{9}$ because $\dfrac{1}{9} \cdot \dfrac{1}{9} = \dfrac{1}{81}$.

9. $\sqrt{256} = 16$ because $16^2 = 256$.

13. $\sqrt{3} \approx 1.732$

17. $\sqrt{14} \approx 3.742$

21. $\sqrt{8} \approx 2.828$

25. $\sqrt{71} \approx 8.426$

29. $a^2 + b^2 = c^2$

$5^2 + 12^2 = c^2$

$25 + 144 = c^2$

$169 = c^2$

$c = \sqrt{169}$

$c = 13$

The length of the hypotenuse is 13 inches.

33.

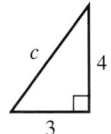

$$a^2 + b^2 = c^2$$
$$3^2 + 4^2 = c^2$$
$$9 + 16 = c^2$$
$$25 = c^2$$
$$c = \sqrt{25}$$
$$c = 5$$

37.

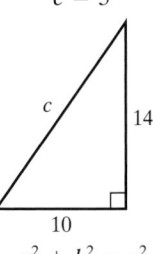

$$a^2 + b^2 = c^2$$
$$10^2 + 14^2 = c^2$$
$$100 + 196 = c^2$$
$$296 = c^2$$
$$c = \sqrt{296}$$
$$c \approx 17.205$$

41.

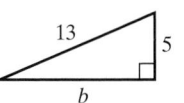

$$a^2 + b^2 = c^2$$
$$5^2 + b^2 = 13^2$$
$$25 + b^2 = 169$$
$$b^2 = 144$$
$$b = 12$$

45.

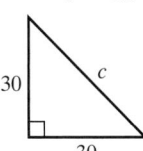

$$a^2 + b^2 = c^2$$
$$30^2 + 30^2 = c^2$$
$$900 + 900 = c^2$$
$$1800 = c^2$$
$$c = \sqrt{1800}$$
$$c \approx 42.426$$

49.
$$a^2 + b^2 = c^2$$
$$100^2 + 100^2 = c^2$$
$$10{,}000 + 10{,}000 = c^2$$
$$20{,}000 = c^2$$
$$c = \sqrt{20{,}000}$$
$$c \approx 141.42$$

The length of the diagonal is about 141.42 yards.

53.
$$a^2 + b^2 = c^2$$
$$300^2 + 160^2 = c^2$$
$$90{,}000 + 25{,}600 = c^2$$
$$115{,}600 = c^2$$
$$c = \sqrt{115{,}600}$$
$$c = 340$$

The length of the diagonal is 340 feet.

57. $\dfrac{24}{60} = \dfrac{2 \cdot 2 \cdot 2 \cdot 3}{2 \cdot 3 \cdot 2 \cdot 5} = \dfrac{2}{5}$

61. $\sqrt{38}$ is between 6 and 7.
$\sqrt{38} \approx 6.164$

CHAPTER 5 TEST

1. 45.092 is forty-five and ninety-two thousandths.

5. $9.83 - 30.25 = 9.83 + (-30.25)$
Subtract the absolute values.
$$\begin{array}{r} 30.25 \\ -\ 9.83 \\ \hline 20.42 \end{array}$$
Attach the sign of the larger absolute value.
$$-20.42$$

9. 0.8623 rounded to the nearest thousandth is 0.862.

13. $-24.73 = -24\dfrac{73}{100}$

17. $\dfrac{0.23 + 1.63}{-0.3} = \dfrac{1.86}{-0.3} = -6.2$

21. $\sqrt{\dfrac{64}{100}} = \dfrac{8}{10} = \dfrac{4}{5}$ since $\left(\dfrac{8}{10}\right)^2 = \dfrac{64}{100}$.

25. Area $= \dfrac{1}{2} \times$ base $\times$ height
$$= 0.5 \times 4.2 \times 1.1$$
$$= 2.31$$
The area is 2.31 square miles.

29.
$$\begin{array}{ll} & 14.2 \quad \text{Estimate:} \quad 14 \\ & 16.1 \qquad\qquad\qquad 16 \\ +\ & 23.7 \qquad\qquad\quad +\ 24 \\ \hline & \qquad\qquad\qquad\qquad\ \ 54 \end{array}$$
The total distance is about 54 miles.

Chapter 6

EXERCISE SET 6.1

1. The ratio of 11 to 14 is $\dfrac{11}{14}$.

5. The ratio of 151 to 201 is $\dfrac{151}{201}$.

9. The ratio of 5 to $7\dfrac{1}{2}$ is $\dfrac{5}{7\frac{1}{2}}$.

13. $\dfrac{16}{24} = \dfrac{2 \cdot 8}{3 \cdot 8} = \dfrac{2}{3}$

17. $\dfrac{4.63}{8.21} = \dfrac{4.63 \times 100}{8.21 \times 100} = \dfrac{463}{821}$

21. $\dfrac{10 \text{ hours}}{24 \text{ hours}} = \dfrac{5 \cdot 2}{12 \cdot 2} = \dfrac{5}{12}$

25. $\dfrac{24 \text{ days}}{14 \text{ days}} = \dfrac{12 \cdot 2}{7 \cdot 2} = \dfrac{12}{7}$

29. $\dfrac{35¢}{\$2} = \dfrac{35¢}{200¢} = \dfrac{35}{200} = \dfrac{7 \cdot 5}{40 \cdot 5} = \dfrac{7}{40}$

33. $\dfrac{\text{length}}{\text{width}} = \dfrac{94 \text{ feet}}{50 \text{ feet}} = \dfrac{2 \cdot 47}{2 \cdot 25} = \dfrac{47}{25}$

37. $\dfrac{\text{calories from fat}}{\text{total calories}} = \dfrac{200 \text{ calories}}{450 \text{ calories}} = \dfrac{4 \cdot 50}{9 \cdot 50} = \dfrac{4}{9}$

41. There are $6000 - 4500 = 1500$ married students.

$$\frac{\text{single students}}{\text{married students}} = \frac{4500}{1500} = \frac{3 \cdot 1500}{1 \cdot 1500} = \frac{3}{1}$$

45. $\dfrac{\text{current minimum wage}}{\text{1938 minimum wage}} = \dfrac{\$5.15}{\$0.25}$

$$= \frac{5.15 \times 100}{0.25 \times 100}$$

$$= \frac{515}{25}$$

$$= \frac{5 \cdot 103}{5 \cdot 5}$$

$$= \frac{103}{5}$$

49. $\dfrac{\text{mountains in Alaska}}{\text{mountains in Colorado}} = \dfrac{19}{57} = \dfrac{1 \cdot 19}{3 \cdot 19} = \dfrac{1}{3}$

53. $3.7\overline{)0.555}$ Move decimal points 1 place.

$$\begin{array}{r} 0.15 \\ 37\overline{)5.55} \\ -37 \\ \hline 185 \\ -185 \\ \hline 0 \end{array}$$

57. $\dfrac{2\frac{1}{2}}{5\frac{3}{4}} = \dfrac{\frac{5}{2}}{\frac{23}{4}}$

$$= \frac{4 \cdot \frac{5}{2}}{4 \cdot \frac{23}{4}}$$

$$= \frac{10}{23}$$

61. a. $\dfrac{\text{states with helmet laws}}{\text{total number of states}} = \dfrac{19}{50}$

b. There are $50 - 19 = 31$ states without mandatory helmet laws.

$$\frac{\text{states with helmet laws}}{\text{states without helmet laws}} = \frac{19}{31}$$

c. No; answers may vary.

EXERCISE SET 6.2

1. $\dfrac{5 \text{ shrubs}}{15 \text{ feet}} = \dfrac{1 \text{ shrub}}{3 \text{ feet}}$

5. $\dfrac{8 \text{ phone lines}}{36 \text{ employees}} = \dfrac{2 \text{ phone lines}}{9 \text{ employees}}$

9. $\dfrac{6 \text{ flight attendants}}{200 \text{ passengers}} = \dfrac{3 \text{ flight attendants}}{100 \text{ passengers}}$

13. $\dfrac{375 \text{ riders}}{5 \text{ subway cars}} = \dfrac{75 \text{ riders}}{1 \text{ subway car}} = 75 \text{ riders/car}$

17. $\dfrac{5400 \text{ wingbeats}}{1 \text{ min}} = \dfrac{5400 \text{ wingbeats}}{60 \text{ sec}} = \dfrac{90 \text{ wingbeats}}{1 \text{ sec}}$

$= 90 \text{ wingbeats/sec}$

21. $\dfrac{26{,}600 \text{ miles}}{20 \text{ days}} = \dfrac{1330 \text{ miles}}{1 \text{ day}} = 1330 \text{ mi/day}$

25. $\dfrac{12{,}000 \text{ good}}{40 \text{ defective}} = \dfrac{300 \text{ good}}{1 \text{ defective}} = 300 \text{ good/defective}$

29. $\dfrac{\$28{,}600{,}000}{500 \text{ species}} = \dfrac{\$57{,}200}{1 \text{ species}}$ or $\$57{,}200$/species

33. a. Greer: $\dfrac{250 \text{ computer boards}}{8 \text{ hours}}$

$= 31.25 \text{ computer boards/hour}$

b. Lamont: $\dfrac{400 \text{ computer boards}}{12 \text{ hours}} \approx 33.3 \text{ computer boards/hour}$

c. Since $33.3 > 31.25$, Lamont can assemble computer boards faster.

37. $\dfrac{\$1.19}{7 \text{ bananas}} = \dfrac{\$0.17}{1 \text{ banana}}$ or $\$0.17$ per banana

41. 6-ounce size: $\dfrac{\$0.69}{6 \text{ ounces}} = \$0.115$ per ounce

16-ounce size: $\dfrac{\$1.69}{16 \text{ ounces}} \approx \$0.106$ per ounce

The 16-ounce size costs less per ounce, thus it is the better buy.

45. 100-pack size: $\dfrac{\$0.89}{100 \text{ napkins}} \approx \$0.009$ per napkin

180-pack size: $\dfrac{\$1.49}{180 \text{ napkins}} \approx \$0.008$ per napkin

The package of 180 napkins cost less per napkin, thus it is the better buy.

49. $\begin{array}{r} 3.7 \\ \times\ 1.2 \\ \hline 74 \\ 370 \\ \hline 4.44 \end{array}$

53.

Miles Driven	Miles Per Gallon (round to the nearest tenth)
$79{,}543 - 79{,}286 = 257$	$\dfrac{257}{13.4} = 19.2$
$79{,}895 - 79{,}543 = 352$	$\dfrac{352}{15.8} = 22.3$
$80{,}242 - 79{,}895 = 347$	$\dfrac{347}{16.1} = 21.6$

57. $\dfrac{481{,}687 \text{ students}}{22{,}008 \text{ teachers}} \approx 22 \text{ students/teacher}$

EXERCISE SET 6.3

1. $\dfrac{10 \text{ diamonds}}{6 \text{ opals}} = \dfrac{5 \text{ diamonds}}{3 \text{ opals}}$

5. $\dfrac{6 \text{ eagles}}{58 \text{ sparrows}} = \dfrac{3 \text{ eagles}}{29 \text{ sparrows}}$

9. $\dfrac{22 \text{ vanilla wafers}}{1 \text{ cup cookie crumbs}} = \dfrac{55 \text{ vanilla wafers}}{2.5 \text{ cups cookie crumbs}}$

13. $\dfrac{8}{6} \overset{?}{=} \dfrac{9}{7}$

$6 \cdot 9 \overset{?}{=} 8 \cdot 7$

$54 \neq 56$

Since the cross products are not equal, the proportion is false.

17.
$$\frac{5}{8} \overset{?}{=} \frac{625}{1000}$$
$$8 \cdot 625 \overset{?}{=} 5 \cdot 1000$$
$$5000 = 5000$$
Since the cross products are equal, the proportion is true.

21.
$$\frac{4.2}{8.4} \overset{?}{=} \frac{5}{10}$$
$$8.4 \cdot 5 \overset{?}{=} 4.2 \cdot 10$$
$$42 = 42$$
Since the cross products are equal, the proportion is true.

25.
$$\frac{2\frac{2}{5}}{\frac{2}{3}} \overset{?}{=} \frac{\frac{10}{9}}{\frac{1}{4}}$$
$$\frac{2}{3} \cdot \frac{10}{9} \overset{?}{=} 2\frac{2}{5} \cdot \frac{1}{4}$$
$$\frac{20}{27} \overset{?}{=} \frac{12}{5} \cdot \frac{1}{4}$$
$$\frac{20}{27} \overset{?}{=} \frac{3}{5}$$
$$27 \cdot 3 \overset{?}{=} 20 \cdot 5$$
$$81 \neq 100$$
Since the cross products are not equal, the proportion is false.

29.
$$\frac{30}{10} = \frac{15}{y}$$
$$10 \cdot 15 = 30y$$
$$150 = 30y$$
$$\frac{150}{30} = \frac{30y}{30}$$
$$5 = y$$

33.
$$\frac{n}{6} = \frac{8}{15}$$
$$6 \cdot 8 = 15n$$
$$48 = 15n$$
$$\frac{48}{15} = \frac{15n}{15}$$
$$3.2 = n$$

37.
$$\frac{n}{\frac{6}{5}} = \frac{4\frac{1}{6}}{6\frac{2}{3}}$$
$$\frac{6}{5} \cdot 4\frac{1}{6} = 6\frac{2}{3} \cdot n$$
$$\frac{6}{5} \cdot \frac{25}{6} = \frac{20}{3}n$$
$$\frac{150}{30} = \frac{20}{3}n$$
$$5 = \frac{20}{3}n$$
$$\frac{3}{20} \cdot \frac{5}{1} = \frac{3}{20} \cdot \frac{20}{3}n$$
$$\frac{15}{20} = n$$
$$\frac{3}{4} = n$$

41.
$$\frac{\frac{2}{3}}{\frac{6}{9}} = \frac{12}{z}$$
$$\frac{6}{9} \cdot 12 = \frac{2}{3}z$$
$$\frac{3}{2} \cdot \frac{6}{9} \cdot \frac{12}{1} = \frac{3}{2} \cdot \frac{2}{3}z$$
$$\frac{3 \cdot 2 \cdot 3 \cdot 12}{2 \cdot 3 \cdot 3 \cdot 1} = z$$
$$12 = z$$

45.
$$\frac{3.5}{12.5} = \frac{7}{z}$$
$$12.5 \cdot 7 = 3.5 \cdot z$$
$$87.5 = 3.5z$$
$$\frac{87.5}{3.5} = \frac{3.5z}{3.5}$$
$$25 = z$$

49.
$$\frac{z}{5.2} = \frac{0.08}{6}$$
$$5.2 \cdot 0.08 = 6z$$
$$0.416 = 6z$$
$$\frac{0.416}{6} = \frac{6z}{6}$$
$$0.07 \approx z$$

53.
$$\frac{43}{17} = \frac{8}{z}$$
$$17 \cdot 8 = 43z$$
$$136 = 43z$$
$$\frac{136}{43} = \frac{43z}{43}$$
$$3.163 \approx z$$

57. $-2 > -3$

61. $-1\frac{1}{2} > -2\frac{1}{2}$

65. $1.23 > 1.039$

69.
$$\frac{n}{1150} = \frac{588}{483}$$
$$1150 \cdot 588 = 483n$$
$$676{,}200 = 483n$$
$$\frac{676{,}200}{483} = \frac{483n}{483}$$
$$1400 = n$$

73. Answers may vary.

EXERCISE SET 6.4

1. Let x be the number of completed passes.
$$\frac{4 \text{ completed}}{9 \text{ attempted}} = \frac{x \text{ completed}}{27 \text{ attempted}}$$
$$9x = 27 \cdot 4$$
$$9x = 108$$
$$\frac{9x}{9} = \frac{108}{9}$$
$$x = 12$$
He completed 12 passes.

5. Let x be the number of students accepted.
$$7x = 2 \cdot 630$$
$$\frac{2 \text{ accepts}}{7 \text{ applicants}} = \frac{x \text{ accepts}}{630 \text{ applicants}}$$
$$7x = 1260$$
$$\frac{7x}{7} = \frac{1260}{7}$$
$$x = 180$$
180 students were accepted.

9. Let x be the amount of floor space.
$$\frac{9 \text{ sq ft}}{1 \text{ student}} = \frac{x \text{ sq ft}}{30 \text{ students}}$$
$$x = 9 \cdot 30$$
$$x = 270$$
270 square feet are required.

13. Let x be the distance from Milan to Rome.
$$\frac{1 \text{ cm}}{30 \text{ km}} = \frac{15 \text{ cm}}{x \text{ km}}$$
$$15 \cdot 30 = 1x$$
$$450 = x$$
It is 450 km from Milan to Rome.

17. Let x be the number of bags of fertilizer.
$$\text{Area of lawn} = 260 \text{ ft} \cdot 180 \text{ ft}$$
$$= 46{,}800 \text{ square feet}$$
$$\frac{1 \text{ bag}}{3000 \text{ sq ft}} = \frac{x}{46{,}800 \text{ sq ft}}$$
$$3000x = 1 \cdot 46{,}800$$
$$3000x = 46{,}800$$
$$\frac{3000x}{3000} = \frac{46{,}800}{3000}$$
$$x = 15.6$$
Since a whole number of bags must be purchased, 16 bags are needed.

21. Let x be the number of hits expected.
$$\frac{3 \text{ hits}}{8 \text{ at bats}} = \frac{x \text{ hits}}{40 \text{ at bats}}$$
$$8x = 3 \cdot 40$$
$$8x = 120$$
$$\frac{8x}{8} = \frac{120}{8}$$
$$x = 15$$
He would be expected to get 15 hits.

25. Let x be the number of weeks it will last.
$$\frac{5 \text{ boxes}}{3 \text{ weeks}} = \frac{144 \text{ boxes}}{x \text{ weeks}}$$
$$3 \cdot 144 = 5x$$
$$432 = 5x$$
$$\frac{432}{5} = \frac{5x}{5}$$
$$86.4 = x$$
The envelopes will last 86 weeks.

29. Let n be the estimated height of Statue of Liberty.
$$\frac{\text{statue}}{\text{student}} = \frac{42 \text{ feet}}{2 \text{ feet}} = \frac{n \text{ feet}}{5\frac{1}{3}\text{feet}}$$
$$2n = 42 \cdot 5\frac{1}{3}$$
$$2n = \frac{42}{1} \cdot \frac{16}{3}$$
$$2n = 224$$
$$\frac{2n}{2} = \frac{224}{2}$$
$$n = 112$$
The estimate is 112 feet. The actual height is 111 feet 1 inch. The difference in the estimated height and the actual height is $112 - 111\frac{1}{12} = \frac{11}{12}$ of a foot or 11 inches.

33. Let x be the number of men that blame their not eating well on fast food.
$$\frac{2 \text{ men}}{5 \text{ men}} = \frac{x \text{ men}}{40 \text{ men}}$$
$$5x = 2 \cdot 40$$
$$5x = 80$$
$$\frac{5x}{5} = \frac{80}{5}$$
$$x = 16$$
16 men blame their not eating well on fast food.

37. Let x be the number of calories in a nine-piece Chicken McNuggets.
$$\frac{190 \text{ calories}}{4 \text{ Chicken McNuggets}} = \frac{x \text{ calories}}{9 \text{ Chicken McNuggets}}$$
$$4x = 190 \cdot 9$$
$$4x = 1710$$
$$\frac{4x}{4} = \frac{1710}{4}$$
$$x = 427.5$$
There are 427.5 calories in a nine-piece Chicken McNuggets.

41. a. Let x be the number of gallons of oil that should be mixed with 5 gallons of gasoline.
$$\frac{50 \text{ gal gas}}{1 \text{ gal oil}} = \frac{5 \text{ gal gas}}{x \text{ gal oil}}$$
$$50x = 1 \cdot 5$$
$$50x = 5$$
$$\frac{50x}{50} = \frac{5}{50}$$
$$x = 0.1$$
Thus, 0.1 gallons of oil should be mixed with 5 gallons of gasoline.

b. Let x be the number of ounces represented by 0.1 gallon.
$$\frac{128 \text{ ounces}}{1 \text{ gal}} = \frac{x \text{ ounces}}{0.1 \text{ gal}}$$
$$1x = 128 \cdot 0.1$$
$$x = 12.8$$
$$x \approx 13$$
About 13 fluid ounces of oil should be mixed with 5 gallons of gasoline.

45. $15 = 3 \cdot 5$

49. $200 = 2 \cdot 100$
$$= 2 \cdot 2 \cdot 50$$
$$= 2 \cdot 2 \cdot 2 \cdot 25$$
$$= 2 \cdot 2 \cdot 2 \cdot 5 \cdot 5$$
$$= 2^3 \cdot 5^2$$

53. $$\frac{15 \text{ mg medicine}}{1 \text{ ml solution}} = \frac{12 \text{ mg medicine}}{x}$$
$$15x = 12 \cdot 1$$
$$15x = 12$$
$$\frac{15x}{15} = \frac{12}{15}$$
$$x = 0.8$$
Thus, 0.8 ml of solution should be administered.

57. $$\frac{\text{first weight}}{\text{second distance}} = \frac{\text{second weight}}{\text{first distance}}$$
$$\frac{40 \text{ pounds}}{n \text{ feet}} = \frac{60 \text{ pounds}}{7 \text{ feet}}$$
$$60n = 40 \cdot 7$$
$$60n = 280$$
$$\frac{60n}{60} = \frac{280}{60}$$
$$n = 4\frac{2}{3}$$
The second distance is $4\frac{2}{3}$ feet.

EXERCISE SET 6.5

1. The triangles are congruent by Side-Side-Side.

5. Since the triangles are similar, we can compare any pair of corresponding sides to find the ratio.

$$\frac{22}{11} = \frac{2}{1}$$

9. $\frac{3}{n} = \frac{6}{9}$

$6n = 27$

$n = 4.5$

13. $\frac{n}{3.75} = \frac{12}{9}$

$9n = 45$

$n = 5$

17. $\frac{n}{3.25} = \frac{17.5}{3.25}$

$3.25n = 56.875$

$n = 17.5$

21. $\frac{34}{n} = \frac{16}{10}$

$16n = 340$

$n = 21.25$

25. $\frac{5}{4} = \frac{n}{48}$

$4n = 240$

$n = 60$

The height of the tree is 60 feet.

29. Average $= \dfrac{14 + 17 + 21 + 18}{4}$

$= \dfrac{70}{4}$

$= 17.5$

33. $\frac{5.2}{n} = \frac{7.8}{12.6}$

$7.8n = 5.2 \cdot 12.6$

$7.8n = 65.52$

$\frac{7.8n}{7.8} = \frac{65.52}{7.8}$

$n = 8.4$

37. Let l be the length of the built deck. Let w be the width of the built deck.

$\dfrac{\frac{1}{4}\text{inch}}{1 \text{ foot}} = \dfrac{3 \text{ inches}}{l \text{ feet}}$

$1 \cdot 3 = \frac{1}{4}l$

$3 = \frac{1}{4}l$

$4 \cdot 3 = 4 \cdot \frac{1}{4}l$

$12 = l$

$\dfrac{\frac{1}{4}\text{inch}}{1 \text{ foot}} = \dfrac{4\frac{1}{2}\text{inches}}{w \text{ feet}}$

$1 \cdot 4\frac{1}{2} = \frac{1}{4}w$

$\frac{9}{2} = \frac{1}{4}w$

$\frac{4}{1} \cdot \frac{9}{2} = 4 \cdot \frac{1}{4}w$

$\frac{36}{2} = w$

$18 = w$

The deck will be 12 feet by 18 feet.

CHAPTER 6 TEST

1. $\dfrac{4500 \text{ trees}}{6500 \text{ trees}} = \dfrac{9 \cdot 500}{13 \cdot 500} = \dfrac{9}{13}$

5. $\dfrac{650 \text{ kilometers}}{8 \text{ hours}} = 81.25$ kilometers/hour

$$\begin{array}{r} 81.25 \\ 8\overline{)650.00} \\ -\underline{64} \\ 10 \\ \underline{-8} \\ 20 \\ \underline{-16} \\ 40 \\ \underline{-40} \\ 0 \end{array}$$

9. 16-ounce size: $\dfrac{\$1.49}{16 \text{ oz}} \approx \$0.0931$ per ounce

24-ounce size: $\dfrac{\$2.39}{24 \text{ oz}} \approx \$0.0996$ per ounce

The 16-ounce size costs less per ounce, thus it is the better buy.

13. $\frac{8}{x} = \frac{11}{6}$

$11 \cdot x = 8 \cdot 6$

$11x = 48$

$\frac{11x}{11} = \frac{48}{11}$

$x = \frac{48}{11} = 4\frac{4}{11}$

17. Let x be the length of time.

$\frac{3}{80} = \frac{x}{100}$

$80x = 300$

$\frac{80x}{80} = \frac{300}{80}$

$x = \frac{15}{4} = 3\frac{3}{4}$

It will take $3\frac{3}{4}$ hours.

21. $\frac{5}{8} = \frac{n}{12}$

$8n = 5 \cdot 12$

$8n = 60$

$\frac{8n}{8} = \frac{60}{8}$

$n = 7.5$

Chapter 7

EXERCISE SET 7.1

1. $\dfrac{81}{100} = 81\%$

5. The largest section of the circle graph represents chocolate chip. Therefore, chocolate chip was the most preferred cookie.

$\dfrac{52}{100} = 52\%$

9. $48\% = 48.(0.01) = 0.48$

13. $100\% = 100.(0.01) = 1.00 = 1$

17. $2.8\% = 2.8(0.01) = 0.028$

21. $300\% = 300.(0.01) = 3.00 = 3$

25. $73.7\% = 73.7(0.01) = 0.737$

29. $11.1\% = 11.1(0.01) = 0.111$

33. $3.1 = 3.10(100\%) = 310\%$

37. $0.003 = 0.003(100\%) = 000.3\% = 0.3\%$

41. $0.056 = 0.056(100\%) = 005.6\% = 5.6\%$

45. $3.00 = 3.00(100\%) = 300.\% = 300\%$

49. $0.10 = 0.10(100\%) = 010.\% = 10\%$

53. $0.38 = 0.38(100\%) = 038.\% = 38\%$

57. $4.5\% = \dfrac{4.5}{100} = \dfrac{45}{1000} = \dfrac{5 \cdot 9}{5 \cdot 200} = \dfrac{9}{200}$

61. $73\% = \dfrac{73}{100}$

65. $6.25\% = \dfrac{6.25}{100} = \dfrac{625}{10,000} = \dfrac{625}{625 \cdot 16} = \dfrac{1}{16}$

69. $22\dfrac{3}{8}\% = \dfrac{22\frac{3}{8}}{100}$

$= \dfrac{\frac{179}{8}}{100}$

$= \dfrac{179}{8} \div 100$

$= \dfrac{179}{8} \cdot \dfrac{1}{100}$

$= \dfrac{179}{800}$

73. $\dfrac{7}{10} = \dfrac{7}{10} \cdot 100\% = \dfrac{700}{10}\% = 70\%$

77. $\dfrac{59}{100} = \dfrac{59}{100} \cdot 100\% = \dfrac{59 \cdot 100}{100}\% = 59\%$

81. $\dfrac{3}{8} = \dfrac{3}{8} \cdot 100\% = \dfrac{300}{8}\% = \dfrac{4 \cdot 75}{4 \cdot 2}\% = \dfrac{75}{2}\% = 37\dfrac{1}{2}\%$

85. $\dfrac{2}{3} = \dfrac{2}{3} \cdot 100\% = \dfrac{200}{3}\% = 66\dfrac{2}{3}\%$

89. $1\dfrac{9}{10} = \dfrac{19}{10} \cdot 100\% = \dfrac{1900\%}{10} = 190\%$

93. $\dfrac{4}{15} = \dfrac{4}{15} \cdot 100\% = \dfrac{400}{15}\% \approx 26.67\%$

$$\begin{array}{r} 26.666 \approx 26.67 \\ 15\overline{)400.000} \\ \underline{-30} \\ 100 \\ \underline{-90} \\ 100 \\ \underline{-90} \\ 100 \\ \underline{-90} \\ 100 \\ \underline{-90} \\ 10 \end{array}$$

97. $\dfrac{11}{12} = \dfrac{11}{12} \cdot 100\% = \dfrac{1100}{12}\% \approx 91.67\%$

$$\begin{array}{r} 91.666 \approx 91.67 \\ 12\overline{)1100.000} \\ \underline{-108} \\ 20 \\ \underline{-12} \\ 80 \\ \underline{-72} \\ 80 \\ \underline{-72} \\ 80 \\ \underline{-72} \\ 8 \end{array}$$

101.

Percent	Decimal	Fraction
40%	0.4	$\dfrac{2}{5}$
$23\dfrac{1}{2}\%$	0.235	$\dfrac{47}{200}$
80%	0.8	$\dfrac{4}{5}$
$33\dfrac{1}{3}\%$	$0.\overline{3}$	$\dfrac{1}{3}$
87.5%	0.875	$\dfrac{7}{8}$
7.5%	0.075	$\dfrac{3}{40}$

105. $\dfrac{17}{200} = \dfrac{17}{200} \cdot 100\%$

$= \dfrac{1700}{200}\%$

$= \dfrac{17}{2}\%$

$= 8.5\%$

$$\begin{array}{r} 8.5 \\ 2\overline{)17.0} \\ \underline{-16} \\ 10 \\ \underline{-10} \\ 0 \end{array}$$

109. $\dfrac{1}{4} = \dfrac{1 \cdot 25}{4 \cdot 25} = \dfrac{25}{100} \cdot 100\%$

$= 25\%$

113. $12\% = \dfrac{12}{100} = \dfrac{3 \cdot 4}{25 \cdot 4} = \dfrac{3}{25}$

117. $3n = 45$

$\dfrac{3n}{3} = \dfrac{45}{3}$

$n = 15$

121. $-6n = -72$

$\dfrac{-6n}{-6} = \dfrac{-72}{-6}$

$n = 12$

125. 4 of the 5 parts of the circle are shaded, or $\dfrac{4}{5}$.

$$\frac{4}{5} \cdot 100\% = \frac{400}{5}\% = \frac{80.5}{5}\% = 80\%$$

129. A fraction written as a percent is greater than 100% when the numerator is greater than the denominator.

133. The longest bar corresponds to computer software engineers and applications, which is predicted to be the fastest-growing occupation.

EXERCISE SET 7.2

1. $15\% \cdot 72 = x$

5. $x \cdot 90 = 20$

9. $x = 9\% \cdot 43$

13. $x = 14\% \cdot 52$
$x = 0.14 \cdot 52$
$x = 7.28$

17. $1.2 = 12\% \cdot x$
$1.2 = 0.12x$
$\dfrac{1.2}{0.12} = \dfrac{0.12x}{0.12}$
$10 = x$

21. $16 = x \cdot 50$
$\dfrac{16}{50} = \dfrac{x \cdot 50}{50}$
$0.32 = x$
$32\% = x$

25. $125\% \cdot 36 = x$
$1.25 \cdot 36 = x$
$45 = x$

29. $2.58 = x \cdot 50$
$\dfrac{2.58}{50} = \dfrac{x \cdot 50}{50}$
$0.0516 = x$
$5.16\% = x$

33. $x \cdot 150 = 67.5$
$\dfrac{x \cdot 150}{150} = \dfrac{67.5}{150}$
$x = 0.45$
$x = 45\%$

37. $\dfrac{27}{n} = \dfrac{9}{10}$
$9 \cdot n = 27 \cdot 10$
$\dfrac{9 \cdot n}{9} = \dfrac{270}{9}$
$n = 30$

41. $\dfrac{17}{12} = \dfrac{n}{20}$

45. $1.5\% \cdot 45,775 = n$
$0.015 \cdot 45,775 = n$
$686.625 = n$

EXERCISE SET 7.3

1. $\dfrac{a}{65} = \dfrac{32}{100}$

5. $\dfrac{70}{200} = \dfrac{p}{100}$

9. $\dfrac{a}{130} = \dfrac{19}{100}$

13. $\dfrac{a}{105} = \dfrac{18}{100}$
$\dfrac{a}{105} = \dfrac{9}{50}$
$a \cdot 50 = 105 \cdot 9$
$\dfrac{a \cdot 50}{50} = \dfrac{945}{50}$
$a = 18.9$
Therefore, 18% of 105 is 18.9.

17. $\dfrac{7.8}{b} = \dfrac{78}{100}$
$\dfrac{7.8}{b} = \dfrac{39}{50}$
$7.8 \cdot 50 = b \cdot 39$
$390 = b \cdot 39$
$\dfrac{390}{39} = \dfrac{b \cdot 39}{39}$
$10 = b$
Therefore, 78% of 10 is 7.8.

21. $\dfrac{14}{50} = \dfrac{p}{100}$
$\dfrac{7}{25} = \dfrac{p}{100}$
$7 \cdot 100 = 25 \cdot p$
$700 = 25 \cdot p$
$\dfrac{700}{25} = \dfrac{25 \cdot p}{25}$
$28 = p$
Therefore, 28% of 50 is 14.

25. $\dfrac{a}{80} = \dfrac{2.4}{100}$
$a \cdot 100 = 80 \cdot 2.4$
$a \cdot 100 = 192$
$\dfrac{a \cdot 100}{100} = \dfrac{192}{100}$
$a = 1.92$
Therefore, 2.4% of 80 is 1.92.

29. $\dfrac{348.6}{166} = \dfrac{p}{100}$
$348.6 \cdot 100 = 166 \cdot p$
$34,860 = 166 \cdot p$
$\dfrac{34,860}{166} = \dfrac{166 \cdot p}{166}$
$210 = p$
Therefore, 210% of 166 is 348.6.

33. $\dfrac{3.6}{8} = \dfrac{p}{100}$
$3.6 \cdot 100 = 8 \cdot p$
$360 = 8 \cdot p$
$\dfrac{360}{8} = \dfrac{8 \cdot p}{8}$
$45 = p$
Therefore, 45% of 8 is 3.6.

37. $\dfrac{11}{16} + \dfrac{3}{16} = \dfrac{11 + 3}{16} = \dfrac{14}{16} = \dfrac{7 \cdot 2}{8 \cdot 2} = \dfrac{7}{8}$

41.
$$\begin{array}{r} \overset{1}{0.41} \\ + \ 0.29 \\ \hline 0.70 \text{ or } 0.7 \end{array}$$

45.
$$\frac{a}{53,862} = \frac{22.3}{100}$$
$$a \cdot 100 = 53,862 \cdot 22.3$$
$$a \cdot 100 = 1,201,122.6$$
$$\frac{a \cdot 100}{100} = \frac{1,201,122.6}{100}$$
$$a = 12,011.226$$

Therefore, 22.3% of 53,862 rounded to the nearest tenth is 12,011.2.

EXERCISE SET 7.4

1. Let n be the number of bolts inspected.
$$1.5\% \cdot n = 24$$
$$0.015 \cdot n = 24$$
$$\frac{0.015 \cdot n}{0.015} = \frac{24}{0.015}$$
$$n = 1600$$
1600 bolts were inspected.

5. Let x be the percent of income spent on food.
$$\$300 = x \cdot \$2000$$
$$\frac{300}{2000} = \frac{x \cdot 2000}{2000}$$
$$0.15 = x$$
$$0.15 \cdot 100\% = x$$
$$15\% = x$$
She spends 15% of her monthly income on food.

9. Let x be the percent of members that attended a community college.
$$73 = x \cdot 535$$
$$\frac{73}{535} = \frac{x \cdot 535}{535}$$
$$0.136 \approx x$$
$$0.136 \cdot 100\% = x$$
$$13.6\% = x$$
13.6% of the members attended a community college.

13. Let x be the percent of calories from fat.
$$35 = x \cdot 120$$
$$\frac{35}{120} = \frac{x \cdot 120}{120}$$
$$0.292 \approx x$$
$$0.292 \cdot 100\% = x$$
$$29.2\% = x$$
29.2% of the food's total calories is from fat.

17. 20% of $170 is what number?
$$20\% \cdot \$170 = n$$
$$0.2 \cdot \$170 = n$$
$$\$34.00 = n$$
$$\text{new bill} = \$170 - \$34$$
$$= \$136$$
Their new bill is $136.

21. 80% of 35 million is what number?
$$80\% \cdot 35 \text{ million} = n$$
$$0.80 \cdot 35,000,000 = n$$
$$28,000,000 = n$$
Projected population:
35 million + 28 million = 63 million
The increase is 28 million. The projected population is 63 million.

25.

Original Amount	New Amount	Amount of Increase	Percent Increase
85	187	$187 - 85 = 102$	$\frac{102}{85} = 1.2 = 120\%$

29.

Original Amount	New Amount	Amount of Decrease	Percent Decrease
160	40	$160 - 40 = 120$	$\frac{120}{160} = 0.75 = 75\%$

33.
$$\begin{aligned}
\text{percent increase} &= \frac{23.7 - 19.5}{19.5} \\
&= \frac{4.2}{19.5} \\
&\approx 0.215 \\
&= 21.5\%
\end{aligned}$$

37.
$$\begin{aligned}
\text{percent decrease} &= \frac{52.1 - 6.7}{52.1} \\
&= \frac{45.4}{52.1} \\
&\approx 0.871 \\
&= 87.1
\end{aligned}$$

41.
$$\begin{aligned}
\text{percent increase} &= \frac{29,000 - 16,000}{16,000} \\
&= \frac{13,000}{16,000} \\
&= 0.8125 \\
&= 81.25\%
\end{aligned}$$
or 81.3% when rounded to the nearest tenth.

45.
$$\begin{aligned}
\text{percent increase} &= \frac{110,000,000 - 16,000,000}{16,000,000} \\
&= \frac{94,000,000}{16,000,000} \\
&= 5.875 \\
&= 587.5\%
\end{aligned}$$

49.
$$\begin{array}{r}
0.12 \\
\times\ 38 \\
\hline
96 \\
360 \\
\hline
4.56
\end{array}$$

53.
$$\begin{array}{r}
78.00 \\
-\ 19.46 \\
\hline
58.54
\end{array}$$

57. Answers may vary.

EXERCISE SET 7.5

1.
$$\begin{aligned}
\text{tax} &= 5\% \cdot \$150.00 \\
&= 0.05 \cdot \$150.00 \\
&= \$7.50
\end{aligned}$$
The sales tax is $7.50.

5.
$$\$54 = r \cdot \$600$$
$$\frac{\$54}{\$600} = r$$
$$0.09 = r$$
$$9\% = r$$
The sales tax rate is 9%.

9. $\text{tax} = 6.5\% \cdot \$1800$
$= 0.065 \cdot \$1800$
$= \$117$
$\text{total} = \$1800 + \$117 = \$1917$
The total price is \$1917.

13. $\text{total purchase} = \$90 + \$15 = \$105$
$\text{tax} = 7\% \cdot \$105$
$= 0.07 \cdot \$105$
$= \$7.35$
$\text{total} = \$105 + \$7.35 = \$112.35$
The total price is \$112.35.

17. $\text{commission} = 4\% \cdot \$1,236,856$
$= 0.04 \cdot \$1,236,856$
$= \$49,474.24$
Her commission was \$49,474.24.

21. $\text{commission} = 1.5\% \cdot \$125,900$
$= 0.015 \cdot \$125,900$
$= \$1888.50$
His commission is \$1888.50.

25.

Original Price	Discount Rate	Amount of Discount	Sale Price
\$68.00	10%	$\$68.00 \cdot 10\%$	$\$68.00 - \$6.80$
		$= \$68.00 \cdot 0.10$	$= \$61.20$
		$= \$6.80$	

29.

Original Price	Discount Rate	Amount of Discount	Sale Price
\$215.00	35%	$\$215.00 \cdot 35\%$	$\$215.00 - \$75.25$
		$= \$215.00 \cdot 0.35$	$= \$139.75$
		$= \$75.25$	

33. $\text{discount} = 15\% \cdot \$300$
$= 0.15 \cdot \$300$
$= \$45$
$\text{sale price} = \$300 - \$45$
$= \$255$

37.

Original Price	Discount Rate	Amount of Discount	Sale Price
\$120	$\dfrac{\$39.60}{\$120} = 0.33$	\$39.60	$\$120 - 39.60$
	$= 33\%$		$= \$80.40$

41. $2000 \cdot 0.3 \cdot 2 = 600 \cdot 2 = 1200$

45. $600 \cdot 0.04 \cdot \dfrac{2}{3} = 24 \cdot \dfrac{2}{3} = \dfrac{3 \cdot 8}{1} \cdot \dfrac{2}{3} = \dfrac{16}{1}$ or 16

49. $60\% \cdot \$50 = 0.60 \cdot \$50 = \$30$,
so the sale price is $\$50 - \$30 = \$20$
$30\% \cdot \$50 = 0.30 \cdot \$50 = \$15$
$\$50 - \$15 = \$35$
$35\% \cdot \$35 = 0.35 \cdot \$35 = \$12.25$
$\$35 - \$12.25 = \$22.75$
A discount of 60% is better.

EXERCISE SET 7.6

1. $I = P \cdot R \cdot T$
$= \$200 \cdot 8\% \cdot 2$
$= \$200 \cdot (0.08) \cdot 2$
$= \$32$

5. $I = P \cdot R \cdot T$
$= \$5000 \cdot 10\% \cdot 1\dfrac{1}{2}$
$= \$5000 \cdot (0.10) \cdot 1.5$
$= \$750$

9. $I = P \cdot R \cdot T$
$= \$2500 \cdot 16\% \cdot \dfrac{21}{12}$
$= \$2500 \cdot (0.16) \cdot 1.75$
$= \$700$

13. $I = P \cdot R \cdot T$
$= \$5000 \cdot 9\% \cdot \dfrac{15}{12}$
$= \$5000 \cdot (0.09) \cdot 1.25$
$= \$562.50$
$\text{total} = \$5000 + \$562.50 = \$5562.50$
The total amount received is \$5562.50.

17. $\text{total} = \text{principal} \cdot \text{compound interest factor}$
$= \$6150(7.61226)$
$= \$46,815.399$
The total amount is \$46,815.40.

21. $\text{total} = \text{principal} \cdot \text{compound interest factor}$
$= \$10,000(5.81636)$
$= \$58,163.60$
The total amount is \$58,163.60.

25. $\text{total} = \text{principal} \cdot \text{compound interest factor}$
$= \$2000(1.46933)$
$= \$2938.66$
$\text{compound interest} = \text{total} - \text{principal}$
$= \$2938.66 - \$2000$
$= \$938.66$

29. $\text{monthly payment} = \dfrac{\text{principal} + \text{interest}}{\text{number of payments}}$
$= \dfrac{\$1500 + \$61.88}{6}$
$= \dfrac{\$1561.88}{6}$
$\approx \$260.31$
The monthly payment is \$260.31.

33. $-5 + (-24) = -29$

37. $\dfrac{7 - 10}{3} = \dfrac{-3}{3} = -1$

41. Answers may vary.

CHAPTER 7 TEST

1. $85\% = 85.(0.01) = 0.85$

5. $6.1 = 6.10 = 6.10(100\%) = 610\%$

9. $0.2\% = \dfrac{0.2}{100} = \dfrac{2}{1000} = \dfrac{1}{500}$

13. $\dfrac{1}{5} = \dfrac{1}{5} \cdot 100\% = \dfrac{100}{5}\% = 20\%$

17.
$$567 = x \cdot 756$$
$$\frac{567}{756} = \frac{x \cdot 756}{756}$$
$$0.75 = x$$
$$75\% = x$$
Therefore, 75% of 756 is 567.

21. percent increase $= \dfrac{26,460 - 25,200}{25,200}$
$$= \frac{1260}{25,200}$$
$$= 0.05$$
$$= 5\%$$
The population increased 5%.

25. $I = P \cdot R \cdot T$
$$= \$2000 \cdot 9.25\% \cdot 3\frac{1}{2}$$
$$= \$2000 \cdot (0.0925) \cdot 3.5$$
$$= \$647.50$$

Chapter 8

EXERCISE SET 8.1

1. The year 2002 has the most cars, so the greatest number of cars were was manufactured in 2002.

5. Compare each year with the previous year to see that the production of automobiles decreased from the previous year in 1998, 1999, and 2003.

9. The year 1997 has seven and a half chickens and each chicken represents 3 ounces of chicken, so approximately $7.5 \cdot 3 = 22.5$ ounces of chicken were consumed per person per week in 1997.

13. There is one more chicken for 2001 than for 1995, so there was an increase of approximately $1 \cdot 3 = 3$ ounces per person per week.

17. The tallest bar corresponds to the month of April, so April has the most tornado-related deaths.

21. Look for bars that extend above the horizontal line at 5. The months of February, March, April, May, and June have more than 5 tornado-related deaths.

25. The two U.S. cities are New York City and Los Angeles, New York City is the largest, with an estimated population of 21.4 million or 21,400,000.

29.

Fiber Content of Selected Foods

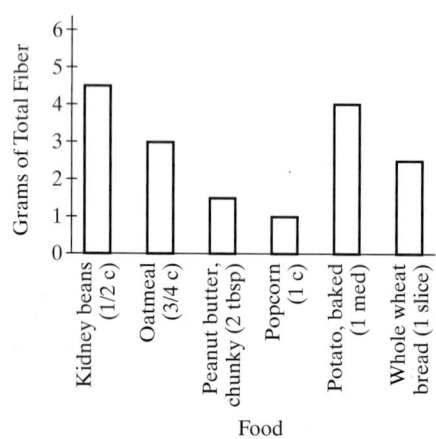

Food

33. The height of the bar representing 100–149 miles is 15. Therefore, 15 adults drive 100–149 miles per week.

37. There are two bars that fit this description, namely 100–149 miles and 150–199 miles. Therefore, $15 + 9 = 24$ adults drive 100–199 miles per week.

41. $\dfrac{\text{number of adults who drive 150–199 miles per week}}{\text{total number of adults}}$
$$= \frac{9}{100}$$

45. The height of the bar representing 55–64 years old is 17. Therefore, 17 million householders will be 55–64 years old.

49. Answers may vary.

53.

Class Intervals (Scores)	Tally	Class Frequency (Number of Games)
90–99	⊪ ⫼⫼⫼	8

57.

Class Intervals (Account Balances)	Tally	Class Frequency (Number of People)
$200–$299	⊪ ⫼	6

61.

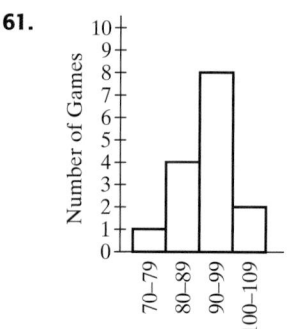

Golf Scores

65. The highest point on the line graph corresponds to 1982. Thus, the average number of goals per game was highest in 1982.

69.
$$30\% \cdot 12 = n$$
$$0.30 \cdot 12 = n$$
$$3.6 = n$$
Thus, 30% of 12 is 3.6.

73. $\dfrac{1}{4} \cdot 100\% = \dfrac{1}{4} \cdot \dfrac{100}{1}\% = \dfrac{100}{4}\% = 25\%$

77. 83°F

81. Look for the day of the week where the lines of the graph are farthest apart. This day is Tuesday, when the high was 86°F and the low was 73°F for a difference of $86 - 73 = 13$ or 13°F.

85. Answers may vary.

EXERCISE SET 8.2

1. The largest sector, or 320 students, corresponds to where most college students live. Most college students live at a parent or guardian's home.

5. $\dfrac{\text{students living in campus housing}}{\text{students living at home}} = \dfrac{180}{320} = \dfrac{9 \cdot 20}{16 \cdot 20} = \dfrac{9}{16}$

9. 30% + 7% = 37%

13. 5% of 57,000,000

= (0.05)(57,000,000)

= 2,850,000

The area of Australia is approximately 2,850,000 square miles.

17. The second-largest sector is 25% which corresponds to "Nonfiction." Nonfiction is the second-largest category of books.

21. 22% of 125,600

= (0.22)(125,600)

= 27,632

There are 27,632 books in the children's fiction category.

25.

Country of Origin	Percent	Degrees in Sector
United States	75%	75% × 360° = 270°
Japan	11%	11% × 360° = 39.6° ≈ 40°
Germany	6%	6% × 360° = 21.6° ≈ 22°
Other Countries	8%	8% × 360° = 28.8° ≈ 29°

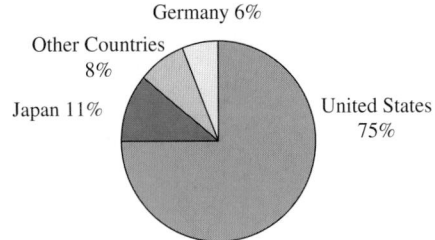

29. $20 = 2 \cdot 10 = 2 \cdot 2 \cdot 5 = 2^2 \cdot 5$

33. $85 = 5 \cdot 17$

37. Pacific Ocean: 49% · 264,489,800 = 129,600,002 square kilometers

41. 21.5% · 2800 = 602 respondents

45. $\dfrac{\text{number of respondents who spend } \$0 - \$15}{\text{number of respondents who spend } \$15 - \$175}$

$= \dfrac{602}{1674} = \dfrac{2 \cdot 301}{2 \cdot 837} = \dfrac{301}{837}$

EXERCISE SET 8.3

1.

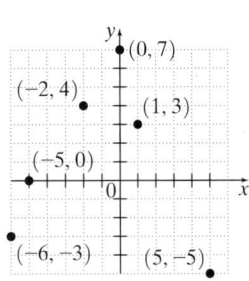

5. Point A has coordinates $(0, 0)$.

Point B has coordinates $\left(3\dfrac{1}{2}, 0\right)$.

Point C has coordinates $(3, 2)$.

Point D has coordinates $(-1, 3)$.

Point E has coordinates $(-2, -2)$.

Point F has coordinates $(0, -1)$.

Point G has coordinates $(2, -1)$.

9. $x - y = 3$

$1 - 2 = 3$

$-1 = 3$ False

No, $(1, 2)$ is not a solution of $x - y = 3$.

13. $y = -4x$

$-8 = -4 \cdot 2$

$-8 = -8$ True

Yes, $(2, -8)$ is a solution of $y = -4x$.

17. $x - 5y = -1$

$3 - 5(1) = -1$

$3 - 5 = -1$

$-2 = -1$ False

No, $(3, 1)$ is not a solution of $x - 5y = -1$.

21. $x = 5y$; $(5, 1), (0, 0), (-5, -1)$

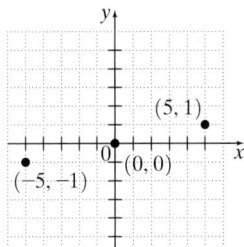

25. $y = 8x$

$y = 8(1)$

$y = 8$

The solution is $(1, 8)$.

$y = 8x$

$y = 8(0)$

$y = 0$

The solution is $(0, 0)$, the origin.

$y = 8x$

$-16 = 8x$

$\dfrac{-16}{8} = \dfrac{8x}{8}$

$-2 = x$

The solution is $(-2, -16)$.

29. $y = x + 5$

$y = 1 + 5$

$y = 6$

The solution is $(1, 6)$.

$y = x + 5$

$7 = x + 5$

$7 - 5 = x + 5 - 5$

$2 = x$

The solution is $(2, 7)$.

$y = x + 5$

$y = 3 + 5$

$y = 8$

The solution is $(3, 8)$.

33.
$$y = -x$$
$$0 = -x$$
$$\frac{0}{-1} = \frac{-x}{-1}$$
$$0 = x$$
The solution is $(0, 0)$, the origin.
$$y = -x$$
$$y = -(2)$$
$$y = -2$$
The solution is $(2, -2)$.
$$y = -x$$
$$2 = -x$$
$$\frac{2}{-1} = \frac{-x}{-1}$$
$$-2 = x$$
The solution is $(-2, 2)$.

37. $5.6 - 3.9 = 1.7$

41. $(0.236)(-100) = -23.6$

45. To plot the point $(0, b)$, we start at the origin and move 0 units to the right and b units up. Our point will be on the y-axis. Thus $(0, b)$ lies on the y-axis is a true statement.

49. To plot $(-a, b)$, we start at the origin and move a units to the left and b units up. Our point will be in quadrant II. Thus $(-a, b)$ lies in quadrant III is a false statement.

53.

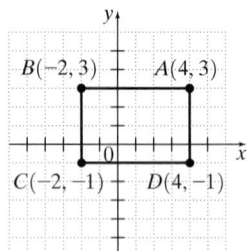

Exercise Set 8.4

1. $x + y = 6$
Find any 3 ordered pair-solutions.
Let $x = 0$.
$$x + y = 6$$
$$0 + y = 6$$
$$y = 6$$
$(0, 6)$
Let $x = 3$.
$$x + y = 6$$
$$3 + y = 6$$
$$3 + y - 3 = 6 - 3$$
$$y = 3$$
$(3, 3)$
Let $x = 6$.
$$x + y = 6$$
$$6 + y = 6$$
$$6 + y - 6 = 6 - 6$$
$$y = 0$$
$(6, 0)$
Plot $(0, 6)$, $(3, 3)$ and $(6, 0)$. Then draw the line through them.

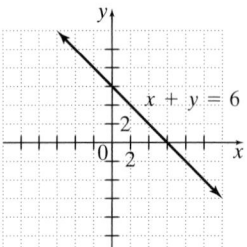

5. $y = 4x$
Find any 3 ordered-pair solutions.
Let $x = 0$.
$$y = 4x$$
$$y = 4(0)$$
$$y = 0$$
$(0, 0)$
Let $x = 1$.
$$y = 4x$$
$$y = 4(1)$$
$$y = 4$$
$(1, 4)$
Let $x = -1$.
$$y = 4x$$
$$y = 4(-1)$$
$$y = -4$$
$(-1, -4)$
Plot $(0, 0)$, $(1, 4)$ and $(-1, -4)$. Then draw the line through them.

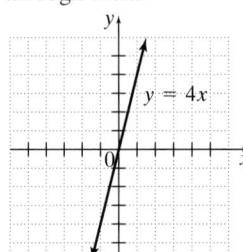

9. $x = 5$
No matter what y-value we choose, x is always 5.

x	y
5	-4
5	0
5	4

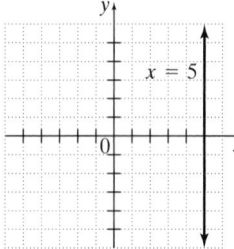

13. $x = 0$
No matter what y-value we choose, x is always 0.

x	y
0	-3
0	0
0	3

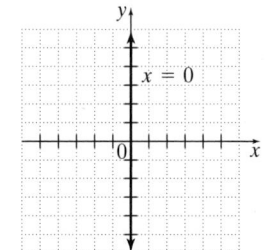

17. $y = -2$

No matter what x-value we choose, y is always -2.

x	y
-4	-2
0	-2
4	-2

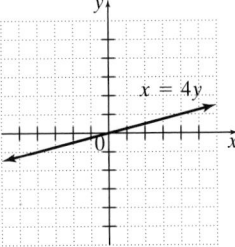

21. $x = 6$

No matter what y-value we choose, x is always 6.

x	y
6	-5
6	0
6	5

25. $x = y - 4$

Find any 3 ordered-pair solutions.

Let $y = 0$.

$x = y - 4$

$x = 0 - 4$

$x = -4$

$(-4, 0)$

Let $y = 4$.

$x = y - 4$

$x = 4 - 4$

$x = 0$

$(0, 4)$

Let $y = 6$.

$x = y - 4$

$x = 6 - 4$

$x = 2$

$(2, 6)$

Plot $(-4, 0)$, $(0, 4)$, and $(2, 6)$. Then draw the line through them.

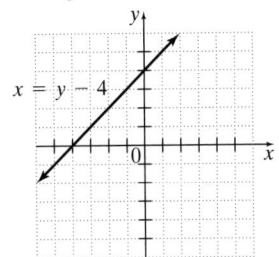

29. $x = 4y$

Find any 3 ordered-pair solutions.

Let $y = -1$.

$x = 4y$

$x = 4(-1)$

$x = -4$

$(-4, -1)$

Let $y = 0$.

$x = 4y$

$x = 4(0)$

$x = 0$

$(0, 0)$

Let $y = 1$.

$x = 4y$

$x = 4(1)$

$x = 4$

$(4, 1)$

Plot $(-4, -1)$, $(0, 0)$ and $(4, 1)$. Then draw the line through them.

33. $y = 4x + 2$

Find any 3 ordered-pair solutions.

Let $x = -1$.

$y = 4x + 2$

$y = 4(-1) + 2$

$y = -4 + 2$

$y = -2$

$(-1, -2)$

Let $x = 0$.

$y = 4x + 2$

$y = 4(0) + 2$

$y = 0 + 2$

$y = 2$

$(0, 2)$

Let $x = 1$.

$y = 4x + 2$

$y = 4(1) + 2$

$y = 4 + 2$

$y = 6$

$(1, 6)$

Plot $(-1, -2)$, $(0, 2)$ and $(1, 6)$. Then draw the line through them.

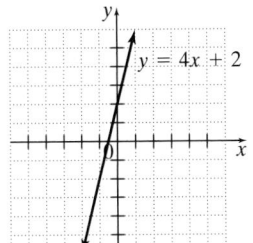

37. $x = -3.5$
No matter what y-value we choose, x is always -3.5.

x	y
-3.5	-3
-3.5	0
-3.5	3

$x = -3.5$

41. $\dfrac{86 + 94}{2} = \dfrac{180}{2} = 90$

45. $\dfrac{30 + 22 + 23 + 33}{4} = \dfrac{108}{4} = 27$

49. Answers may vary.

Exercise Set 8.5

1. Mean:
$\dfrac{21 + 28 + 16 + 42 + 38}{5} = \dfrac{145}{5} = 29$
Median: Write the numbers in order.
$16, 21, 28, 38, 48$
The middle number is 28.
Mode: There is no mode, since each number occurs once.

5. Mean:
$\dfrac{0.2 + 0.3 + 0.5 + 0.6 + 0.6 + 0.9 + 0.2 + 0.7 + 1.1}{9}$
$= \dfrac{5.1}{9} \approx 0.6$
Median: Write the numbers in order.
$0.2, 0.2, 0.3, 0.5, 0.6, 0.6, 0.7, 0.9, 1.1$
The middle number is 0.6.
Mode: Since 0.2 and 0.6 occur twice, there are two modes, 0.2 and 0.6.

9. Mean:
$\dfrac{1483 + 1483 + 1450 + 1381 + 1283}{5} = \dfrac{7080}{5} = 1416 \text{ feet}$

13. Answers may vary.

17. $\text{gpa} = \dfrac{4 \cdot 3 + 4 \cdot 3 + 3 \cdot 4 + 3 \cdot 1 + 3 \cdot 2}{3 + 3 + 4 + 1 + 2} = \dfrac{45}{13} \approx 3.46$

21. Mode: 6.9 since this time appears twice.

25. Mean:
$\dfrac{\text{sum of 15 rates}}{15} = \dfrac{1095}{15} = 73$

29. There are 9 rates lower than the mean. They are $66, 68, 71, 64, 71, 70, 65, 70,$ and 72.

33. $\dfrac{18}{30} = \dfrac{2 \cdot 3 \cdot 3}{5 \cdot 2 \cdot 3} = \dfrac{3}{5}$

37. Since the mode is 35, the number 35 must occur (at least) twice in the list. Since the median is 37 and there are 5 numbers in the list, the number 37 must occur in the list. Let x be the remaining number. Since the mean is 38, we have the following equation.
$\dfrac{35 + 35 + 37 + 40 + x}{5} = 38$
$147 + x = 5 \cdot 38$
$147 + x = 190$
$x = 43$
The missing numbers are $35, 35, 37,$ and 43.

Exercise Set 8.6

1.

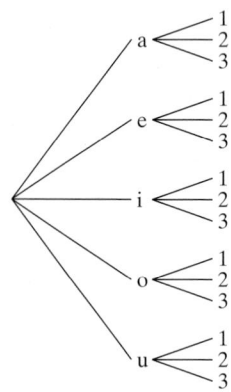

There are 15 outcomes.

5.

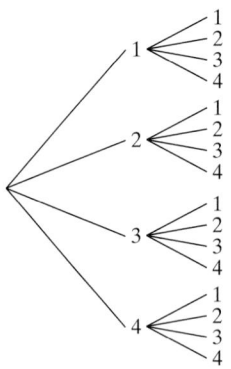

There are 16 outcomes.

9.

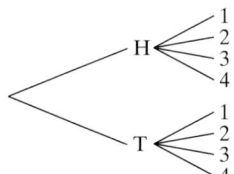

There are 8 outcomes.

13. probability $= \dfrac{2}{6} = \dfrac{1}{3}$

17. probability $= \dfrac{1}{3}$

21. probability $= \dfrac{1}{7}$

25. probability $= \dfrac{38}{200} = \dfrac{19}{100}$

29. $\dfrac{1}{2} + \dfrac{1}{3} = \dfrac{1 \cdot 3}{2 \cdot 3} + \dfrac{1 \cdot 2}{3 \cdot 2} = \dfrac{3}{6} + \dfrac{2}{6} = \dfrac{5}{6}$

33. $5 \div \dfrac{3}{4} = \dfrac{5}{1} \cdot \dfrac{4}{3} = \dfrac{5 \cdot 4}{1 \cdot 3} = \dfrac{20}{3}$ or $6\dfrac{2}{3}$

37. There are four kings.
probability $= \dfrac{4}{52} = \dfrac{1}{13}$

41.   Sum

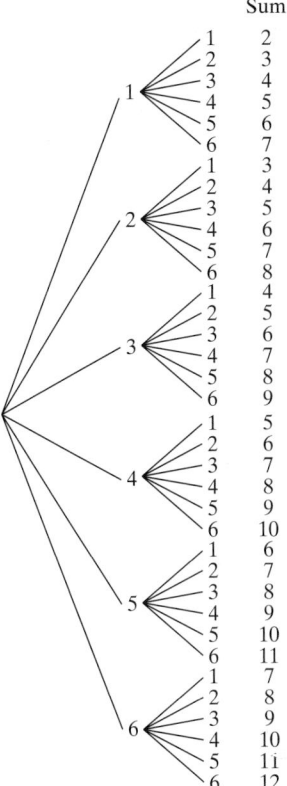

$$\text{probability} = \frac{3}{36} = \frac{1}{12}$$

45. Answers may vary.

CHAPTER 8 TEST

1. The second week has $4\frac{1}{2}$ bills and each bill represents $50.

So, $4.5 \cdot \$50 = \$225$ was collected during the second week.

5. The shortest bar is above the month of February. February has the least amount rainfall with 3 centimeters.

9. $\dfrac{62}{44} = \dfrac{2 \cdot 31}{2 \cdot 22} = \dfrac{31}{22}$

13. There are 5 students $5'0''-5'3''$ tall and 6 students $5'4''-5'7''$ tall, so there are 11 students $5'7''$ or shorter.

17. The point B is 3 units down from the origin on the y-axis. The coordinates of point B are $(0, -3)$

21. $y = 7x - 4$
$y = 7(2) - 4$
$y = 10$
The solution is $(2, 10)$.
$y = 7x - 4$
$y = 7(-1) - 4$
$y = -11$
The solution is $(-1, -11)$.
$y = 7x - 4$
$y = 7(0) - 4$
$y = -4$
The solution is $(0, -4)$.

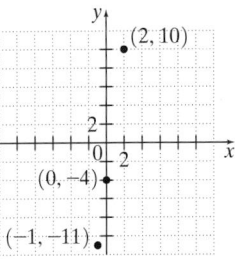

25. $x = 5$
No matter what y-value we choose, x is always 5.

x	y
5	-2
5	0
5	2

29. $\text{mean} = \dfrac{8 + 10 + 16 + 16 + 14 + 12 + 12 + 13}{8}$

$= \dfrac{101}{8} = 12.625$

Median: Write the numbers in order.
8 10 12 12 13 14 16 16

$\text{median:} \dfrac{12 + 13}{2} = \dfrac{25}{2} = 12.5$

mode: 12 and 16

33. $\dfrac{1}{10}$

Chapter 9

EXERCISE SET 9.1

1. The figure extends indefinitely in two directions. It is line yz, or $\overleftrightarrow{yz}$

5. The figure has two endpoints. It is line segment PQ, or $\overline{PQ}$.

9. $\angle ABC = 15°$

13. $\angle DBA = 50° + 15°$
$= 65°$

17. $90°$

21. $\angle S$ is a straight angle.

25. $\angle Q$ is an obtuse angle. It measures between $90°$ and $180°$.

29. The complement of an angle that measures $17°$ is an angle that measures $90° - 17° = 73°$.

33. The complement of a $48°$ angle is an angle that measures $90° - 48° = 42°$.

37. $\angle MNP$ and $\angle RNO$; are complementary angles since $60° + 30° = 90°$. Also, $\angle PNQ$ and $\angle QNR$ are complementary angles since $52° + 38° = 90°$.

41. The supplement of an angle that measures $125.2°$ is an angle that measures $180° - 125.2° = 54.8°$.

45. $\angle x = 90° - 15°$
$= 75°$

49. $\angle x$ and the 103° angle are adjacent angles, so
$\angle x = 180° - 103° = 77°$, $\angle y$ and the 103° angle are
vertical angles, so $\angle y = 103°$, and $\angle z$ and $\angle x$ are
vertical angles so $\angle z = 77°$.

53. $\angle x$ and the 46° angle are adjacent angles so
$\angle x = 180° - 46° = 134°$, $\angle y$ and the 46° angle are
corresponding angles so $\angle y = 46°$, and $\angle z$ and $\angle y$ are
adjacent angles so $\angle z = 180° - 46° = 134°$.

57. $\dfrac{7}{8} \cdot \dfrac{1}{4} = \dfrac{7 \cdot 1}{8 \cdot 4} = \dfrac{7}{32}$

61. $3\dfrac{1}{3} \div 2\dfrac{1}{2} = \dfrac{10}{3} \div \dfrac{5}{2}$

$\qquad = \dfrac{10}{3} \cdot \dfrac{2}{5}$

$\qquad = \dfrac{2 \cdot 5 \cdot 2}{3 \cdot 5}$

$\qquad = \dfrac{4}{3}$ or $1\dfrac{1}{3}$

65. Let $\angle x$ be an angle such that $\angle x + \angle x = 90°$. Then
$2\angle x = 90°$ so $\angle x = 45°$.

EXERCISE SET 9.2

1. $60 \text{ inches} = 60 \text{ inches} \cdot \dfrac{1 \text{ foot}}{12 \text{ inches}} = 5 \text{ feet}$

5. $42,240 \text{ feet} = 42,240 \text{ feet} \cdot \dfrac{1 \text{ mile}}{5280 \text{ feet}} = 8 \text{ miles}$

9. $10 \text{ feet} = 10 \text{ feet} \cdot \dfrac{1 \text{ yard}}{3 \text{ feet}} = 3\dfrac{1}{3} \text{yards}$

13. $40 \text{ feet} \div 3 = 13$ with remainder 1 so $40 \text{ feet} = 13$ yards
1 foot

17. $10,000 \text{ feet} \div 5280 = 1$ with remainder 4720 so
$10,000 = 1$ mile 4720 feet

21. $5 \text{ yards} = 5 \text{ yards} \cdot \dfrac{3 \text{ feet}}{1 \text{ yard}} = 15 \text{ feet}$

$5 \text{ yards } 2 \text{ feet} = 15 \text{ feet} + 2 \text{ feet} = 17 \text{ feet}$

25. $\quad\;\;5 \text{ feet } 8 \text{ inches}$
$\underline{+\; 6 \text{ feet } 7 \text{ inches}}$
$11 \text{ feet } 15 \text{ inches} = 11 \text{ feet} + 1 \text{ foot } 3 \text{ inches}$
$\qquad\qquad\qquad\qquad = 12 \text{ feet } 3 \text{ inches}$

29. $\quad\;\; 24 \text{ feet } 8 \text{ inches}$
$\underline{- \; 16 \text{ feet } 3 \text{ inches}}$
$\qquad 8 \text{ feet } 5 \text{ inches}$

33. $\quad\;\; 3 \text{ feet } 4 \text{ inches}$
$2\overline{)6 \text{ feet } 8 \text{ inches}}$
$\underline{-6 \text{ feet}}$
$\quad\;\; 0 \qquad 8 \text{ inches}$
$\underline{-\qquad\quad\; 8 \text{ inches}}$
$\qquad\qquad\quad\; 0$

37. $\quad\;\; 6 \text{ feet } 10 \text{ inches}$
$\underline{+3 \text{ feet}\;\;\; 8 \text{ inches}}$
$\quad 9 \text{ feet } 18 \text{ inches} = 9 \text{ feet} + 1 \text{ foot } 6 \text{ inches}$
$\qquad\qquad\qquad\qquad = 10 \text{ feet } 6 \text{ inches}$
The bamboo is 10 feet 6 inches tall.

41. $\quad\;\; 1 \text{ foot } 9 \text{ inches}$
$\underline{\qquad\qquad\quad \times\; 9}$
$9 \text{ feet } 81 \text{ inches} = 9 \text{ feet} + 6 \text{ feet } 9 \text{ inches}$
$\qquad\qquad\qquad\quad = 15 \text{ feet } 9 \text{ inches}$
9 stacks would extend 15 feet 9 inches from the wall.

45. $P = 2L + 2W$
$\quad = 2(24 \text{ feet } 9 \text{ inches}) + 2(18 \text{ feet } 6 \text{ inches})$
$\quad = 48 \text{ feet } 18 \text{ inches} + 36 \text{ feet } 12 \text{ inches}$
$\quad = 84 \text{ feet } 30 \text{ inches}$
$\quad = 84 \text{ feet} + 2 \text{ feet } 6 \text{ inches}$
$\quad = 86 \text{ feet } 6 \text{ inches}$
She must buy 86 feet 6 inches of fencing material.

49. 40 meters to centimeters
40 meters = 4000 centimeters
Move the decimal 2 places to the right.

53. 300 meters to kilometers
300 meters = 0.300 kilometer or 0.3 kilometer
Move the decimal 3 places to the left.

57. 1500 centimeters to meters
1500 centimeters = 15 meters
Move the decimal 2 places to the left.

61. 20.1 millimeters to decimeters
20.1 millimeters = 0.201 decimeter
Move the decimal 2 places to the left.

65. $\quad\;\; 8.6 \;\text{ meters}$
$\underline{+\; 0.34 \text{ meters}}$
$\quad\;\; 8.94 \text{ meters}$

69. $24.8 \text{ millimeters} - 1.19 \text{ centimeters}$
$\quad = 2.48 \text{ centimeters} - 1.19 \text{ centimeters}$
$\quad = 1.29 \text{ centimeters or } 12.9 \text{ millimeters}$

73. $18.3 \text{ meters} \times 3 = 54.9 \text{ meters}$

77. $3.4 \text{ meters} + 5.8 \text{ meters} - 8 \text{ centimeters}$
$\quad = 3.4 \text{ meters} + 5.8 \text{ meters} - 0.08 \text{ meter}$
$\quad = 9.12 \text{ meters}$
The tied ropes are 9.12 meters long.

81. $25(1 \text{ meter} + 65 \text{ centimeters})$
$\quad = 25(100 \text{ centimeters} + 65 \text{ centimeters})$
$\quad = 25(165 \text{ centimeters})$
$\quad = 4125 \text{ centimeters or } 41.25 \text{ meters}$
4125 centimeters or 41.25 meters of wood must be
ordered

85. $5.988 \text{ kilometers} + 21 \text{ meters}$
$\quad = 5988 \text{ meters} + 21 \text{ meters}$
$\quad = 6009 \text{ meters or } 6.009 \text{ kilometers}$
The elevation is 6009 meters or 6.009 kilometers.

89. $0.21 = 21\%$

93. $\dfrac{1}{4} = 0.25 = 25\%$

97. A square has 4 equal sides.

$\quad\quad\;\; 6.575$
$4\overline{)26.300}$
$\underline{-24}$
$\quad\; 23$
$\underline{-20}$
$\quad\;\; 30$
$\underline{-28}$
$\qquad 20$
$\underline{-20}$
$\qquad\; 0$

Each side will be 6.575 meters long.

Exercise Set 9.3

1. $P = 2l + 2w$
$= 2(17 \text{ feet}) + 2(15 \text{ feet})$
$= 34 \text{ feet} + 30 \text{ feet}$
$= 64 \text{ feet}$

5. $P = a + b + c$
$= 5 \text{ inches} + 7 \text{ inches} + 9 \text{ inches}$
$= 21 \text{ inches}$

9. $P = 10 \text{ feet} + 8 \text{ feet} + 8 \text{ feet} + 15 \text{ feet} + 7 \text{ feet}$
$= 48 \text{ feet}$

13. $P = 5 \text{ feet} + 3 \text{ feet} + 2 \text{ feet} + 7 \text{ feet} + 4 \text{ feet}$
$= 21 \text{ feet}$

17. $P = 2l + 2w$
$= 2(120 \text{ yards}) + 2(53 \text{ yards})$
$= 240 \text{ yards} + 106 \text{ yards}$
$= 346 \text{ yards}$

21. cost $= 3(22)$
$= \$66$

25. $P = 4s$
$= 4(7 \text{ inches})$
$= 28 \text{ inches}$

29. The missing lengths are:
28 meters $-$ 20 meters $=$ 8 meters
20 meters $-$ 17 meters $=$ 3 meters
$P = 8 \text{ meters} + 3 \text{ meters} + 20 \text{ meters} + 20 \text{ meters}$
$\quad + 28 \text{ meters} + 17 \text{ meters}$
$= 96 \text{ meters}$

33. The missing lengths are:
12 miles $+$ 10 miles $=$ 22 miles
8 miles $+$ 34 miles $=$ 42 miles
$P = 12 \text{ miles} + 34 \text{ miles} + 10 \text{ miles} + 8 \text{ miles}$
$\quad + 22 \text{ miles} + 42 \text{ miles}$
$= 128 \text{ miles}$

37. $C = 2\pi r$
$= 2\pi \cdot 8 \text{ miles}$
$= 16\pi \text{ miles}$
$\approx 16 \cdot 3.14 \text{ miles}$
$= 50.24 \text{ miles}$

41. $C = 2\pi r$
$= 2\pi \cdot 5 \text{ feet}$
$= 10\pi \text{ feet}$
$\approx 10 \cdot \dfrac{22}{7} \text{ feet}$
$= \dfrac{220}{7} \text{ feet or } 31\dfrac{3}{7} \text{ feet}$

45. $5 + 6 \cdot 3 = 5 + 18 = 23$

49. $(18 + 8) - (12 + 4) = 26 - 16 = 10$

53. **a.** Circumference of the smaller circle:
$C = 2\pi r$
$= 2\pi \cdot 10 \text{ meters}$
$= 20\pi \text{ meters}$
$\approx 20 \cdot 3.14 \text{ meters}$
$= 62.8 \text{ meters}$
Circumference of the larger circle:
$C = 2\pi r$
$= 2\pi \cdot 20 \text{ meters}$
$= 40\pi \text{ meters}$
$\approx 40 \cdot 3.14 \text{ meters}$
$= 125.6 \text{ meters}$
b. Yes, the circumference is doubled.

57. $P = 2l + 2w$
30 feet $= 2(9 \text{ feet}) + 2w$
30 feet $= 18 \text{ feet} + 2w$
12 feet $= 2w$
6 feet $= w$
The width of the rectangle is 6 feet.

Exercise Set 9.4

1. $A = lw$
$= 3.5 \text{ meters} \cdot 2 \text{ meters}$
$= 7 \text{ square meters}$

5. $A = \dfrac{1}{2}bh$
$= \dfrac{1}{2} \cdot 6 \text{ yards} \cdot 5 \text{ yards}$
$= \dfrac{1}{2} \cdot 6 \cdot 5 \text{ square yards}$
$= 15 \text{ square yards}$

9. $A = bh$
$= 7 \text{ feet} \cdot 5.25 \text{ feet}$
$= 36.75 \text{ square feet}$

13. $A = \dfrac{1}{2}(b + B)h$
$= \dfrac{1}{2} \cdot (4 \text{ yards} + 7 \text{ yards}) \cdot 4 \text{ yards}$
$= \dfrac{1}{2} \cdot 11 \text{ yards} \cdot 4 \text{ yards}$
$= \dfrac{1}{2} \cdot 11 \cdot 4 \text{ square yards}$
$= 22 \text{ square yards}$

17. $A = bh$
$= 5 \text{ inches} \cdot 4\dfrac{1}{2} \text{ inches}$
$= 5 \cdot \dfrac{9}{2} \text{ square inches}$
$= \dfrac{45}{2} \text{ or } 22\dfrac{1}{2} \text{ square inches}$

21.

Area of rectangle 1 $= bh$
$= 5 \text{ miles} \cdot 10 \text{ miles}$
$= 50 \text{ square miles}$
Area of rectangle 2 $= bh$
$= 12 \text{ miles} \cdot 3 \text{ miles}$
$= 36 \text{ square miles}$
Total area $= 50 \text{ square miles} + 36 \text{ square miles}$
$= 86 \text{ square miles}$

25. $A = \pi r^2$
$= \pi(6 \text{ inches})^2$
$= 36\pi$ square inches
$\approx 36 \cdot \dfrac{22}{7}$ square inches
$= 113\dfrac{1}{7}$ square inches

29. $V = s^3$
$= (8 \text{ centimeters})^3$
$= 512$ cubic centimeters

33. $V = \dfrac{4}{3}\pi r^3$
$= \dfrac{4}{3} \cdot \pi \left(\dfrac{10 \text{ inches}}{2}\right)^3$
$= \dfrac{4}{3} \cdot 125 \cdot \pi$ cubic inches
$= \dfrac{500}{3}\pi$ cubic inches
$\approx \dfrac{500}{3} \cdot \dfrac{22}{7}$ cubic inches
$= \dfrac{11{,}000}{21}$ cubic inches $= 523\dfrac{17}{21}$ cubic inches

37. $V = s^3$
$= \left(1\dfrac{1}{3}\text{ inches}\right)^3$
$= \left(\dfrac{4}{3}\right)^3$ cubic inches
$= \dfrac{64}{27}$ cubic inches $= 2\dfrac{10}{27}$ cubic inches

41. $A = bh$
$= 505 \text{ feet} \cdot 225 \text{ feet}$
$= 505 \cdot 225$ square feet
$= 113{,}625$ square feet

45. $V = \dfrac{1}{3}s^2 h$
$= \dfrac{1}{3} \cdot (12 \text{ centimeters})^2 \cdot 20 \text{ centimeters}$
$= \dfrac{2880}{3}$ cubic centimeters
$= 960$ cubic centimeters

49. Area of shaded part $= \dfrac{1}{2}(b + B)h$
$= \dfrac{1}{2} \cdot (25 \text{ feet} + 36 \text{ feet}) \cdot 12\dfrac{1}{2}\text{feet}$
$= \dfrac{1}{2} \cdot 61 \text{ feet} \cdot \dfrac{25}{2}\text{feet}$
$= \dfrac{1}{2} \cdot \dfrac{61}{1} \cdot \dfrac{25}{2}$ square feet
$= \dfrac{1525}{4}$ square feet $= 381\dfrac{1}{4}$ square feet

The area of the shaded part, to the nearest square foot is 381 square feet.

53. $V = l \cdot w \cdot h$
$= (2 \text{ feet}) \cdot \left(2\dfrac{1}{2}\text{feet}\right) \cdot \left(1\dfrac{1}{2}\text{feet}\right)$
$= 2 \cdot \dfrac{5}{2} \cdot \dfrac{3}{2}$ cubic feet
$= \dfrac{30}{4}$ or $7\dfrac{1}{2}$ cubic feet

57. $r = \dfrac{d}{2} = \dfrac{6 \text{ inches}}{2} = 3$ inches
$V = \dfrac{4}{3} \cdot \pi \cdot r^3 = \dfrac{4}{3} \cdot \pi \cdot (3 \text{ inches})^3$
$= \dfrac{4}{3} \cdot \pi \cdot 27$ cubic inches
$= 36\pi$ cubic inches
$\approx 36 \cdot 3.14$ cubic inches
$= 113.04$ cubic inches

61. $5^2 = 5 \cdot 5 = 25$

65. $1^2 + 2^2 = 1 \cdot 1 + 2 \cdot 2 = 1 + 4 = 5$

69. The fence for a yard goes around the perimeter of the yard.

73. Paint covers the surface of a wall, so you need to know the area of the wall.

77. Area of 12-inch pizza:
$A = \pi r^2$
$= \pi \left(\dfrac{12}{2}\text{inches}\right)^2$
$= 36\pi$ square inches
Area of one 8-inch pizza:
$A = \pi r^2$
$= \pi \left(\dfrac{8}{2}\text{inches}\right)^2$
$= 16\pi$ square inches
Area of two 8-inch pizzas $= 2 \cdot 16\pi$ square inches
$= 32\pi$ square inches
Cost per inch:
12-inch pizza: $\dfrac{\$10}{36\pi \text{ square inches}} \approx \$0.088$ per square inch
two 8-inch pizzas: $\dfrac{\$9}{32\pi \text{ square inches}} \approx \$0.090$ per square inch
The 12-inch pizza is the better buy.

81. Total Area $=$ Area of rectangle $+$ Area of both semicircles
Area of rectangle $= l \cdot w$
$= (22 \text{ m}) \cdot (5 \text{ m} + 5 \text{ m})$
$= (22 \text{ m}) \cdot (10 \text{ m})$
$= 220$ sq m
Area of both semicircles $=$ Area of circle
$= \pi r^2$
$= \pi \cdot (5 \text{ m})^2$
$= 25\pi$ sq m
≈ 78.5 sq m
Total Area $= 220$ sq m $+ 78.5$ sq m
$= 298.5$ sq m

85. $V = \dfrac{1}{3}s^2 h$
$= \dfrac{1}{3} \cdot (344 \text{ meters})^2 \cdot (65.5 \text{ meters})$
$\approx 2{,}583{,}669$ cubic meters

89. Answers may vary.

EXERCISE SET 9.5

1. $2 \text{ pounds} = 2 \text{ pounds} \cdot \dfrac{16 \text{ ounces}}{1 \text{ pound}} = 32 \text{ ounces}$

5. $12{,}000 \text{ pounds} = 12{,}000 \text{ pounds} \cdot \dfrac{1 \text{ ton}}{2000 \text{ pounds}}$
$= 6 \text{ tons}$

9. $3500 \text{ pounds} = 3500 \text{ pounds} \cdot \dfrac{1 \text{ ton}}{2000 \text{ pounds}}$

$= \dfrac{7}{4} \text{tons}$

$= 1\dfrac{3}{4} \text{tons}$

13. $4.9 \text{ tons} = 4.9 \text{ tons} \cdot \dfrac{2000 \text{ pounds}}{1 \text{ ton}}$

$= 9800 \text{ pounds}$

17. $2950 \text{ pounds} = 2950 \text{ pounds} \cdot \dfrac{1 \text{ ton}}{2000 \text{ pounds}}$

$= \dfrac{2950 \text{ tons}}{2000}$

$= 1.475 \text{ tons}$

$\approx 1.5 \text{ tons}$

21.
```
   6 tons 1540 pounds
 +2 tons  850 pounds
```
$8 \text{ tons } 2390 \text{ pounds} = 8 \text{ tons} + 1 \text{ ton } 390 \text{ pounds}$

$= 9 \text{ tons } 390 \text{ pounds}$

25.
```
 12 pounds  4 ounces      11 pounds 20 ounces
- 3 pounds  9 ounces     -3 pounds  9 ounces
                          8 pounds 11 ounces
```

29.
```
        1 ton     700 pounds
 5)6 tons 1500 pounds
  -5 tons
    1 ton = 2000 pounds
            3500 pounds
           -3500 pounds
                      0
```

33.
```
 64 pounds  8 ounces      63 pounds 24 ounces
-28 pounds 10 ounces     -28 pounds 10 ounces
                          35 pounds 14 ounces
```
Her zucchini was 35 pounds 14 ounces below the record.

37.
```
 55 pounds  4 ounces      54 pounds 20 ounces
 -2 pounds  8 ounces      -2 pounds  8 ounces
                          52 pounds 12 ounces
```

```
 52 pounds 12 ounces
              × 4
208 pounds 48 ounces = 208 pounds + 3 pounds
                     = 211 pounds
```

41. 500 grams to kilograms   Move the decimal 3 places to the left.

500 grams = 0.5 kilograms

45. 25 kilograms to grams   Move the decimal 3 places to the right.

25 kilograms = 25,000 grams

49. 6.3 grams to kilograms   Move the decimal 3 places to the left.

6.3 grams = 0.0063 kilograms

53. 4.01 kilograms to grams   Move the decimal 3 places to the right.

4.01 kilograms = 4010 grams

57. 205 mg = 0.205 g or 5.61 g = 5610 mg
```
  0.205 g        205 mg
+ 5.610 g      + 5610 mg
  5.815 g        5815 mg
```

61. 1.61 kg = 1610 g or 250 g = 0.250 kg
```
   1610 g          1.610 kg
 -  250 g        - 0.250 kg
   1360 g          1.360 kg
```

65. 17 kilograms ÷ 8 = 2.125 kilograms

69. 0.09 grams − 60 milligrams
= 90 milligrams − 60 milligrams = 30 milligrams
The extra-strength tablet contains 30 milligrams more medication.

73. $3 \cdot 16 \cdot 3 \text{ milligrams} = 144 \text{ milligrams}$
3 cartons contain 144 milligrams of preservatives.

77. 0.3 kilograms + 0.15 kilograms + 400 grams
= 0.3 kilograms + 0.15 kilograms + 0.4 kilograms
= 0.85 kilograms or 850 grams
The package weighs 0.85 kilograms or 850 grams.

81. $\dfrac{1}{4} = \dfrac{1 \cdot 25}{4 \cdot 25} = \dfrac{25}{100} = 0.25$

85. $\dfrac{7}{8} = \dfrac{7 \cdot 125}{8 \cdot 125} = \dfrac{875}{1000} = 0.875$

EXERCISE SET 9.6

1. $32 \text{ fluid ounces} = 32 \text{ fluid ounces} \cdot \dfrac{1 \text{ cup}}{8 \text{ fluid ounces}}$

$= 4 \text{ cups}$

5. $10 \text{ quarts} = 10 \text{ quarts} \cdot \dfrac{1 \text{ gallon}}{4 \text{ quarts}} = 2\dfrac{1}{2} \text{gallons}$

9. $2 \text{ quarts} = 2 \text{ quarts} \cdot \dfrac{4 \text{ cups}}{1 \text{ quart}} = 8 \text{ cups}$

13. 6 gallons

$= 6 \text{ gallons} \cdot \dfrac{4 \text{ quarts}}{1 \text{ gallon}} \cdot \dfrac{2 \text{ pints}}{1 \text{ quart}} \cdot \dfrac{2 \text{ cups}}{1 \text{ pint}} \cdot \dfrac{8 \text{ fluid ounces}}{1 \text{ cup}}$

$= 768 \text{ fluid ounces}$

17. $2\dfrac{3}{4} \text{gallons} = \dfrac{11}{4} \text{gallons} \cdot \dfrac{4 \text{ quarts}}{1 \text{ gallon}} \cdot \dfrac{2 \text{ pints}}{1 \text{ quart}} = 22 \text{ pints}$

21.
```
   1 c  5 fl oz
 + 2 c  7 fl oz
   3 c 12 fl oz = 3 c + 1 c 4 fl oz = 4c 4 fl oz
```

25.
```
 3 gal 1 qt           2 gal 4 qt 2 pt
-      1 qt 1 pt     -      1 qt 1 pt
                      2 gal 3 qt 1 pt
```

29.
```
 8 gal 2 qt
        × 2
16 gal 4 qt = 17 gal
```

33. $1\dfrac{1}{2} \text{quarts} = \dfrac{3}{2} \text{quarts} \cdot \dfrac{2 \text{ pints}}{1 \text{ quart}} \cdot \dfrac{2 \text{ cups}}{1 \text{ pint}} \cdot \dfrac{8 \text{ fluid ounces}}{1 \text{ cup}}$

$= 48 \text{ fluid ounces}$

37.
```
   5 pints 1 cup
 + 2 pints 1 cup
   7 pints 2 cups
```
Since 8 pints = 1 gallon, the fruit punch can be poured into the container.

41. $12 \text{ fluid ounces} \cdot 24 = 288 \text{ fluid ounces}$

$= 288 \text{ fluid ounces} \cdot \dfrac{1 \text{ cup}}{8 \text{ fluid ounces}} \cdot \dfrac{1 \text{ pint}}{2 \text{ cups}} \cdot \dfrac{1 \text{ quart}}{2 \text{ pints}}$

$= 9 \text{ quarts}$

45. 4500 ml to liters   Move the decimal 3 places to the left.
4500 ml = 4.5 liters

49. 64 ml to liters   Move the decimal 3 places to the left.
64 ml = 0.064 liters

53. 3.6 L to milliliters   Move the decimal 3 places to the right.
3.6 L = 3600 milliliters

57. 2.9 L + 19.6 L = 22.5 L

61. 8.6 L = 8600 ml or 190 ml = 0.190 L

$$\begin{array}{cc} 8600 \text{ ml} & 8.60 \text{ L} \\ -\ 190 \text{ ml} & -\ 0.19 \text{ L} \\ \hline 8410 \text{ ml} & 8.41 \text{ L} \end{array}$$

65. 480 ml × 8 = 3840 ml

69. 2 L − 410 ml = 2 L − 0.41 L = 1.59 L
1.59 L remains in the bottle.

73. $14.00 ÷ 44.3 L ≈ $0.316
The cost is about $0.316 per liter.

77. $0.7 = \dfrac{7}{10}$

81. $0.006 = \dfrac{6}{1000} = \dfrac{3}{500}$

85. B points to $1\frac{1}{2}$ cc or 1.5 cc

89. Each tick mark measures 2 units. B points to 2 tick marks past 50 u, or 54 u. Since 1 u = 0.01 cc, then 54 u = 54(0.01 cc) = 0.54 cc.

93. $2 \text{ pints} \cdot \dfrac{128 \text{ fluid drams}}{1 \text{ pint}}$
$= 256 \text{ fluid drams}$

EXERCISE SET 9.7

1. $578 \text{ ml} \approx \dfrac{578 \text{ ml}}{1} \cdot \dfrac{1 \text{ fl oz}}{29.57 \text{ ml}} = \dfrac{578}{29.57} \text{ fl oz} \approx 19.55 \text{ fl oz}$

5. $1000 \text{ g} \approx \dfrac{1000 \text{ g}}{1} \cdot \dfrac{0.04 \text{ oz}}{1 \text{ g}} = 40 \text{ oz}$

9. $14.5 \text{ L} \approx \dfrac{14.5 \text{ L}}{1} \cdot \dfrac{0.26 \text{ gal}}{1 \text{ L}} = 3.77 \text{ gal}$

13. $10 \text{ cm} = \dfrac{10 \text{ cm}}{1} \cdot \dfrac{1 \text{ in.}}{2.54 \text{ cm}} \approx 3.94 \text{ in.}$

17. $200 \text{ mg} \approx \dfrac{200 \text{ mg}}{1} \cdot \dfrac{1 \text{ g}}{1000 \text{ mg}} \cdot \dfrac{0.04 \text{ oz}}{1 \text{ g}} = 0.008 \text{ oz}$

21. $12 \text{ fl oz} \approx \dfrac{12 \text{ fl oz}}{1} \cdot \dfrac{29.57 \text{ ml}}{1 \text{ fl oz}} = 354.84 \text{ ml}$, so 380 ml is larger.

25. 1.5 lb − 1.25 lb = 0.25 lb
$0.25 \text{ lb} \approx \dfrac{0.25 \text{ lb}}{1} \cdot \dfrac{0.45 \text{ kg}}{1 \text{ lb}} \cdot \dfrac{1000 \text{ g}}{1 \text{ kg}} = 112.5 \text{ g}$

29. $8 \text{ m} \approx \dfrac{8 \text{ m}}{1} \cdot \dfrac{3.28 \text{ ft}}{1 \text{ m}} = 26.24 \text{ ft}$

33. Since 24 ÷ 4 = 6, every 4 hours means 6 doses a day or
6 · 7 = 42 doses a week. Thus 5 ml × 42 = 210 ml for
the week, $210 \text{ ml} \approx \dfrac{210 \text{ ml}}{1} \cdot \dfrac{1 \text{ fl oz}}{29.57 \text{ ml}} \approx 7.1$, rounded up
is 8 fl oz.

37. 1 L = 1.06 qt, thus a liter has greater capacity than a quart. The answer is B.

41. A glass of water is about 8 fluid ounces. Since each fluid ounce is about 30 ml, then 8 fl oz is about 8 · 30 or 240 ml. The correct answer is D.

45. $\begin{aligned} -6 \cdot 4 + 5 \div (-1) &= -24 + (-5) \\ &= -29 \end{aligned}$

49. $\begin{aligned} 3 + 5(17 - 19) - 8 &= 3 + 5(-2) - 8 \\ &= 3 + (-10) + (-8) \\ &= -15 \end{aligned}$

53. $\text{BSA} = \sqrt{\dfrac{(90)(182)}{3600}} \approx 2.13 \text{ sq m}$

57. 60 in. = 152.4 cm; 150 lb ≈ 67.5 kg
$\text{BSA} = \sqrt{\dfrac{(67.5)(152.4)}{3600}} \approx 1.70 \text{ sq m}$

61. $\begin{aligned} \text{Area} &= l \cdot w \\ &= (40 \text{ m})(20 \text{ m}) = 800 \text{ sq m} \end{aligned}$
$\text{Area} \approx (131.2 \text{ ft})(65.6 \text{ ft}) = 8606.72 \text{ sq ft}$

EXERCISE SET 9.8

1. $\begin{aligned} C &= \dfrac{5}{9} \cdot (41 - 32) \\ &= \dfrac{5}{9} \cdot (9) \\ &= 5 \end{aligned}$
41°F = 5°C

5. $\begin{aligned} F &= \dfrac{9}{5} \cdot (60) + 32 \\ &= 9(12) + 32 \\ &= 108 + 32 \\ &= 140 \end{aligned}$
60°C = 140°F

9. $\begin{aligned} C &= \dfrac{5}{9} \cdot (62 - 32) \\ &= \dfrac{5}{9} \cdot (30) \\ &= \dfrac{150}{9} \\ &\approx 16.7 \end{aligned}$
62°F ≈ 16.7°C

13. $\begin{aligned} F &= 1.8(92) + 32 \\ &= 165.6 + 32 \\ &= 197.6 \end{aligned}$
92°C = 197.6°F

17. $\begin{aligned} C &= \dfrac{5}{9} \cdot (122 - 32) \\ &= \dfrac{5}{9} \cdot (90) \\ &= 50 \end{aligned}$
122°F = 50°C

21. $\begin{aligned} C &= \dfrac{5}{9} \cdot (70 - 32) \\ &= \dfrac{5}{9} \cdot (38) \\ &= \dfrac{190}{9} \\ &\approx 21.1 \end{aligned}$
70°F ≈ 21.1°C

25.
$$F = 1.8(118) + 32$$
$$= 212.4 + 32$$
$$= 244.4$$
$$188°C = 244.4°F$$

29.
$$C = \frac{5}{9} \cdot (864 - 32)$$
$$= \frac{5}{9} \cdot (832)$$
$$= \frac{4160}{9}$$
$$\approx 462.2$$
$$864°F \approx 462.2°C$$

33. $P = 3 \text{ feet} + 3 \text{ feet} + 3 \text{ feet} + 3 \text{ feet} + 3 \text{ feet}$
$$= 15 \text{ feet}$$

37.
$$C = \frac{5}{9} \cdot (9010 - 32)$$
$$= \frac{5}{9} \cdot (8978)$$
$$\approx 4988$$
$$9010° \text{ F is} \approx 4988° \text{ C}$$

CHAPTER 9 TEST

1. The complement of a 78° angle is $90° - 78° = 12°$.

5. $\angle x$ and the 73° angle are vertical angles, so
$\angle x = 73°$, $\angle y$ and the 73° angle are corresponding
angles, so $\angle y = 73°$, $\angle z$ and $\angle y$ are vertical angles, so
$\angle z = 73°$.

9. $P = 2l + 2w$
$$= 2(7 \text{ yd}) + 2(5.3 \text{ yd})$$
$$= 14 \text{ yd} + 10.6 \text{ yd}$$
$$= 24.6 \text{ yd}$$
$A = l \cdot w$
$$= (7 \text{ yd})(5.3 \text{ yd})$$
$$= 37.1 \text{ sq. yd}$$

13. $P = 4s$
$$= 4(4 \text{ in.})$$
$$= 16 \text{ in.}$$

17. $2\frac{1}{2}\text{gal} = \frac{5}{2}\text{gal} \cdot \frac{4 \text{ qt}}{1 \text{ gal}} = \frac{20 \text{ qt}}{2} = 10 \text{ qt}$

21. 3.6 cm to millimeters   Move the decimal one
3.6 cm = 36 mm   place to the right.

25. 8.5 in. $= \frac{8.5 \text{ in.}}{1} \cdot \frac{2.54 \text{ cm}}{1 \text{ in.}} = 21.59 \text{ cm}$

29.
$$\begin{array}{r} 20 \text{ gal} \\ - 15 \text{ gal } 1 \text{ qt} \end{array} \text{ or } \begin{array}{r} 19 \text{ gal } 4 \text{ qt} \\ - 15 \text{ gal } 1 \text{ qt} \\ \hline 4 \text{ gal } 3 \text{ qt} \end{array}$$
4 gal 3 qt remains in the drum.

33. $5 \text{ g} \approx \frac{5 \text{ g}}{1} \cdot \frac{0.04 \text{ oz}}{1 \text{ g}} = 0.2 \text{ oz}$

Chapter 10

EXERCISE SET 10.1

1. $(2x + 3) + (-7x - 27) = (2x - 7x) + (3 - 27)$
$$= -5x + (-24)$$
$$= -5x - 24$$

5. $(12y - 20) + (9y^2 + 13y - 20)$
$$= 9y^2 + (12y + 13y) + (-20 - 20)$$
$$= 9y^2 + 25y - 40$$

9. $(5a - 6) - (a + 2) = (5a - 6) + (-a - 2)$
$$= (5a - a) + (-6 - 2)$$
$$= 4a - 8$$

13. $(10y^2 - 7) - (20y^3 - 2y^2 - 3)$
$$= (10y^2 - 7) + (-20y^3 + 2y^2 + 3)$$
$$= (-20y^3) + (10y^2 + 2y^2) + (-7 + 3)$$
$$= -20y^3 + 12y^2 - 4$$

17.
$$\begin{array}{r} 13y^2 - 6y - 14 \\ -(5y^2 + 4y - 6) \end{array} \qquad \begin{array}{r} 13y^2 - 6y - 14 \\ + -5y^2 - 4y + 6 \\ \hline 8y^2 - 10y - 8 \end{array}$$

21. $(4y + 4) - (3y + 8) = (4y + 4) + (-3y - 8)$
$$= (4y - 3y) + (4 - 8)$$
$$= y - 4$$

25. $(10x + 4.5) + (-x - 8.6) = (10x - x) + (4.5 - 8.6)$
$$= 9x - 4.1$$

29. $(21y - 4.6) - (36y - 8.2)$
$$= (21y - 4.6) + (-36y + 8.2)$$
$$= (21y - 36y) + (-4.6 + 8.2)$$
$$= -15y + 3.6$$

33. $(b^3 - 2b^2 + 10b + 11) + (b^2 - 3b - 12)$
$$= b^3 + (-2b^2 + b^2) + (10b - 3b) + (11 - 12)$$
$$= b^3 - b^2 + 7b - 1$$

37. $\left(3z + \frac{6}{7}\right) - \left(3z - \frac{3}{7}\right)$
$$= \left(3z + \frac{6}{7}\right) + \left(-3z + \frac{3}{7}\right)$$
$$= (3z - 3z) + \left(\frac{6}{7} + \frac{3}{7}\right)$$
$$= \frac{9}{7}$$

41. $x^2 - 6x + 3$
Let $x = 2$.
$(2)^2 - 6(2) + 3 = 4 - 12 + 3 = -5$

45. $2x + 10$
Let $x = 5$.
$2(5) + 10 = 10 + 10 = 20$

49. $2x^2 + 4x - 20$
Let $x = 5$.
$2(5)^2 + 4(5) - 20 = 2(25) + 20 - 20 = 50$

53. $3000 + 20x$
Let $x = 10$.
$3000 + 20(10) = 3000 + 200 = 3200$
It costs $3200 to manufacture 10 file cabinets.

57. $3^4 = 3 \cdot 3 \cdot 3 \cdot 3 = 81$

61. $x \cdot x \cdot x = x^3$

65. $(2x + 1) + (x + 11) + (5x - 10)$
$$= (2x + x + 5x) + (1 + 11 - 10)$$
$$= 8x + 2$$
The perimeter is $(8x + 2)$ inches.

69.
$$3x^2 + \_x - \_$$
$$\underline{+ \_x^2 - 6x + 2}$$
$$5x^2 + 14x - 4$$
$$\left[\begin{array}{l}(3 + \_)x^2 = 5x \\ (3 + 2)x^2 = 5x^2\end{array}\right]$$
$$\left[\begin{array}{l}(\_ - 6)x = 14x \\ (20 - 6)x = 14x\end{array}\right]$$
$$\left[\begin{array}{l}(\_ + 2) = -4 \\ (-6 + 2) = -4\end{array}\right]$$
$$3x^2 + 20x - 6$$
$$\underline{+2x^2 - 6x + 2}$$
$$5x^2 + 14x - 4$$

73. When $t = 8$ seconds:
$$\begin{aligned}1053 - 16t^2 &= 1053 - 16(8)^2 \\ &= 1053 - 16 \cdot 64 \\ &= 1053 - 1024 \\ &= 29\end{aligned}$$
The height of the object above the river after 8 seconds is 29 feet.
When $t = 9$ seconds:
$$\begin{aligned}1053 - 16t^2 &= 1053 - 16(9)^2 \\ &= 1053 - 16 \cdot 81 \\ &= 1053 - 1296 \\ &= -243\end{aligned}$$
The height of the object above the river after 9 seconds is -243 feet. The object hits the water between 8 and 9 seconds.

EXERCISE SET 10.2

1. $x^5 \cdot x^9 = x^{5+9} = x^{14}$

5. $3z^3 \cdot 5z^2 = (3 \cdot 5)(z^3 \cdot z^2) = 15z^5$

9. $(-5x^2y^3)(-5x^4y) = (-5)(-5)(x^2 \cdot x^4)(y^3 \cdot y)$
$$= 25x^6y^4$$

13. $2x \cdot 3x \cdot 7x = (2 \cdot 3 \cdot 7)(x \cdot x \cdot x) = 42x^3$

17. $(x^5)^3 = x^{5 \cdot 3} = x^{15}$

21. $(b^7)^6 \cdot (b^2)^{10} = b^{7 \cdot 6} \cdot b^{2 \cdot 10}$
$$\begin{aligned}&= b^{42} \cdot b^{20} \\ &= b^{42+20} \\ &= b^{62}\end{aligned}$$

25. $(a^{11}b^8)^3 = a^{11 \cdot 3}b^{8 \cdot 3} = a^{33}b^{24}$

29. $(-3y)(2y^7)^3 = (-3y) \cdot 2^3(y^7)^3$
$$\begin{aligned}&= (-3y) \cdot 8y^{21} \\ &= (-3)(8)(y^1 \cdot y^{21}) \\ &= -24y^{22}\end{aligned}$$

33. $7(x - 3) = 7x - 21$

37. $9(x + 2y - 3) = 9x + 18y - 27$

41. area $= \dfrac{1}{2}bh$
$$= \dfrac{1}{2} \cdot (6a^3b^4) \cdot (4ab)$$
$$= \left(\dfrac{1}{2} \cdot 6 \cdot 4\right)(a^3 \cdot a)(b^4 \cdot b)$$
$$= 12a^4b^5$$
The area is $12a^4b^5$ square meters.

45. $(8.1x^{10})^5 = 8.1^5(x^{10})^5 = 34{,}867.84401x^{50}$

49. $(a^{20}b^{10}c^5)^5 \cdot (a^9b^{12})^3$
$$\begin{aligned}&= a^{100}b^{50}c^{25} \cdot a^{27}b^{36} \\ &= a^{127}b^{86}c^{25}\end{aligned}$$

EXERCISE SET 10.3

1. $3x(9x^2 - 3) = 3x \cdot 9x^2 + 3x \cdot (-3)$
$$\begin{aligned}&= (3 \cdot 9)(x \cdot x^2) + (3)(-3)(x) \\ &= 27x^3 + (-9x) \\ &= 27x^3 - 9x\end{aligned}$$

5. $7x^2(6x^2 - 5x + 7)$
$$\begin{aligned}&= (7x^2)(6x^2) + (7x^2)(-5x) + (7x^2)(7) \\ &= (7 \cdot 6)(x^2 \cdot x^2) + (7)(-5)(x^2 \cdot x) + (7 \cdot 7)x^2 \\ &= 42x^4 - 35x^3 + 49x^2\end{aligned}$$

9. $(2x - 6)(x + 4) = 2x(x + 4) - 6(x + 4)$
$$\begin{aligned}&= 2x \cdot x + 2x \cdot 4 - 6 \cdot x - 6 \cdot 4 \\ &= 2x^2 + 8x - 6x - 24 \\ &= 2x^2 + 2x - 24\end{aligned}$$

13. $(a + 6)(a^2 - 6a + 3)$
$$\begin{aligned}&= a(a^2 - 6a + 3) + 6(a^2 - 6a + 3) \\ &= a \cdot a^2 + a(-6a) + a \cdot 3 + 6 \cdot a^2 + 6(-6a) + 6 \cdot 3 \\ &= a^3 - 6a^2 + 3a + 6a^2 - 36a + 18 \\ &= a^3 - 33a + 18\end{aligned}$$

17. $(x^3 + 2x + x^2)(3x + 1 + x^2)$
$$\begin{aligned}&= x^3(3x + 1 + x^2) + 2x(3x + 1 + x^2) \\ &\quad + x^2(3x + 1 + x^2) \\ &= x^3 \cdot 3x + x^3 \cdot 1 + x^3 \cdot x^2 + 2x \cdot 3x + 2x \cdot 1 \\ &\quad + 2x \cdot x^2 + x^2 \cdot 3x + x^2 \cdot 1 + x^2 \cdot x^2 \\ &= 3x^4 + x^3 + x^5 + 6x^2 + 2x + 2x^3 \\ &\quad + 3x^3 + x^2 + x^4 \\ &= x^5 + 4x^4 + 6x^3 + 7x^2 + 2x\end{aligned}$$

21. $-2y^2(3y + y^2 - 6)$
$$\begin{aligned}&= -2y^2 \cdot 3y + (-2y^2) \cdot y^2 + (-2y^2)(-6) \\ &= -6y^3 - 2y^4 + 12y^2\end{aligned}$$

25. $(2a + 3)(2a - 3)$
$$\begin{aligned}&= 2a(2a - 3) + 3(2a - 3) \\ &= 2a \cdot 2a + 2a(-3) + 3 \cdot 2a + 3(-3) \\ &= 4a^2 - 6a + 6a - 9 \\ &= 4a^2 - 9\end{aligned}$$

29. $\left(b + \dfrac{3}{5}\right)\left(b + \dfrac{4}{5}\right) = b\left(b + \dfrac{4}{5}\right) + \dfrac{3}{5}\left(b + \dfrac{4}{5}\right)$
$$= b^2 + \dfrac{4}{5}b + \dfrac{3}{5}b + \dfrac{3}{5} \cdot \dfrac{4}{5}$$
$$= b^2 + \dfrac{7}{5}b + \dfrac{12}{25}$$

33. $(7x + 5)^2 = (7x + 5)(7x + 5)$
$$\begin{aligned}&= 7x(7x + 5) + 5(7x + 5) \\ &= 49x^2 + 35x + 35x + 25 \\ &= 49x^2 + 70x + 25\end{aligned}$$

37. $(2x^2 - 3)(4x^3 + 2x - 3)$
$$\begin{aligned}&= 2x^2(4x^3 + 2x - 3) - 3(4x^3 + 2x - 3) \\ &= 8x^5 + 4x^3 - 6x^2 - 12x^3 - 6x + 9 \\ &= 8x^5 - 8x^3 - 6x^2 - 6x + 9\end{aligned}$$

41.
$$2z^2 - z + 1$$
$$\underline{\times \quad 5z^2 + z - 2}$$
$$-4z^2 + 2z - 2$$
$$2z^3 - z^2 + z$$
$$\underline{10z^4 - 5z^3 + 5z^2}$$
$$10z^4 - 3z^3 \qquad + 3z - 2$$

45. $72 = 2 \cdot 2 \cdot 2 \cdot 3 \cdot 3 = 2^3 \cdot 3^2$

49. $(y - 6)(y^2 + 3y + 2)$
$= y(y^2 + 3y + 2) - 6(y^2 + 3y + 2)$
$= y^3 + 3y^2 + 2y - 6y^2 - 18y - 12$
$= y^3 - 3y^2 - 16y - 12$
The area is $(y^3 - 3y^2 - 16y - 12)$ square feet.

53. Answers may vary.

Exercise Set 10.4

1. $48 = 2 \cdot 2 \cdot 2 \cdot 2 \cdot 3$
$15 = 3 \cdot 5$
GCF $= 3$

5. $12 = 2 \cdot 2 \cdot 3$
$20 = 2 \cdot 2 \cdot 5$
$36 = 2 \cdot 2 \cdot 3 \cdot 3$
GCF $= 2 \cdot 2 = 4$

9. $y^7 = y^2 \cdot y^5$
$y^2 = y^2$
$y^{10} = y^2 \cdot y^8$
GCF $= y^2$

13. $x^3 y^2 = x \cdot x^2 \cdot y^2$
$xy^2 = x \cdot y^2$
$x^4 y^2 = x \cdot x^3 \cdot y^2$
GCF $= x \cdot y^2 = xy^2$

17. $2 = 2$
$14 = 2 \cdot 7$
$18 = 2 \cdot 3 \cdot 3$
GCF $= 2$
$z^3 = z^3$
$z^5 = z^3 \cdot z^2$
$z^3 = z^3$
GCF $= z^3$
GCF $= 2z^3$

21. $10a^6 = 5a^6 \cdot 2$
$5a^8 = 5a^6 \cdot a^2$
GCF $= 5a^6$
$10a^6 - 5a^8 = 5a^6 \cdot 2 - 5a^6 \cdot a^2$
$= 5a^6(2 - a^2)$

25. $z^7 = z^5 \cdot z^2$
$6z^5 = z^5 \cdot 6$
GCF $= z^5$
$z^7 - 6z^5 = z^5 \cdot z^2 - z^5 \cdot 6 = z^5(z^2 - 6)$

29. $12a^5 = 12a^5$
$36a^6 = 12a^5 \cdot 3a$
GCF $= 12a^5$
$12a^5 - 36a^6 = 12a^5 \cdot 1 - 12a^5 \cdot 3a$
$= 12a^5(1 - 3a)$

33. $80\% = \dfrac{80}{100} = \dfrac{4}{5}$

37. area on the left: $x \cdot x = x^2$
area on the right: $2 \cdot x = 2x$
total area: $x^2 + 2x$
Notice that $x(x + 2) = x^2 + 2x$.

Chapter 10 Test

1. $(11x - 3) + (4x - 1) = (11x + 4x) + (-3 - 1)$
$= 15x + (-4)$
$= 15x - 4$

5. $x^2 - 6x + 1 = (8)^2 - 6(8) + 1$
$= 64 - 48 + 1$
$= 17$

9. $(6a^3)(-2a^7) = (6)(-2)(a^3 \cdot a^7) = -12a^{10}$

13. $-2y(y^3 + 6y^2 - 4)$
$= -2y \cdot y^3 - 2y \cdot 6y^2 - 2y \cdot (-4)$
$= -2y^4 - 12y^3 + 8y$

17. area:
$(x + 7)(5x - 2) = x(5x - 2) + 7(5x - 2)$
$= 5x^2 - 2x + 35x - 14$
$= 5x^2 + 33x - 14$
The area is $(5x^2 + 33x - 14)$ square inches.
Perimeter: $2(2x) + 2(5x - 2) = 4x + 10x - 4 = 14x - 4$
The perimeter is $(14x - 4)$ inches.

21. $10a^2 = 2a \cdot 5a$
$12a = 2a \cdot 6$
GCF $= 2a$
$10a^2 + 12a = 2a \cdot 5a + 2a \cdot 6$
$= 2a(5a + 6)$

SUBJECT INDEX

A

Absolute value, 113
Acute angle, 681
Addend, 21, 46, 795
Addition
  associative property of, 22-23
  on calculator, 26
  commutative property of, 22
  of decimals, 375-88
    on calculator, 380
    simplifying expressions, 378
    solving problems by, 378-79
  estimating, 46-48
  of fractions, 273-98
    on calculators, 292
    given fractional replacement
      values, 275
    least common denominator for,
      277-79
    like fractions, 273-75, 276-77
    unlike fractions, 287-98
  of integers, 121-28, 130
    on calculator, 124
    solving problems by, 124
  of mixed numbers, 320-22
  modeling equation solving with, 188
  of polynomials, 795-96
  properties of, 172-73
  repeated, 55
  of whole numbers, 21-26
Addition property, 181-88
  of equality, 182
  solving linear equations in one
    variable using, 201-2
  of zero, 22
Additive inverse, 129
Adjacent angles, 684
Algebra, 89
Algebraic expressions, 89-95
  evaluating, 89-91, 123-24, 130-31
    given integer replacement values,
      142
  translating phrases into, 91-92
Alternate interior angles, 685
American Red Cross, 260
American Savings Education Council,
  560
Angles, 679-90
  adjacent, 684
  alternate interior, 685
  classifying, 680-82
  complementary, 682
  corresponding, 685
  exterior, 685
  interior, 685
  measures of, 683-86
  sides of, 680
  supplementary, 682
  vertical, 683-84
Angle-Side-Angle (ASA) congruence,
  495
Annual compounding, 572
Approximating. See Estimating;
  Rounding
Area, 175, 715-18
  of circle, 718
  formulas of common geometric
    figures, 715

of rectangle, 58-59
of square, 84
Area model of fractions, 240
Area problems, 424
Associative property
  of addition, 22-23
  of multiplication, 56
Automobile manufacturing industry,
  515
Average (mean), 74, 151, 641-42

B

Babylonians, ancient, 314
Balance, equal-arm, 208
Bar graphs, 37, 593-96, 608
  containing negative numbers, 115
Bartholdi, Frederic Auguste, 677
Base, solving percent proportions for,
  542
Base of exponent, 81
Binomials, 795, 796
  multiplying, 813-14
Blood, 260
Body dimensions, 478
Borrowing, subtraction with, 34-35
British Thermal Units (BTU), A12

C

Calculator(s)
  addition on, 26
    of decimals, 380
    of fractions, 292
    of integers, 124
  checking equations with, 204
  compound interest factor on, 574
  converting between mixed-number
    and fraction notation, 328
  division on, 74
  exponents on, 84
  graphing, 292
  multiplication on, 60
  order of operations on, 84
  simplifying expressions on, 152
  simplifying fractions on, 254
  square roots on, 440
  subtraction on, 38
    of decimals, 380
    of fractions, 292
Calorie (cal), A13
Capacity, 745-54
  metric system units of, 747-50
  U.S. system units of, 745-47
Carpal tunnel syndrome (CTS), 591
Cartesian coordinate system. See
  Rectangular coordinate system
Celsius, Anders, 763
Celsius scale, 763-65
Centimeter, 695
Central tendency, measures of, 641-46
  mean, 74, 151, 641-42
  median, 642-43
  mode, 643-44
Circle, A5
  area of, 718
  circumference of, 392
Circle graphs, 608, 609-16
  drawing, 610-12

ratios from, 460
reading, 609-10
Circular cylinder, 719, A5, A7
Circumference, 392, 708-9
Class frequency, 596
Class interval, 596
Closing costs, 554
Coefficient, numerical, 171
College degree, earning power and, 44
Combining like terms, 171-73, 174
Commissions, 562-63
Common denominator, 273-74
  least (LCD), 277-79
    to eliminate fractions in
      equations, 311
    simplifying complex fractions
      using, 302-3
Commutative property
  of addition, 22
  of multiplication, 56
Comparison model for subtraction, 38
Complementary angles, 682
Complex fractions, 301-8
Composite number, 249
*Composition with Gray and Light
  Brown* (Mondrian), 430
Compound interest, 572-73
Compound interest factor, 574
Compound interest table, 573, A3
Cone, A7
  right circular, A5
  volume of, 719
Congruent triangles, 495-96
Constant, 171
Convex solid, A6
Corresponding angles, 685
Counters, 147
Counting, 647-48
Counting numbers (natural numbers), 9
Cube, 719, A7
Cubes, 800
Cubic units, 718, A6
Cubit, 766
Cylinder, circular, 719, A5, A7

D

Decimals, 359-456, A1
  adding and subtracting, 375-88
    on calculator, 380
    simplifying expressions, 378
    solving problems by, 378-79
  comparing, 365-66
  decimal notation, 361-63
  dividing, 401-10
    by powers of 10, 403
    solving problems by, 404
  equations containing, 431-33
  estimating operations on, 413-14
  fractions and, 421-29
    area problems containing, 424
    comparing, 422-23
    writing fractions as decimals,
      421-22
  multiplying, 389-92
    modeling, 393
    by powers of 10, 390-91
    solving problems by, 393
  as percents, 519, 522

PHOTO CREDITS

READ THIS LICENSE CAREFULLY BEFORE OPENING THIS PACKAGE. BY OPENING THIS PACKAGE, YOU ARE AGREEING TO THE TERMS AND CONDITIONS OF THIS LICENSE. IF YOU DO NOT AGREE, DO NOT OPEN THE PACKAGE. PROMPTLY RETURN THE UNOPENED PACKAGE AND ALL ACCOMPANYING ITEMS TO THE PLACE YOU OBTAINED THEM. *THESE TERMS APPLY TO ALL LICENSED SOFTWARE ON THE DISK EXCEPT THAT THE TERMS FOR USE OF ANY SHAREWARE OR FREEWARE ON THE DISKETTES ARE AS SET FORTH IN THE ELECTRONIC LICENSE LOCATED ON THE DISK:*

Single PC Site License

1. **GRANT OF LICENSE and OWNERSHIP:** The enclosed computer programs and any data ("Software") are licensed, not sold, to you by Pearson Education, Inc. publishing as Pearson Prentice Hall ("We" or the "Company") in consideration of your adoption of the accompanying Company textbooks and/or other materials, and your agreement to these terms. You own only the disk(s) but we and/or our licensors own the Software itself. This license allows instructors and students enrolled in the course using the Company textbook that accompanies this Software (the "Course") to use and display the enclosed copy of the Software on an unlimited number of computers, for academic use only, so long as you comply with the terms of this Agreement. You may make one copy for back up only. We reserve any rights not granted to you.

2. **USE RESTRICTIONS:** You may not sell or license copies of the Software or the Documentation to others. You may not transfer, distribute or make available the Software or the Documentation. You may not reverse engineer, disassemble, decompile, modify, adapt, translate or create derivative works based on the Software or the Documentation. You may be held legally responsible for any copying or copyright infringement that is caused by your failure to abide by the terms of these restrictions.

3. **TERMINATION:** This license is effective until terminated. This license will terminate automatically without notice from the Company if you fail to comply with any provisions or limitations of this license. Upon termination, you shall destroy the Documentation and all copies of the Software. All provisions of this Agreement as to limitation and disclaimer of warranties, limitation of liability, remedies or damages, and our ownership rights shall survive termination.

4. **DISCLAIMER OF WARRANTY: the Company and its licensors make no warranties about the SOFTWARE, which is provided "AS-IS." IF THE DISK IS DEFECTIVE IN MATERIALS OR WORKMANSHIP, YOUR ONLY REMEDY IS TO RETURN IT TO THE COMPANY WITHIN 30 DAYS FOR REPLACEMENT UNLESS THE COMPANY DETERMINES IN GOOD FAITH THAT THE DISK HAS BEEN MISUSED OR IMPROPERLY INSTALLED, REPAIRED, ALTERED OR DAMAGED. THE COMPANY DISCLAIMS ALL WARRANTIES, EXPRESS OR IMPLIED, INCLUDING WITHOUT LIMITATION, THE IMPLIED WARRANTIES OF MERCHANTABILITY AND FITNESS FOR A PARTICULAR PURPOSE. THE COMPANY DOES NOT WARRANT, GUARANTEE OR MAKE ANY REPRESENTATION REGARDING THE ACCURACY, RELIABILITY, CURRENTNESS, USE, OR RESULTS OF USE, OF THE SOFTWARE.**

5. **LIMITATION OF REMEDIES AND DAMAGES: IN NO EVENT, SHALL THE COMPANY OR ITS EMPLOYEES, AGENTS, LICENSORS OR CONTRACTORS BE LIABLE FOR ANY INCIDENTAL, INDIRECT, SPECIAL OR CONSEQUENTIAL DAMAGES ARISING OUT OF OR IN CONNECTION WITH THIS LICENSE OR THE SOFTWARE, INCLUDING, WITHOUT LIMITATION, LOSS OF USE, LOSS OF DATA, LOSS OF INCOME OR PROFIT, OR OTHER LOSSES SUSTAINED AS A RESULT OF INJURY TO ANY PERSON, OR LOSS OF OR DAMAGE TO PROPERTY, OR CLAIMS OF THIRD PARTIES, EVEN IF THE COMPANY OR AN AUTHORIZED REPRESENTATIVE OF THE COMPANY HAS BEEN ADVISED OF THE POSSIBILITY OF SUCH DAMAGES.** SOME JURISDICTIONS DO NOT ALLOW THE LIMITATION OF DAMAGES IN CERTAIN CIRCUMSTANCES, SO THE ABOVE LIMITATIONS MAY NOT ALWAYS APPLY.

6. **GENERAL:** THIS AGREEMENT SHALL BE CONSTRUED IN ACCORDANCE WITH THE LAWS OF THE UNITED STATES OF AMERICA AND THE STATE OF NEW YORK, APPLICABLE TO CONTRACTS MADE IN NEW YORK, AND SHALL BENEFIT THE COMPANY, ITS AFFILIATES AND ASSIGNEES. This Agreement is the complete and exclusive statement of the agreement between you and the Company and supersedes all proposals, prior agreements, oral or written, and any other communications between you and the company or any of its representatives relating to the subject matter. If you are a U.S. Government user, this Software is licensed with "restricted rights" as set forth in subparagraphs (a)-(d) of the Commercial Computer-Restricted Rights clause at FAR 52.227-19 or in subparagraphs (c)(1)(ii) of the Rights in Technical Data and Computer Software clause at DFARS 252.227-7013, and similar clauses, as applicable.

Should you have any questions concerning this agreement or if you wish to contact the Company for any reason, please contact in writing: Customer Service Pearson Prentice Hall, 200 Old Tappan Road, Old Tappan NJ 07675.

Minimum System Requirements

Windows	Macintosh
Pentium II 300 MHz processor	Power PC G3 233 MHz or better
Windows 98 or later	Mac OS 9.x or 10.x
64 MB RAM	64 MB RAM
800 x 600 resolution	800 x 600 resolution
8x or faster CD-ROM drive	8x or faster CD-ROM drive
QuickTime 6.0 or later	QuickTime 6.0 or later